Dissipative Quantum Mechanics of Nanostructures

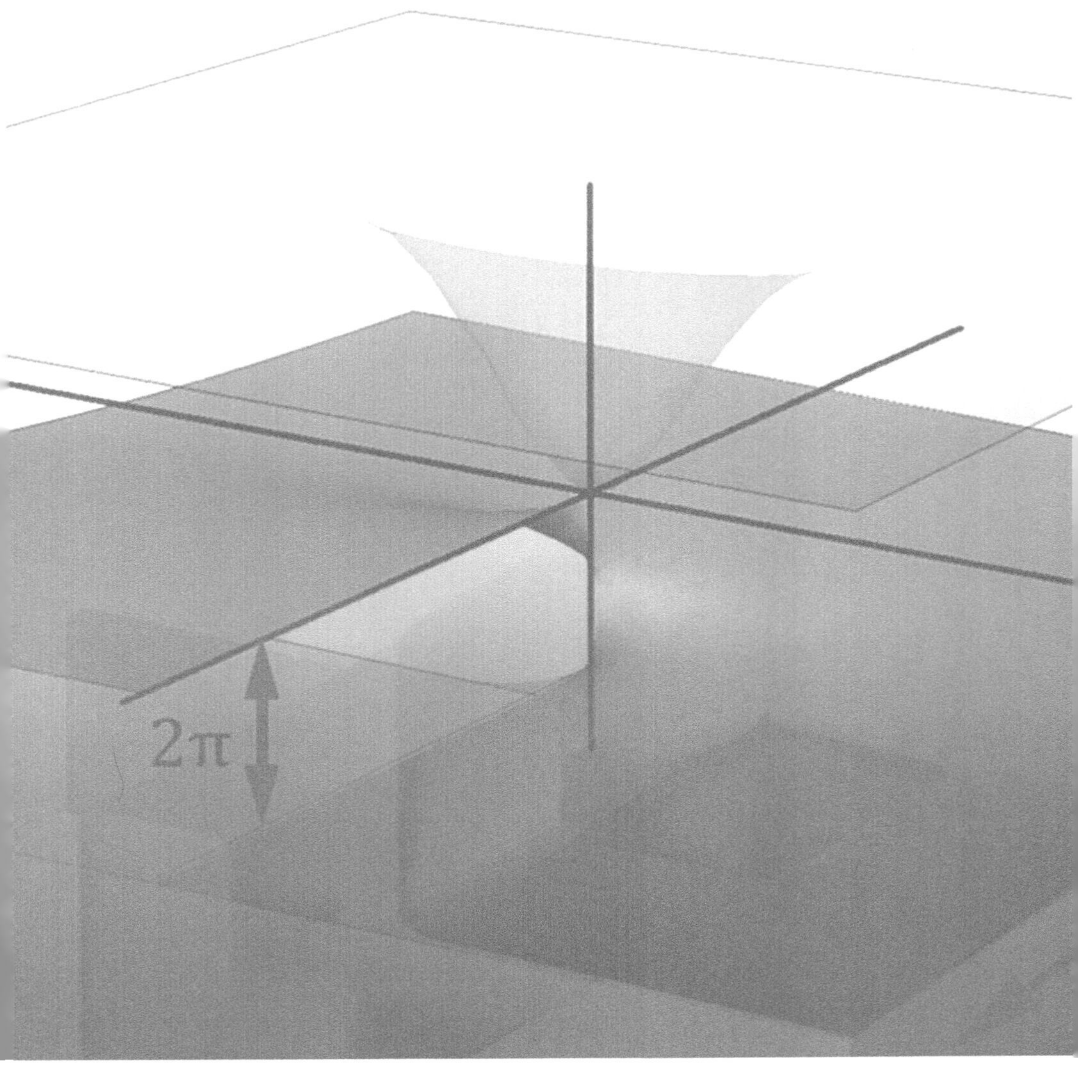

Dissipative Quantum Mechanics of Nanostructures

Electron Transport, Fluctuations, and Interactions

Andrei D. Zaikin
Dmitry S. Golubev

Jenny Stanford
Publishing

Published by

Jenny Stanford Publishing Pte. Ltd.
Level 34, Centennial Tower
3 Temasek Avenue
Singapore 039190

Email: editorial@jennystanford.com
Web: www.jennystanford.com

British Library Cataloguing-in-Publication Data
A catalogue record for this book is available from the British Library.

Dissipative Quantum Mechanics of Nanostructures:
Electron Transport, Fluctuations, and Interactions

ISBN 978-981-4774-50-5 (Hardcover)
ISBN 978-0-429-29823-3 (eBook)

To my father and teacher

—Andrei D. Zaikin

Contents

Preface xvii
Introduction xxi

PART A

1 Quantum Mechanics with Dissipation: Influence Functional Theory 3
1.1 Introduction 3
1.2 Influence Functionals 6
1.3 Influence Functional from a Harmonic Oscillator 12
1.4 Caldeira–Leggett Model 15
1.4.1 Influence Functional 15
1.4.2 Quasiclassical Langevin Equation 18
1.5 Perturbation Theory and Keldysh Diagram Technique 20
1.5.1 Time-Dependent Perturbation Theory 20
1.5.2 Influence Functional and Keldysh Diagrams 22
1.6 Caldeira–Leggett Action in Imaginary Time 28

2 Dissipative Quantum Mechanics of Superconducting Junctions 31
2.1 From Classical Dynamics to Quantum Mechanics: Phenomenology 33
2.2 Effective Action for a Josephson Junction 38
2.2.1 Hubbard–Stratonovich Transformation and Effective Hamiltonian 38
2.2.2 Current Operator 43
2.2.3 Effective Action 47
2.2.4 Electric Current 56

2.2.5 Adiabatic Limit 58
2.2.6 Action for Normal Tunnel Junctions 60
2.3 Langevin Equation 62
2.4 Imaginary-Time Action for Josephson Junctions 64
2.4.1 Adiabatic Limit 68
2.4.2 Normal Metal Limit 69
2.5 Beyond the Tunneling Limit 70
2.5.1 Contacts with Arbitrary Transmissions 73
2.5.2 Normal Ohmic Shunt 78

3 Quantum Particle in a Dissipative Environment 81
3.1 Free Particle Interacting with a Bath of Oscillators 81
3.1.1 Caldeira–Leggett Model: Exact Results 82
3.1.2 Discussion 85
3.2 Particle on a Ring 90
3.2.1 Real-Time Dynamics 90
3.2.2 Decoherence in the Ground State 97
3.3 Damped Harmonic Oscillator 100

4 Quantum Tunneling with Dissipation 107
4.1 Quantum Decay of a Metastable State: Free Energy Approach 108
4.2 Instanton Technique for Decay Rate 111
4.3 Decay of Supercurrent in Josephson Weak Links 120
4.3.1 Macroscopic Quantum Tunneling 122
4.3.2 Thermally Assisted Decay 131
4.3.3 Tunneling through a Non-quasiclassical Potential 141
4.3.4 MQT Experiments 149

5 Macroscopic Quantum Coherence and Dissipation 155
5.1 Thermodynamics of Double-Well Systems with Ohmic Dissipation 157
5.1.1 Instanton Technique 159
5.1.2 Effect of Linear Ohmic Dissipation, Localization Phase Transition 165
5.1.3 Combined Effect of Dissipation and External Bias 171

5.2 Real-Time Dynamics 176
5.2.1 Incoherent Regime 177
5.2.2 Quantum Coherence and Damping: Spin-Boson Model 178
5.2.3 Non-interacting-Blip Approximation 184
5.2.4 Coherent Oscillations and Decoherence 186

6 Quantum Dynamics of Phase and Charge in Josephson Junctions 195
6.1 Quantum Correction to the Critical Current 196
6.2 Incoherent Dynamics of the Phase 199
6.3 Quantum Dissipative Phase Transition and Duality 201
6.4 Phase and Charge Representations 209
6.5 Quasicharge as a Quantum Variable 213
6.6 Phase-Charge Duality and Quantum Tunneling of the Charge 221
6.7 Real-Time Dynamics and Relation for Phase and Charge Mobilities 228
6.8 Quantum Dynamics in the Weak Dissipation Regime 236
6.8.1 Quantum Corrections to Resistance 236
6.8.2 I–V Curve in the Weak Dissipation Limit 238
6.9 I–V Curve at Stronger Dissipation 250
6.9.1 Low-Temperature Limit 250
6.9.2 Higher Temperatures 254
6.10 Including Quasiparticle Tunneling 256

7 Coulomb Effects in Metallic Tunnel Junctions 261
7.1 Weak Electron Tunneling Regime 264
7.1.1 Master Equation Approach 264
7.1.2 Path Integral Analysis 267
7.1.3 Current–Voltage Characteristics 271
7.2 Langevin Equation Analysis 273
7.3 Weak Electron Tunneling in the Presence of an Environment 278
7.4 Thermodynamic Properties: Perturbation Theory 286
7.5 Non-perturbative Effects 296
7.5.1 Diagrammatic Analysis 296
7.5.2 Renormalization Group Equations 306

7.5.3 Instantons 309
7.5.4 Numerical Results 315

PART B

8 Quantum Particle in a Diffusive Electron Gas 323
8.1 Influence Functional from an Electron Gas 324
8.2 Particle on a Ring: Persistent Current and Decoherence 328
8.2.1 Effective Action 329
8.2.2 Persistent Current in the Absence of Interactions 332
8.2.3 Perturbation Theory 338
8.2.4 Non-perturbative Effects: No Periodic Potential 339
8.2.5 Non-perturbative Effects: Strong Periodic Potential 343
8.2.6 Real-Time Analysis 348
8.2.7 Some Concluding Remarks 350
8.3 Persistent Current Noise 352
8.3.1 General Considerations 352
8.3.2 Free Energy and Generating Functional 354
8.3.3 Particle in a Periodic Potential 357
8.3.4 Particle in a Dissipative Environment 361

9 Influence Functional for Interacting Electrons in Disordered Metals 369
9.1 Effective Action: Formal Considerations 371
9.2 Effective Action Expressed via Green Functions 376
9.3 Effective Action Expressed via Evolution Operator 378
9.4 Influence Functional for Electric Potentials 383
9.5 Time Evolution of a Single-Electron Density Matrix 387
9.6 Evolution Operators and Conductivity 391
9.7 Influence Functional for an Interacting Fermi Gas 393
9.8 Discussion 396
9.8.1 Some Applications 397
9.8.2 Quantum Kinetic Equation 398
9.8.3 Keldysh Diagrams 400
9.8.4 Quasiclassical Langevin Equation 403

10 Effective Action for Coherent Scatterers **405**
10.1 Motivation and Choice of the Model 405
10.2 Scattering States and Scattering Matrix 409
10.3 Scattering Matrix and Tunneling Hamiltonian 411
10.4 Evolution Operators and Density Matrix 414
10.5 Green Functions of Normal Reservoirs 417
10.6 Effective Action for a Short Coherent Conductor 418
10.7 Effective Action and Full Counting Statistics 424
10.8 Average Current, Noise, and Third Cumulant 429
10.9 Interaction with Adiabatic Environment 434

11 Coulomb Effects in Short Coherent Conductors **441**
11.1 Weak Coulomb Effects in Metallic Conductors 442
11.1.1 Quasiclassical Expansion for Effective Action 442
11.1.2 Shot Noise and Weak Coulomb Blockade of Electron Transport 443
11.1.3 Interaction Correction to Shot Noise and Relation to Third Cumulant 449
11.2 Coulomb Effects in Almost Transparent Conductors 455
11.2.1 Effective Action 455
11.2.2 Electron Transport and Shot Noise 457
11.3 Coulomb Effects beyond Perturbation Theory 462
11.3.1 Renormalization Group Analysis 462
11.3.2 Weak Coulomb Blockade and Instantons 468
11.3.3 Single-Channel Limit: Exact Solution 471

12 Charging Effects in Metallic Quantum Dots **477**
12.1 Introduction 477
12.2 Strong Coulomb Blockade 481
12.2.1 Sequential Tunneling: Master Equation Analysis 481
12.2.2 Path Integral Formulation 488
12.2.3 Cotunneling and Resonant Tunneling 491
12.3 Coulomb Blockade beyond Perturbation Theory: Inelastic Limit 496
12.3.1 Interaction Correction 497
12.3.2 Gate Effect 500

12.4 Coulomb Blockade in the Presence of Quantum Coherence 508
12.4.1 Scattering Matrix and Langevin Equations 508
12.4.2 Large Metallic Dots 517
12.4.3 Effect of External Impedance 522
12.4.4 Boltzmann–Langevin Approach 525

13 Coulomb Blockade in Quantum Dot Chains and Metallic Wires 533
13.1 Introduction and Basic Model 533
13.2 Strong Inelastic Relaxation Limit 536
13.2.1 Highly Resistive Tunnel Barriers 536
13.2.2 Low Resistive Barriers 545
13.3 Weak Inelastic Relaxation Limit 550
13.3.1 Kinetic Equation Analysis 550
13.3.2 Composite Fano Factor 557
13.3.3 Two Quantum Dots in Series 558
13.3.4 Chains of Identical Quantum Dots 562
13.3.5 Effect of External Environment 567
13.3.6 Weakly Disordered Metallic Wires 568
13.3.7 Discussion 570

14 Weak Localization and Electron Dephasing in Disordered Conductors I: Metallic Limit 573
14.1 Introduction 573
14.2 Weak Localization of Non-interacting Electrons 575
14.3 Electron Dephasing by Classical Fields 584
14.4 Dephasing by Electron–Electron Interactions 590
14.4.1 Exponent 595
14.4.2 Pre-exponent 605
14.4.3 Relation to Perturbation Theory 611
14.5 Relation to $P(E)$-Theory of Electron Tunneling 615

15 Weak Localization and Electron Dephasing in Disordered Conductors II: Beyond Quasiclassics 619
15.1 Introduction 619
15.2 Dephasing in Quantum Dots: Qualitative Arguments 621

15.3 Weak Localization in a System with a Tunnel Barrier 623
15.3.1 Effective Action and Weak Localization 623
15.3.2 Non-interacting Limit 627
15.3.3 Effect of Electron–Electron Interactions 630
15.3.4 Dephasing 634
15.3.5 Coulomb Blockade of Weak Localization Correction 638
15.4 Weak Localization in Arrays of Quantum Dots 641
15.4.1 Non-interacting Limit 641
15.4.2 Electron Dephasing by Coulomb Interactions 644
15.4.3 Good Metals and Strongly Disordered Conductors 648
15.4.4 Comparison with Experiments 651
15.5 Further Extensions: Aharonov–Bohm Effect 660
15.5.1 Ring with Two Quantum Dots 661
15.5.2 Ring Composed of a Quantum Dot Chain 665
15.5.3 Diffusive Nanorings 671
15.6 Short Summary 672

16 Electron Transport, Fluctuations, and Coulomb Effects in Normal-Superconducting Hybrids 675
16.1 Introduction 675
16.2 Shot Noise and Coulomb Blockade of Andreev Reflection 678
16.2.1 Effective Action 678
16.2.2 Shot Noise and Interaction Correction 682
16.3 Combined Effect of Disorder and Interactions 684
16.3.1 Andreev Effective Action 686
16.3.2 Andreev Conductance Affected by Disorder and Interactions 691
16.3.3 Shot Noise 703
16.4 Non-local Electron Transport, Shot Noise, and Interactions 705
16.4.1 Crossed Andreev Reflection 705
16.4.2 The Model and Effective Action 709
16.5 Langevin Equations and Shot Noise 718

16.6 Interaction Correction to the Current 719
16.7 Brief Summary 722

17 Superconducting Contacts beyond the Tunneling Limit 725
17.1 Introduction 725
17.2 Effective Action and Phase Fluctuations 727
17.3 Equilibrium Supercurrent Noise 734
17.4 Capacitance Renormalization 737
17.5 Andreev Level Spectroscopy 741
17.6 Supercurrent Affected by Electron–Electron Interactions 748
17.7 Andreev Levels as an Effective Environment 754
17.8 Supercurrent Decay in Highly Transparent Contacts 760

18 Effective Action and Superconducting Fluctuations 771
18.1 Introduction 771
18.2 Effective Action 773
18.2.1 Perturbation Theory 775
18.2.2 Gaussian Fluctuations in Dirty Superconductors 780
18.3 Langevin Equations 785
18.4 Time-Dependent Ginzburg–Landau Equation 787
18.5 Thin Metallic Wires 789
18.5.1 Propagating Modes 789
18.5.2 Gaussian Fluctuations of the Order Parameter 790
18.6 Matsubara Effective Action 792

19 Thermal and Quantum Phase Slips in Superconducting Nanowires 799
19.1 Introduction 799
19.2 General Background 800
19.3 Thermally Activated Phase Slips 803
19.3.1 Activation Exponent 804
19.3.2 Pre-exponent 805
19.3.3 Comparison with McCumber–Halperin Result 810
19.3.4 Temperature-Dependent Resistance and Noise 811

19.4 Theory of Quantum Phase Slips in Superconducting Nanowires 813
19.4.1 QPS Action 813
19.4.2 QPS Tunneling Amplitude 819
19.4.3 QPS Interactions and Quantum Phase Transitions 823
19.4.4 Wire Resistance at Low Temperatures 828
19.4.5 Discussion 830
19.4.6 Phase–Charge Duality 835
19.4.7 QPS Noise 840
19.5 Key Experiments 848

20 Persistent Currents in Superconducting Nanorings 855
20.1 Persistent Currents in Bulk Superconducting Rings 855
20.2 Quantum Phase Slip Rings 860
20.2.1 Non-interacting QPS: Partition Function 862
20.2.2 Persistent Current and Noise 865
20.2.3 Discussion 868
20.2.4 Including Interactions between QPS 869
20.3 Parity Effect and Persistent Currents 875
20.3.1 Parity Projection Formalism 876
20.3.2 Homogeneous Superconducting Rings 878
20.3.3 Ballistic Superconductor–Normal–Superconductor (SNS) Rings 883
20.3.4 Resonant and Diffusive SNS Rings 892

Bibliography 899
Index 917

Preface

This book is a largely extended and updated version of the lecture course that one of us (A. D. Z.) taught at I. E. Tamm Department of Theoretical Physics of P. N. Lebedev Physical Institute in Moscow in 2007–2008. The purpose of this book is to provide a systematic and comprehensive coverage of physical ideas, theoretical methods, and key results in a rapidly developing field of quantum transport and macroscopic quantum phenomena in a wealth of available nanoscale structures. Better understanding of the whole scope of such phenomena is extremely important both from fundamental point of view as well as for fabrication of high-quality nanodevices with controlled parameters involved in a variety of applications.

In nanostructures elementary charge carriers—electrons—are usually confined to a limited volume interacting both with each other and with lattice ions and simultaneously suffering multiple scattering events on impurities, barriers, surface imperfections, and other defects. Electron interaction with other degrees of freedom generally yields two major consequences—*quantum dissipation* and *quantum decoherence*. In other words, even at very low temperatures interacting electrons can lose both their energy and their ability to quantum interference. These two different (though not unrelated) processes are at the heart of all quantum phenomena discussed in this book.

A perturbative in the interaction approach—intuitively appealing due to its relative simplicity—often fails to adequately account for quantum dynamics of interacting many-body systems, particularly in the low energy limit. In such cases one has no choice but to employ nonperturbative techniques. Path integral formulation of quantum theory is particularly well suited for this purpose and it is intensively exploited throughout our book. Most conveniently, path integral methods allow to implement the central idea of tracing out all the degrees of freedom of an interacting system on which no direct measurements are being performed and, at the same time, keeping complete information about the influence of those on the remaining variable(s) of interest to be measured. In this way we arrive at the concept of effective actions and Feynman-Vernon influence functionals. This approach is elaborated and employed for the analysis of quantum transport and fluctuations in mesoscopic and nanostructured conductors in the presence of interactions.

This book is useful for a possibly broader readership ranging from masters and PhD students to post docs and senior researchers. We assume that the reader has basic knowledge of quantum mechanics and statistical physics (within the frames of the standard university courses) and is also familiar with the Feynman path integrals on an introductory level. The book is divided into two parts. Its first part covers relatively more conventional materials and can also be used as a textbook at the master student level. The second part of the book deals with a number of topics which, although also quite well established by now, may be viewed by some readers as somewhat more advanced. At the same time, this is by no means "two books in one," all 20 chapters are united within the same general concept and, as the reader will observe, are intimately related to each other in many respects. In this sense, the book division into two parts can also be disregarded, if one so wishes.

We have made a substantial effort to make the book self-contained. In most cases we provide the reader with sufficient details and explanations for understanding the basic physics behind the result and learning how this result can technically be reproduced without digging into a lot of extra literature. It is by no means our goal here to present an extensive coverage of the experiment; only in a few exceptional cases we give a brief account of the experimental situation in the corresponding sub-field. At the same time, we provide sufficiently rich bibliography that allows to overview all topics touched upon in this book. While selecting these topics we were merely guided by our research interests and experience. Some other topics are inevitably left behind, which does not necessarily reflect their secondary importance for the field.

In this book we express our views on the field which have been formed in the course of our own research and—not to a lesser extent—as a result of our continuous collaborations and valuable discussions with our many colleagues and friends who shared their interest on this beautiful piece of physics. It is a pleasure to thank all of them for the numerous fruitful interactions we had with them over decades of our research.

Andrei D. Zaikin
Dmitry S. Golubev
Spring 2019

Introduction

Quantum physics of nanostructures is a rich and rapidly developing field. Most of the fundamental problems of modern condensed matter physics involve various aspects of quantum transport and fluctuation phenomena at the nanoscale. In addition, continuing miniaturization of devices, gadgets, and electronic schemes, together with quickly growing number of nanotechnological applications, also demand a profound understanding of the underlying physics.

What are the main specific features of nanostructures? On one hand, these systems are *low-dimensional* by definition. As physics in reduced dimension may differ quite substantially from that in the bulk, theoretical tools should be adjusted accordingly in order to properly respond to the above challenges. On the other hand, the number of relevant degrees of freedom, such as electrons and magnetic impurities. still remains *macroscopically large* despite the small system size. In this situation it is essential to properly account for quantum many-body effects which occur in the presence of interactions. Hence, one generally needs to simultaneously describe both quantum mechanical and quantum statistical evolution of the system. In other words, in order to determine the behavior of relevant observables it is necessary not only to perform the corresponding quantum averages but also to determine the quantum state with respect to which these averages are being performed.

General and elegant ways to deal with this situation were proposed in sixties in two seminal papers by Keldysh [220] and by Feynman and Vernon [119]. Approaching the problem from somewhat different prospectives these authors formulated the central idea to describe quantum evolution of an interacting many-body system in terms of a single particle density matrix. As quantum dynamics of the latter is determined by the product of the two corresponding matrix elements one naturally arrives at the concept of real-time evolution along the so-called *Keldysh contour* which consists of forward and backward time branches. Interactions generally mix up quantum variables defined on each of these branches making it impossible to consider one of them independently of the other. This idea allowed Keldysh to formulate the diagram technique fully describing quantum dynamics of non-equilibrium interacting many-body systems.

Feynman and Vernon took a somewhat different route. They proposed to trace out all "non-interesting" degrees of freedom already at the very beginning of the calculation and then to deal with the reduced density matrix which only accounts for the observable(s) of interest. Technically this approach is most conveniently implemented with the aid of Feynman path integrals and yields the concept of *Feynman–Vernon influence functional,* which in general contains complete information about the influence of the averaged out degrees of freedom on the variable(s) of interest. The equivalence of Keldysh and Feynman–Vernon techniques is verified by expanding the reduced density matrix derived within the latter approach in powers of the corresponding influence functional. In this way one formally recovers the full diagrammatic series of the Keldysh technique.

This book deals with a broad scope of applications of the influence functional technique in condensed matter physics. It consists of 20 chapters and is divided into two parts. The first part (Chapters 1 to 7) describes the basics of the influence functional theory and several most well-known applications of this powerful technique. In Chapter 1 we introduce Feynman–Vernon influence functionals and discuss the Caldeira–Leggett effective action approach to quantum dissipation. This approach yields Gaussian influence functional which captures a number of generic features of quantum dissipative environments and, at the same time, remains simple enough to deal with practical calculations. We also illustrate the relation between the Feynman–Vernon-Caldeira–Leggett path integral technique and several other methods, such as quasiclassical Langevin equation, quantum mechanical time dependent perturbation theory, and Keldysh technique. In Chapter 2 we extend Feynman–Vernon idea of the influence functional, applying it to junctions between two bulk normal metals or superconductors. Of primary importance here is the case of tunnel (Josephson) junctions. Starting from the microscopic Hamiltonian for such systems we trace out all electron degrees of freedom (playing the role of an effective quantum dissipative bath) and derive the so-called Ambegaokar–Eckern–Schön (AES) effective action formulated in terms of a single collective variable—the Josephson phase—which accounts for quantum dissipative dynamics of the system. We also demonstrate how to extend this microscopic approach to metallic junctions with arbitrary transmissions of their conducting channels and to spatially extended metallic conductors. We emphasize a direct link between these results and the phenomenological Caldeira–Leggett effective action derived in Chapter 1.

In Chapters 3–7 we discuss several important examples illustrating key advantages of the influence functional technique developed in Chapters 1 and 2. In Chapter 3 we analyze real-time dynamics of a quantum particle interacting with a dissipative Caldeira–Leggett bath and introduce one of the most fundamental concepts in quantum theory—quantum decoherence. Provided this particle is moving in the absence of any external potential (or in a parabolic potential) one can obtain an exact solution of the problem which also allows to address the issue of quantum decoherence without employing any approximation. In many physical situations, however, nonlinear effects can also play a substantial role. In those cases no exact solution is anymore possible, and one should reside to various approximations. One example of that kind—a dissipative quantum particle on a ring described by the action equivalent to that of AES—is addressed in Chapter 3.

In Chapter 4 we consider another important physical example—quantum dissipative decay of a metastable state. Yet another example, quantum dynamics of a damped particle in a double-well potential, is discussed in Chapter 5. Both these problems are of a special interest in connection with fundamentally important experiments on macroscopic quantum tunneling (MQT) and macroscopic quantum coherence (MQC) demonstrating macroscopic quantum behavior of Josephson junctions, as also indicated in Chapter 2.

The situation of "tunneling into continuum" addressed in Chapter 4 essentially implies incoherent decay since the probability for a quantum particle to return back to its initial state is negligibly small due to both the potential profile and the effect of dissipation. On the contrary, in the case of a double-well potential with (almost) degenerate minima considered in Chapter 5 the quantum particle can tunnel back and forth between these minima gradually losing its coherence in the presence of weak dissipation. Obviously such process cannot anymore be described in terms of incoherent decay and the probability to find the particle in one of the wells exhibits damped coherent oscillations in time. In practice this situation can be realized, e.g., in a Josephson junction embedded in a superconducting loop in which case coherent oscillations of the magnetic flux inside the loop can be observed. Such systems can also be used for practical implementation of superconducting qubits.

Chapter 6 continues our analysis of macroscopic quantum dynamics of ultrasmall Josephson junctions. Provided the bias current is weak, many minima of the Josephson potential in the phase space have almost the same energy and quantum dynamics of the phase variable gets substantially modified as compared to that studied in Chapter 5. According to quantum mechanics, in this case the wave function of a "Josephson particle" describes Bloch states. Dissipation can significantly alter this picture. In Chapter 6 we demonstrate that nontrivial interplay between quantum coherence, dissipation and Coulomb effects yields a rich variety of properties which can be accessed not only by means of the Josephson phase variable but also through the canonically conjugate quantum variable of the junction charge. In the presence of linear Ohmic dissipation these two variables turn out to be strictly dual to each other. This remarkable property allows one to proceed much further in understanding of this problem.

It is rather obvious that quantum fluctuations, dissipation and Coulomb effects should persist not only in superconducting but also in normal tunnel junctions provided their size remains sufficiently small. In Chapter 7 we specifically address quantum behavior of tunnel nanojunctions between two normal metallic electrodes. Transport properties of such structures turn out to be rather complicated provided Coulomb interaction between electrons is taken into account. Coulomb interaction tends to block the process of electron tunneling between metallic electrodes and, hence, to reduce the current flowing through a tunnel barrier. On the other hand, electron tunneling gives rise to charge fluctuations which in turn tend to diminish Coulomb blockade effects. This nontrivial trade-off in small size tunnel junctions is analyzed in details in Chapter 7.

In the second part of this book we further develop the idea of the influence functional applying it to a number of advanced problems of modern condensed matter theory. With this analysis we in general go quite far both from the simple Caldeira–Leggett model of Chapter 1 and from the AES approach of Chapter 2. Following this route we discover plenty of new physics not contained within the Caldeira–Leggett type of models.

Chapters 8 and 9 deal with the problem of a quantum particle propagating in a diffusive electron gas. Provided this particle (e.g., a muon) is distinguishable from electrons in a disordered metal the corresponding influence functional can be constructed relatively easily just by taking into account the effect of quantum electromagnetic fields produced by fluctuating electrons forming an effective environment. In Chapter 8 we implement this program and arrive at a non-Gaussian influence functional for a quantum particle, which is then employed to describe equilibrium properties of this particle in the presence of interactions and disorder. In particular, addressing the effect of persistent currents we again approach the problem of interaction-induced quantum decoherence at low temperatures. In contrast to Chapter 3, however, we do not anymore restrict our analysis to the phenomenological Caldeira–Leggett model but rather employ a microscopically derived influence functional for a diffusive electron gas.

Turning now to the problem of a degenerate interacting electron gas one immediately observes a conceptual problem: Direct application of the same influence functional formalism as in Chapter 8 fails in this case simply because electrons are indistinguishable. Each electron could simultaneously be a particle of interest and a part of an effective environment for this particle. Hence, in order to apply the Feynman–Vernon strategy to interacting Fermi systems it is necessary to keep track of the Pauli principle. This task is accomplished in Chapter 9. The key step here is to derive a formally exact equation for the density matrix of a single electron propagating in the environment formed by quantum electromagnetic fields produced by fluctuating electrons. This equation depends on the electron distribution function and, hence, explicitly accounts for the Pauli principle. Afterward one can integrate out the electromagnetic fields and construct the influence functional for interacting electrons in disordered metallic conductors, which includes Fermi statistics. We also consider several applications of this influence functional technique, including quantum kinetic equation approach, Keldysh technique for interacting fermions, and quasiclassical Langevin equation.

The path integral technique worked out in Chapter 9 usually allows for a straightforward non-perturbative treatment of the problem with disorder and interactions, provided one is able to properly identify relevant quasiclassical electron paths. Technically this is possible, e.g., for weakly disordered spatially extended conductors. On the other hand, for spatially restricted and strongly non-uniform structures, such as quantum dots and granular metals the description in terms of quasiclassical electron trajectories may fail. In such cases electron scattering on disorder should be treated on more general footing without employing any quasiclassics. A proper tool for that is the scattering matrix formalism. The main idea here is to "marry" this formalism with a non-perturbative description of electron–electron interactions. Following this route one can quite generally model a disordered conductor by an array of (metallic) quantum dots connected via junctions (scatterers). A clear advantage of this model is that it allows one to treat weakly and strongly disordered conductors in both spatially restricted and spatially extended geometries within the same theoretical framework. Detailed implementation of these ideas is contained in Chapters 10–13 of this book.

In Chapter 10 we introduce the concept of the scattering matrix and derive the influence functional for short coherent conductors which accounts for the effect of electron–electron interactions. In Chapter 11 we apply this formalism to the problem of (weak) Coulomb blockade in metallic conductors. In particular, we establish a fundamental relation between shot noise and interaction effects in electron transport, extend these ideas further to higher correlators and also address non-perturbative interaction effects. We further extend our scattering matrix analysis of charging effects applying it to single metallic quantum dots (Chapter 12) and their arrays (Chapter 13), eventually building a bridge to describe interaction effects in disordered metallic wires.

Chapters 14 and 15 are devoted to the analysis of weak localization and interaction-induced quantum dephasing of electrons in disordered conductors at low temperatures. While in Chapters 3, 5, and 8 quantum decoherence in (or close to) the ground state of an interacting system is addressed without including Fermi statistics, in Chapters 14 and 15 the Pauli principle is fully accounted for. In Chapter 14 we apply the influence functional formalism developed in Chapter 9 for weakly disordered conductors where quantum dynamics of electrons can be described in terms of their semiclassical trajectories. This formalism is well suited in order to proceed non-perturbatively to all orders in the interaction and also to establish direct links with models considered in Chapters 3 and 8 as well as with the $P(E)$-theory elaborated in Chapter 7. In Chapter 15 we address the same problem employing the combination of the scattering matrix formalism and the path integral influence functional technique of Chapters 10–13. This approach allows to extend our non-perturbative analysis of interaction-induced dephasing of electrons, e.g., to quantum dots and strongly disordered metals and to construct a unified description of this fundamentally important phenomenon embracing a broad range of disordered conductors.

In the last five chapters of this book we develop some of the above ideas further applying them to various superconducting structures. In Chapter 16 we consider hybrid metallic structures composed of normal and superconducting electrodes. Low-energy electron transport in such system is due to the mechanism of Andreev reflection which involves conversion of an incident electron into a Cooper pair and a hole propagating in the opposite direction. As a result, twice the electron charge is being transferred in the course of such electron-hole reflection. In Chapter 16 we analyze various aspects of this charge transport and fluctuation effects in normal-superconducting metallic hybrids and generalize fundamental relation between these effects and electron–electron interactions already established in Chapters 10 and 11 for normal junctions. Remarkably, it turns out that the relation between shot noise and Coulomb interaction effects in such hybrid structures can also be extended to non-local phenomena, such as crossed Andreev reflection.

Chapter 17 is devoted to the analysis of Coulomb effects and fluctuations in superconducting weak links not containing any tunnel barrier, such as superconducting quantum point contacts or superconductor-normal-metal-superconductor junctions. Unlike in the first part of this book (Chapters 2 and 4–6), in Chapter 17 we employ the influence functional technique in order to examine a nontrivial interplay between fluctuation and charging effects in superconducting contacts beyond the tunneling limit. In such systems quasiparticles with energies below the superconducting gap are confined inside the weak link forming a discrete set of Andreev levels that essentially contribute to the supercurrent. At the same time, under certain conditions these low- energy bound states can be viewed as an effective quantum dissipative environment which may become particularly important at higher barrier transmissions.

In Chapter 18 we elaborate a detailed microscopic derivation of the effective action for spatially extended superconductors. We again follow the basic idea of the Feynman–Vernon influence functional technique and integrate out the electron degrees of freedom. This task is now by far more complicated than in the case of various superconducting junctions treated in Chapters 2, 16, and 17. It is not anymore possible to express the effective action in terms of only one quantum variable because (*a*) the superconducting phase is now a quantum field rather than just one degree of freedom, (*b*) fluctuations of the absolute value of the order parameter both in space and in time are important, (*c*) fluctuating electromagnetic potentials cannot in general be integrated out and (*d*) superconducting fluctuations and other dynamical processes may drive the quasiparticle distribution function far from equilibrium. The key idea here is to make use of the Ward identities which help to construct a manifestly gauge invariant effective action for a superconductor. The real-time Keldysh version of this action allows to account for diverse dynamical and fluctuation effects and under certain conditions reduces to a set of Langevin equations which provide a full description within the Gaussian approximation for the fluctuating fields.

An important application of the above general theory concerns quantum dynamical effects in ultrathin superconducting wires. Provided the thickness of such wires enters the range below $\sim$100 nm non-Gaussian fluctuations of the superconducting order parameter—the so-called thermal and quantum phase slips (TAPS and QPS)—start playing an important role. For example, proliferation of QPS in superconducting nanowires drives such systems into a resistive state which may even turn insulating in the zero temperature limit. A detailed theory of this and other TAPS- and QPS-related phenomena in superconducting nanowires is developed in Chapter 19. Chapter 20—the last one in this book—is devoted to quantum effects in superconducting nanorings. In this chapter we analyze the effect of quantum phase slips on the ground properties of these nanorings, such as persistent currents and related phenomena. We also apply our path integral formalism in order to address the superconducting parity effect and its impact on persistent currents in superconducting nanorings with various types of weak links.

In this book an effort is made to maintain a subtle balance between describing theoretical methods and techniques and displaying a rich landscape of physical phenomena accessed by these methods. On one hand, seemingly very different problems and ideas can be united and conveniently compared to each other if analyzed within the same general theoretical approach. On the other hand, the approach itself demonstrates its enormous power if successfully applied in rather different sub-fields of condensed matter physics and beyond. A nice illustration for that is provided by the material covered in this book.

Part A

Chapter 1

Quantum Mechanics with Dissipation: Influence Functional Theory

1.1 Introduction

In classical mechanics, real-time dynamics of a particle with mass m and coordinate q is fully described by the Newton equation of motion

$$m\frac{d^2q}{dt^2} = F(q) \tag{1.1}$$

where F is a total external force acting on a particle. This force can be due to the presence of an external potential profile $U(q)$ in which case F is defined by the potential gradient as $F = -\partial U/\partial q$. In addition, the particle can interact with other degrees of freedom in such a way that the particle energy is dissipated during its motion. This situation is described by an additional dissipative force that, in many cases, is simply proportional to the particle velocity dq/dt, i.e., $F = -\eta dq/dt$.

Note that energy is dissipated only on average, while in reality, interaction with external degrees of freedom results in a random time sequence of energy gains and losses. In order to account for these random energy changes, yet another (stochastic) force

Dissipative Quantum Mechanics of Nanostructures: Electron Transport, Fluctuations, and Interactions
Andrei D. Zaikin and Dmitry S. Golubev

ISBN 978-981-4774-50-5 (Hardcover), 978-0-429-29823-3 (eBook)
www.jennystanford.com

$F = \xi(t)$ needs to be introduced. While the time average of $\xi(t)$ equals to zero, it can differ from zero at any given time moment, thus effectively describing noise produced by external degrees of freedom. Collecting all the above contributions to the external force and substituting them into Eq. (1.1) we arrive at the following Langevin equation:

$$m\ddot{q} + \eta\dot{q} + \frac{\partial U(q)}{\partial q} = \xi(t) \tag{1.2}$$

Here and below we use the standard notations $\dot{q} \equiv dq/dt$ and $\ddot{q} \equiv d^2q/dt^2$. It is usually assumed that $\xi(t)$ and $\xi(t')$ are uncorrelated at different time moments $t \neq t'$, i.e., the noise is "white." Assuming also that a collection of external degrees of freedom interacting with the particle is in thermodynamic equilibrium at temperature T, for the noise correlator $\langle\xi(t)\xi(0)\rangle$ we have

$$\langle\xi(t)\xi(0)\rangle = 2\eta T\delta(t) \tag{1.3}$$

This form of the noise correlation function is consistent with the fluctuation–dissipation theorem in the classical limit. Note that throughout this book, we will always express the temperature in energy units, i.e., we will put the Boltzmann constant equal to unity ($k_{\mathrm{B}} = 1$).

While the above analysis applies for classical systems, it obviously cannot be used for quantum ones; hence, in the latter case, a proper quantization procedure should be constructed. In the absence of dissipation $\eta \to 0$, this procedure is well known from textbooks on quantum mechanics. One first constructs the Lagrange function

$$\mathcal{L} = \frac{m\dot{q}^2}{2} - U(q) \tag{1.4}$$

which yields the Newton equation (1.1) by setting the variation of $\mathcal{L}$ equal to zero,

$$\delta\mathcal{L} = 0$$

for any small variation δq in the vicinity of the classical trajectory. Then with the aid of the Legendre transformation

$$\mathcal{L} = p\dot{q} - H[p, q], \tag{1.5}$$

one introduces the momentum variable p and the Hamilton function

$$H[p, q] = \frac{p^2}{2m} + U(q) \tag{1.6}$$

Finally, one defines the coordinate momentum operators $q \to \hat{q} = q$, $p \to \hat{p} = -i\hbar\partial/\partial q$ as well as the Hamilton operator $\hat{H} = H[\hat{p}, \hat{q}]$ and writes down the Schrödinger equation

$$i\hbar\frac{\partial\psi}{\partial t} = \hat{H}\psi \tag{1.7}$$

for the particle wave function $\psi(q)$.

The question arises if any reasonable generalization of this quantization procedure can be formulated in order to describe quantum systems with dissipation. An attempt to formally follow the same route in this case fails, leading to physically meaningless results. This failure is, of course, easily understandable as quantum behavior of non-conservative systems includes qualitatively new features, which would be entirely absent should energy conservation be maintained. Hence, a different route should be chosen in order to quantize systems with dissipation.

The basic idea was pioneered by Caldeira and Leggett [80, 81] and can be briefly summarized as follows. In order to describe quantum dynamics of any dissipative system, one should first make "one step back" and explicitly define both the "quantum bath" where the energy goes to and the interaction between the system of interest and the bath. In other words, in the process of quantization, it is suggested to deal with a more complicated system than the initial one. An obvious gain, however, is that our new object "system+bath" is already described by the Hamiltonian (the *total* energy is conserved!) and can be quantized in a usual manner. After this quantization is accomplished, extra information is eliminated by averaging the result over the bath degrees of freedom and one arrives at a quantum description of the dissipative system of interest.

Note that dissipation appears at the last step when the bath degrees of freedom are eliminated. This step is, therefore, central for the whole procedure, and it needs to be carried out with sufficient accuracy and care. Fortunately, Feynman and Vernon [118, 119] formulated an approach that allows to tackle this problem in a most general and elegant manner. The goal is accomplished by deriving the so-called influence functional, which fully accounts for the influence of the bath degrees of freedom on the system of interest. In fact, this technique of influence functionals will play a major role in almost all physical examples considered in this book. This method is described in details below in this chapter.

1.2 Influence Functionals

Consider a quantum mechanical particle (or a more general quantum system) with coordinate $q(t)$ (which can, in general, consist of a number of degrees of freedom) and the Hamiltonian $\hat{H}_0$. We will assume that this particle interacts with another quantum system, which will be denoted below as "environment." This latter system is described by coordinate $Q(t)$ and the Hamiltonian $\hat{H}_{\text{env}}$. The total Hamiltonian for the system "particle+environment" then reads

$$\hat{H} = \hat{H}_0 + \hat{H}_{\text{env}} + \hat{H}_{\text{int}} \tag{1.8}$$

where the last term $\hat{H}_{\text{int}}$ describes interaction between the particle and the environment.

At this stage, both subsystems q and Q enter on equal footing, and thus denoting the first one as "particle" and the second one as "environment" is just a matter of convention. This convention becomes meaningful if we assume—as we indeed do below—that all measurements are carried out on subsystem q only, while no measurements on subsystem Q are performed at all. In other words, we will only be interested in quantum dynamics of the variable q but not Q. This situation is rather common, and it is encountered in many physical systems. For instance, if one measures the electric current across some conductor, one is merely interested in the dynamics of electrons in this conductor, while no detailed information, say, about phonon subsystem is needed in this case. Of importance is only the influence of phonons on the electronic states and not quantum dynamics of the phonon subsystem itself. In this example, electrons play the role of a subsystem of interest q, while phonons form an effective environment Q. Similarly, one can study muon scattering and consider muons propagating through some electron gas, e.g., in a metal. As there exists Coulomb interaction between muons and electrons, muon dynamics will be clearly affected by the electron gas. Again, if measurements are only carried out with muons, one is not interested in describing the electron subsystem itself but only its influence on the muon dynamics. Hence, here already electrons play the role of an effective environment Q for the muon subsystem of interest q. The same approach can be applied, e.g., if one studies an atom (q) interacting with an external electromagnetic field (Q) as well as in numerous other physical situations.

Suppose we are interested in the probability for the particle q to have the coordinate $q = q_\mathrm{f}$ at a time t provided at $t = 0$, it had the coordinate $q = q_\mathrm{i}$. By definition, this probability W is given by the square of a quantum mechanical transition amplitude K, i.e., $W = |K|^2$. In the absence of interaction, the particle dynamics does not depend on that of the environment and is fully described by the Schrödinger equation (1.7) for the particle wave function $\psi(q)$. However, at the moment the interaction is "turned on," there appears an additional force acting on a particle q. This force depends on Q, which is itself a quantum variable. Thus, we cannot anymore restrict ourselves to the dynamics of q, but rather should deal with the total Hamiltonian for the system "$q + Q$" (1.8). Of course, also in this case, the probability W remains equal to the square of the corresponding transition amplitude K, which now depends on both q and Q. Since no measurements of Q are performed, we should sum over *all* possible final states of Q. Then the probability W takes the form

$$W = \sum_{Q_\mathrm{f}} K(q_\mathrm{f}, Q_\mathrm{f}, t; q_\mathrm{i}, Q_\mathrm{i}, 0) K^*(q_\mathrm{f}, Q_\mathrm{f}, t; q_\mathrm{i}, Q_\mathrm{i}, 0) \tag{1.9}$$

where

$$K(q_\mathrm{f}, Q_\mathrm{f}, t; q_\mathrm{i}, Q_\mathrm{i}, 0) = \langle q_\mathrm{i}, Q_\mathrm{i} | \hat{U} | q_\mathrm{f}, Q_\mathrm{f} \rangle \tag{1.10}$$

and

$$\hat{U} = T \exp(-\mathrm{i}\hat{H} t/\hbar) \tag{1.11}$$

is the standard quantum mechanical evolution operator for our system. Each of the terms in the sum (1.9) represents a probability for our system to come into the final state $(q_\mathrm{f}, Q_\mathrm{f})$. Since the subsystem Q can be in *any* final state Q_f, in order to find the total probability W, we have to add all these probabilities together.

One can slightly generalize the problem and describe the evolution of the density matrix of the system $\rho(q, q')$ from some initial to some final state. This evolution is described by the following equation:

$$\rho(q_\mathrm{f}, q_\mathrm{f}') = \int \mathrm{d}q_\mathrm{i}\mathrm{d}q_i' J\,(q_\mathrm{f}, q_\mathrm{f}', t; q_\mathrm{i}, q_i', 0)\rho_\mathrm{i}(q_\mathrm{i}, q_i') \qquad (1.12)$$

where $\rho_\mathrm{i}(q_\mathrm{i}, q_i')$ is the initial density matrix for the particle q and we introduced the kernel

$$J = \sum_Q U\,|Q\rangle\langle Q|U^+, \qquad (1.13)$$

which is again given by the product of two transition amplitudes with the sum running over all possible states of the subsystem Q. For the sake of simplicity, in what follows we will assume that there is no interaction between q and Q before $t = 0$. Then the total initial density matrix can be factorized as $\rho_\mathrm{i}(q_\mathrm{i}, q_i')\tilde{\rho}_\mathrm{i}(Q_\mathrm{i}, Q_i')$, where $\tilde{\rho}_\mathrm{i}(Q_\mathrm{i}, Q_i')$ is the density matrix describing the initial state of the environment Q.

It will be convenient for us to express the transition amplitude K via path integral as

$$K(q_\mathrm{f}, Q_\mathrm{f}, t; q_\mathrm{i}, Q_\mathrm{i}, 0) = \int_{q_\mathrm{i}}^{q_\mathrm{f}} \mathrm{D}q \int_{q_\mathrm{i}}^{Q_\mathrm{f}} \mathrm{D}Q \exp\left(\frac{\mathrm{i}}{\hbar}S[q, Q]\right) \qquad (1.14)$$

where

$$S[q, Q] = S_0[q] + S_{\mathrm{env}}[Q(t)] + S_{\mathrm{int}}[q(t), Q(t)] \qquad (1.15)$$

is the total action for our system in which the terms S_0, S_{env}, and S_{int} describe, respectively, the particle, the environment, and the interaction between q and Q. Similarly, the kernel J can be represented in terms of a double path integral

$$J = \int_{q_\mathrm{i}}^{q_\mathrm{f}} \mathrm{D}q_1 \int_{q_i'}^{q_\mathrm{f}'} \mathrm{D}q_2 \exp\left(\frac{\mathrm{i}}{\hbar}S_0[q_1(t)] - \frac{\mathrm{i}}{\hbar}S_0[q_2(t)]\right) F\,[q_1(t), q_2(t)], \qquad (1.16)$$

where F is the Feynman–Vernon *influence functional*, which describes the total effect of the subsystem Q on the particle q. This functional can, in turn, be represented as a double path integral

$$F[q_1(t), q_2(t)] = \sum_{\mathrm{f}} \int \mathrm{d}Q_{\mathrm{i}} \int \mathrm{d}Q_i' \int_{Q_{\mathrm{i}}}^{Q_{\mathrm{f}}} \mathrm{D}Q_1 \int_{Q_i'}^{Q_{\mathrm{f}}'} \mathrm{D}Q_2 \tilde{\rho}(Q_{\mathrm{i}}, Q_i')$$
$$\times \exp\left(\frac{\mathrm{i}}{\hbar} S_{\mathrm{env}}[Q_1] - \frac{\mathrm{i}}{\hbar} S_{\mathrm{env}}[Q_2] + \frac{\mathrm{i}}{\hbar} S_{\mathrm{int}}[q_1, Q_1] - \frac{\mathrm{i}}{\hbar} S_{\mathrm{int}}[q_2, Q_2]\right), \tag{1.17}$$

where $\sum_{\mathrm{f}}$ is equivalent to the integral

$$\sum_{\mathrm{f}} \to \int \mathrm{d}Q_{\mathrm{f}} \int \mathrm{d}Q_{\mathrm{f}}' \delta(Q_{\mathrm{f}} - Q_{\mathrm{f}}') \tag{1.18}$$

implying that after one is no longer interested in the interaction effects, one should set $Q_{\mathrm{f}} = Q_{\mathrm{f}}'$ and integrate over all values of the coordinate Q_{f} of the environment.

We would like to emphasize that no approximations have been done so far, i.e., the above equations are *exact*. Following Feynman and Hibbs [118], we conclude that "F contains the entire effect of the environment, including the change in behavior of the environment resulting from reaction with q. In the classical analogue, F would correspond to knowing not only what the force is as a function of time, but also what it would be for every possible motion $q(t)$ of the object. The force for a given environmental system depends, in general, on the motion of q, of course, since the environmental system is affected by interaction with the system of interest q." Thus, all back-action effects, i.e., all changes of the state of the environment resulting from the interaction with a dynamical variable q, are automatically taken into account within the Feynman–Vernon influence functional formalism.

Before turning to concrete examples, let us formulate some important general properties of the influence functionals that follow directly from Eqs. (1.17) and (1.18). An obvious property is

$$F[q_1, q_2] = F^*[q_2, q_1]. \tag{1.19}$$

Representing the influence functional in the form

$$F[q_1(t), q_2(t)] = \exp\left(-\frac{\mathrm{i}}{\hbar} S_{\mathrm{R}}[q_1, q_2] - \frac{1}{\hbar} S_{\mathrm{I}}[q_1, q_2]\right) \tag{1.20}$$

from Eq. (1.19), one finds

$$S_{\mathrm{R}}[q_1, q_2] = -S_{\mathrm{R}}[q_2, q_1], \quad S_{\mathrm{I}}[q_1, q_2] = S_{\mathrm{I}}[q_2, q_1] \tag{1.21}$$

Another important property of the influence functional is

$$F[q_1, q_1] \equiv 1, \tag{1.22}$$

or, equivalently, $S_{\mathrm{R}}[q_1, q_1] = S_{\mathrm{I}}[q_1, q_1] = 0$. This property stems from the fact that for $q_1(t) \equiv q_2(t)$, the influence functional (1.17) just reduces to the total probability for the system Q described by the action $S_{\mathrm{env}}[Q]$ to make a transition from some initial state to *all* possible final states under the influence of a *classical* force generated by the term $S_{\mathrm{int}}[q_1, Q]$, where $q_1(t)$ is any given function of time. Obviously, this total probability identically equals to unity for any quantum system.

If the subsystem q simultaneously interacts with N different environments $Q^{(1)}, \ldots, Q^{(N)}$ and if all these environments are completely uncorrelated, one can evaluate the influence functional $F^{(n)}$ from the n-th environment disregarding all other $Q^{(m\neq n)}$ and then obtain the total influence functional F as a simple product of influence functionals $F^{(n)}$ from all N environments, i.e.,

$$F = \prod_{n=1}^{N} F^{(n)} \tag{1.23}$$

It is also obvious from our definition of the influence functional that physically different environments described by identical influence functionals produce exactly the same effect on the subsystem q, i.e., these environments are indistinguishable as far as quantum dynamics of the particle q is concerned.

There are also other general requirements that should be obeyed by any influence functional. *Causality* is among the most important of them. In addition, the *fluctuation–dissipation theorem* (FDT) should be justified by an influence functional provided the corresponding environment was in its thermodynamic equilibrium state at the initial moment $t = 0$ when the interaction between q and Q was turned on.

All the above general properties of the influence functionals will be illustrated in our subsequent analysis for a number of specific physical examples. Before turning to such examples, we mention that the prominent role is played by *Gaussian* influence functionals, which in many cases provide sufficiently accurate description of environmental effects. Gaussian influence functionals are exact provided the environment consists of an arbitrary number of harmonic oscillators coupled linearly to the system of interest q. These functionals are also obtained if in the course of its evolution, the coordinate q does not change much as compared to some relevant scale, and one can expand both $S_R[q_1, q_2]$ and $S_I[q_1, q_2]$ in powers of q keeping only the first- and the second-order terms. The corresponding examples will be provided and explicitly analyzed further below.

Consider the general exponential functional

$$F[q_1(t), q_2(t)] = \exp(-S^{(1)}[q_1(t), q_2(t)]/\hbar - S^{(2)}[q_1(t), q_2(t)]/\hbar) \tag{1.24}$$

with, respectively, linear and quadratic in $q_1(t)$ and $q_2(t)$ functionals in the exponent:

$$S^{(1)}[q_1(t), q_2(t)] = \mathrm{i}\int dt(a(t)q_1(t) + b(t)q_2(t)) \tag{1.25}$$

and

$$S^{(2)}[q_1(t), q_2(t)] = \int dt \int^t dt'[\alpha(t, t')q_1(t)q_1(t') + \beta(t, t')q_1(t)q_2(t') + \gamma(t, t')q_2(t)q_1(t') + \delta(t, t')q_2(t)q_2(t')] \tag{1.26}$$

where a and b are arbitrary complex functions of t while α, β, γ, and δ are arbitrary complex functions of t and t'. We also note that the functions α, β, γ, and δ may differ from zero only for $t > t'$ and vanish identically otherwise. This fact just illustrates the requirement of causality: The state of the particle q at time t may only be affected by quantum dynamics of the environment Q at times $t' < t$.

With the aid of the property (1.19), we immediately obtain

$$b(t) = -a^*(t), \quad \delta(t, t') = \alpha^*(t, t'), \quad \gamma(t, t') = \beta^*(t, t') \tag{1.27}$$

These equations allow to exclude the functions $b(t)$, $\gamma(t, t')$, and $\delta(t, t')$ and express the influence functional (1.24) only in terms of the remaining three functions $a(t)$, $\alpha(t, t')$, and $\beta(t, t')$. Setting $q_1(t) = q_2(t)$ and $q_1(t') = q_2(t')$ in the resulting expression for F

and making use of the property (1.22), we find that the function $a(t)$ is purely real, i.e., $a(t) = a^*(t)$, and

$$\beta(t, t') = -\alpha(t, t') \tag{1.28}$$

Introducing real and imaginary parts of the function $\alpha(t, t') = \mathrm{Re}\alpha(t, t') + \mathrm{i}\mathrm{Im}\alpha(t, t')$ with the aid of Eq. (1.28), we finally arrive at the influence functional in the form (1.20) with

$$S_{\mathrm{R}}[q_1, q_2] = \int dt q_-(t) \left[a(t) + 2 \int^t dt' \mathrm{Im}\alpha(t, t') \right] q_+(t') \tag{1.29}$$

$$S_{\mathrm{I}}[q_1, q_2] = \int dt \int^t dt' q_-(t) \mathrm{Re}\alpha(t, t') q_-(t') \tag{1.30}$$

where, as before, we define

$$q_-(t) = q_1(t) - q_2(t) \text{ and } q_+(t) = (q_1(t) + q_2(t))/2.$$

Equations (1.20), (1.29), and (1.30) define the most general Gaussian influence functional. As we will see below, each of the terms in these expressions has its own physical meaning: The function $a(t)$ plays the role of a classical time-dependent force acting on the particle q while the terms containing $\mathrm{Re}\alpha(t, t')$ and $\mathrm{Im}\alpha(t, t')$ are responsible, respectively, for noise and dissipation due to interaction between the particle and the environment. Provided the average properties of the latter do not depend on time, the function α depends only on the time difference $t - t'$, i.e.,

$$\alpha(t, t') = \alpha(t - t') \tag{1.31}$$

If, on top of it, the environment remains in thermodynamic equilibrium, the function α obeys the condition

$$\mathrm{i}(\mathrm{Re}\alpha)_\omega = \coth\left(\frac{\hbar\omega}{2T}\right) (\mathrm{Im}\alpha)_\omega \tag{1.32}$$

which is a direct consequence of FDT.

1.3 Influence Functional from a Harmonic Oscillator

Let us assume that the environment consists of a single harmonic oscillator Q, which has a unit mass and a frequency ω. Then for the action $S_{\mathrm{env}}[Q]$, we have

$$S_{\mathrm{env}}[Q] = \int dt \left(\frac{1}{2} \dot{Q}^2 - \frac{\omega^2}{2} Q^2 \right) \tag{1.33}$$

For simplicity, the interaction is chosen bilinear with respect to both the particle and the oscillator coordinates q and Q, i.e.,

$$S_{\text{int}}[q, Q] = c \int dt\,(qQ) \tag{1.34}$$

where c is a constant that governs the strength of interaction. We will also assume that the oscillator is initially kept in thermodynamic equilibrium at temperature T, i.e., the probability to occupy its k-th energy level $E_k = \hbar\omega(k+1/2)$ is defined by the standard Boltzmann factor

$$w_k = \exp(-k\hbar\omega/T)(1 - \exp(-\hbar\omega/T)) \tag{1.35}$$

and the oscillator initial density matrix can be chosen as

$$\tilde{\rho}(Q_{\text{i}}, Q_i') = \sum_{k=0}^{\infty} w_k \psi_k(Q_{\text{i}})\psi_k^*(Q_{\text{i}}') \tag{1.36}$$

where $\psi_k(Q)$ represents the full set of the oscillator eigenfunctions

$$\psi_k(Q) = (2^k k!)\left(\frac{\omega}{\pi\hbar}\right)^{1/4} H_k\left(Q\sqrt{\frac{\omega}{\hbar}}\right)\exp\left(-\frac{\omega Q^2}{2\hbar}\right) \tag{1.37}$$

and $H_k(x) = (-1)^k e^{x^2} d^k e^{x^2}/dx^k$ are the Hermite polynomials. These expressions yield the equilibrium density matrix for a harmonic oscillator in the following well-known form

$$\tilde{\rho}(Q_{\text{i}}, Q_i') = \sqrt{\frac{\omega}{2\pi\hbar\sinh(\hbar\omega/T)}} \tag{1.38}$$
$$\times \exp\left[-\frac{\omega}{2\hbar\sinh(\hbar\omega/T)}\left((Q_{\text{i}}^2 + (Q_{\text{i}}')^2)\cosh\left(\frac{\hbar\omega}{T}\right) - 2Q_{\text{i}}Q_i'\right)\right]$$

Turning to the influence functional, with the aid of its definition (1.17), we write

$$F = \int dQ_{\text{f}} \int dQ_{\text{i}} \int dQ_i' K_1(Q_{\text{f}}, t; Q_{\text{i}}, 0) K_2^*(Q_{\text{f}}, t; Q_{\text{i}}, 0)\tilde{\rho}(Q_{\text{i}}, Q_i') \tag{1.39}$$

where $K_{1(2)}$ represents the quantum mechanical amplitude for an oscillator Q to pass from the initial state with coordinate q_{i} to the final state with coordinate Q_{f} during time t under the influence of external time-dependent force

$$f_{1(2)}(t) = cq_{1(2)}(t) \tag{1.40}$$

acting on the oscillator coordinate Q, i.e.,

$$K_{1,2}(Q_{\mathrm{f}}, t; Q_{\mathrm{i}}, 0) = \int_{Q_{\mathrm{i}}}^{Q_{\mathrm{f}}} \mathrm{D}Q \exp\left(\frac{\mathrm{i}}{\hbar} \int dt \left(\frac{1}{2}\dot{Q}^2 - \frac{\omega^2}{2} Q^2 + cQq_{1,2}\right)\right) \tag{1.41}$$

The path integral (1.41) is Gaussian and, hence, can be performed directly with the result

$$K_{1,2}(Q_{\mathrm{f}}, t; Q_{\mathrm{i}}, 0) = \sqrt{\frac{\omega}{2\pi \mathrm{i}\hbar \sin \omega t}} \exp\left(\frac{\mathrm{i}}{\hbar} S[q_{1,2}(t)]\right) \tag{1.42}$$

where

$$\begin{aligned} S[q(t)] = \frac{\omega}{2 \sin \omega t} \Bigg[& (Q_{\mathrm{f}}^2 + Q_{\mathrm{i}}^2) \cos \omega t - 2Q_{\mathrm{f}}Q_{\mathrm{i}} \\ & + \frac{2c}{\omega} \int_0^t dt' q(t')(Q_{\mathrm{f}} \sin \omega t + Q_{\mathrm{i}} \sin \omega(t - t')) \\ & - \frac{2c^2}{\omega^2} \int_0^t dt' \int_0^{t'} dt'' q(t') q(t'') \sin \omega t'' \sin \omega(t - t') \Bigg]. \end{aligned} \tag{1.43}$$

What remains is to substitute Eqs. (1.38), (1.42), and (1.43) into Eq. (1.39) and to perform Gaussian integration over initial and final oscillator coordinates $Q_{i,f}$. Then one immediately arrives at the influence functional in the form (1.20) with

$$S_R = \frac{c^2}{\omega} \int_0^t dt' \int_0^{t'} dt'' q_-(t') \sin(\omega(t' - t'')) q_+(t'') \tag{1.44}$$

$$S_I = \frac{c^2}{2\omega} \coth\left(\frac{\hbar\omega}{2T}\right) \int_0^t dt' \int_0^{t'} dt'' q_-(t') \cos(\omega(t' - t'')) q_-(t'') \tag{1.45}$$

where we defined the "center-of-mass" coordinate $q_+ = (q_1 + q_2)/2$ as well as the coordinate difference $q_- = q_1 - q_2$. Equations (1.20)–(1.45) summarize the complete effect of interaction with the oscillator Q on quantum dynamics of the particle q.

We also note that the same result for the influence functional from a harmonic oscillator can also be derived in a slightly different manner. Notice that the functional (1.39) can be identically rewritten in the form

$$F = \sum_{k,m=0}^{\infty} w_k \langle k|\hat{U}_1|m\rangle \langle m|\hat{U}_2^+|k\rangle \tag{1.46}$$

where $\hat{U}_{1(2)}$ is the evolution operator for a harmonic oscillator subject to external force (1.40) and $\langle k|\hat{U}_{1(2)}|m\rangle$ is the transition matrix element between the harmonic oscillator eigenstates $|k\rangle$ and $|m\rangle$ in the presence of the force (1.40). The expression for this matrix element can be taken from the textbooks on quantum mechanics. It reads

$$\langle k|\hat{U}_1|m\rangle = \frac{\langle 0|\hat{U}_1|0\rangle}{\sqrt{m!k!}} \sum_{p=0}^{\max(k,m)} \frac{1}{p!} \frac{(\mathrm{i} f_{1\omega})^{k-p} k!}{(k-p)!} \frac{(\mathrm{i} f_{1\omega}^{*})^{m-p} m!}{(m-p)!} \tag{1.47}$$

where

$$\langle 0|\hat{U}_1|0\rangle = \exp\left[-\frac{1}{2\omega\hbar}\int_0^t dt' \int_0^{t'} dt'' f_1(t') f_1(t'') e^{-\mathrm{i}\omega(t'-t'')}\right] \tag{1.48}$$

is the ground-state matrix element and

$$f_{1\omega} = \frac{1}{\sqrt{2\omega}}\int_0^t dt' f_1(t') e^{-\mathrm{i}\omega t'} \tag{1.49}$$

Plugging Eq. (1.47) together with a similar expression for $\langle m|\hat{U}_2^{+}|k\rangle$ into Eq. (1.46) and performing somewhat tedious summation over k and m with equilibrium occupation probability w_k (1.35) of initial oscillator states, we arrive at the final expression for the influence functional

$$F = \langle 0|\hat{U}_1|0\rangle\langle 0|\hat{U}_2^{+}|0\rangle \exp\left[f_{1\omega}^{*} f_{2\omega} - \frac{(f_{1\omega} - f_{2\omega})(f_{1\omega}^{*} - f_{2\omega}^{*})}{e^{\hbar\omega/T} - 1}\right] \tag{1.50}$$

It is easy to verify that this result is equivalent to Eqs. (1.20), (1.44), and (1.45).

1.4 Caldeira–Leggett Model

1.4.1 Influence Functional

Now let us make one more step and assume that the bath is formed by many—rather than one—harmonic oscillators that do not interact with each other but only with the particle q. As before, for simplicity we will choose this interaction in the bilinear form in both particle and oscillator coordinates. In this case, the environment and interaction contributions to the Hamiltonian can be written in the form

$$\hat{H}_{\text{env}} + \hat{H}_{\text{int}} = \sum_n \left[\frac{\hat{P}_n^2}{2M_n} + \frac{M_n\omega_n^2}{2}\left(Q_n - \frac{c_n q}{M_n\omega_n^2}\right)^2 \right] \tag{1.51}$$

where we included the standard counter term $\sum_n c_n^2 q^2/(2M_n\omega_n^2)$ in order to compensate for unphysical renormalization of the particle Hamiltonian $\hat{H}_0$ originating from a specific form of the interaction between q and the bath of oscillators.

Assuming that initially all the bath oscillators were in thermodynamic equilibrium at temperature T and proceeding in exactly the same way as in the previous section, we establish the influence functionals $F^{(n)}$ for each of the oscillators separately, which again take the form defined in Eqs. (1.20), (1.44), and (1.45). Then with the aid of the rule (1.23), we evaluate the influence functional for the whole bath as a product of the influence functionals for each of the oscillators. In this way, we arrive at Eqs. (1.20), (1.29), and (1.30) with $a(t) = 0$ and

$$\text{Re}\alpha(t) = \sum_n \frac{c_n^2}{2M_n\omega_n} \coth\left(\frac{\hbar\omega_n}{2T}\right) \cos\omega_n t, \tag{1.52}$$

$$\text{Im}\alpha(t) = \sum_n \frac{c_n^2}{2M_n\omega_n} \sin\omega_n t \tag{1.53}$$

What remains is to specify the frequency spectrum of our bath of oscillators. For the reasons which will be clear soon, we will choose

$$\frac{\pi}{2}\sum_n \frac{c_n^2}{M_n\omega_n}\delta(\omega - \omega_n) = \eta\omega, \quad 0 \le \omega \le \omega_c \tag{1.54}$$

i.e., we assume *linear* in frequency spectrum of bath oscillators at frequencies up to ω_c. It is also implied that there exist no oscillators with frequencies $\omega > \omega_c$ in our environment, i.e., ω_c plays the role of a high-frequency cutoff in our problem. Equations (1.51) and (1.54) define the celebrated *Caldeira–Leggett model* for a bath of oscillators [80, 81].

The influence functional from this Caldeira–Leggett bath is obtained by combining Eqs. (1.54) with Eqs. (1.52) and (1.53). One easily finds

$$\mathrm{Re}\alpha(t) = \eta \int \frac{d\omega}{2\pi} \omega \coth\left(\frac{\hbar\omega}{2T}\right) e^{-\mathrm{i}\omega t}, \tag{1.55}$$

$$\mathrm{Im}\alpha(t) = \mathrm{i}\eta \int \frac{d\omega}{2\pi} \omega e^{\mathrm{i}\omega t} \tag{1.56}$$

It is obvious that the latter function is proportional to the derivative of the δ-function, $\mathrm{Im}\alpha(t) = \eta\delta'(t)$.

Equations (1.55) and (1.56)—together with Eq. (1.20)—define the influence functional for the Caldeira–Leggett bath of oscillators, which exactly accounts for the total impact of the environment on quantum dynamics of the particle q. In agreement with our general considerations, this impact is twofold. First, interaction of the particle with oscillators leads to dissipation of energy. This effect is described by the term S_{R}, which depends on $\mathrm{Re}\alpha$ (1.55). Second, in addition to a dissipative force, fluctuations in the environment produce stochastic force acting on the particle q. This noise term is accounted for by the term S_{I} containing the function $\mathrm{Im}\alpha$ (1.56). As we already discussed, these two terms are not independent. Provided initially the environment was in thermodynamic equilibrium at temperature T, as it is the case here, the Fourier components $(\mathrm{Re}\alpha)_\omega$ and $(\mathrm{Im}\alpha)_\omega$ are linked to each other by Eq. (1.32), which is just a consequence of FDT. For the model in question, this FDT relation is directly verified from Eqs. (1.55) and (1.56).

We also remark that as soon as we integrate out the bath degrees of freedom Q and turn to the description in terms of the influence functional, we obviously loose some information about one part of our system (the bath). Hence, after this stage, our system is no longer described by any Hamiltonian. Time irreversibility and dissipation appear exactly at this point. On the other hand, no information is lost about the subsystem of interest, q. All expectation values of all quantum operators related to q are now evaluated with the aid of reduced density matrix $\rho(q, q')$ and the results exactly coincide with those that could be obtained by solving the Schrödinger equation for the total system "particle+bath." This is an important advantage of the method, since from the very beginning, one does not need to deal with excess information about a large number of quantum variables though retaining full information about their influence on the variable of interest q.

1.4.2 Quasiclassical Langevin Equation

Let us describe the particle q by means of the standard Hamiltonian

$$\hat{H}_0 = \frac{\hat{p}^2}{2m} + U(q) \tag{1.57}$$

As before, representing the kernel J for the product of two evolution operators (1.13) in terms of the path integral, after integration over the bath degrees of freedom, we arrive at Eq. (1.16) where now

$$S_0[q] = \int_0^t dt' \left(\frac{m\dot{q}^2}{2} - U(q) \right) \tag{1.58}$$

defines the action of a non-interacting particle and the influence functional F accounts for the effect of interaction between the particle and the Caldeira–Leggett bath. Integrating the kinetic energy term in Eq. (1.58) by parts and introducing again the coordinates $q_\pm$, we obtain

$$J = \int_{q_{+i}}^{q_{+f}} \mathrm{D}q_+ \int_{q_{-i}}^{q_{-f}} \mathrm{D}q_- \exp\left(\frac{\mathrm{i}}{\hbar} \tilde{S}_0[q_+, q_-] - \frac{\mathrm{i}}{\hbar} S_\mathrm{R}[q_+, q_-] - \frac{1}{\hbar} S_\mathrm{I}[q_-] \right) \tag{1.59}$$

where we introduced

$$q_{-i(f)} = q_{i(f)} - q'_{i(f)}, \quad q_{+i(f)} = (q_{i(f)} + q'_{i(f)})/2 \tag{1.60}$$

and defined

$$\begin{aligned} \tilde{S}_0[q_+, q_-] &\equiv S_0[q_1] - S_0[q_2] \\ &= -\int_0^t dt' \, [mq_-\ddot{q}_+ + U(q_+ + q_-/2) - U(q_+ + q_-/2)] \end{aligned} \tag{1.61}$$

and

$$S_\mathrm{R} = \eta \int_0^t dt' q_- \dot{q}_+, \quad S_\mathrm{I} = \frac{1}{2} \int_0^t dt' \int_0^t dt'' q_-(t') \mathrm{Re}\alpha(t' - t'') q_-(t'') \tag{1.62}$$

Performing the Hubbard–Stratonovich transformation of the term with S_I in (1.59), one can identically rewrite Eq. (1.58) in the form

$$J = \int \mathrm{D}\xi \, W[\xi] \int_{q_{+i}}^{q_{+f}} \mathrm{D}q_+ \int_{q_{-i}}^{q_{-f}} \mathrm{D}q_- \exp(\mathrm{i}S[q_+, q_-, \xi]) \tag{1.63}$$

where the functional $W[\xi]$ is Gaussian,

$$W[\xi] = \exp\left(-\frac{1}{2\hbar}\int dt'\int dt''\xi(t')[\mathrm{Re}\alpha(t'-t'')]^{-1}\xi(t'')\right), \quad (1.64)$$

and the effective action $S[q_+, q_-, \xi]$ is purely real and local in time

$$S[q_+, q_-, \xi] = \int_0^t dt'\Big[q_-(\xi - m\ddot{q}_+ - \eta\dot{q}_+)$$
$$+U\left(q_+ + \frac{q_-}{2}\right) - U\left(q_+ + \frac{q_-}{2}\right)\Big] \quad (1.65)$$

At this stage, our equations are still exact and fully describe all quantum mechanical properties of the system. Under certain conditions, however, it might be reasonable to assume that quantum effects, although present, are rather weak, and one can resort to a quasiclassical approximation. As usually, the latter is constructed by means of finding the saddle point of the action $S[q_+, q_-, \xi]$ (1.65) that dominates the path integral (1.63). This saddle point trajectory is determined by the condition

$$\frac{\delta S[q_+, q_-, \xi]}{\delta q_-}|_{q_-=0} = 0 \quad (1.66)$$

which immediately yields equation of motion for the center-of-mass coordinate q_+

$$m\ddot{q}_+ + \eta\dot{q}_+ + \frac{\partial U(q_+)}{\partial q_+} = \xi(t) \quad (1.67)$$

where ξ is the stochastic force with the correlator

$$\langle\xi(t)\xi(0)\rangle = \hbar\eta\int\frac{d\omega}{2\pi}e^{i\omega t}\omega\coth\left(\frac{\hbar\omega}{2T}\right) \quad (1.68)$$

Equation (1.67) together with Eq. (1.68) defines quasiclassical Langevin equation for a quantum particle interacting with Caldeira–Leggett bath of oscillators. In such a form, this equation was derived by Schmid [348]. Note that the Langevin equation (1.67) becomes exact for parabolic or linear external potentials $U(q)$ or if this potential is absent at all.

As we already discussed in the beginning of this chapter, the Langevin equation defined in Eqs. (1.67) and (1.68) describes both linear Ohmic dissipation and noise produced by the quantum environment. If we identify q_+ with the particle coordinate q and restrict ourselves to the (classical) limit of high temperatures or low frequencies $T \gg \hbar\omega$, Eqs. (1.67) and (1.68) just reduce to the classical Langevin equation (1.2), (1.3). This observation clearly demonstrates that the Caldeira–Leggett influence functional, on one hand, fully contains the standard dissipative equation of motion for a classical particle (1.2) and, on the other hand, goes much further describing both quantum dissipation and quantum noise, which become important at sufficiently low temperatures or high enough frequencies $T \lesssim \hbar\omega$.

The above results clarify the key reason behind our choice of the bath oscillator spectrum in Eq. (1.54): It enables one to reproduce linear Ohmic friction in the classical limit. We also stress that particular physical nature of the bath does not play any important role here. In a way, to employ the Caldeira–Leggett model of a bath of harmonic oscillators can be considered just as a physical trick helping to derive the influence functional in the case of linear Ohmic damping. Regardless its physical nature, any other linear Ohmic environment would yield exactly the same influence functional as one derived above. Various examples illustrating this important property will be considered in the next chapters.

1.5 Perturbation Theory and Keldysh Diagram Technique

1.5.1 Time-Dependent Perturbation Theory

Let us now learn how to apply our influence functional technique to standard quantum mechanical tasks, such as perturbative evaluation of the transition probabilities between the initial $|i\rangle$ and final $|f\rangle$ states of the particle q. Assuming that these are two different (and, hence, orthogonal) eigenstates of the non-interacting Hamiltonian $\hat{H}_0$ with respective energies E_i and E_f, one obviously has

$$|\langle i|i\rangle| = 1, \quad \langle f|i\rangle = 0 \tag{1.69}$$

In the presence of interactions with the bath, the transition matrix element between $|i\rangle$ and $|f\rangle$ already differs from zero because these states are not anymore the eigenstates of the total Hamiltonian (1.8) for the system "particle+bath." In order to evaluate the transition probability $W_{\rm if}$ between these states, one should also take into account all possible simultaneous transitions between the states of the environment. Fortunately, we have already carried out this part of the work above in Section 1.2 in a general form and in Section 1.3 for the particular example of a harmonic oscillator. Thus, in order to accomplish our goal, we can simply make use of Eqs. (1.12) and (1.16).

Taking the initial density matrix in the form $\rho_{\rm i} = |i\rangle\langle i|$ and assuming that interaction is weak, one can expand the influence functional F (1.20), (1.29), and (1.30) to the first order in $\alpha(t', t'')$. Setting the external force equal to zero, $a(t) \equiv 0$, one easily recovers the following contributions to the transition probability during time interval t:

$$\begin{aligned} W_{\rm if}(t) = &\int_0^t dt' \int_0^{t'} dt''\alpha(t', t'')\left[\langle f|q(t'')|i\rangle\langle f|q(t')|i\rangle^* \right.\\ &\left. - \langle f|q(t')q(t'')|i\rangle\langle f|i\rangle^*\right] \\ &+ \int_0^t dt' \int_0^{t'} dt''\alpha^*(t', t'')\left[\langle f|q(t')|i\rangle\langle f|q(t'')|i\rangle^* \right.\\ &\left. - \langle f|i\rangle\langle f|q(t')q(t'')|i\rangle^*\right] \end{aligned} \quad (1.70)$$

Making use of the second Eq. (1.69) and observing that

$$\langle f|q(t)|i\rangle = \langle f|q|i\rangle e^{-{\rm i}(E_{\rm f}-E_{\rm i})t/\hbar} \quad (1.71)$$

from Eq. (1.70) we arrive at the result

$$\begin{aligned} W_{\rm if}(t) = &\, 2|\langle f|q|i\rangle|^2 \int_0^t dt' \int_0^{t'} dt''[{\rm Re}\alpha(t', t'')\cos[(E_{\rm f}-E_{\rm i})(t'-t'')/\hbar] \\ &+ 2|\langle f|q|i\rangle|^2 \int_0^t dt' \int_0^{t'} dt''{\rm Im}\alpha(t', t'')\sin[(E_{\rm f}-E_{\rm i})(t'-t'')/\hbar]] \end{aligned} \quad (1.72)$$

Considering now the case of a single oscillator and combining Eq. (1.72) with

$${\rm Re}\alpha(t) = \frac{c^2}{2\omega}\coth\left(\frac{\hbar\omega}{2T}\right)\cos\omega t, \quad (1.73)$$

$${\rm Im}\alpha(t) = \frac{c^2}{2\omega}\sin\omega t \quad (1.74)$$

in the low temperature limit $kT \ll \hbar\omega$, we obtain

$$W_{\mathrm{if}}(t) = 2c^2|\langle i|q|f\rangle|^2 \frac{\sin^2((E_{\mathrm{f}} - E_{\mathrm{i}} + \hbar\omega)t/2\hbar)}{\omega(E_{\mathrm{f}} - E_{\mathrm{i}} + \hbar\omega)^2} \tag{1.75}$$

This is the standard quantum mechanical formula for the transition probability derived within the time-dependent perturbation theory. The transition rate dW_{if}/dt oscillates in time and eventually approaches the delta-function $\delta(E_{\mathrm{f}} - E_{\mathrm{i}} + \hbar\omega)$ in the long time limit $t \to \infty$. This delta-function just assures energy conservation illustrating that for an infinitely long time interval, the transition between the states $|i\rangle$ and $|f\rangle$ is only possible provided the energy difference $E_{\mathrm{i}} - E_{\mathrm{f}}$ exactly equals to $\hbar\omega$, i.e., the oscillator is excited from its ground state to the upper energy level during this transition.

1.5.2 Influence Functional and Keldysh Diagrams

Path integrals appearing in both quantum mechanics and quantum field theory may be viewed as a formal representation of the exact infinite series of Feynman diagrams (see, e.g., the textbook [334]). In other words, path integral and diagram techniques are formally equivalent to each other. Here we will demonstrate the equivalence between Feynman–Vernon influence functional approach and Keldysh diagram technique [220, 274] developed to describe non-equilibrium phenomena in quantum many-body systems with interactions.

The way to recover all Keldysh diagrams from the path integral expression for the evolution kernel J is to expand the integral in Eq. (1.16) in powers of the coupling constant between the particle and the environment. To begin with, let us consider a free particle and set $F[q_1(t), q_2(t)] = 1$ in Eq. (1.16). The path integral is then factorized into the product of two independent integrals:

$$J^{(0)}(t; q_{\mathrm{f}}, q_{\mathrm{i}}; q_{\mathrm{f}}', q_i') = G^R(t, q_{\mathrm{f}}, q_{\mathrm{i}})G^A(t, q_{\mathrm{f}}', q_i') \tag{1.76}$$

where

$$G^R(t, q_{\mathrm{f}}, q_{\mathrm{i}}) = \int_{q_{\mathrm{i}}}^{q_{\mathrm{f}}} \mathrm{D}q_1 \exp\left(\frac{\mathrm{i}}{\hbar} S_0[q_1(t)]\right), \tag{1.77}$$

$$G^A(t, q_{\mathrm{f}}', q_i') = \int_{q_i'}^{q_{\mathrm{f}}'} \mathrm{D}q_2 \exp\left(-\frac{\mathrm{i}}{\hbar} S_0[q_2(t)]\right) \tag{1.78}$$

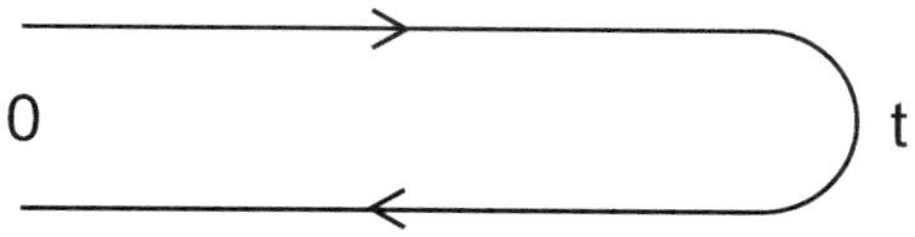

Figure 1.1 Keldysh contour consisting of forward and backward time branches of length t.

represent, respectively, retarded and advanced Green functions for a free particle. These Green functions serve as the main building blocks of the Keldysh diagram technique. In essence, they are just the matrix elements of the free evolution operator, i.e.,

$$G^R(t, q_\mathrm{f}, q_\mathrm{i}) = \langle q_\mathrm{f}|T\,\exp(-\mathrm{i}\hat{H}_0 t/\hbar)|q_\mathrm{i}\rangle$$
$$G^A(t, q'_\mathrm{f}, q'_i) = \langle q'_i|T\,\exp(-\mathrm{i}\hat{H}_0 t/\hbar)|q'_\mathrm{f}\rangle^* \qquad (1.79)$$

If one takes a trace over the final particle states in Eq. (1.76), one can also view the kernel $J^{(0)}$ as a single matrix element

$$\int dq_\mathrm{f} dq'_\mathrm{f} J^{(0)}(t; q_\mathrm{f}, q_\mathrm{i}; q'_\mathrm{f}, q'_i)\delta(q_\mathrm{f} - q'_\mathrm{f}) = \langle q'_i|T_\mathrm{K}\exp(-\mathrm{i}\hat{H}_0 t/\hbar)|q_\mathrm{i}\rangle \qquad (1.80)$$

describing evolution of the particle along the so-called Keldysh contour, which consists of the forward and backward time branches, see Fig. 1.1. Accordingly, one can define the corresponding evolution operator $\hat{U}_\mathrm{K} = T_\mathrm{K}\exp(-\mathrm{i}\hat{H}_0 t/\hbar)$, where T_K is the time-ordering operator along the Keldysh contour.

Note that in this chapter, we only deal with a single degree of freedom (particle), introducing interactions with a bath of harmonic oscillators. Hence, here we do not need to account for any restrictions imposed by Fermi or Bose statistics on the particle wave function. As a result, the definitions of the Green functions (1.77) and (1.78) are simpler than, say, those for electrons in a metal or in any other system where many-body correlations play a significant role. Extensions to the situations where one needs to account, e.g., for the Pauli principle will be considered in the second part of this book.

According to general rules of the diagram technique, the free kernel (1.80) is represented by two disconnected solid lines, which denote the Green functions (1.77) and (1.78), see Fig. 1.2. These two lines also correspond to two branches of the Keldysh contour,

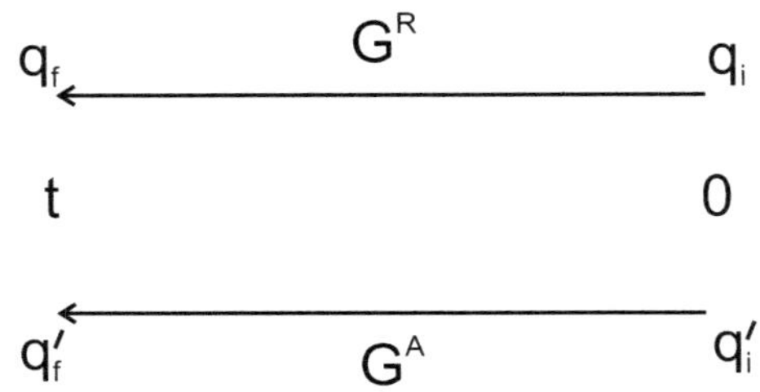

Figure 1.2 Graphical representation of the free evolution kernel (1.76). The Keldysh Green functions (1.77) and (1.78) are shown by solid lines directed from the initial time zero to the final time t.

which can be identified in every diagram in all orders of the perturbation theory. One of these branches represents the forward time evolution of the particle wave function and contains only the retarded particle Green functions G^R, while the other branch is attributed to the backward time evolution accounted for by advanced Green functions G^A.

In order to illustrate a direct relation between the Feynman–Vernon influence functional approach and Keldysh diagram technique, we will take an example of the Caldeira–Leggett model. As we have already observed above, this model leads to the quadratic influence functional of the form (1.24) and (1.26) with the vanishing first-order action, $S^{(1)}[q_1(t), q_2(t)] = 0$. Let us expand the evolution kernel J (1.16) in powers of $S^{(2)}[q_1(t), q_2(t)]$ and retain only the first-order terms in this expansion. Then we arrive at the path integral

$$J = J^{(0)} - \int_{q_\mathrm{i}}^{q_\mathrm{f}} \mathrm{D}q_1 \int_{q_i'}^{q_\mathrm{f}'} \mathrm{D}q_2 \exp\left(\frac{\mathrm{i}}{\hbar} S_0[q_1(t)] - \frac{\mathrm{i}}{\hbar} S_0[q_2(t)]\right)$$
$$\times\left(\int_0^t dt' \int_0^{t'} dt'' \left[\alpha(t', t'')q_1(t')q_1(t'') - \alpha(t', t'')q_1(t')q_2(t'')\right.\right.$$
$$\left.\left. -\alpha^*(t', t'')q_2(t')q_1(t'') + \alpha^*(t', t'')q_2(t')q_2(t'')\right]\right) \quad (1.81)$$

Thus, within the first-order perturbation theory, we obtain four contributions containing the products of coordinates $q_1(t')q_1(t'')$, $q_1(t')q_2(t'')$, $q_2(t')q_1(t'')$, and $q_2(t')q_2(t'')$, respectively.

Let us consider the first of these contributions in more details. As an intermediate step, we need to evaluate the integral

$$\tilde{J}^{(1)}(t', t'') = \int_{q_i}^{q_f} Dq_1 \int_{q_i'}^{q_f'} Dq_2 \times q_1(t')q_1(t'') \exp\left(\frac{i}{\hbar}S_0[q_1(t)] - \frac{i}{\hbar}S_0[q_2(t)]\right) \quad (1.82)$$

At this stage, we make use of the fact that the free particle action $S_0[q_1(t)]$ is local in time. This property allows us to split it into three parts:

$$\begin{aligned} S_0[q_1(t)] &= \int_0^t dt \left[\frac{m\dot{q}_1^2}{2} - U(q_1)\right] \\ &= \int_{t'}^t dt \left[\frac{m\dot{q}_1^2}{2} - U(q_1)\right] + \int_{t''}^{t'} dt \left[\frac{m\dot{q}_1^2}{2} - U(q_1)\right] \\ &\quad + \int_0^{t''} dt \left[\frac{m\dot{q}_1^2}{2} - U(q_1)\right] \end{aligned} \quad (1.83)$$

Then the path integrals over the coordinate q_1 within three time intervals $[t', t]$, $[t'', t']$, and $[0, t']$ may be evaluated independently, resulting in the product of three retarded Green functions. The path integrals over intermediate coordinates $q_1(t')$ and $q_1(t'')$ are then reduced to conventional integrals over coordinates q' and q'', which replace them. Finally, the path integral over $q_2(t)$ gives the advanced Green function. Thus, the integral (1.82) takes the form

$$\tilde{J}^{(1)}(t', t'') = \int dq'dq'' G^R(t - t', q_f, q')\, q'\, G^R(t' - t'', q', q'') \times q''\, G^R(t'', q'', q_i)\, G^A(t, q_f', q_i') \quad (1.84)$$

The remaining integrals in the expression (1.81) are evaluated analogously and the kernel J acquires the form

$$J = J^{(0)} + \delta J_1 + \delta J_2 + \delta J_3 + \delta J_4, \quad (1.85)$$

where

$$\delta J_1 = -\int_0^t dt' \int_0^{t'} dt'' \int dq'dq''\, \alpha(t', t'')\, G^R(t - t', q_f, q') \times q'\, G^R(t' - t'', q', q'')\, q''\, G^R(t'', q'', q_i)\, G^A(t, q_f', q_i') \quad (1.86)$$

$$\delta J_2 = \int_0^t dt' \int_0^{t'} dt'' \int dq'dq''\, \alpha(t', t'')\, G^R(t - t', q_f, q') \times q'\, G^R(t', q', q_i)\, G^A(t - t'', q_f', q'')\, q''\, G^A(t'', q'', q_i') \quad (1.87)$$

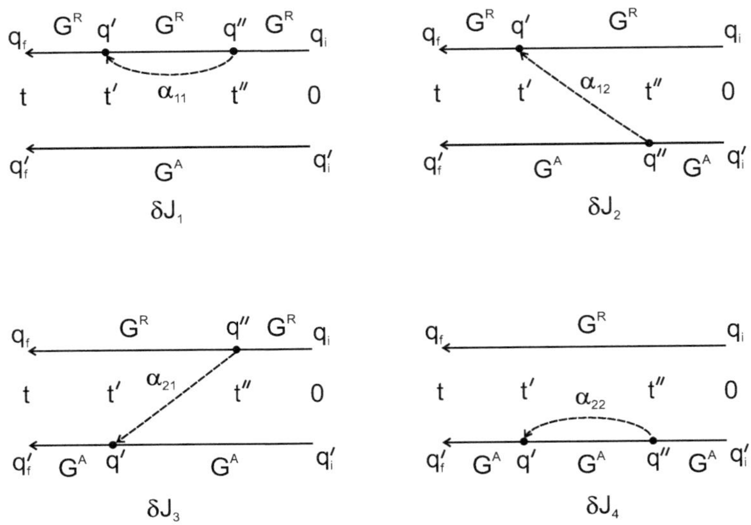

Figure 1.3 Graphical representation of the first-order corrections to the evolution kernel J. Each diagram corresponds to one of the integrals δJ_1, δJ_2, δJ_3, and δJ_4 defined in Eqs. (1.86–1.89).

$$\delta J_3 = \int_0^t dt' \int_0^{t'} dt'' \int dq' dq'' \, \alpha^*(t', t'') \, G^R(t - t'', q_f, q'') \times q'' \, G^R(t'', q'', q_i) \; G^A(t - t', q_f', q') \, q' \, G^A(t', q', q_i') \quad (1.88)$$

$$\delta J_4 = -\int_0^t dt' \int_0^{t'} dt'' \int dq' dq'' \, \alpha^*(t', t'') \, G^R(t, q_f, q_i) \times G^A(t-t', q_f', q') \, q' \, G^A(t'-t'', q', q'') \, q'' \, G^A(t'', q'', q_i) \quad (1.89)$$

Similarly to the zeroth-order contribution to the evolution kernel $J^{(0)}$, the four integrals (1.86)–(1.89) can be illustrated in the form of the diagrams, see Fig. 1.3. As before, the Green functions G^R and G^A are represented by solid lines, while dashed lines stand for the elements of the matrix Keldysh Green function of the environment

$$\alpha_{ij}(t', t'') = \begin{pmatrix} -\alpha(t', t'') & \alpha(t', t'') \\ \alpha^*(t', t'') & -\alpha^*(t', t'') \end{pmatrix}, \quad j = 1, 2 \qquad (1.90)$$

In other words, dashed lines describe interactions, which in the diagrammatic language can be viewed as the process of emitting

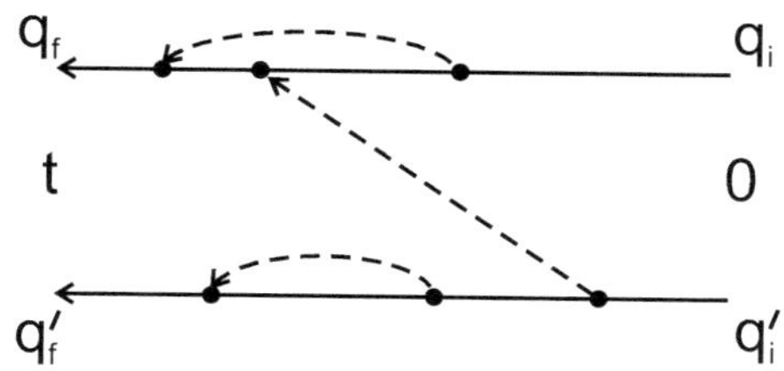

Figure 1.4 Higher-order diagram.

and absorbing a "photon," which in our particular example is just an excitation in the Caldeira–Leggett environment.

The diagrams δJ_1 and δJ_4 are often called "self-energy diagrams." They contain the corrections to the functions G^R and G^A and do not violate the factorized form of the zeroth-order kernel (1.76). The diagrams δJ_2, δJ_3 describe mixing between the two branches of the Keldysh contour and are usually called "vertex diagrams."

Expanding the evolution kernel J in powers of $\alpha(t', t'')$ to higher orders, one encounters a quickly growing number of diagrams of increasing complexity (see Fig. 1.4 for an example). In this way, one recovers all Keldysh diagrams for the problem in question. This observation illustrates the equivalence between Feynman–Vernon and Keldysh approaches to investigation of real-time dynamics of interacting quantum systems.

The main drawback of the diagram technique lies in enormous technical difficulties in handling higher orders of the perturbation series, which usually cannot be evaluated in a closed form. As a consequence, in the vast majority of cases, non-equilibrium quantum dynamics of interacting systems can be treated diagrammatically only within the lowest nontrivial order of the perturbation theory, e.g, taking into account the diagrams depicted in Fig. 1.3. This approach is convenient provided such diagrams are sufficient to capture all essential physics of the problem under consideration. If, however, non-perturbative analysis is required, the diagram technique might be of little help. On the contrary, path integral formulation frequently allows for such non-perturbative analysis. This analysis can further be simplified within the Feynman–Vernon influence functional technique, which allows to trace out all but one interacting degrees of freedom, thereby reducing the effect of interactions to that of some effective environment. In certain classes of problems, this step allows one to perform calculations in a strongly non-perturbative regime, which would be technically impossible to probe with the aid of diagrams.

1.6 Caldeira–Leggett Action in Imaginary Time

Provided the system is in (or close to) thermodynamic equilibrium at some temperature T, instead of real-time analysis, it is frequently more convenient to employ the imaginary time Matsubara technique. The equilibrium density matrix $\rho(q_1, q_2; Q_1, Q_2)$ for the total system "particle+bath" can be expressed as a path integral

$$\rho(q_1, q_2; Q_1, Q_2) = \int_{q_1}^{q_2} \mathrm{D}q \int_{Q_1}^{Q_2} \mathrm{D}Q \exp\left(-\frac{1}{\hbar}\int_0^{\hbar\beta} d\tau\, H[q, Q]\right) \tag{1.91}$$

where, as before, $H[q, Q]$ is the total Hamiltonian of our system (1.8) and $\beta = 1/T$ is the inverse temperature. The double path integral in Eq. (1.91) essentially represents the imaginary time transition matrix element between the system states with coordinates $q_1; Q_1$ and $q_2; Q_2$. Of interest for us is also the grand partition function

$$\mathcal{Z} = \mathrm{Sp}e^{-\beta\hat{H}} = \int dq_0 \int dQ_0 \rho(q_0, q_0; Q_0, Q_0) \tag{1.92}$$

The basic idea is again to get rid of the bath degrees of freedom Q by integrating them out and at the same time to keep their full influence on the quantum variable q. For the Caldeira–Leggett bath of harmonic oscillators, this integration over Q is Gaussian and, hence, can be handled exactly. Substituting the Hamiltonian defined in Eqs. (1.8), (1.51), and (1.57) into Eqs. (1.91) and (1.92) after performing Gaussian integrations over $Q(\tau)$ and Q_0, we arrive at the grand partition function in the form

$$\mathcal{Z} = \int_{q_0}^{q_0} \mathrm{D}q \exp\left(-S_{\mathrm{eff}}[q(\tau)]/\hbar\right) \tag{1.93}$$

where

$$S_{\text{eff}} = \int_0^{\hbar\beta} d\tau \left[\frac{m\dot{q}^2}{2} + U(q)\right] + \frac{\eta}{4\pi}\int_0^{\hbar\beta} d\tau \int_0^{\hbar\beta} d\tau' \frac{(\pi/\hbar\beta)^2 (q(\tau) - q(\tau'))^2}{\sin^2(\pi(\tau - \tau')/\hbar\beta)} \tag{1.94}$$

is the imaginary time effective action for a quantum particle with linear Ohmic dissipation [80]. The first two terms in this equation represent the non-interacting particle action and the third, non-local in time, term defines the imaginary time version of the influence functional for the Caldeira–Leggett model. This term has a particularly simple form in the Fourier representation:

$$\frac{\eta}{2}\sum_{\omega_n} |\omega_n||q_{\omega_n}|^2, \tag{1.95}$$

where $\omega_n = 2\pi n/\hbar\beta$, $n = 0, \pm 1, \pm 2, \ldots$ is the Matsubara frequency. By varying this dissipative contribution to the action with respect to q, we obtain the term $|\omega_n| q_{\omega_n}$, which just coincides with analytic continuation of the dissipative term $-i\omega q_\omega$ in the particle equation of motion (1.67). The action (1.94) will play a prominent role in our future considerations.

Chapter 2

Dissipative Quantum Mechanics of Superconducting Junctions

In this chapter, we will further develop Feynman–Vernon idea of the influence functional applying it to relatively simple, but very important systems–tunnel junctions between two bulk normal metals or superconductors. Such junctions (usually called Josephson junctions in the case of superconducting electrodes) serve as key building blocks of almost every mesoscopic or nanoscale device. These systems are, in many respects, different from those considered in the previous chapter. At the same time, also here we are dealing with quantum dissipative dynamics of one degree of freedom—the Josephson phase—interacting with an effective bath of electrons in metallic conductors. Interestingly, this Josephson phase variable does not even exist in the initial microscopic Hamiltonian of the system; it emerges "from interaction" as a collective variable describing the current across the junction. The influence functional approach works efficiently also in this case: By tracing out all other degrees of freedom (electrons), one can formulate a relatively simple and compact effective action for the system under consideration, which in many important cases turns out to be similar (or even exactly identical) to the Caldeira–Leggett

Dissipative Quantum Mechanics of Nanostructures: Electron Transport, Fluctuations, and Interactions
Andrei D. Zaikin and Dmitry S. Golubev

ISBN 978-981-4774-50-5 (Hardcover), 978-0-429-29823-3 (eBook)
www.jennystanford.com

action already derived in Chapter 1. This observation manifests again the generality of the influence functional technique enabling one to describe complicated problems with physically different types of interactions within the same theoretical framework.

The effective action for a tunnel junction was initially derived from a microscopic theory by Ambegaokar, Eckern, and Schön (AES) [15, 107] and then extensively discussed in a review paper by Schön and one of the present authors [345]. Despite its relative simplicity, the AES action successfully describes a wide range of different phenomena, such as macroscopic quantum tunneling (MQT) and macroscopic quantum coherence (MQC) in Josephson junctions and SQUIDs, Coulomb blockade of single electron, and Cooper pair tunneling. Due to tremendous progress in technology during the last decades, all these phenomena became the subject of increasingly intensive experimental and theoretical studies. They will also be addressed in details in the next several chapters of this book.

More recently, the interest in low-temperature transport properties of ultrasmall tunnel junctions was boosted even further owing to the emergence of a new field in meso- and nanophysics: the so-called quantum computing. Josephson junctions are considered possible building blocks for future quantum computers since, on one hand, they demonstrate quantum mechanical behavior on a macroscopic scale and, on the other hand, they are much easier to control than, say, atoms or molecules. Under certain conditions, a Josephson junction may be reduced to a two-level quantum mechanical system, which may exhibit the whole range of quantum interference effects like "Schrödinger cat" states and Rabi oscillations. While potential applications of Josephson junctions in quantum computing may still be a matter of distant future, all these macroscopic quantum effects have already been successfully demonstrated in real experiments.

In this chapter, we will display and discuss the derivation of the effective action for a Josephson tunnel junction. This derivation will also be helpful for us later when we address quantum properties of more complicated systems.

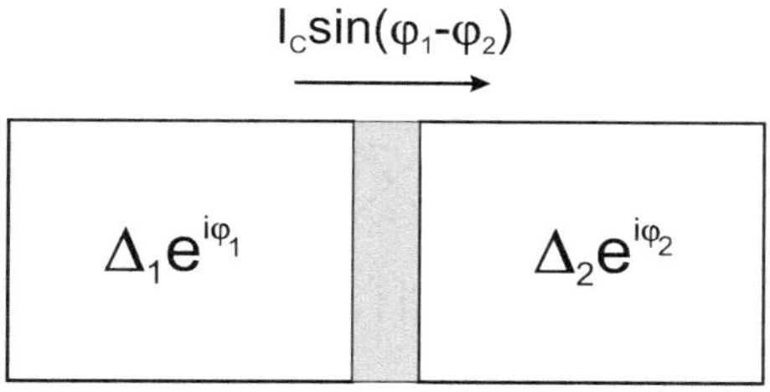

Figure 2.1 Josephson junction, i.e., a tunnel junction between two bulk superconductors with the order parameters $\Delta_1 e^{i\varphi_1}$ and $\Delta_2 e^{i\varphi_2}$.

2.1 From Classical Dynamics to Quantum Mechanics: Phenomenology

We begin with a brief discussion of classical dynamics of Josephson junctions. Consider two bulk superconducting electrodes with the superconducting order parameters $\Delta_1 e^{i\varphi_1}$ and $\Delta_2 e^{i\varphi_2}$ coupled to each other via a tunnel junction (see Fig. 2.1). The current I and the voltage V across the junction are linked to the phase difference $\varphi = \varphi_1 - \varphi_2$ by means of the celebrated Josephson relations

$$I = I_C \sin\varphi, \qquad V = \hbar\dot{\varphi}/2e \tag{2.1}$$

Here I_C is the critical current of the junction, i.e., the maximum value of the supercurrent that can flow through the system at $V = 0$.

Let us introduce a simple dynamical model of a small-area Josephson junction with a capacitance C shunted by an arbitrary linear impedance $Z_S(\omega)$, as shown in Fig. 2.2. Assuming that the system is biased by an external current I_x and making use of the Josephson relations (2.1), one can write down the current balance equation in the form

$$C\frac{\hbar\ddot{\varphi}}{2e} + I_{qp}\left(\frac{\hbar\dot{\varphi}}{2e}\right) + \int dt' Z_S^{-1}(t-t')\frac{\hbar\dot{\varphi}(t')}{2e} + I_C \sin\varphi = I_x + \xi_S(t) + \xi_t(t) \tag{2.2}$$

In the left-hand side of Eq. (2.2), we included four different contributions: the displacement current $C\dot{V}$, the quasiparticle current $I_{\text{qp}}(V)$, the current flowing through an external impedance

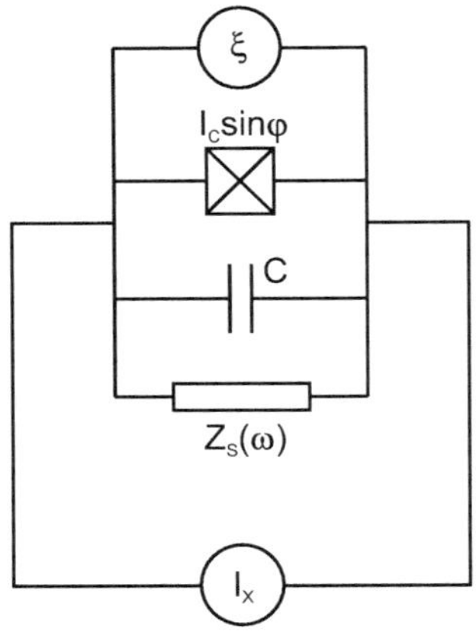

Figure 2.2 The equivalent circuit for a Josephson junction dynamics of which is described by Eq. (2.2). The noise variable ξ is defined as $\xi = -\xi_t - \xi_S$.

Z_S, and the Josephson current $I_C \sin\varphi$. The sum of these terms equals the net current I_x plus the current noise terms $\xi_S(t)$ and $\xi_t(t)$ produced, respectively, by the shunt and the tunnel junction. The current $I_{qp}(V)$ across the junction occurs due to the presence of quasiparticles above the superconducting gap and is defined by the standard formula

$$I_{qp}(V) = \frac{1}{eR_N}\int dE\, N_1(E - eV)N_2(E)\big[f_1(E - eV) - f_2(E)\big] \quad (2.3)$$

where R_N is the normal state junction resistance,

$$N_r(E) = \frac{|E|\theta(|E| - \Delta_r)}{\sqrt{E^2 - \Delta_r^2}}, \quad r = 1, 2 \quad (2.4)$$

are the normalized densities of states in superconducting electrodes, and $f_1(E)$ and $f_2(E)$ are the quasiparticle distribution functions in these electrodes. In equilibrium, these functions just coincide with the Fermi function

$$f_1(E) = f_2(E) = \frac{1}{e^{E/T} + 1} \quad (2.5)$$

The non-local kernel $Z_S^{-1}(t - t')$ in Eq. (2.2) is defined as

$$Z_S^{-1}(t - t') = \int \frac{d\omega}{2\pi}\frac{e^{-i\omega(t-t')}}{Z_S(\omega)} \quad (2.6)$$

According to the general properties of response functions, this kernel differs from zero only for $t > t'$ to obey causality.

The current noise of the shunt impedance is described by Gaussian stochastic variable $\xi_S(t)$ with zero average $\langle \xi_S(t) \rangle = 0$ and with the pair correlator defined in accordance with FDT

$$\langle \xi_S(t)\xi_S(t') \rangle = \int \frac{d\omega}{2\pi} e^{-i\omega(t-t')} \operatorname{Re}\left(\frac{\hbar}{Z_S(\omega)}\right) \omega \coth\frac{\hbar\omega}{2T} \quad (2.7)$$

The junction current noise is also characterized by zero average $\langle \xi_t(t) \rangle = 0$ and by the pair correlator [95]

$$\langle \xi_t(t)\xi_t(t') \rangle = \frac{e}{2} \int \frac{d\omega}{2\pi} e^{-i\omega(t-t')} \left[I_{qp}\left(\frac{\hbar\omega}{e} + V\right) \coth\frac{\hbar\omega + eV}{2T} + I_{qp}\left(\frac{\hbar\omega}{e} - V\right) \coth\frac{\hbar\omega - eV}{2T} \right] \quad (2.8)$$

Equation (2.2) represents the classical equation of motion for the Josephson phase $\varphi(t)$ and serves as a basis for studying classical dynamics of Josephson systems. It describes a great variety of nonlinear effects and provides very accurate description of numerous experiments [39, 275].

Assume now that both temperature and voltage remain well below the superconducting gap. In this case, both the quasiparticle current I_{qp} and the junction shot noise $\xi_t(t)$ are small and can be neglected. In addition, let us assume the external impedance to be purely Ohmic, i.e., $Z_S(\omega) = R_S$. Then the Langevin equation (2.2) reduces to a much simpler one:

$$C\frac{\hbar\ddot{\varphi}}{2e} + \frac{1}{R_S}\frac{\hbar\dot{\varphi}(t')}{2e} + I_C \sin\varphi = I_x + \xi_S(t), \quad (2.9)$$

where

$$\langle \xi_S(t)\xi_S(0) \rangle = \frac{\hbar}{R_S} \int \frac{d\omega}{2\pi} e^{-i\omega t} \omega \coth\frac{\hbar\omega}{2T} \quad (2.10)$$

Taking the classical limit $\hbar\omega \ll T$ in Eq. (2.10), one trivially finds

$$\langle \xi_S(t)\xi_S(0) \rangle = \frac{2}{R_S} T\,\delta(t) \quad (2.11)$$

Then Eq. (2.9) exactly coincides with the Langevin equation (1.2) for a classical particle with coordinate $\hbar\varphi/2e \to q$, effective mass $C \to m$ moving in a tilted washboard potential

$$U = -(\hbar I_C/2e)\cos\varphi - \hbar I_x \varphi/2e$$

under the influence of a linear friction force with effective viscosity $1/R_s \to \eta$ and Gaussian stochastic force $\xi_S(t) \to \xi(t)$ described by the white noise correlator (1.3). Thus, classical dynamics of a resistively shunted Josephson junction is exactly equivalent to that of a linearly damped classical particle. This observation explains why the term "Josephson particle" is frequently used in the literature.

Can this Josephson particle behave quantum mechanically?

On one hand, this question might look paradoxical. Indeed, the "coordinate" φ of this particle is just the difference of the phases of the wave functions of two superconducting condensates. In this respect, the idea to quantize the Josephson phase (i.e., the phase of the wave function!) might appear meaningless.

On the other hand, one can just employ the above formal analogy between Eqs. (2.9), (2.11) and (1.2), (1.3) and proceed along the lines of Chapter 1. Let us, for a moment, ignore dissipation by sending $1/R_S \to 0$. In this case, classical dynamics of the junction can also be defined by introducing its classical Hamilton function

$$H = \frac{Q^2}{2C} - \frac{\hbar I_C}{2e}\cos\varphi - \frac{\hbar I_x}{2e}\varphi \tag{2.12}$$

Here Q is the charge accumulated at the junction capacitance C. It is proportional to the classical "momentum" P, i.e., $Q = 2eP/\hbar$. The classical equations of motion for the junction read

$$\begin{aligned}\dot{\varphi} &= \frac{\partial H}{\partial P} = \frac{4e^2 P}{\hbar^2 C} = \frac{2eQ}{\hbar C},\\ \dot{P} &= \frac{\hbar \dot{Q}}{2e} = -\frac{\partial H}{\partial \varphi} = -\frac{\hbar I_C}{2e}\sin\varphi + \frac{\hbar I_x}{2e}.\end{aligned} \tag{2.13}$$

Eliminating the momentum P, we observe that these two equations are equivalent to Eq. (2.9) with $1/R_S = 0$, i.e., to the Newton equation for the undamped Josephson junction. Note that the Hamilton function (2.12), despite being classical, contains the Planck's constant $\hbar$. It originates from the Josephson relation between the phase and voltage (2.1) and by itself does not yet imply quantum mechanical behavior of the Josephson phase φ.

Now, according to general rules of quantum mechanics in order to quantize the system, one should replace the classical Hamilton function by the Hamiltonian operator. Defining the momentum/charge operator and the corresponding operator commutation relation as

$$\hat{P} = \frac{\hbar \hat{Q}}{2e} \to -\mathrm{i}\hbar \frac{\partial}{\partial \varphi}, \quad [\hat{P}, \hat{\varphi}] = \hat{P}\hat{\varphi} - \hat{\varphi}\hat{P} = -\mathrm{i}\hbar, \tag{2.14}$$

one finds

$$\hat{H} = \frac{\hat{Q}^2}{2C} - \frac{\hbar I_C}{2e} \cos \varphi - \frac{\hbar I}{2e} \varphi, \tag{2.15}$$

where $\hat{Q}^2 = -\partial^2/\partial(\varphi/2e)^2$. Quantum dynamics of the Josephson junction can now be described by the Schrödinger equation

$$\mathrm{i}\hbar \frac{\partial}{\partial t} \Psi(t, \varphi) = \hat{H} \Psi(t, \varphi) \tag{2.16}$$

for the junction wave function $\Psi(t, \varphi)$. One can proceed further and also take both dissipation and noise into account. This task can be accomplished by constructing the corresponding Feynman–Vernon influence functional exactly as it was already done in Chapter 1. For example, linear Ohmic dissipation can be accounted for by coupling the Josephson phase to the Caldeira–Leggett bath of harmonic oscillators with properly chosen parameters. Integrating out these bath oscillators, we immediately arrive at the Caldeira–Legget action for the Josephson phase, which, being reduced back to the classical level, would again yield the Langevin equation defined by Eqs. (2.9) and (2.11).

We note that our choice of the oscillator bath is by no means restrictive since according to general properties of the influence functionals, the physical nature of different baths does not matter as long as their influence to the particle dynamics remains identical. Hence, there is no need to identify the microscopic nature of linear Ohmic dissipation in this case; our procedure should automatically yield the correct quantum description of the effect of an external Ohmic resistor.

Still, on a conceptual level, the possibility to treat a *macroscopic* variable—the Josephson phase—as an ordinary quantum degree of freedom definitely needs further justification based on the microscopic theory of superconductivity. In addition, a more elaborate model is necessary to account for both nonlinear quasiparticle dissipation and shot noise in systems under consideration. We now turn to a rigorous microscopic derivation of the effective action for a Josephson junction, which will allow us to shed light on all the above issues.

2.2 Effective Action for a Josephson Junction

The starting point of our analysis is the Hamiltonian of interacting electrons in a metal

$$\hat{H}_{\text{full}} = \int d^3\boldsymbol{r}\left[\sum_{\alpha=\uparrow,\downarrow} \hat{\Psi}^\dagger_{r\alpha}\left(-\frac{\hbar^2\nabla^2_{\boldsymbol{r}}}{2m} + U(\boldsymbol{r}) - \mu\right)\hat{\Psi}_{r\alpha}\right]$$
$$+ \sum_{\alpha,\beta=\uparrow,\downarrow} \int d^3\boldsymbol{r}d^3\boldsymbol{r}'\ \hat{\Psi}^\dagger_{r\alpha}\hat{\Psi}^\dagger_{r'\beta}\left[\frac{e^2}{2|\boldsymbol{r}-\boldsymbol{r}'|} - \frac{\lambda}{2}\delta(\boldsymbol{r}-\boldsymbol{r}')\delta_{\alpha,\bar{\beta}}\right]\hat{\Psi}_{r'\beta}\hat{\Psi}_{r\alpha}$$
$$(2.17)$$

Here $\hat{\Psi}_{r\alpha}$ is an operator that annihilates an electron with spin α at point $\boldsymbol{r}$, and $\hat{\Psi}^\dagger_{r\alpha}$ is the corresponding creation operator; m is the electron effective mass; the potential of randomly distributed impurities in a metal, $U(\boldsymbol{r})$, also includes the potential of a tunnel barrier between two bulk electrodes; μ is the electron chemical potential. In Eq. (2.17), we included two types of interactions: The term $\propto e^2/2|\boldsymbol{r}-\boldsymbol{r}'|$ accounts for repulsive Coulomb interaction between electrons, while the last term describes attractive BCS interaction with an effective coupling constant λ. The index $\bar{\beta}$ in this term labels the spin direction opposite to one of β.

For simplicity, here we ignored the fluctuating vector potential as well as the energy of the magnetic field in our system. This simplification is appropriate for sufficiently small junctions and in any case will have no effect on our final result for the effective action. If needed, magnetic field effects can easily be included into our consideration. For the corresponding generalization, we refer the reader to the review [345]. We will also return to this issue on more general grounds in Chapter 18 of this book.

2.2.1 Hubbard–Stratonovich Transformation and Effective Hamiltonian

The evolution of the time-dependent density matrix of the system $\boldsymbol{\rho}(t)$ is determined by the standard formula

$$\boldsymbol{\rho}(t) = e^{-\mathrm{i}\hat{H}_{\text{full}}t/\hbar}\boldsymbol{\rho}(0)e^{\mathrm{i}\hat{H}_{\text{full}}t/\hbar}. \qquad (2.18)$$

As in Chapter 1, in order to describe real-time evolution, we will employ the Keldysh time contour, which consists of the forward

(F) and backward (B) branches corresponding, respectively, to the operators $e^{-i\hat{H}_{\text{full}}t}$ and $e^{i\hat{H}_{\text{full}}t}$. We will proceed in a standard manner and decouple both interaction terms in the Hamiltonian (2.17) by means of two Hubbard–Stratonovich transformations introducing complex superconducting order parameter field $\Delta(t, \boldsymbol{r})$ and scalar potential $V(t, \boldsymbol{r})$. In this way, one reduces the problem of an interacting electron gas to that of non-interacting electrons moving in fluctuating potentials $\Delta(t, \boldsymbol{r})$ and $V(t, \boldsymbol{r})$. In Eq. (2.18), we perform the Hubbard–Stratonovich transformations separately for the operators $e^{-i\hat{H}_{\text{full}}t/\hbar}$ and $e^{i\hat{H}_{\text{full}}t/\hbar}$. Thus we generate two pairs of fluctuating fields Δ_{F}, Δ_{B} and V_{F}, V_{B} defined, respectively, of the forward and backward branches of the Keldysh contour. The corresponding transformations read

$$e^{-i\hat{H}_{\text{full}}t/\hbar} = \frac{\int \mathcal{D}^2\Delta_{\text{F}}\,\mathcal{D}V_{\text{F}}\left[\mathcal{T}\,e^{-i\hat{\int}_0^t dt'\,\hat{H}_{\text{eff}}[\Delta_{\text{F}},V_{\text{F}}]}\right]e^{iS_0[\Delta_{\text{F}},V_{\text{F}}]}}{\int \mathcal{D}^2\Delta_{\text{F}}\,\mathcal{D}V_{\text{F}}\,e^{iS_0[\Delta_{\text{F}},V_{\text{F}}]}} \tag{2.19}$$

$$e^{i\hat{H}_{\text{full}}t/\hbar} = \frac{\int \mathcal{D}^2\Delta_{\text{B}}\,\mathcal{D}V_{\text{B}}\left[\mathcal{T}\,e^{-i\int_0^t dt'\,\hat{H}_{\text{eff}}[\Delta_{\text{B}},V_{\text{B}}]/\hbar}\right]^{\dagger}e^{-iS_0[\Delta_{\text{B}},V_{\text{B}}]/\hbar}}{\int \mathcal{D}^2\Delta_{\text{B}}\,\mathcal{D}V_{\text{B}}\,e^{-iS_0[\Delta_{\text{B}},V_{\text{B}}]/\hbar}} \tag{2.20}$$

Here we have defined the free action of the fluctuating fields

$$S_0[\Delta, V] = \frac{1}{2}\int_0^t dt'\int d^3\boldsymbol{r}\left(\frac{|\Delta|^2}{\lambda} + \frac{\nabla^2 V}{4\pi}\right) \tag{2.21}$$

introduced the time-ordered operator exponents indicated by the symbol $\mathcal{T}$ and defined an effective Hamiltonian for electrons interacting with the fluctuating potentials

$$\hat{H}_{\text{eff}}[\Delta, V] = \int d^3\boldsymbol{r}\Bigg[\sum_{\alpha=\uparrow,\downarrow}\hat{\Psi}^{\dagger}_{r\alpha}\left(-\frac{\hbar^2\nabla_{\boldsymbol{r}}^2}{2m_r} + U(\boldsymbol{r}) - \mu - eV(t,\boldsymbol{r})\right)\hat{\Psi}_{r\alpha}$$
$$+\Delta(t,\boldsymbol{r})\hat{\Psi}^{\dagger}_{r\uparrow}\hat{\Psi}^{\dagger}_{r\downarrow} + \Delta^*(t,\boldsymbol{r})\hat{\Psi}_{r\downarrow}\hat{\Psi}_{r\uparrow}\Bigg] \tag{2.22}$$

Thus, having performed the exact transformations (2.19) and (2.20), we reformulated the initial complicated problem with interactions in terms of the effective Hamiltonian (2.22), which is quadratic in the electron operators. Now comes the next important step: We will consider these electron degrees of freedom as an effective bath for the collective variables $\Delta(t, \boldsymbol{r})$ and $V(t, \boldsymbol{r})$. Just in the spirit of the Feynman–Vernon approach, we can trace out all the bath degrees of freedom in Eq. (2.18) and construct the influence functional for fluctuating fields. An important difference with the corresponding procedure described in Chapter 1 is that instead of a collection of harmonic oscillators now we have a bath of fermions. However, tracing out these fermionic degrees of freedom can be handled exactly also in this case because the Hamiltonian (2.22) is quadratic. After this step, we arrive at the influence functional F formulated only in terms of the fields $\Delta(t, \boldsymbol{r})$ and $V(t, \boldsymbol{r})$, dynamics of which is fully determined by the electronic bath. The reduced density matrix $\mathrm{tr}_\psi \boldsymbol{\rho}(t)$ is then expressed as a path integral

$$\mathrm{tr}_\psi \boldsymbol{\rho}(t) \sim \int \mathcal{D}^2\Delta_{\mathrm{F}}\, \mathcal{D}V_{\mathrm{F}} \mathcal{D}^2\Delta_{\mathrm{B}}\, \mathcal{D}V_{\mathrm{B}} \times e^{(\mathrm{i}/\hbar)(S_0[\Delta_{\mathrm{F}}, V_{\mathrm{F}}]-\mathrm{i}S_0[\Delta_{\mathrm{B}}, V_{\mathrm{B}}])} F[\Delta_{\mathrm{F}}, V_{\mathrm{F}}, \Delta_{\mathrm{B}}, V_{\mathrm{B}}] \quad (2.23)$$

This program constitutes an important development of the Feynman–Vernon influence functional technique applied to fermionic systems. Throughout this book, we will repeatedly follow this route demonstrating remarkable efficiency of this approach in various physical situations. Here we implement the above program for a small Josephson junction depicted in Fig. 2.1. The following approximations are in order in this case.

Assuming that both superconducting electrodes are sufficiently large, one can neglect fluctuations of the absolute values of the order parameter fields in these electrodes, i.e., $\Delta_{\mathrm{F,B}}$ identically coincide with the coordinate independent BCS (mean field) order parameter Δ_1 (Δ_2) in the first (second) superconductor. This approximation is secured by the Ginzburg criterion [138] demonstrating that fluctuations of the absolute value of BCS order parameter are negligible with a very high accuracy, except in extremely small superconductors (see also Chapters 18–20 for further discussion). Thus, one has $\Delta_{\mathrm{F,B}} = \Delta_1 e^{\mathrm{i}\varphi_1^{\mathrm{F,B}}(t)}$ in the first electrode and $\Delta_{\mathrm{F,B}} = \Delta_2 e^{\mathrm{i}\varphi_2^{\mathrm{F,B}}(t)}$ in the second one. In contrast, phase fluctuations are, in general, not restricted owing to the presence of the soft mode related to the degeneracy of the BCS ground state.

In addition, we will ignore the coordinate dependence of the fluctuating electric potentials inside the bulk electrodes keeping only their time dependence, i.e., $V_{\mathrm{F,B}} = V_1^{\mathrm{F,B}}(t)$ in the first electrode and $V_{\mathrm{F,B}} = V_2^{\mathrm{F,B}}(t)$ in the second one. The electric potentials and the phases are, in general, not independent. They are linked to each other by means of the Josephson relations

$$V_1^{\mathrm{F,B}}(t) = \frac{\hbar\dot{\varphi}_1^{\mathrm{F,B}}(t)}{2e}, \quad V_2^{\mathrm{F,B}}(t) = \frac{\hbar\dot{\varphi}_2^{\mathrm{F,B}}(t)}{2e} \tag{2.24}$$

Again, these relations remain very accurate as long the electrodes are sufficiently large and involved frequencies are small as compared to the plasma frequency of the metals. In spatially restricted superconductors, such as ultrathin superconducting wires, the Josephson relations can be violated. Rigorous derivation of these relations as well their accuracy will be discussed on more general grounds in Chapter 18.

As a result of the above approximations, the path integral over six quantum fields in Eq. (2.23) gets reduced to that over two phase fields

$$\int \mathcal{D}^2\Delta_{\mathrm{F}}\,\mathcal{D}V_{\mathrm{F}}\mathcal{D}^2\Delta_{\mathrm{B}}\,\mathcal{D}V_{\mathrm{B}}\delta(\Delta_{\mathrm{F}} - \Delta_{1,2})\delta(\Delta_{\mathrm{B}} - \Delta_{1,2})$$
$$\times\,\delta\left(V_{\mathrm{F}} - \frac{\hbar\dot{\varphi}_{\mathrm{F}}}{2e}\right)\delta\left(V_{\mathrm{B}} - \frac{\hbar\dot{\varphi}_{\mathrm{B}}}{2e}\right) \to \int \mathcal{D}\varphi_{\mathrm{F}}\mathcal{D}\varphi_{\mathrm{B}}$$

where we defined $V_{\mathrm{F,B}} = V_1^{\mathrm{F,B}} - V_2^{\mathrm{F,B}}$ and $\varphi_{\mathrm{F,B}} = \varphi_1^{\mathrm{F,B}} - \varphi_2^{\mathrm{F,B}}$. Furthermore, the terms containing $|\Delta|^2/\lambda$ in the forward and backward parts of the action S_0 cancel out exactly in the exponent of Eq. (2.23), while the terms $\nabla^2 V/4\pi$ differ from zero only in the immediate vicinity of the Josephson junction since no voltage drop occurs elsewhere in the system. Introducing the junction capacitance C and making use of Eqs. (2.24), we may approximate

$$S_0[\Delta_{\mathrm{F}}, V_{\mathrm{F}}] - S_0[\Delta_{\mathrm{B}}, V_{\mathrm{B}}] \to \int_0^t dt'\left(\frac{C\hbar^2\dot{\varphi}_{\mathrm{F}}^2}{8e^2} - \frac{C\hbar^2\dot{\varphi}_{\mathrm{B}}^2}{8e^2}\right) \tag{2.25}$$

Let us further specify the effective Hamiltonian (2.22) for the Josephson junction depicted in Fig. 2.1. In order to describe electron tunneling between superconducting electrodes, instead of the potential $U(\boldsymbol{r})$ in the junction area, we introduce the standard tunneling Hamiltonian $\hat{H}_T$ with the tunneling amplitude $t(\boldsymbol{r})$, which differs from zero only at the junction surface. Since properties of our tunnel junctions are not sensitive to disorder in the electrodes, the potential term $U(\boldsymbol{r})$ can be neglected inside these electrodes. Furthermore, it will be convenient for us to perform the gauge transformation multiplying the electron operators in the electrodes by the phase factors, i.e., for the forward (backward) branch of the Keldysh contour in the first electrode, we replace $\hat{\Psi}_1 \rightarrow \hat{\Psi}_1 e^{\mathrm{i}\varphi_{\mathrm{F(B)}}/2}$, $\hat{\Psi}_1^\dagger \rightarrow \hat{\Psi}_1^\dagger e^{-\mathrm{i}\varphi_{\mathrm{F(B)}}/2}$ and similarly in the second electrode. As a result of this transformation, the voltages $V_{1,2}^{\mathrm{F,B}}$ drop out from the effective Hamiltonians for the electrodes, while the tunneling amplitude acquires an additional phase factor, $t(\boldsymbol{r}) \rightarrow t(\boldsymbol{r})e^{-\mathrm{i}\varphi_{\mathrm{F,B}}/2}$. After all these steps, one finds

$$\hat{H}_{\mathrm{eff}} \rightarrow \hat{H}_1 + \hat{H}_2 + \hat{H}_T \tag{2.26}$$

where

$$\hat{H}_j = \int_j d^3\boldsymbol{r} \left[\sum_{\alpha=\uparrow,\downarrow} \hat{\Psi}_{r\alpha}^\dagger \left(-\frac{\hbar^2 \nabla_{\boldsymbol{r}}^2}{2m_j} - \mu \right) \hat{\Psi}_{r\alpha} + \Delta_j \hat{\Psi}_{r\uparrow}^\dagger \hat{\Psi}_{r\downarrow}^\dagger + \Delta_j \hat{\Psi}_{r\downarrow} \hat{\Psi}_{r\uparrow} \right] \tag{2.27}$$

are the Hamiltonians of two superconducting electrodes ($j = 1, 2$), m_j and Δ_j are, respectively, electron effective mass and the absolute value of the superconducting order parameter in the lead j and

$$\hat{H}_{\mathrm{T}} = \sum_{\alpha=\uparrow,\downarrow} \int_{\mathcal{A}} d^2\boldsymbol{r} \left[t(\boldsymbol{r})e^{-\mathrm{i}\hat{\varphi}/2} \hat{\Psi}_{1\alpha}^\dagger \hat{\Psi}_{2\alpha} + t^*(\boldsymbol{r})e^{\mathrm{i}\hat{\varphi}/2} \hat{\Psi}_{2\alpha}^\dagger \hat{\Psi}_{1\alpha} \right] \tag{2.28}$$

Here the integration runs over the junction area $\mathcal{A}$. According to Eq. (2.24), the operators of the phase $\hat{\varphi}$ and the voltage $\hat{V}$ obey the Josephson relation:

$$d\hat{\varphi}/dt = 2e\hat{V}/\hbar \tag{2.29}$$

Finally, we would like to account for the presence of an external impedance $Z_S(\omega)$. This impedance is effectively equivalent to the electromagnetic environment that includes, e.g., shunt resistance and stray capacitance of the wires. Assuming that this environment is linear, as we already discussed in Chapter 1, it can be very generally modeled by a set of harmonic oscillators with the masses M_n, frequencies Ω_n, coordinates and momenta $\hat{X}_n$ and $\hat{P}_n$, respectively, which are linearly coupled to the junction phase $\hat{\varphi}$.

Then the Hamiltonian of this environment takes the form

$$H_{\text{env}} = \sum_n \left(\frac{\hat{P}_n^2}{2M_n} + \frac{M_n \Omega_n^2 (\hat{X}_n - c_n \hat{\varphi})^2}{2} \right) \tag{2.30}$$

The distribution of the environment oscillators over the frequencies is chosen to emulate the impedance "seen" by the tunnel junction $Z_S(\omega)$:

$$\sum_n \frac{2ec_n^2 M_n \Omega_n^2 \, \mathrm{i}\omega}{(\omega + \mathrm{i}0)^2 - \Omega_n^2} = \frac{1}{Z_S(\omega)} \tag{2.31}$$

2.2.2 Current Operator

Having defined the total effective Hamiltonian for our system

$$\hat{H}_{\text{tot}} = \hat{H}_1 + \hat{H}_2 + \hat{H}_T + \frac{C\hat{V}^2}{2} + \hat{H}_{\text{env}} \tag{2.32}$$

we now turn to the current operator. It is defined as a time derivative of the operator of the electron number in the first lead $\hat{N}_1$:

$$\hat{I} = e\frac{d\hat{N}_1}{dt} = e\frac{d}{dt} \sum_{\alpha=\uparrow,\downarrow} \int d^3\boldsymbol{r} \, \hat{\Psi}_{1\alpha}^\dagger \hat{\Psi}_{1\alpha} \tag{2.33}$$

where e is the absolute value of the electron charge. Following the standard rules of quantum mechanics, we rewrite this operator in the form

$$\begin{aligned} \hat{I} &= \frac{\mathrm{i}e}{\hbar} \sum_{\alpha=\uparrow,\downarrow} \int d^3\boldsymbol{r} \left[\hat{H}_T, \hat{\Psi}_{1\alpha}^\dagger \hat{\Psi}_{1\alpha} \right] \\ &= -\frac{\mathrm{i}e}{\hbar} \sum_{\alpha=\uparrow,\downarrow} \int_{\mathcal{A}} d^2\boldsymbol{r} \left[t(\boldsymbol{r}) e^{-\mathrm{i}\hat{\varphi}/2} \hat{\Psi}_{1\alpha}^\dagger \hat{\Psi}_{2\alpha} - t^*(\boldsymbol{r}) e^{\mathrm{i}\hat{\varphi}/2} \hat{\Psi}_{2\alpha}^\dagger \hat{\Psi}_{1\alpha} \right] \end{aligned} \tag{2.34}$$

Comparing this expression with the tunneling Hamiltonian (2.28), we notice that the current operator may be written in a simple symbolic form

$$\hat{I} = \frac{2e}{\hbar} \frac{\partial}{\partial \hat{\varphi}} \hat{H}_{\mathrm{T}} \tag{2.35}$$

Let us now derive a formal expression for the current across the junction in terms of a path integral over the fluctuating phases φ_F and φ_B. To this end, we assume that at time $t < 0$, the density matrix of the system $\boldsymbol{\rho}(0)$ is factorized into the product of equilibrium density matrices for two superconductors and the environment:

$$\boldsymbol{\rho}(0) = \frac{e^{-\hat{H}_1/T}}{\mathrm{tr}\left(e^{-\hat{H}_1/T}\right)} \frac{e^{-\hat{H}_2/T}}{\mathrm{tr}\left(e^{-\hat{H}_2/T}\right)} \frac{e^{-\hat{H}_{\mathrm{env}}/T}\Big|_{C_n=0}}{\mathrm{tr}\left(e^{-\hat{H}_{\mathrm{env}}/T}\Big|_{C_n=0}\right)} \qquad (2.36)$$

Here by T we denote the temperature of both the electrodes and the environment. "Turning on" electron tunneling between the superconductors as well as coupling between the junction phase and the environment at the time moment $t = 0$, we define the current I across the junction at time t

$$I(t) = \frac{2e}{\hbar}\,\mathrm{tr}\left[\frac{\partial \hat{H}_{\mathrm{T}}}{\partial \hat{\varphi}} e^{-\mathrm{i}\hat{H}t/\hbar} \boldsymbol{\rho}(0) e^{\mathrm{i}\hat{H}t/\hbar}\right] \qquad (2.37)$$

This formula can also be rewritten in a symmetric form

$$I(t) = \frac{e}{\hbar}\,\mathrm{tr}\left[\frac{\partial \hat{H}_{\mathrm{T}}}{\partial \hat{\varphi}} e^{-\mathrm{i}\hat{H}t/\hbar} \boldsymbol{\rho}(0) e^{\mathrm{i}\hat{H}t/\hbar}\right] + \frac{e}{\hbar}\mathrm{tr}\left[e^{-\mathrm{i}\hat{H}t/\hbar} \boldsymbol{\rho}(0) e^{\mathrm{i}\hat{H}t/\hbar} \frac{\partial \hat{H}_{\mathrm{T}}}{\partial \hat{\varphi}}\right] \qquad (2.38)$$

The trace (2.38) can be expressed in terms of a path integral over both Grassmann fermionic fields ψ_α, $\psi_\alpha^\dagger$ and bosonic fields φ, X_n. Equation (2.38) contains the operator $e^{-\mathrm{i}\hat{H}t}$, describing the evolution along the forward branch of the Keldysh contour, as well as the inverse operator $e^{\mathrm{i}\hat{H}t/\hbar}$ responsible for evolution along the backward branch. Accordingly, we introduce two sets of fluctuating fields and get

$$\begin{aligned} I(t) = \frac{e}{\hbar} \int \mathcal{D}\varphi^F \mathcal{D}\varphi^B \int \mathcal{D}X_n^F \mathcal{D}X_n^B \int \mathcal{D}\psi_\alpha^F \mathcal{D}\psi_\alpha^B \mathcal{D}\psi_\alpha^{\dagger F} \mathcal{D}\psi_\alpha^{\dagger B} \\ \times \left(\frac{\partial H_T\left(\varphi_F(t), \psi_\alpha^{\dagger F}(t), \psi_\alpha^F(t)\right)}{\partial \varphi_F(t)} + \frac{\partial H_T\left(\varphi_B(t), \psi_\alpha^{\dagger B}(t), \psi_\alpha^B(t)\right)}{\partial \varphi_B(t)} \right) \\ \times\ e^{\mathrm{i}S[\varphi_F, X_n^F, \psi_\alpha^F, \psi_\alpha^{\dagger F}]/\hbar - \mathrm{i}S[\varphi_B, X_n^B, \psi_\alpha^B, \psi_\alpha^{\dagger B}]/\hbar}\Big|_{\varphi_F(t)=\varphi_B(t)} \end{aligned} \qquad (2.39)$$

where the action S is defined by the sum of electronic and environmental contributions:

$$S[\varphi, X_n, \psi_\alpha, \psi_\alpha^\dagger] = S_{\mathrm{el}}[\varphi, \psi_\alpha, \psi_\alpha^\dagger] + S_{\mathrm{env}}[\varphi, X_n] \qquad (2.40)$$

According to general principles of quantum field theory, the electronic part of the action has the form

$$S_{\text{el}} = \sum_{j=1,2}\sum_{\alpha=\uparrow,\downarrow}\int dt'\left[\mathrm{i}\hbar\psi^{\dagger}_{j\alpha}\frac{\partial\psi_{j\alpha}}{\partial t'} - H\left(\varphi,\psi^{\dagger}_{j,\alpha},\psi_{j,\alpha}\right)\right]$$

which yields

$$\begin{aligned}S_{\text{el}} = &\int_0^t dt'\int d^3\boldsymbol{r}\Bigg[\sum_{\alpha=\uparrow,\downarrow}\psi^{\dagger}_{1\alpha}\left(\mathrm{i}\hbar\frac{\partial}{\partial t'}+\frac{\hbar^2\nabla^2_{\boldsymbol{r}}}{2m_1}+\mu\right)\psi_{1\alpha} - \Delta_1\psi^{\dagger}_{1\uparrow}\psi^{\dagger}_{1\downarrow}\\ &-\Delta_1\psi_{1\downarrow}\psi_{1\uparrow}\Bigg]+\int_0^t dt'\int d^3\boldsymbol{r}\Bigg[\sum_{\alpha=\uparrow,\downarrow}\psi^{\dagger}_{2\alpha}\left(\mathrm{i}\hbar\frac{\partial}{\partial t'}+\frac{\hbar^2\nabla^2_{\boldsymbol{r}}}{2m_2}+\mu\right)\psi_{2\alpha}\\ &-\Delta_2\psi^{\dagger}_{2\uparrow}\psi^{\dagger}_{2\downarrow}-\Delta_2\psi_{2\downarrow}\psi_{2\uparrow}\Bigg]\\ &-\sum_{\alpha=\uparrow,\downarrow}\int_{\mathcal{A}} d^2\boldsymbol{r}\left[t(\boldsymbol{r})e^{\mathrm{i}\varphi/2}\psi^{\dagger}_{1\alpha}\psi_{2\alpha}+t^*(\boldsymbol{r})e^{-\mathrm{i}\varphi/2}\psi^{\dagger}_{2\alpha}\psi_{1\alpha}\right]\end{aligned}$$

The remaining environmental part of the action (which also includes the charging term) reads

$$S_{\text{env}} = \int_0^t dt'\left[\frac{C\hbar^2\dot{\varphi}^2}{8e^2}+\sum_n\left(\frac{M_n\dot{X}_n^2}{2}-\frac{M_n\Omega_n^2(X_n-c_n\varphi)^2}{2}\right)\right] \quad (2.41)$$

This is essentially the Caldeira–Leggett action already discussed in Chapter 1. The above expressions for the effective action can also be interpreted as a path integral formulation of the Hubbard–Stratonovich transformation (2.19) and (2.20).

It is easy to observe that the total action (2.40) is quadratic both in the oscillator coordinates $X_n^{\mathrm{F,B}}$ and in the Grassmann fields $\psi_\alpha^{\mathrm{F,B}}$. Hence, these variables can be integrated out exactly. To integrate out the bosonic coordinates X_n, we just repeat the standard procedure outlined in Chapter 1. As for the Grassmann fields, we apply the following rule of integration (see, e.g., the book [334]):

$$\int \mathcal{D}\boldsymbol{\psi}^{\dagger}\mathcal{D}\boldsymbol{\psi}\; e^{\mathrm{i}\boldsymbol{\psi}^{\dagger T}M\boldsymbol{\psi}} = A\det[M] \quad (2.42)$$

Here $\boldsymbol{\psi}$ and $\boldsymbol{\psi}^{\dagger}$ are the two N-dimensional vectors composed of Grassmann variables, M is the $N\times N$ matrix, and A is an unimportant numerical prefactor. Uniting all available time and coordinate dependent fermionic fields, we define two 8-dimensional vectors

$$\boldsymbol{\psi}^T = \left(\psi^F_{1\uparrow}, \psi^B_{1\uparrow}, \psi^{\dagger F}_{1\downarrow}, \psi^{\dagger B}_{1\downarrow}, \psi^F_{2\uparrow}, \psi^B_{2\uparrow}, \psi^{\dagger F}_{2\downarrow}, \psi^{\dagger B}_{2\downarrow}\right) \qquad (2.43)$$

$$\boldsymbol{\psi}^{\dagger T} = \left(\psi^{\dagger F}_{1\uparrow}, \psi^{\dagger B}_{1\uparrow}, \psi^F_{1\downarrow}, \psi^B_{1\downarrow}, \psi^{\dagger F}_{2\uparrow}, \psi^{\dagger B}_{2\uparrow}, \psi^F_{2\downarrow}, \psi^B_{2\downarrow}\right) \qquad (2.44)$$

Accordingly the matrix M has the dimensions 8×8 with each of its matrix elements being the operator in the 4-dimensional space of time and coordinates. In fact, the matrix M just coincides with the inverse Keldysh Green function of the system

$$M \equiv \check{G}^{-1} = \begin{pmatrix} \check{G}_1^{-1} & \check{t} \\ \check{t}^\dagger & \check{G}_2^{-1} \end{pmatrix}. \qquad (2.45)$$

The 4×4 sub-blocks of this matrix read

$$\check{G}_j^{-1} = \begin{pmatrix} i\hbar\frac{\partial}{\partial t} + \frac{\hbar^2\nabla_r^2}{2m_j} + \mu & 0 & -\Delta_j & 0 \\ 0 & -i\hbar\frac{\partial}{\partial t} - \frac{\hbar^2\nabla_r^2}{2m_j} - \mu & 0 & \Delta_j \\ -\Delta_j & 0 & i\hbar\frac{\partial}{\partial t} - \frac{\hbar^2\nabla_r^2}{2m_j} - \mu & 0 \\ 0 & \Delta_j & 0 & -i\hbar\frac{\partial}{\partial t} + \frac{\hbar^2\nabla_r^2}{2m_j} + \mu \end{pmatrix} \qquad (2.46)$$

$$\check{t} = \begin{pmatrix} -t(\boldsymbol{r})e^{-i\varphi_F/2} & 0 & 0 & 0 \\ 0 & t(\boldsymbol{r})e^{-i\varphi_B/2} & 0 & 0 \\ 0 & 0 & t(\boldsymbol{r})e^{i\varphi_F/2} & 0 \\ 0 & 0 & 0 & -t(\boldsymbol{r})e^{i\varphi_B/2} \end{pmatrix} \qquad (2.47)$$

The matrices $\check{G}_{1,2}^{-1}$ are the inverse Keldysh Green functions of isolated normal electrodes and the matrix $\check{t}$ originates from the tunneling Hamiltonian (2.28).

After these manipulations, the current (2.39) takes the form

$$I(t) = -\frac{e}{\hbar}\int \mathcal{D}\varphi_{F,B} \left(\frac{\delta S_{\rm el}}{\delta\varphi_F(t)} - \frac{\delta S_{\rm el}}{\delta\varphi_B(t)}\right)\Bigg|_{\varphi_F(t)=\varphi_B(t)} e^{iS_{\rm el}/\hbar + iS_{\rm env}/\hbar}. \qquad (2.48)$$

The action now depends only on two remaining fluctuating variables: φ^F and φ^B. Our next task is to derive an explicit expression for this action.

2.2.3 Effective Action

The environmental part of the action S_{env} is recovered immediately. After integration over the oscillator coordinates X_n, one finds

$$S_{\text{env}} = \int dt_1 \frac{C\hbar^2\dot{\varphi}^+\dot{\varphi}^-}{4e^2} - \int dt_1 dt_2 \frac{\hbar\varphi^-(t_1)}{2e} Z_S^{-1}(t_1 - t_2)\frac{\hbar\dot{\varphi}^+(t_2)}{2e}$$
$$+ \frac{\mathrm{i}}{2}\int dt_1 dt_2 \frac{\hbar\varphi^-(t_1)}{2e} K_S(t_1 - t_2)\frac{\hbar\varphi^-(t_2)}{2e}, \tag{2.49}$$

where, as in Chapter 1, we introduced the so-called "classical" and "quantum" phases

$$\varphi^+ = (\varphi_F + \varphi_B)/2, \quad \varphi^- = \varphi_F - \varphi_B, \tag{2.50}$$

The inverse impedance kernel $Z_S^{-1}(t)$ is defined in Eq. (2.6) being just the response function of the environment to the fluctuating voltage across the junction. The function $K_S(t)$ defines the noise correlation function of the environment

$$K_S(t) = \langle \xi_S(t)\xi_S(0)\rangle$$
$$= \int \frac{d\omega}{2\pi} \operatorname{Re}\left(\frac{1}{Z_S(\omega)}\right) \hbar\omega \coth\frac{\hbar\omega}{2T}\, e^{-\mathrm{i}\omega t} \tag{2.51}$$

Now we turn to the electronic part of the action, which is our main goal here. As we already discussed, it is defined by a formally exact expression

$$\mathrm{i}S_{\text{el}} = \hbar \ln\det\left(\check{G}^{-1}\right) = \hbar\,\mathrm{tr}\,\ln\left(\check{G}^{-1}\right) \tag{2.52}$$

where $\check{G}^{-1}$ is specified in Eqs. (2.45)–(2.47). In order to simplify Eq. (2.52), we recall that we are dealing with a tunnel junction characterized by a small value of the tunneling amplitude $t(\boldsymbol{r})$. Having defined the zeroth order Green function $\check{G}_0^{-1}$ by setting $\check{t} = 0$ in Eq. (2.45) as well as a small correction to it $\delta\check{G}^{-1} = \check{G}^{-1} - \check{G}_0^{-1} \propto t(\boldsymbol{r})$, we expand the action in powers of $\delta\check{G}^{-1}$ and get

$$\mathrm{i}S_{\text{el}} = \hbar\,\mathrm{tr}\,\ln\left(\check{G}_0^{-1} + \delta\check{G}^{-1}\right) = \hbar\,\mathrm{tr}\,\ln\left(\check{G}_0^{-1}\right) + \mathrm{tr}\,\ln\left(1 + \check{G}_0\delta\check{G}^{-1}\right)$$
$$= \hbar\,\mathrm{tr}\,\ln\left(\check{G}_0^{-1}\right) + \hbar\,\mathrm{tr}\left(\check{G}_0\delta\check{G}^{-1}\right) - \frac{\hbar}{2}\mathrm{tr}\left(\check{G}_0\delta\check{G}^{-1}\check{G}_0\delta\check{G}^{-1}\right) + \mathcal{O}(t^4(\boldsymbol{r}))$$
$$\tag{2.53}$$

The first term in Eq. (2.53), $\hbar\,\mathrm{tr}\,\ln\left(\check{G}_0^{-1}\right)$, describes the contribution of isolated superconducting electrodes and contains information about the bias current I_x. It may be cast to the form

$$\hbar\,\mathrm{tr}\,\ln\left(\check{G}_0^{-1}\right) = \mathrm{i}\int dt' \frac{\hbar I_x}{2e}\varphi^-(t') \tag{2.54}$$

In order to derive this formula, it is necessary to take into account the spatial dependence of the phase of the superconducting order parameter in the electrodes since its gradient is proportional to the current, $\nabla\varphi_j \propto I_x$. Further details of this derivation can be revealed from a more general analysis, which we postpone to Chapter 18.

The second term in Eq. (2.53) identically equals to zero, i.e.,

$$\hbar\,\mathrm{tr}\,\left(\check{G}_0\delta\check{G}^{-1}\right) \equiv 0.$$

Thus, the lowest-order nontrivial contribution to the action (2.53) reads

$$S_{\mathrm{el}} = \int dt' \frac{\hbar I_x}{2e}\varphi^-(t') + \frac{\mathrm{i}\hbar}{2}\mathrm{tr}\,\left(\check{G}_0\delta\check{G}^{-1}\check{G}_0\delta\check{G}^{-1}\right) \tag{2.55}$$

Taking into account the block structure of the matrices $\check{G}_0$ and $\delta\check{G}^{-1}$, we transform the last term of this formula and rewrite the action as

$$S_{\mathrm{el}} = \int dt' \frac{\hbar I_x}{2e}\varphi^-(t') + \mathrm{i}\hbar\,\mathrm{tr}\,\left(\check{G}_1\check{t}\check{G}_2\check{t}^{\dagger}\right) \tag{2.56}$$

where $\check{G}_1$ and $\check{G}_2$ are 4×4 Keldysh Green functions of isolated superconducting electrodes, which contain information about their temperature and the density of states. Explicitly, the Green function of the first superconductor reads

$$\check{G}_1 = \begin{pmatrix} \breve{G}_1 & \breve{F}_1 \\ \breve{F}_1^+ & \breve{G}_1^+ \end{pmatrix} \tag{2.57}$$

where 2×2 sub-blocks are defined in terms of the operators $\hat{\Psi}_\alpha$ and $\hat{\Psi}_\alpha^\dagger$,

$$\breve{G}_1(X, X') = \frac{1}{\mathrm{i}}\begin{pmatrix} \langle\mathcal{T}\hat{\Psi}_{1\uparrow}(X)\hat{\Psi}_{1\uparrow}^{\dagger}(X')\rangle & -\langle\hat{\Psi}_{1\uparrow}^{\dagger}(X')\hat{\Psi}_{1\uparrow}(X)\rangle \\ -\langle\hat{\Psi}_{1\uparrow}(X)\hat{\Psi}_{1\uparrow}^{\dagger}(X')\rangle & -\langle\mathcal{T}^{-1}\hat{\Psi}_{1\uparrow}(X)\hat{\Psi}_{1\uparrow}^{\dagger}(X')\rangle \end{pmatrix} \tag{2.58}$$

$$\breve{F}_1(X, X') = \frac{1}{\mathrm{i}}\begin{pmatrix} \langle\mathcal{T}\hat{\Psi}_{1\uparrow}(X)\hat{\Psi}_{1\downarrow}(X')\rangle & -\langle\hat{\Psi}_{1\downarrow}(X')\hat{\Psi}_{1\uparrow}(X)\rangle \\ -\langle\hat{\Psi}_{1\uparrow}(X)\hat{\Psi}_{1\downarrow}(X')\rangle & -\langle\mathcal{T}^{-1}\hat{\Psi}_{1\uparrow}(X)\hat{\Psi}_{1\downarrow}(X')\rangle \end{pmatrix} \tag{2.59}$$

$$\breve{F}_1^+(X, X') = \frac{1}{\mathrm{i}}\begin{pmatrix} -\langle\mathcal{T}\hat{\Psi}_{1\downarrow}^{\dagger}(X)\hat{\Psi}_{1\uparrow}^{\dagger}(X')\rangle & \langle\hat{\Psi}_{1\uparrow}^{\dagger}(X')\hat{\Psi}_{1\downarrow}^{\dagger}(X)\rangle \\ \langle\hat{\Psi}_{1\downarrow}^{\dagger}(X)\hat{\Psi}_{1\uparrow}^{\dagger}(X')\rangle & \langle\mathcal{T}^{-1}\hat{\Psi}_{1\downarrow}^{\dagger}(X)\hat{\Psi}_{1\uparrow}^{\dagger}(X')\rangle \end{pmatrix} \tag{2.60}$$

$$\check{G}_1^{+}(X, X') = \frac{1}{\mathrm{i}} \begin{pmatrix} -\langle \mathcal{T}\hat{\Psi}_{1\downarrow}^{\dagger}(X)\hat{\Psi}_{1\downarrow}^{\dagger}(X')\rangle & \langle \hat{\Psi}_{1\downarrow}^{\dagger}(X')\hat{\Psi}_{1\downarrow}^{\dagger}(X)\rangle \\ \langle \hat{\Psi}_{1\downarrow}^{\dagger}(X)\hat{\Psi}_{1\downarrow}^{\dagger}(X')\rangle & \langle \mathcal{T}^{-1}\hat{\Psi}_{1\downarrow}^{\dagger}(X)\hat{\Psi}_{1\downarrow}^{\dagger}(X')\rangle \end{pmatrix} \tag{2.61}$$

Here we introduced 4-dimensional vectors $X = (t, \boldsymbol{r})$, $X' = (t', \boldsymbol{r}')$ as well as the time ordering operator $\mathcal{T}$.

Let us now evaluate the Keldysh Green functions (2.58)–(2.61). The electron annihilation and creation operators can be expressed in a standard form

$$\hat{\Psi}_{1\uparrow}(X) = \frac{1}{\sqrt{\mathcal{V}}}\sum_{\boldsymbol{k}} e^{\mathrm{i}\boldsymbol{kr}}\left(u(\boldsymbol{k})e^{-\frac{\mathrm{i}\varepsilon(\boldsymbol{k})t}{\hbar}}\hat{\gamma}_{1\uparrow,\boldsymbol{k}} - v(\boldsymbol{k})e^{\frac{\mathrm{i}\varepsilon(\boldsymbol{k})t}{\hbar}}\hat{\gamma}_{1\downarrow,-\boldsymbol{k}}^{\dagger}\right),$$

$$\hat{\Psi}_{1\uparrow}^{\dagger}(X) = \frac{1}{\sqrt{\mathcal{V}}}\sum_{\boldsymbol{k}} e^{-\mathrm{i}\boldsymbol{kr}}\left(u(\boldsymbol{k})e^{\frac{\mathrm{i}\varepsilon(\boldsymbol{k})t}{\hbar}}\hat{\gamma}_{1\uparrow,\boldsymbol{k}}^{\dagger} - v(\boldsymbol{k})e^{-\frac{\mathrm{i}\varepsilon(\boldsymbol{k})t}{\hbar}}\hat{\gamma}_{1\downarrow,-\boldsymbol{k}}\right),$$

$$\hat{\Psi}_{1\downarrow}(X) = \frac{1}{\sqrt{\mathcal{V}}}\sum_{\boldsymbol{k}} e^{\mathrm{i}\boldsymbol{kr}}\left(u(\boldsymbol{k})e^{-\frac{\mathrm{i}\varepsilon(\boldsymbol{k})t}{\hbar}}\hat{\gamma}_{1\downarrow,\boldsymbol{k}} + v(\boldsymbol{k})e^{\frac{\mathrm{i}\varepsilon(\boldsymbol{k})t}{\hbar}}\hat{\gamma}_{1\uparrow,-\boldsymbol{k}}^{\dagger}\right),$$

$$\hat{\Psi}_{1\downarrow}^{\dagger}(X) = \frac{1}{\sqrt{\mathcal{V}}}\sum_{\boldsymbol{k}} e^{-\mathrm{i}\boldsymbol{kr}}\left(u(\boldsymbol{k})e^{\frac{\mathrm{i}\varepsilon(\boldsymbol{k})t}{\hbar}}\hat{\gamma}_{1\downarrow,\boldsymbol{k}}^{\dagger} + v(\boldsymbol{k})e^{-\frac{\mathrm{i}\varepsilon(\boldsymbol{k})t}{\hbar}}\hat{\gamma}_{1\uparrow,-\boldsymbol{k}}\right) \tag{2.62}$$

Here $\mathcal{V}$ is the volume of a superconductor, $\varepsilon(\boldsymbol{k}) = \hbar^2\boldsymbol{k}^2/2m_1 - \mu$ is the electron energy,

$$u(\boldsymbol{k}) = \frac{1}{\sqrt{2}}\sqrt{1 + \frac{\varepsilon(\boldsymbol{k})}{\sqrt{\varepsilon^2(\boldsymbol{k}) + \Delta_1^2}}}, \quad v(\boldsymbol{k}) = \frac{1}{\sqrt{2}}\sqrt{1 - \frac{\varepsilon(\boldsymbol{k})}{\sqrt{\varepsilon^2(\boldsymbol{k}) + \Delta_1^2}}} \tag{2.63}$$

are BCS coherence factors, and $\hat{\gamma}_{1\alpha,\boldsymbol{k}}^{\dagger}$, $\hat{\gamma}_{1\alpha,\boldsymbol{k}}$ are creation and annihilation operators for quasiparticles with momentum $\boldsymbol{k}$ and spin α. Assuming that the superconductor is in equilibrium at temperature T, we may apply the standard rules of averaging

$$\langle \hat{\gamma}_{1\alpha,\boldsymbol{k}}^{\dagger}\hat{\gamma}_{1\alpha',\boldsymbol{k}'}\rangle = n_{\boldsymbol{k}}\delta_{\boldsymbol{kk}'}\delta_{\alpha\alpha'}, \quad \langle \hat{\gamma}_{1\alpha,\boldsymbol{k}}\hat{\gamma}_{1\alpha',\boldsymbol{k}'}^{\dagger}\rangle = (1 - n_{\boldsymbol{k}})\delta_{\boldsymbol{kk}'}\delta_{\alpha\alpha'}$$

where the occupation probability for quasiparticle states with momentum $\boldsymbol{k}$ is given by the Fermi function

$$n_{\boldsymbol{k}} = \frac{1}{1 + e^{\sqrt{\varepsilon^2(\boldsymbol{k}) + \Delta_1^2}/T}}. \tag{2.64}$$

The Green function (2.57) then takes the form

$$\check{G}_1(X, X') = \int \frac{dE\, d^3\boldsymbol{k}}{(2\pi)^4\hbar}\, e^{-\mathrm{i}E(t-t')/\hbar + \mathrm{i}\boldsymbol{k}(\boldsymbol{r}-\boldsymbol{r}')}\check{\mathcal{G}}_1(E, \boldsymbol{k}), \tag{2.65}$$

where

$$\check{\mathcal{G}}_1(E, \boldsymbol{k}) = \left(\hat{G}_1^R(E, \boldsymbol{k}) + \hat{G}_1^A(E, \boldsymbol{k})\right) \otimes \frac{\hat{\sigma}_z}{2} + \left(\hat{G}_1^R(E, \boldsymbol{k}) - \hat{G}_1^A(E, \boldsymbol{k})\right) \otimes \frac{\hat{Q}_1(E)\hat{\sigma}_z}{2} \quad (2.66)$$

Here the 2×2 retarded and advanced matrix Green functions $\hat{G}_1^R$ and $\hat{G}_1^A$ read

$$\hat{G}_1^{R,A} \equiv \begin{pmatrix} G_1^{R,A} & F_1^{R,A} \\ F_1^{+R,A} & G_1^{+R,A} \end{pmatrix} = \frac{1}{(E \pm \mathrm{i}0)^2 - \varepsilon^2(\boldsymbol{k}) - \Delta_1^2} \begin{pmatrix} E + \varepsilon(\boldsymbol{k}) & \Delta_1 \\ \Delta_1 & E - \varepsilon(\boldsymbol{k}) \end{pmatrix} \quad (2.67)$$

Next, the 2×2 matrix $\hat{Q}_1(E)$ obeys the normalization condition $\hat{Q}_1^2(E) = 1$ and is defined as

$$\hat{Q}_1(E) = \begin{pmatrix} 1 - 2n(E) & 2n(E) \\ 2 - 2n(E) & -1 + 2n(E) \end{pmatrix} \quad (2.68)$$

where $n(E)$ is the quasiparticle distribution function. Provided large superconducting electrodes remain in equilibrium, $n(E)$ coincides with the Fermi function, i.e., $n(E) = 1/(1 + e^{E/T})$. Finally, $\hat{\sigma}_z$ is one of the three 2×2 Pauli matrices,

$$\hat{\sigma}_x = \begin{pmatrix} 0 & 1 \\ 1 & 0 \end{pmatrix}, \quad \hat{\sigma}_y = \begin{pmatrix} 0 & -\mathrm{i} \\ \mathrm{i} & 0 \end{pmatrix}, \quad \hat{\sigma}_z = \begin{pmatrix} 1 & 0 \\ 0 & -1 \end{pmatrix} \quad (2.69)$$

Note that the 4×4 matrix Green function (2.66) contains direct products of the 2×2 matrices. The matrices (2.67) depend on Δ_1 and account for superconducting properties of the first electrode. They can be interpreted as operators acting in the so-called "Nambu space." In contrast, the matrices $\hat{Q}_1(E)$ and $\hat{\sigma}_j$ do not depend on Δ_1 and act in "Keldysh space." Thus, Eqs. (2.65)–(2.68) define the Green function $\check{G}_1$. The Keldysh Green function of the second superconductor, $\check{G}_2$, is defined by interchanging the indices 1 and 2 in these equations.

Turning back to the electronic part of the action $S_{\rm el}$ (2.56), we transform it as follows

$$
\begin{aligned}
S_{\rm el} &= \int dt' \frac{\hbar I_x}{2e}\varphi^-(t') \\
&+ \mathrm{i}\hbar \int_{\mathcal{A}} d^2\boldsymbol{r} \int dt_1 dt_2 \,\mathrm{tr}\left(\check{G}_1(t_1, \boldsymbol{r}, t_2, \boldsymbol{r})\check{t}(t_2, \boldsymbol{r})\check{G}_2(t_2, \boldsymbol{r}, t_1, \boldsymbol{r})\check{t}^\dagger(t_1, \boldsymbol{r})\right) \\
&= \int dt' \frac{\hbar I_x}{2e}\varphi^-(t') \\
&+ \mathrm{i} \int_{\mathcal{A}} d^2\boldsymbol{r} \int dt_1 dt_2 \int \frac{dE_1 dE_2}{(2\pi\hbar)^2} \int \frac{d^3\boldsymbol{k}_1 d^3\boldsymbol{k}_2}{(2\pi)^6} e^{-\mathrm{i}(E_1-E_2)(t_1-t_2)/\hbar} \\
&\times \mathrm{tr}\left(\check{\mathcal{G}}_1(E_1, \boldsymbol{k}_1)\check{t}(t_2, \boldsymbol{r})\check{\mathcal{G}}_2(E_2, \boldsymbol{k}_2)\check{t}^\dagger(t_1, \boldsymbol{r})\right) \qquad (2.70)
\end{aligned}
$$

In order to proceed further and also for later purposes, it is instructive to introduce the so-called quasiclassical Green–Eilenberger matrix functions

$$
\check{g}_j(E_j) = \frac{\mathrm{i}}{\pi\nu_j} \int \frac{d^3\boldsymbol{k}_j}{(2\pi)^3} (\hat{\tau}_z \otimes \hat{1})\, \check{\mathcal{G}}_j(E_j, \boldsymbol{k}_j)\, (\hat{1} \otimes \hat{\sigma}_z), \quad (j = 1, 2) \qquad (2.71)
$$

where $\nu_j = m_j\sqrt{2m_j\mu}/2\pi^2\hbar^3$ is the normal state density of states per spin in the j-th electrode and $\hat{\tau}_z$ is the Pauli matrix equal to $\hat{\sigma}_z$. The reason to choose different notations for these formally identical Pauli matrices $\hat{\sigma}_z$ and $\hat{\tau}_z$ is to emphasize that these matrices are defined, respectively, in Keldysh space and in Nambu space, the latter playing a prominent role in the theory of superconductivity.

The quasiclassical functions are normalized by the condition

$$
\check{g}_j^2(E) = 1 \qquad (2.72)
$$

and obey the Eilenberger equations [51, 110], which will be specified below in Section 2.5.

As usually, we will employ the quasiclassical approximation applicable provided the Fermi energy in the electrodes exceeds all other relevant energy scales in our problem. Technically this approximation amounts to replacing

$$
\int \frac{d^3\boldsymbol{k}_j}{(2\pi)^3} \to \nu_j \int d\varepsilon_j \qquad (2.73)
$$

Then for uniform bulk superconducting electrodes, we get

$$\check{g}_j(E) = -\mathrm{i}\frac{\theta(\Delta_j - |E|)}{\sqrt{\Delta_j^2 - E^2}}\begin{pmatrix} E & \Delta_j \\ -\Delta_j & -E \end{pmatrix} \otimes \hat{1}$$
$$+\frac{\theta(|E| - \Delta_j)}{\sqrt{E^2 - \Delta_j^2}}\frac{E}{|E|}\begin{pmatrix} E & \Delta_j \\ -\Delta_j & -E \end{pmatrix} \otimes \hat{Q}_j(E) \qquad (2.74)$$

With the aid of the quasiclassical Green functions (2.71), we can rewrite the action (2.70) in the form

$$S_{\mathrm{el}} = \int dt' \frac{\hbar I_x}{2e}\varphi^-(t')$$
$$-\pi^2 \mathrm{i}\nu_1\nu_2\hbar \int_{\mathcal{A}} d^2\boldsymbol{r} \int dt_1 dt_2 \int \frac{dE_1 dE_2}{(2\pi\hbar)^2} e^{-\mathrm{i}(E_1-E_2)(t_1-t_2)/\hbar}\,\mathrm{tr}\Big[(\hat{\tau}_z \otimes \hat{1})$$
$$\times\, \check{g}_1(E_1)(\hat{1}\otimes\hat{\sigma}_z)\check{t}(t_2,\boldsymbol{r})(\hat{\tau}_z\otimes\hat{1})\check{g}_2(E_2)(\hat{1}\otimes\hat{\sigma}_z)\check{t}^\dagger(t_1,\boldsymbol{r})\Big] \qquad (2.75)$$

Substituting now Eq. (2.74) into Eq. (2.75), we arrive at the final expression for the AES effective action [107], which reads

$$S_{\mathrm{el}} = S_{\mathrm{R}} + \mathrm{i}S_{\mathrm{I}} \qquad (2.76)$$

where

$$S_{\mathrm{R}} = \int dt' \frac{\hbar I_x}{2e}\varphi^-(t')$$
$$-4\hbar\int dt_1 dt_2 \left[\alpha_R(t_1-t_2)\sin\frac{\varphi^+(t_1)-\varphi^+(t_2)}{2}\right.$$
$$\left.-\beta_R(t_1-t_2)\sin\frac{\varphi^+(t_1)+\varphi^+(t_2)}{2}\right]\sin\frac{\varphi^-(t_1)}{4}\cos\frac{\varphi^-(t_2)}{4} \qquad (2.77)$$

$$S_{\mathrm{I}} = 2\hbar\int dt_1 dt_2 \left[\alpha_I(t_1-t_2)\cos\frac{\varphi^+(t_1)-\varphi^+(t_2)}{2}\right.$$
$$\left.+\beta_I(t_1-t_2)\cos\frac{\varphi^+(t_1)+\varphi^+(t_2)}{2}\right]\sin\frac{\varphi^-(t_1)}{4}\sin\frac{\varphi^-(t_2)}{4} \qquad (2.78)$$

The kernels $\alpha_{R,I}$ and $\beta_{R,I}$ are expressed via the Eilenberger functions of the bulk electrodes (2.74), as we will specify further below.

The expression $F = \exp(\mathrm{i}S_{\mathrm{el}})$ defines the influence functional for the phase variable interacting with an effective environment formed by electrons in the superconducting banks. We observe that the structure of this influence functional is, in many respects, similar to that already discussed in Chapter 1. As before, the total action consists of two parts, real S_{R} and imaginary $\mathrm{i}S_{\mathrm{I}}$. Classical dynamics of the Josephson phase is encoded in the action S_{R}, while the current noise is accounted for by the term S_{I}. Both these contributions to the action are non-local in time. The latter feature is due to retardation effects and dissipative nature of the quasiparticle current.

On the other hand, there are also important differences with the influence functionals considered in Chapter 1. For example, at $I_x = 0$, the action derived here is 4π-periodic in the phase. As we will see in Chapter 7, this property implies that electric charge is transferred across the junction in units of e.

Let us now specify the kernels $\alpha_{R,I}(t)$ and $\beta_{R,I}(t)$ in Eqs. (2.77), (2.78). For this purpose, we introduce the functions

$$\mathcal{N}_j(E) = \frac{E}{\sqrt{\Delta_j^2 - E^2}}\theta(\Delta_j - |E|),$$

$$F_j(E) = \frac{E}{|E|}\frac{\Delta_j}{\sqrt{E^2 - \Delta_j^2}}\theta(|E| - \Delta_j),$$

$$\mathcal{F}_j(E) = \frac{\Delta_j}{\sqrt{\Delta_j^2 - E^2}}\theta(\Delta_j - |E|) \tag{2.79}$$

and define the normal state resistance of our tunnel barrier

$$R_N = \left[\frac{4\pi e^2 \nu_1 \nu_2}{\hbar}\int_{\mathcal{A}} d^2\boldsymbol{r}|t(\boldsymbol{r})|^2\right]^{-1} \tag{2.80}$$

Then the kernels $\alpha_{R,I}(t)$ and $\beta_{R,I}(t)$ can be expressed in the form

$$\alpha_R(t) = -\frac{\hbar}{4e^2 R_N}\int \frac{dE_1 dE_2}{2\pi\hbar^2} e^{-\mathrm{i}(E_1-E_2)t/\hbar}\Big\{\left[\mathcal{N}_1(E_1)+\mathrm{i}N_1(E_1)\right]N_2(E_2)$$
$$\times \tanh\frac{E_2}{2T} + N_1(E_1)\left[\mathcal{N}_2(E_2) - \mathrm{i}N_2(E_2)\right]\tanh\frac{E_1}{2T}\Big\} \tag{2.81}$$

$$\alpha_I(t) = \frac{\hbar}{2e^2 R_N}\int \frac{dE_1 dE_2}{2\pi\hbar^2} e^{-\mathrm{i}(E_1-E_2)t/\hbar} N_1(E_1)N_2(E_2)$$
$$\times \left[1 - \tanh\frac{E_1}{2T}\tanh\frac{E_2}{2T}\right], \tag{2.82}$$

$$\beta_R(t) = -\frac{\hbar}{4e^2R_N}\int\frac{dE_1dE_2}{2\pi\hbar^2}e^{-i(E_1-E_2)t/\hbar}\Big\{\big[\mathcal{F}_1(E_1)+iF_1(E_1)\big]F_2(E_2)$$
$$\times\tanh\frac{E_2}{2T}+F_1(E_1)\big[\mathcal{F}_2(E_2)-iF_2(E_2)\big]\tanh\frac{E_1}{2T}\Big\} \quad (2.83)$$
$$\beta_I(t) = \frac{\hbar}{2e^2R_N}\int\frac{dE_1dE_2}{2\pi\hbar^2}e^{-i(E_1-E_2)t/\hbar}F_1(E_1)F_2(E_2)$$
$$\times\left[1-\tanh\frac{E_1}{2T}\tanh\frac{E_2}{2T}\right]. \quad (2.84)$$

where $N_j(E)$ define the normalized densities of states in the superconducting electrodes (2.4).

Let us discuss the properties of the above functions in more details. Due to causality, the functions $\alpha_R(t)$ and $\beta_R(t)$ differ from zero only at positive times, while at $t < 0$, one finds

$$\alpha_R(t) = \beta_R(t) \equiv 0. \quad (2.85)$$

Technically the requirement of causality is secured by the identities

$$\int dE\big[\mathcal{N}_1(E)+iN_1(E)\big]e^{-i\frac{Et}{\hbar}} = \int dE\big[\mathcal{N}_2(E)-iN_2(E)\big]e^{i\frac{Et}{\hbar}} = 0,$$
$$\int dE\big[\mathcal{F}_1(E)+iF_1(E)\big]e^{-i\frac{Et}{\hbar}} = \int dE\big[\mathcal{F}_2(E)-iF_2(E)\big]e^{i\frac{Et}{\hbar}} = 0$$
$$(2.86)$$

which apply at $t < 0$ and can be verified with the aid of the following integrals

$$\int\frac{dE}{2\pi\hbar}N_r(E)e^{-iEt/\hbar} = \delta(t)-\frac{\Delta_r}{2\hbar}J_1\left(\frac{\Delta_r|t|}{\hbar}\right), \quad (2.87)$$

$$\int\frac{dE}{2\pi\hbar}\mathcal{N}_r(E)e^{-iEt/\hbar} = -i\frac{\Delta_r}{2\hbar}J_1\left(\frac{\Delta_r t}{\hbar}\right),$$
$$\int\frac{dE}{2\pi\hbar}F_r(E)e^{-iEt/\hbar} = -i\frac{\Delta_r}{2\hbar}J_0\left(\frac{\Delta_r t}{\hbar}\right)\text{sign}\,t,$$
$$\int\frac{dE}{2\pi\hbar}\mathcal{F}_r(E)e^{-iEt/\hbar} = \frac{\Delta_r}{2\hbar}J_0\left(\frac{\Delta_r t}{\hbar}\right), \quad (2.88)$$

where $J_{0,1}(x)$ are the Bessel functions.

Let us emphasize that the causality requirement is an important one. For instance, the current value I at time t may depend only on the phase values $\varphi(t')$ at earlier times $t' \leq t$. This is indeed the case, as it is demonstrated below, e.g., by Eq. (2.97).

The kernels $\alpha_{R,I}(t)$ and $\beta_{R,I}(t)$ may also be expressed in an alternative form revealing their close relation to transport properties of Josephson junctions in the classical regime [256, 404]. Let us remind the reader that in this regime, the current across a voltage-biased Josephson junction consists of three different contributions. One of them is the Josephson current $I_C \sin\varphi$. Another one is a dissipative current produced by quasiparticles

$$I_{qp}(V) = \frac{1}{2eR_N}\int dE\, N_1\left(E+\frac{eV}{2}\right) N_2\left(E-\frac{eV}{2}\right) \times\left[\tanh\left(\frac{E}{2T}+\frac{eV}{4T}\right)-\tanh\left(\frac{E}{2T}-\frac{eV}{4T}\right)\right] \quad (2.89)$$

Yet another contribution is represented by the so-called interference current between quasiparticles and Cooper pairs, which also has a dissipative nature and can be written in the form [256, 404]

$$I_{\cos}(V) = \frac{1}{2eR_N}\int dE\, F_1\left(E+\frac{eV}{2}\right) F_2\left(E-\frac{eV}{2}\right) \times\left[\tanh\left(\frac{E}{2T}+\frac{eV}{4T}\right)-\tanh\left(\frac{E}{2T}-\frac{eV}{4T}\right)\right] \quad (2.90)$$

In order to relate the currents (2.89), (2.90) and the kernels (2.81)–(2.84) to each other, we make use of the integrals (2.88) and derive the identities

$$\int \frac{dE_1}{2\pi\hbar}\left[\mathcal{N}_1(E_1)+\mathrm{i}N_1(E_1)\right]e^{-iE_1t/\hbar} = 2\mathrm{i}\theta(t)\int \frac{dE_1}{2\pi\hbar}N_1(E_1)e^{-\mathrm{i}E_1t/\hbar},$$
$$\int \frac{dE_2}{2\pi\hbar}\left[\mathcal{N}_2(E_2)-\mathrm{i}N_2(E_2)\right]e^{iE_2t/\hbar} = -2\mathrm{i}\theta(t)\int \frac{dE_2}{2\pi\hbar}N_2(E_2)e^{\mathrm{i}E_2t/\hbar},$$
$$\int \frac{dE_1}{2\pi\hbar}\left[\mathcal{F}_1(E_1)+\mathrm{i}F_1(E_1)\right]e^{-iE_1t/\hbar} = 2\mathrm{i}\theta(t)\int \frac{dE_1}{2\pi\hbar}F_1(E_1)e^{-\mathrm{i}E_1t/\hbar},$$
$$\int \frac{dE_2}{2\pi\hbar}\left[\mathcal{F}_2(E_2)-\mathrm{i}F_2(E_2)\right]e^{iE_2t/\hbar} = -2\mathrm{i}\theta(t)\int \frac{dE_2}{2\pi\hbar}F_2(E_2)e^{\mathrm{i}E_2t/\hbar}$$

where $\theta(t)$ is the Heaviside theta-function taking care of the causality principle. Then, after some manipulations and with the aid of the identity

$$1-\tanh\frac{E_1}{2T}\tanh\frac{E_2}{2T} = \left[\tanh\frac{E_1}{2T}-\tanh\frac{E_2}{2T}\right]\coth\frac{E_1-E_2}{2T}, \quad (2.91)$$

we arrive at the following relations

$$
\begin{aligned}
\alpha_R(t) &= \frac{\theta(t)}{e} \int \frac{d\omega}{2\pi} I_{qp}\left(\frac{\hbar\omega}{e}\right) \sin\omega t, \\
\alpha_I(t) &= \frac{1}{e} \int \frac{d\omega}{2\pi} I_{qp}\left(\frac{\hbar\omega}{e}\right) \coth\frac{\hbar\omega}{2T} \cos\omega t, \\
\beta_R(t) &= \frac{\theta(t)}{e} \int \frac{d\omega}{2\pi} I_{\cos}\left(\frac{\hbar\omega}{e}\right) \sin\omega t, \\
\beta_I(t) &= \frac{1}{e} \int \frac{d\omega}{2\pi} I_{\cos}\left(\frac{\hbar\omega}{e}\right) \coth\frac{\hbar\omega}{2T} \cos\omega t,
\end{aligned}
\tag{2.92}
$$

which demonstrate that the kernels $\alpha_R(t)$ and $\beta_R(t)$ are simply proportional to the Fourier transformed currents for $I_{qp}(V)$ and $I_{\cos}(V)$, respectively, while the expressions $\alpha_I(t)$ and $\beta_I(t)$ also contain an extra factor $\coth(\hbar\omega/2T)$. The latter observation indicates that—in the agreement with FDT—these kernels are proportional to the current noise.

To complete this section, let us briefly discuss the validity range of the effective action (2.77) and (2.78) derived here. As in the course of our derivation we proceeded perturbatively in the tunneling Hamiltonian, it is necessary to require the tunneling transmission amplitude $t(\boldsymbol{r})$ to be sufficiently small. In order to quantify this requirement, it is necessary to expand the exact action to higher orders in $t(\boldsymbol{r})$ and compare the resulting expressions with the second-order terms derived here. We postpone this expansion and further details to Section 2.5 and here only quote the corresponding validity condition for the AES action (2.77) and (2.78). It reads

$$
\frac{e^2 R_N}{\pi\hbar} \frac{k_F^2 \mathcal{A}}{4\pi^2} \gg 1 \tag{2.93}
$$

This condition is equivalent to requirement of small transmissions for all conducting channels of our junction. Obviously, it allows for large values of the junction total conductance $\pi\hbar/e^2 R_N > 1$. The latter inequality is often referred to as strong tunneling limit.

2.2.4 Electric Current

Let us now express the electric current across the Josephson junction as a path integral over the fluctuating phases $\varphi^{\pm}$. Combining the definition (2.48) with the explicit form of the action (2.77) and (2.78), we arrive at the following result:

$$I(t) = \int \mathcal{D}\varphi^+ \mathcal{D}\varphi^- \, \tilde{I}[\varphi^+, \varphi^-] \, e^{iS_{el}[\varphi^+,\varphi^-]/\hbar + iS_{env}[\varphi^+,\varphi^-]/\hbar} \tag{2.94}$$

Here the current depending on the instantaneous phase values

$$\tilde{I}[\varphi^+, \varphi^-] = -\frac{2e}{\hbar} \frac{\delta S_{el}}{\delta \varphi^-(t)}\bigg|_{\varphi^-(t)=0} \tag{2.95}$$

reads

$$\begin{aligned}
\tilde{I}[\varphi^+, \varphi^-] = 2e \int dt' \bigg[& \alpha_R(t-t') \sin\frac{\varphi^+(t) - \varphi^+(t')}{2} \\
& - \beta_R(t-t') \sin\frac{\varphi^+(t) + \varphi^+(t')}{2} \bigg] \cos\frac{\varphi^-(t')}{4} \\
- 2ie \int dt' \bigg[& \alpha_I(t-t') \cos\frac{\varphi^+(t) - \varphi^+(t')}{2} \\
& + \beta_I(t-t') \cos\frac{\varphi^+(t) + \varphi^+(t')}{2} \bigg] \sin\frac{\varphi^-(t')}{4}.
\end{aligned} \tag{2.96}$$

Due to peculiar properties of the action (2.77) and (2.78), the phase-dependent current (2.96) turns out to be a complicated function of both time and the phases. Indeed, it depends on $\varphi^\pm$ in a nonlinear manner, and, on top of that, it also depends on the prehistory of their time evolution. Let us first consider the classical limit and put $\varphi^- = 0$. Formally, this limit is achieved provided the environment impedance remains small, i.e., $Z_S(\omega) \to 0$. In this limit, the environment action S_{env} (2.49) acquires a very large negative real part for any $\varphi^- \neq 0$, thus fully suppressing fluctuations of the quantum phase φ^-. Hence, the junction operates in the regime of time-dependent voltage bias $V(t) = I_x(t) Z_s(0)$ and the current reduces to

$$\begin{aligned}
I(t) = \tilde{I}[\varphi, 0] = 2e \int_{-\infty}^{t} dt' \bigg[& \alpha_R(t-t') \sin\frac{\varphi^+(t) - \varphi^+(t')}{2} \\
& - \beta_R(t-t') \sin\frac{\varphi^+(t) + \varphi^+(t')}{2} \bigg]
\end{aligned} \tag{2.97}$$

with $\dot{\varphi}^+(t) = 2eV(t)/\hbar$. In particular, for a constant voltage bias $\varphi^+(t) = 2eVt/\hbar$, the current takes the form [256, 404]

$$I(t) = I_{qp}(V) + I_C(V) \sin[2eVt/\hbar] + I_{\cos}(V) \cos[2eVt/\hbar] \tag{2.98}$$

The terms $I_{qp}(V)$ and $I_{\cos}(V)$ are defined, respectively, in Eqs. (2.89) and (2.90), while $I_C(V)$ equals to

$$I_C(V) = \int \frac{dV'}{2\pi} I_{\cos}(V') \frac{2V'}{V^2 - V'^2} \tag{2.99}$$

Equations (2.97) and (2.98) form a basis for theoretical description of classical dynamics of Josephson junctions.

Turning back to the general expression (2.96), we note that it can be split into two parts. The first part comes from the action S_R and contains the kernels $\alpha_R(t)$ and $\beta_R(t)$. This part can be interpreted as a direct generalization of the classical formula (2.97) to the quantum regime with nonzero phase φ^-. The second part containing the kernels $\alpha_I(t)$ and $\beta_I(t)$ originates from S_I and vanishes in the classical limit. This part can be interpreted as a correction to the current due to, respectively, shot noise (α_I-terms) and fluctuations of the Josephson current (β_I-terms).

2.2.5 Adiabatic Limit

Although the effective action (2.76)–(2.78) is already much simpler than the initial expression (2.52), it still remains sufficiently complicated to deal with. Further simplifications can be achieved in the low-energy limit, i.e., at subgap temperatures and voltages $T, eV \ll \Delta_{1,2}$. In this regime, the quasiparticle current is exponentialy small, $I_{qp} \sim \exp[-\min\{\Delta_1, \Delta_2\}/T]$. Hence, in the leading approximation, we can put $I_{qp} = 0$, and $\alpha_R(t) = \alpha_I(t) = 0$. Similar considerations show that we may also put $\beta_I(t) = 0$. Thus, the only remaining nonzero kernel is $\beta_R(t)$ (2.83). In the limit of slow voltage fluctuations, it can be reduced to a simple expression

$$\beta_R(t) = -\frac{I_C}{2e}\delta(t), \tag{2.100}$$

where I_C is the Josephson critical current given by the integral

$$I_C = \frac{1}{eR_N} \int dE \{\mathcal{F}_1(E) F_2(E) + F_1(E)\mathcal{F}_2(E)\} \tanh \frac{E}{2T} \tag{2.101}$$

This integral can be evaluated in several limiting cases. In particular, if one considers the junction between two identical superconductors with $\Delta_1 = \Delta_2 = \Delta$, the critical current takes the form [14]

$$I_C = \frac{\pi\Delta}{2eR_N} \tanh \frac{\Delta}{2T} \tag{2.102}$$

At $T \to 0$ and for $\Delta_1 < \Delta_2$, one finds [14]

$$I_C = \frac{\Delta_1}{eR_N} K\left(\sqrt{1 - \frac{\Delta_1^2}{\Delta_2^2}}\right) \tag{2.103}$$

where $K(x)$ is the complete elliptic integral of the first kind.

Employing Eq. (2.100), we arrive at the adiabatic form of the action $S = S_{\text{env}} + S_{\text{el}}$, which reads

$$S = \int dt_1 \left(\frac{C\hbar^2\dot{\varphi}^+\dot{\varphi}^-}{4e^2} - \frac{\hbar I_C}{e}\sin\varphi^+ \sin\frac{\varphi^-}{2} + \frac{\hbar I_x}{2e}\varphi^-\right) + S_{\text{diss}}[\varphi^+, \varphi^-], \tag{2.104}$$

where the dissipative part of the action,

$$S_{\text{diss}}[\varphi^+, \varphi^-] = -\int dt_1 dt_2 \frac{\hbar\varphi^-(t_1)}{2e} Z_S^{-1}(t_1 - t_2)\frac{\hbar\dot{\varphi}^+(t_2)}{2e} + \frac{\mathrm{i}}{2}\int dt_1 dt_2 \frac{\hbar\varphi^-(t_1)}{2e} K_S(t_1 - t_2)\frac{\hbar\varphi^-(t_2)}{2e} \tag{2.105}$$

differs from the total action of the environment (2.49) due to the absence of the junction capacitance. Rewriting the adiabatic action in terms of the Keldysh phases φ_F and φ_B, we find

$$S = S_0[\varphi_F] - S_0[\varphi_B] + S_{\text{diss}}[(\varphi_F + \varphi_B)/2, \varphi_F - \varphi_B] \tag{2.106}$$

with

$$S_0[\varphi] = \int dt' \left(\frac{C}{2}\frac{\hbar^2\dot{\varphi}^2}{4e^2} + E_J \cos\varphi + \frac{\hbar I_x}{2e}\varphi\right) \tag{2.107}$$

where we defined the so-called Josephson coupling energy $E_J = \hbar I_C/2e$. Equation (2.107) just represents the action of a particle with mass $m^* = C\hbar^2/4e^2$ moving in a tilted washboard potential

$$U(\varphi) = -\frac{\hbar I_C}{2e}\cos\varphi - \frac{\hbar I_x}{2e}\varphi. \tag{2.108}$$

Taking the limit of vanishing dissipation $Z_S(\omega) \to \infty$ and switching from the path integral formulation of quantum mechanics to the Hamilton one, we conclude that in the low-energy limit, quantum dynamics of a Josephson junction is indeed described by the Hamilton operator (2.15), as we already anticipated in the beginning of this chapter.

More generally, we have demonstrated that low-energy quantum dynamics of the Josephson junction is equivalent to that of a quantum particle moving in a tilted Josephson potential (2.108) and linearly coupled to an oscillator bath described by the action (2.105) in the spirit of the Caldeira–Leggett model. The bath spectrum should be properly adjusted to emulate the frequency dependence of the impedance of the electromagnetic environment $Z_S(\omega)$.

The formal expression for the current (2.94) is also greatly simplified in the adiabatic regime. Combining Eqs. (2.96) and (2.106), one arrives at the result

$$I(t) = \int \mathcal{D}\varphi_F \mathcal{D}\varphi_B \, I_C \sin\varphi_F(t) \times e^{iS_0[\varphi_F]/\hbar - iS_0[\varphi_B]/\hbar + iS_{\rm diss}[\varphi_F,\varphi_B]/\hbar}\Big|_{\varphi_F(t)=\varphi_B(t)} \quad (2.109)$$

Assuming that characteristic frequencies ω of voltage or phase fluctuations remain much smaller than the superconducting gap, $\omega \ll \Delta$, one can expand the exact expression for the effective action in powers of ω/Δ and find corrections to the adiabatic form of the action due to retardation effects. For example, it is straightforward to demonstrate that both kernels α_R and β_R generate corrections $\propto \omega^2/\Delta^2$, which can be absorbed in the capacitive term in the action since it has exactly the same dependence on ω. As a result, we again arrive at the action (2.104), where the capacitance C should be replaced by its renormalized value

$$C^*(\varphi^+) = C + \frac{3\pi\hbar}{32\Delta R_N}\left(1 - \frac{\cos\varphi^+}{3}\right) \quad (2.110)$$

The first term inside the brackets in Eq. (2.110) originates from the kernel α_R. It does not depend on φ^+ and describes the effect of capacitance renormalization due to quasiparticle tunneling [345]. The second term containing $\cos\varphi^+$ emerges from the kernel β_R. This term is frequently neglected in the literature, though it is in general of the same order as the first term and, hence, should also be accounted for within the accuracy adopted here.

2.2.6 Action for Normal Tunnel Junctions

Although in the above analysis it was always implied that both metallic electrodes are superconducting, particular values of their superconducting gaps Δ_1 and Δ_2 were not really important for our general derivation. Thus, this derivation remains valid even if at least one such gap equals to zero or, in other words, at least one of the electrodes is normal. In this case, the kernels $\beta_R(t)$ and $\beta_I(t)$ vanish identically, while the kernels $\alpha_R(t)$ and $\alpha_I(t)$ become simpler. Of particular interest is to consider the behavior of normal tunnel junctions in which case one should set $\Delta_1 = \Delta_2 = 0$. Then the quasiparticle current (2.89) between two normal metals reduces to Ohmic $I_{qp} = V/R_N$. Evaluating the integrals in Eqs. (2.92), one finds

$$\alpha_R(t) = -\frac{\hbar}{e^2 R_N}\theta(t)\frac{d}{dt}\delta(t),$$

$$\alpha_I(t) = \frac{1}{e^2 R_N}\int\frac{d\omega}{2\pi}\hbar\omega\coth\frac{\hbar\omega}{2T}\cos\omega t = \frac{\hbar}{e^2 R_N}K_N(t) \qquad (2.111)$$

where

$$K_N(t) = \int\frac{d\omega}{2\pi}\omega\coth\frac{\hbar\omega}{2T}e^{-i\omega t} = -\frac{\pi T^2}{\hbar^2\sinh^2[\pi T t/\hbar]} \qquad (2.112)$$

The action (2.77), (2.78) for the system "junction plus environment" then acquires the form

$$S[\varphi^+,\varphi^-] = \int dt_1\left(\frac{C\hbar^2\dot\varphi^+\dot\varphi^-}{4e^2} - \frac{\hbar^2}{2e^2 R_N}\dot\varphi^+\sin\frac{\varphi^-}{2} + \frac{\hbar I_x}{2e}\varphi^-\right)$$
$$+\frac{2i\hbar^2}{e^2 R_N}\int dt_1 dt_2 K_N(t_1-t_2)\cos\frac{\varphi^+(t_1)-\varphi^+(t_2)}{2}$$
$$\times\sin\frac{\varphi^-(t_1)}{4}\sin\frac{\varphi^-(t_2)}{4} + S_{\rm diss}[\varphi^+,\varphi^-] \qquad (2.113)$$

Accordingly, the current across normal tunnel junctions is given by the path integral

$$I(t) = \frac{1}{R_N}\int\mathcal{D}\varphi^+\mathcal{D}\varphi^-\, e^{iS[\varphi^+,\varphi^-]/\hbar}\left(\frac{\hbar\dot\varphi^+(t)}{2e}\cos\frac{\varphi^-(t)}{2}\right.$$
$$-\frac{2i\hbar}{e}\int dt' K_N(t-t')\cos\frac{\varphi^+(t)-\varphi^+(t')}{2}$$
$$\left.\times\cos\frac{\varphi^-(t)}{4}\sin\frac{\varphi^-(t')}{4}\right) \qquad (2.114)$$

Note that in the normal case, the variables φ_F and φ_B cannot anymore be interpreted as fluctuating superconducting phases across the Josephson junction. On the other hand, these variables can still be defined with the aid of Eq. (2.24), i.e., in the normal limit they are simply proportional to the time integral of fluctuating voltages across the tunnel junction. They remain convenient variables also in this case.

2.3 Langevin Equation

Let us now turn back to the case of superconducting junctions. As we have already discussed, in the limit of very strong dissipation $Z_S(\omega) \to 0$, fluctuations of the quantum phase φ^- are completely suppressed and the path integral expression for the current (2.94), (2.96) reduces to the classical result (2.97). Let us now keep an external impedance small but finite. In this case, dissipation remains strong, although not anymore strong enough to fully suppress quantum fluctuations. Under the condition

$$\frac{2\pi\hbar}{e^2}\frac{dI_{qp}(V)}{dV} + \frac{2\pi\hbar}{e^2 Z_S(eV/\hbar)} \gg 1 \tag{2.115}$$

fluctuations of the quantum phase φ^- remain weak, $\langle(\varphi^-)^2\rangle \ll \pi^2$, and it suffices to expand the action (2.76)–(2.78) in powers of φ^- to the second order. Then we obtain

$$\begin{aligned} S_{\mathrm{R}} = &\int dt_1 \frac{\hbar I_x}{2e}\varphi^-(t_1) - \hbar\int dt_1 dt_2 \left[\alpha_R(t_1-t_2)\sin\frac{\varphi^+(t_1)-\varphi^+(t_2)}{2}\right. \\ &\left. - \beta_R(t_1-t_2)\sin\frac{\varphi^+(t_1)+\varphi^+(t_2)}{2}\right]\varphi^-(t_1) \end{aligned} \tag{2.116}$$

$$\begin{aligned} S_{\mathrm{I}} = &\frac{\hbar}{8}\int dt_1 dt_2 \left[\alpha_I(t_1-t_2)\cos\frac{\varphi^+(t_1)-\varphi^+(t_2)}{2}\right. \\ &\left. + \beta_I(t_1-t_2)\cos\frac{\varphi^+(t_1)+\varphi^+(t_2)}{2}\right]\varphi^-(t_1)\varphi^-(t_2) \end{aligned} \tag{2.117}$$

Let us now employ the same idea as in Chapter 1 and introduce, in the spirit of the Hubbard–Stratonovich transformation, the noise current of the junction $\xi_t(t_1, \varphi^+(t_1))$ as an extra quantum variable.

Then the part of the action S_I can be rewritten in the form

$$\exp\left\{-\frac{S_I[\varphi^+,\varphi^-]}{\hbar}\right\} = \int \mathcal{D}\xi(t)\exp\left\{\frac{\mathrm{i}}{\hbar}\int dt_1 \frac{\hbar\varphi^-(t_1)}{2e}\xi_t(t_1,\varphi^+(t_1))\right\}$$
$$\times\exp\left\{-\frac{1}{2}\int dt_1 dt_2 \xi_t(t_1,\varphi^+(t_1))\left(\langle\xi_t(t_1,\varphi^+(t_1))\xi_t(t_2,\varphi^+(t_2))\rangle^{-1}\right)\right.$$
$$\left.\times\,\xi_t(t_2,\varphi^+(t_2))\right\} \qquad (2.118)$$

The pair correlator for Gaussian noise $\xi_t(t_1,\varphi^+)$ is defined as

$$\langle\xi_t(t_1,\varphi(t_1))\xi_t(t_2,\varphi(t_2))\rangle = e^2\left[\alpha_I(t_1-t_2)\cos\frac{\varphi(t_1)-\varphi(t_2)}{2}\right.$$
$$\left.+\,\beta_I(t_1-t_2)\cos\frac{\varphi(t_1)+\varphi(t_2)}{2}\right] \qquad (2.119)$$

Provided the voltage bias remains constant in time, one has $\varphi(t) = 2eVt/\hbar$ and Eq. (2.119) reduces to the standard formula (2.8). The correlator (2.119) serves as a direct generalization of Eq. (2.8) to arbitrarily time-dependent voltages across the junction. The symbolic expression $\langle\xi_t(t_1,\varphi(t_1))\xi_t(t_2,\varphi(t_2))\rangle^{-1}$ in Eq. (2.118) stands for the inverse operator of the noise correlator, which is formally defined by the identity

$$\int dt'\langle\xi_t(t_1,\varphi(t_1))\xi_t(t',\varphi(t'))\rangle\langle\xi_t(t',\varphi(t'))\xi_t(t_2,\varphi(t_2))\rangle^{-1} = \delta(t_1-t_2).$$

The explicit form of the inverse operator is not important for our considerations.

Performing the same transformation with the imaginary part of the environment action (2.49) containing the kernel $K_S(t)$, we introduce the environment noise $\xi_S(t)$ with the pair correlator (2.7). After that, the full action $S = S_{\mathrm{el}} + S_{\mathrm{env}}$ contains only the terms that either are linear in the quantum phase φ^- or do not depend on this phase at all (quadratic terms in noise variables). The path integral over φ^- then becomes trivial and gives the functional delta-function symbolically denoted as $\delta[\delta S/\delta\varphi^-]$. Further evaluation of the path integral over the classical phase φ^+ just requires solving the equation $\delta S/\delta\varphi^- = 0$. This equation has the form of the Langevin equation [107, 146]

$$C\frac{\hbar\ddot{\varphi}}{2e}+\int dt' Z_S^{-1}(t-t')\frac{\hbar\dot{\varphi}}{2e}+2e\int dt'\alpha_R(t-t')\sin\frac{\varphi(t)-\varphi(t')}{2}$$
$$-2e\int dt'\beta_R(t-t')\sin\frac{\varphi(t)+\varphi(t')}{2}=I_x+\xi_t(t,\varphi(t))+\xi_S(t) \tag{2.120}$$

Here, for simplicity, we omitted the superscript for the phase variable, i.e., we set $\varphi^+ \to \varphi$.

Comparing now Eqs. (2.120) and (2.2), we observe that they differ at two important points. First, the expressions for the quasiparticle and Josephson currents in Eq. (2.120) are non-local in time, unlike the corresponding expressions in Eq. (2.2). This difference could be expected from the very beginning. It could be eliminated should we employ a more accurate non-local in time expression for the tunneling current (2.97) in Eq. (2.2). Second, in Eq. (2.120), the junction noise $\xi_t(t)$ depends on the phase $\varphi(t)$, while such dependence is absent in Eq. (2.2). The origin of this difference is less obvious. As we will demonstrate in Chapter 7, this dependence accounts for Coulomb effects, e.g., for the Coulomb blockade correction to the current across the junction. We also note that the junction current noise $\xi_t(t,\varphi(t))$ is fully determined by the imaginary part of the action S_I (2.117).

2.4 Imaginary-Time Action for Josephson Junctions

The Keldysh action or, equivalently, the Feynman–Vernon influence functional for a Josephson junction derived above fully accounts for real-time non-equilibrium quantum dynamics of the Josephson phase. There is, however, an important class of problems where equilibrium thermodynamic properties of Josephson junctions are of interest. In this case, it can be more appropriate to analyze the behavior of the grand partition function of the system or its equilibrium density matrix. As we already discussed in Chapter 1, in such situations, it is convenient to employ the imaginary-time Matsubara technique, which allows to recover the imaginary-time (Euclidean) version of effective action for our system, $S[\varphi(\tau)]$. With the aid of this action, one can directly approach various thermodynamic properties of the system, such as free energy and specific heat. On top of that, within the imaginary-time technique, one can also analyze transport properties of the system in the linear regime. It order to access this information, it is necessary to evaluate the so-called Matsubara susceptibility of the system [273] with its subsequent analytic continuation to real time. In this way, one can recover, e.g., the system linear conductance by applying the Kubo formula. Yet another important application of the imaginary-time technique is the analysis of both thermal and quantum decay rates for metastable states. All these issues will be extensively addressed in the next chapters of this book.

Let us express the junction grand canonical partition function $\mathcal{Z}$ in the form of the path integral over the fluctuating phase $\varphi(\tau)$. It reads

$$\mathcal{Z} = \mathrm{tr}\left[e^{-\hat{H}/T}\right] = \int \mathcal{D}\varphi(\tau)\, e^{-S[\varphi(\tau)]/\hbar} \tag{2.121}$$

where $\hat{H}$ is the junction Hamiltonian (2.32). The phase variable $\varphi(\tau)$ is now defined as a function of imaginary time τ restricted to the interval $0 < \tau < \hbar\beta$ where, as before, we introduced the inverse temperature $\beta = 1/T$. The phase $\varphi(\tau)$ obeys the Josephson relation

$$\hbar\frac{\partial\varphi}{\partial\tau} = 2eV(\boldsymbol{r}, t) \tag{2.122}$$

which links it to the fluctuating voltage V across the junction.

Proceeding similarly to our previous real-time analysis, we split the action of the junction into two parts describing, respectively, an external electromagnetic environment (S_{env}) and electrons tunneling through the junction (S_{el}), i.e.,

$$S[\varphi] = S_{\mathrm{env}}[\varphi] + S_{\mathrm{el}}[\varphi] \tag{2.123}$$

The action of the environment is given by the Euclidean action of the Caldeira–Leggett bath with the spectral density (2.31) and reads (cf. Section 1.6):

$$\begin{aligned} S_{\mathrm{env}}[\varphi] = &\int_0^{\hbar\beta} d\tau \frac{C}{2}\left(\frac{\hbar\dot{\varphi}}{2e}\right)^2 \\ &+ \frac{1}{4}\int_0^{\hbar\beta} d\tau \int_0^{\hbar\beta} d\tau' \tilde{\alpha}_S(\tau-\tau')\left[\frac{\hbar\varphi(\tau)}{2e} - \frac{\hbar\varphi(\tau')}{2e}\right]^2 \end{aligned} \tag{2.124}$$

where

$$\tilde{\alpha}_S(\tau) = -\frac{T}{\hbar}\sum_{\Omega_n}\frac{|\Omega_n|}{Z_S(\mathrm{i}|\Omega_n|)}\cos(\Omega_n\tau) \tag{2.125}$$

and $\Omega_n = 2\pi T n/\hbar$ are bosonic Matsubara frequencies.

In order to derive the electronic part of the action, we will follow the same route as in the previous section. The first step of the derivation is the integration over the Grassmann electronic fields ψ, $\psi^\dagger$. The corresponding integral is Gaussian, and as before, it is expressed via the determinant of the inverse Green function, cf. Eqs. (2.42), (2.45). Accordingly, the electronic part of the action acquires the form analogous to that of the real-time action (2.52). One finds

$$S_{\rm el}[\varphi] = -\hbar\,\mathrm{tr}\left(\hat{\mathcal{G}}^{-1}\right), \tag{2.126}$$

where the imaginary-time Green function has the block structure

$$\hat{\mathcal{G}}^{-1} = \begin{pmatrix} \hat{\mathcal{G}}_1^{-1} & \hat{t} \\ \hat{t}^\dagger & \hat{\mathcal{G}}_2^{-1} \end{pmatrix} \tag{2.127}$$

Here $\mathcal{G}_1^{-1}$ and $\mathcal{G}_2^{-1}$ are the inverse Matsubara Green functions of the electrodes defined as 2×2 matrices in the Nambu space

$$\hat{\mathcal{G}}_j^{-1} = \begin{pmatrix} -\hbar\frac{\partial}{\partial\tau} + \frac{\hbar^2\nabla^2}{2m_j} + \mu & -\Delta_j \\ -\Delta_j & -\hbar\frac{\partial}{\partial\tau} - \frac{\hbar^2\nabla^2}{2m_j} - \mu \end{pmatrix} \tag{2.128}$$

while the operator $\hat{t}$ describing tunneling of electrons between the electrodes reads

$$\hat{t} = \begin{pmatrix} t(\boldsymbol{r})e^{\mathrm{i}\varphi(\tau)/2} & 0 \\ 0 & -t(\boldsymbol{r})e^{-\mathrm{i}\varphi(\tau)/2} \end{pmatrix} \tag{2.129}$$

Next we expand the trace in Eq. (2.126) in powers of $\hat{t}$ to the second order,

$$S_{\rm el}[\varphi] = -\hbar\,\mathrm{tr}\left(\hat{\mathcal{G}}_0^{-1}\right) - \hbar\,\mathrm{tr}\left(\hat{\mathcal{G}}_0^{-1}\delta\hat{G}^{-1}\right) + \frac{\hbar}{2}\,\mathrm{tr}\left(\hat{\mathcal{G}}_0^{-1}\delta\hat{G}^{-1}\hat{\mathcal{G}}_0^{-1}\delta\hat{G}^{-1}\right) \tag{2.130}$$

where

$$\hat{\mathcal{G}}_0^{-1} = \begin{pmatrix} \hat{\mathcal{G}}_1^{-1} & 0 \\ 0 & \hat{\mathcal{G}}_2^{-1} \end{pmatrix}, \quad \delta\hat{\mathcal{G}}^{-1} = \begin{pmatrix} 0 & \hat{t} \\ \hat{t}^\dagger & 0 \end{pmatrix} \tag{2.131}$$

As before, the lowest-order term in this expansion can be expressed in the form (cf. Eq. (2.54))

$$-\hbar\,\mathrm{tr}\left(\hat{\mathcal{G}}_0^{-1}\right) = -\int_0^\beta d\tau \frac{\hbar I_x}{2e}\varphi(\tau) \tag{2.132}$$

It is straightforward to verify that the first-order term vanishes,

$$-\hbar\,\mathrm{tr}\left(\hat{\mathcal{G}}_0^{-1}\delta\hat{G}^{-1}\right) = 0.$$

Evaluating the second-order term, we arrive at the imaginary-time version of the AES action [107]

$$\begin{aligned} &S_{\mathrm{el}}[\varphi] \\ &= -\int_0^\beta d\tau \frac{\hbar I_x}{2e}\varphi(\tau) - \hbar\int_0^\beta d\tau \int_0^\beta d\tau' \left[\alpha(\tau-\tau')\cos\frac{\varphi(\tau)-\varphi(\tau')}{2}\right. \\ &\left. -\beta(\tau-\tau')\cos\frac{\varphi(\tau)+\varphi(\tau')}{2}\right] \end{aligned} \tag{2.133}$$

Non-local kernels $\alpha(\tau)$ and $\beta(\tau)$ in Eq. (2.133) are fixed by the relations

$$\begin{aligned} &\alpha(\tau) \\ &= -\frac{T^2}{2\pi\hbar^3 e^2 \nu_1\nu_2 R_N}\sum_{\omega_\mu,\omega_\nu}\int \frac{d^3\boldsymbol{k}_1 d^3\boldsymbol{k}_2}{(2\pi)^6}\mathcal{G}_1(\omega_\nu,\boldsymbol{k}_1)\mathcal{G}_2(\omega_\mu,\boldsymbol{k}_2)e^{\mathrm{i}(\omega_\mu-\omega_\nu)\tau} \\ &\beta(\tau) \\ &= -\frac{T^2}{2\pi\hbar^3 e^2 \nu_1\nu_2 R_N}\sum_{\omega_\mu,\omega_\nu}\int \frac{d^3\boldsymbol{k}_1 d^3\boldsymbol{k}_2}{(2\pi)^6}\mathcal{F}_1(\omega_\nu,\boldsymbol{k}_1)\mathcal{F}_2(\omega_\mu,\boldsymbol{k}_2)e^{\mathrm{i}(\omega_\mu-\omega_\nu)\tau} \end{aligned} \tag{2.134}$$

where

$$\mathcal{G}_j(\omega_\mu,\boldsymbol{k}) = -\hbar\frac{\mathrm{i}\hbar\omega_\nu + \frac{\hbar^2k^2}{2m_j}-\mu}{\hbar^2\omega_\nu^2+\left(\frac{\hbar^2k^2}{2m_j}-\mu\right)^2+\Delta_j^2},$$

$$\mathcal{F}_j(\omega_\mu,\boldsymbol{k}) = \frac{\hbar\Delta_j}{\hbar^2\omega_\nu^2+\left(\frac{\hbar^2k^2}{2m_j}-\mu\right)^2+\Delta_j^2}$$

are Matsubara Green functions of superconducting electrodes and $\omega_\nu = \pi(2\nu+1)T/\hbar$, $\nu = 0, \pm1, \pm2, \ldots$ are fermionic Matsubara frequencies.

Performing the momentum integration over $\boldsymbol{k}_1$, $\boldsymbol{k}_2$ in Eqs. (2.134) in the constant density of states approximation (2.73), we arrive at the final expressions

$$\alpha(\tau) = \frac{\pi\hbar}{2e^2 R_N}\frac{T^2}{\hbar^2}\sum_{\omega_\mu,\omega_\nu}\frac{\hbar\omega_\nu\,\hbar\omega_\mu e^{\mathrm{i}(\omega_\mu-\omega_\nu)\tau}}{\sqrt{\Delta_1^2+\hbar^2\omega_\nu^2}\sqrt{\Delta_2^2+\hbar^2\omega_\mu^2}} \tag{2.135}$$

$$\beta(\tau) = -\frac{\pi\hbar}{2e^2 R_N}\frac{T^2}{\hbar^2}\sum_{\omega_\mu,\omega_\nu}\frac{\Delta_1\Delta_2 e^{\mathrm{i}(\omega_\mu-\omega_\nu)\tau}}{\sqrt{\Delta_1^2+\hbar^2\omega_\nu^2}\sqrt{\Delta_2^2+\hbar^2\omega_\mu^2}} \tag{2.136}$$

In the zero temperature limit, the sums over Matsubara frequencies are replaced by the integrals and can be handled exactly. In this way at $T = 0$, we obtain

$$\alpha(\tau) = \frac{\hbar}{2\pi e^2 R_N}\frac{\Delta_1\Delta_2}{\hbar^2}K_1\left(\frac{\Delta_1\tau}{\hbar}\right)K_1\left(\frac{\Delta_2\tau}{\hbar}\right) \tag{2.137}$$

$$\beta(\tau) = -\frac{\hbar}{2\pi e^2 R_N}\frac{\Delta_1\Delta_2}{\hbar^2}K_0\left(\frac{\Delta_1\tau}{\hbar}\right)K_0\left(\frac{\Delta_2\tau}{\hbar}\right) \tag{2.138}$$

Here $K_0(x)$ and $K_1(x)$ are the modified Bessel functions of the second kind. These relations complete our derivation of the imaginary time effective action for a Josephson junction.

2.4.1 Adiabatic Limit

In the low-temperature limit, $T \ll \Delta_1, \Delta_2$, the phase variable becomes a slow function of time. In this case, one can neglect the quasiparticle term by setting $\alpha(\tau) = 0$ and approximate $\beta(\tau) = -I_C\delta(\tau)/2e$. Then the action gets significantly simplified and takes the form

$$S[\varphi] = \int_0^{\hbar\beta} d\tau\left[\frac{C}{2}\left(\frac{\hbar\dot\varphi}{2e}\right)^2 - E_J\cos\varphi - \frac{\hbar I_x}{2e}\varphi\right] + \frac{1}{4}\int_0^{\hbar\beta} d\tau\int_0^{\hbar\beta} d\tau'\tilde\alpha_S(\tau-\tau')\left[\frac{\hbar\varphi(\tau)}{2e} - \frac{\hbar\varphi(\tau')}{2e}\right]^2 \tag{2.139}$$

As before, taking into account retardation effects yields effective capacitance renormalization, i.e., in Eq. (2.139) one should replace C by its renormalized value $C^*(\varphi)$ (2.110).

As one could anticipate from our previous considerations, we just arrived at the Euclidean action of a Josephson particle moving in a tilted washboard potential (2.108). The action (2.139) is particularly useful, e.g., for the analysis of quantum decay of metastable current states in Josephson junctions, quantum coherent properties of the ground state of such junctions as well as superconducting rings involving such junctions. In all these problems, the condition $T \ll \Delta_{1,2}$ is usually well satisfied and, hence, the action (2.139) provides a good starting point for their analysis. Employing this imaginary-time action, we will provide a detailed discussion of these phenomena in Chapters 4–6 of this book.

2.4.2 Normal Metal Limit

Let us now again turn to normal tunnel junctions by setting $\Delta_1 = \Delta_2 = 0$. As before, it follows immediately that the kernel $\beta(\tau) = 0$ in this case, while the kernel $\alpha(\tau)$ reduces to

$$\alpha(\tau) = \frac{\hbar}{2\pi e^2 R_N} \frac{(\pi T/\hbar)^2}{\sin^2(\pi T \tau/\hbar)} \tag{2.140}$$

Accordingly, the imaginary-time AES action takes the form

$$\begin{aligned} S[\varphi] = & \int_0^{\hbar\beta} d\tau \left[\frac{C}{2} \left(\frac{\hbar\dot{\varphi}}{2e} \right)^2 - \frac{\hbar I_x}{2e} \varphi \right] \\ & - \hbar \int_0^{\hbar\beta} d\tau \int_0^{\hbar\beta} d\tau' \alpha(\tau - \tau') \cos \frac{\varphi(\tau) - \varphi(\tau')}{2} \\ & + \frac{1}{4} \int_0^{\hbar\beta} d\tau \int_0^{\hbar\beta} d\tau' \tilde{\alpha}_S(\tau - \tau') \left[\frac{\hbar\varphi(\tau)}{2e} - \frac{\hbar\varphi(\tau')}{2e} \right]^2 \end{aligned} \tag{2.141}$$

Here again the variable φ does not have a meaning of the Josephson phase but is simply proportional to the time integral of the fluctuating voltage across the junction, cf. Eq. (2.122).

We will explicitly use this form of the action, e.g., in Chapter 7 while discussing Coulomb effects in normal tunnel junctions. Also, as it will be demonstrated in the next chapter, in the limit $I_x \to 0$ and $\tilde{\alpha}_S(\tau) \to 0$, the action (2.141) for normal tunnel junctions is identical to that for a quantum particle on a ring interacting with the Caldeira–Leggett bath of oscillators. This equivalence offers a possibility to map these two physically different problems onto each other. We will make use of this observation in Chapters 3 and 7.

2.5 Beyond the Tunneling Limit

In the above analysis, we directly evaluated the real-time effective action for a tunnel junction $S_{\rm el}$ (2.52) by expanding tr ln $\left(\check{G}^{-1}\right)$ in powers of the tunneling amplitude $t(r)$ in Eq. (2.53). The same kind of expansion (2.130) was also used while dealing with the imaginary-time version of the action (2.126). This approach works well provided it suffices to treat electron tunneling as a small perturbation. What if electron transport between metallic electrodes cannot anymore be handled perturbatively? For instance, one can think of a small-area contact (constriction) with highly transmitting conducting channels between two superconducting electrodes. Alternatively, one can connect these two electrodes by a normal metallic conductor (shunt) assuming high transmissions of metallic interfaces. In both these examples, no electron tunneling takes place and, hence, in order to construct the corresponding expression for the effective action, it appears inevitable to evaluate the full trace of the logarithm in formal expressions for $S_{\rm el}$ (2.52) and (2.126).

Note that in the case of an Ohmic shunt, we avoided this complication by treating the effect of a shunt phenomenologically as that of a bath of Caldeira–Leggett oscillators coupled linearly to the phase variable φ, cf. Eq. (2.30). The oscillator spectrum (2.31) was chosen in a way to mimic the effect of an external impedance $Z_S(\omega)$. Provided the resulting influence functional remains Gaussian (and, hence, unique), this procedure is sufficient due to a general property of Feynman–Vernon influence functionals (see Chapter 1) stating that the effect of physically different baths is indistinguishable provided they yield identical influence functionals. In other words, tracing out the Caldeira–Leggett oscillators with the spectrum (2.31) should yield exactly the same Gaussian influence functional for an Ohmic shunt as tracing out electronic degrees of freedom. Still, even in this case, it would be desirable to verify this equivalence by means of a microscopic calculation without employing any phenomenological arguments.

Below we will illustrate another method initially suggested by Zaikin and Panyukov [420] and then extensively used [344,

345, 411] in order to microscopically derive the effective action describing a variety of different metallic structures. Here we restrict our consideration to the imaginary-time analysis. Its real-time version on the Keldysh contour can be constructed in a completely analogous manner [345].

Consider again a general metallic structure described by the Hamiltonian (2.17), which includes both Coulomb and BCS interactions between electrons. As before, for simplicity we ignore the vector potential, which can be easily restored, if necessary. Our first steps are the same as above. Performing the Hubbard–Stratonovich transformations, we arrive at the effective Hamiltonian (2.22). Then, neglecting fluctuations of the absolute value of the order parameter field Δ, we express the grand partition function of the system $\mathcal{Z}$ in terms of the path integral over electronic degrees of freedom as well as over the superconducting phase and voltage variables $\varphi(\boldsymbol{r}, t)$ and $V(\boldsymbol{r}, t)$. Since the effective Hamiltonian (2.22) is quadratic in the electronic variables, they can be traced out exactly. Then one finds

$$\mathcal{Z} = \int \mathcal{D}\varphi(\boldsymbol{r}, \tau)\mathcal{D}V(\boldsymbol{r}, \tau)\, e^{-S_{\rm el}[\varphi, V]/\hbar} \tag{2.142}$$

where the effective action

$$S_{\rm el} = -\hbar \ln \mathrm{tr}_\psi \left[e^{-\hat{H}_{\rm eff}/T}\right] = -\hbar\, \mathrm{tr}\left(\hat{\mathcal{G}}^{-1}\right) \tag{2.143}$$

is expressed in terms of the inverse Green function operator for our system

$$\hat{\mathcal{G}}^{-1} = \begin{pmatrix} G^{-1} & -|\Delta|e^{-\mathrm{i}\varphi(\boldsymbol{r},\tau)} \\ -|\Delta|e^{\mathrm{i}\varphi(\boldsymbol{r},\tau)} & \tilde{G}^{-1} \end{pmatrix} \tag{2.144}$$

with

$$G^{-1} = -\hbar\frac{\partial}{\partial\tau} + \frac{\hbar^2\nabla^2}{2m} + \mu - \mathrm{i}eV(\boldsymbol{r}, t) - U(\boldsymbol{r}), \tag{2.145}$$

$$\tilde{G}^{-1} = -\hbar\frac{\partial}{\partial\tau} - \frac{\hbar^2\nabla^2}{2m} - \mu + \mathrm{i}eV(\boldsymbol{r}, t) + U(\boldsymbol{r}) \tag{2.146}$$

Our next step is to perform a gauge transformation under the trace, which keeps the result invariant but removes the superconducting phase from the order parameter field. We define

$$\hat{\mathcal{G}}^{-1} \to \hat{U}\hat{\mathcal{G}}^{-1}\hat{U}^{-1}, \quad \hat{U} = \exp(\mathrm{i}\varphi(\boldsymbol{r}, t)\hat{\tau}_z/2) \tag{2.147}$$

where, as before, $\hat{\tau}_z$ is the Pauli matrix. Employing the Josephson relation (2.122), we get

$$\hat{\mathcal{G}}^{-1} = \begin{pmatrix} G^{-1} + i\hbar \boldsymbol{v}_s \nabla - m v_s^2/2 & -|\Delta| \\ -|\Delta| & \tilde{G}^{-1} + i\hbar \boldsymbol{v}_s \nabla + m v_s^2/2 \end{pmatrix} \quad (2.148)$$

where $\boldsymbol{v}_s = -\hbar\nabla\varphi/2m$ is the superconducting velocity of the condensate. Note that the Josephson relation also removes the path integral over the scalar potential V in Eq. (2.142).

At the first glance, it appears that after all these manipulations, we still reached little progress in evaluating the action $S_{\rm el}$ in Eq. (2.142). Note, however, that strictly speaking, we are not interested in the exact expression for $\mathrm{tr}\left(\hat{\mathcal{G}}^{-1}\right)$, which contains a lot of excessive information we will make no use of. Rather, we are only interested to recover the contribution to this expression, which explicitly depends on the fluctuating phase $\varphi(\boldsymbol{r}, t)$. To this end, we introduce a "coupling constant" λ multiplying $\nabla\varphi(\boldsymbol{r}, \tau)$ in Eq. (2.148). In this way, we define $\hat{\mathcal{G}}_\lambda^{-1}$ and

$$\mathcal{Z}_\lambda = \int \mathcal{D}\varphi(\boldsymbol{r}, \tau)\mathcal{D}\, e^{-S_\lambda/\hbar}, \quad S_\lambda = -\hbar\,\mathrm{tr}\left(\hat{\mathcal{G}}_\lambda^{-1}\right) \quad (2.149)$$

Taking the derivative of S_λ with respect to λ and employing Eq. (2.148), one readily finds

$$\frac{\partial S_\lambda}{\partial\lambda} = \frac{1}{8em}\int_0^{\hbar\beta} d\tau \int d\boldsymbol{r}\mathrm{tr}\left[i\nabla\varphi(\boldsymbol{r}, \tau)(\nabla_r - \nabla_{r'}) + \lambda(\nabla\varphi(\boldsymbol{r}, \tau))^2\right]$$
$$\times\, \hat{\mathcal{G}}_\lambda(\boldsymbol{r}', \boldsymbol{r}; \tau, \tau)_{r'=r} \quad (2.150)$$

where the trace in the Nambu space is implied. Defining now $\hat{U}_\lambda = \exp(i\lambda\varphi\hat{\tau}_3/2)$ and performing the inverse transformation $\hat{\mathcal{G}}_\lambda \to \hat{U}_\lambda^{-1}\hat{\mathcal{G}}_\lambda\hat{U}_\lambda$ in Eq. (2.150), we remove the term containing $(\nabla\varphi)^2$ and obtain

$$\frac{\partial S_\lambda}{\partial\lambda} = \frac{1}{2e}\int_0^{\hbar\beta} d\tau \int d\boldsymbol{r}\nabla\varphi(\boldsymbol{r}, \tau)\boldsymbol{j}_\lambda(\boldsymbol{r}, \tau) \quad (2.151)$$

where

$$\boldsymbol{j}_\lambda(\boldsymbol{r}, \tau) = \frac{i}{2m}(\nabla_r - \nabla_{r'})G_\lambda(\boldsymbol{r}', \boldsymbol{r}; \tau, \tau)_{r'=r} \quad (2.152)$$

is the current density expressed in a standard way via the Green function G_λ. What remains is to perform the integration of $\partial S_\lambda/\partial\lambda$ over λ, which yields

$$S_{\rm el}[\varphi] = -\hbar\,\mathrm{tr}\left(\hat{\mathcal{G}}_{\nabla\varphi=0}^{-1}\right) + \int_0^1 d\lambda \frac{\partial S_\lambda}{\partial\lambda}, \quad (2.153)$$

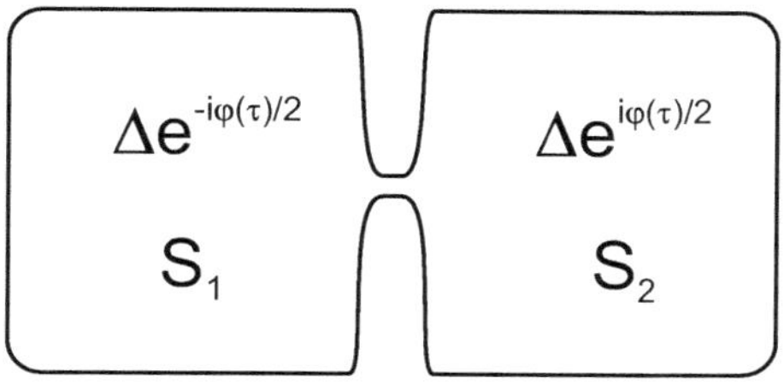

Figure 2.3 A contact between two bulk superconductors S_1 and S_2 with arbitrary transmission of conducting channels.

We finally arrived at a desired result, which allows to bypass a complicated calculation of the whole action S_{el} in Eq. (2.142) and recover only its part that explicitly depends on the phase gradient. This part is given by the second term in Eq. (2.153).

Another important advantage of this representation is that in order to derive the phase-dependent contribution to S_{el}, we need to evaluate the Green function G or the current density $\boldsymbol{j}$ only in the space region where the phase gradient differs from zero. This is an enormous simplification of our tasks. Actually, already the AES action for tunnel junctions can be recovered in this way [345], since the general expression for the current across such junctions (2.96)–(2.97) is known for arbitrary phase values, cf. Eqs. (2.96), (2.97). It is, therefore, quite natural to observe that the AES action and the current are closely related to each other, as it was indeed demonstrated, e.g., in Eqs. (2.92).

We will now proceed to other examples where electron transport is provided by non-tunneling mechanisms.

2.5.1 Contacts with Arbitrary Transmissions

Let us first evaluate the effective action for a contact between two superconductors characterized by arbitrary transmissions of their conducting channels (see Fig. 2.3). In principle, one could try to proceed within the same framework as in Section 2.4. In this case, one should formally extend the expansion (2.130) to all orders in the tunneling amplitude $t(\boldsymbol{r})$ evaluating all these contributions term by term and summing them up without assuming $t(\boldsymbol{r})$ to be small. This is a formidable task that requires a lot of calculational efforts. Below we will follow the analysis [411], which is based on Eqs. (2.151)–(2.153). It provides a much shorter and more elegant way to the desired effective action.

Let us choose the x-axis normal to the contact area and assume that the contact between two superconducting electrodes is located at the point $x = 0$. Since the contact area $\mathcal{A}$ is sufficiently small, the superconductors are little affected by the current flowing through the system. Hence, the order parameter field $\Delta(\boldsymbol{r}, \tau)$ and the voltage $V(\boldsymbol{r}, \tau)$ can be chosen in a simple form

$$\Delta(\boldsymbol{r}, \tau) = |\Delta| \left[e^{i\varphi(\tau)/2}\theta(x) + e^{-i\varphi(\tau)/2}\theta(-x)\right],$$
$$V(\boldsymbol{r}, \tau) = \frac{1}{2}V(\tau)\mathrm{sign}x, \tag{2.154}$$

where $\varphi(\tau)$ and $V(\tau)$ define, respectively, the fluctuating phase and voltage across the contact again related to each other by means of the Josephson relation (2.122), i.e., both phase and voltage gradients differ from zero only in the contact area.

According to the above arguments, our main goal is to recover the general expression for the current across the contact. To this end, we will employ the formalism of quasiclassical energy integrated matrix Green functions $\hat{g}(\boldsymbol{v}_F, \boldsymbol{r}, \tau, \tau')$, which we already defined in Eq. (2.71). These functions are 2×2 matrices in the Nambu space obeying the Eilenberger equations [51, 110]

$$\begin{aligned}
&i\hat{\tau}_z\frac{\partial}{\partial\tau}\hat{g}(\tau, \tau') + i\frac{\partial}{\partial\tau'}\hat{g}(\tau, \tau')\hat{\tau}_z + \boldsymbol{v}_F\nabla\hat{g}(\tau, \tau') \\
&+ (eV(\boldsymbol{r}, \tau)\hat{1} - i\hat{\Delta}(\boldsymbol{r}, \tau))\hat{g}(\tau, \tau') - \hat{g}(\tau, \tau')(eV(\boldsymbol{r}, \tau')\hat{1} - i\hat{\Delta}(\boldsymbol{r}, \tau')) \\
&+ \frac{v_F}{2\ell}\int d\tau_1 \left[\langle\hat{g}(\tau, \tau_1)\rangle_{\boldsymbol{v}_F}\hat{g}(\tau_1, \tau') - \hat{g}(\tau, \tau_1)\langle g(\boldsymbol{v}_F, \boldsymbol{r}, \tau_1, \tau')\rangle_{\boldsymbol{v}_F}\right] = 0
\end{aligned} \tag{2.155}$$

where in the expressions for $\hat{g}$, we omitted the arguments $\boldsymbol{v}_F$ and $\boldsymbol{r}$ for the sake of brevity, denoted averaging over directions of the Fermi velocity vector $\boldsymbol{v}_F$ by angular brackets $\langle\ldots\rangle_{\boldsymbol{v}_F}$ and defined

$$\hat{\Delta}(\boldsymbol{r}, \tau), = \begin{pmatrix} 0 & \Delta(\boldsymbol{r}, \tau) \\ -\Delta^*(\boldsymbol{r}, \tau) & 0 \end{pmatrix}, \quad \hat{1} = \begin{pmatrix} 1 & 0 \\ 0 & 1 \end{pmatrix} \tag{2.156}$$

In addition, the Eilenberger functions obey the normalization condition

$$\int d\tau_1\hat{g}(\tau, \tau_1)\hat{g}(\tau_1, \tau') = \hat{1}\delta(\tau - \tau') \tag{2.157}$$

Evaluating $\hat{g}(\boldsymbol{v}_F, \boldsymbol{r}, \tau, \tau')$ from Eqs. (2.155) in the vicinity of the contact, one immediately obtains the current density

$$j[\varphi(\tau)] = \pi \mathrm{i} e N_0 \mathrm{tr} \hat{\tau}_z \langle \boldsymbol{v}_F \hat{g}(\boldsymbol{v}_F, \boldsymbol{r}, \tau, \tau) \rangle_{\boldsymbol{v}_F} \tag{2.158}$$

and recovers the action from the formula (cf. Eqs. (2.151)–(2.153))

$$S_{\mathrm{el}}[\varphi] = \frac{\mathcal{A}}{2e} \int_0^1 d\lambda \int_0^{\hbar\beta} d\tau \varphi(\tau) j[\lambda\varphi(\tau)]. \tag{2.159}$$

Let us now implement this program for the system under consideration. We first obtain the solution to Eqs. (2.155) in superconducting electrodes. The Eilenberger matrix function can be found in the form [345]

$$\hat{g}(\boldsymbol{v}_F, \boldsymbol{r}, \tau, \tau') = \hat{A}_{1(2)}(\tau, \tau') + \hat{B}_{1(2)}(\boldsymbol{v}_F, \boldsymbol{r}, \tau, \tau') \tag{2.160}$$

where the index 1 (2) corresponds to the left (right) electrode located at $x < 0$ ($x > 0$),

$$\hat{A}_j = \exp\left[(-1)^{j+1} \frac{\mathrm{i}\hat{\tau}_z \varphi(\tau)}{4}\right] \hat{g}_0(\tau - \tau') \exp\left[(-1)^j \frac{\mathrm{i}\hat{\tau}_z \varphi(\tau')}{4}\right], \; j = 1, 2 \tag{2.161}$$

and

$$\hat{g}_0(\tau) = T \sum_{\omega} e^{-\mathrm{i}\omega\tau} \left[\frac{\omega\hat{\tau}_z + |\Delta| \mathrm{i}\hat{\tau}_y}{\sqrt{\omega^2 + |\Delta|^2}}\right] \tag{2.162}$$

is the Eilenberger matrix for a bulk superconductor.

The matrices $\hat{B}_{1(2)}$ essentially depend on the type of the electron trajectories. For trajectories that either do not pass through the contact or get reflected from it, we have $\hat{B}_{1(2)} \equiv 0$. Electrons following these trajectories always remain in one electrode and, hence, do not contribute to the current across the contact. Of interest for us are the trajectories that start in one electrode, pass through the contact, and end in another electrode. For such trajectories, one has

$$\hat{B}_j(\boldsymbol{v}_F, \boldsymbol{r}, \tau, \tau') = \exp\left[-\hat{K}_j \boldsymbol{r} \boldsymbol{v}_F / v_F^2\right] \hat{C}(\boldsymbol{v}_F) \exp\left[\hat{K}_j \boldsymbol{r} \boldsymbol{v}_F / v_F^2\right], \; j = 1, 2, \tag{2.163}$$

where

$$\hat{K}_j = \exp\left[(-1)^{j+1} \frac{\mathrm{i}\hat{\tau}_z \varphi(\tau)}{4}\right] \hat{K}(\tau - \tau') \exp\left[(-1)^j \frac{\mathrm{i}\hat{\tau}_z \varphi(\tau')}{4}\right] \tag{2.164}$$

and

$$\hat{K}_\omega = \omega\hat{\tau}_z + |\Delta|\mathrm{i}\hat{\tau}_y + (v_F/2\ell)\hat{g}_{0\omega} \tag{2.165}$$

Far from the contact, the matrices $\hat{B}_j$ should vanish since deep inside the superconducting electrodes, the matrix $\hat{g}$ should coincide with its bulk values $\hat{A}_j$ (2.161). This requirement is equivalent to the following relations:

$$\hat{A}_j\hat{C}_j(\boldsymbol{v}_F) = (-1)^{j+1}\hat{C}_j(\boldsymbol{v}_F)\mathrm{sign}v_x,\ \ \hat{C}_j(\boldsymbol{v}_F)\hat{A}_j = (-1)^j\hat{C}_j(\boldsymbol{v}_F)\mathrm{sign}v_x \tag{2.166}$$

which fully determine the general solution to Eqs. (2.155) in both electrodes. Here and below, $v_x = v_F\theta$ is the x-component of the Fermi velocity.

What remains is to match these solutions at the contact between two superconductors, i.e., at $x = 0$. This matching can be performed with the aid of Zaitsev's boundary conditions describing transfer of electrons across the barrier with arbitrary transmission coefficient $\mathcal{T}(v_x/v_F) \equiv \mathcal{T}(\theta)$, which explicitly depends on the direction of the Fermi velocity $\boldsymbol{v}_F$ for incident electrons and may take values from zero (impenetrable interface) to one (full transmission). These boundary conditions read [428]

$$\hat{g}_a[(1-\mathcal{T}(\theta))\hat{g}_{s+}^2 + \hat{g}_{s-}^2] = \mathcal{T}(\theta)\hat{g}_{s-}\hat{g}_{s+}, \tag{2.167}$$

where we defined

$$\hat{g}_a = \frac{1}{2}[\hat{g}_1(v_x,0) - \hat{g}_1(-v_x,0)] = \frac{1}{2}[\hat{g}_2(v_x,0) - \hat{g}_2(-v_x,0)], \tag{2.168}$$

$$\hat{g}_{s\pm} = \frac{1}{4}[\hat{g}_2(v_x,0) + \hat{g}_2(-v_x,0) \pm (\hat{g}_1(v_x,0) + \hat{g}_1(-v_x,0))] \tag{2.169}$$

In Eq. (2.167) and below, an integration over the internal time variable is implied for the product of matrices $\hat{g}$.

Combining the above solution to the Green–Eilenberger functions in the electrodes (2.160)–(2.166) with boundary conditions (2.167) after a simple algebra, we obtain

$$\hat{g}_a(v_x) = \mathrm{sign}v_x\mathcal{T}(\theta)\hat{g}_-\hat{g}_+[\hat{1} - \mathcal{T}(\theta)\hat{g}_-^2]^{-1} \tag{2.170}$$

where we defined

$$\hat{g}_\pm = (\hat{A}_2 \pm \hat{A}_1)/2. \tag{2.171}$$

Then with the aid of Eqs. (2.170) and (2.158), we arrive at the expression for the current across the contact [411]

$$j[\varphi(\tau)] = -\frac{iep_F^2}{2\pi}\mathrm{tr}\hat{\tau}_z \int_0^1 d\theta \mathcal{T}(\theta)\hat{g}_-\hat{g}_+[\hat{1} - \mathcal{T}(\theta)\hat{g}_-^2]^{-1} \quad (2.172)$$

What remains is to employ the relation (2.159) between the current density and the effective action. Substituting Eq. (2.172) into Eq. (2.159), expanding the inverse operator in Eq. (2.172) in powers of $\mathcal{T}(\theta)$ and integrating over λ, after a proper symmetrization we recover the final expression for the effective action [344, 345, 411]

$$S_{\mathrm{el}}[\varphi] = \frac{\mathcal{A}p_F^2}{4\pi\hbar}\int_0^1 \theta d\theta \mathrm{tr}\sum_{n=1}^{\infty}\frac{\mathcal{T}^n(\theta)}{n}\int_0^{\hbar\beta} d\tau_1 \int_0^{\hbar\beta} d\tau_2 \int_0^{\hbar\beta} d\tau_3 \ldots$$

$$\int_0^{\hbar\beta} d\tau_{2n} \times \hat{g}_-(\tau_1 - \tau_2)\hat{g}_-(\tau_2 - \tau_3)\ldots\hat{g}_-(\tau_{2n} - \tau_1) \quad (2.173)$$

We observe that the resulting action takes the form of an infinite series in powers of the contact transmission $\mathcal{T}(\theta)$. The n-th term in this series corresponds to $2n$ traverses of an electron through the contact, and the sum in Eq. (2.173) can schematically be represented as a collection of diagrams in Fig. 2.4 Making use of Eqs. (2.161), (2.171) and defining the normal state contact resistance R_N by means of the relation

$$\frac{1}{R_N} = \frac{e^2\mathcal{A}p_F^2}{2\pi^2\hbar}\int_0^1 \theta d\theta \mathcal{T}(\theta) \quad (2.174)$$

it is easy to verify that the term $n = 1$ in Eq. (2.173) yields the AES action for a tunnel barrier derived above. Other terms with $n \geq 2$ describe higher-order processes in the contact transmission. At small transmissions $\mathcal{T}(\theta) \ll 1$ or, equivalently, under the condition (2.93), these terms remain small and can be neglected. On the other hand, for higher transmissions, the terms with $n \geq 2$ in Eq. (2.173) become important and need to be fully accounted for.

One can also perform a formal summation of the series in Eq. (2.173) and rewrite the effective action S_{el} in a much more compact symbolic form

$$S_{\mathrm{el}}[\varphi] = \frac{\mathcal{A}p_F^2}{4\pi\hbar}\mathrm{tr}\langle \boldsymbol{v}_F \ln[\hat{1} + \mathcal{T}(\theta)\hat{g}_-^2]\rangle_{\boldsymbol{v}_F} \quad (2.175)$$

where, as before, the trace is taken in the Nambu space. Here we also note that averaging over directions of the Fermi velocity

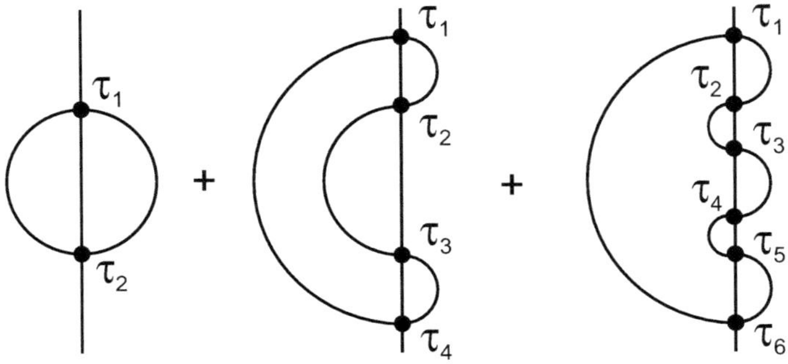

Figure 2.4 Diagrammatic representation of the first three terms of the series (2.173).

is equivalent to a summation over the conducting channels (see Chapters 10 and 11).

Finally, let us specify the expression for the effective action in the case of normal electrodes. By setting $|\Delta| = 0$ in Eq. (2.162), one finds $\hat{g}_0(\tau) = -\mathrm{i}T\,\hat{\tau}_z/\sin(\pi T\tau/\hbar)$ and, hence,

$$\hat{g}_-(\tau - \tau') = T\,\frac{\sin[(\varphi(\tau) - \varphi(\tau'))/4]}{\sin[\pi T(\tau - \tau')/\hbar]}\,\hat{\tau}_z \tag{2.176}$$

Combining this expression with Eq. (2.173), we arrive at the effective action for normal contacts with arbitrary transmission:

$$S_{\rm el}[\varphi] = \frac{\mathcal{A}p_F^2}{2\pi\hbar}\int_0^1 \theta d\theta \sum_{n=1}^{\infty} \frac{\mathcal{T}^n(\theta)T^{2n}}{n} \int_0^{\hbar\beta} d\tau_1 \int_0^{\hbar\beta} d\tau_2 \ldots \int_0^{\hbar\beta} d\tau_{2n}$$
$$\times \frac{\sin[(\varphi(\tau_1) - \varphi(\tau_2))/4]}{\sin[\pi T(\tau_1 - \tau_2)/\hbar]} \cdots \frac{\sin[(\varphi(\tau_{2n}) - \varphi(\tau_1))/4]}{\sin[\pi T(\tau_{2n} - \tau_1)/\hbar]} \tag{2.177}$$

We will explicitly make use of this form of the action later in Chapter 11 when we will discuss Coulomb blockade effects in short conductors with arbitrary transmissions of their conducting channels.

2.5.2 Normal Ohmic Shunt

Let us now turn to another example and consider a normal metallic conductor (shunt) of length L and cross section $\mathcal{A}$ connecting two superconducting electrodes with the phase difference $\varphi(\tau)$. We

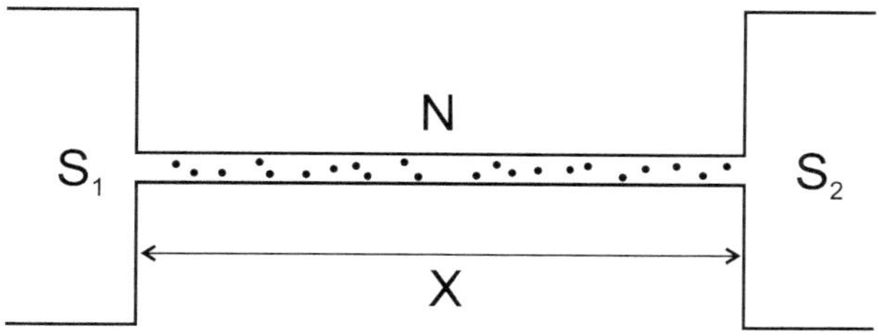

Figure 2.5 Two superconductors S_1 and S_2 connected by a normal disordered metallic wire of length X.

will assume this shunt resistor to be long enough and ignore the proximity effect, i.e., we neglect the anomalous Green functions inside the normal conductor. Then the Eilenberger matrix function becomes diagonal $\hat{g} = g\hat{\tau}_z$ and the Eilenberger equations (2.155) reduce to a single equation for the function $g(\boldsymbol{v}_F, \boldsymbol{r}, \tau, \tau')$.

Performing the gauge transformation

$$\tilde{g}(\boldsymbol{v}_F, \boldsymbol{r}, \tau, \tau') = \exp(-i\varphi(\boldsymbol{r}, \tau)/2) g(\boldsymbol{v}_F, \boldsymbol{r}, \tau, \tau') \exp(i\varphi(\boldsymbol{r}, \tau')) \tag{2.178}$$

and choosing the function $\varphi(\boldsymbol{r}, \tau)$ inside the normal conductor to obey the Josephson relation (2.122) (with $V(\boldsymbol{r}, \tau)$ being the fluctuating voltage inside the shunt), with the aid of (2.155), we obtain

$$\left[i\left(\frac{\partial}{\partial\tau} + \frac{\partial}{\partial\tau'}\right) + \boldsymbol{v}_F\nabla + (i\boldsymbol{v}_F/2)(\nabla\varphi(\boldsymbol{r}, \tau) - \nabla\varphi(\boldsymbol{r}, \tau'))\right] \tilde{g}(\tau, \tau')$$
$$+ \frac{v_F}{2\ell}\int d\tau_1 \left[\langle\tilde{g}(\tau, \tau_1)\rangle_{\boldsymbol{v}_F}\tilde{g}(\tau_1, \tau') - \tilde{g}(\tau, \tau_1)\langle\tilde{g}(\tau_1, \tau')\rangle_{\boldsymbol{v}_F}\right] = 0 \tag{2.179}$$

Provided our metal is sufficiently dirty, i.e., the electron elastic mean free path ℓ is rather small, the electron motion is diffusive. In this limit, the Eilenberger function of our metal is almost isotripic. Performing the Fourier transformation with respect to time, we have [383]

$$\tilde{g}_{\omega,\omega'}(\boldsymbol{v}_F, \boldsymbol{r}) = \delta_{\omega,\omega'}\mathrm{sign}\omega + \frac{\boldsymbol{v}_F}{v_F}\boldsymbol{g}_{\omega,\omega'}(\boldsymbol{v}_F, \boldsymbol{r}) \tag{2.180}$$

where the vector function $\boldsymbol{g}$ identically equals to zero provided the frequencies ω and ω' have the same sign, whereas in the opposite case $\omega\omega' \leq 0$, the nontrivial solution reads

$$\boldsymbol{g}_{\omega,\omega'}(\boldsymbol{v}_F, \boldsymbol{r}) = \mathrm{i}\ell\nabla\varphi_{\omega-\omega'}(\boldsymbol{r}). \tag{2.181}$$

Performing the inverse Fourier transformation, we obtain

$$\boldsymbol{g}(\boldsymbol{v}_F, \boldsymbol{r}, \tau, \tau) = \frac{\mathrm{i}\ell T}{2\pi\hbar}\sum_{\omega} |\omega| e^{\mathrm{i}\omega\tau}\nabla\varphi_{\omega}(\boldsymbol{r}) \tag{2.182}$$

Combining now Eqs. (2.180) and (2.182) with the expression for the current (2.158) after averaging over the Fermi velocity directions, we arrive at the result

$$\boldsymbol{j}(\boldsymbol{r}, \tau) = \pi e N_0 D T^2 \int_0^{\hbar\beta} d\tau' \frac{\nabla\varphi(\boldsymbol{r}, \tau) - \nabla\varphi(\boldsymbol{r}, \tau')}{\sin^2[\pi T(\tau - \tau')/\hbar]} \tag{2.183}$$

where $D = v_F\ell/3$ is the diffusion coefficient of a normal metal. Assuming that this metal is uniform at length scales exceeding ℓ, we can approximate the phase gradient by a constant in space, i.e., $\nabla\varphi(\boldsymbol{r}, \tau) = \varphi(\tau)/L$. Subsituting this relation into Eq. (2.183) and making use of Eq. (2.159), we recover the effective action for the normal Ohmic shunt

$$S_{\mathrm{el}} = \frac{\pi T^2}{16e^2 R_S}\int_0^{\hbar\beta} d\tau \int_0^{\hbar\beta} d\tau' \frac{[\varphi(\tau) - \varphi(\tau')]^2}{\sin^2[\pi T(\tau - \tau')/\hbar]} \tag{2.184}$$

where $1/R_S = 2e^2 N_0 D\mathcal{A}/L$ is the shunt conductance. This result is, of course, not surprising. It exactly coincides with the action S_{env} for the external Ohmic environment of Caldeira–Leggett oscillators defined, e.g., by Eqs. (2.124), (2.125) with $C \to 0$ and $Z_S \equiv R_S$. Here, however, the same effective action was derived microscopically by tracing out electronic degrees of freedom rather than those of fictitious harmonic oscillators.

In the next chapters, we will employ the effective actions derived here in order to describe a variety of macroscopic quantum effects in small-size metallic junctions.

Chapter 3

Quantum Particle in a Dissipative Environment

Let us now turn to several examples illustrating key advantages of the influence functional technique, which allows to obtain important physical results in a simple and elegant manner.

3.1 Free Particle Interacting with a Bath of Oscillators

We start from the most simple case of a freely propagating quantum particle with mass m and coordinate q. In the absence of any interaction, the probability W to find the particle in the state with some coordinate $q = q_f$ at a time t provided initially it had a coordinate $q(0) = q_i$ is given by the product of two transition amplitudes K,

$$W = K(q_f, t; q_i, 0)K^*(q_f, t; q_i, 0) \tag{3.1}$$

where

$$K(q_f, t; q_i, 0) = \int_{q_i}^{q_f} \exp\left(\frac{\mathrm{i}}{\hbar}\int_0^t dt' \frac{m\dot{q}^2}{2}\right) \tag{3.2}$$

Dissipative Quantum Mechanics of Nanostructures: Electron Transport, Fluctuations, and Interactions
Andrei D. Zaikin and Dmitry S. Golubev

ISBN 978-981-4774-50-5 (Hardcover), 978-0-429-29823-3 (eBook)
www.jennystanford.com

This path integral is dominated by the classical saddle point trajectory

$$\tilde{q}(t') = q_i + (q_f - q_i)t'/t \tag{3.3}$$

Substituting this trajectory into Eq. (3.2) and integrating over fluctuations of q around this trajectory, we recover the standard quantum mechanical expression

$$K(q_f, t; q_i, 0) = \sqrt{m/(2\pi i t\hbar)}\exp(iS_0[\tilde{q}]/\hbar),\ S_0[\tilde{q}] = m(q_f - q_i)^2)/t \tag{3.4}$$

It is clear that, while the transition amplitude (3.4) explicitly depends on the phase iS_0, the probability W does not depend on this phase. It simply cancels out in Eq. (3.1), which yields $W \propto 1/t$.

Now let us turn on the interaction between our particle q and a dissipative bath described, as before, by a collection of coordinates Q. According to Eq. (1.12), the reduced density matrix of the particle, $\rho(q_f, q'_f)$, at a final moment of its evolution is fully determined for any initial density matrix $\rho_i(q_i, q'_i)$, provided we evaluate the kernel J of the product of the evolution operators (1.13). Applying the influence functional technique, we again arrive at Eq. (1.16) with $S_0[q] = \int dt m\dot{q}^2/2$.

3.1.1 Caldeira–Leggett Model: Exact Results

In order to fix the influence functional (1.20), we need to specify the model for a dissipative bath. As an important example here, we will consider the Caldeira–Leggett bath of oscillators. The influence functional for this model was already evaluated in Chapter 1 and is defined by Eqs. (1.29), (1.30) together with Eqs. (1.55), (1.56). For practical calculations, it is convenient to again introduce the coordinates $q_\pm$ and evaluate the kernel J in the form (1.59)–(1.62) with $U(q) \equiv 0$. The problem can be solved exactly, as in this case both path integrals in Eq. (1.59) are Gaussian. Integrating first over q_+, we obtain the functional delta-function

$$\delta(m\ddot{q}_- - \eta\dot{q}_-)$$

which trivially resolves the second integral over q_-. As a result, we arrive at the exact expression for the kernel J:

$$J = \frac{\eta}{2\pi\hbar(1-\mathrm{e}^{-\gamma t})}\exp\left[\mathrm{i}\frac{\eta}{\hbar}\frac{q_{+f}q_{-f}+q_{+i}q_{-i}-\mathrm{e}^{\gamma t}q_{+f}q_{-i}-q_{-f}q_{+i}}{\mathrm{e}^{\gamma t}-1}\right.$$
$$\left.-\frac{m}{\hbar}f_1(t)q_{-i}^2-\frac{m}{\hbar}f_2(t)(q_{-f}-q_{-i})^2-\frac{m}{\hbar}f_3(t)q_{-i}(q_{-f}-q_{-i})\right] \tag{3.5}$$

Here we introduced the parameter $\gamma = \eta/m$ and defined the functions $f_{1,2,3}(t)$, which essentially determine the dependence of J on time. The first of these functions reads

$$f_1(t) = \frac{\gamma}{2}\int_0^t \mathrm{d}s\int_0^t \mathrm{d}s'\int_{-\omega_c}^{\omega_c}\frac{\mathrm{d}\omega}{2\pi}\omega\coth\frac{\hbar\omega}{2T}\mathrm{e}^{-\mathrm{i}\omega(s-s')}$$
$$= \frac{\gamma T t}{\hbar}+\gamma\ln\frac{1-\mathrm{e}^{-2\pi Tt/\hbar}}{2\pi(T/\hbar\omega_c)} \tag{3.6}$$

Clearly, the long-time behavior of this function at any nonzero temperature $f_1(t) \simeq \gamma Tt/\hbar$ differs considerably from $f_1(t) = \gamma\ln(\omega_c t)$ at $T = 0$. The functions $f_2(t)$ and $f_3(t)$ tend to constant values in the (interesting for us) long-time limit. In this limit, we obtain

$$f_2 = \gamma\int_0^{\omega_c}\frac{\mathrm{d}\omega}{2\pi}\frac{\omega\coth\frac{\hbar\omega}{2T}}{\omega^2+\gamma^2} \simeq \frac{\gamma}{2\pi}\ln\frac{\omega_c}{\gamma}+\frac{T}{\pi\hbar}\arctan\frac{T}{\hbar\gamma} \tag{3.7}$$

and $f_3 = T/\hbar + 2f_2$.

Let us analyze the above exact results and make several important observations. To begin with, we note that at sufficiently long times, the function $f_1(t)$ becomes large, $f_1(t) \gg f_2,\ f_3$, and the kernel (3.5) effectively reduces to

$$J = \frac{1}{2}\sqrt{\frac{\eta\gamma}{\pi\hbar f_1(t)}}\mathrm{e}^{-mf_2q_{-f}^2/\hbar}\delta(q_{-i}), \tag{3.8}$$

Substituting this result into Eq. (1.12), we immediately conclude that for any initial conditions after some interaction with a dissipative bath, the particle density matrix tends to the same universal form

$$\rho_{\mathrm{eq}}(q_1, q_2) = \frac{1}{L}\exp(-mf_2(q_1-q_2)^2/\hbar) \tag{3.9}$$

which is not sensitive to the initial phase and is totally independent on the initial density matrix ρ_i. Equation (3.9) represents the equilibrium density matrix for the particle q in the Caldeira–Leggett bath at a temperature T. We also introduced L as the system size, which provides correct normalization of ρ_{eq}. We observe that off-diagonal elements of the density matrix are exponentially suppressed beyond some length L_φ, which will be called *decoherence length* (or *dephasing length*) further below. It follows from Eq. (3.7) that at high temperatures $T \gg \hbar\gamma \ln(\omega_c/\gamma)$, this length increases with decreasing T as $L_\varphi \propto \hbar/\sqrt{T}$ and then saturates to a temperature-independent value

$$L_\varphi \sim \sqrt{\hbar/\eta \ln(\omega_c/\gamma)} \tag{3.10}$$

at $T \lesssim \hbar\gamma \ln(\omega_c/\gamma)$, i.e., the dephasing length stays finite down to $T = 0$. This is a fundamentally important effect of *quantum decoherence* produced by interactions, in this case interactions between the particle and the Caldeira–Leggett bath of oscillators.

Our analysis also allows to determine characteristic time scales at which interactions bring the particle in equilibrium with the bath. For example, according to Eqs. (3.5) and (3.6), the decay of the off-diagonal elements of the initial density matrix is exponential at any nonzero T with the characteristic *decoherence time* (or *dephasing time*)

$$\tau_\varphi = \hbar^2/\left(\eta T q_{-i}^2\right). \tag{3.11}$$

Note that although this time diverges in the low temperature limit, Eq. (3.11) by no means implies the absence of dephasing by interactions at $T \to 0$. Rather, it implies the absence of an exponential decay in this limit. Indeed, at $T = 0$, the off-diagonal elements decay as a power law

$$\rho_i(q_{-i}) \propto (t\omega_c)^{-\eta q_{-i}^2/\hbar} \tag{3.12}$$

but also in this case, the information about the initial phase is practically lost in the long-time limit and the particle will not show any coherent properties beyond the length scale (3.10) set by interactions.

3.1.2 Discussion

Although the Caldeira–Leggett model is solved exactly leaving no room for doubts in the validity of the above results, the statement about interaction-induced dephasing of a quantum particle at $T \to 0$ in (or close to) its ground state sometimes causes certain concerns and/or misunderstanding. Therefore, for pedagogical purposes, it would be appropriate to analyze the issue in more details.

First of all, the very idea about quantum decoherence in the ground state could sometimes be naively considered contradicting quantum mechanics. Of course, in reality no such contradiction exists. The wave function of a closed system "particle+bath" remains fully coherent at all times. The above results, however, apply not to the whole system, but to its part, namely to an interacting particle the quantum state of which can be described only by some reduced density matrix rather than by the wave function. In other words, even at $T = 0$, the particle always remains in the mixed state, so in general, its decoherence should, by no means, be taken as a surprise. If measurements are performed with the interacting particle, the effect of decoherence in its behavior could be observed even in or close to equilibrium at $T = 0$ no matter that the wave function of the total system "particle+bath" remains coherent.

For the model in question, one can exactly diagonalize the Hamiltonian and introduce a new set of independent (and, therefore, fully coherent) modes. In this case, the wave function for the total system is factorized into the product of wave functions of all such individual modes. Performing the calculation in this new basis of coherent states [187], one arrives at the same exact results for the evolution kernel J (3.5) and the equilibrium density matrix $\rho_{\rm eq}(q, q')$ (3.9), though with much more efforts and with loss of physical transparency. The message here is that quantum decoherence may occur even from fully coherent modes provided measurements are performed not with each of these modes separately, but with the variable (in our case q), which is in general expressed as a combination of *infinite* number of these new modes. In this case, the information about initial phase is lost since the particle will not be able to return to its initial state after any finite time.

It is sometimes believed that if initially all the bath oscillators are in their ground states and the particle energy is close to zero, no energy exchange between the particle and the oscillators would be possible because the particle has no energy to excite the oscillators and the latter in turn cannot transfer their zero-point energy to the particle. Hence, the particle should always stay in its initial coherent state. If so, no zero-temperature decoherence would be possible on these general grounds.

In order to identify the main drawbacks of this argument, let us consider a quantum particle q scattered on a harmonic oscillator Q with a frequency ω. Since the above argument is formulated as a scattering problem operating with well-defined in- and out-scattering states, we will need to assume that interaction exists only in a certain space region $-q_0/2 < q < q_0/2$ and vanishes outside this region. As before, we will choose the interaction in the form $\hat{H}_{\text{int}} = CqQ$ (1.34), where we now set

$$C = C_0(\Theta(q + q_0/2) - \Theta(q - q_0/2)) \tag{3.13}$$

where $q_0 < L$ and C_0 is some constant. As requested by the above argument, before scattering, the oscillator is assumed to be in its ground state and the initial particle energy E_i is assumed to be small.

Let us evaluate the probability $W(t)$ for a particle q to leave the initial state ψ_i after time t. In fact we have already treated this problem in Chapter 1 when we evaluated the transition probability perturbatively in the interaction (see Section 1.5.1). Collecting transition probabilities into all possible final states ψ_f, from Eq. (1.75) we obtain

$$W(t) = 2\sum_{f\neq i} |\langle i|Cq|f\rangle|^2 \frac{\sin^2(\Delta E_{\text{if}}t/2\hbar)}{\omega(\Delta E_{\text{if}})^2} \tag{3.14}$$

where $\Delta E_{\text{if}} = \hbar\omega + E_f - E_i$. This result implies $dW/dt \propto \sin(\Delta E_{\text{if}}t/\hbar)/\Delta E_{\text{if}}$, i.e., at large t, the transition rate experiences fast oscillations and approaches $\delta(\Delta E_{\text{if}})$ at $t \to \infty$. If $E_f > E_i \to 0$ for all f, the transitions are highly improbable *at very long times*. Note, however, that the time average of the escape probability W (3.14) differs from zero already in this case. This is a direct consequence of energy–time uncertainty.

In the case of scattering on a bath of oscillators with frequencies ω_n, Eq. (3.14) is trivially modified by changing $\omega \to \omega_n$, $C_0 \to C_n$ and by summing up the resulting expression over all n. For the Ohmic spectrum, this summation is carried out by means of Eq. (1.54). What remains is to evaluate the transition matrix elements between the states ψ_i and ψ_f. Without interaction, the eigenstates of the problem are plane waves, i.e.,

$$\psi_{i/f}(q,t) = (L)^{-1/2}\exp\left[\frac{\mathrm{i}}{\hbar}(p_{i/f}q - E_{i/f}t)\right] \tag{3.15}$$

where $E_{i/f} = p_{i/f}^2/2m$. With this in mind, we observe that an important contribution comes from $|p_f - p_i| \lesssim \hbar/q_0$, and we get

$$\langle \mathrm{i}|q(\Theta(q+q_0/2) - \Theta(q-q_0/2))|f\rangle|^2 \sim q_0^6(p_f-p_i)^2/\hbar^2 L^2 \tag{3.16}$$

Employing Eq. (3.14) and carrying out the summation over the oscillator frequencies with the aid of Eq. (1.54) and over the final states $\sum_f \to L\int dp_f$, we obtain

$$W(t) \sim \frac{\eta q_0^6}{\hbar^4 L}\int_0^{\omega_c}\omega d\omega \int_{-\hbar/q_0}^{\hbar/q_0} dp_f (p_f - p_i)^2 \frac{\sin^2(\Delta E_{\mathrm{if}}t/2\hbar)}{(\Delta E_{\mathrm{if}})^2} \tag{3.17}$$

Employing further the required condition $E_i \to 0$ after simple integrations, we obtain

$$W(t) \sim \frac{\eta q_0^3}{\hbar L}\ln(\omega_c \min(t, mq_0^2/\hbar)) \tag{3.18}$$

We observe that—in contrast to the above naive expectations—the escape probability $W(t)$ from the initial state is not zero and, moreover, it is not necessarily small. Although the contribution of each oscillator to $W(t)$ (3.14) is small, it differs from zero at any finite t due to the energy–time uncertainty. The sum of a large number of such small contributions from all bath oscillators yields the result (3.18).

Hence, the main problem with the above argument against zero temperature decoherence is that it *violates the energy–time uncertainty* by implicitly taking the limit $t \to \infty$ and approximating

$$\sin(\Delta E_{\mathrm{if}}t/\hbar)/\Delta E_{\mathrm{if}} \to \delta(\Delta E_{\mathrm{if}}) \tag{3.19}$$

already at the very first step of the calculation. This approximation may be sufficient at high enough temperatures, but not in the limit $T \to 0$, as we observed in our rigorous analysis. For instance, this approximation allows to correctly recover exponential decay of quantum correlations described by the first term $\propto Tt$ in Eq. (3.6). However, it completely misses the second term, which accounts for the power-law decay (3.12) dominating the system behavior at $T \to 0$. The reason for this failure is obvious: Employing Eq. (3.19) in Eq. (3.14) for $\Delta E_{\rm if} > 0$, one finds $dW/dt \equiv 0$ at all times in contradiction with rigorous results.

Another useful conclusion that can be drawn from the above analysis is that the interaction range can also play an important role. For example, by choosing the parameter q_0 sufficiently small, $q_0^3 \ll L/\eta$, one can also keep the escape probability $W(t)$ from the particle's initial state small. Thus, in the case of short-range interactions, with the dominating probability the particle could remain in its initial state also after scattering. If so, quantum dephasing of the particle wave function can be neglected.

For long-range interactions, the situation is entirely different. In order to model such interactions and to observe the relation between (3.18) and our previous results, we now set q_0 to be of the order of the system size $q_0 \sim L$. Then requiring the escape from the initial state to be complete $W(t, L \sim L_\varphi) \sim 1$ from Eq. (3.18), we immediately recover our previous result (3.10) for the decoherence length L_φ. Thus, in this case for sufficiently large $L > L_\varphi$, quantum dephasing by interactions dominates even exactly at $T = 0$ and any nonzero interaction strength η.

Actually for truly long-range interactions, in- and out-scattering states cannot be defined at all and the scattering problem cannot even be formulated. Let us evaluate the average value of the kinetic energy of the particle $\langle E_{\rm kin}\rangle = \langle m\dot{q}^2/2\rangle$ in the Caldeira–Leggett bath. Resolving the Langevin equation (1.67), we find

$$\langle E_{\rm kin}\rangle = \frac{m}{2}\int \frac{d\omega}{2\pi}\frac{\langle \xi_\omega \xi_{-\omega}\rangle}{m^2\omega^2 + \eta^2} \tag{3.20}$$

Combining this result with Eq. (1.68), we immediately obtain that $\langle E_{\rm kin}\rangle$ exactly coincides with the expression for f_2 (3.7) and, hence,

$$\langle E_{\rm kin}\rangle = \frac{\hbar\gamma}{2\pi}\ln\frac{\omega_c}{\gamma} + \frac{T}{\pi}\arctan\frac{T}{\hbar\gamma} \tag{3.21}$$

Thus, at sufficiently long times, the average value of the particle kinetic energy is not zero (and in general not small) even at $T = 0$ irrespective of its initial energy. At high temperatures

$T \gg \hbar\gamma \ln(\omega_c/\gamma)$, the energy is given by its classical value $T/2$ in accordance with the well-known equipartition theorem, but at lower T, its value is determined by the interaction strength γ and the high-frequency cutoff ω_c. The exact result (3.21) demonstrates that the above naive scattering argument effectively assumes the energy of the system to be *lower than its ground-state energy in the interacting state*, which can never be realized in nature. It also follows from Eq. (3.21) that the particle q can exchange energy with *all* oscillators, including high-frequency ones, up to $\omega \sim \omega_c$. Indeed, in the presence of long-range interactions, the oscillators are not anymore in their non-interacting energy states implying that all their energy levels acquire a finite width. Hence, these oscillators can exchange energy in arbitrarily small portions without being excited to higher energy levels.

Likewise, there also exists momentum exchange between the particle and the oscillators. As a result, the particle momentum p fluctuates and the expectation value $\langle \hat{p}^2 \rangle$ differs from zero down to $T = 0$. Employing again the Langevin equation analysis, we easily obtain $\langle \hat{p}^2 \rangle = 2m\langle E_{\text{kin}} \rangle$. Hence, at $T = 0$, we have

$$\langle \hat{p}^2 \rangle = \frac{\hbar\eta}{\pi} \ln \frac{\omega_c}{\gamma} \tag{3.22}$$

also implying that the dephasing length L_φ (3.10) is essentially determined by the momentum fluctuations as

$$L_\varphi \sim \frac{\hbar}{\sqrt{\langle \hat{p}^2 \rangle}} \tag{3.23}$$

Employing the above exact result (3.5), it is also straightforward to see how the initial momentum state decays as a result of interaction between the particle and the Caldeira–Leggett bath. Let us choose $\psi \sim \exp(ip_i q/\hbar)$ as the initial state of the particle q and evaluate the transition probability $W_{p_i p_f}$ to the state with momentum p_f. In the absence of interactions, the particle momentum is a good quantum number, i.e., one simply has $W_{p_i p_f} = \delta_{p_i p_f}$. For nonzero η at $T = 0$ and in the long-time limit from Eqs. (1.12) and (3.5), we obtain

$$W_{p_i p_f} \sim \frac{1}{\sqrt{\chi_1 \chi_2}} \exp\left(-\frac{1}{\hbar\eta} \left(\frac{p_i^2}{\chi_1} + \frac{p_f^2}{\chi_2} \right) \right) \tag{3.24}$$

where $\chi_1 \simeq 4\ln(\omega_c t)$ and $\chi_2 \simeq (2/\pi)\ln(\omega_c/\gamma)$. The result (3.24) demonstrates that in the limit of very long times, information about the initial momentum p_i is lost and the particle momentum distribution slowly relaxes to the universal one. We also note that Eq. (3.24) cannot be recovered in any finite order of the perturbation theory in the interaction strength η.

3.2 Particle on a Ring

Having understood the behavior of a free particle in a dissipative environment, we now proceed to our second example. We will now assume that the particle motion is confined to a ring of radius R, i.e., our quantum particle can move freely along the ring without having a possibility to leave it [154, 181]. In this case, it will be convenient for us to choose the origin in the center of the ring and to parameterize all particle trajectories by some angle variable $\theta(t)$, i.e., the particle position will be described by a two-dimensional vector $\boldsymbol{r}$ with components $(R\cos\theta,\ R\sin\theta)$ in the ring plane.

3.2.1 Real-Time Dynamics

As before, we would like to calculate the transition probability W, but now we have to choose both initial and final particle coordinates on the ring. Let us set initial and final values of $\boldsymbol{r}$ equal, i.e., $\boldsymbol{r}_i = \boldsymbol{r}_f = (0, -R)$ (see Fig. 3.1). With this choice, W defines the probability for a particle to return to the same point located in the bottom of the ring. As we will see below in Chapter 14, this return probability plays an important role, e.g., in the problem of weak localization, giving a great deal of information about quantum coherence of electrons moving in a disordered conductor. Since the latter problem is rather complex, here we will use a simpler model that, however, captures essential physics, teaching us a lot about the effect of quantum decoherence for a particle interacting with a dissipative environment.

In the absence of interactions, the return probability is evaluated trivially. In order to return to the initial point during the time interval

t, the particle should move along one of the classical trajectories

$$\tilde{\theta}^{(n)}(t') = 2\pi n t'/t - \pi/2, \quad 0 \leq t' \leq t, \quad n = 0, \pm 1, \pm 2, \ldots \quad (3.25)$$

describing $|n|$ full clockwise ($n > 0$) or counterclockwise ($n < 0$) circles around the ring. Substituting the trajectories (3.25) into the particle action

$$S[\theta] = \int_0^t \frac{mR^2\dot{\theta}^2(t_1)}{2} dt_1 \quad (3.26)$$

we get the classical action for the n-th trajectory equal to

$$S_n = S_{-n} = 2\pi^2 m R^2 n^2/t = \pi m v R n^2 \quad (3.27)$$

where $v = 2\pi R/t$ is the particle velocity.

Again, in the absence of interaction, the two amplitudes K and K^* in Eq. (1.9) for the transition probability decouple and can be evaluated separately. For simplicity, we assume that the action on the classical trajectory (3.25) with $n = \pm 1$ is large, $S_{\pm 1} \gg \hbar$, i.e.,

$$mvR \gg \hbar. \quad (3.28)$$

This condition allows to evaluate the transition amplitude

$$K = \int \mathrm{D}\theta \exp(\mathrm{i}S[\theta]/\hbar) \quad (3.29)$$

semiclassically. We have

$$K = \sum_n K_n, \quad K_n \sim \exp(\mathrm{i}S_n/\hbar) \quad (3.30)$$

and similarly for K^*. Evaluating now the return probability $W = KK^*$, we get the sum of terms $K_n K_{n'}^* \sim \exp(\mathrm{i}(S_n - S_{n'}/\hbar))$ for all integer n and n'. Since the action in the exponent is large, terms with $n' \neq \pm n$ contain large phase factors, which oscillate rapidly with changing $n - n'$ and t and effectively cancel out in the course of summation over n and n'. What remains is the contribution from all terms with $n = \pm n'$. These terms are special since large phases $\mathrm{i}S_n/\hbar$ and $-\mathrm{i}S_{n'}/\hbar$ coming, respectively, from K and K^* cancel each other exactly in the exponent. Hence, these terms and their sum do not vanish and the probability W can be written in the form

$$W = W_1 + W_2, \quad W_1 = \sum_n K_n K_n^*, \quad W_2 = \sum_{n \neq 0} K_n K_{-n}^* \quad (3.31)$$

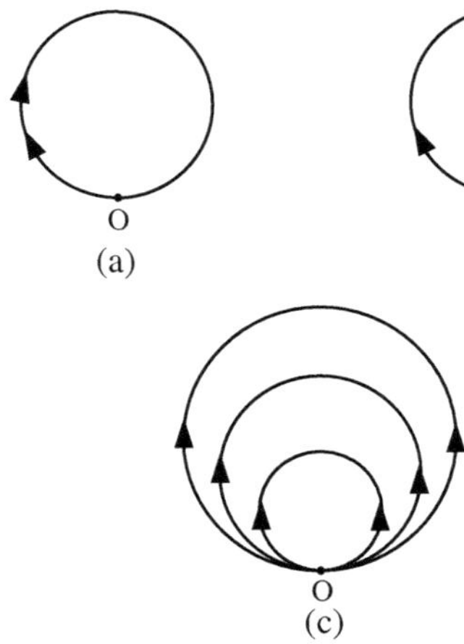

Figure 3.1 A particle on a ring: (a) classical return paths, (b) time-reversed return paths, and (c) time-reversed return paths of different sizes.

The term W_1 contains all contributions with $n = n'$, i.e., it is determined by a pair of *equivalent* clockwise (or counterclockwise) paths $\theta_1(t') = \theta_2(t') = \tilde{\theta}^{(n)}(t')$ (Fig. 3.1a). This contribution does not vanish in the classical limit and just reduces to the classical return probability. The term W_2 originates from the contributions with $n = -n'$. Equivalently, one can say that this term comes from a pair of time-reversed paths $\theta_1(t') = \theta_2(t - t') = \tilde{\theta}^{(n)}(t')$, i.e., one of the paths is going clockwise and another one, counterclockwise (Fig. 3.1b). The term W_2 describes specifically quantum contribution to the return probability. For that reason, this term is very sensitive to the presence of the phase coherence in our system and it vanishes identically in the classical limit. Without interaction, the two terms W_1 and W_2 are related to each other simply as $W_2 = W_1 - |K_0|^2$.

Now let us analyze the effect of interactions. For that purpose, we will assume that our ring is immersed in a bath of harmonic oscillators and, as before, consider the interaction between the particle and the oscillators in the form of a bilinear function in their coordinates. Integrating out the bath oscillators, we again arrive at a Gaussian influence functional $F[\mathbf{r}_F(t), \mathbf{r}_B(t)]$. Following the arguments of Chapter 1, we can write down this functional in the following general form:

$$F = \exp(-S^{(2)}[\boldsymbol{r}_F(t), \boldsymbol{r}_B(t)]/\hbar), \tag{3.32}$$

where

$$S^{(2)} = \int_0^t dt' \int_0^{t'} dt''(\boldsymbol{r}_F(t') - \boldsymbol{r}_B(t')(\alpha(t', t'')\boldsymbol{r}_F(t'') - \alpha^*(t', t'')\boldsymbol{r}_B(t'')) \tag{3.33}$$

Expressing this formula in terms of the angle variable $\theta(t)$, we get

$$\begin{aligned} S^{(2)} = R^2 \int_0^t dt' \int_0^{t'} & dt''[\alpha(t', t'')(\cos(\theta_1(t') - \theta_1(t'')) \\ & - \cos(\theta_2(t') - \theta_1(t''))) + \alpha^*(t', t'')(\cos(\theta_2(t') - \theta_2(t'')) \\ & - \cos(\theta_1(t') - \theta_2(t'')))]. \end{aligned}$$

Defining again $S^{(2)} = \mathrm{i}S_\mathrm{R} + S_\mathrm{I}$, after trivial algebra, we obtain

$$S_\mathrm{R} = 8R^2 \int_0^t \mathrm{d}t' \int_0^{t'} \mathrm{d}t'' \mathrm{Im}\alpha(t' - t'') \sin(\theta_+(t') - \theta_+(t'')) \times \sin\frac{\theta_-(t')}{2} \cos\frac{\theta_-(t'')}{2} \tag{3.34}$$

$$S_\mathrm{I} = 4R^2 \int_0^t \mathrm{d}t' \int_0^{t'} \mathrm{d}t'' \mathrm{Re}\alpha(t' - t'') \cos(\theta_+(t') - \theta_+(t'')) \times \sin\frac{\theta_-(t')}{2} \sin\frac{\theta_-(t'')}{2} \tag{3.35}$$

where $\theta_+(t) = (\theta_F(t) + \theta_B(t))/2$ and $\theta_-(t) = \theta_F(t) - \theta_B(t)$. We note that the dependence of both real and imaginary parts of the action (3.34) and (3.35) on the angle variable θ is identical to that on the phase variable $\varphi/2$ established in Chapter 2 for normal tunnel barriers, cf. Eqs. (2.77), (2.78) with $I_x \to 0$ and $\beta_{R,I} \to 0$.

For pedagogical purposes, we first consider a bath consisting of a single oscillator, in which case the kernels $\mathrm{Re}\alpha(t)$ and $\mathrm{Im}\alpha(t)$ are defined, respectively, in Eqs. (1.73) and (1.74). Provided the classical paths (3.25) are weakly perturbed by the interaction with an oscillator, i.e., for

$$S_{R,I}[\theta^{(n)}(t)] \ll S_n \tag{3.36}$$

and provided, as before, the action S_n is large, $S_{n\neq 0} \gg \hbar$, the effect of the interaction is trivially captured within the semiclassical approximation just by substituting the trajectories (3.25) directly into the influence functional (1.20). Evaluating the contribution W_1 to the return probability, we immediately conclude that within our approximation this contribution is insensitive to the interaction. This is because W_1 is determined by the pairs of equivalent paths $\theta_F(t) = \theta_B(t)$, for which one has (cf. Eq. (1.19))

$$F[\theta_F(t), \theta_F(t)] \equiv 1 \tag{3.37}$$

and, hence, W_1 remains unaffected.

In contrast to W_1, the quantum contribution W_2 to the return probability determined by the pairs of time-reversed paths *is* affected by the interaction. Substituting these paths into Eqs. (3.34), (3.35) after a simple integration, we find

$$S_{Rn} = 0, \quad S_{\text{In}}(t) = \frac{4R^2C^2}{\omega} \coth\left(\frac{\hbar\omega}{2T}\right)\left(\frac{t}{2\pi n}\right)^2 \frac{\sin^2(\omega t/2)}{(1 - \omega^2 t^2/4\pi^2 n^2)^2} \tag{3.38}$$

We emphasize that the term S_R identically equals to zero for any pair of time-reversed paths simply because the combination $\theta_+(t') - \theta_+(t'') = 0$ for such paths. This is a generic property of S_R related to its particular symmetry. As we will see later, this property is also preserved in other models. The term S_I, on the contrary, differs from zero on any pair of time-reversed paths. It is proportional to T at high temperatures $T \gg \omega$ and becomes temperature independent in the opposite limit $T \ll \omega$.

As a result, in the presence of the interaction between the particle and a harmonic oscillator with frequency ω, the return probability W gets modified as

$$W = W_1 + W_2, \quad W_1 = \sum_n K_n K_n^*, \quad W_2 = \sum_{n \neq 0} K_n K_{-n}^* e^{-S_{\text{In}}/\hbar} \tag{3.39}$$

where S_{In} is defined in Eq. (3.38). We see that the values $S_{\text{In}}(t)$ and, hence, W_2 and W oscillate in time with a period $2\pi/\omega$. After each such period, the quantum contribution to the return probability W_2 restores its "non-interacting" value, while at all intermediate times, the value W_2 is smaller than in the non-interacting case. Furthermore, in the long-time limit, we have $S_{\text{In}} \propto n^2/t^2$, i.e., the oscillations of W_2 should eventually die out in this limit.

The situation changes drastically if we add up the effect from many oscillators in Eq. (3.38). Considering an infinite bath of Caldeira–Leggett oscillators and summing up all their contributions with the Ohmic spectrum (1.54), we again recover Eqs. (3.34) and (3.35) where the kernels $\mathrm{Re}\alpha(t)$ and $\mathrm{Im}\alpha(t)$ are now defined, respectively, in Eqs. (1.55) and (1.56). In this case, not only the angle dependence, but the whole influence functional turns out to be exactly identical to that derived in Chapter 2 for normal tunnel barriers if we identify

$$\theta(t) \leftrightarrow \varphi(t)/2, \qquad \eta R^2 \leftrightarrow \hbar^2/e^2 R_N \tag{3.40}$$

Hence, we conclude that the effective action for normal tunnel barriers is fully equivalent to that for a quantum particle on a ring interacting with a bath of Caldeira–Leggett oscillators.

Integrating the result (3.38) over ω and employing again Eq. (1.54), we obtain

$$S_{\mathrm{In}} = \frac{2\eta R^2}{\pi} \int \frac{zdz}{n^2} \frac{\sin^2(\pi z)}{(1-(z/n)^2)^2} \coth\left(\frac{\pi\hbar z}{Tt}\right) \tag{3.41}$$

This integral can be easily handled with the result

$$S_{\mathrm{In}} \sim \begin{cases} \eta R^2, & T \to 0 \\ \eta T R^2 t/n\hbar, & Tt/\hbar \gg n \end{cases} \tag{3.42}$$

Combining Eqs. (3.39) and (3.42), we arrive at the final result for the return probability W for a quantum particle on a ring interacting with the Caldeira–Leggett bath of oscillators. We observe from Eq. (3.39) that if S_{In} is small, the influence functional $F \simeq 1$ and interactions do not lead to any significant modifications of the return probability. The decoherence effect of the quantum dissipative bath can be ignored in this limit. However, in the opposite limit of large $S_{\mathrm{In}} \gg \hbar$, the interference of the time-reversed paths is exponentially suppressed and the quantum contribution to the return probability W_2 vanishes. This is strong quantum dephasing by interactions. The borderline between these two regimes is obtained simply by setting $S_{\mathrm{In}} \sim \hbar$. This condition defines a typical ring perimeter $2\pi R \sim L_\varphi$ beyond which the effect of quantum decoherence becomes important. In our particular example, the characteristic dephasing length L_φ at $T \to 0$ becomes [154]

$$L_\varphi \sim \sqrt{\hbar/\eta} \tag{3.43}$$

Hence, quantum coherence of a particle is suppressed by interactions provided the ring perimeter $2\pi R$ strongly exceeds the length scale (3.43). It is also satisfactory to observe that the dephasing length (3.43) derived semiclassically is fully consistent with the result (3.10) obtained for the exactly solvable model studied above.

As we already discussed, the validity of our semiclassical approximation is limited by Eq. (3.28). In addition, Eq. (3.36) should be obeyed, which allows us to ignore perturbation of classical trajectories (3.25) by interactions. In fact, it suffices to verify this condition only at $R \lesssim L_\varphi$, since for bigger values of R, quantum coherence would already be strongly suppressed anyway. For $R \sim \sqrt{\hbar/\eta}$, Eq. (3.42) at $T = 0$ obviously yields $S_{\rm In} \sim \hbar$, and hence the inequality (3.36) is obeyed automatically provided Eq. (3.28) is fulfilled.

The above semiclassical results allow to qualitatively understand the effect of interaction-induced quantum decoherence in a number of physical situations. For instance, one can consider conduction electrons in a metal propagating with almost constant (in absolute value) velocities of order Fermi one, $v \simeq v_F$. In order to account for this feature in our consideration, it is sufficient to keep the velocity of a particle v fixed, implying that the radius R for a classical return path should increase linearly with time t. In other words, we can slightly modify our model allowing our particle to "choose" the ring with a proper $R \propto t$ for each time t (see Fig. 3.1c). Then Eq. (3.28) remains justified also in the long-time limit and we observe that $S_{\rm In}$ (3.41) grows in time as $S_{\rm In} \propto t^2$ at $T = 0$ and $S_{\rm In} \propto t^3$ at $Tt \gg \hbar$, i.e., in this case, the decay of the quantum interference term W_2 in time is faster than exponential.

Let us also recall that in a disordered metal, although the length of the electron trajectory increases linearly with time, its dynamics is diffusive and the effective loop size grows as $R^2 \sim Dt$, where D is the electron diffusion coefficient. Substituting this expression into Eq. (3.41) for $n \sim 1$, one finds

$$F = \begin{cases} \exp(-t/\tau_\varphi), & \tau_\varphi \sim \hbar/D\eta, \qquad\qquad Tt \ll \hbar, \\ \exp(-(t/\tau_\varphi)^2), & \tau_\varphi \sim \hbar/(D\eta T)^{1/2}, \ Tt \gg \hbar \end{cases} \tag{3.44}$$

Although for a number of reasons the above simple model cannot be directly applied to disordered metals, it demonstrates several important properties, which will be also observed in a rigorous calculation, which we postpone to Chapter 14 of this book. One such property is that the suppression of quantum interference between time-reversed paths increases if the size of the loop grows in time. Another property is that oscillators with $\hbar\omega \gg T$ may give the dominating contribution to the particle dephasing. This property was already observed and explained in Section 3.1 when we dealt with the exact solution for a free particle in the Caldeira–Leggett bath. In the case of a particle on a ring, frequencies $\omega \sim 1/t$ give the maximum contribution, but if the spectrum of the problem is different (as, e.g., in a disordered metal, see Chapters 8 and 9), the environmental modes with higher frequencies up to the cutoff one $\sim \omega_c$ also become important.

The above analysis also clearly demonstrates at which step of our calculation the effect of quantum decoherence occurs. Interaction of the particle with one harmonic oscillator leads to the oscillations of the interference term W_2 with the oscillator frequency ω. These quantum coherent oscillations occur simply because the initial state is not an eigenstate of the system. Although the time-averaged quantum contribution to the return probability is smaller than in the non-interacting case, after each period of these oscillations, the particle returns to its initial state. Yet no dephasing occurs. In the case of interaction with several oscillators with incommensurate frequencies, the return probability remains decreased at any finite time, but approaches its non-interacting value in the long-time limit. If, finally, coupling to *infinitely many* oscillators with different frequencies is introduced, the probability W_2 will always be suppressed, the system will never return to its initial state, and, provided the ring perimeter is large enough, $2\pi R \gg L_\varphi$, the information about its initial phase will be washed out at any bath temperature, including $T = 0$.

3.2.2 Decoherence in the Ground State

Although the above quasiclassical description is sufficient for particles with sufficiently high velocities, it might fail at lower energies. For instance, keeping the ring radius R constant in time, we inevitably decrease the particle velocity $v = 2\pi R/t$ with increasing t, in which case the inequality (3.28) is violated in the long-time limit. Do we have to expect any significant modifications of our results and conclusions in this case? In other words, does the interaction-induced decoherence persist also in the ground state of an interacting system? Below we will demonstrate that the answer to the last question is positive and our result for the zero-temperature decoherence length L_φ (3.43) remains valid also in this case.

In order to investigate the ground-state properties of our interacting system consisting of the particle on a ring and the Caldera–Leggett bath of oscillators, it will be convenient to analyze its grand partition function $\mathcal{Z}$. This partition function can be represented as a path integral (1.93), (1.94) modified in order to account for the ring geometry. In the effective action (1.94), one should set $U(q) \to 0$ and $q(\tau) - q(\tau') \to 2R\sin(\theta(\tau)/2 - \theta(\tau')/2)$. Then one finds

$$
\begin{aligned}
S_{\text{eff}} &= \int_0^{\hbar\beta} d\tau \frac{\dot{\theta}^2}{4E_C} \\
&+ \frac{\eta R^2}{\pi} \int_0^{\hbar\beta} d\tau \int_0^{\hbar\beta} d\tau' \frac{(\pi/\hbar\beta)^2 \sin^2[(\theta(\tau) - \theta(\tau'))/2]}{\sin^2(\pi(\tau - \tau')/\hbar\beta)}
\end{aligned}
\tag{3.45}
$$

where the parameter $E_C = 1/2mR^2$. Comparing this expression with Eq. (2.141) (where $\alpha(\tau)$ is defined in Eq. (2.140) and $\tilde{\alpha}_S(\tau) \to 0$) and identifying the parameters according to Eq. (3.40), we again observe the equivalence of the actions describing our model and that of a normal tunnel barrier, now in their imaginary time formulations.

The boundary conditions in the path integral (1.93) should also be modified having in mind that the angle variable θ is compact (i.e., defined on a ring) and, hence, the particle returns to the same state (the same position) after m winds around the ring. The contributions from all these winding numbers m should be included into the expression for $\mathcal{Z}$. We skip a formal discussion of this point here and postpone it to Chapters 6 and 7 (see Sections 6.4, 6.5, and 7.4) where exactly the same physical situations of compact and extended phase variables are analyzed in details.

Finally, we will assume that our particle has the electric charge e and that an external magnetic flux Φ_x is inserted inside the ring. In this case, the term $E_C(\hat{\Phi} + \Phi_x)^2/\Phi_0^2$ should be added to the

Hamiltonian of our system, where $\hat{\Phi} = -i\Phi_0\partial/\partial\theta$ is the magnetic flux operator, and $\Phi_0 = 2\pi\hbar c/e$ is the flux quantum.

As a result, the grand partition function for our system takes the form (see also Chapter 8)

$$\mathcal{Z} = \sum_{m=-\infty}^{\infty} \int_0^{2\pi} d\theta_0 \int_0^{2\pi m} \mathcal{D}\theta \exp(i2\pi m\phi_x - S_{\text{eff}}[\theta]/\hbar), \quad (3.46)$$

where we defined $\phi_x = \Phi_x/\Phi_0$.

The effect of quantum decoherence within the above model can be conveniently investigated by analyzing, e.g., the Aharonov–Bohm effect [181] or persistent current circulating around the ring [143]. Both these properties are sensitive to quantum coherence of the particle being suppressed, provided this coherence is lost. For instance, persistent current is defined as a derivative of the free energy $F = -T \ln \mathcal{Z}$ with respect to the flux

$$I = -\frac{e}{2\pi\hbar}\frac{\partial F}{\partial \phi_x} \quad (3.47)$$

In the absence of interactions $\eta \to 0$ and in the limit $T \to 0$, this formula yields (see also Chapter 8)

$$I = (eE_C/\pi\hbar)\frac{\partial}{\partial\phi_x}\min_k\frac{(\phi_x - k)^2}{2}, \quad (3.48)$$

i.e., persistent current is proportional to the parameter E_C and is periodic in the magnetic flux with the period equal to the flux quantum Φ_0.

In order to find out in which way this simple result is modified in the presence of interactions, it is in general necessary to work out a detailed non-perturbative analysis of the grand partition function defined in Eqs. (3.45) and (3.46). Fortunately, due to exact equivalence between the partition functions for our model and that of the so-called single-electron box studied below in Chapter 7, it is not necessary to perform this analysis here and postpone it to Section 7.4. The equivalence between these two models is observed if we identify ϕ_x with the dimensionless gate charge n_g in Chapter 7 and also employ Eq. (3.40). Hence, we can directly translate the non-perturbative results derived in Chapter 7 to our present model. For the persistent current at $T \to 0$, we obtain

$$I \simeq \frac{eE_C}{\pi\hbar}\left(\phi_x - \frac{\frac{R^2}{L_\varphi^2\pi^2}\ln\frac{1+2\phi_x}{1-2\phi_x}}{2 + \frac{2R^2}{L_\varphi^2\pi^2}\ln\frac{1+2\phi_x}{1-2\phi_x}}\right) \quad (3.49)$$

for $R \ll L_\varphi$ and $0 \le \phi_g \le 1/2$ (cf. Eq. (7.146)) and

$$I \propto (eE_C/\hbar)\exp(-R^2/L_\varphi^2) \tag{3.50}$$

for $R \gg L_\varphi$ (cf. Eq. (7.169)). In these equations, we defined the dephasing length $L_\varphi = \sqrt{\hbar/\pi\eta}$, which matches with the result (3.43) derived from our semiclassical analysis.

Equation (3.50) demonstrates exponential suppression of the persistent current for ring perimeters exceeding L_φ (3.43) exactly in the ground state of an interacting system. This is yet one more clear manifestation of zero-temperature interaction-induced decoherence encountered within the Caldeira–Leggett model. We will return to this issue again in Chapters 8, 14, and 15 when we consider more complicated models describing interacting electrons in realistic disordered conductors.

3.3 Damped Harmonic Oscillator

Let us now discuss another important physical situation of a linearly damped harmonic oscillator. The kernel J, which accounts for quantum evolution of the oscillator density matrix, is given by the double path integral (1.59) with the boundary conditions (1.60). The corresponding action is defined by Eq. (1.61) with potential energy $U(q) = m\omega_0^2 q^2/2$ (where ω_0 is the oscillator frequency) and the action of the environment (1.62). With this choice of the potential, the path integral is Gaussian and can be handled exactly. The whole procedure repeats that for a free particle. As the first step, one performs the integration over the variable q_+. Since the action contains only linear q_+ terms, this integration just yields the δ−function in the functional space. Thus, the path integral for the kernel J reduces to the form

$$J \propto \int \mathcal{D}q_-\, \delta(m\ddot{q}_- - \eta\dot{q}_- + m\omega_0^2 q_-) e^{im[q_{+f}\dot{q}_-(t) - q_{+i}\dot{q}_-(0)]/\hbar} e^{-S_I[q_-]/\hbar}$$

In order to evaluate this integral, it suffices to solve the linear differential equation ($\gamma = \eta/m$)

$$\ddot{q}_- - \gamma\dot{q}_- + \omega_0^2 q_- = 0 \tag{3.51}$$

with the boundary conditions $q_-(0) = q_{-i}$, $q_-(t) = q_{-i}$ and then substitute the corresponding solution both into the action $S_I[q_-]$

and into the boundary terms $im[q_{+f}\dot{q}_-(t) - q_{+i}\dot{q}_-(0)]/\hbar$. Note that Eq. (3.51) just coincides with the time-reversed version of the classical equation of motion of a linearly damped harmonic oscillator.

The procedure outlined above is straightforward but rather tedious. In particular, one should pay special attention to the boundary terms in the action. Here we omit the corresponding details of the calculation referring the reader to the review paper [176]. Having performed this calculation, one arrives at the final result:

$$\begin{aligned} J = & \frac{m}{2\pi\hbar G_+(t)} \exp\left\{ \frac{im}{\hbar}\frac{\dot{G}_+(t)}{G_+(t)}(q_{+f}q_{-f} + q_{+i}q_{-i}) \right. \\ & - \frac{im}{\hbar}\left(\frac{q_{+f}q_{-i}}{G_+(t)} + \frac{q_{+i}q_{-f}}{G_-(t)}\right) \\ & \left. - \frac{m[q_{-i}^2 R^{++}(t) + 2q_{-i}q_{-f}R^{+-}(t) + q_{-f}^2 R^{--}(t)]}{2\hbar} \right\} \end{aligned} \tag{3.52}$$

Here we defined the functions

$$\begin{aligned} G_+(s) &= \int \frac{d\omega}{2\pi} \frac{e^{-\mathrm{i}\omega s}}{\omega_0^2 - \omega^2 - \mathrm{i}\gamma\omega}, \\ G_-(s) &= \frac{G_+(t-s)\dot{G}_+(t) - G_+(t)\dot{G}_+(t-s)}{G_+(t)\ddot{G}_+(t) - \dot{G}_+^2(t)}, \\ R^{++}(t) &= \gamma \int_0^t ds \int_0^t ds' \int_{-\omega_c}^{\omega_c} \frac{d\omega}{2\pi}\omega \coth\frac{\hbar\omega}{2T} e^{-\mathrm{i}\omega(s-s')} \\ &\quad \times \frac{G_+(t-s)}{G_+(t)}\frac{G_+(t-s')}{G_+(t)}, \\ R^{+-}(t) &= \gamma \int_0^t ds \int_0^t ds' \int_{-\omega_c}^{\omega_c} \frac{d\omega}{2\pi}\omega \coth\frac{\hbar\omega}{2T} e^{-\mathrm{i}\omega(s-s')} \\ &\quad \times \frac{G_+(t-s)}{G_+(t)}\frac{G_-(s')}{G_-(t)}, \\ R^{--}(t) &= \gamma \int_0^t ds \int_0^t ds' \int_{-\omega_c}^{\omega_c} \frac{d\omega}{2\pi}\omega \coth\frac{\hbar\omega}{2T} e^{-\mathrm{i}\omega(s-s')} \frac{G_-(s)}{G_-(t)}\frac{G_-(s')}{G_-(t)} \end{aligned} \tag{3.53}$$

The function $G_+(s)$ obeys the equation

$$\ddot{G}_+ + \gamma\dot{G}_+ + \omega_0^2 G_+ = 0 \tag{3.54}$$

with the boundary conditions $G_+(0) = 0$ and $\dot{G}_+(0) = 1$. It may be interpreted as a response function of the oscillator to an external force of the form $\delta(t)$. In the limit $\omega_0 \to 0$, the oscillator becomes a free particle. In this case, $G_+(s) = (1 - e^{-\gamma t})/\gamma$ and the kernel of the evolution operator (3.52) reduces to that defined in Eq. (3.5). For an underdamped oscillator with $\gamma < 2\omega_0$, one finds

$$G_\pm(s) = e^{\mp\frac{\gamma s}{2}} \frac{\sin\left(\sqrt{\omega_0^2 - \frac{\gamma^2}{4}}\, s\right)}{\sqrt{\omega_0^2 - \frac{\gamma^2}{4}}} \tag{3.55}$$

while in the overdamped regime $\gamma > 2\omega_0$, the functions $G_\pm(s)$ take the form

$$G_\pm(s) = e^{\mp\frac{\gamma s}{2}} \frac{\sinh\left(\sqrt{\frac{\gamma^2}{4} - \omega_0^2}\, s\right)}{\sqrt{\frac{\gamma^2}{4} - \omega_0^2}} \tag{3.56}$$

Consider a Gaussian initial density matrix

$$\rho_0(q_i, q_i') = \sqrt{\frac{1}{2\pi\,\delta q_0^2}} \exp\left\{-\frac{(q_{+i} - q_0)^2}{2\,\delta q_0^2} - \frac{\delta p_0^2 q_{-i}^2}{2\hbar^2} + \mathrm{i}\frac{p_0 q_{-i}}{\hbar}\right\} \tag{3.57}$$

The task at hand is to describe its evolution in real time.

To begin with, let us recollect the relations between the coordinates q_i, q_i' and q_{+i}, q_{-i}. They read $q_i = q_{+i} + q_{-i}/2$ and $q_i' = q_{+i} - q_{-i}/2$. It is convenient to define the Wigner function, which determines the probability distribution for both coordinate and momentum variables

$$\begin{aligned} w(q_{+i}, p_i) &= \int \frac{dq_{-i}}{2\pi\hbar} \rho_0\left(q_{+i} + \frac{q_{-i}}{2}, q_{+i} - \frac{q_{-i}}{2}\right) e^{-\mathrm{i}p_i q_{-i}/\hbar} \\ &= \frac{1}{2\pi\,\delta p_0\,\delta q_0} \exp\left\{-\frac{(q_{+i} - q_0)^2}{2\,\delta q_0^2} - \frac{(p_i - p_0)^2}{2\,\delta p_0^2}\right\} \end{aligned} \tag{3.58}$$

Thus, the initial state (3.57) corresponds to the wave packet centered around the coordinate q_0 and momentum p_0 with mean square coordinate and momentum fluctuations δq_0^2 and δp_0^2, respectively. The density matrix at time t,

$$\rho(t, q_{+f}, q_{-f}) = \int dq_{+i} dq_{-i}\, J(t, q_{+f}, q_{-f}, q_{+i}, q_{-i}) \rho_0 \left(q_{+i} + \frac{q_{-i}}{2}, q_{+i} - \frac{q_{-i}}{2}\right)$$

can be evaluated exactly since both integrals over q_{+i} and q_{-i} are Gaussian. One finds [176]

$$\rho(t, q_{+f}, q_{-f}) = \frac{1}{\sqrt{2\pi\sigma_q(t)}} \exp\left\{ -\frac{\left(q_{+f} - q_{\text{cl}}(t)\right)^2}{2\sigma_q(t)} - \frac{1}{2\hbar^2}\left(\sigma_p(t) - \frac{\sigma_{pq}^2(t)}{\sigma_q(t)}\right) q_{-f}^2 + \frac{\mathrm{i}}{\hbar}\left(p_{\text{cl}}(t) + \frac{\sigma_{pq}(t)}{\sigma_q(t)}\left(q_{+f} - q_{\text{cl}}(t)\right)\right) q_{-f} \right\}. \tag{3.59}$$

Here we have introduced the classical time-dependent coordinate and momentum,

$$q_{\text{cl}}(t) = \frac{p_0}{m} G_+(t) + q_0 \dot{G}_+(t), \tag{3.60}$$

$$p_{\text{cl}}(t) = m\dot{q}_{\text{cl}}(t) = p_0 \dot{G}_+(t) + m q_0 \ddot{G}_+(t) \tag{3.61}$$

the corresponding time-dependent mean square fluctuations,

$$\sigma_q(t) = \delta q_0^2 \dot{G}_+^2(t) + \left(\frac{\hbar}{m} R^{++}(t) + \frac{\delta p_0^2}{m^2}\right) G_+^2(t), \tag{3.62}$$

$$\sigma_p(t) = \delta p_0^2 \dot{G}_+^2(t) + m^2 \delta q_0^2 \ddot{G}_+^2(t) + m\hbar\left[R^{--}(t) + 2R^{+-}(t)\dot{G}_+(t) + R^{++}(t)\dot{G}_+^2(t)\right], \tag{3.63}$$

and the cross-correlation of the coordinate and momentum defined as

$$\sigma_{pq}(t) = \frac{m}{2}\frac{d\sigma_q(t)}{dt} = m\delta q_0^2 \dot{G}_+(t)\ddot{G}_+(t) + \hbar R^{+-}(t) G_+(t) + \left(\hbar R^{++}(t) + \frac{\delta p_0^2}{m}\right) \dot{G}_+(t) G_+(t) \tag{3.64}$$

These parameters determine the expectation values of the coordinate and momentum operators $\hat{q}(t)$ and $\hat{p}(t)$. Namely, one can verify that

$$\langle \hat{q}(t)\rangle = q_{\text{cl}}(t), \quad \langle \hat{p}(t)\rangle = p_{\text{cl}}(t),$$

$$\sigma_q(t) = \langle \hat{q}^2(t)\rangle - \langle \hat{q}(t)\rangle^2, \quad \sigma_p(t) = \langle \hat{p}^2(t)\rangle - \langle \hat{p}(t)\rangle^2,$$

$$\sigma_{pq}(t) = \frac{\langle \hat{q}(t)\hat{p}(t) + \hat{p}(t)\hat{q}(t)\rangle}{2} - \langle \hat{q}(t)\rangle\langle \hat{p}(t)\rangle \tag{3.65}$$

The classical path (3.60) indicating the position of the maximum for the probability distribution $\rho(t, q_{+f}, 0)$ is the solution to the classical equation of motion

$$\ddot{q}_{\mathrm{cl}} + \gamma \dot{q}_{\mathrm{cl}} + \omega_0^2 q_{\mathrm{cl}} = 0 \tag{3.66}$$

with initial conditions $q_{\mathrm{cl}}(0) = q_0$ and $\dot{q}_{\mathrm{cl}}(0) = p_0/m + q_0 \ddot{G}_+(0)$. From Eqs. (3.55) and (3.56), one finds $\ddot{G}_+(0) = -\gamma$. On the other hand, taking into account the high-frequency cutoff of the bath spectrum ω_c, one can demonstrate that $\ddot{G}_+(0) \equiv 0$. Thus, the initial speed of the particle is set to be $\dot{q}_{\mathrm{cl}}(0) = p_0/m$ in full agreement with classical mechanics. However, due to the coupling between the oscillator and the bath after a very short time $\sim\omega_c^{-1}$, this derivative approaches the value $\dot{q}_{\mathrm{cl}}(t > \omega_c^{-1}) = p_0/m - \gamma q_0$ consistent with Eqs. (3.55) and (3.56). The presence of such a short initial evolution period of initially factorized density matrix is typical for "bath + environment" models.

In the opposite long-time limit, the density matrix of the damped oscillator (3.59) tends to relax to its equilibrium form. In this limit, one finds $q_{\mathrm{cl}} = 0$, $p_{\mathrm{cl}} = 0$ and the average coordinate and momentum values vanish. The mean coordinate dispersion $\sigma_q(t)$ approaches the equilibrium value σ_q^{eq},

$$\begin{aligned}\lim_{t=\infty} \sigma_q(t) &\equiv \sigma_q^{\mathrm{eq}} \\ &= \frac{\hbar\gamma}{m} \int_0^\infty ds ds' \int_{-\omega_c}^{\omega_c} \frac{d\omega}{2\pi} \omega \coth \frac{\hbar\omega}{2T} e^{-i\omega(s-s')} G_+(s) G_+(s') \\ &= \frac{\gamma}{m} \int_{-\omega_c}^{\omega_c} \frac{d\omega}{2\pi} \frac{\hbar\omega \coth \frac{\hbar\omega}{2T}}{(\omega^2 - \omega_0^2)^2 + \gamma^2\omega^2}\end{aligned} \tag{3.67}$$

Similarly, for the momentum dispersion $\sigma_p(t)$ and for the cross-correlation term $\sigma_{pq}(t)$, one finds, respectively

$$\lim_{t=\infty} \sigma_p(t) \equiv \sigma_p^{\mathrm{eq}} = \gamma m \int_{-\omega_c}^{\omega_c} \frac{d\omega}{2\pi} \frac{\hbar\omega^3 \coth \frac{\hbar\omega}{2T}}{(\omega^2 - \omega_0^2)^2 + \gamma^2\omega^2} \tag{3.68}$$

and

$$\lim_{t=\infty} \sigma_{pq}(t) \equiv 0 \tag{3.69}$$

Note that due to the presence of dissipation, the values σ_q^{eq}, σ_p^{eq}, and $\sigma_{pq}^{\mathrm{eq}}$ deviate from those for a free harmonic oscillator,

$$\sigma_q^{(0)} = \frac{\hbar}{2m\omega_0} \coth \frac{\hbar\omega_0}{2T}, \quad \sigma_p^{(0)} = \frac{\hbar m\omega_0}{2} \coth \frac{\hbar\omega_0}{2T} \tag{3.70}$$

which can be achieved only in the limit $\eta \to 0$. In the opposite limit of strong dissipation, $\gamma \gg \omega_0$, and at $T = 0$, the integrals (3.67) and (3.68) may be evaluated approximately, and we get

$$\sigma_q^{\rm eq} = \frac{2\hbar}{\pi\eta} \ln \frac{\eta}{m\omega_0}, \qquad \sigma_p^{\rm eq} = \frac{\hbar\eta}{\pi} \ln \frac{\omega_c}{\omega_0} \tag{3.71}$$

Comparing these results with the corresponding expressions for an undamped oscillator (3.70), we conclude that dissipation tends to suppress coordinate fluctuations and simultaneously enhances momentum fluctuations.

Finally, it is worth pointing out that the exact solution to a quantum mechanical problem of a damped harmonic oscillator can also be derived with the aid of quasiclassical Langevin equation (1.67) combined with the noise correlator defined in the quantum limit (1.68). This property is a direct consequence of the identity

$$U\left(q_+ + \frac{q_-}{2}\right) - U\left(q_+ - \frac{q_-}{2}\right) \equiv \frac{dU(q_+)}{dq_+} q_- \tag{3.72}$$

which holds for any parabolic potential $U(q)$. For instance, Eqs. (3.62)–(3.68) can also be easily recovered with the aid of the Langevin equation (1.67). In general, one can formulate the following rule: In order to find the time evolution of an arbitrary quantum initial state, one should first construct its Wigner function and then describe its time evolution by means of Eq. (1.67) as if it would be the classical coordinate and momentum distribution function. This procedure yields the exact solution to the quantum mechanical problem of a linearly damped harmonic oscillator.

Chapter 4

Quantum Tunneling with Dissipation

In Chapter 3, we discussed real-time quantum dynamics of a quantum particle interacting with a dissipative Caldeira–Leggett bath. Provided this particle is moving in the absence of any external potential, the problem is Gaussian and, hence, allows for its exact solution. For the same reason, one can also obtain an exacted solution to the problem of linearly damped harmonic oscillator, i.e., linearly damped quantum particle moving in a parabolic external potential. In the majority of physical situations, however, anharmonicity of an external potential and/or other nonlinear effects play a substantial role. In those cases, no exact description of the system dynamics can usually be obtained, and one should reside to various approximations. In this chapter, we will consider one important physical example of that kind, quantum dissipative decay of a metastable state. Another example, quantum dynamics of a damped particle in a double-well potential, will be discussed in Chapter 5. Both these problems are of special interest in connection with fundamentally important experiments demonstrating macroscopic quantum behavior of Josephson junctions, as we have already discussed in Chapter 2.

Dissipative Quantum Mechanics of Nanostructures: Electron Transport, Fluctuations, and Interactions
Andrei D. Zaikin and Dmitry S. Golubev

ISBN 978-981-4774-50-5 (Hardcover), 978-0-429-29823-3 (eBook)
www.jennystanford.com

4.1 Quantum Decay of a Metastable State: Free Energy Approach

Consider a particle with coordinate q in a potential with a local minimum, as shown in Fig. 4.1. On a classical level, the particle trapped inside this potential well will stay there forever provided it does not have enough energy to overcome the potential barrier and to escape. Quantum mechanics makes this escape possible even for low-energy particles. The particle can simply tunnel under the barrier, thus going through the classically forbidden region and reappearing on the other side of the barrier either with the same or with lower energy (see Fig. 4.1). The latter situation occurs if the particle is coupled to a dissipative environment. Then the particle energy loss is accompanied by excitation of environmental modes, and in this way, the total energy of the system "particle+environment" remains conserved in the course of tunneling.

Provided the potential barrier is very high, the particle remains trapped for a long time and can occupy one of the discrete energy levels near the bottom of the potential well. As long as tunneling

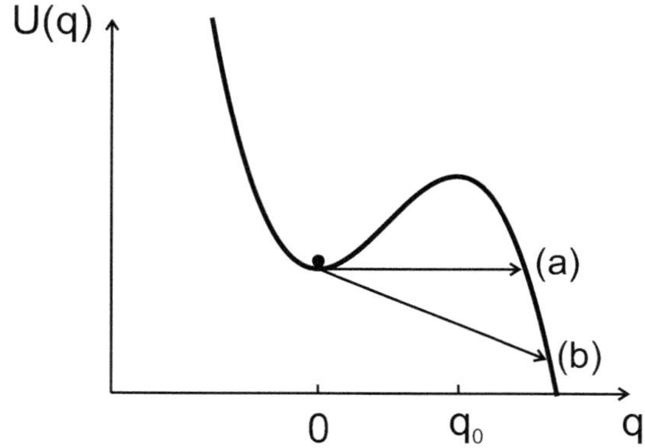

Figure 4.1 Quantum tunneling of a particle out of its metastable state: (a) the particle energy is conserved during the whole process (no dissipation included) and (b) the particle loses energy due to interaction with a dissipative bath.

effect can be ignored, the particle wave function can be written in the form

$$\psi_k(q,t) = \psi_k(q)\exp(-\mathrm{i}E_k t/\hbar) \tag{4.1}$$

where $\psi_k(q)$ obeys the stationary Schrödinger equation

$$\hat{H}\,\psi_k(q) = E_k\psi_k(q) \tag{4.2}$$

and defines the wave function of the k-th eigenstate of the system with energy E_k.

The probability density to find the particle in this eigenstate with coordinate q is given by the product $\psi_k(q,t)\psi_k^*(q,t)$, which, in the absence of tunneling, does not depend on time t and equals to $\psi_k(q)\psi_k^*(q)$. Including (weak) tunneling of the particle across the barrier implies that the k-th energy level acquires a finite width

$$E_k \to E_k - \mathrm{i}\hbar\Gamma_k/2 \tag{4.3}$$

where $\hbar\Gamma_k \ll E_k$. Defining the probability W_k for the particle to occupy the k-th energy level

$$W_k = \int dq\,\psi_k(q,t)\psi_k^*(q,t) \tag{4.4}$$

and substituting Eq. (4.3) into Eq. (4.1), we immediately obtain

$$\frac{dW_k}{dt} = -\Gamma_k W_k \tag{4.5}$$

i.e., the k-th level decays with the rate Γ_k due to tunneling.

It follows from the above consideration that, e.g., the decay rate of the lowest energy state $k = 0$ in the potential well is given by a simple formula

$$\Gamma = -\frac{2}{\hbar}\mathrm{Im}E_0 \tag{4.6}$$

As in the low-temperature limit $T \to 0$ only this lowest energy state can initially be occupied, this formula defines the quantum decay rate of a metastable state at $T = 0$.

At nonzero T, the occupation probabilities cannot be neglected anymore and the problem of finding the decay rate Γ becomes more complicated since different energy states in general decay with different rates Γ_k. Thus, the population of different states evolves differently in time. In order to correctly account for this fact, it is necessary to develop a self-consistent quantum kinetic analysis of the problem. However, if the decay is very slow and thermal equilibrium in the well is maintained at all moments of time (e.g., due to interaction between the particle and some large external reservoir/bath), one universally has $W_k(t) \propto \exp(-E_k/T)$ and the time evolution of the probability $W(t) = \sum_k W_k(t)$ to find the system in the metastable state is controlled by the following equations:

$$\frac{dW}{dt} = -\Gamma W, \qquad \Gamma = \frac{\sum_k \Gamma_k e^{-E_k/T}}{\sum_k e^{-E_k/T}} \tag{4.7}$$

On the other hand, defining the "grand partition function" for our metastable system as

$$\mathcal{Z} = \sum_k \exp\left(-\frac{E_k - \mathrm{i}\hbar\Gamma_k/2}{T}\right) \tag{4.8}$$

and assuming the decay rates to be very small, $\hbar\Gamma_k \ll T$, we arrive at the expression for the system "free energy"

$$F = -T \ln \mathcal{Z} \simeq -T \ln \sum_k \exp\left(-\frac{E_k}{T}\right) - \frac{\mathrm{i}}{2} \frac{\sum_k \Gamma_k e^{-E_k/T}}{\sum_k e^{-E_k/T}} \tag{4.9}$$

Comparing the expression for Γ in Eq. (4.7) with the last term in Eq. (4.9), we immediately identify the decay rate as

$$\Gamma(T) = -\frac{2}{\hbar}\mathrm{Im}F \tag{4.10}$$

This relation generalizes Eq. (4.6) to the case of nonzero temperatures and smoothly crosses over to the latter in the limit $T \to 0$.

Originally, the "ImF" method of calculating the decay rate of a metastable state was proposed by Langer [251, 253] in order to obtain the nucleation rate governing first-order phase transitions. Later it was developed and extensively used by different authors in order to describe quantum statistical metastability in a variety of physical systems. An important advantage of this "ImF" method is that it can be directly and conveniently applied to systems with arbitrary number of degrees of freedom, in which case other approaches such as the standard WKB method may be hard to employ. This method turns out to be particularly useful if combined with the Feynman–Vernon influence functional technique, and it was widely used in order to describe quantum decay of a metastable state in the presence of dissipation [400].

It is also necessary to point out that Eq. (4.10) for the decay rate Γ remains correct only provided the decay is dominated by the process of quantum tunneling. At sufficiently high temperatures, thermal activation takes over and a somewhat different formula for Γ should be employed. Affleck [2] argued that in the regime of thermal activation, the formula

$$\Gamma(T) = -\frac{2}{\hbar}\frac{T_0}{T}\mathrm{Im}F \tag{4.11}$$

should be used instead of Eq. (4.10). Here T_0 is the temperature at which the crossover between the regimes of quantum tunneling and thermal activation is expected to occur. This temperature essentially depends on the model under consideration, and it should also be self-consistently derived from the calculation of the decay rate. We will consider this issue in Section 4.3.2.

4.2 Instanton Technique for Decay Rate

Let us demonstrate how to evaluate the quantum decay rate of a metastable state for a quantum particle in the presence of dissipation. For definiteness and simplicity, we will assume linear Ohmic dissipation and describe the system by means of the Caldeira–Leggett effective action (1.94), which can be rewritten in the form

$$S_{\mathrm{eff}} = \int_{-\hbar\beta/2}^{\hbar\beta/2} d\tau \left[\frac{m\dot{q}^2}{2} + U(q) + \frac{\eta}{4\pi}\int_{-\infty}^{\infty} d\tau' \left(\frac{q(\tau)-q(\tau')}{\tau-\tau'}\right)^2\right] \tag{4.12}$$

Here $q(\tau)$ is periodically continued outside the interval $-\hbar\beta/2 \leq \tau < \hbar\beta/2$, and $U(q)$ is a metastable potential with local minimum at $q = 0$ and local maximum at $q = q_0$ (see Fig. 4.1). At this stage, an explicit functional form of this potential is not important for our analysis. For instance, one can choose $U(q)$ in the form

$$U(q) = \frac{\kappa}{2}q^2 - \frac{\lambda}{6}q^3 \tag{4.13}$$

which is realized in the problem of quantum decay of supercurrent in ultra-small Josephson junctions to be considered below. Also other potential profiles can be considered. Here it is only important to assume that the potential barrier U_0 is sufficiently high in order to have exponentially small decay rate Γ. The corresponding condition reads

$$U_0 \equiv U(q_0) \gg \hbar\omega_0 \tag{4.14}$$

where q_0 is the coordinate at which the potential maximum is reached and ω_0 is the oscillation frequency near the bottom of the potential well $q = 0$. For instance, for the potential of the form (4.13), one has

$$q_0 = \frac{2\kappa}{\lambda}, \quad U_0 = \frac{2\kappa^3}{3\lambda^2}, \quad \omega_0 = \sqrt{\frac{\kappa}{m}} \tag{4.15}$$

Under the condition (4.14), quantum tunneling across the barrier can be treated semiclassically.

As in Eq. (1.91), let us express the grand partition function in terms of the path integral

$$\mathcal{Z} \sim \mathrm{D}q \exp\left(-S_{\mathrm{eff}}[q(\tau)]/\hbar\right) \tag{4.16}$$

now having in mind that for a metastable potential $U(q)$ of Fig. 4.1, $\mathcal{Z}$ should acquire an imaginary part. In fact, it is this imaginary part that is mainly of interest for us here, since it determines the quantum decay rate via Eq. (4.10).

Evaluating the path integral in Eq. (4.16) semiclassically, we should first find all relevant saddle point trajectories obeying the equation

$$\delta S_{\mathrm{eff}}/\delta q = 0 \tag{4.17}$$

together with the boundary conditions $q(\pm\hbar\beta/2) = q_E$, where the initial and final particle coordinate q_E is located inside the potential well and is determined by the condition $U(q(E)) = E$, where E is the particle energy. For the action (4.12), Eq. (4.17) yields

$$m\frac{d^2q}{d\tau^2} = \frac{\partial U}{\partial q} + \frac{\eta}{\pi}\int_{-\infty}^{\infty} d\tau' \frac{q(\tau) - q(\tau')}{(\tau - \tau')^2} \tag{4.18}$$

In the absence of dissipation $\eta \to 0$, Eq. (4.18) just coincides with the classical Newton equation of motion for a particle of mass m

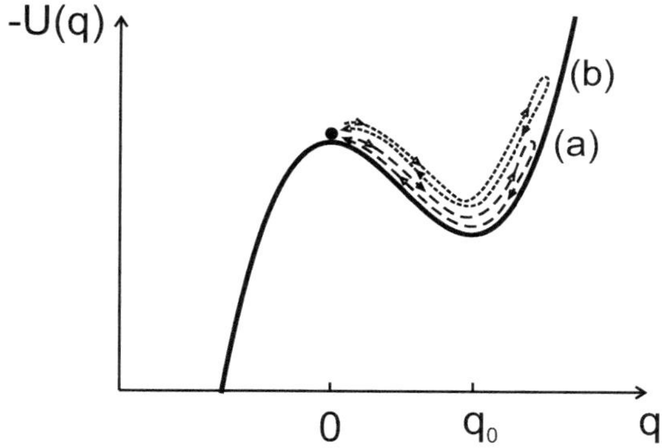

Figure 4.2 Quantum decay of a metastable state expressed in terms of classical dynamics of a particle in the inverted potential (cf. Fig. 4.1): (a) bounce trajectory in the presence of energy conservation and (b) bounce trajectory in the presence of dissipation.

in the external potential $-U(q)$, i.e., as compared to Fig. 4.1 the potential is turned upside down (see Fig. 4.2). The reason for that is, of course, clear: The imaginary time variable τ plays the role of a "real" time in Eq. (4.18), thus giving extra minus sign in front of $U(q)$. The last—non-local in time—term in Eq. (4.18) represents an extra force acting on our classical particle due to its interaction with the Caldeira–Leggett bath.

Let us first assume temperature to be very low ($T \to 0$) and find all relevant solutions to Eq. (4.18). A trivial solution is

$$q(\tau) = 0 \tag{4.19}$$

i.e., the particle just stays at rest on the very top of the "hill" and never moves down. Another trivial solution is

$$q(\tau) = q_0 \tag{4.20}$$

i.e., the particle just remains in the minimum of the inverted potential. Below we shall see that this solution will play an important role at sufficiently high temperatures.

Yet another (nontrivial) solution $\tilde{q}(\tau)$ describes the following process: The particle first moves downhill from the top point $q(-\infty) = 0$ gaining the kinetic energy, passes the potential minimum at $q = q_0$, climbs up and reaches the highest possible point

$q = q_{\rm ex}$ at $\tau = 0$. Then it goes back along the same path, eventually reaching the initial point $q = 0$ on the top of the hill at $\tau = \infty$. This is a specific instanton trajectory, which is also called "bounce" (since the particle is bounced back at $\tau = 0$). This bounce trajectory $\tilde{q}(\tau)$ accounts for particle tunneling across the barrier (Fig. 4.2). In the dissipativeless case (a) in Figs. 4.1 and 4.2, the total particle energy is conserved during the whole process, while in the presence of dissipation (the process (b) in Figs. 4.1 and 4.2), the particle loses energy during tunneling. One can also define an effective bounce time

$$\tau_B = \int_{-\infty}^{\infty} d\tau \tilde{q}(\tau)/q_0 \tag{4.21}$$

indicating for how long the tunneling particle "lives" under the barrier.

In a number of limits, the bounce trajectory $\tilde{q}(\tau)$ can be found analytically. For example, in the important case (4.13) and in the absence of dissipation, the exact solution to Eq. (4.18) reads

$$\tilde{q}(\tau) = \frac{3q_0/2}{\cosh^2(\omega_0\tau/2)} \tag{4.22}$$

Substituting (4.22) into Eq. (4.21), one finds $\tau_B = 6/\omega_0$. Some other examples of analytic solutions to Eq. (4.18) will be considered below.

We also note that at $T \to 0$, the bounce trajectory always starts and ends at $q = 0$. At nonzero temperatures, the "time interval" $-\hbar\beta/2 \leq \tau \leq \hbar\beta/2$ is finite and $\tilde{q}(\pm\hbar\beta/2) = q_E$ becomes positive, $q_E > 0$, and grows with temperature, eventually reaching the top of the barrier $q_E = q_0$ at some $T = T_0$. In the latter case, the bounce trajectory reduces to a trivial one $\tilde{q}(\tau) = q_0$. For even higher temperatures $T > T_0$, thermal activation prevails over quantum tunneling and Eq. (4.10) should be modified in order to correctly evaluate thermal decay rate. This regime of thermally activated decay of a metastable state will be addressed separately below in this chapter.

Turning back to the low-temperature regime of quantum tunneling, our next step is to account for paths that deviate from the above saddle point trajectories. As usually, within the semiclassical approximation, it suffices to expand our exact action to the second order in such deviations, since higher terms may only give small

($\sim\hbar$) corrections to the leading terms and, hence, can safely be neglected. Evaluating the contribution of Gaussian fluctuations around the trivial saddle point (4.19), one finds

$$\int \mathrm{D}q \exp\left(-\frac{1}{2\hbar}\int_{-\hbar\beta/2}^{\hbar\beta/2} d\tau \int_{-\hbar\beta/2}^{\hbar\beta/2} d\tau' G_0^{-1}(\tau-\tau')q(\tau)q(\tau')\right)$$
$$= N(\det G_0^{-1})^{-1/2} \tag{4.23}$$

where N is the normalization constant, which we do not specify here. The Fourier transform of G_0^{-1} reads

$$G_{0\omega}^{-1} = m\omega^2 + \eta|\omega| + \kappa \tag{4.24}$$

and, hence,

$$(\det G_0^{-1})^{-1/2} = \prod_n \Lambda_{0n}^{-1/2}, \quad \Lambda_{0n} = m\omega_n^2 + \eta|\omega_n| + \kappa \tag{4.25}$$

where $\kappa = U''(0)$ and $\omega_n = 2\pi nT$ with $n = 0, \pm1, \pm2, \ldots$ are the Matsubara frequencies.

Analogously, expanding the action to the second order in fluctuations around the bounce trajectory $\delta q(\tau) \equiv q(\tau) - \tilde{q}(\tau)$, we obtain

$$\int \mathrm{D}\delta q \exp\left(-\frac{S_{\mathrm{eff}}[\tilde{q}(\tau)]}{\hbar}\right.$$
$$\left. -\frac{1}{2\hbar}\int_{-\hbar\beta/2}^{\hbar\beta/2} d\tau \int_{-\hbar\beta/2}^{\hbar\beta/2} d\tau' G^{-1}(\tau,\tau')\delta q(\tau)\delta q(\tau')\right) \tag{4.26}$$

where

$$G^{-1}(\tau,\tau') = \frac{\delta^2 S_{\mathrm{eff}}}{\delta\tilde{q}(\tau)\delta\tilde{q}(\tau')} \tag{4.27}$$

At the first sight, the path integral (4.26) can be evaluated in exactly the same way as that in Eq. (4.23) with the result expressed as $(\det G^{-1})^{-1/2}$. More caution is required at this point, however.

Let us express fluctuations $\delta q(\tau)$ around the bounce $\tilde{q}(\tau)$ in terms of the following Fourier series:

$$\delta q(\tau) = \sum_n a_n q_n(\tau) \tag{4.28}$$

where the functions $q_n(\tau)$ are chosen to obey the conditions

$$\int_{-\hbar\beta/2}^{\hbar\beta/2} d\tau q_n(\tau)q_{n'}(\tau) = \delta_{nn'}, \quad q_n(\pm\hbar\beta/2) = 0 \tag{4.29}$$

and the equation

$$\left(\frac{\delta^2 S_{\text{eff}}}{\delta q^2}|_{q=\tilde{q}(\tau)}\right) q_n(\tau) = \Lambda_n q_n(\tau) \tag{4.30}$$

i.e., Λ_n and $q_n(\tau)$ define, respectively, the eigenvalues and the complete set of (vanishing at the boundaries) orthonormal eigenfunctions of the operator $\delta^2 S_{\text{eff}}/\delta q^2|_{q=\tilde{q}}$. With the aid of Eq. (4.28), the path integral (4.26) can be rewritten as an infinite product of ordinary integrals

$$N \exp\left(-\frac{S_{\text{eff}}[\tilde{q}(\tau)]}{\hbar}\right) \prod_{n=-1}^{\infty} \left[\int \frac{da_n}{\sqrt{2\pi\hbar}} \exp\left(-\frac{\Lambda_n a_n^2}{2\hbar}\right)\right] \tag{4.31}$$

where we choose to assign the integer numbers $n \geq -1$ to all eigenmodes of our problem. It turns out that two of the integrals in this product—ones with $n = -1$ and $n = 0$—formally diverge and need to be regularized. This is because the corresponding eigenvalues are, respectively, negative $\Lambda_{-1} < 0$ and equal to zero $\Lambda_0 = 0$.

A negative eigenvalue always signals instability of the system, which tends to decay to the state with lower energy. Physically, the corresponding divergence in the integral over a_{-1} implies that a purely real partition function of a metastable state simply cannot be defined. This divergence is handled by an analytic continuation, which provides nonzero imaginary part of the partition function. For this purpose, the integration contour over a_{-1} needs to be distorted into the complex plane [82]. As a result, the imaginary part of the integral over a_{-1} takes the form

$$\frac{i}{2}\int_{-\infty}^{\infty} \frac{da_0}{\sqrt{2\pi\hbar}} \exp\left(-\frac{|\Lambda_{-1}|a_{-1}^2}{2\hbar}\right) = \frac{i}{2\sqrt{|\Lambda_{-1}|}} \tag{4.32}$$

We note that the prefactor i/2 in front of this integral appears as a result of the deformation of the integration contour at the saddle point in the steepest descent direction.

The presence of the eigenvalue $\Lambda_0 = 0$ is related to the fact that Eq. (4.18) has not just one solution $\tilde{q}(\tau)$ but a family of them: The periodic (with period $\hbar\beta$) function $\tilde{q}(\tau - \tau_0)$ also represents a solution to Eq. (4.18) for any $-\hbar\beta/2 < \tau_0 \leq \hbar\beta/2$ and, furthermore, the action evaluated for all these solutions does not depend on τ_0, i.e., $S_{\text{eff}}[\tilde{q}(\tau + \tau_0)] = A$ for any τ_0. In other words, there exists a direction in the functional space in which the bounce trajectory can be distorted without changing the action.

In order to handle this zero mode, we note that upon changing the corresponding Fourier coefficient a_0 by Δa_0 (and keeping all other a_n unchanged), the function $\delta q(\tau)$ (4.28) changes by

$$\Delta\delta q(\tau) = q_0(\tau)\Delta a_0 \tag{4.33}$$

On the other hand, in this case, we obviously have

$$\Delta\delta q(\tau) = \tilde{q}(\tau + \Delta\tau_0) - \tilde{q}(\tau) = \frac{d\tilde{q}(\tau)}{d\tau}\Delta\tau_0 \tag{4.34}$$

What remains is to identify the expression for the eigenfunction $q_0(\tau)$ corresponding to the eigenvalue $\Lambda_0 = 0$. From the condition

$$S_{\text{eff}}[\tilde{q}(\tau + \Delta\tau_0)] - S_{\text{eff}}[\tilde{q}(\tau)] = 0$$

we have

$$\left(\frac{\delta^2 S_{\text{eff}}}{\delta q^2}|_{q=\tilde{q}(\tau)}\right)\frac{d\tilde{q}(\tau)}{d\tau} = 0 \tag{4.35}$$

Comparing this equation with Eq. (4.30), we conclude that $q_0(\tau)$ is simply proportional to $d\tilde{q}(\tau)/d\tau$. Including the correct normalization factor, we obtain

$$q_0(\tau) = \left[\int_{-\hbar\beta/2}^{\hbar\beta/2} d\tau \left(\frac{d\tilde{q}}{d\tau}\right)^2\right]^{-1/2} \frac{d\tilde{q}}{d\tau} \tag{4.36}$$

Combining now Eqs. (4.33) and (4.34) with Eq. (4.36), we arrive at the desired relation

$$da_0 = \left[\int_{-\hbar\beta/2}^{\hbar\beta/2} d\tau \left(\frac{d\tilde{q}}{d\tau}\right)^2\right]^{1/2} d\tau_0 \tag{4.37}$$

This relation implies that the integral over a_0 in Eq. (4.31) can be converted into that over τ_0, which, in turn, can trivially be performed.

The remaining eigenvalues Λ_n with $n \geq 1$ are all positive. Hence, the integrals over the corresponding Fourier coefficients a_n in Eq. (4.31) are handled without problems. Collecting the contributions from all integrals, we derive the final expression for the combination in Eq. (4.31). It reads

$$\frac{\mathrm{i}N}{2\sqrt{2\pi\hbar}}\exp\left(-\frac{S_{\text{eff}}[\tilde{q}(\tau)]}{\hbar}\right)\left[\int_{-\hbar\beta/2}^{\hbar\beta/2} d\tau \left(\frac{d\tilde{q}}{d\tau}\right)^2\right]^{1/2}$$
$$\times\, (\det{}' G^{-1})^{-1/2} \int_{-\hbar\beta/2}^{\hbar\beta/2} d\tau_0 \tag{4.38}$$

where we denoted

$$(\det{}' G^{-1})^{-1/2} = \frac{1}{|\Lambda_{-1}|^{1/2}} \prod_{n=1}^{\infty} \frac{1}{\Lambda_n^{1/2}} \tag{4.39}$$

i.e., $\det{}' G^{-1}$ is essentially the absolute value of the determinant of G^{-1} with excluded zero mode.

What remains is to bring together two contributions to the partition function (4.23) and (4.38) and to evaluate the free energy

$$F = -T \ln \mathcal{Z} \simeq -T \ln \Big[N (\det G_0^{-1})^{-1/2} - \frac{i\sqrt{\hbar}}{2\sqrt{2\pi}} \left[\int_{-\hbar\beta/2}^{\hbar\beta/2} d\tau \left(\frac{d\tilde{q}}{d\tau} \right)^2 \right]^{1/2} \left(\frac{\det G_0^{-1}}{\det{}' G^{-1}} \right)^{1/2} \exp\left(-\frac{S_{\mathrm{eff}}[\tilde{q}(\tau)]}{\hbar} \right) \Big] \tag{4.40}$$

Combining this result with Eq. (4.10), we immediately identify the quantum statistical decay rate of a metastable state in the form

$$\Gamma = B \exp(-A) \tag{4.41}$$

where

$$A = \frac{S_{\mathrm{eff}}[\tilde{q}(\tau)]}{\hbar}, \quad B = \frac{1}{\sqrt{2\pi\hbar}} \left[\int_{-\hbar\beta/2}^{\hbar\beta/2} d\tau \left(\frac{d\tilde{q}}{d\tau} \right)^2 \right]^{1/2} \left(\frac{\det G_0^{-1}}{\det{}' G^{-1}} \right)^{1/2} \tag{4.42}$$

Note that in our derivation of the decay rate (4.41), (4.42), we made use of the condition $\hbar\Gamma \ll T$, which allowed us to retain only the contribution of one bounce. Due to smallness of Γ, this condition is well satisfied except at exponentially low temperatures. In the limit $T \to 0$, the above derivation should be slightly extended in order to include configurations with more than one bounce into our consideration. The point here is that at $T \to 0$, a sum of k bounces

$$\tilde{q}(\tau - \tau_1) + \tilde{q}(\tau - \tau_2) + \ldots + \tilde{q}(\tau - \tau_k) \tag{4.43}$$

also represents an approximate solution to Eq. (4.17) for any $k \geq 1$. This is because each bounce has a finite size $\Delta\tau$ and is centered around the point $\tau = \tau_l$, $l = 1, 2, \ldots, k$. Beyond this size, i.e., for $|\tau - \tau_l| \gg \Delta\tau$, with a good accuracy one has $\tilde{q}(\tau - \tau_l) \simeq 0$ and, hence, at $T \to 0$ within the interval $-\hbar\beta/2 < \tau < \hbar\beta/2$, one can "pack" as many bounces with non-overlapping cores as one wishes. In a way, we can think of the analogy with one-dimensional gas of k non-interacting hard particles with size τ_B and center-of-mass coordinates $\tau_1, \tau_2, \ldots, \tau_k$.

Substituting the solution (4.43) into the effective action, one gets $S_{\text{eff}} = kA$, i.e., naturally the action of k bounces is k times bigger than that of one bounce. In order to account for fluctuations around the saddle point path (4.43), we identify k negative and k zero modes, which are treated analogously to the one bounce case. Integration over the remaining modes is again trivial. As a result, we get

$$N(\det G_0^{-1})^{-1/2}\left(\frac{\mathrm{i}B}{2}\right)^k \exp(-kA)\int_{-\hbar\beta/2}^{\hbar\beta/2} d\tau_k \int_{-\hbar\beta/2}^{\tau_k} d\tau_{k-1}\ldots\int_{-\hbar\beta/2}^{\tau_2} d\tau_1 \tag{4.44}$$

where A and B are defined in Eq. (4.42). The integrals over all center-of-mass coordinates in Eq. (4.44) obviously yield the factor $(\hbar\beta)^k/k!$. Collecting now the contributions from all bounce configurations, we arrive at the following expression for the partition function at $T \to 0$

$$\begin{aligned}\mathcal{Z} &= N(\det G_0^{-1})^{-1/2}\left(1+\sum_{k=1}^{\infty}\left(\frac{\mathrm{i}}{2}\right)^k \frac{(\hbar\beta B)^k}{k!}\exp(-kA)\right)\\ &\sim \exp\left(-\frac{E_0-(\mathrm{i}\hbar B/2)e^{-A}}{T}\right)\end{aligned} \tag{4.45}$$

Identifying the decay rate of a metastable state at $T \to 0$ with the aid of Eqs. (4.6) and (4.45), we again reproduce the result (4.41), (4.42). We conclude that these equations can be employed for evaluation of the quantum statistical decay rate Γ at any temperature below a certain value T_0, which marks the crossover between quantum tunneling and thermal activation and which will be explicitly defined in the next section.

Before we turn to concrete calculations, we would like to make one remark. Although the above treatment of zero modes is pretty standard and commonly used, an attentive reader surely already noticed its formal inconsistency. Indeed, in the beginning of our calculation, we assumed that all fluctuations $\delta q(\tau)$ around the bounce trajectories should remain small. This assumption enabled us to expand the effective action only up to quadratic in $\delta q(\tau)$ terms. On the other hand, in the course of our analysis it turned out that at least in one direction in the functional space, such fluctuations are not at all small: The bounce center-of-mass coordinate can be freely shifted within the interval from $-\hbar\beta/2$ to $\hbar\beta/2$. Fortunately, one can develop more rigorous approaches [258, 421], which—being free from the above formal inconsistency—yield exactly the same results (4.41), (4.42) for the decay rate. By making use of the interpretation of zero modes in terms of collective coordinates here, we intentionally sacrificed the rigor in favor of physical transparency. This interpretation will also be extensively employed in a number of other problems analyzed below in this book.

4.3 Decay of Supercurrent in Josephson Weak Links

As we already demonstrated in Chapter 2, macroscopic quantum dynamics of small superconducting junctions can be described in terms of one collective variable—the Josephson phase difference φ—playing the role of a coordinate for an effective quantum "particle" with mass $m = C^*/(4e^2)$ moving in a tilted (by external bias current I) periodic potential

$$U(\varphi) = -\hbar I\varphi/2e - E_J \cos\varphi \tag{4.46}$$

and interacting with a dissipative environment formed by electronic degrees of freedom in the system. Provided an external bias current I remains smaller than the critical current of the junction $I_C = 2eE_J/\hbar$ and in the absence of fluctuations—thermal or quantum—of the Josephson phase φ, the junction sustains non-dissipative dc current

$$I = I_C \sin\varphi \tag{4.47}$$

implying that the "Josephson particle" remains at rest in one of the potential minima $\varphi = \arcsin(I/I_C) + 2\pi n$, where n is any integer number. In the presence of phase fluctuations, however, such current states become unstable and decay simply because the "particle" can escape from the local potential minimum either by thermal activation or by quantum tunneling. After that the particle starts sliding down in the washboard potential (4.46) with some "velocity" $d\varphi/dt$. Hence, in accordance with the second Josephson relation (2.1), a nonzero voltage $V = (2e/\hbar)d\varphi/dt$ appears across the junction, implying its switching to a resistive state. Thus, both thermal and quantum fluctuations tend to destroy superconductivity in Josephson junctions as well as in other types of weak links already for $I < I_C$.

While this conclusion formally holds for any nonzero I and any system parameters, in a large number of cases, an effective potential barrier for a "Josephson particle" turns out to be very high, thus making fluctuation-induced supercurrent decay a very slow process, which can safely be neglected. On the other hand, for sufficiently large I approaching I_C, the barrier becomes lower and the supercurrent decays much faster. In this case, such decay in sufficiently small Josephson junctions can be detected experimentally [205, 350, 389, 390, 395].

In order to evaluate the decay rate Γ of macroscopic supercurrent states in small Josephson junctions, we will make use of Eqs. (4.41), (4.42). In what follows we will assume that the current I is close to the critical one,

$$\chi \equiv \frac{I_C - I}{I_C} \ll 1 \tag{4.48}$$

in which case the phase variable φ remains close to $\varphi_C = \pi/2$ and does not change significantly in the course of tunneling. Hence, it suffices to expand an effective potential for the "Josephson particle" in powers of the fluctuating phase. Introducing the variable $q = \varphi - \arcsin(I/I_C)$, expanding (4.46) up to the third order in q, and making use of Eq. (4.48), we arrive at the effective potential (4.13) with

$$\kappa = \sqrt{\frac{\lambda(I_C - I)}{e}}, \quad \lambda = -\frac{1}{2e}\frac{\partial^2 I(\varphi)}{\partial\varphi^2}|_{I=I_C} \tag{4.49}$$

where the condition (4.48) implies $\kappa \ll \lambda$. For example, for junctions with sinusoidal current-phase relation $I(\varphi) = I_C \sin\varphi$, we have $\lambda = I_C/2e$, $\kappa = I_C(2\chi)^{1/2}/2e$, and $q_0 = 2(2\chi)^{1/2}$.

Note that Eqs. (4.13) and (4.49) apply not only for Josephson junctions with purely sinusoidal current-phase relation, but also for some other types of superconducting weak links described by smooth current-phase relations such as diffusive superconducting contacts (constrictions) with zero-temperature relation [420]

$$I(\varphi) = \frac{\pi\Delta_0}{2eR_N}\ln\left[\frac{1+\sin(\varphi/2)}{1-\sin(\varphi/2)}\right]\cos(\varphi/2) \tag{4.50}$$

and the critical current $I_C \equiv I(\varphi_c)$ reached at $\varphi_c \simeq 1.25\pi/2$. On the other hand, there also exist superconducting weak links, e.g., short ballistic constrictions or long SNS junctions, where at $T \to 0$ the critical current I_C is reached at $\varphi_c = \pi$ and $I(\varphi)$ experiences a jump at this point. In such cases, the smooth potential of the form (4.13) is insufficient even provided the condition (4.48) is fulfilled. For instance, in the case of ballistic superconducting contacts with $I(\varphi) = I_C\sin(\varphi/2)$ (here we assume $T = 0$ and $-\pi < \varphi \leq \pi$), one has

$$U(q) = \left(\frac{\kappa}{2}q^2 - \frac{\lambda}{6}q^3\right)\theta(q_0 - q) + \frac{\gamma}{2}(\tilde{q}_0 - q)\theta(q - q_0) \tag{4.51}$$

where $q = \varphi - 2\arcsin(I/I_C)$, $\kappa = I_C(2\chi)^{1/2}/4e$, $\lambda = I_C/8e$, $\gamma = 2I_C/e$, $q_0 = 2(2\chi)^{1/2} = \kappa/\lambda$ and $\tilde{q}_0 = q_0[1 + 2\kappa q_0/(3\gamma)]$. In Section 4.3.3, we will demonstrate that some features of the supercurrent decay in the non-quasiclassical potential of the type (4.51) differ from those in the case of a smooth potential (4.13).

Let us now turn to evaluation of the decay rate Γ.

4.3.1 Macroscopic Quantum Tunneling

We first consider the limit $T \to 0$ in which case the supercurrent decay occurs due to macroscopic quantum tunneling (MQT) of the Josephson phase. As it was already discussed above, we will describe our system by the effective action defined in Eqs. (4.12) and (4.13). In order to evaluate the factor A in the expression for Γ, we need to obtain the solution to Eq. (4.18) with boundary conditions $q(\pm\infty) = 0$.

In the absence of dissipation $\eta \to 0$, this solution has the form of a bounce (4.22) and the first Eq. (4.42) yields the well-known result

$$A = \frac{36U_0}{5\hbar\omega_0} \tag{4.52}$$

where the potential barrier height U_0 and the frequency ω_0 are defined in Eq. (4.15). The result (4.52) remains applicable in the semiclassical limit $A \gg 1$, cf. Eq. (4.14). Of course, Eq. (4.52) can also be recovered by means of the standard semiclassical formula

$$W(E) = 2\int_{q_E}^{q_{\text{ex}}} dq\sqrt{2m(U(q) - E)} \qquad (4.53)$$

where, as before, q_E is defined from the condition $U(q_E) = E$ and q_{ex} is the particle exit point, i.e., the coordinate at which the particle "re-appears" from under the barrier. Obviously, in the dissipativeless limit, the particle energy is conserved during tunneling. As in our case this energy equals to zero ($E = 0$), one readily finds $q_E = 0$ and $q_{\text{ex}} = 3\kappa/\lambda \equiv 3q_0/2$. Then for the potential energy of the form (4.13), Eq. (4.53) with $A = W(0)/\hbar$ yields the result (4.52).

Let us now include dissipation into our consideration. We first assume that the viscosity η remains sufficiently small, in which case the dissipative term in Eq. (4.12) can be treated perturbatively. Making use of the bounce trajectory (4.22), from the first Eq. (4.42), we again recover the term (4.52) as well as the leading order dissipative correction [80]

$$A \simeq \frac{36U_0}{5\hbar\omega_0}\left(1 + \frac{45\zeta(3)}{\pi^3}\alpha\right) \qquad (4.54)$$

where the parameter

$$\alpha = \frac{\eta}{2m\omega_0} \qquad (4.55)$$

controls effective dissipation strength and $\zeta(3) \simeq 1.202$ is the Riemann number. We observe that for nonzero η, the value A increases, i.e., dissipation *decreases* the quantum decay rate Γ (4.41). This result can easily be understood since in the course of tunneling, the particle can excite environmental modes, thereby loosing energy and "traveling" a longer path under the barrier (process (b) in Fig. 4.1). Hence, dissipation *increases* the supercurrent lifetime, thereby helping to maintain superconductivity in our system. This conclusion, though perhaps looking somewhat paradoxical at the first glance, is in fact natural since damping tends to suppress fluctuations of the Josephson phase, making the variable φ "more classical" and, hence, supporting the supercurrent across the junction.

The result (4.52) is applicable as long as dissipation remains sufficiently weak, i.e., for

$$\eta \ll m\omega_0 \qquad (4.56)$$

or, equivalently, for $\alpha \ll 1$. In this case, the dissipative contribution to A remains small as compared to the leading non-dissipative term. Note that already in this underdamped limit, dissipation can yield exponentially strong extra suppression of the decay rate Γ. As follows from Eq. (4.54), this is the case for

$$\frac{\eta q_0^2}{\hbar} \gg 1 \tag{4.57}$$

As soon as one goes beyond the underdamped limit (4.56), dissipative effects cannot anymore be treated perturbatively. In the opposite limit of strong dissipation

$$\eta \gg m\omega_0 \tag{4.58}$$

one can again derive an analytic expression for the factor A. In this overdamped limit, the mass term in Eq. (4.18) remains small as compared to the dissipative one. Setting $m \to 0$ in this equation, one arrives at the bounce solution for a massless particle

$$\tilde{q}(\tau) = \frac{4\kappa/\lambda}{1 + (m\omega_0^2/\eta)^2\tau^2} \tag{4.59}$$

Substituting this function into Eq. (4.21), we determine the characteristic bounce time in the overdamped limit $\tau_B = 4\pi\alpha/\omega_0$. This time is parametrically longer than that for the underdamped particle ($\sim\omega_0^{-1}$), implying that dissipation indeed increases effective tunneling time $\sim\tau_B$ (i.e., time the particle "spends" in the classically forbidden region under the barrier). Setting $\tau = 0$ in Eq. (4.59), we find the coordinate of the exit point $q_{\text{ex}} = 4\kappa/\lambda$ and the energy dissipated by the bath in the course of particle tunneling equal to $\Delta E = -U(q_{\text{ex}}) = 8\kappa^3/(3\lambda^2)$. Remarkably, in the limit of strong dissipation q_{ex} and, hence, ΔE turn out to be independent of the viscosity η.

Substituting the bounce trajectory (4.59) into the action (4.12), we obtain [81]

$$A \simeq 6\pi\alpha\frac{U_0}{\hbar\omega_0} = \frac{\pi}{2}\frac{\eta q_0^2}{\hbar} \tag{4.60}$$

This result demonstrates that in the overdamped limit, the supercurrent decay rate Γ is suppressed much stronger than in the underdamped one.

At intermediate viscosity values $\eta \sim m\omega_0$, both the solution to Eq. (4.18) and the factor A can be found numerically [88, 179]. The latter quantity turns out to increase with η almost linearly for all values of η. Thus, one can approximate $A(\alpha)$ by a simple formula

$$A(\alpha) \simeq \frac{U_0}{\hbar\omega_0}\left(\frac{36}{5} + 4\pi\alpha s(\alpha)\right) \tag{4.61}$$

where $s(\alpha)$ is a smooth function growing monotonously from $s(0) \simeq 1$ to $s(\infty) = 1.5$. From the exact solution [336] established at the particular value $\alpha = 1.175$, one also has $s(1.175) \simeq 1.22$. All this information is sufficient to conclude that Eq. (4.61) provides a good estimate for A at all values of the viscosity η.

Now we turn to evaluation of the pre-exponent B in the expression for the decay rate Γ. The task here is to obtain the ratio of fluctuation determinants in the second Eq. (4.42), which in turn requires finding all eigenvalues of the operators (4.24) and (4.27) for the potential $U(q)$ defined in Eq. (4.13). The eigenvalues of the first operator Λ_{0n} are defined in Eq. (4.25), while those of the second operator should be determined from Eq. (4.30). This task will be handled analytically in the limits of weak and strong dissipation.

First let us stick to the underdamped limit (4.56). In this limit, the effect of dissipation on the pre-exponent B is unimportant and on the leading approximation can be neglected. Setting $\eta \to 0$, one can rewrite Eq. (4.30) in the form

$$\left(-m\frac{d^2}{d\tau^2} + \kappa - \frac{3\kappa}{\cosh^2(\omega_0\tau/2)}\right) q_n(\tau) = \Lambda_n q_n(\tau) \tag{4.62}$$

This equation formally coincides with the Schrödinger equation for a particle with coordinate τ in the inverse $\cosh^2$-potential. Fortunately, the energy spectrum of this quantum mechanical problem is well studied [250]. There exist three eigenvalues corresponding to discrete energy levels:

$$\Lambda_{-1} = -\frac{5}{4}\kappa, \quad \Lambda_0 = 0, \quad \Lambda_1 = \frac{3}{4}\kappa \tag{4.63}$$

The remaining eigenvalues Λ_n correspond to the continuous energy spectrum. The delocalized solution to Eq. (4.62) reads

$$q_n(\tau) = C\psi_{\leftarrow}(\tau) + D\psi_{\rightarrow}(\tau) \tag{4.64}$$

where C and D are complex constants, while $\psi_{\leftarrow}(\tau)$ and $\psi_{\rightarrow}(\tau)$ denote, respectively, the left and right moving scattering states. These states have the following asymptotic:

$$\psi_{\leftarrow}(\tau) = \begin{cases} t\,e^{-\mathrm{i}\sqrt{(\Lambda_n-\kappa)/m}\tau}, & \tau < 0,\ |\tau| \gg \frac{1}{\omega_0}, \\ e^{-\mathrm{i}\sqrt{(\Lambda_n-\kappa)/m}\,\tau} - (r^* t/t^*)e^{\mathrm{i}\sqrt{(\Lambda_n-\kappa)/m}\,\tau}, & \tau > 0,\ |\tau| \gg \frac{1}{\omega_0}, \end{cases} \tag{4.65}$$

$$\psi_{\rightarrow}(\tau) = \begin{cases} e^{\mathrm{i}\sqrt{(\Lambda_n-\kappa)/m}\,\tau} + r\,e^{-\mathrm{i}\sqrt{(\Lambda_n-\kappa)/m}\,\tau} & \tau < 0, \quad |\tau| \gg \frac{1}{\omega_0}, \\ t\,e^{\mathrm{i}\sqrt{(\Lambda_n-\kappa)/m}\,\tau}, & \tau > 0, \quad |\tau| \gg \frac{1}{\omega_0}. \end{cases} \tag{4.66}$$

Here we defined both transmission (t) and reflection (r) amplitudes, which read [250]

$$r = 0,$$
$$t = -\frac{\left(3 - 2\mathrm{i}\sqrt{\frac{\Lambda_n}{\kappa} - 1}\right)\left(2 - 2\mathrm{i}\sqrt{\frac{\Lambda_n}{\kappa} - 1}\right)\left(1 - 2\mathrm{i}\sqrt{\frac{\Lambda_n}{\kappa} - 1}\right)}{\left(3 + 2\mathrm{i}\sqrt{\frac{\Lambda_n}{\kappa} - 1}\right)\left(2 + 2\mathrm{i}\sqrt{\frac{\Lambda_n}{\kappa} - 1}\right)\left(1 + 2\mathrm{i}\sqrt{\frac{\Lambda_n}{\kappa} - 1}\right)} \tag{4.67}$$

The boundary condition for the eigenfunctions provides us with the following equation:

$$C\,t\,e^{\frac{\mathrm{i}\hbar\sqrt{(\Lambda_n-\kappa)/m}}{2T}} + De^{-\frac{\mathrm{i}\hbar\sqrt{(\Lambda_n-\kappa)/m}}{2T}} = Ae^{-\frac{\mathrm{i}\hbar\sqrt{(\Lambda_n-\kappa)/m}}{2T}} + D\,t\,e^{\frac{\mathrm{i}\hbar\sqrt{(\Lambda_n-\kappa)/m}}{2T}} \tag{4.68}$$

Furthermore, by construction the eigenfunctions $q_n(\tau)$ should be real, implying $Ct = D^*$ and $Dt = C^*$. With this in mind, we conclude that a nontrivial solution to Eq. (4.68) may only exist provided the condition

$$te^{\frac{\mathrm{i}\hbar\sqrt{(\Lambda_n-\kappa)/m}}{2T}} = e^{-\frac{\mathrm{i}\hbar\sqrt{(\Lambda_n-\kappa)/m}}{2T}} \tag{4.69}$$

is fulfilled. This condition yields the dispersion relation

$$-\frac{(3 - 2\mathrm{i}u_n)(2 - 2\mathrm{i}u_n)(1 - 2\mathrm{i}u_n)}{(3 + 2\mathrm{i}u_n)(2 + 2\mathrm{i}u_n)(1 + 2\mathrm{i}u_n)} = e^{-\mathrm{i}u_n\hbar\omega_0/T}, \quad u_n = \sqrt{\Lambda_n/\kappa - 1} \tag{4.70}$$

which implicitly defines the eigenvalues Λ_n.

With the aid of Eqs. (4.25) and (4.63) for the ratio of fluctuation determinants, we obtain

$$\left(\frac{\det G_0^{-1}}{\det' G^{-1}}\right)^{1/2} = \frac{4}{\sqrt{15}}\kappa^{1/2}\sqrt{\frac{\prod_n(\Lambda_{0n}/\kappa)}{\prod_n(\Lambda_n/\kappa)}} \tag{4.71}$$

In the low-temperature limit considered here, we can replace the summation over n by the frequency integration:

$$\sqrt{\frac{\prod_n(\Lambda_{0n}/\kappa)}{\prod_n(\Lambda_n/\kappa)}} \to \exp\left[\frac{1}{2}\frac{\hbar}{2\pi T}\int d\omega \ln\frac{1+\omega^2/\omega_0^2}{\Lambda(\omega)/\kappa}\right] \tag{4.72}$$

where $\Lambda(\omega)/\kappa = 1 + u^2$ and u obeys the equation

$$u - \frac{2T}{\hbar\omega_0}\left(\arctan\frac{2u}{3} + \arctan u + \arctan 2u - \frac{3\pi}{2}\right) = \frac{\omega}{\omega_0} \tag{4.73}$$

which follows directly from Eq. (4.70). Resolving Eq. (4.73) in the lowest order in T, we obtain

$$\frac{\Lambda(\omega)}{\kappa} = 1+\frac{\omega^2}{\omega_0^2}+\frac{4T\omega}{\hbar\omega_0^2}\left(\arctan\frac{2\omega}{3\omega_0}+\arctan\frac{\omega}{\omega_0}+\arctan\frac{2\omega}{\omega_0}-\frac{3\pi}{2}\right) \tag{4.74}$$

Combining this result with Eq. (4.72) and performing the frequency integral, from Eq. (4.71), we get

$$\left(\frac{\det G_0^{-1}}{\det' G^{-1}}\right)^{1/2} = 2\sqrt{15}\kappa^{1/2} \tag{4.75}$$

What remains is to obtain the value of the normalization factor entering in the expression for the zero-mode eigenfunction $q_0(\tau)$ (4.36) and in the second equation (4.42). Employing the bounce trajectory (4.22), one readily finds

$$\int_{-\infty}^{\infty} d\tau\left(\frac{d\tilde{q}}{d\tau}\right)^2 = \frac{24}{5}\frac{\omega_0\kappa^2}{\lambda^2} = \frac{A}{m}, \tag{4.76}$$

where A is defined in Eq. (4.52). Substituting Eqs. (4.75) and (4.76) into the second equation (4.42), we finally get

$$B = 6\sqrt{\frac{6}{\pi}}\sqrt{\frac{U_0\omega_0}{\hbar}} \tag{4.77}$$

This is a well-known quantum mechanical result valid in the absence of dissipation. Weak dissipation can be accounted for perturbatively. As a result, in the limit (4.56), one finds [400]

$$B \simeq 6\sqrt{\frac{6}{\pi}}\sqrt{\frac{U_0\omega_0}{\hbar}}(1+2.86\alpha) \tag{4.78}$$

i.e., in this case the effect of a dissipative bath on B remains small.

Now let us carry out the same calculation in the more interesting overdamped limit (4.58). Our goal is again to find the eigenvalues of the operator (4.27). For this purpose, let us specify Eq. (4.30) employing the bounce trajectory (4.59). We obtain

$$\left(-m\frac{d^2}{d\tau^2}+\kappa-\frac{4\kappa}{1+(m\omega_0^2/\eta)^2\tau^2}\right)q_n(\tau)$$
$$-\frac{\eta}{2\pi}\int_{-\infty}^{\infty}d\tau'\frac{\partial q(\tau')}{\partial\tau'}\left(\frac{1}{\tau'-\tau+\mathrm{i}0}+\frac{1}{\tau'-\tau-\mathrm{i}0}\right)=\Lambda_n q_n(\tau) \tag{4.79}$$

Note that for the reasons that will become clear shortly, in Eq. (4.79) we keep the mass term even though we are now working in the limit of strong dissipation (4.58). The eigenfunctions $q_n(\tau)$ obeying this equation can be expanded in the Fourier series

$$q_n(\tau)=\sum_{k=-\infty}^{\infty}c_k^{(n)}\exp(\mathrm{i}2\pi k\tau/\hbar\beta) \tag{4.80}$$

Here and below, it is convenient for us to keep the inverse temperature β large but finite. Substituting Eq. (4.80) into Eq. (4.79), we arrive at the equation for the Fourier coefficients

$$\left(\frac{(2\pi k)^2 m}{\hbar^2\beta^2}+\frac{2\pi|k|\eta}{\hbar\beta}+\kappa-\Lambda_n\right)c_k^{(n)}$$
$$=\frac{4\pi\eta}{\hbar\beta}\sum_{l=-\infty}^{\infty}c_l^{(n)}\exp\left(-\frac{2\pi|k+l|\eta}{\hbar\beta\kappa}\right) \tag{4.81}$$

It is easy to observe that in the limit (4.58) and for Matsubara frequencies $\omega_k = 2\pi kT/\hbar$ with sufficiently small numbers k, the mass term in Eq. (4.81) is irrelevant and can be neglected. Seeking the solutions to Eq. (4.81) with $m \to 0$ in the form [258]

$$c_k^{(n)} = (|k|+C_n)\exp\left(\frac{2\pi|k|\eta}{\hbar\beta\kappa}\right) \tag{4.82}$$

we obtain

$$C_n = C_n^2 \frac{2\pi\eta}{\hbar\beta\kappa} - \frac{\hbar\beta\kappa}{2\pi\eta} = -\frac{\hbar\beta}{2\pi\eta}\Lambda_n \tag{4.83}$$

Resolving these equations and taking the limit $\beta \to \infty$, we easily determine the two first nonzero eigenvalues

$$\Lambda_{\pm 1} = -\kappa \frac{1 \mp \sqrt{5}}{2} \tag{4.84}$$

In accordance with our general considerations, the lowest eigenvalue Λ_{-1} is negative, as it should be in the problem with metastability. In addition, there also exists one zero mode $\Lambda_0 = 0$ corresponding to the eigenfunction $d\tilde{q}/d\tau$.

In order to find all other eigenvalues, it is necessary to keep the kinetic term in Eq. (4.81). Without going into further details of the calculation, here we only present the result [258]:

$$\Lambda_n = m(2\pi n T/\hbar)^2 + 2\pi\eta T|n-2|/\hbar + \kappa \tag{4.85}$$

where $|n| \geq 2$. Equations (4.84) and (4.85) account for all nonzero eigenvalues of the operator G^{-1} in the limit of strong dissipation (4.58). We can now evaluate the ratio of determinants

$$\left(\frac{\det G_0^{-1}}{\det' G^{-1}}\right)^{1/2} = \frac{\kappa^{1/2}}{\sqrt{|\Lambda_{-1}|\Lambda_1}} \frac{\prod_{n=1}^{\infty} \Lambda_{0n}}{\prod_{n=2}^{\infty} \Lambda_n} \tag{4.86}$$

The products in Eq. (4.86) are evaluated explicitly. As a result, we get

$$\frac{\prod_{n=1}^{\infty} \Lambda_{0n}}{\prod_{n=2}^{\infty} \Lambda_n} = (\kappa - 2\pi\eta T/\hbar) \frac{\Gamma(1+\hbar\tilde{\omega}_+/(2\pi T))\Gamma(1+\hbar\tilde{\omega}_-/(2\pi T))}{\Gamma(1+\hbar\omega_+/(2\pi T))\Gamma(1+\hbar\omega_-/(2\pi T))} \tag{4.87}$$

where $\Gamma(x)$ is the Euler gamma function and we defined

$$\omega_\pm = \omega_0(\alpha \pm \sqrt{\alpha^2 - 1}), \quad \tilde{\omega}_\pm = \omega_0(\alpha \pm \sqrt{\alpha^2 - 1 + 8\pi\alpha T/\hbar\omega_0}) \tag{4.88}$$

In the limit $T \to 0$ and $\alpha \gg 1$, this combination reduces to a much simpler form. In this limit from Eqs. (4.86), (4.84), and (4.87), we obtain

$$\left(\frac{\det G_0^{-1}}{\det' G^{-1}}\right)^{1/2} = 16\alpha^4\kappa^{1/2} \tag{4.89}$$

To complete our preparations, we again establish the normalization factor, which is easily evaluated employing Eq. (4.59). We obtain

$$\int_{-\infty}^{\infty} d\tau \left(\frac{d\tilde{q}}{d\tau}\right)^2 = \frac{4\pi\kappa^3}{\lambda^2\eta} = \frac{6\pi U_0}{\eta} \tag{4.90}$$

Combining Eqs. (4.89) and (4.90) with the second Eq. (4.42), we arrive at the final result [258]

$$B = 8\sqrt{6}\alpha^{7/2}\sqrt{\frac{U_0\omega_0}{\hbar}} \tag{4.91}$$

As compared to the result (4.77) derived in the dissipative-less limit, the pre-exponent B (4.91) contains an extra factor $\sim\alpha^{7/2} \gg 1$. This enhancement of the pre-exponent originates from the contribution of high-frequency modes making the ratio of fluctuation determinants to strongly increase at sufficiently large α, cf. Eqs. (4.75) and (4.89). In fact, these high-frequency fluctuations have nothing to do with the process of quantum tunneling. In Chapter 6, we will demonstrate that exactly the same fluctuations produced by a dissipative environment yield modifications in the current-phase relation $I(\varphi)$ of the Josephson junction and partial suppression of the critical current I_C. In other words, an effective potential (4.13) for the Josephson particle is renormalized by quantum fluctuations. In particular, in the overdamped limit, the renormalized value of the critical current I_C^* reads [422]

$$I_C^* = I_C - \frac{2e\lambda\hbar}{\pi\eta}\ln\alpha \tag{4.92}$$

The last term (defined within the logarithmic accuracy) just represents the leading quantum correction to the critical current. Equation (4.92) is valid provided $I_C - I_C^* \ll I_C$. It implies that the effective potential barrier "seen" by the Josephson particle becomes lower because of quantum fluctuations in our dissipative bath. Hence, it is the renormalized current I_C^* (4.92)—not the bare one I_C—that represents the physical value of the critical current, which can be, e.g., detected experimentally in the overdamped limit (4.58). Indeed, the supercurrent I flowing across the junction cannot exceed I_C^*. Hence, the bigger value I_C becomes physically irrelevant as it simply cannot be reached.

Turning back to the problem of quantum tunneling, one can rewrite our results (4.41), (4.60), and (4.91) for the decay rate in the form

$$\Gamma = \frac{\sqrt{2}\eta^{7/2}}{\lambda m^2} \exp\left(-\frac{2\pi\eta}{e\lambda\hbar}(I_C - I)\right) \tag{4.93}$$

This expression grows with decreasing mass m and formally diverges in the limit $m \to 0$. Expressing the result (4.93) in terms of the renormalized critical current I_C^*, with the aid of Eq. (4.92), one easily finds

$$\Gamma = \frac{\sqrt{2}\kappa^2}{\eta^{1/2}\lambda} \exp\left(-\frac{2\pi\eta}{e\lambda\hbar}(I_C^* - I)\right) \tag{4.94}$$

We observe that now the mass drops out from the pre-exponent, indicating that the dependence of the decay rate (4.93) on m is solely due to renormalization of the critical current (4.92) by quantum fluctuations in a dissipative environment. This dependence disappears provided Γ is expressed in terms of the measurable value of the critical current I_C^*. Hence, Eq. (4.94)—rather than Eq. (4.93)—should be used for a comparison between theoretical predictions and experimental data for the decay rate Γ. Below we shall see that Eq. (4.94) was indeed found to be in a very good agreement with the experimental results in the overdamped limit [350].

4.3.2 Thermally Assisted Decay

Our previous analysis was restricted to the limit $T \to 0$ in which case a metastable state can decay only due to quantum tunneling. At nonzero temperatures in addition to tunneling, there also exists another decay mechanism: The system can overcome the potential barrier and escape from a metastable state by thermal activation. The corresponding escape rate remains exponentially small under the condition

$$T \ll U_0 \tag{4.95}$$

which we only consider further below. Thus, there is a trade-off between two decay mechanisms: quantum tunneling and thermal activation. In the case of a smooth potential barrier, like one defined in Eq. (4.13), the former mechanism dominates at low enough T, while at higher temperatures, the latter takes over. On a qualitative level, the corresponding physics can be understood as follows.

In the regime of quantum tunneling, the state with energy E inside the potential well decays with the rate $\propto \exp(-W(E)/\hbar)$, with $W(E)$ defined in Eq. (4.53). Bearing in mind that in the (quasi-) equilibrium situation this state is populated with the probability $\propto \exp(-E/T)$, one easily concludes that the total quantum decay rate is proportional to the integral

$$\Gamma(T) \propto \int_0^{U_0} dE \exp\left(-\frac{E}{T} - \frac{W(E)}{\hbar}\right) \tag{4.96}$$

Assuming the combination in the exponent to be large and evaluating this integral by the steepest descent method, from the saddle point condition $\partial W(E)/\partial(E) = -\hbar/T$, one obtains the energy $E = E_T$, which contributes most to the decay rate at a given temperature T. Then within the exponential accuracy, the quantum decay rate is defined as

$$\Gamma(T) \propto \exp\left(-\frac{E_T}{T} - \frac{W(E_T)}{\hbar}\right) \tag{4.97}$$

At $T = 0$, one has $E_T \to 0$ and we are back to the situation considered in Section 4.3.1. As temperature grows, E_T increases, while $W(E_T)$ (4.53) decreases since the potential barrier becomes effectively lower and "thinner" with increasing energy. Accordingly, the quantum decay rate $\Gamma(T)$ increases with increasing temperature. Eventually at some temperature $T = T_0$, the energy E_T reaches the value $E_{T_0} = U_0$, which simultaneously implies that $W(E_{T_0}) \to 0$. At all $T > T_0$, the decay of a metastable state will be due to thermal activation, and in this regime, the decay rate will increase with temperature as $\Gamma(T) \propto \exp(-U_0/T)$.

In the absence of dissipation, the above consideration allows to easily estimate (with exponential accuracy) the quantum decay rate (4.97) as a function of temperature as well as to establish the crossover temperature T_0 between the regimes of quantum tunneling and thermal activation. In the presence of dissipation, however, the above simple arguments are in general insufficient and more advanced analysis becomes necessary. It is also desirable to go beyond the exponential accuracy and to determine both A and B in the expression for the decay rate (4.41). These goals can be achieved by the "ImF" method combined with the instanton technique, as outlined above.

We first consider sufficiently low temperatures in which case the decay is due to quantum tunneling. In order to determine the factor A in the exponent, one again has to find the bounce trajectory from Eq. (4.18). In the low-temperature limit, it suffices to evaluate the leading T-dependent correction to the zero-temperature result. While in the dissipativeless limit, this correction is exponentially small, at nonzero dissipation it grows with temperature as $\propto T^2$. In order to explicitly obtain this correction, we will make use of the zero-temperature bounce $\tilde{q}(\tau)$ and proceed perturbatively in temperature. Then we get [179, 400]

$$A(T) \simeq A_0 - \frac{\pi}{6}\frac{\eta q_0^2}{\hbar}\left(\frac{\tau_B T}{\hbar}\right)^2 \tag{4.98}$$

where $A_0 \equiv A(T = 0)$ and τ_B is the bounce time defined in Eq. (4.21). Provided the temperature is low enough, the result (4.98) universally holds for any dissipation strength. The next-to-leading order temperature correction turns out to scale with temperature as $\propto T^4$. In the underdamped and overdamped limits, Eqs. (4.56) and (4.58), one should substitute, respectively, $\tau_B = 6/\omega_0$ and $\tau_B = 4\pi\alpha/\omega_0$ into Eq. (4.98).

At not too low temperatures, the perturbative in T result (4.98) can become insufficient. In this case, it is necessary to determine the bounce $\tilde{q}(\tau)$ as a solution to Eq. (4.18) at nonzero T. Since this solution is periodic (with period $\hbar\beta$), it may be written as a Fourier series

$$\tilde{q}(\tau) = \sum_{n=-\infty}^{\infty} b_n \exp(\mathrm{i}\omega_n\tau) + q_a \tag{4.99}$$

where q_a is an arbitrary constant. The choice of this constant is just a matter of our convenience. It turns out that in the most interesting overdamped limit (4.58), one can obtain an exact analytic solution of Eq. (4.18) at any T. Choosing $q_a = 0$, one finds the Fourier coefficients b_n in Eq. (4.99) in the following form [257]:

$$b_n = (T\tau_B q_0/\hbar)\exp(-b|n|), \qquad \tanh b = T\tau_B/\hbar \tag{4.100}$$

where, as before, $\tau_B = 4\pi\alpha/\omega_0$. Evaluating the effective action for this bounce trajectory, one again arrives at Eq. (4.98), thus proving that in the overdamped limit (4.58), this result applies not only at low T but for the whole temperature range where quantum tunneling remains the dominating decay mechanism in our problem.

The pre-exponent B in the expression for the decay rate (4.41) is evaluated similarly to the zero-temperature limit considered above. At low enough T, thermal fluctuations can only produce a small correction to $B_0 \equiv B(T = 0)$, which can be safely neglected. Remarkably, in the overdamped limit $\alpha \gg 1$, no T-dependent correction to B_0 is found at any temperature. Indeed, repeating the whole analysis of fluctuations around the bounce trajectory (4.99) and (4.100), we again recover the eigenvalues (4.85) for $|n| \geq 2$, while instead of Eq. (4.84), we now have

$$\Lambda_{\pm 1} = -\frac{\kappa}{2}\left[1 \mp \left(1 + \frac{4}{\cosh b}\right)^{1/2}\right] \tag{4.101}$$

where b is defined in Eq. (4.100). Finally, with the aid of Eqs. (4.99) and (4.100), we evaluate the normalization factor, which now reads

$$\int_{-\infty}^{\infty} d\tau \left(\frac{d\tilde{q}}{d\tau}\right)^2 = \frac{6\pi U_0}{\eta}\left(1 - \frac{T^2\tau_B^2}{\hbar^2}\right) \tag{4.102}$$

Combining Eqs. (4.101), (4.85), and (4.102) with the general expression for B (4.42), we observe that the temperature-dependent terms contained in Eqs. (4.101) and (4.102) cancel out exactly and, hence, B remains temperature independent in the limit $\alpha \gg 1$.

Summarizing our results derived for the regime of thermally assisted quantum tunneling, we have

$$\frac{\Gamma(T)}{\Gamma_0} = \begin{cases} \exp\left(72\pi\alpha\frac{U_0}{\hbar\omega_0}\frac{T^2}{\hbar^2\omega_0^2}\right), & \alpha \ll 1, \\ \exp\left(32\pi^3\alpha^3\frac{U_0}{\hbar\omega_0}\frac{T^2}{\hbar^2\omega_0^2}\right), & \alpha \gg 1, \end{cases} \tag{4.103}$$

where $\Gamma_0 \equiv \Gamma(T = 0)$ is the decay rate at zero temperature. In the underdamped limit $\alpha \ll 1$, Eq. (4.103) applies at $T \ll \hbar\omega_0/2\pi$, while in the opposite limit of strong dissipation $\alpha \gg 1$, this result remains valid at any temperature $T < \hbar\omega_0/(4\pi\alpha)$.

Note that at $T \to \hbar/\tau_B$, the normalization factor (4.102) tends to zero, meaning that at even higher temperature, the above consideration is not anymore applicable. A formal reason for that is obvious: At $T = \hbar\omega_0/(4\pi\alpha)$, our bounce solution (4.99), (4.100) reduces to a trivial one $\tilde{q}(\tau) = q_0$ at all values of τ. Physically, it implies that at this value of temperature, thermally assisted quantum tunneling crosses over to thermal activation, which is the dominant decay mechanism at all higher T.

Let us denote this crossover temperature as T_0. It turns out that at temperatures sufficiently close to T_0, one can construct the full analytic solution to our problem for any dissipation strength and for any smooth potential [178, 258]. The reason is that at such temperatures, all significant values of $q(\tau)$ remain in the vicinity of q_0, i.e., it suffices to operate not far from the barrier top where the potential is close to parabolic. Here, as before, we restrict our analysis to the potential (4.13), which we rewrite in the form

$$U(q) = U_0 - \frac{\kappa}{2}(q - q_0)^2 - \frac{\lambda}{6}(q - q_0)^3 \tag{4.104}$$

At T very close to T_0, the bounce trajectory is again represented by the Fourier series (4.99) where now it is convenient to choose $q_a = q_0$ and $\tilde{q}(\tau) = \tilde{q}(-\tau)$ or, equivalently $b_n = b_{-n}$. For $|\tilde{q}(\tau) - q_0| \ll q_0$ in the leading approximation, one can retain only the principal harmonics with $|n| = 1$ in the Fourier series (4.99). Then, substituting the bounce $\tilde{q}(\tau) = q_0 + 2b_1 \cos(2\pi T\tau/\hbar)$ into the action (4.12) and for a moment disregarding the cubic nonlinearity in Eq. (4.104), we obtain

$$S[\tilde{q}(\tau)] \simeq U_0/T + \Lambda_1 b_1^2 \tag{4.105}$$

where Λ_1 is one of the eigenvalues

$$\Lambda_n = m(2\pi Tn/\hbar)^2 + 2\pi\eta|n|T/\hbar - \kappa \tag{4.106}$$

of the operator $\delta^2 S_{\text{eff}}/\delta q^2_{q=q_0}$. Since at $T = T_0$ the bounce reduces to $\tilde{q}(\tau) = q_0$, we obviously must have $S[q_0] = U_0/T_0$. This is generally achieved provided the eigenvalue Λ_1 vanishes. Thus, the condition $\Lambda_1(T_0) = 0$ defines the crossover temperature

$$T_0 = \frac{\hbar\omega_0}{2\pi}\left(\sqrt{1+\alpha^2} - \alpha\right) \tag{4.107}$$

As one could already expect from our previous consideration, in the overdamped limit T_0 reduces to $T = \hbar/\tau_B = \hbar\omega_0/(4\pi\alpha)$. In the opposite dissipativeless limit, Eq. (4.107) just yields $T_0 = \hbar\omega_0/(2\pi)$. Thus, damping decreases the crossover temperature between the regimes of quantum tunneling and thermal activation.

It follows from Eq. (4.105) that at all $T \geq T_0$, the minimal action value $S = U_0/T$ is provided for $b_1(T) = 0$, i.e., $\tilde{q}(\tau) = q_0$ remains the only relevant saddle point above T_0. On the other hand, at $T < T_0$, nonzero Fourier coefficient $b_1(T) \neq 0$ corresponds to lower action value, which is achieved on the bounce trajectory $\tilde{q}(\tau)$ deviating from q_0, i.e., $S[\tilde{q}(\tau)] < S[q_0] = U_0/T$. In order to evaluate both $\tilde{q}(\tau)$ and the action $S[\tilde{q}(\tau)]$ at temperatures slightly below T_0,

$$\varepsilon = \frac{T_0 - T}{T_0} \ll 1 \tag{4.108}$$

we can proceed perturbatively in ε. Substituting Eq. (4.99) into Eq. (4.18), for the potential energy of the form (4.104), we obtain

$$b_{\pm 1} = q_0\sqrt{\varepsilon d(\alpha)}, \quad b_0 = -2q_0\varepsilon d(\alpha), \quad b_{\pm 2} = q_0\varepsilon d(\alpha)\kappa/\Lambda_2 \tag{4.109}$$

where the eigenvalue Λ_2 is found from Eq. (4.106) and

$$d(\alpha) = \frac{(1+\alpha^2 - \alpha\sqrt{1+\alpha^2})(3+4\alpha^2 - 4\alpha\sqrt{1+\alpha^2})}{(5+8\alpha^2 - 8\alpha\sqrt{1+\alpha^2})} \tag{4.110}$$

The Fourier coefficients b_n with $|n| \geq 3$ are proportional to higher powers of ε and, therefore, can be omitted in the limit (4.108). Keeping only terms with $|n| \leq 2$ in Eq. (4.99) and making use of the results (4.109), we evaluate the effective action on the bounce trajectory $\tilde{q}(\tau)$. In the leading order in $\varepsilon \ll 1$, we get [178, 257]

$$A(T) = \frac{U_0}{T} - \frac{12\pi U_0}{\hbar\omega_0}\left(\frac{T_0 - T}{T_0}\right)^2 d(\alpha)\sqrt{1+\alpha^2} \tag{4.111}$$

This is a general expression for the exponential factor entering the quantum decay rate (4.41) at temperatures slightly below T_0 and at any dissipation strength α. In the dissipativeless ($\alpha \to 0$) and overdamped ($\alpha \gg 1$) limits, we have, respectively, $d \to 3/5$ at $d \to 1/2$ and Eq. (4.111) matches with the results [2] and the second Eq. (4.103) in the corresponding limits close to T_0.

At $T \to T_0$, the bounce merges with a trivial solution $q(\tau) = q_0$, which remains the only saddle point with the action $S_{\text{eff}}[q_0] = \hbar U_0/T$ at all $T \geq T_0$. As before, in order to determine the decay rate $\Gamma(T)$, we now need to properly evaluate the fluctuation determinants and find the imaginary part of the free energy. At $T \geq T_0$, it is necessary to account for fluctuations around the trivial saddle point (4.20), while at temperatures below T_0, one should rather take care about fluctuations around the bounce $\tilde{q}(\tau)$. In general, these two situations should be analyzed separately. However, since at T just slightly below T_0 (4.108) the difference between the bounce and (4.20) is small, $|\tilde{q}(\tau)-q_0| \ll 1$, it is possible to cover the whole temperature range both above and slightly below T_0 within the same procedure [258] to be discussed below.

Let us express fluctuations $\delta q(\tau) = q(\tau) - q_0$ as a Fourier series (cf. Eq. (4.28)):

$$\delta q(\tau) = T^{1/2} \sum_{n=-\infty}^{\infty} a_n \exp(\mathrm{i}\omega_n \tau), \quad a_n = a^*_{-n} \tag{4.112}$$

Substituting Eq. (4.112) into the action and expanding in powers of the Fourier coefficients a_n, we get

$$S_{\text{eff}}[q(\tau)] = \frac{\hbar U_0}{T} + \frac{\hbar}{2} \sum_{n=-\infty}^{\infty} \Lambda_n |a_n|^2 + \frac{\hbar\lambda T^{1/2}}{2}(2a_0|a_1|^2 + a_2 a_{-1}^2 + a_{-2}a_1^2) \tag{4.113}$$

where the eigenvalues Λ_n are defined in Eq. (4.106). Note that here the action is expanded up to second-order terms in all a_n with $n \neq 1$, while terms containing $a_{\pm 1}$ are kept up to the third order since the eigenvalue Λ_1 is small at temperatures close to T_0. In general terms $\propto |a_{\pm 1}|^4$ should also be included into our expansion (4.113), but these terms vanish for the cubic potential (4.104) considered here.

With the aid of Eq. (4.112), the path integral (4.26) can be rewritten as an infinite product of ordinary integrals

$$N \prod_{n=-\infty}^{\infty} \left[\int \frac{da_n}{\sqrt{2\pi\hbar}} \exp\left(-\frac{S_{\text{eff}}[q(\tau)]}{\hbar} \right) \right] \tag{4.114}$$

As before, one of the eigenvalues (which we now denote as Λ_0) is negative since we are dealing with metastability. The integral over a_0 is again handled by distortion of the integration contour in the complex plane, which defines an imaginary part of the free energy, cf. Eq. (4.32). Gaussian integrals over all a_n with $n \geq 2$ are trivially performed. As result, from Eq. (4.114), we obtain

$$\frac{\mathrm{i}N}{2} \exp\left(-\frac{U_0}{T} \right) (\det' G^{-1})^{-1/2} \int \frac{d^2 a_1}{2\pi\hbar} \exp(-\Lambda_1 |a_1|^2 - Q|a_1|^4) \tag{4.115}$$

where we defined

$$Q = \frac{\lambda T}{q_0}\left(1 - \frac{\kappa}{2\Lambda_2}\right) \tag{4.116}$$

and also denoted

$$(\det{}' G^{-1})^{-1/2} = \frac{1}{|\Lambda_0|^{1/2}} \prod_{n=2}^{\infty} \frac{1}{\Lambda_n} \tag{4.117}$$

i.e., in this case, $\det' G^{-1}$ coincides with the absolute value of the determinant of $G^{-1}_{q=q_0}$ with excluded eigenvalues $\Lambda_{\pm 1}$. This is in contrast to Eq. (4.39) where one zero mode is excluded. Here exact zero modes are lacking, but instead there exist two soft modes that are accounted for by the remaining non-Gaussian integral in Eq. (4.115). This integral can be expressed in terms of the error integral $\mathrm{erfc}(x) = (2/\pi^{1/2}) \int_x^{\infty} dy \exp(-y^2)$. As a result, we obtain

$$F = -T \ln \mathcal{Z} \simeq -T \ln \left[N(\det G_0^{-1})^{-1/2}\right] - \frac{iT}{2}\left(\frac{\det G_0^{-1}}{\det' G^{-1}}\right)^{1/2}$$
$$\times \frac{\sqrt{\pi((U_0/T) - A(T))}}{|\Lambda_1|} erfc\left(-\frac{\epsilon}{|\epsilon|}\sqrt{(U_0/T) - A(T)}\right) \exp(-A(T)) \tag{4.118}$$

where $A(T)$ and $\det' G^{-1}$ are defined, respectively, in Eqs. (4.111) and (4.117), while $\det G_0^{-1}$, as before, is given by the second Eq. (4.25). What remains is to evaluate the ratio of determinants in Eq. (4.118). This task is easily accomplished, and the result is again expressed in terms of the Euler gamma functions

$$\left(\frac{\det G_0^{-1}}{\det' G^{-1}}\right)^{1/2} = m(2\pi T/\hbar)^2 \frac{\Gamma(2 + \hbar\Omega_+/(2\pi T))\Gamma(2 + \hbar\Omega_-/(2\pi T))}{\Gamma(1 + \hbar\omega_+/(2\pi T))\Gamma(1 + \hbar\omega_-/(2\pi T))} \tag{4.119}$$

where $\omega_\pm$ is defined in the first Eq. (4.88) and

$$\Omega_\pm = \omega_0(\alpha \pm \sqrt{\alpha^2 + 1}) \tag{4.120}$$

We have completed all necessary preparations, and now we are able to evaluate the decay rate Γ. Let us remind the reader that the validity range of our calculation and the resulting Eqs. (4.118) and (4.119) embraces all temperatures above T_0 and also includes the temperature interval (4.108) slightly below T_0. For $T \geq T_0$, the path $q(\tau) = q_0$ is the only saddle point, while in the case (4.108), the actual saddle point path $\tilde{q}(\tau)$ remains close to q_0. We also note that at $T < T_0$, one could proceed differently, and directly evaluate the fluctuation contribution around the true bounce $\tilde{q}(\tau)$. Then, two soft modes $\Lambda_{\pm 1}$ split into one nonzero and one zero mode, which can be handled in a standard manner as described in Section 4.2. The corresponding calculation [179, 400] again yields Eqs. (4.118) and (4.119), thus reconfirming their validity at temperatures (4.108) slightly below T_0.

Although the imaginary part of the free energy in Eq. (4.118) looks the same both below and above T_0, the decay rate Γ should be extracted from it somewhat differently. As we already discussed in Section 4.1, for $T < T_0$, Eq. (4.10) applies, while at $T > T_0$, Eq. (4.11) should rather be employed. The formal difference between these two cases lies in the presence of the exact zero mode at $T < T_0$ and its absence at $T \geq T_0$. Accordingly, while integration over the collective coordinate τ_0 (corresponding to the zero mode) is performed providing the factor $1/T$ in the expression for $\mathrm{Im}F$ at $T < T_0$, cf. Eq. (4.38), no such integration is carried out for $T > T_0$. Following [2], at $T > T_0$ it is necessary to "correct" for that by multiplying the expression for the quantum decay rate (4.10) by the (dimensionless) factor $\propto 1/T$. Since at $T = T_0$ one should get Eq. (4.10) back, the correct value of this factor is T_0/T. In this way, one arrives at Eq. (4.11) describing thermally activated decay of a metastable state in the presence of quantum fluctuations.

Needless to say, the above considerations serve merely as plausibility arguments rather than a rigorous derivation of Eq. (4.11). However, one can independently verify by other methods that this formula indeed yields correct results for the decay rate Γ at $T \geq T_0$. For more details on this subject, we refer the reader to the book [400].

Combining Eqs. (4.118) and (4.119) with the formula (4.11), at $T \geq T_0$ we obtain

$$\Gamma = \frac{T_0}{\hbar}\mathcal{F}(T)\sqrt{\pi((U_0/T) - A(T))}\,\mathrm{erfc}\left(\sqrt{(U_0/T) - A(T)}\right) \times \exp(-A(T)) \tag{4.121}$$

where

$$\mathcal{F}(T) = \frac{\Gamma(1+\hbar\Omega_+/(2\pi T))\Gamma(1+\hbar\Omega_-/(2\pi T))}{\Gamma(1+\hbar\omega_+/(2\pi T))\Gamma(1+\hbar\omega_-/(2\pi T))} \tag{4.122}$$

The general result (4.121) can further be simplified outside immediate vicinity of T_0, i.e., for $T - T_0 \gg T_0(T/U_0)^{1/2}$. In this case with the aid of the asymptotic form $erfc(x) \approx (1/\pi^{1/2}x)\exp(-x^2)$ for $x \gg 1$ from Eq. (4.121), one finds [300, 407]

$$\Gamma = \frac{T_0}{\hbar}\mathcal{F}(T)\exp(-U_0/T) \tag{4.123}$$

In the high-temperature limit $T \gg T_0$, Eq. (4.123) further reduces to the classical thermal activation decay rate [239]

$$\Gamma_{\rm cl} = \frac{\omega_0}{2\pi}\left(\sqrt{1+\alpha^2}-\alpha\right)\exp(-U_0/T) \tag{4.124}$$

The combination in the pre-exponent in Eq. (4.124) just represents the classical attempt frequency for a particle in the presence of linear Ohmic dissipation. Equation (4.123) demonstrates that as temperature is lowered and becomes comparable to T_0, the classical activation rate (4.124) gets enhanced by quantum fluctuations by the factor $\mathcal{F}(T)$. This effect is twofold. On one hand, quantum fluctuations increase the particle energy as compared to its classical value. On the other hand, at energies slightly below U_0, the process of quantum tunneling (not included in the classical result (4.124)) yields a contribution comparable to that of thermal activation and further increases the decay rate Γ.

With the aid of Eqs. (4.123) and (4.124), we can further analyze this quantum enhancement effect in some limiting cases. For example, for $T \gg T_0$ and $\alpha \gg 1$, we obtain

$$\mathcal{F}(T) \equiv \frac{\Gamma}{\Gamma_{\rm cl}} = \exp\left[\frac{2T_0}{T}\left(\psi\left(1+\frac{\hbar\eta}{2\pi mT}\right)-\psi(1)\right)\right] \tag{4.125}$$

which further reduces to

$$\frac{\Gamma}{\Gamma_{\rm cl}} = \exp\left(\frac{\hbar^2\omega_0^2}{12T^2}\right) \tag{4.126}$$

in the limit $T \gg \hbar\eta/(2\pi m)$ and to

$$\Gamma/\Gamma_{\rm cl} = \left(4\alpha^2 T_0/T\right)^{2T_0/T} \tag{4.127}$$

at $T_0 \ll T \ll \hbar\eta/(2\pi m)$. We observe that even at temperatures substantially above the crossover one T_0, the enhancement of the classical activation decay rate can be large, in particular in the overdamped limit $\alpha \gg 1$. We also remark that Eq. (4.126) actually applies at any damping provided $T \gg T_0$.

Very close to the crossover temperature, i.e., at $T - T_0 \lesssim T_0(T/U_0)^{1/2}$, Eq. (4.123) becomes inaccurate and a more general formula (4.121) should be employed. We also note that, as we already argued above, our result (4.118) remains valid not only above the crossover temperatures but also at temperatures below (though not far from) T_0, i.e., under the condition (4.108). Combining Eqs. (4.118) and (4.10), in the temperature interval (4.108), we obtain

$$\Gamma = \frac{T}{\hbar}|\mathcal{F}(T)|\sqrt{\pi((U_0/T) - A(T))} \times \text{erfc}\left(-\sqrt{(U_0/T) - A(T)}\right)\exp(-A(T)) \quad (4.128)$$

This result simplifies further outside the immediate vicinity of the crossover temperature $(T/U_0)^{1/2} \ll \epsilon \ll 1$. In this case (bearing in mind that $\text{erfc}(-x) \approx 2$ for $x \gg 1$) from (4.128), we get

$$\Gamma = \frac{2T}{\hbar}|\mathcal{F}(T)|\sqrt{\pi((U_0/T) - A(T))}\exp(-A(T)) \quad (4.129)$$

As in this equation one should keep $T \approx T_0$, it is easy to check that in the overdamped limit $\alpha \gg 1$, the combination in the pre-exponent is temperature independent and reduces to that defined in Eq. (4.91). It is, therefore, satisfactory to observe that the decay rate (4.129), (4.111) evaluated just below T_0 exactly coincides (under the condition (4.108)) with that defined by the second Eq. (4.103). Hence, in the strong dissipation limit, we established the complete analytical description of the decay rate for all temperatures.

We conclude this part of our analysis by pointing out that for experimental verification of the above results, it is in general necessary, similarly to the case $T = 0$, to take into account fluctuation renormalization of the Josephson critical current. The corresponding general expressions are worked out in the literature [422]; however, we will avoid presenting them here.

4.3.3 Tunneling through a Non-quasiclassical Potential

As we already pointed out above, there exist superconducting weak links where the low-temperature current-phase relation strongly

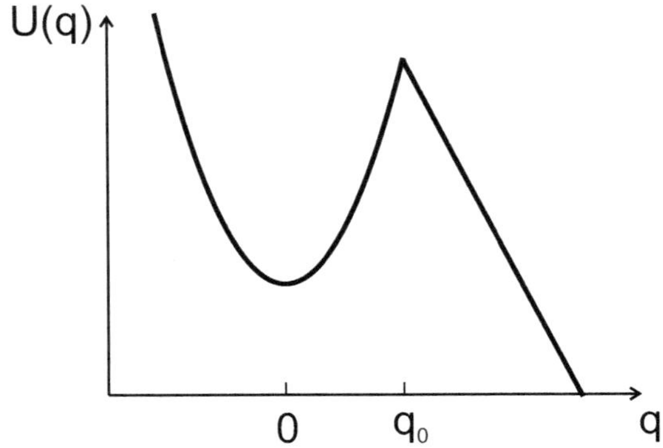

Figure 4.3 A non-quasiclassical metastable potential defined in Eq. (4.130).

deviates from the standard sinusoidal form. Provided the value of the critical current is reached at $\varphi = \pi$ and $I(\varphi)$ experiences a jump at this point, under the condition $\chi \ll 1$ (4.48) the effective potential $U(q)$ for the "Josephson particle" is determined by Eq. (4.51). This potential—in contrast to Eq. (4.13)—is not smooth on the top, i.e., it cannot be reduced to the parabolic form at $q \sim q_0$. Below we shall demonstrate that this feature can yield substantial modifications of the decay rate Γ as compared to that evaluated in the case (4.13).

For simplicity in what follows, we will consider the potential

$$U(q) = \frac{\kappa}{2} q^2 + \frac{\gamma}{2}(q_0 - q)\theta(q - q_0) \qquad (4.130)$$

which also demonstrates non-quasiclassical behavior at $q \approx q_0$ (see also Fig. 4.3). This "parabolic plus linear" potential describes, e.g., long ballistic SNS junctions (in which case $\kappa = I_C/2\pi e$, $\gamma = 2I_C/e$ and $q_0 = \pi\chi$) and can also be interesting in the context of quantum tunneling in some chemical and biological systems.

Let us first stick to the zero-temperature limit $T \to 0$. In order to determine the bounce trajectory, we again consider classical dynamics of a particle with mass m in the inverted potential $-U(q)$. As before, we need to solve Eq. (4.18), which now turns out to be linear at any $q(\tau)$ except for the point $q = q_0$. In the course of its motion, the particle crosses this point twice at the time moments $\tau = \pm\tau_0$. Resolving Eq. (4.18) with the potential (4.130) at all $\tau \neq \pm\tau_0$ and making use of the condition $q(\pm\tau_0) = q_0$, we obtain the bounce trajectory for arbitrary dissipation strength η. It reads

$$\tilde{q}_\omega = \frac{\gamma}{\omega} G_{0\omega} \sin \omega\tau_0, \qquad q_0 = \int \frac{d\omega}{2\pi} \tilde{q}_\omega e^{-i\omega\tau_0} \tag{4.131}$$

Here $\tilde{q}_\omega$ is the Fourier component of the bounce $\tilde{q}(\tau)$, $G_{0\omega}$ is defined in Eq. (4.24), and $\tau = \pm\tau_0$ are the imaginary time values at which the system coordinate $q(\tau)$ crosses the point $q = q_0$. Substituting this bounce solution into the effective action, from the first equation (4.42), we get [420]

$$A = \frac{\gamma q_0 \tau_0}{\hbar} - \frac{\gamma^2}{2\hbar} \int \frac{d\omega}{2\pi} G_{0\omega} \frac{\sin^2 \omega\tau_0}{\omega^2} \tag{4.132}$$

Now let us evaluate the pre-exponential factor B. For this purpose, let us first determine the expression for G^{-1} (4.27) in the case of the potential (4.130). Making use of Eq. (4.131), we obtain

$$G_g^{-1}(\tau, \tau') = \int \frac{d\omega d\omega'}{(2\pi)^2} G^{-1}_{\omega,\omega'} e^{-i\omega\tau + i\omega'\tau'},$$

$$G^{-1}_{\omega,\omega'} = 2\pi G_{0\omega}\delta(\omega - \omega') - g\left[e^{i(\omega-\omega')\tau_0} + e^{-i(\omega-\omega')\tau_0}\right] \tag{4.133}$$

where the parameter g is defined by the following relations

$$g = [f(0) - f(2\tau_0)]^{-1}, \qquad f(\tau) = \int \frac{d\omega}{2\pi} G_{0\omega} \cos \omega\tau \tag{4.134}$$

Our main goal is to evaluate the path integral

$$K(g) = \int \mathrm{D}\delta q \exp\left(-\frac{1}{2\hbar} \int d\tau \int d\tau' G_g^{-1}(\tau, \tau')\delta q(\tau)\delta q(\tau')\right) \tag{4.135}$$

where $\delta q(\tau) = q(\tau) - \tilde{q}(\tau)$ denotes fluctuations around the bounce trajectory (4.131). As we already know, this integral formally diverges in the presence of metastability and needs to be analytically continued to provide an imaginary contribution to the partition function. Further calculation can be performed similarly to that carried out above for the potential (4.13). Here we will illustrate another method of evaluating B [421], which appears more convenient in the case of "parabolic plus linear" potential (4.130).

Let us introduce $\tilde{g} = g - \epsilon$ and choose $\mathrm{Re}\epsilon$ to be positive and large enough to have the path integral for $K(\tilde{g})$ well defined and purely real. In this case, one obviously has

$$K(\tilde{g}) = N\left[\prod_{n=0}^{\infty} \Lambda_n(\epsilon)\right]^{-1/2} \tag{4.136}$$

where $\Lambda_n(\epsilon)$ are the eigenvalues of the operator (4.133) with $g \to \tilde{g}$. These eigenvalues are determined from the equation

$$1 = \tilde{g} \int \frac{d\omega}{2\pi} \frac{1 + (-1)^n \cos(2\omega\tau_0)}{G_{0\omega}^{-1} - \Lambda_n(\epsilon)} \tag{4.137}$$

The quantity $K(\tilde{g})$ can also be evaluated in a different manner. Differentiating $\ln K(\tilde{g})$ with respect to $\tilde{g}$, we get

$$\frac{\partial \ln K(\tilde{g})}{\partial g} = \frac{1}{2\hbar} \langle (\delta q(\tau_0))^2 + (\delta q(-\tau_0))^2 \rangle = G_{\tilde{g}}(\tau_0, \tau_0) \tag{4.138}$$

Here $G_{\tilde{g}}(\tau, \tau')$ is the inverse of the Fourier-transformed function (4.133) with $g \to \tilde{g}$. For $\mathrm{Re}\tilde{g} < [f(0) + f(2\tau_0)]^{-1}$, this function is found explicitly. We obtain

$$G_{\tilde{g}}(\tau, \tau') = f(\tau - \tau') + \frac{\tilde{g}}{2} \left(\sum_{\pm} \frac{[f(\tau + \tau_0) \pm f(\tau - \tau_0)][f(\tau' + \tau_0) \pm f(\tau' - \tau_0)]}{1 - \tilde{g}[f(0) \pm f(2\tau_0)]} \right) \tag{4.139}$$

Combining Eqs. (4.138) and (4.139) and integrating the first of them over $\tilde{g}$ from zero to $\tilde{g}$, we arrive at an important formula

$$K(\tilde{g}) = \frac{K(0)}{[1 - \tilde{g}(f(0) + f(2\tau_0)][1 - \tilde{g}(f(0) - f(2\tau_0)]} \tag{4.140}$$

This result allows to avoid solving the transcendental Eq. (4.137), finding all eigenvalues Λ_n with subsequent evaluation of the ratio of two fluctuation determinants in the second Eq. (4.42). In fact, Eq. (4.140) already provides us with a formal expression for the needed fluctuation determinants ratio $K(\tilde{g})/K(0)$.

What remains is to perform analytic continuation by setting $\epsilon \to 0$. As a result of this step, as usually, one eigenvalue becomes negative $\Lambda_{-1} < 0$ due to the presence of metastability and another one tends to zero $\Lambda_0(\epsilon \to 0) \to 0$. We have already learned how to deal with both these eigenvalues. The negative eigenvalue just yields a nonzero imaginary part in the expression (4.136). Indeed, one can directly observe that the combination (4.140) becomes imaginary for sufficiently small $\mathrm{Re}\epsilon$. The zero mode should be extracted from Eq. (4.136). As a result, making use of the second Eq. (4.42), we obtain

$$B = \frac{1}{(2\pi\hbar)^{1/2}} \left[\int d\tau \left(\frac{\partial \tilde{q}}{\partial \tau} \right)^2 \right]^{1/2} \left[\frac{|K(g-\epsilon)| \Lambda_0^{1/2}(\epsilon)}{K(0)} \right]_{\epsilon \to 0} \tag{4.141}$$

Now we just need to evaluate $\Lambda_0(\epsilon)$. From Eq. (4.137) in the limit $\epsilon \to 0$, we get

$$\Lambda_0(\epsilon) = \epsilon\gamma^2 [f(0) - f(2\tau_0)]^2 \left[2 \int d\tau \left(\frac{\partial \tilde{q}}{\partial \tau} \right)^2 \right]^{-1} \tag{4.142}$$

Substituting Eq. (4.142) into Eq. (4.141) and making use of Eq. (4.140), we eventually derive the general formula for the pre-exponential factor

$$B = \frac{\gamma}{(2\pi\hbar)^{1/2}} \frac{f(0) - f(2\tau_0)}{2\sqrt{f(2\tau_0)}} \tag{4.143}$$

The combination of Eqs. (4.132) and (4.143)—together with (4.131) and (4.134)—fully determines the quantum decay rate Γ (4.41) of a metastable state in the potential (4.130).

Let us now recall that while deriving the result (4.143), we restricted our analysis to Gaussian fluctuations around the bounce (4.131), i.e., expanded our action to the second order in $\delta q(\tau)$. Since the action is quadratic in $q(\tau)$ for $q \neq q_0$, this expansion (4.135) is exact for all $\tau \neq \pm\tau_0$. On the other hand, at $q = q_0$ our action is strongly nonlinear; therefore, the applicability of Gaussian approximation (4.135) at $\tau = \pm\tau_0$ needs to be verified separately.

Let us define fluctuations around the bounce with excluded zero mode as

$$\delta' q(\tau) = \sum_{n \neq 0} a_n q_n(\tau) \tag{4.144}$$

These fluctuations remain small in the "non-quasiclassical" point $\tau = \tau_0$ provided the inequality

$$\sqrt{|\langle (\delta' q(\tau_0))^2 \rangle|} \ll q_0 \tag{4.145}$$

is satisfied. In order to get an estimate for the left-hand side of Eq. (4.145), let us consider the correlator

$$\langle \delta' q(\tau_0) \delta' q(\tau_0) \rangle / \hbar = \sum_{n \neq 0} q_n(\tau) q_n(\tau') \Lambda_n^{-1} \tag{4.146}$$

which can also be rewritten in the form

$$\langle \delta' q(\tau_0) \delta' q(\tau_0) \rangle / \hbar = \lim_{\epsilon \to 0} [G(\tau, \tau') - q_0(\tau) q_0(\tau') \Lambda_0(\epsilon)] \tag{4.147}$$

With the aid of Eqs. (4.139) and (4.142) combined with Eqs. (4.36) and (4.131) at $\tau = \tau' = \tau_0$, one readily finds

$$\frac{\langle (\delta' q(\tau_0))^2 \rangle}{\hbar} = -\frac{(f(0) - f(2\tau_0))^2}{4 f(2\tau_0)} \tag{4.148}$$

The minus sign in the right-hand side of Eq. (4.148) just implies that the correlator (4.146) is dominated by the negative mode with the eigenvalue Λ_{-1}. Combining Eqs. (4.148) and (4.145) and making use of the result (4.143), we arrive at its validity condition

$$B \ll \frac{\gamma q_0}{(2\pi)^{1/2} \hbar} \tag{4.149}$$

Let us analyze the behavior of Γ at the bias current values close to the critical one (4.48). Specifically, below we will consider the limit

$$\xi = 2\pi \kappa q_0 / \gamma \ll 1 \tag{4.150}$$

In this limit, Eq. (4.132) yields

$$A = \begin{cases} 2\pi r \frac{U_0}{\hbar \omega_0} \sqrt{|1 - \alpha^2|}, & \alpha^2 \lesssim 1/\xi, \\ \frac{\pi p(1 - p/2)}{2} \frac{\eta q_0^2}{\hbar}, & \alpha^2 \gtrsim 1/\xi \end{cases} \tag{4.151}$$

Here we defined $U_0 = \kappa q_0^2 / 2$,

$$r^{-1} = \begin{cases} \pi - 2 \arctan \left(\frac{\alpha}{\sqrt{1 - \alpha^2}} \right), & \alpha < 1, \\ \ln \left[\frac{\alpha + \sqrt{\alpha^2 - 1}}{\alpha - \sqrt{\alpha^2 - 1}} \right], & \alpha > 1 \end{cases} \tag{4.152}$$

and the parameter p is determined by the equation

$$p(1 - \gamma_0 - \ln(\xi p)) = 1 \tag{4.153}$$

where $\gamma_0 \simeq 0.577$ is the Euler constant. Under the condition (4.150), the above equations fully determine the exponential factor A at all values of α. We observe that at any dissipation strength, this factor behaves quite similarly to that derived for the smooth potential (4.13), cf., e.g., Eq. (4.61).

Evaluating Eq. (4.143) in the limit (4.150), we get

$$B = \begin{cases} \pi r^{3/2} \sqrt{\frac{2 U_0 \omega_0}{\hbar}} |1 - \alpha^2|^{3/4}, & \alpha^2 \lesssim 1/\xi, \\ \frac{\gamma \eta}{2\pi \hbar^{1/2}} \sqrt{\frac{p}{2(1-p)}} \left[\ln \left(\frac{2\pi p q_0 \eta^2}{m \gamma} \right) + \gamma_0 \right], & \alpha^2 \gtrsim 1/\xi \end{cases} \tag{4.154}$$

These results for the pre-exponential factor B turn out to be qualitatively similar to those for the smooth potential (4.13) only in the underdamped limit, while for $\alpha \gg 1$, this factor is parametrically different from that defined in Eq. (4.91).

Finally, let us briefly address the case of nonzero temperatures restricted by the condition (4.95). As we have already learned, thermal fluctuations enhance the decay rate Γ. The whole calculation can be repeated at nonzero T and the results for A and B are again presented, respectively, by Eqs. (4.132) and (4.143) where one should substitute $\omega \to \omega_n = 2\pi n T/\hbar$ and replace frequency integrals by summands over Matsubara frequencies. Equations (4.131), (4.134), and others should be modified in the same manner.

We again consider the limit (4.150). In the case of not too strong dissipation

$$\alpha^2 \ll \frac{1 + 4\pi\alpha T/\hbar\omega_0}{\xi} \tag{4.155}$$

the temperature-dependent exponential factor reads [419]

$$A = 2\pi \frac{U_0}{\hbar\omega_0} \left[\frac{\psi(\hbar\omega_+/2\pi T) - \psi(\hbar\omega_-/2\pi T)}{\sqrt{\alpha^2 - 1}} - \frac{2\pi T}{\hbar\omega_0} \right]^{-1} \tag{4.156}$$

where the frequencies $\omega_\pm$ are defined in Eq. (4.88) and $\psi(x)$ is the digamma function. In the limit $T \to 0$, this result reduces to Eq. (4.151). At nonzero but sufficiently small T, Eq. (4.156) yields negative correction to the zero-temperature result for A. This correction turns out to be exponentially small in the dissipativeless limit and proportional to T^2 for $\eta \neq 0$.

Increasing the temperature further, one observes that both the above bounce solution and Eq. (4.156) remain applicable at any $T \ll U_0$, i.e., in contrast to the case of a smooth potential (4.13); here no crossover between quantum tunneling and classical thermal activation occurs at all. In other words, no decay by thermal activation is realized for the non-quasiclassical potential (4.130); thermally assisted quantum tunneling always dominates in this case. At sufficiently high temperatures from Eq. (4.156), we find

$$A = \begin{cases} \frac{U_0}{T}\left[1 - \frac{\hbar\omega_0}{2\pi\alpha T}\ln\left(\frac{\hbar\omega_0\alpha}{\pi T}\right)\right], & \frac{\hbar\omega_0}{4\pi\alpha} \ll T \ll \frac{\hbar\omega_0\alpha}{\pi}, \\ \frac{U_0}{T}\left[1 - \frac{\hbar^2\omega_0^2}{12T^2}\right], & \frac{\hbar\omega_0}{2\pi}(\alpha+1) \ll T \ll U_0 \end{cases} \tag{4.157}$$

i.e., upon increasing temperature, $A(T)$ gradually approaches the classical value $A = U_0/T$.

The pre-exponential factor B is also affected by thermal fluctuations. While at low enough temperatures this factor gets slightly enhanced as compared to Eq. (4.154), at higher $T \gg (\hbar\omega_0/2\pi)(\alpha + 1)$, we obtain

$$B \simeq \frac{\omega_0^2 U_0^{1/2} \hbar^{1/2}}{2\sqrt{\pi} T^{3/2}} \tag{4.158}$$

i.e., in this regime the pre-exponent B decreases with increasing T.

Note that the result (4.157) coincides with that obtained for the parabolic potential that is abruptly "cut" at $q = q_0$ [177], i.e., for $\gamma \to \infty$. Hence, under the condition (4.155) and for sufficiently large γ, the value A is insensitive to γ, implying that the particle energy loss during tunneling is sufficiently small and the difference between q_0 and $q_{\rm ex}$ can be neglected. However, if the condition (4.155) is violated, this difference becomes non-negligible. For example, in the opposite strong dissipation limit

$$\alpha^2 \gg \frac{1 + 4\pi\alpha T/\hbar\omega_0}{\xi} \tag{4.159}$$

and for $T \ll \hbar\omega_0/4\pi\alpha$, we obtain

$$A = \frac{\pi p}{2} \frac{\eta q_0^2}{\hbar} \left[1 - \frac{p}{2} - \frac{16\pi^2\alpha^2 p}{3} \left(\frac{T}{\hbar\omega_0} \right)^2 \right] \tag{4.160}$$

while B acquires a small positive correction $\propto T^2$.

Summarizing our analysis, we can conclude that quantum decay of a metastable state through a non-quasiclassical potential of the type (4.130) is in several important aspects different from that in the potential (4.13) analyzed in previous sections. Perhaps the most important difference between these two situations is the absence of the quantum-to-classical crossover in the non-quasiclassical case (4.130). In the latter case, the decay remains "quantum" at any temperature (of course, restricted by the condition (4.95)), meaning that the nontrivial bounce solution does not disappear with increasing temperature and the dominating decay mechanism always remains quantum tunneling under the barrier. At higher values of T, the barrier becomes lower and the suppression factor $A(T)$ in the exponent gradually approaches the classical limit

$A(T) = U_0/T$ (see Eq. (4.157)). Another important difference is that the pre-exponential factors in the expressions for the decay rate Γ may strongly differ for the potentials (4.13) and (4.130), in particular in the overdamped limit $\alpha \gg 1$. All these differences are essentially due to the fact that the potential (4.130) remains strictly parabolic up to energies $E \approx U_0$ and, accordingly, discrete energy levels remain equidistant up to barrier top. In contrast, in the case (4.13), anharmonic terms play an important role at such energies and the inter-level distance shrinks with increasing E.

4.3.4 MQT Experiments

Quantum tunneling of the Josephson phase φ considered above serves as one of the most important manifestations of applicability of quantum mechanics to macroscopic objects, such as Josephson junctions and SQUIDs. As we have already demonstrated in Chapter 2, the validity of quantum mechanical analysis of a collective variable φ describing macroscopically large number of electron states can be proven theoretically. Experimental observation of MQT phenomenon in Josephson structures and quantitative verification of the above theoretical predictions for the supercurrent decay rate Γ would also be of fundamental importance.

MQT experiments in Josephson junctions are usually performed in the following way. One applies a small current bias I to the junction and slowly increases I, keeping it below I_C. In the absence of phase fluctuations (thermal or quantum), the system would remain in the zero-voltage state since only supercurrent $I = I_C \sin\varphi$ would flow across the junction. In the presence of phase fluctuations, however, this current state can decay, i.e., Josephson particle can either overcome the barrier by thermal activation or tunnel through the classically forbidden region, as shown in Fig. 4.1. After that the system slides down in the tilted washboard potential (4.46) with a nonzero velocity $\dot{\varphi}(t)$, implying switching to a resistive state with nonzero voltage $V = \hbar\dot{\varphi}/2e$. The current value $I = I_1$ at which this spontaneous switching takes place is recorded. Then the system is returned to a superconducting state, and the whole cycle is repeated giving the current switching value $I = I_2$. Repeating this experiment many times, one can present the results in the form of a histogram, which provides the probability distribution of the switching events w as a function of the bias current I. Usually it is considered sufficient to perform $10^4 - 10^5$ runs in order to achieve a good accuracy of the experimental data.

In order to extract the decay rate Γ from this probability distribution, let us keep in mind that the probability $P(t)$ that switching to a resistive state has already taken place increases with time as

$$\frac{dP}{dt} = \Gamma(I)(1 - P) \qquad (4.161)$$

Then for the decay rate, one trivially finds

$$\Gamma(I) = \frac{v_I w(I)}{1 - \int_0^I w(I)dI} \qquad (4.162)$$

where $v_I = dI/dt$ is the current sweeping rate and $w(I) = dP/dI$ is the probability distribution of switching events. This formula allows to directly reconstruct the decay rate Γ from the experimentally obtained probability density $w(I)$ for further comparison with theoretical predictions. A similar analysis of MQT data can be performed in SQUIDS where one varies an external flux Φ_x threading a superconducting loop. Typical experimental histogram is presented in Fig. 4.4.

Classical process of thermally activated decay of supercurrent in Josephson junctions was first demonstrated in 1974 [124], and the results turned out to be in a good agreement with theoretical predictions. First unambiguous experimental evidence of macroscopic quantum tunneling in these systems was reported in 1981 [205, 389, 390]. Later experiments [92, 100, 350, 395] allowed to quantitatively verify a number of important theoretical predictions, e.g., ones concerning the effect of temperature and damping on the quantum decay rate Γ. For instance, the dependence $\ln(\Gamma(T)/\Gamma_0) \propto T^2$ (4.103) was convincingly demonstrated in these experiments. Both the predicted magnitude and the functional form of the exponent $A(T)$ evaluated in the quantum regime in the strong damping limit were quantitatively confirmed by the experimental data [92, 350], as it is illustrated, e.g., in Fig. 4.5. Thus, it is now firmly established both in theory and in experiment that dissipation yields suppression of quantum decay of supercurrent thereby helping to maintain superconductivity in Josephson structures.

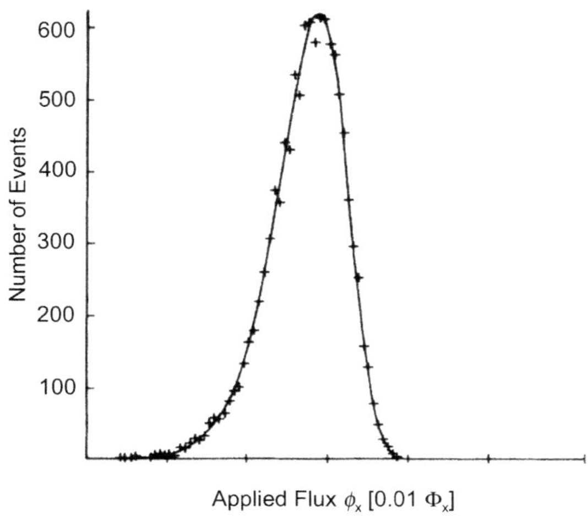

Figure 4.4 Typical experimental histogram representing the probability distribution for tunneling events as a function of the applied magnetic flux. Reprinted with permission from Ref. [350], Copyright 1985, by the American Physical Society.

At the same time, a significant discrepancy was found between the theoretical expression for the pre-exponential factor B in the strong damping limit (4.91) and its experimentally measured value [350]. For the parameters employed in these experiments, the measured value B turned out to be $4 \div 5$ orders of magnitude smaller than predicted by Eq. (4.91). However, this discrepancy disappears completely if we take into account the effect quantum renormalization of the critical current (4.92).

Indeed, in order to correctly compare experimental data and theoretical predictions for the decay rate Γ, the value of the junction critical current should be measured independently. If the experimental accuracy is sufficient to resolve quantum corrections to the exponent A (of order $\sim \hbar \ln B$) in the expression for the decay rate Γ, the same accuracy should also allow to identify reduction in the critical current due to quantum fluctuations (4.92). In other words, it is in general the renormalized value I_C^* (not the bare one I_C) that is independently measured in experiments. Hence, in order to perform the correct comparison with theory, it is necessary to

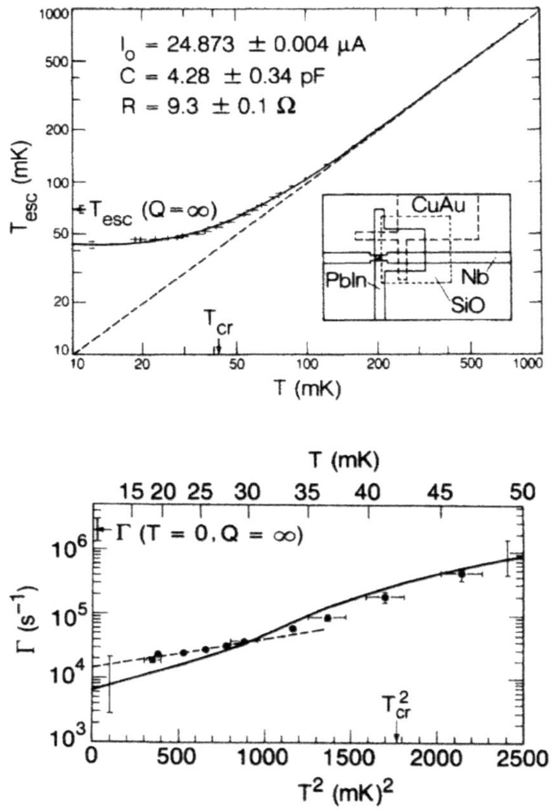

Figure 4.5 Experimental data for the decay rate $\Gamma(T)$ [92] and their comparison with theory. Top panel: The so-called escape temperature $T_{\text{esc}}(T)$ defined from $\Gamma(T) = (\omega_0/2\pi)\exp(-U_0/T_{\text{esc}})$. Experimental data are presented by crosses, and the solid line represents theoretical results discussed above. Bottom panel: $\Gamma(T)$ versus T^2. Solid circles are the data points [92]; the solid line indicates theoretical results [178]; and the dashed line is a result of a least-squares linear fit of the data to the dependence $\ln\Gamma(T) = a + bT^2$ with a and b being constants. For more details, we refer the reader to the work [92]. Adapted with permission from Ref. [92], Copyright 1988, by the American Physical Society.

express the decay rate Γ in terms of I_C^* exactly as we have done in Eq. (4.94). In contrast to Eq. (4.91), the latter expression does not contain the particle mass m in the pre-exponent and yields the results [422] in a very good agreement with experimental findings [350].

We also remark that this renormalization effect remains substantial not only at $T \to 0$ but also at higher temperatures, including the regime $T > T_0$ where smaller values of B as compared to those predicted theoretically are also recovered. For more details on this issue, we refer the reader to the work [422].

Chapter 5

Macroscopic Quantum Coherence and Dissipation

In Chapter 4, we analyzed "tunneling into continuum" assuming that there exists no noticeable probability for a particle to return to its initial state. This property is due to both the potential profile considered and the effect of dissipation. As it was explained, in Josephson junctions, this potential serves as a good approximation provided the bias current is sufficiently close to the critical one, I_C. In this case, in order to describe the process of quantum tunneling from a metastable state, it suffices to take into account only one local minimum of the potential (4.46) ignoring all others. However, if the bias current I is considerably smaller than I_C, in order to properly describe quantum dynamics of the system, it is in general necessary to account for other potential minima in Eq. (4.46) (see also Fig. 5.1a). Qualitatively, the main difference with the situation considered in Chapter 4 is that now the particle can jump from one potential minimum to the other and then return back. Provided quantum coherence is not significantly destroyed by interaction with an external environment, such processes cannot anymore be described as incoherent decay. For example, in the absence of any dissipation and in the limit of vanishing bias current

Dissipative Quantum Mechanics of Nanostructures: Electron Transport, Fluctuations, and Interactions
Andrei D. Zaikin and Dmitry S. Golubev

ISBN 978-981-4774-50-5 (Hardcover), 978-0-429-29823-3 (eBook)
www.jennystanford.com

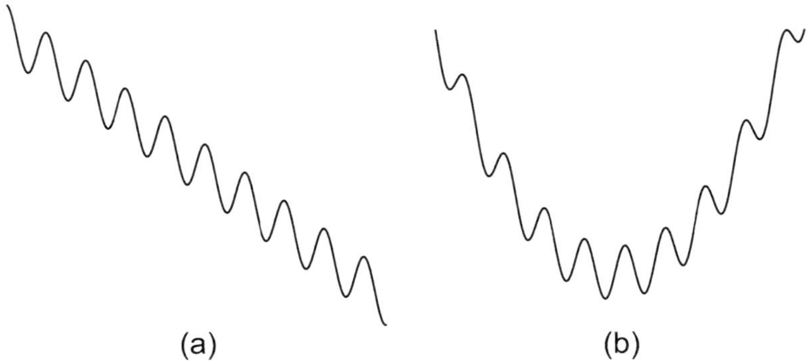

Figure 5.1 Potential energy of a current-biased Josephson junction (a) and of a dc SQUID (b).

$I \to 0$, the Josephson junction is equivalent to a quantum particle with coordinate φ in a 2π-periodic potential $U(\varphi) = -E_J \cos\varphi$. Hence, according to standard quantum mechanics, in this case the particle wave function describes Bloch states and the correct energy states are Bloch energy bands. In other words, the coordinate of the "Josephson particle" φ is not localized in space and can take any value from minus to plus infinity. Dissipation can significantly alter this picture and yield a variety of novel effects to be discussed in Chapter 6.

One can also consider a somewhat different situation of a SQUID, i.e., of a Josephson junction embedded in a superconducting ring of inductance L (see Fig. 5.2). In this case, the potential energy $U(\varphi)$ consists of two terms: the Josepson energy $-E_J \cos\varphi$ and the magnetic energy of a ring $(\Phi - \Phi_x)^2/2L$, where Φ is the magnetic flux inside the ring produced by circulating current $I = I_C \sin\varphi$ and Φ_x is an external flux. The magnetic energy obviously violates periodicity of the Josephson potential, and one arrives at the potential profile depicted in Fig. 5.1b. By properly choosing the system parameters, one can adjust $U(\varphi)$ in a way to make the energies of the two lowest potential wells very close to each other and the energies of all other wells considerably higher. In this case, we obtain an effective realization of a double-well potential in a SQUID. Such systems are of fundamental importance since they can be conveniently used in order to experimentally verify the idea of macroscopic quantum

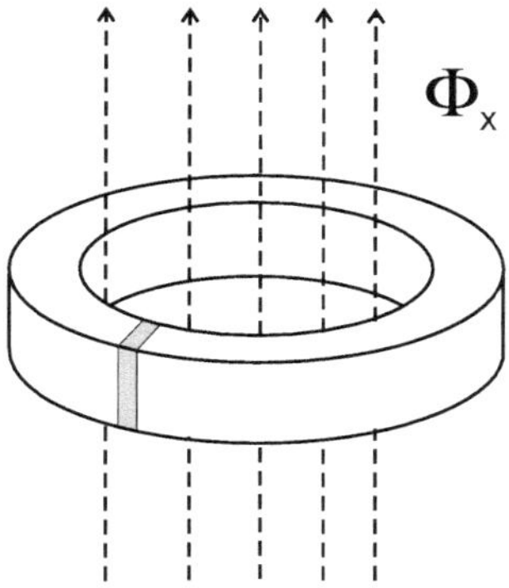

Figure 5.2 The schematics of a dc SQUID threaded by the magnetic flux Φ_x.

coherence (MQC) [263], e.g., by observing coherent oscillations of the magnetic flux inside the SQUID loop, as well as to study the effect of dissipation on quantum coherence in or close to the ground state. In addition, the possibility to further reduce the ring behavior to that of a two-level system opens unique opportunities for practical implementation of superconducting qubits, so-called Josephson flux qubits [291]. The issue of dephasing due to interaction with an external (dissipative) environment is absolutely crucial for such structures.

This chapter will be devoted to the analysis of the effect of dissipation on quantum properties of double-well systems.

5.1 Thermodynamics of Double-Well Systems with Ohmic Dissipation

Let us consider a superconducting ring interrupted by a Josephson junction, as shown in Fig. 5.2. As before, we will assume that the junction is shunted by an Ohmic resistor and, hence, one can again describe quantum dynamics of the system by means of the Caldeira–Leggett effective action (1.94)

$$S_{\text{eff}} = \int_{-\hbar\beta/2}^{\hbar\beta/2} d\tau \left[\frac{m\dot{\varphi}^2}{2} + U(\varphi) + \frac{\eta}{4\pi} \int_{-\infty}^{\infty} d\tau' \left(\frac{\varphi(\tau) - \varphi(\tau')}{\tau - \tau'} \right)^2 \right] \tag{5.1}$$

where the Josephson phase variable φ needs to be periodically continued outside the interval $-\hbar\beta/2 \leq \tau < \hbar\beta/2$. The potential energy $U(\varphi)$ now includes the Josephson and magnetic contributions

$$U(\varphi) = -E_J \cos\varphi + \frac{\Phi_0^2}{8\pi^2 L}(\varphi - \varphi_x)^2 \tag{5.2}$$

where, as before, $\Phi_0 = hc/2e$ is the superconducting magnetic flux quantum and we defined the phase $\varphi_x = 2\pi\Phi_x/\Phi_0$ proportional to the external magnetic flux Φ_x piercing the superconducting ring with inductance L. While at most values of φ_x the ground state of the system is non-degenerate, it is obvious from Eq. (5.2) that for

$$\gamma = 4\pi^2 E_J L/\Phi_0^2 > 1 \tag{5.3}$$

and particular values $\varphi_x = \pi(2n+1)$, the potential $U(\varphi)$ describes doubly degenerate ground state (see Fig. 5.3a). Below in this chapter, we will be interested only in the situation of (almost) degenerate double-well potential. Provided the parameter γ remains just slightly bigger than unity, the potential (5.2) has only two minima at $\varphi = \varphi_\pm$ determined from the equation

$$\varphi_\pm - \varphi_x + \gamma \sin\varphi_\pm = 0 \tag{5.4}$$

with $\varphi_x = \pi(2n+1) \neq \varphi_\pm$. For $\gamma \gg 1$, the total number of minima depends on a particular value of γ and, in any case, is bigger than two. However, also in this case, an effective double-well potential remains a good approximation provided one is merely interested in low-temperature and low-energy properties of our quantum system. In this case, only low-energy levels play a significant role, while high-energy levels will be of little importance. In the limit

$$E_J \gg E_C \tag{5.5}$$

(only which will be considered below in this chapter), the potential barrier between two neighboring potential minima is high, and each potential well remains parabolic sufficiently close to its bottom. Hence, the low lying energy levels in these wells are those of a harmonic oscillator with frequency $\omega_0 \simeq \sqrt{E_J E_C}/\hbar$. Then under the condition

$$T, \hbar\omega_0 \ll \Phi_0^2/L = 4\pi^2 E_J/\gamma \tag{5.6}$$

all energy levels in higher potential wells can be neglected, and the potential (5.2) effectively reduces to the double-well one again.

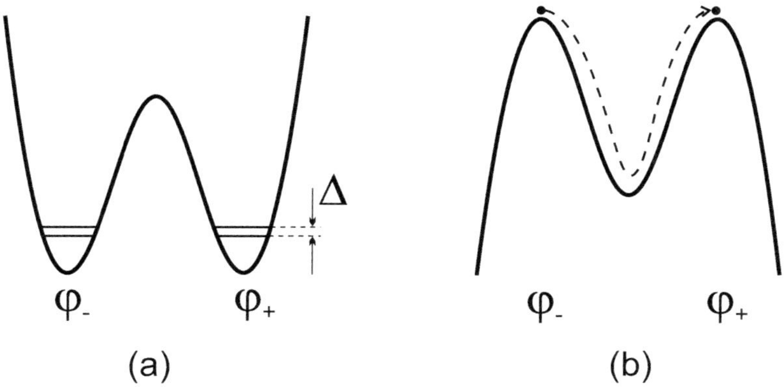

Figure 5.3 (a) Splitting of the energy levels in the double-well potential. (b) Instanton trajectory in the inverted double-well potential.

5.1.1 Instanton Technique

Let us first adopt the instanton technique already developed in Chapter 4 to the case of a degenerate double-well potential with minima located at $\varphi = \varphi_{\pm}$. For definiteness, we will choose $\varphi_x = \pi$. Then, e.g., for $\gamma \gg 1$ in the leading approximation in $1/\gamma$, we have $\varphi_- \simeq 0$ and $\varphi_+ \simeq 2\pi$.

As before, let us represent the grand partition function of our system via the path integral

$$\mathcal{Z} \sim \int d\varphi_0 \int_{\varphi_0}^{\varphi_0} \mathrm{D}\varphi \exp\left(-S_{\text{eff}}[\varphi(\tau)]/\hbar\right) \tag{5.7}$$

where S_{eff} is defined in Eqs. (5.1) and (5.2). Under the condition (5.5), this path integral can be evaluated semiclassically. For this purpose, one should again determine all relevant saddle point trajectories from the equation

$$\delta S_{\text{eff}}/\delta\varphi = 0 \tag{5.8}$$

evaluate the ratio of fluctuation determinants for each of these paths, and then sum up all these contributions to the partition function $\mathcal{Z}$.

We first demonstrate how to accomplish this calculation in the absence of dissipation, i.e., in the limit $\eta \to 0$. Just like in our previous examples, in this limit Eq. (5.8) simply coincides with the classical

Newton equation of motion for a particle in the inverted double-well potential (see Fig. 5.3b), i.e.,

$$\frac{1}{E_C}\frac{d^2\varphi}{d\tau^2} = E_J(\sin\varphi + (\varphi - \pi)/\gamma) \tag{5.9}$$

A trivial solution is

$$\varphi(\tau) = \varphi_- \tag{5.10}$$

i.e., the Josephson particle always remains on the top of the left "hill" and never moves down. Evaluating the contribution of Gaussian fluctuations around this trivial saddle point, as before, we arrive at the following contribution to the partition function $\mathcal{Z}$:

$$N(\det G_0^{-1})^{-1/2} = N\prod_n [\omega_n^2/E_C + \tilde{E}_J]^{-1/2}, \quad \tilde{E}_J = E_J(\cos\varphi_- + 1/\gamma) \tag{5.11}$$

where N is a normalization constant and $\omega_n = 2\pi nT/\hbar$ with $n = 0, \pm 1, \pm 2, \ldots$ are Bose–Matsubara frequencies. In fact, the combination (5.11) is identified as the diagonal element $\rho(\varphi_-, \varphi_-)$ of the density matrix for a harmonic oscillator with frequency $\omega_0 = \sqrt{8\tilde{E}_J E_C}/\hbar$. The full expression for this density matrix is well known and was already encountered in Chapter 1. For our later purposes, here we recollect the diagonal part of this matrix. In our present notations, it reads

$$\rho(\varphi_0, \varphi_0) = \sqrt{\frac{m\omega_0}{2\pi\hbar\sinh(\hbar\omega_0/T)}} \times \exp\left[-\frac{m\omega_0}{\hbar}(\varphi_0 - \varphi_-)^2\tanh\left(\frac{\hbar\omega_0}{T}\right)\right] \tag{5.12}$$

Obviously, for $\varphi_0 = \varphi_-$, this expression simplifies and yields $\rho(\varphi_-, \varphi_-)$, which in a way can also be considered a definition of the normalization factor N.

A nontrivial solution to Eq. (5.9) $\tilde{\varphi}(\tau)$ describes tunneling of a particle between two potential minima. Similarly to the case of a quantum decay considered in the previous chapter, the particle first moves downhill from the top point $\tilde{\varphi}(-\infty) = \varphi_-$ gaining the kinetic energy, passes the potential minimum $\tilde{\varphi} = \pi$ at $\tau = 0$, and then it climbs uphill reaching the highest point $\tilde{\varphi} = \varphi_+$ at $\tau = \infty$. An important difference between this instanton or kink solution $\tilde{\varphi}(\tau)$ and the bounce trajectory discussed in Chapter 4 is that here the particle is not bounced back immediately after it reaches the top of the second "hill." Rather it can remain at rest there for a long time. Substituting the instanton trajectory $\tilde{\varphi}(\tau)$ into the action, we immediately conclude that the corresponding contribution is proportional to the factor $\exp(-S_{\text{eff}}[\tilde{\varphi}(\tau)]/\hbar)$, which remains exponentially small in the semiclassical approximation controlled by the inequality (5.5).

In order to analyze the effect of fluctuations, we again expand the action around the kink trajectory to the second order in $\delta\varphi(\tau) \equiv \varphi(\tau) - \tilde{\varphi}(\tau)$. Then we obtain

$$\int D\delta\varphi \exp\left(-\frac{S_{\text{eff}}[\tilde{\varphi}(\tau)]}{\hbar} - \frac{1}{2\hbar}\int_{-\hbar\beta/2}^{\hbar\beta/2} d\tau \int_{-\hbar\beta/2}^{\hbar\beta/2} d\tau' G^{-1}(\tau, \tau')\delta\varphi(\tau)\delta\varphi(\tau')\right) \tag{5.13}$$

where

$$G^{-1}(\tau, \tau') = \frac{\delta^2 S_{\text{eff}}}{\delta\tilde{\varphi}(\tau)\delta\tilde{\varphi}(\tau')} \tag{5.14}$$

Further calculation is carried out analogously to that in Chapter 4 with only one important difference. Evaluating the eigenvalues Λ_n of the operator (5.14), we observe that the lowest eigenvalue equals to zero, $\Lambda_0 = 0$, while all other eigenvalues $\Lambda_{n>0}$ are positive. Thus, in contrast to the decay problem analyzed in Chapter 4, here the operator (5.14) does not posses a negative eigenvalue. This fact is easily understandable since the tunneling process between two degenerate potential minima does not correspond to any decay. The meaning of the zero mode remains exactly the same as before: It corresponds to shifts of the instanton trajectory in time without any change in the effective action value $S_{\text{eff}}[\tilde{\varphi}(\tau)]$. Turning again the integral over Λ_0 into that over the collective coordinate τ_0 (coordinate of the instanton "center of mass") and performing a standard Gaussian integration over all remaining eigenvalues, we arrive at the following result for the integral (5.13):

$$\rho(\varphi_-, \varphi_-)\frac{\Delta}{2\hbar}\int_{-\hbar\beta/2}^{\hbar\beta/2} d\tau_0 \tag{5.15}$$

where

$$\frac{\Delta}{2} = \sqrt{\frac{S_{\text{eff}}[\tilde{\varphi}(\tau)]}{2\pi\hbar}}\left(\frac{\det G_0^{-1}}{\det' G^{-1}}\right)^{1/2} \exp\left(-\frac{S_{\text{eff}}[\tilde{\varphi}(\tau)]}{\hbar}\right) \tag{5.16}$$

defines the tunneling amplitude between two potential wells and

$$(\det' G^{-1})^{-1/2} = \prod_{n=1}^{\infty} \frac{1}{\Lambda_n^{1/2}} \tag{5.17}$$

is the determinant of G^{-1} with excluded zero mode. Note that in Eq. (5.16), we made use of the expression for the square of the normalization factor

$$\int_{-\hbar\beta/2}^{\hbar\beta/2} d\tau \left(\frac{d\tilde{\varphi}}{d\tau}\right)^2 = S_{\text{eff}}[\tilde{\varphi}(\tau)]/2m \tag{5.18}$$

which holds only in the dissipativeless limit.

In some cases, the tunneling amplitude (5.16) can be evaluated analytically in a closed form. For example, in the limit $\gamma^{-1} \to 0$, the instanton trajectory takes the form

$$\tilde{\varphi}(\tau) = 4 \arctan \exp(\omega_0 \tau) \tag{5.19}$$

and yields $S_{\text{eff}}[\tilde{\varphi}(\tau)]/\hbar = (8E_J/E_C)^{1/2}$. The ratio of determinants in (5.16) can also be found analytically in this case. The corresponding calculation is completely analogous to that carried out in a dissipativeless case in Chapter 4. Here we quote only the final result

$$\Delta = 16 \left(\frac{E_J E_C}{\pi}\right)^{1/2} \left(\frac{E_J}{2E_C}\right)^{1/2} \exp\left[-\left(\frac{8E_J}{E_C}\right)^{1/2}\right] \tag{5.20}$$

For finite values of γ, the expression for Δ also depends on this parameter, but remains qualitatively similar to Eq. (5.20). The explicit form of the tunneling amplitude will not be significant for our further analysis in this chapter; of importance is only that under the condition (5.5) Δ remains exponentially small and, hence, the inequality

$$\Delta \ll \hbar\omega_0 \tag{5.21}$$

is fulfilled throughout the whole calculation.

Having evaluated the fluctuation contribution around the saddle point path $\tilde{\varphi}(\tau)$, we realize that the corresponding term (5.15) cannot contribute to the partition function $\mathcal{Z}$ because of the boundary conditions in the path integral (5.7): All trajectories contributing to this integral should start and end in one point $\varphi = \varphi_0$, but $\tilde{\varphi}(\tau)$ starts at $\varphi = \varphi_-$ and ends at $\varphi = \varphi_+$. In other words, the partition function $\mathcal{Z}$ is expressed in terms of the matrix element $\langle\varphi_-|e^{-\beta\hat{H}}|\varphi_-\rangle$, while the term (5.15) contributes to a different matrix element $\langle\varphi_-|e^{-\beta\hat{H}}|\varphi_+\rangle$ describing tunneling between the wells. Although the latter matrix element can also be evaluated with the aid of our instanton technique, here we will concentrate on the partition function $\mathcal{Z}$.

In order to construct the simplest configuration that describes tunneling between the wells and contributes to the path integral (5.7), it is necessary to consider two tunneling events in a row: the particle first tunnels from the left well to the right one and then tunnels back to the left well. This process corresponds to two instantons or, better to say, a "kink–anti-kink" pair

$$\tilde{\varphi}(\tau-\tau_1)-\tilde{\varphi}(\tau-\tau_2),\qquad \tau_2>\tau_1 \tag{5.22}$$

forming a kind of bounce with two zero modes with collective coordinates τ_1 and τ_2. This is in contrast to the bounce trajectory analyzed in Chapter 4, which describes quantum decay of a metastable state and has one negative and one zero mode. Here no metastability exists, and the particle returns to the initial well after making a detour to a neighboring one. Having in mind that the bounce (5.22) is not "rigid," i.e., the time difference between two tunneling events $\tau_2-\tau_1$ can take any value between zero and $\hbar\beta$, and repeating the same analysis of fluctuations, we arrive at the term

$$\rho(\varphi_-,\varphi_-)\left(\frac{\Delta}{2\hbar}\right)^2\int_{-\hbar\beta/2}^{\hbar\beta/2}d\tau_2\int_{-\hbar\beta/2}^{\tau_2}d\tau_1 \tag{5.23}$$

which contributes to the partition function $\mathcal{Z}$. Likewise, trajectories describing $2k$ tunneling events back and forth between the wells (or k bounces with $2k$ zero modes) should also contribute to $\mathcal{Z}$. Collecting all such contributions, we finally obtain

$$\begin{aligned}\langle\varphi_-|\cdots|\varphi_-\rangle &= \rho(\varphi_-,\varphi_-)\sum_{k=0}^{\infty}\left(\frac{\Delta}{2\hbar}\right)^{2k}\\ &\times\int_{-\hbar\beta/2}^{\hbar\beta/2}d\tau_{2k}\int_{-\hbar\beta/2}^{\tau_{2k}}d\tau_{2k-1}\ldots\int_{-\hbar\beta/2}^{\tau_2}d\tau_1\end{aligned} \tag{5.24}$$

The integrals over all center-of-mass coordinates in the k-th term in the sum (5.24) obviously yield the factor $(\hbar\beta)^{2k}/2k!$, and hence the sum just reduces to $\cosh(\Delta/2T)$.

Finally, we should bear in mind that the evaluated matrix element (5.24) is not yet equivalent to the grand partition function (5.7): In order to obtain the latter, one should evaluate $\langle\varphi_0|e^{-\beta\hat{H}}|\varphi_0\rangle$ also for $\varphi_0 \neq \varphi_-$ and then integrate over φ_0. Fortunately, under the condition (5.5) and in the low-temperature limit

$$T \ll \hbar\omega_0 \tag{5.25}$$

the latter steps can be easily accomplished since the particle density matrix (5.12) is sharply peaked at $\varphi_0 = \varphi_-$. In this case, the contribution from fluctuations near the bottom of the well with a good accuracy gets factorized from that describing tunneling to a neighboring well. Then in order to find $\mathcal{Z}$, it suffices to substitute $\rho(\varphi_-, \varphi_-) \to \rho(\varphi_0, \varphi_0)$ in Eq. (5.24). Integrating over φ_0 and employing Eq. (5.12), one readily finds

$$\mathcal{Z} \approx \mathcal{Z}_0 \cosh(\Delta/2T) \tag{5.26}$$

where

$$\mathcal{Z}_0 = \int d\varphi_0 \rho(\varphi_0, \varphi_0) = \left[2\sinh\left(\frac{\hbar\omega_0}{2T}\right)\right]^{-1} \tag{5.27}$$

is the grand partition function of a harmonic oscillator with frequency ω_0. In fact, under the condition $T \ll \hbar\omega_0$, only the lowest energy level of this oscillator contributes to $\mathcal{Z}_0$ and we obtain

$$\mathcal{Z} \approx \frac{1}{2}\left[\exp\left(-\frac{\hbar\omega_0 - \Delta}{2T}\right) + \exp\left(-\frac{\hbar\omega_0 + \Delta}{2T}\right)\right] \tag{5.28}$$

This is the standard quantum mechanical result demonstrating that in the low-temperature limit, the partition function for a particle in a degenerate double-well potential is determined by contributions from two states with energies $\hbar\omega_0/2 \pm \Delta/2$. In other words, at low enough T, our problem can effectively be reduced to that of a two-level system. Level splitting Δ is determined by quantum mechanical tunneling, which transfers the wave function from one well to another and "mixes" the ground states in these wells. As a result, instead of two independent states with ground-state energies $\hbar\omega_0/2$ and wave functions $\Psi_-(\varphi)$ and $\Psi_+(\varphi)$ localized in each of the two wells, one obtains a unique ground state with lower energy $\hbar\omega_0/2 - \Delta/2$ described by symmetric wave function $(\Psi_-(\varphi) + \Psi_+(\varphi))/\sqrt{2}$ and the first excited state with energy $\hbar\omega_0/2 + \Delta/2$ and antisymmetric wave function $(\Psi_-(\varphi) - \Psi_+(\varphi))/\sqrt{2}$.

5.1.2 Effect of Linear Ohmic Dissipation, Localization Phase Transition

Now let us see how these results get modified provided the particle is coupled to a dissipative bath. In what follows we will again restrict ourselves to the case of linear Ohmic dissipation and assume effective viscosity η to be sufficiently small, meaning that dissipation does not significantly affect the form of instantons determined above. In other words, we will treat the dissipative part of the action (5.1) perturbatively in which case its effect is captured simply by substituting all possible (multi-)instanton trajectories into the dissipative term. Repeating the same analysis as in Section 5.1.1, we arrive at the following expression for the grand partition function [66, 85]

$$\mathcal{Z} \sim \sum_{k=0}^{\infty} \left(\frac{y}{2}\right)^{2k} \int_{-\hbar\beta/2}^{\hbar\beta/2} \frac{d\tau_{2k}}{\tau_c} \int_{-\hbar\beta/2}^{\tau_{2k}-\tau_c} \frac{d\tau_{2k-1}}{\tau_c} \cdots \int_{-\hbar\beta/2}^{\tau_2-\tau_c} \frac{d\tau_1}{d\tau_c} \times \exp\left[2K \sum_{i>j} (-1)^{i-j} \Lambda(\tau_i - \tau_j)\right] \tag{5.29}$$

where we defined $y = \Delta\tau_c/\hbar$,

$$\Lambda(\tau) = \ln\left(\frac{|\sin(\pi T \tau)|}{\pi T \tau_c}\right) \tag{5.30}$$

and $\tau_c \sim \omega_0^{-1}$ is the instanton core size. The parameter K is proportional to viscosity η, i.e.,

$$K = \frac{\eta(\varphi_+ - \varphi_-)^2}{2\pi\hbar} \tag{5.31}$$

In the absence of dissipation ($K \to 0$), Eq. (5.29) again yields standard quantum mechanical results considered above.

The partition function (5.29) is formally equivalent to that for a one-dimensional Coulomb gas of logarithmically interacting hard particles with size τ_c and fugacity y. The parameter τ_c also defines an effective cutoff of the interaction between charges. The situation is illustrated in Fig. 5.4. Figure 5.4a shows a typical multi-instanton trajectory consisting of kinks and anti-kinks, which correspond to successive tunneling between the two wells. Under the conditions (5.21) and (5.25) adopted here, the average distance

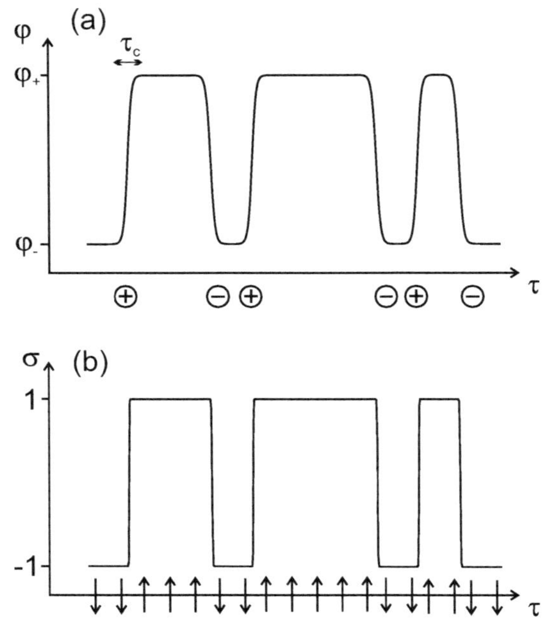

Figure 5.4 (a) A multi-instanton trajectory and the corresponding topological charges. (b) Spin representation of a multi-instanton trajectory.

between instantons strongly exceeds their typical size $\sim \tau_c$, i.e., in this case, we can talk about dilute gas of instantons. We can also identify each instanton (anti-instanton) with a positively (negatively) charged particle, which are ordered as shown in Fig. 5.4a. As each configuration that contributes to the partition function $\mathcal{Z}$ contains equal number of instantons and anti-instantons, here we will only have to deal with neutral charge configurations. In addition, according to Eq. (5.29), the charges interact logarithmically in a way that opposite charges attract each other while identical charges repel each other. It follows immediately that in the presence of interaction, the energy of each charge configuration reaches its minimum provided the neighboring (opposite) charges form closed pairs. This effect becomes more pronounced with increasing interaction strength K. On the other hand, in the absence of interaction, the energy of any configuration of charges does not depend on the distance between them.

Already this simple qualitative observation suggests that interaction with a dissipative bath can significantly alter physical properties of a quantum mechanical particle in a double-well potential. Indeed, while for $K \to 0$ both wells are equivalent and the particle tunnels back and forth with equal probability, possible formation of kink–anti-kink pairs at nonzero dissipation would effectively imply breaking of the symmetry between the wells. This is because in this case, each tunneling event out of one well into the other is accompanied by immediate tunneling back after which the particle keeps staying in the initial well for a long time. Hence, the probability to find the particle in one of the wells is bigger than in the other one implying symmetry breaking between them.

Quantitative analysis of the many-body problem (5.29) is, in general, a complicated task. Fortunately, a powerful approach allows to capture essential properties of the system without explicit evaluation of the partition function $\mathcal{Z}$. Before turning to this approach, let us note that the partition function of a logarithmically interacting Coulomb gas (5.29) is known to describe not only the problem discussed here but also a number of other important physical problems, such as anisotropic Kondo model and one-dimensional Ising model with inverse-square-law interactions. In addition, as we will discuss a bit later in this chapter, Eq. (5.29) applies for the so-called spin-boson model, i.e., the model describing quantum dynamics of a spin interacting with a phonon bath. A detailed discussion of the relation between all these models can be found elsewhere [264, 400]; here we only schematically illustrate the equivalence of these models in Fig. 5.4b. Of importance is that in order to analyze our model, one can make use of the technique pioneered before in a different physical context [18].

The essence of the method is that one can split the integral over each collective coordinate τ_i in Eq. (5.29) into two, namely from $-\hbar\beta/2$ to $\tau_{i+1} - (\tau_c + \delta\tau_c)$ and from $\tau_{i+1} - (\tau_c + \delta\tau_c)$ to $\tau_{i+1} - \tau_c$. Assuming $\delta\tau_c$ to be sufficiently small, one can explicitly perform the second of these integrals for all τ_i. After this step, provided $y \ll 1$, one arrives at the partition function of the same form (5.29) but with modified parameters $y \to \tilde{y} = y + \delta y$, $K \to \tilde{K} = K + \delta K$ and $\tau_c \to \tilde{\tau}_c = \tau_c + \delta\tau_c$. Repeating the same procedure again, one generates the partition function (5.29) with yet again modified

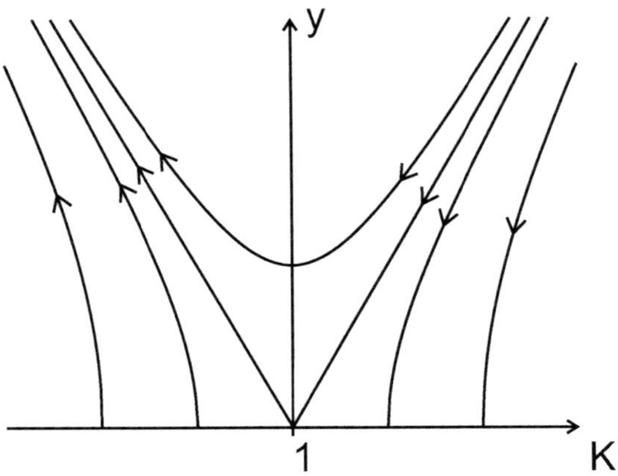

Figure 5.5 Renormalization group flow as predicted by Eqs. (5.32).

fugacity, interaction strength, and cutoff. Having repeated the above procedure many times, one arrives at the so-called renormalization group (RG) recursion relations [18]

$$d\tilde{y}/d\ln\tilde{\tau}_c = \tilde{y}(1-\tilde{K})+O(\tilde{y}^3), \quad d\tilde{K}/d\ln\tilde{\tau}_c = -\tilde{K}\tilde{y}^2+O(\tilde{y}^4) \quad (5.32)$$

demonstrating how the initial charge fugacity y and the interaction strength K get modified upon changing the cutoff parameter τ_c. One can also define the renormalized level splitting

$$\Delta_r = \hbar\tilde{y}/\tilde{\tau}_c \quad (5.33)$$

which sets an important energy scale in our problem.

Equations (5.32) generate RG flows in the $(\tilde{y}, \tilde{K})$ plane schematically depicted in Fig. 5.5. It is easy to observe that there exists a separatrix $\tilde{y}^2/2 = \tilde{K} - 1 - \ln\tilde{K}$ between flows toward a line of stable fix points with $\tilde{K} \geq 1$ and $\tilde{y} = 0$ and flows with $\tilde{K} < 1$ corresponding to increasing $\tilde{y}$ in the course of renormalization (i.e., with increasing cutoff τ_c) eventually driving the parameters outside the applicability range of RG equations (5.32). In this latter regime, renormalization should eventually be stopped as soon as $\tilde{y}$ reaches the value of order one beyond which Eqs. (5.32) become inaccurate. At not very low temperatures, renormalization should be stopped in both these regimes provided the renormalized cutoff reaches the value $\tilde{\tau}_c \sim \hbar\beta$.

Let us first consider the limit $T \to 0$. In this case for $K < 1$, an essential scale can be identified by neglecting the renormalization of K, i.e., simply by dropping the second Eq. (5.32) while solving the first one. Then we obtain

$$\tilde{y}/y = (\tilde{\tau}_c/\tau_c)^{1-K} \tag{5.34}$$

Setting $\tilde{y} \sim 1$ in Eq. (5.34), we estimate the cutoff value $\tilde{\tau}_c \sim \tau_c y^{1/(K-1)}$ at which renormalization should be stopped. Substituting these parameters into Eq. (5.33), we arrive at the following estimate for the renormalized level splitting

$$\Delta_r = \Delta \left(\frac{\Delta}{\hbar\omega_0} \right)^{\frac{K}{1-K}}, \quad K < 1 \tag{5.35}$$

It is easy to verify that taking into account Eq. (5.32) does not qualitatively alter this estimate.

Equation (5.35) is a nontrivial result, which sets new energy scale in our problem, showing that interaction with a dissipative environment reduces an effective tunneling amplitude between the wells. This reduction effect increases with increasing K. Most remarkably, for $K \to 1$, the value Δ_r (5.35) tends to zero, indicating that inter-well particle motion should be prohibited at any $K \geq 1$. This conclusion is fully consistent with the RG flow, which follows from Eqs. (5.32): As we already pointed out, for $K \geq 1$ and $T \to 0$, the fugacity $\tilde{y}$ gets renormalized to zero, meaning that

$$\Delta_r = 0, \quad K \geq 1 \tag{5.36}$$

Equations (5.35) and (5.36) demonstrate that Ohmic dissipation can qualitatively change the ground-state properties of a degenerate double-well system. As long as dissipation remains sufficiently weak, $K < 1$, the particle keeps traveling between the wells, though with a reduced tunneling amplitude Δ_r. Hence, the symmetry between the left and the right wells remains preserved, i.e., the particle can be found in each of the wells with equal probability. For stronger dissipation, $K > 1$, however, no particle transfer between the wells is anymore possible implying *spontaneous symmetry breaking* between them. In other words, at $T \to 0$, Ohmic dissipation yields quantum phase transition (QPT) in our system, which occurs at the particular value of the interaction parameter $K = 1$ [66, 85].

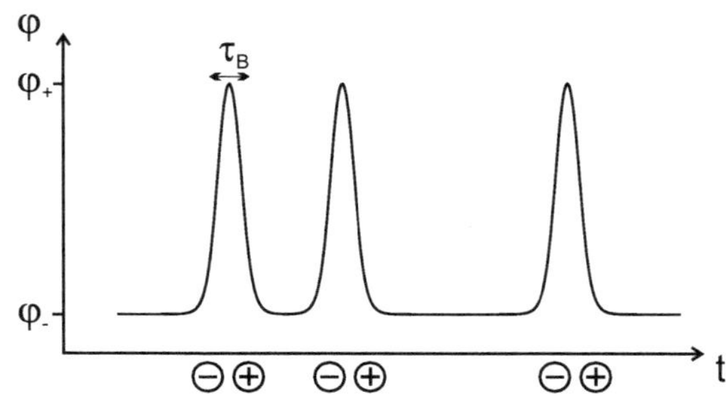

Figure 5.6 Multi-instanton trajectory with bound kink–anti-kink pairs (bounces).

This QPT implies particle localization: Being initially put into one of the wells, at $T = 0$ and $K \geq 1$, the particle will stay there forever with no chance to move to another well.

In order to understand underlying physics behind this remarkable result, we first point out that the above effect of particle localization can, by no means, be naively interpreted as a consequence of "too strong friction," which simply does not let the particle move out of the well. Just on the contrary, under the condition (5.5), the critical values $K \approx 1$ are reached while effective viscosity still remains sufficiently low $\eta \ll m\omega_0$. In this case, dissipation can only weakly affect the particle motion within one well. Also it would not be correct to say that tunneling processes are fully suppressed for $K > 1$. Rather what happens is that for such values of K, each tunneling event from one well to another is accompanied by immediate tunneling back. After that the particle can stay in the initial well for a long time until the next tunneling event occurs, which is again accompanied by immediate tunneling event in the opposite direction, i.e., back to the initial well. As a result, the particle stays in this well essentially all time, while the probability to find it in another well is negligible.

This situation is illustrated in Fig. 5.6. For $K > 1$, attractive interaction between kinks and anti-kinks becomes strong enough to bind them in close pairs. All other configurations are strongly suppressed by dissipation. In a way in this case, we are again dealing with rigid bounces now formed due to interaction with a dissipative

Caldeira–Leggett bath. In contrast to the situation considered in Chapter 4, here the contribution from these bounce configurations to the partition function is purely real, i.e., no decay occurs. The latter observation, as we already discussed, just follows from the fact that our double-well potential is degenerate, i.e., its two minima are reached at exactly the same energy.

We also emphasize that the above physical picture of a quantum dissipative phase transition applies only in the zero-temperature limit $T \to 0$. At nonzero temperature, an effective tunneling amplitude between the wells Δ_r differs from zero at any dissipation strength K. In order to estimate the value Δ_r in this regime, we again ignore renormalization of K and resolve only the first RG equation (5.32) now stopping renormalization at the cutoff value $\tilde{\tau}_c \sim \hbar\beta$. Then from Eqs. (5.33) and (5.34), we obtain

$$\Delta_r(T) = \Delta \left(\frac{T}{\hbar\omega_0}\right)^K \tag{5.37}$$

i.e., also at nonzero T, an effective inter-well tunneling amplitude gets reduced by dissipation. As long as $T > 0$, tunneling between the wells persists making the equilibrium probability to find the particle in each of the two wells equal to 1/2. For $K \geq 1$, Eq. (5.37) remains valid down to zero temperature, illustrating again full suppression of Δ_r at $T \to 0$. On the other hand, for $K < 1$ Eq. (5.37) applies only at sufficiently high temperatures $T \gtrsim \Delta_r$ (with Δ_r defined in Eq. (5.35)), while at lower $T \lesssim \Delta_r$, the renormalized level splitting reduces to the zero-temperature result (5.35).

5.1.3 Combined Effect of Dissipation and External Bias

It is clear that symmetry between the wells can also be broken in a different (trivial) way simply by slightly shifting the external flux value Φ_x away from $\Phi_0(n+1/2)$. In this case, the degeneracy is lifted and the two potential minima acquire some energy difference, which will be denoted as ϵ (see Fig. 5.7).

In what follows, we will restrict our consideration to small values

$$\epsilon \ll \hbar\omega_0 \tag{5.38}$$

In this case, distortion of our double-well potential remains weak and we can repeat the above instanton analysis of the grand partition

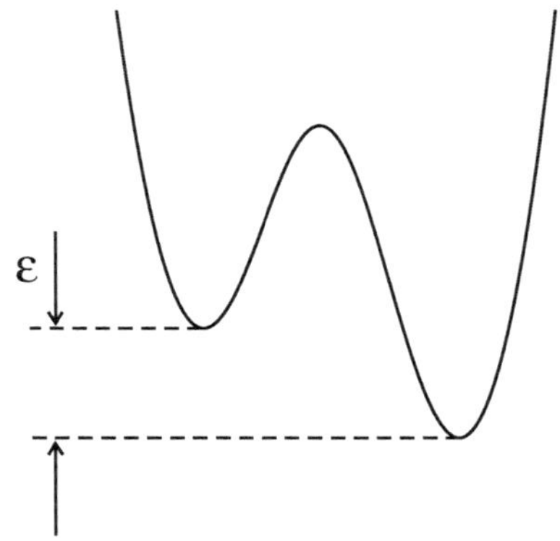

Figure 5.7 Tilted double-well potential.

function $\mathcal{Z}$ now also taking into account a small energy mismatch ϵ between the two wells. As a result, we obtain

$$\mathcal{Z} \sim \sum_{k=0}^{\infty} \left(\frac{y}{2}\right)^{2k} \int_{-\hbar\beta/2}^{\hbar\beta/2} \frac{d\tau_{2k}}{\tau_c} \int_{-\hbar\beta/2}^{\tau_{2k}-\tau_c} \frac{d\tau_{2k-1}}{\tau_c} \cdots \int_{-\hbar\beta/2}^{\tau_2-\tau_c} \frac{d\tau_1}{d\tau_c}$$
$$\times \exp\left[2K\sum_{i>j}(-1)^{i-j}\Lambda(\tau_i-\tau_j)+\epsilon\sum_i(-1)^i\tau_i\right] \quad (5.39)$$

Obviously, this expression reduces to Eq. (5.29) in the limit $\epsilon \to 0$. For nonzero values of ϵ, additional terms appear in the exponent describing extra interactions in our effective one-dimensional Coulomb gas of charges. In this case, the system behavior, in general, turns out to be entirely different from that discussed above for a degenerate situation.

In order to proceed, one can again employ RG analysis of the Coulomb gas partition function (5.39). We again arrive at RG Eqs. (5.32), which should now be supplemented by the third equation describing renormalization of the parameter ϵ. Keeping only the leading orders in $\tilde{y}$, one finds [85]

$$d\tilde{\epsilon}/d\ln\tilde{\tau}_c = \tilde{\epsilon}(1-\tilde{y}^2/2) \quad (5.40)$$

where we defined $\tilde{\epsilon} = \epsilon\tau_c$. Initially, this parameter is small, $\tilde{\epsilon} \ll 1$, due to the condition (5.38). According to Eq. (5.40), $\tilde{\epsilon}$ grows upon scaling and one can define a new scale τ_ϵ set by the condition $\tilde{\epsilon} \sim 1$.

Assuming that $\tilde{y}$ remains much smaller than unity, one readily finds $\tau_\epsilon \sim 1/\epsilon$. At $\tau \sim \tau_\epsilon$, our scaling breaks down and should be stopped. As a result, similar to Eq. (5.37), we arrive at the renormalized tunneling amplitude

$$\Delta_r(\epsilon) = \Delta \left(\frac{\epsilon}{\hbar\omega_0} \right)^K \tag{5.41}$$

For $K > 1$, this result is valid as long as $\epsilon > T$, while for $K < 1$, one should also verify that τ_ϵ remains smaller than the scale $\sim\tau_c y^{1/(K-1)}$ at which the renormalized fugacity $\tilde{y}$ becomes of order one. The latter condition is equivalent to $\epsilon > \Delta_r$ (with Δ_r defined in Eq. (5.35)). At smaller $\epsilon \lesssim \Delta_r$ and $K < 1$, the renormalized level splitting is again set by Eq. (5.35) and the energy difference ϵ has little effect on the properties of our system.

Physically non-vanishing effective tunneling amplitude (5.41) implies that even at $T \to 0$, the system eventually reaches its non-degenerate ground state in the lower minimum no matter from which well it initially started. In other words, the state in the upper well is metastable and decays to the lower well. In order to evaluate the corresponding decay rate, we notice that for nonzero ϵ the formal expression for the free energy F defined by the partition function (5.39) diverges and acquires an imaginary part. Hence, we can employ the same method as in Chapter 4 and find the decay rate by means of Eq. (4.10).

Let us formally represent all instanton configurations in Eq. (5.39) as a collection of kink–anti-kink pairs (bounces), as shown in Fig. 5.6. If we assume that the average size of each such bounce τ_B is small as compared to the average distance between them, we can neglect inter-bounce interactions and keep only logarithmic interactions inside each bounce. Within this approximation in Eq. (5.39), one can trivially integrate over "center-of-mass" coordinates of all bounces and get

$$\mathcal{Z} = \exp(-F/T), \quad F = \frac{\Delta^2}{4\hbar} \int_0^\infty d\tau \exp\left[\epsilon\tau/\hbar - 2K\Lambda(\tau)\right] \tag{5.42}$$

It is obvious that in general, the expression (5.42) for the free energy F diverges. This formal divergence, as we already know, signals instability and decay of a metastable state. As before, the imaginary part of F can be defined by means of a proper analytic continuation procedure [400, 401]. To this end, let us distort the contour of integration at the saddle point

$$\tau = (\hbar/\pi T)\text{arccot}(\epsilon/2\pi KT) \tag{5.43}$$

into the upper half-plane along the direction of the steepest descent. Making use of the analytic properties of the exponent in the second Eq. (5.42), we define

$$\text{Im}F = \frac{\text{i}\Delta^2}{8\hbar}\int_C d\tau \exp\left[\epsilon\tau/\hbar - 2K\Lambda(\tau)\right] \tag{5.44}$$

where the contour C starts at $\tau = -\infty - \text{i}0$, follows the negative part of the real axis, circles the origin, and returns to $\tau = -\infty + \text{i}0$ along the positive part of the real axis. The integral in Eq. (5.44) can be evaluated in a straightforward manner. Substituting the result into Eq. (4.10), we obtain [400, 401]

$$\Gamma = \frac{\Delta^2}{4\hbar\omega_0}\left(\frac{\hbar\omega_0}{2\pi T}\right)^{1-2K} \exp\left(\frac{\epsilon}{2T}\right)\frac{|\Gamma(K + \text{i}\epsilon/(2\pi T))|^2}{\Gamma(2K)} \tag{5.45}$$

where, as before, $\Gamma(x)$ is the Euler gamma function. Note that this general expression applies both for positive and negative values of ϵ thereby defining the decay rates for upper and lower energy wells, respectively, Γ_+ and Γ_-. Obviously, these two rates should obey the detailed balance relation

$$\Gamma_- = \exp\left(-\frac{\epsilon}{T}\right)\Gamma_+ \tag{5.46}$$

which trivially follows from Eq. (5.45).

At $T \to 0$ and $\epsilon > 0$, Eq. (5.45) reduces to the expression

$$\Gamma_+ = \frac{\pi}{2\Gamma(2K)}\frac{\Delta^2}{\hbar\omega_0}\left(\frac{\epsilon}{\hbar\omega_0}\right)^{2K-1} \tag{5.47}$$

which defines the decay rate from upper to lower well in the zero-temperature limit. This decay is possible due to the presence of dissipation: The particle lowers its energy and the released energy is taken by a dissipative bath. Obviously, tunneling in the opposite direction (from lower to upper well) is not possible at $T = 0$ because in this case, the bath cannot supply the particle with extra energy. Hence, in the zero-temperature limit, one has $\Gamma_- = 0$, as also follows from the above equations.

At nonzero temperatures, tunneling can go in both directions and both rates $\Gamma_{\pm}$ differ from zero. In the limit $\epsilon \to 0$ from Eq. (5.45), one finds

$$\Gamma_+ = \Gamma_- = \frac{\sqrt{\pi}\Gamma(K)}{4\Gamma(K+1/2)} \frac{\Delta^2}{\hbar\omega_0} \left(\frac{\pi T}{\hbar\omega_0}\right)^{2K-1} \tag{5.48}$$

Let us now recall that in the beginning of our derivation, we assumed the average bounce size τ_B to remain much smaller than the average distance between them $\sim 1/\Gamma_{\pm}$. Estimating τ_B from Eq. (5.42) as

$$\tau_B \sim \max(\hbar/\epsilon, \hbar/(2\pi K T)) \tag{5.49}$$

and making use of the above results for the decay rates, we arrive at the inequality

$$\max(\epsilon, T) \gg b(K)\Delta_r \tag{5.50}$$

where Δ_r is defined in Eqs. (5.35) and (5.36) and $b(K)$ is represented by some combination of gamma functions. Obviously, for $K \geq 1$, Eq. (5.50) is obeyed at all nonzero values of T and ϵ. At $K \lesssim 1$, the function $b(K)$ is of order unity and the condition (5.50) just coincides with that already encountered within the RG analysis above. Finally, in the weak dissipation limit $K \ll 1$ in the leading approximation from Eq. (5.50), one finds

$$\epsilon \gg K^{1/2}\Delta_r, \quad T \gg \Delta_r/K \tag{5.51}$$

The physical meaning of the condition (5.50) is simple: It controls the regime of incoherent tunneling between the wells. Provided this condition is violated, coherent effects should be taken into account and the analysis of the problem becomes much more involved. Various approximate methods can be employed in order to evaluate the system partition function $\mathcal{Z}$ in a number of limiting cases, e.g., the weak dissipation limit and high-temperature limit. The corresponding calculations are well documented in the literature [264, 400] and will not be discussed here.

In a special case $K = 1/2$, it is possible to carry out an exact summation in Eqs. (5.29) and (5.39) and to obtain the exact solution to the problem. This special case is also known as the so-called Toulouse limit in the corresponding anisotropic Kondo model [264, 400]. The key simplification is achieved by transforming the logarithmic interaction terms in our Coulomb gas of alternating charges in the following manner. Consider, e.g., the configuration consisting of two kinks and two anti-kinks (or two bounces). For $K = 1/2$, the corresponding interaction term for this configuration can be rewritten with the aid of the identity

$$\begin{aligned}
&\exp[-\Lambda(\tau_2-\tau_1)-\Lambda(\tau_4-\tau_3)-\Lambda(\tau_3-\tau_2)-\Lambda(\tau_4-\tau_1)\\
&\quad+\Lambda(\tau_3-\tau_1)+\Lambda(\tau_4-\tau_2)]\\
&=\exp[-\Lambda(\tau_2-\tau_1)-\Lambda(\tau_4-\tau_3)]+\exp[-\Lambda(\tau_3-\tau_2)-\Lambda(\tau_4-\tau_1)]
\end{aligned}$$

which enables one to exactly perform all time integrations in the corresponding term Eq. (5.29). Employing the same decomposition in all other charge configurations and taking into account the contributions to the partition function from both wells in the case of a symmetric double-well system, we arrive at the following result:

$$\mathcal{Z} = 2\exp\left[\sum_{n=1}^{\infty}\frac{(-1)^{n-1}}{n!}\frac{\Delta^{2n}}{(4\hbar\omega_0 T)^n}\int_{\frac{\pi T}{\hbar\omega_0}}^{\infty} dz\frac{z^{n-1}}{\sinh z}\right] \tag{5.52}$$

which yields the partition function in the form

$$\mathcal{Z} = 2\exp\left[\int_{\frac{\pi T}{\hbar\omega_0}}^{\infty} dz\frac{1-e^{-\Delta^2 z/4\hbar\omega_0 T}}{z\sinh z}\right] \tag{5.53}$$

With some more efforts, one can also extend the procedure to the case of nonzero values of ϵ and obtain

$$\mathcal{Z} = 2\exp\left[\int_{\frac{\pi T}{\hbar\omega_0}}^{\infty} dz\frac{1-e^{-\Delta^2 z/4\hbar\omega_0 T}\cos(\epsilon z/\pi T)}{z\sinh z}\right] \tag{5.54}$$

Equation (5.54) is a direct generalization of the exact result (5.53) to the case of nonzero bias. We will make use of these results in Chapter 8.

5.2 Real-Time Dynamics

We can now turn to the description of real-time dynamics of the double-well systems in the presence of linear Ohmic dissipation, which is the primary goal of this chapter.

5.2.1 Incoherent Regime

We first assume that dissipation yields total loss of coherence and, hence, real-time dynamics of the system remains incoherent at all times. In this case, off-diagonal elements of the density matrix of our Josephson particle are suppressed, and in the low-temperature limit (5.25), it suffices to operate with time-dependent probabilities to find the particle near the bottom of the left and right wells, respectively, $W_L(t)$ and $W_R(t)$. These probabilities are formally defined via diagonal elements of the density matrix integrated in the vicinity of the bottom of the corresponding well, i.e.,

$$W_{L,R} = \int_{L,R} d\varphi \rho(\varphi, \varphi)$$

These probabilities can change in time only as a result of tunneling of the Josephson particle between the wells. Hence, we arrive at the following master equations:

$$\frac{dW_L}{dt} = -\Gamma_+ W_L + \Gamma_- W_R, \quad \frac{dW_R}{dt} = -\Gamma_- W_R + \Gamma_+ W_L \qquad (5.55)$$

which should be solved together with the normalization condition $W_L + W_R = 1$. Here $\Gamma_\pm$ are incoherent tunneling rates between the wells that we have already evaluated above; see Eqs. (5.45)–(5.48).

The solution to Eqs. (5.55) is trivially obtained. With the aid of Eq. (5.46), the probability $W_L(t)$ to find the particle in the left well is derived in the form

$$W_L(t) = \left[W_L(0) - \frac{1}{1 + e^{\epsilon/T}} \right] e^{-(\Gamma_+ + \Gamma_-)t} + \frac{1}{1 + e^{\epsilon/T}} \qquad (5.56)$$

where $W_L(0)$ is the occupation probability of the left well in the initial time moment $t = 0$.

Equation (5.56) fully describes incoherent dissipative dynamics of our double-well system at sufficiently low temperatures (5.25), demonstrating that the equilibrium state is always reached at times substantially exceeding the inverse rate $1/(\Gamma_+ + \Gamma_-)$.

According to our analysis in Section 5.1, the regime of incoherent tunneling applies for sufficiently strong dissipation $K \geq 1$, as well as for weaker dissipation $K < 1$ provided either temperature or external bias ϵ remains sufficiently large; see Eq. (5.50). If, however, the latter inequality is violated, quantum coherence comes into play and may essentially influence the system dynamics. Obviously, this regime cannot be described by simple master equations (5.55) and more sophisticated techniques need to be elaborated.

5.2.2 Quantum Coherence and Damping: Spin-Boson Model

As we already observed in our previous analysis, under the conditions (5.21), (5.25), and (5.38), only the ground states of the two wells are involved, while all higher energy states can be safely neglected. In other words, an effective Hilbert space of our problem is two dimensional: The two flux states (or, equivalently, the Josephson particle states in the two wells) $\varphi = \varphi_-$ and $\varphi = \varphi_+$ are well separated and are only weakly coupled by means of the tunneling amplitude Δ. Such two-state system can also be described by the Hamiltonian for a fictitious particle with spin 1/2. Its "spin-up" and "spin-down" states, $|\uparrow\rangle$ and $|\downarrow\rangle$, correspond to the states localized, respectively, in the right and left wells of our double-well potential. A convenient and frequently used representation of the two-state system Hamiltonian reads

$$\hat{H}_{TSS} = -\frac{1}{2}\Delta\hat{\sigma}_x - \frac{1}{2}\epsilon\hat{\sigma}_z \tag{5.57}$$

where $\hat{\sigma}_x$ and $\hat{\sigma}_z$ are the two Pauli matrices (2.69). The Hamiltonian (5.57) has two eigenstates with eigenvalues E_g and E_e, and the level splitting is

$$\delta E = E_e - E_g = \sqrt{\Delta^2 + \epsilon^2} \tag{5.58}$$

It is convenient to choose the basis formed by the states $|\uparrow\rangle$ and $|\downarrow\rangle$, which are eigenstates of $\hat{\sigma}_z$ with eigenvalues ± 1. In this basis, the density matrix is 2×2 matrix of the form

$$\hat{\rho} = \begin{pmatrix} \rho_{\uparrow\uparrow} & \rho_{\uparrow\downarrow} \\ \rho_{\downarrow\uparrow} & \rho_{\downarrow\downarrow} \end{pmatrix}. \tag{5.59}$$

Obviously, its diagonal elements are identical to probabilities to find the system in the right and left wells, i.e., $\rho_{\uparrow\uparrow}(t) \equiv W_R(t)$ and $\rho_{\downarrow\downarrow} \equiv W_L(t)$. Unlike in the incoherent regime considered above, here off-diagonal elements $\rho_{\uparrow\downarrow}(t)$ and $\rho_{\downarrow\uparrow}(t)$ differ from zero and should be accounted for while describing the system dynamics.

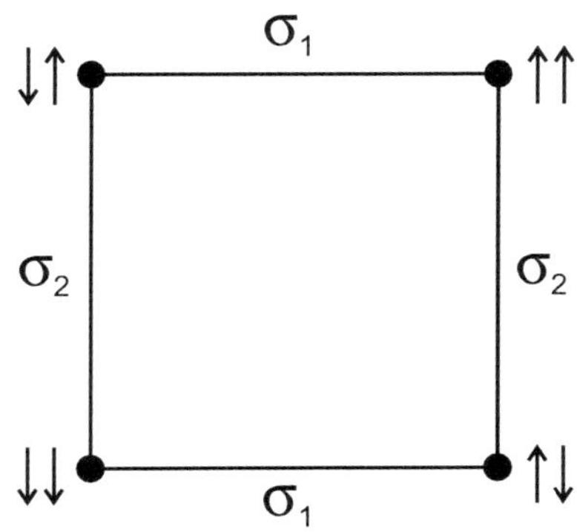

Figure 5.8 Four possible states on the Keldysh contour.

Let us first consider the unbiased limit $\epsilon = 0$ and assume that in the initial time moment $t = 0$, the system was in the state $|\uparrow\uparrow\rangle$, i.e., $\rho_{\uparrow\uparrow}(0) = 1$. At sufficiently short times, the system dynamics can be described perturbatively in Δ. Let us, for instance, inspect the probability to be in the state $|\downarrow\downarrow\rangle$ at a time t. In the lowest non-vanishing order in Δ, two spin flip processes should be taken into account: $\uparrow\uparrow\rightarrow\downarrow\uparrow\rightarrow\downarrow\downarrow$ and $\uparrow\uparrow\rightarrow\uparrow\downarrow\rightarrow\downarrow\downarrow$. The transition probability for each of these processes is the product of two transition amplitudes

$$\frac{-\mathrm{i}\Delta}{2\hbar}\int_0^t dt_2 \frac{\mathrm{i}\Delta}{2\hbar}\int_0^{t_2} dt_1 = \frac{\Delta^2}{4\hbar^2}\frac{t^2}{2}$$

i.e., the total transition probability at short times is simply $\rho_{\downarrow\downarrow}(t) = (\Delta t/2\hbar)^2$. Accordingly, the probability $\rho_{\uparrow\uparrow}(t)$ to remain in the initial state is

$$\rho_{\uparrow\uparrow}(t) = 1 - \rho_{\downarrow\downarrow}(t) = 1 - \frac{\Delta^2}{4\hbar^2}t^2 \tag{5.60}$$

where the last term can also be interpreted as the contribution from the following two spin flip processes: $\uparrow\uparrow\rightarrow\downarrow\uparrow\rightarrow\uparrow\uparrow$ and $\uparrow\uparrow\rightarrow\uparrow\downarrow\rightarrow\uparrow\uparrow$. These processes can be visualized graphically as detours from the upper right angle of the square along the sidelines, respectively, to the upper left and lower right angles and followed by returns back to the initial state along the same paths (see Fig. 5.8).

It is clear that the above expressions for $\rho_{\uparrow\uparrow}(t)$ and $\rho_{\downarrow\downarrow}(t)$ remain valid as long as $\Delta t/\hbar \ll 1$, while at longer times t, the contributions from all higher-order spin flip processes should also be taken into account. Consider, e.g., the probability $\rho_{\uparrow\uparrow}(t)$. In order to identify all relevant paths, we again use the graphical representation in Fig. 5.8 and count all possible ways along the square sidelines a walker can choose if he/she wants to start from the upper right corner and eventually return to the initial point after the time interval t. Obviously, going along the horizontal and vertical lines corresponds to the amplitude per unit time equal to, respectively, $-\mathrm{i}\Delta/2\hbar$ and $\mathrm{i}\Delta/2\hbar$. Then summing up over all possible configurations and integrating over all spin flip times, one readily finds

$$\rho_{\uparrow\uparrow}(t) = 1 - \rho_{\downarrow\downarrow}(t) = \frac{1 + P(t)}{2} \tag{5.61}$$

where

$$\begin{aligned} P(t) \equiv \langle \sigma_z \rangle_t &= \sum_{n=0}^{\infty} (-1)^n (\Delta/\hbar)^{2n} \int_0^t dt_{2n} \int_0^{t_{2n}} dt_{2n-1} \ldots \int_0^{t_2} dt_1 \\ &= \cos(\Delta t/\hbar) \end{aligned} \tag{5.62}$$

As it could be anticipated from the very beginning, we arrived at the well-known result, according to which the probabilities to occupy the right and left wells oscillate as $W_R(t) = \cos^2(\Delta t/2\hbar)$ and $W_L(t) = \sin^2(\Delta t/2\hbar)$. The frequency of these coherent oscillations equals to $\Delta/\hbar$.

The evolution of the density matrix can formally be described by the double path integral

$$\hat{\rho}(t) \sim \int \mathrm{D}\sigma_1(t) \int \mathrm{D}\sigma_2(t) \mathcal{A}[\sigma_1(t)] \mathcal{A}^*[\sigma_2(t)] \tag{5.63}$$

where the path integrals are taken over variables $\sigma_{1,2}(t)$ that switch between $+1$ and -1 in all possible combinations and $\mathcal{A}[\sigma_{1,2}(t)]$ are the corresponding transition amplitudes. Employing the graphical representation of Fig. 5.8, we can say that $\sigma_1(t)$ and $\sigma_2(t)$ are responsible for walks along, respectively, horizontal and vertical sidelines of the square. Introducing new variables

$$\sigma_{\pm}(t) = \frac{\sigma_1(t) \pm \sigma_2(t)}{2} \tag{5.64}$$

we represent all relevant trajectories with $2n$ transitions contributing to the path integral (5.63) in the form

$$\sigma_+(t') = \sum_{j=0}^{n} \eta_j[\theta(t' - t_{2j}) - \theta(t' - t_{2j-1})],$$
$$\sigma_-(t') = \sum_{j=1}^{n} \xi_j[\theta(t' - t_{2j-1}) - \theta(t' - t_{2j})], \qquad (5.65)$$

where $t_0 \equiv 0$ and $t_{2n+1} \equiv t$. It was suggested [264] to call the time periods $t_{2j} < t < t_{2j+1}$ the system spends in the diagonal states $\uparrow\uparrow$ and $\downarrow\downarrow$ *sojourns* and the time intervals $t_{2j-1} < t < t_{2j}$ it remains in off-diagonal states $\uparrow\downarrow$ and $\downarrow\uparrow$ *blips*. It is obvious from Eq. (5.65) that the function $\sigma_+(t)$ equals to zero during any sojourn, while $\sigma_-(t)$ vanishes during any blip. Two sojourn states $\uparrow\uparrow$ and $\downarrow\downarrow$ are labeled, respectively, by $\eta_j = +1$ and $\eta_j = -1$. Similarly, two different blip states $\downarrow\uparrow$ and $\uparrow\downarrow$ correspond to, respectively, $\xi_j = +1$ and $\xi_j = -1$. For later purposes, we also define the sojourn and blip lengths, respectively, as

$$s_j = t_{2j+1} - t_{2j}, \quad b_j = t_{2j} - t_{2j-1} \qquad (5.66)$$

The above analysis can be easily modified in order to take nonzero bias ϵ into account. It is sufficient to notice that, while the amplitude to remain in a sojourn does not change being equal to unity, the amplitude to be in the j-th blip state with label ξ_j and length τ_j is $\exp(-i\epsilon\xi_j b_j/\hbar)$. With this in mind instead of Eq. (5.62), we obtain

$$P(t) = \sum_{n=0}^{\infty}\left(-\frac{1}{2}\right)^n\left(\frac{\Delta}{\hbar}\right)^{2n}\sum_{\xi_j}\int_0^t dt_{2n}\int_0^{t_{2n}} dt_{2n-1}\ldots\int_0^{t_2} dt_1$$
$$\times \cos\left[\sum_{j=1}^{n}\epsilon\xi_j b_j/\hbar\right] \qquad (5.67)$$

where the sum over all possible combinations ξ_j is included. Evaluating Eq. (5.67), one finds

$$P(t) = \frac{\epsilon^2 + \Delta^2\cos(\sqrt{\Delta^2+\epsilon^2}\,t/\hbar)}{\Delta^2+\epsilon^2} \qquad (5.68)$$

We observe that also in this case, the frequency of coherent oscillations coincides with the value of level splitting (5.58).

Obviously, in the limit $\epsilon \to 0$, Eqs. (5.67) and (5.68) reduce to the result (5.62).

Let us now make a crucial step and include dissipation. For this purpose, the above model Hamiltonian (5.57) should be modified by adding interactions to the Caldeira–Leggett bath of harmonic oscillators, exactly as we have already done in Chapter 1. Employing bilinear coupling between the particle position operator $\sim \hat{\sigma}_z$ and the coordinates of the bath oscillators Q_n (cf. Eq. (1.51)), we arrive at the Hamiltonian of the so-called spin-boson model

$$\hat{H}_{SB} = -\frac{1}{2}\Delta\hat{\sigma}_x - \frac{1}{2}\epsilon\hat{\sigma}_z + \sum_n \left[\frac{\hat{P}_n^2}{2M_n} + \frac{M_n\omega_n^2 Q_n^2}{2} - \frac{\varphi_+ - \varphi_-}{2}\hat{\sigma}_z C_n Q_n\right] \tag{5.69}$$

which is frequently used in the literature in various contexts. One of the important applications of this model is superconducting qubits [291].

The remaining steps are standard and exactly repeat those already performed in Chapter 1. Representing the time-dependent density matrix of the system as a double path integral over both the particle spin variable $\sigma(t)$ and the oscillator coordinates Q_n (all defined on the Keldysh contour), integrating out all Q_n and choosing the linear Ohmic spectrum of oscillators (1.54) with the cutoff frequency $\omega_c \sim \omega_0$, we arrive at the modified path integral for the density matrix

$$\hat{\rho}(t) \sim \int \mathrm{D}\sigma_1(t) \int \mathrm{D}\sigma_2(t)\mathcal{A}[\sigma_1(t)]\mathcal{A}^*[\sigma_2(t)]F[\sigma_1(t), \sigma_2(t)] \tag{5.70}$$

where

$$F[\sigma_1(t), \sigma_2(t)] = \exp\left(-\frac{\mathrm{i}}{\hbar}S_\mathrm{R}[\sigma_1, \sigma_2] - \frac{1}{\hbar}S_\mathrm{I}[\sigma_1, \sigma_2]\right), \tag{5.71}$$

is the Feynman–Vernon influence functional for the σ-variable and

$$S_\mathrm{R}[\sigma_1, \sigma_2] = 4\int dt \int^t dt'\sigma_-(t)\mathrm{Im}\alpha(t, t')\sigma_+(t'), \tag{5.72}$$

$$S_\mathrm{I}[\sigma_1, \sigma_2] = 4\int dt \int^t dt'\sigma_-(t)\mathrm{Re}\alpha(t, t')\sigma_-(t'). \tag{5.73}$$

For the Ohmic bath spectrum, we have

$$\mathrm{Re}\alpha(t) = K\int d\omega^2 \coth\left(\frac{\hbar\omega}{2kT}\right)e^{-\mathrm{i}\omega t}, \tag{5.74}$$

$$\mathrm{Im}\alpha(t) = \pi K\delta'(t) \tag{5.75}$$

What remains is to evaluate the integral (5.70) over the same collection of paths as in the absence of dissipation. Substituting the trajectories (5.65) into Eqs. (5.71)–(5.75), we observe that the term S_{I} (5.73) in the exponent of Eq. (5.71) gives rise to the contribution to F of the form

$$\exp\left[-\sum_{j}\mathcal{Q}(b_j)-\sum_{j>k}\sum_{k}\xi_j\xi_k\Lambda_{jk}\right] \tag{5.76}$$

while the term S_R (5.72) yields the factor

$$\exp\left[\mathrm{i}\sum_{k=0}^{n-1}\sum_{j=k+1}^{n}\eta_k\xi_j X_{jk}-(\mathrm{i}\epsilon/\hbar)\sum_{j=1}^{n}\xi_j(b_j)\right] \tag{5.77}$$

where b_j is defined in Eq. (5.66).

Here we introduce the notations

$$\begin{aligned}\Lambda_{jk} &= \mathcal{Q}(t_{2j}-t_{2k+1})+\mathcal{Q}(t_{2j-1}-t_{2k})-\mathcal{Q}(t_{2j}-t_{2k})\\ &\quad-\mathcal{Q}(t_{2j-1}-t_{2k-1}) \qquad (5.78)\\ X_{jk} &= \mathcal{R}(t_{2j}-t_{2k-1})+\mathcal{R}(t_{2j-1}-t_{2k})-\mathcal{R}(t_{2j}-t_{2k})\\ &\quad-\mathcal{R}(t_{2j-1}-t_{2k+1}) \qquad (5.79)\end{aligned}$$

and

$$\begin{aligned}\mathcal{Q}(t) &= 2K\ln\left[\frac{\hbar}{\pi T t}\sinh\left(\frac{\pi T t}{\hbar}\right)\right]+K\ln(1+\omega_0^2t^2),\\ \mathcal{R}(t) &= 2K\arctan(\omega_0 t) \qquad (5.80)\end{aligned}$$

Note that in the resulting expression for the influence functional, one should carry out the summation over all possible trajectories (5.65), i.e., over all $\eta_k=\pm1$ and $\xi_j=\pm1$. The summation over η_j in Eq. (5.77) is carried out explicitly, giving the term

$$2^{n-1}\prod_{k=1}^{n-1}\cos\left(\sum_{j=k+1}\xi_j X_{jk}\right) \tag{5.81}$$

Collecting all these contributions, we arrive at the final result [264]

$$\begin{aligned}P(t) &= \sum_{n=0}^{\infty}(-1)^n\frac{\Delta^{2n}}{2^n}\sum_{\{\xi_j=\pm1\}}\int_0^t dt_{2n}\int_0^{t_{2n}}dt_{2n-1}\dots\int_0^{t_2}dt_1\\ &\quad\times F_n[t_1,\dots t_2n;\xi_1,\dots\xi_n;\epsilon] \qquad (5.82)\end{aligned}$$

where

$$F_n = F_n^{(1)} F_n^{(2)} \tag{5.83}$$

$$F_n^{(1)} = \exp\left[-\sum_{j=1}^{n} \mathcal{Q}(b_j) - \sum_{j=2}^{n}\sum_{k=1}^{n-1} \xi_j \xi_k \Lambda_{jk}\right] \tag{5.84}$$

$$F_n^{(2)} = \cos\left[\sum_{j=1}^{n} \epsilon \xi_j (b_j/\hbar - X_{j0})\right] \prod_{k=1}^{n-1} \cos\left[\sum_{i=k+1}^{n} \xi_i X_{ik}\right] \tag{5.85}$$

Equations (5.82)–(5.85) together with Eqs. (5.78)–(5.80) and Eq. (5.61) fully determine real-time dynamics of diagonal elements of the density matrix for the spin-boson model with linear Ohmic dissipation. Actually, the specifics of the Ohmic bath enter only in Eqs. (5.80). With proper modifications of the kernels $\mathcal{Q}(t)$ and $\mathcal{R}(t)$, this general result is equally applicable for non-Ohmic spectra of the bath oscillators. Here, however, we restrict ourselves to the most important regime of Ohmic dissipation.

5.2.3 Non-interacting-Blip Approximation

The result for $P(t)$ (5.82), though formally exact, still looks rather cumbersome. Therefore, in order to reduce the above equations to a more tractable form, it is necessary to employ some approximations. Drastic simplifications occur if one assumes that the average time the system remains in a state described by the diagonal density matrix strongly exceeds the time it spends in the off-diagonal state. In other words, the average lengths of sojourns $\langle s_j \rangle$ are all much longer than those of blips $\langle b_j \rangle$. This approximation is usually called non-interacting-blip approximation. Technically, it consists of two simple prescriptions [264]: (i) set all interblip interaction terms Λ_{jk} in the expression for $F_n^{(1)}$ (5.84) equal to zero, i.e., $\Lambda_{jk} \to 0$ and (ii) set all sojourn-blip correlation terms X_{jk} in the expression for $F_n^{(2)}$ (5.85) equal to zero $X_{jk} = 0$ whenever $j \neq k+1$ and put $X_{jk} = \mathcal{R}(t_{2k+2} - t_{2k+1})$ otherwise (i.e., for $j = k+1$). After that the terms (5.84) and (5.85) reduce, respectively, to

$$F_n^{(1)} = \prod_{j=1}^{n} \exp(-\mathcal{Q}(b_j)) \tag{5.86}$$

and

$$F_n^{(2)} = \cos\left[\sum_{j=1}^{n} \epsilon \xi_j b_j/\hbar - \xi_1 \mathcal{R}(b_1)\right] \prod_{j=2}^{n} \cos(\mathcal{R}(b_j)) \qquad (5.87)$$

Combining Eqs. (5.86) and (5.87) with Eqs. (5.82) and (5.83) and summing up over all ξ_j-configurations, we obtain [264]

$$P(t) = \sum_{n=0}^{\infty} (-1)^n (\Delta/\hbar)^{2n} \int_0^t dt_{2n} \int_0^{t_{2n}} dt_{2n-1} \ldots \int_0^{t_2} dt_1$$
$$\times \left[(g(b_1) + h(b_1)) \prod_{j=2}^{n} g(b_j)\right] \qquad (5.88)$$

where

$$g(t) = \cos(\epsilon t/\hbar) \cos(\mathcal{R}(t)) \exp(-Q(t)), \qquad (5.89)$$
$$h(t) = \sin(\epsilon t/\hbar) \sin(\mathcal{R}(t)) \exp(-Q(t)) \qquad (5.90)$$

Equations (5.88)–(5.90) define the function $P(t)$ within the non-interacting-blip approximation. The validity of this approximation can be rigorously justified in a number of limiting cases. Specifically, in the case of Ohmic dissipation considered here, the above prescription (i) is strictly valid in the important weak coupling limit $K \ll 1$ and zero bias $\epsilon \to 0$ (since the second term in the exponent in Eq. (5.84) after the summation over $\{\xi_j\}$ may only generate terms of higher order $\propto K^2$ as compared to the leading ones) as well as in the opposite strong damping limit or sufficiently large bias ϵ or at nonzero T. The prescription (ii) for the case of Ohmic dissipation is correct in the leading order in the small parameter $\Delta/\hbar\omega_0$, i.e., it is exact in the so-called scaling limit (implying that one sends the high-frequency cutoff ω_0 to infinity formally keeping the ratio $\Delta/\hbar\omega_0$ constant). Both prescriptions (i) and (ii) can also be justified under certain conditions [264, 400] for non-Ohmic dissipative baths.

It is important to mention that the non-interacting-blip approximation remains qualitatively correct also outside the above parameter domains where it can be rigorously justified. At the same time, at zero temperature $T \to 0$, it is known to fail in the limit of very long times $t \to \infty$ [400]. This latter observation, however, remains of a limited significance if one is aiming at describing the decoherence effect of dissipation. This effect usually becomes important already at intermediate times when the non-interacting-blip approximation still remains physically meaningful.

5.2.4 Coherent Oscillations and Decoherence

Turning to concrete results, let us first consider the unbiased case $\epsilon = 0$. It is convenient to perform the Laplace transformation of the function $P(t)$, i.e., define

$$P_\lambda = \int_0^\infty e^{-\lambda t} P(t) dt \tag{5.91}$$

Since for $\epsilon = 0$ the function $h(t)$ (5.90) vanishes, Eq. (5.88) only contains the product of g-functions, and we obtain

$$\begin{aligned} P_\lambda &= \sum_{n=0}^{\infty} (-1)^n (\Delta/\hbar)^{2n} \int_0^\infty dt \int_0^\infty dt_1 \ldots \\ &\times \int_0^\infty dt_{2n} e^{-\lambda(t_1+t_2+\ldots+t_{2n})} \prod_{j=1}^{n} g(t_{2j}) \\ &= \sum_{n=0}^{\infty} (-1)^n (\Delta/\hbar)^{2n} \frac{g_\lambda^n}{\lambda^{n+1}} = \frac{1}{\lambda + \Delta^2 g_\lambda/\hbar^2} \end{aligned} \tag{5.92}$$

where g_λ is the Laplace transform of the function $g(t)$ (5.89) taken at $\epsilon = 0$, i.e.,

$$g_\lambda = \int_0^\infty dt \cos(\mathcal{R}(t)) \exp(-\lambda t - Q(t)) \tag{5.93}$$

Inverting now the Laplace transformation (5.91) and employing Eq. (5.92), we arrive at the final result for $P(t)$:

$$P(t) = \frac{1}{2\pi \mathrm{i}} \int_C \frac{\exp(\lambda t) d\lambda}{\lambda + \Delta^2 g_\lambda/\hbar^2} \tag{5.94}$$

where C defines the standard Bromwich contour running from $-\mathrm{i}\infty$ to $+\mathrm{i}\infty$ in such a way that all singularities of P_λ remain to the left from this contour. The form (5.94) turns out to be very convenient for further investigations.

Consider first the most important zero-temperature limit $T \to 0$. According to our previous results, in this case for $K \geq 1$, the particle is localized in one of the wells and, hence, no real-time dynamics can be expected. For $K < 1$ in the scaling limit, we obtain

$$P_\lambda = \frac{1}{\lambda[1 + (\Delta_{\rm eff}/\hbar\lambda)^{2(1-K)}]} \tag{5.95}$$

where we defined

$$\Delta_{\rm eff} = [\Gamma(1 - 2K)\cos(\pi K)]^{1/2(1-K)}\Delta_r \tag{5.96}$$

with Δ_r defined in Eq. (5.35). The function of the form (5.95) is known to be the Laplace transform of the Mittag–Leffler function $E_{2-2K}(z)$, i.e., we obtain

$$P(t) = E_{2-2K}(z(t)), \quad z(t) = -(\Delta_{\rm eff}t/\hbar)^{2-2K} \tag{5.97}$$

In the limit $K \to 0$, Eq. (5.97) reduces to the result $P(t) = \cos(\Delta t/\hbar)$ (5.62) describing coherent oscillations, while at $K = 1/2$, it yields purely exponential decay

$$P(t) = \exp(-\pi\Delta^2 t/2\hbar^2\omega_0) \tag{5.98}$$

At $\Delta_{\rm eff}t \ll \hbar$, Eq. (5.97) provides a simple formula

$$P(t) \simeq 1 - (\Delta_{\rm eff}t/\hbar)^{2-2K}/\,\Gamma(3 - 2K) \tag{5.99}$$

which describes the system behavior exactly in the short-time limit.

Let us now further concentrate on the case $K < 1/2$. In this case, the function P_λ (5.95) has a complex conjugate pair of simple poles on the principal sheet at $\lambda = (\Delta_{\rm eff}/\hbar)\exp[\pm i\pi/(2 - 2K)]$ and a branch point at $\lambda = 0$. Integrating along the Bromwich contour from the poles, one recovers the term describing damped coherent oscillations, while the branch point leads to an additional incoherent cut contribution. Collecting both these contributions, we obtain

$$P(t) = \frac{1}{1 - K}\cos(\omega_r t)\exp(-\gamma_\varphi t) + \delta P(t) \tag{5.100}$$

where

$$\omega_r = (\Delta_{\rm eff}/\hbar)\cos[\pi K/2(1 - K)], \quad \gamma_\varphi = (\Delta_{\rm eff}/\hbar)\sin[\pi K/2(1 - K)] \tag{5.101}$$

and the incoherent contribution $\delta P(t)$ equals to

$$\delta P(t) \simeq -\frac{\sin(2\pi K)\Gamma(2 - 2K)}{\pi(\Delta_{\rm eff}t/\hbar)^{2-2K}} f(t) \tag{5.102}$$

in the long-time limit $\Delta_{\rm eff}t/\hbar \gg 1$. Within the non-interacting-blip approximation, the function $f(t)$ is close to unity $f(t) \simeq 1$. In the opposite limit of short times $\Delta_{\rm eff}t/\hbar \ll 1$, the function $\delta P(t)$ is trivially recovered by combining Eqs. (5.100) and (5.99). Overall, the incoherent contribution $\delta P(t)$ is negative; it does not oscillate and monotonously decays to zero at $t \to \infty$. In addition, this contribution vanishes identically for $K = 0$ and $K = 1/2$.

Of more significance is the first term in Eq. (5.100), which describes coherent oscillations with the amplitude exponentially decaying in time. Thus, interaction with a dissipative Caldeira–Leggett environment yields two effects: (a) The frequency of coherent oscillations ω_r (as well as renormalized level splitting Δ_r) decreases with increasing dissipation strength K and (b) these coherent oscillations get suppressed on a timescale $\sim 1/\gamma_\varphi$, i.e., the system cannot demonstrate quantum coherent properties provided time exceeded the dephasing time $\tau_\varphi = 1/\gamma_\varphi$. The latter effect manifests zero-temperature decoherence by interaction, which we already encountered and discussed in details in Chapter 3 in the case of a free quantum particle interacting with a dissipative bath. While the physical nature of this effect remains exactly the same, some qualitative features differ. For instance, in the case of a spin-boson model considered here, the off-diagonal elements of the reduced density matrix decay exponentially in time already at zero temperature, while for a free particle, there is a power-law decay at $T = 0$, which turns exponential at $T > 0$, cf., e.g., Eqs. (3.5)–(3.7).

One can introduce the quality factor $Q = \omega_r/\gamma_\varphi$ describing damped coherent oscillations (5.100). It follows from Eq. (5.101) that this quality factor is independent of $\Delta_{\rm eff}$ and reads

$$Q = \cot[\pi K/2(1 - K)] \qquad (5.103)$$

In other words, coherent oscillations are only pronounced in the weak dissipation limit $K \ll 1$ (corresponding to $Q \gg 1$) and vanish completely at $K \to 1/2$ in which case purely exponential decay of correlations (5.98) sets in. For $1/2 < K < 1$, no coherent oscillations take place either. In this case, the poles of P_λ are not on a principal sheet and within the non-interacting-blip approximation, $P(t)$ is determined solely by the branch-cut contribution (5.102). Note, however, that this approximation turns out to be insufficient in order to establish quantitatively correct behavior of $P(t)$ in the long-time limit. In this case, one should employ more sophisticated techniques.

Exact results can be derived in the vicinity of the Toulouse limit, i.e., for the values K sufficiently close to $1/2$. As we already indicated, at $K = 1/2$ the decay of correlations is purely exponential and $P(t)$ is defined by Eq. (5.98). For $0 < 1/2 - K \ll 1$, one arrives at Eqs. (5.100)–(5.102) with $f(t) = \exp(-\Delta_{\text{eff}} t/2\hbar)$ [400], i.e., incoherent decay of $P(t)$ at long times is exponential rather than algebraic predicted by the non-interacting-blip approximation. Conformal field theory arguments [267] allow to conclude that the decay of $P(t)$ is universally exponential at long times

$$P(t) \propto \exp(-\gamma_\varphi t), \quad t \gg \hbar/\Delta_r \tag{5.104}$$

with the zero-temperature decoherence rate

$$\gamma_\varphi = \frac{1}{\sqrt{\pi}} \sin^2\left(\frac{\pi K}{2-2K}\right) \frac{\Gamma\left(\frac{K}{2-2K}\right)}{\Gamma\left(\frac{1}{2-2K}\right)} \left(\frac{\Gamma(1-K)}{2^K}\right)^{\frac{1}{1-K}} \frac{\Delta_r}{\hbar} \tag{5.105}$$

The behavior of the function $P(t)$ for symmetric double-well system at $T \to 0$ is also illustrated in Fig. 5.9 for different values of the dissipation parameter K.

It follows from the above consideration that coherent oscillations in a double-well system in the presence of Ohmic dissipation can only be observed in the limit $K \ll 1$. As measurements are performed at nonzero temperatures, it is important to find out in which way these coherent oscillations are affected by thermal fluctuations. Employing again the non-interacting-blip approximation results (5.92)–(5.94) at nonzero T and in the, interesting for us, weak dissipation limit $K \ll 1$, we again arrive at Eq. (5.100), where now [400]

$$\omega_r(T) = \frac{\Delta_{\text{eff}}}{\hbar}\left[1 + K\left(\text{Re}\psi\left(\frac{i\Delta_{\text{eff}}}{2\pi T}\right) - \ln\left(\frac{\Delta_{\text{eff}}}{2\pi T}\right)\right)\right], \tag{5.106}$$

$$\gamma_\varphi(T) = \frac{\pi K \Delta_{\text{eff}}}{2\hbar} \coth\left(\frac{\Delta_{\text{eff}}}{2T}\right) \tag{5.107}$$

and the term $\delta P(t)$ is small in the measure of K. For $T \ll \Delta_{\text{eff}}$, the corresponding quality factor is $Q \simeq 2/\pi K \gg 1$ in the agreement with Eq. (5.103) in the limit of small K. At higher temperatures $T \gg \Delta_{\text{eff}}$, one finds $Q \simeq \Delta_r/(\pi K T)$. Then the inequality $Q \gg 1$ translates into

$$T \ll \Delta_r/\pi K, \quad K \ll 1 \tag{5.108}$$

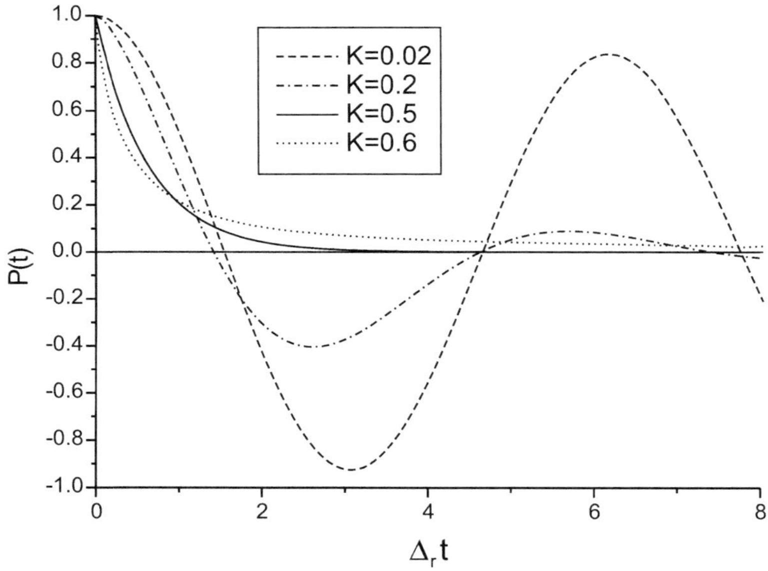

Figure 5.9 Function $P(t)$ (determined by Eq. (5.97)) at $T = 0$ and different values of the dissipation parameter K.

This condition should be fulfilled in order to observe coherent oscillations in degenerate double-well systems under consideration.

To complete our analysis, let us briefly discuss the case of a nonzero bias $\epsilon \neq 0$. One can again employ non-interacting-blip approximation in which case from Eq. (5.88) one readily finds

$$P_\lambda = \frac{1 - \Delta^2 h_\lambda/\hbar^2\lambda}{\lambda + \Delta^2 g_\lambda/\hbar^2} \tag{5.109}$$

where, as before, g_λ and h_λ are, respectively, the Laplace transforms of the functions (5.89) and (5.90). Inverting the Laplace transformation and employing Eq. (5.109) in the long-time limit, one finds the result $P(t \to \infty) = -\tanh(\epsilon/2T)$, which implies symmetry breaking and the particle localization in the lower well at $T = 0$ and any nonzero bias. This formula, in general, *does not* agree with the standard quantum mechanical result for the occupation probabilities both wells in equilibrium at a temperature T in the weak damping limit:

$$P(t \to \infty) \equiv P_\infty = -\frac{\epsilon}{\sqrt{\Delta^2 + \epsilon^2}} \tanh\left(\frac{\sqrt{\Delta^2 + \epsilon^2}}{2T}\right) \tag{5.110}$$

The latter result demonstrates that at low enough temperature, the particle is confined to the lower well only provided the bias ϵ strongly exceeds the level splitting Δ. The above disagreement again indicates that the non-interacting-blip approximation fails to correctly describe the long-time behavior of the system in the limit of low temperature, low bias, and low damping. On the other hand, it remains valid in the incoherent regime where, by inverting the Laplace transformation from Eq. (5.109), one easily obtains

$$P(t) = -\tanh(\epsilon/2T) + (1 + \tanh(\epsilon/2T))e^{-(\Gamma_+ + \Gamma_-)t} \tag{5.111}$$

where the decay rates $\Gamma_\pm$ are defined in Eqs. (5.45)–(5.48). Equation (5.111) fully agrees with the master equation result (5.56). Thus, we have demonstrated the equivalence of Eq. (5.55) and the non-interacting-blip approximation in the incoherent regime, which applies under the condition (5.50) reducing to Eq. (5.51) in the weak dissipation limit $K \ll 1$.

Provided both inequalities (5.51) are violated, the regime of damped coherent oscillations sets in for $K \ll 1$. In order to correctly describe this regime, it is in general necessary to go beyond the non-interacting-blip approximation and employ the exact form of $P(t)$ (5.82). Tractable results can be obtained, e.g., by working out a systematic perturbation theory in the dissipation strength K. The corresponding calculation can be found elsewhere [400]; here we only present the final result for $P(t)$. It reads

$$\begin{aligned} P(t) &= [\epsilon^2/\hbar^2\omega_r^2 - P_\infty]e^{-\gamma_r t} + P_\infty \\ &+ \left[\frac{\Delta_{\text{eff}}^2}{\hbar^2\omega_r^2}\cos\omega_r t + \left(\frac{\gamma_r\epsilon^2 + \gamma_\varphi\Delta_{\text{eff}}^2}{\hbar^3\omega_r^3} - \frac{\gamma_r P_\infty}{\omega_r}\right)\sin\omega_r t\right] e^{-\gamma_\varphi t} \end{aligned} \tag{5.112}$$

where we defined

$$P_\infty = -\frac{\epsilon}{\Delta_\text{b}} \tanh\left(\frac{\Delta_\text{b}}{2T}\right), \quad \Delta_\text{b} = \sqrt{\Delta_{\text{eff}}^2 + \epsilon^2} \tag{5.113}$$

and

$$\omega_r^2 = \frac{\Delta_b^2}{\hbar^2} + 2K\left[\mathrm{Re}\psi\left(\frac{i\Delta_{\mathrm{eff}}}{2\pi T}\right) - \ln\left(\frac{\Delta_{\mathrm{eff}}}{2\pi T}\right)\right], \quad (5.114)$$

$$\gamma_r = \frac{\pi K \Delta_{\mathrm{eff}}^2}{\hbar\Delta_b}\coth\left(\frac{\Delta_b}{2T}\right), \quad (5.115)$$

$$\gamma_\varphi = \gamma_r/2 + 2\pi KT(\epsilon^2/\hbar\Delta_b^2). \quad (5.116)$$

The above equations remain applicable at sufficiently low temperatures $T \lesssim \Delta_b$.

As expected, in the weak dissipation limit, Eq. (5.113) reduces to the standard quantum mechanical result (5.110). We also note that exponential relaxation in Eq. (5.112) is governed by two different relaxation rates: γ_r and γ_φ. The rate γ_r describes energy relaxation. The corresponding contribution to $P(t)$ in Eq. (5.112) remains nonzero only for non-degenerate systems $\epsilon \neq 0$. The rate $\gamma_\varphi = 1/\tau_\varphi$ accounts for loss of phase coherence in our system. We observe that this dephasing rate consists of two different terms: the relaxation one $\gamma_r/2$ and the so-called "pure dephasing" term, which in the case of Ohmic dissipation vanishes in the limit $T \to 0$. In the zero bias limit $\epsilon \to 0$ from Eqs. (5.112)–(5.116), we again recover our previous results (5.100), (5.106), and (5.107) in which we assume $K \ll 1$. This observation concludes our discussion of real-time behavior of a two-state system with Ohmic dissipation. More details on quantum dissipative dynamics of spin-boson models (including the properties of off-diagonal elements of the reduced density matrix) can be found elsewhere [264, 400].

As we already pointed out in the beginning of this chapter, experimental realization of coherent oscillations in superconducting rings interrupted by a Josephson junction (flux qubits [291]) implies observation of Macroscopic Quantum Coherence [263], since in this case a macroscopic variable—magnetic flux Φ inside the ring (equivalently, the Josephson phase $\varphi = 2\pi\Phi/\Phi_0$ or persistent current $I = I_C \sin\varphi$ circulating inside the ring) describing a large number of microscopic (electron) degrees of freedom in the system—oscillates coherently between macroscopically distinguishable flux states $\varphi = \varphi_\pm$ defined in Eq. (5.4). Such coherent oscillations in the flux qubit were first observed by Mooij and coworkers [89]. Shortly before that, the first successful MQC experiment in the so-called charge qubits was performed [312]. In the latter case, coherent oscillations of the charge were observed. This collective charge variable is canonically conjugate to that of the phase (or flux), thus also describing quantum states of a macroscopically large number of electron degrees of freedom. In the next chapter, we will learn more about the relation between the phase and the charge variables in Josephson junctions.

Chapter 6

Quantum Dynamics of Phase and Charge in Josephson Junctions

In this chapter, we will continue our analysis of macroscopic quantum dynamics of small-size Josephson junctions.

In Chapter 4, we already considered an important quantum effect—macroscopic quantum tunneling of the Josephson phase—which yields instability and decay of supercurrent in such systems. While dealing with this effect, we intentionally ignored quantum coherence, i.e., considered incoherent decay of supercurrent states. This is appropriate as long as the Josephson particle escapes from the potential well and cannot return due to the presence of a sufficiently large bias current tilting the Josephson potential and/or due to strong enough dissipation.

At small current bias and weak dissipation, quantum coherence, on the contrary, may play an important role and significantly influence quantum dynamics of the system. In Chapter 5, we addressed this issue restricting our consideration to a special case of a Josephson particle moving in a double-well potential, which can be realized provided a Josephson junction is inserted into a superconducting ring with sufficiently large inductance L obeying the condition (5.3). In this case, the particle can tunnel

Dissipative Quantum Mechanics of Nanostructures: Electron Transport, Fluctuations, and Interactions
Andrei D. Zaikin and Dmitry S. Golubev

ISBN 978-981-4774-50-5 (Hardcover), 978-0-429-29823-3 (eBook)
www.jennystanford.com

back and forth between (almost) degenerate potential minima, gradually losing its coherence due to the presence of (relatively weak) dissipation.

Consider now the current-biased configuration schematically shown, e.g., in Fig. 2.2. In this case, the double-well potential model does not apply anymore, and the effective potential for the Josephson particle $U(\varphi)$ is defined in Eq. (2.108). Provided the current bias I_x is small, many potential minima of $U(\varphi)$ have almost the same energy and quantum dynamics of the phase variable gets substantially modified as compared to that studied in Chapter 5. Below in this chapter, we will address this situation. We will demonstrate that nontrivial interplay between quantum coherence, dissipation, and Coulomb effects yields a rich variety of properties, which can be accessed not only by means of the phase variable φ but also through the canonically conjugate quantum variable of the junction charge Q. In fact, it turns out that in the presence of linear Ohmic dissipation, these two variables are strictly dual to each other. As we will see, this remarkable property allows one to proceed much further in understanding of quantum properties of ultrasmall Josephson junctions.

We also note that the effective action that accounts for quantum dissipative dynamics of a Josephson "particle" can be exactly mapped onto those describing a number of other important problems of modern condensed matter theory. Among these, we mention the problem of quantum transport through a coherent one-channel conductor in a resistive electromagnetic environment [338] (see also Chapter 11), the problem of a quantum impurity in a Luttinger liquid [215], and the problem of tunneling of edge currents through constrictions in fractional quantum-Hall systems [402]. Thus, although in this chapter we focus our attention only on resistively shunted Josephson junctions, the results presented below actually find much broader applications.

6.1 Quantum Correction to the Critical Current

We start by considering the situation when the behavior of the Josephon phase φ is almost classical, i.e., the effect of quantum fluctuations is weak and can be considered a small perturbation. An external bias I is set to be lower than the Josephson critical current I_C. This condition means that classical motion of the Josephson particle between different potential minima is forbidden. At this stage, we will also neglect the effect of quantum tunneling of the phase out of the potential well, implying that the supercurrent decay rate Γ is vanishingly small at least for not-too-long time scales. As follows from the analysis in Chapter 4, this condition requires either large Josephson coupling energy E_J (as compared to the charging energy E_C, cf., e.g., Eq. (5.5)) or strong enough dissipation or both. Under such assumptions, the Josephson particle should not move at all and remain close to the bottom of one of the potential wells.

In the absence of quantum fluctuations, the value of the phase is fixed by the Josephson relation (first Eq. (2.1)), i.e., $\varphi = \arcsin(I/I_C)$. We will now demonstrate that already weak quantum fluctuations yield both modifications in the current-phase relation and reduction of the critical current I_C. To this end, let us consider the system free energy

$$F = -T \ln \int \mathcal{D} \exp(-S_{\text{eff}}/\hbar) \tag{6.1}$$

where the effective action is set by Eq. (5.1) with $U(\varphi)$ defined in Eq. (2.108). In order to evaluate this free energy, we will follow the analysis [322] and employ the so-called self-consistent harmonic approximation (SCHA). This approximation amounts to substituting a non-harmonic potential by an effective harmonic one with the renormalized parameters. For the problem under consideration, we set

$$\begin{aligned} E_J \cos\varphi \to\ & E_J \exp(-\langle(\delta\varphi)^2\rangle/2) \\ & \times [\cos\langle\varphi\rangle(2 - (\delta\varphi)^2 + \langle(\delta\varphi)^2\rangle)/2 - \delta\varphi \sin\langle\varphi\rangle] \end{aligned} \tag{6.2}$$

where $\delta\varphi = \varphi - \langle\varphi\rangle$ and the expectation value of the phase is defined by the condition

$$\delta F[\varphi]/\delta\varphi_{\varphi=\langle\varphi\rangle} = 0 \tag{6.3}$$

We also remark that the results obtained within SCHA are essentially equivalent to those derived within the so-called one-loop approximation, which in our case would also allow to recover the first-order quantum correction to the classical value of the critical current I_C.

Substituting Eq. (6.2) into the action S_{eff} in Eq. (6.1), evaluating the Gaussian path integral over φ, and combining the result with Eq. (6.3), we obtain [322]

$$I = I_C^* \sin\langle\varphi\rangle, \qquad I_C^* = I_C \exp(-\langle(\delta\varphi)^2\rangle/2) \tag{6.4}$$

where

$$\langle(\delta\varphi)^2\rangle = \frac{\psi(a_+(\kappa)) - \psi(a_-(\kappa))}{\sqrt{K^2/4 - \pi^2\kappa/2E_C}}, \qquad \kappa = \frac{\hbar I_C}{2e}\sqrt{1 - \frac{I^2}{I_C^2}} \tag{6.5}$$

$\psi(a)$ is the digamma function and

$$a_\pm(\kappa)) = \frac{1}{2} + \frac{2E_C}{\pi T}\left[\frac{K}{2\pi} \pm \sqrt{\frac{K^2}{4\pi^2} - \frac{\kappa}{2E_C}}\right] \tag{6.6}$$

Here and below in this chapter, the parameter K is identical to that introduced in Eq. (5.31), i.e.,

$$K = 2\pi\eta/\hbar = \pi\hbar/(2e^2 R_S) \tag{6.7}$$

In the limit of strong dissipation $K \gg 1$, $\sqrt{\kappa/E_C}$ and at $T \to 0$ from Eqs. (6.4)–(6.6), we obtain

$$I = I_C^{\frac{K}{K-1}}\left(\frac{R_S C}{8e}\right)^{\frac{1}{K-1}} \sin\langle\varphi\rangle(\cos\langle\varphi\rangle)^{\frac{1}{K-1}} \tag{6.8}$$

This result demonstrates that quantum fluctuations modify the current-phase relation of small Josephson junctions as compared to the classical dependence $I = I_C \sin\varphi$.

The strength of this effect depends on the value of K. Provided K is large and, hence, the effect of quantum fluctuations is weak, $\langle(\delta\varphi)^2\rangle \ll 1$, one can expand the exponent in Eq. (6.4) and find

$$I_C^* = I_C\left[1 - \frac{1}{K}\ln\left(\frac{2K^2 E_C}{\pi^2\kappa}\right)\right] \tag{6.9}$$

which agrees with Eq. (4.92) within the logarithmic accuracy.

At sufficiently small dissipation, $K < 1$, SCHA yields $\exp(-\langle(\delta\varphi)^2\rangle/2) = 0$, implying $I_C^* = 0$, i.e., the critical current vanishes due to strong quantum fluctuations. In other words, at $K < 1$ the Josephson junction cannot sustain any supercurrent. Below we will re-derive this result with the aid of more rigorous analysis.

6.2 Incoherent Dynamics of the Phase

Now let us "turn on" quantum tunneling, i.e., allow the Josephson particle to move "under the barrier" between the potential wells. If the potential wells are high enough, i.e., the condition (5.5) is satisfied, the particle spends most of the time near the bottom of the well and occasionally tunnels from one well to another. Here we will assume that tunneling between the wells is incoherent. As we have already learned in Chapter 5, this regime is realized provided either dissipation is strong enough or temperature and/or external bias are sufficiently high. In this case, the Josephson particle may only hop between nearest wells of the tilted washboard potential (2.108), whereas tunneling beyond adjacent wells is strongly suppressed and can be disregarded. Such incoherent dynamics of the system was already considered for the double-well potential in Section 5.2.1 where the occupation probabilities (i.e., diagonal elements of the particle density matrix) for the two wells were evaluated from the master equations (5.55). Here we will proceed analogously with only one difference that now the master equations should be formulated for the infinite number of probabilities to occupy each of the potential wells (2.108).

Let us denote the probability to occupy the k-th well $W_k(t)$. Provided tunneling is incoherent, this probability can decrease due to quantum decay into neighboring wells $k \pm 1$ and increase due to tunneling back from these two wells. Taking all these processes into account, we obtain [401]

$$\frac{dW_k(t)}{dt} = \Gamma_+ W_{k-1}(t) + \Gamma_- W_{k-1}(t) - (\Gamma_+ + \Gamma_-)W_k(t) \qquad (6.10)$$

where $\Gamma_\pm$ are the tunneling rates from the k-th well to the well $k \pm 1$. The solution to Eqs. (6.10) can be easily constructed. For example, assuming that initially the particle occupied one of the wells, i.e., $W_{k=0}(t=0) = 1$ and $W_{k\neq 0}(t=0) = 0$, at $t > 0$ we get

$$W_k(t) = \exp\left[\frac{\pi I \hbar k}{2eT} - (\Gamma_+ + \Gamma_-)t\right] I_k\left(\frac{(\Gamma_+ + \Gamma_-)t}{\cosh(\pi\hbar I/2eT)}\right) \qquad (6.11)$$

where $I_k(x)$ is the modified Bessel function. Then for the expectation value for the Josephson phase, we derive

$$\langle\varphi(t)\rangle = 2\pi(\Gamma_+ + \Gamma_-)t\tanh(\pi\hbar I/2eT) \qquad (6.12)$$

and with the aid of the Josephson relation between the phase and the voltage, we immediately recover the effective resistance R_{eff} of our device consisting of a Josephson junction shunted by an Ohmic resistor, as it is shown, e.g., in Fig. 2.2. We find

$$R_{\text{eff}} = \frac{\langle V \rangle}{I} = \frac{\pi\hbar}{eI}(\Gamma_+ + \Gamma_-)\tanh\left(\frac{\pi\hbar I}{2eT}\right) \tag{6.13}$$

Provided both temperature and the energy difference between the nearest potential minima $\epsilon = \pi\hbar I/e$ remain much smaller than the inter-level distance $\hbar\omega_0$, cf. Eq. (5.38), the tunneling rates $\Gamma_\pm$ have already been evaluated in Section 5.1.3., cf. Eqs. (5.45)–(5.48). Combining these equations with Eq. (6.13), we arrive at the current–voltage characteristics of our system in the limit of small temperature and current bias. In the limit $I \to 0$ from Eqs. (6.13) and (5.48), we obtain the linear resistance [401]

$$R_{\text{eff}} = R_Q \frac{\pi^{5/2}\Gamma(K)}{2\Gamma(K+1/2)}\left(\frac{\Delta}{\hbar\omega_0}\right)^2\left(\frac{\pi T}{\hbar\omega_0}\right)^{2K-2} \tag{6.14}$$

where $R_Q = \pi\hbar/2e^2$ is the "superconducting" quantum resistance unit and the expression for the tunneling amplitude Δ is defined in Eq. (5.20). We observe that in the limit $T \to 0$, the effective resistance vanishes provided $K > 1$. Physically, it implies that the Josephson particle remains localized in the initial well and does not move to other wells, which, in turn, means that the junction remains in a superconducting state. This observation emphasizes a crucial role played by dissipation, which tends to decrease fluctuations of the Josephson phase, thereby helping to maintain superconductivity.

At $K < 1$, however, dissipation is not strong enough and the Josephson particle gets delocalized being able to move to other wells of the washboard potential. Accordingly, the effective resistance (6.14) *increases* with decreasing T. Note, however, that the validity of Eq. (6.14) is restricted to not-too-low temperatures, i.e., to the applicability range of Eq. (5.48), cf. Eqs. (5.50) and (5.51). At $K < 1$ and very low temperatures, incoherent tunneling approximation breaks down and quantum coherence essentially influences the particle dynamics. This issue will be addressed below in this chapter.

From Eq. (6.13), one can also recover the nonlinear effective resistance, which depends on the external current bias I. At $T = 0$, the particle can only move down in energy. Then with the aid of Eq. (5.47), we find [401, 423, 424]

$$R_{\mathrm{eff}} = R_Q \frac{\pi^2 \Gamma(2K)}{\Gamma(K+1/2)} \left(\frac{\Delta}{\hbar\omega_0}\right)^2 \left(\frac{\pi I}{e\omega_0}\right)^{2K-2} \tag{6.15}$$

Again for $K > 1$ and $I \to 0$, the Josephson particle remains trapped in the initial well and the effective resistance vanishes, hence implying superconductivity. In contrast, for $K < 1$, the resistance (6.15) grows with decreasing I, implying delocalization of the Josephson particle and full suppression of superconductivity by quantum fluctuations of the phase. As before, the applicability range of Eq. (6.15) is restricted to not-too-low bias values (5.50) and (5.51), while for very low I, quantum coherence should be taken into account.

To complete our analysis of incoherent dynamics of the Josephson phase, we would like to point out that master equations (6.10) also allow to recover higher correlators of the phase. For example, for the expectation value $\langle\varphi^2\rangle$, one easily gets

$$\langle\varphi^2(t)\rangle = \langle\varphi(t)\rangle^2 + Dt \tag{6.16}$$

where the first term in the right-hand side of this equation is determined by Eq. (6.12), while the second term accounts for phase diffusion. The phase diffusion coefficient D reads

$$D = 4\pi^2(\Gamma_+ + \Gamma_-) \tag{6.17}$$

In the linear regime $I \to 0$, one finds $D \propto (T/\hbar\omega_0)^{2K-1}$, i.e., for $K < 1/2$, the coefficient D increases as temperature gets lower.

6.3 Quantum Dissipative Phase Transition and Duality

Let us now set an external bias equal to zero, $I = 0$. In this case, the Josephson particle is moving in a strictly 2π-periodic potential $U(\varphi) = -E_J \cos\varphi$. As the reader already knows from standard courses in quantum mechanics, in the absence of dissipation, the wave functions of a quantum particle in a periodic potential are delocalized forming Bloch bands. In simple terms, this implies that the particle will eventually occupy all potential wells even if initially it would have stayed only in one well. In the so-called tight-binding limit $E_J \gg E_C$ (5.5), the wells are deep and the particle hopping between the wells (which leads to its delocalization) is due to quantum tunneling. In the case of a double-well potential, we have already developed the corresponding instanton analysis of this process (see Section 5.1.1). Exactly the same treatment can be repeated in the case of a periodic potential with the only difference that now one should account for instanton trajectories describing tunneling between all potential wells (in all possible combinations) rather than just between two neighboring wells. In other words, in addition to instanton trajectories depicted in Fig. 5.4a, we should include many others going beyond adjacent wells. For instance, the configuration with $2k$ instantons has the form

$$\varphi^{(2k)}(\tau) = \sum_{j=1}^{2k} \nu_j \tilde{\varphi}(\tau - \tau_j), \qquad \nu_j = \pm 1 \tag{6.18}$$

where $\tilde{\varphi}(\tau)$ is defined in Eq. (5.19). Examples of such trajectories are presented in Fig. 6.1.

The corresponding problem can again be reformulated in terms of a one-dimensional gas of positive and negative unity charges ν_i. As suggested by Eq. (6.18), now we should account for all possible charge configurations (see Fig. 6.1), whereas in Section 5.1.1, we only dealt with the case of alternating positive and negative charges $\nu_{i\pm1} = -\nu_i$ (cf. Fig. 5.4a). In both cases, the gas of charges is neutral, i.e., the total number of positive and negative charges (equivalently, instantons and anti-instantons) should be the same and, hence,

$$\sum_i \nu_i = 0 \tag{6.19}$$

Formally, this restriction is dictated by the boundary condition $\varphi(0) = \varphi(\hbar\beta)$ in the path integral for the partition function, cf., e.g., Eq. (5.7), implying that the particle imaginary-time trajectory should always return to its initial point.

Proceeding in exactly the same way as in Section 5.1.1 and summing up over all relevant instanton configurations, we evaluate

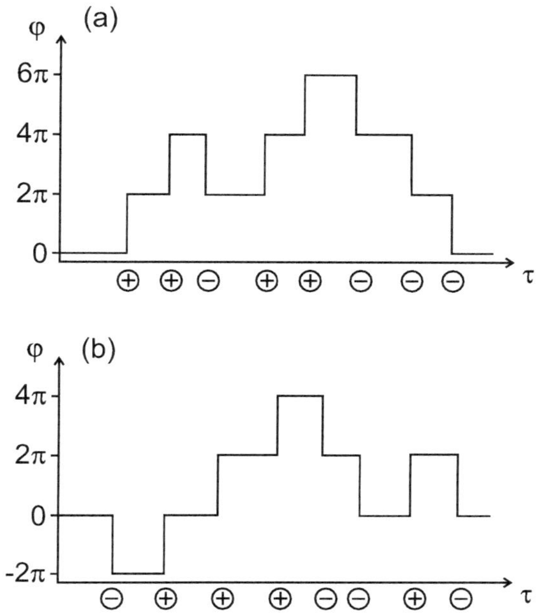

Figure 6.1 Two different instanton configurations contributing to the partition function of a quantum particle in a tight-binding periodic potential.

the grand partition function $\mathcal{Z}$ and reproduce the well-known result for the system ground-state energy equal to $\hbar\omega_0/2 + E_0(q)$, where

$$E_0(q) = -\Delta \cos(\pi q/e) \tag{6.20}$$

defines the lowest Bloch band in the tight-binding limit. Here q is the so-called (quasi-)charge, which is fully analogous to the quasi-momentum for a quantum particle in a periodic potential. We will return to this issue below in this chapter.

Let us now include linear Ohmic dissipation and, as before in Section 5.1.2, assume that dissipation remains sufficiently weak and does not significantly affect instanton trajectories (5.19). Treating the dissipative part of the action (5.1), perturbatively we again substitute all possible instanton trajectories into this term and arrive at the following expression for the grand partition function

[349]

$$\mathcal{Z} \sim \sum_{k=0}^{\infty} \left(\frac{y}{2}\right)^{2k} \frac{1}{2k!} \int_0^{\hbar\beta} \frac{d\tau_{2k}}{\tau_c} \int_0^{\hbar\beta} \frac{d\tau_{2k-1}}{\tau_c} \cdots \int_0^{\hbar\beta} \frac{d\tau_1}{\tau_c} \times \sum_{\nu_i=\pm 1} \exp\left[-K \sum_{i \neq j} \nu_i \Lambda(\tau_i - \tau_j)\nu_j\right] \quad (6.21)$$

where the sum $\sum_{\nu_i=\pm 1}$ is taken over all neutral charge configurations (6.19), the logarithmic interaction between charges is defined by the kernel $\Lambda(\tau \gtrsim \tau_c)$ (5.30), $y = \Delta\tau_c/\hbar$ and $\tau_c \sim \omega_0^{-1}$ are again, respectively, the charge fugacity and the instanton core size. In the absence of dissipation ($K \to 0$), Eq. (6.21) again yields the standard quantum mechanical results discussed above.

In order to analyze the problem of a one-dimensional neutral Coulomb gas of logarithmically interacting charges described by the partition function (6.21), we will make use of the renormalization group treatment similar to that already employed in Chapter 5. We again progressively increase the cutoff parameter $\tau_c \to \tau_c + \delta\tau_c$ integrating out configurations with close instanton pairs exactly as we did in Section 5.1.2. Provided $\delta\tau_c$ remains sufficiently small and $y \ll 1$, after each of such steps, we recover the partition function (6.21) with modified parameters $y \to \tilde{y} = y + \delta y$, $K \to \tilde{K} = K + \delta K$, and $\tau_c \to \tilde{\tau}_c = \tau_c + \delta\tau_c$. Repeating this procedure many times we arrive at the following RG equations [75, 182, 349]:

$$d\tilde{y}/d\ln\tilde{\tau}_c = \tilde{y}(1 - \tilde{K}), \quad d\tilde{K}/d\ln\tilde{\tau}_c = 0 \quad (6.22)$$

The first of these RG equations is similar to the analogous equation (5.32), demonstrating that upon increasing the cutoff parameter τ_c, the charge fugacity y increases for $K < 1$ and decreases for $K > 1$. The first type of behavior corresponds to delocalization of the quantum particle in a periodic potential, while the second type implies particle localization in the limit $T \to 0$. This particle localization-by-dissipation is fully consistent with our observations made above in this chapter and is, to a large extent, analogous to that discussed in Section 5.1.2 in the case of a double-well potential. As before, one can also define the renormalized energy band

$$\Delta_r = \hbar\tilde{y}/\tilde{\tau}_c \quad (6.23)$$

and along the lines with our previous analysis identify

$$\Delta_r = \Delta\left(\frac{\Delta}{\hbar\omega_0}\right)^{\frac{K}{1-K}}, \quad K < 1 \tag{6.24}$$

and

$$\Delta_r = 0, \quad K \geq 1 \tag{6.25}$$

We again observe that Ohmic dissipation qualitatively changes the ground-state properties of a quantum particle in a periodic potential causing localization phase transition at $K = 1$ and $T \to 0$. The basic physics behind this QPT is essentially the same as in the double-well case already considered in Chapter 5. At the same time, the RG flow is not identical to that illustrated in Fig. 5.5. An important difference between these two situations is that now the effective interaction parameter K does not change in the course of renormalization.

Note that the above analysis does not yet allow to fully describe the junction behavior, since it only applies provided the fugacity y remains small, i.e., for $E_J \gg E_C$. If the latter condition is violated, no useful information can be extracted from the RG equations (6.22). In order to extend our description to smaller values of E_J, let us split the effective action (5.1) into two parts

$$S_{\text{eff}}[\varphi] = S_G[\varphi] + S_J[\varphi] \tag{6.26}$$

where

$$S_G[\varphi] = \frac{TK}{4}\sum_{\omega_l}\tilde{\Lambda}^{-1}_{\omega_l}|\varphi_{\omega_l}|^2, \quad \tilde{\Lambda}^{-1}_{\omega_l} = \frac{\omega_l^2}{4KE_C} + \frac{|\omega_l|}{\pi} \tag{6.27}$$

and

$$S_J[\varphi] = -E_J\int_0^{\hbar\beta} d\tau\cos\varphi \tag{6.28}$$

The main problem in evaluating the path integral for the grand partition function (5.7) is the presence of non-Gaussian term S_J in the action. This problem can be cured by formally expanding the term $\exp(-S_J/\hbar)$ in infinite series [349]

$$\exp(-S_J/\hbar) = \sum_{n=0}^{\infty}\frac{(E_J/2\hbar)^n}{n!}\int_0^{\hbar\beta}d\tau_n\int_0^{\hbar\beta}d\tau_{n-1}\dots\int_0^{\hbar\beta}d\tau_1$$
$$\times\sum_{\varepsilon_j=\pm1}\exp\left[\mathrm{i}\int_0^{\hbar\beta}d\tau\sum_{j=1}^{n}\epsilon_j\delta(\tau-\tau_j)\varphi(\tau)\right] \tag{6.29}$$

where the sum $\sum_{\varepsilon_j=\pm 1}$ is taken only over neutral "charge" configurations

$$\sum_j \epsilon_j = 0 \tag{6.30}$$

The action for all other configurations is infinite, and, hence, such configurations do not contribute to the grand partition function at all. Substituting the expansion (6.29) into the path integral (5.7) together with Eqs. (6.26) and (6.27), we arrive at the infinite series of Gaussian path integrals that can be handled exactly with the result

$$\begin{aligned}\mathcal{Z} \sim \sum_{n=0}^{\infty} \frac{1}{2n!}\left(\frac{E_J}{2\hbar}\right)^{2n} \int_0^{\hbar\beta} d\tau_{2n} \int_0^{\hbar\beta} d\tau_{2n-1} \ldots \int_0^{\hbar\beta} d\tau_1 \\ \times \sum_{\varepsilon_j=\pm 1} \exp\left[-\frac{1}{K}\int_0^{\hbar\beta} d\tau \int_0^{\hbar\beta} d\tau' \rho_n(\tau)\tilde{\Lambda}(\tau-\tau')\rho_n(\tau')\right]\end{aligned} \tag{6.31}$$

where we defined the "charge density"

$$\rho_n(\tau) = \sum_{j=1}^{2n} \epsilon_j \delta(\tau - \tau_j) \tag{6.32}$$

and the sum $\sum_{\varepsilon_j=\pm 1}$ is again taken only over configurations obeying the neutrality condition (6.30). Since the second equation (6.27) yields

$$\tilde{\Lambda}(\tau) \simeq \begin{cases} -2KE_C|\tau|, & |\tau| \ll 1/KE_C, \\ -\ln(KE_C|\tau|), & |\tau| \gg 1/KE_C \end{cases} \tag{6.33}$$

it follows immediately that the partition function (6.31) again describes one-dimensional gas of logarithmically interacting charges similarly to Eq. (6.21) derived by means of the instanton analysis. Moreover, we observe that Eqs. (6.31) and (6.21) can be exactly mapped onto each other if we substitute

$$K \leftrightarrow 1/K, \quad \frac{E_J}{8KE_C} \leftrightarrow \frac{\Delta}{\hbar\omega_0}, \quad \epsilon_j \leftrightarrow \nu_j \tag{6.34}$$

In other words, the representations (6.21) and (6.31) for the same partition function are dual to each other under the transformation (6.34).

This duality feature [349] plays a very important role as it allows to obtain complementary information about the system properties.

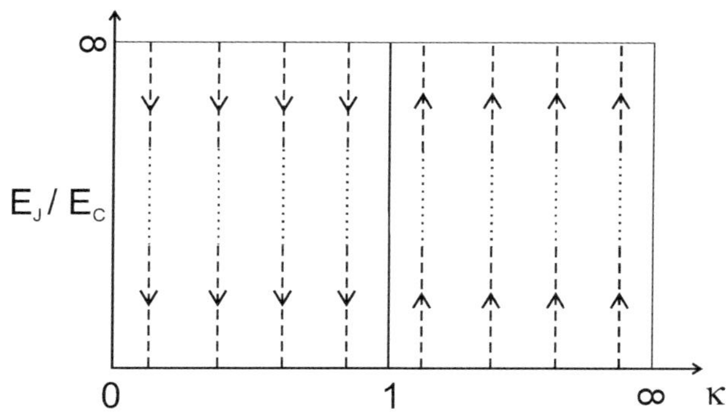

Figure 6.2 Schmid diagram for a linearly damped quantum particle in a periodic potential at $T = 0$. Dashed lines with arrows indicate the direction of RG flow in both delocalized ($K < 1$) and localized ($K > 1$) phases.

For instance, with the aid of Eqs. (6.34) and (6.22), we immediately recover the RG equations [75, 122, 182, 349]

$$d\tilde{z}/d\ln\tilde{\tau}_c = \tilde{z}(1 - 1/\tilde{K}), \quad d(1/\tilde{K})/d\ln\tilde{\tau}_c = 0 \tag{6.35}$$

where we now define the fugacity $z = E_J/KE_C$, renormalized fugacity $\tilde{z}$ and set the initial cutoff parameter to be $\tau_c \sim 1/KE_C$. As before, these equations apply as long as the renormalized fugacity remains small $\tilde{z} \ll 1$, i.e., for small values of E_J. In contrast, Eqs. (6.22) work well at large values of E_J. The RG flows obtained from two sets of RG Eqs. (6.22) and (6.35) can be combined in the phase diagram presented in Fig. 6.2. It is often called Schmid diagram.

We observe that for $K < 1$, the potential barrier $\sim E_J$ for the Josephson particle always decreases in the course of renormalization. Indeed, at large E_J, the fugacity $\tilde{y}$ increases with increasing parameter $\tilde{\tau}_c$, implying decreasing E_J, and at small E_J, the fugacity $\tilde{z}$ decreases implying the same trend for E_J. As a result, at $T \to 0$, the Josephson coupling energy is renormalized to zero, i.e., the quantum particle effectively "feels" no potential at all and, hence, remains delocalized for *any* E_J. Formally, it implies that the expectation value $\langle[\varphi(\tau) - \varphi(0)]^2\rangle$ diverges in the long-time limit. On the contrary, for $K > 1$, the Josephson coupling energy E_J always increases in the course of renormalization, as follows from Eqs. (6.35) at small E_J and Eqs. (6.22) at large E_J. Eventually at $T \to 0$, an effective potential barrier for the Josephson particle becomes very large, i.e., the particle remains localized without any chance to escape from an infinitely deep potential well. Hence, the expectation value $\langle[\varphi(\tau)-\varphi(0)]^2\rangle$ always stays finite and never diverges.

We arrive at the conclusion that for the model under consideration, dissipative quantum phase transition takes place at the dissipation strength $K = 1$ *at all values* E_J/E_C, as indicated by the vertical solid line in the diagram in Fig. 6.2. This line separates two different phases: $K < 1$ where the particle is delocalized no matter how strong the potential is and $K > 1$ where the particle is localized and cannot move even for very small E_J. Note that strictly speaking, the above RG analysis does not yet allow to draw any definite conclusion for the parameter region $E_J \sim E_C$ where both Eqs. (6.22) and (6.35) reach the borderline of their applicability. In this region, the RG treatment was complemented by quantum Monte Carlo (QMC) analysis [193], which confirmed that the localization phase transition indeed occurs at $K = 1$ also for $E_J \sim E_C$. Experimental evidence consistent with the above physical picture of a dissipative QPT in single Josephson junctions was reported in a number of studies [327, 329, 367, 408].

The above analysis also allows to establish a useful relation between the phase–phase and "density–density" correlators $\langle\varphi(\tau)\varphi(\tau')\rangle$ and $\langle\rho(\tau)\rho(\tau')\rangle$. For this purpose, let us define the generating functional

$$\mathcal{Z}[\eta] \sim \int \mathcal{D}\exp\left[-\frac{S_{\text{eff}}[\varphi]}{\hbar} + \mathrm{i}\int d\tau \int_0^{\hbar\beta} \eta(\tau)\varphi(\tau)\right] \tag{6.36}$$

Obviously, the phase–phase correlation function is determined by means of the following relation:

$$\langle\varphi(\tau)\varphi(\tau')\rangle = -\frac{\delta^2\mathcal{Z}[\eta]}{\delta\eta(\tau)\delta\eta(\tau')}_{\eta=0} \tag{6.37}$$

On the other hand, combining Eqs. (6.29) and (6.36) and, exactly as before, performing Gaussian path integrals in each term of the series, we arrive at the expression for the generating functional $\mathcal{Z}[\eta]$ defined by Eq. (6.31) with $\rho(\tau) \to \rho(\tau) + \eta(\tau)$. Then making use of Eq. (6.37) and performing the Fourier transformation with respect to time, we obtain

$$\langle\varphi\varphi\rangle_\omega = \frac{2}{K}\tilde{\Lambda}_\omega\left(1 - \frac{2}{K}\tilde{\Lambda}_\omega\langle\rho\rho\rangle_\omega\right) \tag{6.38}$$

This relation is important since evaluating the phase–phase correlation function, one can immediately determine the system resistance. Indeed, turning to real times, one can write down the standard Kubo formula for the voltage response to the external current bias

$$\langle V(t)\rangle = \frac{\hbar}{2e}\langle\dot{\varphi}\rangle = \frac{\hbar}{4e^2}\int_{-\infty}^{t}\langle[\dot{\varphi}(t), \varphi(t')]\rangle I(t') \tag{6.39}$$

where $[a, b] = ab - ba$ is the commutator. Thus, by calculating the imaginary-time correlation function $|\omega|\langle\varphi\varphi\rangle_\omega$ after the standard procedure of its analytic continuation [273], one obtains the real-time voltage response $\langle V(t)\rangle$ to the external current bias $I(t)$. In the low-frequency limit, one gets the effective resistance

$$R_{\text{eff}} = \frac{\hbar}{4e^2}|\omega|\langle\varphi\varphi\rangle_{\omega\to 0} \tag{6.40}$$

Employing the analogy between a Josephson junction and a quantum particle, one can also define the mobility μ_φ as a ratio between the "particle velocity" $\langle\dot{\varphi}\rangle$ and the "external force" I, i.e.,

$$\mu_\varphi(\omega) = \langle V\rangle_\omega / I_\omega \tag{6.41}$$

Obviously, at low frequencies, this mobility is just proportional to the system resistance. Making use of this definition together with Eqs. (6.38) and (6.39), at sufficiently low frequencies, we can write

$$\mu_\varphi(\omega) = (4e^2 R_S/\hbar)[1 - \mu_q(\omega)/2e] \tag{6.42}$$

where we formally defined the quantity $\mu_q = (4e/K)\tilde{\Lambda}_\omega\langle\rho\rho\rangle_\omega$. We will later demonstrate that the physical meaning of this quantity is the mobility in the quasicharge space. Hence, Eq. (6.42) can be interpreted as the relation between the mobilities in the phase and charge spaces. Bearing in mind the duality between these two spaces (see below) and depending on the system parameters, it will be more convenient to analyze only one of these mobilities, either μ_φ or μ_q, and then reconstruct the other one by means of Eq. (6.42).

6.4 Phase and Charge Representations

Until now we mainly described quantum properties of Josephson junctions in terms of the phase variable φ, which is a relevant quantum degree of freedom in this case, as we have demonstrated in Chapter 2. However, the phase representation of the problem may be not the only convenient one. Below we will argue that useful and sometimes complementary information can be obtained if one considers quantum dynamics of another collective variable—the junction charge.

Let us for a moment ignore dissipation. Then, as it was already proven in Chapter 2, in the low frequency and energy limit and in the absence of external bias quantum properties of ultrasmall Josephson junctions are described by the Hamiltonian

$$H = \frac{\hat{Q}^2}{2C} - E_J \cos\varphi, \qquad \hat{Q} = -\mathrm{i}\hbar\frac{d}{d(\hbar\varphi/2e)} + Q_x \tag{6.43}$$

which is identical to that for a quantum particle in a periodic potential. Here $\hat{Q}$ is the operator of the junction charge canonically conjugated to that of the phase. This operator is fully analogous to that of the momentum of a quantum particle. Standard quantum mechanical considerations immediately lead us to the picture of Bloch states

$$\psi_{n,Q_x}(\varphi + 2\pi) = \exp(2\pi\mathrm{i}Q_x/2e)\psi_{n,Q_x}(\varphi) \tag{6.44}$$

and Bloch energy bands $E(Q_x)$, which depend on the quasicharge Q_x playing exactly the same role as the quasi-momentum for a quantum particle in a periodic potential.

Of conceptual importance is the question whether the states φ and, say, $\varphi + 2\pi$ are physically equivalent or not. In other words, the question is whether we should treat the phase φ (i) as a compact variable defined on a unit circle, i.e., $0 \le \varphi < 2\pi$ or (ii) as an extended variable defined for $-\infty < \varphi < \infty$. The answer to this question depends on a concrete physical realization of our Josephson system [345].

At the first sight, the phase states differing by 2π should be considered identical since they describe the same state of the superconducting condensate. In this case, the choice (i) applies. This consideration is indeed sufficient provided our junction is not attached to any external circuit/environment. The situation changes, however, if the junction is embedded in some external environment. Examples of that kind were already considered, e.g., in Chapters

4 and 5. One can attach the leads and bias the junction by the current I_x. In this case, the energies of the states φ and $\varphi + 2\pi$ differ by $\pi\hbar I_x/e$ (see also Fig. 5.1a), which obviously implies that these states are physically distinguishable. Similar arguments allow to distinguish these states in the case of a SQUID (see Fig. 5.1b). In both these cases, the system Hamiltonian is clearly not invariant under the phase shift transformation $\varphi \to \varphi+2\pi$. Even if no external bias is applied to the junction, but it is connected to a (dissipative) environment, the states φ and $\varphi + 2\pi$ are different because the transition between these states is accompanied by a change in the quantum state of the environment. In all these cases, the choice (ii) applies provided the states φ and $\varphi + 2\pi$ can be distinguished.

Having learned that the physical difference between compact and non-compact phase φ lies in the absence or presence of an external environment, it is easy to realize that (i) and (ii) just correspond to different choices of the junction charge states [345]. Namely, if the junction is decoupled from any external circuit/environment (choice (i)), only discrete charge states $Q_x + 2em$ with integer m are allowed. In other words, (i) corresponds to conserved quasicharge Q_x and the junction charge may differ from Q_x only due to tunneling of m Cooper pairs from one superconducting electrode to another. If, on the contrary, the junction is coupled to external leads, its charge may vary continuously, i.e., in the case (ii), all charge states are allowed. Thus, in order to choose between compact and extended phase states, it is necessary to determine if discrete or continuous charge states are adequate for the corresponding physical situation.

In both cases (i) and (ii), one can establish a transparent relation between phase and charge representations of the grand partition function $\mathcal{Z}$ for our system. We begin with choice (i), implying that phase states differing by 2π are indistinguishable. In this case, the partition function of the Josephson junction described by the Hamiltonian (6.43) can be written in the form of the path integral

$$\mathcal{Z}_{Q_x} = \sum_{k=-\infty}^{\infty} \exp(2\pi i k Q_x/2e) \int d\varphi_0 \int_{\varphi_0}^{\varphi_0+2\pi k} \mathcal{D}\varphi \exp(-S_0[\varphi]/\hbar) \tag{6.45}$$

where

$$S_0[\varphi] = \int_0^{\hbar\beta} d\tau \left[\frac{C}{2}\left(\frac{\hbar\dot{\varphi}}{2e}\right)^2 - E_J \cos\varphi\right] \tag{6.46}$$

In Eq. (6.45), we included the contribution from all trajectories $\varphi(\tau)$, which wind around the circle k times by explicitly summing up over this winding number k. One can demonstrate [345] that the same partition function in the charge representation reads

$$\mathcal{Z}_{Q_x} = \sum_m \exp\left(-\frac{E_m(Q_x)}{T}\right) \tag{6.47}$$

where $E_m(Q_x)$ are Bloch energy bands derived from the Hamiltonian (6.43).

The equivalence between the expressions for the partition function (6.45) and (6.47) can be easily verified in the limit $E_C \gg E_J$. In this case, in the main approximation, one can neglect the Josephson term in the action (6.46). After that the path integral in Eq. (6.45) becomes Gaussian and, hence, can be handled exactly. Then from Eq. (6.45), one finds

$$\begin{aligned}\mathcal{Z}_{Q_x} &= \sqrt{\frac{\pi T}{4E_C}} \sum_{k=-\infty}^{\infty} \exp\left(i\frac{Q_x}{2e}2\pi k - \frac{T}{4E_C}\pi^2 k^2\right) \\ &= \sum_{p=-\infty}^{\infty} \exp\left(-\frac{(Q_x + 2ep)^2}{2CT}\right)\end{aligned} \tag{6.48}$$

where in the last equation, we employed an extension of Poisson resummation theorem related to properties of Jacoby theta-function. Equation (6.48) demonstrates the equivalence of the two representations for the partition function (6.45) and (6.47) in the limit $E_C \gg E_J$ since in this case, the energy bands are parabolic $E(Q) \simeq Q^2/2C$ with $Q = Q_x + 2ep$.

Note that the equivalence of Eqs. (6.45) and (6.47) holds for any E_C and E_J, and it can also be proven analytically in the limit $E_J \gg E_C$ by means of the instanton technique. For the corresponding calculation, we refer the reader to the review [345].

Let us now turn to the situation (ii) with distinguishable states φ and $\varphi + 2\pi$. In this case, we do not have to sum up over the winding numbers, but rather we should integrate over all trajectories in the phase space, i.e.

$$\mathcal{Z} = \int d\varphi_0 \int_{\varphi_0}^{\varphi_0} \mathcal{D}\varphi \exp(-S[\varphi]/\hbar) \tag{6.49}$$

As we already discussed, choice (ii) implies continuous charge states and, hence, the charge representation of the partition function (6.49) reads

$$\mathcal{Z} = \int dQ \sum_m \exp\left(-\frac{E_m(Q)}{T}\right) \tag{6.50}$$

Again the equivalence between phase and charge representations holds in a general case, but it can most easily be verified in the limit of large $E_C \gg E_J$. Substituting the action (6.46) into Eq. (6.49), disregarding the Josephson term and performing Gaussian path integral over $\varphi(\tau)$, we observe the equivalence of the resulting expression to the partition function

$$\mathcal{Z} = \int_{-\infty}^{\infty} \frac{dQ}{2e} \exp\left(-\frac{Q^2}{2CT}\right) \tag{6.51}$$

6.5 Quasicharge as a Quantum Variable

In the above analysis, the Josephson phase φ is considered a quantum variable, whereas quasicharge remains essentially classical. This is only possible in certain physical limits. In this respect, it is necessary to bear in mind that the phase and the charge are canonically conjugate variables in a complete analogy to, e.g., the coordinate and momentum variables in quantum mechanics. This observation implies that both these collective variables are quantum, i.e., also the (quasi-)charge dynamics (along with that of the phase) should in general be described quantum mechanically.

In order to illustrate this statement, let us consider a simple $JCRL$ circuit shown in Fig. 6.3. This circuit consists of a Josephson junction (J), which has a capacitance C and which is connected in series to an Ohmic resistor R_S and an inductance L. In the absence of the Josephson element, the circuit obviously reduces to a standard CRL one which can be quantized in a usual manner. For this purpose,

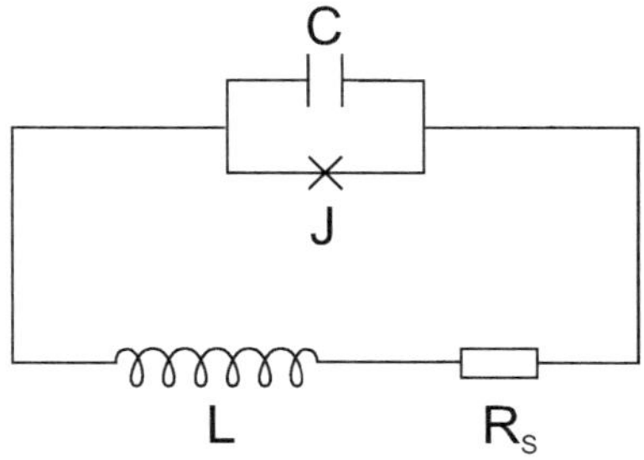

Figure 6.3 The JCRL circuit described by the Hamiltonian (6.52).

one can introduce the charge variable q, which time derivative just defines the current in the circuit, i.e., $I = \dot{q}$. The energies of a capacitor and an inductor are, respectively, $q^2/2C$ and $L\dot{q}^2/2$, while the contribution of a resistor can generally be described by introducing the Caldeira–Leggett bath of oscillators interacting with the charge variable q following exactly the same line of reasoning as in Chapter 1. What remains is to combine the corresponding Hamiltonian for the CRL circuit with that for the Josephson element (6.43), where the role of an external charge Q_x is now played by the collective charge variable q.

Collecting all relevant contributions, we arrive at the Hamiltonian [23]

$$\hat{H} = \frac{1}{2C}\left(-i\hbar\frac{\partial}{\partial(\hbar\varphi/2e)} + q\right)^2 - E_J\cos\varphi - \frac{\hbar^2}{2L}\frac{\partial^2}{\partial q^2} + \sum_i\left[\frac{P_i^2}{2M_i} + \frac{M_i\Omega_i^2}{2}\left(X_i + \frac{c_i}{\Omega_i^2}q\right)^2\right] \qquad (6.52)$$

which describes the $JCRL$ circuit of Fig. 6.3. In order to properly account for an Ohmic resistor R_S, the bath frequency spectrum and the coupling constants c_i should obey Eq. (1.54), which now takes the form

$$\frac{\pi}{2}\sum_i\frac{c_i^2}{M_i\Omega_i}\delta(\omega - \Omega_i) = \omega R_S \qquad (6.53)$$

Our next step is pretty standard. We write down the partition function of the system in terms of the path integral over all quantum variables entering the Hamiltonian (6.52) and trace out the oscillator coordinates X_i. After this exact procedure (which we already described in details in Chapter 1), one arrives at the effective action

$$S[\varphi, q] = S_0[\varphi] + \int_0^{\hbar\beta} d\tau \left(\frac{L}{2}\dot{q}^2 + \mathrm{i}\frac{\hbar}{2e}\dot{\varphi} q + \frac{\pi T^2 R_S}{4\hbar^2} \int_0^{\hbar\beta} d\tau' \frac{(q(\tau) - q(\tau'))^2}{\sin^2[\pi T(\tau - \tau')/\hbar]} \right), \tag{6.54}$$

where $S_0[\varphi]$ is defined in Eq. (6.46). We observe that the effective action (6.54) depends on the two quantum variables—$\varphi(\tau)$ and $q(\tau)$—and that dissipation is accounted for by the non-local in time Caldeira–Leggett term, which now involves the q-variable (in contrast to the analogous term involving the φ-variable, as in Chapters 2, 4, 5 and in the beginning of this chapter). The phase and charge variables are coupled by the term $\sim\dot{\varphi}q$, and, hence, their quantum dynamics is not independent.

In order to write down the path integral expression for the partition function, it is necessary to define proper boundary conditions for both quantum variables φ and q. In our present formulation, the phase variable φ describes only the properties of the junction and not those of the external circuit. Hence, φ should be considered compact, i.e., defined on a unit circle. In this case, the boundary conditions analogous to those in Eq. (6.45) apply. As for the charge variable (playing the role of an external charge for the Josephson junction), it can vary from $-\infty$ to $+\infty$, i.e., it should be treated as extended. Accordingly, the grand partition function takes the form

$$\mathcal{Z} = \int dq_0 \int_{q_0}^{q_0} \mathcal{D}q \sum_{k=-\infty}^{\infty} \int d\varphi_0 \int_{\varphi_0}^{\varphi_0+2\pi k} \mathcal{D}\varphi \exp(-S[\varphi, q]/\hbar) \tag{6.55}$$

Note that since the action (6.54) is quadratic in q, one can integrate out the charge variable exactly. For pedagogical purposes, it is worthwhile to carry out this integration in a detailed manner. Expanding the trajectories $q(\tau)$ and $\varphi(\tau)$ in the Fourier series, i.e.,

$$q(\tau) = q_0 + \sum_{n\neq 0} q_n \exp(-\mathrm{i}\omega_n \tau), \quad \omega_n = 2\pi n T \tag{6.56}$$

and similarly for $\varphi(\tau)$, we rewrite the q-integrals in Eq. (6.55) as

$$\int \prod_{n=-\infty}^{\infty} dq_n \exp\left[2\pi \mathrm{i} k q_0/2e - \sum_{n\neq 0}\left((L\omega_n^2 + R_S|\omega_n|)\frac{|q_n|^2}{2\hbar} + \mathrm{i}\frac{\omega_n}{2e}\varphi_n q_n^*\right)\right] \tag{6.57}$$

Integration over q_0 eliminates the sum over winding numbers k leaving only the term with $k = 0$. This is just an illustration of "decompactification" of the phase variable already discussed above: In the presence of an external circuit (leads) attached to the Josephson junction, its charge becomes effectively continuous and the phase φ becomes a non-compact (extended) variable defined on the whole axis $-\infty < \varphi < \infty$ rather than on a unit circle. Gaussian integrals over all other Fourier coefficients $q_{n\neq 0}$ in Eq. (6.57) can be easily handled. As a result, we obtain the partition function in the "phase-only" representation

$$\mathcal{Z} = \int d\varphi_0 \int_{\varphi_0}^{\varphi_0} \mathcal{D}\varphi \exp(-S[\varphi]/\hbar) \tag{6.58}$$

where

$$S[\varphi] = S_0[\varphi] + \frac{\hbar^2 T}{16e^2 R_S} \times \sum_{n\neq 0}\int_0^{\hbar\beta} d\tau \int_0^{\hbar\beta} d\tau' \frac{|\omega_n| \cos[\omega_n(\tau-\tau')][\varphi(\tau)-\varphi(\tau')]^2}{1+(L/R_S)|\omega_n|} \tag{6.59}$$

In the zero inductance limit $L \to 0$, the action (6.59) reduces to the standard Caldeira–Leggett form, while in the limit $R_S \to 0$, Eq. (6.59) yields $S[\varphi] = S_0[\varphi] + S_L[\varphi]$, where

$$S_L[\varphi] = \sum_{n\neq 0} \frac{\Phi_0^2}{2L}\frac{|\varphi_n|^2}{(2\pi)^2} \tag{6.60}$$

defines the inductance contribution.

We may say that the derivation of the effective action from the Hamiltonian (6.52) is very similar to one already employed, e.g., in Chapters 1 and 2 with only one important difference: Here we introduced an additional collective variable—the quasicharge q—which mediates interaction between the phase variable φ and the bath oscillators. In a way, we effectively extracted the q-variable from an infinite number of the bath degrees of freedom and traced out the bath variables in two steps: first by integrating out all oscillator coordinates X_i (which yields the action (6.54) in the mixed phase-charge representation) and then performing an additional integration over the collective variable q.

It is also clear that the quasicharge q may formally be introduced even without employing the Hamiltonian (6.52) simply as a result of the Hubbard–Stratonovich transformation, which "decouples" the Caldeira–Leggett dissipative term in the action. In other words, one can go in the "opposite" direction starting from the Caldeira–Leggett action derived microscopically in Section 2.5.2 and demonstrating the equivalence of the corresponding partition function to that defined in Eqs. (6.54) and (6.55), which contain an extra path integral over $q(\tau)$.

In a number of physical situations, one might be interested only in describing quantum dynamics of the charge q, not that of the phase φ. In such situations, it can be convenient to integrate out $\varphi(\tau)$ in Eqs. (6.54) and (6.55) and derive the "charge-only" representation of the partition function. The problem here is that the corresponding path integral over the phase variable is non-Gaussian and, hence, cannot in general be performed exactly. However, provided the time scale of variation of $q(\tau)$ is slow as compared to that of φ, the latter variable can be integrated out.

Let us rewrite Eq. (6.55) in the form

$$\mathcal{Z} = \int dq_0 \int_{q_0}^{q_0} \mathcal{D}q \exp(-\tilde{S}_G[q]/\hbar)\mathcal{F}[q(\tau)] \tag{6.61}$$

where

$$\tilde{S}_G[q] = \int_0^{\hbar\beta} d\tau \frac{L}{2}\dot{q}^2 + \frac{R_S}{4\pi}\int_0^{\hbar\beta} d\tau \int_0^{\hbar\beta} d\tau' \frac{\pi^2 T^2 (q(\tau) - q(\tau'))^2/\hbar^2}{\sin^2[\pi T(\tau - \tau')/\hbar]} \tag{6.62}$$

and

$$\mathcal{F}[q(\tau)] = \sum_{k=-\infty}^{\infty} \int d\varphi_0 \int_{\varphi_0}^{\varphi_0 + 2\pi k} \mathcal{D}\varphi \exp\left(-\frac{S_0[\varphi]}{\hbar} + \mathrm{i}\int_0^{\hbar\beta} d\tau \dot{\varphi} q/2e\right) \tag{6.63}$$

Transforming the charging term in the action $S_0[\varphi]$ in this equation by means of the identity [345]

$$\exp\left[-\int_0^{\hbar\beta}\frac{d\tau}{\hbar}\frac{C}{2}\left(\frac{\hbar}{2e}\frac{\partial\varphi}{\partial\tau}\right)^2\right]$$
$$=\int\mathcal{D}Q\exp\left[-\int_0^{\hbar\beta}\frac{d\tau}{\hbar}\left(\frac{Q^2(\tau)}{2C}+\mathrm{i}\frac{\partial Q(\tau)}{\partial\tau}\frac{\hbar\varphi(\tau)}{2e}\right)\right.$$
$$\left.+\mathrm{i}\frac{Q(\beta)\varphi(\beta)-Q(0)\varphi(0)}{2e}\right] \tag{6.64}$$

setting $Q(\beta)=Q(0)\equiv Q_0$, $\varphi(\beta)=\varphi(0)+2\pi k$, expanding the resulting partition function in powers of E_J (cf. Eq. (6.29)) and performing Gaussian path integrals in each term of this expansion, we obtain

$$\mathcal{F}[q(\tau)]=\int dQ_0\int_{Q_0}^{Q_0}\mathcal{D}Q\sum_{n=0}^{\infty}\frac{1}{2n!}\left(\frac{E_J}{2}\right)^{2n}$$
$$\times\int_0^{\hbar\beta}d\tau_{2n}\int_0^{\hbar\beta}d\tau_{2n-1}\ldots\int_0^{\hbar\beta}d\tau_1$$
$$\times\sum_{\varepsilon_j=\pm1}\delta\left[\dot{Q}(\tau)-\dot{q}(\tau)-\dot{Q}_n(\tau)\right]\exp\left[-\int_0^{\hbar\beta}\frac{d\tau}{\hbar}\frac{Q^2(\tau)}{2C}\right] \tag{6.65}$$

with $\dot{Q}_n(\tau)=2e\rho_n(\tau)$ and $\rho_n(\tau)$ being the "charge density" defined in Eq. (6.32). Here, as before, the sum $\sum_{\varepsilon_j=\pm1}$ is taken only over neutral "charge" configurations $\sum_j\epsilon_j=0$ (cf. Eq. (6.30)).

Assuming that the variable $q(\tau)$ changes adiabatically, one would be able to separate the characteristic time scales of the slow and the fast variables, respectively, $q(\tau)$ and $Q_n(\tau)$. Then the sum over all possible charge configurations in Eq. (6.65) can be easily handled, since in this case the combination of series in Eq. (6.65) exactly coincides with the grand partition function of the system, which in the low-temperature limit reduces to [410]

$$\mathcal{F}[q(\tau)]=\exp\left(-\frac{1}{\hbar}\int_0^{\hbar\beta}d\tau\,E_0(q)\right) \tag{6.66}$$

Here, as before, $E_0(q)$ is the $2e$-periodic charge-dependent ground-state energy (lowest Bloch band) corresponding to the Hamiltonian (6.43). Equation (6.66) extends the equivalence between the two representations of the partition function (6.45) and (6.47) to the case of (slowly) fluctuating charge variable $q(\tau)$.

Combining Eqs. (6.61) and (6.66), we immediately arrive at the "charge-only" representation of the partition function

$$\mathcal{Z} = \int dq_0 \int_{q_0}^{q_0} \mathcal{D}q \exp(-S[q]/\hbar) \tag{6.67}$$

where $S[q]$ is the adiabatic charge-dependent effective action

$$S[q] = \tilde{S}_G[q] + \int_0^{\hbar\beta} d\tau E_0(q) \tag{6.68}$$

One can also apply an external voltage bias V_x in which case one should also account for an extra term in the action [35]

$$S[q, V_x] = S[q] - \int_0^{\hbar\beta} d\tau q V_x \tag{6.69}$$

which yields nonzero expectation value of the quasicharge obtained from the equation

$$V_x = \partial E_0(q)/\partial q \tag{6.70}$$

What remains is to explicitly formulate the adiabaticity condition under which the effective action (6.68) for the quasicharge has been derived. As we already explained, this action remains applicable provided the characteristic time scale ω_q^{-1} for the variable $q(\tau)$ remains much longer than that of

$$Q_n(\tau) = 2e \sum_{j=1}^{2n} \epsilon_j \theta(\tau - \tau_j) \tag{6.71}$$

which describes a sequence of virtual tunneling events of a Cooper pair across the junction. After each of these events corresponding to jumps between neighboring energy bands, the junction charge changes by $\pm 2e$.

Consider first the limit of small Josephson coupling energy $E_J \ll E_C$ in which case the Bloch bands are nearly parabolic $E(q) \simeq (q - 2en)^2/2C$ except close to the edges of the Brillouin zones $q = 2e(n + 1/2)$ where the interband gaps open up (see Fig. 6.4). Provided temperature is sufficiently low $T \ll E_C$, the system spends most of the time in the lowest energy band $E_0(q)$, but from time to time, it undergoes virtual transitions to the higher band $E_1(q)$ and

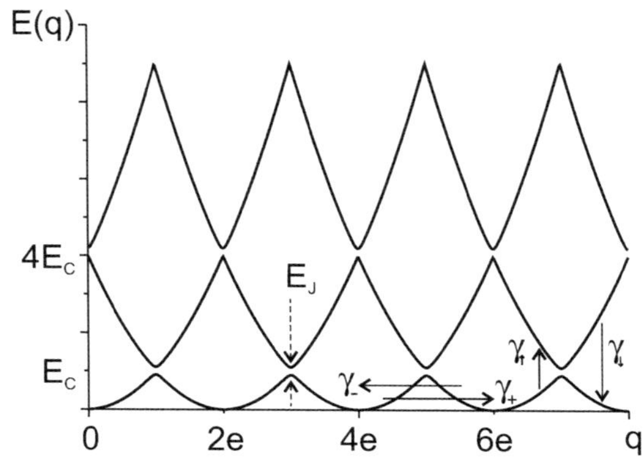

Figure 6.4 The bandstructure $E(q)$ in the limit $E_J \ll E_C$. Quasicharge tunneling rates $\gamma_\pm$ between the neighboring wells within the lowest zone $E_0(q)$ are indicated by horizontal arrows, while Zener tunneling rate from the lowest to the first Brillouin zone $\gamma_\uparrow$ as well as inelastic relaxation rate $\gamma_\downarrow$ in the opposite direction are indicated by vertical arrows.

returns after a typical time interval $\sim (E_1(q) - E_0(q))^{-1}$. Hence, for q not very close to $2e(n+1/2)$, the applicability condition for the action (6.68) reads $\hbar\omega_q \ll E_C|q/e - 2n - 1| \sim E_C$. In the vicinity of the points $q = 2e(n+1/2)$, this condition changes since in this case, $E_1(e) - E_0(e) = E_J$ and, on top of that, for $\dot{q} \neq 0$, it is necessary to bear in mind the real process of Zener tunneling between the lowest and the first excited energy states. Below we will demonstrate that the corresponding Zener tunneling rate is defined as

$$\gamma_\uparrow = \omega_q \exp\left(-\frac{\pi}{16}\frac{E_J^2}{\hbar\omega_q E_C}\right) \tag{6.72}$$

Hence, in order to ignore Zener transitions from the lowest to higher bands at $q \approx 2e(n+1/2)$, one should require $\hbar\omega_q \ll E_J^2/E_C$. Combining this condition with that found above, one gets

$$\hbar\omega_q \ll \max(E_C|q/e - 2n - 1|, E_J^2/E_C) \tag{6.73}$$

At low temperatures $T \ll E_C$, this condition controls the applicability of Eq. (6.68) at all values $q(\tau)$ and in the limit $E_J \ll E_C$.

In the opposite limit of large Josephson coupling energies $E_J \gg E_C$, a similar analysis can be developed with the only difference that in this case $E_1(q) - E_0(q) \simeq \hbar\omega_0 = (8E_J E_C)^{1/2}$ and, hence, in the low-temperature limit $T \ll \hbar\omega_0$ instead of (6.73), one should request

$$\hbar\omega_q \ll (8E_J E_C)^{1/2} \tag{6.74}$$

Perhaps we also note that in the limit $E_J \gg E_C$, one might consider a formal expansion of $\mathcal{F}[q(\tau)]$ in powers of large Josephson coupling energy E_J (6.65) somewhat inconvenient. An alternative and more appropriate way to proceed in this limit is to employ an instanton technique [410]. Similarly to our analysis carried out in the beginning of Section 6.3, we evaluate the path integral over φ in Eq. (6.63) semiclassically with the aid of instanton trajectories (6.18). Then we arrive at the result

$$\mathcal{F}[q] = \sum_{k=0}^{\infty} \left(\frac{\Delta}{2}\right)^{2k} \frac{1}{2k!} \int_0^{\hbar\beta} d\tau_{2k} \int_0^{\hbar\beta} d\tau_{2k-1} \ldots \int_0^{\hbar\beta} d\tau_1$$
$$\times \sum_{\nu_i=\pm 1} \exp\left[i \sum_{j=1}^{2n} \int_0^{\hbar\beta} d\tau \nu_j \frac{q(\tau)}{2e} \frac{d\tilde{\varphi}(\tau-\tau_j)}{d\tau} \right] \tag{6.75}$$

where the sum $\sum_{\nu_i=\pm 1}$ again includes only neutral charge configurations (6.19) and $\tilde{\varphi}(\tau)$ is defined in Eq. (5.19). Provided the characteristic frequency ω_q is smaller than that of the phase variable $\omega_\varphi \sim \omega_0$, we can approximate

$$\frac{d\tilde{\varphi}(\tau-\tau_j)}{d\tau} \approx 2\pi\delta(\tau-\tau_j),$$

in which case, the sum in Eq. (6.75) is easily evaluated and yields Eq. (6.66) with $E_0(q)$ defined in Eq. (6.20). Thus, we reproduce the same applicability condition (6.74) also with the aid of the instanton analysis.

6.6 Phase-Charge Duality and Quantum Tunneling of the Charge

The above analysis reveals *exact duality* between effective actions in the charge and phase representations in the limit of large Josephson coupling energies $E_J \gg E_C$. This is a remarkable property of our system, which is observed if we perform the transformation

$$\varphi \leftrightarrow \frac{\pi q}{e}, \; E_J \leftrightarrow \Delta, \; K \leftrightarrow 1/K, \quad E_C \leftrightarrow E_L = \frac{\pi^2}{8e^2 L}, \; \frac{I_x}{2e} \leftrightarrow \frac{eV_x}{\pi} \tag{6.76}$$

translating each of the terms in the action $S[q]$ defined by Eqs. (6.68), (6.69), (6.62), and (6.20) into the corresponding term in the action $S[\varphi]$ extensively used in our previous considerations, cf., e.g., Eqs. (2.139) and (2.125) (with $Z_S(\Omega_n) = R_S$) or equivalent expressions elsewhere in Chapters 2–6. This duality property formulated on the level of effective actions is similar to one established in Section 6.3 for the grand partition function (cf. Eq. (6.34)), though it differs from the latter in several aspects, in particular related to its applicability range.

Most importantly, it demonstrates that all physical phenomena—both classical and quantum—established and analyzed in the phase space have their exact replicas in the dual charge space. For instance, we may give an additional interpretation to the dissipative localization phase transition and the Schmid diagram discussed above. Namely, this quantum phase transition not only implies localization of the phase at $K > 1$ but also simultaneous delocalization of the charge at such values of K. Vice versa, for $K < 1$, the charge is localized while the phase is delocalized. In the dissipativeless limit $K \to 0$, the phase variable φ behaves purely quantum mechanically (meaning that its properties are described by the wave function $\psi(\varphi)$ determined from the Schrödinger equation with the Hamiltonian (6.43)), while the charge q remains a good quantum number (so-called c-number variable). In the opposite very strong dissipation limit $K \to \infty$, already the phase variable behaves classically, while quantum dynamics of the charge is determined from the Schrödinger equation with the Hamiltonian

$$\hat{H} = \frac{\hat{\Phi}^2}{2L} - \Delta \cos(\pi q/e) \tag{6.77}$$

where $\hat{\Phi} = -i\hbar d/dq$ is the flux operator proportional to that of the phase $\hat{\varphi} = 2\pi \hat{\Phi}/\hat{\Phi}_0$.

Provided the phase φ remains a classical variable, its dynamics is controlled by the Josephson relations (2.1) and the corresponding Langevin equations (see, e.g., Eqs. (2.9) and (2.10)). Likewise, should the charge q behave classically, one can immediately write down the relations strictly dual to the Josephson ones, i.e.,

$$V = V_C \sin(\pi q/e), \qquad I = \dot{q} \tag{6.78}$$

where V_C is the critical voltage, which equals to $V_C = \pi\Delta/e$ in the limit $E_J \gg E_C$. We observe that, as compared to the Josephson relations (2.1), one should simply interchange voltage and current (along with phase and charge). In the case of a circuit dual to one of Fig. 2.2 (cf., e.g., Figs. 6.7a and 6.7b below), classical dynamics of the charge variable is controlled by the Langevin equation

$$L\ddot{q} + R_S\dot{q} + \partial E_0(q)/\partial q = V_x + \tilde{\xi}_S(t) \tag{6.79}$$

where $\partial E_0(q)/\partial q = V_C \sin(\pi q/e)$ for $E_J \gg E_C$ and the correlator

$$\langle \tilde{\xi}_S(t)\tilde{\xi}_S(0)\rangle = R_S \int \frac{d\omega}{2\pi} e^{-i\omega t}\,\omega \coth\frac{\hbar\omega}{2T} \tag{6.80}$$

defines the charge noise produced by the resistor R_S. As before, taking the classical limit $\hbar\omega \ll T$ in Eq. (6.80), we recover the white noise

$$\langle \tilde{\xi}_S(t)\tilde{\xi}_S(0)\rangle = 2R_S T\delta(t) \tag{6.81}$$

Equations (6.79) and (6.80) (or (6.81)) are exactly dual to the Langevin equations defined in Eqs. (2.9) and (2.10) (or (2.11)). While the latter equations for the phase are applicable for $R_S \ll R_Q$, the Langevin equations for the charge (6.79), (6.80) hold, provided $R_S \gg R_Q$. In this limit, it follows immediately that for $V_x < V_C$ and low enough temperatures, the classical charge variable remains constant $q = (e/\pi)\arcsin(V_x/V_C)$ and, hence, $I = \dot{q} = 0$, i.e., the current cannot flow through the junction at all. This is the well-known phenomenon of *Coulomb blockade* of Cooper pairs [30, 345], which is suppressed tunneling across the junction at low enough voltages and temperatures, since it would yield an increase in the total energy of the system.

At $V_x > V_C$, the current $I = \dot{q}$ is not anymore equal to zero, Cooper pairs periodically tunnel across the junction, and, hence, the voltage $V(t)$ oscillates with the fundamental frequency $\nu_B \equiv \omega_B/2\pi = I/2e$. This is the phenomenon of Bloch oscillations [36, 276], which is exactly dual to that of Josephson oscillations of the current with the frequency $\omega_J = 2eV/\hbar$. Similarly, in the presence of an external ac signal depending on which variable behaves classically—$\varphi(t)$ or $q(t)$—there appear either Shapiro steps or Bloch steps on the $I-V$ curves of the system. For more details, we refer the reader to the review papers [30, 345].

It is important to emphasize that the above physical picture of dual phenomena in the phase and charge spaces remains applicable not only in the limit $E_J \gg E_C$ when the actions $S[\varphi]$ and $S[q]$ are strictly dual under the transformation (6.76), but also for smaller values of the Josephson coupling energy $E_J \lesssim E_C$. In the latter case, the ground-state energy E_0 deviates from a sinusoidal form (6.20) but remains $2e$-periodic. Hence, all our considerations hold also for such values of E_J, except in Eq. (6.78) one should substitute the term $V_C \sin(\pi q/e)$ by $\partial E_0(q)/\partial q$.

Also the duality for the effective actions $S[\varphi]$ and $S[q]$ can be easily extended to the case $E_J \lesssim E_C$ if we assume that the current-phase relation for the Josephson element deviates from a sinusoidal one. For instance, in the limit $E_J \ll E_C$, strict duality between the actions in the phase and charge spaces can be achieved if instead of a tunnel barrier between two superconductors, we consider a clean long SNS junction, which at low T is characterized by 2π-periodic saw-tooth current-phase relation [203, 241]. The duality between $S[\varphi]$ and $S[q]$ becomes obvious in this case if we recall that for $E_J \ll E_C$, the ground-state energy reduces to a set of parabolas $E_0(q) = \min_n(q - 2en)^2/2C$ (except in the immediate vicinity of the points $q = 2e(n + 1/2)$) and, hence, its derivative with respect to q also takes the saw-tooth form. The critical voltage then equals to $V_C = e/C$.

The above duality arguments are also useful if both canonically conjugate variables φ and q simultaneously demonstrate quantum behavior. For instance, making use of our analysis of macroscopic quantum tunneling of the phase performed in Chapter 4 and employing the duality transformation (6.76), one immediately arrives at the conclusion that even at $V_x < V_C$, the junction charge states in general become metastable and decay due to macroscopic quantum tunneling of the charge down to $T = 0$. Provided an externally applied voltage V_x is only slightly smaller than the critical one V_C, explicit expressions for both quantum and thermal decay rates of metastable charge states can be directly recovered simply by applying the duality transformation (6.76) to the corresponding results for the phase decay rates derived in Chapter 4.

Likewise, at small values of $V_x \ll V_C$, the quantum tunneling rates for the charge between adjacent wells of the $2e$-periodic potential $E_0(q)$ are recovered from the "dual" phase decay rates derived in Chapter 5, cf. Eqs. (5.45)–(5.48). Provided interaction with a dissipative Caldeira–Leggett bath destroys macroscopic quantum coherence of the charge states, quantum dynamics of the charge can be described by means of the master equations

$$\frac{dw_k(t)}{dt} = \gamma_+ w_{k-1}(t) + \gamma_- w_{k-1}(t) - (\gamma_+ + \gamma_-) w_k(t) \tag{6.82}$$

where $w_k(t)$ is the time-dependent occupation probability of the k-th potential well of $E_0(q)$ and $\gamma_\pm$ are tunneling rates between adjacent potential wells or, in other words, Cooper pair tunneling rates across the junction. Clearly, the master equations (6.82) are dual to Eqs. (6.10), and the rates $\gamma_\pm$ are derived by applying the duality transformation (6.76) to Eqs. (5.45) and (5.46).

Let us define the parameters $\tilde{\omega}_0 = (\pi/e)(\Delta/L)^{1/2}$, $E_{2e} = 2\Delta$ for $E_J \gg E_C$ and $\tilde{\omega}_0 = (1/LC)^{1/2}$, $E_{2e} = E_C$ for $E_J \ll E_C$. Assuming both temperature and external bias to be low,

$$T, eV_x \ll \tilde{\omega}_0 \ll E_{2e} \tag{6.83}$$

we immediately recover the total decay rate of the charge state

$$\gamma = \gamma_+ + \gamma_- = \frac{\tilde{\Delta}^2}{2\hbar\tilde{\omega}_0} \left(\frac{\hbar\tilde{\omega}_0}{2\pi T}\right)^{1-2/K} \cosh\left(\frac{eV_x}{2T}\right) \frac{|\Gamma\left(\frac{1}{K} + i\frac{eV_x}{\pi T}\right)|^2}{\Gamma(2/K)} \tag{6.84}$$

where we also defined the charge tunneling amplitude

$$\tilde{\Delta} = 16 \left(\frac{\Delta E_L}{\pi}\right)^{1/2} \left(\frac{\Delta}{2E_L}\right)^{1/4} \exp\left[-\left(\frac{8\Delta}{E_L}\right)^{1/2}\right] \tag{6.85}$$

for $E_J \gg E_C$ and

$$\tilde{\Delta} \sim E_C^{3/4} E_L^{1/4} \exp\left[-a\left(\frac{E_C}{E_L}\right)^{1/2}\right], \quad a \sim 1 \tag{6.86}$$

for $E_J \ll E_C$. The applicability conditions for the expression (6.84) can also be established by duality arguments. Repeating the corresponding analysis of Section 5.1.3, we arrive at the conclusion that under the condition (6.83), the above expression for γ applies for $K < 1$. Equation (6.84) also holds for $K > 1$ provided an additional requirement

$$\max(eV_x, T) \gg b(1/K)\tilde{\Delta}_r \tag{6.87}$$

is fulfilled (cf. Eq. (5.50)), where

$$\tilde{\Delta}_r = \tilde{\Delta}\left(\frac{\tilde{\Delta}}{\tilde{\omega}_0}\right)^{\frac{1}{1-K}} \tag{6.88}$$

is the corresponding renormalized bandwidth dual to that defined in Eq. (5.35).

At this stage, we would like to remind the reader that it is essential to obey the adiabaticity conditions (6.73) or (6.74) for the charge variable q in order to assure the phase-charge duality at the level of the effective action analysis. These conditions can be fulfilled only provided the inductance L remains not very low. In the opposite case, the characteristic time scale ω_q^{-1} for the charge variable becomes short, and the above adiabaticity conditions (6.73), (6.74) are violated. In this limit, Eq. (6.66) is not anymore accurate and it is necessary to employ the exact Eqs. (6.63) or (6.65). Still, also in this case, the duality considerations for the partition function in the spirit of those outlined in Section 6.3 remain applicable and allow to easily reconstruct, e.g., the Cooper pair tunneling rates $\gamma_\pm$ also in the limit of low inductances.

Setting $L \to 0$ in Eq. (6.62), substituting it into Eq. (6.61) together with Eq. (6.65) and performing Gaussian integrals over both variables, we again arrive at the partition function represented in terms of a gas of interacting charges $\rho_n(\tau)$ (6.32), cf., e.g., Eq. (6.31). In the limit of small $E_J \ll E_C$, one can proceed perturbatively in E_J, which amounts to taking into account only the contribution of pairs of alternating charges ϵ_j. In this case with the aid of the duality transformation (6.34), the problem is essentially reduced to that already analyzed in Chapter 5. In particular, at low enough temperatures and small values of E_J, the free energy of the state with the junction charge equal to $Q_x = CV_x$ takes the form [345]

$$F(Q_0) = \frac{Q_x^2}{2C} - \frac{E_J^2}{4\hbar}\int_0^{\hbar\beta} d\tau \exp\left(\frac{2e|Q_x|\tau}{C} - \frac{2}{K}\tilde{\Lambda}(\tau)\right) \tag{6.89}$$

where $\tilde{\Lambda}(\tau)$ is defined in Eq. (6.33). For any $K > 0$, this expression for the free energy is dual to that defined in Eq. (5.42). As before, the expression for $F(Q_x)$ formally diverges, signaling instability of the charge state Q_x. Proceeding in exactly the same manner as in Chapter 5, we define the imaginary part of the free energy, which yields the Cooper pair tunneling rates

$$\gamma_{\pm} = \frac{E_J^2 R_S C}{4\hbar} (2\pi T R_S C)^{2/K-1} \exp\left(\pm\frac{eQ_x}{CT}\right) \frac{|\Gamma\left(\frac{1}{K} + \mathrm{i}\frac{eQ_x}{\pi T C}\right)|^2}{\Gamma(2/K)} \tag{6.90}$$

Thus, the total charge decay rate $\gamma = \gamma_+ + \gamma_-$ obtained in this case is in many respects similar to Eq. (6.90) derived for sufficiently large values of L. We finally note that in the limit $K \to 0$, the last term in the exponent in Eq. (6.89) just reduces to $(2/K)\tilde{\Lambda}(\tau) \to 2e^2\tau/C$. The same procedure of finding the imaginary part of the free energy for $0 \leq q \leq 2e$ in this case yields [418]

$$\gamma(q) = \frac{\pi}{8}\frac{eE_J^2}{\hbar E_C}\delta(q-e) \tag{6.91}$$

We observe that in the absence of dissipation, the Cooper pair tunneling rate is peaked at $q = e$ where the parabolas $q^2/2C$ and $(q - 2e)^2/2C$ cross, i.e., the energies of the charge states q and $q - 2e$ coincide exactly. This result also implies that a Cooper pair can be transferred such as to keep the energy of the system low (i.e., confined to the lowest Brillouin zone). Indeed, let us sweep the quasicharge across the lowest band $E_0(q)$ with some speed $\dot{q}$. The probability $P(t)$ that no Cooper pair gets transferred (i.e., the system occupies the parabola $q^2/2C$ and does not jump to the parabola $(q-2e)^2/2C$) until the time moment t is obtained from the equation

$$dP/dt = -\gamma(q(t))P(t) \tag{6.92}$$

Combining Eqs. (6.91) and (6.92) and integrating the latter from t_1 to t_2 chosen such that $q(t_1) < e < q(t_2)$, we easily find

$$P(t_2)/P(t_1) = \exp\left(-\frac{\pi}{8}\frac{eE_J^2}{\hbar\dot{q}E_C}\right) \tag{6.93}$$

implying that for very slow sweep rates $\dot{q}$, the probability $P(t_2)$ tends to zero, i.e., a Cooper pair is transferred across the junction and, accordingly, the system stays in the lowest band. Should the sweep rate $\dot{q}$ or, equivalently, the frequency $\omega_q = \dot{q}/2e$ increase, the probability $P(t)$ for a Cooper pair $P(t)$ not to be transferred becomes nonzero. This probability (6.93) divided by the time interval $1/\omega_q$ defines the Zener tunneling rate $\gamma_\uparrow$ (6.72).

6.7 Real-Time Dynamics and Relation for Phase and Charge Mobilities

Above in this chapter, we merely discussed some (quasi-) equilibrium properties of ultrasmall Josephson junctions, which can be adequately accounted for by evaluating the grand partition function $\mathcal{Z}$ within the imaginary-time path integral formalism. We also observed that making use of the "ImF method," one can extend this formalism in order to find decay rates of the quantum states and, hence, to obtain some information about real-time dynamics of the system. In general, however, the imaginary-time technique may be insufficient and/or inconvenient to properly describe real-time evolution of Josephson junctions in the presence of macroscopic quantum coherence and dissipation. As we already demonstrated, e.g., in Chapter 5, the real-time path integral technique based on the Feynman–Vernon influence functional theory appears to be the most efficient tool for this purpose.

Let us introduce a reduced density matrix of a superconducting junction in the phase representation $\rho(t;\varphi,\varphi')$, which is obtained from the exact density matrix (2.18) by tracing out all electronic degrees of freedom in the system (see Chapter 2). Following general considerations of Chapter 1, we can write down the equation describing the time evolution of this reduced density matrix in the form (cf. Eq. (1.12))

$$\rho(t;\varphi,\varphi') = \int d\varphi_i \int d\varphi_i' J(\varphi,\varphi',t;\varphi_i,\varphi_i',0)\langle\varphi_i|\rho(0)|\varphi_i'\rangle \quad (6.94)$$

where $\rho(0)$ is the density matrix at the initial time moment $t = 0$, J represents the kernel of the evolution operator on the Keldysh contour, which can be expressed as a double path integral (cf. Eq. (1.16))

$$J(\varphi, \varphi', t; \varphi_i, \varphi_i', 0) = \int_{\varphi_i}^{\varphi} \mathcal{D}\varphi_F \int_{\varphi_i'}^{\varphi'} \mathcal{D}\varphi_B \exp(\mathrm{i}S[\varphi_F, \varphi_B]/\hbar) \quad (6.95)$$

and the system action $S[\varphi_F, \varphi_B]$ is defined, e.g., in Eqs. (2.105)–(2.107). For simplicity below, we again stick to the limit of linear Ohmic dissipation, in which case the kernels $Z_S^{-1}(t)$ and $K_S(t)$ entering the action S_{diss} (2.105) are defined, respectively, by Eqs. (2.6) and (2.51) with $Z_S(\omega) \equiv R_S$.

It is also convenient for our analysis to introduce the reduced density matrix in the quasicharge representation $\rho_{q,q'}(t)$ and the generating functional $X(t; \chi, \xi)$ [345, 423, 424]. The first of these quantities is defined by the Fourier transformation of $\rho(t; \varphi, \varphi')$,

$$\rho_{q,q'}(t) = \sum_{\varphi,\varphi'} \rho(t; \varphi, \varphi') \exp(\mathrm{i}q\varphi/2e - \mathrm{i}q'\varphi'/2e) \quad (6.96)$$

while the second one has the form

$$X(t; \chi, \xi) = \sum_{\phi} \rho(t; \phi + \chi/2, \phi - \chi/2) \exp(-\mathrm{i}\xi\phi/2) \quad (6.97)$$

It is also easy to observe that the quantities (6.96) and (6.97) are related to each other as

$$X(t; \chi, 0) = \sum_{q} \rho_{q,q}(t) \exp(\mathrm{i}q\chi/2e) \quad (6.98)$$

Evaluating the generating functional (6.97), one would be able to immediately recover various expectation values for the phase variable simply by differentiating $X(t; 0, \xi)$ with respect to ξ. A formal expression for this generating functional can be derived with the aid of Eqs. (6.94) and (6.95). In order to handle the only non-Gaussian Josephson term in the action, we again perform its expansion in formal series on both forward and backward parts of the Keldysh contour, e.g.,

$$\exp\left(\frac{\mathrm{i}E_J}{\hbar}\int_0^t dt' \cos\varphi_F(t')\right)$$
$$= \sum_{m=0}^{\infty} \frac{(\mathrm{i}E_J/2\hbar)^m}{m!} \int_0^t dt_{m,F} \int_0^t d\tau_{m-1,F} \ldots \int_0^t d\tau_{1,F}$$
$$\times \sum_{\varepsilon_{jF}=\pm 1} \exp\left[\mathrm{i}\int_0^{\hbar\beta} d\tau \sum_{j=1}^{m} \epsilon_{jF}\delta(\tau - \tau_{j,F})\varphi(\tau)\right] \quad (6.99)$$

and similarly for $\exp(-\frac{iE_J}{\hbar}\int_0^t dt' \cos\varphi_B(t'))$, cf. also Eq. (6.29). Here we introduce the "charge" densities on both contours

$$\rho_{m,F}(\tau) = \sum_{j=1}^{m} \epsilon_{jF}\delta(\tau - \tau_{j,F}), \quad \rho_{n,B}(\tau) = \sum_{i=1}^{n} \epsilon_{iB}\delta(\tau - \tau_{i,B}) \quad (6.100)$$

Substituting Eq. (6.99) and a similar expansion for the backward contour into the path integral (6.95), we arrive at an infinite series of Gaussian integrals. Each of them can be handled exactly. One possible way to do it is to proceed in the same manner as, e.g., in Chapter 3 performing first the integration over the sum of the phases $\varphi_+ = (\varphi_F + \varphi_B)/2$, which yields the functional δ-function eliminating the remaining path integral over the phase difference $\varphi_- = \varphi_F - \varphi_B$. For further details of this procedure, we can also refer the reader to the analysis [122]. Alternatively one may parameterize the trajectories in the phase space in a way effectively equivalent to deforming the Keldysh contour to imaginary time. This trick also allows to proceed in a simple and straightforward manner [321, 345, 423, 424]. Both approaches yield the same results, which are briefly summarized below.

It is easy to observe that in the limit $\xi \to 0$, only charge configurations obeying the generalized neutrality condition

$$\sum_{j=1}^{m} \epsilon_{jF} + \sum_{i=1}^{2n-m} \epsilon_{iB} = 0 \quad (6.101)$$

give non-vanishing contributions to the path integrals. For these charge configurations, it might be convenient to omit the forward and backward indices and label the charges simply as $\epsilon_{jF} \equiv \epsilon_j$ for $1 \leq j \leq m$ and $\epsilon_{iB} \equiv \epsilon_{j=i+m}$, i.e., in the latter case, one has $m+1 \leq j \leq 2n$. Then the condition (6.101) reads

$$\sum_{j=1}^{2n} \epsilon_j = 0 \quad (6.102)$$

These configurations and their ordering in real time are schematically depicted in Fig. 6.5. We also note that in general no charge neutrality condition separately for the forward and backward parts of the Keldysh contour should be imposed.

Setting first $\xi = 0$ in Eq. (6.97), under the condition

$$\max(T, I_x/2eK) \ll 1/R_S C, \quad t \gg R_S C \quad (6.103)$$

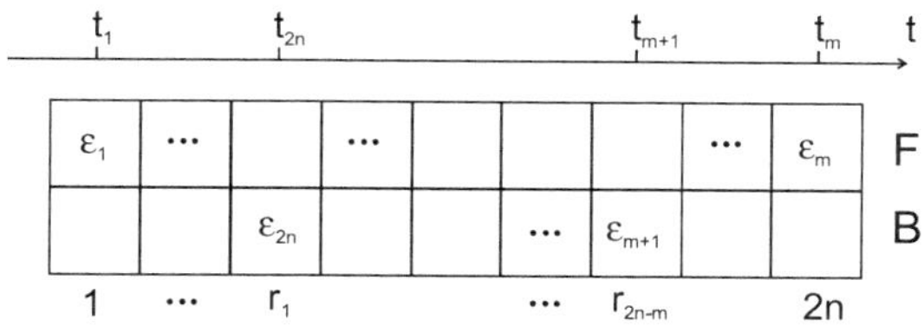

Figure 6.5 Relevant configurations of topological charges on the Keldysh contour and their ordering in real time. The phase factor $\Phi[\epsilon_i]$ (6.107) is determined by the following rules: (1) $\Phi = \pi n/K$ if all $2n$ charges remain on the forward branch of the Keldysh contour. (2) The remaining configurations can be obtained by moving the charges from the cell $(F, 2n)$ to the cell $(B, 2n)$. During each such move, the factor $\Phi[\epsilon_i]$ remains unchanged. (3) With every change in relative positions of charges ϵ_{iF} and ϵ_{jB} from $t_i < t_j$ to $t_i > t_j$, the phase factor Φ changes by $\epsilon_{iF}\epsilon_{jB}2\pi/K$.

for the generating functional $X(t;\chi,0)$, we obtain [321, 345, 423, 424]

$$X(t;\chi,0) = \exp(\mathrm{i}I_x R_S C\chi/2e)\sum_{k=-\infty}^{\infty} w_k(t)\exp(\mathrm{i}k\chi) \tag{6.104}$$

where

$$\begin{aligned} w_k(t) &= \sum_{n=k}^{\infty}(-1)^{n+k}\left(\frac{E_J}{2\hbar}\right)^{2n}\sum_{\epsilon_j}\int\prod_{j=1}^{2n}dt_j \\ &\times\exp\left[(2/K)\sum_{i,j=1,i>j}^{2n}\epsilon_i\epsilon_j\tilde{\Lambda}(t_i-t_j) + \mathrm{i}\frac{\pi I_x}{eK}\sum_{i=1}^{2n}\epsilon_i t_i + \mathrm{i}\Phi[\epsilon_i]\right] \end{aligned} \tag{6.105}$$

Here we denote

$$\tilde{\Lambda}(t) = \ln\left(\frac{\sinh(\pi T t/\hbar)}{\pi T R_S C/\hbar}\right) \tag{6.106}$$

$$\begin{aligned} \Phi[\epsilon_i] &= \frac{\pi n}{K} \\ &+\frac{2\pi}{K}\left(\sum_{i,j=m+1}^{2n}\epsilon_i\epsilon_j + \epsilon_{2n}\sum_{i=r_1+1}^{m}\epsilon_i + \ldots + \epsilon_{m+1}\sum_{i=r_{2n-m}+1}^{m}\epsilon_i\right) \end{aligned} \tag{6.107}$$

and the sum is performed over charge configurations obeying the condition

$$\sum_{j=1}^{m} \epsilon_j - \sum_{j=m+1}^{2n} \epsilon_j = 2k \tag{6.108}$$

along with the neutrality condition (6.102).

Comparing Eqs. (6.98) and (6.104), we immediately observe that

$$w_k(t) = \rho_{q,q}(t), \quad q = 2ek + Q_x, \quad Q_x = I_x R_S C \tag{6.109}$$

and $k = 0, \pm 1, \pm 2, \ldots$ is an integer number. This equation implies that the quantities $w_k(t)$ introduced in a purely formal way have a clear physical meaning being identical to the diagonal elements of the density matrix in the quasicharge space $\rho_{qq}(t)$, which, in turn, represent the probabilities that k Cooper pairs carrying the total charge $2ek$ have tunneled across the Josephson junction within the time interval t. Note that this process, by no means, implies the junction charge accumulation, since tunneling of each Cooper pair is accompanied by the compensating current through the shunt in the opposite direction (see Fig. 6.6). Hence, physical charge of the Josephson junction remains unchanged after this process being equal to Q_x. If so, one may wonder if the transitions in the quasicharge space are detectable. The answer to this question is positive. Indeed, the quasicharge states labeled by different values of k are physically distinguishable because they correspond to different quantum states of the junction environment (shunt). Hence, transitions between such states may, in principle, be registered and single Cooper pair tunneling events may be studied experimentally.

Yet another remarkable observation that follows from Eq. (6.109) is that under the condition (6.103), diagonal elements of the density matrix in the q-representation $\rho_{q,q}(t)$ differ from zero only for quantized values $q = 2ek + q_x$, i.e., in the long-time limit $t \gg R_S C$, the junction quasicharge inevitably relaxes to one of these discrete values also in the presence of an Ohmic shunt.

As we already mentioned above, our analysis also allows to recover various expectation values of the phase variable $\varphi(t)$. Here we are merely interested in the average

$$\langle \varphi(t) \rangle = \mathrm{i} \frac{\partial X(t; \chi, \xi)}{\partial \xi} |_{\xi=0} \tag{6.110}$$

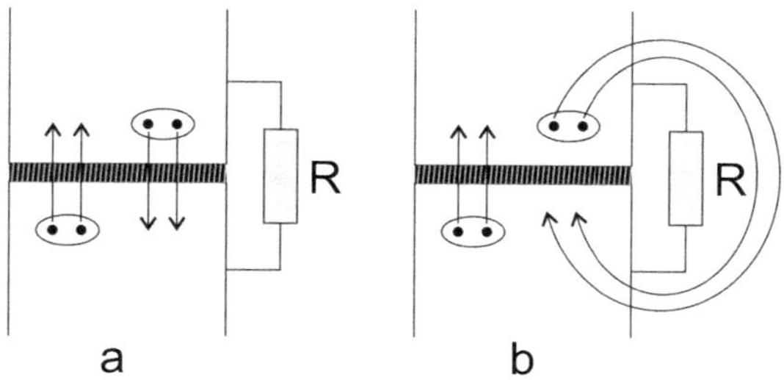

Figure 6.6 (a) Virtual tunneling of Cooper pairs between superconducting electrodes and (b) tunneling event of a Cooper pair accompanied by the compensating normal current through the shunt. Electric charge across the junction is conserved in both processes (a) and (b), while the junction quasicharge is conserved only in the process (a) and changes by $2e$ in the process (b).

In order to evaluate this expectation value, it suffices to establish the generating functional (6.97) keeping only linear in ξ terms. With trivial modifications, the above analysis can be repeated for nonzero values of ξ. As a result, we obtain

$$X(t;0,\xi) = X(t;0,0) - \mathrm{i}\xi\left[\frac{\pi}{K}\sum_{k=-\infty}^{\infty} kw_k(t) - 2eI_x R_S t\right] + \mathcal{O}(\xi^2) \tag{6.111}$$

Substituting this result into Eq. (6.110), one immediately arrives at the formula

$$\langle\varphi(t)\rangle = 2eI_x R_S t - \frac{\pi}{K}\sum_{k=-\infty}^{\infty} kw_k(t) \tag{6.112}$$

Introducing now the mobilities in both phase and quasicharge spaces

$$\mu_\varphi = \frac{2e}{I_x}\langle\dot{\varphi}(t)\rangle, \quad \mu_q = \frac{2e}{I_x}\langle\dot{q}(t)\rangle \equiv \frac{2e}{I_x}\sum_{k=-\infty}^{\infty} k\dot{w}_k(t) \tag{6.113}$$

and making use of Eq. (6.112), we recover the relation between these mobilities [122, 345, 423, 424]

$$\mu_\varphi = (4e^2 R_S/\hbar)[1 - \mu_q/2e] \tag{6.114}$$

which generalizes Eq. (6.42), established earlier in the linear in the external bias regime and at temperatures $T \to 0$. Here we demonstrated that Eq. (6.114) is, in fact, valid also at nonzero T and in the nonlinear in I_x regime (both μ_φ and μ_q may, in general, depend on I_x).

Note that initially the above relation for mobilities was interpreted as a formal consequence of the duality transformation (6.34) involving a fictitious particle in an effective tight-binding lattice. Our present derivation of Eq. (6.114) is not based on the duality arguments at all and, hence, is unrelated to the transformation (6.34). Furthermore, Eq. (6.114) was shown to apply also in a more general case when quasiparticle tunneling is present along with that of Cooper pairs [321]. In the latter case (which we do not consider here), no duality between the phase and quasicharge spaces exists at all.

In fact, the relation (6.114) has a clear physical meaning being just the current balance equation in the circuit depicted in Fig. 6.7a. Combining the definition (6.113) with the Josephson relation $2eV = \hbar\langle\dot{\varphi}\rangle$ and having in mind that $\langle\dot{q}\rangle$ defines the Copper pair tunneling current across the Josephson element, one immediately observes that Eq. (6.114) is equivalent to a simple current balance condition

$$I_x = V/R_S + \langle\dot{q}\rangle \tag{6.115}$$

Alternatively, one can consider the circuit where a voltage source V_x, the shunt resistor, and the Josephson element are all switched in series (see Fig. 6.7b). In this case, the charge mobility μ_q defines the current in the circuit as

$$\langle\dot{q}\rangle = \frac{\mu_q}{2e}\frac{V_x}{R_S} \tag{6.116}$$

where the expression for μ_q is derived from Eq. (6.114) with the substitution $I_x = V_x/R_S$.

Although the relation for mobilities (6.114) is not directly based on the duality arguments, this relation can be conveniently used in combination with such arguments. In order to proceed in this direction, as before, let us consider the tight-binding limit $E_J \gg E_C$ and assume both temperature and current bias to be small $T/\hbar$, $I_x/e \ll \omega_0$. Under these conditions, the Josephson particle φ is located near the bottoms of the potential wells $\varphi_k = 2\pi k + \hbar I_x/2eE_J$ so that the density matrix effectively reduces to a discrete one

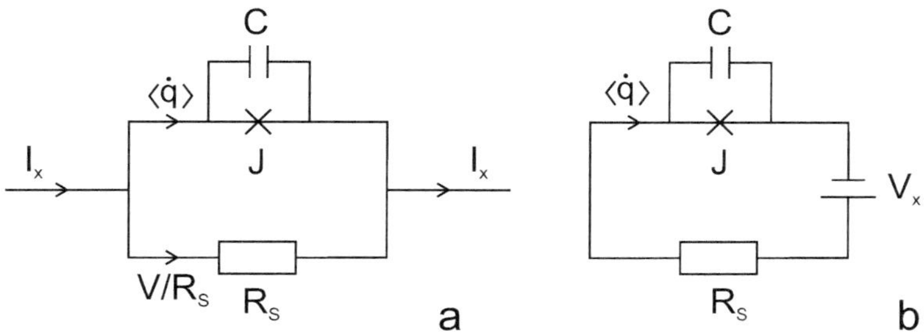

Figure 6.7 Current-biased (a) and voltage-biased (b) circuits involving a Josephson junction with a capacitance C and Ohmic resistance R_S.

$\rho(\varphi_k, \varphi_k')$. Actually, this situation was already considered in Section 6.2 in the incoherent regime, which can be handled with the aid of the master equations (6.10). In order to include coherent effects into consideration, it is necessary to properly account for real-time dynamics of all elements of the density matrix, not just the diagonal ones. This goal can be accomplished either by generalizing the real-time blip approach (see Chapter 5) to the multiple well case [122] or by employing real-time quantum tunneling technique [345, 424]. As in the tight-binding limit the density matrix changes in time only due to tunneling between the wells, it is necessary to take into account all possible tunneling events to all orders in the tunneling amplitude Δ. The corresponding analysis is straightforward, but somewhat tedious. For this reason, we do not give further details here. The interested reader can find them in the papers [345, 424].

Defining the occupation probabilities of the wells as $W_k(t) = \rho(\varphi_k, \varphi_k)$ and assuming for definiteness that initially the Josephson particle occupied one of the wells $W_k(t = 0) = \delta_{k,0}$, we obtain [345, 424]

$$W_k(t) = \sum_{n=k}^{\infty}(-1)^{n+k}\left(\frac{\Delta}{2\hbar}\right)^{2n}\sum_{\nu_j}\int\prod_{j=1}^{2n}dt_j$$
$$\times \exp\left[2K\sum_{i,j=1,i>j}^{2n}\nu_i\nu_j\Lambda(t_i - t_j) + i\frac{\pi I_x}{e}\sum_{i=1}^{2n}\nu_i t_i + i\Phi[\nu_i]\right] \tag{6.117}$$

where we defined

$$\Lambda(t) = \ln\left(\frac{\sinh(\pi T t/\hbar)}{\pi T \tau_c/\hbar}\right), \qquad \tau_c \sim \omega_0^{-1} \tag{6.118}$$

and

$$\Phi[\nu_i] = \pi n K + 2\pi K \left(\sum_{i,j=m+1}^{2n} \nu_i \nu_j + \nu_{2n} \sum_{i=r_1+1}^{m} \nu_i + \ldots + \nu_{m+1} \sum_{i=r_{2n-m}+1}^{m} \nu_i \right) \tag{6.119}$$

It is easy to observe that the occupation probabilities in the phase and charge spaces $W_k(t)$ and $w_k(t)$ defined, respectively, by Eqs. (6.117)–(6.119) and (6.105)–(6.107) are exactly dual to each other under the transformation

$$\Delta \leftrightarrow E_J, \quad K \leftrightarrow 1/K, \quad \omega_0 \leftrightarrow 1/R_S C, \quad I_x \leftrightarrow I_x/K, \quad \nu_j \leftrightarrow \epsilon_j \tag{6.120}$$

Hence, the mobility in the phase space μ_φ (evaluated in the tight-binding limit) is also dual to that in the charge space μ_q. As these two quantities are related to each other by means of Eq. (6.114), the analysis of quantum dynamics of resistively shunted Josephson junctions gets significantly simplified, since it is necessary to evaluate only one of these mobilities to be able to reconstruct another one without any calculation. We will make use of this property below in this chapter.

6.8 Quantum Dynamics in the Weak Dissipation Regime

6.8.1 Quantum Corrections to Resistance

General expressions for occupation probabilities both in the phase and charge spaces established above formally allow to fully describe quantum dissipative dynamics of the Josephson particle for any value of the dissipation strength K. Although these expressions are rather cumbersome, in the majority of situations, significant simplifications can be achieved. One of such situations was already analyzed in Section 6.2 where it was demonstrated that for $E_J \gg E_C$ and sufficiently large dissipation $K > 1$, the Josephson particle behaves nearly classically, i.e., for most of the time, it remains trapped in one of the potential wells and from time to time incoherently tunnels between the wells. This quantum tunneling yields non-vanishing effective resistance of our device $R_{\rm eff}$ defined in Eqs. (6.13)–(6.15) as well as the particle diffusion in the Josephson potential. Both $R_{\rm eff}$ and the corresponding diffusion coefficient D (see Eq. (6.17)) remain small, being proportional to the square of the inter-well tunneling amplitude Δ^2 and, on top of that, vanish, provided both temperature T and external bias I_x tend to zero. This corresponds to quantum localization of the phase φ at $K > 1$, as discussed in Section 6.3. In this regime, the quasicharge q is, in contrast, delocalized, its mobility μ_q is close to $2e$, and, hence, $\langle\dot{q}\rangle = I_x - O(\Delta^2)$ (cf. Eq. (6.114)). Physically, these results imply that, loosely speaking, in the circuit depicted in Fig. 6.7a, practically all external current I_x passes through the Josephson junction in a non-dissipative manner (dc Josephson effect) and only its very small fraction ($\propto \Delta^2$) flows across the shunting resistance, giving rise to nonzero $R_{\rm eff}$.

Note that both this physical picture and the formal expressions for $R_{\rm eff}$ (6.13)–(6.15) remain applicable also for $K < 1$, provided either temperature or external bias (or both) remains sufficiently large. The corresponding validity conditions are defined in Eqs. (5.50) and (5.51). In the limit of very small T and I_x, the situation becomes entirely different due to intensive quantum fluctuations of the phase φ, which delocalize the Josephson particle in the periodic potential $-E_J \cos\varphi$. At the same time, the dual quasicharge variable behaves nearly classically for $K < 1$ being localized at $T \to 0$, $I_x \to 0$, and moving incoherently in the effective $2e$-periodic potential for nonzero T and I_x. At low enough temperature and bias, the quasicharge tunnels incoherently between the wells with the rates $\gamma_\pm$ and the occupation probabilities of different quasicharge states $w_k(t)$ are obtained from the master equations (6.82). With this in mind, one easily recovers the quasicharge mobility in this regime. It reads

$$\mu_q = \frac{4e^2\gamma}{I_x}\tanh\left(\frac{\pi\hbar I_x}{2eKT}\right) \tag{6.121}$$

or, equivalently,

$$\langle\dot{q}\rangle = 2e(\gamma_+ - \gamma_-) \tag{6.122}$$

Here the rate $\gamma = \gamma_+ + \gamma_-$ is defined in Eqs. (6.84) and (6.90). As this rate remains small (meaning $\mu_q \ll 2e$ or $\langle \dot{q} \rangle \ll I_x$) and in the leading approximation can be neglected, from Eq. (6.114) one immediately finds the voltage V across the junction equal to $V \simeq I_x R_S$, i.e., $R_{\text{eff}} \simeq R_S$. This result implies that almost all current I_x flows through the shunt R_S, while tunneling of Cooper pairs across the junction is suppressed by Coulomb effects. As we already discussed in Section 6.6, this is a manifestation of Coulomb blockade of Cooper pair tunneling, which is equivalent to quasicharge localization in the limit of zero temperature and bias.

At nonzero T and/or I_x, the Cooper pair current $\langle \dot{q} \rangle$ differs from zero and, according to Eq. (6.114), causes a small reduction in the effective resistance R_{eff} below R_S. For definiteness, let us consider the limit $E_J \ll E_C$. At $\hbar I_x/eK \ll T \ll \hbar/R_S C$ from Eqs. (6.114), (6.121), and (6.90) for the system effective linear resistance R_{eff}, one finds [122, 345]

$$R_{\text{eff}}(T) = R_S \left(1 - \frac{\pi^{9/2}\Gamma(1/K)(E_J/E_C)^2}{32K^3\Gamma(1/2+1/K)}(\pi T R_S C/\hbar)^{2/K-2}\right) \tag{6.123}$$

Analogously at $T \ll \hbar I_x/eK \ll \hbar/R_S C$ for the nonlinear resistance $R_{\text{eff}} \equiv \langle V \rangle / I_x$, one gets

$$R_{\text{eff}}(I_x) = R_S \left(1 - \frac{\pi^4 (E_J/E_C)^2}{16K^3\Gamma(2/K)}(\pi I_x R_S C/eK)^{2/K-2}\right) \tag{6.124}$$

In both these cases, the negative correction to R_S is due to quantum tunneling of Cooper pairs across the Josephson junction. In the considered limit, this correction remains small in the parameter $\sim E_J^2/E_C^2 \ll 1$ and, on top of that, vanishes for $K < 1$ and at small T and I_x as $\propto \; [\max(T, I_x/e)]^{2/K-2}$. The latter feature remains applicable also for $E_J > E_C$. For further details related to the behavior of the effective resistance R_{eff} in the Coulomb blockade regime, we refer the reader to the review paper [345].

6.8.2 *I–V* Curve in the Weak Dissipation Limit

Below we will specifically address the system behavior in the weak dissipation limit $K \ll 1$ in which case the phase φ fluctuates strongly, while the quasicharge only weakly deviates from its classical value, so that at low enough temperatures, one has $\langle\delta q^2\rangle/e^2 \ll 1$. Neglecting for simplicity weak quantum tunneling effects (by setting $\gamma_\pm \to 0$), one can only consider quasicharge fluctuations within one potential well and approximate $\partial E_0(q)/\partial q \simeq q/C$. Then one easily resolves Eqs. (6.79) and (6.80) and finds

$$\langle\delta q^2\rangle = R_S C^2 \int \frac{d\omega}{2\pi} \frac{\omega \coth(\hbar\omega/2T)}{1+\omega^2 R_S^2 C^2}$$
$$= \begin{cases} \frac{2e^2 K}{\pi^2} \ln(\omega_c R_S C), & T \ll E_C, \\ e^2 T/2E_C, & T \gg K E_C, \end{cases} \tag{6.125}$$

where $\omega_c \sim R_S/L$ defines the high-frequency cutoff. It follows immediately that for $K \ll 1$, the condition $\langle\delta q^2\rangle/e^2 \ll 1$ remains justified, provided $T \ll E_C$. Both the result (6.125) and its validity condition apply for not very large $E_J \lesssim E_C$, whereas in the limit $E_J \gg E_C$ in Eq. (6.125), one should simply substitute $C \to C_{\text{eff}} = e^2/\pi^2\Delta$ and $E_C \to \pi^2\Delta/2$.

With this in mind in the leading approximation, we can treat the quasicharge dynamics classically and conclude that provided the current I_x does not exceed some threshold value I_{th}, the quasicharge q will remain at rest in the bottom of one of the potential wells formed by $2e$-periodic ground-state energy $E_0(q)$. In the limit $E_J \gg E_C$, we have $E_0(q) = -\delta\cos(\pi q/e)$ and, as we already discussed, the situation is strictly dual to dc Josephson effect (cf. Eq. (6.78)), with I_{th} being the exact analogue of the Josephson critical current. For $E_J \gg E_C$, this current is trivially defined from the first Eq. (6.78) by setting $V = I_{\text{th}} R_S = V_C$. In the opposite limit $E_J \ll E_C$, the ground-state energy is close to the set of parabolas $E_0(q) = \min_n(q - 2en)^2/2C$ and the critical voltage equals to $V_C = e/C$. Hence, we obtain

$$I_{\text{th}} = \begin{cases} e/R_S C, & E_J \ll E_C, \\ \pi\Delta/eR_S, & E_J \gg E_C \end{cases} \tag{6.126}$$

As long as I_x remains below I_{th}, Cooper pair tunneling is blocked by Coulomb interaction, implying that in this regime, the voltage across the junction increases linearly with I_x as $V = R_S I_x$. Should, however, the bias current I_x exceed I_{th}, the quasicharge q starts moving across the lowest Brillouin zone $E_0(q)$, thereby causing voltage oscillations. In the low-temperature limit, the classical dynamics of q is controlled by the following equation (cf. Eq. (6.79)):

$$\dot{q} + \frac{1}{R_S}\frac{\partial E_0(q)}{\partial q} = I_x \tag{6.127}$$

As we already discussed in Section 6.6, in the limit $E_J \gg E_C$, these Bloch oscillations are strictly dual to the Josephson ones, i.e., $V(t) = V_C \sin(\pi q(t)/e)$. Accordingly, the corresponding $I-V$ curve turns out to be exactly dual to one obtained within the resistively shunted junction (RSJ) model [375]. In the opposite limit $E_J \gg E_C$, one finds $V(t) = (q(t) - 2en)/C$ for $2e(n-1/2) < q(t) < 2e(n+1/2)$. Employing Eq. (6.127), for the average in time voltage (which we denote as V_0), we obtain [36, 276]

$$V_0/R_S = \begin{cases} I_x - 2I_{\text{th}} \ln^{-1}\left(\frac{I_x + I_{\text{th}}}{I_x - I_{\text{th}}}\right), & E_J \ll E_C, \\ I_x - (I_x^2 - I_{\text{th}}^2)^{1/2}, & E_J \gg E_C \end{cases} \tag{6.128}$$

These relations define the $I-V$ curve of a current-biased resistively shunted Josephson junction in the Bloch oscillation regime $I_x > I_{\text{th}}$. We observe that with increasing I_x, the average voltage (6.128) *decreases* and in the limit $I_x \gg I_{\text{th}}$ becomes $V_0 = E_C/R_S C I_x$ for $E_J \ll E_C$ and $V_0 = \pi^2\Delta^2/2e^2 R_S I_x$ for $E_J \gg E_C$. As a result, the Bloch oscillation frequency $\nu_B = (I_x - V_0/R_S)/2e$ increases with increasing I_x and approaches the fundamental value $\nu_B = I_x/2e$ in the limit of large current bias. The $I-V$ curve describing both Coulomb blockade and Bloch oscillations regimes is depicted in Fig. 6.8. It has a typical form of a "Bloch nose."

This $I-V$ curve is stable as long as the system remains in the lowest Brillouin zone and does not tunnel to higher zones. As we shall see below, this is the case provided the current bias I_x does not exceed some crossover current value I_{cr}. In order to evaluate I_{cr}, it is necessary to properly account for two competing processes: Zener tunneling from the lowest band $E_0(q)$ to higher ones with the rate $\gamma_\uparrow$ and inelastic relaxation back to the ground state with the rate $\gamma_\downarrow$ (see Fig. 6.4). Both these rates depend on the external bias I_x. Provided the inelastic relaxation rate exceeds Zener tunneling rate $\gamma_\downarrow > \gamma_\uparrow$, the system always remains in the lowest zone and has no chance to climb up to higher zones. Should, however, Zener tunneling be more efficient than inelastic relaxation, i.e., $\gamma_\uparrow > \gamma_\downarrow$, the system quickly

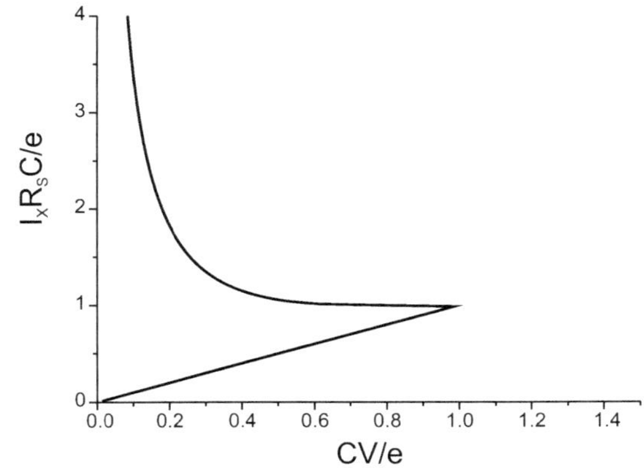

Figure 6.8 Bloch-nose-type $I-V$ curve for a current-biased Josephson junction with Ohmic dissipation at low enough external bias I_x when Zener tunneling to higher bands can be neglected. One distinguishes two regimes of Coulomb blockade $I_x < I_{\rm th}$ and Bloch oscillations $I_x > I_{\rm th}$.

gets to higher bands, the Bloch oscillation regime gets destroyed, and the $I-V$ curve switches to a dissipative branch. Obviously, the crossover between these two regimes occurs, provided both rates become equal, i.e.,

$$\gamma_\uparrow \approx \gamma_\downarrow \tag{6.129}$$

As both these rates depend on the external bias I_x, it is clear that this condition can be reached at some value $I_x \approx I_{\rm cr}$, thereby defining the crossover current $I_{\rm cr}$.

In the absence of dissipation (i.e., in the limit $1/R_S \to 0$), the relaxation rate obviously vanishes $\gamma_\downarrow \to 0$, while Zener tunneling rate $\gamma_\uparrow$ in the limit $E_J \ll E_C$ can be easily recovered from Eq. (6.72) by setting $\omega_q = I_x/2e$. We have

$$\gamma_\uparrow = \frac{I_x}{2e}\exp\left(-I_Z/I_x\right), \quad I_Z = \frac{\pi}{8}\frac{eE_J^2}{\hbar E_C} \tag{6.130}$$

Judging from this result, one could assume that Zener tunneling becomes important only provided the current I_x becomes of order of the Zener breakdown current I_Z (in which case, $I_{\rm cr}$ would be of order I_Z as well). Below we will demonstrate that the actual value of the crossover current is in general parametrically different from the Zener breakdown current and the crossover between Bloch oscillations and dissipation-dominated current flow can occur already at much lower current values, i.e., $I_{\rm cr} \ll I_Z$. The task at hand is now to explicitly evaluate both rates $\gamma_\uparrow$ and $\gamma_\downarrow$ in the presence of dissipation [414].

Let us define the reduced density matrix of our system in the phase representation

$$\rho(t;\varphi_1,\varphi_2) = \int dq_f dq_i \int \mathcal{D}q_F \mathcal{D}q_B \tilde{\rho}(t;\varphi_1,\varphi_2;q_F,q_B) \times \exp\left(\mathrm{i}\frac{S_{\rm ext}[q_F,q_B]}{\hbar}\right) \tag{6.131}$$

where the action

$$S_{\rm ext} = \int_0^t q_-(t')(V_x - \dot{q}_+(t')R_S)dt' + (\mathrm{i}R_S/2)\int_0^t dt' \int_0^t dt'' q_-(t')q_-(t'') \int \frac{d\omega}{2\pi} e^{-\mathrm{i}\omega(t'-t'')}\,\omega \coth\frac{\hbar\omega}{2T} \tag{6.132}$$

with $V_x \equiv I_x R_S$ describing the effect of an external circuit containing an Ohmic resistor R_S. Note that both this action and our subsequent analysis are equally applicable to both current- and voltage-biased schemes of Fig. 6.7a,b. In the current-biased circuit, $V_x \equiv I_x R_S$ denotes an externally induced voltage across the junction.

The matrix $\tilde{\rho}$ can be expressed in the form

$$\tilde{\rho}(t;\varphi_1,\varphi_2;q_F,q_B) = \int dp_1 dp_2 \rho(0,p_1,p_2)\psi_{p_1q_F}(t,\varphi_1)\psi^*_{p_2q_F}(t,\varphi_2) \tag{6.133}$$

where the wave functions $\psi_{pq}(t,\varphi)$ obey the Schrödinger equation

$$\mathrm{i}\hbar\frac{\partial\psi_{pq}(t,\varphi)}{\partial t} = \hat{H}_J[q]\psi_{pq}(t,\varphi), \quad \psi_{pq}(0,\varphi) = \frac{1}{\sqrt{2\pi}}\exp(\mathrm{i}p\varphi/2e) \tag{6.134}$$

with the Hamiltonian

$$\hat{H}_J[q] = -\frac{\hbar^2}{2C}\frac{\partial^2}{\partial(\hbar\varphi/2e)^2} - E_J\cos\varphi - \frac{\dot{q}(t)\varphi}{2e} \tag{6.135}$$

describing the Josephson junction in the presence of a varying in time current $\dot{q}(t)$. The solution to Eq. (6.134) can be expressed in the form

$$\psi_{pq}(t,\varphi) = \sum_n \frac{c_n(t)}{\sqrt{2\pi}}\exp[\mathrm{i}(p+q(t)-q(0)-2en)\varphi/2e] \tag{6.136}$$

with $c_0(0) = 1$ and $c_n(0) = 0$ for all $n \neq 0$. Combining Eqs. (6.134) and (6.136), we arrive at the equation

$$i\hbar\frac{dc_n}{dt} = \frac{(p + q(t) - q(0) - 2en)^2}{2C}c_n + \frac{E_J^2}{2}(c_{n-1} - c_{n+1}) \quad (6.137)$$

Then, making use of Eqs. (6.133) and (6.136), after a shift $p_1 + q_F(t) - q_-(t)/2 \rightarrow q_F(t)$ and $p_2 + q_B(t) + q_-(t)/2 \rightarrow q_B(t)$, we obtain

$$\rho(t; \varphi_1, \varphi_2) = \sum_{n,m} \int \frac{dq_1 dq_2}{2\pi} \rho_{nm}(t; q_1, q_2)$$
$$\times \exp[i(q_1 - 2en)\varphi_1/2e] \exp[-i(q_2 - 2em)\varphi_2/2e] \quad (6.138)$$

where

$$\rho_{nm}(t; q_1, q_2) = \int dp_1 dp_2 \rho(0, p_1, p_2)$$
$$\times \int_{p_1}^{q_1} \mathcal{D}q_F \int_{p_2}^{q_2} \mathcal{D}q_B c_n[q_F] c_m^*[q_B] \exp[iS_{\text{ext}}[q_F - q_1, q_B - q_2]/\hbar) \quad (6.139)$$

In order to proceed, we restrict ourselves to the limit $E_J \ll E_C$, in which interband Zener tunneling plays the most significant role. In this case, as we already discussed, the band structure is close to a set of parabolas (see Fig. 6.4) and the gaps between the bands n and $n - 1$ are defined by a simple formula

$$\delta E_{n,n-1} \simeq E_C (E_J/E_C)^n / n^{n-1} \quad (6.140)$$

This formula implies that the gap between the two lowest Brillouin zones $\delta E_{1,0}$ parametrically (in the parameter $E_C/E_J \gg 1$) exceeds all bandgaps between higher zones, which can, therefore, simply be neglected. In other words, in our further analysis, we will take into account only the gap between the lowest Brillouin zones $\delta E_{1,0} = E_J$ ignoring all higher gaps. This approximation brings remarkable simplifications because in this case only two amplitudes c_0 and c_1 in Eqs. (6.136), (6.137), and (6.139) differ from zero. These amplitudes correspond to two parabolas $E(q) = q^2/2C$ ($n = 0$) and $E(q) = (q - 2e)^2/2C$ ($n = 1$), which can only be included into our consideration. With this in mind, we now proceed with the calculation of the tunneling rates $\gamma_\uparrow$ and $\gamma_\downarrow$.

We first consider Zener transitions between the two lowest bands and evaluate the tunneling rate $\gamma_\uparrow$. Without loss of generality, the initial state of the system at $t = 0$ is chosen to be in the lowest Brillouin zone and the density matrix $\rho(0; p_1, p_2)$ is assumed to be nonzero only for $-e < p_{1,2} < e$. In other words, the quasicharge q is assumed to initially stay on the parabola $n = 0$. Then the rate $\gamma_\uparrow$ to tunnel to higher bands is defined by the probability for the quasicharge q to remain on *the same* parabola at times exceeding one period of Bloch oscillations $t > 1/\nu_B$, whereas the jump to the parabola $n = 1$ would imply that no Zener tunneling has occurred up to such times (see also our analysis toward the end of Section 6.6). The probability density for the quasicharge q to stay on the parabola $n = 0$ at time t is defined by $\rho_{00}(t; q, q)$ and, hence,

$$\gamma_\uparrow = \nu_B \int \rho_{00}(t > 1/\nu_B; q, q) dq \tag{6.141}$$

It is convenient for us to introduce new variables u_0 and u_1 with the aid of the relations

$$c_0(t) = \exp\left(-\frac{\mathrm{i}}{\hbar}\int_0^t \frac{q^2}{2C} dt'\right) u_0(t),$$
$$c_1(t) = \exp\left(-\frac{\mathrm{i}}{\hbar}\int_0^t \frac{(q-2e)^2}{2C} dt'\right) u_1(t)$$

and rewrite Eqs. (6.137) in the form

$$\mathrm{i}\hbar\frac{du_0}{dt} = \frac{E_J^2}{2}\exp\left(\mathrm{i}\frac{2e}{\hbar C}\int_0^t (q-e)dt'\right) u_1,$$
$$\mathrm{i}\hbar\frac{du_1}{dt} = \frac{E_J^2}{2}\exp\left(-\mathrm{i}\frac{2e}{\hbar C}\int_0^t (q-e)dt'\right) u_0 \tag{6.142}$$

Expressing the rate (6.141) in terms of these new variables, we find (cf. Eq. (6.139)

$$\gamma_\uparrow/\nu_B = \langle u_0[q_F]u_0^*[q_B]\rangle = \int dq \int dp_1 dp_2 \rho(0, p_1, p_2)$$
$$\times \int_{p_1}^{q} \mathcal{D}q_F \int_{p_2}^{q} \mathcal{D}q_B u_0[q_F]u_0^*[q_B] \exp\left[\frac{\mathrm{i}}{\hbar}\left(S_{\mathrm{ext}}[q_F, q_B] - \int_0^t \frac{q_+ q_-}{C} dt'\right)\right] \tag{6.143}$$

Proceeding perturbatively in E_J and making use of the condition $\gamma_\uparrow \ll E_C/\hbar$, it is straightforward to solve Eqs. (6.142) and derive

$\gamma_{\uparrow}/\nu_B = \langle \exp(-f) \rangle$ with

$$f[q_F, q_B] = \frac{E_J^2}{4\hbar^2} \int_0^t dt_2 \int_0^{t_2} dt_1 \left[\exp\left(\mathrm{i}\frac{2e}{\hbar C} \int_{t_1}^{t_2} (q_F - e)dt' \right) + \exp\left(-\mathrm{i}\frac{2e}{\hbar C} \int_{t_1}^{t_2} (q_B - e)dt' \right) \right] \quad (6.144)$$

Evaluating the average in the expression for $\gamma_{\uparrow}$, we make use of the following formula

$$\langle \exp(-f) \rangle = \exp(-\langle f \rangle)[1 - (\langle f^2 \rangle - \langle f \rangle^2)/2 + \ldots] \quad (6.145)$$

Under the condition $K \ll 1$ employed here, fluctuations of trajectories $q_{F,B}$ near the least action path $\tilde{q}(t)$ remain small (cf. Eq. (6.125)). Hence, it suffices to replace the combination in the square brackets by unity and simply perform averaging of the function f in the exponent. Then one finds

$$\langle f \rangle = \frac{E_J^2}{2\hbar^2} \mathrm{Re} \int_0^t dt_2 \int_0^{t_2} dt_1 \exp\left(-\mathcal{K}(t_2 - t_1) + \mathrm{i}\frac{2e}{\hbar C} \int_{t_1}^{t_2} \tilde{q}(t')dt' \right) \quad (6.146)$$

where

$$\mathcal{K}(t) = \frac{1}{K} \int \frac{d\omega}{1 + \omega^2 R_S^2 C^2} \left[\coth\left(\frac{\hbar\omega}{2T} \right) \frac{1 - \cos(\omega t)}{\omega} + \mathrm{i}\frac{\sin(\omega t)}{\omega} \right] \quad (6.147)$$

and the least action path $\tilde{q}(t)$ obeys the classical equation of motion (6.127) with $\partial E_0(q)/\partial q = q/C$. Making use of the inequality $\langle \delta q^2 \rangle \ll e^2$, from the above equations, we obtain [414]

$$\gamma_{\uparrow} = \begin{cases} 0, & I_x \leq I_{\mathrm{th}}, \\ \nu_B \exp\left[-\frac{I_Z}{I_x - I_{\mathrm{th}}} \left(1 + \frac{\langle \delta q^2 \rangle}{(I_x R_S C - e)^2} \right) \right], & I_x > I_{\mathrm{th}} \end{cases} \quad (6.148)$$

This result defines the Zener tunneling rate in the presence of dissipation, which turns out to be smaller than that evaluated in a dissipativeless limit (6.130). It is clear that at low T, Zener tunneling should be totally suppressed in the Coulomb blockade regime $I_x < I_{\mathrm{th}}$, since in this case, Bloch oscillations are absent, the energy $\hbar\omega_B$ tends to zero, and the system cannot gain enough energy to overcome the interband gap E_J. In the regime of Bloch oscillations $I_x > I_{\mathrm{th}}$, the rate $\gamma_{\uparrow}$ (6.148) is also reduced by dissipation because the photon energy $\hbar\omega_B$ absorbed by the system is smaller than $\pi\hbar I_x/e$ and also because of non-vanishing junction charge fluctuations $\langle\delta q^2\rangle$.

We now turn to the analysis of the process of inelastic relaxation and evaluate the rate $\gamma_\downarrow$. In order to extract this rate, we need to consider the time derivative of the occupation probability $\rho_{00}(t; q, q)$ for the parabola $n = 0$ at $q > e$. In general, there are two different charge relaxation processes in our system that contribute to this derivative. One of them is due to continuous charge flow through the Ohmic resistor R_S and another one is due to Cooper pair tunneling across the junction. The first process can be neglected, provided I_x is larger than (and not too close to) $I_{\rm th}$ and $\gamma_\downarrow \ll 1/R_S C$. Under these conditions without loss of generality, one can assume that the initial system state is close to the equilibrium one. Accordingly, no charge relaxation via an external circuit would occur, and one can choose

$$\rho(0; p, p) = \frac{1}{\sqrt{2\pi\langle\delta q^2\rangle}} \exp\left[-\frac{(CV_x - p)^2}{2\langle\delta q^2\rangle}\right] \tag{6.149}$$

With this choice, the only contribution to $\partial\rho_{00}/\partial t$ is due to the process of Cooper pair tunneling between the states q and $q - 2e$. Combining Eqs. (6.149) and (6.139), one readily finds

$$\begin{aligned}\frac{\partial\rho_{00}(t; q, q)}{\partial t} &= \int dq \int dp_1 dp_2 \rho(0, p_1, p_2) \int_{p_1}^{q} \mathcal{D}q_F \int_{p_2}^{q} \mathcal{D}q_B \\ &\times \exp\left[\frac{\mathrm{i}}{\hbar}\left(S_{\rm ext}[q_F, q_B] - \int_0^t \frac{q_+ q_-}{C} dt'\right)\right] \frac{\partial(u_0[q_F]u_0^*[q_B])}{\partial t}\end{aligned} \tag{6.150}$$

Substituting the solution to Eqs. (6.142) into Eq. (6.150), after a straightforward calculation, we obtain

$$\begin{aligned}\frac{\partial\rho_{00}(t; q, q)}{\partial t} &= -\frac{E_J^2}{4\hbar^2\sqrt{2\pi\langle\delta q^2\rangle}} \\ &\times \int dt \exp\left[-\frac{(q - CV_x + e - \mathrm{i}C\dot{\mathcal{K}}(t)/2e)^2}{2\langle\delta q^2\rangle} - \mathcal{K}(t) + 2\mathrm{i}eV_x t\right]\end{aligned} \tag{6.151}$$

This expression defines the relaxation rate of the initial charge state q due to tunneling of a Cooper pair into the final charge state $q - 2e$.

Integrating this rate over all values of q, we obtain the total Cooper pair tunneling rate in the presence of dissipation

$$\gamma(V_x) = \int dq \frac{\partial \rho_{00}(t; q, q)}{\partial t} = \frac{1}{2}\mathrm{Re} \int_0^{\infty} dt \exp[-\mathcal{K}(t) + 2ieV_x t] \tag{6.152}$$

Multiplying this expression by $2e$, we recover the Cooper pair current $I = 2e\gamma(V_x)$ that flows across the Josephson junction [34]. In the low bias limit, Eq. (6.152) reduces to the result (6.90), while at higher bias values and $T \gtrsim KE_C$, one gets

$$\gamma(V_x) = \frac{\pi E_J^2}{8\hbar E_C} \frac{e}{\sqrt{2\pi \langle \delta q^2 \rangle}} \exp\left[-\frac{(CV_x - e)^2}{2\langle \delta q^2 \rangle}\right] \tag{6.153}$$

i.e., the Cooper pair tunneling rate $\gamma(q)$ is peaked around the quasicharge value $q = e$. Provided the charge fluctuations decrease, this peak becomes narrower, and in the limit $\langle \delta q^2 \rangle \to 0$, the expression (6.153) reduces to Eq. (6.91). We also point out that at very low temperatures $T \to 0$ and for $q > e$, the exact expression for the Cooper pair decay rate deviates from the Gaussian form (6.153) showing a power-law dependence $\gamma(q) \propto (q - e)^{-3}$ in the limit of large q.

In order to obtain the rate $\gamma_\downarrow$, it is necessary to bear in mind that here we are interested in the process of relaxation to the ground-state $E_0(q)$, i.e., the final charge state is restricted by the condition $-e < q < e$. Since this relaxation process involves tunneling of exactly one Cooper pair, the initial state should be in the first excited zone corresponding to $e < q < 3e$. Should the initial charge state be outside this interval, relaxation to the ground state could only be possible as a result of more than one Cooper pair tunneling events, which yield small (in the parameter E_J/E_C) corrections to $\gamma_\downarrow$ to be safely ignored here. Hence, in the leading approximation, the relaxation rate $\gamma_\downarrow$ is defined as

$$\gamma_\downarrow = \int_e^{3e} dq \frac{\partial \rho_{00}(t; q, q)}{\partial t} \tag{6.154}$$

Substituting the expression (6.151) into this formula and evaluating the integrals, we get [414]

$$\gamma_\downarrow = \frac{K^{3/2} E_J^2}{4\pi^{5/2}\hbar E_C} \sqrt{\ln(\omega_c R_S C)} \left(\frac{e}{CV_x}\right)^4 \exp\left[-\frac{\pi^2 (CV_x - 3e)^2}{4e^2 K \ln(\omega_c R_S C)}\right] \tag{6.155}$$

for $T \ll KE_C$ and

$$\gamma_{\downarrow} = \frac{\pi e E_J^2}{8\hbar E_C \sqrt{2\pi C T}} \exp\left[-\frac{(CV_x - e)^2}{2CT}\right] \tag{6.156}$$

for $T \gg KE_C$. We note that the expression in the pre-exponent in Eq. (6.155) is applicable for $CV_x \gg e$.

Now we are ready to construct the full I–V curve of our system. As we already discussed, provided $\gamma_{\uparrow} \ll \gamma_{\downarrow}$, the system remains in the lowest zone. In this case, the I–V curve $V_0(I_x)$ describes the "Bloch nose" (defined by $V_0 = R_S I_x$ for $I \leq I_{\rm th}$ and by Eq. (6.128) for $I > I_{\rm th}$). On the other hand, for $\gamma_{\uparrow} \gg \gamma_{\downarrow}$, the system easily tunnels to higher zones and ends up in the charge state $q = CV_x - 2e\gamma R_s$ (the last term in this equation accounts for decharging of the capacitor C by the "leakage" current of Cooper pairs through the Josephson junction). Accordingly, in this regime, the I–V curve is defined as

$$V_1(I_x) = R_S(I_x - 2e\gamma) \tag{6.157}$$

Provided the crossover between these two regimes is sufficiently sharp, the full I–V curve $V(I_x)$ can be described by a simple superposition of $V_0(I_x)$ and $V_1(I_x)$ as [414, 418]

$$V = \frac{V_0(I_x)}{1 + \gamma_{\uparrow}\gamma_{\downarrow}} + \frac{V_1(I_x)}{1 + \gamma_{\downarrow}\gamma_{\uparrow}} \tag{6.158}$$

Assuming that $I_{\rm th} \ll I_Z$ (which requires $K \ll (E_J/E_C)^2$) and substituting the expressions for the rates $\gamma_{\uparrow}$ (6.148) and $\gamma_{\downarrow}$ (6.155), (6.156) into the crossover condition (6.129), we determine $I_{\rm cr}$. At not very low temperatures $KE_C \ll T \ll E_C$, this condition yields within the exponential accuracy

$$I_{\rm cr} = I_{\rm th} + (\pi E_J^2 T / 2eR_S^2)^{1/3} \tag{6.159}$$

whereas at $T \ll KE_C$ from Eq. (6.129), we obtain

$$I_{\rm cr} - I_{\rm th} \sim eK(E_J^2 E_C)^{1/3} \tag{6.160}$$

In the, interesting for us, parameter range $K \ll (E_J/E_C)^2$ and at $T \ll KE_C$, the value $I_{\rm cr}$ turns out to be only slightly higher than $I_{\rm th}$, thus leaving very little room for Bloch oscillations. On the other hand, with increasing temperature, the crossover current $I_{\rm cr}$ increases and at $T \gg KE_C^3/E_J^2$ (but still $T \ll E_C$) may strongly exceed $I_{\rm th}$. We conclude, therefore, that at higher T, the system stays

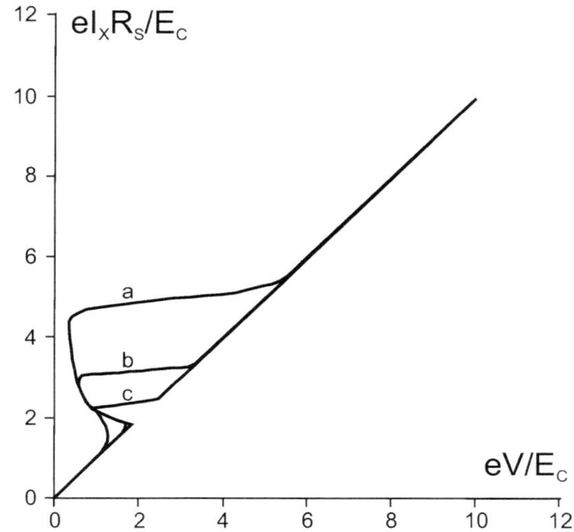

Figure 6.9 I–V curves for a current-biased Josephson junction with $E_J/E_C = 0.2$ and weak Ohmic dissipation $K = 0.01$. Zener tunneling to higher bands is included into consideration and temperature is chosen as $T/E_C = 0.2$ (a), 0.05 (b), and 0 (c).

in the lowest zone up to higher values I_x and Bloch oscillations may occur within a larger current interval $I_{\text{th}} < I_x < I_{\text{cr}}(T)$. This (at the first sight counterintuitive) result just follows from the observation that the inelastic relaxation rate $\gamma_\downarrow$ increases with temperature, while the rate $\gamma_\uparrow$ only weakly depends on T. Hence, with increasing temperature, the balance between these rates shifts to higher current values $I_x \approx I_{\text{cr}}(T)$. The corresponding I–V curves are depicted in Fig. 6.9.

We can add that the whole theoretical picture developed above has received a convincing experimental confirmation. Without going into details, let us only mention some key experiments on the subject. For example, Bloch oscillations and Bloch nose on the I–V curve in small-area Josephson junctions were demonstrated in experiments [245]. Incoherent tunneling of Cooper pairs with the rate γ (6.152), (6.153) peaked around $q = e$ was observed in the work [246]. The effect of Zener tunneling to higher zones in the presence of dissipation, which causes the crossover on the

$I{-}V$ curves, was also studied experimentally [247], and a good agreement between the above expression for the crossover current I_{cr} (6.159) and experimental data was demonstrated.

In this context, it is also worth pointing out that in some experiments—depending on the system parameters—it may be problematic to resolve the Bloch nose structure on the $I{-}V$ curve, in which case the shape of the latter may strongly resemble that for a classical Josephson junction. In other words, there exists a danger to visually confuse the regimes of Coulomb blockade and Bloch oscillations $I_x < I_{\text{cr}}$ in the quantum limit (strong phase fluctuations and nearly classical charge) with a zero-voltage branch (dc Josephson effect) $I_x < I_C = 2eE_J/\hbar$ in the classical limit (classical phase variable and intensive charge fluctuations). This similarity is, of course, superficial and by no means allows to interpret I_{cr} as an analogue of the superconducting critical current I_C. These two currents are parametrically different and, e.g., in experiments [247] the former was found to be by more than an order of magnitude lower than the latter.

6.9 *I–V* Curve at Stronger Dissipation

Now let us go beyond the underdamped limit considered in Section 6.8 and establish the $I{-}V$ curve for a resistively shunted Josephson junction at not-too-small values of the parameter K. Provided quantum fluctuations of the phase can be treated as incoherent, a simple master equation approach of Section 6.2 is sufficient. However, in a general situation, it is necessary to make use of a more elaborate analysis developed in Section 6.7.

6.9.1 Low-Temperature Limit

The average voltage V across the Josephson junction embedded in a circuit depicted in Fig. 6.7a defined as

$$V = R_S \left(I_x - I_C \int d\varphi \, \sin\varphi \, \rho(\varphi, \varphi) \right) \tag{6.161}$$

where the density matrix $\rho(\varphi, \varphi')$ is given by Eq. (6.94) in the long-time limit $t \to \infty$. Performing again the expansion in powers of E_J (6.99) and evaluating the path integrals in the same way as in Section 6.7, we rewrite Eq. (6.161) in the form of a series

$$V = I_x R_S - 2eR_S \sum_{n=1}^{\infty}(-1)^n \left(\frac{E_J}{2\hbar}\right)^{2n} \sum_{\epsilon_j=\pm1} \sum_{\eta_k=\pm1}$$
$$\times \int_0^{\infty} dt_1 \int_0^{t_1} dt_2 \dots \int_0^{t_{2n-2}} dt_{2n-1}$$
$$\times \exp\left[\frac{2}{K}\sum_{k=1}^{2n-1}\sum_{l=0}^{k-1} \epsilon_k\epsilon_l \tilde{\Lambda}\left(\eta_k(t_k - t_l)\right) + \mathrm{i}\frac{\pi I_x}{eK}\sum_{k=0}^{2n-1}\epsilon_k t_k\right], \tag{6.162}$$

where

$$\tilde{\Lambda}(t) = 2\int_0^{\infty} d\omega \frac{\coth\frac{\hbar\omega}{2T}(1 - \cos\omega t) + \mathrm{i}\sin\omega t}{\omega(1 + \omega^2 R_S^2 C^2)} \tag{6.163}$$

In the zero-temperature limit $T \to 0$ and at sufficiently low bias

$$\frac{I_x}{I_C} \lesssim \frac{4K^2 E_C}{\pi^2 E_J} \tag{6.164}$$

one can approximate the function $\tilde{\Lambda}(t)$ as

$$\tilde{\Lambda}(t) = \ln\frac{|t|}{R_S C} + \gamma + \mathrm{i}\frac{\pi}{2}\,\mathrm{sign}(t) \tag{6.165}$$

Within this approximation, the time integrals in Eq. (6.162) can be handled exactly and we obtain [201]

$$V = I_x R_S \left(1 - \sum_{n=1}^{\infty} c_n(K)\left(\frac{I_x}{I_C^*}\right)^{2\frac{1-K}{K}n}\right) \tag{6.166}$$

where we defined both the coefficients

$$c_n(K) = (-1)^{n-1}\frac{\Gamma\left(1 + \frac{n}{K}\right)\Gamma\left(\frac{3}{2}\right)}{\Gamma(1+n)\Gamma\left(\frac{3}{2} + \left(\frac{1}{K} - 1\right)n\right)} \tag{6.167}$$

and the renormalized critical current

$$I_C^* = \left(\Gamma\left(1 + \frac{1}{K}\right)\right)^{\frac{K}{1-K}}\left(\frac{e^{\gamma}K^2 E_C}{\pi^2 E_J}\right)^{\frac{1}{1-K}} I_C \tag{6.168}$$

which is essentially identical to that already introduced in Eq. (6.4). In particular, it is easy to observe that in the strong damping limit $K \gg 1$, Eq. (6.168) reduces to our previous result (6.9).

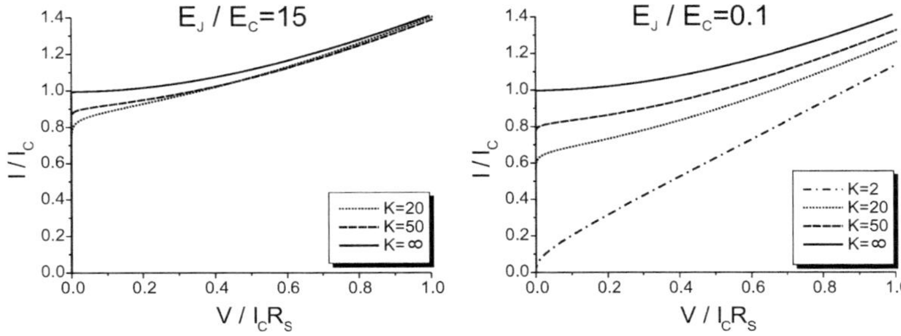

Figure 6.10 $I{-}V$ curves of an overdamped Josephson junction at $T = 0$ for two different values of the ratio E_J/E_C. The condition (6.170) imposes the following restrictions on the dissipation strength: $K \gtrsim 12$ for $E_J/E_C = 15$ and $K \gtrsim 1$ for $E_J/E_C = 0.1$.

Provided dissipation remains weak, i.e., $K < 1$, the condition (6.164) restricts the applicability range of the result (6.166)–(6.168) to sufficiently small currents I_x. In this case, the first term in the series in Eq. (6.166) gives the main contribution and the $I{-}V$ curve matches with that already defined in Eq. (6.124). At higher values of the damping parameter $K > 1$, the series (6.166) converges only at sufficiently large current values $|I_x| > I_{cr}$, where

$$I_{cr} = I_C^* K^{\frac{K}{2(1-K)}} \sqrt{|1-K|} \tag{6.169}$$

In order to derive the $I{-}V$ curve for $K > 1$ and smaller currents, one can make use of an alternative description in terms of MQT of the Josephson phase between different minima of the tilted washboard potential (4.46). This physical picture was already discussed in details in Chapters 4 and 5 and employed, e.g., in Section 6.2 of this chapter. At $T \to 0$ in the tight-binding limit, $E_J \gg E_C$ and for not-too-strong dissipation $1 < K \ll (E_J/E_C)^{1/2}$ the MQT description yields the result (6.15). In the overdamped limit $K \gg 1$ or, more precisely

$$K \gg \max(\sqrt{E_J/E_C}, \ln(E_C/E_J)) \tag{6.170}$$

the same physical picture applies, but the MQT rate for the phase should be modified taking into account the influence of strong dissipation on the corresponding instanton trajectories. For current values I_x in the vicinity of the critical one $I_C \simeq I_C^* \simeq I_{cr}$, the MQT rate in the overdamped limit was already analyzed in details in Chapter 4 of this book (see Section 4.3).

A generalization of this analysis to arbitrary values of I_x can also be worked out. Repeating the analysis of Section 6.2 and employing now the expression for the MQT rate [232] instead of Eq. (5.47), from Eq. (6.13) in the limit $T \to 0$ we get

$$R_{\rm eff}(I_x) \sim \frac{\hbar}{(2e)^2} \frac{K^{7/2}}{\pi^{5/2}} \frac{E_C^2}{E_J^2} \left(\frac{I_x}{I_C}\right)^{2K-2} \qquad (6.171)$$

This equation defines the effective nonlinear resistance $R_{\rm eff}(I_x) = V/I_x$ of a strongly overdamped Josephson junction for current values I_x lower than (and not too close to) $I_C^* \simeq I_{cr}$. It demonstrates that in the overdamped case, the leading power-law dependence of V on I_x remains the same as in Eq. (6.15). We also note that the inequality (6.170) does not require the condition $E_J \gg E_C$ to be satisfied, i.e., for $K \gg 1$, the instanton analysis [232] and the result (6.171) also apply for $E_J \lesssim E_C$.

On the other hand, for very small E_J/E_C, the condition eventually breaks down. Nevertheless, also in this case, one can demonstrate that the dependence $V \propto I_x^{2K-1}$ remains applicable at sufficiently small currents. This is achieved by combining the present analysis with the RG arguments developed in Section 6.3. Indeed, in the limit $K \gg 1$, we can start from an arbitrarily small ratio E_J/E_C, but scaling yields bigger and bigger values of this ratio. Eventually, the condition (6.170) is satisfied for the renormalized parameter E_J/E_C, and we again arrive at the dependence (6.171).

In order to restore the $I-V$ curve at all currents $|I_x| < I_{cr}$, it is worthwhile to employ, e.g., the formal equivalence between the problem studied here and that of a single channel coherent conductor in a resistive electromagnetic environment [338]. Translating the exact results [116, 399] derived for the latter problem within the thermodynamic Bethe ansatz technique to that of a resistively shunted Josephson junction, one obtains [201]

$$V = I_x R_S \sum_{n=1}^{\infty} c_n(1/K) \left(\frac{I_x}{I_C^*}\right)^{2(K-1)n} \qquad (6.172)$$

The first term $n = 1$ in the series (6.172) is consistent with Eq. (6.171) and all other terms with $n \geq 2$ can be interpreted as contributions from higher-order tunneling processes.

The two series (6.166) and (6.172) match at $|I_x| = I_{cr}$. Thus, they describe the I–V curve in the entire range of currents $0 < |I_x| < 4I_C K^2 E_C/\pi^2 E_J$. Due to the condition (6.170), the upper bound of this interval significantly exceeds I_C and, hence, Eqs. (6.166) and (6.172) cover the whole range of currents within which the I–V curve demonstrates a nonlinear behavior. At even higher currents, $|I_x| > 4I_C K^2 E_C/\pi^2 E_J$, one can still use the result (6.166) because in this case, it essentially reduces to the Ohmic form $V = I_x R_S$, which also follows from the exact solution in this limit.

The zero-temperature I–V curves for a resistively shunted Josephson junction are depicted in Fig. 6.10 at different values of the dissipation parameter K. In the strong dissipation limit $K \to \infty$, one finds $I_C^* = I_{cr} = I_C$. In this case, Eqs. (6.166) and (6.172) just reduce to the classical result

$$
\begin{aligned}
V &= 0, \quad |I_x| < I_C, \\
V &= R_S\sqrt{I_x^2 - I_C^2}, \quad |I_x| > I_C
\end{aligned}
\tag{6.173}
$$

At finite values of K, dissipation visibly suppresses the critical current and leads to smearing of the I–V curve.

We also note that some of the above results can also be extended to nonzero (though sufficiently low) temperatures. For instance, in the strong dissipation limit (6.170) after the replacement $\hbar I_x/e \to T$, Eq. (6.171) describes the temperature dependence of the effective linear resistance R_{eff} at $T > \hbar I_x/e$.

6.9.2 Higher Temperatures

Let us now turn to the case of higher temperatures, which can be analyzed by means of the quasiclassical Langevin equation (2.9). It turns out that under the condition $K \gg (E_J/E_C)^{1/2}$, one can derive relatively simple analytical expressions for the I–V curve in the wide range of parameters covering both classical and quantum regimes.

In the high-temperature limit $T \gtrsim \hbar/R_S C$, the phase dynamics can be treated as classical. The condition $K \gg (E_J/E_C)^{1/2}$ allows one to put $C = 0$ in the Langevin equation (2.9) and obtain

$$
\frac{1}{R_S}\frac{\hbar\dot{\varphi}}{2e} + I_C \sin\varphi = I_x + \xi_S(t) \tag{6.174}
$$

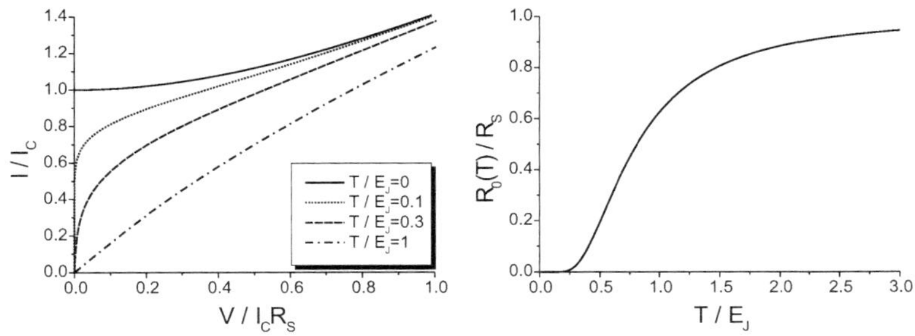

Figure 6.11 $I-V$ curves of an overdamped Josephson junction at $T \gtrsim \hbar/R_S C$, i.e., in the classical high-temperature regime. Left panel—voltage (6.178) at various temperatures; right panel—zero bias resistance (6.179).

where the stochastic variable $\xi_S(t)$ is described by the white noise correlator (2.11). The solution to Eq. (6.174) [16, 204] yields the average voltage across the junction in the form

$$V = I_x R_S - I_C \int_{-\pi}^{\pi} d\varphi \, \sin\varphi \, \sigma(\varphi) \tag{6.175}$$

where $\sigma(\varphi)$ is the stationary distribution function of the phase normalized by the condition $\int_{-\pi}^{\pi} d\varphi\sigma(\varphi) = 1$. This function is derived from the Smoluchowski phase diffusion equation

$$\frac{\partial\sigma}{\partial t} + \frac{2eR_S}{\hbar}\frac{\partial}{\partial\varphi}\big[(I_x - I_C\sin\varphi)\sigma\big] - \frac{4e^2R_S T}{\hbar^2}\frac{\partial^2\sigma}{\partial\varphi^2} = 0 \tag{6.176}$$

in the long-time limit $t \to \infty$. It reads

$$\sigma(\varphi) = \frac{\hbar V}{4\pi e R_S T}\int_{\varphi}^{\infty} d\varphi' e^{\hbar I_x(\varphi-\varphi')/2eT} e^{(\cos\varphi - \cos\varphi')E_J/T} \tag{6.177}$$

Combining this result with Eq. (6.175), we find

$$V = \frac{\frac{T\,eR_S}{\hbar}\sinh\frac{\pi\hbar I_x}{2eT}}{\int_0^{\pi/2} d\varphi \cosh\left(\frac{\hbar I_x\varphi}{eT}\right) I_0\left(\frac{\hbar I_C\cos\varphi}{eT}\right)} \tag{6.178}$$

where $I_0(x)$ is the modified Bessel function of the first kind. In the limit $T \ll E_J$, this result again reduces to the form (6.173), while at low bias $I_x \ll \hbar/eT$, one gets $V = R_0(T)I_x$, where

$$R_0(T) = R_S/I_0^2(E_J/T) \tag{6.179}$$

defines the zero bias resistance.

One can also extend the above analysis to include quantum effects [21]. In the case of an overdamped Josephson junction, the corresponding quantum Smoluchowski equation reads [22]

$$\frac{\partial \sigma}{\partial t} + \frac{2eR_S}{\hbar}\frac{\partial}{\partial \varphi}\left[\left(I_x - I_C\left(1 - \frac{\Lambda}{2}\right)\sin\varphi\right)\sigma\right]$$
$$- \frac{4e^2 R_S T}{\hbar^2}\frac{\partial^2}{\partial \varphi^2}\frac{\sigma}{1 - \Lambda E_J \cos\varphi / T} = 0 \quad (6.180)$$

where

$$\Lambda = \frac{2}{K}\left[\gamma + \frac{2\pi^2 T}{K E_C} + \Psi\left(\frac{K E_C}{2\pi^2 T}\right)\right] \quad (6.181)$$

is the dimensionless parameter characterizing the strength of quantum effects. The quantum equation (6.180) is handled in exactly the same way as the classical one (6.176). Its stationary solution reads [22]

$$\sigma(\varphi) = A\left(1 - \Lambda\frac{E_J\cos\varphi}{T}\right)\int_{\varphi}^{\infty} d\varphi' \exp\left[\frac{\hbar I_x}{2eT}(\varphi - \varphi')\right.$$
$$+ \frac{E_J}{T}\left(1 - \frac{\Lambda}{2}\right)(\cos\varphi - \cos\varphi') - \Lambda\frac{E_J}{T}\frac{\hbar I_x}{2eT}(\sin\varphi - \sin\varphi')$$
$$\left. - \frac{\Lambda}{4}\frac{E_J^2}{T^2}\left(1 - \frac{\Lambda}{2}\right)(\cos 2\varphi - \cos 2\varphi')\right] \quad (6.182)$$

Here the normalization constant A is determined from the normalization condition $\int_{-\pi}^{\pi} d\varphi\sigma(\varphi) = 1$, and the voltage across the junction V is again defined by Eq. (6.175).

The analysis based on the quantum Smoluchowski equation (6.180) remains valid at not very low temperatures $T \gg (eI_C R_S)^2 R_S C/\hbar$ and, in addition, the conditions $K \gg (E_J/E_C)^{1/2}$ and (6.164) should be satisfied. Thus, at $K \gtrsim 1$, the I–V curve (6.175) combined with the distribution function of the phase (6.182) covers most of the gap between the high-temperature result (6.178) and the low-temperature one defined by Eqs. (6.166) and (6.172).

6.10 Including Quasiparticle Tunneling

Throughout this chapter—as well as earlier in Chapters 4 and 5—we on purpose restricted our analysis to the situation assuming that the Josephson junction is attached to the leads with linear Ohmic resistance R_S, i.e., dissipation produced by an external environment can be described by the Caldeira–Leggett term in the effective action. As we already argued in Chapters 1 and 2, this term is in fact very general and captures a lot of physics related to the effect of dissipation on quantum dynamics of ultrasmall Josephson junctions. The whole analysis of this chapter is trivially generalized to external circuits characterized by an arbitrary linear impedance $Z_S(\omega)$, which should be employed instead of R_S. For instance, in Eqs. (6.125) and (6.147), one should substitute

$$\frac{R_S}{1+\omega^2 R_S^2 C^2} \to \frac{\mathrm{Re} Z_S(\omega)}{|1-\mathrm{i}\omega C Z_S(\omega)|^2}.$$

In addition to dissipation produced by an external circuit, there exists yet another source of dissipation, which occurs intrinsically due to quasiparticle tunneling across the junction. In Chapter 2, we have already analyzed this source of dissipation microscopically. We have evaluated the corresponding contribution to the effective action controlled by the kernel $\alpha_R(t)$, which is in turn directly related to the quasiparticle current I_{qp} across the junction, cf. the first Eq. (2.92). Since at low temperatures $T \ll \Delta$ and subgap frequencies and voltages, the number of quasiparticles above the gap is exponentially small, I_{qp} is negligible, and this mechanism of dissipation is inefficient. If, however, the relevant energies are comparable with and/or higher than Δ or, alternatively, there exist sufficiently many subgap states inside the tunnel junction, the current I_{qp} is not small and dissipation due to quasiparticle tunneling needs to be taken into account. The same is true if a Josephson junction is shunted by yet another tunnel junction, which remains in the normal state.

In all these cases, the corresponding dissipative contribution to the effective action is described by nonlinear in the phase terms containing the Ohmic α-kernel, e.g., in Eqs. (2.111)–(2.113) (real time) or Eqs. (2.140) and (2.141) (imaginary time). For small values of the phase difference $\delta\varphi \ll 2\pi$, these nonlinear expressions can be expanded in powers of $\delta\varphi$ and in the leading approximation again yield the Caldeira–Leggett action studied above. Hence, the difference between linear Ohmic dissipation and that produced by quasiparticle tunneling (in the gapless case) is negligible for processes involving only small changes of the phase φ. Examples of such processes were already considered in Chapter 4 of this book, where we analyzed both quantum and thermally activated decay of macroscopic current states in small-area Josephson junctions in the vicinity of the critical current. In contrast, in Chapter 5, we already dealt with larger variations of the phase $\delta\varphi \sim 2\pi$, in which case no expansion of the nonlinear dissipative terms in the action in powers of $\delta\varphi$ is in general possible. Nevertheless, at least some physics of double-well systems considered in Chapter 5 (for instance, that of the dissipative localization phase transition discussed in Section 5.1.2) remains essentially unchanged also in the case of quasiparticle dissipation.

Qualitatively new effects emerge if one considers quantum dynamics of a Josephson particle in its 2π-periodic potential $U(\varphi) = -E_J \cos\varphi$ subject to quasiparticle Ohmic dissipation instead of linear Ohmic one (as discussed above in this chapter). For instance, the dissipation-driven quantum phase transition occurs also in this case, but now it implies localization of the Josephson particle in *all* even (or all odd) potential wells [183, 321] (thus implying no real localization) rather than in just one well as in the case of linear Ohmic dissipation (Section 6.3). Analogously to Section 6.8, the $I-V$ curve of ultrasmall Josephson junction is strongly affected by Coulomb interaction also in the presence of quasiparticle dissipation; however, the shape of the Bloch nose is entirely different in this case [27, 425] as well as the expression for the crossover current I_{cr} [384, 418]. Further interesting effects emerge in the presence of both linear Ohmic and quasiparticle dissipation [345].

Note that there exists a profound physical reason for all these and other differences between the cases of linear Ohmic dissipation and that produced by quasiparticle tunneling. On a formal level, it manifests itself by different symmetry properties of the corresponding contributions to the effective action: The quasiparticle term is invariant under $\varphi \to \varphi + 4\pi$ (i.e., is 4π-periodic in φ), while the Caldeira–Leggett term is quadratic in φ and, hence, explicitly violates the translational symmetry of the action in the phase space (cf., e.g., the last two terms in Eq. (2.141)), thus making the states φ and $\varphi + 4\pi$ distinguishable. Accordingly, quasiparticle dissipation preserves discreteness of the charge, making the quasicharge space e-periodic, whereas the presence of a linear Ohmic resistor makes the charge states continuous, as we already discussed in Sections 6.4 and 6.5. For an extended discussion of all these issues, we refer the reader to the review paper [345].

We finally note that, while in this chapter we only considered quantum behavior and charging effects in ultrasmall superconducting junctions, it is also interesting to analyze Coulomb interaction effects in the absence of superconductivity, i.e., in normal tunnel junctions where only single-electron tunneling between metallic electrodes is present. This will be the subject of the next chapter.

Chapter 7

Coulomb Effects in Metallic Tunnel Junctions

In previous chapters, we described the physics of small-size Josephson junctions, where quantum fluctuations and Coulomb effects play a prominent role, and demonstrated the power of the influence functional path integral approach to such systems. It is quite obvious that such effects should persist not only in superconducting but also in normal tunnel junctions, provided their size remains sufficiently small. Although a tunnel junction between two normal electrodes appears to be one of the simplest mesoscopic structures, its transport properties turn out to be sufficiently complicated, provided Coulomb interaction between electrons is taken into account.

In this chapter, we will focus on charging effects in tunnel junctions between normal metallic electrodes. We will demonstrate that also in this case the influence functional path integral formalism is very useful and provides a unified description of electron transport and other properties of such systems. Our analysis will also serve as a basis for further generalizations aiming at better understanding of electron–electron interaction effects in more complicated systems, such as metallic quantum dots and diffusive

Dissipative Quantum Mechanics of Nanostructures: Electron Transport, Fluctuations, and Interactions
Andrei D. Zaikin and Dmitry S. Golubev

ISBN 978-981-4774-50-5 (Hardcover), 978-0-429-29823-3 (eBook)
www.jennystanford.com

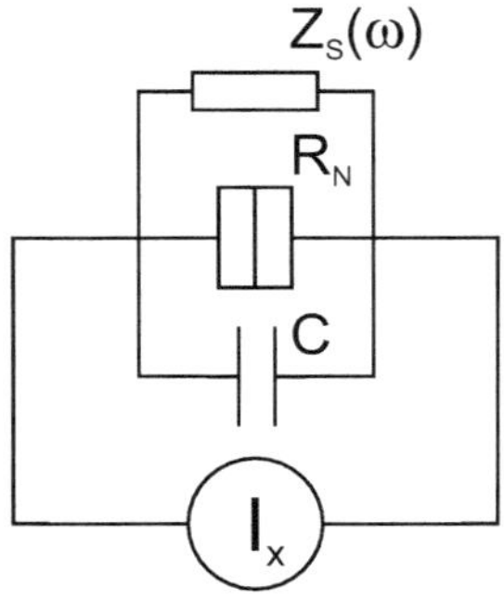

Figure 7.1 Schematics of a tunnel junction with a resistance R_N and capacitance C embedded in the electromagnetic environment characterized by the impedance $Z_S(\omega)$. The junction is biased by a current source I_x.

wires. These systems will be addressed in details in the second part of this book.

Intuitively, it is quite clear that Coulomb interaction tends to block the process of electron tunneling between metallic electrodes and, hence, reduce the current flowing through a tunnel barrier. This Coulomb blockade has to do with a simple observation that normal electrodes connected by a tunnel barrier effectively form a capacitor. Tunneling electrons charge this capacitor, thereby generating an electric force in the direction opposite to that of electron tunneling. In other words, Coulomb interaction creates an extra potential barrier for electron tunneling, and, hence, the probability of this process gets reduced.

Having in mind this simple physical picture, let us consider a circuit depicted in Fig. 7.1. An important element of this circuit is a tunnel junction characterized by its normal resistance $R_{\rm N}$ and capacitance C. We also assume that this junction is electromagnetically coupled to external circuit elements, such as wires and transmission lines. All these elements form an effective electromagnetic environment with a frequency-dependent impedance $Z_{\rm S}(\omega)$. In the simplest case of a purely Ohmic environment, this frequency dependence can be disregarded. Then we have $Z_{\rm S}(\omega) = R_{\rm S}$, where $R_{\rm S}$ is the shunt resistance.

In the absence of interactions, the junction differential conductance is $dI/dV = 1/R_N$. Since Coulomb interaction tends to block electron tunneling, this relation should generally be modified as

$$\frac{dI(V)}{dV} = \frac{1}{R_N} - \delta G(T, V), \tag{7.1}$$

where $-\delta G(V)$ is a negative interaction correction to the differential conductance, which depends on both temperature T and the average voltage V across the junction. Both T and eV should be compared with the charging energy $E_C = e^2/2C$, which defines electrostatic energy of a capacitor C charged by an electron charge e. In the high-energy limit $\max\{T, eV\} \gg E_C$, charging effects are relatively weakly pronounced and the interaction correction $\delta G(T, V)$ remains small. On the other hand, in the opposite limit $\max\{T, eV\} \ll E_C$, one can expect strong charging effects and large values $\delta G(T, V)$.

Below we will demonstrate that this is indeed the case, provided both resistances R_N and R_S strongly exceed the normal quantum resistance unit

$$R_q = 2\pi\hbar/e^2 \approx 25.8\text{k}\Omega \tag{7.2}$$

If, however, at least one of the two dimensionless conductances

$$g = \frac{R_q}{R_N}, \quad g_S = \frac{R_q}{R_S} \tag{7.3}$$

is not small, charge fluctuations in the system become more pronounced and tend to smear Coulomb blockade effects even at relatively low temperatures and voltages $\max\{T, eV\} \ll E_C$. In this case, Coulomb blockade remains strong only at exponentially low energies.

To summarize, below we will show that the Coulomb interaction correction $\delta G(T, V)$ to the conductance of a tunnel barrier may become large (i.e., comparable with $1/R_N$ in the following two cases:

$$\begin{aligned} &\max\{T, eV\} \lesssim E_C, \quad \text{for } g_+ g_S \ll 1, \\ &\max\{T, eV\} \lesssim (g_N + g_S)E_C \exp[-(g + g_S)/2], \quad \text{for } g + g_S \gtrsim 1 \end{aligned} \tag{7.4}$$

Outside this parameter region, the Coulomb blockade is weak and $\delta G(T, V) \ll 1/R_N$.

In this chapter, with the aid of the path integral influence functional technique, we will analyze electron transport across tunnel junctions affected by Coulomb interaction. We will demonstrate that by employing a number of complementary approximations, it is possible to analytically cover practically all experimentally relevant parameter regimes of electron transport. In addition, we will also discuss nontrivial thermodynamic properties of tunnel junctions in the presence of electron–electron interactions.

7.1 Weak Electron Tunneling Regime

7.1.1 Master Equation Approach

Let us begin our analysis from the weak tunneling limit $R_N \gg R_q$ or, equivalently, $g \ll 1$. First we consider an unshunted junction, i.e., we set $1/Z_S \to 0$. In this case, the junction is directly biased by the current I_x. In the absence of electron tunneling, the charge Q across the capacitor C would grow linearly with time as $Q_x = I_x t$. Single-electron tunneling results in rapid charge jumps in units of e, i.e., only discrete charge values $Q(t) = Q_x + pe$ with p integer are allowed.

Since electron tunneling is a stochastic process, it is convenient to introduce the probability $W(Q, t)$ for a junction to be in the state with charge Q at time t. This probability distribution obeys a simple master equation [29]

$$\frac{\partial W(Q,t)}{\partial t} = -I_x \frac{\partial W(Q,t)}{\partial Q} - [\overrightarrow{\Gamma}(Q) + \overleftarrow{\Gamma}(Q)]W(Q,t) + \overrightarrow{\Gamma}(Q-e)W(Q-e,t) + \overleftarrow{\Gamma}(Q+e)W(Q-e,t), \tag{7.5}$$

which accounts both for the current bias I_x and for single-electron tunneling. Here $\overrightarrow{\Gamma}(Q)$ and $\overleftarrow{\Gamma}(Q)$ are the electron tunneling rates from the charge state Q, respectively, to the states $Q - e$ and $Q + e$. In the limit $g \ll 1$, these tunneling rates can be evaluated perturbatively in $1/R_N$ and are given by the standard expression proportional to the energy-integrated product of the electron and hole distribution functions in the left and right electrodes. Should an electron tunnel from the left electrode to the right one, the electrostatic energy of our capacitor changes from $Q^2/2C$ to $(Q - e)^2/2C$. Likewise, if an electron tunnels in the opposite direction, the energy changes from $Q^2/2C$ to $(Q + e)^2/2C$. Bearing in mind that the total electron energy E should be conserved during tunneling and representing E as a sum of its kinetic energy E_{kin} and the electrostatic energy of the capacitor, we immediately conclude that before and after tunneling, we have, respectively, $E_{\text{kin}} = E - Q^2/2C$ and $E_{\text{kin}} = E - (Q \pm e)^2/2C$ depending on the tunneling direction. With this in mind, we readily obtain

$$\begin{aligned}\overrightarrow{\Gamma}(Q) &= \frac{1}{eR_{\text{N}}}\int dE\, f_{\text{L}}\left(E - \frac{Q^2}{2C}\right)\left[1 - f_{\text{R}}\left(E - \frac{(Q-e)^2}{2C}\right)\right],\\ \overleftarrow{\Gamma}(Q) &= \frac{1}{eR_{\text{N}}}\int dE\left[1 - f_{\text{L}}\left(E - \frac{(Q+e)^2}{2C}\right)\right] f_{\text{R}}\left(E - \frac{Q^2}{2}\right)\end{aligned} \tag{7.6}$$

Here f_{L} and f_{R} are the electron distribution functions in the two leads. In equilibrium, they are equal to the Fermi function $f(E) = 1/(1 + e^{E/T})$. The latter has a useful property

$$\int dE\, f(E - E_1)\left[1 - f(E - E_2)\right] = \frac{E_1 - E_2}{1 - e^{(E_2 - E_1)/T}}, \tag{7.7}$$

which allows us to express the rates (7.6) in the form

$$\overrightarrow{\Gamma}(Q) = \frac{1}{R_{\text{N}}}\frac{(e/2 - Q)/C}{e^{(E_{\text{C}} - eQ/C)/T} - 1}, \quad \overleftarrow{\Gamma}(Q) = \frac{1}{R_{\text{N}}}\frac{(e/2 + Q)/C}{e^{(E_{\text{C}} + eQ/C)/T} - 1} \tag{7.8}$$

Note that the same expressions can also be derived by other means, for example, by evaluating an imaginary part of the system free energy, as we already demonstrated in Chapters 4–6. We will return to this point in Section 7.4.

It follows from Eq. (7.5) that within the time scale $\sim R_{\text{N}}C$, an arbitrary initial probability distribution relaxes to a universal form

$$W(Q)\sum_{p=-\infty}^{\infty}\delta(Q - Q_x(t) - pe) \tag{7.9}$$

where $W(Q)$ is a stationary envelope distribution. At low temperatures and small bias $I_x \ll e/R_\mathrm{N}C$, one finds [29]

$$W(Q) = \begin{cases} \exp\left(-\dfrac{(Q-e/2)^2}{2eI_x R_\mathrm{N}C}\right), & Q \geq e/2, \\ 1-\exp\left(-\dfrac{(Q+e/2)^2}{2eI_x R_\mathrm{N}C}\right), & -e/2 \leq Q \leq e/2, \\ 0, & Q \leq -e/2 \end{cases} \tag{7.10}$$

In the opposite large bias limit $I_x \gg e/R_\mathrm{N}C$ and $T \to 0$, a detailed solution to the master equation (7.5) was also worked out [290]. Close to its maximum, the distribution is Gaussian and reads

$$W(Q) = \frac{e}{\pi I_x R_\mathrm{N}C}\exp\left(-\frac{(Q-\langle Q\rangle)^2}{eI_x R_\mathrm{N}C}\right) \quad \text{for } |Q-\langle Q\rangle| \ll I_x R_\mathrm{N}C \tag{7.11}$$

where in the above limit the average charge is $\langle Q\rangle = I_x R_\mathrm{N}C + e/2$. At nonzero T, this distribution gets broadened further due to thermal fluctuations.

Evaluating the average time-dependent charge across the junction

$$\langle Q(t)\rangle = \sum_{p=-\infty}^{\infty} (Q_x(t)+pe)W(Q_x(t)+pe) \tag{7.12}$$

at low T and I_x, one finds

$$\langle Q(t)\rangle = I_x t + e\left[\exp(-I_x t^2/2eR_\mathrm{N}C) - 1/2\right] \tag{7.13}$$

within the time interval $0 \leq t \leq e/I_x$. Outside this interval, the above function should be continued periodically. Hence, at small enough values of both temperature and external bias, the average junction charge oscillates in time with the frequency $\nu_\mathrm{SET} = I_x/e$. These single-electron tunneling oscillations [29, 52] get suppressed with increasing I_x and T. Averaging the function $\langle Q(t)\rangle$ (7.12) over time, we obtain the junction $I-V$ curve in the form

$$V = \begin{cases} \sqrt{\pi e I_x R_\mathrm{N}/2C}, & I_x R_\mathrm{N}C \ll e \\ I_x R_\mathrm{N} + e/2C, & I_x R_\mathrm{N}C \gg e \end{cases} \tag{7.14}$$

We observe that in the single-electron tunneling oscillations regime of low current bias, the $I-V$ curve is highly nonlinear and the junction differential resistance dV/dI_x diverges in the zero-bias limit $I_x \to 0$. This is a clear signature of strong Coulomb blockade. Yet another effect of Coulomb blockade is the offset of the Ohmic $I-V$ curve by the value $V_{\rm off} = e/2C$ in the limit of large I_x, as it is displayed by Eq. (7.14).

Now let us shunt the junction by an external impedance $Z_{\rm S}(\omega)$, as shown in Fig. 7.1. In this case, the total current I_x splits between the junction and the shunt. Just like in the case of a superconducting junction, we have

$$I_x = I + V/R_{\rm S} \tag{7.15}$$

where I is the current through the junction and V is the average voltage value across the shunt and also across the junction. Here we will assume that $Z_{\rm S}(0) = R_{\rm S}$ is large as compared to the quantum resistance unit (7.2) and, hence,

$$g + g_{\rm S} \ll 1 \tag{7.16}$$

Under this condition, electron tunneling can be treated perturbatively, and we immediately obtain the junction current in the form

$$I = e\left(\overrightarrow{\Gamma}(CV) - \overleftarrow{\Gamma}(CV)\right) \tag{7.17}$$

where $\overrightarrow{\Gamma}(Q)$ and $\overleftarrow{\Gamma}(Q)$ are the decay rates (7.6) of the charge state $Q = CV$ into the states $Q - e$ and $Q + e$, respectively. Note that this equation is exactly analogous to Eq. (6.122) defining the current across the Josephson junction due to incoherent tunneling of Cooper pairs in the presence of a resistive shunt. The equivalence between Eqs. (6.122) and (7.17) becomes obvious if we establish the correspondence between the elementary charges $2e \leftrightarrow e$ and the tunneling rates $\gamma_+ \leftrightarrow \overrightarrow{\Gamma}$ and $\gamma_- \leftrightarrow \overleftarrow{\Gamma}$.

Substituting the rates (7.8) into Eq. (7.17), we arrive at the $I-V$ curve for a resistively shunted normal tunnel junction in the strong Coulomb blockade regime

$$I = \frac{1}{eR_{\rm N}}\left(\frac{eV + E_{\rm C}}{1 - e^{(E_{\rm C}+eV)/T}} - \frac{E_{\rm C} - eV}{1 - e^{(E_{\rm C}-eV)/T}}\right) \tag{7.18}$$

7.1.2 Path Integral Analysis

Before investigating the result (7.18) in various limiting cases, let us re-derive it from a more rigorous path integral analysis. This derivation will also serve as an illustration of our method, which will later be applied to more complicated problems.

In order to evaluate the current in our system, we will make use of a general expression (2.114) already derived in Chapter 2. For the sake of convenience, here we will stick to the same definition of the superconducting phase variable φ as in Chapter 2 despite the fact that in this chapter, we will not deal with superconductivity at all. In the second part of this book, we will, however, rescale the phase by the factor 1/2.

In order to proceed, let us perform a shift of the classical phase in Eq. (2.114)

$$\varphi^+(t) \to 2eVt/\hbar + \phi^+(t) \tag{7.19}$$

where, as before, V is the average voltage across the junction. Combining Eqs. (7.15) and (7.19), we can eliminate the linear in I_x term in the action (2.113). After that we explicitly make use of the condition (7.16) and keep only the charging term in the action (2.113), disregarding both the environment contribution $S_{\text{diss}}[\varphi+, \varphi^-]$ and all the remaining after the shift (7.19) terms $\sim 1/R_{\text{N}}$. Then the current (2.114) across the junction acquires the form

$$\begin{aligned} I = \frac{1}{R_{\text{N}}} \int \mathcal{D}\phi^+ \mathcal{D}\varphi^- \exp\left[\frac{\mathrm{i}}{\hbar}\int dt_1 \frac{C\hbar^2\dot{\phi}^+\dot{\varphi}^-}{4e^2}\right] \\ \times \left(\left(V + \frac{\hbar\dot{\phi}^+(t)}{2e}\right)\cos\frac{\varphi^-(t)}{2} - \frac{2\mathrm{i}\hbar}{e}\int dt' K_N(t-t') \right. \\ \left. \times \cos\left[\frac{eV(t-t')}{\hbar} + \frac{\phi^+(t)-\phi^+(t')}{2}\right]\cos\frac{\varphi^-(t)}{4}\sin\frac{\varphi^-(t')}{4}\right) \end{aligned} \tag{7.20}$$

where the kernel $K_{\text{N}}(t)$ is defined in Eq. (2.112).

We observe that the path integrals in Eq. (7.20) are Gaussian and can be handled in a straightforward manner. For pedagogical purposes, let us illustrate the key points of this procedure. We first evaluate the integral over the classical phase ϕ^+. For that purpose, let us rewrite Eq. (7.20) in an equivalent form

$$I = \frac{1}{R_N} \int \mathcal{D}\varphi^- \left(V \cos \frac{\varphi^-(t)}{2} B[\varphi^-, 0] \right.$$
$$- \mathrm{i}\frac{\hbar}{e} \lim_{t' \to t} \frac{B[\varphi^-, 1] - B[\varphi^-, 0]}{t - t'} \cos \frac{\varphi^-(t)}{2}$$
$$- \frac{\mathrm{i}\hbar}{e} \int dt' K_N(t - t') \left[e^{\mathrm{i}eV(t-t')/\hbar} B[\varphi^-, 1] + e^{-\mathrm{i}eV(t-t')/\hbar} B[\varphi^-, -1] \right]$$
$$\left. \times \cos \frac{\varphi^-(t)}{4} \sin \frac{\varphi^-(t')}{4} \right), \tag{7.21}$$

where we defined

$$B[\varphi^-, \nu] = \int \mathcal{D}\phi^+ \exp \left[\frac{\mathrm{i}}{\hbar} \int dt_1 \frac{C\hbar^2 \dot{\phi}^+ \dot{\varphi}^-}{4e^2} + \mathrm{i}\nu \frac{\phi^+(t) - \phi^+(t')}{2} \right] \tag{7.22}$$

As the combination in the exponent in the latter integral is a linear function of ϕ^+, this integral reduces to the δ-function in the functional space,

$$B[\varphi^-, \nu] = \Omega \, \delta \left(\hat{L}\varphi^- - f \right) \tag{7.23}$$

where the operator $\hat{L}$ and the function f are defined as follows:

$$\hat{L} = -C \frac{\hbar}{4e^2} \frac{d^2}{dt^2}, \quad f(\tau) = -\frac{\nu}{2} \delta(\tau - t) + \frac{\nu}{2} \delta(\tau - t') \tag{7.24}$$

The prefactor Ω should ensure the correct normalization of the whole expression,

$$\int \mathcal{D}\varphi^- B[\varphi^-, \nu = 0] \equiv 1 \tag{7.25}$$

Utilizing the properties of the δ-function, we cast the integral $B[\varphi, \nu]$ to the form

$$B[\varphi, \nu] = \frac{\Omega}{\det \hat{L}} \delta(\varphi^- - \hat{L}^{-1} f) \tag{7.26}$$

The normalization condition (7.25) implies that $\Omega / \det \hat{L} = 1$. The combination $\hat{L}^{-1} f$ can be found by solving the second-order linear differential equation

$$\hat{L}x = f \tag{7.27}$$

which yields

$$\hat{L}^{-1} f = \frac{2e^2}{\hbar C} \nu \left[\theta(t - \tau)(t - \tau) - \theta(t' - \tau)(t' - \tau) \right] \tag{7.28}$$

An attentive reader may have already noticed that the solution (7.28) is, strictly speaking, not a general one. Indeed, since $\hat{L}(a\tau + b) = 0$, the function $a\tau + b$ with arbitrary a and b can also be added to the right-hand side of Eq. (7.28). However, as we will see shortly below, this ambiguity is eliminated by employing the requirement of causality.

Combining all the above expressions, we arrive at the final result for the integral $B[\varphi^-, \nu]$,

$$B[\varphi^-, \nu] = \delta\left(\varphi^-(\tau) - \frac{2e^2}{\hbar C}\nu\left[\theta(t-\tau)(t-\tau) - \theta(t'-\tau)(t'-\tau)\right]\right) \quad (7.29)$$

Substituting this functional δ-function into Eq. (7.21), we eliminate the remaining integral over the quantum phase φ^- and obtain

$$\begin{aligned} I(t) &= \frac{1}{R_N}\left(V + \frac{\hbar}{e}\int dt' K_N(t-t')\cos\left[\frac{e^2}{2\hbar C}\theta(t'-t)(t-t')\right]\right. \\ &\quad \left.\times \sin\left[\frac{e^2}{2\hbar C}\theta(t-t')(t-t')\right]\sin\frac{eV(t-t')}{\hbar}\right) \\ &= \frac{V}{R_N} + \frac{\hbar}{eR_N}\int_{-\infty}^{t} dt' K_N(t-t')\sin\left[\frac{E_C(t-t')}{\hbar}\right]\sin\frac{eV(t-t')}{\hbar} \end{aligned} \quad (7.30)$$

We note that causality is explicitly maintained in Eq. (7.30) due to the presence of the Heaviside step functions $\theta(t)$, cf. Eq. (7.28). As a result, the current at time t depends only on the voltage values at earlier times $t' < t$, as it is evident from Eq. (7.30). On the other hand, any additional term of the form $a\tau + b$ with nonzero a and b in Eq. (7.28) would violate causality. Hence, only the solution (7.28) is physically meaningful.

Combining Eqs. (7.30) and (2.112), we arrive at the final expression for the current across the tunnel junction in the presence of Coulomb effects. This expression remains valid under the condition (7.16) and reads

$$I = \frac{V}{R_N} - \frac{1}{e\hbar R_N}\int_0^{\infty} dt \frac{\pi T^2}{\sinh^2[\pi T t/\hbar]}\sin\frac{E_C t}{\hbar}\sin\frac{eVt}{\hbar} \quad (7.31)$$

Evaluating the time integral in Eq. (7.31), one reduces this result to the form (7.18). Thus, our rigorous path integral analysis fully justifies the validity of a simple approach of Section 7.1.1 in the limit (7.16).

7.1.3 Current–Voltage Characteristics

Let us now analyze the expression for the current (7.18). This expression is plotted in Fig. 7.2 (left panel) for some typical parameter values. We observe that, in the agreement with our initial expectations, the current is strongly suppressed as long as both temperature and the bias voltage remain small, i.e., $T, e|V| \lesssim E_C$. This is a clear manifestation of strong Coulomb blockade of electron tunneling.

Coulomb blockade manifests itself also at higher voltages. In this case, electrons have enough energy to overcome the Coulomb barrier and, hence, the current across the junction is not anymore small. In the large voltage limit $eV \gg E_C$, one finds

$$I = \frac{1}{R_N}\left(V - \frac{e}{2C}\right) \tag{7.32}$$

i.e., just like in the case of an unshunted junction (7.14), the $I-V$ curve shows a negative offset as compared to the non-interacting Ohmic form. By measuring the corresponding offset voltage $V_{off} = e/2C$ in the limit of large V, one can conveniently determine the capacitance C of a tunnel barrier. We also note that at $T \to 0$, Eq. (7.32) applies not only at large voltages, but at any value of V down to $V \to e/2C$. At even lower voltages $0 < V < e/2C$ and $T = 0$, both electron tunneling rates $\overrightarrow{\Gamma}(CV)$ and $\overleftarrow{\Gamma}(CV)$ (7.17) vanish, i.e., electron tunneling across the junction is not possible due to Coulomb blockade and, hence, $I = 0$. For this reason, the offset voltage V_{off} is also referred to as the Coulomb gap.

From Eq. (7.18), one can also recover the temperature dependence of the zero-bias conductance $G_0 = dI/dV|_{V=0}$. It reads

$$G_0 = \frac{1}{R_N}\left(1 - \coth\frac{E_C}{2T} + \frac{E_C}{2T}\frac{1}{\sinh^2\frac{E_C}{2T}}\right) \tag{7.33}$$

This dependence is also plotted in Fig. 7.2 (right panel). At $T \ll E_C$, the conductance is strongly suppressed showing a pronounced

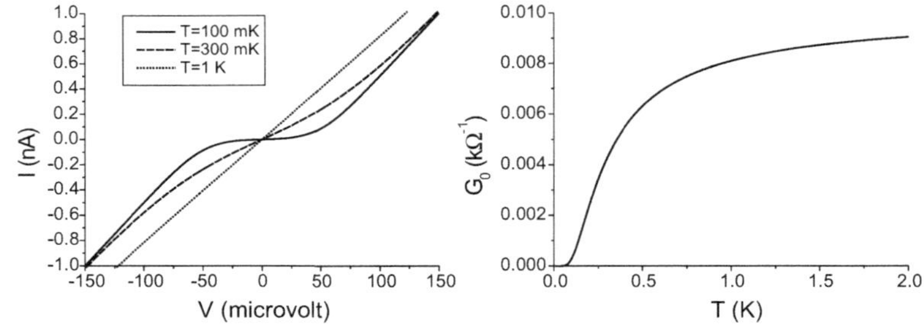

Figure 7.2 The current (left panel) and zero-bias conductance (right panel) for a tunnel junction in the strong Coulomb blockade regime. The parameters are chosen to be $E_C = 50$ μeV and $R_N = 100$ kΩ.

activation behavior,

$$G_0 \approx \frac{2}{R_N}\left(\frac{E_C}{T} - 1\right) e^{-E_C/T} \tag{7.34}$$

At high temperatures $T \gg E_C$, on the contrary, the junction conductance is only weakly influenced by Coulomb effects. In this limit, one finds

$$G_0 = \frac{1}{R_N}\left(1 - \frac{E_C}{3T} + \frac{E_C^3}{90T^3} + \mathcal{O}\left(\frac{E_C^5}{T^5}\right)\right) \tag{7.35}$$

The result (7.18) also yields an explicit expression for the differential conductance of a tunnel barrier in the strong Coulomb blockade regime:

$$\frac{dI}{dV} = \frac{1}{R_N}\left[1 + \frac{\coth(x-u)}{2} - \frac{\coth(x+u)}{2} - \frac{x-u}{2\sinh^2(x-u)} + \frac{x+u}{2\sinh^2(x+u)}\right], \tag{7.36}$$

where we defined $x = eV/2T$ and $u = E_C/2T$. At $T \gg E_C$, one can further simplify the expression (7.36) by expanding it in powers of u. Keeping only the lowest-order terms, one gets

$$\frac{dI}{dV} = \frac{1}{R_N} - 2\frac{u}{R_N}\frac{x\cosh x - \sinh x}{\sinh^3 x} \tag{7.37}$$

The second term in this expression describes a pronounced dip in the differential conductance due to Coulomb blockade. Since the voltage dependence of the interaction correction in Eq. (7.37) occurs only via the combination $x = eV/2T$, the half-width of this dependence is determined only by temperature. This remarkable property was used to construct an absolute thermometer [137]. In practice, it turns out to be more convenient to fabricate such Coulomb blockade thermometers as arrays consisting of several connected in series tunnel junctions. This way is preferable as it allows to avoid using high Ohmic leads, which otherwise would be necessary to achieve the strong Coulomb regime (7.16). Later we will see that Eq. (7.37) with some simple modifications remains applicable also in the case of tunnel junction arrays.

7.2 Langevin Equation Analysis

From our previous consideration, we can conclude that in the high resistance limit (7.16), Coulomb effects remain sufficiently weak only at high energies $\max\{T, eV\} \gg E_C$. However, if one relaxes the condition (7.16), charge fluctuations become more pronounced and the Coulomb blockade gets weaker. Accordingly, one can perform a perturbative analysis of the interaction effects down to much lower values of T and eV. In this section, we will elaborate a detailed treatment of the weak Coulomb blockade regime in which the interaction correction to the junction conductance remains much smaller than the standard non-interacting Ohmic term. Specifically, we will address the two limiting cases:

$$\begin{aligned} &\max\{T, eV\} \gg E_C, \ \text{at } g + g_S \ll 1, \\ &\max\{T, eV\} \gg (g + g_S)E_C \exp[-(g + g_S)/2], \ \text{at } g + g_S \gtrsim 1 \end{aligned} \tag{7.38}$$

These conditions are complementary to those defined in Eq. (7.4) for the strong Coulomb blockade. Also, the first of the two inequalities (7.38) corresponds to the regime already considered in the previous section. Accordingly, one can also expect partial overlap between the results obtained here and in Section 7.1.

To begin with, let us make an intuitively obvious observation (to be verified in our subsequent analysis, see Eqs. (7.80), (7.81)) that under the conditions (7.38), fluctuations of the quantum phase φ^- remain small and, hence, the path integral (2.114) can be simplified by performing a quasiclassical expansion of the action $S[\varphi^+, \varphi^-]$ in powers of φ^-. Expanding the action to the second order in φ^- and proceeding analogously to Section 2.3, we arrive at the following Langevin equation (cf. Eq. (2.120))

$$C\frac{\hbar\ddot{\varphi}}{2e} + \frac{1}{R_N}\frac{\hbar\dot{\varphi}}{2e} + \int_{-\infty}^{t} dt' Z_S^{-1}(t-t')\frac{\hbar\dot{\varphi}(t')}{2e} = I_x + \xi_S(t) + \xi_1(t)\cos\frac{\varphi}{2} + \xi_2(t)\sin\frac{\varphi}{2} \quad (7.39)$$

describing real-time dynamics of the classical phase φ^+ (for brevity replaced by φ here and below) in the system depicted in Fig. 7.1. Here we defined the environment response function

$$Z_S^{-1}(t) = \int \frac{d\omega}{2\pi}\frac{e^{-i\omega t}}{Z_S(\omega)} \quad (7.40)$$

the stochastic variable $\xi_S(t)$ with the correlator

$$\langle \xi_S(t)\xi_S(t')\rangle = \int \frac{d\omega}{2\pi}\hbar\omega\coth\frac{\hbar\omega}{2T}\,\mathrm{Re}\left(\frac{1}{Z_S(\omega)}\right) \quad (7.41)$$

which accounts for the current noise produced by the environment and two more stochastic variables ξ_1 and ξ_2 characterized by the correlators

$$\langle \xi_1(t)\xi_1(t')\rangle = \langle \xi_2(t)\xi_2(t')\rangle = \frac{1}{R_N}\int \frac{d\omega}{2\pi}\hbar\omega\coth\frac{\hbar\omega}{2T},$$
$$\langle \xi_1(t)\xi_2(t')\rangle = 0 \quad (7.42)$$

The latter two variables describe the current noise produced by electron tunneling across the junction.

Since within the approximation adopted here the above quasiclassical Langevin equation is equivalent to the path integral formulation of our problem, instead of employing Eq. (2.114) in order to evaluate the interaction correction to the current, it is sufficient simply to solve Eq. (7.39) for the classical phase φ. In doing so, below we will essentially follow the analysis [146].

Let us perform averaging of Eq. (7.39) over the noise variables $\xi_S(t)$ and $\xi_{1,2}(t)$ and over time. Observing that $\langle \hbar\dot{\varphi}/2e\rangle$ defines the average voltage across the junction V, setting for simplicity $Z_S(\omega) = R_S$ and combining Eqs. (7.15) and (7.39), we arrive at the following expression for the current through this junction

$$I = \frac{V}{R_N} - \left\langle \xi_1(t)\cos\frac{\varphi(t)}{2} + \xi_2(t)\sin\frac{\varphi(t)}{2}\right\rangle \quad (7.43)$$

We emphasize that although $\langle \xi_{1,2} \rangle = 0$, the average in the right-hand side of Eq. (7.43) differs from zero because it contains the phase φ. We will evaluate this average perturbatively in the noise variables $\xi_{1,2}$ and ξ_S.

In the absence of any current noise, Eq. (7.39) has a trivial solution $\varphi(t) = 2eVt/\hbar$, where

$$V = \frac{R_N R_S}{R_N + R_S} I_x \tag{7.44}$$

Noise terms in Eq. (7.39) cause deviations of the phase from the above value, i.e., $\varphi(t) = 2eVt/\hbar + \delta\varphi$. Proceeding perturbatively in $\xi_{1,2}$ and ξ_S, in the leading order from Eq. (7.39), we obtain

$$C\frac{\hbar\delta\ddot{\varphi}}{2e} + \frac{1}{R_N}\frac{\hbar\delta\dot{\varphi}}{2e} + \int_{-\infty}^{t} dt' Z_S^{-1}(t-t')\frac{\hbar\delta\dot{\varphi}(t')}{2e}$$
$$= \xi_S(t) + \xi_1(t)\cos\frac{eVt}{\hbar} + \xi_2(t)\sin\frac{eVt}{\hbar} \tag{7.45}$$

Solving this equation, we recover the fluctuating part of the phase in the form

$$\delta\varphi = \int_{-\infty}^{t} dt' \frac{K(t-t')}{e}\{\xi_S(t') + \xi_1(t)\cos[eVt'] + \xi_2(t)\sin[eVt']\} \tag{7.46}$$

where

$$K(t) = \frac{e^2}{\hbar}\int \frac{d\omega}{2\pi}\frac{1}{-\mathrm{i}\omega C + R_N^{-1} + Z_S^{-1}(\omega)}\frac{e^{-\mathrm{i}\omega t}}{-\mathrm{i}\omega + 0} \tag{7.47}$$

defines the phase response function to a short current pulse. Transforming the average in Eq. (7.43) as

$$\left\langle \xi_1(t)\cos\frac{\varphi(t)}{2} + \xi_2(t)\sin\frac{\varphi(t)}{2}\right\rangle \approx \left\langle \xi_1(t)\cos\frac{eVt}{\hbar} + \xi_2(t)\sin\frac{eVt}{\hbar}\right\rangle$$
$$+ \left\langle\left(- \xi_1(t)\sin\frac{eVt}{\hbar} + \xi_2(t)\cos\frac{eVt}{\hbar}\right)\delta\varphi(t)\right\rangle, \tag{7.48}$$

we observe that the first term in the right-hand side of this equation vanishes, while the second one is evaluated with the aid of Eq. (7.46) and yields

$$\left\langle\left(- \xi_1(t)\sin\frac{eVt}{\hbar} + \xi_2(t)\cos\frac{eVt}{\hbar}\right)\delta\varphi(t)\right\rangle$$
$$= - \int_{-\infty}^{t} dt' \frac{K(t-t')}{e}\langle \xi_1(t)\xi_1(t')\rangle \sin\frac{eV(t-t')}{\hbar} \tag{7.49}$$

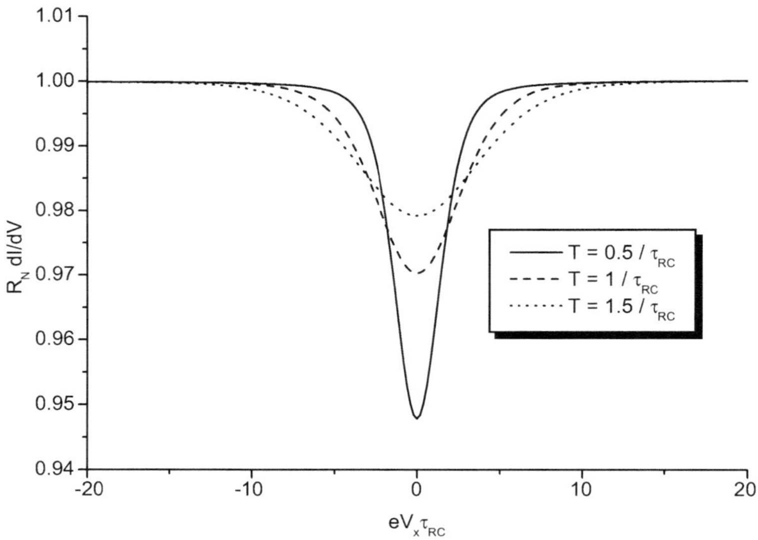

Figure 7.3 Differential conductance of a tunnel junction in the weak Coulomb blockade regime evaluated with the aid of Eq. (7.52) at three different temperatures and $R_N \gg R_S = 860\ \Omega$.

Substituting the average defined in Eqs. (7.48), (7.49) into Eq. (7.43) and making use of Eqs. (7.42), we arrive at the final expression for the current [146, 319]

$$I = \frac{V}{R_N} - \frac{1}{e\hbar R_N}\int_0^\infty dt\, \frac{\pi T^2}{\sinh^2[\pi T t/\hbar]} K(t) \sin\frac{eVt}{\hbar} \quad (7.50)$$

In the important Ohmic limit $Z_S(\omega) = R_S$, the function $K(t)$ (7.47) takes the form

$$K(t) = \frac{e^2}{\hbar}\frac{R_N R_S}{R_N + R_S}\left(1 - e^{-t/\tau_{RC}}\right), \quad \tau_{RC} = \frac{R_N R_S C}{R_N + R_S} \quad (7.51)$$

Combining this expression with Eq. (7.50) and evaluating the remaining time integral, we obtain

$$I = \frac{V}{R_N} - \frac{e}{\pi\hbar}\frac{R_S}{R_S + R_N}\,\mathrm{Im}\left[\left(\frac{\hbar}{\tau_{RC}} + \mathrm{i}eV\right)\Psi\left(1 + \frac{\hbar}{2\pi T\tau_{RC}} + \mathrm{i}\frac{eV}{2\pi T}\right) - ieV\Psi\left(1 + \mathrm{i}\frac{eV}{2\pi T}\right)\right] \quad (7.52)$$

where $\Psi(x)$ is the digamma function.

The differential conductance derived from Eq. (7.52) is plotted in Fig. 7.3. We observe that Coulomb interaction yields a shallow dip in dI/dV centered at zero voltage. As the temperature increases, the dip becomes weaker and wider.

The result (7.52) can further be simplified in several limits. At high bias voltages $eV \gg \hbar/\tau_{\mathrm{RC}}$, the $I-V$ curve evaluated from this equation approaches the form (7.32) and exhibits the offset $-e/2C$. Thus, in the high voltage limit, the current offset occurs at any value of junction resistance and also for any type of the environment, as one can demonstrate from the more general Eq. (7.50). In practice, however, this offset can reliably be detected mainly in tunnel junctions with sufficiently small g, whereas for $g \gg 1$, it gets shifted to very high voltages where other effects, such as the frequency dispersion of $Z_{\mathrm{S}}(\omega)$, tunnel barrier suppression, or heating, dominate the system behavior [328, 391, 392].

At high temperatures, one can expand the result (7.52) to the first order in $\hbar/2\pi T\tau_{\mathrm{RC}}$ and again recover Eq. (7.37) which, therefore, remains applicable for any value of the conductances g_{N} and g_{S}. Of course, this overlap between the results derived in the limit of high voltages and/or temperatures here and in the previous section could be expected already prior to our calculation.

In the zero-temperature limit $T \to 0$ and at not very high voltages $eV \lesssim \hbar\tau_{\mathrm{RC}}$, Eq. (7.52) yields the differential conductance in the form

$$\frac{dI}{dV} = \frac{1}{R_{\mathrm{N}}}\left(1 - \frac{1}{g+g_{\mathrm{S}}}\ln\left(1 + \frac{\hbar^2}{e^2V^2\tau_{\mathrm{RC}}^2}\right)\right) \qquad (7.53)$$

demonstrating that the leading order interaction correction to the conductance grows logarithmically with decreasing V. This result remains valid as long as the second of the conditions (7.38) is satisfied, i.e., at

$$eV \gg (g+g_{\mathrm{S}})E_{\mathrm{C}}\exp(-(g+g_{\mathrm{S}})/2) \qquad (7.54)$$

At even lower voltages, the Coulomb correction becomes large and cannot anymore be treated perturbatively. In this case, the system again enters the strong Coulomb blockade regime characterized by a pronounced suppression of the system conductance. We will return to this issue below in Section 7.4 and also within a broader context in Chapters 11–13.

From Eq. (7.52), one can also derive a general expression for the temperature-dependent zero-bias differential conductance:

$$G_0 = \frac{1}{R_N} - \frac{e^2}{\pi\hbar}\frac{R_S}{R_S + R_N}$$
$$\times \left[\Psi\left(1 + \frac{\hbar}{2\pi T \tau_{RC}}\right) + \frac{\hbar}{2\pi T \tau_{RC}}\Psi'\left(1 + \frac{\hbar}{2\pi T \tau_{RC}}\right) + \gamma\right] \tag{7.55}$$

where $\gamma = 0.5772\ldots$ is the Euler constant. In the temperature interval $(g + g_S)E_C \exp(-(g + g_S)/2 \ll T \lesssim \hbar/2\pi\tau_{RC}$, this equation reduces to

$$G_0 = \frac{1}{R_N}\left(1 - \frac{2}{g + g_S}\ln\frac{\hbar}{2\pi T \tau_{RC}}\right) \tag{7.56}$$

In the opposite high-temperature limit $T \gg \hbar/2\pi\tau_{RC}$, one finds

$$G_0 = \frac{1}{R_N}\left(1 - \frac{E_C}{3T} + \frac{3\zeta(3)}{2\pi^4}\frac{(g + g_S)E_C^2}{T^2} + \mathcal{O}\left(\frac{(g + g_S)^2 E_C^3}{T^3}\right)\right) \tag{7.57}$$

where $\zeta(x)$ is the zeta-function. Comparing this result with Eq. (7.35), we observe that the leading order high-temperature interaction correction to the conductance is universal, i.e., it turns out to be independent of both g and g_S. These parameters enter only in the sub-leading terms.

In order to establish the validity conditions for the results derived in this section, it is necessary to require the corrections in Eqs. (7.52)–(7.56) to be smaller than the leading terms. In this way, one arrives at the inequalities (7.38), just as it was expected.

7.3 Weak Electron Tunneling in the Presence of an Environment

Let us now consider highly resistive junctions with $g \ll 1$ interacting with an arbitrary electromagnetic environment described by a linear impedance $Z_S(\omega)$. This physical situation overlaps in part with both regimes (7.4) and (7.38), but at the same time, it differs from them in some important aspects, mainly because the effect of an external environment is now accounted for without any approximations.

In this weak tunneling limit, one can proceed perturbatively in g, which amounts to dropping the terms $\sim 1/R_N$ in the expression for the effective action (2.113) in the path integral (2.114). Then we obtain

$$I = \frac{1}{R_N}\int \mathcal{D}\varphi^+\mathcal{D}\varphi^- \, e^{iS[\varphi^+,\varphi^-]/\hbar}\left(\frac{\hbar\dot{\varphi}^+(t)}{2e}\cos\frac{\varphi^-(t)}{2}\right.$$
$$+\frac{2i\hbar}{e}\int dt' \frac{\pi T^2}{\hbar^2\sinh^2[\pi T(t-t')/\hbar]}\cos\frac{\varphi^+(t)-\varphi^+(t')}{2}$$
$$\left.\times\cos\frac{\varphi^-(t)}{4}\sin\frac{\varphi^-(t')}{4}\right), \qquad (7.58)$$

where

$$iS[\varphi^+,\varphi^-] = i\int dt_1\left(\frac{C\hbar^2\dot{\varphi}^+\dot{\varphi}^-}{4e^2}+\frac{\hbar I_x}{2e}\varphi^-\right)$$
$$-i\int dt_1 dt_2 \frac{\hbar\varphi^-(t_1)}{2e}Z_S^{-1}(t_1-t_2)\frac{\hbar\dot{\varphi}^+(t_2)}{2e}$$
$$-\frac{1}{2}\int dt_1 dt_2\frac{\hbar\varphi^-(t_1)}{2e}K_S(t_1-t_2)\frac{\hbar\varphi^-(t_2)}{2e} \qquad (7.59)$$

and the function $K_S(t)$ is defined in Eq. (2.51).

The path integral in Eq. (7.58) is Gaussian and, hence, can be evaluated exactly. We will proceed analogously to Section 7.1 and define an auxiliary path integral

$$A(t,t',\kappa,\mu,\nu) = \frac{\int \mathcal{D}\varphi^\pm \, e^{i\kappa\frac{\varphi^+(t)-\varphi^+(t')}{2}} \, e^{i\frac{\mu\varphi^-(t)+\nu\varphi^-(t')}{4}} \, e^{iS[\varphi^\pm]/\hbar}}{\int \mathcal{D}\varphi^\pm \, e^{iS[\varphi^\pm]/\hbar}} \qquad (7.60)$$

Integration over the classical phase φ^+ yields the functional δ-function and we find

$$A = \int \mathcal{D}\varphi^-(\tau)\, e^{i[\mu\varphi^-(t)+\nu\varphi^-(t')]/4}$$
$$\times\delta\left(\frac{\kappa}{2}\delta(\tau-t)-\frac{\kappa}{2}\delta(\tau-t')-C\frac{\hbar\ddot{\varphi}^-}{4e^2}+\frac{\hbar}{4e^2}\frac{d}{d\tau}\int_\tau^t d\tau'\varphi^-(\tau')Z_S^{-1}(\tau'-\tau)\right)$$
$$\times\exp\left[i\frac{I_x}{2e}\int_{-\infty}^t dt_1\,\varphi^-(t_1)-\frac{\hbar}{8e^2}\int_{-\infty}^t dt_1\int_{-\infty}^t dt_2\varphi^-(t_1)K_S(t_1-t_2)\varphi^-(t_2)\right] \qquad (7.61)$$

The δ-function in the above equation implies that the quantum phase φ^- should obey the differential equation

$$-C\frac{\hbar\ddot{\varphi}^-}{4e^2}+\frac{\hbar}{4e^2}\frac{d}{d\tau}\int_\tau^t d\tau'\varphi^-(\tau')Z_S^{-1}(\tau'-\tau) = -\frac{\kappa}{2}\delta(\tau-t)$$
$$+\frac{\kappa}{2}\delta(\tau-t')$$

with the solution

$$\varphi^-(\tau) = -2\kappa K(t' - \tau) + 2\kappa K(t - \tau) \tag{7.62}$$

where

$$K(t) = \frac{e^2}{\hbar} \int \frac{d\omega}{2\pi} \frac{1}{-\mathrm{i}\omega C + Z_S^{-1}(\omega)} \frac{e^{-\mathrm{i}\omega t}}{-\mathrm{i}\omega + 0} \tag{7.63}$$

is the environment response function, which obeys the requirement of causality $K(t < 0) \equiv 0$ and matches with the function defined in Eq. (7.47) in the limit $1/R_N \to 0$.

With the solution (7.62), the path integral (7.61) is evaluated trivially with the result

$$A(t, t', \kappa, \mu, \nu) = e^{-\kappa^2 F(t-t')} e^{-\mathrm{i}\frac{\mu\kappa}{2} K(t'-t) + \mathrm{i}\frac{\nu\kappa}{2} K(t-t')} e^{\mathrm{i}\kappa eV(t-t')} \tag{7.64}$$

Here we again defined the voltage $V = I_x Z_S(0)$ and introduced the function

$$F(t) = e^2 \int \frac{d\omega}{2\pi} \mathrm{Re}\left(\frac{1}{-\mathrm{i}\omega C + Z_S^{-1}(\omega)}\right) \omega \coth \frac{\omega}{2T} \frac{1 - \cos\omega t}{\omega^2} \tag{7.65}$$

which coincides with the phase correlation function

$$F(t) = \frac{1}{8} \left\langle [\varphi^+(t) - \varphi^+(0)]^2 \right\rangle \tag{7.66}$$

It is straightforward to verify that the path integral for the current (7.58) can be expressed as

$$\begin{aligned}
I = \frac{1}{eR_N} \lim_{t' \to t; \kappa \to 0} & \left[\frac{A(t, t', \kappa, 2, 0) + A(t, t', \kappa, -2, 0)}{2} \right. \\
& \left. - \frac{A(t, t', 0, 2, 0) + A(t, t', 0, -2, 0)}{2} \right] \\
& - \frac{1}{4eR_N} \int_0^t dt' \alpha(t - t') \Big[A(t, t', 1, 1, 1) - A(t, t', -1, -1, -1) \\
& - A(t, t', 1, -1, -1) + A(t, t', -1, 1, 1) - A(t, t', 1, 1, -1) \\
& + A(t, t', -1, -1, 1) + A(t, t', 1, -1, 1) - A(t, t', -1, 1, -1) \Big]
\end{aligned}$$

Combining this expression with Eq. (7.64), after a simple algebra, we arrive at the final result

$$I = \frac{V}{R_N} - \frac{2}{e\hbar R_N} \int_0^\infty dt \frac{\pi T^2}{\sinh^2[\pi T t/\hbar]} e^{-F(t)} \sin\left[\frac{K(t)}{2}\right] \sin\frac{eVt}{\hbar} \tag{7.67}$$

As before, the second term in the right-hand side of this equation defines the Coulomb blockade contribution to the current across the tunnel junction.

It is instructive to compare Eq. (7.67) with the corresponding expressions for the current derived in Sections 7.1 and 7.2. In the high environment impedance limit $Z_{\rm S}(\omega) \gg R_q$ in the main approximation, one can ignore $Z_{\rm S}^{-1}(\omega)$ in Eqs. (7.63), (7.65). Then one has $K(t) = 2E_{\rm C}t/\hbar$ and $F(t) \to 0$ implying that Eq. (7.67) reduces to the result (7.31) derived in the strong Coulomb blockade regime. On the other hand, in the low environment impedance limit $Z_{\rm S}(\omega) \ll R_q$, one can again set $F(t) \to 0$ and the result (7.67) then matches with the formula (7.50) although the response functions $K(t)$ differ in these two cases.

It follows immediately that the expression for the current (7.67) remains accurate not only in the initially imposed weak tunneling limit $g \ll 1$, but covers essentially all limiting cases considered in Sections 7.1, 7.2, and 7.3 provided the function $K(t)$ is defined by Eq. (7.47). Equation (7.67) remains qualitatively correct also for $g + g_{\rm S} \sim 1$ and low voltages and temperatures in which case it can be considered a good interpolation. Note that in the absence of an external environment, the expression of the form (7.67) was first established by means of a different approach [319] exploiting a close similarity between the Coulomb blockade problem and the polaron model [117].

Turning back to the weak tunneling limit $g \ll 1$, let us again consider a particular example of the Ohmic environment with the impedance $Z_{\rm S}(\omega) = R_{\rm S}$. In the limit of large voltages and/or temperatures $\max\{eV, T\} \gg g_{\rm S}E_{\rm C}$ from Eq. (7.67), we recover the same results (7.32) for $eV \gg T$ and (7.37) for $eV \ll T$. In the opposite limit $\max\{eV, T\} \ll g_{\rm S}E_{\rm C}$, Eq. (7.67) yields

$$I = \frac{2T}{eR_{\rm N}}\left(\frac{2\pi e^{-\gamma}T\tau_{\rm RC}}{\hbar}\right)^{\frac{2}{g_{\rm S}}}\frac{\left|\Gamma\left(1+\frac{1}{g_{\rm S}}+\frac{ieV}{2\pi T}\right)\right|^2}{\Gamma\left(2+\frac{2}{g_{\rm S}}\right)}\sinh\frac{eV}{2T} \tag{7.68}$$

From this result, one easily derives both the differential conductance of the tunnel junction at low temperatures $T \ll eV \ll \hbar/\tau_{\rm RC}$

$$\frac{dI}{dV} = \frac{1}{R_{\rm N}}\frac{\left(e^{-\gamma}eV\tau_{\rm RC}\right)^{2/g_{\rm S}}}{\Gamma(1+2/g_{\rm S})} \tag{7.69}$$

and the temperature-dependent conductance at low voltages $eV \ll T \ll \hbar/\tau_{RC}$

$$G_0 = \frac{1}{R_N} \frac{\sqrt{\pi}\Gamma\left(1+\frac{1}{g_S}\right)}{2\Gamma(3/2+1/g_S)} \left(\frac{\pi e^{-\gamma} T \tau_{RC}}{\hbar}\right)^{2/g_S} \tag{7.70}$$

The above results (7.68)–(7.70) were derived within both real-time [321] and imaginary-time [425] techniques (see also the review [345]). These expressions clearly demonstrate the role of an external Ohmic environment in Coulomb blockade suppression of electron tunneling at low-enough energies. For instance, Eq. (7.69) defines the so-called Coulomb zero-bias anomaly (ZBA) in the differential conductance, which depends on the applied voltage as $dI/dV \propto V^{2/g_S}$. According to this result, the conductance is very strongly suppressed for highly resistive environment with $g_S \ll 1$, whereas in the opposite limit $g_S \gg 1$, this suppression is much weaker because of strong charge fluctuations in the shunt resistor R_S. In the limit $g_S \gg 1$ and for not-too-low voltages, one can expand the result (7.69) in powers of $1/g_S$ and within the logarithmic accuracy recover Eq. (7.53) with $g \to 0$. Note, however, that—in contrast to Eq. (7.53)—the expression (7.69) remains valid also in the zero voltage limit $V \to 0$. Similarly, Eq. (7.70) shows the power-law dependence of the zero-bias conductance on temperature $G_0 \propto T^{2/g_S}$, which demonstrates strong Coulomb suppression of G_0 at large g_S and matches with Eq. (7.56) in the limit $g_S \gg 1$ and at not-too-low T.

Zero-temperature $I–V$ curves evaluated from Eq. (7.67) for different values of the environment dimensionless conductance g_S are displayed in Fig. 7.4. In accordance with the above analytic expressions, we observe that with increasing g_S in the considered voltage interval, the $I–V$ curve gradually crosses over from $IR_N = (V - e/2C)\theta(eV - E_C)$ at $g_S \to 0$ to the standard Ohmic form $IR_N = V$ at $g_S \to \infty$.

The above results demonstrate that interaction of tunneling electrons with an external environment plays a prominent role and—depending on the strength of this interaction (controlled by the dimensionless parameter $1/g_S$)—may essentially influence electron transport in the Coulomb blockade regime. The physical reason for that is clear: Tunneling electrons exchange energies

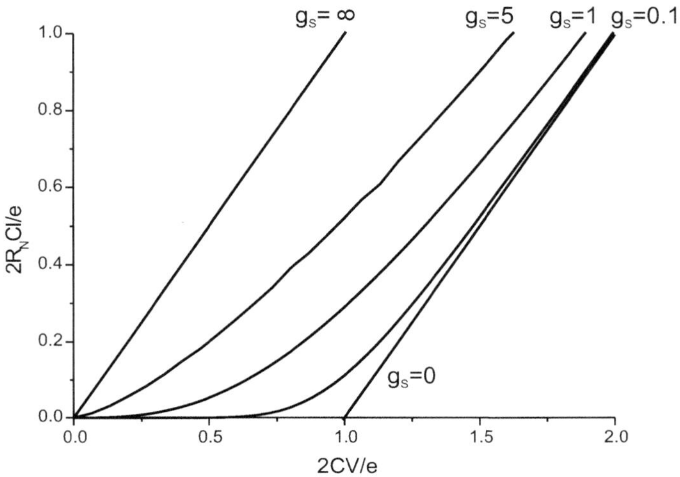

Figure 7.4 Zero-temperature $I-V$ curves (7.67) for a highly resistive tunnel junction in an Ohmic environment with different values of g_S.

with the environmental modes. This exchange causes dissipation of electron energy and, hence, influences the probability of electron tunneling. In order to quantify this physical picture, let us define the probability $P(E)$ for a tunneling electron to excite an environmental mode with energy E. It reads

$$P(E) = \int \frac{dt}{2\pi\hbar} e^{\mathrm{i}Et/\hbar} e^{-F(t)-\mathrm{i}\frac{K(t)}{2}+\mathrm{i}\frac{K(-t)}{2}}$$

$$= \int \frac{dt}{2\pi\hbar} \exp\left[\mathrm{i}\frac{Et}{\hbar} + \frac{e^2}{\pi\hbar}\int_0^{\infty} \frac{d\omega}{\omega} \mathrm{Re}\left(\frac{1}{-\mathrm{i}\omega C + Z_S^{-1}(\omega)}\right)\right.$$

$$\left.\times \left([\cos\omega t - 1]\coth\frac{\hbar\omega}{2T} - \mathrm{i}\sin\omega t\right)\right], \qquad (7.71)$$

With the aid of this function, one can rewrite the result (7.67) in the form [202]

$$I(V) = \frac{1}{eR_\mathrm{N}} \int dE dE' \{ f(E)[1 - f(E')] P(E - E' + eV)$$

$$- [1 - f(E)] f(E') P(E' - E - eV) \} \qquad (7.72)$$

where, as before, $f(E) = 1/(1 + e^{E/T})$ is the Fermi function.

The function $P(E)$ obeys the detailed balance condition

$$P(-E) = e^{-E/T} P(E) \qquad (7.73)$$

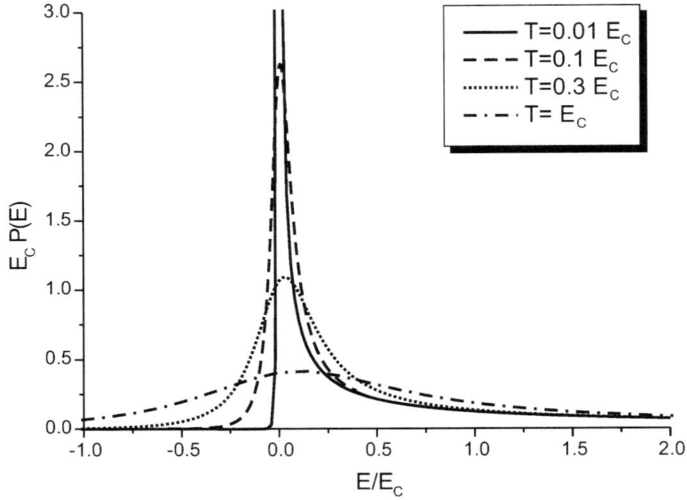

Figure 7.5 The function $P(E)$ (7.78) for an Ohmic environment with the dimensionless conductance $g_S = 10$.

implying that the probability of energy absorption by the environment is higher than the probability to emit energy. With the aid of Eq. (7.73), one can further transform the expression (7.72) as

$$I(V) = \frac{1}{eR_N} \int dE\, E \frac{1 - e^{-eV/T}}{1 - e^{-E/T}} P(eV - E). \tag{7.74}$$

Equations (7.72) and (7.74) clearly illustrate the importance of energy exchange between electrons and the environment.

Let us consider several important limits. In the absence of interactions $Z_S(\omega) \to 0$, one finds

$$P(E) = \delta(E),$$

effectively implying energy conservation for tunneling electrons. In this limit, the charge accumulated in the capacitor after an electron tunneling event quickly relaxes and does not create any additional potential barrier. Obviously the I–V curve reduces to $I = V/R_N$ in this case.

In the opposite limit of a highly resistive environment or, equivalently, very strong interactions $Z_S(\omega) \to \infty$, we obtain

$$F(t) + \mathrm{i}\frac{K(t)}{2} - \mathrm{i}\frac{K(-t)}{2} = \mathrm{i}E_C t \tag{7.75}$$

and the function $P(E)$ takes the form

$$P(E) = \delta(E - E_C). \tag{7.76}$$

This is the strong Coulomb blockade regime described by Eq. (7.18).

Let us also note that the function $P(E)$ obeys the following useful identity:

$$\int dE\, E\, P(E) = \mathrm{i}\hbar \frac{\partial}{\partial t} e^{-F(t) - \mathrm{i}\frac{K(t)}{2} + \mathrm{i}\frac{K(-t)}{2}}\bigg|_{t=0} = E_C \tag{7.77}$$

which holds for an arbitrary impedance $Z_S(\omega)$. Combining this identity with Eq. (7.74), we immediately observe that in the limit of very large V, the I–V curve always exhibits the offset and approaches the form (7.32).

Finally, let us evaluate the function $P(E)$ for a special case of the Ohmic environment with $Z_S(\omega) = R_S$. At low-enough energies $E \ll \hbar/\tau_{RC}$, from Eq. (7.71) one finds [202]

$$P(E) = \frac{e^{E/2T}}{4\pi^2 T \Gamma(2/g_S)} \left(\frac{2\pi e^{-\gamma} T \tau_{RC}}{\hbar}\right)^{2/g_S} \left|\Gamma\left(\frac{1}{g_S} + \frac{\mathrm{i}E}{2\pi T}\right)\right|^2 \tag{7.78}$$

For $g_S \gg 1$, this dependence covers essentially all relevant energies for the problem of electron transport. Substituting the result (7.78) into Eq. (7.74), we again recover the result (7.68).

The function $P(E)$ (7.78) is plotted in Fig. 7.5. It has a form of an asymmetric peak centered around $E = 0$. The peak width grows linearly with temperature.

At the end of this section, let us return to our approach based on the quasiclassical Langevin equation (7.39). As we already discussed, this approach is applicable as long as the quantum phase φ^- remains small, i.e. $|\varphi^-| \ll \pi/2$. Hence, making use of Eq. (7.62), which determines the saddle point value of φ^-, one can easily specify the validity range of the Langevin equation (7.39). With the aid of Eqs. (7.62), (7.63) (derived here in the limit $g \ll 1$, but straightforwardly extended to arbitrary g), we obtain

$$|\varphi^-| \leq \begin{cases} 4\pi/(g + g_S), & \text{if } |t - t'| \gg \tau_{RC}, \\ 2E_C|t - t'|/\hbar, & \text{if } |t - t'| \ll \tau_{RC} \end{cases} \tag{7.79}$$

Here we have put $Z_S(\omega) = R_S$ for simplicity. The time difference $|t - t'|$ can be estimated from the path integral expressions for the current (2.114), (7.58) as $|t - t'| \sim \min\{\hbar/T, \hbar/eV\}$. It follows immediately that the quasiclassial Langevin equation (7.39) remains accurate provided either one of the conditions

$$g + g_S \lesssim 1, \quad \text{but} \quad \max\{eV, T\} \gg E_C, \tag{7.80}$$

or

$$g + g_S \gtrsim 1 \tag{7.81}$$

is satisfied. Equation (7.80) coincides with the first condition (7.38). At the same time, the inequality (7.81) is more general than the second condition (7.38). The latter, more restrictive, condition not only ensures the validity of the Langevin equation, but also allows one to solve this equation perturbatively in the noise.

7.4 Thermodynamic Properties: Perturbation Theory

So far we merely discussed electron transport in small tunnel junctions in the presence of electron–electron interactions. Of interest are also thermodynamic properties of such junctions. Below in this chapter, we will demonstrate that such properties may be essentially influenced by Coulomb interaction. In experiment the effects under consideration can be conveniently studied with the aid of the so-called single-electron box setup schematically shown in Fig. 7.6. This single-electron box consists of a small normal metal island connected to a big metallic electrode via a small-size tunnel barrier and, in addition, capacitively coupled to another electrode. As before, the tunnel junction has the resistance R_N and the total capacitance of our device is denoted by C.

It is obvious that no dc current can flow through the structure of Fig. 7.6. However, by changing the gate voltage V_g, one can tune the box energy and the average number of extra electrons on the metallic grain $\langle n \rangle$, which can be directly measured in modern experiments. In the weak tunneling limit $g \ll 1$, the latter quantity can easily be evaluated. In this case, the electrostatic energy of the system corresponding to n extra electrons in the box is

$$E_n = \frac{(CV_g - en)^2}{2C} \tag{7.82}$$

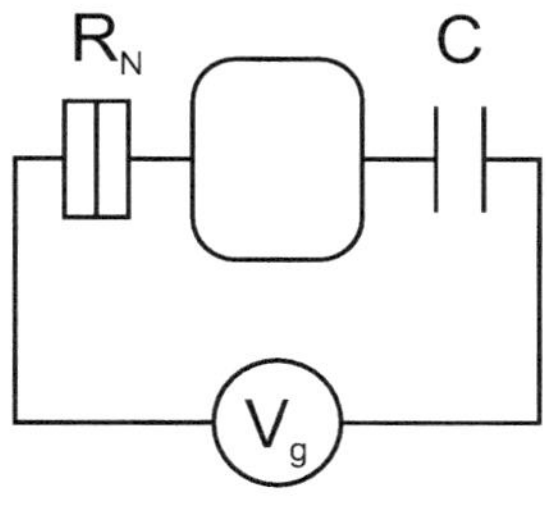

Figure 7.6 Single-electron box is a metallic island coupled to a big electrode by a tunnel junction with the resistance R_{N}. Its energy may be tuned by applying the gate voltage V_{g} via the capacitor C.

i.e., the energy states are simply defined by a set of parabolas, as shown in Fig. 7.7a. In equilibrium, the occupation probabilities for these states are defined as

$$p_n = \frac{e^{-E_n/T}}{\mathcal{Z}} \tag{7.83}$$

where

$$\mathcal{Z} = \sum_n e^{-E_n/T} \tag{7.84}$$

is the grand partition function of our structure. Then for the average number of electrons $\langle n \rangle$, we obtain

$$\langle n \rangle = \sum_n n p_n = \sum_n \left(\frac{C V_{\mathrm{g}}}{e} - \frac{1}{e}\frac{\partial E_n}{\partial V_{\mathrm{g}}} \right) \frac{e^{-E_n/T}}{\mathcal{Z}} = \frac{C V_{\mathrm{g}}}{e} - \frac{1}{e}\frac{\partial F}{\partial V_{\mathrm{g}}} \tag{7.85}$$

where $F = -T \ln \mathcal{Z}$ is the free energy of our system.

The average number of electrons $\langle n \rangle$ is plotted in Fig. 7.7b. At low temperatures $T \ll E_{\mathrm{C}} = e^2/2C$, the function $\langle n(V_{\mathrm{g}}) \rangle$ is represented by a set of sharp steps occurring at half-integer values of the polarization charge $C V_{\mathrm{g}}$. Whenever $C V_{\mathrm{g}}$ approaches the value $ne + 1/2$ where the two neighboring parabolas cross, one more electron is pulled into the box in order to minimize its energy. At higher temperatures comparable to E_{C}, the sharp steps get smeared since several states with higher energies can also be occupied in this case. Finally, at $T \gg E_{\mathrm{C}}$, many states with different n are simultaneously occupied and the average number of electrons approaches a simple linear dependence $\langle n(V_{\mathrm{g}}) \rangle = C V_{\mathrm{g}}/e$.

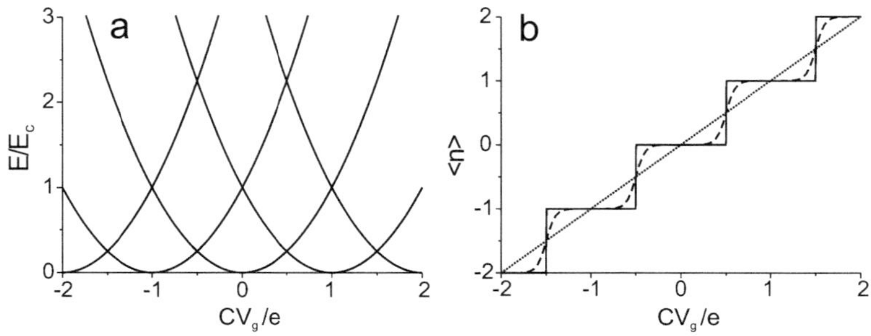

Figure 7.7 (a) Energies of a single-electron box (corresponding to different number of excess electrons n on the island) as a function of the gate voltage. (b) Average number of excess electrons $\langle n \rangle$ for $R_N \gg R_q$ and $T \ll E_C$ (solid line), for $R_N \lesssim R_q$ and/or $T \sim E_C$ (dashed line) and for $R_N \ll R_q$ and/or $T \gg E_C$ (dotted line).

The above simple physical picture relies on the assumption of vanishingly small junction conductance g. With increasing g, virtual processes of electron tunneling into and out of the box gain importance and tend to smear the Coulomb staircase $\langle n(V_g) \rangle$ even at $T = 0$. In order to carry out a quantitative analysis of this effect, let us consider the grand partition function

$$\mathcal{Z} = \sum_m e^{i2\pi C V_g m/e} \int_0^{4\pi} d\varphi_0 \int_{\varphi(0)=\varphi_0}^{\varphi(\hbar\beta)=\varphi_0+4\pi m} \mathcal{D}\varphi \, e^{-S[\varphi]/\hbar} \quad (7.86)$$

where $S[\varphi]$ is the imaginary-time effective action for our structure. This action was already derived in Chapter 2. It is represented by the sum of Coulomb and tunneling terms and is defined by Eq. (2.141), in which one should now set $I_x = 0$ and $\tilde{\alpha}_S = 0$. Then one finds

$$S[\varphi] = \frac{\hbar^2}{16E_C} \int_0^{\hbar\beta} d\tau \dot{\varphi}^2 - \frac{g\hbar}{4\pi^2} \int_0^{\hbar\beta} d\tau \int_0^{\hbar\beta} d\tau' \frac{(\pi T/\hbar)^2}{\sin^2[\pi T(\tau - \tau')/\hbar]} \cos\frac{\varphi(\tau) - \varphi(\tau')}{2} \quad (7.87)$$

where, as before, $\beta = 1/T$.

As we have already learned in Chapter 6, it is also important to define proper boundary conditions for the path integral (7.86). The main question here is whether one needs to consider the phase as a compact (i.e., defined on a ring) or an extended variable. In the case of Josephson junctions, we have already demonstrated in Sections 6.4 and 6.5 that this choice essentially depends on that for the proper charge states, which can be either discrete (in units of the Cooper pair charge $2e$) or continuous. Here we have a similar situation, except the charge changes in units of e rather than $2e$.

Let us for a moment ignore electron tunneling by putting $g = 0$ and evaluate the Gaussian path integral

$$\rho_0(\varphi_f, \varphi_i) = \int_{\varphi(0)=\varphi_i}^{\varphi(\hbar\beta)=\varphi_f} \mathcal{D}\varphi \exp\left[-\frac{\hbar}{16E_C}\int_0^{\hbar\beta} d\tau \dot{\varphi}^2\right] \quad (7.88)$$

which in fact defines the equilibrium density matrix of the system. The least action trajectory in this case is just the straight line $\varphi_0(\tau) = \varphi_i + (\varphi_f - \varphi_i)\tau/\hbar\beta$. Shifting the integration parameter, $\varphi \to \varphi_0(\tau) + \varphi$, we find

$$\rho_0(\varphi_f, \varphi_i) = B \exp\left[-\frac{(\varphi_f - \varphi_i)^2}{16E_C\beta}\right] \quad (7.89)$$

where the pre-exponent B is equal the integral (7.88) with the boundary conditions $\varphi(0) = \varphi(\hbar\beta) = 0$. The latter is easily recovered from the normalization condition for the density matrix $\lim_{\beta\to 0} = \delta(\varphi_f - \varphi_i)$. In this way, we obtain

$$B = \frac{1}{4\sqrt{\pi E_C \beta}}. \quad (7.90)$$

After a proper redefinition of the parameters, the density matrix (7.89) coincides with that for a free quantum particle. If one considers the particle coordinate φ as an extended variable, the grand partition function becomes

$$\mathcal{Z} = \int d\varphi \rho_0(\varphi, \varphi) = \frac{X}{4\sqrt{\pi E_C \beta}} \quad (7.91)$$

where X is an effective system size. This expression differs from the correct one (Eq. (7.84)) for our physical situation. As in Chapter 6, this inconsistency is eliminated if we consider the phase as a compact variable employing the boundary conditions appropriate for a quantum particle on a ring threaded by a magnetic flux. Specifically, we impose the boundary conditions

$$\psi(\varphi + 4\pi m) = \psi(\varphi) \exp\left[i\frac{2\pi C V_g}{e} m\right] \quad (7.92)$$

for the particle wave function $\psi(\varphi)$. Similar to our analysis in Chapter 6, the parameter m fixes the number of winds made by the particle around the ring. With these boundary conditions, the partition function for a free quantum particle reads

$$\mathcal{Z} = \sum_m \int_0^{4\pi} d\varphi_0\, \rho_0(\varphi_0 + 4\pi m, \varphi_0) \exp\left[\mathrm{i}\frac{2\pi C V_\mathrm{g}}{e} m\right] \quad (7.93)$$

Employing Eq. (7.89) and making use of Poisson's resummation formula, one can bring the partition function to the correct form (7.84). Thus, similar to the case of Josephson junctions, here the boundary conditions in Eq. (7.86) properly account for discreteness of the electric charge accumulated in the box. It is also clear that due to their topological nature, these boundary conditions apply not only for $g = 0$ but also for any nonzero value of the junction conductance. Indeed, single-electron tunneling may only yield changes of the junction charge in units of e and, hence, does not alter the discrete nature of the charge states. We also point out that instead of the 2π-periodicity of the boundary conditions in the case of Josephson junctions, here we encounter the 4π-periodicity, cf. Eqs. (6.45) and (7.86). This is a direct consequence of the e-periodicity in the reciprocal charge space as compared to the $2e$-periodicity in the case of superconducting tunnel junctions.

As in Chapter 6, by expanding the partition function (7.86), (7.87) in formal series in powers of the dimensionless conductance g, employing the identity (6.64) and performing Gaussian integrals over the phase variable φ in each term of this expansion, we arrive at the "charge-only" representation for the partition function corresponding to the junction charge state $Q = Q_0$ [345]

$$\mathcal{Z}(Q_0) = \sum_{p=-\infty}^{\infty} \int_{Q_0+pe}^{Q_0+pe} \mathcal{D}Q \sum_{n=0}^{\infty} \frac{1}{n!}\left(\frac{g\hbar}{4\pi^2}\right)^n \int_0^{\hbar\beta} d\tau_1 \ldots \int_0^{\hbar\beta} d\tau_n$$
$$\times \int_0^{\hbar\beta} d\tau_1' \ldots \int_0^{\hbar\beta} d\tau_n' \frac{(\pi T/\hbar)^2}{\sin^2[\pi T(\tau_1 - \tau_1')/\hbar]} \cdots \frac{(\pi T/\hbar)^2}{\sin^2[\pi T(\tau_n - \tau_n')/\hbar]}$$
$$\times \delta\left[\dot{Q}(\tau) - \dot{Q}_n(\tau)\right] \exp\left[-\int_0^{\hbar\beta} \frac{d\tau}{\hbar} \frac{Q^2(\tau)}{2C}\right] \quad (7.94)$$

where

$$Q_n(\tau) = e\sum_{k=1}^{n} (\theta(\tau - \tau_k') - \theta(\tau - \tau_k)) \quad (7.95)$$

In the weak tunneling limit $g \ll 1$, one can restrict the analysis to the lowest-order correction to the partition function $\mathcal{Z}$. Expanding the partition function (7.86), (7.87) to the first order in g, we obtain $\mathcal{Z} = \mathcal{Z}_0 + \delta\mathcal{Z}_1$, where

$$\delta\mathcal{Z}_1 = \frac{g\hbar}{4\pi^2}\int_0^{\hbar\beta} d\tau \int_0^{\hbar\beta} d\tau' \frac{(\pi T/\hbar)^2}{\sin^2[\pi T(\tau-\tau')/\hbar]} \\ \times \sum_m e^{i2\pi C V_g m/e} \int d\varphi_0 \int_{\varphi(0)=\varphi_0}^{\varphi(\hbar\beta)=\varphi_0+4\pi m} \mathcal{D}\varphi \\ \times \cos\frac{\varphi(\tau)-\varphi(\tau')}{2} \exp\left[-\frac{\hbar}{16E_C}\int_0^{\hbar\beta} ds\dot{\varphi}^2(s)\right] \quad (7.96)$$

This expression may be recast to the form

$$\delta\mathcal{Z}_1 = \frac{g\hbar}{8\pi^2}\int_0^{\hbar\beta} d\tau \int_0^{\hbar\beta} d\tau \frac{(\pi T/\hbar)^2}{\sin^2[\pi T(\tau-\tau')/\hbar]} \\ \times \sum_m e^{i2\pi C V_g m/e}\left[B_m(1,\tau,\tau') + B_m(-1,\tau,\tau')\right] \quad (7.97)$$

where we defined the integral

$$B_m(\nu,\tau,\tau') = \int_0^{4\pi} d\varphi_0 \int_{\varphi(0)=\varphi_0}^{\varphi(\hbar\beta)=\varphi_0+4\pi m} \mathcal{D}\varphi \\ \times \exp\left[-\frac{\hbar}{16E_C}\int_0^{\hbar\beta} ds\dot{\varphi}^2(s) - i\nu\frac{\varphi(\tau)-\varphi(\tau')}{2}\right] \quad (7.98)$$

In order to evaluate this integral, let us determine the saddle point path satisfying the above boundary conditions. We have

$$\varphi_0(s) = \varphi_0 + \frac{4\pi m}{\hbar\beta}s \\ + i\nu\frac{4E_C}{\hbar}\left(\frac{\tau-\tau'}{\hbar\beta}s + \theta(s-\tau)(s-\tau) - \theta(s-\tau')(s-\tau')\right) \quad (7.99)$$

Integrating over the fluctuations around this saddle point path, we get

$$B_m(\nu,\tau,\tau') = B\int_0^{4\pi} d\varphi_0 \exp\left[-\frac{\hbar}{16E_C}\int_0^{\hbar\beta} ds\dot{\varphi}_0^2(s) - i\nu\frac{\varphi_0(\tau)-\varphi_0(\tau')}{2}\right]$$

and with the aid of Eq. (7.99), we obtain

$$B_m(\nu,\tau,\tau') = \sqrt{\frac{\pi}{E_C\beta}}\exp\left[-\frac{\pi^2 m^2}{E_C\beta} - 2\pi i\nu\frac{\tau-\tau'}{\hbar\beta}m + \nu^2\frac{E_C(\tau-\tau')^2}{\hbar^2\beta} - \nu^2\frac{E_C|\tau-\tau'|}{\hbar}\right] \qquad (7.100)$$

Then the first-order correction to the partition function (7.97) reads

$$\delta\mathcal{Z}_1 = \frac{g\hbar}{4\pi^2}\sum_n e^{-E_C\beta(n_g-n)^2}\int_0^{\hbar\beta}d\tau\int_0^{\hbar\beta}d\tau'\frac{(\pi T/\hbar)^2}{\sin^2[\pi T(\tau-\tau')/\hbar]} \times \exp\left[-\frac{E_C|\tau-\tau'|}{\hbar}\right]\cosh\left[2(n_g-n)\frac{E_C(\tau-\tau')}{\hbar}\right] \qquad (7.101)$$

where $n_g = CV_g/e$ is the dimensionless polarization charge induced by the gate voltage. Here we again employed Poisson's resummation formula in order to pass from the summation over the winding numbers m to that over the number of excess electrons in the box n.

Now we are in a position to derive the first-order correction to the ground-state energy of the system $E_0(n_g)$ due to single-electron tunneling across the junction. For polarization charges within the interval $-0.5 < n_g < 0.5$, we obtain

$$E_0(n_g) = \lim_{T\to 0}(-T\ln\mathcal{Z}) = \lim_{T\to 0}\left(-T\ln\mathcal{Z}_0 - T\frac{\delta\mathcal{Z}_1}{\mathcal{Z}_0}\right) + \mathcal{O}(g^2)$$

$$= E_C n_g^2 - \frac{g\hbar}{2\pi^2}\int_{\tau_c}^{+\infty}\frac{d\tau}{\tau^2}\exp\left[-\frac{E_C\tau}{\hbar}\right]\cosh\left[2n_g\frac{E_C\tau}{\hbar}\right] + \mathcal{O}(g^2)$$

where the parameter τ_c provides the short time cutoff for a formally diverging integral. Evaluating the integral over τ, we arrive at the system energy in the form

$$E_0(n_g) = E_C n_g^2 - \frac{gE_C}{4\pi^2}\left[(1-2n_g)\ln\frac{(1-2n_g)E_C\tau_c}{\hbar} + (1+2n_g)\ln\frac{(1+2n_g)E_C\tau_c}{\hbar} + 2\gamma + \frac{2\hbar}{E_C\tau_c}\right] \qquad (7.102)$$

We observe that the last term in the square brackets does not depend on n_g and, hence, can be safely ignored as a physically irrelevant constant. Furthermore, this unphysical divergence can be totally eliminated by performing a shift in the effective action [149]

$$S[\varphi] \to S[\varphi] + \frac{g\hbar}{4\pi^2}\int_0^{\hbar\beta} d\tau \int_0^{\hbar\beta} d\tau' \frac{(\pi T/\hbar)^2}{\sin^2[\pi T(\tau - \tau')/\hbar]} \times \exp[-E_C(\tau - \tau')] \quad (7.103)$$

This regularization of our effective action just implies an energy shift by a constant value, which does not influence any physical quantity evaluated with the aid of this action. Dropping the last term in Eq. (7.102) and taking the derivative of the energy over V_g, we determine the average number of extra electrons in the box:

$$\langle n\rangle = n_g - \frac{1}{e}\frac{\partial E_0}{\partial V_g} = n_g - \frac{1}{2E_C}\frac{\partial E_0}{\partial n_g} \quad (7.104)$$

In combination with Eq. (7.102), this result yields

$$\langle n\rangle = \frac{g}{4\pi^2}\ln\frac{1+2n_g}{1-2n_g}, \quad -0.5 < n_g < 0.5 \quad (7.105)$$

At this point let us remind the reader that in the limit $g \to 0$, one has $\langle n\rangle = 0$ for the whole interval $-0.5 < n_g < 0.5$. The formula (7.105) demonstrates that already weak electron tunneling smears the sharp Coulomb staircase $\langle n(n_g)\rangle$ even at $T = 0$.

At small $n_g \ll 1$ from the above equations, we have $\langle n\rangle \simeq g n_g/\pi^2$ and

$$E_0(n_g) = E_C^* n_g^2 \quad (7.106)$$

where

$$E_C^* = E_C(1 - g/\pi^2 + O(g^2)) \quad (7.107)$$

is the so-called renormalized charging energy of our device. Equations (7.106), (7.107) demonstrate that single-electron tunneling yields partial screening of the gate charge n_g and, hence, reduces the effective charging energy E_C^* as compared to its bare value E_C. This effect can easily be understood since even for $-0.5 < n_g < 0.5$ and $T = 0$, electrons can virtually tunnel to the island and back. As a result, the system "spends" part of the time in the charge states $CV_g \pm e$, $CV_g \pm 2e$ etc., thereby effectively reducing the absolute value of the capacitor charge below $|CV_g|$.

With increasing V_g, such virtual tunneling processes become progressively more important and for $|n_g| \to 1/2$, the first order in

g contribution to $\langle n \rangle$ formally diverges, as it is seen in Eq. (7.105). This kind of divergencies are routinely encountered within the perturbation theory applied to systems with a degenerate ground state. In our case, the states corresponding, e.g., to the gate charge $n_{\rm g}$ and $n_{\rm g} - 1$ have exactly the same energy provided $n_{\rm g} = 1/2$ (see also Fig. 7.7a). Hence, tunneling between these two states becomes strong even for very small g and cannot anymore be treated perturbatively. Below we will demonstrate that in this case, the situation can be repaired if one properly accounts for higher-order electron tunneling processes.

Note that Eq. (7.102) derived for the system energy within the interval $-0.5 < n_{\rm g} < 0.5$ can also be used outside this interval provided one performs an analytical continuation of the corresponding logarithms. As a result, the energy also acquires an imaginary part, indicating metastability of the charge state with $n = 0$ and $|n_{\rm g}| > 0.5$. As we have already learned in Chapter 4, there exists a simple relation between the imaginary part of the energy and the zero temperature decay rate of a metastable state $\Gamma = -(2/\hbar)\,{\rm Im}\,E(n_{\rm g})$. More generally, also at nonzero temperature, one can relate the decay rate Γ of any charge state Q to the imaginary part of the free energy $F(Q)$ by means of the formula (4.10). In order to evaluate the leading order correction to the free energy, we again expand the grand partition function (7.86), (7.87) to the first order in g and perform the path integral over the phase, like it was already done above while evaluating $\delta\mathcal{Z}_1$. As a result, one finds [418]

$$F(Q_0) = \frac{Q^2}{2C} - \frac{g\hbar}{4\pi^2\tau_0^2}\int_{\tau_c}^{\infty} d\tau\, \exp\left[\frac{eQ\tau}{C} - \alpha \ln\left(\frac{\sin^2(\pi T\tau/\hbar)}{(\pi T\tau_0/\hbar)^2}\right)\right] \tag{7.108}$$

where $\alpha = 1$. We also introduced the parameter τ_0, particular value of which is unimportant here since for $\alpha = 1$, it obviously cancels out from the expression (7.108). Nevertheless, the presence of this parameter in Eq. (7.108) makes no harm to our calculation and we keep τ_0 for our later purposes.

The imaginary part of $F(Q)$ (7.108) is evaluated in exactly the same way as, e.g., it was done in Chapter 5. Separating the decay rate into two,

$$\overrightarrow{\Gamma}(Q) = -\frac{2}{\hbar}\,{\rm Im}\,F(Q), \quad \overleftarrow{\Gamma}(Q) = -\frac{2}{\hbar}\,{\rm Im}\,F(-Q) \tag{7.109}$$

employing Eq. (7.108) and proceeding along the same lines as in Section 5.1.3, we reproduce the expressions (7.8), which in combination with Eq. (7.17) again yield the result (7.18). In particular, at $T = 0$ and $Q > 0$, we have $\overleftarrow{\Gamma}(Q) = 0$ and

$$\overrightarrow{\Gamma}(Q) = \begin{cases} \frac{Q/e-1/2}{R_N C}, & Q > e/2C \\ 0, & Q < e/2C \end{cases} \tag{7.110}$$

Identifying $Q = VC$, we trivially reproduce the $I-V$ curve (7.32).

One can also shunt our tunnel junction by an Ohmic resistor R_S and again perform the calculation of the first order in g correction to the free energy. The calculation remains the same as above, except in Eq. (7.86) along with the action (7.87), we should now also include the action for the Ohmic shunt (2.184). Then we again arrive at the result (7.108) where now $\alpha = 1 + 1/g_S$ and $\tau_0 = R_S C$. Performing the analytic continuation of the free energy yet again and having in mind Eq. (7.109), we obtain [345, 425]

$$\overrightarrow{\Gamma}(Q) = \frac{g}{4\pi^2}\left(\frac{2\pi T \tau_0}{\hbar}\right)^{2\alpha-1} \frac{\left|\Gamma\left(\alpha + \frac{ieQ}{2\pi CT}\right)\right|^2}{\Gamma(2\alpha)} \exp\left(\frac{eQ}{2CT}\right) \tag{7.111}$$

which in turn yields $T \ll eV \ll \hbar/R_S C$

$$\overrightarrow{\Gamma}(Q) = \frac{g}{2\pi R_S C} \frac{(eQR_S)^{1+2/g_S}}{\Gamma(2+2/g_S)} \tag{7.112}$$

for $T \to 0$ and $0 < Q \ll eg_S$ and

$$\overrightarrow{\Gamma}(Q) = \frac{g}{4\pi^{3/2} R_S C} \frac{\Gamma\left(1+\frac{1}{g_S}\right)}{\Gamma(3/2+1/g_S)} \left(\frac{\pi T \tau_{RC}}{\hbar}\right)^{1+2/g_S} \tag{7.113}$$

in the limit $Q \to 0$ and $T \ll g_S E_C$. According to Eq. (7.109), the rate $\overleftarrow{\Gamma}(Q)$ is trivially recovered from the results (7.111)–(7.113) by means of the condition

$$\overleftarrow{\Gamma}(Q) = \overrightarrow{\Gamma}(Q) \exp(-eQ/CT) \tag{7.114}$$

which also follows from detailed balance. Combining Eqs. (7.111)–(7.114) with Eq. (7.17), we again arrive at the results (7.68)–(7.70) describing the $I-V$ curve of a tunnel barrier in the presence of both Coulomb interaction and the Ohmic environment. Thus, we managed to recover important nonlinear transport properties of our system from its thermodynamics.

Note that so far we restricted our perturbative analysis to the lowest-order terms in the junction conductance g. Outside an immediate vicinity of the points $n_{\rm g} = \pm 1/2$, this approximation is usually sufficient provided g remains small. For larger values of g, it might be necessary to include also higher-order tunneling processes. The calculation of higher-order terms becomes progressively more and more tedious, but in some cases, it substantially improves the accuracy of the results even at large values of g. For instance, the renormalized charging energy $E_{\rm C}^*$ was evaluated up to the third-order terms in g with the result [174]

$$\frac{E_{\rm C}^*}{E_{\rm C}} = 1 - 4\frac{g}{4\pi^2} + c_2\left(\frac{g}{4\pi^2}\right)^2 - c_3\left(\frac{g}{4\pi^2}\right)^3 + O(g^4) \quad (7.115)$$

with $c_2 \simeq 5.066$ and $c_3 \simeq 1.457$. It turns out that Eq. (7.115) remains accurate even if the junction conductance becomes as large as $g \approx 10$. This is actually not surprising if we recall that, as follows, e.g., from Eq. (7.94), the actual expansion parameter here is $g/4\pi^2$ rather than g. In other situations, it is not sufficient to retain only several lowest-order terms of the perturbation theory in g. In these cases, one should perform an effective summation of the whole perturbation theories. This goal can be accomplished by means of various non-perturbative methods to be discussed in the next section.

7.5 Non-perturbative Effects

7.5.1 Diagrammatic Analysis

As we already demonstrated, the first-order perturbative expression for the average extra charge $\langle n\rangle$ (7.105) diverges logarithmically as $n_{\rm g}$ approaches the values $\pm 1/2$. This divergence signals insufficiency of the first-order perturbation theory and the necessity to take into account higher-order terms. For instance, evaluating the average extra charge to the second order in g, we obtain [149]

$$\begin{aligned} < n > = {} & \frac{g}{4\pi^2}\ln(a(n_{\rm g})) + 2\left(\frac{g}{4\pi^2}\right)^2 [(\ln(a(Q_x)))^2 \\ & + 3\ln(a(Q_x))] + \ldots \end{aligned} \quad (7.116)$$

where we defined $a(n_g) = 1/2 - n_g \ll 1$. It is straightforward to show that the third-order correction to $< n >$ contains the main logarithmic contribution $4(g/4\pi^2)^3(\ln a(n_g))^3$ as well as lower powers of $\ln a$. Analogously the n-th order term of the perturbation expansion for $< n >$ should contain all powers of $\ln a$ from one to n. It is possible to effectively sum up our perturbation expansion in all orders in g and to evaluate the energy spectrum of our problem for the values of the gate charge close to the degeneracy point $n_g = 1/2$. This goal will be accomplished by means of the diagrammatic approach [149], which will be worked out below.

Let us introduce the density matrix

$$\rho(\tau, \varphi) = \int_0^{\varphi} D\varphi \exp\{-S[\varphi]/\hbar\}. \tag{7.117}$$

Making use of Poisson's resummation theorem, after a trivial algebra, one can rewrite the expression for $\mathcal{Z}(Q_g)$ in the form

$$\mathcal{Z} = \frac{A}{4\pi} \sum_n Z(Q_g - en), \quad A = \int d\varphi \tag{7.118}$$

where—in contrast to $\mathcal{Z}(Q_g)$—the function

$$Z(Q_g) = \int_{-\infty}^{\infty} \rho(\varphi) \exp(\mathrm{i}Q_g\varphi/2e)d\varphi \tag{7.119}$$

is not e-periodic in $Q_g = CV_g$. Let us furthermore define the function $Z(\tau, Q_g)$, which coincides with $Z(Q_g)$ (7.119) at $\tau = \hbar\beta$. The function $Z(\tau, Q_g)$ will play the role of a propagator in the diagram technique to be developed below.

Expanding $Z(\tau, Q_g)$ in powers of g, we get

$$\begin{aligned} Z(\tau, Q_g) &= e^{-\frac{Q_g^2}{2C}\tau} \\ &+ \sum_{n=1}^{\infty} \frac{1}{n!} \int_0^{\tau} \cdots \int_0^{\tau} d\tau_1 d\tau_1' \ldots d\tau_n d\tau_n' \alpha(\tau_1 - \tau_1') \ldots \alpha(\tau_n - \tau_n') \\ &\times \exp\{-\frac{1}{2C}\int_0^{\tau} dt (Q_g - e\sum_{j=1}^{n}(\theta(t-\tau_j) - \theta(t-\tau_j')))^2\} \end{aligned} \tag{7.120}$$

where we defined

$$\alpha(\tau) = \frac{g}{4\pi^2}\left(\frac{\pi T/\hbar}{\sin(\pi T\tau/\hbar)}\right)^2, \tag{7.121}$$

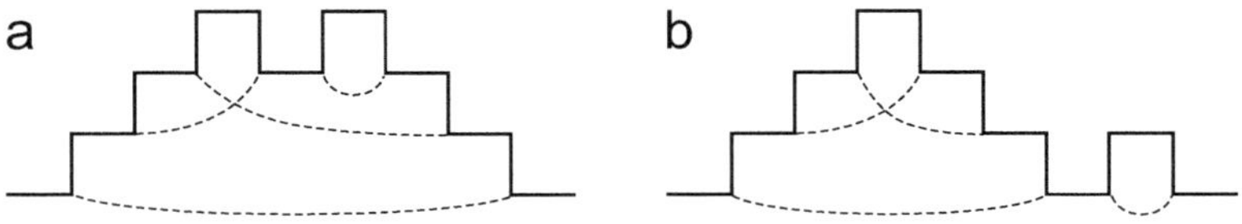

Figure 7.8 Typical connected (a) and disconnected (b) diagrams illustrating terms of the perturbative expansion for Z (7.122). Dashed lines correspond to logarithmic interaction between pairs of electron tunneling events.

introduced a Heaviside step function $\theta(x)$, and set $0 < \tau_j, \tau_j' < \tau$ for all j. Let us order the times τ_j, τ_j' and re-denote them according to their ordering: $\tau_1 < \tau_2 < \ldots < \tau_{2n-1} < \tau_{2n}$. Then we have

$$Z(\tau, Q_x) = e^{-\frac{Q_x^2}{2C}\tau} + \sum_{n=1}^{\infty} \sum_{\epsilon_r=\pm 1} \sum_{\text{pairs}} \int_0^{\tau} d\tau_{2n} \int_0^{\tau_{2n}} d\tau_{2n-1} \ldots \int_0^{\tau_2} d\tau_1 \times \prod_{n=1}^{n} \alpha(\tau_{i_k} - \tau_{j_k}) \exp\{-\frac{1}{2C} \int_0^{\tau} dt (Q_g - e \sum_{r=1}^{n} \epsilon_r \theta(t - \tau_r))^2\} \tag{7.122}$$

The sum $\sum_{\epsilon_r=\pm 1}$ is taken over neutral charge configurations $\sum_r \epsilon_r = 0$, whereas the sum $\sum_{\text{pairs}}$ is taken over all possible pairs i_k and j_k with $\epsilon_{i_k} = 1, \epsilon_{j_k} = -1$.

It is convenient to express the expansion (7.122) graphically (see Fig. 7.8). Each term of this expansion corresponds to an even number of successive electron tunneling events, interaction between each pair of these events τ_{i_k} and τ_{j_k} introduced by the kernel $\alpha(\tau_{i_k} - \tau_{j_k})$ is depicted by a dashed line. We shall distinguish between connected and disconnected diagrams. Connected diagrams (Fig. 7.8a) cannot be divided into two parts without crossing of at least one dashed line. Disconnected diagrams (Fig. 7.8b) have parts without dashed lines. The total charge corresponding to these parts is always equal to Q_g because connected blocks are neutral.

As in the standard Feynman diagram technique, let us introduce the sum of all connected diagrams $\Sigma(\tau, Q_g)$ and then rewrite the diagram series in terms of a Dyson equation

$$Z(\tau, Q_x) = e^{-Q_g^2\tau/2C} + \int_0^{\tau} d\tau_2 \int_0^{\tau_2} d\tau_1 Z(\tau - \tau_2)\Sigma(\tau_2 - \tau_1) e^{-Q_g^2\tau_1/2C} \tag{7.123}$$

Figure 7.9 Graphical representation of the Dyson equation (7.123).

which can also be represented graphically as shown in Fig. 7.9. Making use of the Laplace transformation, we find

$$Z_p(Q_g) = \int_0^\infty d\tau \exp(-p\tau) Z(\tau, Q_g) = 1/(p + Q_g^2/2C - \Sigma_p(Q_g)) \tag{7.124}$$

where p is the Laplace variable. Below we shall use this equation in order to evaluate the energy spectrum of a single-electron box.

The self-energy Σ in the Dyson equation defined by Eqs. (7.123), (7.124) can be identified in terms of the diagram series displayed in Fig. 7.10. These series allow one to express Σ in terms of $Z(\tau, Q_g)$ and to obtain the closed system of equations for Z.

Equivalently, one can rewrite the Dyson equation introducing the vertex parts $\Gamma(p, Q_g; p', Q_g \pm e)$

$$\begin{aligned}\frac{1}{Z(p, Q_g)} &= p + \frac{Q_g^2}{2C} + \frac{1}{2\pi \mathrm{i}} \\ &\times \int_{-\mathrm{i}\infty}^{\mathrm{i}\infty} dp' \alpha(p - p') \{\Gamma(p, Q_g; p', Q_g + e) Z(p', Q_g + e) \\ &+ \Gamma(p, Q_x; p', Q_x - e) Z(p', Q_x - e)\}\end{aligned} \tag{7.125}$$

and express Γ diagrammatically as shown in Fig. 7.11.

At this stage, it is worth pointing a useful analogy to the so-called "zero charge problem," which is well known in quantum electrodynamics, see, e.g., [197]. In order to observe this analogy, it suffices to identify solid lines in Fig. 7.11 with electron propagators, while dashed lines would then denote photon propagators. The only

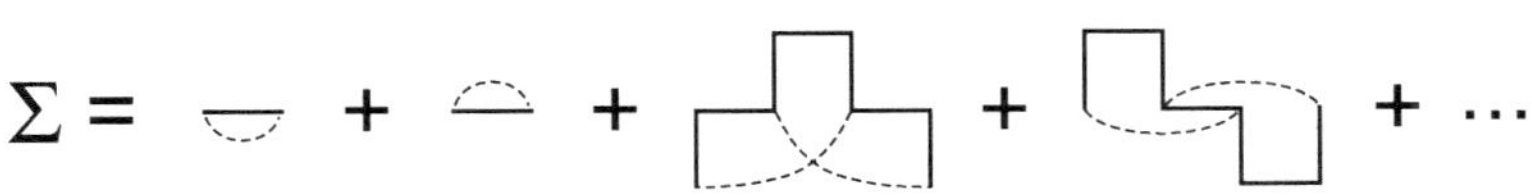

Figure 7.10 Diagrammatic representation of the perturbative expansion for the self-energy Σ.

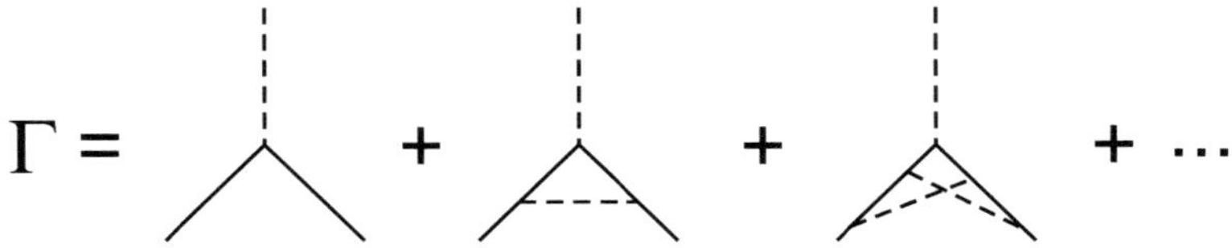

Figure 7.11 Diagrammatic representation of the perturbative expansion for the vertex part Γ.

important difference between the Coulomb blockade problem considered here and the "zero charge" one in quantum electrodynamics is that there exists no effect of vacuum polarization in our problem and the expression for the "photon propagator" $\alpha(\tau)$ remains fixed.

The function $Z_p(Q_g)$ contains complete information about the energy spectrum of our problem. Indeed for $-e/2 < Q_g < e/2$, this function defines the junction ground-state energy $E_0(Q_g) = -T \ln Z(Q_g)|_{T\to 0}$. From the inverse Laplace transformation

$$Z(\tau, Q_g) = \frac{1}{2\pi i}\int_{-i\infty}^{+i\infty} dp Z_p(Q_g)\exp(p\tau) \tag{7.126}$$

one can easily find $Z(\tau, Q_g) \sim \exp(p_0\tau)$ for $\tau \to \infty$. Here p_0 is the pole of the function $Z_p(Q_g)$ in the complex plane p with the largest value $\mathrm{Re}\, p_0$ (all poles of the function Z_p are in the halfplane $\mathrm{Re}\, p < 0$). Therefore, the ground-state energy $E_0(Q_g)$ coincides with the smallest positive solution to the equation

$$Z_p^{-1}(Q_g) = 0 \tag{7.127}$$

or, equivalently,

$$E_0(Q_g) = \frac{Q_g^2}{2C} - \Sigma_p(Q_g)|_{p=-E_0(Q_g)}. \tag{7.128}$$

Making use of analytic properties of the function $Z_p(Q_g)$, one can also obtain information about higher energy states. Indeed, provided $Z_p(Q_g)$ is an analytic function of the parameters p and Q_g, the function $E_0(Q_g)$ is also an analytic function of Q_g. This observation allows to reconstruct the energy of excited states by means of an analytic continuation of the function $Z_p(Q_g)$ from the interval $|Q_g| < e/2$ to $|Q_g| > e/2$.

In order to proceed, let us again regularize the action by means of the shift (7.103). It is straightforward to verify [149] that this regularization eliminates all ultraviolet divergencies in all orders of the perturbation theory. After that one can rewrite the Dyson equation (7.125) in the form

$$\frac{1}{Z_p(Q_g)} = p + \frac{Q_g^2}{2C} - \frac{g}{4\pi^2}\int_p^{+\infty} dp'(p'-p)(\Gamma(p, Q_g; p', Q_g - e)Z_{p'}(Q_g - e)$$
$$+\Gamma(p, Q_g; p', Q_x + e)Z_{p'}(Q_g + e) - 2(p, Q_g; p' + E_C, Q_g)Z_{p'+E_C}(Q_g)). \tag{7.129}$$

Equations (7.123), (7.124) are modified accordingly. In order to solve these equations in the limit of small junction conductances $g \ll 1$, we proceed perturbatively in g and take into account only two first diagrams in the series depicted in Fig. 7.10. This approximation corresponds to a summation of a subsequence of diagrams with non-crossing dashed lines. It is equivalent to the standard approximation of main logarithms routinely employed, e.g., in the Kondo or zero charge problems. Analogously, only the first diagram in the series for the vertex part Γ (Fig. 7.11) has to be taken into account. Within the framework of this approximation, the self-energy reads

$$\Sigma(\tau, Q_g) = \alpha(\tau)[Z(\tau, Q_g - e) + Z(\tau, Q_g + e) - 2\exp(-E_C\tau)Z(\tau, Q_g)] \tag{7.130}$$

and the vertex part equals to unity $\Gamma = 1$. Then with the aid of Eq. (7.130), one can rewrite the equation for Z (7.129) in the form

$$\frac{1}{Z_p(Q_g)} = p + \frac{Q_g^2}{2C} - \frac{g}{4\pi^2}\int_p^{+\infty} dp'(p'-p)$$
$$\times [Z_{p'}(Q_g - e) + Z_{p'}(Q_g + e) - 2Z_{p'+E_C}(Q_g)] \tag{7.131}$$

Taking the second derivative of Eq. (7.131) with respect to p, we get

$$\frac{d^2}{dp^2}\frac{1}{Z_p(Q_g)} = -\frac{g}{4\pi^2}[Z_p(Q_g - e) + Z_p(Q_g + e) - 2Z_{p+E_C}(Q_g)] \tag{7.132}$$

So far we have not yet made use of the fact that in the regime interesting for us here the value of the gate charge Q_g is close to $e/2$. In this limit, the diagrams describing tunneling between the charge states Q_g and $Q_g{-}e$ give the main contribution and, hence, one can reduce the problem to the effective two-state one. Taking the Laplace parameter p close to the poles of the functions $Z_p(Q_g)$ and

$Z_p(Q_g - e)$, for $|e/2 - Q_g| \ll e$ and $g \ll 1$, one can estimate the functions Z_p in the right-hand side of Eq. (7.130) as

$$Z_p^{-1}(Q_g - e) \sim \frac{(Q_g - e)^2}{2C} - \frac{Q_g^2}{2C} \equiv \Delta(Q_g),$$

$$Z_p^{-1}(Q_g + e) \sim Z_p(Q_g - e) \sim E_C.$$

Then for $p + Q_g^2/2C \ll E_C$ and $\Delta(Q_g) \ll E_C$, Eq. (7.132) reduces to

$$\ddot{x}(p) = -\frac{g}{4\pi^2 y(p)}, \quad \ddot{y}(p) = -\frac{g}{4\pi^2 x(p)} \tag{7.133}$$

where we introduced the notations $x(p) = Z_p^{-1}(Q_g)$, $y(p) = Z_p^{-1}(Q_g - e)$ and denoted the derivatives over p by overdots. In order to establish the initial conditions for these equations, we rewrite the first of Eqs. (7.133) in the form of the integral equation

$$x(p) = -\frac{g}{4\pi^2} \int_p^0 dp'(p' - p)y^{-1}(p') + x_0 + \dot{x}_0 p \tag{7.134}$$

and compare it with Eq. (7.131) for $p + Q_g^2/2C \ll E_C$, i.e., in the vicinity of the pole of the function $Z_p(Q_g)$. Then we arrive at the desired initial conditions:

$$x_0 = \frac{Q_g^2}{2C} - \frac{g}{4\pi^2} \int_0^\infty dp' p' Z_{p'}(Q_g - e) - \frac{g}{4\pi^2} \int_{-Q_g^2/2C}^\infty dp' p' [Z_{p'}(Q_g + e) - 2Z_{p'+E_C}(Q_g)] \tag{7.135}$$

and

$$\dot{x}_0 = 1 + \frac{g}{4\pi^2} \int_0^\infty dp' Z_{p'}(Q_g - e) - \frac{g}{4\pi^2} \int_{-Q_g^2/2C}^\infty dp' [Z_{p'}(Q_g + e) - 2Z_{p'+E_C}(Q_g)] \tag{7.136}$$

Initial conditions for y_0 and y_0' can be derived analogously. As integration over p' in Eqs. (7.135) and (7.136) runs far from the poles of the function Z_p, with a sufficient accuracy one can put $Z_p(Q_g) = (p + Q_g^2/2C)^{-1}$. Then for $Q_g \simeq e/2$, we obtain

$$x_0 = Q_g^2/2C + \frac{g}{4\pi^2} E_C((1 - 7\ln 2)/4 + (1/4 + 5\ln 2)a(Q_g)) \tag{7.137}$$

$$y_0 = (Q_g - e)^2/2C + \frac{g}{4\pi^2} E_C((1 - 7\ln 2)/4 - (1/4 + 5\ln 2)a(Q_g)) \tag{7.138}$$

$$\dot{x}_0 + \frac{3g}{4\pi^2} a(Q_g) = \dot{y}_0 - \frac{3g}{4\pi^2} a(Q_g) = 1 + \frac{g}{4\pi^2} \ln 2 \qquad (7.139)$$

In order to resolve Eqs. (7.133) with the initial conditions (7.137)–(7.139), it is convenient to introduce the function $u(p) = x(p)y(p)$ obeying the equation

$$\ddot{u} = 2\dot{x}_0\dot{y}_0 - \frac{g}{2\pi^2}(1 + \ln(u/x_0y_0)) \qquad (7.140)$$

with a solution

$$p = \int_{x_0y_0}^{u} du[(x_0\dot{y}_0 - \dot{x}_0y_0)^2 + 4\dot{x}_0\dot{y}_0 - (gu/\pi^2)\ln(u/x_0y_0)]^{-1/2} \qquad (7.141)$$

Combining this solution with Eq. (7.128), we arrive at the final result for the ground-state energy

$$E_0(Q_g) = \frac{1}{2}\left(\frac{x_0y_0}{\dot{x}_0\dot{y}_0}\right)^{1/2} \int_0^1 dt \left(\frac{(x_0\dot{y}_0 - \dot{x}_0y_0)^2}{4x_0y_0\dot{x}_0\dot{y}_0} + t - \frac{gt\ln t}{\dot{x}_0\dot{y}_0}\right)^{-1/2} \qquad (7.142)$$

Within the accuracy of our calculation, this expression yields

$$E_0(Q_g) = E_0(e/2) - E_C a(Q_g)\left[\frac{1 - gb}{L(Q_g)} + \frac{g(b - 1/2\pi^2)}{L^2(Q_g)}\right] \qquad (7.143)$$

where we defined

$$E_0(e/2) = \frac{E_C}{4}\left(1 - \frac{g}{4\pi^2}\right), \quad b = (1 + 4\ln 2)/4\pi^2 \qquad (7.144)$$

and

$$L(Q_g) = 1 - \frac{g}{4\pi^2}\left[\ln a(Q_g) + \ln\left(\frac{E_0(e/2) - E_0(Q_g)}{E_C}\right)\right] \qquad (7.145)$$

The above equations fully determine the system ground-state energy in the vicinity of the degeneracy point $Q_g = e/2$. In particular, exactly at this point, we have $\partial E_0/\partial Q_g|_{Q_g=e/2} = 0$, i.e., the band flattens off due to intensive electron tunneling between the charge states $e/2$ and $-e/2$.

We also point out that the result defined by Eqs. (7.143)–(7.145) derived by a summation of non-crossing diagrams actually goes beyond the approximation of the main logarithms. This is because the two-state problem diagrams with crossing dashed lines give a small (of order g^2) contribution to the self-energy. Hence, it is

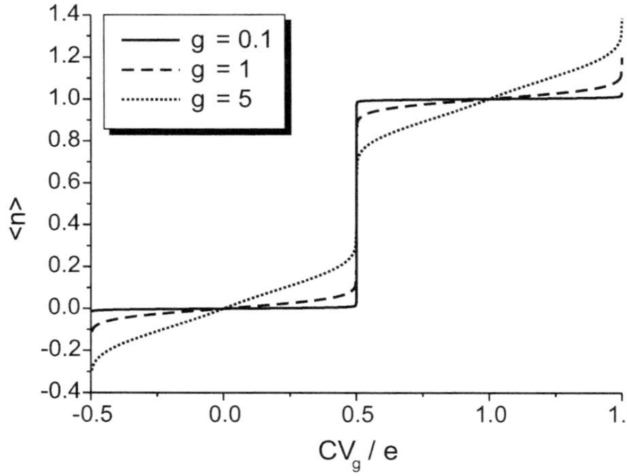

Figure 7.12 The average excess number of electrons $\langle n \rangle$ versus n_g (7.146) for different values of the junction dimensionless conductance g.

possible also to recover the next order term in $g/(1 - 2g \ln a)$ in the expression for $\langle n \rangle$. On the other hand, the approximation of non-crossing two-state diagrams does not allow to correctly reproduce the prefactor in front of the term $\sim g^2 \ln a$ in Eq. (7.116) because diagrams describing tunneling to higher energy states also contribute to this term. This effect might at most lead to a small unimportant renormalization of g in Eqs. (7.143)–(7.145).

Combining Eqs. (7.143)–(7.145) and (7.104) and keeping only the dominant logarithmic terms, we obtain [149, 295, 343]

$$\langle n \rangle \simeq \frac{1}{2}\left(1 - \frac{1}{1 + \frac{g}{2\pi^2}\ln\frac{1+2n_g}{1-2n_g}}\right), \quad 0 \le n_g \le 1/2 \quad (7.146)$$

This dependence is plotted in Fig. 7.12 for different values of g. In contrast to the first-order result (7.105), the non-perturbative expression (7.146) does not diverge as n_g approaches 1/2 and yields $\langle n \rangle|_{n_g=1/2} = 1/2$. We also note that the difference between the results (7.105) and (7.146) becomes significant only in a very narrow region in the vicinity of the Coulomb blockade threshold

$$1/2 - n_g \lesssim \exp(-2\pi^2/g) \quad (7.147)$$

As we already explained above, our method allows to evaluate not only the ground-state energy $E_0(Q_g)$ but also the energy of excited states. The latter can be obtained by means of the analytic continuation of $E_0(Q_g)$ to the values $|Q_g| > e/2$. Let us, for instance, reconstruct the energy of the first excited state $E_1(Q_g)$ in the vicinity of the point $Q_x = e/2$. The analytic continuation procedure in Eq. (7.143) is completely straightforward and yields

$$E_1(Q_g) = E_0(e/2) + E_C a(Q_g)\left[\frac{1-gb}{L(Q_g)-ig/2\pi} + \frac{g(b-1/2\pi^2)}{(L(Q_g)-ig/2\pi)^2}\right] \tag{7.148}$$

Combining this result with Eq. (7.143), we immediately recover the level splitting $\Delta_{10}(Q_g) = \mathrm{Re}E_1(Q_g) - E_0(Q_g) \simeq 2[E_0(e/2) - E_0(Q_g)]$ in the vicinity of the point $Q_g = e/2$. For $n_g \to 1/2$, we obtain

$$\Delta_{10}(n_g) \simeq -\frac{4\pi^2 E_C|1/2-n_g|}{g\ln|1/2-n_g|} \tag{7.149}$$

As usually, the presence of an imaginary part in the expression for $E_1(Q_g)$ (7.148) signals instability, i.e., the system decays to its ground state with energy $E_0(Q_g)$. At $T \to 0$, the corresponding decay rate $\Gamma(Q)$ of the first excited state is again defined as $-(2/\hbar)\mathrm{Im}E_1(Q)$. Then from Eq. (7.148) in the main approximation, we get

$$\Gamma(Q) \simeq \frac{g}{2\pi}\theta(Q-e/2)\frac{eQ/C - E_C}{[1-(g/2\pi^2)\ln|a(Q)|]^2} \tag{7.150}$$

This expression accounts for the effect of strong quantum fluctuations of the charge in the vicinity of the point $Q_x = e/2$.

From the above analysis, one can also conclude that at $T \to 0$ and in the immediate vicinity of the Coulomb blockade threshold $V \equiv Q/C = e/2C$, the I–V curve of a tunnel junction should deviate from Eq. (7.32). Two effects need to be taken into account. One of them is the reduction of the single-electron tunneling rate (7.150) by the factor $1/L^2(Q)$ below its standard perturbative value (7.110). Another effect is the charge screening (7.146) according to which at $Q \approx e/2$ after each electron tunneling event, the junction charge changes not by e but only by $q^* = e/L(Q)$. If one combines both effects within a phenomenological formula for the current $I = q^*(VC)\Gamma(VC)$, one immediately finds that at $\delta V = V - e/2C \to 0$, the current goes to zero as

$$I = \frac{\delta V}{R_N}\frac{1}{[(g/2\pi^2)\ln(C\delta V/e)]^3} \tag{7.151}$$

instead of $I = \delta V/R_N$.

7.5.2 Renormalization Group Equations

The above diagrammatic approach enabling one to sum up the leading logarithms to all orders of the perturbative expansion in g remains efficient provided the junction dimensionless conductance is sufficiently small $g \ll 1$. In this limit, the leading logarithms can also be summed up with the aid of the renormalization group approach. Employing, e.g., Eq. (7.94), restricting the analysis to the effective two-state problem (i.e., one which includes only the lowest and the first excited charge states) and making use of the standard scaling arguments similar to ones already encountered in Chapters 5 and 6, one arrives at the RG equations for the renormalized values of the ground-state energy $\tilde{E}_0$, level splitting $\tilde{\Delta}_{10}$, and dimensionless conductance $\tilde{g}$ [113]

$$\frac{d\tilde{E}_0}{d\ln\omega_c} = -\frac{\tilde{g}}{4\pi^2}\tilde{\Delta}_{10}, \quad \frac{d\tilde{\Delta}_{10}}{d\ln\omega_c} = \frac{\tilde{g}}{2\pi^2}\tilde{\Delta}_{10}, \quad \frac{d\tilde{g}}{d\ln\omega_c} = \frac{\tilde{g}^2}{8\pi^4} \tag{7.152}$$

which leave the free energy of our single-electron box invariant under the procedure of successive decrease in the high-frequency cutoff ω_c. We note that the renormalized junction conductance g decreases upon scaling, i.e., the system is driven toward lower and lower conductance values where the perturbation theory works better and better.

It is also worth pointing out that the RG equations (7.152) are rather similar to ones derived in Chapter 5 for the two-level system with linear Ohmic dissipation (cf. Eqs. (5.32)). This similarity is, by no means, surprising since both problems can be mapped onto that of logarithmically interacting one-dimensional Coulomb gas. An important formal difference between these two problems is that here only pairs of charges interact with each other, whereas in the case of a dissipative two-level system considered in Chapter 5, the interaction exists between all charges.

An alternative real-time RG scheme was developed in the work [236]. This scheme allows for a partial summation of diagrams in all orders in g, though neglects higher-order vertex corrections. The corresponding RG equations may be integrated numerically and yield the results that remain applicable at sufficiently large g. While the explicit validity range of this RG analysis is difficult to establish analytically, the corresponding information can be extracted from a comparison to the numerical results discussed in Section 7.5.4. This comparison reveals differences between the real-time RG approach and the results of quantum Monte Carlo (QMC) simulations for $g \gtrsim 12$.

Yet another RG approach was developed, which allows to obtain important information about the properties of our system in the strong tunneling limit $g \gg 1$. In this case, one can follow the method [234] developed to analyze the critical behavior of one-dimensional XY models with isotropic long-range interactions. The same approach was applied to tunnel junctions in the strong tunneling regime $g \gg 1$ [76, 183, 184]. In general, this analysis also allows to include superconductivity, i.e., to simultaneously treat both tunneling of Cooper pairs and single electrons. Here we only consider the case of normal tunnel barriers.

Following the same route as in our previous scaling considerations, in the expression for the partition function (7.86) and (7.87), one can split the fluctuating phase variable φ into "fast" (with frequencies $\omega_{\mathrm{c}} > \omega > \omega_{\mathrm{c}} - \delta\omega$) and "slow" (with frequencies $\omega_{\mathrm{c}} - \delta\omega > \omega$) parts and integrate over the "fast" phase fluctuations. Making use of the fact that in the limit $g \gg 1$ phase fluctuations remain weak, one can expand the action to the second order in the "fast" fluctuations and integrate them out. Then one again arrives at the partition function (7.86) and (7.87) where the action $S[\varphi]$ depends only on "slow" phase variable and the effective dimensionless conductance $\tilde{g}$. Repeating this procedure many times and proceeding perturbatively in $1/\tilde{g}$, one finds

$$\frac{d\tilde{g}}{d\ln\omega_{\mathrm{c}}} = 2 + \frac{4}{\tilde{g}} + O(\tilde{g}^{-2}) \tag{7.153}$$

The main term in the right-hand side of this equation was established within the analysis [76, 183, 184], while the leading ($\sim 1/\tilde{g}$) correction to it was derived later by means of the two-loop RG calculation [194].

Equation (7.153) demonstrates that—in a qualitative similarity to the weak tunneling limit described by Eqs. (7.152)—the renormalized dimensionless conductance $\tilde{g}$ *decreases* upon decreasing the cutoff frequency ω_{c}. Starting renormalization from the highest frequency $\omega_{\mathrm{c}} \sim 1/R_{\mathrm{N}}C \sim gE_{\mathrm{C}}/\hbar$, integrating Eq. (7.153), and setting the final value of the dimensionless conductance to be of order one $\tilde{g} \sim 1$, we arrive at a new energy scale

$$\hbar\omega_{\text{c}} \sim E^{*} \sim g^{2} E_{\text{C}} \exp(-g/2) \tag{7.154}$$

Renormalization should be stopped at this scale, since the perturbation theory in $1/\tilde{g}$ ceases to be valid and the RG equation (7.153) becomes inaccurate. Combining RG equations (7.153) and (7.152), one can introduce the two-stage scaling procedure [113], which allows to cover essentially the whole range of the junction conductances.

Equation (7.154) constitutes an important energy scale in our problem demonstrating that at exponentially low energies (temperatures, voltages) of order or below E^{*}, Coulomb blockade becomes strong even in the limit of large junction conductances $g \gg 1$. Accordingly, for $T \to 0$ and $V \to 0$, one can expect electron tunneling across the junction to be fully suppressed by electron–electron interactions even at very large values of g.

We also note that the energy scale similar to E^{*} also emerges in our previous real-time analysis of Coulomb effects on transport properties of our system in the limit of large conductances (cf., e.g., Eq. (7.54)). In the limit $g \gg g_{\text{S}}$, the corresponding energy scale differs from E^{*} (7.154) only by one power of g in the pre-exponent, which is not a significant effect as compared to the exponential factor $\sim \exp(-g/2)$. Nevertheless, it is worth discussing how the prefactor $\sim g^{2}$ in Eq. (7.154) comes about. One power of g is simply due to the chosen high-frequency cutoff $\omega_{\text{c}} \sim gE_{\text{C}}/\hbar$, which also naturally appears in our real-time calculations. An extra power of g in Eq. (7.154) emerges from the sub-leading term $\sim 1/g$ in the right-hand side of the RG equation (7.153). Should this two-loop correction be ignored (which would correspond to the accuracy of our real-time calculations), this extra power of g would be absent and the scales obtained here and, e.g., in Eq. (7.54) would coincide already at the level of the pre-exponential factors. Finally, we would like to mention that the non-perturbative scaling analysis yields additional non-renormalizable interactions not captured by the RG equation (7.154). Hence, the dependence of the pre-exponent on g in Eq. (7.154), although most likely accurate, could be the subject of further minor modifications.

7.5.3 Instantons

Although the above RG analysis already yields an important information about the non-perturbative ground-state properties of our structure in the strong tunneling limit $g \gg 1$, it does not explicitly account for charge discreteness and, hence, cannot be applied, e.g., for the evaluation of the ground-state energy as a function of the gate charge $E_0(n_g)$. On the other hand, the grand partition function (7.86), (7.87) fully accounts for the discrete nature of charge states. The task at hand is to directly evaluate this grand partition function in the large conductance limit $g \gg 1$ and at low temperature $T \to 0$. This information will then allow to establish $E_0(n_g)$.

In order to proceed, we first notice that at large junction conductance $g \gg 1$, the second term in the action (7.87) becomes large and, hence, phase fluctuations should remain weak. Therefore, the path integral (7.86) can be handled semiclassically within the saddle point approximation. As usually, the first step is to establish the saddle point trajectories from the least action condition $\delta S[\varphi]/\delta\varphi = 0$. Provided the action is defined by Eq. (7.87), this condition takes the form of the following equation

$$-C\frac{\hbar\ddot{\varphi}}{2e} + \frac{\hbar}{\pi e R_N}\int_0^{\hbar\beta} d\tau' \frac{(\pi T/\hbar)^2}{\sin^2[\pi T(\tau - \tau')/\hbar]} \sin\frac{\varphi(\tau) - \varphi(\tau')}{2} = 0 \tag{7.155}$$

Note that this saddle point equation has certain similarities, e.g., with Eqs. (4.18) and (5.9). On the other hand, an important difference of Eq. (7.155) from the above equations considered in Chapters 4 and 5 is that the time-independent potential energy term is now absent, and the only remaining nonlinearity is due to the non-local in-time friction term proportional to $1/R_N$. In a way, this term plays the role of a time-dependent potential for the phase variable.

Along with a trivial solution $\varphi(\tau) = \varphi_0$ corresponding to the topological sector $m = 0$, there exist nontrivial ones that describe tunneling of the phase variable in the non-local in-time potential between different topological sectors. The simplest solution of that kind is the so-called "straight line"

$$\varphi(\tau) = \varphi_0 + 4\pi m T\tau/\hbar \tag{7.156}$$

corresponding to the tunneling of the phase between the topological sectors $m = 0$ and m. Substituting these straight-line solutions into the action (7.87), we obtain

$$S_m/\hbar = \pi^2 m^2 T/E_\mathrm{C} + |m|g/2 \tag{7.157}$$

Note that both Eqs. (7.156) and (7.157) remain applicable for any value of g, i.e., at this point, g need not be necessarily large. Restricting our analysis to these semiclassical trajectories, we immediately arrive at the following approximate expression for the partition function

$$\mathcal{Z} \sim \sum_{m=-\infty}^{\infty} e^{\mathrm{i}2\pi n_\mathrm{g} m - S_m/\hbar} \tag{7.158}$$

In the limit $g \to 0$ analogously to Eq. (6.48), this expression can be transformed to

$$\mathcal{Z} \sim \sum_{p=-\infty}^{\infty} \exp\left[\frac{E_\mathrm{C}}{T}(n_\mathrm{g} + p)^2\right] \tag{7.159}$$

i.e., we again recover the standard parabolic dependence of the energy on the gate charge depicted in Fig. 7.7a. At high temperatures $T \gg E_\mathrm{C}$ and/or large $g \gg 1$, the action S_m (7.157) remains large for any nonzero m. In this case, the sum over m in Eq. (7.158) can be restricted to the terms with $m = 0$ and $m = \pm 1$ and the n_g-dependent part of the free energy $F = -T \ln \mathcal{Z}$ is easily evaluated. Employing Eq. (7.85), for the average value of the excess charge, one finds

$$\langle n \rangle \approx n_\mathrm{g} - \frac{2\pi T}{E_\mathrm{C}} \exp(-\pi^2 T/E_C - \mathrm{g}/2)\sin(2\pi n_\mathrm{g}) \tag{7.160}$$

This result holds at high-enough temperatures $T \gtrsim E_\mathrm{C}$ and any value of g. We observe that the gate effect persists even in the strong tunneling regime $g \gg 1$, although in this case its magnitude gets additionally suppressed by the factor $\exp(-g/2)$. One can also define the temperature-dependent renormalized charging energy

$$E_\mathrm{C}^*(T) = \frac{1}{2}\frac{\partial^2 F}{\partial n_\mathrm{g}^2} = E_\mathrm{C}\left(1 - \frac{\partial\langle n\rangle}{\partial n_\mathrm{g}}|_{n_\mathrm{g}=0}\right) \tag{7.161}$$

Then from Eq. (7.160), one gets

$$E_C^*(T) = 4\pi^2 T \exp(-\pi^2 T/E_C - g/2) \qquad (7.162)$$

Although this high-temperature result cannot be directly used in the low-temperature limit, it nevertheless allows to correctly estimate the dominant effect of strong electron tunneling on E_C^* even at low T. Indeed, at the borderline of applicability of Eq. (7.162) $\pi^2 T \sim E_C$ and for $g \gg 1$, we obtain

$$E_C^*/E_C \propto \exp(-g/2) \qquad (7.163)$$

Since there is no other energy scale in our problem below E_C, it would be reasonable to conjecture that this exponential suppression of the renormalized charging energy should persist down to $T = 0$. We also note that in this case, the renormalized charging energy E_C^* (7.163) within the exponential accuracy matches with the energy scale E^* (7.154) derived from the RG analysis.

The key limitation of the above simple semiclassical consideration is related to the fact that fluctuations of the phase around the straight-line trajectories (7.156) were totally neglected so far. One could try to repair this drawback in a usual manner by integrating out small Gaussian fluctuations of φ. It turns out, however, that even in the limit $g \gg 1$, there exist trajectories that may strongly deviate from the straight lines (7.156) without significantly increasing the value of the action $S[\varphi]$. In order to account for these nontrivial soft modes in our problem, we will make use of the observation that in the limit $C \to 0$ and $T \to 0$, the saddle point equation (7.155) has an exact solution of the form [231]

$$\tilde{\varphi}_1(\tau) = \varphi_0 + 2\pi + 4\arctan(\Omega_1(\tau - \tau_1)) \qquad (7.164)$$

where Ω_1 is an arbitrary frequency and $-\infty < \tau_1 < \infty$. Hence, similar to the trajectory (7.156) with $m = 1$, it also describes quantum tunneling of the phase φ between the topological sectors $m = 0$ and $m = 1$. Actually, in the zero frequency limit $\Omega_1 \to 0$, the solution (7.164) reduces to the form (7.156) with $m = 1$ and $T \to 0$.

Substituting this solution into the action (7.87), we obtain

$$\frac{S[\tilde{\varphi}_1(\tau)]}{\hbar} = \frac{\pi}{2}\frac{\hbar\Omega_1}{E_C} + \frac{g}{2} \qquad (7.165)$$

Thus, we observe that for all frequencies $\Omega_1 \lesssim E_C/\hbar$, the action (7.165) for the path (7.164) matches with that in Eq. (7.157) for $m = 1$ and $T \ll E_C$. The trajectory (7.164) represents an instanton, which accounts for tunneling of the phase variable between the states $\varphi = \varphi_0$ and $\varphi = \varphi_0 + 4\pi$ with the tunneling amplitude $\propto \exp(-g/2)$. Similar to Chapters 4 and 5, we can develop an instanton technique [321, 323] and evaluate both the grand partition function (7.86) and the ground-state energy $E_0(n_g)$.

As before, in order to evaluate the effect of fluctuations around the trajectory $\tilde{\varphi}_1(\tau)$, we need to establish the eigenvalues of the operator $\hat{M}_1 = \delta^2 S/\delta\tilde{\varphi}_1^2$. For $C \to 0$, there exist two zero modes with collective coordinates τ_1 and $1/\Omega_1$ corresponding, respectively, to a shift of the instanton position along the imaginary-time axis and to a change of a typical instanton size. These two zero modes have the eigenfunctions $\psi_1(\tau) = -(8\pi\Omega)^{-1/2}\partial\tilde{\varphi}_1/\partial\tau_1$ and $\psi_2(\tau) = (8\pi/\Omega^3)^{-1/2}\partial\tilde{\varphi}_1/\partial\Omega_1$. All remaining eigenvalues of the operator $\hat{M}_1$ are positive. Following the strategy outlined in Chapters 4 and 5, we integrate out phase fluctuations around $\tilde{\varphi}_1(\tau)$ and recover the standard expression (cf. Eqs. (5.15), (5.16)) consisting of the integrals over the two collective variables τ_1 and $1/\Omega_1$ times $\exp(-g/2)$ times the square root of the ratio of the two determinants $(\det\hat{M}_0/\det'\hat{M}_1)^{1/2}$, where the operator $\hat{M}_0 = \delta^2 S/\delta\varphi^2$ is defined on the trajectory $\varphi_\tau = \varphi_0$ and $\det'\hat{M}_1$ is the determinant with two zero modes excluded. In the presence of a nonzero capacitance C, one zero mode with the collective coordinate $1/\Omega_1$ turns to a soft mode, meaning that fluctuations with frequencies $\Omega_1 \gtrsim E_C/\hbar$ are suppressed stronger than those with lower frequencies. This observation effectively restricts the integral over Ω_1 to values below $E_C/\hbar$ (cf. Eq. (7.165)).

The next step is to consider the contribution that accounts for the phase tunneling between the topological sectors $m = 0$ and m by means of N instantons

$$\tilde{\varphi}_N(\tau) = \varphi_0 + 2\pi m + \sum_{k=1}^{N} \epsilon_k 4\arctan(\Omega_k(\tau - \tau_k)), \quad \sum_{k=1}^{N} \epsilon_k = m \tag{7.166}$$

where $\epsilon_k = \pm 1$ are, as before, the topological charges of instantons. This configuration has $2N$ zero modes. Proceeding in exactly the same way as above, we arrive at the following expression

$$4^N \int d\tau_N \int \frac{d\Omega_N}{\Omega_N} \ldots \int d\tau_1 \int \frac{d\Omega_1}{\Omega_1} \left(\frac{\det \hat{M}_0}{\det' \hat{M}_N} \right)^{1/2}$$
$$\times \exp\left(-\frac{S[\tilde{\varphi}_N]}{\hbar} + 2\pi i m n_g \right) \tag{7.167}$$

where we ordered the instantons as $\tau_N > \tau_{N-1} > \ldots > \tau_1$ and defined $\det' \hat{M}_N$ as the determinant with excluded $2N$ zero modes of the operator $\hat{M}_N = \delta^2 S / \delta \tilde{\varphi}_N^2$.

What remains is to evaluate the ratio of the determinants, perform integrals over all collective coordinates, and sum up all contributions (7.167) with all values of m and N. Unfortunately, this program cannot be carried out rigorously, merely because the usual approximation of a dilute instanton gas does not work well in this case. Provided all $\Omega_k \lesssim E_C/\hbar$, one finds

$$\frac{S[\tilde{\varphi}_N(\tau)]}{\hbar} \simeq \frac{Ng}{2} + \frac{S_{\text{int}}}{\hbar} \tag{7.168}$$

where the term S_{int} accounts for the power-law inter-instanton interactions. The analysis [321, 323] reveals that instantons with the same topological charges do not interact with each other, while oppositely charged instantons at a distance τ interact as $\sim 1/\tau^2$. At inter-instanton distances τ comparable with the instanton width $\sim 1/\Omega_k$, the term S_{int} is not small; hence, the dilute instanton gas approximation may become insufficient, in particular in the low-temperature limit.

Different authors addressed this problem within various approximate schemes. Without going into further details, we stress that presently there is a general agreement that for $g \gg 1$ and small values of the gate charge $n_g \ll 1$, the ground-state energy of the system has the form (7.106) with E_C^* defined in Eq. (7.163). The predictions concerning an explicit dependence of the pre-exponential factor on g in the expression for E_C^* differ. Here we specify the result for the renormalized charging energy derived in the form [323]

$$E_C^*/E_C = g^2 f \exp(-g/2) \tag{7.169}$$

with $f \simeq \exp(\gamma) \simeq 1.781$. Although this pre-exponential factor f is not exact and in general $f = f(g, T)$, it turns out that the formula (7.169) yields a fairly good estimate for E_C^* in the interesting range of g. We also note that this result matches with the energy scale E^* (7.154) obtained from the RG analysis. Such an agreement between E^* and E_C^* could not be a priori expected, since these two quantities are derived within different approaches, which may in general yield different energy scales. In addition, we remark that some other authors predict somewhat stronger dependence of the pre-exponent in the expression for E_C^* on g, e.g., $\propto g^3$ [289, 394] or even stronger [236]. For instance, it was argued [289] that with the aid of quantum field theoretical methods, one can accurately evaluate the function $f(g, T)$ in the pre-exponent of Eq. (7.169) with the asymptotic result

$$f(g, 0) = g - 5\ln g - 5\gamma - \ln 2 - 3\pi^2/4 + o(1), \quad g \to \infty. \tag{7.170}$$

Also the predictions for an explicit form of $E_0(n_g)$ at n_g close to $e/2$ differ ranging from a completely flat band [323] to a nearly parabolic dependence [236].

All these minor discrepancies are, by no means, significant for us here. Of primary importance is the conclusion that Coulomb effects in normal tunnel junctions persist even at large junction conductances $g \gg 1$ and that Coulomb blockade becomes strong at energy scales comparable to (and lower than) E_C^*. From a fundamental viewpoint, this conclusion appears natural since charge discreteness in our system remains preserved even in the large conductance limit. It is also worth pointing out that charging effects in tunnel junctions were systematically observed experimentally [90, 211, 277] up to effective junction conductances as high as $g \simeq 33$. We will return to this important issue in Chapter 12.

From a practical point of view, exponential suppression of E_C^* (7.169) implies that in order to observe substantial charging effects in the limit of very large g, it is necessary to lower the temperature down to exponentially small values $T \lesssim E_C^*$. This condition naturally restricts the range of conductances at which the energy scale E_C^* remains physically meaningful. Of course, weak Coulomb blockade can also be observed at higher $T \gg E_C^*$, but in this case, the parameter E_C^* becomes irrelevant. Since experimental values of E_C realized in devices with highly conducting tunnel barriers are typically of order $E_C \sim 1 \div 4$ K, it follows immediately that the renormalized charging energy suppressed below $E_C^*/E_C \sim 10^{-3}$ would be of little experimental relevance because such low temperatures are hard to achieve in modern experiments. With the aid of Eq. (7.169), we conclude that for practical purposes it suffices to evaluate $E_C^*(g)$ at most up to $g \sim 30 \div 35$. At the same time, the asymptotic result (7.170) should be reliable only at much higher values of $g \gg 4\pi^2$.

Below we will demonstrate that this gap can be filled by means of numerical calculations. Combining both perturbative (7.115) and non-perturbative (7.169) results with those obtained numerically, one arrives at an accurate estimate for the renormalized charging energy covering the whole interesting range of junction conductances.

7.5.4 Numerical Results

Both the renormalized charging energy $E_C^*(g)$ and the dependence of the ground-state energy on the gate charge $E_0(n_g)$ were intensively studied in the literature by numerical methods, which allow to simultaneously address both perturbative and non-perturbative regimes. König and Schoeller [236] numerically integrated their real-time RG equations, while numerous other authors [71, 113, 175, 192, 194, 289, 393, 403] approached the problem by means of QMC simulations. Despite minor differences in certain details, all these numerical calculations fully support our key conclusions reached within the above analytic considerations. Here we will present the results of an extended QMC analysis [192], which also contains the comparison with the results of other authors.

Combining Eq. (7.161) with Eqs. (7.86) and (7.87), we immediately observe that the effective charging energy $E_C^*(T)$ can be expressed in terms of the average

$$E_C^*(T) = 2\pi^2 T \langle m^2 \rangle_{n_g=0} , \tag{7.171}$$

Taking the limit $T \to 0$ in this relation, one recovers the value E_C^* evaluated above. QMC simulations for this and other relevant physical quantities are carried out by the standard discretization of the quantum paths into N (Trotter number) imaginary-time slices. As the temperature is reduced, the number of time-slices N has to increase as $1/T$ in order to preserve sufficient accuracy of the calculation. It was argued [192] that the value $N = 4E_C/T$ should typically be sufficient in order to reach convergence of QMC data in the interesting-for-us parameter range. The partition function was sampled for temperatures down to $T = E_C/500$. Such values of T provide the low-temperature limit $T \ll E_C^*(g)$ for all studied values of g except for the last point $g \approx 24$ in which case one has $T \lesssim E_C^*$. For more details, we refer the reader to the original literature on the subject; here we only present the corresponding results.

At high temperatures $T \gg E_C$, the grand partition function can be approximated semiclassically by means of Eq. (7.158) and the corresponding temperature-dependent charging energy $E_C^*(T)$ is defined in Eq. (7.162). Introducing the function $\tilde{E}_C(T) = \pi^2 T/\ln(4\pi^2 T/E_C^*(T))$ and employing Eq. (7.162), we obtain [394]

$$\tilde{E}_C(T) = \frac{E_C}{1 + (g/2\pi^2)E_C/T} \tag{7.172}$$

As the temperature is lowered, quantum fluctuations become more significant and $\tilde{E}_C(T)$ decreases. In Fig. 7.13, the QMC data points [192] are compared with the expression (7.172) (continuous lines) for several values of g. At all values of g, a good agreement between QMC and Eq. (7.172) results is observed up to $T \sim 5E_C$, thus confirming the validity of the semiclassical expression for $E_C^*(T)$ (7.162).

At lower temperatures, Eq. (7.172) becomes insufficient, dropping below the QMC results. Going to lower and lower values of T, one eventually observes saturation of the QMC data for $E_C^*(T)$ to a temperature-independent value $E_C^*(0)$, which defines the zero-temperature renormalized charging energy E_C^*. The corresponding QMC results for $E_C^*(0)$ [192] are displayed in Fig. 7.14 for different values of g and compared with the results of other authors.

Figure 7.14 demonstrates that the QMC results [192] (black symbols) and [194] (open triangles) are in a good agreement with each other, whereas the data points [393] (open squares) are systematically higher. The QMC results [192] for $E_C^*(0)$ also agree well with the results of the perturbation expansion (7.115) [174] and the real-time RG approach [236] for conductance values

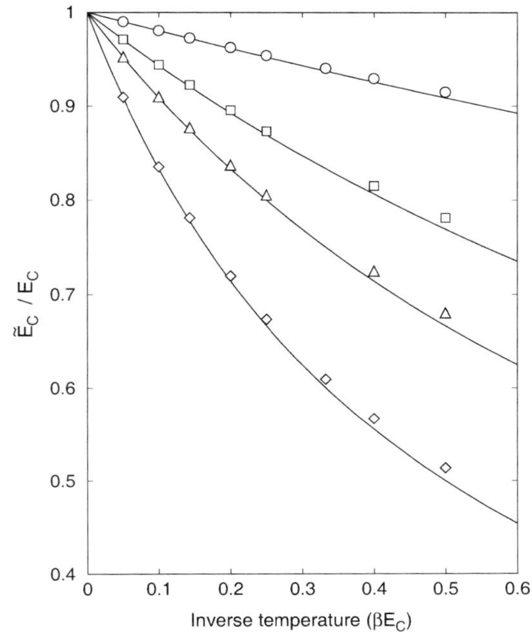

Figure 7.13 The QMC data [192] for $\tilde{E}_C(T)$ obtained at high temperatures and for several values of the parameter $\alpha_t = g/4\pi^2 = 0.1$ (circles), 0.3 (squares), 0.5 (triangles), and 1 (diamonds). Error bars are smaller than the symbol size. Solid lines correspond to Eq. (7.172).

$g \lesssim 10$. For larger conductances, the QMC data points deviate from these results and approach the instanton result [323] defined in Eq. (7.169).

The QMC method also allows us to determine the free energy of the system F as a function of the gate charge n_g. The QMC data [192] demonstrate that at high temperatures, $F(n_g)$ approaches the function $F(n_g) \simeq E_C^*(T)[1-\cos(2\pi n_g)]/(2\pi^2)$, while at lower T, the free energy band is closer to a parabolic shape, with some deviations mostly pronounced close to $n_g = 1/2$. The average number of excess electrons in the island $\langle n \rangle$ derived from Eq. (7.85) is displayed in Fig. 7.15 as a function of the gate charge n_g. In Fig. 7.15a, this function is presented at several temperatures for a dimensionless parameter $\alpha_t \equiv g/4\pi^2 = 0.1$. As expected, at high temperature, the

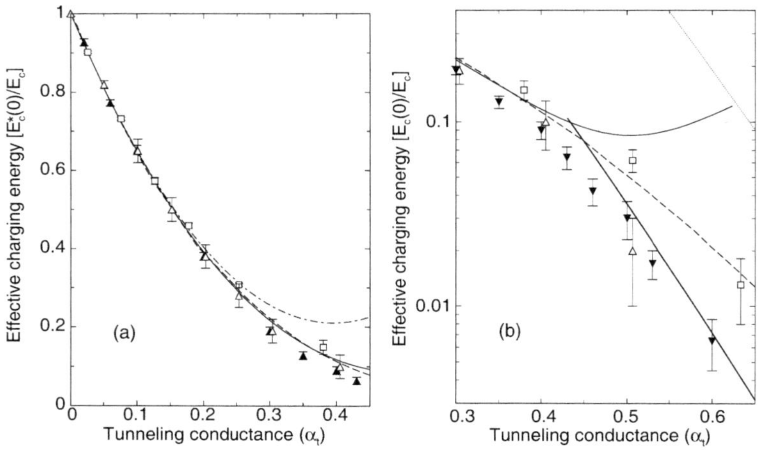

Figure 7.14 Low-temperature renormalized charging energy $E_C^*(0)$: (a) in the conductance region up to $\alpha_t = 0.4$ and (b) at higher conductances (semilogarithmic scale). Black symbols: QMC results [192], open symbols: triangles [194] and squares [393]. The dashed-dotted and continuous lines correspond to second- and third-order perturbation theory in g, respectively (see Eq. (7.115)). The dashed line represents the real-time RG results [236]. The bold and dotted lines in (b) correspond to non-perturbative results derived from instanton calculations [323] (Eq. (7.169)) and [394], respectively.

average charge follows closely the result (7.160), while for lower T, deviations from this dependence become more pronounced. These QMC results are consistent with those of real-time RG calculations [236].

In Fig. 7.15b, we present the dependence $\langle n \rangle$ versus n_g for several values of the tunneling conductance, as obtained from QMC simulations [192] at $T = 0.03E_C$. We observe that at this temperature for the largest $g \approx 12$ shown, the dependence is close to a linear one. These finite-temperature results compare well with those of QMC simulations in the charge representation [175] performed at a lower temperature $T = 10^{-4}E_C$.

For completeness, let us also mention about an earlier QMC calculation [71] investigating the correlator $\langle \cos(\varphi(\tau) - \varphi(0)) \rangle$ as a function of temperature at different values of g. A sharp crossover was observed at a temperature $T^*(g)$, which should be close to

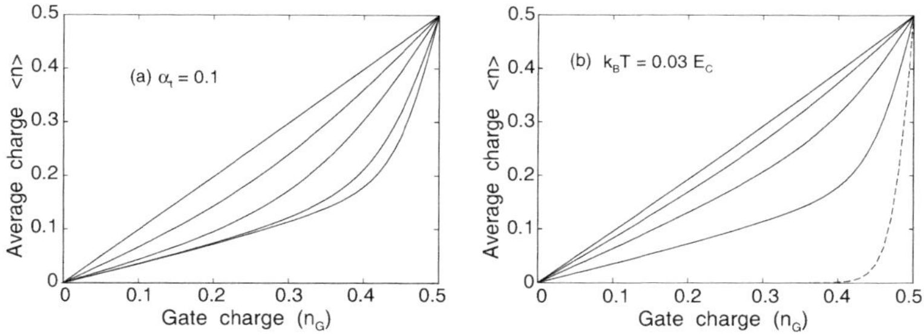

Figure 7.15 The QMC results [192] for the average number of excess electrons $\langle n \rangle$: (a) For $\alpha_t = 0.1$ at various temperatures $T/E_C = 0.6, 0.2, 0.1, 0.04$, and 0.03 (from top to bottom); (b) At $T = 0.03 E_C$ for several values of $\alpha_t = 0.4, 0.3, 0.2$, and 0.1 (from top to bottom). For comparison, the average charge for $\alpha_t = 0$ is also presented (dashed line).

the renormalized charging energy, i.e., $T^*(g) \sim E_C^*$. A comparison between these two quantities [426] revealed a very good agreement of the data for $T^*(g)$ [71] with the result of the instanton calculation [323] (7.169) and, hence, also with the QMC data [192] in the moderately strong tunneling regime.

More recently, a detailed QMC analysis of the problem was carried out [289, 403] with the emphasis put on higher conductance values. These data allow to verify the validity of the exponential dependence of the renormalized charging energy on g (7.169) down to $E_C^*/E_C \sim 10^{-34}$ and to numerically establish the pre-exponential function $f(g, T)$ at higher values of g as compared to earlier calculations. At the same time, the QMC data points [289, 403] corresponding to dimensionless conductances $g \gtrsim 30 \div 35$ appear to be merely of a theoretical significance, since temperatures $T \lesssim 10^{-3} E_C$ are hardly attainable in modern experiments and the renormalized charging energy E_C^* at such high values of g is suppressed too strongly to be measured.

From the above analysis, we may conclude that the combination of the perturbative expansion (7.115), the instanton result (7.169), and the available numerical QMC data provide complete information about the renormalized charging energy E_C^* at low temperatures covering all relevant values of the junction conductance g. The QMC results for the average excess charge in the island $\langle n \rangle$ as a function of the gate charge $0 \leq n_{\mathrm{g}} \leq 0.5$ cover most of the temperature region actually studied in the experiments.

Finally, we would like to remind the reader that the non-perturbative results established here for the Coulomb blockade problem also have fundamental implications for the equivalent model of a quantum particle on a ring interacting with the Caldeira–Leggett bath, which we discussed in Chapter 3. In particular, exponential reduction of E_{C}^{*} (7.169) for $g \gg 1$ can be translated into exponential suppression of quantum coherent properties of the particle (3.50) for the ring radii R exceeding the decoherence length L_{φ} (3.43), which is set by interaction and does not depend on temperature at all.

Part B

Chapter 8

Quantum Particle in a Diffusive Electron Gas

The Caldeira–Leggett model considered in Chapter 1 yields the influence functional, which captures a number of generic features of quantum dissipative environments. In addition, the Caldeira–Leggett influence functional is Gaussian, thus being relatively easy to deal with in practical calculations. On the other hand, there exist many physical situations where Gaussian influence functionals become insufficient. Typically this is the case provided nonlinear effects are significant and yield new physics not contained within the Caldeira–Leggett-type of models.

The systems of this kind are diverse. For instance, in Chapter 2 we already went beyond the simple phenomenological model employing harmonic oscillators and carried out a microscopic derivation of the effective action for a Josephson tunnel junction. In that case, the collective quantum degree of freedom—the Josephson phase—was extracted "from interaction" and the role of an effective environment was played by electrons in metallic electrodes forming the junction. The resulting AES effective action serves as an example of non-Gaussian influence functional, which indeed describes new

Dissipative Quantum Mechanics of Nanostructures: Electron Transport, Fluctuations, and Interactions
Andrei D. Zaikin and Dmitry S. Golubev

ISBN 978-981-4774-50-5 (Hardcover), 978-0-429-29823-3 (eBook)
www.jennystanford.com

physical effects, such as the effect of Coulomb blockade in tunnel barriers considered in the previous chapter.

On the other hand, the microscopic action for tunnel barriers derived in Chapter 2 may also be related to the phenomenological Caldeira–Leggett type of models: In Chapter 3, we demonstrated a formal equivalence between the action for a normal tunnel barrier and that for a quantum particle on a ring in a bath of harmonic oscillators. Below in this book, we will further develop the idea of the influence functional by applying it to various problems of modern condensed matter theory. The corresponding analysis will, in general, require going much further from the simple Caldeira–Leggett model than we already did in the case of metallic tunnel barriers. However, it appears to be a reasonable price to pay for learning a lot of new physics and taking advantage of a unified and elegant approach of the influence functional.

8.1 Influence Functional from an Electron Gas

We will start by considering an electron gas, e.g., in a metallic conductor. Electrons in this gas fluctuate and cause random fluctuations of the electric potential V in the system. As the system remains neutral (negative electron charge is compensated by positively charged lattice ions), on average this potential vanishes, $\langle V \rangle = 0$. At the same time, the voltage–voltage correlator $\langle V V \rangle$ remains nonzero. In equilibrium, this correlator can be related to the dielectric function of the system $\epsilon(\omega, \boldsymbol{k})$ by means of the standard formula (see, e.g., [273])

$$\langle V V \rangle_{\omega,\mathbf{k}} = -\coth\frac{\omega}{2T}\mathrm{Im}\frac{4\pi}{\boldsymbol{k}^2\epsilon(\omega,\boldsymbol{k})} \tag{8.1}$$

Consider now a charged particle, e.g., a muon, propagating through this electron gas. Obviously, the fluctuating electric potential $V(t, \boldsymbol{r})$ will result in an additional random force acting on this particle. In this way, the electronic environment affects the particle dynamics. Provided the environment and, hence, this random force can be considered classical, Eq. (8.1) would be sufficient for a complete description of environmental effects. Here, however, we are generally aiming at dealing with quantum environments. Quantum mechanical description requires more efforts, since in this case the random electric potential V should be treated as a quantum operator. As before, the influence functional technique allows to formulate fully quantum mechanical description of the problem in a straightforward and elegant manner.

Let us introduce the total Hamiltonian for our system. It reads

$$\hat{H} = \hat{H}_0 + \hat{H}_{\rm el} + \hat{H}_{\rm int} \tag{8.2}$$

where

$$\hat{H}_0 = \frac{\hat{\boldsymbol{p}}^2}{2m} + U(\hat{\boldsymbol{r}}) \tag{8.3}$$

is the particle Hamiltonian (as usually, $\hat{\boldsymbol{p}}$ and $\hat{\boldsymbol{r}}$ are, respectively, momentum and coordinate operators), $\hat{H}_{\rm el}$ is the standard Hamiltonian for electrons in a metallic conductor, and

$$\hat{H}_{\rm int} = e\hat{V} \tag{8.4}$$

is the Coulomb term that describes interaction between the particle with charge e and fluctuating electrons in the environment. As before, we are interested in real-time evolution of the particle density matrix, which is described as a product of two transition amplitudes (1.13). The corresponding evolution operator kernel J is again expressed as a double path integral

$$J = \int \mathcal{D}\mathbf{r_F}\mathcal{D}\mathbf{r_B} \exp\left(\mathrm{i}S_0[\boldsymbol{r}_F] - \mathrm{i}S_0[\boldsymbol{r}_B]\right) F[\boldsymbol{r}_F, \boldsymbol{r}_B] \tag{8.5}$$

where

$$S_0 = \int_0^t dt' \left(\frac{m\dot{\boldsymbol{r}}^2}{2} - U(\boldsymbol{r})\right) \tag{8.6}$$

is the action of the particle described by the non-interacting part of the Hamiltonian (8.3) and

$$F[\boldsymbol{r}_F, \boldsymbol{r}_B] = \left\langle \exp\left[\mathrm{i}\int_0^t dt' \left(eV_F(t', \mathbf{r}_F) - eV_B(t', \boldsymbol{r}_B)\right)\right]\right\rangle_{V_F, V_B} \tag{8.7}$$

represents the influence functional, which accounts for the effect of the electron gas. As before, in contrast to the classical case, the *quantum* field on the forward part of the time contour $V_F(t, \boldsymbol{r})$ is to be distinguished from the one on the backward part $V_B(t, \boldsymbol{r})$. We also note that for the sake of simplicity starting from this chapter and throughout the rest of this book, we will set Planck's constant equal to unity $\hbar = 1$.

In order to evaluate the average over the quantum fields in Eq. (8.7), it is necessary to establish their correlators. We postpone this task to Chapter 9 where this issue will be addressed in details (see Eqs. (9.64)–(9.66)). Here we only present the final result, which takes the form

$$\langle V_i(t, \boldsymbol{r}) V_j(0, 0)\rangle = I(t, \boldsymbol{r}) + \frac{\mathrm{i}}{2}[(-1)^j R(t, \boldsymbol{r}) + (-1)^i R(-t, -\boldsymbol{r})] \tag{8.8}$$

where $i, j = F, B, (-1)^F = +1, (-1)^B = -1$, and

$$R(t, \boldsymbol{r}) = \int \frac{d\omega d^3k}{(2\pi)^4} \frac{4\pi}{k^2\epsilon(\omega, \boldsymbol{k})} e^{-\mathrm{i}\omega t + \mathrm{i}\boldsymbol{k}\boldsymbol{r}} \tag{8.9}$$

$$I(t, \boldsymbol{r}) = \int \frac{d\omega d^3k}{(2\pi)^4} \mathrm{Im}\left(\frac{-4\pi}{k^2\epsilon(\omega, \boldsymbol{k})}\right) \coth\frac{\omega}{2T} e^{-\mathrm{i}\omega t + \mathrm{i}\boldsymbol{k}\boldsymbol{r}} \tag{8.10}$$

These equations extend Eq. (8.1) to the quantum case.

For our purposes, it is sufficient to consider fluctuations of the fields $V_{F,B}$ Gaussian. Then averaging over these fields in Eq. (8.7) can easily be accomplished. With the aid of Eqs. (8.8), we again reproduce the general structure of the influence functional

$$F[\boldsymbol{r}_F, \boldsymbol{r}_B] = \exp(-\mathrm{i}S_{\mathrm{R}}[\boldsymbol{r}_F, \boldsymbol{r}_B] - S_{\mathrm{I}}[\boldsymbol{r}_F, \boldsymbol{r}_B]) \tag{8.11}$$

with

$$S_{\mathrm{R}} = \frac{e^2}{2}\int_0^t \big[R(t' - t'', \boldsymbol{r}'_F - \boldsymbol{r}''_F) - R(t' - t'', \boldsymbol{r}'_B - \boldsymbol{r}''_B) + R(t' - t'', \boldsymbol{r}'_F - \boldsymbol{r}''_B) - R(t' - t'', \boldsymbol{r}'_B - \boldsymbol{r}''_F)\big]\, dt'dt'', \tag{8.12}$$

$$S_{\mathrm{I}} = \frac{e^2}{2}\int_0^t \big[I(t' - t'', \mathbf{r}'_F - \boldsymbol{r}''_F) + I(t' - t'', \mathbf{r}'_B - \mathbf{r}''_B) - I(t' - t'', \boldsymbol{r}'_F - \boldsymbol{r}''_B) - I(t' - t'', \boldsymbol{r}'_B - \boldsymbol{r}''_F)\big]\, dt'dt'', \tag{8.13}$$

where we employed the short-hand notation $\boldsymbol{r}'_{F,B} = \mathbf{r}_{F,B}(t')$.

We observe that the above influence functional describing the environment formed by electron gas is considerably more complicated as compared to that for a bath of Caldeira–Leggett oscillators. In particular, both S_{R} and S_{I} are now essentially nonlinear in the particle coordinates $\boldsymbol{r}_{F,B}$. On the other hand, on a qualitative level, there exist clear similarities between the models. For instance, the function $\mathrm{Im}(-1/\epsilon(\omega, \boldsymbol{k}))$ generalizes the spectral density (1.54) in the Caldeira–Leggett model. In addition, from Eqs. (8.9), (8.10), it is easy to observe that the real and imaginary parts of the action (8.12) and (8.13) describing, respectively, dissipation and noise are again linked to each other by the FDT relation

$$I_{\omega,\boldsymbol{k}} = -\coth\frac{\omega}{2T}\mathrm{Im}\,R_{\omega,\boldsymbol{k}} \tag{8.14}$$

The above results for the influence functional are rather general and can be applied to electronic environments described by different dielectric functions $\epsilon(\omega, \boldsymbol{k})$. Of particular interest for us here is the model of the so-called "dirty electron gas," i.e., the gas formed by conduction electrons in a diffusive conductor. In this case, the system is described by the Drude model with dielectric function

$$\epsilon(\omega, \boldsymbol{k}) = 1 + \frac{4\pi\sigma_D}{-\mathrm{i}\omega + D\boldsymbol{k}^2} \tag{8.15}$$

where $\sigma_\mathrm{D} = 2e^2 N_0 D$ is the Drude conductivity of a metal, $N_0 = mp_\mathrm{F}/2\pi^2$ is the electron density of states at the Fermi level, $D = v_\mathrm{F}\ell/3$ is the electron diffusion coefficient, and ℓ is the electron elastic mean free path. Equation (8.15) applies at frequencies $\omega \ll v_\mathrm{F}/\ell$, where v_F is the Fermi velocity for electrons. Hence, the frequency $\sim v_\mathrm{F}/\ell$ effectively provides the high-frequency cutoff for the dirty electron gas model similarly to that set by ω_c (1.54) in the Caldeira–Leggett model.

Finally, we should point out one significant limitation of the above derivation. Namely, the particle propagating in the electronic environment should not be identical to electrons forming this environment. The reason for this limitation is clear: Our derivation does not account for Pauli exclusion principle and, hence, in general cannot apply to electrons propagating in the electron gas. Is it possible to generalize our derivation in order to formulate yet "more clever" influence functional that would automatically keep track of Fermi statistics? Fortunately, the answer to this question is positive. We will postpone a detailed analysis of this important issue to the next chapter.

8.2 Particle on a Ring: Persistent Current and Decoherence

It is well known that quantum coherence is one of the most fundamental properties of quantum mechanical systems. In meso- and nanostructures at sufficiently low temperatures, quantum coherence of electron wave functions may persist at distances comparable with the system size, thus making the system properties markedly different from those observed at higher temperatures.

According to general principles, quantum coherence of the wave function is maintained during its evolution and it cannot be destroyed due to elastic scattering on a static external potential. On the other hand, as we have already demonstrated in Chapter 3, interactions with a quantum dissipative environment may dephase the particle wave function even at $T \to 0$, i.e., if the whole interacting system is in its true quantum mechanical ground state. While this conclusion was obtained for the particular example of Caldeira–Leggett type of environment, the question remains if a similar effect can be observed in other models. Here we address this issue for a dissipative environment formed by a diffusive electron gas already discussed in Section 8.1. Coulomb interaction between a charged particle and such dissipative environment to a large extent mimics electron–electron interactions in realistic disordered conductors and, hence, allows to understand a number of important features of the latter systems within a relatively simple theoretical framework.

It is obvious that for a system in thermodynamic equilibrium, quantum dephasing cannot be associated with any kind of decay or relaxation in real time; rather, it manifests itself via suppression of off-diagonal elements of the electron density matrix beyond a certain length L_φ, as we already demonstrated in Chapter 3. Provided there exists nonzero electron dephasing due to its interaction with quantum environment even at $T \to 0$, this dephasing length L_φ should stay finite down to zero temperature. This, in turn, would imply that effects sensitive to quantum coherence, such as persistent currents and Aharonov–Bohm oscillations in meso- and nanorings, should be suppressed by interactions at any temperature, including $T = 0$, if the perimeter of the ring exceeds L_φ. In order

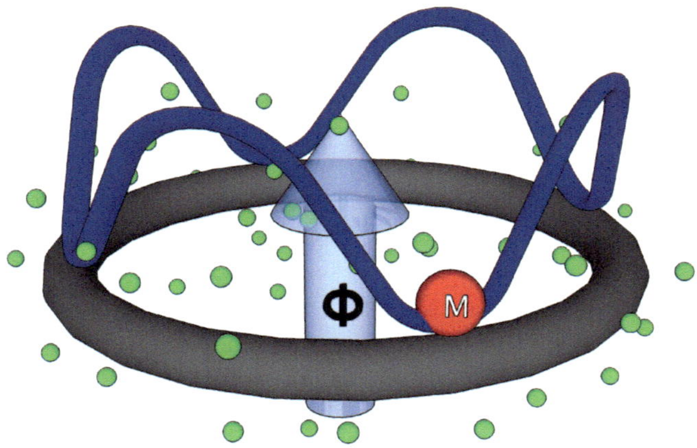

Figure 8.1 The system under consideration: A particle on a ring in the presence of a periodic potential. The ring is pierced by the magnetic flux and the particle interacts with an effective environment formed by a diffusive electron gas.

to better understand the effect of decoherence by electron–electron interactions, Guinea [181] proposed to study these effects within the framework of the model, which we will explicitly specify below.

8.2.1 Effective Action

Similar to Chapter 3, let us again consider a quantum particle with mass m and electric charge e on a ring with radius R threaded by external magnetic flux Φ_x (see Fig. 8.1). We will assume that the ring is embedded in the environment of the "dirty electron gas" described by the Drude-type dielectric function (8.15). In this way, we will to a large extent mimic the effect of Coulomb interactions between electrons in a diffusive conductor: The particle on a ring "feels" exactly the same fluctuating electromagnetic potential $V(\mathbf{r})$ as any other electron with the coordinate $\mathbf{r}$ in this conductor.

Here we are merely interested in equilibrium properties of this system. Therefore, we will also assume that the whole system "particle+environment" is in equilibrium at a temperature T. The derivation of the influence functional for the environment model was already performed in the previous section. Here we will essentially follow the same track with two minor modifications. First, it will be convenient for us to reformulate our derivation within the equilibrium imaginary-time technique defining the grand partition function, which is better suited for calculation of equilibrium quantities, like persistent current. Second, we will account for the fact that the particle motion is confined to a ring by describing the particle position by a vector $\boldsymbol{r} = (R\cos\theta,\ R\sin\theta)$. As before, in what follows the angle θ will play the role of the quantum variable of interest. As compared to our previous treatment in Chapter 3, here we will assume that our particle moves in a periodic potential, which—just for the sake of definiteness—is chosen in the form $U(\theta) = U_0(1 - \cos(\kappa\theta))$. Here κ is the total number of periods of the potential $U(\theta)$, which the particle should pass before it makes one full circle along the ring. Accordingly, in this case the Hamiltonian (8.3) can be written in the form

$$\hat{H}_0 = \frac{E_C(\hat{\Phi} + \Phi_x)^2}{\Phi_0^2} + U_0(1 - \cos(\kappa\theta)) \tag{8.16}$$

where $\hat{\Phi} = -\mathrm{i}\Phi_0\partial/\partial\theta$ is the magnetic flux operator, $E_C = 1/(2mR^2)$ and $\Phi_0 = 2\pi c/e$ is the flux quantum.

The grand partition function $\mathcal{Z}$ of our system can be expressed via the path integral over the θ-variable

$$\begin{aligned} \mathcal{Z} &\equiv \sum_{n=-\infty}^{\infty} e^{2\pi \mathrm{i} n\phi_x}\,\mathcal{Z}_n \\ &= \sum_{n=-\infty}^{\infty} \int_0^{2\pi} d\theta_0 \int_0^{2\pi n} \mathcal{D}\theta \exp(\mathrm{i}2\pi n\phi_x - S_0[\theta] - S_{\mathrm{int}}[\theta]) \end{aligned} \tag{8.17}$$

Here we defined $\phi_x = \Phi_x/\Phi_0$ and the winding-number-projected partition functions $\mathcal{Z}_n$.

The first term in the exponent containing ϕ_x takes care of the magnetic flux. Note that this topological term is exactly equivalent to that already encountered in Chapter 6 in the case of Josephson junctions if we identify the normalized flux ϕ_x with the normalized external charge $Q_x/2e$ (cf., e.g., Eq. (6.45)). The second term

$$S_0[\theta] = \int_0^{\beta} d\tau \left[\frac{1}{4E_C}\left(\frac{\partial\theta}{\partial\tau}\right)^2 + U_0(1 - \cos(\kappa\theta))\right] \tag{8.18}$$

defines the action for a particle in the absence of the environment. For $\kappa = 1$, $U_0 \leftrightarrow E_J$, and $\theta \leftrightarrow \varphi$, the action (8.18) is identical to one for the Josephson junction where E_C would then play the role of the "superconducting" charging energy $E_C \rightarrow (2e)^2/2C$. In Chapter 20, we will also demonstrate that the same action (8.18) with $\kappa = 1$ also describes superconducting nanorings in the presence of quantum phase slips.

Finally, the term S_{int} accounts for the effect of interaction between the particle and the electromagnetic field V produced by fluctuating electron gas. Averaging the interacting term (8.4) over this fluctuating field, one finds

$$S_{\text{int}} = -\ln\left\langle \exp\left(-\mathrm{i}\int_0^\beta d\tau eV(\tau, \theta(\tau))\right)\right\rangle_V \tag{8.19}$$

This equation represents the imaginary-time analogue of the influence functional (8.7). Provided fluctuations of the field V are sufficiently well described within the Gaussian approximation, we obtain

$$S_{\text{int}} = \frac{e^2}{2}\int_0^\beta d\tau \int_0^\beta d\tau' \langle V(\tau, \theta(\tau))V(\tau', \theta(\tau'))\rangle \tag{8.20}$$

where the correlator $\langle VV\rangle$ can be expressed via the dielectric susceptibility of the environment $\epsilon(\omega, \boldsymbol{k})$ as:

$$\langle VV\rangle = T\sum_{\omega_l}\int \frac{d^3k}{(2\pi)^3}\frac{4\pi}{k^2\epsilon(\mathrm{i}|\omega_l|, \boldsymbol{k})}e^{-\mathrm{i}\omega(\tau-\tau')+\mathrm{i}\boldsymbol{k}X} \tag{8.21}$$

where, as before, $\omega_l = 2\pi l T$ is the Matsubara frequency and $\boldsymbol{X} = \boldsymbol{r}(\tau) - \boldsymbol{r}(\tau')$. Employing the 3D diffusive electron gas model and combining Eqs. (8.15) and (8.21), after the integration we observe that the term $\sim Dk^2$ yields only an X-independent energy shift, which can be safely omitted. Integrating the remaining contribution $\sim|\omega_l|$ over $k \lesssim 1/\ell$ in Eq. (8.21), we obtain the result $\propto \min(1/X, 1/\ell) \approx 1/\sqrt{X^2+\ell^2}$. Defining $X = 2R\sin[(\theta(\tau)-\theta(\tau'))/2]$ and summing over ω_l, from Eqs. (8.20) and (8.21), we get [143]

$$S_{\text{int}}[\theta] = \alpha\int_0^\beta d\tau\int_0^\beta d\tau' \frac{\pi^2T^2K(\theta(\tau)-\theta(\tau'))}{\sin^2[\pi T(\tau-\tau')]} \tag{8.22}$$

$$K(\theta) = 1 - \frac{1}{\sqrt{4r^2\sin^2(\theta/2)+1}} \tag{8.23}$$

where we introduced $\alpha = 3/(8k_F^2\ell^2)$ and $r = R/\ell$. The integral in Eq. (8.22) is understood as a principal value. The divergence at $\tau = \tau'$ is then regularized in a standard manner by requiring $K(0) = 0$, which explains the origin of the first term in (8.23). Note that it can also be convenient to rewrite the function K in terms of the Fourier series [181]

$$K = \sum_k a_k \sin^2\left[\frac{k(\theta(\tau) - \theta(\tau'))}{2}\right] \tag{8.24}$$

with Fourier coefficients $a_k \sim (2/\pi r)\ln(r/k)$ for $1 \leq k \lesssim r$ and $a_k \approx 0$ otherwise. We would also like to point out that the term with $k = 1$ has exactly the same form as the analogous dissipative contribution in the Caldeira–Leggett model or the part of the action for normal tunnel barriers. Thus, in a certain sense, the above effective action can be viewed as a generalization of the above models.

What remains is to establish the validity conditions for our model. The Gaussian approximation (8.20) in a 3D diffusive conductor is known to work sufficiently well provided this conductor is not too strongly disordered, i.e., if the electron elastic mean free path remains not too short, $k_F\ell \gg 1$. This condition also implies that our effective interaction constant α is small,

$$\alpha \ll 1 \tag{8.25}$$

i.e., Coulomb interaction remains weak. In addition, we should request the distance between the neighboring potential minima for the particle on a ring to be much larger than the electron mean free path in a diffusive conductor, i.e., $2\pi R/\kappa \gg \ell$. Accordingly, the parameter $r = R/\ell$ obeys the inequality

$$r \gg \kappa/2\pi \tag{8.26}$$

8.2.2 Persistent Current in the Absence of Interactions

It is well known that as soon as nonzero magnetic flux is applied to the ring, the system develops a circulating current I, which does not decay in time. The presence of such persistent current in the ring is a fundamental consequence of quantum coherence. This effect was intensively studied both theoretically and experimentally in a large number of metallic and semiconductor rings. For a brief review of this very interesting field, we could refer the reader, e.g., to the paper [108]. Our aim here is not to discuss various details of this phenomenon in non-interacting systems but specifically address the effect of Coulomb interaction within the model outlined above in this chapter. We also note that we already briefly dealt with the phenomenon of persistent currents in Chapter 3 when we considered a quantum particle on a ring interacting with the Caldeira–Leggett bath. Here we will present a more detailed treatment of this effect for a more general model of the fluctuating electromagnetic environment.

Persistent current can be determined from the general formula

$$I = -\frac{eT}{2\pi}\frac{\partial}{\partial\phi_x}\ln\mathcal{Z} \tag{8.27}$$

which always yields *diamagnetic* current response for the problem in question.

Let us first briefly address the non-interacting limit $S_{\text{int}} \to 0$ and evaluate persistent current in two specific cases: (a) vanishingly small external potential $U_0 \to 0$ and (b) the opposite limit of strong external potential

$$U_0 \gg \kappa^2 E_C \tag{8.28}$$

Note that provided U_0 differs from zero and the parameter κ does not depend on R, the condition (8.28) is automatically satisfied for sufficiently large values of R simply because $E_C \propto 1/R^2$.

In the case (a), S_0 is quadratic in θ and the partition function (8.17) is trivially evaluated. Just as in Eq. (6.48), we obtain

$$\mathcal{Z} = \sqrt{\frac{\pi T}{E_C}}\sum_{n=-\infty}^{\infty}\exp\left(\mathrm{i}2\pi n\phi_x - \frac{\pi^2 n^2 T}{E_C}\right) = \sum_{p=-\infty}^{\infty}\exp(-E_p(\phi_x)/T) \tag{8.29}$$

where

$$E_p(\phi_x) = E_C(p+\phi_x)^2 \tag{8.30}$$

For the flux-dependent part of the ground-state energy at $-0.5 < \phi_x \leq 0.5$, one finds $E_0(\phi_x) = E_C\phi_x^2$. In the limit $T \to 0$, the value of persistent current is trivially obtained from the formula

$$I = \frac{e}{2\pi}\frac{\partial E_0(\phi_x)}{\partial\phi_x} \tag{8.31}$$

which yields (cf. Eq. (3.48))

$$I = I_{C0}\frac{\partial}{\partial\phi_x}\min_n\frac{(\phi_x - n)^2}{2}, \qquad I_{C0} = eE_C/\pi, \tag{8.32}$$

i.e., at $T = 0$ the flux dependence of I is described by a periodic saw-tooth function with the period equal to the flux quantum.

At nonzero temperatures not only the lowest, but also higher energy levels contribute to the current. The amplitude of persistent current decreases with increasing T and the function $I(\phi_x)$ becomes smoother. In the limit $T \gg E_C/\pi^2$ in Eq. (8.29), it suffices to keep only the terms with $n = 0, \pm 1$, which give the dominant contribution. Then from Eq. (8.27), we obtain

$$I = I_{C0}(T)\sin(2\pi\phi_x), \tag{8.33}$$

where

$$I_{C0}(T) = 2eT\exp\left(-\frac{\pi^2 T}{E_C}\right) \tag{8.34}$$

Let us now turn to the tight binding limit (8.28). In this case, at sufficiently low temperatures, the particle is located at the bottom of one of the potential wells (see Fig. 8.1), i.e., in the vicinity of the points $\theta = 2\pi p/\kappa$, where $0 < p \le \kappa$ is an integer number. Accordingly, in Eq. (8.17) one should substitute

$$\int_0^{2\pi} d\theta_0 \to \sum_{p=1}^{\kappa}\int d\theta_0\delta(\theta_0 - 2\pi p/\kappa) \tag{8.35}$$

The particle can move around the ring only by hopping between the neighboring minima $\theta = 2\pi p/\kappa$ and $\theta = 2\pi(p\pm 1)/\kappa$ of the periodic potential $U(\theta)$. As we already discussed in Chapter 5, each of these tunneling events is described by the instanton (kink) trajectory

$$\tilde{\theta}(\tau) = \frac{4}{\kappa}\arctan(\exp(\omega_0\tau)), \tag{8.36}$$

which corresponds to the tunneling amplitude $\Delta/2$, where

$$\Delta = 8\left(\frac{\kappa U_0 E_C}{\pi}\right)^{1/2}\left(\frac{2U_0}{E_C}\right)^{1/4}\exp\left(-\frac{4}{\kappa}\sqrt{\frac{2U_0}{E_C}}\right) \tag{8.37}$$

and $\omega_0 = \kappa\sqrt{2U_0E_C}$. In order to evaluate the grand partition function

$$\mathcal{Z} \sim \sum_{p=1}^{\kappa}\langle p|e^{-\beta\hat{H}_0}|p\rangle \tag{8.38}$$

it is necessary to sum over all possible tunneling events of the particle between all potential minima to all orders in Δ. The minimum number of such hops should be equal to $n\kappa$ for any trajectory corresponding to the winding number n. Taking into account that effective duration of each tunneling event is $\sim\omega_0^{-1}$ and that the total imaginary-time span equals to $\beta \equiv 1/T$, we can distinguish two different limits. In the limit $n\kappa\omega_0^{-1} \ll \beta$, the average distance between instantons is large as compared to their typical size, i.e., in this case we are dealing with a dilute instanton gas. In the opposite limit $n\kappa\omega_0^{-1} \gg \beta$, instantons are very close to each other and essentially merge forming a single trajectory [352].

Here we will merely be interested in the dilute instanton gas limit $T \ll \omega_0/\kappa$. To proceed, let us consider multi-instanton trajectories

$$\Theta(\tau) = \frac{2\pi p_1}{\kappa} + \sum_j \nu_j \tilde{\theta}(\tau - \tau_j), \tag{8.39}$$

where $\nu_j = \pm 1$ and τ_j are, respectively, the topological charges and collective coordinates of instantons and $\tilde{\theta}(\tau)$ is defined in Eq. (8.36). The trajectory (8.39) describes the particle between the states $|p\rangle$ after n winds around the ring, provided we fix

$$\sum_j \nu_j = n_2 - n_1 = n\kappa \tag{8.40}$$

i.e., we consider configurations containing totally $n_1 + n_2$ instantons corresponding to n_1 hops clockwise and n_2 hops counterclockwise. Taking into account all possible tunneling events restricted by the condition (8.40) and summing over all winding numbers n, we obtain

$$\langle p|e^{-\beta \hat{H}_0}|p\rangle = \left(\frac{\omega_0}{\pi}\right)^{1/2} \sum_{n=-\infty}^{\infty} e^{2\pi i n\phi_x - \frac{\beta\omega_0}{2}} \sum_{n_1, n_2=0}^{\infty} \frac{(\beta\Delta/2)^{n_1+n_2}}{n_1! n_2!} \delta_{n_2-n_1, n\kappa}$$

Making use of the integral representation for the Kronecker symbol

$$\delta_{j,k} = \int_0^{2\pi} \frac{dy}{2\pi} e^{iy(j-k)} \tag{8.41}$$

after performing the summation over n with the aid of Poisson's resummation formula

$$\sum_{n=-\infty}^{\infty} e^{2\pi i n x} = \sum_{k=-\infty}^{\infty} \delta(x - k) \tag{8.42}$$

we get

$$\mathcal{Z} \sim \left(\frac{\omega_0}{\pi\kappa^2}\right)^{1/2} e^{-\frac{\beta\omega_0}{2}} \sum_{j=1}^{\kappa} e^{\beta\Delta\cos\left(\frac{2\pi(j+\phi_x)}{\kappa}\right)} \tag{8.43}$$

The ground-state energy of the particle, $E_0(\phi_x)$, is obtained by taking the limit $T \to 0$ in Eq. (8.43), which yields

$$E_0(\phi_x) = \frac{\omega_0}{2} - \Delta\cos\left(\frac{2\pi\phi_x}{\kappa}\right) \tag{8.44}$$

for $-1/2 < \phi_x < 1/2$. Equation (8.44) should be continued periodically outside this interval. In the limit $\kappa \gg 1$, this dependence reduces to a set of parabolas $(2\pi^2\Delta/\kappa^2)\min_n(\phi_x - n)^2$.

Combining Eqs. (8.43) and (8.27), we obtain

$$I = \frac{e\Delta}{\kappa}\frac{\sum_{j=1}^{\kappa}\sin\left(\frac{2\pi(j+\phi_x)}{\kappa}\right)e^{\beta\Delta\cos\left(\frac{2\pi(j+\phi_x)}{\kappa}\right)}}{\sum_{j=1}^{\kappa}e^{\beta\Delta\cos\left(\frac{2\pi(j+\phi_x)}{\kappa}\right)}} \tag{8.45}$$

This expression fully determines the value of persistent current in the limit (8.28) at temperatures $T \ll \omega_0/\kappa$.

In the low-temperature limit $T \ll \Delta/\kappa^2$, Eq. (8.45) reduces to a simple formula

$$I = \frac{e\Delta}{\kappa}\sin\left(\frac{2\pi\phi_x}{\kappa}\right), \qquad -1/2 < \phi_x \leq 1/2 \tag{8.46}$$

which also trivially follows from Eq. (8.44). This formula demonstrates that at $T = 0$, persistent current is proportional to Δ and its flux dependence deviates from the simple sinusoidal form for all $\kappa > 1$. In particular, in the limit $\kappa \gg 1$, this dependence approaches a saw-tooth one (8.32) with

$$I_{C0} = 2\pi e\Delta/\kappa^2 \tag{8.47}$$

At higher temperatures $T \gg \Delta/\kappa^2$, the leading contribution to the partition function is defined by configurations with minimal number of instantons. Hence, in the sum over winding numbers in Eq. (8.17), it is sufficient to keep only the terms with $n = 0, \pm 1$. For $n = 0$, it is necessary to sum over all configurations, whereas at $n = \pm 1$ only configurations with κ instantons contribute. After some algebra, we find

$$\mathcal{Z}_0 = \kappa e^{-\beta\omega_0/2}\mathrm{I}_0(\beta\Delta), \qquad \mathcal{Z}_{\pm 1} = \kappa\frac{(\beta\Delta)^{\kappa}}{2^{\kappa}\kappa!}e^{-\beta\omega_0/2} \tag{8.48}$$

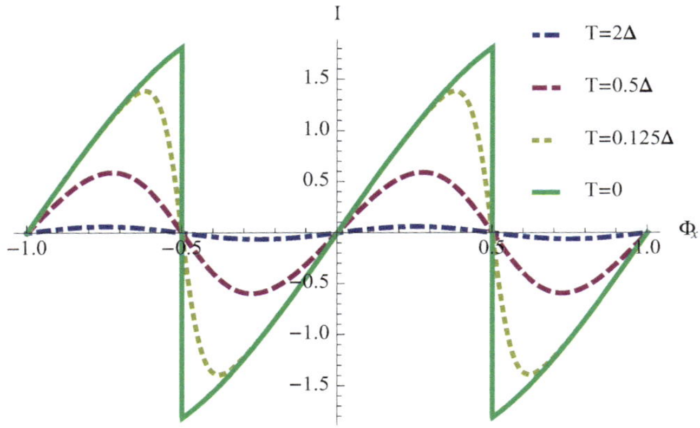

Figure 8.2 Persistent current I (measured in units $e\Delta/2\pi$) as a function of the magnetic flux ϕ_x for $\kappa = 3$ and different temperatures.

where $\mathrm{I}_0(x)$ is the modified Bessel function of imaginary argument. As a result, we again arrive at Eq. (8.33), where now

$$I_{C0}(T) = \frac{e\Delta^{\kappa}}{(2T)^{\kappa-1}\kappa!\mathrm{I}_0\left(\Delta/T\right)} \tag{8.49}$$

The current-flux dependencies at different values of T are depicted in Fig. 8.2.

Comparing now both limits (a) and (b), we conclude that the main physical difference between them is the presence of two distinct energy scales—ω_0 and Δ—in the case (b) instead of one energy scale E_C in the case (a). Otherwise, all significant features are the same in both limits. Indeed, at higher temperatures, the current–flux dependence is sinusoidal with period Φ_0 and its amplitude decreases with increasing T (cf. Eqs. (8.34) and (8.49)). In the low-temperature limit, the dependence $I(\phi_x)$ strongly deviates from a sinusoidal one (except for a special case $\kappa = 1$), Eqs. (8.32) and (8.46). The dependence of I_{C0} on the ring size is $I_{C0} \propto 1/R^2$ in the limit (a), while in the limit (b), Eqs. (8.47) and (8.37) yield exponential decay of I_{C0} with increasing R (since $E_C \propto 1/R^2$) provided the parameter κ does not change with R. However, if one chooses to keep the distance between the adjacent potential minima unchanged while increasing R, one should set $\kappa \propto R$. Under this condition, one would get $I_{C0} \propto 1/R^2$ exactly as in the case (a). Due to Eq. (8.28), this regime can apply only at not-too-large values of R.

8.2.3 Perturbation Theory

Let us now analyze the effect of Coulomb interactions. With Eq. (8.25) in mind, one could try to argue that interaction effects should be sufficiently weak and, hence, could be treated perturbatively in the effective coupling constant α. Following this conjecture, it would then be sufficient to expand the partition function (8.17) to the first order in $S_{\rm int}$ (8.22). After that in the absence of the periodic potential (case (a)), the calculation would reduce to a simple Gaussian integration over the θ-variable in each term of the series. Performing this integration and summing over n, one easily evaluates the first-order interaction correction to the ground-state energy $E_0(\phi_x)$ and finds

$$I = I_{C0}\left[\phi_x - \frac{\alpha}{2}\sum_{n=1}^{r} na_n \ln\left(\frac{n+2\phi_x}{n-2\phi_x}\right)\right] \qquad (8.50)$$

where $-0.5 < \phi_x \leq 0.5$. The last term in Eq. (8.50) represents the first-order Coulomb correction to the current at $T = 0$. This correction is negative, i.e., Coulomb interaction tends to suppress persistent current even in the ground state. For $\phi_x \ll 1$, the interaction correction in Eq. (8.50) turns linear in ϕ_x and small in the parameter $\alpha\sum_{n=1}^{r} a_n \sim \alpha \ll 1$ as compared to the non-interacting term. However, in the immediate vicinity of half-integer flux values $\phi_x = \pm 1/2, \pm 3/2, \ldots$, this correction becomes large. The logarithmic divergence in Eq. (8.50) originates from the fact that at such values of ϕ_x, the two lowest energy levels become very close to each other and the perturbation theory in the interaction fails already in the first order even at small values of the effective coupling constant $\alpha \ll 1$.

A similar behavior is observed in the strong potential limit (b), Eq. (8.28). In this case, the calculation is somewhat more tedious and for that reason is omitted here. At $T \to 0$, the result reads [352]

$$I = I_C\left[\sin\left(\frac{2\pi\phi_x}{\kappa}\right) - \alpha\sum_{n=-r}^{r} a_n \sin\left(\frac{\pi n}{\kappa}\right)\cos\left(\frac{2\pi\phi_x + \pi n}{\kappa}\right)\right.$$
$$\left.\times \ln\left|\sin\left(\frac{\pi n}{\kappa}\right)\sin\left(\frac{2\pi\phi_x + \pi n}{\kappa}\right)\right|\right] \qquad (8.51)$$

We again observe that the first-order α terms in Eq. (8.51) become singular at half-integer values of ϕ_x, signaling the failure of the first-order perturbation theory, at least for such values of ϕ_x.

Higher-order terms of the perturbation theory in the interaction may contain similar (or even stronger) singularities and, on top of that, may grow with increasing r. Yet another effect is the r-dependent renormalization of Δ and, hence, the current I_{C0}. This effect will be considered below. Here we only quote the result obtained within the first-order perturbation theory in α:

$$I_C \simeq I_{C0}(1 - 4\pi\alpha r/\kappa) \tag{8.52}$$

This result also clearly indicates insufficiency of a perturbative treatment of Coulomb interaction at large values of r.

All these considerations allow us to conclude that—contrary to our initial expectations—perturbation theory in α cannot, in general, be trusted, particularly in the low-temperature limit and at sufficiently large r. Hence, even in the limit of small effective coupling constants α, it is necessary to go beyond the perturbative treatment and to include non-perturbative effects into our consideration.

8.2.4 Non-perturbative Effects: No Periodic Potential

We start from the case (a), i.e., in Eq. (8.18) we set $U_0 = 0$. Let us first consider the regime of sufficiently high temperatures. In this case, the partition function (8.17) can be evaluated semiclassically. The classical trajectory providing the minimum of the total action $S_0 + S_{\text{int}}$ for a winding number n is represented by the so-called "straight line" path $\theta_{\text{cl}}(\tau) = 2\pi n T \tau$. This trajectory corresponds to n complete rotations of the particle around the ring. Substituting the paths $\theta_{\text{cl}}(\tau)$ into Eq. (8.17), one gets

$$\mathcal{Z} \sim \sum_{n=-\infty}^{\infty} \exp\left(\mathrm{i}2\pi n\phi_x - \frac{\pi^2 n^2 T}{E_C} - 4\pi|n|\alpha r\right) \tag{8.53}$$

As before, in the limit $T \gg E_C/\pi^2$, it suffices to keep only the terms with $n = 0, \pm 1$. Then from Eq. (8.27), one obtains

$$I = I_C(T)\sin(2\pi\phi_x) \tag{8.54}$$

where

$$I_C(T) = I_{C0}(T)e^{-4\pi\alpha r} \tag{8.55}$$

and $I_{C0}(T)$ is defined in Eq. (8.34). We observe that in this temperature range in addition to the effect of thermal fluctuations, Coulomb interaction yields an extra T-independent term, which provides exponential suppression of I for ring perimeters $2\pi R$ exceeding the value

$$L_\varphi \sim \ell/\alpha \sim \ell(k_F\ell)^2 \tag{8.56}$$

Equation (8.56) defines the dephasing length for the problem in question. This length does not depend on temperature and is controlled by the effective interaction strength α.

As it is clear from Eq. (8.53), the above semiclassical analysis is well justified at sufficiently high temperatures, whereas at lower $T \lesssim E_C/\pi^2$, fluctuations around the classical paths $\theta_{\text{cl}}(\tau)$ become important and our treatment needs to be modified. Provided the suppression of persistent current is sufficiently strong, one can employ the instanton technique similar to one we already made use of in Chapter 7. As before, nontrivial saddle points $\tilde{\theta}(\tau)$, which describe quantum tunneling of the variable θ between different topological sectors, e.g., between the states $\theta = 0$ and $\theta = 2\pi$, are defined as solutions to the equation $\delta S/\delta\tilde{\theta} = 0$. For the action (8.22), (8.23), this equation is approximately satisfied for a wide class of sufficiently smooth functions obeying the boundary conditions $\tilde{\theta}(0) = 0$ and $\tilde{\theta}(\beta) = 2\pi$. Defining the typical instanton frequency Ω, one finds that for $\Omega \lesssim E_C$ and $T \ll E_C$, the total action $S[\tilde{\theta}(\tau)]$ does not depend on E_C and is equal to $S_{\text{int}}[\tilde{\theta}(\tau)] = 4\pi\alpha r$, while for $\Omega > E_C$, the instanton contribution gets suppressed by the kinetic energy term S_0. For $S_{\text{int}}[\tilde{\theta}(\tau)] \gg 1$, the tunneling amplitude between the states $\theta = 2\pi m$ and $\theta = 2\pi(m \pm 1)$ is exponentially reduced as $\propto \exp(-S[\tilde{\theta}(\tau)])$. Hence, in the limit of small $\phi_x \ll 1$, we obtain

$$I = I_C\phi_x, \quad I_C/I_{C0} = A(T)\exp(-4\pi\alpha r) \tag{8.57}$$

This result is justified provided $4\pi\alpha r \gg 1$ and the ratio I_C/I_{C0} is sufficiently small. It demonstrates that also at low $T \ll E_C$, persistent current is strongly suppressed by Coulomb interaction provided the ring perimeter $2\pi R$ exceeds the dephasing length L_φ

(8.56). Equation (8.57) indicates that at low T interaction-induced decoherence in our model is controlled by the parameter [143]

$$\alpha r \sim \alpha \sum_{k=1}^{r} k a_k \tag{8.58}$$

rather than by α. Hence, for $r \gg 1$, the decoherence effect of Coulomb interactions is indeed *much stronger* than it could naively be expected from first-order perturbation theory (Eq. (8.52)).

While the above instanton analysis is applicable for $4\pi\alpha\, r \gg 1$, it can hardly provide a quantitative description of the problem for moderate $\alpha r \sim 1$. Furthermore, even for large rings, rigorous analytic evaluation of the pre-exponent $A(T)$ in Eq. (8.57) is difficult because the action (8.22), (8.23) is strongly non-Gaussian and the standard techniques are hard to employ. On a qualitative level, one can expect that many non-Gaussian quasi-zero modes should yield a large pre-exponent $A(T)$, which increases upon decreasing T. Hence, eventually the instanton technique—at least in the form employed here—may become insufficient, and the result (8.57) may get modified in the limit of very low temperatures $T \to 0$.

In order to get additional information about the system behavior at arbitrary values of αr and at very low temperatures $T \to 0$, it is desirable to complement the above analysis with some independent consideration. Similar to the Coulomb blockade problem discussed in Chapter 7, one can also carry out extensive quantum Monte Carlo (QMC) simulations [143], which allow to directly compute persistent current as a function of α and r for the problem in question. The corresponding results are summarized in Fig. 8.3. In order to emphasize the analogy to the Coulomb blockade problem, similar to Chapter 7, these results are presented for the renormalized "charging energy" E_C^* related to persistent current as $E_C^* = \pi I_C/e$. At relatively high temperatures $T \gtrsim \alpha r E_C$, these QMC data fully confirm our semiclassical result (8.55). Also at lower T, one observes dramatic suppression of E_C^* with increasing r. At sufficiently large r, this suppression turns exponential and at $E_C/(4\pi\alpha r) \lesssim T \lesssim E_C$ is well described by Eq. (8.57) with $A(T) \approx \exp(cE_C/T)$ and $c \sim 1$. Thus, the QMC data displayed in Fig. 8.3 (left panel) confirm the validity of Eq. (8.57) also at temperatures well below E_C being fully consistent with our instanton results.

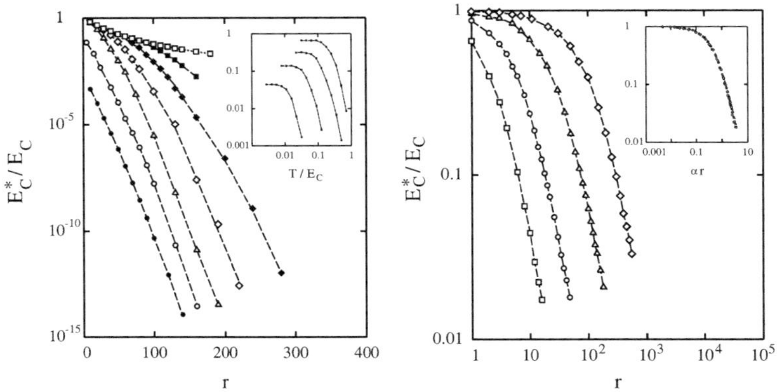

Figure 8.3 Left panel: The value $E_C^*/E_C = I_C/I_{C0}$ as a function of r for $\alpha = 0.019$ and different T. From bottom to top: $T/E_C = 1$, 0.5, 0.14, 0.06, 0.03, 0.016, and 0. The inset: E_C^*/E_C as a function of T. From top to bottom: $r = 10$, 30, 60, and 120. Right panel: E_C^*/E_C as a function of r at $T = 0$ (bottom to top: $\alpha = 0.21$, 0.07, 0.019, and 0.005). The inset: the same data for different α collapse onto one curve if plotted versus αr.

Note, however, that at very low temperatures, roughly at $T \lesssim E_C/4\pi\alpha r \ll E_C$ (assuming, of course, $4\pi\alpha r \gg 1$), the QMC data indicate that persistent current I_C decays slower with increasing r than it is suggested by Eq. (8.57). This observation is also in line with our expectations possibly indicating that very large number of quasi-zero modes in our problem makes the instanton result (8.57) unreliable at very low T.

The QMC data corresponding to the true zero-temperature limit $T = 0$ are displayed in Fig. 8.3 (right panel) for different values of the coupling constant α. At relatively small r, we observe very weak suppression of E_C^* in agreement with our perturbative result (8.50). However, at $\alpha r \gtrsim 1$, the ratio E_C^*/E_C rapidly decreases approximately as $E_C^*/E_C \propto r^{-\gamma}$ with the power γ numerically close to 2. Thus, the QMC data demonstrate strong suppression of persistent current by Coulomb interactions even exactly at $T = 0$ in rings with $2\pi R > L_\varphi$ as well as the failure of perturbation theory for such values of R. In other words, though exactly at $T = 0$, Eq. (8.57) is not supported by the QMC analysis, this analysis confirms that L_φ (8.56) remains the relevant scale down to zero temperature. It is also remarkable that the zero-temperature QMC data taken at different values of α essentially collapse onto one single curve if re-plotted as a function of αr instead of r; see the inset in the right panel of Fig. 8.3. This observation also demonstrates that at least for sufficiently large values of r, the actual parameter that controls the system behavior in the presence of interactions is indeed Eq. (8.58) rather than α.

Although the above physical picture already looks consistent and reliable, it should be pointed out that some other approaches, such as RG analysis [181] and variational calculation [196], yield weaker suppression of persistent current by Coulomb interaction than that established above. Without dwelling into further details, which could possibly account for such differences, we note that it would be desirable to verify the above results by yet another independent approach, which will be elaborated below.

8.2.5 Non-perturbative Effects: Strong Periodic Potential

Let us now consider the limit of a strong periodic potential (8.28). We have already learned that in the absence of interaction, key features of persistent current in this limit remain essentially the same as in the case (a). It is also clear that in both physical limits (a) and (b), the only possible source of the particle decoherence is its interaction with the fluctuating electromagnetic field $V(\boldsymbol{r})$ produced by a dissipative environment. Thus, for the same environment, it is natural to expect a similar decoherence effect regardless of the presence of any periodic potential for a particle on a ring. If so, similar interaction-induced suppression of persistent current should also be expected in both limits (a) and (b).

In order to evaluate the path integral in Eq. (8.17) in the case (8.28) and for $T \ll \omega_0/\kappa$, we will employ the same instanton technique as in the absence of interactions. As compared to the instanton analysis of the problem with $U_0 = 0$, in the limit (8.28), there exist no technical problems with taking into account fluctuations around the saddle point trajectory (8.36) at low temperatures, and the whole calculation remains well controlled down to $T = 0$.

Let us substitute the trajectory (8.39) describing quantum tunneling of the particle between different potential minima into the full effective action $S_0[\theta] + S_{\text{int}}[\theta]$. As before, we fix the winding number equal to n and again arrive at configurations of totally k instantons restricted by the condition (8.40). Evaluating the interaction term S_{int} (8.22) on multi-instanton trajectories (8.39) after some algebra, we obtain

$$
\begin{aligned}
S_{\text{int}}[\Theta(\tau)] &= 4\pi\alpha rk/\kappa \\
&- 2\alpha \sum_{a,b=1,\ a<b}^{k} \nu_a \nu_b g(\varphi_{ab}) \ln\left[\frac{\sin[\pi T(\tau_b - \tau_a)]}{\pi T \omega_0^{-1}}\right]
\end{aligned}
\tag{8.59}
$$

where

$$
g(\varphi) = K(\varphi + 2\pi/\kappa) + K(\varphi - 2\pi/\kappa) - 2K(\varphi) \tag{8.60}
$$

and

$$
\varphi_{ab} = \frac{2\pi}{\kappa}\sum_{j=a}^{b} \nu_j - \frac{\pi}{\kappa}(\nu_b + \nu_a) \tag{8.61}
$$

We observe that the interaction term (8.59) consists of two different contributions. One of them, $4\pi\alpha rk/\kappa$, describes interaction-induced suppression of quantum tunneling of the angle variable θ between different potential minima. This term yields effective r-dependent renormalization of the tunneling amplitude

$$
\Delta \to \Delta_r = \Delta e^{-4\pi\alpha r/\kappa} \tag{8.62}
$$

The remaining contribution in Eq. (8.59) describes logarithmic interaction between different instantons, which occurs for $\kappa \geq 2$ due to the presence of a dissipative environment. This logarithmic interaction is absent for $\kappa = 1$ in which case $g(\varphi) \equiv 0$.

At not-too-low temperatures $T \gg \Delta_r/\kappa^2$, inter-instanton interactions just provide further renormalization of the tunneling amplitude

$$
\Delta_r \to \Delta_r(1 + 2\alpha K(2\pi/\kappa)\ln(2\pi T/\omega)) \tag{8.63}
$$

One can also write down the corresponding RG equation

$$
\frac{d\Delta_r}{d\ln\omega} = 2\alpha K(2\pi/\kappa)\Delta_r \tag{8.64}
$$

which yields both r- and T-dependent renormalized tunneling amplitude of the form

$$
\Delta_R = \Delta_r\left(\frac{T}{\omega}\right)^{2\alpha K(2\pi/\kappa)} \tag{8.65}
$$

At even lower temperatures $T \ll \Delta_r/\kappa^2$, interactions again yield renormalization of the tunneling amplitude, which now becomes $\Delta_r \to \Delta_r \left(1 + 2\alpha K(2\pi/\kappa) \ln \frac{2\Delta_r e^{\gamma}}{\omega}\right)$. The RG equation (8.64) remains the same also in this case but the relevant energy scale now becomes

$$\Delta_R = \Delta_r \left(\frac{\Delta_r}{\omega}\right)^{\frac{2\alpha K(2\pi/\kappa)}{1-2\alpha K(2\pi/\kappa)}} \tag{8.66}$$

Equation (8.66), together with Eq. (8.62), defines the renormalized tunneling amplitude in the limit $T \to 0$ and for $\kappa \geq 2$. In the particular case $\kappa = 2$, our results reduce to those established for the spin-boson model with Ohmic dissipation, which we already addressed in details in Chapter 5.

We also note that Eq. (8.66) is formally applicable for $\alpha < 1/2K(2\pi/\kappa)$, while for larger values of the interaction strength α and at $T = 0$, the tunneling amplitude is renormalized to zero, $\Delta_R = 0$. This is the consequence of the quantum dissipative phase transition, which we have already encountered in Chapters 5 and 6. In the model under consideration, it occurs at the critical interaction strength $\alpha_c = 1/2K(2\pi/\kappa) \simeq 1/2$ and, as before, implies full suppression of tunneling between adjacent potential wells and localization of a quantum particle in one of these wells at any α exceeding the critical value α_c. Accordingly, no persistent current could flow at $T = 0$ and $\alpha \geq 1/2K(2\pi/\kappa)$. This conclusion, though formally correct, appears to be of little physical significance for us here, since the applicability of our model is restricted to small values of $\alpha \ll 1$.

Due to the absence of logarithmic interaction between instantons for $\kappa = 1$, the only effect of interactions in this case is r-dependent renormalization of the tunneling amplitude (8.62). Hence, for $\kappa = 1$ at all $T \ll \omega_0$, we arrive at Eq. (8.54) with temperature-independent

$$I_C = e\Delta e^{-4\pi\alpha r} \tag{8.67}$$

For $\kappa \geq 2$, instantons interact logarithmically, and in addition to Eq. (8.62), at not-too-low temperatures, the tunneling amplitude is renormalized according to Eq. (8.65). Combining these two equations and substituting the renormalized tunneling amplitude Δ_R into Eq. (8.49) instead of Δ, we again arrive at Eq. (8.54) where now

$$I_C(T) = I_{C0}(T)\left(\frac{T}{\omega_0}\right)^{2\alpha\kappa K(2\pi/\kappa)} e^{-4\pi\alpha r} \tag{8.68}$$

and $I_{C0}(T)$ is defined in Eq. (8.49). We again observe the same exponential suppression of persistent current at ring perimeters $2\pi R \gtrsim L_\varphi$. At the same time, the pre-exponent in the expression for I_C depends on temperature as a power law $I_C(T) \propto T^{-\mu}$ with $\mu = \kappa(1 - 2\alpha K(2\pi/\kappa)) - 1$. For $\kappa \geq 2$ and $\alpha \ll 1$, we have $\mu > 0$, i.e., $I_C(T)$ grows with decreasing temperature. This growth, though somewhat weaker than in the non-interacting case (8.49) (since $\mu < \kappa - 1$), implies that Eq. (8.68) can be trusted only at $T \gtrsim \Delta_R/\kappa^2$, whereas at even lower temperatures, one expects a crossover to a different regime where the renormalized tunneling amplitude Δ_R becomes temperature independent and is defined in Eqs. (8.66), (8.62).

It follows from these equations that exactly at $T = 0$ and at sufficiently large ring perimeters, persistent current is exponentially suppressed as

$$I_C \propto \exp\left(-\frac{4\pi\alpha r}{\kappa(1 - 2\alpha K(2\pi/\kappa))}\right) \tag{8.69}$$

i.e., in this case for any fixed value κ, we can define an effective zero-temperature dephasing length

$$\tilde{L}_\varphi = L_\varphi\kappa(1 - 2\alpha K(2\pi/\kappa)) \tag{8.70}$$

or simply $\tilde{L}_\varphi \approx l/\alpha\kappa$. Should, however, we assume that $\kappa \propto r$, no finite dephasing length could be formally defined from Eq. (8.69), although also in this case persistent current can suffer exponentially strong suppression by electron–electron interactions due to the condition (8.26).

In addition, the result (8.51) demonstrates that at $T \to 0$, the effect of electron–electron interactions on PC does not just reduce to renormalization of the tunneling amplitude $\Delta \to \Delta_R$. We observe that Eq. (8.51) also contains extra r-dependent terms evaluated here within the first-order perturbation theory in α. These extra terms become singular at half-integer values of ϕ_x and are not captured by the RG equation (8.64). Such terms, evaluated to all orders in α, might yield an additional dependence of persistent current on both the ring radius r and the flux ϕ_x not accounted for in the expression for Δ_R. This statement is illustrated by the exact solution presented below.

As we already pointed out, for $\kappa = 2$ our problem is exactly mapped onto the well-known spin-boson model with Ohmic dissipation. In order to non-perturbatively evaluate persistent current at all values of r and at all temperatures including $T = 0$, we can make use of the exact solution known for the particular value of the interaction strength $\alpha = 1/4$, the so-called Toulouse limit discussed toward the end of Section 5.1.3. Employing the exact expression for the grand partition function (5.54), from Eq. (8.27) one readily finds

$$I = \frac{e\Delta_r^2}{2\omega_0}\sin(2\pi\phi_x)\int\limits_{\frac{\pi T}{\omega_0}}^{\infty} dx\frac{e^{-\frac{x\Delta_r^2}{\omega_0 T}\cos^2(\pi\phi_x)}}{\sinh x} \tag{8.71}$$

This is the exact expression for persistent current in our problem established for $\kappa = 2$ and $\alpha = 1/4$. At not-too-low temperatures $\Delta_r \ll T \ll \omega_0$, we reproduce Eq. (8.54) with

$$I_C(T) = \frac{e\Delta^2}{2\omega_0}e^{-\pi r}\ln\frac{2\omega_0}{\pi T} \tag{8.72}$$

In low-temperature limit $T \to 0$, one finds

$$I = \frac{e\Delta^2}{2\omega_0}e^{-\pi r}\left[\pi r + \ln\left(\frac{\omega_0^2 e^{-\gamma}}{\pi\Delta^2\cos^2(\pi\phi_x)}\right)\right]\sin(2\pi\phi_x) \tag{8.73}$$

Both at nonzero temperatures and exactly at $T = 0$, the above exact expressions yet again confirm exponential suppression of persistent current at ring perimeters exceeding $L_\varphi = \tilde{L}_\varphi$ (these two length scales coincide for $\kappa = 2$ and $\alpha = 1/4$; see Eq. (8.70)). In addition, the result (8.73) demonstrates that at $T = 0$, the dependence of persistent current on r deviates from purely exponential while its dependence on ϕ_x deviates from purely sinusoidal and contains the logarithmic singularity at half-integer values of ϕ_x (cf. also Eq. (8.51)).

Thus, our main observations can be summarized as follows: (i) Coulomb interaction yields R-dependent renormalization of the tunneling amplitude (8.62) which, in turn, results in exponential suppression of persistent current at large-enough R; (ii) logarithmic interaction between instantons yields additional renormalization of the tunneling amplitude described by the RG equation (8.64); and (iii) Coulomb interactions generate yet additional contributions not captured by Eqs. (8.62) and (8.64) (cf. Eqs. (8.51) and (8.73)). These contributions may become particularly important at $T \to 0$ indicating the failure of the naive perturbation theory in α at sufficiently large R. We conclude that—similar to the limit $U_0 \to 0$—in the case (8.28), the decoherence effect of the "dirty electron gas" is controlled by the parameter αr (8.58) and yields exponential suppression of persistent current down to $T \to 0$ at ring perimeters exceeding the dephasing length defined in Eqs. (8.56) and (8.70).

8.2.6 Real-Time Analysis

In order to complete our analysis of the effect of the "dirty electron gas" environment on a quantum particle on a ring, let us now turn to the real-time calculation and evaluate the return probability for a pair of classical paths (3.25), which encircle the ring and return to the initial point after time t. As in Chapter 3, we distinguish two probabilities W_1 and W_2 corresponding, respectively, to two identical and two time-reversed paths. The quantity W_1 is the standard classical return probability, and W_2 describes quantum interference between time-reversed paths. The latter quantity is sensitive to quantum coherence and vanishes in the classical limit. Therefore, it would be of our particular interest to investigate the effect of the electronic environment on real-time dynamics on W_2.

Our further analysis closely follows that developed in Chapter 3. Without interactions, the time-reversal symmetry yields $W_1 = W_2 - K_0^2$. The effect of interactions with the diffusive electron gas is again accounted for by the influence functional (8.11). The form of this functional for a particle on a ring is readily established. Employing Eqs. (8.12), (8.13) and introducing the θ-variables, after a simple algebra we obtain

$$S_R[\theta_+, \theta_-] = \alpha \sum_{k=1}^{\infty} a_k k \int_0^t dt' \dot{\theta}_+(t') \sin(n\theta_-(t')) \qquad (8.74)$$

and

$$S_I[\theta_+, \theta_-] = -2\alpha \sum_{k=1}^{\infty} a_k \int_0^t dt' \int_0^t dt'' \frac{\pi T^2}{\sinh^2(\pi T(t' - t''))}$$
$$\times \cos(k(\theta_+(t') - \theta_+(t''))) \sin \frac{k\theta_-(t')}{2} \sin \frac{k\theta_-(t'')}{2} \qquad (8.75)$$

where $\theta_\pm$ were already defined after Eq. (3.35).

Proceeding semiclassically, we can restrict our attention to classical paths (3.25), which fully encircle the ring, exactly as we have already done in Chapter 3. By doing so, we will neglect the effect of the periodic potential (in case it exists) on these paths. This approximation amounts to assuming the particle kinetic energy to be sufficiently high, $mv^2/2 \gg U_0$. Substituting classical trajectories (3.25) into Eqs. (8.74) and (8.75), for two identical paths $\theta_F(t') = \theta_B(t') = \tilde{\theta}^{(n)}(t')$, we again find $S_{\mathrm{R}} = S_{\mathrm{I}} = 0$, i.e., within our semiclassical approximation, the probability W_1 is not affected by interactions at all. For a pair of time-reversed paths, one also has $S_R = 0$; however, the action S_I is now positive, $S_I > 0$. Inserting the time-reversed paths $\theta_F(t') = \theta_B(t - t') = \tilde{\theta}^{(n)}(t')$ into the action (8.75), we obtain

$$S_{In} = \frac{2\eta R^2}{\pi} \int \frac{zdz}{n^2} \frac{\sin^2(\pi z)}{(1 - (z/n)^2)^2} \coth\left(\frac{\pi z}{Tt}\right) \qquad (8.76)$$

At $T \to 0$, this expression yields

$$S_{In} \sim 4\pi\alpha r \qquad (8.77)$$

As in Chapter 3, the total return probability W is defined in Eq. (3.39). Combining this equation with our result (8.77), we conclude that at $T \to 0$, the probability W_2 is suppressed by interactions beyond the dephasing length (8.56) due to interaction between the particle and the bath of fluctuating electrons. This conclusion is fully consistent with the results derived within the imaginary-time analysis above in this chapter.

As it was already observed in the case of the Caldeira–Leggett model, dephasing of the particle wave function is produced by the environmental noise (accounted for by the term S_I in the action), whereas dissipation (described by the term S_R) turns out to be irrelevant for this effect, at least within the semiclassical analysis employed here. Loosely speaking, the wave function of a moving particle accumulates extra stochastic phase produced by the fluctuating electromagnetic field V. As a result, after some time, the information about its initial phase gets entirely lost. We also emphasize that this dephasing mechanism does not assume any energy losses for our particle; its average energy remains essentially conserved during the whole process.

Our real-time analysis also allows us to define the particle dephasing time τ_φ. For ballistic particle motion along the ring considered here, one can identify τ_φ with the time it takes the particle to cover the distance equal to the dephasing length, i.e., $\tau_\varphi = L_\varphi/v$. If, furthermore, we identify our particle with an electron moving in a 3D metal with the velocity close to the Fermi one, $v \approx v_F$, we get

$$\tau_\varphi \sim \tau_e (k_F l)^2 \tag{8.78}$$

where $\tau_e = \ell/v_F$ is the elastic scattering time for electrons. Of course, so far our latter assumption looks completely unjustified because in our consideration, we deliberately ignored Pauli exclusion principle meaning that our particle on a ring is distinguishable from electrons in a 3D diffusive conductor forming our effective environment. This important issue will be addressed in the next chapter.

8.2.7 Some Concluding Remarks

Combining the results of different approaches employed above, we can conclude that Coulomb interaction between the particle on a ring and fluctuating electrons in a diffusive conductor yields exponential suppression of persistent current at large-enough values of the ring perimeter $2\pi R$ exceeding the dephasing length L_φ (8.56). This length is inversely proportional to the effective coupling constant α of our problem, i.e., L_φ remains finite down to $T = 0$. Despite certain quantitative differences, there are clear physical similarities between the model studied here and that of the Caldeira–Leggett bath analyzed in Chapters 3 and 7. In both cases, the most important effect of fluctuations in a dissipative bath is manifested as suppression of quantum coherence of the particle wave functions not only at finite temperatures but also in the limit

$T \to 0$ (cf. Eqs. (8.56) and (3.43)). In both cases, this effect cannot be adequately described perturbatively in the interaction and requires more sophisticated non-perturbative approaches.

At this stage, it is also important to note that exponential suppression of persistent current with the ring radius $I \propto \exp(-AR)$ by itself does not yet necessarily imply decoherence. For instance, even in the absence of interactions, the current $I \propto \Delta$ (8.46) can decrease exponentially with increasing R, provided the parameter κ is fixed to be independent of the ring radius. Obviously, quantum coherence remains fully preserved in this case. In Chapter 20, we will demonstrate that exponential reduction of persistent current with increasing R at $T \to 0$ can also occur in superconducting nanorings due to proliferation of quantum phase slips. Also in that case, the dependence $I \propto \exp(-AR)$ can be interpreted just as a nontrivial coordinate-dependent renormalization effect.

How to discriminate between pure renormalization and decoherence? An important qualitative difference between the two last examples and nanorings in a dissipative regime considered here is that in our problem, dissipation explicitly *violates time-reversal symmetry* (thus causing genuine decoherence of a quantum particle), while no such symmetry is violated in superconducting nanorings or in normal rings in the absence of dissipation. Hence, quantum coherence remains fully preserved in the last two cases despite exponential suppression of persistent current at large R.

In practical terms, one can discriminate between decoherence and pure renormalization in a number of ways. For instance, one can weakly drive the system out of equilibrium and investigate its phase relaxation by means of a real-time analysis. If this analysis reveals real-time decay of off-diagonal elements of the particle density matrix, one can unambiguously identify quantum decoherence. Examples of this kind are provided in Chapter 3 and above in this chapter.

Another possible way [84] amounts to analyzing equilibrium current fluctuations in the ground state of an interacting system. For instance, one can study the current–current correlation function $\langle(\hat{I} - \langle\hat{I}\rangle)^2\rangle$, where $\hat{I}$ is the current operator whose expectation value $\langle\hat{I}\rangle$ defines persistent current in the ground state. Within the Caldeira–Leggett model, it was demonstrated [84] that while the average current $\langle \hat{I} \rangle$ decreases with increasing interaction strength, its fluctuations *increase*, thus implying genuine decoherence rather than pure renormalization. Below we will demonstrate that a similar situation occurs within the model of a quantum particle on a ring studied here.

8.3 Persistent Current Noise

The above arguments suggest that the analysis of persistent current fluctuations can provide an additional useful information about the ground-state properties of interacting quantum systems. Unlike persistent current itself, its fluctuations have so far received rather limited attention in the literature. Gefen and Entin-Wohlman [136] investigated persistent current fluctuations in disordered metallic rings in the absence of electron–electron interactions. Thermal fluctuations of persistent current were also addressed by Moskalets [309] within the framework of the Luttinger model.

Below we will outline a general approach [353, 355] that allows to analyze persistent current correlators at any frequencies and temperatures, including the most interesting limit $T \to 0$. For the sake of definiteness, we will again stick to the model of a quantum particle on a ring in a dissipative environment already introduced in the previous section. At the same time, the main observations established within this model are of a general nature and, hence, should also survive in other models describing mesoscopic rings in the presence of disorder and interactions.

8.3.1 General Considerations

For the model system of a quantum particle on a ring described in Section 8.2 (see also Fig. 8.1), it is straightforward to define the current operator $\hat{I}$. In the Schrodinger representation, it reads

$$\hat{I} = \frac{e}{2\pi}\dot{\hat{\theta}} = \frac{ie}{2\pi}[\hat{H}, \hat{\theta}] = \frac{e(\hat{\phi} + \phi_x)}{2\pi MR^2} \tag{8.79}$$

where $\hat{H}$ is the total Hamiltonian of the system "particle+bath." We start from the imaginary-time technique. Switching to the Matsubara representation $\hat{I}_M(\tau) = e^{\tau\hat{H}}\hat{I}e^{-\tau\hat{H}}$, we define the current–current correlation function

$$\Pi(\tau) = \langle\langle \mathcal{T}\hat{I}_M(\tau)\hat{I}_M(0)\rangle\rangle = T\sum_{n=-\infty}^{\infty}\Pi_{\mathrm{i}\omega_n}e^{-\mathrm{i}\omega_n\tau} \quad (8.80)$$

which describes equilibrium current noise. Here $\mathcal{T}$ is the time-ordering operator, $\omega_n = 2\pi nT$ are Bose–Matsubara frequencies, and the symbol $\langle\langle\ldots\rangle\rangle$ stands for irreducible correlators (or cumulants), e.g., $\langle\langle\hat{I}(\tau)\hat{I}(0)\rangle\rangle = \langle\hat{I}(\tau)\hat{I}(0)\rangle - \langle\hat{I}(\tau)\rangle\langle\hat{I}(0)\rangle$.

Employing the full set of eigenstates $\hat{H}|m\rangle = \varepsilon_m(\phi_x)|m\rangle$ after straightforward algebra, we obtain

$$\Pi(\tau) = \mathcal{P} + \tilde{\Pi}(\tau) \quad (8.81)$$

where $\mathcal{P}$ does not depend on imaginary time and reads

$$\mathcal{P} = \frac{1}{\mathcal{Z}}\sum_m |\langle m|\hat{I}|m\rangle|^2 e^{-\beta\varepsilon_m} - \langle\hat{I}\rangle^2 \quad (8.82)$$

while the Fourier components of $\Pi(\tau)$ are defined as

$$\tilde{\Pi}_{i\omega_k} = \frac{1}{\mathcal{Z}}\sum_{m\neq n}|\langle m|\hat{I}|n\rangle|^2\frac{e^{-\beta\varepsilon_n} - e^{-\beta\varepsilon_m}}{\mathrm{i}\omega_k + \varepsilon_m - \varepsilon_n} \quad (8.83)$$

Turning now to the real-time technique, let us establish the relation between the imaginary-time correlator (8.80) and the current noise power. For this purpose, let us define the Heisenberg operators $\hat{I}(t) = e^{\mathrm{i}t\hat{H}}\hat{I}e^{-\mathrm{i}t\hat{H}}$ and the Keldysh Green function

$$S(t) = \langle\langle\hat{I}(t)\hat{I}(0) + \hat{I}(0)\hat{I}(t)\rangle\rangle = \int\frac{d\omega}{2\pi}S_\omega e^{-\mathrm{i}\omega t} \quad (8.84)$$

Decomposing $S(t)$ into time-independent and time-dependent contributions, we find

$$S(t) = 2\mathcal{P} + \tilde{S}(t) \quad (8.85)$$

where $\mathcal{P}$ is again defined in Eq. (8.82) and the Fourier components of $\tilde{S}(t)$ take the form

$$\tilde{S}_\omega = \frac{2\pi}{\mathcal{Z}}\sum_{m\neq n}|\langle m|\hat{I}|n\rangle|^2\left(e^{-\beta\varepsilon_m} + e^{-\beta\varepsilon_n}\right)\delta(\omega + \varepsilon_m - \varepsilon_n) \quad (8.86)$$

Comparing now Eqs. (8.83) and (8.86), we arrive at the fluctuation-dissipation relation

$$\tilde{S}_\omega = 2\coth\frac{\beta\omega}{2}\mathrm{Im}\tilde{\Pi}_{\omega+\mathrm{i}0} \quad (8.87)$$

which allows to immediately recover the current noise power from the correlator [(8.81)–(8.83)].

We would like to point out that the above general relations also allow to analyze the linear current response to the time-dependent flux inside the ring and to formally define the ac conductance of the system. According to the Kubo formula, this ac conductance is expressed in terms of the commutator of the current operators, unlike the noise spectrum defined by the anticommutator of these operators (8.84).

The above exact relations allow to observe that as long as the current operator $\hat{I}$ commutes with the Hamiltonian of the system, the Fourier components (8.83) and (8.86) vanish identically together with the matrix elements $\langle m|\hat{I}|n\rangle$ with $m \neq n$, while the time-independent term $\mathcal{P}$ (8.82) tends to zero only in the zero-temperature limit. Thus, in this case, no current fluctuations can occur at $T \to 0$, and at nonzero temperatures, persistent current noise does not vanish only in the zero-frequency limit, $S_\omega = 4\pi\mathcal{P}\delta(\omega)$.

If, however, the operators $\hat{I}$ and $\hat{H}$ do not commute with each other (as is the case for a vast majority of realistic nanorings), the matrix elements $\langle m|\hat{I}|n\rangle$ in general remain nonzero for any pair of eigenstates. Thus, we arrive at an important conclusion that in this case, fluctuations of persistent current do not vanish down to $T = 0$. As we will see below, analyzing zero-temperature fluctuations of persistent current in nanorings, one may obtain useful additional information about the ground-state properties of such systems.

8.3.2 Free Energy and Generating Functional

Although current correlators can be evaluated directly from Eqs. (8.80)–(8.86) employing the representation of eigenstates, for practical calculations it is sometimes more convenient to make use of alternative techniques, which are illustrated as follows.

To begin with, we notice that in both limits of zero Matsubara frequency and zero imaginary time, the current–current correlator can be conveniently related to the free energy of the system $F = -T \ln \mathcal{Z}$. Making use of Eq. (8.27), which defines the current expectation value $I = \langle \hat{I} \rangle$ and employing the identity

$$\frac{\partial (e^{-\beta \hat{H}})}{\partial \phi_x} = -\int_0^\beta d\tau e^{-(\beta-\tau)\hat{H}} \frac{\partial \hat{H}}{\partial \phi_x} e^{-\tau \hat{H}} \quad (8.88)$$

for the second derivative of the free energy with respect to the flux, we obtain

$$\frac{e^2}{4\pi^2} \frac{\partial^2 F}{\partial \phi_x^2} = \frac{e^2}{4\pi M R^2} - \int_0^\beta d\tau \frac{\mathrm{tr}(\hat{I} e^{-(\beta-\tau)\hat{H}} \hat{I} e^{-\tau \hat{H}})}{\mathrm{tr}(e^{-\beta \hat{H}})} + \beta \langle \hat{I} \rangle^2 \quad (8.89)$$

From this equation, one finds

$$\int_0^\beta d\tau \Pi(\tau) \equiv \Pi_0 = \frac{e^2}{4\pi^2 M R^2} - \frac{e^2}{4\pi^2} \frac{\partial^2 F}{\partial \phi_x^2} \quad (8.90)$$

On the other hand, with the aid of the identity

$$\frac{1}{R^2} \frac{\partial}{\partial M} \mathrm{tr}\left(e^{-\beta \hat{H}}\right) = \frac{2\pi^2}{e^2 T} \mathrm{tr}\left(\hat{I}^2 e^{-\beta \hat{H}}\right) \quad (8.91)$$

we get

$$\langle \hat{I}^2 \rangle = -\frac{e^2}{2\pi^2 R^2} \frac{\partial F}{\partial M} \quad (8.92)$$

and, hence,

$$\Pi(0) = -\frac{e^2}{2\pi^2 R^2} \frac{\partial F}{\partial M} - \frac{e^2}{4\pi^2} \left(\frac{\partial F}{\partial \phi_x} \right)^2 \quad (8.93)$$

The free energy of the system can be readily evaluated, and Eqs. (8.90), (8.93) provide a great deal of information about persistent current noise. Further simplifications may appear in the zero-temperature limit since in this case, the free energy reduces to the ground-state energy $F(T \to 0) = \varepsilon_0(\phi_x)$. For example, for a free particle on a ring, one has $\varepsilon_0(\phi_x) = \phi^2/(2MR^2)$. Hence, in this case in the limit $T \to 0$ from Eqs. (8.90) and (8.93), one trivially finds

$$\Pi_0 = \Pi(0) = 0 \quad (8.94)$$

in agreement with our general considerations stating that at $T = 0$, the correlator (8.80) should vanish identically at any time provided the current operator commutes with the total Hamiltonian of the system.

If, however, the particle on a ring is moving in an external potential (e.g., $U_0 \neq 0$ in Eq. (8.16)) or interacting with an external environment, these two operators do not commute anymore and, hence, persistent current fluctuations do not vanish even at very low T. This conclusion can be reached, e.g., from Eq. (8.90) without any additional calculation. Indeed, for $U_0 \neq 0$, the ground-state energy $\varepsilon_0(\phi_x)$ deviates from $\phi^2/(2MR^2)$ and, hence, $\Pi_0 \neq 0$ down to $T = 0$.

In order to fully account for persistent current fluctuations, it might be convenient to employ the generating functional technique. Consider, e.g., the particle on a ring described by the Hamiltonian

$$\hat{H} = \frac{(\phi + \phi_x)^2}{2MR^2} + U(\theta) \tag{8.95}$$

and define the generating functional as a path integral

$$\mathcal{Z}[\eta] = \int \mathcal{D}\phi \mathcal{D}\theta e^{\int_0^\beta d\tau \left(i\phi\dot{\theta} - \frac{(\phi+\phi_x)^2}{2MR^2} - U(\theta) - \eta\phi \right)} \tag{8.96}$$

where $\eta(\tau)$ is the source field for the flux variable ϕ. Performing the integration over ϕ, we obtain

$$\mathcal{Z}[\eta] \sim \int \mathcal{D}\theta e^{-\int_0^\beta d\tau \left(\frac{MR^2(\dot{\theta}+i\eta)^2}{2} + i\phi_x(\dot{\theta}+i\eta) + U(\theta) \right)} \tag{8.97}$$

Taking then the variational derivative of $F[\eta] = -T \ln \mathcal{Z}[\eta]$ over the source field $\eta(\tau)$ and setting this field equal to zero afterward, we derive the relation between the expectation values for the current and the particle "velocity" $\dot{\theta}$:

$$\langle I(\tau) \rangle = \frac{ie}{2\pi} \langle \dot{\theta}(\tau) \rangle. \tag{8.98}$$

Similarly, the second derivative of $F[\eta]$ with respect to $\eta(\tau)$ yields the second current cumulant:

$$\langle\langle I(\tau_1) I(\tau_2) \rangle\rangle = \frac{e^2}{4\pi^2 MR^2} \delta(\tau_1 - \tau_2) - \frac{e^2}{4\pi^2} \langle\langle \dot{\theta}(\tau_1) \dot{\theta}(\tau_2) \rangle\rangle \tag{8.99}$$

Analogously one can establish the relations between higher current and velocity cumulants. Up to some unimportant δ-functions at coinciding times (which cancel out in the final result as we will see below), the latter cumulants, in turn, are evaluated from the relation

$$\langle\langle \dot{\theta}(\tau_1) \ldots \dot{\theta}(\tau_N) \rangle\rangle = (-i)^N \frac{\delta^N \ln \mathcal{Z}[\zeta]}{\delta\zeta(\tau_1) \ldots \delta\zeta(\tau_N)} |_{\zeta=0} \tag{8.100}$$

where

$$\mathcal{Z}[\zeta] = \int_0^{2\pi} d\theta_0 \sum_{m=-\infty}^{\infty} e^{2\pi \mathrm{i} m\phi_x} \int_{\theta_0}^{\theta_0+2\pi m} \mathcal{D}\theta e^{-\int_0^\beta d\tau \left(\frac{MR^2\dot{\theta}^2}{2}+U(\theta)-\mathrm{i}\zeta\dot{\theta}\right)} \tag{8.101}$$

The above general expressions allow to conveniently evaluate all current cumulants in our problem. If necessary, these relations can be straightforwardly generalized to systems described by more complicated Hamiltonians.

8.3.3 Particle in a Periodic Potential

In order to demonstrate the advantages of the above method, let us analyze fluctuations of persistent current in the system described by the Hamiltonian (8.16). For the sake of definiteness, we will stick to the tight binding limit (8.28) in which case one can evaluate the generating functional quasiclassically with the aid of the instanton technique already developed in Section 8.2.2.

Substituting multi-instanton trajectories (8.39) into Eq. (8.101) and performing Gaussian integrals, we get

$$\mathcal{Z}[\zeta] = \kappa \sum_{n=0}^{\infty} \sum_{\nu_1=\pm 1} .. \sum_{\nu_n=\pm 1} \left(\frac{\Delta}{2}\right)^n \int_0^\beta d\tau_1 \int_{\tau_1}^\beta d\tau_2 \ldots \int_{\tau_{n-1}}^\beta d\tau_n \sum_{m=-\infty}^{\infty} e^{2\pi \mathrm{i} m\phi_x}$$
$$\times e^{\mathrm{i}\sum_j \nu_j \int_0^\beta \zeta(\tau)\dot{\tilde{\theta}}(\tau-\tau_j)d\tau} \mathcal{Z}_n[\zeta]\delta_{\sum_i \nu_i, m\kappa} \tag{8.102}$$

where the terms

$$\mathcal{Z}_n[\zeta] = e^{-\frac{1}{2}\int \zeta(\tau)G(\tau,\tau')\zeta(\tau')d\tau d\tau'} \tag{8.103}$$

are set by Gaussian fluctuations around n-instanton trajectories $\Theta(\tau)$ (8.39). The correlator $G(\tau, \tau') = \langle \delta\dot{\theta}(\tau)\delta\dot{\theta}(\tau')\rangle$ can easily be evaluated for a dilute instanton gas provided both times τ and τ' are outside the instanton cores, i.e., $|\tau-\tau_j|$, $|\tau'-\tau_j| \gg \omega_0^{-1}$ for every j. In this case, $\mathcal{Z}_n[\zeta]$ reduces to the generating functional for a harmonic oscillator $\mathcal{Z}_0[\zeta]$ defined by Eq. (8.103) with

$$G(\tau, \tau') \approx \frac{\omega_0^3}{2\kappa^2 U_0} e^{-\omega_0|\tau-\tau'|} + \frac{1}{MR^2}\delta(\tau - \tau') \tag{8.104}$$

where the last expression remains valid for $\beta\omega_0 \gg 1$ and $|\tau - \tau'| \ll \beta$. Proceeding analogously to Section 8.2.2 and employing Poisson's resummation formula (8.42), we obtain

$$\mathcal{Z}[\zeta] = \mathcal{Z}_0[\zeta] \sum_{k=1}^{\kappa} e^{\Delta \int_0^\beta d\tau \cos\left(\frac{2\pi(\phi_x - k)}{\kappa} + \int_0^\beta \zeta(\tau_1)\dot{\tilde{\theta}}(\tau - \tau_1)d\tau_1\right)} \tag{8.105}$$

In the limit $\zeta \to 0$, Eq. (8.105) reduces to the partition function $\mathcal{Z}$ (8.43) and the average value of persistent current derived from Eqs. (8.105) and (8.100) with $N = 1$ coincides with our result (8.45).

The same equations for $N = 2$ yield the second current cumulant

$$\Pi(\tau) = \mathcal{P} + \mathcal{P}_{osc}(\tau) + \mathcal{P}_{osc}(\beta - \tau) + \tilde{\mathcal{P}}(f(\tau) + f(\beta - \tau)) \tag{8.106}$$

where for $\omega_0^{-1} \leq \tau \leq \beta - \omega_0^{-1}$, we find

$$\mathcal{P} = \frac{e^2\Delta^2}{\kappa^2} \frac{\sum_{k=1}^{\kappa} \sin^2\left(\frac{2\pi(\phi_x - k)}{\kappa}\right) e^{\beta\Delta\cos\left(\frac{2\pi(\phi_x - k)}{\kappa}\right)}}{\sum_{k=1}^{\kappa} e^{\beta\Delta\cos\left(\frac{2\pi(\phi_x - k)}{\kappa}\right)}} - \frac{e^2\Delta^2}{\kappa^2}\left(\frac{\sum_{k=1}^{\kappa} \sin\left(\frac{2\pi(\phi_x - k)}{\kappa}\right) e^{\beta\Delta\cos\left(\frac{2\pi(\phi_x - k)}{\kappa}\right)}}{\sum_{k=1}^{\kappa} e^{\beta\Delta\cos\left(\frac{2\pi(\phi_x - k)}{\kappa}\right)}}\right)^2 \tag{8.107}$$

and

$$\tilde{\mathcal{P}} = -\frac{e^2\Delta}{\kappa^2} \frac{\sum_{k=1}^{\kappa} \cos\left(\frac{2\pi(\phi_x - k)}{\kappa}\right) e^{\beta\Delta\cos\left(\frac{2\pi(\phi_x - k)}{\kappa}\right)}}{\sum_{k=1}^{\kappa} e^{\beta\Delta\cos\left(\frac{2\pi(\phi_x - k)}{\kappa}\right)}} \tag{8.108}$$

In Eq. (8.106) for $\tau \gg \omega_0^{-1}$, we also defined

$$f(\tau) = \frac{\kappa^2}{4\pi^2} \int_{-\infty}^{\infty} \dot{\tilde{\theta}}(\tau_1 - \tau)\dot{\tilde{\theta}}(\tau_1)d\tau_1 \simeq \frac{4\Omega^2\tau}{\pi^2} e^{-\omega_0\tau} \tag{8.109}$$

and

$$\mathcal{P}_{osc}(\tau) = \frac{e^2\omega_0^3}{8\pi^2\kappa^2 U_0} e^{-\omega_0\tau} \tag{8.110}$$

In accordance with our general analysis in terms of the exact eigenstates, the result (8.106) consists of two different (time-independent and time-dependent) contributions. The meaning of each of these terms can be identified with the aid of Eqs. (8.81)–(8.83).

As we already discussed, at zero temperature, the time-independent part $\mathcal{P}$ should vanish, $\mathcal{P} = 0$, as is indeed observed from Eq. (8.107) in the limit $T \to 0$. At nonzero temperatures, however, the term $\mathcal{P}$ does not vanish. At $\Delta/\kappa \ll T \ll \omega_0$ from Eq. (8.107), we get

$$\mathcal{P} = \frac{e^2\Delta^2}{2\kappa}\left(I_0(\beta\Delta) - I_2(\beta\Delta)\right) - \frac{e^2(\kappa-1)\Delta^\kappa}{2^{\kappa-1}\kappa! T^{\kappa-2}}\cos(2\pi\phi_x) \tag{8.111}$$

where, as before, $I_i(x)$ are the Bessel functions.

Let us now turn to time-dependent contribution to $\Pi(\tau)$. With the aid of Eqs. (8.81), (8.83), and (8.106), we identify

$$\mathcal{P}_{osc} + \tilde{\mathcal{P}} f(\tau) = \frac{1}{\mathcal{Z}}\sum_{m>n} |\langle m|\hat{I}|n\rangle|^2 e^{-\beta\varepsilon_n} e^{-\tau(\varepsilon_m-\varepsilon_n)} \tag{8.112}$$

In order to proceed, let us bear in mind that there exist κ low-lying quantum levels with energies $\omega_0/2 - \Delta\cos\left(2\pi(\phi_x - k)/\kappa\right)$ in our problem. These states originate from tunneling depletion of the ground-state energy level $\omega_0/2$ in each of κ potential wells. Below we will label these states as $|0k\rangle$ with $k = 1, \ldots, \kappa$. Due to the rotation symmetry of our model, all the matrix elements between these states vanish and, hence, do not contribute to the current–current correlation function.

Next κ energy levels $|1l\rangle$ with $l = 1, \ldots, \kappa$ occur due to depletion of the first excited state $3\omega_0/2$. These states are characterized by the energies $3\omega_0/2 + \tilde{\Delta}\cos\left(2\pi(\phi_x - l)/\kappa\right)$, where the parameter $\tilde{\Delta}$ is to be defined below. With the aid of the symmetry arguments, one can again demonstrate that the matrix elements of the current operator between the states $|1l\rangle$ with different l vanish, while the matrix elements between the states $|0k\rangle$ and $|1l\rangle$ remain nonzero, provided $k = l$. In order to evaluate these matrix elements, it suffices to consider the instanton contribution small by setting $(\Delta+\tilde{\Delta})\tau \ll 1$ in Eq. (8.112) and to expand the right-hand side of this equation in powers of $\Delta\tau$ and $\tilde{\Delta}\tau$. Comparing the first two terms of this expansion with $\mathcal{P}_{osc}$ and $\tilde{P} f$ [(8.108)–(8.110)], in the limit $U_0 \gg \omega_0$ considered here, we identify

$$\langle 0k|\hat{I}|1l\rangle|^2 \approx \delta_{lk}\frac{e^2\omega_0^3}{8\pi^2\kappa^2 U_0}, \qquad \tilde{\Delta} \approx \frac{32U_0}{\omega_0}\Delta \tag{8.113}$$

Finally, let us evaluate the real-time current noise power S_ω defined in Eqs. (8.85) and (8.86). Employing the above results, at $T \ll \omega_0$, we obtain

$$S_\omega = 4\pi \mathcal{P}\delta(\omega) + \frac{e^2\omega_0^3}{4\pi\kappa^2 U_0 \mathcal{Z}} \sum_{k=1}^{\kappa} e^{\beta\Delta\cos\left(\frac{2\pi(\phi_x - k)}{\kappa}\right)} \times \left(\delta(\omega - \omega_0 - \epsilon_k) + \delta(\omega + \omega_0 + \epsilon_k)\right) \quad (8.114)$$

where we defined

$$\epsilon_k = \frac{32U_0\Delta}{\omega_0}\cos\left(\frac{2\pi(\phi_x - k)}{\kappa}\right) \quad (8.115)$$

We observe that—in agreement with our general analysis—persistent current noise power has the form of sharp peaks at frequencies equal to the distance between the energy levels with nonzero matrix elements of the current operator plus an additional peak at zero frequency. In the zero-temperature limit $T \to 0$, the amplitude of this peak tends to zero along with the terms related to transitions to higher energy levels and Eq. (8.114) reduces to

$$S_\omega = \frac{e^2\omega_0^3}{4\pi\kappa^2 U_0}\left(\delta(\omega - \omega_0 - \epsilon_0) + \delta(\omega + \omega_0 + \epsilon_0)\right) \quad (8.116)$$

where $\epsilon_0(\phi_x) = \max_k \epsilon_k(\phi_x)$. This result demonstrates that current fluctuations indeed persist down to $T = 0$ and that S_ω differs from zero even at zero external flux $\phi_x = 0$ when the average value of persistent current is zero. As we will see below, in the presence of dissipation due to interaction of the particle with other (quantum) degrees of freedom, the energy levels acquire a finite width, the peaks get broadened, and the noise power should differ from zero in a wider range of frequencies.

Similarly, broadening of such peaks inevitably occurs in ensembles of rings or individual rings with many conducting channels. Within our model, this broadening can be illustrated by considering an ensemble of rings with the parameter U_0 uniformly distributed within a certain energy interval. In this case, the total persistent current noise produced by the system is given by the sum of a large number of very close peaks (8.116) effectively resulting in a much smoother and broader noise spectrum, as it is shown in Fig. 8.4.

A specific feature of persistent current noise is the dependence of S_ω on the external magnetic flux ϕ_x. This dependence occurs due

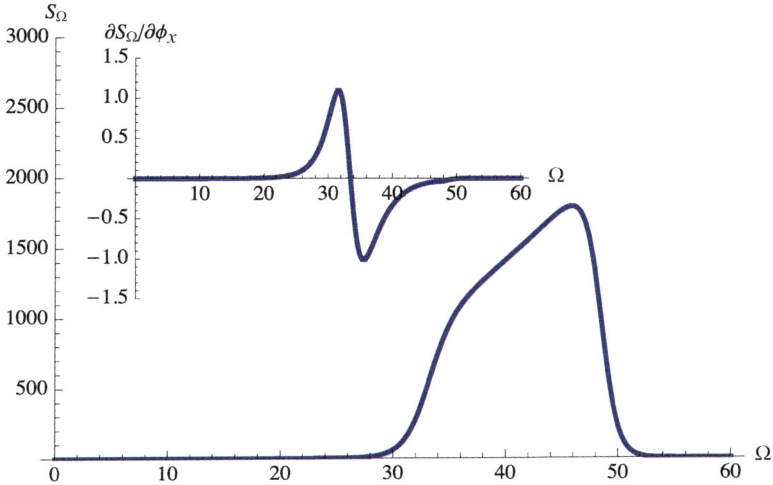

Figure 8.4 Persistent current noise spectrum S_ω (arbitrary units) and its derivative with respect to the flux $\partial S_\omega/\partial\phi_x$ (arbitrary units) at $T = 0$ as functions of ω (measured in units of $1/2MR^2$) for an ensemble of rings (or for a ring containing many independent channels) with U_0 uniformly distributed within the interval from $30/MR^2$ to $65/MR^2$.

to the presence of quantum coherence in the system and disappears if this coherence gets destroyed. Hence, such sensitivity of the noise spectrum to the flux can be used as a measure of quantum coherence in our system. Taking the derivative of S_ω with respect to the flux, we obtain

$$\partial S_\omega/\partial\phi_x \propto \sin(2\pi\phi_x/\kappa) \tag{8.117}$$

Typical dependence of $\partial S_\omega/\partial\phi_x$ on ω is also illustrated in Fig. 8.4.

8.3.4 Particle in a Dissipative Environment

Now let us include dissipation into our consideration. As in Section 8.2, we will describe the effect of a dissipative environment within the model of a dirty electron gas, which yields the effective action $S_{\text{int}}[\theta]$ (8.22), (8.23). For simplicity, we will also assume that the particle moves along the ring in the absence of any external potential, i.e., $U_0 = 0$. In other words, the situation is exactly the same as the one considered in Section 8.2.4.

To begin with, let us proceed perturbatively in the effective coupling constant α. We will employ the diagrammatic approach already developed in Section 7.5.1 for the Coulomb blockade problem. Here the structure of the effective action is exactly the same except the interaction part $S_{\rm int}[\theta]$ is now somewhat more complicated than that analyzed in Chapter 7. However, technically this difference between these two problems is not significant and can be easily accounted for.

Following the approach of Section 7.5.1, we expand the partition function $\mathcal{Z}$ in powers of α and express the result via the diagrammatic series, which can in turn be rewritten in terms of a Dyson equation (see Fig. 7.9). Then we obtain

$$\mathcal{Z} = \sum_{n=-\infty}^{\infty} Z[\beta, \phi_x - n] \tag{8.118}$$

where the Laplace transform of Z is defined as

$$Z_p[q] = \int_0^{\infty} d\tau e^{-p\tau} Z[\tau, q] = \frac{1}{p + q^2/(2MR^2) - \Sigma_p[q]} \tag{8.119}$$

(cf. Eqs. (7.118)–(7.124)). Here $\Sigma_p[q]$ is the self-energy defined as a sum of all irreducible diagrams depicted in Fig. 7.10. At this stage, we will proceed perturbatively taking into account only the simplest one-loop diagrams, i.e., two first diagrams in Fig. 7.10. We also restrict our analysis to zero-temperature limit. Employing the Fourier representation (8.24) and performing the same calculation as in Section 7.5.1, within the logarithmic accuracy for the self-energy, we get

$$\Sigma_p[q] = \frac{\alpha}{2} \sum_{n=1}^{r} a_n \left[\sum_{s=\pm 1} (p + (q+sn)^2/(2MR^2)) \ln\left(p + \frac{(q+sn)^2}{(2MR^2)} \right) \right.$$
$$\left. -(2p + (q^2+n^2)/(MR^2)) \ln\left(p + \frac{(q^2+n^2)}{(2MR^2)} \right) \right] \tag{8.120}$$

Making use of this expression, it is straightforward to derive the current–current correlator. Neglecting vertex corrections, we obtain

$$\langle \dot\theta(0)\dot\theta(\tau)\rangle = -\frac{1}{(MR^2)^2} \frac{\sum\limits_{n=-\infty}^{\infty} (n-\phi_x)^2 Z[\tau, n-\phi_x] Z[\beta-\tau, n-\phi_x]}{\sum\limits_{n=-\infty}^{\infty} Z[\beta, n-\phi_x]} \tag{8.121}$$

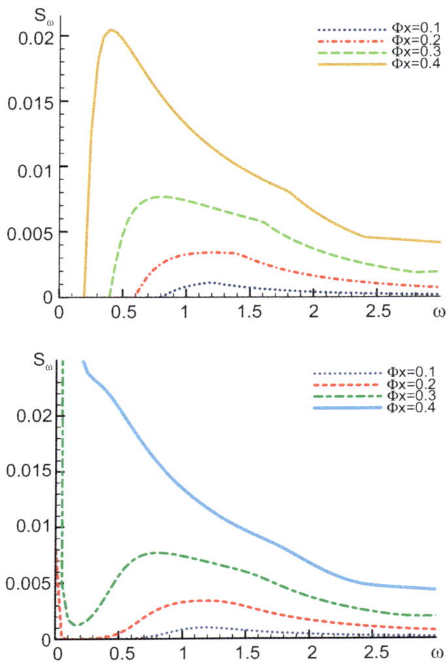

Figure 8.5 Persistent current noise power for $\pi\alpha = 0.05$, $r = 5$ and different flux values at $T = 0$ (top panel) and $T = 0.05E_C$ (bottom panel). Frequency ω and noise power S_ω are normalized by, respectively, E_C and $e^2E_C/(4\pi^2)$.

After analytic continuation to real time, we arrive at the following expression for zero-temperature persistent current noise power:

$$S_\omega = -\frac{2\phi_x^2}{(MR^2)^2}\left(\rho_{p+\omega}[-\phi_x] + \rho_{p-\omega}[-\phi_x]\right)|_{p=\phi_x^2/(2MR^2)} \quad (8.122)$$

where we defined $\rho_p[q] = \mathrm{Im}Z_{-p+i0}[q]$. This expression can be evaluated numerically, and the result is displayed in Fig. 8.5 (top panel).

One observes that at $T = 0$, the noise power S_ω remains zero at frequencies smaller than the inter-level distance $\omega < \delta E(\phi_x)$ and becomes nonzero otherwise. The noise power strongly depends on the magnetic flux ϕ_x, illustrating the coherent nature of persistent current fluctuations in the perturbative α regime. Persistent current noise grows with increasing ϕ_x and diverges as the flux approaches the point $\phi_x = 0.5$. In this limit, the distance between the two lowest energy levels $\delta E(\phi_x) = E_C(1 - 2|\phi_x|)$ becomes small, and the system undergoes rapid transitions between these energy states. As these levels correspond to different persistent current values, such transitions, in turn, yield strong current fluctuations.

We also note that in the lowest ($\sim\alpha$) order of the perturbation theory, zero-temperature persistent current noise vanishes at $\phi_x = 0$. However, beyond the first-order perturbation theory, this feature does not hold, as it can be observed, e.g., from the general expressions derived in Section 8.3.1 as well as from the non-perturbative in α analysis to be presented below.

The above perturbative analysis can be easily extended to nonzero temperatures [355]. Here we will not go into corresponding technical details and only display the main results in Fig. 8.5 (bottom panel). One observes that at nonzero T, there appears additional zero-frequency noise power peak. This peak grows rapidly with increasing temperature and eventually assimilates all other peaks. As a result, at sufficiently high temperatures, only a wide hump remains, and persistent current noise becomes flux independent, i.e., incoherent. In this limit, one can also derive analytic results for S_ω, as we will demonstrate below.

Let us now proceed to the limit of strong interactions $4\pi\alpha r \gg 1$ (or high temperatures) in which case the effect of a dissipative environment on the particle motion becomes large, substantially reducing fluctuations of the angle variable θ. In this limit and provided temperature is not too low, it suffices to employ the semiclassical approximation and to expand the effective action (8.74), (8.75) up to quadratic in θ_- terms. As we demonstrated before, e.g., in Chapters 1, 2, and 7, the resulting effective action can be exactly rewritten in terms of the quasiclassical Langevin equation for the "center-of-mass" variable θ_+. We obtain

$$-\frac{1}{2E_C}\ddot{\theta}_+(t) - \frac{\gamma}{2}\dot{\theta}_+(t) = \sum_{n=1}^{\infty}(\xi_n(t)\cos(n\theta_+(t)) + \lambda_n(t)\sin(n\theta_+(t))) \tag{8.123}$$

where we defined the parameter

$$\gamma = 2\pi\alpha\sum_{n=1}^{\infty} a_n n^2 = 4\pi\alpha r^2 \tag{8.124}$$

and introduced Gaussian stochastic fields $\xi_n(t)$ with the correlators

$$\langle \xi_n(t)\xi_m(t')\rangle_{\xi,\lambda} = \langle \lambda_n(t)\lambda_m(t')\rangle_{\xi,\lambda} = -\delta_{m,n}\pi\alpha a_n n^2 \frac{\pi T^2}{\sinh^2(\pi T(t-t'))}$$

$$\langle \xi_n(t)\lambda_m(t')\rangle_{\xi,\lambda} = 0 \qquad (8.125)$$

In the high-temperature limit, these correlators reduce to those describing the white noise

$$\langle \xi_n(t)\xi_m(t')\rangle_{\xi,\lambda} = 2\delta_{m,n}\pi\alpha a_n n^2 T\delta(t-t') \qquad (8.126)$$

and the Langevin equation can be solved exactly. As a result, we arrive at the high-temperature noise power

$$S_\omega = \frac{e^2\gamma T E_C^2}{\pi^2(\omega^2 + (\gamma E_C)^2)} \qquad (8.127)$$

It is important to emphasize that this result remains applicable at all values of the coupling constant α, thus covering both perturbative and non-perturbative regimes at sufficiently high temperatures.

At lower temperatures, the white noise approximation (8.126) becomes inaccurate and Eqs. (8.125), (8.125) should be employed. Similar to our analysis in Chapter 7, in this case noise terms in the Langevin equation can be treated perturbatively. Keeping only the zeroth and the first-order contributions, one gets the solution to Eq. (8.123) in the form

$$\theta_+(t) = \theta_+^{(0)} + \theta_+^{(1)}(t) \qquad (8.128)$$

where $\theta_+^{(0)}$ is an arbitrary (and physically irrelevant) constant and $\theta_+^{(1)}(t)$ obeys the equation

$$-\frac{1}{2E_C}\ddot{\theta}_+^{(1)}(t) - \frac{\gamma}{2}\dot{\theta}_+^{(1)}(t) = \sum_{n=1}^{\infty} \xi_n(t) \qquad (8.129)$$

Resolving this equation, one immediately arrives at the noise power in the form

$$S_\omega = \frac{e^2\gamma E_C^2}{2\pi^2(\omega^2 + (\gamma E_C)^2)}\omega \coth\frac{\omega}{2T} \qquad (8.130)$$

which again reduces to Eq. (8.127) in the high-temperature limit $T \gg \omega$. For $\omega \ll \gamma E_C$, the parameter E_C drops out and the noise power becomes

$$S_\omega = \frac{e^2\omega}{2\pi^2\gamma}\coth\frac{\omega}{2T} \qquad (8.131)$$

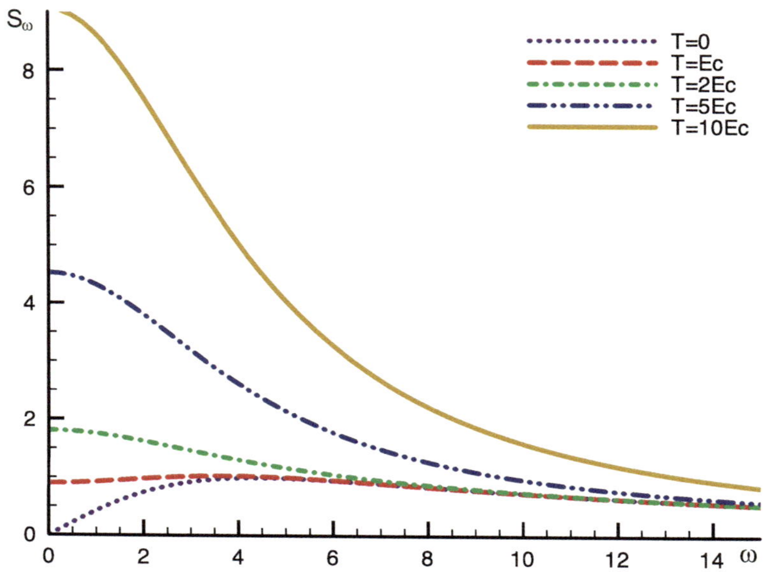

Figure 8.6 Noise power at different temperatures for $\pi\alpha = 0.05$ and $r = 25$. Units are the same as in Fig. 8.5.

For $\omega \to 0$, this expression further reduces to $S_0 \propto T/\gamma$. The noise power S_ω (8.130) is depicted in Fig. 8.6 at different values of T.

Comparing Eq. (8.130) with the above perturbative in the interaction results, we observe a striking difference between them in the low-temperature limit: While at weak interactions, persistent current noise is sensitive to the externally applied magnetic flux ϕ_x, in the opposite limit of strong interactions, the noise power S_ω turns out to be essentially independent of ϕ_x (cf. Figs. 8.5 and 8.6). These observations imply that fluctuations of persistent current remain coherent only in sufficiently small rings, while in larger rings with $R > L_\varphi$, quantum coherence of the particle is suppressed down to $T \to 0$ in full agreement with our results and conclusions in Section 8.2. In both coherent and incoherent regimes, the key role is played by interaction between the particle and the dissipative environment, in particular at $T \to 0$ where persistent current fluctuations vanish in the limit $\alpha \to 0$.

We also point out that while in the perturbative limit, S_ω tends to increase with the coupling constant α, in the non-perturbative regime, the dependence of the noise power on α (8.130) becomes somewhat more complicated. In particular, at sufficiently low frequencies, we find $S_\omega \propto 1/(\alpha R^2)$, i.e., in this regime, the noise power decreases with increasing both α and the ring radius R.

Finally, we note that Eq. (8.130) defines only the dominating contribution to the noise power. In addition there also exist small corrections to this result, which do depend on the external flux ϕ_x piercing the ring. At low temperatures, the leading flux-dependent correction to Eq. (8.130) is estimated as $\delta S_\omega(\phi_x) \propto \exp(-4\pi\alpha r)\cos(2\pi\phi_x)$ [355].

To conclude, the above analysis clearly demonstrates that fluctuations of persistent current can provide additional useful information about quantum properties of interacting systems. For instance, both quantum coherence and its suppression by interactions in meso- and nanorings can be experimentally investigated not only by detecting the average persistent current but also by measuring its noise and extracting its dependence on the external magnetic flux.

Chapter 9

Influence Functional for Interacting Electrons in Disordered Metals

The problem of an interacting electron gas in disordered metallic conductors is one of the most important and complicated problems in modern condensed matter physics. The complexity of this problem is primarily due to a necessity to simultaneously take into account electron scattering on impurities, long-range electron–electron interactions, and Fermi statistics. While there exists a well-established technique [1] enabling one to reliably describe scattering of electrons on impurities in metals, Coulomb interaction between electrons is usually accounted for within some perturbative schemes. Although such schemes might give an adequate description in some cases, in general perturbative analysis of long-range Coulomb interaction can be insufficient, as we already demonstrated, e.g., in Chapter 8. Hence, the task at hand is to construct a more general approach, which would allow one to non-perturbatively treat the interaction effects.

As we have already learned in previous chapters, the Feynman–Vernon influence functional formalism appears to be the most adequate tool for this purpose. However, a direct application of the same technique to a degenerate interacting electron gas

Dissipative Quantum Mechanics of Nanostructures: Electron Transport, Fluctuations, and Interactions
Andrei D. Zaikin and Dmitry S. Golubev

ISBN 978-981-4774-50-5 (Hardcover), 978-0-429-29823-3 (eBook)
www.jennystanford.com

immediately faces a conceptual problem. Indeed, in order to construct the influence functional, it is necessary (i) to separate the quantum variable(s) of interest from all other degrees of freedom (forming an effective environment for such variable(s)) and (ii) to integrate out these environmental degrees of freedom. This is exactly the program that was implemented, e.g., in Chapters 1 and 2 as well as in Chapter 8 where we intentionally disregarded Fermi statistics by assuming that the propagating particle can be distinguished from those forming its dissipative environment. Now we are in trouble: Already, step (i) cannot be performed in the case of a degenerate electron gas simply because electrons are indistinguishable. Hence, it appears totally unclear in which way one can split one electron from their fellows in the Fermi sea. In other words, each electron could simultaneously be a particle of interest and a part of an effective environment for this particle. This complication indicates that in order to apply the Feynman–Vernon strategy to interacting Fermi systems, it is necessary to work out its generalization that would allow to explicitly account for the Pauli principle.

Fortunately, this generalization can be constructed [155] and the main idea is as follows. To begin with, we will express the exact time-dependent density matrix for the system of interacting electrons in terms of the path integral and extract (by virtue of the proper Hubbard–Stratonovich transformation) the collective variables describing fluctuating quantum electromagnetic fields produced by fluctuating electrons. This step allows us, on one hand, to reduce the problem of interacting electrons to that of a single electron interacting with these quantum electromagnetic fields (defined on the Keldysh contour) and, on the other hand, by integrating out all electron degrees of freedom to obtain the influence functional or, equivalently, the effective action for these fluctuating fields. This part of our program is somewhat similar to that already implemented in Chapter 2. As the next step, we will derive a formally exact equation for the density matrix of a single electron propagating in the environment formed by these quantum electromagnetic fields fluctuating both in space and in time. It is important that this equation will depend on the electron distribution function and, hence, will explicitly account for the Pauli principle. Then what remains is to take the last step, namely, to again express the electron density matrix in terms of a path integral and to integrate out the fluctuating electromagnetic fields. This procedure yields the desired influence functional for interacting electrons.

Before we can proceed with this program, we will need to make some formal preparations that will constitute the subject of the next three sections of this chapter.

9.1 Effective Action: Formal Considerations

As we already pointed out, the main conceptual difference between the bath of electrons and, e.g., that of Caldeira–Leggett oscillators is the necessity to account for the Fermi statistics in the former case. Pauli principle imposes specific restrictions on the properties of the wave function of interacting electrons. Namely, the wave function of N electrons $\Psi(\boldsymbol{r}_1, \ldots, \boldsymbol{r}_N)$ should change its sign if two arbitrary chosen coordinates get interchanged. For instance, the wave function should change its sign provided we interchange $\boldsymbol{r}_1 \leftrightarrow \boldsymbol{r}_2$, i.e.,

$$\Psi(\boldsymbol{r}_1, \boldsymbol{r}_2, \boldsymbol{r}_3, \ldots, \boldsymbol{r}_N) = -\Psi(\boldsymbol{r}_2, \boldsymbol{r}_1, \boldsymbol{r}_3, \ldots, \boldsymbol{r}_N) \tag{9.1}$$

In order to account for the Pauli principle, we will employ the standard second quantization formalism describing the electrons by means of creation and annihilation field operators $\hat{\Psi}^{\dagger}_{\alpha}(\boldsymbol{r})$ and $\hat{\Psi}_{\alpha}(\boldsymbol{r})$, which depend on both the coordinate $\boldsymbol{r}$ and the spin index $\alpha = \uparrow, \downarrow$. These operators satisfy the canonical commutation relations

$$\hat{\Psi}_{\alpha}(\boldsymbol{r})\hat{\Psi}^{\dagger}_{\beta}(\boldsymbol{r}') + \hat{\Psi}^{\dagger}_{\beta}(\boldsymbol{r}')\hat{\Psi}_{\alpha}(\boldsymbol{r}) = \delta_{\alpha\beta}\delta(\boldsymbol{r} - \boldsymbol{r}') \tag{9.2}$$

The Hamiltonian of non-interacting electrons reads

$$\hat{H}_0 = \sum_{\alpha=\uparrow,\downarrow} \int d\boldsymbol{r}\, \hat{\Psi}^{\dagger}_{\alpha}(\boldsymbol{r}) \left[-\frac{\nabla^2}{2m} + U(\boldsymbol{r}) \right] \hat{\Psi}_{\alpha}(\boldsymbol{r}) \tag{9.3}$$

where m is the electron mass and $U(\boldsymbol{r})$ is the potential energy, which originates, e.g., from impurities in a metal. We assume that at negative times $t < 0$, the interaction between electrons was absent, and the Hamiltonian of the system is defined by Eq. (9.3). Then at

$t = 0$, the Coulomb interaction is "turned on" and, hence, at $t > 0$, the system Hamiltonian acquires the form

$$\hat{H} = \hat{H}_0 + \sum_{\alpha,\beta=\uparrow,\downarrow} \frac{1}{2} \int d\boldsymbol{r} \int d\boldsymbol{r}' \, \hat{\Psi}^\dagger_\alpha(\boldsymbol{r}) \hat{\Psi}^\dagger_\beta(\boldsymbol{r}') \frac{e^2}{|\boldsymbol{r} - \boldsymbol{r}'|} \hat{\Psi}_\beta(\boldsymbol{r}') \hat{\Psi}_\alpha(\boldsymbol{r}) \tag{9.4}$$

We will further assume that at $t < 0$ the system of N electrons was prepared in a mixed state characterized by the many-body density matrix $\rho_0(\boldsymbol{r}_1, \ldots, \boldsymbol{r}_N; \boldsymbol{r}'_1, \ldots, \boldsymbol{r}'_N)$.

Let us derive a formal path integral expression for the density of electric charge at point $\boldsymbol{r}$ and time t

$$\rho(t, \boldsymbol{r}) = -e \sum_{\alpha=\uparrow,\downarrow} \mathrm{tr} \left[\hat{\Psi}^\dagger_\alpha(\boldsymbol{r}) \hat{\Psi}_\alpha(\boldsymbol{r}) \mathrm{e}^{-\mathrm{i}\hat{H}t} \tilde{\rho}_0 \mathrm{e}^{\mathrm{i}\hat{H}t} \right] \tag{9.5}$$

Following the standard procedure, we perform the Hubbard–Stratonovich transformation and decouple the interaction term in the Hamiltonian (9.4), introducing the path integral over the fluctuating electric potential $V(t, \boldsymbol{r})$:

$$\mathrm{e}^{-\mathrm{i}\hat{H}t} = \frac{\int \mathcal{D}V(t', \boldsymbol{r}) \, \hat{\boldsymbol{U}}[t, 0; V(t, \boldsymbol{r})] \exp\left[\frac{\mathrm{i}}{2} \int_0^t dt' \int d\boldsymbol{r} \, \frac{(\nabla V(t', \boldsymbol{r}))^2}{4\pi} \right]}{\int \mathcal{D}V(t', \boldsymbol{r}) \, \exp\left[\frac{\mathrm{i}}{2} \int_0^t dt' \int d\boldsymbol{r} \, \frac{(\nabla V(t', \boldsymbol{r}))^2}{4\pi} \right]} \tag{9.6}$$

where

$$\hat{\boldsymbol{U}}[t_1, t_2; V] = \mathcal{T} \exp\left[-\mathrm{i} \int_{t_1}^{t_2} dt' \hat{H}_{\mathrm{eff}}[t'; V] \right] \tag{9.7}$$

is the time-ordered quantum mechanical evolution operator for the system of non-interacting electrons moving in an external potential $U_{\mathrm{imp}}(\boldsymbol{r})$ and being subject to the time- and coordinate-dependent electric field $\boldsymbol{E}(t, \boldsymbol{r}) = -\nabla V(t, \boldsymbol{r})$. This system is described by the Hamiltonian

$$\hat{H}_{\mathrm{eff}}[t; V] = \sum_{\alpha=\uparrow,\downarrow} \int d\boldsymbol{r} \, \hat{\Psi}^\dagger_\alpha(\boldsymbol{r}) \left[-\frac{\nabla^2}{2m} + U(\boldsymbol{r}) - eV(t, \boldsymbol{r}) \right] \hat{\Psi}_\alpha(\boldsymbol{r}) \tag{9.8}$$

Turning back to the charge density (9.5), we observe that the transformation (9.6) has to be performed separately in the operators $\mathrm{e}^{-\mathrm{i}\hat{H}t}$ and $\mathrm{e}^{\mathrm{i}\hat{H}t}$ corresponding, respectively, to the forward and backward branches of the Keldysh contour. As before, the corresponding quantities will be indicated by the subscripts F and B. The charge density (9.5) can now be expressed in the following form

$$\rho(t, \mathbf{r}) = -e\frac{\int \mathcal{D}V_F(t', \mathbf{r}) \int \mathcal{D}V_B(t', \mathbf{r})\, P(t, \mathbf{r}; V_F, V_B)\, e^{\mathrm{i}S_{\mathrm{em}}[V_F, V_B]}}{\int \mathcal{D}V_F(t', \mathbf{r}) \int \mathcal{D}V_B(t', \mathbf{r})\, e^{\mathrm{i}S_{\mathrm{em}}[V_F, V_B]}} \tag{9.9}$$

where we defined an effective, voltage-dependent, charge density

$$P(t, \mathbf{r}; V_F, V_B) = \sum_{\alpha=\uparrow,\downarrow} \mathrm{tr}\left[\hat{\boldsymbol{U}}^{\dagger}[t, 0; V_B]\hat{\Psi}_{\alpha}^{\dagger}(\mathbf{r})\hat{\Psi}_{\alpha}(\mathbf{r})\hat{\boldsymbol{U}}[t, 0; V_F]\tilde{\boldsymbol{\rho}}_0\right] \tag{9.10}$$

and the free electric field action

$$S_{\mathrm{em}}[V_F, V_B] = \int_0^t dt' \int d\mathbf{r}\, \frac{(\nabla V_F(t', \mathbf{r}))^2 - (\nabla V_B(t', \mathbf{r}))^2}{8\pi} \tag{9.11}$$

Let us now introduce an effective action S_{el} for a non-interacting electron gas subject to fluctuating electric potentials. It is given by the following formal expression:

$$\mathrm{i}S_{\mathrm{el}}[V_F, V_B] = \ln\left\{\mathrm{tr}\left[\hat{\boldsymbol{U}}^{\dagger}[t, 0; V_B]\hat{\boldsymbol{U}}[t, 0; V_F]\tilde{\boldsymbol{\rho}}_0\right]\right\} \tag{9.12}$$

The action (9.12) depends only on the fluctuating electric potentials V_F and V_B and not on the coordinates of individual electrons. We will now demonstrate that the charge density (9.9) can be expressed in terms of the action (9.12).

Let us add a small perturbation to the electric potential, $V \to V + \delta V$, and find a linear in δV correction to the time-ordered exponent (9.7). It reads

$$\delta\hat{\boldsymbol{U}}[t, 0; V] = \mathrm{i}e \sum_{\alpha=\uparrow,\downarrow} \int_0^t dt' \int d\mathbf{r}'\, \hat{\boldsymbol{U}}[t, t'; V]\delta V(t', \mathbf{r}')\hat{\Psi}_{\alpha}^{\dagger}(\mathbf{r}')\hat{\Psi}_{\alpha}(\mathbf{r}') \times \hat{\boldsymbol{U}}[t', 0; V].$$

Having found this correction, we can evaluate the functional derivative of the evolution operator

$$\frac{\delta\hat{\boldsymbol{U}}[t, 0; V_F]}{\delta V_F(t', \mathbf{r}')} = \mathrm{i}e \sum_{\alpha=\uparrow,\downarrow} \hat{\boldsymbol{U}}[t, t'; V_F]\hat{\Psi}_{\alpha}^{\dagger}(\mathbf{r}')\hat{\Psi}_{\alpha}(\mathbf{r}')\hat{\boldsymbol{U}}[t', 0; V_F] \tag{9.13}$$

In exactly the same way, we obtain

$$\frac{\delta \hat{\boldsymbol{U}}^\dagger[t,0;V_B]}{\delta V_B(t',\boldsymbol{r}')} = -\mathrm{i}e \sum_{\alpha=\uparrow,\downarrow} \hat{\boldsymbol{U}}^\dagger[t',0;V_B]\hat{\Psi}^\dagger_\alpha(\boldsymbol{r}')\hat{\Psi}_\alpha(\boldsymbol{r}')\hat{\boldsymbol{U}}^\dagger[t,t';V_B] \tag{9.14}$$

With the aid of these identities, we rewrite the charge density in the form

$$\rho(t,\boldsymbol{r}) = -\frac{1}{2}\frac{\lim_{t'\to t}\int \mathcal{D}V_F \int \mathcal{D}V_B \left(\frac{\delta S_{\rm el}}{\delta V_F(t',\boldsymbol{r})} - \frac{\delta S_{\rm el}}{\delta V_B(t',\boldsymbol{r})}\right) e^{\mathrm{i}S_{\rm el}+\mathrm{i}S_{\rm em}}}{\int \mathcal{D}V_F \int \mathcal{D}V_B\, e^{\mathrm{i}S_{\rm em}}} \tag{9.15}$$

At this point, exactly as we already did in earlier chapters, we again introduce both "classical" and "quantum" components of the fluctuating electric potential, respectively, $V^+ = (V_F + V_B)/2$ and $V^- = V_F - V_B$. Then we get

$$\rho(t,\boldsymbol{r}) = -\frac{\lim_{t'\to t}\int \mathcal{D}V^+ \int \mathcal{D}V^- \frac{\delta S_{\rm el}}{\delta V^-(t',\boldsymbol{r})}\, e^{\mathrm{i}S_{\rm el}+\mathrm{i}S_{\rm em}}}{\int \mathcal{D}V^+ \int \mathcal{D}V^-\, e^{\mathrm{i}S_{\rm em}}} \tag{9.16}$$

This equation demonstrates that the charge density $\rho(t,\boldsymbol{r})$ (9.5) can indeed be expressed as a path integral over the fluctuating potentials, which only involves the action (9.12). It is also worth pointing out that this expression contains the variation of the action over the quantum potential V^-.

In the same way, one can derive the path integral representation for the correlation functions of the charge density operators

$$\langle\hat{\rho}(t_1,\boldsymbol{r}_1)\hat{\rho}(t_2,\boldsymbol{r}_2)\rangle =$$
$$= e^2 \sum_{\alpha,\beta=\uparrow,\downarrow} \mathrm{tr}\left[e^{\mathrm{i}\hat{H}t_1}\hat{\Psi}^\dagger_\alpha(\boldsymbol{r}_1)\hat{\Psi}_\alpha(\boldsymbol{r}_1)e^{-\mathrm{i}\hat{H}(t_1-t_2)}\hat{\Psi}^\dagger_\beta(\boldsymbol{r}_2)\hat{\Psi}_\beta(\boldsymbol{r}_2)e^{-\mathrm{i}\hat{H}t_2}\tilde{\boldsymbol{\rho}}_0\right]$$

Let us first assume that $t_1 > t_2$ and perform the Hubbard–Stratonovich transformation in the operators $e^{\mathrm{i}\hat{H}t_1}$, $e^{-\mathrm{i}\hat{H}(t_1-t_2)}$, and $e^{-\mathrm{i}\hat{H}t_2}$. Having in mind that for such time ordering both $e^{-\mathrm{i}\hat{H}t_2}$ and $e^{-\mathrm{i}\hat{H}(t_1-t_2)}$ belong to the forward branch of the Keldysh contour, we obtain

$$\langle\hat{\rho}(t_1,\boldsymbol{r}_1)\hat{\rho}(t_2,\boldsymbol{r}_2)\rangle = e^2 \sum_{\alpha,\beta=\uparrow,\downarrow} \int \mathcal{D}V_F\,\mathcal{D}V_B\, e^{\mathrm{i}S_{\rm em}} \times$$
$$\frac{\mathrm{tr}\left[\hat{\boldsymbol{U}}^\dagger(t_1,0;V_B)\hat{\Psi}^\dagger_\alpha(\boldsymbol{r}_1)\hat{\Psi}_\alpha(\boldsymbol{r}_1)\hat{\boldsymbol{U}}(t_1,t_2;V_F)\hat{\Psi}^\dagger_\beta(\boldsymbol{r}_2)\hat{\Psi}_\beta(\boldsymbol{r}_2)\hat{\boldsymbol{U}}(t_2,0;V_F)\tilde{\rho}_0\right]}{\int \mathcal{D}V_F\,\mathcal{D}V_B\, e^{\mathrm{i}S_{\rm em}}}$$

This expression can be cast to the form

$$\langle \hat{\rho}(t_1, \boldsymbol{r}_1)\hat{\rho}(t_2, \boldsymbol{r}_2)\rangle = -\frac{\int \mathcal{D}V_{F,B}\, e^{\mathrm{i}S_{\mathrm{em}}}\, \mathrm{tr}\left[\hat{\boldsymbol{U}}^{\dagger}(t_1, 0; V_B)\frac{\delta^2 \hat{\boldsymbol{U}}(t,0;V_F)}{\delta V_F(t_1,\boldsymbol{r}_1)\delta V_F(t_2,\boldsymbol{r}_2)}\tilde{\boldsymbol{\rho}}_0\right]}{\int \mathcal{D}V_F\, \mathcal{D}V_B\, e^{\mathrm{i}S_{\mathrm{em}}}}$$

Comparing this formula with the effective action (9.12), we arrive at the following result

$$\langle \hat{\rho}(t_1, \boldsymbol{r}_1)\hat{\rho}(t_2, \boldsymbol{r}_2)\rangle = -\frac{\int \mathcal{D}V_F\, \mathcal{D}V_B\, e^{\mathrm{i}S_{\mathrm{em}}}\frac{\delta^2\, e^{\mathrm{i}S_{\mathrm{el}}}}{\delta V_F(X_1)\delta V_F(X_2)}}{\int \mathcal{D}V_F\, \mathcal{D}V_B\, e^{\mathrm{i}S_{\mathrm{em}}}} \tag{9.17}$$

where, as usual, $X_1 = (t_1, \boldsymbol{r}_1)$ and $X_2 = (t_2, \boldsymbol{r}_2)$ are the four component vectors. In the same way at $t_1 < t_2$, we obtain

$$\langle \hat{\rho}(t_1, \boldsymbol{r}_1)\hat{\rho}(t_2, \boldsymbol{r}_2)\rangle = -\frac{\int \mathcal{D}V_F\, \mathcal{D}V_B\, e^{\mathrm{i}S_{\mathrm{em}}}\frac{\delta^2\, e^{\mathrm{i}S_{\mathrm{el}}}}{\delta V_B(X_1)\delta V_B(X_2)}}{\int \mathcal{D}V_F\, \mathcal{D}V_B\, e^{\mathrm{i}S_{\mathrm{em}}}} \tag{9.18}$$

It is now straightforward to find the symmetrized correlation function of the charge densities. It reads

$$\frac{1}{2}\langle \hat{\rho}(t_1, \boldsymbol{r}_1)\hat{\rho}(t_2, \boldsymbol{r}_2) + \hat{\rho}(t_2, \boldsymbol{r}_2)\hat{\rho}(t_1, \boldsymbol{r}_1)\rangle$$
$$= \frac{\int \mathcal{D}V^{\pm}\, e^{\mathrm{i}S_{\mathrm{em}}}\left(\frac{1}{4}\frac{\delta^2}{\delta V^+(t_1,\boldsymbol{r}_1)\delta V^+(t_2,\boldsymbol{r}_2)} + \frac{\delta^2}{\delta V^-(t_1,\boldsymbol{r}_1)\delta V^-(t_2,\boldsymbol{r}_2)}\right)e^{\mathrm{i}S_{\mathrm{el}}}}{\int \mathcal{D}V_F\, \mathcal{D}V_B\, e^{\mathrm{i}S_{\mathrm{em}}}} \tag{9.19}$$

Proceeding analogously, one can also derive the corresponding path integral representations for higher-order correlation functions of the charge density. In addition, employing the continuity equation

$$\frac{\partial \hat{\rho}}{\partial t} + \nabla \hat{\boldsymbol{j}} = 0 \tag{9.20}$$

one can also find the correlation functions of the current densities $\hat{\boldsymbol{j}}$. Hence, having established the effective action (9.12), one can evaluate essentially any quantity that can be measured in transport experiments. For this reason, it is highly desirable to transform the formal expression (9.12) to a more tractable form, which could be useful for practical calculations.

Below we will illustrate two different ways of reaching this goal. The first one involves electron Keldysh Green functions and is frequently used in the literature. This approach is useful if one is interested, e.g., in the effect of weakly fluctuating electromagnetic potentials in a metal. An alternative and fully equivalent way is to express the action via the evolution operators for electrons propagating in the fluctuating electromagnetic potentials. This approach turns out to be particularly convenient for the analysis of transport properties of various spatially confined conductors, such as tunnel junctions and quantum dots. Both these approaches are outlined as follows.

9.2 Effective Action Expressed via Green Functions

We begin with more traditional representation of the effective action (9.12) involving electron Keldysh Green functions. In order to formulate this representation, let us rewrite the charge density via the path integral over anticommuting Grassmann fields $\psi_{\alpha,F,B}$ and $\psi^{\dagger}_{\alpha,F,B}$:

$$\rho(t,\boldsymbol{r}) = -e\int \mathcal{D}\psi_{\alpha,F,B}(t,\boldsymbol{r})\mathcal{D}\psi^{\dagger}_{\alpha,F,B}(t,\boldsymbol{r}) \sum_{\alpha=\uparrow,\downarrow} \psi^{\dagger}_{\alpha,F}(t,\boldsymbol{r})\psi_{\alpha,F}(t,\boldsymbol{r})$$
$$\times \exp\left[\sum_{\alpha=\uparrow,\downarrow}\int dt \int d\boldsymbol{r}\left(\mathrm{i}\psi^{\dagger}_{\alpha,F}\frac{\partial\psi_{\alpha,F}}{\partial t} - H[\psi^{\dagger}_{\alpha,F},\psi_{\alpha,F}]\right)\right]$$
$$\times \exp\left[-\sum_{\alpha=\uparrow,\downarrow}\int dt \int d\boldsymbol{r}\left(\mathrm{i}\psi^{\dagger}_{\alpha,B}\frac{\partial\psi_{\alpha,B}}{\partial t} - H[\psi^{\dagger}_{\alpha,B},\psi_{\alpha,B}]\right)\right] \tag{9.21}$$

Here the Hamiltonian of the electron gas reads

$$H[\psi^{\dagger}_{\alpha},\psi_{\alpha}] = \sum_{\alpha=\uparrow,\downarrow}\int d\boldsymbol{r}\left[\psi^{\dagger}_{\alpha}\left(-\frac{\nabla^2}{2m}+U(\boldsymbol{r})\right)\psi_{\alpha}\right]$$
$$+\sum_{\alpha,\beta=\uparrow,\downarrow}\frac{1}{2}\int d\boldsymbol{r}\int d\boldsymbol{r}'\,\psi^{\dagger}_{\alpha}(\boldsymbol{r})\psi^{\dagger}_{\beta}(\boldsymbol{r}')\frac{e^2}{|\boldsymbol{r}-\boldsymbol{r}'|}\psi_{\beta}(\boldsymbol{r}')\psi_{\alpha}(\boldsymbol{r}) \tag{9.22}$$

We now decouple the Coulomb terms in the Hamiltonian by means of the Hubbard–Stratonovich transformation giving rise to the fields $V_{F,B}(t,\boldsymbol{r})$. The integral over the electron variables $\psi_{\alpha,F,B}$, $\psi^{\dagger}_{\alpha,F,B}$ then becomes Gaussian and can be handled exactly. One can verify that as a result of this procedure, the charge density acquires the form (9.16) with

$$\mathrm{i}S_{\mathrm{el}} = \ln \int \mathcal{D}\psi_{F,B}\mathcal{D}\psi^{\dagger}_{F,B} \exp\left\{\mathrm{i}\tilde{S}[\psi_F, \psi^{\dagger}_F, V_F] - \mathrm{i}\tilde{S}[\psi_B, \psi^{\dagger}_B, V_B]\right\} \tag{9.23}$$

Here we defined the action

$$\mathrm{i}\tilde{S}[\psi, \psi^{\dagger}, V] = \sum_{\alpha=\uparrow,\downarrow} \int dt \int d\boldsymbol{r}\, \psi^{\dagger}_{\alpha}\left(\mathrm{i}\frac{\partial}{\partial t} + \frac{\nabla^2}{2m} - U(\boldsymbol{r}) + eV(t, \boldsymbol{r})\right)\psi_{\alpha}$$

Performing the path integrals over the fields ψ, $\psi^{\dagger}$ in Eq. (9.23), we arrive at the standard "trace of logarithm" formula for the action

$$\mathrm{i}S_{\mathrm{el}} = \mathrm{tr}\ \ln\left(\check{G}_V^{-1}\right) \tag{9.24}$$

where $\check{G}_V^{-1}$ is the inverse electron Keldysh Green function

$$\check{G}_V^{-1} = \begin{pmatrix} \mathrm{i}\frac{\partial}{\partial t} + \frac{\hbar^2\nabla^2}{2m} - U(\boldsymbol{r}) + eV_F & 0 \\ 0 & -\mathrm{i}\frac{\partial}{\partial t} - \frac{\hbar^2\nabla^2}{2m} + U(\boldsymbol{r}) - eV_B \end{pmatrix} \tag{9.25}$$

Equation (9.24) serves as a convenient starting point for various perturbative expansions. For instance, one can expand the exact action up to the second order in the fluctuating potentials V_F, V_B, thereby providing complete information about Gaussian fluctuations of these fields. This program will be accomplished in Section 9.4. Alternatively, one can expand the action in some other small parameter of the problem in question, e.g., in the effective transmission of a tunnel barrier between two metallic electrodes. This expansion has already been carried out in Chapter 2 while deriving the AES action for a Josephson junction.

In order to proceed with this perturbative expansion, one splits the inverse Green function (9.25) into an exactly solvable part $\check{G}_0^{-1}$ and the perturbation $\delta\check{G}^{-1}$. The part $\check{G}_0^{-1}$ can be inverted to give the zeroth-order Keldysh Green function of the system, which is expressed in terms of the averages of electron creation and annihilation operators

$$\begin{aligned}\check{G}_0 &\equiv \begin{pmatrix} G_{FF} & -G_{FB} \\ G_{BF} & -G_{BB} \end{pmatrix} \\ &= -\mathrm{i}\begin{pmatrix} \langle \mathcal{T}\hat{\Psi}_{\uparrow}(t_1, \boldsymbol{r}_1)\hat{\Psi}^{\dagger}_{\uparrow}(t_2, \boldsymbol{r}_2)\rangle & -\langle \hat{\Psi}^{\dagger}_{\uparrow}(t_2, \boldsymbol{r}_2)\hat{\Psi}_{\uparrow}(t_1, \boldsymbol{r}_1)\rangle \\ \langle \hat{\Psi}_{\uparrow}(t_1, \boldsymbol{r}_1)\hat{\Psi}^{\dagger}_{\uparrow}(t_2, \boldsymbol{r}_2)\rangle & \langle \tilde{\mathcal{T}}\hat{\Psi}_{\uparrow}(t_1, \boldsymbol{r}_1)\hat{\Psi}^{\dagger}_{\uparrow}(t_2, \boldsymbol{r}_2)\rangle \end{pmatrix}\end{aligned} \tag{9.26}$$

The symbols $\mathcal{T}$ and $\tilde{\mathcal{T}}$ stand, respectively, for the operator of time ordering and time anti-ordering of the fermionic operators, i.e.,

$$
\begin{aligned}
\langle \mathcal{T}\hat{\Psi}_\uparrow(t_1, \boldsymbol{r}_1)\hat{\Psi}^\dagger_\uparrow(t_2, \boldsymbol{r}_2)\rangle &= \theta(t_1 - t_2)\langle \hat{\Psi}_\uparrow(t_1, \boldsymbol{r}_1)\hat{\Psi}^\dagger_\uparrow(t_2, \boldsymbol{r}_2)\rangle \\
&\quad - \theta(t_2 - t_1)\langle \hat{\Psi}^\dagger_\uparrow(t_2, \boldsymbol{r}_2)\hat{\Psi}_\uparrow(t_1, \boldsymbol{r}_1)\rangle, \\
\langle \tilde{\mathcal{T}}\hat{\Psi}_\uparrow(t_1, \boldsymbol{r}_1)\hat{\Psi}^\dagger_\uparrow(t_2, \boldsymbol{r}_2)\rangle &= -\theta(t_1 - t_2)\langle \hat{\Psi}^\dagger_\uparrow(t_2, \boldsymbol{r}_2)\hat{\Psi}_\uparrow(t_1, \boldsymbol{r}_1)\rangle \\
&\quad + \theta(t_2 - t_1)\langle \hat{\Psi}_\uparrow(t_1, \boldsymbol{r}_1)\hat{\Psi}^\dagger_\uparrow(t_2, \boldsymbol{r}_2)\rangle
\end{aligned}
\tag{9.27}
$$

For simplicity here we consider only one spin orientation.

The next step is to expand the action (9.24) in powers of $\delta\check{G}^{-1}$,

$$
\mathrm{i}S_{\mathrm{el}} = \mathrm{tr}\,\left(\check{G}_0^{-1}\right) + \mathrm{tr}\,\left(\check{G}_0\,\delta\check{G}^{-1}\right) - \frac{1}{2}\,\mathrm{tr}\,\left(\check{G}_0\,\delta\check{G}^{-1}\check{G}_0\,\delta\check{G}^{-1}\right) + \ldots
\tag{9.28}
$$

enabling one to perturbatively evaluate the effective action term by term. This method is widely appreciated and frequently employed in the literature. Its main drawback, however, is that progressively more complicated integrals are encountered as one proceeds to higher orders of the perturbation theory. In many cases, it becomes hard to go beyond the second-order term in Eq. (9.28).

9.3 Effective Action Expressed via Evolution Operator

Let us now turn to an alternative and somewhat less standard representation of the effective action via the system evolution operators. For the sake of simplicity, we will initially assume that at $t < 0$, the system of N electrons was prepared in a pure state with the wave function $\Psi_0(\boldsymbol{r}_1, \ldots, \boldsymbol{r}_N)$, which we choose in the form of a Slater determinant constructed from N orthogonal single-electron wave functions $\psi_{k_j}(\boldsymbol{r})$, $j = 1 \ldots N$, where the indices k_j also include spin orientation and indicate the occupied states chosen from a full infinite set of orthogonal states $\psi_k(\boldsymbol{r})$. The wave function in the form of the Slater determinant is fully antisymmetric in accordance with the Pauli principle. It can be conveniently expressed with the aid of electron creation operators $a^\dagger_{k_j}$ in the states $\psi_{k_j}(\boldsymbol{r})$,

$$
\Psi_0(\boldsymbol{r}_1, \ldots, \boldsymbol{r}_N) = a^\dagger_{k_N} \ldots a^\dagger_{k_1}|0\rangle
\tag{9.29}
$$

The effective action (9.12) then acquires the form

$$\mathrm{i}S_{\mathrm{el}} = \left\langle 0 \left| a_{k_1} \ldots a_{k_N} \hat{\boldsymbol{U}}^{\dagger}[t, 0; V_B] \hat{\boldsymbol{U}}[t, 0; V_F] a^{\dagger}_{k_N} \ldots a^{\dagger}_{k_1} \right| 0 \right\rangle \quad (9.30)$$

In order to proceed, let us define the two sets of orthogonal wave functions $\psi_k^F(t)$ and $\psi_k^B(t)$, representing the solutions to the Schrödinger equations

$$\mathrm{i}\frac{\partial}{\partial t}\psi_k^{(j)}(t, \boldsymbol{r}) = \left[-\frac{\nabla^2}{2m} + U(\boldsymbol{r}) - eV_j(t, \boldsymbol{r})\right]\psi_k^{(j)}(t, \boldsymbol{r}), \quad j = F, B \quad (9.31)$$

with the initial conditions $\psi_k^F(0) = \psi_k^B(0) = \psi_k$. Equations (9.31) account for quantum evolution of a single electron interacting with fluctuating electric potentials $V_F(t)$ and $V_B(t)$. Having solved these equations, we describe the time evolution of the N-particle Slater determinants:

$$\begin{aligned} \hat{\boldsymbol{U}}(t, 0; V_F) a^{\dagger}_{k_N} \ldots a^{\dagger}_{k_1} \left| 0 \right\rangle &= \left(a^F_{k_N}(t)\right)^{\dagger} \ldots \left(a^F_{k_1}(t)\right)^{\dagger} \left| 0 \right\rangle, \\ \left\langle 0 \right| a_{k_1} \ldots a_{k_N} \hat{\boldsymbol{U}}^{\dagger}(t, 0; V_B) &= \langle 0 | a^B_{k_1}(t) \ldots a^B_{k_N}(t) \end{aligned} \quad (9.32)$$

where $a^B_{k_s}(t)$ and $\left(a^F_{k_s}(t)\right)^{\dagger}$ are, respectively, annihilation and creation operators for an electron in the states $\psi^B_{k_s}(t)$ and $\psi^F_{k_s}(t)$. Then the effective action (9.30) takes the form

$$\mathrm{i}S_{\mathrm{el}} = \ln \left\langle 0 \left| a^B_{k_1}(t) \ldots a^B_{k_N}(t) \left(a^F_{k_N}(t)\right)^{\dagger} \ldots \left(a^F_{k_1}(t)\right)^{\dagger} \right| 0 \right\rangle \quad (9.33)$$

In order to evaluate the action (9.33), it is necessary to consider the product of two Slater determinants constructed with the "forward" and "backward" wave functions $\psi^F_{k_s}(t)(\boldsymbol{r}_s)$ and $\psi^{B*}_{k_s}(t)(\boldsymbol{r}_s)$. Integrating this product over all electron coordinates, we recover the determinant of $N \times N$ matrix $A_{ij}(t)$ defined by its matrix elements

$$A_{ij}(t) = \int d\boldsymbol{r}\, \psi^{B*}_{k_i}(t, \boldsymbol{r}) \psi^F_{k_j}(t, \boldsymbol{r}) = \langle \psi^B_{k_i}(t) | \psi^F_{k_j}(t) \rangle \quad (9.34)$$

i.e.,

$$\mathrm{i}S_{\mathrm{el}} = \ln\left(\det[A_{ij}(t)]\right) = \mathrm{tr}\ln\left[A_{ij}(t)\right] \quad (9.35)$$

Let us now rewrite the action (9.35) in a somewhat more general form. First, we encode information about the initial distribution of electrons over the wave functions $\psi_k(\boldsymbol{r})$ in a single-particle density matrix

$$\boldsymbol{\rho}_0 = \sum_k n_k \left| \psi_k \right\rangle \left\langle \psi_k \right| \quad (9.36)$$

where $n_k = 1$ if the state k is initially occupied and $n_k = 0$ otherwise. Second, we define a single-particle evolution operator $\hat{\boldsymbol{u}}(t, t'; V_j)$ as a solution to Eq. (9.31) with the initial condition $\hat{\boldsymbol{u}}(t \to t', t'; V_j) = \delta(\boldsymbol{r} - \boldsymbol{r}')$. The matrix elements (9.34) can now be written in the form

$$A_{ij} = \langle \psi_{k_i} | \hat{\boldsymbol{u}}^\dagger(t, 0; V_B) \hat{\boldsymbol{u}}(t, 0; V_F) | \psi_{k_j} \rangle \tag{9.37}$$

One can verify that the matrix element of the operator

$$\boldsymbol{A} = \boldsymbol{\rho}_0 \hat{\boldsymbol{u}}^\dagger(t, 0; V_B) \hat{\boldsymbol{u}}(t, 0; V_F) \boldsymbol{\rho}_0$$

between two initially occupied states $\langle \psi_{k_i} |$ and $| \psi_{k_j} \rangle$ equals to A_{ij} defined in Eq. (9.37). The same matrix element equals to zero if at least one of these states remains unoccupied. Therefore, we can treat the operator $\boldsymbol{A}$ as an extension of the $N \times N$ matrix (9.34) to the full Hilbert space of wave functions with an infinite number of dimensions. Our next step is to construct the operator

$$\boldsymbol{B} = 1 + \boldsymbol{\rho}_0 \left(\hat{\boldsymbol{u}}^\dagger(t, 0; V_B) \hat{\boldsymbol{u}}(t, 0; V_F) - 1 \right) \boldsymbol{\rho}_0$$

with $\det \boldsymbol{B} = \det[A_{ij}]$. Then the action (9.35) can be cast to the form

$$\mathrm{i}S_{\mathrm{el}} = \mathrm{tr} \ln \left[1 + \boldsymbol{\rho}_0 \left(\hat{\boldsymbol{u}}^\dagger(t, 0; V_B) \hat{\boldsymbol{u}}(t, 0; V_F) - 1 \right) \boldsymbol{\rho}_0 \right] \tag{9.38}$$

The last step is to perform the formal series expansion of this expression in powers of the second term under the logarithm:

$$\mathrm{i}S_{\mathrm{el}} = \sum_{m=1}^{+\infty} \mathrm{tr} \left[\left(\boldsymbol{\rho}_0 \left(\hat{\boldsymbol{u}}^\dagger(t, 0; V_B) \hat{\boldsymbol{u}}(t, 0; V_F) - 1 \right) \boldsymbol{\rho}_0 \right)^m \right] \tag{9.39}$$

Since the initial state is pure, i.e., it is described by the wave function in the form of a Slater determinant, the corresponding density matrix satisfies the identity $\boldsymbol{\rho}_0^2 = \boldsymbol{\rho}_0$. With the aid of this identity, we can rewrite the action in the form

$$\mathrm{i}S_{\mathrm{el}} = \sum_{m=1}^{+\infty} \mathrm{tr} \left[\left(\left(\hat{\boldsymbol{u}}^\dagger(t, 0; V_B) \hat{\boldsymbol{u}}(t, 0; V_F) - 1 \right) \boldsymbol{\rho}_0 \right)^m \right] \tag{9.40}$$

and, finally,

$$\mathrm{i}S_{\mathrm{el}} = \mathrm{tr} \ln \left[1 + \left(\hat{\boldsymbol{u}}^\dagger(t, 0; V_B) \hat{\boldsymbol{u}}(t, 0; V_F) - 1 \right) \boldsymbol{\rho}_0 \right] \tag{9.41}$$

Equation (9.41) provides the desired expression for the effective action in terms of the evolution operators. Depending on the problem, this representation for the action may have certain advantages over Eq. (9.24). For instance, Eq. (9.41) is more convenient if there exists an explicit and relatively simple expression for the evolution operator $\hat{\boldsymbol{u}}$. If this is not the case, it may be inevitable to employ the perturbation theory, in which case Eq. (9.24) may, on the contrary, be more convenient than Eq. (9.41).

At this point, let us remind the reader that—for the sake of transparency—Eq. (9.41) was so far derived under the assumption of pure initial states described by the Slater determinant (9.29). It turns out, however, that the formula (9.41) is actually more general, i.e., it also applies for the case of mixed initial states corresponding to the equilibrium grand canonical ensemble of electrons with temperature T and chemical potential μ. The corresponding initial many-body density matrix reads

$$\begin{aligned}\tilde{\boldsymbol{\rho}}_0 &= \sum_N \sum_{k_1,\dots,k_N} W(N, k_1, \dots, k_N) \\ &\times \Psi_0(k_1, \dots, k_N, \boldsymbol{r}_1, \dots, \boldsymbol{r}_N)\Psi_0^*(k_1, \dots, k_N, \boldsymbol{r}'_1, \dots, \boldsymbol{r}'_N) \quad (9.42)\end{aligned}$$

where $\Psi_0(k_1, \dots, k_N, \boldsymbol{r}_1, \dots, \boldsymbol{r}_N)$ is the Slater determinant defined in Eq. (9.29) and

$$W(N, k_1, \dots, k_N) = \frac{e^{-E(k_1,\dots,k_N)/T}\, e^{\mu N/T}}{\sum_N \sum_{k_1,\dots,k_N} e^{-E(k_1,\dots,k_N)/T}\, e^{\mu N/T}} \quad (9.43)$$

is the occupation probability of the corresponding state. Here

$$E(k_1, \dots, k_N) = \sum_k \epsilon_k n_k \quad (9.44)$$

is the energy of the initial state (9.29), ϵ_k is the energy of k-th single-electron state, the occupation number equals to $n_k = 1$ ($n_k = 0$) for occupied (empty) states. Repeating the same procedure as in the case of pure initial states, for the action (9.12), we obtain

$$\begin{aligned}\mathrm{i}S_{\mathrm{el}}[V_F, V_B] = \ln \Bigg\{ &\sum_N \sum_{k_1,\dots,k_N} W(N, k_1, \dots, k_N) \\ &\times \langle 0|a_{k_1} \dots a_{k_N} \hat{\boldsymbol{u}}^\dagger[t, 0; V_B]\hat{\boldsymbol{u}}[t, 0; V_F]a^\dagger_{k_N} \dots a^\dagger_{k_1}|0\rangle \Bigg\} \quad (9.45)\end{aligned}$$

Next we apply a usual recipe of statistical physics and replace the double sum running over both N and all possible combinations of the occupied states $k_1, \dots, k_N$ by the sum over all possible combinations of the occupation numbers n_k equal to either 0 or 1. Then the action (9.45) acquires the form

$$
\begin{aligned}
iS_{\rm el}[V_F, V_B] = \ln\Bigg\{ \sum_{n_k=0,1} \left(\prod_k \frac{e^{-(\epsilon_k-\mu)n_k/T}}{1+e^{-(\epsilon_k-\mu)/T}} \right) \\
\times \det\left[1 + \left(\hat{\boldsymbol{u}}^\dagger[t,0;V_B]\hat{\boldsymbol{u}}[t,0;V_F] - 1\right)\sum_k n_k|\psi_k\rangle\langle\psi_k|\right]\Bigg\}
\end{aligned}
\tag{9.46}
$$

Defining the equilibrium occupation probability of k-th energy level $\bar{n}_k = 1/(1+e^{(\epsilon_k-\mu)/T})$, we rewrite the product in Eq. (9.46) as a determinant

$$
\prod_k \frac{e^{-(\epsilon_k-\mu)n_k/T}}{1+e^{-(\epsilon_k-\mu)/T}} = \det\left[\sum_k ((1-\bar{n}_k)(1-n_k) + \bar{n}_k n_k)|\psi_k\rangle\langle\psi_k|\right]
\tag{9.47}
$$

and get

$$
\begin{aligned}
iS_{\rm el}[V_F, V_B] = \ln\Bigg\{ \sum_{n_k=0,1} \det\Bigg[1 + \left(\hat{\boldsymbol{u}}^\dagger[t,0;V_B]\hat{\boldsymbol{u}}[t,0;V_F] - 1\right) \\
\times \sum_k n_k|\psi_k\rangle\langle\psi_k|\Bigg] \\
\times \det\left[\sum_k ((1-\bar{n}_k)(1-n_k) + \bar{n}_k n_k)|\psi_k\rangle\langle\psi_k|\right]\Bigg\}
\end{aligned}
\tag{9.48}
$$

Next we employ the property $\det A \det B = \det(AB)$ and rewrite the action in the form

$$
\begin{aligned}
iS_{\rm el}[V_F, V_B] = \ln\Bigg\{ \sum_{n_k=0,1} \det\Bigg[\sum_k ((1-\bar{n}_k)(1-n_k) + \bar{n}_k n_k)|\psi_k\rangle\langle\psi_k| \\
+ \left(\hat{\boldsymbol{u}}^\dagger[t,0;V_B]\hat{\boldsymbol{u}}[t,0;V_F] - 1\right)\sum_k \bar{n}_k n_k|\psi_k\rangle\langle\psi_k|\Bigg]\Bigg\}
\end{aligned}
\tag{9.49}
$$

Finally, we make use of the determinant additivity with respect to its columns. Specifically, denoting the $j-$th column of the matrix by bold letters $\boldsymbol{a}_j$, one finds

$$\det[\boldsymbol{a}_1, \boldsymbol{a}_2, \dots, \boldsymbol{a}_j + \boldsymbol{b}_j, \dots, \boldsymbol{a}_N] = \det[\boldsymbol{a}_1, \boldsymbol{a}_2, \dots, \boldsymbol{a}_j, \dots, \boldsymbol{a}_N] + \det[\boldsymbol{a}_1, \boldsymbol{a}_2, \dots, \boldsymbol{b}_j, \dots, \boldsymbol{a}_N] \tag{9.50}$$

Applying this property to each column, we arrive at the result

$$\det[\boldsymbol{a}_1 + \boldsymbol{b}_1, \boldsymbol{a}_2 + \boldsymbol{b}_2, \dots, \boldsymbol{a}_j + \boldsymbol{b}_j, \dots, \boldsymbol{a}_N + \boldsymbol{b}_N] = = \sum_{n_k=0,1} \det[\boldsymbol{a}_1(1-n_1) + \boldsymbol{b}_1 n_1, \boldsymbol{a}_2(1-n_2) + \boldsymbol{b}_2 n_2, \dots, \boldsymbol{a}_j(1-n_j) + \boldsymbol{b}_j n_j, \dots, \boldsymbol{a}_N(1-n_N) + \boldsymbol{b}_N n_N] \tag{9.51}$$

We observe that the right-hand side of Eq. (9.49) takes the same form as that of Eq. (9.51) in the basis $|\psi_k\rangle$. Hence, the action (9.49) can be recast to the form similar to that of the left-hand side of Eq. (9.51). Performing this operation, we again arrive at the action (9.41) with the initial density matrix

$$\boldsymbol{\rho}_0 = \sum_k \bar{n}_k |\psi_k\rangle\langle\psi_k| \tag{9.52}$$

This is nothing but the equilibrium single-electron density matrix of an ideal electron gas at temperature T.

Thus, we verified the validity of Eq. (9.41) for a much wider class of initial free electron gas states. In fact, one can also prove that the formula (9.41) also applies for non-equilibrium initial distributions with arbitrary occupation probabilities $\bar{n}_k$ of the single-particle states $|\psi_k\rangle$. However, we will not dwell into the corresponding analysis here.

9.4 Influence Functional for Electric Potentials

Having gone through the above formal exercises, we are now ready for practical calculations. As it was already announced, our first example will be the calculation of the effective action (or the influence functional) for the fluctuating electric potentials $V^{\pm}$. Below we will restrict ourselves to the so-called random phase approximation (RPA), which is widely used to describe electromagnetic properties of metals as well as many other materials. This approximation amounts to expanding the exact effective action up to quadratic terms in $V^{\pm}$, i.e., the action is determined by Eq. (9.28) terminated after the third term. RPA is formally justified only at sufficiently high electron concentrations. In practice, however, it provides a sufficiently accurate description for almost all metals and semiconductors.

Let us now explicitly employ Eq. (9.28), evaluating its right-hand side, term by term. The zero-order Keldysh Green function $\check{G}_0$ (9.26) just coincides with that for non-interacting electron gas, i.e., it should be evaluated at $V^+ = V^- = 0$. Actually, we have already performed this calculation in Chapter 2 in the case of a superconductor (cf. Eqs. (2.65)–(2.68)). The corresponding expressions for normal metals are obtained from these equations simply by setting the superconducting order parameter Δ equal to zero.

The correction $\delta\check{G}^{-1}$ reads

$$\delta\check{G}^{-1} = \begin{pmatrix} eV_F & 0 \\ 0 & -eV_B \end{pmatrix} \tag{9.53}$$

Substituting both $\check{G}_0$ and $\delta\check{G}^{-1}$ into Eq. (9.28), we observe that the first two terms in the right-hand side of this equation vanish identically since the Green function $\check{G}_0$ describes the system state with no electric current and zero charge density. Evaluating the third term in the right-hand side of Eq. (9.28), we obtain

$$\begin{aligned} iS_{\rm el} = -e^2\mathrm{tr}\Big[& G_{FF}V^+G_{FF}V^+ - G_{FB}V^+G_{BF}V^+ - G_{BF}V^+G_{FB}V^+ \\ & + G_{BB}V^+G_{BB}V^+ + G_{FF}V^+G_{FF}V^- - G_{FB}V^+G_{BF}V^- \\ & + G_{BF}V^+G_{FB}V^- - G_{BB}V^+G_{BB}V^- + \frac{1}{4}G_{FF}V^-G_{FF}V^- \\ & + \frac{1}{4}G_{FB}V^-G_{BF}V^- + \frac{1}{4}G_{BF}V^-G_{FB}V^- + \frac{1}{4}G_{BB}V^-G_{BB}V^-\Big] \end{aligned} \tag{9.54}$$

This expression can be simplified with the aid of the identities

$$G_{FF} = G_{FB} + G^R = G_{BF} + G^A, \qquad G_{BB} = G_{FB} - G^A = G_{BF} - G^R \tag{9.55}$$

where G^R and G^A are the retarded and advanced Green functions, which allow to exclude the functions G_{FF}, G_{BB}, G_{FB} and G_{BF} from the action (9.54). The terms containing V^+V^+ are reduced to

$$\mathrm{tr}(G^RV^+G^RV^+ + G^AV^+G^AV^+) \tag{9.56}$$

Making use of Eqs. (9.26), (9.27), and (9.55) and rewriting the traces (9.56) in the time–space representation, we immediately observe that the product $\theta(t_1 - t_2)\theta(t_2 - t_1) = 0$ appears under the integral and, hence, the whole combination (9.56) vanishes. This observation implies that the terms V^+V^+ give no contribution to the action (9.54). The latter observation is a direct consequence of the causality principle, which manifests itself through the presence of the Heaviside step functions $\theta(t_1 - t_2)$ and $\theta(t_2 - t_1)$ in the retarded and advanced Green functions. Collecting all other terms, we obtain [155]

$$\begin{aligned}
S_{\text{field}} &= S_{\text{em}} + S_{\text{el}} \\
&= \int dt \int d^3\boldsymbol{r} \frac{\nabla V^+(t,\boldsymbol{r})\,\nabla V^-(t,\boldsymbol{r})}{4\pi} \\
&\quad - \int dt_1 dt_2 \int d^3\boldsymbol{r}_F d^3\boldsymbol{r}_B\, V^-(t_1,\boldsymbol{r}_F)\chi(t_1,t_2,\boldsymbol{r}_F,\boldsymbol{r}_B)V^+(t_2,\boldsymbol{r}_B) \\
&\quad + \frac{\mathrm{i}}{2}\int dt_1 dt_2 \int d^3\boldsymbol{r}_F d^3\boldsymbol{r}_B\, V^-(t_1,\boldsymbol{r}_F)\eta(t_1,t_2,\boldsymbol{r}_F,\boldsymbol{r}_B)V^-(t_2,\boldsymbol{r}_B)
\end{aligned} \tag{9.57}$$

where

$$\begin{aligned}
\chi(t_1,t_2,\boldsymbol{r}_F,\boldsymbol{r}_B) = -\mathrm{i}e^2\big[&G^R(t_1,t_2,\boldsymbol{r}_F,\boldsymbol{r}_B)\big(G_{BF}(t_2,t_1,\boldsymbol{r}_B,\boldsymbol{r}_F) \\
&+ G_{FB}(t_2,t_1,\boldsymbol{r}_B,\boldsymbol{r}_F)\big) + \big(G_{FB}(t_1,t_2,\boldsymbol{r}_F,\boldsymbol{r}_B) \\
&+ G_{BF}(t_1,t_2,\boldsymbol{r}_F,\boldsymbol{r}_B)\big)G^A(t_2,t_1,\boldsymbol{r}_B,\boldsymbol{r}_F)\big],
\end{aligned} \tag{9.58}$$

$$\begin{aligned}
\eta(t_1,t_2,\boldsymbol{r}_F,\boldsymbol{r}_B) = \frac{e^2}{2}\big[&G_{FB}(t_1,t_2,\boldsymbol{r}_F,\boldsymbol{r}_B)G_{BF}(t_2,t_1,\boldsymbol{r}_B,\boldsymbol{r}_F) \\
&+ G_{BF}(t_1,t_2,\boldsymbol{r}_F,\boldsymbol{r}_B)G_{FB}(t_2,t_1,\boldsymbol{r}_B,\boldsymbol{r}_F)\big]
\end{aligned} \tag{9.59}$$

The expressions (9.57)–(9.59) define the influence functional for the field V in terms of the Green–Keldysh functions for (in general non-equilibrium) electron subsystems. It is easy to check that the above expressions satisfy the requirement of causality: in the $V^-(t_1)V^+(t_2)$ terms, the time t_1 is always larger than t_2 due to analytic properties of retarded and advanced Green functions. It is also straightforward to verify that in thermodynamic equilibrium, the kernels $\eta(\omega)$ and $\mathrm{Im}\,\chi(\omega)$ (Fourier transformed with respect to the time difference $t_1 - t_2$) satisfy the equation [155]

$$\eta(\omega, \boldsymbol{r}_F, \boldsymbol{r}_B) = -\frac{1}{2}\coth\left(\frac{\omega}{2T}\right)\mathrm{Im}\chi(\omega, \boldsymbol{r}_F, \boldsymbol{r}_B) \tag{9.60}$$

which is in agreement with FDT.

In homogeneous metals and at sufficiently large scales, one finds that the functions χ and η depend only on the differences of their arguments $t_1 - t_2$ and $\boldsymbol{r}_1 - \boldsymbol{r}_2$. Then one can express them via the dielectric susceptibility of the system $\epsilon(\omega, \boldsymbol{k})$,

$$\chi = -\int \frac{d^3k}{(2\pi)^3}\frac{k^2\epsilon(\omega, \boldsymbol{k})}{4\pi} e^{-\mathrm{i}\omega(t_1-t_2)+\mathrm{i}\boldsymbol{k}(\boldsymbol{r}_F-\boldsymbol{r}_B)}, \tag{9.61}$$

$$\eta = \int \frac{d^3k}{(2\pi)^3}\frac{k^2\,\mathrm{Im}\,[\epsilon(\omega, \boldsymbol{k})]}{4\pi}\coth\frac{\hbar\omega}{2T} e^{-\mathrm{i}\omega(t_1-t_2)+\mathrm{i}\boldsymbol{k}(\boldsymbol{r}_F-\boldsymbol{r}_B)} \tag{9.62}$$

In the case of ballistic electron gas, its dielectric susceptibility is defined by means of the exact Lindhard formula [283]

$$\epsilon(\omega, \boldsymbol{k}) = 1 + \frac{e^2}{\pi^2k^2}\int d^3q\,\frac{n(\xi_{\boldsymbol{q}+\boldsymbol{k}}) - n(\xi_{\boldsymbol{q}})}{\omega - \xi_{\boldsymbol{q}+\boldsymbol{k}} + \xi_{\boldsymbol{q}} + \mathrm{i}0} \tag{9.63}$$

where $n(\xi) = 1/(1 + e^{\xi/T})$ is the Fermi function, $\xi_{\boldsymbol{q}} = \hbar^2q^2/2m - \mu$ is the energy of an electron with a momentum $\hbar\boldsymbol{q}$, and μ is the chemical potential of the electron gas. In the diffusive limit, the function $\epsilon(\omega, \boldsymbol{k})$ is defined by the Drude formula (8.15), which remains accurate at wave vectors smaller than the inverse elastic mean free path $k \lesssim 1/\ell$ and at small frequencies $\omega \lesssim v_F/\ell$. Note that for (nearly) uniform in space ($k \approx 0$) fluctuations of the V-fields, the above effective action reduces to the Caldeira–Leggett form and exactly coincides with the action of a metallic shunt already derived in Section 2.5.2 by means of the quasiclassical Eilenberger equations. Hence, in a disordered metal, the effect of electron–electron interactions is equivalent to that of an effective *dissipative* environment with the correspondent effective viscosity governed by the Drude conductivity σ_D.

It is also important to point out that the effect of lattice ions (phonons) as well as other possibly existing interactions can be easily incorporated into our analysis simply by proper generalization of the dielectric function $\epsilon(\omega, \boldsymbol{k})$ for our problem. However, we will refrain from specifying the corresponding expressions here.

Having derived the action (9.57), we are now in a position also to evaluate the pair correlators of the fluctuating potentials. We obtain (cf. also Eqs. (8.8)–(8.10)):

$$\left\langle V^{+}(t_1, \boldsymbol{r}_1)V^{+}(t_2, \boldsymbol{r}_2)\right\rangle \equiv I(t_1 - t_2, \boldsymbol{r}_1 - \boldsymbol{r}_2)$$
$$= -\int \frac{d\omega d^3k}{(2\pi)^4}\frac{4\pi}{k^2}\,\mathrm{Im}\left(\frac{\coth\frac{\hbar\omega}{2T}}{\epsilon(\omega, k)}\right) e^{-\mathrm{i}\omega(t_1-t_2)+\mathrm{i}\boldsymbol{k}(\boldsymbol{r}_1-\boldsymbol{r}_2)} \quad (9.64)$$

$$\left\langle V^{+}(t_1, \boldsymbol{r}_1)V^{-}(t_2, \boldsymbol{r}_2)\right\rangle \equiv \mathrm{i}R(t_1 - t_2, \boldsymbol{r}_1 - \boldsymbol{r}_2)$$
$$= \mathrm{i}\int \frac{d\omega d^3k}{(2\pi)^4}\frac{4\pi}{k^2\epsilon(\omega, k)} e^{-\mathrm{i}\omega(t_1-t_2)+\mathrm{i}\boldsymbol{k}(\boldsymbol{r}_1-\boldsymbol{r}_2)} \quad (9.65)$$

$$\left\langle V^{-}(t_1, \boldsymbol{r}_1)V^{-}(t_2, \boldsymbol{r}_2)\right\rangle \equiv 0 \quad (9.66)$$

The above averages also define the Keldysh Green functions of photons frequently used in the literature. For example, the function $R(t_1 - t_2, \boldsymbol{r}_1 - \boldsymbol{r}_2)$ in Eq. (9.65) coincides with the retarded photon Green function. One can show that due to the causality principle and the corresponding analytic properties of $\epsilon(\omega, k)$, the average value $\left\langle V^{+}(t_1, \boldsymbol{r}_1)V^{-}(t_2, \boldsymbol{r}_2)\right\rangle$ differs from zero at $t_1 > t_2$ and vanishes identically at $t_1 < t_2$.

9.5 Time Evolution of a Single-Electron Density Matrix

Let us now turn back to quantum dynamics of interacting electrons and derive a formally exact evolution equation for their density matrix. This equation will then play a crucial role in deriving the influence functional for an interacting Fermi gas.

In what follows, we will employ the standard definition for the single-electron density matrix

$$\rho(t, \boldsymbol{r}_1, \boldsymbol{r}_2) = \mathrm{tr}\left[\hat{\Psi}^{\dagger}(\boldsymbol{r}_2)\hat{\Psi}(\boldsymbol{r}_1)e^{-\mathrm{i}Ht}\tilde{\rho}_0 e^{\mathrm{i}Ht}\right] \quad (9.67)$$

Repeating the analysis of Section 9.1, we again apply the Hubbard–Stratonovich transformations to the operators $e^{-\mathrm{i}Ht}$ and $e^{\mathrm{i}Ht}$ and express $\boldsymbol{\rho}$ as a path integral

$$\rho(t, \boldsymbol{r}_1, \boldsymbol{r}_2) = \frac{\int \mathcal{D}V_F\mathcal{D}V_B\, \rho_V(t, \boldsymbol{r}_1, \boldsymbol{r}_2)\, e^{\mathrm{i}S_{\mathrm{em}}+\mathrm{i}S_{\mathrm{el}}}}{\int \mathcal{D}V_F\mathcal{D}V_B\, e^{\mathrm{i}S_{\mathrm{em}}}} \quad (9.68)$$

where we introduced the voltage-dependent density matrix of a single electron

$$\boldsymbol{\rho}_V(t, \boldsymbol{r}_1, \boldsymbol{r}_2) = \frac{\mathrm{tr}\left[\hat{\boldsymbol{U}}^\dagger(t, 0; V_B)\hat{\Psi}^\dagger(\boldsymbol{r}_2)\hat{\Psi}(\boldsymbol{r}_1)\hat{\boldsymbol{U}}(t, 0; V_F)\tilde{\boldsymbol{\rho}}_0\right]}{\mathrm{tr}\left[\hat{\boldsymbol{U}}^\dagger(t, 0; V_B)\hat{\boldsymbol{U}}(t, 0; V_F)\tilde{\boldsymbol{\rho}}_0\right]} \tag{9.69}$$

Let us derive the time-dependent Schrödinger equation for the density matrix $\boldsymbol{\rho}_V$. For convenience, we express the Hamiltonians (9.8) in the form

$$\hat{H}_{\mathrm{eff}}[t; V_j] = \int d\boldsymbol{r}' d\boldsymbol{r}''\, \hat{\Psi}^\dagger(\boldsymbol{r}') H_j(t, \boldsymbol{r}', \boldsymbol{r}'')\hat{\Psi}(\boldsymbol{r}''), \quad j = F, B \tag{9.70}$$

and also suppress the spin index by absorbing it into the electron coordinates. Employing obvious relations

$$\mathrm{i}\partial_t \hat{\boldsymbol{U}}(t, 0; V_F) = \hat{H}_{\mathrm{eff}}[t; V_F]\hat{\boldsymbol{U}}(t, 0; V_F),$$
$$\mathrm{i}\partial_t \hat{\boldsymbol{U}}^\dagger(t, 0; V_B) = -\hat{\boldsymbol{U}}^\dagger(t, 0; V_B)\hat{H}_{\mathrm{eff}}[t; V_B]$$

we obtain

$$\begin{aligned}
\mathrm{i}\frac{\partial \boldsymbol{\rho}_V}{\partial t} = &-\frac{\mathrm{tr}\left[\hat{\boldsymbol{U}}^\dagger(t, 0; V_B)\hat{H}_{\mathrm{eff}}[t; V_B]\hat{\Psi}^\dagger(\boldsymbol{r}_2)\hat{\Psi}(\boldsymbol{r}_1)\hat{\boldsymbol{U}}(t, 0; V_F)\tilde{\boldsymbol{\rho}}_0\right]}{\mathrm{tr}\left[\hat{\boldsymbol{U}}^\dagger(t, 0; V_B)\hat{\boldsymbol{U}}(t, 0; V_F)\tilde{\boldsymbol{\rho}}_0\right]} \\
&+\frac{\mathrm{tr}\left[\hat{\boldsymbol{U}}^\dagger(t, 0; V_B)\hat{\Psi}^\dagger(\boldsymbol{r}_2)\hat{\Psi}(\boldsymbol{r}_1)\hat{H}_{\mathrm{eff}}[t; V_F]\hat{\boldsymbol{U}}(t, 0; V_F)\tilde{\boldsymbol{\rho}}_0\right]}{\mathrm{tr}\left[\hat{\boldsymbol{U}}^\dagger(t, 0; V_B)\hat{\boldsymbol{U}}(t, 0; V_F)\tilde{\boldsymbol{\rho}}_0\right]} \\
&-\frac{\mathrm{tr}\left[\hat{\boldsymbol{U}}^\dagger(t, 0; V_B)\hat{\Psi}^\dagger(\boldsymbol{r}_2)\hat{\Psi}(\boldsymbol{r}_1)\hat{\boldsymbol{U}}(t, 0; V_F)\tilde{\boldsymbol{\rho}}_0\right]}{\mathrm{tr}\left[\hat{\boldsymbol{U}}^\dagger(t, 0; V_B)\hat{\boldsymbol{U}}(t, 0; V_F)\tilde{\boldsymbol{\rho}}_0\right]} \\
&\times\frac{\mathrm{tr}\left[\hat{\boldsymbol{U}}^\dagger(t, 0; V_B)\left(\hat{H}_{\mathrm{eff}}[t; V_F] - \hat{H}_{\mathrm{eff}}[t; V_B]\right)\hat{\boldsymbol{U}}(t, 0; V_F)\tilde{\boldsymbol{\rho}}_0\right]}{\mathrm{tr}\left[\hat{\boldsymbol{U}}^\dagger(t, 0; V_B)\hat{\boldsymbol{U}}(t, 0; V_F)\tilde{\boldsymbol{\rho}}_0\right]}
\end{aligned}$$

This equation for the density matrix ρ_V may also be rewritten in an equivalent form

$$\begin{aligned}
\mathrm{i}\frac{\partial \boldsymbol{\rho}_V}{\partial t} = &\int d\boldsymbol{r}' d\boldsymbol{r}''\, H_F(t, \boldsymbol{r}', \boldsymbol{r}'')\left\langle \hat{\Psi}^\dagger(\boldsymbol{r}_2)\hat{\Psi}(\boldsymbol{r}_1)\hat{\Psi}^\dagger(\boldsymbol{r}')\hat{\Psi}(\boldsymbol{r}'')\right\rangle \\
&-\int d\boldsymbol{r}' d\boldsymbol{r}''\, H_B(t, \boldsymbol{r}', \boldsymbol{r}'')\left\langle \hat{\Psi}^\dagger(\boldsymbol{r}')\hat{\Psi}(\boldsymbol{r}'')\hat{\Psi}^\dagger(\boldsymbol{r}_2)\hat{\Psi}(\boldsymbol{r}_1)\right\rangle \\
&-\int d\boldsymbol{r}' d\boldsymbol{r}''\, H_F(t, \boldsymbol{r}', \boldsymbol{r}'')\left\langle \hat{\Psi}^\dagger(\boldsymbol{r}_2)\hat{\Psi}(\boldsymbol{r}_1)\right\rangle\left\langle \hat{\Psi}^\dagger(\boldsymbol{r}')\hat{\Psi}(\boldsymbol{r}'')\right\rangle \\
&-\int d\boldsymbol{r}' d\boldsymbol{r}''\, H_F(t, \boldsymbol{r}', \boldsymbol{r}'')\left\langle \hat{\Psi}^\dagger(\boldsymbol{r}')\hat{\Psi}(\boldsymbol{r}'')\right\rangle\left\langle \hat{\Psi}^\dagger(\boldsymbol{r}_2)\hat{\Psi}(\boldsymbol{r}_1)\right\rangle
\end{aligned} \tag{9.71}$$

where we denote

$$\langle \ldots \rangle \equiv \frac{\mathrm{tr}\left[\hat{\boldsymbol{U}}^{\dagger}(t,0;V_B)\ldots\hat{\boldsymbol{U}}(t,0;V_F)\tilde{\boldsymbol{\rho}}_0\right]}{\mathrm{tr}\left[\hat{\boldsymbol{U}}^{\dagger}(t,0;V_B)\hat{\boldsymbol{U}}(t,0;V_F)\tilde{\boldsymbol{\rho}}_0\right]} \tag{9.72}$$

An important advantage of the adopted formulation is that the operators $\hat{\boldsymbol{U}}[t, 0; V_{F,B}]$ describe evolution of *non-interacting* electrons propagating in external fluctuating electric fields corresponding to potentials $V_{F,B}$. Hence, in order to handle averaging of the products of creation and annihilation operators, one can make use of the standard Wick's theorem. According to this theorem, one has

$$\begin{aligned}\left\langle\hat{\Psi}^{\dagger}(\boldsymbol{r}_2)\hat{\Psi}(\boldsymbol{r}_1)\hat{\Psi}^{\dagger}(\boldsymbol{r}')\hat{\Psi}(\boldsymbol{r}'')\right\rangle &= \left\langle\hat{\Psi}^{\dagger}(\boldsymbol{r}_2)\hat{\Psi}(\boldsymbol{r}_1)\right\rangle\left\langle\hat{\Psi}^{\dagger}(\boldsymbol{r}')\hat{\Psi}(\boldsymbol{r}'')\right\rangle\\ &+\left\langle\hat{\Psi}^{\dagger}(\boldsymbol{r}_2)\hat{\Psi}(\boldsymbol{r}'')\right\rangle\left\langle\hat{\Psi}(\boldsymbol{r}_1)\hat{\Psi}^{\dagger}(\boldsymbol{r}')\right\rangle,\end{aligned} \tag{9.73}$$

$$\begin{aligned}\left\langle\hat{\Psi}^{\dagger}(\boldsymbol{r}')\hat{\Psi}(\boldsymbol{r}'')\hat{\Psi}^{\dagger}(\boldsymbol{r}_2)\hat{\Psi}(\boldsymbol{r}_1)\right\rangle &= \left\langle\hat{\Psi}^{\dagger}(\boldsymbol{r}')\hat{\Psi}(\boldsymbol{r}'')\right\rangle\left\langle\hat{\Psi}^{\dagger}(\boldsymbol{r}_2)\hat{\Psi}(\boldsymbol{r}_1)\right\rangle\\ &+\left\langle\hat{\Psi}^{\dagger}(\boldsymbol{r}')\hat{\Psi}(\boldsymbol{r}_1)\right\rangle\left\langle\hat{\Psi}(\boldsymbol{r}'')\hat{\Psi}^{\dagger}(\boldsymbol{r}_2)\right\rangle\end{aligned} \tag{9.74}$$

Employing now the definition of the electron density matrix $\boldsymbol{\rho}_V$ (9.69), we obtain

$$\begin{aligned}\left\langle\hat{\Psi}^{\dagger}(\boldsymbol{r}_2)\hat{\Psi}(\boldsymbol{r}'')\right\rangle &= \rho_V(t,\boldsymbol{r}'',\boldsymbol{r}_2),\\ \left\langle\hat{\Psi}(\boldsymbol{r}_1)\hat{\Psi}^{\dagger}(\boldsymbol{r}')\right\rangle &= \delta(\boldsymbol{r}_1-\boldsymbol{r}')-\rho_V(t,\boldsymbol{r}_1,\boldsymbol{r}'),\\ \left\langle\hat{\Psi}^{\dagger}(\boldsymbol{r}')\hat{\Psi}(\boldsymbol{r}_1)\right\rangle &= \rho_V(t,\boldsymbol{r}_1,\boldsymbol{r}')\\ \left\langle\hat{\Psi}(\boldsymbol{r}'')\hat{\Psi}^{\dagger}(\boldsymbol{r}_2)\right\rangle &= \delta(\boldsymbol{r}''-\boldsymbol{r}_2)-\rho_V(t,\boldsymbol{r}'',\boldsymbol{r}_2)\end{aligned} \tag{9.75}$$

With the aid of the above relations, the Schrödinger equation (9.71) can be rewritten as

$$\begin{aligned}&\mathrm{i}\frac{\partial}{\partial t}\rho_V(t,\boldsymbol{r}_1,\boldsymbol{r}_2)\\ &=\int d\boldsymbol{r}'d\boldsymbol{r}''[\delta(\boldsymbol{r}_1-\boldsymbol{r}')-\rho_V(t,\boldsymbol{r}_1,\boldsymbol{r}')]H_F(t,\boldsymbol{r}',\boldsymbol{r}'')\rho_V(t,\boldsymbol{r}'',\boldsymbol{r}_2)\\ &-\int d\boldsymbol{r}'d\boldsymbol{r}''\,\rho_V(t,\boldsymbol{r}_1,\boldsymbol{r}')H_B(t,\boldsymbol{r}',\boldsymbol{r}'')[\delta(\boldsymbol{r}''-\boldsymbol{r}_2)-\rho_V(t,\boldsymbol{r}'',\boldsymbol{r}_2)]\end{aligned}$$

or in an equivalent symbolic form [155]:

$$\mathrm{i}\frac{\partial\boldsymbol{\rho}_V}{\partial t} = (\mathbf{1}-\boldsymbol{\rho}_V)\hat{H}_F\boldsymbol{\rho}_V - \boldsymbol{\rho}_V\hat{H}_B(\mathbf{1}-\boldsymbol{\rho}_V) \tag{9.76}$$

where H_F and H_B are the Hamiltonians of a single electron propagating in the "forward" or "backward" fluctuating electric field created by all other electrons, i.e.

$$\hat{H}_{F,B} = \hat{H}_0 - eV_{F,B}(t, \boldsymbol{r}) \tag{9.77}$$

and

$$\hat{H}_0 = -\frac{\nabla^2}{2m} + U(\boldsymbol{r}) \tag{9.78}$$

is, as before, the Hamiltonian of non-interacting electrons.

Equation (9.76) constitutes an important result. This equation is *exact* and remains formally applicable at any interaction strength. It clearly illustrates the importance of the Pauli principle for quantum dynamics of an interacting Fermi gas. In order to make the effect of Fermi statistics completely transparent, let us identically rewrite Eq. (9.76) as

$$\mathrm{i}\frac{\partial \boldsymbol{\rho}_V}{\partial t} = [\hat{H}_0 - eV^+, \boldsymbol{\rho}_V] - (1-\boldsymbol{\rho}_V)\frac{eV^-}{2}\boldsymbol{\rho}_V - \boldsymbol{\rho}_V\frac{eV^-}{2}(1-\boldsymbol{\rho}_V) \tag{9.79}$$

where $\hat{H}_0$ is the Hamiltonian of non-interacting electrons (9.3) and, as before, $[a, b] = ab - ba$ denotes the commutator. We observe that the classical and quantum components of the fluctuating potential V^+ and V^- enter differently in Eq. (9.79). The classical potential component $V^+(t, \boldsymbol{r})$ plays the same role as an external potential. All electrons move collectively in this field; its presence is essentially equivalent to local fluctuations of the Fermi energy $\mu \rightarrow \mu + eV^+(t, \boldsymbol{r})$. Accordingly, the Pauli principle is irrelevant for this field.

The terms containing the quantum field V^- in Eq. (9.79) are, on the contrary, very sensitive to the Pauli principle. These terms demonstrate that the electron scattering rate is proportional to the concentration of electrons multiplied by the concentration of unoccupied states (or holes). The structure of these terms is standard for Fermi systems. It appears, e.g., in the electron–phonon collision integral of the Boltzmann kinetic equation. The terms containing V^- are also responsible for damping of an electron that propagates in a metal. Corresponding energy losses can be only due to electron transitions into lower energy states. At $T = 0$ in equilibrium, all such states are already occupied by other electrons; therefore, such processes are forbidden and the electron energy remains unchanged due to the Pauli principle.

9.6 Evolution Operators and Conductivity

Let us define the evolution operators

$$\hat{\boldsymbol{U}}_{F,B}(t,t') = \mathcal{T}\exp\left[-\mathrm{i}\int_{t'}^{t} d\tau\,(\hat{\mathcal{H}}_{F,B}(\tau)-\mu)\right] \tag{9.80}$$

where

$$\begin{aligned}\hat{\mathcal{H}}_F(t) &= \hat{H}_0 - e\hat{V}^+(t) - \frac{1}{2}[1-2\hat{\rho}_V(t)]e\hat{V}^-(t),\\ \hat{\mathcal{H}}_B(t) &= \hat{H}_0 - e\hat{V}^+(t) + \frac{1}{2}e\hat{V}^-(t)[1-2\hat{\rho}_V(t)]\end{aligned} \tag{9.81}$$

Here and below in this section for brevity, we explicitly indicate only the time dependence of the operators and suppress the coordinate and momentum variables.

As it was demonstrated above, the density matrix $\rho_V(t)$ obeys Eq. (9.79). Averaging of $\rho_V(t)$ over $V^{\pm}$ yields the exact single-electron density matrix in the presence of disorder and Coulomb interactions. Before turning to the path integral representation, let us perform some manipulations and derive exact relations involving the evolution operators. We first note that the solution to Eq. (9.79) with the initial condition $\hat{\rho}_V(0) = \hat{\rho}_0 \equiv [1+\mathrm{e}^{(\hat{H}_0-\mu)/T}]^{-1}$ can be expressed in the following exact form

$$\hat{\rho}_V(t) = \left[1 + \hat{\boldsymbol{u}}_B(t,0)\mathrm{e}^{(\hat{H}_0-\mu)/T}\hat{\boldsymbol{u}}_F(0,t)\right]^{-1} \tag{9.82}$$

where

$$\hat{\boldsymbol{u}}_{F,B}(t,t') = \mathcal{T}\exp\left[-\mathrm{i}\int_{t'}^{t} d\tau\,(\hat{H}_{F,B}(\tau)-\mu)\right] \tag{9.83}$$

and the Hamiltonians $\hat{H}_{F,B}$ (9.77) are rewritten as

$$\hat{H}_{F,B} = \hat{H}_0 - e\hat{V}^+(t) \mp e\hat{V}^-(t)/2 \tag{9.84}$$

The operators $\hat{\boldsymbol{U}}_{F,B}$ satisfy the Schrödinger equation

$$\begin{aligned}\mathrm{i}\frac{\partial}{\partial t}\hat{\boldsymbol{U}}_F(t,t') &= \hat{\mathcal{H}}_F(t)\hat{\boldsymbol{U}}_F(t,t'), \quad \hat{\boldsymbol{U}}_F(t',t') = \hat{1},\\ \mathrm{i}\frac{\partial}{\partial t}\hat{\boldsymbol{U}}_B(t',t) &= \hat{\boldsymbol{U}}_B(t',t)\hat{\mathcal{H}}_B(t), \quad \hat{\boldsymbol{U}}_B(t',t') = \hat{1}\end{aligned} \tag{9.85}$$

The solutions to Eq. (9.85) can be found exactly. They read

$$\begin{aligned}\hat{\boldsymbol{U}}_F(t,t') &= [1-\hat{\rho}_V(t)]\hat{\boldsymbol{u}}_F(t,t')[1-\hat{\rho}_V(t')]^{-1},\\ \hat{\boldsymbol{U}}_B(t',t) &= [1-\hat{\rho}_V(t')]^{-1}\hat{\boldsymbol{u}}_B(t',t)[1-\hat{\rho}_V(t)]\end{aligned} \tag{9.86}$$

Combining these expressions with Eq. (9.82), one can rewrite the operators (9.86) in the following identical form [162]:

$$\hat{\boldsymbol{U}}_F(t,t') = [(1-\hat{\boldsymbol{\rho}}_V(t'))\hat{\boldsymbol{u}}_F(t',t) + \hat{\boldsymbol{\rho}}_V(t')\hat{\boldsymbol{u}}_B(t',t)]^{-1},$$
$$\hat{\boldsymbol{U}}_B(t',t) = [\hat{\boldsymbol{u}}_B(t,t')(1-\hat{\boldsymbol{\rho}}_V(t')) + \hat{\boldsymbol{u}}_F(t,t')\hat{\boldsymbol{\rho}}_V(t')]^{-1} \quad (9.87)$$

Equations (9.87) establish the exact relations between the evolution operators $\hat{\boldsymbol{U}}_{F,B}$ (9.80) containing the density matrix operator $\hat{\boldsymbol{\rho}}_V$ and the operators $\hat{\boldsymbol{u}}_{F,B}$ (9.83) not involving this density matrix at all. It is also important to note that the operators $\hat{\boldsymbol{U}}_{F,B}$ (9.87) depend on the density matrix $\hat{\boldsymbol{\rho}}_V$ taken at the initial time t' only but not at later time moments. The above relations can be convenient for implementing various approximations. We will explicitly make use of Eqs. (9.87) in Chapter 14.

It is also instructive to express the general quantum mechanical expression for the linear conductivity σ of our system in terms of the evolution operators $\hat{\boldsymbol{U}}_{F,B}$ (9.80). The corresponding expression reads

$$\sigma = \frac{e}{3\mathrm{i}} \int_{-\infty}^{t} dt' \left\langle \mathrm{tr}\left(\hat{\boldsymbol{j}}(\boldsymbol{r})\hat{\boldsymbol{U}}_F(t,t')[\hat{\boldsymbol{r}}, \hat{\boldsymbol{\rho}}_V(t')]\hat{\boldsymbol{U}}_B(t',t)\right)\right\rangle_V \quad (9.88)$$

where the current density operator is defined via

$$\langle \boldsymbol{r}_1|\hat{\boldsymbol{j}}(\boldsymbol{r})|\boldsymbol{r}_2\rangle = \frac{e}{\mathrm{i}m}\left[\nabla_{\boldsymbol{r}_1}\delta_{\boldsymbol{r}_1,\boldsymbol{r}}\delta_{\boldsymbol{r}_2,\boldsymbol{r}} - \delta_{\boldsymbol{r}_1,\boldsymbol{r}}\nabla_{\boldsymbol{r}_2}\delta_{\boldsymbol{r}_2,\boldsymbol{r}}\right] \quad (9.89)$$

In Eq. (9.88), we implicitly assume averaging over the coordinate $\boldsymbol{r}$ wherever the current density is evaluated.

Equations (9.88) and (9.89) together with Eq. (9.87) provide a formally exact expression for the linear conductivity of an arbitrary disordered conductor in the presence of electron–electron interactions. Hence, all diagrams of the perturbation theory in all orders in the interaction are fully contained in the above results and can be recovered by expanding Eq. (9.87) in $V^{\pm}$ with subsequent averaging over these fields. For instance, expanding Eq. (9.87) to the second order in $V^{\pm}$, after averaging one arrives at contributions of the type $\langle V^+V^+\rangle$ and $\langle V^+(1-2\hat{\rho})V^-\rangle$. We will return to this issue below in this chapter as well as in Chapter 14.

9.7 Influence Functional for an Interacting Fermi Gas

We are now in a position to construct the influence functional for interacting electrons. As in the future we will merely be interested in describing transport properties of disordered conductors, we will assume that a weak external potential $V_x(t, \boldsymbol{r})$ is applied to our metallic conductor, thereby modifying the Hamiltonian (9.78) simply by substituting $U \to U + eV_x(t, \boldsymbol{r})$. Provided this potential is non-uniform in space, an external electric field is created, which forces electrons to diffuse across the conductor. Accordingly, the electron density matrix is driven out of equilibrium, $\boldsymbol{\rho}_V = \boldsymbol{\rho}_V^{\rm eq} + \delta\boldsymbol{\rho}_V$. The correction $\delta\boldsymbol{\rho}_V$ to the equilibrium density matrix $\boldsymbol{\rho}_V^{\rm eq}$ can be found from the equation

$$\mathrm{i}\frac{\partial\delta\boldsymbol{\rho}_V}{\partial t} = \hat{\mathcal{H}}_F\delta\boldsymbol{\rho}_V - \delta\boldsymbol{\rho}_V\hat{\mathcal{H}}_B - [eV_x(t, \hat{\boldsymbol{r}}), \boldsymbol{\rho}_V^{\rm eq}] \qquad (9.90)$$

derived from Eq. (9.76) by expanding it to the first order in V_x and $\delta\boldsymbol{\rho}_V$. Here the Hamiltonians $\hat{\mathcal{H}}_{F,B}$ are defined in Eq. (9.81) with $\boldsymbol{\rho}_V \to \boldsymbol{\rho}_V^{\rm eq}$. The formal solution to Eq. (9.90) reads

$$\delta\boldsymbol{\rho}_V(t) = \mathrm{i}\int_{-\infty}^{t} dt'\hat{\boldsymbol{U}}_F(t, t')[eV_x(t', \hat{\boldsymbol{r}}), \boldsymbol{\rho}_V^{\rm eq}(t')]\hat{\boldsymbol{U}}_B(t', t) \quad (9.91)$$

where $\hat{\boldsymbol{U}}_{F,B}(t_f, t_i)$ are defined in Eq. (9.80). The kernels of these evolution operators may be expressed via the path integrals over both coordinate and momentum as

$$\langle\boldsymbol{r}_f|\hat{\boldsymbol{U}}_{F,B}(t_f, t_i)|\boldsymbol{r}_i\rangle = \int_{\boldsymbol{r}(t_i)=\boldsymbol{r}_i}^{\boldsymbol{r}(t_f)=\boldsymbol{r}_f} \mathcal{D}\boldsymbol{r}\int\mathcal{D}\boldsymbol{p}\exp\left[\mathrm{i}\int_{t_i}^{t_f} dt'\big(\boldsymbol{p}\dot{\boldsymbol{r}} - \mathcal{H}_{F,B}(\boldsymbol{p}, \boldsymbol{r})\big)\right] \quad (9.92)$$

Here $\mathcal{H}_{F,B}(\boldsymbol{p}, \boldsymbol{r})$ are defined by Eq. (9.81) with the momentum and coordinate operators replaced by the classical variables $\boldsymbol{p}, \boldsymbol{r}$. Care is needed at this point since the operators $\hat{\boldsymbol{p}}, \hat{\boldsymbol{r}}$ and, hence, the operators $V^-(t, \hat{\boldsymbol{r}})$ and $\boldsymbol{\rho}_V$ do not commute with each other. Accordingly, one should additionally specify the rules of integration in Eq. (9.92) considering it as a limit of the product of ordinary integrals defined for a large number of small time slices. Explicitly, the functions $\mathcal{H}_{F,B}(\boldsymbol{p}, \boldsymbol{r})$ read

$$\mathcal{H}_{F,B}(\boldsymbol{p}, \boldsymbol{r}) = \frac{\boldsymbol{p}^2}{2m} + U(\boldsymbol{r}) - eV^+(t, \boldsymbol{r}) \mp \frac{1}{2}\left[1 - 2n_V(\boldsymbol{p}, \boldsymbol{r})\right] eV^-(t, \boldsymbol{r}), \tag{9.93}$$

where $n_V(\boldsymbol{p}, \boldsymbol{r})$ is the coordinate and momentum-dependent electron distribution function, which also, generally speaking, depends on the potentials $V^\pm$.

Integrating Eq. (9.92) over the fluctuating potentials $V^\pm$, we define the correction to the density matrix $\delta\boldsymbol{\rho} \equiv \int \mathcal{D}V^\pm e^{\mathrm{i}S_{\mathrm{em}}+\mathrm{i}S_{\mathrm{el}}}\delta\boldsymbol{\rho}_V$, which reads

$$\begin{aligned}\delta\boldsymbol{\rho}(t, \boldsymbol{r}_{Ff}, \boldsymbol{r}_{Bf}) = \mathrm{i}e \int_{-\infty}^{t} dt' \int d\boldsymbol{r}_{Fi} d\boldsymbol{r}_{Bi}\, J(t, t'; \boldsymbol{r}_{Ff}, \boldsymbol{r}_{Bf}; \boldsymbol{r}_{Fi}, \boldsymbol{r}_{Bi}) \\ \times \left(V_x(t', \boldsymbol{r}_{Fi}) - V_x(t', \boldsymbol{r}_{Bi})\right)\boldsymbol{\rho}(t, \boldsymbol{r}_{Fi}, \boldsymbol{r}_{Bi})\end{aligned} \tag{9.94}$$

where the evolution kernel J is formally expressed as a double path integral over the two coordinates and two momenta corresponding to forward and backward branches of the Keldysh contour

$$\begin{aligned}J(t, t'; \boldsymbol{r}_{Ff}, \boldsymbol{r}_{Bf}; \boldsymbol{r}_{Fi}, \boldsymbol{r}_{Bi}) &= \int_{\boldsymbol{r}_F(t')=\boldsymbol{r}_{Fi}}^{\boldsymbol{r}_F(t)=\boldsymbol{r}_{Ff}} \mathcal{D}\boldsymbol{r}_F \int_{\boldsymbol{r}_B(t')=\boldsymbol{r}_{Bi}}^{\boldsymbol{r}_B(t)=\boldsymbol{r}_{Bf}} \mathcal{D}\boldsymbol{r}_B \int \mathcal{D}\boldsymbol{p}_F \mathcal{D}\boldsymbol{p}_B \\ &\times \exp\{\mathrm{i}S_0[\boldsymbol{p}_F, \boldsymbol{r}_F] - \mathrm{i}S_0[\boldsymbol{p}_B, \boldsymbol{r}_B]\} \int \mathcal{D}V^\pm \\ &\times \exp\left\{\mathrm{i}\int_{t'}^{t} dt'' \int d\boldsymbol{r}''\left(\rho^-(t'', \boldsymbol{r}'')V^+(t'', \boldsymbol{r}'') + \rho^+(t'', \boldsymbol{r}'')V^-(t'', \boldsymbol{r}'')\right)\right\} \\ &\times \exp\left\{\mathrm{i}S_{\mathrm{em}}[V^+, V^-] + \mathrm{i}S_{\mathrm{el}}[V^+, V^-]\right\}\end{aligned} \tag{9.95}$$

Here

$$\begin{aligned}\rho^-(t'', \boldsymbol{r}'') &= e\delta\left(\boldsymbol{r}'' - \boldsymbol{r}_F(t'')\right) - e\delta\left(\boldsymbol{r}'' - \boldsymbol{r}_B(t'')\right), \\ \rho^+(t'', \boldsymbol{r}'') &= \frac{e}{2}\left[1 - 2n_V(\boldsymbol{p}_F(t''), \boldsymbol{r}_F(t''))\right]\delta\left(\boldsymbol{r}'' - \boldsymbol{r}_F(t'')\right) \\ &\quad + \frac{e}{2}\left[1 - 2n_V(\boldsymbol{p}_B(t''), \boldsymbol{r}_B(t''))\right]\delta\left(\boldsymbol{r}'' - \boldsymbol{r}_B(t'')\right)\end{aligned} \tag{9.96}$$

are effective charge densities associated with the propagation of a single electron in a metal and

$$S_0[\boldsymbol{p}, \boldsymbol{r}] = \int_{t'}^{t} dt''\left(\boldsymbol{p}\dot{\boldsymbol{r}} - \frac{\boldsymbol{p}^2}{2m} - U(\boldsymbol{r})\right) \tag{9.97}$$

is the free electron action. Note that while deriving Eq. (9.94) for the sake of simplicity, we disregarded weak correlation effects between the kernel J and the equilibrium density matrix $\boldsymbol{\rho}^{\rm eq} \equiv \int \mathcal{D}V^{\pm} e^{{\rm i}S_{\rm em}+{\rm i}S_{\rm el}} \boldsymbol{\rho}_V^{\rm eq}$. Such correlations may only be important in materials with strongly correlated electrons.

Although the general structure of Eqs. (9.94), (9.95) resembles that of the evolution kernel for a particle coupled to a bath of oscillators, there remains an important difference: Eq. (9.95) contains an additional path integral over the potentials $V^{\pm}$. This integral can be easily handled, provided Coulomb interaction between electrons remains weak, which is the case for practically all metallic conductors. In this case (i) the distribution function n_V may be replaced by the Fermi function

$$n_V(\boldsymbol{p}, \boldsymbol{r}) \to n_0(\boldsymbol{p}, \boldsymbol{r}) \equiv \frac{1}{1 + \exp\left[\frac{\frac{p^2}{2m} + U(\boldsymbol{r}) - \mu}{T}\right]} \tag{9.98}$$

and (ii) the field action ${\rm i}S_{\rm em} + {\rm i}S_{\rm el}$ may be taken in the RPA form (9.57). Employing Eqs. (9.57)–(9.62), we obtain [155]

$$J(t, t'; \boldsymbol{r}_{Ff}, \boldsymbol{r}_{Bf}; \boldsymbol{r}_{Fi}, \boldsymbol{r}_{Bi}) = \int_{\boldsymbol{r}_F(t')=\boldsymbol{r}_{Fi}}^{\boldsymbol{r}_F(t)=\boldsymbol{r}_{Ff}} \mathcal{D}\boldsymbol{r}_F \int_{\boldsymbol{r}_B(t')=\boldsymbol{r}_{Bi}}^{\boldsymbol{r}_B(t)=\boldsymbol{r}_{Bf}} \mathcal{D}\boldsymbol{r}_B \int \mathcal{D}\boldsymbol{p}_F \mathcal{D}\boldsymbol{p}_B$$
$$\exp\left\{{\rm i}S_0[\boldsymbol{p}_F, \boldsymbol{r}_F] - {\rm i}S_0[\boldsymbol{p}_B, \boldsymbol{r}_B] - {\rm i}S_{\rm R}[\boldsymbol{p}_F, \boldsymbol{r}_F, \boldsymbol{p}_B, \boldsymbol{r}_B] - S_{\rm I}[\boldsymbol{r}_F, \boldsymbol{r}_B]\right\}, \tag{9.99}$$

where

$$\begin{aligned} S_{\rm R} = \frac{e^2}{2} \int_{t'}^{t} dt_1 \int_{t'}^{t} dt_2 \{ & R(t_1 - t_2, \boldsymbol{r}_F(t_1) - \boldsymbol{r}_F(t_2))[1 - 2n_0(\boldsymbol{p}_F(t_2), \boldsymbol{r}_F(t_2))] \\ & - R(t_1 - t_2, \boldsymbol{r}_B(t_1) - \boldsymbol{r}_B(t_2))[1 - 2n_0(\boldsymbol{p}_B(t_2), \boldsymbol{r}_B(t_2))] \\ & + R(t_1 - t_2, \boldsymbol{r}_F(t_1) - \boldsymbol{r}_B(t_2))[1 - 2n_0(\boldsymbol{p}_B(t_2), \boldsymbol{r}_B(t_2))] \\ & - R(t_1 - t_2, \boldsymbol{r}_B(t_1) - \boldsymbol{r}_F(t_2))[1 - 2n_0(\boldsymbol{p}_F(t_2), \boldsymbol{r}_F(t_2))]\} \end{aligned} \tag{9.100}$$

and

$$\begin{aligned} S_{\rm I} = \frac{e^2}{2} \int_{t'}^{t} dt_1 \int_{t'}^{t} dt' \{ & I(t_1 - t_2, \boldsymbol{r}_F(t_1) - \boldsymbol{r}_F(t_2)) \\ & + I(t_1 - t_2, \boldsymbol{r}_B(t_1) - \boldsymbol{r}_B(t_2)) - I(t_1 - t_2, \boldsymbol{r}_F(t_1) - \boldsymbol{r}_B(t_2)) \\ & - I(t_1 - t_2, \boldsymbol{r}_B(t_1) - \boldsymbol{r}_F(t_2))\} \end{aligned} \tag{9.101}$$

Equations (9.99)–(9.101) constitute the final result of our consideration. They determine the kernel J of the evolution operator (defined on the Keldysh contour) for a single electron interacting with other electrons forming a degenerate Fermi gas.

The general structure of this result is essentially identical to that already encountered in previous chapters while describing quantum dynamics of a particle interacting with a dissipative environment: The terms S_0 (9.97) in the exponent of Eq. (9.99) describe propagation of a free electron in a disordered metal, while the remaining exponent $e^{-iS_R-S_I}$ defines the Feynman–Vernon influence functional for an electron gas.

It is instructive to compare these results with those for the influence functional for a charged particle (e.g., muon) interacting with an electron gas derived in Chapter 8 (cf. Eqs. (8.11)–(8.13)). We observe that the terms S_I, which account for the environmental noise, are identical in both cases (cf. Eqs. (8.13) and (9.101)). Hence, the noise contribution to the influence functional of the electron gas S_I is not sensitive to the Pauli exclusion principle at all. The dissipative contribution to the effective action S_R is, in contrast, sensitive to Fermi statistics. Comparing Eqs. (8.12) and (9.100), we observe that having a similar structure they nevertheless differ by the combination $1 - 2n_0$, which accounts for the Pauli principle in the latter case. The presence of this factor in the influence functional has a clear physical meaning: It demonstrates that electrons in a metal cannot infinitely decrease their energy since this factor (and, hence, the whole dissipative term S_R) vanishes identically at the Fermi energy ($n_0 = 1/2$) and turns negative below this energy. On the other hand, at energies well above the Fermi energy, n_0 becomes small and the effect of Fermi statistics loses unimportance meaning that the results for S_R (8.12) and (9.100) coincide in this limit. In other words, the constructed influence functional (9.99)–(9.101) is "clever" enough to assure that quantum dynamics of electrons with initial energies above μ is dissipative until they reach the Fermi level and, on the contrary, holes with initial energies below μ are pushed up to the Fermi surface.

9.8 Discussion

The above formalism of the influence functional for a degenerate Fermi gas is a powerful tool which can be—and has already been—employed in a number of problems of modern condensed matter physics where long-range interactions play an important role. For instance, this formalism was extensively applied in the calculation of the weak localization correction to conductance and in the analysis of interaction-induced quantum decoherence of electrons in disordered conductors [155, 157, 162]. These issues will be addressed in details in Chapters 14 and 15 of this book. This formalism was also employed in the problem of quantum transport of composite fermions in the fractional quantum Hall regime [111].

9.8.1 Some Applications

In order to successfully apply the above influence functional approach, it is also useful to establish its connection to other techniques frequently used in the literature. First of all, let us reproduce the standard calculations of the conductivity for disordered conductors. Making use of the definition of the current density

$$\boldsymbol{j}(t,\boldsymbol{r}) = \frac{\mathrm{i}e}{m}\left(\nabla_{r_1}\delta\rho(t,\boldsymbol{r}_1,\boldsymbol{r}_2) - \nabla_{r_2}\delta\rho(t,\boldsymbol{r}_1,\boldsymbol{r}_2)\right)\Big|_{\boldsymbol{r}_1=\boldsymbol{r}_2=\boldsymbol{r}} \qquad (9.102)$$

combining it with Eq. (9.94) and assuming that the external electric field is constant both in space and in time, $V_x = -\boldsymbol{E}\boldsymbol{r}$, we arrive at the expression for the system conductivity

$$\sigma = \frac{e^2}{3m}\int_{-\infty}^{t} dt' \int d\boldsymbol{r}_{i1} d\boldsymbol{r}_{i2} \left(\nabla_{r_1} - \nabla_{r_2}\right)\Big|_{\boldsymbol{r}_1=\boldsymbol{r}_2} J\left(t,t';\boldsymbol{r}_1,\boldsymbol{r}_2;\boldsymbol{r}_{1i},\boldsymbol{r}_{2i}\right)$$
$$\times\,(\boldsymbol{r}_{1i}-\boldsymbol{r}_{2i})\rho_0(\boldsymbol{r}_{1i},\boldsymbol{r}_{2i}) \qquad (9.103)$$

Here we have shifted the initial time to $-\infty$ (cf. also Eqs. (9.88), (9.89)).

In order to establish a connection between Eq. (9.103) and other conductivity calculations, it is convenient to introduce the Wigner function $n(\boldsymbol{p},\boldsymbol{r}) = \int d\boldsymbol{r}^- e^{-\mathrm{i}\boldsymbol{p}\boldsymbol{r}^-}\rho(\boldsymbol{r}+\boldsymbol{r}^-/2, \boldsymbol{r}-\boldsymbol{r}^-/2)$ instead of the density matrix. For homogeneous systems, it does not depend on $\boldsymbol{r}$ at the scales exceeding the mean free path. With this in mind, we obtain

$$\sigma = -\frac{2e^2}{3m}\int_{-\infty}^{t} dt' \int \frac{d^3\mathbf{p}}{(2\pi)^3}\mathbf{p}\hat{W}(t,t')\frac{\partial n(\mathbf{p})}{\partial \mathbf{p}} \qquad (9.104)$$

where $\hat{W}(t, t')$ is the evolution operator for the Wigner function:

$$n(t, \mathbf{p}, \mathbf{r}) = \hat{W}(t, t')n(t', \mathbf{p}, \mathbf{r}) \tag{9.105}$$

The kernel of this operator and that in Eq. (9.99) are related by means of the Fourier transformation with respect to the difference of coordinates at two branches of the Keldysh contour $\mathbf{r}_1 - \mathbf{r}_2$. In equilibrium, one has $n(\mathbf{p}) = 1/(\exp(\xi_{\mathbf{p}}/T) + 1)$. Hence, at small T, one finds

$$\frac{\partial n(\mathbf{p})}{\partial \mathbf{p}} \simeq -\mathbf{v}_F \delta(\xi)$$

and arrives at the well-known expression for the conductivity

$$\sigma = \frac{2e^2 N_0}{3} \int_{-\infty}^{t} dt' \langle \mathbf{v}(t)\mathbf{v}(t')\rangle \tag{9.106}$$

Employing the standard form of the velocity–velocity correlator for conduction electrons in a diffusive metal

$$\langle \mathbf{v}(t)\mathbf{v}(t')\rangle = v_F^2 \exp(-v_F(t - t')/\ell) \tag{9.107}$$

with the aid of Eq. (9.106), we immediately reconstruct the classical Drude formula for the conductivity of such metals

$$\sigma_D = 2e^2 N_0 D \tag{9.108}$$

Note that in addition to this classical contribution to the system conductivity, Eq. (9.103) also contains both interaction and quantum correction terms, which in the case of diffusive metallic conductors will be addressed in Chapters 13–15.

9.8.2 Quantum Kinetic Equation

Let us now demonstrate how the usual quantum kinetic description can be derived from our analysis. We start from the equation for the density matrix $\boldsymbol{\rho}_V$ (9.79). Rewriting this equation in the "interaction representation," i.e., substituting $\boldsymbol{\rho}_V \to e^{-\mathrm{i}H_0 t}\boldsymbol{\rho}_V e^{\mathrm{i}H_0 t}$, we find

$$\begin{aligned} i\frac{\partial \boldsymbol{\rho}_V}{\partial t} &= -e\hat{V}^+(t)\boldsymbol{\rho}_V + \boldsymbol{\rho}_V e\hat{V}^+(t) \\ &\quad -\frac{e}{2}\left((1 - \boldsymbol{\rho}_V)\hat{V}^-(t)\boldsymbol{\rho}_V + \boldsymbol{\rho}_V \hat{V}^-(t)(1 - \boldsymbol{\rho}_V)\right) \end{aligned} \tag{9.109}$$

where $\hat{V}^{\pm}(t) = e^{\mathrm{i}H_0 t} V^{\pm}(t) e^{-\mathrm{i}H_0 t}$. Let us integrate this equation over time, substitute the resulting expression for ρ_V into the right-hand side of Eq. (9.109), and average over $V^{\pm}$. If the Coulomb interaction is sufficiently weak, one can proceed perturbatively in V and neglect the dependence of the density matrix $\boldsymbol{\rho}_V$ on this field in the right-hand side of the resulting equation. Then the result of averaging can be expressed in terms of the correlation functions $\langle V V \rangle$. We obtain

$$\frac{\partial \boldsymbol{\rho}}{\partial t} = e^2 \int_{-\infty}^{t} dt' \Big\langle - \hat{V}^+(t)\hat{V}^+(t')\boldsymbol{\rho}(t') + \hat{V}^+(t)\boldsymbol{\rho}(t')\hat{V}^+(t')$$
$$+ \hat{V}^+(t')\boldsymbol{\rho}(t')\hat{V}^+(t) - \boldsymbol{\rho}(t')\hat{V}^+(t')\hat{V}^+(t)$$
$$- \frac{1}{2}\hat{V}^+(t)(1-\boldsymbol{\rho}(t'))\hat{V}^-(t')\boldsymbol{\rho}(t') - \frac{1}{2}\hat{V}^+(t)\boldsymbol{\rho}(t')\hat{V}^-(t')(1-\boldsymbol{\rho}(t'))$$
$$+ \frac{1}{2}(1-\boldsymbol{\rho}(t'))\hat{V}^-(t')\boldsymbol{\rho}(t')V^+(t) + \frac{1}{2}\boldsymbol{\rho}(t')\hat{V}^-(t')(1-\boldsymbol{\rho}(t'))V^+(t) \Big\rangle_{V^+,V^-} \tag{9.110}$$

where $\boldsymbol{\rho}(t) = \boldsymbol{\rho}_{V=0}(t)$. The averaging over the fluctuating potentials V^+ and V^- should be performed via path integration with the quadratic RPA action (9.57). As a result of this procedure, one arrives at the integrals containing the averages $\langle V^+V^+\rangle$ and $\langle V^+V^-\rangle$. In a homogeneous metal, the latter averages are given by Eqs. (9.64), (9.65); the density matrix takes a simple form

$$\boldsymbol{\rho}(\boldsymbol{r}_1 - \boldsymbol{r}_2) = \int \frac{d^3p}{(2\pi)^3} n_{\boldsymbol{p}} e^{\mathrm{i}\boldsymbol{p}(\boldsymbol{r}_1 - \boldsymbol{r}_2)}$$

and the operator $e^{-\mathrm{i}H_0 t}$ reduces to $e^{-\mathrm{i}\xi_{\boldsymbol{p}} t}$. After Fourier transformation over the coordinates, the kinetic equation (9.110) reduces to that for the electron distribution function n_p, i.e.,

$$\frac{dn_{\boldsymbol{p}}}{dt} = \frac{e^2}{\pi^2}\int d\omega d^3k \mathrm{Im}\left(\frac{-1}{k^2\epsilon(\omega,k)}\right)\delta(\omega + \xi_{p-k} - \xi_p)$$
$$\times\left[\coth\left(\frac{\omega}{2T}\right)(n_{\boldsymbol{p}-\boldsymbol{k}} - n_{\boldsymbol{p}}) - n_{\boldsymbol{p}}(1-n_{\boldsymbol{p}-\boldsymbol{k}}) - n_{\boldsymbol{p}-\boldsymbol{k}}(1-n_{\boldsymbol{p}})\right] \tag{9.111}$$

The right-hand side of this equation represents the standard electron–electron collision integral, which vanishes in equilibrium, i.e., for $n_{\boldsymbol{p}} = 1/(\exp(\xi_{\boldsymbol{p}}/T)+1)$.

The above kinetic equation can also be generalized to the case of disordered conductors, in which case the electron momentum is not anymore conserved due to elastic scattering on impurities. The corresponding derivation is much more involved in this case and will not be displayed here. In the diffusive limit, the result of this derivation is roughly equivalent to Eq. (9.111) in which one should now perform a replacement

$$\delta(\omega + \xi_{\boldsymbol{p}-\boldsymbol{k}} - \xi_{\boldsymbol{p}}) \to \mathrm{Re}\left[\frac{1}{\mathrm{i}\omega + Dk^2}\right].$$

9.8.3 Keldysh Diagrams

The effect of electron–electron interactions on transport properties of disordered conductors is often analyzed with the aid of the Keldysh diagrammatic technique [220]. Let us demonstrate that the approach based on the modified evolution equation for the density matrix (9.76) is equivalent to the Keldysh technique.

For this purpose, we first express the Keldysh Green functions of free electrons (9.26) in terms of the single-electron density matrix $\boldsymbol{\rho}_0$ and the single-electron evolution operator $\hat{\boldsymbol{u}}_0$. The former is defined as

$$\rho_0(\boldsymbol{r}_1, \boldsymbol{r}_2) = \sum_p \frac{\psi_p(\boldsymbol{r}_1)\psi_p^*(\boldsymbol{r}_2)}{1 + e^{(\epsilon_p - \mu)/T}} \tag{9.112}$$

where $\psi_p(\boldsymbol{r})$ and ϵ_p are, respectively, the eigenfunctions and the energy eigenvalues of the free electron Hamiltonian (9.78), i.e.,

$$\hat{H}_0|\psi_p\rangle = \epsilon_p|\psi_p\rangle \tag{9.113}$$

The evolution operator of non-interacting electrons is $\hat{\boldsymbol{u}}_0 = e^{-\mathrm{i}\hat{H}_0 t}$. Accordingly, its matrix elements read

$$u_0(t, \boldsymbol{r}_1, \boldsymbol{r}_2) = \langle \boldsymbol{r}_1|\hat{\boldsymbol{u}}_0(t)|\boldsymbol{r}_2\rangle = \sum_p e^{-\mathrm{i}\epsilon_p t}\psi_p(\boldsymbol{r}_1)\psi_p^*(\boldsymbol{r}_2) \tag{9.114}$$

It is straightforward to verify that the components of the Keldysh Green function (9.26) may be expressed in the form

$$G_{FF}(X_1, X_2) = -\mathrm{i}\int d\boldsymbol{r}' d\boldsymbol{r}'' u_0(t_1, \boldsymbol{r}_1, \boldsymbol{r}')\big[\theta(t_1 - t_2)\delta(\boldsymbol{r}' - \boldsymbol{r}'')$$
$$-\rho_0(\boldsymbol{r}', \boldsymbol{r}'')\big] \times u_0(-t_2, \boldsymbol{r}'', \boldsymbol{r}_2),$$

$$G_{BB}(X_1, X_2) = -\mathrm{i} \int d\boldsymbol{r}' d\boldsymbol{r}'' u_0(t_1, \boldsymbol{r}_1, \boldsymbol{r}') \big[\theta(t_2 - t_1)\delta(\boldsymbol{r}' - \boldsymbol{r}'') \\ -\boldsymbol{\rho}_0(\boldsymbol{r}', \boldsymbol{r}'')\big] \times u_0(-t_2, \boldsymbol{r}'', \boldsymbol{r}_2),$$

$$G_{BF}(X_1, X_2) = -\mathrm{i} \int d\boldsymbol{r}' d\boldsymbol{r}'' u_0(t_1, \boldsymbol{r}_1, \boldsymbol{r}') \big[\delta(\boldsymbol{r}' - \boldsymbol{r}'') - \boldsymbol{\rho}_0(\boldsymbol{r}', \boldsymbol{r}'')\big] \\ \times u_0(-t_2, \boldsymbol{r}'', \boldsymbol{r}_2),$$

$$G_{FB}(X_1, X_2) = \mathrm{i} \int d\boldsymbol{r}' d\boldsymbol{r}'' u_0(t_1, \boldsymbol{r}_1, \boldsymbol{r}') \boldsymbol{\rho}_0(\boldsymbol{r}', \boldsymbol{r}'') u_0(-t_2, \boldsymbol{r}'', \boldsymbol{r}_2) \tag{9.115}$$

Let us now assume that at $t = 0$ the density matrix $\boldsymbol{\rho}_V$ coincides with $\boldsymbol{\rho}_0$ and solve the evolution equation (9.76) perturbatively in the fluctuating potentials V_F, V_B keeping only the terms up to the second order. Then we obtain

$$\boldsymbol{\rho}_V(t) = \boldsymbol{\rho}_0(t) + \delta\boldsymbol{\rho}_V^{(1)}(t) + \delta\boldsymbol{\rho}_V^{(2)}(t) \tag{9.116}$$

Here the zeroth-order term describes the evolution of free electrons,

$$\boldsymbol{\rho}_0(t, \boldsymbol{r}_1, \boldsymbol{r}_2) = \int d\boldsymbol{r}' d\boldsymbol{r}'' u_0(t, \boldsymbol{r}_1, \boldsymbol{r}') \boldsymbol{\rho}_0(\boldsymbol{r}', \boldsymbol{r}''), u_0(-t, \boldsymbol{r}'', \boldsymbol{r}_2) \tag{9.117}$$

the first-order correction $\delta\boldsymbol{\rho}_V^{(1)}(t) \propto V_F$, V_B vanishes upon averaging over the fluctuating potentials, and the second-order correction to the density matrix reads

$$\delta\boldsymbol{\rho}_V^{(2)}(t, \boldsymbol{r}_1, \boldsymbol{r}_2) = \mathrm{i}e^2 \int_0^t dt' \int_0^t dt'' \int d\boldsymbol{r}' d\boldsymbol{r}'' \\ \times \big\{ - G_{FF}(X_1, X')V_F(X')G_{FF}(X', X'')V_F(X'')G_{FB}(X'', X_2) \\ - G_{FB}(X_1, X')V_B(X')G_{BB}(X', X'')V_B(X'')G_{BB}(X'', X_2) \\ + G_{FF}(X_1, X')V_F(X')G_{FB}(X', X'')V_B(X'')G_{BB}(X'', X_2) \\ + G_{FB}(X_1, X')V_B(X')G_{BF}(X', X'')V_F(X'')G_{FB}(X'', X_2)\big\} \tag{9.118}$$

Here we already employed the relation between the Green function components and the density matrix $\boldsymbol{\rho}_0$ (9.115). We also defined the four-dimensional vectors $X_1 = (t, \boldsymbol{r}_1)$, $X_2 = (t, \boldsymbol{r}_2)$, $X' = (t', \boldsymbol{r}')$, and $X'' = (t'', \boldsymbol{r}'')$.

Averaging over the fluctuating potentials V_F and V_B in Eq. (9.118), we introduce Keldysh Green functions of photons

$$\mathcal{D}_{\alpha\beta}(X', X'') = \langle V_\alpha(X')V_\beta(X'')\rangle \tag{9.119}$$

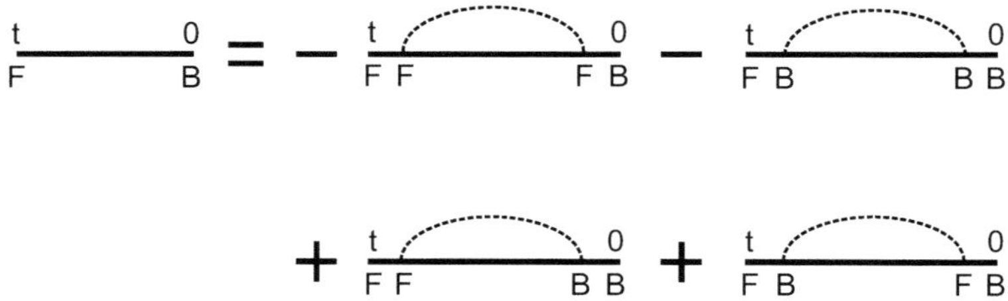

Figure 9.1 Diagrammatic representation of Eq. (9.120). The solid lines stand for the electron Green functions (9.115), while the dashed lines represent the photon Green functions (9.119).

where the indices α and β label the Keldysh contour branches F and B. Averaging in Eq. (9.118) is performed with the aid of the action (9.57), and the Green functions $\mathcal{D}_{\alpha\beta}(X', X'')$ may be expressed via the retarded and Keldysh components of the photon Green functions $R(X', X'')$ and $I(X', X'')$ defined in Eqs. (9.64), (9.65).

Finally, we note that the density matrix is related to the component G_{FB} of the Green function matrix, namely $\boldsymbol{\rho}(t, \boldsymbol{r}_1, \boldsymbol{r}_2) = -\mathrm{i}G_{FB}(X_1, X_2)$. After all these operations, Eq. (9.118) takes the form

$$\begin{aligned}
\delta G_{FB}(X_1, X_2) &= e^2 \int_0^t dt' \int_0^t dt'' \int d\boldsymbol{r}' d\boldsymbol{r}'' \\
&\times \big\{ - G_{FF}(X_1, X')G_{FF}(X', X'')G_{FB}(X'', X_2)\mathcal{D}_{FF}(X', X'') \\
&- G_{FB}(X_1, X')G_{BB}(X', X'')G_{BB}(X'', X_2)\mathcal{D}_{BB}(X', X'') \\
&+ G_{FF}(X_1, X')G_{FB}(X', X'')G_{BB}(X'', X_2)\mathcal{D}_{FB}(X', X'') \\
&+ G_{FB}(X_1, X')G_{BF}(X', X'')G_{FB}(X'', X_2)\mathcal{D}_{BF}(X', X'')\big\} \qquad (9.120)
\end{aligned}$$

which defines the lowest-order electron–electron interaction correction to the component G_{FB} of the Green function matrix. Equation (9.120) also follows from the standard diagrammatic technique and the correction to G_{FB} can be conveniently represented as the sum of four diagrams displayed in Fig. 9.1.

Thus, our analysis demonstrates that evaluating the second-order correction to the density matrix, one just reproduces the lowest-order diagrams, which determine the interaction correction to the Green function G_{FB}. Analogously, one can take into account higher-order terms and verify the equivalence between our approach and the Keldysh diagram technique in all higher orders of the perturbation theory.

It is important to emphasize that all Keldysh diagrams can also be generated by expanding the evolution operator J (9.99)–(9.101) in powers of $iS_R + S_I$. In order to verify the exact equivalence between the influence functional for interacting electrons derived here and the Keldysh diagram technique for such electrons, one can proceed analogously to Section 1.5.2 and take a proper account on Fermi statistics. As we already explained above, the latter step requires keeping track of non-commutativity of the operators $\hat{\boldsymbol{p}}$ and $\hat{\boldsymbol{r}}$ and their proper ordering while performing the integrals over the corresponding variables. An example of the perturbation theory in the electron–electron interactions is presented in Section 14.4.3 where we discuss the conductance σ of an arbitrary disordered conductor. For more details and some practical calculations, we also refer the reader to the papers [157, 162].

9.8.4 Quasiclassical Langevin Equation

To complete this section, let us derive the quasiclassical Langevin equation describing electron dynamics in a disordered metal. For that purpose, we will follow the same procedure as that already employed, e.g., in Chapters 1 and 2.

Consider only electron paths for which the values $\boldsymbol{r}^- = \boldsymbol{r}_F - \boldsymbol{r}_B$ and $\boldsymbol{p}^- = \boldsymbol{p}_F - \boldsymbol{p}_B$ remain small. For such trajectories, we can expand the effective action in the exponent of Eq. (9.99) in powers of $\boldsymbol{r}^-$ and $\boldsymbol{p}^-$ keeping only quadratic terms. The action then becomes Gaussian, and the integrals in Eq. (9.99) are dominated by the saddle points $\delta S/\delta \boldsymbol{p}^- = 0$ and $\delta S/\delta \boldsymbol{r}^- = 0$. The first equation coincides with one without dissipation: $\dot{\boldsymbol{r}} = \boldsymbol{p}/m$. With the aid of this equation, we exclude the electron momentum and find

$$m\ddot{\boldsymbol{r}} + \nabla U(\boldsymbol{r}) + e^2\left[1 - 2n(\boldsymbol{r}, m\dot{\boldsymbol{r}})\right] \times \int_{-\infty}^{t} dt' \nabla_r R(t-t', \boldsymbol{r}(t) - \boldsymbol{r}(t')) = -e\boldsymbol{E}(t, \boldsymbol{r}) \quad (9.121)$$

Here $\boldsymbol{E}(t, \boldsymbol{r})$ represents the fluctuating electric field produced by fluctuating electrons. Equilibrium fluctuations of this field are described by the correlator

$$\langle E_i(t_1, \boldsymbol{r}_1) E_j(t_2, \boldsymbol{r}_2) \rangle = 4\pi \delta_{ij} \int \frac{d\omega d^3k}{(2\pi)^4} \times \mathrm{Im} \left(\frac{-\coth \frac{\omega}{2T}}{\epsilon(\omega, \boldsymbol{k})} \right) e^{-\mathrm{i}\omega(t_1 - t_2) + \mathrm{i}\boldsymbol{k}(\boldsymbol{r}_1 - \boldsymbol{r}_2)} \tag{9.122}$$

The Langevin equation (9.121) satisfies the requirement of causality and semiclassically describes dissipative dynamics of electrons in a disordered metal. The physical origin of the damping term (which contains the function $R(t, \boldsymbol{r})$) in Eq. (9.121) is quite transparent: An electron (or a hole) propagating in a metal produces the screened electric potential due to the presence of other electrons and "feels" this potential itself. In this sense, Eq. (9.121) is to a certain extent similar to that describing dynamics of a high-energy particle (e.g., a muon) in a metal. The important difference between these two cases, however, lies in the factor $1 - 2n$, which enters in the dissipative term and reflects the effect of the Pauli exclusion principle. In the case of a muon, this factor would be absent. The fluctuating electric field $\boldsymbol{E}$ in the right-hand side of Eq. (9.121) is, on the contrary, not affected by the Pauli principle at all.

Chapter 10

Effective Action for Coherent Scatterers

10.1 Motivation and Choice of the Model

In Chapter 9, we extended the Feynman–Vernon–Caldeira–Leggett path integral influence functional technique to fermionic systems with disorder and interactions and we arrived at a formulation that allows for a formally exact treatment of quantum dynamics of electrons in disordered conductors in the presence of interaction effects. This formulation operates with the kernel of the evolution operator defined as the integral over all electron paths (9.99) affected both by dissipation (9.100) and noise (9.101) produced by an effective electronic environment. As in general the path integral (9.99) is essentially non-Gaussian, it cannot be handled exactly. Hence, in practice, one usually needs to resort to some approximations, for instance, by properly selecting electron trajectories (saddle point paths), which give dominant contributions to the path integral. Provided such quasiclassical trajectories can be identified, as is the case, e.g., for weakly disordered spatially extended metals, this formalism allows for a straightforward non-perturbative treatment of the problem with disorder and interactions. An important example of this kind will be discussed in Chapter 14.

Dissipative Quantum Mechanics of Nanostructures: Electron Transport, Fluctuations, and Interactions
Andrei D. Zaikin and Dmitry S. Golubev

ISBN 978-981-4774-50-5 (Hardcover), 978-0-429-29823-3 (eBook)
www.jennystanford.com

On the other hand, for some spatially restricted and strongly non-uniform structures, such as quantum dots and granular metals, it can be rather difficult to directly evaluate the path integral (9.99) for a number of reasons. One of them is that the description in terms of quasiclassical electron trajectories may become insufficient in the above cases, and electron scattering on disorder should be treated on more general footing without employing any quasiclassics. Another, purely technical, reason has to do with averaging over disorder. Within the approach developed in Chapter 9, it is usually convenient to postpone disorder averaging until the last stage of the calculation. In some systems, like ones mentioned above, it might be, on the contrary, more appropriate to perform disorder averaging already in the beginning of the whole consideration. In addition, it is desirable to deal with the model that would embrace various types of conductors with well-defined properties both in the long and short wavelength limits. This feature will help to construct a fully self-contained theory free of any divergencies and cutoff parameters.

In order to construct such a theory, one can follow a different route and, e.g., adopt a model for a disordered conductor consisting of an array of (metallic) quantum dots connected via junctions (scatterers) with arbitrary transmission distribution of their conducting channels. This model would allow one to easily crossover between the limits of a granular metal and that with point-like impurities and to treat spatially restricted and spatially extended conductors within the same theoretical framework. This route will be chosen in Chapters 13 and 15.

To summarize, the task at hand is to develop a formalism that accounts for each electron scattering event in a possibly most general manner and then to incorporate electron–electron interactions into this scheme. In order to proceed with the first part of this task, we will employ the scattering matrix formalism [44, 77], which is proven to be a very efficient tool to describe both electron transport and fluctuation effects. We will then demonstrate how to generalize this formalism to include the effect of electron–electron interactions.

Electrons propagating in a metal get scattered every time they approach any kind of obstacle, which can be, e.g., an impurity, a

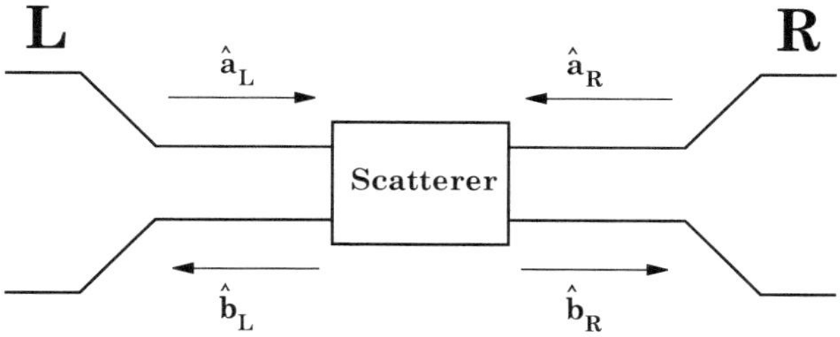

Figure 10.1 A short coherent scatterer connecting left (L) and right (R) reservoirs with electric potentials V_L and V_R. The amplitudes of the wave functions for incoming and outgoing electrons are denoted, respectively, by $\hat{a}_L$, $\hat{a}_R$ and $\hat{b}_L$, $\hat{b}_R$.

boundary imperfection, and a tunnel barrier. In order to describe this process, we will adopt a model of a coherent scatterer placed in between two big normal metallic reservoirs. This system is schematically depicted in Fig. 10.1. The scatterer is indicated by a box, which couples the left and the right reservoirs (leads). Electrons coming, e.g., from the left lead can either be transmitted through the scatterer into the right lead or get reflected back into the left one. The probabilities of both these processes depend on the properties of the scatterer encoded in the unitary scattering matrix

$$\hat{S} = \begin{pmatrix} \hat{r} & \hat{t}' \\ \hat{t} & \hat{r}' \end{pmatrix} \tag{10.1}$$

Here $\hat{r}$ and $\hat{t}$ are, respectively, the reflection and transmission matrices. The eigenvalues T_n of the matrix $\hat{t}^\dagger\hat{t}$ define transmission probabilities of the corresponding conducting channels.

If electric potentials V_L and V_R are applied, respectively, to the left and the right leads, the current I proportional to the voltage difference $V = V_L - V_R$ will flow across the scatterer, i.e.,

$$I = GV \tag{10.2}$$

which is, of course, just the standard Ohm's law. The conductance of the scatterer G is defined by the famous Landauer formula. Provided the scatterer is small enough, implying that the corresponding electron flight time τ_{fl} between the reservoirs is shorter than all other important timescales of the problem, such as the inverse temperature/voltage or the typical inelastic relaxation time τ_{in}:

$$\tau_{\text{fl}} \ll 1/T, \quad 1/eV, \quad \text{or} \quad \tau_{\text{in}} \tag{10.3}$$

one can neglect the dependence of the scattering matrix on the energy of incoming electrons. In this case, the Landauer formula reads

$$G = \frac{e^2}{2\pi}\,\mathrm{tr}\,\left[\hat{t}^\dagger\hat{t}\right] = \frac{e^2}{2\pi}\left(\sum_n T_n\right) \tag{10.4}$$

The combination in front of the trace is the normal quantum conductance unit $G_q \equiv 1/R_q = e^2/h$ (see also Chapter 7), where $h = 2\pi$ in the employed system of units with $\hbar = 1$.

Note that here and below, we distinguish the conductance channels with opposite spin directions, i.e., both the trace and the sum in Eq. (10.4) include also the summation over the spin index. Provided there exists spin degeneracy in the problem, one can count only channels with one spin direction and multiply the corresponding sum by the factor 2. This way of channel counting is also frequently used in the literature. Of course, both ways lead to identical final results in any spin degenerate situation. Provided spin degeneracy is lifted, it is in general necessary to distinguish between channel transmissions with opposite spin directions. In this case, the way of channel counting adopted here should be used.

Current fluctuations in the system can also be conveniently treated within the scattering formalism. For instance, the low-frequency current noise defined by the symmetrized current–current correlator

$$\mathcal{S} = \frac{1}{2}\int dt\langle I(t)I(0) + I(0)I(t)\rangle \tag{10.5}$$

takes the form (see, e.g., [60]):

$$\begin{aligned}\mathcal{S} &= \frac{e^2}{2\pi}\,\mathrm{tr}\,\left[\left(\hat{t}^\dagger\hat{t}\right)^2\right]2T + \frac{e^2}{2\pi}\,\mathrm{tr}\,\left[\hat{t}^\dagger\hat{t}(\hat{1} - \hat{t}^\dagger\hat{t})\right]eV\coth\frac{eV}{2T}\\ &= \frac{e^2}{2\pi}\left(\sum_n T_n^2\right)2T + \frac{e^2}{2\pi}\left(\sum_n T_n(1 - T_n)\right)eV\coth\frac{eV}{2T}\end{aligned} \tag{10.6}$$

In what follows, employing Eqs. (9.24) and (9.41), we will systematically derive the action for a short coherent conductor, which explicitly depends on fluctuating voltages (phases) and, hence, can be used to analyze the effect of electron–electron interactions. We start our derivation by evaluating a single particle evolution operator $\hat{\mathbf{u}}(t, 0; V)$ and the electron Green functions in the reservoirs (9.26). This will be the subject of the next several sections.

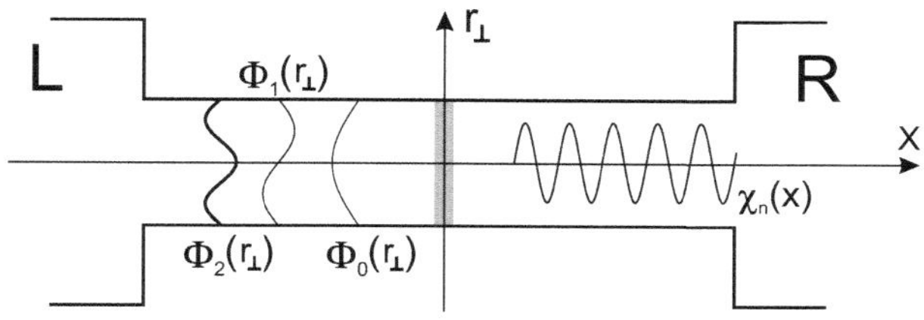

Figure 10.2 The waveguide geometry.

10.2 Scattering States and Scattering Matrix

To begin with, let us introduce the concept of transmission channels and scattering states. For simplicity, we assume that normal reservoirs far from the scatterer are ballistic (i.e., no additional scattering is involved) and, on top of that, our structure has an ideal waveguide geometry (see Fig. 10.2). Both these assumptions do not affect our final results but significantly simplify our consideration. For instance, the latter assumption implies that the potential $W(\boldsymbol{r})$ describing both the potential barrier between the leads and their boundaries can be effectively factorized:

$$W(\boldsymbol{r}) = W_0(\boldsymbol{r}_\perp) W_{\text{barrier}}(x) \tag{10.7}$$

Here the coordinate x describes the electron motion in the direction perpendicular to the barrier, $\boldsymbol{r}_\perp$ accounts for its transverse motion, the potential $W_{\text{barrier}}(x)$ determines the barrier profile, and $W_0(\boldsymbol{r}_\perp)$ originates from the walls of the waveguide.

Let us consider the Schrödinger equation

$$-\frac{\nabla^2}{2m}\psi(\boldsymbol{r}) + W(\boldsymbol{r})\psi(\boldsymbol{r}) = (E + \epsilon_F)\psi(\boldsymbol{r}) \tag{10.8}$$

where ϵ_F is the electron Fermi energy. Adopting the assumption (10.7), one can also factorize the solution to Eq. (10.8) as

$$\psi(\boldsymbol{r}) = \sum_n c_n \Phi_n(\boldsymbol{r}_\perp)\chi_n(x) \tag{10.9}$$

The transverse wave functions $\Phi_n(\boldsymbol{r}_\perp)$ satisfy the equation

$$-\frac{\nabla_\perp^2}{2m}\Phi_n(\boldsymbol{r}_\perp) + W_0(\boldsymbol{r}_\perp)\Phi_n(\boldsymbol{r}_\perp) = E_n\Phi(\boldsymbol{r}_\perp) \tag{10.10}$$

where the subscript n enumerates the transmission channels (we consider only channels with $E_n < \epsilon_F$). The function $\chi_n(x)$ outside the scatterer region obeys the equation

$$-\frac{1}{2m}\frac{d^2}{dx^2}\chi_n(x) = (E + \epsilon_F - E_n)\chi_n(x) \tag{10.11}$$

In what follows, we will treat electron dynamics quasiclassically. This is appropriate since both electrodes are good metals and, hence, electronic states with energies sufficiently close to ϵ_F provide the dominant contribution. Let us define the electron velocity in the n-th channel $v_n = \sqrt{2(\epsilon_F - E_n)/m}$. Then the wave functions can be expressed as

$$\begin{aligned}\chi_n(x) &= e^{imv_n x} f_n^{in}(x) + e^{-imv_n x} f_n^{out}(x), \text{ left reservoir.}\\ \chi_n(x) &= e^{imv_n x} g_n^{out}(x) + e^{-imv_n x} g_n^{in}(x), \text{ right reservoir.}\end{aligned} \tag{10.12}$$

Consider first the left reservoir. The functions $f_n^{in}(x)$ and $f_n^{out}(x)$ satisfy the following quasiclassical equations:

$$\begin{aligned}-iv_n\frac{d}{dx} f_n^{in}(x) &= E f_n^{in}(x),\\ iv_n\frac{d}{dx} f_n^{out}(x) &= E f_n^{out}(x)\end{aligned} \tag{10.13}$$

with the solutions

$$f_n^{in}(x) = \frac{e^{iEx/v_n}}{\sqrt{v_n}}, \qquad f_n^{out}(x) = \frac{e^{-iEx/v_n}}{\sqrt{v_n}} \tag{10.14}$$

Analogously, for the right reservoir we obtain

$$g_m^{in}(x) = \frac{e^{-iEx/v_m}}{\sqrt{v_m}}, \qquad g_n^{out}(x) = \frac{e^{iEx/v_m}}{\sqrt{v_m}} \tag{10.15}$$

Note that all these functions are normalized in such a way that they carry the same current.

With the aid of the above expressions, the eigenfunction of Eq. (10.8) with energy E in the left and the right reservoirs may be expressed, respectively, as

$$\psi_E(\mathbf{r}) = \sum_n \left[a_{Ln} e^{imv_n x} f_n^{in}(x) + b_{Ln} e^{-imv_n x} f_n^{out}(x)\right] \Phi_n(\mathbf{r}_\perp) \tag{10.16}$$

and

$$\psi_E(\mathbf{r}) = \sum_k \left[b_{Rk} e^{imv_k x} g_k^{out}(x) + a_{Rk} e^{-imv_k x} g_k^{in}(x)\right] \Phi_k(\mathbf{r}_\perp) \tag{10.17}$$

The amplitudes of both outgoing and incoming states $b_{L,R}$ and $a_{L,R}$ (cf. Fig. 10.1) are related to each other via the scattering matrix $\hat{S}(E)$

$$\begin{pmatrix} b_{L1} \\ \dots \\ b_{LN_L} \\ b_{R1} \\ \dots \\ b_{RN_R} \end{pmatrix} = \hat{S}(E) \begin{pmatrix} a_{L1} \\ \dots \\ a_{LN_L} \\ a_{R1} \\ \dots \\ a_{RN_R} \end{pmatrix} \tag{10.18}$$

Now it will be convenient for us to introduce a new coordinate y and combine both incident and outgoing wave functions $f_n^{in}(x)$ and $f_n^{out}(x)$ belonging to the same channel into one wave function $\psi_n(y)$. Then in the left reservoir $(x < 0)$, we set

$$\psi_n(y) = \begin{cases} f_n^{in}(y), & y < 0, \\ f_n^{out}(-y), & y > 0, \end{cases} \tag{10.19}$$

while for the right reservoir $(x > 0)$, we have

$$\psi_m(y) = \begin{cases} f_m^{in}(-y), & y < 0, \\ f_m^{out}(y), & y > 0. \end{cases} \tag{10.20}$$

These functions are defined in the range $y \in [-\infty, +\infty]$ and read

$$\psi_j(y) = \frac{e^{\mathrm{i}Ey/v_j}}{\sqrt{v_j}} \tag{10.21}$$

It is important to emphasize that here the index j enumerates all conducting channels first in the left reservoir and then in the right one.

10.3 Scattering Matrix and Tunneling Hamiltonian

It is instructive to establish the relation between the scattering matrix $\hat{S}$ (10.1) and the tunneling Hamiltonian

$$H_T = \sum_{nk} \left(\tau_{nk} a_n^\dagger b_k + \tau_{nk}^* b_k^\dagger a_n \right) \tag{10.22}$$

which is widely used in the literature in order to describe tunneling between the electrodes. We have already employed the

Hamiltonian (10.22) in Chapter 2 of this book. As it is clear from the above formula, this Hamiltonian explicitly describes creation and annihilation of electrons in the leads, e.g., a_n is the annihilation operator for an electron in the n-th state of the left electrode, $b_k^\dagger$ is the creation operator for an electron in the k-th state of the right electrode, and τ_{nk} is the tunneling matrix element between these two states.

Turning back to the scattering problem, we note that in the basis of the scattering states

$$\tilde{\psi}_n(y) = e^{\mathrm{i}Ey/v_n} \tag{10.23}$$

and for sufficiently small energies $|E| \ll \epsilon_F$, the Hamiltonian describing the electron motion through the barrier,

$$\hat{H} = -\frac{\nabla^2}{2m} + W(\boldsymbol{r}) \tag{10.24}$$

takes a simple matrix form

$$H_{nk} = -\mathrm{i}v_n \delta_{nk} \frac{\partial}{\partial y} + W_{nk}\delta(y) \tag{10.25}$$

where the matrix $W_{nk}\delta(y)$ accounts for electron scattering at $y = 0$. It has the block structure

$$\hat{W} = \begin{pmatrix} \hat{W}^{LL} & \hat{W}^{LR} \\ \hat{W}^{RL} & \hat{W}^{RR} \end{pmatrix} \tag{10.26}$$

with four sub-blocks describing scattering both between the leads ($\hat{W}^{LR}$, $\hat{W}^{RL}$) and within the leads ($\hat{W}^{LL}$, $\hat{W}^{RR}$). Comparing the Hamiltonians (10.22) and (10.25), we arrive at the relation

$$\tau_{nk} = W_{nk}^{LR} \tag{10.27}$$

In order to obtain the relation between the matrix $\hat{W}$ and the scattering matrix, it is necessary to resolve the Schrödinger equation

$$-\mathrm{i}v_n \frac{\partial \psi_n(y)}{\partial y} + \sum_k W_{nk}\delta(y)\psi_k(0) = E\psi_n(y) \tag{10.28}$$

Following the procedure outlined in the previous section, we express the solution to this equation in terms of incoming (a_n) and outgoing (b_n) amplitudes,

$$\psi_n(y) = \left[a_n\theta(-y) + b_n\theta(y)\right]\frac{e^{\mathrm{i}Ey/v_n}}{\sqrt{v_n}} \tag{10.29}$$

and apply the boundary condition

$$\psi_n(+0) - \psi_n(-0) = \frac{1}{\mathrm{i}} \sum_k \frac{W_{nk}}{v_n} \psi_k(0) \tag{10.30}$$

which follows directly from Eq. (10.28). We immediately observe that $\psi_n(-0) = a_n$, $\psi_n(+0) = b_n$ and $\psi_k(0) = (a_k + b_k)/2$. Combining all these results and making use of the definition of the scattering matrix $\hat{S}$ (10.18) we arrive at the following result (see, for example, the review [44], where a similar formula for a chaotic cavity is discussed)

$$\hat{S} = \left[1 + \mathrm{i}\hat{W}'\right]^{-1} \left[1 - \mathrm{i}\hat{W}'\right] \tag{10.31}$$

Here $\hat{W}'$ is a Hermitian matrix with the elements $W'_{nk} = W_{nk}/2\sqrt{v_n v_k}$.

Equation (10.31) defines the relation between the tunneling Hamiltonian and the scattering matrix we were aiming at, since the matrix $\hat{W}'$ is directly expressed via the tunneling matrix elements (cf. Eq. (10.27)). In order to explicitly relate the channel transmissions T_m (defined as the eigenvalues of the matrix $\hat{t}^\dagger\hat{t}$) and the tunneling matrix elements in the Hamiltonian (10.22) to each other, let us rotate the channel basis in such a way that the matrix $\hat{t}^\dagger\hat{t}$ and also the matrices $\hat{t}$, $\hat{r}$, and $\hat{W}'^{LR}$ become diagonal with t_m, r_m, and W'_m being the corresponding eigenvalues. In addition, we set $\hat{W}^{RR} = \hat{W}^{LL} = 0$, since scattering processes within the same electrode are not described by the Hamiltonian H_T. Resolving Eq. (10.31) for t_m and evaluating the transmission $T_m = |t_m|^2$, we get

$$T_m = \frac{4|W'_m|^2}{\left(1 + |W'_m|^2\right)^2} = \frac{|\tau_m|^2}{v_m^L v_m^R \left(1 + \frac{|\tau_m|^2}{4 v_m^L v_n^R}\right)^2} \tag{10.32}$$

where τ_m is the tunneling matrix element of the Hamiltonian in the new channel basis, while v_m^L and v_m^R are the corresponding channel velocities, respectively, in the left and in the right electrodes. Equation (10.32) concludes our analysis demonstrating the equivalence between the scattering matrix and the tunneling Hamiltonian approaches. Both these techniques can generally be employed in order to describe electron transport at arbitrary channel transmissions.

10.4 Evolution Operators and Density Matrix

Our next step is to find the single-electron evolution operator $\boldsymbol{u}(t, 0, V)$, which appears in the action (9.41). In order to accomplish this goal, it is necessary to find the solution to the Schrödinger equation

$$\mathrm{i}\frac{\partial}{\partial t}\psi_n(t, y) = \sum_k H_{nk}(y)\psi_k(t, y) \tag{10.33}$$

with the Hamiltonian

$$H_{nk} = -\mathrm{i}v_n\delta_{nk}\frac{\partial}{\partial y} - eV_n(t)\delta_{nk} + W_{nk}\delta(y) \tag{10.34}$$

Let us assume that at initial time t_1, the wave function in the n-th channel was equal to $\psi_n(t_1, y_1)$ and find the wave function $\psi_n(t_2, y_2)$ at time $t_2 > t_1$. To begin with, we note that the evolution equation (10.33) in the absence of scattering imposes the restriction on the particle coordinates

$$y_2 - y_1 = v_n(t_2 - t_1) \tag{10.35}$$

which simply implies classical propagation of the electron wave packets.

We begin with two simple cases: (i) $y_2 < 0$, implying that electrons do not reach the barrier during the time interval $t_2 - t_1$, and (ii) $y_2 > v_n(t_2 - t_1)$, $y_1 > 0$, meaning that electrons move away from the barrier all the time. In both cases, the scattering does not occur and the wave function is given by a simple formula

$$\psi_n(t_2, y_2) = e^{\mathrm{i}[\varphi_n(t_2)-\varphi_n(t_1)]}\psi_n(t_1, y_2 - v_n(t_2 - t_1)) \tag{10.36}$$

which describes dispersionless electron propagation and which depends on the phase of the n-th channel

$$\varphi_n(t) = \int_0^t dt' eV_n(t') \tag{10.37}$$

where V_n is the electric potential in this channel.

Let us now consider the case $0 < y_2 < v_n(t_2 - t_1)$. According to Eq. (10.35), in this case electrons hit the scatterer at time

$$t_{\mathrm{sc}} = t_2 - y_2/v_n \tag{10.38}$$

and the wave function takes the form

$$\begin{aligned}\psi_n(t_2, y_2) &= \sum_k e^{\mathrm{i}[\varphi_n(t_2)-\varphi_k(t_1)]-\mathrm{i}[\varphi_n(t_{sc})-\varphi_k(t_{sc})]} \\ &\quad\times s_{nk}\sqrt{\frac{v_k}{v_n}}\psi_k\left(t_1, -v_k(t_{sc}-t_1)\right) \\ &= \sum_k e^{\mathrm{i}[\varphi_n(t_2)-\varphi_k(t_1)]-\mathrm{i}\left[\varphi_n\left(t_2-\frac{y_2}{v_n}\right)-\varphi_k\left(t_2-\frac{y_2}{v_n}\right)\right]} \\ &\quad\times s_{nk}\sqrt{\frac{v_k}{v_n}}\psi_k\left(t_1, \frac{v_k}{v_n}y_2 - v_k(t_2-t_1)\right) \qquad (10.39)\end{aligned}$$

where s_{nk} are the matrix elements of the $\hat{S}$-matrix.

We are now in a position to derive the expression for the evolution operator $\boldsymbol{u}_{nk}(t_2, t_1; y_2, y_1)$. According to its definition, we have

$$\psi_n(t_2, y_2) = \sum_k \int dy_1\ \boldsymbol{u}_{nk}(t_2, t_1; y_2, y_1)\psi_k(t_1, y_1).$$

Comparing this expression with Eqs. (10.36) and (10.39), we find

$$\begin{aligned}\boldsymbol{u}_{nk}(t_2, t_1; y_2, y_1) &= e^{\mathrm{i}\varphi_n(t_2)}\left\{\frac{\delta_{nk}}{v_n}\delta\left(\frac{y_2-y_1}{v_n} - t_2 + t_1\right)\right. \qquad (10.40) \\ &\quad + \theta(y_2)\theta(v_n(t_2-t_1)-y_2)e^{-\mathrm{i}\varphi_n\left(t_2-\frac{y_2}{v_n}\right)}[s_{nk}-\delta_{nk}] \\ &\quad\times\left.\sqrt{\frac{v_k}{v_n}}\ e^{\mathrm{i}\varphi_k\left(t_2-\frac{y_2}{v_n}\right)}\delta\left(\frac{v_k}{v_n}y_2 - y_1 - v_k(t_2-t_1)\right)\right\}e^{-\mathrm{i}\varphi_k(t_1)}.\end{aligned}$$

This expression can be simplified if we introduce the flight times $\tau = y/v_n$. Specifically, instead of the wave function with the channel components $\psi_n(y)$, we introduce the functions $\eta_n(\tau) = \sqrt{v_n}\psi_n(y/v_n)$. Provided the two functions are related to each other by means of a linear operator

$$\psi_n^{(2)}(y) = \sum_k \int dy'\ K_{nk}(y, y')\psi_k^{(1)}(y') \qquad (10.41)$$

the corresponding wave functions $\eta^{(2)}$ and $\eta^{(1)}$ obey the relation:

$$\eta_n^{(2)}(\tau) = \sum_k \int d\tau'\ \tilde{K}_{nk}(\tau, \tau')\eta_k^{(1)}(\tau') \qquad (10.42)$$

where

$$\tilde{K}_{nk}(\tau, \tau') = \sqrt{v_n}K_{nk}(v_n\tau, v_k\tau')\sqrt{v_k} \qquad (10.43)$$

Applying this transformation to the evolution operator (10.40) and expressing the result in the matrix form, we obtain [159]

$$\hat{\boldsymbol{u}}(t_2, t_1; \tau_2, \tau_1) = \delta(\tau_2 - \tau_1 - t_2 + t_1)e^{i\hat{\varphi}(t_2)}\Big\{\hat{1} + \theta(\tau_2)\theta(-\tau_1) \times e^{-i\hat{\varphi}(t_2-\tau_2)}[\hat{S} - \hat{1}]e^{i\hat{\varphi}(t_1-\tau_1)}\Big\}e^{-i\hat{\varphi}(t_1)} \quad (10.44)$$

where we introduced the diagonal phase matrix $\hat{\varphi}$ with the elements $\hat{\varphi}_{ik} = \varphi_i\delta_{ik}$.

The conjugate operator $\hat{\boldsymbol{u}}^\dagger$ is evaluated analogously. We find

$$\hat{\boldsymbol{u}}^\dagger(t_2, t_1; \tau_2, \tau_1) = \delta(\tau_1 - \tau_2 - t_2 + t_1)e^{i\hat{\varphi}(t_1)}\Big\{\hat{1} + \theta(\tau_1)\theta(-\tau_2) \times e^{-i\hat{\varphi}(t_1-\tau_2)}[\hat{S}^+ - \hat{1}]e^{i\hat{\varphi}(t_2-\tau_1)}\Big\}e^{-i\hat{\varphi}(t_2)}. \quad (10.45)$$

Equations (10.44) and (10.45) apply for $t_2 > t_1$. In the opposite case $t_2 < t_1$, the evolution operators are constructed from these equations employing the property

$$\hat{\boldsymbol{u}}(t_2, t_1; \tau_2, \tau_1) = \hat{\boldsymbol{u}}^{-1}(t_1, t_2; \tau_2, \tau_1) = \hat{\boldsymbol{u}}^\dagger(t_1, t_2; \tau_2, \tau_1) \quad (10.46)$$

In addition, the effective action (9.41) depends on the initial electron density matrix ρ_0. Its matrix elements between the channel states read

$$\rho_{0,nk}(y_1, y_2) = \delta_{nk}\int \frac{dp}{2\pi}\frac{e^{ip(y_1-y_2)}}{1+e^{pv_n/T}} = \frac{\delta_{nk}}{2}\delta(y_1 - y_2) - \frac{\delta_{nk}}{2\pi}\frac{\pi i T}{v_n \sinh\left[\frac{\pi T(y_1-y_2)}{v_n}\right]} \quad (10.47)$$

Introducing again the flight times $\tau_{1,2}$ instead of coordinates $y_{1,2}$ and performing the transformation (10.43), we rewrite the density matrix ρ_0 as a diagonal matrix in the channel space

$$\rho_{0,nk}(\tau_1, \tau_2) = f(\tau_1 - \tau_2)\delta_{nk}, \quad (10.48)$$

where the function $f(\tau)$ is the Fourier transform of the Fermi function

$$f(\tau) = \frac{1}{2}\left(\delta(\tau) - \frac{iT}{\sinh \pi T\tau}\right) \quad (10.49)$$

The established equations both for the evolution operators (10.44), (10.45) and for the initial density matrix (10.48) are sufficient to construct the effective action for a coherent scatterer with the aid of Eq. (9.41).

10.5 Green Functions of Normal Reservoirs

In order to employ an alternative form of the effective action (9.24), it is necessary to evaluate the Keldysh Green functions of isolated normal reservoirs defined in Eq. (9.26). Making use of the definitions already employed in Chapter 2 (see, e.g., Eq. (2.58)) and repeating the same procedure as one which has led to Eqs. (2.65)–(2.68) (in which we now set the superconducting order parameter equal to zero), we obtain

$$\check{G}_{nk} = \begin{pmatrix} e^{i\varphi_n(t_1)} G^{FF}_{nk} e^{-i\varphi_k(t_2)} & e^{i\varphi_n(t_1)} G^{FB}_{nk} e^{-i\varphi_k(t_2)} \\ e^{i\varphi_n(t_1)} G^{BF}_{nk} e^{-i\varphi_k(t_2)} & e^{i\varphi_n(t_1)} G^{BB}_{nk} e^{-i\varphi_k(t_2)} \end{pmatrix}, \qquad (10.50)$$

where the functions $G^{\alpha\beta}$ are diagonal matrices in the channel space. They read

$$G^{FB}_{nk}(t_1, t_2, \tau_1, \tau_2) = i\delta_{nk}\, f(\tau_1 - \tau_2 - t_1 + t_2) \qquad (10.51)$$

$$G^{BF}_{nk}(t_1, t_2, \tau_1, \tau_2) = -i\delta_{nk}\left[\delta(\tau_1 - \tau_2 - t_1 + t_2) - f(\tau_1 - \tau_2 - t_1 + t_2)\right] \qquad (10.52)$$

$$G^{FF}_{nk}(t_1, t_2, \tau_1, \tau_2) = \theta(t_1 - t_2) G^{BF}_{nk}(t_1, t_2, \tau_1, \tau_2) + \theta(t_2 - t_1) G^{FB}_{nk}(t_1, t_2, \tau_1, \tau_2) \qquad (10.53)$$

$$G^{BB}_{nk}(t_1, t_2, \tau_1, \tau_2) = \theta(t_1 - t_2) G^{FB}_{nk}(t_1, t_2, \tau_1, \tau_2) + \theta(t_2 - t_1) G^{BF}_{nk}(t_1, t_2, \tau_1, \tau_2) \qquad (10.54)$$

These functions obey the relation

$$G^{FF} + G^{BB} - G^{BF} - G^{FB} = 0 \qquad (10.55)$$

implying that only three of those are independent.

Along with $G^{\alpha\beta}$, one can also define the retarded, advanced, and Keldysh Green functions, respectively, G^R, G^A, and G^K:

$$G^R_{nk}(t_1, t_2, \tau_1, \tau_2) = \theta(t_1 - t_2)\left[G^{BF}_{nk}(t_1, t_2, \tau_1, \tau_2) - G^{FB}_{nk}(t_1, t_2, \tau_1, \tau_2)\right] = -i\delta_{nk}\, \theta(t_1 - t_2)\, \delta(\tau_1 - \tau_2 - t_1 + t_2) \qquad (10.56)$$

$$G^A_{nk}(t_1, t_2, \tau_1, \tau_2) = -\theta(t_2 - t_1)\left[G^{BF}_{nk}(t_1, t_2, \tau_1, \tau_2) - G^{FB}_{nk}(t_1, t_2, \tau_1, \tau_2)\right] = i\delta_{nk}\, \theta(t_2 - t_1)\, \delta(\tau_1 - \tau_2 - t_1 + t_2) \qquad (10.57)$$

$$G^K_{nk}(t_1, t_2, \tau_1, \tau_2) = G^{BF}_{nk}(t_1, t_2, \tau_1, \tau_2) + G^{FB}_{nk}(t_1, t_2, \tau_1, \tau_2)$$
$$= -\mathrm{i}\delta_{nk}\left[\delta(\tau_1 - \tau_2 - t_1 + t_2) - 2f(\tau_1 - \tau_2 - t_1 + t_2)\right] \quad (10.58)$$

Below we will also make use of the quasiclassical Green functions

$$\check{Q}_{nk}(t_1, t_2) = 2\mathrm{i}\delta_{nk} \lim_{\tau_1 \to \tau_2} \begin{pmatrix} G^{FF}_{nn} & -G^{FB}_{nn} \\ G^{BF}_{nn} & -G^{BB}_{nn} \end{pmatrix}$$
$$= \delta_{nk} \begin{pmatrix} \delta(t_2 - t_1) - 2f(t_2 - t_1) & 2f(t_2 - t_1) \\ 2[\delta(t_2 - t_1) - f(t_2 - t_1)] & 2f(t_2 - t_1) - \delta(t_2 - t_1) \end{pmatrix} \quad (10.59)$$

where we employed the regularization rule

$$\lim_{\tau_1 \to \tau_2} \theta(t_1 - t_2)\delta(\tau_1 - \tau_2 - t_1 + t_2) = \frac{1}{2}\delta(t_2 - t_1) \quad (10.60)$$

The quasiclassical function $\check{Q}$ obeys the normalization condition

$$\sum_q \int dt' \check{Q}_{nq}(t_1, t')\check{Q}_{qk}(t', t_2) = \delta_{nk}\delta(t_1 - t_2) \quad (10.61)$$

or, equivalently,

$$\check{Q}^2 = 1, \quad (10.62)$$

which can be verified directly from the above equations. This normalization condition is very useful as it can greatly simplify our calculations.

10.6 Effective Action for a Short Coherent Conductor

We are finally done with our formal preparations and can now proceed directly toward the effective action for a short coherent scatterer S_{sc}.

Substituting the expressions (10.44), (10.45) into Eq. (9.41) and performing the integrals over the corresponding flight times, we arrive at the following expression for the action [127, 142]

$$\mathrm{i}S_{\mathrm{sc}} = \mathrm{tr}\ln\left[\hat{1} + \theta(t - \tau_1)\theta(\tau_1)\left(e^{-\mathrm{i}\hat{\varphi}_B(\tau_1)}\hat{S}^\dagger e^{\mathrm{i}\hat{\varphi}_B(\tau_1)} e^{-\mathrm{i}\hat{\varphi}_F(\tau_1)}\hat{S}e^{\mathrm{i}\hat{\varphi}_F(\tau_1)} - \hat{1}\right)\right.$$
$$\left. \times\, \rho_0(-\tau_1, -\tau_2)\right] \quad (10.63)$$

The operator traces in Eq. (10.63) imply both the sum over all scattering channels and the integrals over the flight times τ. The combination $\theta(t - \tau_1)\theta(\tau_1)$ in front of the second term under the logarithm indicates that the integrals over intermediate flight times τ_j are taken from 0 to t.

As we will see below, the action (10.63) contains complete information about transport properties of a coherent scatterer in the presence of electron–electron interactions. It explicitly depends on both the scattering matrix $\hat{S}$ (10.1) and on the fluctuating phases $\varphi_L^{F,B}$ and $\varphi_R^{F,B}$ for the left and the right leads. The "phase" matrices $e^{i\hat{\varphi}_{F,B}}$ are diagonal and have the block structure in the channel space

$$e^{i\hat{\varphi}_{F,B}} = \begin{pmatrix} e^{i\varphi_L^{F,B}}\,\hat{1}_L & 0 \\ 0 & e^{i\varphi_R^{F,B}}\,\hat{1}_R \end{pmatrix} \tag{10.64}$$

where $\hat{1}_L$ and $\hat{1}_R$ are the unity matrices in the channel subspace, respectively, for the left and the right reservoirs. In the same basis, the initial density matrix reads

$$\rho_0(-\tau_1, -\tau_2) = \begin{pmatrix} \tilde{f}(\tau_1 - \tau_2)\,\hat{1}_L & 0 \\ 0 & \tilde{f}(\tau_1 - \tau_2)\,\hat{1}_R \end{pmatrix} \tag{10.65}$$

where we employed the complex conjugate Fourier transform of the Fermi function (cf. Eq. (10.49))

$$\tilde{f}(\tau) \equiv f^*(\tau) = \int \frac{dE}{2\pi}\frac{e^{-iE\tau}}{1 + e^{E/T}} = \frac{1}{2}\left(\delta(\tau) + \frac{iT}{\sinh \pi T \tau}\right) \tag{10.66}$$

For pedagogical purposes, let us further clarify the meaning of the operator trace in Eq. (10.63). For this purpose, we formally expand the action in powers of ρ_0. Keeping only the first- and the second-order terms $S_{sc} = S_1 + S_2 + \ldots$, we obtain

$$iS_1 = \int_0^t d\tau_1 \lim_{\tau_2 \to \tau_1} \text{tr}\left[\left(e^{-i\hat{\varphi}_B(\tau_1)}\hat{S}^\dagger e^{i\hat{\varphi}_B(\tau_1)} e^{-i\hat{\varphi}_F(\tau_1)}\hat{S} e^{i\hat{\varphi}_F(\tau_1)} - \hat{1}\right)\tilde{f}(\tau_1 - \tau_2)\right]$$

$$iS_2 = -\frac{1}{2}\int_0^t d\tau_1 \int_0^t d\tau_2\, \text{tr}\left[\left(e^{-i\hat{\varphi}_B(\tau_1)}\hat{S}^\dagger e^{i\hat{\varphi}_B(\tau_1)} e^{-i\hat{\varphi}_F(\tau_1)}\hat{S} e^{i\hat{\varphi}_F(\tau_1)} - \hat{1}\right)\right.$$
$$\left. \times\, \tilde{f}(\tau_1 - \tau_2)\left(e^{-i\hat{\varphi}_B(\tau_2)}\hat{S}^\dagger e^{i\hat{\varphi}_B(\tau_2)} e^{-i\hat{\varphi}_F(\tau_2)}\hat{S} e^{i\hat{\varphi}_F(\tau_2)} - \hat{1}\right)\tilde{f}(\tau_2 - \tau_1)\right].$$

Here the integrals over the flight times are explicitly indicated, and the remaining traces just imply summing over the scattering channels. Higher-order terms are constructed in the same way. Hence, a symbolic formula (10.63) can be formally interpreted as the sum of the corresponding infinite series in powers of ρ_0.

Finally, let us identically rewrite the action (10.63) in the form that can be more convenient for practical calculations:

$$\begin{aligned} \mathrm{i}S_{\mathrm{sc}} = \mathrm{tr}\ln\Bigg\{ \begin{pmatrix} \hat{1}_L & 0 \\ 0 & \hat{1}_R \end{pmatrix} + \theta(t-\tau_1)\theta(\tau_1) \Bigg[\begin{pmatrix} e^{-\mathrm{i}\varphi_L^B(\tau_1)}\,\hat{1}_L & 0 \\ 0 & e^{-\mathrm{i}\varphi_R^B(\tau_1)}\,\hat{1}_R \end{pmatrix} \\ \times \begin{pmatrix} \hat{r}^\dagger & \hat{t}^\dagger \\ \hat{t}'^\dagger & \hat{r}'^\dagger \end{pmatrix} \begin{pmatrix} e^{-\mathrm{i}\varphi_L^-(\tau_1)}\,\hat{1}_L & 0 \\ 0 & e^{-\mathrm{i}\varphi_R^-(\tau_1)}\,\hat{1}_R \end{pmatrix} \begin{pmatrix} \hat{r} & \hat{t}' \\ \hat{t} & \hat{r}' \end{pmatrix} \\ \times \begin{pmatrix} e^{\mathrm{i}\varphi_L^F(\tau_1)}\,\hat{1}_L & 0 \\ 0 & e^{\mathrm{i}\varphi_R^F(\tau_1)}\,\hat{1}_R \end{pmatrix} - \begin{pmatrix} \hat{1}_L & 0 \\ 0 & \hat{1}_R \end{pmatrix} \Bigg] \begin{pmatrix} \tilde{f}(\tau_1-\tau_2) & 0 \\ 0 & \tilde{f}(\tau_1-\tau_2) \end{pmatrix} \Bigg\}. \end{aligned} \tag{10.67}$$

Equation (10.67) will be directly employed in our subsequent analysis of transport properties of coherent conductors in the presence of electron–electron interactions.

Let us now elaborate an alternative representation of the action displayed in Eq. (9.24). Making use of the channel representation for the Hamiltonian (10.34), we can rewrite the effective action (9.24) in the form

$$\mathrm{i}S_{\mathrm{sc}} = \mathrm{tr}\ln \begin{pmatrix} \check{G}_L^{-1} & \check{H}_T^{LR} \\ \check{H}_T^{RL} & \check{G}_R^{-1} \end{pmatrix} \tag{10.68}$$

where the inverse Keldysh Green function reads

$$\check{G}_L^{-1} = \begin{pmatrix} \mathrm{i}\frac{\partial}{\partial t} + \mathrm{i}\hat{v}^L\frac{\partial}{\partial y} - \hat{W}^{LL}\delta(y) + eV_F^L & 0 \\ 0 & -\mathrm{i}\frac{\partial}{\partial t} - \mathrm{i}\hat{v}^L\frac{\partial}{\partial y} + \hat{W}^{LL}\delta(y) - eV_B^L \end{pmatrix}$$

for the left electrode and similarly for the right one. Coupling between the leads is defined with the aid of the matrices

$$\check{H}_T^{LR} = \begin{pmatrix} -\hat{W}^{LR}\delta(y) & 0 \\ 0 & \hat{W}^{LR}\delta(y) \end{pmatrix}, \quad \check{H}_T^{RL} = \begin{pmatrix} -\hat{W}^{RL}\delta(y) & 0 \\ 0 & \hat{W}^{RL}\delta(y) \end{pmatrix}$$

We proceed by formally expanding the action (10.68) in powers of $\check{H}_T^{LR}$ and $\check{H}_T^{RL}$. Performing an explicit resummation of the corresponding series, we arrive at the result

$$\mathrm{i}S_{\mathrm{sc}} = \mathrm{tr}\ln\left[\check{1} - \check{G}_L\check{H}_T^{LR}\check{G}_R\check{H}_T^{RL}\right] \tag{10.69}$$

Since the matrices $\check{H}_T^{LR}$ and $\check{H}_T^{RL}$ are proportional to $\delta(y)$, the integrals over coordinates in the trace (10.69) can be handled exactly to all orders. As a result, the series is expressed via the Green functions with coincident coordinates $\check{G}_{L,R}(t_1, t_2, 0, 0)$. The latter, in turn, are expressed through the quasiclassical function $\check{Q}$ defined in Eq. (10.59). Performing all these operations, we get

$$\begin{aligned} \mathrm{i}S_{\mathrm{sc}} &= \mathrm{tr}\ln\left[\check{1} + \frac{1}{4}e^{\mathrm{i}\check{\varphi}_L}\check{Q}e^{-\mathrm{i}\check{\varphi}_L}\hat{v}_L^{-1/2}\hat{W}^{LR}\hat{v}_R^{-1/2}e^{\mathrm{i}\check{\varphi}_R}\check{Q}e^{-\mathrm{i}\check{\varphi}_R}\hat{v}_R^{-1/2}\hat{W}^{RL}\hat{v}_L^{-1/2}\right] \\ &= \mathrm{tr}\ln\left[\check{1} + \frac{\hat{v}_L^{-1/2}\hat{W}^{LR}\hat{v}_R^{-1}\hat{W}^{RL}\hat{v}_L^{-1/2}}{4}e^{\mathrm{i}\check{\varphi}_L}\check{Q}e^{-\mathrm{i}\check{\varphi}_L}e^{\mathrm{i}\check{\varphi}_R}\check{Q}e^{-\mathrm{i}\check{\varphi}_R}\right] \end{aligned} \quad (10.70)$$

Here we defined the phase matrices in the Keldysh space

$$\check{\varphi}_L = \begin{pmatrix} \varphi_L^F\,\hat{1}_L & 0 \\ 0 & -\varphi_L^B\,\hat{1}_L \end{pmatrix}, \quad \check{\varphi}_R = \begin{pmatrix} \varphi_R^F\,\hat{1}_R & 0 \\ 0 & -\varphi_R^B\,\hat{1}_R \end{pmatrix} \quad (10.71)$$

It is now convenient to diagonalize the matrix $\hat{v}_L^{-1/2}\hat{W}^{LR}\hat{v}_R^{-1}\hat{W}^{RL}\hat{v}_L^{-1/2}$ by properly transforming the channel basis. Then, making use of the relation between the matrix elements of the tunnel Hamiltonian $W'_m = W_m/2\sqrt{v_m^L v_m^R}$ and the channel transmissions T_m (10.32), we can cast the action (10.70) to a much more compact form

$$\mathrm{i}S_{\mathrm{sc}} = \sum_m \mathrm{tr}\ln\left[\check{1} + \frac{1-\sqrt{1-T_m}}{1+\sqrt{1-T_m}}\,e^{\mathrm{i}\check{\varphi}}\check{Q}e^{-\mathrm{i}\check{\varphi}}\check{Q}\right] \quad (10.72)$$

where $\check{\varphi} = \check{\varphi}_L - \check{\varphi}_R$. Employing the properties of the trace, we observe that a replacement $e^{\mathrm{i}\check{\varphi}}\check{Q}e^{-\mathrm{i}\check{\varphi}}\check{Q} \to \check{Q}e^{\mathrm{i}\check{\varphi}}\check{Q}e^{-\mathrm{i}\check{\varphi}}$ under the logarithm leaves the action (10.72) unchanged. Hence, this action can be rewritten as

$$\begin{aligned} \mathrm{i}S_{\mathrm{sc}} &= \frac{1}{2}\sum_m \mathrm{tr}\ln\left[\check{1} + \frac{1-\sqrt{1-T_m}}{1+\sqrt{1-T_m}}\,e^{\mathrm{i}\check{\varphi}}\check{Q}e^{-\mathrm{i}\check{\varphi}}\check{Q}\right] \\ &+ \frac{1}{2}\sum_m \mathrm{tr}\ln\left[\check{1} + \frac{1-\sqrt{1-T_m}}{1+\sqrt{1-T_m}}\,\check{Q}e^{\mathrm{i}\check{\varphi}}\check{Q}e^{-\mathrm{i}\check{\varphi}}\right] \end{aligned} \quad (10.73)$$

This expression, in turn, may further be transformed with the aid of the normalization condition $\check{Q}^2 = 1$. As a result, we obtain

$$\mathrm{i}S_{\mathrm{sc}} = \frac{1}{2}\sum_m \operatorname{tr}\ln\left[\check{1} + \frac{T_m}{4}\left(\{e^{\mathrm{i}\check{\varphi}}\check{Q}e^{-\mathrm{i}\check{\varphi}}, \check{Q}\} - 2\right)\right] \qquad (10.74)$$

$$- 2\sum_m \ln\frac{1+\sqrt{1-T_m}}{2}.$$

The combination in the second line of this expression does not depend on the fluctuating phases and, hence, only yields an unimportant constant that can be dropped in our further analysis of interaction effects. The first term in Eq. (10.74) provides the alternative form of the action for a short coherent conductor, which is exactly equivalent to the action (10.67) derived above. Both these forms originate from the same Eq. (9.12).

Note that an imaginary time version of Eq. (10.74) for a coherent scatterer with arbitrary channel transmissions was also derived in Chapter 2 by means of yet another technique [411] (see Eqs. (2.175), (2.177)). A detailed derivation of the real-time formula (10.74) is also presented in the paper [368]. In the next chapters, we will extensively use the effective action defined in Eqs. (10.67) and (10.74) in order to analyze electron transport and fluctuations in the presence of electron–electron interactions.

If expanded to the first order in the barrier transmissions T_n, both actions (10.67) and (10.74) reduce to one derived in Chapter 2 for a normal tunnel barrier (cf. Eq. (2.113)). One can also proceed further and expand the effective action (10.67) up to terms $\sim (\hat{t}^\dagger\hat{t})^2$ keeping the complete nonlinear dependence on the fluctuating phases $\varphi^\pm$. It is easy to see that in order to recover all such terms, it is necessary to expand the logarithm in Eq. (10.67) up to the fourth order in the term containing the function $\tilde{f}$. Higher-order terms of this expansion can be omitted within the required accuracy since they do not contain contributions proportional to the first and second powers of the eigenvalues of the matrix $\hat{t}^\dagger\hat{t}$, i.e., terms $\sim T_n$ and $\sim T_n^2$. The whole calculation is straightforward, although requires some care. The final result reads [142]

$$
\begin{aligned}
\mathrm{i}S_{\mathrm{sc}} = & -\frac{\mathrm{i}}{2\pi}\int_0^t dx\, \dot{\varphi}^+(x)\sin\varphi^-(x)\left[\mathrm{tr}[\hat{t}^\dagger\hat{t}] + \frac{2}{3}\mathrm{tr}[(\hat{t}^\dagger\hat{t})^2]\sin^2\frac{\varphi^-(x)}{2}\right] \\
& -\frac{1}{\pi}\int_0^t dxdy\, \alpha(x-y)\sin\frac{\varphi^-(x)}{2}\sin\frac{\varphi^-(y)}{2} \\
& \times\left\{\mathrm{tr}[\hat{t}^\dagger\hat{t}(1-\hat{t}^\dagger\hat{t})]\cos[\varphi^+(x)-\varphi^+(y)]\right. \\
& \left. +\,\mathrm{tr}[(\hat{t}^\dagger\hat{t})^2]\cos\frac{\varphi^-(x)-\varphi^-(y)}{2}\right\} \\
& +\frac{2\mathrm{i}}{3}\mathrm{tr}[(\hat{t}^+\hat{t})^2]\int_0^t dxdydz \\
& \times\frac{T^3\sin\frac{\varphi^-(x)}{2}\sin\frac{\varphi^-(y)}{2}\sin\frac{\varphi^-(z)}{2}}{\sinh[\pi T(y-x)]\sinh[\pi T(x-z)]\sinh[\pi T(z-y)]} \\
& \times\left\{\sin[\varphi^+(y)-\varphi^+(x)]\cos\frac{\varphi^-(z)}{2} + \sin[\varphi^+(z)-\varphi^+(y)]\cos\frac{\varphi^-(x)}{2}\right. \\
& \left. +\sin[\varphi^+(x)-\varphi^+(z)]\cos\frac{\varphi^-(y)}{2}\right\} \\
& -8\mathrm{tr}[(\hat{t}^\dagger\hat{t})^2]\int_0^t dxdydzdw\; f(y-x)f^*(x-w)f(w-z)f^*(z-y) \\
& \times\sin\frac{\varphi^-(x)}{2}\sin\frac{\varphi^-(y)}{2}\sin\frac{\varphi^-(z)}{2}\sin\frac{\varphi^-(w)}{2} \\
& \times\cos\left[\varphi^+(x)-\varphi^+(y)+\varphi^+(z)-\varphi^+(w)\right]
\end{aligned}
\tag{10.75}
$$

where the function $f(x)$ is defined in Eq. (10.49), $\varphi^\pm$ are the classical and quantum phase differences across the scatterer defined in Eq. (2.50) and

$$
\alpha(x) = \int \frac{d\omega}{2\pi} e^{-\mathrm{i}\omega x}\omega\coth\frac{\omega}{2T} = -\frac{\pi T^2}{\sinh^2\pi T x} \tag{10.76}
$$

Equation (10.75) represents the complete expression for the effective action valid up to the second order in the transmissions T_n. This expression involves no further approximations and fully accounts for the nonlinear dependence on fluctuating phases $\varphi^\pm$.

Note that the above expansion of the full effective action in powers of T_n works well, provided the channel transmissions remain small and represents only one possible type of approximation. In Chapter 11, we will also demonstrate that one can successfully employ other approximations, e.g., expanding the effective action in powers of the reflection coefficients $R_n = 1 - T_n$ in the case of weakly reflecting scatterers or in powers of the quantum phase φ^- (quasiclassical approximation).

10.7 Effective Action and Full Counting Statistics

Let us establish a useful link between the effective action defined by Eqs. (10.67) and (10.74) and full counting statistics (FCS) of electron transport in coherent conductors. The concept of FCS plays an important role in modern theory of quantum transport. On one hand, there exists a relatively simple and elegant theoretical formulation of FCS in terms of the cumulant-generating function [269], which allows to fully account for fundamental issues such as quantum coherent electron scattering and Pauli principle. On the other hand, first several current cumulants can be reliably detected in modern experiments thereby enabling one a detailed comparison between experimental results and theoretical predictions.

Let us introduce the probability $P(t, N)$ for N electrons to be transferred through the junction during the time t. Provided a complete distribution of probabilities $P(t, N)$ is established, one can fully determine the FCS of electron transport. Unfortunately, it appears difficult to directly measure the function $P(t, N)$ in experiments because typical values of the electron tunneling frequency I/e are usually too high. Hence, it is instructive to search for some other quantities that may characterize this distribution and, at the same time, can be accessed in experiment. These parameters are current cumulants.

To proceed let us also define the cumulant-generating function

$$\mathcal{F}(t, \chi) = \ln\left(\sum_N P(t, N)\, e^{-\mathrm{i}\chi N}\right) \tag{10.77}$$

where χ is the so-called counting field. The functions $P(t, N)$ and $\mathcal{F}(t, \chi)$ contain the same information about the statistics of electron transport, i.e., it is sufficient to evaluate $\mathcal{F}(t, \chi)$ in order to recover $P(t, N)$. The n-th current cumulant C_n is formally defined with the aid of the function $\mathcal{F}(t, \chi)$ by means of the following equation:

$$C_n = \lim_{t\to\infty} \frac{(\mathrm{i}e)^n}{t} \left.\frac{\partial^n \mathcal{F}(t, \chi)}{\partial \chi^n}\right|_{\chi=0} \tag{10.78}$$

Evaluating the first three current cumulants, we find

$$\mathcal{C}_1 = -\lim_{t\to\infty} \frac{e}{t} \sum_N N P(t, N),$$

$$\mathcal{C}_2 = \lim_{t\to\infty} \frac{e^2}{t} \sum_N (N - \mathcal{C}_1 t)^2 P(t, N),$$

$$\mathcal{C}_3 = -\lim_{t\to\infty} \frac{e^3}{t} \sum_N (N - \mathcal{C}_1 t)^3 P(t, N) \qquad (10.79)$$

It is easy to observe that the first cumulant $\mathcal{C}_1$ just coincides with the average charge transferred through the conductor per one second, i.e., it simply equals to the average current value, $\mathcal{C}_1 = \langle I \rangle$. The second cumulant $\mathcal{C}_2$ measures the width of the charge distribution $P(t, N)$ and coincides with zero-frequency noise power $\mathcal{S}$:

$$\mathcal{C}_2 \equiv \mathcal{S} = \int dt \left[\langle I(t)I(0)\rangle - \mathcal{C}_1^2\right] \qquad (10.80)$$

Here we used the relation between the number of transferred electrons N and the fluctuating current,

$$-eN = \int_0^t dt' \, I(t') \qquad (10.81)$$

and assumed that the time integral in Eq. (10.80) converges implying that the correlator $\langle I(t)I(0)\rangle$ should decay in time quickly enough. The third cumulant is proportional to the skewness of the charge distribution $P(t, N)$ and can be expressed in the form

$$\mathcal{C}_3 = \int dt_1 dt_2 \left\langle \big(I(t_1) - \mathcal{C}_1\big)\big(I(t_2) - \mathcal{C}_1\big)\big(I(0) - \mathcal{C}_1\big)\right\rangle \qquad (10.82)$$

The cumulant (10.82) characterizes deviations of transport statistics from a simple Gaussian form. While the cumulants $\mathcal{C}_1$ and $\mathcal{C}_2$ can be routinely measured, the cumulant $\mathcal{C}_3$ is much more difficult to detect. It has only become possible during recent years [261, 335, 376].

Similar to the first three cumulants, the n-th cumulant $\mathcal{C}_n$ can be related to the n-th order current correlator integrated over all intermediate times. Explicit expressions for $\mathcal{C}_n$ get increasingly complicated with increasing n. The cumulants $\mathcal{C}_n$ further characterize the distribution of probabilities $P(t, N)$. For example, in the simplest case of a Gaussian distribution

$$P(t, N) = \frac{\exp\left[-\frac{(N-\langle I\rangle t/e)^2}{2\langle \delta N^2\rangle}\right]}{\sqrt{2\pi \langle \delta N^2\rangle}} \tag{10.83}$$

one obtains

$$\mathcal{C}_1 = \langle I\rangle, \quad \mathcal{C}_2 = \frac{e^2 \langle \delta N^2\rangle}{t} \tag{10.84}$$

while all higher cumulants vanish identically. In this case, the cumulant-generating function reads

$$\mathcal{F}_{\text{Gauss}}(t, \chi) = t\left(-i\mathcal{C}_1\chi - \frac{\mathcal{C}_2}{2}\chi^2\right) + \mathcal{O}(t) \tag{10.85}$$

where $\mathcal{O}(t)$ stands for the terms that do not grow in the long-time limit $t \to \infty$. Another example is given by the Poisson distribution of transferred electrons

$$P(t, N) = e^{-\langle I\rangle t/e}\frac{\left(\langle I\rangle t/e\right)^N}{N!} \tag{10.86}$$

In this case, the cumulant-generating function has the form

$$\mathcal{F}_{\text{Poisson}} = \frac{\langle I\rangle t}{e}\left(e^{-\mathrm{i}\chi} - 1\right) \tag{10.87}$$

All current cumulants can be evaluated exactly with the result

$$\mathcal{C}_n = e^{n-1}\langle I\rangle \tag{10.88}$$

One can also investigate FCS for a perfectly voltage-biased conductor characterized by a set of transmissions T_n and demonstrate that in this case, the cumulant-generating function can be formally expressed in terms of the current operator in the Heisenberg representation $\hat{I}(t)$ as [269]

$$\mathcal{F}(\chi) = \ln\left\langle \left(\mathcal{T}\, e^{\mathrm{i}\frac{\chi}{2e}\int_0^t dt'\,\hat{I}(t')}\right)^\dagger \mathcal{T}\, e^{-\mathrm{i}\frac{\chi}{2e}\int_0^t dt'\,\hat{I}(t')}\right\rangle \tag{10.89}$$

This expression can be evaluated exactly. In the long-time limit, $t \to \infty$, one gets

$$\mathcal{F}(t, \chi) = t\sum_n \int \frac{dE}{2\pi} \ln\left[1 + T_n(e^{\mathrm{i}\chi} - 1)\left(1 - f_L(E)\right) f_R(E)\right.$$
$$\left. + T_n(e^{-\mathrm{i}\chi} - 1) f_L(E)\left(1 - f_R(E)\right)\right] \tag{10.90}$$

Here f_L and f_R are the electron distribution functions in the corresponding reservoirs. In equilibrium, they are

$$f_r(E) = \frac{1}{1 + e^{(E - eV_r)/T}}, \quad r = L, R \tag{10.91}$$

The formula (10.90) becomes simpler in the shot noise regime $eV \gg T$. Then the integral over energies is evaluated exactly, and one finds

$$\mathcal{F}(t, \chi) = \frac{eVt}{2\pi} \sum_n \ln\left[1 + T_n(e^{-i\chi} - 1)\right] \tag{10.92}$$

Combining Eqs. (10.78) and (10.92), we immediately recover the first three cumulants:

$$I \equiv \mathcal{C}_1 = \frac{e^2 V}{2\pi} \sum_n T_n, \tag{10.93}$$

$$\mathcal{S} \equiv \mathcal{C}_2 = \frac{e^3 V}{2\pi} \sum_n T_n(1 - T_n), \tag{10.94}$$

$$\mathcal{C}_3 = \frac{e^4 V}{2\pi} \sum_n T_n(1 - T_n)(1 - 2T_n) \tag{10.95}$$

Higher cumulants can be derived in the same manner.

We will now demonstrate that the effective action (10.67) contains the full information about FCS. Let us consider a perfectly biased conductor. In this limit, phase fluctuations are suppressed and one has $\varphi^+ = eVt$, where V is the voltage bias. In addition, we identify $\varphi^- = \chi$. Then the action (10.67) takes the form

$$\begin{aligned} iS_{\mathrm{sc}} = t \int \frac{dE}{2\pi} \ln\det\Bigg\{ &\begin{pmatrix} \hat{1}_L & 0 \\ 0 & \hat{1}_R \end{pmatrix} \\ &+ \begin{pmatrix} \hat{t}^\dagger \hat{t}(e^{i\chi} - 1) & 2i\hat{t}^\dagger \hat{r}' \sin\frac{\chi}{2} \\ 2i\hat{r}'^\dagger \hat{t} \sin\frac{\chi}{2} & \hat{t}'^\dagger \hat{t}'(e^{-i\chi} - 1) \end{pmatrix} \begin{pmatrix} (1 - f_L)\hat{1}_L & 0 \\ 0 & (1 - f_R)\hat{1}_R \end{pmatrix} \Bigg\}, \end{aligned}$$

Evaluating this expression and comparing the result with Eq. (10.90), we find the relation between the cumulant-generating function and the effective action

$$iS_{\mathrm{sc}}[\varphi^+ = eVt, \varphi^- = \chi] \equiv \mathcal{F}(t, \chi) \tag{10.96}$$

Finally let us express the action (10.67) in terms of the current operator $\hat{I}(t)$ and derive a formula similar to Eq. (10.89). We start from the general expression for the action (9.12) bearing in mind that fluctuating electric potentials are spatially constants in the left and right electrodes, i.e.,

$$V_{F,B}(t, x) = V_{F,B;L}(t)\theta(-x) + V_{F,B;R}(t)\theta(x) \tag{10.97}$$

where $\theta(x)$ is the Heaviside theta-function. Let us write down the effective Hamiltonian (9.8) as a sum of four terms

$$H_{\text{eff}}[t; V] = \hat{H}_L + \hat{H}_R + \hat{Q}_L V_L(t) + \hat{Q}_R V_R(t) \tag{10.98}$$

where

$$\hat{H}_{L,R} = \int_{L,R} dx\, \hat{\Psi}^\dagger(x) \left[-\frac{\nabla^2}{2m} + U(x) - \mu_{L,R} \right] \hat{\Psi}(x) \tag{10.99}$$

are the Hamiltonians of the left and right leads and

$$\hat{Q}_{L,R} = -e \int_{L,R} dx\, \hat{\Psi}^\dagger(x)\hat{\Psi}(x) \tag{10.100}$$

define the charge operators for these electrodes. Introducing time-dependent Heisenberg operators

$$\hat{Q}_r(t) = e^{\mathrm{i}\hat{H}_r t}\hat{Q}_r e^{-\mathrm{i}\hat{H}_r t}, \quad r = L, R, \tag{10.101}$$

we can formally rewrite the action (9.12) in the form

$$\mathrm{i}S_{\text{sc}} = \ln \left\langle \left(\mathcal{T} e^{-\mathrm{i}\int_0^t dt' [\hat{Q}_L V_{B,L} + \hat{Q}_R V_{B,R}]} \right)^\dagger \mathcal{T} e^{-\mathrm{i}\int_0^t dt' [\hat{Q}_L V_{F,L} + \hat{Q}_{B,R} V_R]} \right\rangle$$

Next we note that the Hamiltonian (9.8) conserves the total electric charge $Q = \hat{Q}_L(t) + \hat{Q}_R(t)$. Hence, without loss of generality, we can choose $Q = 0$ or, equivalently, $\hat{Q}_R(t) = -\hat{Q}_L(t)$. Then the action will only depend on the voltage differences $V_{F,B} = V_{F,B;L} - V_{F,B;R}$ and we obtain

$$\mathrm{i}S_{\text{sc}} = \ln \left\langle \left(\mathcal{T} e^{-\mathrm{i}\int_0^t dt' \hat{Q}_L(t') V_B(t')} \right)^\dagger \mathcal{T} e^{-\mathrm{i}\int_0^t dt' \hat{Q}_L(t') V_F(t')} \right\rangle$$

Integrating the term in the first exponent by parts, we get

$$\int_0^t dt' \hat{Q}_L(t') V_B(t') = \frac{\varphi_B(t)\hat{Q}_L(t)}{e} - \frac{\varphi_B(0)\hat{Q}_L(0)}{e} + \frac{1}{e}\int dt' \hat{I}(t')\varphi_B(t')$$

where

$$\hat{I} = -\frac{d}{dt}\hat{Q}_L, \tag{10.102}$$

is the current operator. Since the dependence of the action on the phases taken at $t' = 0$ and $t' = t$ is not important for our analysis, we can safely put $\varphi_B(0) = \varphi_B(t) = 0$. Repeating the same operation for the forward branch of the Keldysh contour, we arrive at the action

$$\mathrm{i}S_{\mathrm{sc}} = \ln\left\langle \left(\mathcal{T} e^{-\frac{\mathrm{i}}{e}\int_0^t dt' \hat{I}(t')\varphi_B(t')}\right)^{\dagger} \mathcal{T} e^{-\frac{\mathrm{i}}{e}\int_0^t dt' \hat{I}(t')\varphi_F(t')}\right\rangle \qquad (10.103)$$

This result may be interpreted as a direct generalization of Eq. (10.89) to the case of time-dependent counting field and bias voltage. Functional derivatives of the action (10.103) or, equivalently, the action (10.67) over the phases $\varphi_F(t)$ and $\varphi_B(t)$ generate arbitrary time-dependent correlation functions of the current operators. The cumulant-generating function (10.90) is less general since it allows one to generate only zero-frequency current cumulants.

10.8 Average Current, Noise, and Third Cumulant

Let us now apply our effective action formalism to a simple model of a short coherent scatterer embedded in the electromagnetic environment modeled by a linear circuit with an arbitrary frequency-dependent impedance $Z_S(\omega)$. Schematics of the system is shown in Fig. 10.3a. Below we will adopt the model that only includes purely electromagnetic coupling between the two elements of our circuit: the scatterer and the environment. It implies that fluctuations of the current in one element cause fluctuations of electric potentials and in this way affect the charge transport in the second element of the circuit. More subtle effects like, for instance, entanglement of the electron wave functions in these two subsystems as well as possible deviations of the electron distribution function in between the two elements from equilibrium will be disregarded. The latter assumption implies that electrons arriving from the scatterer to the impedance $Z_S(\omega)$ completely relax to their equilibrium states. In addition, we will neglect the resistance of the wire segment in which this electron relaxation occurs.

With these assumptions, the effective action can be expressed as the sum of two contributions:

$$S = S_{\mathrm{sc}} + S_{\mathrm{env}} \qquad (10.104)$$

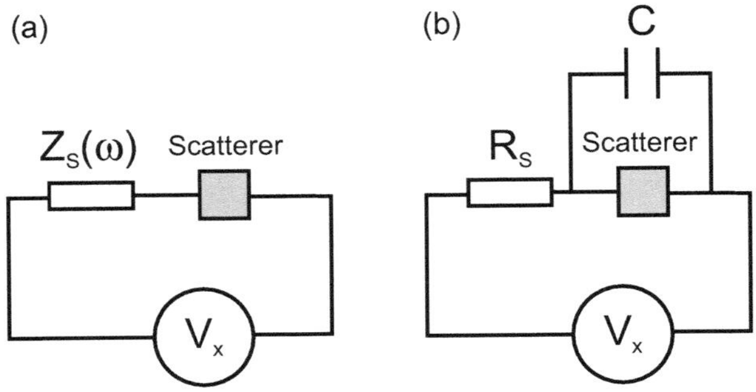

Figure 10.3 (a) Short coherent scatterer embedded in the environment modeled by the impedance $Z_S(\omega)$. (b) Specific model of the environment: Ohmic resistor R_S and junction capacitance C. In this case, one has $Z_S(\omega) = [1/R_S - i\omega C]^{-1}$.

where S_{sc} is the action of a scatterer (10.67) and S_{env} describes the effect of the environment

$$
\begin{aligned}
iS_{env} = &-\frac{i}{e^2}\int_{-\infty}^{t} dt_1 \int_{-\infty}^{t_1} dt_2 \varphi_{env}^{-}(t_1) k_S(t_1 - t_2)\dot{\varphi}_{env}^{+}(t_2) \\
&-\frac{1}{2e^2}\int_{-\infty}^{t} dt_1 \int_{-\infty}^{t} dt_2 \varphi_{env}^{-}(t_1) g_S(t_1 - t_2)\varphi_{env}^{-}(t_2) \quad (10.105)
\end{aligned}
$$

The classical and quantum phases $\varphi_{env}^{\pm}$ are expressed via fluctuating voltages $V_{env}^{\pm}(t')$ across the impedance $Z_S(\omega)$, i.e.,

$$
\varphi_{env}^{\pm} = \int_0^t dt'\, eV_{env}^{\pm}(t') \quad (10.106)
$$

The functions $k_S(t)$ and $g_S(t)$ read

$$
\begin{aligned}
k_S(t) &= \int \frac{d\omega}{2\pi} \frac{e^{-i\omega t}}{Z_S(\omega)}, \\
g_S(t) &= \int \frac{d\omega}{2\pi} \operatorname{Re}\left(\frac{\omega}{Z_S(\omega)}\right) \coth\frac{\omega}{2T} e^{-i\omega t} \quad (10.107)
\end{aligned}
$$

Equation (10.105) provides the most general form of the quadratic effective action obeying the symmetries imposed by unitarity, causality, and FDT.

It is obvious that the following voltage balance equations are satisfied for the equivalent circuit depicted in Fig. 10.3a:

$$
V_x = V^{+} + V_{env}^{+}, \quad 0 = V^{-} + V_{env}^{-} \quad (10.108)
$$

These equations in turn imply that

$$\dot{\varphi}^+_{\rm env} = eV_x - \dot{\varphi}^+, \quad \varphi^-_{\rm env} = -\varphi^- \tag{10.109}$$

and, hence, the action (10.105) takes the form

$$\begin{aligned} {\rm i}S_{\rm env} = &\frac{\rm i}{e^2}\int_{-\infty}^{t} dt_1 \int_{-\infty}^{t_1} dt_2 \varphi^-(t_1) k_S(t_1 - t_2)\left[eV_x - \dot{\varphi}^+(t_2)\right] \\ &- \frac{1}{2e^2}\int_{-\infty}^{t} dt_1 \int_{-\infty}^{t} dt_2 \varphi^-(t_1) g_S(t_1 - t_2)\varphi^-(t_2) \end{aligned} \tag{10.110}$$

Let us derive the formal expressions for the average current, current noise, and the third current cumulant. From Eq. (10.103), we find

$$\begin{aligned} \frac{\delta(S_{\rm sc})}{\delta\varphi^-(t)} &= \frac{1}{2}\left(\frac{\delta(S_{\rm sc})}{\delta\varphi_F(t)} - \frac{\delta(S_{\rm sc})}{\delta\varphi_B(t)}\right) \\ &= -\frac{1}{e}\frac{\left\langle \left(\mathcal{T}e^{-\frac{\rm i}{e}\int_0^t dt' \hat{I}(t')\varphi_B(t')}\right)^\dagger \hat{I}(t)\mathcal{T}e^{-\frac{\rm i}{e}\int_0^t dt' \hat{I}(t')\varphi_F(t')}\right\rangle}{\left\langle \left(\mathcal{T}e^{-\frac{\rm i}{e}\int_0^t dt' \hat{I}(t')\varphi_B(t')}\right)^\dagger \mathcal{T}e^{-\frac{\rm i}{e}\int_0^t dt' \hat{I}(t')\varphi_F(t')}\right\rangle}. \end{aligned} \tag{10.111}$$

On the other hand, the average current is formally defined by the integral

$$\begin{aligned} I &= {\rm tr}\left(e^{{\rm i}Ht}\hat{I}e^{-{\rm i}Ht}\rho_0\right) \\ &= \frac{\int \mathcal{D}\varphi_{F,B}\left\langle \left(\mathcal{T}e^{-\frac{\rm i}{e}\int_0^t dt' \hat{I}(t')\varphi_B(t')}\right)^\dagger \hat{I}(t)\mathcal{T}e^{-\frac{\rm i}{e}\int_0^t dt' \hat{I}(t')\varphi_F(t')}\right\rangle e^{{\rm i}S_{\rm env}[\varphi_F,\varphi_B]}}{\int \mathcal{D}\varphi_{F,B}\, e^{{\rm i}S_{\rm env}[\varphi_F,\varphi_B]}} \end{aligned}$$

Comparing these two expressions, we arrive at the following result

$$I = \frac{{\rm i}e\int \mathcal{D}\varphi^+\mathcal{D}\varphi^- \frac{\delta({\rm i}S_{\rm sc})}{\delta\varphi^-}\, e^{{\rm i}S_{\rm sc}+{\rm i}S_{\rm env}}}{\int \mathcal{D}\varphi^+\mathcal{D}\varphi^-\, e^{{\rm i}S_{\rm sc}+{\rm i}S_{\rm env}}} \tag{10.112}$$

Note that the path integral in the denominator of the above formula—being properly normalized—equals to unity. Nevertheless, we explicitly keep track of this denominator here and below in order to automatically provide correct normalization of the corresponding path integral in the numerator.

A formal expression for the current noise can be derived analogously. Going through a slightly more involved manipulations, we arrive at the formula for the symmetrized current correlator [127]

$$\tilde{\mathcal{S}}(t_1, t_2) = \frac{1}{2}\langle \hat{I}(t_1)\hat{I}(t_2) + \hat{I}(t_2)\hat{I}(t_1)\rangle$$
$$= \frac{-e^2 \int \mathcal{D}\varphi^+\mathcal{D}\varphi^- \left(\frac{1}{4}\frac{\delta^2 e^{iS_{sc}}}{\delta\varphi^+(t_1)\delta\varphi^+(t_2)} + \frac{\delta^2 e^{iS_{sc}}}{\delta\varphi^-(t_1)\delta\varphi^-(t_2)}\right) e^{iS_{env}}}{\int \mathcal{D}\varphi^+\mathcal{D}\varphi^-\, e^{iS_{sc}+iS_{env}}} \tag{10.113}$$

Employing the properties of the total action (10.104), such as $S[-\varphi^+, -\varphi^-] = S[\varphi^+, \varphi^-]$ and $S[\varphi^+, -\varphi^-] = -S^*[\varphi^+, \varphi^+]$, one can demonstrate that the average value of the derivative $\delta^2 e^{iS_{sc}}/\delta\varphi^+(t_1)\delta\varphi^+(t_2)$ in Eq. (10.113) equals to zero. Hence, one finds

$$\tilde{\mathcal{S}}(t_1, t_2) = \frac{-e^2 \int \mathcal{D}\varphi^+\mathcal{D}\varphi^- \frac{\delta^2 e^{iS_{sc}}}{\delta\varphi^-(t_1)\delta\varphi^-(t_2)} e^{iS_{env}}}{\int \mathcal{D}\varphi^+\mathcal{D}\varphi^-\, e^{iS_{sc}+iS_{env}}} \tag{10.114}$$

In order to convert the correlator (10.114) into the noise spectral power (which can be directly measured in experiment), it is necessary to perform the Fourier transformation

$$\mathcal{S}(\omega) = \int dt\, e^{i\omega t}\left(\tilde{\mathcal{S}}(t, 0) - I^2\right) \tag{10.115}$$

Let us now turn to the third cumulant $\mathcal{C}_3$ and define the average of the product of three current operators $\tilde{C}(t_1, t_2, t_3)$. This average is sensitive to specific time ordering of the current operators $\hat{I}(t)$. Below we will use the so-called Keldysh ordering appropriate for certain types of detectors. This ordering corresponds to the definition [128]

$$\tilde{\mathcal{C}}_3(t_1, t_2, t_3) = \frac{(ie)^3 \int \mathcal{D}\varphi^+\mathcal{D}\varphi^- \frac{\delta^3 e^{iS_{sc}}}{\delta\varphi^-(t_1)\delta\varphi^-(t_2)\delta\varphi^-(t_3)} e^{iS_{env}}}{\int \mathcal{D}\varphi^+\mathcal{D}\varphi^-\, e^{iS_{sc}+iS_{env}}} \tag{10.116}$$

Making use of Eq. (10.103), one can demonstrate that this correlator can be rewritten in the equivalent operator form

$$\begin{aligned}\tilde{\mathcal{C}}_3(t_1, t_2, t_3) = \frac{1}{8}\{&\langle \hat{I}(t_1)(\mathcal{T}\hat{I}(t_2)\hat{I}(t_3))\rangle + \langle(\tilde{\mathcal{T}}\hat{I}(t_2)\hat{I}(t_3))\hat{I}(t_1)\rangle \\ &+ \langle \hat{I}(t_2)(\mathcal{T}\hat{I}(t_1)\hat{I}(t_3))\rangle + \langle(\tilde{\mathcal{T}}\hat{I}(t_1)\hat{I}(t_3))\hat{I}(t_2)\rangle \\ &+ \langle \hat{I}(t_3)(\mathcal{T}\hat{I}(t_1)\hat{I}(t_2))\rangle + \langle(\tilde{\mathcal{T}}\hat{I}(t_1)\hat{I}(t_2))\hat{I}(t_3)\rangle \\ &+ \langle \mathcal{T}\hat{I}(t_1)\hat{I}(t_2)\hat{I}(t_3)\rangle + \langle \tilde{\mathcal{T}}\hat{I}(t_1)\hat{I}(t_2)\hat{I}(t_3)\rangle\}\end{aligned} \tag{10.117}$$

We again perform a Fourier transformation over times and obtain

$$\mathcal{C}_3(\omega_1, \omega_2) = \int dt_1 dt_2\, e^{i\omega_1 t_1 + i\omega_2 t_2} \left[\tilde{\mathcal{C}}_3(0, -t_1, -t_2) - I\, \mathcal{S}(-t_1, -t_2) \right.$$
$$\left. - I\, \mathcal{S}(0, -t_2) - I\mathcal{S}(0, -t_1) + 2I^3 \right] \quad (10.118)$$

Here I stands for the average current. One can verify that in zero-frequency limit, this expression reduces to the formula (10.82), which we derived earlier.

Equation (10.118) defines the frequency-dependent third current cumulant $\mathcal{C}_3(\omega_1, \omega_2)$, which can be directly measured in modern experiments, as it was done, for instance, in the experiment [126] with a linear amplifier. On the other hand, since current operators taken at different times do not commute with each other, different ordering of these operators would in general yield different frequency dependencies of the third current cumulant, which could possibly be measured by other types of detectors.

We also point out that in the zero-frequency limit realized in a vast majority of experiments, time ordering of the current operators is unimportant. For instance, let us consider the second-order cumulant in the limit $\omega \to 0$. One has

$$\mathcal{S} = \lim_{t\to\infty} \frac{1}{2t} \int_0^t dt_1 \int_0^t dt_2 \langle \hat{I}(t_1)\hat{I}(t_2) + \hat{I}(t_2)\hat{I}(t_1)\rangle \quad (10.119)$$

Interchanging the integration parameters $t_1 \leftrightarrow t_2$, we get

$$\mathcal{S} = \lim_{t\to\infty} \frac{1}{t} \int_0^t dt_1 \int_0^t dt_2 \langle \hat{I}(t_1)\hat{I}(t_2)\rangle = \lim_{t\to\infty} \frac{1}{t} \int_0^t dt_1 \int_0^t dt_2 \langle \hat{I}(t_2)\hat{I}(t_1)\rangle$$

This equation illustrates our point. The same arguments apply to higher cumulants as well. Hence, while taking the zero-frequency limit, it suffices to treat the time-dependent operator $\hat{I}(t)$ simply as a real-valued function.

As it was already demonstrated above, in the zero-impedance limit $Z_S(\omega) \to 0$, the cumulant-generating function $\mathcal{F}(\chi)$ (10.90) coincides with the action of a short coherent conductor evaluated at $\varphi^+ = eVt$, $\varphi^- = \chi$ (see Eq. (10.96)). At nonzero $Z_S(\omega)$, both phases φ^+ and φ^- fluctuate and the cumulant-generating function $\mathcal{F}(\chi)$ is expressed as a path integral over these phases

$$\mathcal{F}(\chi) = \ln \left[\int \mathcal{D}\varphi^+ \mathcal{D}\varphi^- e^{iS_{sc}[\varphi^+, \chi+\varphi^-] + iS_{env}[eV_x t - \varphi^+, -\varphi^-]} \right] \quad (10.120)$$

Here $S_{\text{env}}[\varphi_{\text{env}}^+, \varphi_{\text{env}}^-]$ is the environment action defined by Eq. (10.105). It is straightforward to verify that by differentiating the function (10.120) with respect to the counting field χ and making use of Eq. (10.78), one reproduces the expressions both for current noise (10.115) and for the third cumulant (10.118) taken in the zero-frequency limit. The same conclusion also applies to higher-order cumulants and, hence, Eq. (10.120) indeed defines the cumulant-generating function for a coherent scatterer in the presence of an electromagnetic environment.

10.9 Interaction with Adiabatic Environment

As a simple application of the above results, let us consider a specific model of the environment illustrated in Fig. 10.3b, i.e., we will assume that the scatterer has a capacitance C and the leads connecting it to the voltage source have a resistance R_S. In this case, one trivially finds

$$\frac{1}{Z_S(\omega)} = \frac{1}{R_S} - \mathrm{i}\omega C \tag{10.121}$$

Despite its simplicity, this model is rather general since Eq. (10.121) should in fact always apply at sufficiently low frequencies ω, being just a low-frequency Taylor expansion of an arbitrary function $1/Z_S(\omega)$.

In what follows, we will also assume that the inverse RC-time is low as compared to all other relevant frequencies, i.e.,

$$\frac{1}{R_S C} \ll T,\, eV,\, I/e. \tag{10.122}$$

In this case, fluctuations of both the quantum phase φ^- and the bias voltage $v = \dot{\varphi}^+/e$ remain slow and the path integral in Eq. (10.120) may be replaced by ordinary integrals over these two parameters. Then the scatterer action reduces to the zero-frequency cumulant-generating function (10.90). On top of that, we will consider the shot noise regime $eV \gg T$ and employ an approximate expression for $\mathcal{F}(t, \chi)$ (10.92). In this case, the cumulant-generating function

(10.120) reduces to

$$\mathcal{F}(t,\chi) = \ln\left[\int dv\, d\varphi^- \exp\left(\frac{evt}{2\pi}\sum_n \ln\left[1+T_n(e^{-i(\chi+\varphi^-)}-1)\right]\right.\right.$$
$$\left.\left.+\,i\frac{e(V_x-v)t}{e^2R_S}\varphi^-\right)\right]. \tag{10.123}$$

Let us evaluate the current cumulants (10.78) determined by the long-time asymptotics of the function (10.123). Taking the limit $t\to\infty$ and performing the integral (10.123) by means of the saddle point approximation, we obtain the following result [224]

$$\mathcal{F}(t,\chi) = \frac{ev_0t}{2\pi}\sum_n \ln\left[1+T_n(e^{-i(\chi+\varphi_0^-)}-1)\right] + i\frac{(V_x-v_0)t}{eR_S}\varphi_0^- \tag{10.124}$$

where the parameters v_0 and φ_0^- depend on the counting field χ and are fixed by the least action conditions

$$\frac{e^2R_S}{2\pi}\sum_n \ln\left[1+T_n(e^{-i(\chi+\varphi_0^-)}-1)\right] - i\varphi_0^- = 0,$$

$$v_0\left(1+\frac{e^2R_S}{2\pi}\sum_n \frac{T_ne^{-i(\chi+\varphi_0^-)}}{1+T_n(e^{-i(\chi+\varphi_0^-)}-1)}\right) = V_x. \tag{10.125}$$

Excluding the saddle point value v_0 from the above equations, we arrive at the final formula for the cumulant-generating function:

$$\mathcal{F}(\chi) = i\frac{V_xt}{eR_S}\varphi_0^-(\chi), \tag{10.126}$$

$$\varphi_0^-(\chi) = -i\frac{e^2R_S}{2\pi}\sum_n \ln\left[1+T_n(e^{-i[\chi+\varphi_0^-(\chi)]}-1)\right] \tag{10.127}$$

In order to recover the first three current cumulants, we expand both sides of Eq. (10.127) in powers of the counting field χ keeping all terms up to the third order. Then we get

$$\varphi_0^{-'}\chi + \frac{\varphi_0^{-''}}{2}\chi^2 + \frac{\varphi_0^{-'''}}{6}\chi^3 = \frac{eR_S}{V_x}\left[-\frac{\mathcal{C}_1^{(0)}}{e}(1+\varphi_0^{-'})\chi\right.$$
$$+\left(-\frac{\mathcal{C}_1^{(0)}}{e}\varphi_0^{-''} + i\frac{\mathcal{C}_2^{(0)}}{e^2}(1+\varphi_0^{-'})^2\right)\frac{\chi^2}{2}$$
$$\left.+\left(-\frac{\mathcal{C}_1^{(0)}}{e}\varphi_0^{-'''} + 3i\frac{\mathcal{C}_2^{(0)}}{e^2}(1+\varphi_0^{-'})\varphi_0^{-''} + \frac{\mathcal{C}_3^{(0)}}{e^3}(1+\varphi_0^{-'})^3\right)\frac{\chi^3}{6}\right].$$

Here $\mathcal{C}_1^{(0)}$, $\mathcal{C}_2^{(0)}$, and $\mathcal{C}_3^{(0)}$ are "non-interacting" current cumulants defined in Eqs. (10.93)–(10.95) with V replaced by V_x. From the above equation, we reconstruct the derivatives of the function $\varphi_0^-(\chi)$,

$$\varphi_0^{-'} = -\frac{R_S \mathcal{C}_1^{(0)}}{V_x + R_S \mathcal{C}_1^{(0)}}, \quad \varphi_0^{-''} = \frac{\mathrm{i}}{e} \frac{R_S \mathcal{C}_2^{(0)} (1 + \varphi_0^{-'})^2}{V_x + R_S \mathcal{C}_1^{(0)}},$$
$$\varphi_0^{-'''} = \frac{R_S}{e^2} \frac{3\mathrm{i}e\mathcal{C}_2^{(0)}(1 + \varphi_0^{-'})\varphi_0^{-''} + \mathcal{C}_3^{(0)}(1 + \varphi_0^{-'})^3}{V_x + R_S \mathcal{C}_1^{(0)}} \tag{10.128}$$

and derive the expressions for the cumulants affected by our adiabatic environment. They read

$$\mathcal{C}_1 = -\frac{V_x}{R_S}\varphi_0^{-'} = \frac{V_x \mathcal{C}_1^{(0)}}{V_x + R_S \mathcal{C}_1^{(0)}}, \tag{10.129}$$

$$\mathcal{C}_2 = -\mathrm{i}\frac{eV_x}{R_S}\varphi_0^{-''} = \frac{V_x^3 \mathcal{C}_2^{(0)}}{\left(V_x + R_S \mathcal{C}_1^{(0)}\right)^3}, \tag{10.130}$$

$$\mathcal{C}_3 = \frac{e^2 V_x}{R_S}\varphi_0^{-'''}$$
$$= \frac{V_x^4}{\left(V_x + R_S \mathcal{C}_1^{(0)}\right)^4} \left(\mathcal{C}_3^{(0)} - 3\frac{R_S \left(\mathcal{C}_2^{(0)}\right)^2}{V_x + R_S \mathcal{C}_1^{(0)}} \right) \tag{10.131}$$

Rewriting these results in terms of bias voltage V_x, we obtain

$$I = \frac{V_x}{R_N + R_S}, \tag{10.132}$$

$$\mathcal{S} = \frac{R_N^3}{(R_N + R_S)^3} \beta \frac{eV_x}{R_N}, \tag{10.133}$$

$$\mathcal{C}_3 = \frac{R_N^4}{(R_N + R_S)^4} \left(\beta - 2\gamma - 3\frac{R_S}{R_N + R_S}\beta^2 \right) \frac{e^2 V_x}{R_N} \tag{10.134}$$

where we defined the conductor resistance

$$R_N = \left[\frac{e^2}{2\pi} \sum_n T_n \right]^{-1} \tag{10.135}$$

its Fano factor

$$\beta = \frac{\sum_n T_n (1 - T_n)}{\sum_n T_n} \tag{10.136}$$

and yet another dimensionless parameter

$$\gamma = \frac{\sum_n T_n^2(1 - T_n)}{\sum_n T_n} \tag{10.137}$$

which characterizes non-Gaussian nature of current fluctuations.

The above expressions determine zero-frequency current cumulants $\mathcal{C}_1$, $\mathcal{C}_2$, and $\mathcal{C}_3$ in the presence of adiabatic voltage fluctuations induced by an external electromagnetic environment.

Let us now discuss a physical meaning of these results for the current cumulants. The expression (10.132) is completely transparent: It just describes the current reduction due to the environment resistance R_S switched in series to our conductor. In order to qualitatively interpret the environment effect on both the second and the third current cumulants in Eqs. (10.133) and (10.134), it will be convenient to provide an alternative (simpler) derivation of these equations.

In order to rederive Eq. (10.133), we employ the system of two coupled equations for the fluctuating current $I(t)$ and instantaneous voltage drop across the scatterer $v(t)$:

$$I(t) = \frac{v(t)}{R_N} + \xi\left(t, v(t)\right), \quad I(t) = \frac{V_x - v(t)}{R_S} \tag{10.138}$$

The first equation just represents the current across the conductor as the sum of Ohmic and stochastic (shot noise) terms, respectively, $v(t)/R_N$ and $\xi\left(t, v(t)\right)$. The latter is characterized by the following pair correlator:

$$\left\langle \xi\left(t_1, v(t_1)\right)\xi\left(t_2, v(t_2)\right)\right\rangle = \beta\frac{ev(t_1)}{R_N}\delta(t_1 - t_2) \tag{10.139}$$

i.e., the noise intensity also fluctuates with $v(t)$. Excluding $v(t)$ from Eqs. (10.138), we find

$$I(t) = \frac{V_x}{R_N + R_S} + \frac{R_N}{R_N + R_S}\xi(t) \tag{10.140}$$

Averaging this result over $\xi(t)$, we again recover Eq. (10.132). The fluctuating contribution $\delta I(t) = I(t) - \langle I\rangle$

$$\delta I(t) = \frac{R_N}{R_N + R_S}\xi(t) \tag{10.141}$$

gives rise to the current noise. Combining Eqs. (10.141) and (10.139), we get

$$\mathcal{S} \equiv |\delta I|_\omega^2 = \frac{R_N^2}{(R_N + R_S)^2}\beta\frac{e\langle v\rangle}{R_N} \tag{10.142}$$

Substituting now the average value of the voltage drop across the conductor

$$\langle v\rangle = R_N V_x/(R_N + R_S) \tag{10.143}$$

into Eq. (10.142), we again recover Eq. (10.133). Hence, we may conclude that the effect of our adiabatic environment on the current noise is due to a reduction in both current fluctuations (10.141) and the average voltage drop across the conductor (10.143).

The same calculation can be performed in order to rederive Eq. (10.134) for the third cumulant. We again employ Eq. (10.141) and express the third current cumulant in the form

$$\langle \delta I(t_1)\delta I(t_2)\delta I(t_3)\rangle = \frac{R_N^3}{(R_N+R_S)^3}\langle \xi\left(t_1, v(t_1)\right)\xi\left(t_2, v(t_2)\right)\xi\left(t_3, v(t_3)\right)\rangle \tag{10.144}$$

The voltage drop across the conductor can be expressed as the sum of the average value and the fluctuating part,

$$v(t) = \langle v\rangle + \delta v(t),$$

where the latter is caused by the current noise and reads

$$\delta v(t) = -R_S\delta I(t) = -\frac{R_N R_S}{R_N + R_S}\xi(t) \tag{10.145}$$

Voltage fluctuations, in turn, modify the current noise. This effect can be approximately accounted for by the following expansion:

$$\xi\left(t, v(t)\right) = \xi\left(t, \langle v\rangle + \delta v(t)\right) \approx \xi(t, \langle v\rangle) + \frac{\partial \xi(t, \langle v\rangle)}{\partial v}\delta v(t) \tag{10.146}$$

Substituting this formula into Eq. (10.144) and keeping only the lowest-order corrections in δv, we get

$$\begin{aligned}\langle \delta I(t_1)\delta I(t_2)\delta I(t_3)\rangle &= \frac{R_N^3}{(R_N+R_S)^3}\langle \xi(t_1, \langle v\rangle)\xi(t_2, \langle v\rangle)\xi(t_3, \langle v\rangle)\rangle \\ &+ \frac{R_N^3}{(R_N+R_S)^3}\left\langle \frac{\partial \xi(t_1, \langle v\rangle)}{\partial v}\delta v(t_1)\xi(t_2, \langle v\rangle)\xi(t_3, \langle v\rangle)\right\rangle \\ &+ \frac{R_N^3}{(R_N+R_S)^3}\left\langle \xi(t_1, \langle v\rangle)\frac{\partial \xi(t_2, \langle v\rangle)}{\partial v}\delta v(t_2)\xi(t_3, \langle v\rangle)\right\rangle \\ &+ \frac{R_N^3}{(R_N+R_S)^3}\left\langle \xi(t_1, \langle v\rangle)\xi(t_2, \langle v\rangle)\frac{\partial \xi(t_3, \langle v\rangle)}{\partial v}\delta v(t_3)\right\rangle\end{aligned}$$

Then for the zero-frequency current cumulant $\mathcal{C}_3$, we obtain

$$\mathcal{C}_3 = \int dt_1 dt_2 \langle \delta I(t_1)\delta I(t_2)\delta I(0)\rangle = \frac{R_N^3}{(R_N + R_S)^3}\mathcal{C}_3^{(0)}(\langle v\rangle)$$
$$-3\frac{R_N^4}{(R_N+R_S)^4}\int dt_1 dt_2 \left\langle \xi(t_1,\langle v\rangle)\xi(t_2,\langle v\rangle)\frac{\partial \xi(0,\langle v\rangle)}{\partial v}\xi(0,\langle v\rangle)\right\rangle$$

Evaluating the average in the last term of the above equation, one can neglect non-Gaussian fluctuations. Hence, one can decompose this average in accordance with the Wick theorem and find

$$\left\langle \xi(t_1,\langle v\rangle)\xi(t_2,\langle v\rangle)\frac{\partial \xi(0,\langle v\rangle)}{\partial v}\xi(0,\langle v\rangle)\right\rangle$$
$$= \langle \xi(t_1,\langle v\rangle)\xi(t_2,\langle v\rangle)\rangle \left\langle \frac{\partial \xi(0,\langle v\rangle)}{\partial v}\xi(0,\langle v\rangle)\right\rangle$$
$$+ \left\langle \xi(t_1,\langle v\rangle)\frac{\partial \xi(0,\langle v\rangle)}{\partial v}\right\rangle \langle \xi(t_2,\langle v\rangle)\xi(0,\langle v\rangle)\rangle$$
$$+ \left\langle \xi(t_2,\langle v\rangle)\frac{\partial \xi(0,\langle v\rangle)}{\partial v}\right\rangle \langle \xi(t_1,\langle v\rangle)\xi(0,\langle v\rangle)\rangle$$

The first average in the right-hand side of this equation vanishes due to causality, i.e.,

$$\left\langle \frac{\partial \xi(0,\langle v\rangle)}{\partial v}\xi(0,\langle v\rangle)\right\rangle = 0 \qquad (10.147)$$

The remaining averages can be expressed as follows

$$\left\langle \xi(t_1,\langle v\rangle)\frac{\partial \xi(0,\langle v\rangle)}{\partial v}\right\rangle = \frac{1}{2}\frac{\partial}{\partial v}\langle \xi(t_1,\langle v\rangle)\xi(0,\bar{v})\rangle \quad (10.148)$$

Collecting now all contributions and performing the time integrals, we finally obtain

$$\mathcal{C}_3 = \frac{R_N^3}{(R_N+R_S)^3}\mathcal{C}_3^{(0)}(\langle v\rangle) - 3\frac{R_N^4}{(R_N+R_S)^4}\mathcal{S}^{(0)}(\langle v\rangle)\frac{\partial \mathcal{S}^{(0)}(\langle v\rangle)}{\partial v} \qquad (10.149)$$

Making use of Eqs. (10.94) and (10.95), one can verify that Eq. (10.149) is fully equivalent to the result (10.134), which was derived rigorously without employing perturbation theory in $\delta v(t)$. On the other hand, our present derivation allows for a transparent physical interpretation of this result. Indeed, it is easy to observe that the pre-factor in front of the first term in Eq. (10.149) just reflects a reduction of current fluctuations (10.141) due to the presence of an extra resistor R_S, while the second term in Eq. (10.149) is due to fluctuations of the shot noise power caused by voltage fluctuations. Depending on the ratio R_S/R_N and the parameters β and γ, the negative terms in Eqs. (10.134) and (10.149) may become large, resulting in the sign change of the third cumulant [224]. This effect was indeed observed in experiments [335].

To conclude, we have demonstrated that the zero-frequency FCS of a short coherent conductor interacting with an adiabatic environment can be evaluated exactly. Provided high-frequency fluctuations gain importance, the environment cannot anymore be considered adiabatic and the situation becomes much more complicated. For instance, Eqs. (10.121) and (10.122) do not apply in this case. Instead, e.g., at high frequencies and/or voltage bias values, one should rather model the environment as a transmission line with distributed capacitance and inductance [328, 391, 392].

In the next chapters, we will demonstrate that high-frequency fluctuations should necessarily be taken into account while describing electron–electron interactions in disordered conductors. An interplay between electron scattering, quantum coherence, and interactions yields new physical effects to be discussed in the following chapters of this book.

Chapter 11

Coulomb Effects in Short Coherent Conductors

In this chapter, we will make use of the general expressions for the effective action derived above for short coherent conductors and apply them in order to analyze Coulomb interaction effects in such conductors. Understanding of an interplay between coherent scattering and electron–electron interactions in such spatially confined system will then allow us to proceed to more complicated structures, including spatially extended disordered conductors, transport properties of which will be addressed in the next few chapters.

In Chapter 7, we have already analyzed charging effects in normal tunnel junctions characterized by small transmissions ($T_n \ll 1$) of all their conducting channels. We demonstrated that at sufficiently low temperatures, electron–electron interactions tend to suppress electron transport across such junctions. This effect is usually called Coulomb blockade of electron tunneling. Assume now that instead of a tunnel barrier between two bulk metallic leads, one would place a short coherent conductor characterized by arbitrary transmission distribution of its conducting channels T_n, as described, e.g., in Section 10.1. Do Coulomb blockade effects persist also in this case? The answer to this question will be presented here.

Dissipative Quantum Mechanics of Nanostructures: Electron Transport, Fluctuations, and Interactions
Andrei D. Zaikin and Dmitry S. Golubev

ISBN 978-981-4774-50-5 (Hardcover), 978-0-429-29823-3 (eBook)
www.jennystanford.com

As before, for simplicity we will assume the transmissions T_n to be energy independent. This model of a coherent scatterer embraces practically all types of conductors, e.g., break junctions, point contacts, carbon nanotubes, and diffusive metallic bridges. The only essential requirement involved in this model is that the electron flight time τ_{fl} across the scatterer should be short enough; see Eq. (10.3). This requirement imposes natural restrictions on the size of our conductor.

11.1 Weak Coulomb Effects in Metallic Conductors

11.1.1 Quasiclassical Expansion for Effective Action

From the analysis developed in Chapter 7 (see, in particular, Eqs. (7.80) and (7.81)), we already learned that provided either the conductance of a tunnel junction $1/R_{\mathrm{N}}$ (or, more generally the sum of $1/R_{\mathrm{N}}$ and the inverse environment impedance $1/Z_{\mathrm{S}}$) strongly exceeds the quantum conductance unit $1/R_{\mathrm{q}} = e^2/h$ or the corresponding bias voltage and/or temperature values are much higher than the junction charging energy E_{C}, Coulomb effects are relatively weakly pronounced and—at least in some parameter regions—can be treated perturbatively in the interaction. Technically, this perturbative analysis is based on the observation that under the above conditions, fluctuations of the quantum phase, φ^-, remain small. Estimating φ^- from Eqs. (7.47) and (7.62), one finds

$$\varphi^- \sim K(t) \sim e^2 \frac{R_{\mathrm{N}} Z_{\mathrm{S}}}{R_{\mathrm{N}} + Z_{\mathrm{S}}} \ll 1 \tag{11.1}$$

in the first limit and

$$\varphi^- \sim K(t \sim 1/eV, 1/T) \sim \frac{E_{\mathrm{C}}}{\max[eV, T]} \ll 1 \tag{11.2}$$

in the second one.

We will now apply the same idea to the case of short coherent scatterers. Namely, we assume that the quantum phase φ^- fluctuates weakly and perform a regular quasiclassical expansion of the effective action (10.67) in powers of φ^-. For our purposes, it will be sufficient to keep all terms up to the third order in φ^-. After a straightforward calculation, we arrive at the result [127]

$$\begin{aligned} \mathrm{i}S_{\mathrm{sc}} = &-\frac{\mathrm{i}}{e^2 R_{\mathrm{N}}} \int_0^t dt_1 \dot{\varphi}^+(t_1) \left(\varphi^-(t_1) - \beta \frac{[\varphi^-(t_1)]^3}{6} \right) \\ &- \frac{1-\beta}{2e^2 R_{\mathrm{N}}} \int_0^t dt_1 dt_2 \varphi^-(t_1) \alpha(t_1 - t_2) \varphi^-(t_2) \\ &- \frac{\beta}{2e^2 R_{\mathrm{N}}} \int_0^t dt_1 dt_2 \varphi^-(t_1) \alpha(t_1 - t_2) \varphi^-(t_2) \cos\left[\varphi^+(t_1) - \varphi^+(t_2)\right] \\ &+ \frac{\pi \mathrm{i} \gamma T^3}{6e^2 R_{\mathrm{N}}} \int_0^t \frac{dt_1 dt_2 dt_3 \ \varphi^-(t_1) \varphi^-(t_2) \varphi^-(t_3)}{\sinh[\pi T(t_3 - t_2)] \sinh[\pi T(t_2 - t_1)] \sinh[\pi T(t_1 - t_3)]} \\ &\times \left\{ \sin[\varphi^+(t_3) - \varphi^+(t_2)] + \sin[\varphi^+(t_2) - \varphi^+(t_1)] \right. \\ &\left. + \sin[\varphi^+(t_1) - \varphi^+(t_3)] \right\} \end{aligned} \tag{11.3}$$

Here, as before, the kernel $\alpha(t)$ is defined in Eq. (10.76). We observe that the action (11.3) explicitly depends on the Fano factor β (10.136) as well as on the parameter γ (10.137), which we already encountered in Chapter 10.

11.1.2 Shot Noise and Weak Coulomb Blockade of Electron Transport

In order to evaluate the current across the conductor, we substitute the action (11.3) into the general expression (10.112). While performing the path integrals, we neglect the terms $\sim (\varphi^-)^3$ in Eq. (11.3). Then we obtain [159]

$$I = \frac{V}{R_{\mathrm{N}}} - \frac{\beta}{eR_{\mathrm{N}}} \int_0^\infty dt \frac{\pi T^2}{\sinh^2[\pi T t]} K(t) \sin[eVt], \tag{11.4}$$

where

$$V = \frac{R_{\mathrm{N}}}{R_{\mathrm{N}} + Z_{\mathrm{S}}(0)} V_x \tag{11.5}$$

is the average voltage across the conductor and

$$K(t) = e^2 \int \frac{d\omega}{2\pi} \frac{R_{\mathrm{N}} Z_{\mathrm{S}}(\omega)}{R_{\mathrm{N}} + Z_{\mathrm{S}}(\omega)} \frac{e^{-\mathrm{i}\omega t}}{-\mathrm{i}\omega + 0} \tag{11.6}$$

It is easy to verify that in the limit $\beta \to 1$, Eq. (11.4) reduces to our previous result (7.50) derived in Chapter 7 for tunnel barriers. In a single-channel limit, the relation equivalent to Eq. (11.4) was also derived by Levy Yeyati et al. [271].

The result (11.4) allows to make two important observations: (i) Coulomb correction to the conductance of short coherent conductors is negative, being determined by the *universal* function of voltage and temperature, i.e., within its applicability range (Eq. (11.4) equally well describes all types of conductors irrespectively to their physical nature) and (ii) the interaction correction to the conductance is proportional to the Fano factor β of the conductor, which also enters into the expression for its shot noise. The second observation hints at an intimate relation between the shot noise and Coulomb effects in disordered conductors. For instance, for given values V and T, the absolute value of the interaction correction to the conductance in Eq. (11.4) reaches its maximum in the tunneling limit $\beta \to 1$ and vanishes for fully open barriers $\beta \to 0$. The conductor shot noise demonstrates exactly the same behavior.

To further illustrate this fundamentally important point, it is instructive to rederive the result (11.4) by means of the equivalent Langevin equation approach, which makes the relation between current noise and electron–electron interaction effects completely transparent. Let us consider the quasiclassical Langevin equation

$$\frac{v(t)}{R_{\rm N}} + \int_{-\infty}^{t} dt'\, z_{\rm S}^{-1}(t-t')v(t') = \frac{V_x}{Z_{\rm S}(0)} + \xi_{\rm S}(t) + \xi(t, v(t)), \quad (11.7)$$

which accounts for charge dynamics in the circuit depicted in Fig. 10.3(a) and generalizes Eq. (10.138) in order to describe fluctuations at arbitrary frequencies. Here

$$z_{\rm S}^{-1}(t) = \int \frac{d\omega}{2\pi} \frac{e^{-i\omega t}}{Z_{\rm S}(\omega)} \quad (11.8)$$

defines the time-dependent admittance of the effective environment, while $\xi_{\rm S}(t)$ and $\xi(t, v(t))$ are stochastic variables describing, respectively, the current noise produced by the environmental impedance $Z_{\rm S}(\omega)$ and the noise produced by the conductor itself. The variable $\xi_{\rm S}(t)$ is Gaussian and its pair correlator can be easily determined from FDT as

$$\langle \xi_{\rm S}(t_1)\xi_{\rm S}(t_2)\rangle = \int \frac{d\omega}{2\pi} e^{-i\omega(t_1-t_2)}\, {\rm Re}\left(\frac{1}{Z_{\rm S}(\omega)}\right) \omega \coth\frac{\omega}{2T} \quad (11.9)$$

The conductor noise $\xi(t)$ is defined by the correlator

$$\langle \xi(t_1, v(t_1))\xi(t_2, v(t_2))\rangle = \left(\frac{1-\beta}{R_{\rm N}} + \frac{\beta}{R_{\rm N}}\cos\left[\varphi^+(t_1) - \varphi^+(t_2)\right]\right) \times \left(-\frac{\pi T^2}{\sinh^2[\pi T(t_1 - t_2)]}\right) \quad (11.10)$$

The quasiclassical Langevin equation in Eqs. (11.7)–(11.10) is exactly equivalent to the effective action (11.3) (with omitted third order in φ^- terms). It can be derived from this action in the same manner as it was done, e.g., in Section 2.3 for Josephson junctions. For instance, the conductor noise correlator (11.10) for an arbitrary time-dependent voltage $v(t)$ is obtained from the action (11.3) by taking the derivatives over the quantum phase

$$\langle \xi(t_1)\xi(t_2)\rangle = -e^2 \frac{\mathrm{i}\delta^2 S_{\rm sc}}{\delta\varphi^-(t_1)\delta\varphi^-(t_2)}\bigg|_{\varphi^-=0} \quad (11.11)$$

and identifying the instantaneous value of the classical phase in Eq. (11.10) with the fluctuating voltage drop across the conductor as $\dot{\varphi}^+(t) = ev(t)$.

In order to derive the average current I in the system, we make use of the general expression for the fluctuating current

$$I(t) = \frac{v(t)}{R_{\rm N}} - \xi(t, v(t)) \quad (11.12)$$

flowing in our conductor and average Eq. (11.12) over the noise. Then we obtain

$$I = \frac{V}{R_{\rm N}} - \langle \xi(t, v(t))\rangle \quad (11.13)$$

where $V = \langle v(t)\rangle$. Comparing Eqs. (11.13) and (11.4), we immediately observe that the interaction correction to the current is identical to the average value $\langle \xi(t, v(t))\rangle$ describing the shot noise in our conductor. Thus, there indeed exists a fundamental relation between shot noise and Coulomb effects in disordered conductors, as it was anticipated above.

It is important to emphasize that in the presence of nonzero voltage $v(t)$, the average value $\langle \xi(t, v(t))\rangle$ differs from zero. In order to see that we use the approach similar to one already employed in Section 10.9., now we consider the correlator (11.10) and treat the noise variable ξ as a function of the time-dependent phase

$\varphi^+(t)$. Splitting this phase into regular and fluctuating parts $\varphi^+(t) = \varphi_0(t) + \delta\varphi^+$ (where $\varphi_0(t) = eVt$) and assuming $\delta\varphi^+$ to be small, we may approximate

$$\langle \xi(t, v(t)) \rangle = \left\langle \frac{\partial \xi(t, \varphi_0(t))}{\partial \varphi_0(t)} \delta\varphi^+(t) \right\rangle \tag{11.14}$$

Fluctuations of the classical phase are described by the Langevin equation (11.7), which yields

$$\begin{aligned} \delta\varphi^+(t) &= e \int_0^t dt'\, \delta v(t') \\ &= \frac{1}{e} \int_0^t dt' \int dt''\, \dot{K}(t' - t'') \left[\xi_S(t'') + \xi(t'', \varphi_0(t'')) \right] \\ &= \frac{1}{e} \int dt'' \left[K(t - t'') - K(-t'') \right] \left[\xi_S(t'') + \xi(t'') \right] \end{aligned} \tag{11.15}$$

Substituting this expression into Eq. (11.14) and bearing in mind that the noise variables $\xi(t)$ and $\xi_S(t)$ are uncorrelated, we get

$$\langle \xi(t, v(t)) \rangle = \frac{1}{e} \int dt'' \left[K(t - t'') - K(-t'') \right] \left\langle \frac{\partial \xi(t, \varphi_0(t))}{\partial \varphi_0(t)} \xi(t'', \varphi_0(t'')) \right\rangle$$

The average under the integral may be easily evaluated from Eq. (11.10):

$$\begin{aligned} \left\langle \frac{\partial \xi(t, \varphi_0(t))}{\partial \varphi_0(t)} \xi(t'', \varphi_0(t'')) \right\rangle &= \frac{\partial}{\partial \varphi_0(t)} \left\langle \xi(t, \varphi_0(t)) \xi(t'', \varphi_0(t'')) \right\rangle \\ &= \frac{\pi T^2}{\sinh^2[\pi T(t - t'')]} \frac{\beta}{R_N} \sin[eV(t - t'')] \end{aligned} \tag{11.16}$$

Combining all the above expressions and taking the limit $t \to \infty$, we again arrive at Eq. (11.4).

This simple Langevin equation analysis clearly demonstrates the physical reason behind the dependence of the interaction correction to the conductance on the Fano factor β. According to Eq. (11.14), the average value of the noise variable ξ differs from zero only provided it depends on the voltage drop across the conductor. The effect of the bias voltage on the noise power implies, in turn, that the conductor produces shot noise. Since the intensity of the latter scales linearly with β, so does the interaction correction to the conductance.

Let us evaluate the expression (11.4) for an important special case of the environment describing the impedance (10.121). In this case, the response function acquires a simple form

$$K(t) = e^2 \frac{R_N R_S}{R_N + R_S} \theta(t) \left[1 - e^{-t/\tau_{RC}} \right] \tag{11.17}$$

where $\tau_{\rm RC} = R_{\rm N}R_{\rm S}C/(R_{\rm N} + R_{\rm S})$ is the charge relaxation time. The current (11.4) can then be evaluated exactly with the result [159]

$$I = \frac{V}{R_{\rm N}} - \frac{e\beta}{\pi}\frac{R_{\rm S}}{R_{\rm N}+R_{\rm S}}\,{\rm Im}\left[\left(\frac{1}{\tau_{\rm RC}} + {\rm i}eV\right)\Psi\left(1+\frac{1}{2\pi T\tau_{\rm RC}} + {\rm i}\frac{eV}{2\pi T}\right)\right.$$
$$\left. -{\rm i}eV\Psi\left(1+{\rm i}\frac{eV}{2\pi T}\right)\right]. \tag{11.18}$$

We emphasize again that Eq. (11.18) describes the current–voltage characteristics of a short coherent conductor with an arbitrary transmission distribution T_n of its conducting channels in the weak Coulomb blockade regime. Equation (11.18) contains exactly the same combination of digamma functions as the corresponding expression (7.52) derived in Chapter 7 for tunnel barriers. Actually the only difference between these two results is the presence of the Fano factor β in front of the interaction correction in Eq. (11.18).

Let us present several useful limiting expressions that follow from Eq. (11.18). As we already did in Chapter 7, we again introduce the dimensionless conductances both for the shunt $g_{\rm S} = R_{\rm q}/R_{\rm S}$ and for the scatterer:

$$g = \frac{R_{\rm q}}{R_{\rm N}} = \sum_n T_n \tag{11.19}$$

In the low temperature limit $T \ll eV$, the digamma functions in Eq. (11.18) reduce to logarithms and the differential conductance of the scatterer takes a remarkably simple form

$$R_{\rm N}\frac{dI}{dV} = 1 - \frac{\beta}{g+g_{\rm S}}\ln\left(1+\frac{1}{e^2V^2\tau_{\rm RC}^2}\right) \tag{11.20}$$

i.e., in this case the magnitude of the interaction correction increases logarithmically with decreasing voltage V. Equation (11.20) also indicates the validity range of our analysis: For not-too-low voltages, the interaction correction remains small in the so-called metallic limit provided at least one of the dimensionless conductances—g or g_S—is much larger than unity, i.e.,

$$g + g_{\rm S} \gg 1 \tag{11.21}$$

At exponentially low voltages, the logarithm in Eq. (11.20) may become large, so that the second term in the right-hand side of Eq. (11.20) may formally reach values of order one or bigger. Obviously, for such low voltages, Coulomb effects become strong and cannot be anymore considered perturbatively. Accordingly, Eqs. (11.18) and (11.20) cease to be valid at such low values of V. In order to estimate the voltage range at which the perturbative result (11.20) remains applicable, it is necessary to require the last term in this equation to be much smaller than unity and, in addition, that higher-order terms remain much smaller than the leading interaction correction. Employing the action (11.3) together with Eq. (10.112), one can evaluate the subleading interaction correction to the current and get [142]

$$I = \frac{V}{R}\left[1 - \frac{2\beta L}{g+g_S} + \frac{2L^2}{(g+g_S)^2}\left(\beta - \frac{\beta^2 R_S}{R_N + R_S} - 2\gamma\right) + \mathcal{O}\left(\frac{L^3}{(g+g_S)^3}\right)\right] \tag{11.22}$$

where we defined

$$L = \frac{1}{2}\ln\left(1 + \frac{1}{e^2V^2\tau_{RC}^2}\right) \tag{11.23}$$

This result demonstrates that in the limit $eV \gg T$, the actual parameter of our perturbative expansion is $L/(g + g_S)$, i.e., the condition (11.21) indeed becomes insufficient at low enough voltages and one should require

$$\frac{2L}{g+g_S} \ll 1 \tag{11.24}$$

for our perturbative approach to remain applicable. In order to find the higher-order terms of the perturbation theory in $L/(g + g_S)$, one needs to retain the contributions to the effective action of order $\sim(\varphi^-)^4$ and higher and then to sum up the corresponding logarithmic series for the current. This summation will be carried out by means of the renormalization group approach in Section 11.3.

On the other hand, in the limit of very high voltages $eV/E_C \gg \max(1, g + g_S)$, Coulomb blockade remains weak even if one abandons the condition (11.21). In this limit from Eq. (11.18), one finds

$$R_N I = V - \beta e/2C \tag{11.25}$$

i.e., at large V, the I–V curve of any short coherent conductor is offset by the value $\beta e/2C$ due to Coulomb effects. For tunnel junctions with $\beta \to 1$, this offset value coincides with one already discussed in Chapter 7.

Let us now analyze the dependence of the Coulomb blockade effects on temperature. In the limit $T \gg eV$, the I–V curve remains linear, i.e., the system conductance $G = dI/dV$ does not depend on the applied voltage. In this case, under the condition (11.21) and provided $T \ll (g + g_S)E_C$ from Eq. (11.18), one obtains

$$GR_N = 1 - \frac{2\beta}{g + g_S}\left[\gamma + 1 + \ln\left(\frac{1}{2\pi T \tau_{RC}}\right)\right] \tag{11.26}$$

Similar to the result (11.20), this equation remains applicable at

$$\frac{2\ln(1/T\tau_{RC})}{g + g_S} \ll 1 \tag{11.27}$$

and fails at exponentially low temperatures when this condition is violated. Finally, in the high temperature limit $T/E_C \gg \max(1, g + g_S)$, one can relax the condition (11.21) and get

$$GR_N = 1 - \beta\left[\frac{E_C}{3T} - \left(\frac{3\zeta(3)(g + g_S)}{2\pi^4} + \frac{1}{15}\right)\left(\frac{E_C}{T}\right)^2\right] \tag{11.28}$$

Thus, we see that weak Coulomb blockade in short coherent conductors shows a number of nontrivial features, which can be observed in different types of systems. These features were indeed observed in various experiments. For instance, the logarithmic behavior of the interaction correction as a function of voltage and/or temperature, Eqs. (11.20) and (11.26), was detected in diffusive metallic bridges [397], multiwalled carbon nanotubes [255, 285, 346, 373], break junctions [7], and other systems. In the high temperature limit, the power-law behavior (11.28) was observed, e.g., in granular chains [114] and metallic wires [63], while the offset feature was found in experiments with carbon nanotubes [372] and granular metallic strips [240]. The dependence of the interaction correction on the Fano factor β was also verified experimentally [7].

11.1.3 Interaction Correction to Shot Noise and Relation to Third Cumulant

Let us now investigate the effect of electron–electron interactions on shot noise in short coherent conductors. The calculation is analogous to that carried out above for the average current, though technically somewhat more complicated. All relevant details can be found in the paper by Galaktionov et al. [127]; here we only present the final result of this calculation. Substituting the action (11.3) into the path integral (10.114) and repeating the same steps as in Section 11.1.2, we arrive at the following expression [127]:

$$\tilde{S}(t_1, t_2) = \frac{\alpha(t_1 - t_2)}{R_N}\{1 - \beta + \beta e^{-\tilde{F}(t_1 - t_2)} \cos[eV(t_1 - t_2)]\}$$
$$- \frac{\pi \gamma T^3}{R_N} \int dt_3 \frac{K(t_2 - t_3)\left[\cos eV(t_1 - t_2) - \cos eV(t_3 - t_2)\right]}{\sinh[\pi T(t_3 - t_2)] \sinh[\pi T(t_2 - t_1)] \sinh[\pi T(t_1 - t_3)]}$$
$$- \frac{\pi \gamma T^3}{R_N} \int dt_3 \frac{K(t_1 - t_3)\left[\cos eV(t_1 - t_3) - \cos eV(t_1 - t_2)\right]}{\sinh[\pi T(t_3 - t_2)] \sinh[\pi T(t_2 - t_1)] \sinh[\pi T(t_1 - t_3)]}$$
$$+ \frac{\beta \dot{K}(0)}{2R_N} \delta(t_1 - t_2) \tag{11.29}$$

Here $\tilde{F}(t_1 - t_2)$ is the correlation function of the classical phases

$$\tilde{F}(t_1 - t_2) = \frac{1}{2}\left\langle (\varphi^+(t_1) - \varphi^+(t_2))^2 \right\rangle \tag{11.30}$$

which reads

$$\tilde{F}(t) = e^2 \int \frac{d\omega}{2\pi} \left| \frac{R_N Z_S(\omega)}{R_N + Z_S(\omega)} \right|^2 \left[\left(\mathrm{Re}\, \frac{1}{Z_S(\omega)} + \frac{1-\beta}{R_N} \right) \omega \coth \frac{\omega}{2T} \right.$$
$$\left. + \frac{\beta}{2R_N} \left((\omega + eV) \coth \frac{\omega + eV}{2T} + (\omega - eV) \coth \frac{\omega - eV}{2T} \right) \right]$$
$$\times \frac{1 - \cos \omega t}{\omega^2} \tag{11.31}$$

As before, Eq. (11.29) was derived perturbatively in the interaction and remains applicable provided the effective dimensionless conductance $g_{\tilde{R}} = R_N Z_S(0)/[R_N + Z_S(0)]$ is much larger than one. Since the function $\tilde{F}(t_1 - t_2) \propto \tilde{R}$, one actually needs to expand the exponent in Eq. (11.29) making a replacement

$$e^{-\tilde{F}(t_1 - t_2)} \to 1 - \tilde{F}(t_1 - t_2) \tag{11.32}$$

The first line of Eq. (11.29) at $\tilde{F} = 0$, being Fourier transformed, yields the noise spectral density of a voltage-biased conductor in the non-interacting limit [222]

$$\mathcal{S}_0(\omega) = \frac{1}{R_N} \left\{ \frac{\beta}{2} \left((\omega + eV) \coth \frac{\omega + eV}{2T} + (\omega - eV) \coth \frac{\omega - eV}{2T} \right) \right.$$
$$\left. + (1 - \beta)\omega \coth \frac{\omega}{2T} \right\} \tag{11.33}$$

Of course, the same result follows from Eq. (11.10). The remaining terms in Eq. (11.29) define the lowest-order interaction correction to the shot noise. Let us briefly analyze the main properties of this interaction correction.

It is straightforward to verify that as long as $Z_S(\omega) \ll R_N$, one can replace the function (11.31) by its value at $V = 0$. Then one finds

$$\tilde{F}(t)|_{V=0} = e^2 \int \frac{d\omega}{2\pi} \mathrm{Re}\left[\frac{R_N Z_S(\omega)}{R_N + Z_S(\omega)}\right] \omega \coth\frac{\omega}{2T} \frac{1-\cos\omega t}{\omega^2} \tag{11.34}$$

Substituting this result into Eq. (11.29) and performing the Fourier transformation, we get

$$\begin{aligned} \mathcal{S}(\omega) &= \mathcal{S}_0(\omega) + e\frac{\gamma}{\beta}\left[\delta I\left(\frac{\omega}{e}+V\right) + \delta I\left(\frac{\omega}{e}-V\right)\right]\coth\frac{\omega}{2T} \\ &+ \frac{e}{2}\frac{\beta-2\gamma}{\beta}\left[\delta I\left(\frac{\omega}{e}+V\right)\coth\frac{\omega+eV}{2T}\right. \\ &\left.+\delta I\left(\frac{\omega}{e}-V\right)\coth\frac{\omega-eV}{2T}\right] \end{aligned} \tag{11.35}$$

where by $\delta I = I - V/R_N$, we denoted the interaction correction to the current defined by Eq. (11.4). In the shot noise limit $eV \gg T$ and at low frequencies $\omega \to 0$, this expression reduces to

$$\mathcal{S}(0) = \beta\frac{eV}{R_N} + \frac{\beta - 2\gamma}{\beta} e\,\delta I(V) \tag{11.36}$$

Since the Coulomb correction to the current is negative ($\delta\ I < 0$), from Eq. (11.36), we conclude that for $\beta > 2\gamma$, the shot noise is suppressed by Coulomb interaction, whereas for $\beta < 2\gamma$, electron–electron interactions *enhance* the shot noise of a short coherent conductor. For example, in the simple case of a one-channel conductor with a transmission T_1, one has $\beta - 2\gamma = (1-2T_1)(1-T_1)$, implying that the shot noise is suppressed by interactions for $T_1 < 1/2$ and enhanced for $T_1 > 1/2$. Thus, for weakly transmitting conductors, we may speak about Coulomb blockade of the shot noise. This effect was indeed observed in experiments with tunnel junctions [8]. In the case of highly transmitting conductors, the shot noise, on the contrary, gets enhanced by Coulomb interaction. In an important case of diffusive metallic conductors, we have $\beta = 1/3$,

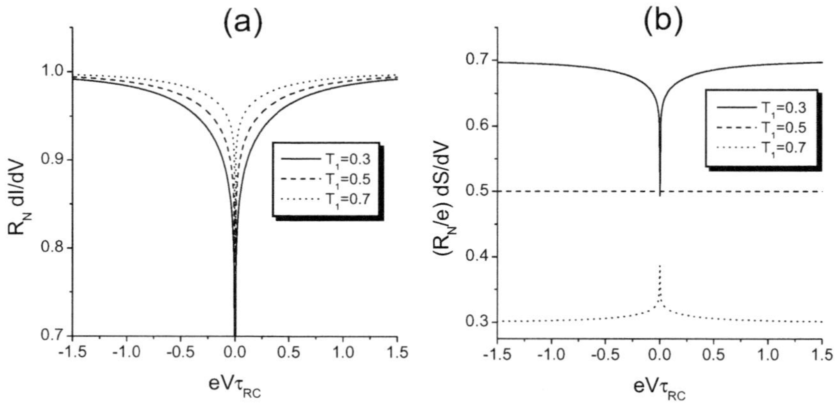

Figure 11.1 Differential conductance dI/dV (a) and the differential Fano factor $(R_{\mathrm{N}}/e)d\mathcal{S}/dV$ (b) of a single-channel conductor with transmission T_1 equal to $T_1 = 0.3$, 0.5, and 0.7. The dimensionless conductance of the environment is $g_{\mathrm{S}} = 30$. dI/dV is defined by Eq. (11.20), while $d\mathcal{S}/dV$ is derived from Eq. (11.37).

$\gamma = 2/15$, and, hence, $\beta - 2\gamma = 1/15 > 0$, implying that the interaction correction to the shot noise (11.36) is negative.

Let us further specify our result (11.36) for the environment with the impedance (10.121). Substituting the expression (11.18) into Eq. (11.35) or (11.36) and taking the limit $eV \gg T$, we obtain

$$\mathcal{S}(0) = \beta\frac{eV}{R_{\mathrm{N}}} - \frac{e^2(\beta-2\gamma)}{\pi}\frac{R_{\mathrm{S}}}{R_{\mathrm{N}}+R_{\mathrm{S}}} \times \left[\frac{eV}{2}\ln\left(1+\frac{1}{e^2V^2\tau_{\mathrm{RC}}^2}\right)+\frac{\arctan(eV\tau_{\mathrm{RC}})}{\tau_{\mathrm{RC}}}\right] \quad (11.37)$$

Similar to Eq. (11.18), this result applies under the condition (11.24), i.e., as long as the interaction correction remains small.

Our results are illustrated in Fig. 11.1, which shows the differential conductance (11.20) and the differential Fano factor $(R_{\mathrm{N}}/e)d\mathcal{S}/dV$ derived from Eq. (11.37) for a single-channel conductor with transmission T_1. The differential conductance is always suppressed by Coulomb interaction, although the magnitude of this effect decreases with increasing transmission T_1. The differential Fano factor is suppressed at $T_1 < 1/2$ and enhanced at $T_1 > 1/2$. We can also add that, in contrast to the shot noise, Gaussian Nyquist noise in the limit $T \gg eV$ is always suppressed by electron–electron interactions, as it is seen in Eq. (11.35). Of course, the same result also directly follows from FDT.

Finally, let us briefly discuss the relation between the interaction correction to the shot noise and the third current cumulant. As we have demonstrated, this interaction correction is proportional to the combination $\beta - 2\gamma$. Remarkably, the third current cumulant $\mathcal{C}_3$ (10.95) is proportional to the same combination. Hence, it is reasonable to conjecture that there exists a close relation between these two quantities.

In order to get more information about the third current cumulant $\mathcal{C}_3$, let us establish its full frequency dependence. Neglecting for a moment the interaction effects by setting $Z_S(\omega) = 0$ and substituting the action (11.3) into Eq. (10.116), we obtain [128]

$$\mathcal{C}_3(\omega_1, \omega_2) = \int dt_1 dt_2\, e^{i\omega_1 t_1 + i\omega_2 t_2} (ie)^3 \frac{\delta^3\, iS_{sc}}{\delta\varphi^-(0)\delta\varphi^-(-t_1)\delta\varphi^-(-t_2)}\Bigg|_{\varphi^-=0}$$
$$= \frac{e^2 V}{R_N}\left(\beta - 2\gamma F_3(V, \omega_1, \omega_2)\right), \tag{11.38}$$

where we defined

$$F_3(V, \omega_1, \omega_2) = \frac{T \sinh^3 \frac{eV}{2T}}{4eV} \int dx \frac{dx}{\chi(x)\chi\left(x - \frac{\omega_1}{2T}\right)\chi\left(x + \frac{\omega_2}{2T}\right)} \tag{11.39}$$

and $\chi(x) = \cosh^2 x + \sinh^2[eV/4T]$. The integral in Eq. (11.39) can be evaluated with the result [134]

$$F_3 = Z(v, -w_1, w_1 + w_2) + Z(v, -w_2, w_1 + w_2) + Z(v, w_1, w_2) \tag{11.40}$$

where $v = V/T$, $w_{1,2} = \omega_{1,2}/2T$, and the function Z is defined as

$$Z(v, x, y) = \frac{\sinh(v/2)}{v \sinh x \sinh y} \times \left(\frac{(x+y)/2 - v/4}{\sinh[x + y - v/2]} + \frac{(x+y)/2 + v/4}{\sinh[x + y + v/2]}\right) \tag{11.41}$$

In the zero-frequency limit $w_{1,2} \to 0$, the function F_3 (11.40) reduces to a much simpler form [270]

$$F_3(V) = 1 + 3\frac{1 - (\sinh v/v)}{\cosh v - 1} \tag{11.42}$$

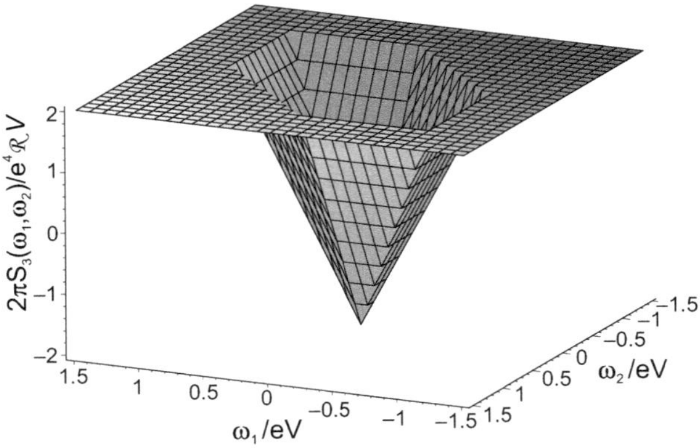

Figure 11.2 Third cumulant of the current operator $\mathcal{C}_3$ at zero temperature as a function of frequencies ω_1 and ω_2. Note that $\mathcal{C}_3(\omega_1, \omega_2)$ changes its sign with increasing frequencies and flattens off at $|\omega_1|, |\omega_2| > eV$.

which in turn reduces to $F_3(V) \to 1$ in the shot noise regime $eV \gg T$, in which case we get back to the result (10.95). For illustration, the function $\mathcal{C}_3(\omega_1, \omega_2)$ determined by Eqs. (11.38)–(11.41) is shown in Fig. 11.2 in the low temperature limit. We also note that Eq. (11.38) defines the symmetrized version of the third cumulant with respect to time ordering of the current operators [128]. Non-symmetrized versions of the third cumulant can also be investigated [341], which may lead to different frequency dependencies.

The above results for the frequency-dependent third current cumulant $\mathcal{C}_3(\omega_1, \omega_2)$ can be verified experimentally [126, 382] and provide a great deal of information about non-Gaussian character of current fluctuations in coherent conductors. In addition, it is indeed possible to confirm the conjecture about a close relation between the third current cumulant and the interaction correction to the shot noise. This relation can be proven, e.g., with the aid of the quasiclassical Langevin equation approach as it was already done while demonstrating the relation between the shot noise and the interaction correction to the current. In Section 11.3, we will show how to relate the third current cumulant and the interaction correction to the shot noise to each other by means of the renormalization group analysis. As we will see, this analysis also allows to proceed further and demonstrate that the interaction correction to the nth current cumulant is proportional to the $1+n$th current cumulant for all values $n \geq 1$. This observation plays a very important role in our understanding of fluctuations and interaction effects in disordered conductors. We will return to this point in Section 11.3.

11.2 Coulomb Effects in Almost Transparent Conductors

Let us consider a specific limit of weakly reflecting scatterers, i.e., we will assume that N conducting channels of a scatterer are all characterized by transmission probabilities close to unity $R_n = 1 - T_n \ll 1$. Possibly the remaining conducting channels are assumed to be closed and, hence, do not participate in the electron transport through the scatterer. It is clear that the Fano factor of such scatterer is small $\beta \ll 1$. According to our previous results, the quantities that scale with β, namely non-interacting shot noise and the interaction correction to the current, will also be small in the measure of β. This observation allows to develop yet one more perturbative approach that relies on the smallness of the Fano factor β but does not require the dimensionless conductances g and g_S to be large.

11.2.1 Effective Action

The key idea is to formally expand the effective action (10.67) in powers of the total reflection probability

$$\mathcal{R} = \sum_{n=1}^{N} (1 - T_n) \tag{11.43}$$

and keep only two lowest-order terms of this expansion, i.e.,

$$S_{\text{sc}} = S_{\text{sc}}^{(0)} + S_{\text{sc}}^{(1)} \tag{11.44}$$

where $S_{\text{sc}}^{(0)} \propto \mathcal{R}^0$ and $S_{\text{sc}}^{(1)} \propto \mathcal{R}$.

Let us first establish the action $S_{\text{sc}}^{(0)}$ for reflectionless barriers. To begin with, we note that in the reflectionless limit, i.e., at $\hat{r}' = \hat{r} = 0$, our problem can be mapped onto the so-called Luttinger model describing one-dimensional gas of interacting electrons; see, e.g., [125]. As the RPA becomes exact for the Luttinger model [104], one can expect that in the limit $\hat{r}' \to 0$, the action (10.67) should be quadratic in $\varphi^{\pm}$, at least for not very large values of the phase. Rigorous analysis [142] demonstrates that this is indeed the case if $|\varphi^-| < \pi$.

Technically, the derivation of the action $S_{\mathrm{sc}}^{(0)}$ is rather complicated and relies on the so-called loop cancelation theorem. The latter theorem states that in the reflectionless limit, the expansion of the action (10.67) in powers of the fluctuating phases $\varphi_R^{\pm}, \varphi_L^{\pm}$ terminates at the second order. The cubic terms $\propto (\varphi^{\pm})^3$ and all higher-order terms identically vanish due to the cancelation of the loop diagrams.

Let us put $\hat{r}' = 0$ in Eq. (10.67). Then the action can be split into two parts:

$$S_{\mathrm{sc}}^{(0)} = NS_0\left[\theta(t-x)\theta(x)(\mathrm{e}^{\mathrm{i}\varphi^-(x)} - 1),\ -\varphi^+\right] + NS_0\left[\theta(t-x)\theta(x)(\mathrm{e}^{-\mathrm{i}\varphi^-(x)} - 1),\ \varphi^+\right], \quad (11.45)$$

where

$$\mathrm{i}S_0[a, \varphi^+] = \mathrm{tr}\ln\left[1 + a(x)\rho_0(y-x)\mathrm{e}^{\mathrm{i}\frac{\varphi^+(y)-\varphi^+(x)}{2}}\right] \quad (11.46)$$

The action (11.46) can be evaluated exactly. For corresponding details, we refer the reader to the work of Golubev et al. [142]; here we only present the final result:

$$\mathrm{i}S_0[a, \varphi^+] = \int dx\left(\rho_0(0) + \frac{\dot{\varphi}^+(x)}{4\pi}\right)\ln[1 + a(x)] + \int dx dy\, \alpha(x-y)\frac{\ln[1 + a(x)]\ln[1 + a(y)]}{4\pi} \quad (11.47)$$

where $\rho_0(0)$ has a meaning of the electron density being equal to a large constant. At a later stage, this term is canceled by the corresponding contribution of ions. Combining this formula with Eq. (11.45), we obtain

$$\mathrm{i}S_{\mathrm{sc}}^{(0)} = -\frac{\mathrm{i}N}{2\pi}\int_0^t dx\, W(\varphi^-(x))\dot{\varphi}^+(x) - \frac{N}{4\pi}\int_0^t dx dy\, W(\varphi^-(x))\alpha(x-y)W(\varphi^-(y)), \quad (11.48)$$

where $W(\varphi)$ is the $2\pi-$periodic function of φ^-, which equals to φ^- in the interval $-\pi < \varphi^- < \pi$. Under certain conditions, the contribution of large phase values ($|\varphi^-| > \pi$) can be disregarded; in which case it suffices to employ the Gaussian action

$$\begin{aligned} iS_{sc}^{(0)} &= -\frac{i}{e^2 R_0}\int_0^t dt_1 \varphi^-(t_1)\dot{\varphi}^+(t_1) \\ &\quad - \frac{1}{2e^2 R_0}\int_0^t dt_1 \int_0^t dt_2\, \varphi^-(t_1)\alpha(t_1 - t_2)\varphi^-(t_2) \end{aligned} \quad (11.49)$$

instead of the exact one (11.48). Here $R_0 = 2\pi/e^2 N$ is the Landauer resistance of the junctions in the absence of any reflection. Within this approximation, the action $S_{sc}^{(0)}$ is quadratic in the fluctuating phases and coincides with the action for a linear Ohmic resistor R_0. It is important to emphasize, however, that this approximation ignores the action periodicity in the phase space and, hence, becomes inadequate as soon as electron charge discreteness turns out to be an important effect.

The derivation of the first-order correction to the action $S_{sc}^{(1)}$ is even more involved since it requires the inversion of the operator appearing under the logarithm in Eq. (10.67) at $\hat{r} = \hat{r}' = 0$. For further details, we again refer the reader to the work of Golubev et al. [142] and present only the final expression here. It reads

$$\begin{aligned} iS_{sc}^{(1)} &= \frac{i\mathcal{R}}{2\pi}\int_0^t dt_1 \dot{\varphi}^+(t_1)\sin\varphi^-(t_1) \\ &\quad + \frac{\mathcal{R}}{2\pi}\int_0^t dt_1 dt_2\, \varphi^-(t_1)\alpha(t_1 - t_2)\sin\varphi^-(t_2) \\ &\quad - \frac{\mathcal{R}}{\pi}\int_0^t dt_1 dt_2\, \sin\frac{\varphi^-(t_1)}{2}\alpha(t_1 - t_2)\sin\frac{\varphi^-(t_2)}{2} e^{i[\varphi^+(t_1)-\varphi^+(t_2)]} \\ &\quad \times e^{\int_0^t dt' \left[T\coth\left(\pi T(t_1 - t')\right) - T\coth\left(\pi T(t_2 - t')\right)\right]\varphi^-(t')} \end{aligned} \quad (11.50)$$

As we have already pointed out, this expression is justified provided all the channels are weakly reflecting, $R_n \ll 1$. At the same time, the parameter $\mathcal{R} = \sum_n R_n$ needs not be small. We also note that different limiting forms of the actions (10.75) and (11.48), (11.50) can be combined if some channels are weakly transmitting $T_n \ll 1$, while the others have small reflection coefficients $R_m \ll 1$.

11.2.2 Electron Transport and Shot Noise

Let us now employ the aforementioned results for the effective action in order to evaluate both the current and shot noise in weakly reflecting scatterers $\mathcal{R} \ll 1$ in the presence of electron–electron interactions. Substituting the action defined in Eqs. (11.44), (11.48), and (11.50) into the general expression for the current (10.112) and evaluating the corresponding Gaussian path integrals, in the limit $\mathcal{R} \ll 1$, we obtain

$$I = \frac{V}{R_{\rm N}} - \frac{e\mathcal{R}}{\pi}\int_0^t dt \frac{\pi T^2 e^{F(t)}}{\sinh^2[\pi T t]} \sin\left[\frac{K(t)}{2}\right] \sin[eVt] \qquad (11.51)$$

where, as before, we defined the voltage drop across the scatterer

$$V = \frac{R_{\rm N}}{R_{\rm N} + Z_{\rm S}(0)} V_x \qquad (11.52)$$

the scatterer resistance $R_{\rm N} = 2\pi/e^2(N - \mathcal{R})$ and the functions

$$K(t) = e^2 \int \frac{d\omega}{2\pi} \frac{R_{\rm N} Z_{\rm S}(\omega)}{R_{\rm N} + Z_{\rm S}(\omega)} \frac{e^{-{\rm i}\omega t}}{-{\rm i}\omega + 0} \qquad (11.53)$$

$$F(t) = e^2 \int \frac{d\omega}{2\pi} {\rm Re}\left[\frac{R_{\rm N} Z_{\rm S}(\omega)}{R_{\rm N} + Z_{\rm S}(\omega)}\right] \omega \coth\frac{\omega}{2T} \frac{1 - \cos\omega t}{\omega^2} \qquad (11.54)$$

As one could expect, the interaction correction to the current (11.51) scales linearly with the parameter $\mathcal{R}$. Overall, the form of this interaction correction (11.51) is similar to that derived in the previous subsection, cf. Eq. (11.4); except instead of the decaying exponent $e^{-F(t)}$ under the time integral, we now have the exponent $e^{F(t)}$, which grows with time. We will get back to this point below.

The current noise can be evaluated analogously. Again employing the action given by Eqs. (11.44), (11.48), and (11.50) and performing the path integrals in the expression for the correlator (10.114), after the Fourier transformation, we recover the spectral density of the current noise [142]

$$\mathcal{S}(\omega) = \frac{e^2\mathcal{R}}{4\pi}\dot{K}(0) + \frac{e^2}{2\pi}(N - 2\mathcal{R})\omega\coth\frac{\omega}{2T}$$
$$+ \frac{e^2\mathcal{R}}{2\pi}\int dt \alpha(t) e^{\Phi(t)} \cos eVt \left[\cos\omega t - 2{\rm i}\coth\frac{\omega}{2T}\sin\omega t\right] \qquad (11.55)$$

Here we defined the function

$$\Phi(t) = F(t) + {\rm i}\frac{K(t)}{2} + {\rm i}\frac{K(-t)}{2} \qquad (11.56)$$

Now we are ready to establish the relation between the noise (11.55) and the current (11.51). As before, we denote the interaction correction to the current as

$$\delta I(V) = I(V) - \frac{V}{R_N} \tag{11.57}$$

and employ the property $\Phi(t - i/T) = \Phi(t)$, which allows to prove the following identity:

$$\int dt\alpha(t)e^{\Phi(t)}\cos\omega t = -\frac{\dot{K}(0)}{2} + \omega\coth\frac{\omega}{2T} + \mathrm{i}\coth\frac{\omega}{2T}\int dt\alpha(t)e^{\Phi(t)}\sin\omega t \tag{11.58}$$

With the aid of this identity, we rewrite the spectral density (11.55) in the form

$$\begin{aligned}\mathcal{S}(\omega) = &\frac{e^2(N-2\mathcal{R})}{2\pi}\omega\coth\frac{\omega}{2T} \\ &+ \frac{e^2\mathcal{R}}{4\pi}\left[(\omega+eV)\coth\frac{\omega+eV}{2T} + (\omega-eV)\coth\frac{\omega-eV}{2T}\right] \\ &- \frac{e}{2}\left[\delta I\left(\frac{\omega}{e}+V\right)\coth\frac{\omega+eV}{2T} + \delta I\left(\frac{\omega}{e}-V\right)\coth\frac{\omega-eV}{2T}\right] \\ &+ e\left[\delta I\left(\frac{\omega}{e}+V\right) + \delta I\left(\frac{\omega}{e}-V\right)\right]\coth\frac{\omega}{2T}\end{aligned} \tag{11.59}$$

The first two lines of this expression represent the noise spectral power in the absence of electron–electron interactions, while the last two lines define an additional contribution due to such interactions. In the zero-frequency limit $\omega \to 0$, the above expression takes a simple form

$$\begin{aligned}\mathcal{S}(0) = &\left[\frac{e^2(N-2\mathcal{R})}{2\pi} + 2\frac{\partial}{\partial V}\delta I(V)\right]2T \\ &+ \left[\frac{e^2\mathcal{R}}{2\pi}eV - e\delta I(V)\right]\coth\frac{eV}{2T}\end{aligned} \tag{11.60}$$

which further reduces to

$$\mathcal{S}(0) = \frac{e^2\mathcal{R}}{2\pi}eV - e\delta I(V) \tag{11.61}$$

in the shot noise regime $eV \gg T$. Since the Coulomb correction to the current δI is negative, the interaction term in the expression for the shot noise is positive, i.e., the shot noise in almost transparent conductors is enhanced by electron–electron interaction, as we already demonstrated above in Section 11.1.3. Note that both non-interacting and Coulomb terms in Eq. (11.61) are proportional to $\mathcal{R}$ and, hence, can be of the same order. Moreover, below we will demonstrate that under certain conditions, the interaction contribution can even dominate the shot noise in weakly reflecting scatterers.

Let us again consider the impedance (10.121) and specify the above general expressions for this case. Evaluating the differential conductance of the scatterer at voltages at

$$T \ll eV \ll (R_S + R_N)/R_S R_N C,$$

we obtain

$$\frac{dI}{dV} = \frac{1}{R_N} - \frac{e^2\mathcal{R}}{2\pi\Gamma(1-2/g_0)}\left(\frac{e^\gamma(R_N+R_S)}{eV\,R_N R_S C}\right)^{2/g_0} \tag{11.62}$$

where we defined the dimensionless conductance

$$g_0 = R_q/R_N + R_q/R_S.$$

Analogously, in the linear response regime

$$T \ll eV \ll (R_S + R_N)/R_S R_N C,$$

we find

$$G = \frac{1}{R_N} - \frac{e^2\mathcal{R}}{4\sqrt{\pi}}\frac{\Gamma(1-2/g_0)}{\Gamma(3/2-1/g_0)}\left(\frac{e^\gamma(R_N+R_S)}{\pi T\,R_N R_S C}\right)^{2/g_0} \tag{11.63}$$

We note that both Eqs. (11.62) and (11.63), being perturbative in $\mathcal{R}$, become inapplicable at voltages/temperatures below the energy scale

$$E^* \sim \frac{R_N + R_S}{R_N R_S C}\left(\frac{\mathcal{R}}{N}\right)^{g_0/2} \tag{11.64}$$

At the same time, both these results *do not* require the dimensionless conductance g_0 to be large and, hence, provide complementary information to the perturbative in $1/g$ expressions derived in Section 11.1.2.

Let us now turn to the current noise. The frequency dependence of the noise spectrum defined in Eq. (11.59) is illustrated in Fig. 11.3. One can see that the noise is enhanced due to interaction at low frequencies $|\omega| \lesssim eV$ and reduced at higher ones $|\omega| \gtrsim eV$.

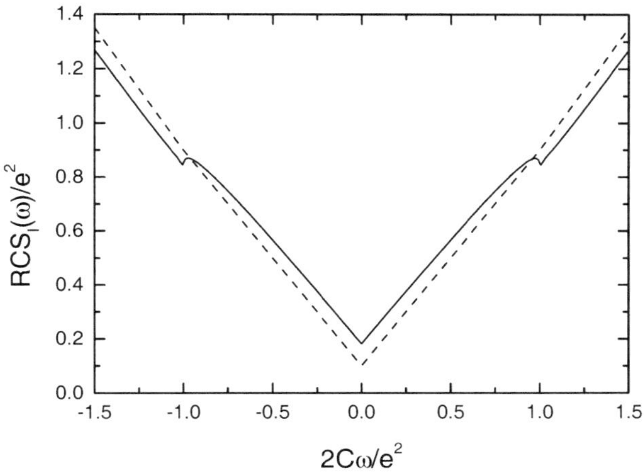

Figure 11.3 Frequency dependence of the noise power spectrum (solid line) at $T = 0$ and in the presence of interactions. For comparison, the noise spectrum in the absence of interactions is shown by the dashed line. Both curves are plotted for $N = 1$, $\mathcal{R} = 0.1$, $g_S = 2$, and $V = e/2C$.

It is interesting to analyze the zero-frequency noise $\mathcal{S}(0)$ in more details. At high voltages, the noise spectrum (11.61) takes the form

$$\mathcal{S}_I(0) = \frac{e^3\mathcal{R}}{2\pi}\left(V + \frac{e}{2C}\right) \tag{11.65}$$

i.e., in this limit, electron–electron interactions induce voltage-independent excess noise with the magnitude that scales linearly with the Coulomb gap $e/2C$. At lower voltages $eVRR_SC/(R + R_S) \ll 1$, one finds

$$\mathcal{S}(0) = \frac{e^2}{2\pi}\mathcal{R}\frac{R_N + R_S}{R_N R_S C}\frac{e^{2\gamma/g_0}}{\Gamma(2 - 2/g_0)}\left(\frac{eV R_N R_S C}{R_N + R_S}\right)^{1-2/g_0} \tag{11.66}$$

The expression (11.66) is valid provided $e|V| > E^*$, where E^* is defined in Eq. (11.64). In the limit $g_0 \to \infty$, Eq. (11.66) reduces to the non-interacting result. For $g_0 \gg 1$, the interaction effects remain weak and Eq. (11.66) agrees with the perturbative results derived in the previous section. For smaller conductance values, the excess noise becomes larger, and for g_0 of order one, the shot noise is *strongly enhanced* by electron–electron interactions. The effect of interactions on the shot noise spectrum $\mathcal{S}(0)$ of highly transmitting coherent scatterers is clearly observed in Fig. 11.4.

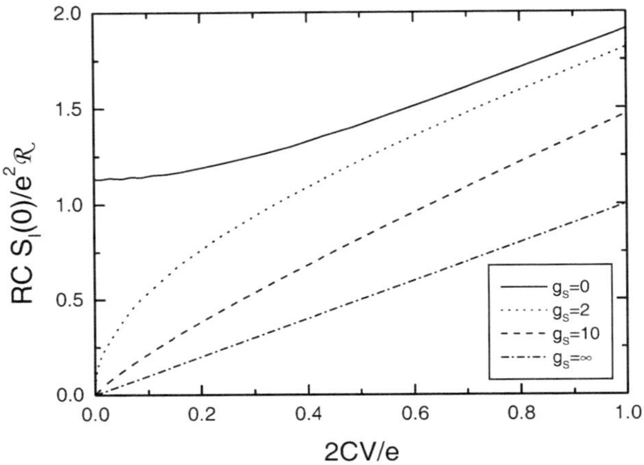

Figure 11.4 The zero-frequency shot noise power spectrum $\mathcal{S}(0)$ normalized by the value $e^2\mathcal{R}/RC$ as a function of voltage V at $T = 0$ for $N = 1$ and small $\mathcal{R} \ll 1$. The conductance g_S effectively controls the interaction strength, i.e., at small g_S, the interaction is strong, while it tends to zero in the limit $g_S \to \infty$. The interaction-induced excess noise is clearly observed even at rather large values of g_S. Our results remain formally applicable at nonzero voltages $eV \gtrsim E^*$, while $\mathcal{S}(0)$ should always vanish in the zero voltage limit $V = 0$. However, the corresponding crossover region may become extremely narrow for $R_q/R_N \to 2$ and very small $\mathcal{R}$, cf. the top curve.

11.3 Coulomb Effects beyond Perturbation Theory

11.3.1 Renormalization Group Analysis

Let us rewrite Eqs. (11.20) and (11.37) in the form

$$\frac{dI}{dV} = \frac{e^2}{2\pi}\sum_n \left(T_n - \frac{2T_n(1-T_n)}{g_S + \sum_n T_n} L \right) \tag{11.67}$$

$$\mathcal{S}(0) = \frac{e^3 V}{2\pi}\sum_n \left(T_n(1-T_n) - \frac{2T_n(1-T_n)(1-2T_n)}{g_S + \sum_n T_n}(L+1) \right) \tag{11.68}$$

where we have assumed $eV\tau_{RC} \ll 1$ and employed the expression for the parameter L defined in Eq. (11.23). In the limit of very low voltages, this parameter becomes large $L \gg 1$, and, hence, one can replace $L + 1$ by L in Eq. (11.68). Then one can express the results (11.67) and (11.68) as

$$\frac{dI}{dV} = \frac{e^2}{2\pi}\sum_n \tilde{T}_n, \quad \mathcal{S}(0) = \frac{e^3 V}{2\pi}\sum_n \tilde{T}_n(1 - \tilde{T}_n) \qquad (11.69)$$

i.e., in the presence of electron–electron interactions, both the scatterer conductance and its shot noise take the same form as in the non-interacting theory if instead of bare channel transmissions T_n, one substitutes their renormalized values

$$\tilde{T}_n = T_n - \frac{2T_n(1 - T_n)}{g_S + \sum_n T_n} L \qquad (11.70)$$

Equations (11.69) and (11.70) are highly suggestive. Namely, on their basis, one can conjecture that the effect of electron–electron interactions in short coherent conductors could be fully accounted for by proper renormalization of their channel transmissions. In order to proceed in this direction, we notice that Eqs. (11.69) and (11.70) hold as long as the transmission renormalization remains weak, i.e., provided $|\tilde{T}_n - T_n| \ll T_n$, or, equivalently, $L/(g_S + \sum_n T_n) \ll 1$. If, however, the effect of renormalization becomes large, $|\tilde{T}_n - T_n| \sim T_n$, it is necessary to also include higher-order terms in the parameter $L/(g_S + \sum_n T_n)$ (cf., e.g., Eq. (11.22)) and to sum up the corresponding logarithmic series for the current and all higher cumulants. This summation can conveniently be employed by means of the renormalization group technique in the same spirit as it was repeatedly done, e.g., in Chapters 5–8. This approach allows to recover the leading logarithmic contributions to all orders.

The main idea is as follows. Expressing the cumulant generating function $\mathcal{F}(\chi)$ as a path integral over fluctuating phases $\varphi^{\pm}$ (see Eq. (10.103)), splitting these phase variables into "fast" (with frequencies $\omega_c > \omega > \omega_c - \delta\omega$) and "slow" (with frequencies $\omega_c - \delta\,\omega > \omega$) ones, and integrating over the "fast" phase fluctuations, we again arrive at Eq. (10.103) where both actions $S_{\rm env}$ and $S_{\rm sc}$ will depend only on "slow" variables, and the latter action will be again defined by Eq. (10.67) or by Eq. (10.74) in which one should substitute $T_n \to \tilde{T}_n = T_n + \delta T_n$ and $\omega_c \to \tilde{\omega}_c - \delta\omega$. Then one performs the same steps again. Repeating this procedure many times, under the condition

$$g_S + \sum_n \tilde{T}_n(L) \gg 1 \tag{11.71}$$

one effectively carries out the summation of the leading logarithms in every order of the perturbation and arrives at the differential RG equations for the channel transmissions [37, 225]

$$\frac{d\tilde{T}_n}{dL} = -\frac{2\tilde{T}_n(1-\tilde{T}_n)}{g_S + \sum_n \tilde{T}_n} \tag{11.72}$$

where $L = \ln(1/\tilde{\omega}_c)$ or, equivalently, L is defined in Eq. (11.23). Loosely speaking, one can say that the RG equations can be reconstructed simply by applying the renormalization rule (11.70) iteratively M times and setting $M \to \infty$.

We note that in the limit $T_n \ll 1$, there exists a direct relation between Eq. (11.72) and the RG equation (7.153) derived in Chapter 7 for tunnel barriers with $g \gg 1$. In order to observe this relation, it suffices to replace the term $1 - T_n$ in Eq. (11.72) by unity, set $g_S \to 0$, identify $\tilde{g} = \sum_n \tilde{T}_n$, and perform a summation over all conducting channels in both sides of this equation. In this way, we reproduce Eq. (7.153) except for the two-loop term $4/\tilde{g}$ in its right-hand side, which apparently goes beyond the accuracy of Eq. (11.72). It is worth pointing out that RG equations (11.72) can also be derived from a more general equation describing renormalization of the full scattering matrix [163].

The RG equations (11.72) demonstrate that any transmission $T_n < 1$ *decreases* monotonously in the course of renormalization, i.e., as one proceeds to lower and lower energies. Eventually the transmissions become small, and one enters the regime of strong Coulomb blockade. On the other hand, fully open channels remain open also in the presence of electron–electron interactions, i.e., channel transmissions $T_n = 1$ are not renormalized at all.

Resolving Eqs. (11.72), we find the renormalized transmissions in the implicit form

$$\tilde{T}_n = \frac{T_n Y}{1 - T_n + T_n Y}, \quad 2L = -g_S \ln Y - \sum_n \ln(1 - T_n + T_n Y) \tag{11.73}$$

It follows from this RG flow that as the voltage decreases, the channels are "turned off" by interactions one by one depending on their transmission values. Most transparent channels $1 - T_n \ll 1$ remain open down to the lowest voltages. Resolving Eqs. (11.73), one can explicitly determine the renormalized (voltage dependent) transmissions $\tilde{T}_n$ and, e.g., substituting $\tilde{T}_n$ into the Landauer formula, derive the expression for the I–V curve.

Furthermore, the above RG approach effectively demonstrates that also all higher-order cumulants in the presence of electron–electron interactions can be derived by substituting the renormalized transmissions into the corresponding non-interacting formulae. This conclusion implies that in the leading approximation, the cumulant generating function (10.120) reduces to the form

$$\mathcal{F}(\chi) = \frac{eVt}{2\pi} \ln\left(1 + \tilde{T}_n\left(e^{-i\chi} - 1\right)\right), \quad eV \gg T \qquad (11.74)$$

which resembles the non-interacting formula (10.92).

In what follows, we will specifically consider two different limits. First, we assume that the shunt conductance is large $g_S \gg \max(1, g = \sum_n T_n)$. In this case, the condition (11.71) is well satisfied and one can make use of the solution (11.73) down to $V \to 0$. Then in the limit $eV\tau_{\mathrm{RC}} \ll 1$, one finds

$$\tilde{T}_n = \frac{T_n(eV\tau_{\mathrm{RC}})^{2/g_S}}{1 - T_n + T_n(eV\tau_{\mathrm{RC}})^{2/g_S}} \qquad (11.75)$$

This result implies that all transmissions $T_n < 1$ tend to zero in the course of renormalization, i.e., at $V \to 0$. Note that in the case of a single-channel quantum point contact, the scaling dependence (11.75) was also verified experimentally [208, 324].

Provided the scatterer has only a few conducting channels with $T_n < 1$, it follows from Eq. (11.75) that its behavior in the low bias limit and at low temperature resembles that of a tunnel junction embedded in the electromagnetic environment (see Section 7.3). If the conductor has many transport channels $N \gg 1$, it can be described by a continuous distribution of the transmission eigenvalues $W(\mathcal{T})$, where the transmission $\mathcal{T}$ is introduced instead T_n. Then, according to Eq. (11.75), at low V and T, the distribution for the renormalized transmission $\tilde{\mathcal{T}}$ reads

$$\tilde{W}(\tilde{\mathcal{T}}) = \frac{(eV\tau_{\mathrm{RC}})^{2/g_{\mathrm{S}}}}{\left(\tilde{\mathcal{T}} + (1-\tilde{\mathcal{T}})(eV\tau_{\mathrm{RC}})^{2/g_{\mathrm{S}}}\right)^2} W\left(\frac{\tilde{\mathcal{T}}}{\tilde{\mathcal{T}} + (1-\tilde{\mathcal{T}})(eV\tau_{\mathrm{RC}})^{2/g_{\mathrm{S}}}}\right) \tag{11.76}$$

Let us consider an important case of a sufficiently short diffusive metallic conductor of length X. In this case, the transmission distribution is defined by a universal function [102]

$$W(\mathcal{T}) = \frac{Nl}{X}\frac{\theta(\mathcal{T} - \mathcal{T}_{\mathrm{min}})}{\mathcal{T}\sqrt{1-\mathcal{T}}} \tag{11.77}$$

Here, as before, l is the electron elastic mean free path and $\mathcal{T}_{\mathrm{min}} \approx 4e^{-X/l}$ is the minimum transmission eigenvalue. Under the transformation (11.76) and in the limit of very small V, this distribution takes the form [225]

$$\tilde{W}(\tilde{\mathcal{T}}) = \frac{Nl}{X}(eV\tau_{\mathrm{RC}})^{1/g_{\mathrm{S}}}\frac{\theta(\tilde{\mathcal{T}} - \tilde{\mathcal{T}}_{\mathrm{min}})}{\tilde{\mathcal{T}}^{3/2}\sqrt{1-\tilde{\mathcal{T}}}} \tag{11.78}$$

This distribution turns out to be identical to that for a chaotic quantum dot coupled to metallic electrodes via two tunnel junctions.

Now let us turn to another important limit of a coherent scatterer with a large dimensionless conductance $g = \sum_n T_n \gg 1$ embedded in an effective environment with a large Ohmic resistance. In this case in the RG equations (11.72), one can set $g_{\mathrm{S}} \to 0$ and observe that for each channel, the contribution of all other channels plays the same role as that of an effective Ohmic environment. The key difference from the former situation is, however, that now the dimensionless conductance of this environment $\tilde{g} = \sum_n \tilde{T}_n$ gets renormalized. As all $\tilde{T}_n$ decrease in the course of renormalization, $\tilde{g}$ decreases too and eventually becomes of order one in which case the applicability condition (11.71) of our RG approach is violated. Hence, the RG procedure has to be stopped at $L \sim L^*$ defined by the condition

$$\sum_n \tilde{T}_n(L^*) \sim 1 \tag{11.79}$$

At even lower values of the parameter $L < L^*$, the Coulomb interaction cannot anymore be treated perturbatively, and the system enters the regime of strong Coulomb blockade. In this respect, the situation is fully analogous to that of tunnel barriers in the strong tunneling limit considered in Section 7.5. Similar to that case, L^* defines an important energy (or voltage) scale in our problem.

In order to find this scale, we employ the RG flow solution (11.73). Since we are only interested in the order-of-magnitude estimate, for simplicity we may replace the condition (11.79) by another one, $\sum_n \tilde{T}_n(L^*) \approx 0$, which yields the same scale L^* but is easier to handle. Combining the latter condition with Eq. (11.73), we obtain

$$L^* = -\frac{1}{2}\sum_n \ln(1 - T_n) \tag{11.80}$$

As in Section 7.5.2, this equation defines the characteristic energy scale $E^* = eV^*$ via the relation $L^* = \ln(1/E^*\tau_{\rm RC})$. As a result, we get [37]

$$E^* \sim \frac{e^{-L^*}}{\tau_{\rm RC}} = \frac{\chi_1}{\tau_{\rm RC}}\sqrt{\prod_n (1 - T_n)} \sim g\chi_1 E_{\rm C} \exp(-ag) \tag{11.81}$$

with $\chi_1 = 1$ and the parameter

$$a = -\frac{\sum_n \ln(1 - T_n)}{2g} \tag{11.82}$$

depending on the particular transmission distribution in the conductor.

Equation (11.81) constitutes an important non-perturbative in the interaction result. It demonstrates that strong Coulomb effects may occur not only in tunnel junctions but also in any coherent conductor even for $g \gg 1$ provided both temperature and voltage remain lower than E^*:

$$T, eV \leq E^* \tag{11.83}$$

Accordingly, in the limit $T, eV \to 0$, one can expect to observe the conductance $G \to 0$, i.e., complete blocking of electron transport by Coulomb effects in our coherent conductor.

Note that this result applies only provided all channel transmissions T_n are smaller than unity. If at least one transport channel is fully open, e.g., $T_1 = 1$, the energy scale (11.81) vanishes identically $E^* = 0$. In this particular case, no complete Coulomb blockade sets in even in the limit $T, eV \to 0$, and the corresponding system conductance G reduces to $e^2/2\pi$ times the number of fully open channels.

We also note that Eq. (11.81) with $\chi_1 = 1$ was derived from the RG equations (11.72), which do not take into account the two-loop corrections. We speculate that the corresponding two-loop terms could alter the prefactor χ_1 in Eq. (11.81). For instance, in the tunneling limit $T_n \ll 1$, Eq. (11.82) yields $a = 1/2$ and the scale E^* (11.81) coincides with that in Eq. (7.154) provided $\chi_1 \sim g$. As we demonstrated in Chapter 7, this extra power of g in the pre-exponent originates from the two-loop term $\sim 1/\tilde{g}$ in Eq. (7.153). One can expect that a similar effect should also occur in the case of arbitrary transmissions considered here.

11.3.2 Weak Coulomb Blockade and Instantons

Following the same line of reasoning as in Chapter 7, we will now analyze the ground-state energy of an arbitrary coherent scatterer characterized by the transmission distribution T_n and the dimensionless conductance $g \gg 1$. For simplicity, we will also ignore the effect of an external impedance by setting $1/Z_S \to 0$. Then the grand partition function of our coherent scatterer $\mathcal{Z}$ takes the form of an imaginary time path integral over the fluctuating phase

$$\mathcal{Z} = \int_0^{2\pi} d\varphi_0 \sum_{l=-\infty}^{\infty} \int_{\varphi_0}^{\varphi_0+2\pi l} \mathcal{D}\varphi \exp\left[\frac{2\pi i Q_x}{e} - S[\varphi]\right] \tag{11.84}$$

where Q_x is an external charge and

$$S[\varphi] = \frac{1}{4E_C}\int_0^{\beta} d\tau \dot{\varphi}^2 + S_{sc}[\varphi] \tag{11.85}$$

is the effective action. Equation (11.84) is essentially identical to Eq. (7.86) except instead of the action for a normal tunnel barrier in Eq. (7.87), we now have the scatterer action $S_{sc}[\varphi]$. In Chapter 10 and earlier in this chapter, we merely dealt with the real-time version of this action. Here we need to employ its imaginary time version $S_{sc} \equiv S_{el}$, which was already derived in Chapter 2 (see Eq. (2.173)).

In what follows it will be convenient for us to replace the angle-averaged transmissions in Eq. (2.173) by the corresponding sum over independent conducting channels, i.e.,

$$\frac{\mathcal{A}p_F^2}{\pi}\int_0^1 \theta d\theta \mathcal{T}^m(\theta) = \sum_n T_n^m \tag{11.86}$$

where $\mathcal{A}$ is the cross section of the scatterer and $\mathcal{T}(\theta)$ is its angle-dependent transparency. Then we obtain

$$S_{\rm sc}[\varphi] = \sum_n \sum_{m=1}^{\infty} \frac{T_n^m}{2m\beta^{2m}} \int_0^\beta d\tau_1 \int_0^\beta d\tau_2 \int_0^\beta d\tau_3 ... \int_0^\beta d\tau_{2m}$$
$$\times \frac{\sin[(\varphi(\tau_1)-\varphi(\tau_2))/2]}{\sin[\pi T(\tau_1-\tau_2)]} \cdots \frac{\sin[(\varphi(\tau_{2m})-\varphi(\tau_1))/2]}{\sin[\pi T(\tau_{2m}-\tau_1)]} \tag{11.87}$$

Technically, charging effects occur due to the nonlinear dependence of the action $S_{\rm sc}$ on the phase, Eq. (11.87). In the large conductance limit $g \gg 1$ and at low T, such nonlinearities give rise to instantons, which are responsible for tunneling of the phase φ between the states $\varphi = 0$ and $\varphi = 2\pi l$. Should the corresponding tunneling amplitude be fully suppressed, charge discreteness will be effectively washed out and the metallic behavior will be restored at all temperatures, including $T = 0$.

At large conductances $g \gg 1$, the remaining integral over φ is evaluated within the saddle point approximation. We are interested in the nontrivial solutions $\tilde{\varphi}_l(\tau)$ of the saddle point equation $\delta S/\delta\varphi(\tau) = 0$, which "connect" topological sectors with different winding numbers l. It is easy to verify that the corresponding solutions are again defined by the straight lines

$$\tilde{\varphi}_l(\tau) = \varphi_0 + 2\pi l T \tau \tag{11.88}$$

cf. Eq. (7.156). Further analysis is completely analogous to that carried out in Section 7.5.3. Substituting the solutions (11.88) into the effective action, we observe that the charging term again gives the contribution $\pi^2 l^2 T/E_C$, which vanishes in the limit $T \to 0$, whereas the scatterer action (11.87) yields

$$S_{\rm sc}[\tilde{\varphi}_l(\tau)] = |l| \sum_n \sum_{m=1}^{\infty} \frac{T_n^m}{2m} = -\frac{|l|}{2}\sum_n \ln R_n \tag{11.89}$$

where, as before, we defined $R_n = 1 - T_n$. In the tunneling limit $T_n \ll 1$, this expression reduces to $|l|g/2$ and exactly coincides with the last term in Eq. (7.157).

Further analysis goes along the same lines as in Section 7.5.3. Taking into account fluctuations around the straight lines (11.88), we observe that there exist directions in the functional space in which trajectories which may strongly deviate from (11.88) without increasing the value of the action. A general expression for such instantons is not needed for our further analysis, and we will omit it here. It is sufficient for us to observe that in a certain limit, this general solution again reduces to Korshunov's instantons [231], i.e.,

$$\tilde{\varphi}_1(\tau) = \varphi_0 + \pi + 2\arctan(\Omega_1(\tau - \tau_1)) \tag{11.90}$$

cf. Eq. (7.164). Substituting Eq. (11.90) into Eq. (11.87), we again recover the result (11.89) with $l = 1$ implying that, exactly as in the case of a tunnel barrier, Ω_1 and τ_1 are collective coordinates of the two zero modes corresponding, respectively, to changes in the instanton width and shifts of its position.

Next we consider the general configuration of N instantons describing tunneling of the phase between topological sectors $l = 0$ and l. The contribution from this configuration to the partition function takes the form analogous to that in Eq. (7.167). We again have to (i) integrate out the zero modes, (ii) evaluate the ratio of the determinants, and (iii) account for interactions between different instantons. As in Chapter 7, Step (iii) is problematic because of a complicated form of the inter-instanton interactions. Fortunately, this can only affect the pre-exponential factor in our final result.

Proceeding similarly to the analysis [323], we recover the ground-state energy of our coherent scatterer $E_0(Q_x)$. For small $Q_x \ll e$, we obtain $E_0(Q_x) = E_C^*(Q_x/e)^2$, where

$$E_C^*/E_C = \chi_2 \left[\prod_n R_n\right]^{1/2} \ln\left[\prod_n R_n\right] \sim ag\chi_2 \exp(-ag) \tag{11.91}$$

This formula is valid for $ag \gg 1$, i.e., either at large conductances $g \gg 1$ or, if $g \sim 1$, for very small values R_n implying $a \gg 1$. It sets an important energy scale at which strong Coulomb blockade is expected in short coherent scatterers with arbitrary transmission distribution of their conducting channels T_n.

Within the exponential accuracy, Eq. (11.91) was derived by Nazarov [315]. The contribution to the pre-exponent in the form $\ln\left[\prod_n R_n\right]$ follows directly from the integration over the two zero modes [161] (Step (i)), while the remaining function χ_2 (which we cannot determine explicitly) emerges from Steps (ii) and (iii).

Let us evaluate the energy scale (11.91) for several specific examples of coherent conductors. Provided all channel transmissions are small, $T_n \ll 1$, the expression (11.91) should reduce to that for the renormalized charging energy E_C^* already derived in Chapter 7 for tunnel barriers. Indeed, in this limit, the parameter (11.82) equals to $a = 1/2$ and Eq. (11.91) coincides with the result (7.169) if we identify $\chi_2 \sim gf$. If at least one (k-th) conducting channel in our conductor is perfectly transmitting ($T_k = 1$), the renormalized charging energy (11.91) vanishes identically, $E_C^* = 0$. In a (spin-degenerate) single-channel limit, Eq. (11.91) yields $E_C^*/E_C \sim R_1 \ln R_1$ in agreement with the result derived in the limit $R_1 \ll 1$ by means of a different approach [296].

Finally, for an important example of a diffusive metallic conductor with the transmission distribution (11.77), the parameter a (11.82) is evaluated with the result $a = \pi^2/8$ [315], and Eq. (11.91) takes the form

$$E_C^*/E_C \sim g\chi_2 e^{-\pi^2 g/8} \tag{11.92}$$

Comparing this result for E_C^* with Eq. (7.169), we observe that in diffusive metallic conductors with $g \gg 1$, strong Coulomb blockade effects occur at much lower energies than in the case of tunnel barriers with the same g and E_C.

To complete our analysis, we point out that—just as in the case of tunnel barriers—the renormalized charging E_C^* (11.91) coincides (up to possible difference in the unknown pre-exponential functions χ_1 and χ_2) with the energy scale E^* (11.81) derived from the RG approach. This agreement is curious, since one generally expects the instanton and RG types of analysis to account for different kinds of fluctuations and, hence, to produce different energy scales. Here these scales turned out to be essentially the same, which is a remarkable feature of the problem under consideration.

11.3.3 Single-Channel Limit: Exact Solution

To complete our analysis, let us consider a specific limit of a conductor with only one transport channel with transmission T_1.

We will assume that the spin degeneracy is lifted by applying a sufficiently strong magnetic field, in which case the Landauer conductance is fully determined by T_1 and reads

$$\frac{1}{R_\mathrm{N}} = \frac{e^2}{2\pi} T_1 \tag{11.93}$$

As before, we assume that the conductor is embedded in an Ohmic electromagnetic environment with the resistance R_S (see Fig. 10.3b).

It was demonstrated [338, 340] that in the limit eV_x, $T \ll 1/\tau_{\mathrm{RC}}$, this problem can be mapped onto that of a single impurity embedded in a Luttinger liquid. In the zero temperature limit, the latter problem can be solved exactly [116, 399]. Adapting these results to our problem, we arrive at the I–V curve for a single-channel conductor. Provided the bias voltage remains lower than a certain critical value, $V_x < V_{cr}^*$, one has

$$I = \frac{e^2}{2\pi} \frac{g_\mathrm{S} V_x}{1+g_\mathrm{S}} \sum_{n=1}^{\infty} (-1)^{n+1} \frac{\Gamma\left(1 + \frac{(1+g_\mathrm{S})n}{g_\mathrm{S}}\right)\Gamma\left(\frac{3}{2}\right)}{\Gamma(n+1)\Gamma\left(\frac{3}{2} + \frac{n}{g_\mathrm{S}}\right)} \left(\frac{eV_x}{E_0}\right)^{\frac{2n}{g_\mathrm{S}}} \tag{11.94}$$

while at higher voltages $V_x > V_{cr}^*$, one finds

$$I = \frac{e^2}{2\pi} \frac{g_\mathrm{S} V_x}{1+g_\mathrm{S}} \left[1 - \sum_{n=1}^{\infty} \frac{(-1)^{n+1}\Gamma\left(1 + \frac{g_\mathrm{S} n}{1+g_\mathrm{S}}\right)\Gamma\left(\frac{3}{2}\right)}{\Gamma(n+1)\Gamma\left(\frac{3}{2} - \frac{n}{1+g_\mathrm{S}}\right)} \left(\frac{E_0}{eV_x}\right)^{\frac{2n}{1+g_\mathrm{S}}}\right] \tag{11.95}$$

Unlike in our previous analysis, here we expressed the current I via the total voltage drop V_x across the whole system "coherent conductor + shunt resistor." The voltage value V_{cr}^* at which the expressions (11.94) and (11.95) coincide is defined as

$$V_{\mathrm{cr}}^* = \frac{E_0}{e\sqrt{g_\mathrm{S}}} \left(1 + \frac{1}{g_\mathrm{S}}\right)^{-\frac{1+g_\mathrm{S}}{2}} \tag{11.96}$$

Here E_0 is the characteristic energy scale of our problem, which depends on both the charging energy of the conductor and its transmission T_1. In our current notations, the exact expression for the parameter E_0 [399] reads

$$E_0 = \frac{2e^\gamma}{\pi} E_\mathrm{C} \frac{(1+g_\mathrm{S})(T_1+g_\mathrm{S})}{g_\mathrm{S}\left[\Gamma\left(\frac{g_\mathrm{S}}{1+g_\mathrm{S}}\right)\right]^{1+g_\mathrm{S}}} \left(\frac{1-T_1}{T_1}\right)^{\frac{1+g_\mathrm{S}}{2}} \tag{11.97}$$

where the inverse RC-time of our circuit is identified as

$$\frac{1}{\tau_{\mathrm{RC}}} = \frac{(T_1 + g_{\mathrm{S}})E_{\mathrm{C}}}{\pi} \tag{11.98}$$

It is important to keep in mind that the requirement $eV_x\tau_{\mathrm{RC}} \ll 1$ effectively restricts the applicability range of the results (11.94) and (11.95) to sufficiently low bias voltages. Since the nonlinearity in the I–V curve occurs at voltages lower than or comparable with E_0/e, in order to justify the applicability of the series (11.94) and (11.95), one should impose the condition $E_0\tau_{\mathrm{RC}} \ll 1$, which roughly translates to

$$T_1 \geq 1/2 \tag{11.99}$$

In the limit $1 - T_1 \ll 1$, the result (11.95) can be interpreted as an expansion of the current in powers of the reflection probability. To the lowest order in $1 - T_1$, this expansion should match with Eq. (11.62). In order to verify the equivalence of these results, it is necessary to account for the difference between V and V_x. These two voltages are related as $V_x = V + I(V_x)R_{\mathrm{S}}$ and, hence, we have

$$\frac{dI}{dV} = \frac{dI/dV_x}{1 - R_{\mathrm{S}}\,dI/dV_x} \tag{11.100}$$

Substituting Eq. (11.95) into the right-hand side of this formula, expanding it to the lowest order in $1 - T_1$, and performing the replacement $V_x \to (1 + g_{\mathrm{S}})V/g_{\mathrm{S}}$, one indeed recovers Eq. (11.62).

It is also straightforward to verify that in the limit $g_{\mathrm{S}} \gg 1$ (in which the difference between V and V_x becomes negligible), Eq. (11.95) yields

$$I = \frac{e^2}{2\pi}\frac{V}{1 + (E_0/eV)^{2/g_{\mathrm{S}}}} \tag{11.101}$$

for any value of T_1. This result is consistent with that of the RG analysis in Eq. (11.75).

In order to test the validity range of another, low bias, series (11.94), let us compare its first term with the expression for the differential conductance (7.69) derived under the assumption $T_1 \ll \min\{1, g_{\mathrm{S}}\}$. The two results agree only in the limit $g_{\mathrm{S}} \gg 1$. Hence, at $T_1 \ll 1$ and $g_{\mathrm{S}} \ll 1$, Eq. (11.94) captures the dependence of the current on V_x, T_1, g_{S}, and E_{C} only qualitatively. This fact is, by no means, surprising in the light of the condition (11.99).

We also note that from the mathematical perspective, the $I-V$ curve for a single-channel coherent conductor defined in Eqs. (11.94) and (11.95) can also be mapped onto the zero temperature $I-V$ curve of a resistively shunted Josephson junction already discussed in Chapter 6, cf. Eqs. (6.166) and (6.172). In order to observe this mapping, one needs to redefine the dissipation parameter as $K \leftrightarrow (1+g_S)/g_S$.

For a specific value of the shunt conductance $g_S = 1$, the series in Eq. (11.94) can be effectively summed up and one arrives at a very simple expression

$$I = \frac{e^2 V_x}{4\pi} - \frac{eE_0}{8\pi} \arctan \frac{2eV_x}{E_0} \tag{11.102}$$

For $g_S = 1$, one can also derive the expression for the current at nonzero temperatures. Making use of the results [116], one finds

$$I = \frac{e^2}{4\pi} V_x \left[1 - \frac{E_0}{2eV_x} \operatorname{Im} \Psi \left(\frac{1}{2} + \frac{E_0}{2\pi T} + \mathrm{i} \frac{eV_x}{4\pi T} \right) \right] \tag{11.103}$$

In addition, the current noise spectrum of the single-mode conductor can be related to its $I-V$ curve (11.103) as [226]

$$\mathcal{S}(0) = 2T \frac{\partial I}{\partial V_x} - \frac{e}{2} \omega_c \coth \frac{eV_x}{2T} \frac{\partial I}{\partial \omega_c} + T\omega_c \frac{\partial I}{\partial V_x \partial \omega_c} \tag{11.104}$$

where $\omega_c = e^{\gamma}/\tau_{RC}$ is the effective cutoff frequency.

The exact solutions (11.94), (11.95) provide a useful test for both RG and instanton analysis developed in Sections 11.3.1 and 11.3.2. To begin with, it reconfirms our conclusion that as long as the transmission probability T_1 differs from unity, the low-energy behavior of the conductor eventually reduces to that of a tunnel barrier. Furthermore, it also allows to estimate the characteristic energy scale E^* in our problem. This energy scale follows directly from Eq. (11.97) provided one accounts for the difference between V and V_x. Taking the limit $g_S \to 0$, we arrive at the estimate for the "bare" energy scale or, equivalently, for the renormalized charging energy of our conductor

$$E^* = \lim_{g_S \to 0} (eV_x - eI(V_x)R_S) \Big|_{V_x = E_0/e} = \frac{2e^{\gamma}}{\pi} E_C \sqrt{T_1(1-T_1)} \tag{11.105}$$

As one could expect, under the condition (11.99), this expression is consistent with the corresponding RG and instanton results (11.81) and (11.91), which also reduce to the form $E_C^* \sim E_C T_1 \sqrt{1 - T_1}$ in the single-channel limit. Finally, the exact solutions (11.94) and (11.95) also demonstrate that there exists only one fundamental energy scale in our problem. This is the energy at which the low bias perturbation theory (operating in the same energy domain as the instanton approach) matches to the high bias perturbation theory, which in a way extends the RG approach.

Chapter 12

Charging Effects in Metallic Quantum Dots

12.1 Introduction

In this chapter, we will analyze the behavior of metallic quantum dots in the presence of electron–electron interactions. The term "quantum dot" is usually employed if one deals with a small object (e.g., cavity) coupled to two big metallic reservoirs, thereby allowing one to apply bias voltage and to drive electric current through the system. Quantum dots play a very important role in modern mesoscopic physics and are being intensively studied both theoretically and experimentally. Quite generally, quantum dots may consist of a large variety of materials, such as metals, semiconductors, superconductors, carbon nanotubes, or even single molecules. The quantum dot size may vary from a few microns to a few Ångstroms. Physical properties of quantum dots are very diverse resulting from a nontrivial interplay between level quantization, electron scattering, quantum coherence, electron–electron interactions, and other effects. Such properties may strongly depend on the quantum dot size, composition, temperature, etc. Here we are not going to provide a detailed overview of the whole subject; rather, we will

Dissipative Quantum Mechanics of Nanostructures: Electron Transport, Fluctuations, and Interactions
Andrei D. Zaikin and Dmitry S. Golubev

ISBN 978-981-4774-50-5 (Hardcover), 978-0-429-29823-3 (eBook)
www.jennystanford.com

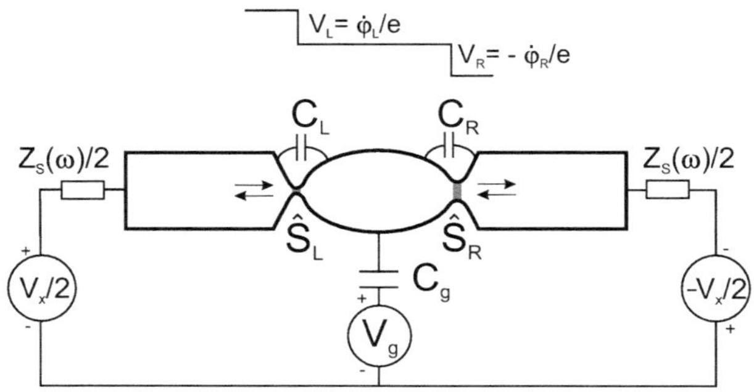

Figure 12.1 Schematics of a metallic quantum dot.

concentrate on just one important example of metallic quantum dots. More precisely, in what follows we will consider sufficiently large normal metallic islands with the electron energy level spacing δ much smaller than any other relevant energy scale in the problem, e.g., $\delta \ll T$ and eV.

Schematics of the system under consideration is displayed in Fig. 12.1. We will consider a normal metallic island coupled to two reservoirs via two junctions (scatterers) characterized by energy-independent S–matrices $\hat{S}_{\rm L}$ and $\hat{S}_{\rm R}$. The bias voltage is applied to the left and the right reservoirs via two identical impedances $Z_{\rm S}(\omega)/2$ playing the role of an effective electromagnetic environment. Electric potentials of both reservoirs and of the central island may fluctuate in time but do not depend on spatial coordinates. The voltage drops across the left and the right junctions, $V_{\rm L}$ and $V_{\rm R}$, are the two relevant degrees of freedom describing quantum dynamics of the system in the presence of Coulomb interaction. They are related to fluctuating phases across the corresponding scatterers by means of the standard relations $\dot{\varphi}_{\rm L} = eV_{\rm L}$, $\dot{\varphi}_{\rm R} = -eV_{\rm R}$. The electrostatic energy of extra electric charge accumulated on the central island is stored in the junction capacitors $C_{\rm L}$ and $C_{\rm R}$ as well as in the gate capacitor $C_{\rm g}$. In addition, the gate voltage $V_{\rm g}$ is applied to the gate electrode and may shift the quantum dot potential via the capacitor $C_{\rm g}$.

Note that although the size of metallic quantum dots may be very small, their properties are, in many respects, different from those of short coherent scatterers analyzed in Chapter 11. In order to highlight this difference, we begin with some general remarks on the scattering matrix formalism employed in our consideration. One of the key parameters characterizing the properties of a quantum dot is the electron dwell time τ_{D}. It is the average time spent by the incoming electrons inside the dot before escaping to one of the leads. Within our analysis in Chapter 11, this parameter was considered to be very small and, hence, was totally ignored. In spatially extended conductors, on the contrary, it needs to be taken into account, since in this case, τ_{D} may exceed the inverse temperature, inverse bias voltage, or some other relevant timescale in the problem, i.e.,

$$\tau_{\mathrm{D}} > \max\{1/T,\ 1/eV\} \tag{12.1}$$

In this case, transport properties of the dot already in the absence of interaction effects are described by the *energy-dependent* scattering matrix

$$\hat{S}(E) = \begin{pmatrix} \hat{r}(E) & \hat{t}'(E) \\ \hat{t}(E) & \hat{r}'(E) \end{pmatrix} \tag{12.2}$$

in contrast to short coherent conductors characterized by scattering matrices independent of energy.

The Landauer–Büttiker scattering formalism should be modified accordingly; see the review by Blanter and Büttiker [60] for further details. In particular, the current across our extended conductor is now expressed as follows:

$$I = \frac{e}{2\pi}\sum_n \int dE\ \mathrm{tr}\left[\hat{t}^\dagger(E)\hat{t}(E)\right]\left[f_{\mathrm{R}}(E+eV_{\mathrm{R}}) - f_{\mathrm{L}}(E+eV_{\mathrm{L}})\right] \tag{12.3}$$

where f_{L} and f_{R} are the electron distribution functions in the left and the right reservoirs, and V_{L}, V_{R} are the corresponding electric potentials. Likewise, zero-frequency noise power in this case is given by the integral

$$\begin{aligned}\mathcal{S}(0) = \frac{e}{2\pi}\sum_n \int dE \Big\{ &\mathrm{tr}\left[\hat{t}^\dagger(E)\hat{t}(E)\right]\big[f_{\mathrm{L}}(E+eV_{\mathrm{L}})(1-f_{\mathrm{R}}(E+eV_{\mathrm{R}})) \\ &+ (1-f_{\mathrm{L}}(E+eV_{\mathrm{R}}))\,f_{\mathrm{R}}(E+eV_{\mathrm{R}})\big] \\ &+ \mathrm{tr}\left[\hat{t}^\dagger(E)\hat{t}(E)\hat{r}^\dagger(E)\hat{r}(E)\right]\left[f_{\mathrm{L}}(E+eV_{\mathrm{L}}) - f_{\mathrm{R}}(E+eV_{\mathrm{R}})\right]^2\Big\}\end{aligned} \tag{12.4}$$

As we will see below, the energy dependence of the S-matrix yields substantial modification of both the formalism and the physical picture as compared to those discussed in Chapter 11.

In addition to the electron dwell time τ_D, there exist two other important timescales in the problem: electron inelastic relaxation time τ_{in} and dephasing time τ_φ. Depending on the relation between these three different timescales, electron transport and fluctuations in quantum dots may be described by a variety of different regimes. In order to account for all these regimes, it is, in general, necessary to obtain full information about the non-equilibrium electron distribution function in the dot, e.g., by solving the corresponding kinetic equation. There exist, however, two limiting cases in which the electron distribution function inside the dot is known and, hence, no kinetic equation is needed. They are: (i) the strong inelastic relaxation limit $\tau_{in} \ll \tau_D$ and (ii) the coherent regime $\tau_\varphi \gg \tau_D$. Only these two regimes will be addressed below in this chapter.

As electrons in a quantum dot are confined to a limited volume, Coulomb interaction between them is usually substantial and, hence, needs to be taken into account. The Coulomb interaction strength in a quantum dot is characterized by its charging energy E_C defined as

$$E_C = \frac{e^2}{2C_\Sigma}, \quad C_\Sigma = C_L + C_R + C_g \tag{12.5}$$

Similar to the case of single-tunnel junctions and short coherent conductors discussed in Chapters 7 and 11, Coulomb effects in quantum dots are pronounced at low enough energies $T, eV < E_C$, while they become much weaker in the opposite limit $T, eV > E_C$.

Finally, the strength of Coulomb effects in quantum dots also strongly depends on the junction conductances connecting the dot and the leads. As before, we define the dimensionless conductances of these junctions as

$$g_L = R_q/R_L, \quad g_R = R_q/R_R \tag{12.6}$$

where the quantum resistance unit R_q is defined in Eq. (7.2). As long as these dimensionless conductances are small, $g_L, g_R \ll 1$, the quantum dot remains in the strong Coulomb blockade regime implying, e.g., exponential suppression of its zero-bias conductance G in the low-temperature limit,

$$G \propto \exp[-E_C/T] \tag{12.7}$$

In this regime, the current through the dot is also sensitive to the gate voltage V_g. This is the so-called gate effect, which constitutes one of the characteristic features of the strong Coulomb blockade. In a way, our system acquires the properties of a transistor. For this reason, the system under consideration depicted in Fig. 12.1 is also known as single-electron tunneling (SET) transistor.

On the other hand, provided at least one of the junction conductances is large,

$$\max\{g_L, g_R\} \gg 1 \tag{12.8}$$

Coulomb blockade effects get substantially weakened by charge fluctuations in a qualitative similarity with the situations described in previous chapters. The gate effect also becomes weak in this case.

Let us now turn to a more detailed analysis of the properties of metallic quantum dots in the presence of electron–electron interactions.

12.2 Strong Coulomb Blockade

Let us first consider the limit of highly resistive tunnel junctions with $R_{L,R}$ strongly exceeding the quantum resistance R_q, i.e.,

$$R_L, R_R \gg R_q \tag{12.9}$$

In addition, we will assume that electron tunneling rates across the junctions are sufficiently low and the electron distribution function inside the dot fully relaxes to its equilibrium form between two successive tunneling events. In other words, we will consider the strong inelastic relaxation regime $\tau_{in} \ll \tau_D$.

12.2.1 Sequential Tunneling: Master Equation Analysis

In order to proceed, we will adapt a simple master equation formalism already discussed in Section 7 to the quantum dot configuration depicted in Fig. 12.1 For simplicity, we will neglect the impedance of the leads and set $Z_S(\omega) \to 0$. The energy for the system consisting of three capacitors C_L, C_R, and C_g obviously reads

$$E_0 = \frac{q_L^2}{2C_L} + \frac{q_R^2}{2C_R} + \frac{q_g^2}{2C_g} \tag{12.10}$$

where q_{L}, q_{R}, and q_{g} are the corresponding electric charges. Let us assume that the total charge of the quantum dot equals to Ne, where N is an integer number. Then we immediately recover the following relation between the charges:

$$-q_{\mathrm{L}} + q_{\mathrm{R}} - q_{\mathrm{g}} = Ne \tag{12.11}$$

The total voltage drop across the quantum dot V equals to

$$V = \frac{q_{\mathrm{L}}}{C_{\mathrm{L}}} + \frac{q_{\mathrm{R}}}{C_{\mathrm{R}}} \tag{12.12}$$

Yet another relation between the charges can be easily established in the form

$$\frac{q_{\mathrm{L}}}{C_{\mathrm{L}}} - \frac{q_{\mathrm{g}}}{C_{\mathrm{g}}} = \frac{V}{2} - V_{\mathrm{g}} \tag{12.13}$$

Resolving the system of the above three equations, we get

$$\begin{aligned} q_{\mathrm{L}}(N) &= -\frac{C_{\mathrm{L}}}{C_{\Sigma}} Ne + q_{\mathrm{L}}^{(0)} \\ q_{\mathrm{R}}(N) &= \frac{C_{\mathrm{R}}}{C_{\Sigma}} Ne + q_{\mathrm{R}}^{(0)} \\ q_{\mathrm{g}}(N) &= -\frac{C_{\mathrm{g}}}{C_{\Sigma}} Ne + q_{\mathrm{g}}^{(0)} \end{aligned} \tag{12.14}$$

where $q_j^{(0)}$ are the offset charges, which do not depend on the number of extra electrons in the quantum dot and, hence, can be disregarded. Combining the above results with Eq. (12.10), we find

$$E_0(N) = E_{\mathrm{C}} N^2 + \tilde{E} \tag{12.15}$$

where $\tilde{E}$ is an unimportant combination of the voltages V and V_{g}.

Let us now include electron tunneling into our consideration. Assume that one electron has tunneled from the left electrode to the quantum dot. Immediately after this tunneling event, the charge C_{L} of the left capacitor changes from $q_{\mathrm{L}}(N)$ to $q_{\mathrm{L}}(N) - e$ and very quick recharging process through the external circuit with vanishing impedance occurs. As a result, the charges take the new values $q_{\mathrm{L}}(N+1)$, $q_{\mathrm{R}}(N+1)$, and $q_{\mathrm{g}}(N+1)$ and, hence, the charges

$$\begin{aligned} q_{\mathrm{L}}(N+1) - q_{\mathrm{L}}(N) + e &= e - \frac{C_{\mathrm{L}}}{C_{\Sigma}} e = \frac{C_{\mathrm{R}} + C_{\mathrm{g}}}{C_{\Sigma}} e \\ q_{\mathrm{R}}(N) - q_{\mathrm{R}}(N+1) &= -\frac{C_{\mathrm{R}}}{C_{\Sigma}} e \\ q_{\mathrm{g}}(N+1) - q_{\mathrm{g}}(N) &= -\frac{C_{\mathrm{g}}}{C_{\Sigma}} e \end{aligned} \tag{12.16}$$

pass, respectively, through the left, right, and gate voltage sources. The presence of the electron charge e in the first of Eqs. (12.16) just indicates that an electron tunnels across the left junction. From these equations, one can easily evaluate the work performed by the voltage sources. It reads

$$W = \frac{V}{2}\left[q_{\mathrm{L}}(N+1) - q_{\mathrm{L}}(N) + e\right] - \frac{V}{2}\left[q_{\mathrm{R}}(N) - q_{\mathrm{R}}(N+1)\right] + V_{\mathrm{g}}\left[q_{\mathrm{g}}(N+1) - q_{\mathrm{g}}(N)\right] = \frac{C_{\mathrm{R}} + C_{\mathrm{g}}/2}{C_{\Sigma}} eV - \frac{C_{\mathrm{g}}}{C_{\Sigma}} eV_{\mathrm{g}} \quad (12.17)$$

The total energy change produced by tunneling of a single electron through the left junction from the lead to the dot $\Delta \overrightarrow{E}_{\mathrm{L}}$ is defined by a sum

$$\Delta \overrightarrow{E}_{\mathrm{L}}(N) = E_0(N) - E_0(N+1) + W = -E_{\mathrm{C}}(2N+1) + \frac{C_{\mathrm{R}} + C_{\mathrm{g}}/2}{C_{\Sigma}} eV - \frac{C_{\mathrm{g}}}{C_{\Sigma}} eV_{\mathrm{g}} \quad (12.18)$$

Considering now three remaining elementary tunneling processes in our system and repeating the same procedure, we arrive at the corresponding energy changes involved in these processes:

$$\Delta \overleftarrow{E}_{\mathrm{L}}(N) = E_{\mathrm{C}}(2N-1) - \frac{C_{\mathrm{R}} + C_{\mathrm{g}}/2}{C_{\Sigma}} eV + \frac{C_{\mathrm{g}}}{C_{\Sigma}} eV_{\mathrm{g}}$$
$$\Delta \overrightarrow{E}_{\mathrm{R}}(N) = E_{\mathrm{C}}(2N-1) + \frac{C_{\mathrm{L}} + C_{\mathrm{g}}/2}{C_{\Sigma}} eV + \frac{C_{\mathrm{g}}}{C_{\Sigma}} eV_{\mathrm{g}}$$
$$\Delta \overleftarrow{E}_{\mathrm{R}}(N) = -E_{\mathrm{C}}(2N+1) - \frac{C_{\mathrm{L}} + C_{\mathrm{g}}/2}{C_{\Sigma}} eV - \frac{C_{\mathrm{g}}}{C_{\Sigma}} eV_{\mathrm{g}} \quad (12.19)$$

Equations (12.18) and (12.19) allow to immediately recover the corresponding electron tunneling rates. Analogously to Eq. (7.8), we obtain

$$\overrightarrow{\Gamma}_{\mathrm{L}}(N) = \frac{1}{e^2 R_{\mathrm{L}}} \frac{\Delta \overrightarrow{E}_{\mathrm{L}}}{1 - e^{-\Delta \overrightarrow{E}_{\mathrm{L}}/T}}, \quad \overleftarrow{\Gamma}_{\mathrm{L}}(N) = \frac{1}{e^2 R_{\mathrm{L}}} \frac{\Delta \overleftarrow{E}_{\mathrm{L}}}{1 - e^{-\Delta \overleftarrow{E}_{\mathrm{L}}/T}},$$
$$\overrightarrow{\Gamma}_{\mathrm{R}}(N) = \frac{1}{e^2 R_{\mathrm{R}}} \frac{\Delta \overrightarrow{E}_{\mathrm{R}}}{1 - e^{-\Delta \overrightarrow{E}_{\mathrm{R}}/T}}, \quad \overleftarrow{\Gamma}_{\mathrm{R}}(N) = \frac{1}{e^2 R_{\mathrm{R}}} \frac{\Delta \overleftarrow{E}_{\mathrm{R}}}{1 - e^{-\Delta \overleftarrow{E}_{\mathrm{R}}/T}}, \quad (12.20)$$

where the argument N indicates the number of excess electrons in the dot before tunneling.

Our next step is to construct the master equations for the occupation probabilities W_N of the quantum dot charge states. This equation reads

$$\begin{aligned}\frac{dW_N}{dt} = &-\left[\overrightarrow{\Gamma}_{\mathrm{L}}(N) + \overleftarrow{\Gamma}_{\mathrm{R}}(N) + \overleftarrow{\Gamma}_{\mathrm{L}}(N) + \overrightarrow{\Gamma}_{\mathrm{R}}(N)\right] W_N \\ &+ \left[\overrightarrow{\Gamma}_{\mathrm{L}}(N-1) + \overleftarrow{\Gamma}_{\mathrm{R}}(N-1)\right] W_{N-1} \\ &+ \left[\overleftarrow{\Gamma}_{\mathrm{L}}(N+1) + \overrightarrow{\Gamma}_{\mathrm{R}}(N+1)\right] W_{N+1}\end{aligned} \tag{12.21}$$

Obviously, the terms in the first line of Eq. (12.21) account for the decay of the charge state N into the states $N-1$ and $N+1$, while the second and the third lines, respectively, describe the transitions $N-1 \to N$, and $N+1 \to N$.

Provided the system remains in its equilibrium stationary state, the probabilities W_N do not depend on time and, hence, in Eq. (12.21), one should set $dW_N/dt = 0$. Then, combining Eqs. (12.21) for different charge states labeled by the numbers from $-\infty$ to N, one finds

$$0 = -\left[\overrightarrow{\Gamma}_{\mathrm{L}}(N) + \overleftarrow{\Gamma}_{\mathrm{R}}(N)\right] W_N + \left[\overleftarrow{\Gamma}_{\mathrm{L}}(N+1) + \overrightarrow{\Gamma}_{\mathrm{R}}(N+1)\right] W_{N+1} \tag{12.22}$$

or, equivalently,

$$W_{N+1} = \frac{\overrightarrow{\Gamma}_{\mathrm{L}}(N) + \overleftarrow{\Gamma}_{\mathrm{R}}(N)}{\overleftarrow{\Gamma}_{\mathrm{L}}(N+1) + \overrightarrow{\Gamma}_{\mathrm{R}}(N+1)} W_N \tag{12.23}$$

These simple recursive relations can be conveniently employed for a numerical evaluation of the probabilities W_N and the current I across the dot. In practice, it is necessary to first truncate the system of equations keeping only the finite number of the occupied charging states ranging from $-M$ to M, where M is chosen to be sufficiently large. Then one should express all probabilities via one of them, say W_0, with the aid of Eq. (12.23). After that one finds the probability W_0 from the normalization condition $\sum_{-M}^{M} W_N = 1$ and evaluate the current I through the dot from the equation

$$I = e\sum_N \left[\overrightarrow{\Gamma}_{\mathrm{L}}(N) - \overleftarrow{\Gamma}_{\mathrm{L}}(N)\right] W_N \tag{12.24}$$

The current (12.24) is a periodic function of the gate voltage V_{g} with the period e/C_{g} corresponding to the change in the induced

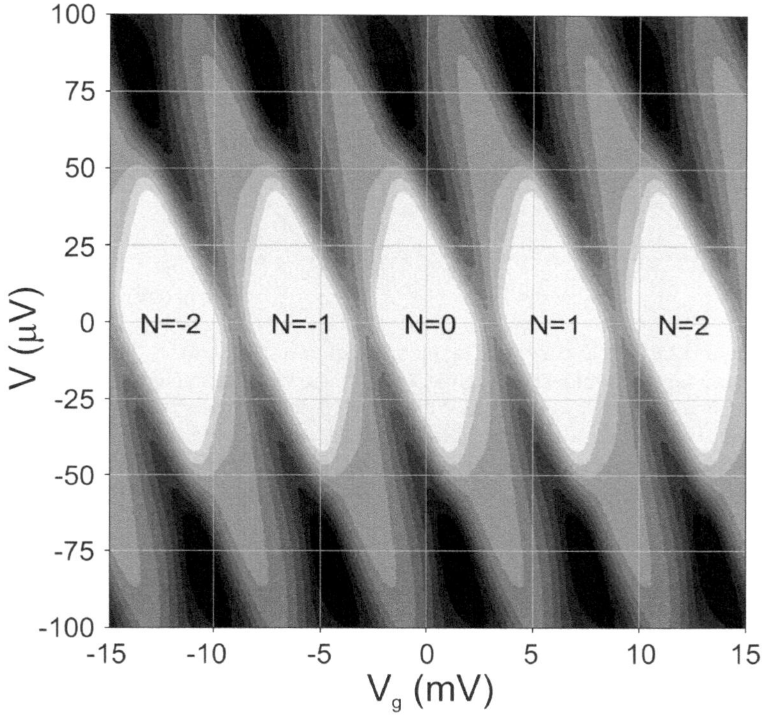

Figure 12.2 Gray-scale plot of the differential conductance $\partial I(V_g, V)/\partial V$ for a SET transistor with typical experimental parameters $E_C = 30$ μV, $R_L = R_R = 50$ kΩ, $C_L = 1.87$ fF, $C_R = 0.77$ fF, $C_g = 0.027$ fF and $T = 35$ mK. Darker areas correspond to higher values of $\partial I/\partial V$. The graph exhibits the so-called "diamond structure." Every white diamond on the V_g, V plane indicates the area of strong Coulomb blockade of electron tunneling. For every diamond, the number of extra electrons accumulated in the dot coincides with a particular value of N.

charge value $Q_g = C_g V_g$ by e. This periodic dependence of the differential conductance on V_g is illustrated in Fig. 12.2. This figure also displays the so-called "diamond structure" observed in the dependence of the differential conductance on the two voltages V_g and V. Such diamond structure is a hallmark of the Coulomb blockade. Every diamond in Fig. 12.2 corresponds to the fixed number N_d of the extra electrons accumulated in the quantum dot and indicates the area where the current is strongly suppressed. The

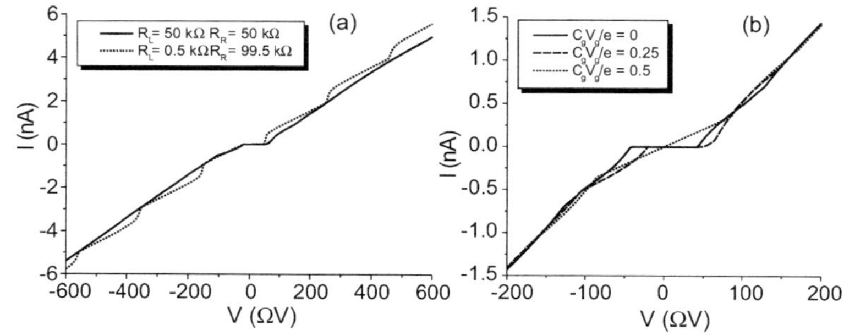

Figure 12.3 $I-V$ curves of a metallic quantum dot at $T \ll E_C$ in the strong Coulomb blockade regime. (a) $I-V$ curve for a strongly asymmetric transistor with $R_L = 0.5$ kΩ, $R_R = 99.5$ kΩ exhibits strong Coulomb staircase, while in a symmetric case $R_L = R_R = 50$ kΩ, the staircase is almost invisible. The gate voltage is chosen to be $V_g = 0.25e/C_g$. (b) $I-V$ curves of a SET transistor at three different values of the gate voltage. Note that at $C_gV_g/e \neq 0; 0.5$, the $I-V$ curve is not symmetric, i.e., $I(-V) \neq -I(V)$.

diamonds are formed by the two sets of parallel straight lines clearly seen in the figure. The first set of lines corresponds to vanishing of the Coulomb potential barrier for electron tunneling across the left junction. This set of lines is determined by the equation $\Delta \vec{E}_L(N) = 0$, which yields

$$\left.\frac{dV}{dV_g}\right|_L = \frac{C_g}{C_R + C_g/2} \tag{12.25}$$

Similarly, the second set of lines has the slope

$$\left.\frac{dV}{dV_g}\right|_R = -\frac{C_g}{C_L + C_g/2} \tag{12.26}$$

determined by the equations $\Delta \vec{E}_R(N) = 0$. Equations (12.25) and (12.26) can be used in order to evaluate the capacitances C_L, C_R, and C_g from the available experimental data. The boundaries of every diamond define the threshold voltage $V_{th}(V_g)$ for the Coulomb blockade. This voltage vanishes, $V_{th}(V_g) = 0$, at the values $V_g = (n + 1/2)e/C_g$ corresponding to zero-bias conductance peaks. The maximum value of the threshold voltage is achieved at $V_g = ne/C_g - e(C_2 - C_g)/2C_\Sigma$ and equals to $V_{th}^{max} = e/C_\Sigma$.

The $I-V$ curves of the quantum dot are displayed in Fig. 12.3. They exhibit two distinct features: (i) Coulomb staircase observed

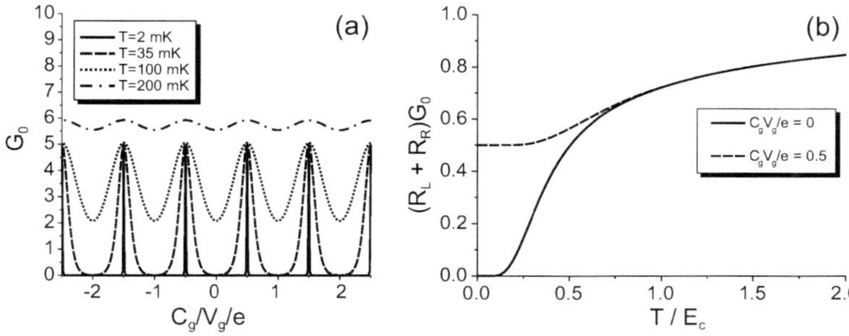

Figure 12.4 Zero-bias conductance G_0 for an SET transistor. (a) Gate voltage dependence of G_0 at different temperatures. The parameters of the system are the same as in Fig. 12.2: $E_\mathrm{C} = 30\ \mu$V, $R_\mathrm{L} = R_\mathrm{R} = 50$ kΩ, $C_\mathrm{L} = 1.87$ fF, $C_\mathrm{R} = 0.77$ fF, $C_\mathrm{g} = 0.027$ fF. (b) Maximum and minimum zero-bias conductance values $G_0(T, 0.5e/C_\mathrm{g})$ and $G_0(T, 0)$ as a function of temperature.

on the $I-V$ curve in a strongly asymmetric case and (ii) asymmetric behavior of the $I-V$ curve with $I(-V) \neq -I(V)$ provided the charge $C_\mathrm{g}V_\mathrm{g}$ differs from 0 or $e/2$.

The zero-bias conductance of the quantum dot $G_0(T)$ can be derived analytically. It reads [243]

$$G_0(T, V_\mathrm{g}) = \frac{E_\mathrm{C}}{T(R_\mathrm{L}+R_\mathrm{R})} \frac{1}{\sum_N e^{-E_\mathrm{C}(N-N_\mathrm{g})^2/T}} \times \sum_N \frac{(1+2N-2N_\mathrm{g})e^{-E_\mathrm{C}(N-N_\mathrm{g})^2/T}}{e^{E_\mathrm{C}(1+2N-2N_\mathrm{g})/T}-1} \quad (12.27)$$

where $N_\mathrm{g} = C_\mathrm{g}V_\mathrm{g}/e$. This function is plotted in Fig. 12.4. Figure 12.4(a) demonstrates that at low temperature $T \ll E_\mathrm{C}$, the zero-bias conductance is strongly suppressed between the conductance peaks (of width $\sim T$) centered around the half-integer values of the gate charge $C_\mathrm{g}V_\mathrm{g} = en + e/2, n = 0, \pm 1, \pm 2, \ldots$. Within one period $-0.5 < N_\mathrm{g} < 0.5$, the conductance is approximately defined as

$$(R_\mathrm{L}+R_\mathrm{R})G_0 \approx \frac{E_\mathrm{C}}{T}\left(\frac{1-2N_\mathrm{g}}{e^{E_\mathrm{C}(1-2N_\mathrm{g})/T}-1} + \frac{1+2N_\mathrm{g}}{e^{E_\mathrm{C}(1+2N_\mathrm{g})/T}-1}\right) \quad (12.28)$$

Close to the peaks, the energy barriers (12.18), (12.19) are low, and the Coulomb blockade is lifted. At $T \ll E_\mathrm{C}$, the conductance peak centered at $N_\mathrm{g} = 1/2$ is described by a simple formula

$$(R_\mathrm{L}+R_\mathrm{R})G_0 \approx \frac{E_\mathrm{C}\left(N_\mathrm{g}-\frac{1}{2}\right)}{T\sinh\left[\frac{2E_\mathrm{C}}{T}\left(N_\mathrm{g}-\frac{1}{2}\right)\right]} \quad (12.29)$$

which yields $G_0 \propto \exp(-E_\mathrm{C}/2T)$ for $N_\mathrm{g} \approx 0$ and $G_0 = 1/2(R_\mathrm{L} + R_\mathrm{R})$ for $N_\mathrm{g} \to 1/2$.

The gate dependence of the conductance weakens with growing temperature and essentially vanishes at $T \sim E_\mathrm{C}$ (see Fig. 12.4(b)). At higher temperatures $T > E_\mathrm{C}$, one finds

$$(R_\mathrm{L} + R_\mathrm{R})G_0 = 1 - \frac{E_\mathrm{C}}{3T} + \frac{E_\mathrm{C}^2}{15T^2} - \frac{2\pi^2 E_\mathrm{C}^2}{3T^2} e^{-\pi^2 T/E_\mathrm{C}} \cos 2\pi N_\mathrm{g} + \mathcal{O}\left(\frac{E_\mathrm{C}^3}{T^3}\right) \quad (12.30)$$

Finally, we also provide the high-temperature expansion of the differential conductance of a symmetric SET transistor with $R_\mathrm{L} = R_\mathrm{R} = R$ and $C_\mathrm{L} = C_\mathrm{R} = C$. At $T \gg E_\mathrm{C}$, we obtain

$$2R\frac{dI}{dV} = 1 - \frac{e^2}{4CT}\,\frac{\frac{eV}{4T}\cosh\frac{eV}{4T} - \sinh\frac{eV}{4T}}{\sinh^3\frac{eV}{4T}} \quad (12.31)$$

This formula resembles Eq. (7.37) derived for a single-tunnel junction embedded in the high Ohmic environment. An important difference between these results, however, lies in the fact that the width of the Coulomb blockade dip in Eq. (12.31) is two times larger than in Eq. (7.37). In fact, one can demonstrate that this width scales proportionally to the number of connected-in-series tunnel junctions in the system. Equation (12.31) serves as a basis for the absolute thermometry [137]: The half-width of the Coulomb dip in the differential conductance is proportional to the absolute temperature T and, hence, the latter can be extracted from accurate measurements of the $I-V$ curve of a SET transistor in the Coulomb blockade regime.

12.2.2 Path Integral Formulation

So far we only considered the lowest-order tunneling processes and evaluated the conductance perturbatively in the junction conductances g_L and g_R, restricting ourselves to the leading order in each of these parameters. According to this physical picture, electrons tunnel sequentially through the left and the right junctions. Since Coulomb blockade yields exponential suppression of this sequential tunneling current at T, $eV \ll E_\mathrm{C}$, higher-order tunneling processes also need to be taken into account even for small values g_{L} and g_{R}.

In order to quantitatively characterize such processes, it is convenient to reformulate our problem via path integrals. Adopting the same model as in Section 12.2.1, in order to describe the Coulomb interaction between electrons, we introduce the electric potential $V(t)$ of the island as a fluctuating macroscopic field. The whole procedure is completely analogous to that already described, e.g., in Chapters 2 and 9. Therefore, it is not necessary to go into further details here. We only point out that interaction between electrons in our system is replaced in this way by an interaction with the collective variable $V(t)$.

Since the electrochemical potentials of the reservoirs are fixed to be $\mu_r = -eV_r$ for $r = L, R$, the only fluctuating field in our problem is the island voltage $V(t)$, which we again relate to the corresponding phase field $\varphi(t)$ as $eV(t) \equiv -\dot{\varphi}(t)$. As we have already repeatedly done before, let us introduce the reduced density matrix $\rho(t; \varphi_1, \varphi_2)$ and express it as a double path integral over the phase variables $\varphi_{\mathrm{F,B}}$ defined on the forward and backward parts of the Keldysh contour. Then we obtain (cf., e.g., Chapter 2):

$$\rho(t_f; \varphi_{1f}, \varphi_{2f}) = \sum_{m_1, m_2} \int_{-\infty}^{\infty} d\varphi_{1i} \int_{-\infty}^{\infty} d\varphi_{2i} \int_{\varphi_{1i}}^{\varphi_{1f}+2\pi m_1} \mathcal{D}\varphi_{\mathrm{F}} \int_{\varphi_{2i}}^{\varphi_{2f}+2\pi m_2} \mathcal{D}\varphi_{\mathrm{B}}$$
$$\times \exp\left(\mathrm{i}S[\varphi_{\mathrm{F}}(t), \varphi_{\mathrm{B}}(t)]\right) \rho_i(t_i; \varphi_{1i}, \varphi_{2i}), \qquad (12.32)$$

This equation fully describes the real-time evolution of the reduced density matrix ρ from the initial state described by the matrix ρ_i at time $t = t_i$ until the final time moment t_f.

Note that in Eq. (12.32), we made use of the fact that the charge on the island is discrete, being quantized in units e. In this case, the phase variable is compact (i.e., it is defined on a circle with the states φ and $\varphi + 2\pi$ being physically equivalent), and it is necessary to sum up over all integer winding numbers $m_1, m_2 = 0, \pm 1, \pm 2, \ldots$. For an extended discussion of this point, we refer the reader to Chapters 6 and 7.

As usually, the two path integrals in Eq. (12.32) can be combined to a single integral over the phase variable $\varphi(t)$ defined on the Keldysh contour K, which runs forward and backward between the times t_i and t_f along the real-time axis, i.e.,

$$\int \mathcal{D}\varphi_{\rm F} \int \mathcal{D}\varphi_{\rm B} \exp\left(\mathrm{i}S[\varphi_{\rm F}(t), \varphi_{\rm B}(t)]\right) \equiv \int \mathcal{D}\varphi \exp\left(\mathrm{i}S[\varphi]\right) \quad (12.33)$$

The effective action for our system consists of two terms:

$$S[\varphi] = S_{\rm ch}[\varphi] + S_{\rm el}[\varphi] \quad (12.34)$$

describing, respectively, the charging energy and electron tunneling between the island and two bulk metallic electrodes across the junctions L and R, cf., e.g., Eq. (2.113). The charging energy term has the standard form

$$S_{\rm ch}[\varphi(t)] = \int_{\rm K} dt \left[\frac{C_\Sigma}{2}\left(\frac{\dot{\varphi}}{e}\right)^2 + \tilde{N}(V_{\rm L}, V_{\rm R}, V_{\rm g})\dot{\varphi}\right] \quad (12.35)$$

where we defined

$$\tilde{N}(V_{\rm L}, V_{\rm R}, V_{\rm g}) = (C_{\rm L}V_{\rm L} + C_{\rm R}V_{\rm R} + C_{\rm g}V_{\rm g})/e \quad (12.36)$$

The tunneling term contains the contributions from both left and right junctions. With the aid of Eqs. (2.111–2.113), these contributions can be rewritten as

$$S_{\rm el}[\varphi] = \frac{\mathrm{i}}{2\pi} \sum_{r=L,R} \int_{\rm K} dt \int_K dt' \alpha_r^K(t,t') e^{\mathrm{i}\varphi(t)} e^{-\mathrm{i}\varphi(t')} \quad (12.37)$$

The kernels $\alpha_r^K(t,t') = \alpha_r^{\pm}(t-t')$ for $t < t'$ $(t > t')$ depend on the time ordering on the Keldysh contour. Their Fourier transforms are

$$\alpha_r^{\pm}(\omega) = \pm g_r \frac{\omega - \mu_r}{\exp[\pm(\omega-\mu_r)/T] - 1} \quad (12.38)$$

As we have already demonstrated above, in the regime of strong Coulomb blockade, the phase variable φ fluctuates strongly, while the conjugate charge variable or the number of excess electrons $N(t)$ underlies only weak fluctuations. For this reason, it may be useful to change from the phase to the charge representation. The time evolution of the density matrix in the charge representation depends on the propagator from N_1 forward to N_1' and from N_2' backward to N_2. It is given by the kernel of the evolution operator

$$U(N_1, N_2, N_1', N_2') = \int d\varphi_1 \int d\varphi_1' \int d\varphi_2' \int d\varphi_2\, e^{\mathrm{i}N_1\varphi_1} e^{-\mathrm{i}N_1'\varphi_1'} e^{\mathrm{i}N_2'\varphi_2'} e^{-\mathrm{i}N_2\varphi_2}$$
$$\times \int_{\varphi_2,\varphi_2'}^{\varphi_1,\varphi_1'} \mathcal{D}\varphi \int \mathcal{D}N \exp\left(-\mathrm{i}S_{\rm ch}[N(t)] + \mathrm{i}S_{\rm el}[\varphi(t)] + \mathrm{i}\int_K dt\, N(t)\dot{\varphi}(t)\right) \quad (12.39)$$

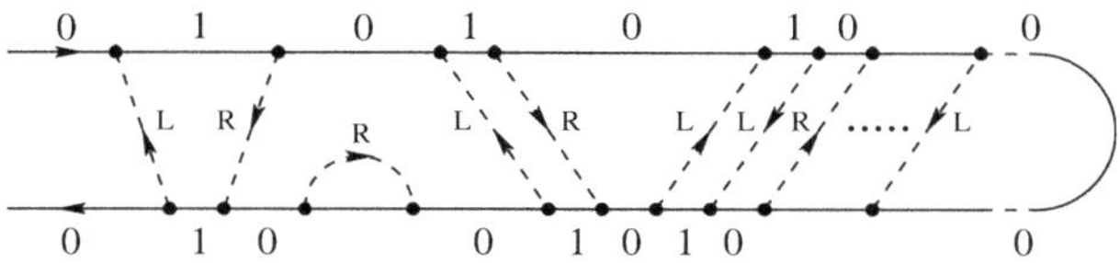

Figure 12.5 A diagram indicating (from left to right): (a) sequential tunneling in both left and right junctions, (b) the term preserving the norm, (c) cotunneling, and (d) resonant tunneling.

In the charge representation, one has

$$S_{\mathrm{ch}}[N(t)] = \int_{\mathrm{K}} dt\, E_{\mathrm{C}} \left[N(t) - \tilde{N}(V_{\mathrm{L}}, V_{\mathrm{R}}, V_{\mathrm{g}})\right]^2 \tag{12.40}$$

where the charging energy E_{C} is defined in Eq. (12.5).

12.2.3 Cotunneling and Resonant Tunneling

Regular expansion of Eq. (12.39) in powers of S_{el} with subsequent integration over φ generates infinite series of terms that can be visualized by diagrams. Several examples of such diagrams are displayed in Fig. 12.5. This diagram technique [237, 343] is similar in the spirit to that developed in Section 7.5.1 for the grand partition function. But, in contrast to the latter, here we are dealing with the real-time Keldysh technique, which allows to also address non-equilibrium problems.

Each diagram can be evaluated with the aid of the rules [237, 343] that can be directly established from our expansion. Each of the exponentials $\exp[\pm i\varphi(t)]$ describes tunneling of an electron at time t. These electron tunneling events occur in pairs in each of the two junctions L and R, and are connected by the tunneling lines $\alpha_r^K(t, t')$, as shown in Fig. 12.5. The propagator $U(N_1, N_2; N_1', N_2') \equiv U(N, N')$ between two diagonal charge states N and N' can be expressed by an irreducible self-energy part $\Sigma_{N,N'}$ defined as the sum of all diagrams in which any vertical line cutting through them crosses at least one tunneling line (cf. Section 7.5.1). Eventually, one arrives at the Dyson equation [145, 237]

$$U(N, N') = U^{(0)}(N)\delta_{N,N'} + \sum_{N_1} U(N, N_1)\, \Sigma_{N_1,N'}\, U^{(0)}(N). \tag{12.41}$$

The term $U^{(0)}(N)$ defines the propagator in the absence of tunneling. The stationary probability for the charge state N is defined as $W_N = \sum_{N'} W_{N'}^{(0)} U(N', N)$, where $W_N^{(0)}$ is the initial probability distribution. Our diagram rules then yield

$$0 = \sum_{N'} (-W_N \Sigma_{N,N'} + W_{N'} \Sigma_{N',N}) \tag{12.42}$$

This result is equivalent to a stationary master equation for the occupation probabilities W_N and the corresponding charge transition rates defined by the self-energies $\Sigma_{N',N}$.

In general, the irreducible self-energy Σ contains the rates of all possible charge tunneling processes. For instance, evaluating the diagrams containing no overlapping tunneling lines, we recover the single-electron tunneling rates (12.20) and, having in mind that in this case tunneling is only possible between the charge states N and $N \pm 1$, from Eq. (12.42) we immediately reconstruct the master equation (12.21) in the stationary limit $dW_N/dt = 0$.

As we already demonstrated, at low enough temperatures and for gate charge values not very close to the resonant points $N_g = \pm 1/2$, the system conductance evaluated within the approximation of sequential electron tunneling can be strongly suppressed by Coulomb interaction. In this case even for small $g_{L,R} \ll 1$, it can be necessary to include higher-order tunneling processes into our consideration. This task can be accomplished within our diagrammatic analysis. It is easy to verify that the most significant second-order contribution to the current originates from the terms $\sim g_L g_R$ describing the so-called cotunneling processes [31, 35]. In the limit of strong inelastic relaxation inside the dot, the process of inelastic cotunneling is described by the diagrams with two tunneling lines overlapping in time, as shown in Fig. 12.5.

Physically this process implies tunneling of one electron to the island through one junction accompanied by (almost) simultaneous tunneling of another electron out of the island through another junction. In contrast to sequential tunneling, this cotunneling process does not imply charging of the island by an extra charge e, i.e., the island charge $N(t) \approx \tilde{N}$ remains practically unchanged during the whole process. Hence, cotunneling does not "cost" a large extra energy $\sim E_C$ (cf. Eq. (12.40)) and its contribution to the current is not exponentially suppressed by Coulomb interaction even at $T \to 0$.

The corresponding diagrams for the self-energy are evaluated in a straightforward manner, and we recover the expressions for the inelastic cotunneling rates [31]

$$\overrightarrow{\gamma}_{\text{in}} = \frac{g_L g_R}{(2\pi)^3} \int dE_1 \dots dE_4\, f(E_1)\,[1 - f(E_2)]\, f(E_3)\,[1 - f(E_4)] \times \left(\frac{1}{E_2 - E_1 + \Delta \overleftarrow{E}_L(1)} + \frac{1}{E_4 - E_3 + \Delta \overleftarrow{E}_R(-1)} \right)^2 \times \delta(eV + E_1 - E_2 + E_3 - E_4), \quad (12.43)$$

$$\overleftarrow{\gamma}_{\text{in}} = \frac{g_L g_R}{(2\pi)^3} \int dE_1 \dots dE_4\, f(E_1)\,[1 - f(E_2)]\, f(E_3)\,[1 - f(E_4)] \times \left(\frac{1}{E_2 - E_1 + \Delta \overrightarrow{E}_R(1)} + \frac{1}{E_4 - E_3 + \Delta \overrightarrow{E}_L(-1)} \right)^2 \times \delta(-eV + E_1 - E_2 + E_3 - E_4) \quad (12.44)$$

and the contribution to the current in the form

$$I_{\text{cot}}^{\text{in}} = e\left(\overrightarrow{\gamma}_{\text{in}} - \overleftarrow{\gamma}_{\text{in}}\right) \quad (12.45)$$

Here, as before, $f(E)$ is the Fermi function, the energy differences $\Delta \overleftarrow{E}_{L,R}(N)$, $\Delta \overrightarrow{E}_{L,R}(N)$ are defined in Eqs. (12.18), (12.19) and the gate voltage is restricted to the interval $-e/2 < C_g V_g < e/2$.

At $T \to 0$, the integrals in Eqs. (12.43) and (12.44) can be evaluated exactly. Combining the corresponding results with Eq. (12.45), in this regime one finds

$$I_{\text{cot}}^{\text{in}} = e^2 \frac{g_L g_R}{(2\pi)^3} V \left[\left(1 + \frac{\Delta \overleftarrow{E}_L(1) \Delta \overleftarrow{E}_R(-1)}{eV E_C} \right) \left(\ln\left(1 + \frac{eV}{\Delta \overleftarrow{E}_L(1)}\right) + \ln\left(1 + \frac{eV}{\Delta \overleftarrow{E}_R(-1)}\right) \right) - 2 \right] \quad (12.46)$$

Finally, in the limit of low voltages and temperatures $eV, T \ll E_C$, we obtain

$$I_{\text{cot}}^{\text{in}} = \frac{e}{48\pi^3} \frac{g_L g_R}{E_C^2} \left(\frac{1}{1 + 2N_g} + \frac{1}{1 - 2N_g} \right)^2 (e^2 V^2 + 4\pi^2 T^2) V \quad (12.47)$$

As it was expected, at $T \to 0$, the cotunneling current does not vanish even for $N_g \to 0$. In this limit, Coulomb interaction yields the zero-bias anomaly $I_{cot}^{in} \propto v^3$.

For completeness, let us recall that provided both electron energy and phase relaxation processes inside the island can be neglected, there exists yet another contribution to the current proportional to the product $g_L g_R$. This contribution is due to the so-called elastic cotunneling process, which can be viewed as tunneling of a single electron between the electrodes through the whole metallic island. Obviously, at low temperatures, this elastic cotunneling contribution to the current I_{cot}^{el} is not suppressed by Coulomb effects either. In the limit $eV, T \to 0$, one finds

$$I_{cot}^{el} = G_{cot}^{el} V \qquad (12.48)$$

where the conductance G_{cot}^{el} reads

$$G_{cot}^{el} = \frac{e^2}{16\pi^3} \frac{g_L g_R \delta}{E_C} \left(\frac{1}{1 + 2N_g} + \frac{1}{1 - 2N_g} \right) \qquad (12.49)$$

where δ is the average energy level spacing in the island. Here we again assumed $-1/2 < N_g < 1/2$.

Collecting all the above contributions in the first and second orders in tunneling, we arrive at the result for the current

$$I(N_g) = I_{seq}(N_g) + I_{cot}^{in}(N_g) + I_{cot}^{el}(N_g) \qquad (12.50)$$

where I_{seq} is the sequential tunneling term (12.24) and both inelastic and elastic cotunneling terms are presented above. This current is periodic in N_g, i.e., outside the interval $-1/2 < N_g < 1/2$, all three terms in the right-hand side of Eq. (12.50) should be continued periodically with the period equal to 1.

We also note that both expressions (12.47) and (12.49) describing, respectively, inelastic and elastic cotunneling contributions formally diverge at resonance $N_g \to \pm 1/2$. As usually, this divergence signals the failure of the second-order perturbation theory in the presence of a degeneracy in which case also higher-order tunneling terms need to be taken into account.

As we already pointed out, elastic cotunneling is sensitive to both dephasing and inelastic relaxation processes inside the dot and, in addition, $G_{cot}^{el} \propto \delta$. Since here we are merely interested in the case of large metallic dots with $\delta \to 0$ and strong inelastic relaxation, in Eq. (12.50) we can safely neglect the term $I_{\text{cot}}^{\text{el}}(N_{\text{g}})$. What remains is to regularize the inelastic cotunneling contribution (12.47) in the vicinity of the resonance points $N_{\text{g}} = \pm 1/2$. This task is accomplished [145, 237] by taking into account higher-order diagrams indicated in Fig. 12.5. The contributions from these diagrams to the self-energy can effectively be summed up within the standard approximation of main logarithms (cf., e.g., Section 7.5.2). As a result, the maximum value of the linear conductance $G_0(N_{\text{g}} \simeq \pm 1/2)$ as well as the width of the conductance peak get effectively renormalized by the factors $L(T)$ and $L^{-1}(T)$, respectively, where

$$L(T) = \frac{1}{1 + (g/2\pi^2)\ln(E_{\text{C}}/2\pi T)}, \quad g = g_{\text{L}} + g_{\text{R}} \tag{12.51}$$

For this reason, as compared to the sequential tunneling result (12.29), in the temperature limit the conductance peak does not reach the value $1/2(R_{\text{L}} + R_{\text{R}})$ and decreases with decreasing T. For $N_{\text{g}} = \pm 1/2$, we obtain [145, 237]

$$(R_{\text{L}} + R_{\text{R}})G_0(T) \approx L(T)\left[\frac{1}{2} - \frac{1}{\pi}\arctan\left(\frac{g^2 L^2(T) - 16\pi^2}{8\pi g L(T)}\right)\right] \tag{12.52}$$

At the same time, the peak width is increased as compared to the lowest-order perturbation theory result (12.29).

Both suppression of the conductance peaks and their broadening were indeed observed in experiments [90, 211]. In Fig. 12.6, we display the temperature-dependent minimum ($N_{\text{g}} = 0$) and maximum ($N_{\text{g}} = \pm 1/2$) conductances for our SET transistor evaluated within the above theory for two different values of the total conductance equal to $g \approx 0.15$ and $g \approx 0.62$. On the same plot, we also present experimental results [211] obtained for SET transistors with these values of g. In both cases, a good agreement between theory and experiment is observed.

It also follows from the above analysis of higher-order diagrams that a simple perturbative treatment of the linear conductance holds provided the factor $L(T)$ (12.51) remains close to unity or, equivalently, provided the condition

$$\frac{g}{2\pi^2}\ln\left(\frac{E_{\text{C}}}{2\pi T}\right) \ll 1 \tag{12.53}$$

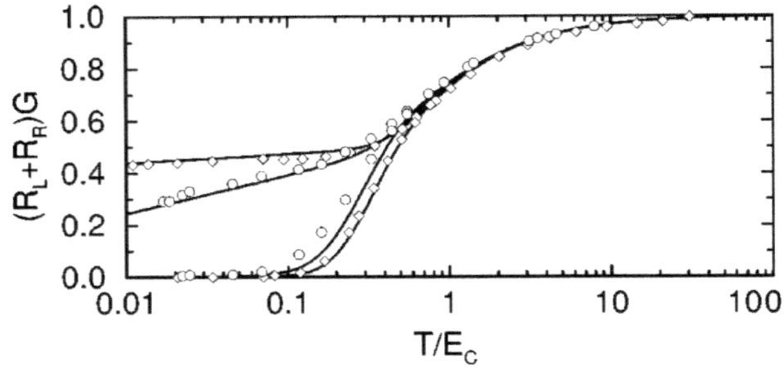

Figure 12.6 The sequential plus cotunneling contribution to the maximal and minimal linear conductance for $E_C = 1.47$ K, $g \approx 0.15$, and for $E_C = 1$ K, $g \approx 0.62$. The data points are taken from the experiment [211] for the two samples with the above parameters. Reprinted with permission from Ref. [431], Copyright 1997, by the American Physical Society.

is fulfilled. Accordingly, for sufficiently small conductance values $g \ll 2\pi^2$ (i.e., for $g \lesssim 1 \div 2$), non-perturbative renormalization effects are significant only at exponentially low temperatures, which are irrelevant for the vast majority of experiments. On the other hand, for higher conductance values (e.g., for $g \sim 10$ or bigger), the approximation of main logarithms becomes clearly insufficient and all classes of diagrams need to be taken into account. For this reason, the diagrammatic analysis is inconvenient for such values of g and one should proceed with other non-perturbative means. A similar situation was already encountered in Chapter 7 for systems containing single-tunnel junctions where a variety of non-perturbative methods were employed in order to account for Coulomb blockade in the strong tunneling regime. In Chapter 11, these techniques were generalized to the case of arbitrary short coherent conductors. Below we will adopt the same methods in order to describe Coulomb effects in metallic quantum dots with large conductances $g \gg 1$.

12.3 Coulomb Blockade beyond Perturbation Theory: Inelastic Limit

Let us now relax the condition (12.9) and assume that at least one of the resistances R_L or R_R is lower than the quantum resistance unit

R_q, i.e.,

$$\frac{R_L R_R}{R_L + R_R} \ll R_q \tag{12.54}$$

or, equivalently, $g \gg 1$. In this limit, we are not anymore restricted to the case of tunnel junctions and will further assume that both conductors R_L and R_R are characterized by arbitrary transmission distributions of their conducting channels. As before, we introduce the Fano factors for these conductors β_L and β_R. In addition, we will again stick to the limit of strong inelastic electron relaxation in the metallic island $\tau_{in} \ll \tau_D$ in which case the two conductors are practically independent of each other.

12.3.1 Interaction Correction

Let us first ignore the gate effect, i.e., we will assume that the gate modulation of the current flowing through the quantum dot is absent. The condition (12.54) implies that fluctuations of the island phase φ remain weak and—similar to Chapters 7 and 11—we can again employ the quasiclassical Langevin equation analysis, which allows to easily recover the Coulomb interaction correction to the system conductance. Applying the results derived in Chapter 11 to each of the two conductors, we obtain their $I-V$ curves in the form

$$I_L(V_L) = \frac{V_L}{R_L} + \delta I_L(V_L), \quad I_R(V_R) = \frac{V_R}{R_R} + \delta I_R(V_R) \tag{12.55}$$

where $\delta I_L(V_L)$ and $\delta I_R(V_R)$ are the corresponding interaction corrections directly related to the shot noise in our system. It is easy to observe that in the main approximation, the effect of the right conductor on the interaction correction $\delta I_L(V_L)$ is equivalent to that of an Ohmic resistance R_R and vice versa. Hence, we can make use of Eq. (11.18) derived for a short coherent in the presence of an Ohmic shunt and immediately recover the interaction correction $\delta I_L(V_L)$ in the form

$$\delta I_L = -\frac{e\beta_L}{\pi}\frac{R_R}{R_L + R_R}\,\mathrm{Im}\left[\left(\frac{1}{\tau_0} + \mathrm{i}eV_L\right)\Psi\left(1 + \frac{1}{2\pi T\tau_0} + \mathrm{i}\frac{eV_L}{2\pi T}\right)\right.$$
$$\left. - \mathrm{i}eV_L\Psi\left(1 + \mathrm{i}\frac{eV_L}{2\pi T}\right)\right], \tag{12.56}$$

where

$$\tau_0 = \frac{R_{\rm L} R_{\rm R} C_\Sigma}{R_{\rm L} + R_{\rm R}} \tag{12.57}$$

is the charge relaxation time in our system. The interaction correction $\delta I_{\rm R}(V_{\rm R})$ is defined by the same expression with interchanged indices $L \leftrightarrow R$.

What remains is to evaluate the total current I through our quantum dot as a function of the total bias voltage V applied to the system. This current is derived from the following set of equations:

$$I = I_{\rm L}(V_{\rm L}), \quad I = I_{\rm R}(V_{\rm R}), \quad V = V_{\rm L} + V_{\rm R} \tag{12.58}$$

Resolving these equations and keeping only the lowest-order terms in $\delta I_{\rm L}$ and $\delta I_{\rm R}$, we obtain

$$I = \frac{V}{R_{\rm L} + R_{\rm R}} + \frac{R_{\rm L}\delta I_{\rm L}(V_{\rm L}) + R_{\rm R}\delta I_{\rm R}(V_{\rm R})}{R_{\rm L} + R_{\rm R}}, \quad V_{\rm L,R} = \frac{R_{\rm L,R} V}{R_{\rm L} + R_{\rm R}} \tag{12.59}$$

where $\delta I_{\rm L,R}$ is defined in Eq. (12.56). As before, in the limit $g \gg 1$, the results of our quasiclassical analysis (12.59), (12.56) apply down to exponentially small temperatures and/or voltages

$$\max(T, eV) \gg E^* \sim g E_{\rm C} \exp(-g/2),$$

cf. Eq. (11.24).

The differential conductance dI/dV for our quantum dot determined from Eq. (12.59) is depicted in Fig. 12.7 (top panel) together with the experimental curves measured at different temperatures in one of the SET transistors [90] with low resistive tunnel barriers ($\beta_{\rm L} = \beta_{\rm R} = 1$). A good agreement between theory and experiment is observed provided $\max(T, eV) \gg E^*$. A similarly good agreement with theoretical predictions was also found for other samples [90]. The differential conductance dI/dV is shown in the bottom panel of Fig. 12.7 for all five samples at the lowest temperature $T \approx 20$ mK. It is remarkable that even for the sample V with $g \sim 33$, Coulomb suppression of the differential conductance at small bias voltages is quite well pronounced.

The zero-bias conductance G_0 of the quantum dot derived from Eqs. (12.59), (12.56) in the limit $V \to 0$ reads

$$(R_{\rm L} + R_{\rm R})G_0(T) = 1 - f(T) \tag{12.60}$$

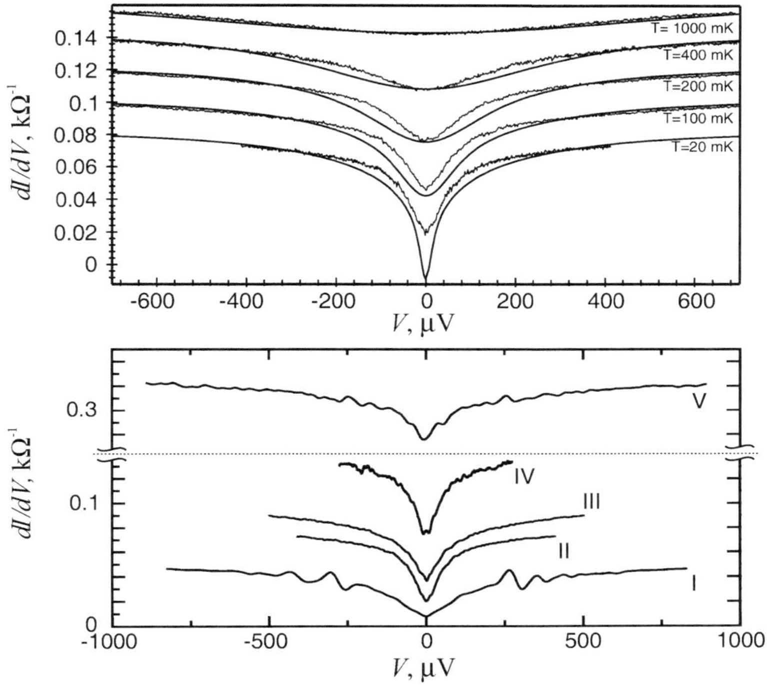

Figure 12.7 Top panel: The differential conductance dI/dV derived from Eqs. (12.59), (12.56) for a symmetric metallic quantum dot with low resistive tunnel barriers (thicker curves) together with the experimental data measured in one of the samples (sample II) (thinner curves). For the sake of clarity, the curves at higher T are shifted vertically with a step 0.02 $k\Omega^{-1}$. No fit parameters are employed. Bottom panel: Experimental curves for all five samples [90] measured at the lowest temperature $T \simeq 20$ mK. The sample parameters are: $E_C \simeq 2.25$ K, $g \simeq 5.9$, $10E^* \sim 0.6$ K (sample I), $E_C \simeq 1.10$ K, $g \simeq 8.6$, $10E^* \sim 0.1$ K (sample II), $E_C \simeq 1.04$ K, $g \simeq 9.9$, $10E^* \sim 0.06$ K (sample III), $E_C \simeq 1.16$ K, $g \simeq 15.9$, $10E^* \sim 6 \times 10^{-3}$ K (sample IV), $E_C \sim 0.3$ K, $g \simeq 33.3$, $10E^* \sim 5 \times 10^{-6}$ K (sample V).

where [159]

$$f(T) = \frac{2(g_R\beta_L + g_L\beta_R)}{(g_L + g_R)^2}\left[\gamma + \Psi\left(1 + \frac{gE_C}{2\pi^2 T}\right) + \frac{gE_C}{2\pi^2 T}\Psi'\left(1 + \frac{gE_C}{2\pi^2 T}\right)\right] \tag{12.61}$$

Comparing this result, e.g., with Eq. (11.26), we observe that in the case of metallic quantum dots, the magnitude of the interaction correction depends on both Fano factors $\beta_{\rm L}$ and $\beta_{\rm R}$, i.e., it is determined by the shot noise in both left and right conductors. Equations (12.60) and (12.61) apply as long as the interaction correction to the system conductance remains much smaller than the non-interacting contribution, i.e., for

$$\frac{2\ln(1/T\tau_0)}{g_{\rm L}+g_{\rm R}} \ll 1 \tag{12.62}$$

cf. Eq. (11.27).

The temperature-dependent zero-bias conductance (12.60), (12.61) is displayed in Fig. 12.8 together with the experimental data [90]. The Coulomb blockade induced suppression of the conductance at low T is clearly seen even for a highly conducting sample IV with $g \sim 16$. We again observe a good agreement between theory and experiment for all samples except at the lowest temperatures where the condition (12.62) is violated.

In the high-temperature limit $2\pi T\tau_0 \gg 1$, Eq. (12.59) yields a simple expression for the differential conductance

$$\frac{dI}{dV} = \frac{1}{R_{\rm L}+R_{\rm R}}\left[1 - \frac{R_{\rm L}\beta_{\rm L}}{R_{\rm L}+R_{\rm R}}\frac{E_{\rm C}}{T}\frac{\frac{eV_{\rm L}}{2T}\cosh\frac{eV_{\rm L}}{2T} - \sinh\frac{eV_{\rm L}}{2T}}{\sinh^3\frac{eV_{\rm L}}{2T}}\right.$$
$$\left. - \frac{R_{\rm R}\beta_{\rm R}}{R_{\rm L}+R_{\rm R}}\frac{E_{\rm C}}{T}\frac{\frac{eV_{\rm R}}{2T}\cosh\frac{eV_{\rm R}}{2T} - \sinh\frac{eV_{\rm R}}{2T}}{\sinh^3\frac{eV_{\rm R}}{2T}}\right] \tag{12.63}$$

In the symmetric case $C_{\rm L} = C_{\rm R}$, $R_{\rm L} = R_{\rm R}$, and $\beta_{\rm L} = \beta_{\rm R} = 1$, this expression coincides with Eq. (12.31) derived for $g_{\rm L}, g_{\rm R} \gg 1$ within the sequential tunneling approximation. Thus, in the high-temperature limit $T/E_{\rm C} \gg \max(1, g)$, the leading order perturbative results of Section 12.2.1 remain quantitatively correct regardless of the dimensionless conductance values $g_{\rm L}$ and $g_{\rm R}$.

12.3.2 Gate Effect

The above results derived from the Langevin equation analysis do not account for charge discreteness in our system and, hence, totally ignore the dependence of the transport current I on the gate charge

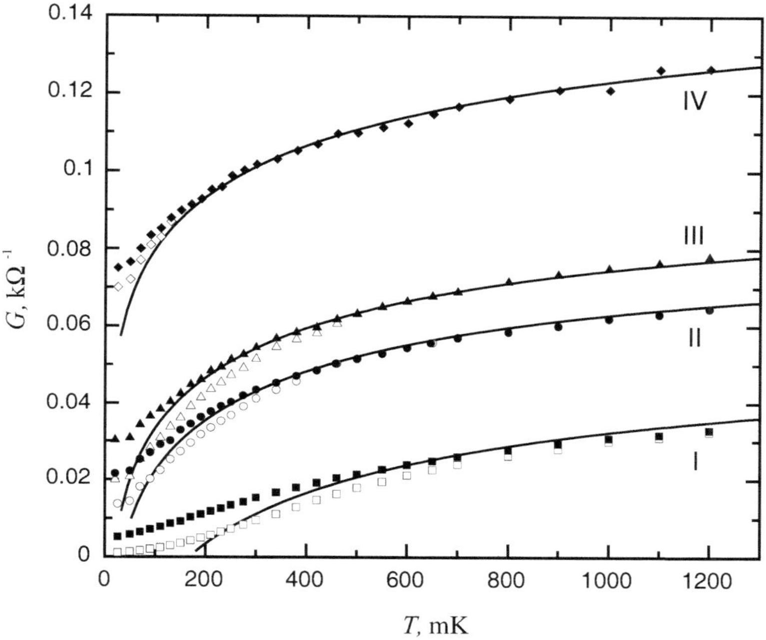

Figure 12.8 The zero-bias conductance $G_0(T)$ defined by Eqs. (12.60) and (12.61) is shown by solid curves together with the experimental data [90] for the samples I–IV (filled symbols). No fit parameters are employed. The sample parameters are the same as in Fig. 12.7. Also the data for the minimum conductance reached at $N_g = 0$ (gate effect) are indicated by open symbols.

N_g. Fortunately, this dependence can be restored by means of the following arguments [152].

Let us again recall that provided the charge of the central island is quantized in units of e, the reduced density matrix of our system $\rho(\varphi, \varphi')$ (12.32) depends on the compact phase variable φ defined on the circle with physically identical states φ and $\varphi + 2\pi$. This property implies

$$\rho(\varphi_1+2\pi m_1, \varphi_2+2\pi m_2) = \exp[2\pi i N_g(m_1-m_2)]\rho(\varphi_1, \varphi_2) \quad (12.64)$$

together with the normalization

$$\int_{-\pi}^{\pi} d\varphi \rho(\varphi, \varphi) = 1 \quad (12.65)$$

On the other hand, since the above Langevin equation analysis effectively treats the charge as a continuous variable, the corresponding reduced density matrix $\rho_c(\varphi, \varphi')$ is non-periodic in φ and obeys the standard normalization condition

$$\int_{-\infty}^{\infty} d\varphi \rho_c(\varphi, \varphi) = 1 \tag{12.66}$$

Let us now introduce a non-periodic in φ density matrix $\tilde{\rho}(\varphi_1, \varphi_2)$ by means of the relation

$$\rho(\varphi_1, \varphi_2) = \sum_{m_1, m_2} \exp[2\pi \mathrm{i} N_{\mathrm{g}}(m_1 - m_2)]\tilde{\rho}(\varphi_1 - 2\pi m_1, \varphi_2 - 2\pi m_2) \tag{12.67}$$

This density matrix is normalized by the condition

$$\sum_m \int_{-\infty}^{\infty} d\varphi \exp(2\pi \mathrm{i} N_{\mathrm{g}} m)\tilde{\rho}(\varphi - 2\pi m, \varphi) = 1 \tag{12.68}$$

The matrix $\tilde{\rho}$ (12.67) obeys the same equation of motion as the density matrix ρ_{c}. Hence, the connection between these two density matrices can be easily established bearing in mind that at sufficiently long times, the solution to a linear equation of motion for a dissipative ergodic system takes a unique form independent of the initial conditions. In other words, in the long time limit, the two solutions of this equation may differ only by a constant, which can be easily determined from the normalization condition. As a result, at sufficiently long times, we obtain

$$\rho_{\mathrm{c}}(t, \varphi_1, \varphi_2) = A\tilde{\rho}(t, \varphi_1, \varphi_2) \tag{12.69}$$

where the normalization constant A is

$$A = \sum_m \int_{-\infty}^{+\infty} d\varphi e^{2\pi \mathrm{i} N_{\mathrm{g}} m} \rho_{\mathrm{c}}(\varphi - 2\pi m, \varphi) \tag{12.70}$$

Combining Eqs. (12.67), (12.69), and (12.70), we arrive at the desired relation between ρ and ρ_{c}:

$$\rho(\varphi, \varphi) = A^{-1} \sum_{m_1, m_2} e^{2\pi \mathrm{i} N_{\mathrm{g}}(m_1 - m_2)} \rho_{\mathrm{c}}(\varphi + 2\pi m_1, \varphi + 2\pi m_2) \tag{12.71}$$

Equation (12.71) helps to establish the connection between the expectation values for the operator of any physical quantity evaluated with discrete and continuous charge distributions. The expectation value of an arbitrary operator $\hat{O}(\hat{\varphi})$ for the system with discrete charges is

$$\langle \hat{O} \rangle = \int_{-\pi}^{\pi} d\varphi O(\varphi)\rho(\varphi, \varphi) = \sum_m \int_{-\infty}^{+\infty} d\varphi O(\varphi) e^{2\pi i N_g m} \rho_c(\varphi - 2\pi m, \varphi)$$

$$= \frac{\sum_m \left\langle \hat{O}(\hat{\varphi}) e^{2\pi i (N_g - \hat{N})m} \right\rangle_c}{\sum_m \left\langle e^{2\pi i (N_g - \hat{N})m} \right\rangle_c}, \tag{12.72}$$

where by $\langle (\ldots) \rangle_c = \int d\varphi (\ldots)\rho_c(\varphi, \varphi)$, we denoted the average performed for the same system with continuous charges and also used an obvious identity

$$\int d\varphi (\ldots)\rho_c(\varphi - 2\pi m, \varphi) = \langle (\ldots) e^{-2\pi i \hat{N} m} \rangle_c$$

Now we are ready to evaluate the current through our metallic quantum dot in the presence of the gate effect. In order to do so, let us first rewrite the general expression for the current (12.59) in an equivalent form

$$I = \frac{V - R_L \langle \xi_L \rangle_c - R_R \langle \xi_R \rangle_c}{R_L + R_R} \tag{12.73}$$

where the stochastic variables $\xi_{L,R}$ define the current shot noise in the left and right conductors described by the correlators

$$\langle \xi_{L,R}(t_1)\xi_{L,R}(t_2) \rangle_c = \left(\frac{1 - \beta_{L,R}}{R_{L,R}} + \frac{\beta_{L,R}}{R_{L,R}} \cos\left[\varphi(t_1) - \varphi(t_2)\right] \right) \times \left(-\frac{\pi T^2}{\sinh^2[\pi T(t_1 - t_2)]} \right) \tag{12.74}$$

In Eqs. (12.73) and (12.74), we explicitly indicated that the averages $\langle \ldots \rangle_c$ are performed with the continuous charge distribution, i.e., in Eq. (12.73), the gate effect is still missing.

This effect is restored if we evaluate the averages in Eq. (12.73) keeping the island charges discrete, i.e., we replace $\langle \ldots \rangle_c \to \langle \ldots \rangle$ and employ the relation (12.72). Treating the island charge fluctuations as Gaussian, one can readily estimate the contribution of the m-th winding number to the expectation value (12.72) as

$$\left\langle \ldots e^{2\pi i \hat{N} m} \right\rangle_c \sim \exp\left(-2\pi^2 \left\langle \delta \hat{N}^2 \right\rangle_c m^2\right) \tag{12.75}$$

Since in the low resistance limit (12.54), the phase fluctuations are weak and, on the contrary, the charge fluctuations are considerable $\langle \delta \hat{N}^2 \rangle_c \gtrsim 1$, in Eq. (12.72) it is sufficient to retain only terms with winding numbers $m = 0, \pm 1$. Within this approximation, we get

$$\langle \xi_{\mathrm{L,R}} \rangle = \frac{\langle \xi_{\mathrm{L,R}} \rangle_c + 2 \langle \xi_{\mathrm{L,R}} \cos(2\pi(\hat{N} - N_{\mathrm{g}})) \rangle_c}{1 + 2 \langle \cos(2\pi(\hat{N} - N_{\mathrm{g}})) \rangle_c} \tag{12.76}$$

Substituting $\langle \ldots \rangle_C \to \langle \ldots \rangle$ in Eq. (12.73), employing Eqs. (12.76) and (12.74) with $\varphi(t) \to eVt$, and proceeding perturbatively in the noise terms, we arrive at the I–V curve for our quantum dot, which now fully accounts for the gate effect [145].

A close inspection of these results demonstrates that at sufficiently high bias voltages V, the gate effect is strongly suppressed and, hence, becomes unimportant. On the other hand, at low voltages, the effect is more pronounced. In this limit for the linear conductance, we obtain [152]

$$(R_{\mathrm{L}} + R_{\mathrm{R}}) G_0(T, N_{\mathrm{g}}) = 1 - f(T) - B(T) e^{-F(T)} \cos(2\pi N_g) \tag{12.77}$$

Here $f(t)$ is again defined in Eq. (12.61), and the last term in the right-hand side of Eq. (12.77) describes the dependence of the conductance on the gate charge. This result demonstrates that at not very low $T \gg E^*$, the gate modulation is universally described by $\cos(2\pi N_{\mathrm{g}})$. The function $F(T) = 2\pi^2 \langle \delta N^2 \rangle$ determines the temperature dependence of the island charge fluctuations (within the Gaussian approximation). It reads

$$F(T) = F(0) + \frac{\pi^2 T}{E_{\mathrm{C}}} + g \left[\ln\left(\frac{g E_{\mathrm{C}}}{2\pi^2 T} \right) - \Psi\left(1 + \frac{g E_{\mathrm{C}}}{2\pi^2 T} \right) \right] \tag{12.78}$$

Note that the value $F(0)$ evaluated within our quasiclassical Langevin approach is defined by the time integral, which diverges logarithmically at high frequencies. This formal divergence just indicates the failure of this approach in the limit $T \to 0$, as we already discussed in Chapter 11. The problem can be cured by observing that in thermodynamic equilibrium ($V \to 0$) within the

Gaussian approximation, we have

$$e^{-F(0)} = \langle \cos(2\pi \hat{N}) \rangle = \frac{\int d\varphi_0 \rho_c^{\text{eq}}(2\pi + \varphi_0, \varphi_0)}{\int d\varphi_0 \rho_c^{\text{eq}}(\varphi_0, \varphi_0)}$$
$$= \int d\varphi_0 \rho_c^{\text{eq}}(2\pi + \varphi_0, \varphi_0) \quad (12.79)$$

This expectation value involving the equilibrium density matrix can be estimated employing the results of Chapter 11. Let us rewrite the expression (12.79) in terms of the imaginary time path integral

$$\int d\varphi_0 \rho_c^{\text{eq}}(2\pi + \varphi_0, \varphi_0) =$$
$$\int d\varphi_0 \int_{\varphi_0+2\pi}^{\varphi_0} \mathcal{D}\varphi \exp\left[-\int d\tau \frac{C_\Sigma \dot{\varphi}^2}{2e^2} - S_{\text{sc}}[\varphi]\right] \quad (12.80)$$

As usually, the term involving the capacitance C_Σ accounts for the charging energy of the dot, while the action $S_{\text{sc}}[\varphi]$ describes electron transfer through both left and right conductors and is defined by Eq. (11.87) where the summation over transport channels includes both these conductors. In the limit of large $g \gg 1$, the path integral (12.80) can be estimated within the saddle point approximation in the same spirit as in Chapter 11. Evaluating the total action on a straight line path $\varphi(\tau) = \varphi_0 + 2\pi(1 - T\tau)$, within the exponential accuracy, we obtain (cf. Eq. (11.91))

$$\int d\varphi_0 \rho_c^{\text{eq}}(2\pi + \varphi_0, \varphi_0) \sim \left[\prod_n (1 - T_n)\right]^{1/2} = \exp(-ag) \quad (12.81)$$

where the parameter a is defined by Eq. (11.82) with the sum taken over all conducting channels in both left and right conductors (as well as the product in Eq. (12.81)). Combining Eqs. (12.79) and (12.81), we arrive at the estimate

$$F(0) \simeq ag \quad (12.82)$$

which disregards unimportant subleading logarithmic terms. Such terms can be recovered only if one accurately evaluates the pre-exponential factor in Eq. (12.81).

The above results demonstrate that the gate effect vanishes identically provided at least one of the transmission channels is fully open. In this case, a tends to infinity and, hence, $\exp(-F) \equiv 0$.

On the other hand, for weakly transmitting tunnel barriers (with $g \gg 1$), one has $a = 1/2$ and one can expect to observe considerable gate modulation of the linear conductance (12.77) at sufficiently low temperatures. In this case at $T \ll gE_C$, from Eqs. (12.78) and (12.82), we find

$$F(T) \simeq g/2 + (T/T_0)^2, \quad T_0 = \sqrt{3g}E_C/\pi^2 \tag{12.83}$$

What remains is to establish the function $B(T)$ in Eq. (12.77). For tunnel barriers with $g \gg 1$, it was explicitly evaluated [145, 152]. In the interesting low-temperature and -voltage limit, this analysis yields $B \approx 11.29 + 4.88/g$.

Now we are ready to compare the predicted gate effect in our highly conducting SET transistors with the experimental results [90]. The dependence of the linear conductance G_0 on the gate charge is displayed in Fig. 12.8, which shows both the minimum conductance $G_{\min} \equiv G_0(N_g = 0)$ (open symbols) and the gate-averaged conductance (filled symbols) as a function of temperature. Note that due to the simple cosine modulation of the conductance, the last term in Eq. (12.77) averages out and the gate-averaged conductance exactly coincides with the result in Eqs. (12.60) and (12.61). For the same reason, the difference between $G_{\max} \equiv G_0$ $(N_g = 1/2)$ and $G_{\min}$ is twice as large as the difference between the gate-averaged conductance and $G_{\min}$.

In the experiments [90], the gate modulation of the current was indeed found to be of a simple cosine form (12.77) for all samples except for a relatively highly resistive sample I at low temperatures $T \lesssim E^*$. The results for the gate-modulated linear conductance of sample II are shown in Fig. 12.9 (inset). The amplitude of modulation increases with decreasing temperature in a qualitative agreement with theory. At low T, the modulation effect is considerable even for sample IV with $g \simeq 16$. At the lowest temperatures, this effect is visible also for sample V. However, in this case the amplitude of modulation was only slightly above the average noise level.

The data for the temperature-dependent amplitude of conductance modulation for samples I–IV are also presented in Fig. 12.9. Solid curves correspond to the best fit with a theoretical

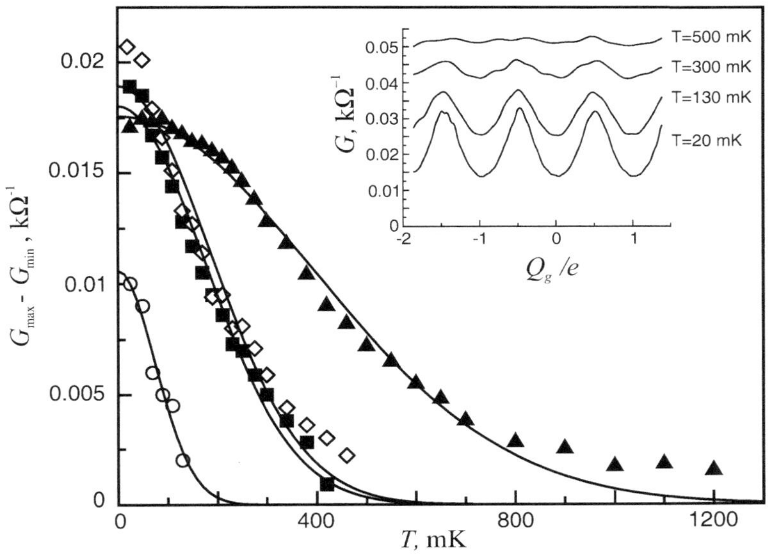

Figure 12.9 Experimental data [90] for the gate modulation amplitude $G_{\max} - G_{\min}$ of a linear conductance as a function of T for the samples I–IV (triangles, squares, diamonds, and circles, respectively). The sample parameters are the same as in Fig. 12.7. Solid lines provide the best fit to the formula $K \exp(-(T/T_0)^2)$. The inset: The linear conductance of sample II as a function of $Q_g = C_g V_g$ measured at different T.

dependence $G_{\max} - G_{\min} = K \exp(-T^2/T_0^2)$, where we denoted $K = 2B \exp(-F(0))$. Also in this case, experimental data points follow closely the above theoretical predictions showing considerable gate effect at $T \lesssim T_0$ and its rapid decay at higher temperatures.

Overall we observe a very good agreement between theory and experiment, demonstrating that at sufficiently low temperatures, SET transistors can operate well even if the resistances of tunnel barriers $R_{L,R}$ are $\sim$ 30 times smaller than the quantum resistance unit R_q. On the other hand, the gate effect gets strongly decreased even at $T \to 0$ if conductors with the same resistance values $R_{L,R}$ but higher channel transmissions T_n are employed instead of the tunnel barriers.

12.4 Coulomb Blockade in the Presence of Quantum Coherence

Up to now we merely addressed the inelastic limit characterized by strong electron energy and phase relaxation inside a large metallic island forming our quantum dot. Now let us analyze the opposite limit, i.e., assume that both the electron inelastic relaxation time τ_{in} and its dephasing time τ_φ inside the quantum dot strongly exceed the electron dwell time τ_{D} in this dot:

$$\tau_{\text{in}} \gg \tau_{\text{D}}, \qquad \tau_\varphi \gg \tau_{\text{D}} \tag{12.84}$$

In this case, the electron motion inside the dot remains coherent and, hence, in the absence of Coulomb effects, transport properties of our system are fully described by its energy-dependent scattering matrix $\hat{S}(E)$.

12.4.1 Scattering Matrix and Langevin Equations

In order to describe the effect of electron–electron interactions in this regime, we again employ the effective action formalism in the same spirit as it was already done, e.g., in Chapters 10 and 11 for short coherent conductors. The whole derivation of the effective action is analogous to that of Eq. (10.63), except one should now explicitly account for the energy dependence of the scattering matrix. For further technical details, we refer the reader to the work [163]. Here we only present the final result of this derivation, which turns out to be very similar to the action for a short coherent conductor (10.63). It reads

$$\begin{aligned} \mathrm{i}S_{\text{qd}} = \operatorname{tr}\ln\Bigg[\hat{1} + \int_0^t d\tau_2 \int_0^t d\tau_3\, \theta(t-\tau_1)\theta(\tau_1)\theta(\tau_3) \\ \times \Bigg(e^{-\mathrm{i}\hat{\varphi}_{\text{B}}(\tau_1)}\hat{\mathbf{S}}^\dagger(\tau_1-\tau_2)e^{\mathrm{i}\hat{\varphi}_{\text{B}}(\tau_2)}e^{-\mathrm{i}\hat{\varphi}_{\text{F}}(\tau_2)}\hat{\mathbf{S}}(\tau_2-\tau_3)e^{\mathrm{i}\hat{\varphi}_{\text{F}}(\tau_3)} \\ -\delta(\tau_1-\tau_2)\,\hat{1}\,\delta(\tau_2-\tau_3)\Bigg)\rho_0(-\tau_3,-\tau_4)\Bigg] \end{aligned} \tag{12.85}$$

where we defined the Fourier-transformed scattering matrix as well as its Hermitian conjugate

$$\hat{\mathbf{S}}(\tau) = \int \frac{dE}{2\pi}\,\hat{S}(E)\,e^{-\mathrm{i}E\tau}, \quad \hat{\mathbf{S}}^\dagger(\tau) = \int \frac{dE}{2\pi}\,\hat{S}^\dagger(E)\,e^{-\mathrm{i}E\tau} \tag{12.86}$$

The matrices involving the phase factors in Eq. (12.85) remain the same as in the case of a short conductor,

$$e^{\mathrm{i}\hat{\varphi}_{\mathrm{F,B}}} = \begin{pmatrix} e^{\mathrm{i}\varphi_L^{\mathrm{F,B}}}\hat{1}_L & 0 \\ 0 & e^{\mathrm{i}\varphi_R^{\mathrm{F,B}}}\hat{1}_R \end{pmatrix} \tag{12.87}$$

where the phases are now related to the fluctuating-in-time electric potentials of the leads, $\dot{\varphi}_j^{\mathrm{F,B}}(\tau) = eV_j^{\mathrm{F,B}}(\tau)$. Note that Eq. (12.85) defines a symbolic expression, which should be interpreted in the same way as Eq. (10.63), i.e., as a formal expansion of the logarithm in powers of the initial density matrix ρ_0.

Let us recollect that one of the crucial points in our derivation of the action for a short coherent scatterer is Eq. (10.38), which establishes the relation between the time at which the classical electron trajectory hits this scatterer and the initial (or final) times/coordinates of the quantum evolution operator. Accordingly, this evolution operator acquires an additional phase factor $\exp[\mathrm{i}(\varphi_L^{\mathrm{F,B}} - \varphi_R^{\mathrm{F,B}})]$ picked up by an electron while crossing the barrier. Inspecting classical electron paths in the quantum dot, one easily concludes that these paths may cross the barriers connecting the dot and the leads only twice, i.e., at the times an electron enters and leaves the dot. The corresponding crossing times are again related to the initial and final coordinates similar to Eq. (10.38), thereby explaining the appearance of the phase factors in Eq. (12.85).

The same arguments demonstrate that the action of the form (12.85) cannot be directly applied to more complicated systems, such as quantum dot arrays, despite the particular form of the $S-$matrix in this equation is not yet specified. Indeed, such systems contain three or more barriers at which the fluctuating electric potential drops. As electron trajectories may cross the inner barrier(s) an arbitrary number of times, numerous phase factors emerge in the quantum evolution operator and, hence, in this case, the action cannot be reduced to the simple form (12.85).

In order to proceed with our analysis, we again employ the quasiclassical approximation for the fluctuating phases and expand the action (12.85) to the second order in φ_L^- and φ_R^-. Then we again recover the standard structure of the effective action

$$\mathrm{i}S_{\mathrm{qd}} = \mathrm{i}S_{\mathrm{R}} - S_{\mathrm{I}} \tag{12.88}$$

where the real and imaginary parts of the action are, respectively, linear and quadratic in the quantum phases, $S_{\rm R} \propto \varphi^-$ and $S_{\rm I} \propto (\varphi^-)^2$. A straightforward algebra yields [163]

$$\begin{aligned} iS_{\rm R} = i \int_0^t d\tau_1 d\tau_2 d\tau_3 \lim_{\tau_4 \to \tau_1} \operatorname{tr}\big\{ \big[\delta(\tau_1 - \tau_2)\, \hat{\varphi}^-(\tau_2)\, \delta(\tau_2 - \tau_3) \\ - \hat{\boldsymbol{S}}^\dagger(\tau_1 - \tau_2)\hat{\varphi}^-(\tau_2)\hat{\boldsymbol{S}}(\tau_2 - \tau_3)\big] e^{i\hat{\varphi}^+(\tau_3)} \tilde{f}(\tau_3 - \tau_4) e^{-i\hat{\varphi}^+(\tau_4)} \big\} \end{aligned} \quad (12.89)$$

$$\begin{aligned} S_{\rm I} = \frac{1}{2} \int_0^t d\tau_1 \ldots d\tau_6 \\ \times \operatorname{tr}\big\{ \big[\delta(\tau_1 - \tau_2)\, \hat{\varphi}^-(\tau_2)\, \delta(\tau_2 - \tau_3) - \hat{\boldsymbol{S}}^\dagger(\tau_1 - \tau_2)\hat{\varphi}^-(\tau_2)\hat{\boldsymbol{S}}(\tau_2 - \tau_3)\big] \\ \times e^{i\hat{\varphi}^+(\tau_3)} \tilde{f}(\tau_3 - \tau_4) e^{-i\hat{\varphi}^+(\tau_4)} \\ \times \big[\delta(\tau_4 - \tau_5)\hat{\varphi}^-(\tau_5)\delta(\tau_5 - \tau_6) - \hat{\boldsymbol{S}}^\dagger(\tau_4 - \tau_5)\hat{\varphi}^-(\tau_5)\hat{\boldsymbol{S}}(\tau_5 - \tau_6)\big] \\ \times e^{i\hat{\varphi}^+(\tau_6)} \big[\delta(\tau_6 - \tau_1) - \tilde{f}(\tau_6 - \tau_1)\big] e^{-i\hat{\varphi}^+(\tau_1)} \big\} \end{aligned} \quad (12.90)$$

Here we introduced the matrices

$$\hat{\varphi}^- = \begin{pmatrix} \varphi_{\rm L}^- \hat{1}_L & 0 \\ 0 & \varphi_R^- \hat{1}_R \end{pmatrix}, \quad e^{i\hat{\varphi}^+} = \begin{pmatrix} e^{i\varphi_{\rm L}^+} \hat{1}_L & 0 \\ 0 & e^{i\varphi_{\rm R}^+} \hat{1}_R \end{pmatrix} \quad (12.91)$$

and employed the Fourier-transformed Fermi function $\tilde{f}(\tau)$ defined in Eq. (10.49).

Let us formulate the Langevin equations describing the currents flowing through the left and right conductors forming our quantum dot. It will be convenient for us to introduce the matrix $\hat{\boldsymbol{\zeta}}(\tau_1, \tau_2)$ with the elements $\zeta_{\alpha\beta}(\tau_1, \tau_2)$, where the indices α, β label the conducting channels. These matrix elements are Gaussian stochastic variables with the pair correlators

$$\big\langle \zeta_{\alpha\beta}(\tau_1, \tau_2)\zeta_{\gamma\delta}(\tau_3, \tau_4)\big\rangle = \delta_{\alpha\delta}\delta_{\beta\gamma}\, \tilde{f}(\tau_1 - \tau_4)\big[\delta(\tau_3 - \tau_2) - \tilde{f}(\tau_3 - \tau_2)\big] \quad (12.92)$$

In the spirit of a path integral, one can interpret the matrix element $\zeta_{\alpha\beta}(\tau_1, \tau_2)$ in terms of the operator expression involving electron creation and annihilation operators at the energy E in the conducting channels α and β, respectively, $a_\alpha^\dagger(E)$ and $a_\beta(E)$. We identify

$$\zeta_{\alpha\beta}(\tau_1, \tau_2) \leftrightarrow \int \frac{dE_1 dE_2}{(2\pi)^2} e^{-iE_1\tau_1 + iE_2\tau_2} \big[a_\alpha^\dagger(E_2)a_\beta(E_1) - \langle a_\alpha^\dagger(E_2)a_\beta(E_1)\rangle\big]$$

Let us formally decouple the imaginary part of the action by means of the following Hubbard–Stratonovich transformation:

$$\exp\left\{\mathrm{i}S_{\mathrm{R}}[\varphi_{\mathrm{L}}^{\pm},\varphi_{\mathrm{R}}^{\pm}]-S_{\mathrm{I}}[\varphi_{\mathrm{L}}^{\pm},\varphi_{\mathrm{R}}^{\pm}]\right\}=\frac{\int\mathcal{D}\zeta_{\alpha\beta}\exp\left\{\mathrm{i}\tilde{S}_{\mathrm{R}}[\varphi_{\mathrm{L}}^{\pm},\varphi_{\mathrm{R}}^{\pm},\zeta_{\alpha\beta}]\right\}P[\zeta_{\alpha\beta}]}{\int\mathcal{D}\zeta_{\alpha\beta}P[\zeta_{\alpha\beta}]}$$

where we introduced the Gaussian distribution function for the fluctuating noise terms $P[\zeta_{\alpha\beta}]$ characterized by the pair correlators (12.92). Here we also defined

$$\begin{aligned}\tilde{S}_{\mathrm{R}}=&\int_0^t d\tau_1 d\tau_2 d\tau_3\lim_{\tau_4\to\tau_1}\mathrm{tr}\Big\{\big[\delta(\tau_1-\tau_2)\,\hat{\varphi}^-(\tau_2)\,\delta(\tau_2-\tau_3)\\&-\hat{\mathbf{S}}^\dagger(\tau_1-\tau_2)\hat{\varphi}^-(\tau_2)\hat{\mathbf{S}}(\tau_2-\tau_3)\big]\\&\times e^{\mathrm{i}\hat{\varphi}^+(\tau_3)}\big[\tilde{f}(\tau_3-\tau_4)+\hat{\boldsymbol{\zeta}}(\tau_3,\tau_4)\big]e^{-\mathrm{i}\hat{\varphi}^+(\tau_4)}\Big\}\end{aligned}\qquad(12.93)$$

Note that Eq. (12.93) allows us to interpret the stochastic variables $\zeta_{\alpha\beta}$ as fluctuating parts of the distribution function for incident electrons.

In addition to the action (12.93), we should also specify the contribution S_{env} from the electromagnetic environment, which in our case consists of three capacitors C_{L}, C_{R}, C_{g} and the impedance $Z_{\mathrm{S}}(\omega)$. Assuming for simplicity $C_{\mathrm{g}}\ll C_{\mathrm{L}}, C_{\mathrm{R}}$, we obtain

$$\begin{aligned}\mathrm{i}S_{\mathrm{env}}=&\frac{\mathrm{i}}{e^2}\int_{-\infty}^t dt_1\left[C_{\mathrm{L}}\dot{\varphi}_{\mathrm{L}}^+\dot{\varphi}_{\mathrm{L}}^-+C_{\mathrm{R}}\dot{\varphi}_{\mathrm{R}}^+\dot{\varphi}_{\mathrm{R}}^-\right]\\&-\frac{\mathrm{i}}{e^2}\int_{-\infty}^t dt_1\int_{-\infty}^{t_1}dt_2\varphi_{\mathrm{env}}^-(t_1)k_{\mathrm{S}}(t_1-t_2)\dot{\varphi}_{\mathrm{env}}^+(t_2)\\&-\frac{1}{2e^2}\int_{-\infty}^t dt_1\int_{-\infty}^t dt_2\varphi_{\mathrm{env}}^-(t_1)g_{\mathrm{S}}(t_1-t_2)\dot{\varphi}_{\mathrm{env}}^-(t_2)\end{aligned}\qquad(12.94)$$

(cf Eq. (2.49)), with the functions $k_{\mathrm{S}}(t)$ and $g_{\mathrm{S}}(t)$ defined in Eq. (10.107). We again decouple the last term of the action (12.94), introducing the Gaussian current noise of the external circuit $\xi_{\mathrm{S}}(t)$ with the pair correlator

$$\langle\xi_{\mathrm{S}}(t_1)\xi_{\mathrm{S}}(t_2)\rangle=g_{\mathrm{S}}(t_1-t_2)\qquad(12.95)$$

Then the action of the environment takes the form

$$\begin{aligned}\tilde{S}_{\mathrm{env}}=&\int_{-\infty}^t dt_1\left[C_{\mathrm{L}}\frac{\dot{\varphi}_{\mathrm{L}}^+\dot{\varphi}_{\mathrm{L}}^-}{e^2}+C_{\mathrm{R}}\frac{\dot{\varphi}_{\mathrm{R}}^+\dot{\varphi}_{\mathrm{R}}^-}{e^2}+\frac{\xi_{\mathrm{S}}\varphi_{\mathrm{env}}^-}{e}\right]\\&-\frac{1}{e^2}\int_{-\infty}^t dt_1\int_{-\infty}^{t_1}dt_2\varphi_{\mathrm{env}}^-(t_1)k_{\mathrm{S}}(t_1-t_2)\dot{\varphi}_{\mathrm{env}}^+(t_2)\end{aligned}\qquad(12.96)$$

Finally, for the circuit depicted in Fig. 12.1, the balance of voltages across its elements yields

$$\frac{\dot\varphi_{\rm L}^+}{e} - \frac{\dot\varphi_{\rm R}^+}{e} + \frac{\dot\varphi_{\rm env}^+}{e} = V_x, \quad \dot\varphi_{\rm L}^- - \dot\varphi_{\rm R}^- + \dot\varphi_{\rm env}^- = 0 \qquad (12.97)$$

These two equations allow one to exclude the phases $\varphi_{\rm env}^{\pm}$ from Eq. (12.96), rewriting it in terms of the phase variables $\varphi_{\rm L,R}^{\pm}$.

As a result of all these manipulations, we arrive at the total effective action of our system $S_{\rm tot} = \tilde S_{\rm R} + \tilde S_{\rm env}$, which is linear in the phases $\varphi_{\rm L,R}^{\pm}$. Expressing the kernel of the corresponding Keldysh evolution operator as a path integral over $\varphi_{\rm L,R}^{\pm}$ and performing the integrals over $\varphi_{\rm L,R}^{-}$, one gets two functional delta-functions symbolically denoted as $\delta[\delta S_{\rm tot}/\delta\varphi_{\rm L,R}^{-}]$. Hence, the remaining path integrals over $\varphi_{\rm L,R}^{+}$ are equivalent to solving the equations $\delta S_{\rm tot}/\delta\varphi_{\rm L,R}^{-} = 0$. These equations read

$$C_{\rm L}\frac{\ddot\varphi_{\rm L}^+}{e} + I_{\rm L}(\varphi_{\rm L}^+, \varphi_{\rm R}^+) + \int_{-\infty}^{t} dt' k_{\rm S}(t-t')\left(\frac{\dot\varphi_{\rm L}^+}{e} - \frac{\dot\varphi_{\rm R}^+}{e}\right)$$
$$= \frac{V_x}{Z_{\rm S}(0)} - \xi_{\rm L}(\varphi_{\rm L}^+, \varphi_{\rm R}^+) - \xi_{\rm S},$$
$$C_{\rm R}\frac{\ddot\varphi_{\rm R}^+}{e} + I_{\rm R}(\varphi_{\rm L}^+, \varphi_{\rm R}^+) - \int_{-\infty}^{t} dt' k_{\rm S}(t-t')\left(\frac{\dot\varphi_{\rm L}^+}{e} - \frac{\dot\varphi_{\rm R}^+}{e}\right)$$
$$= -\frac{V_x}{Z_{\rm S}(0)} - \xi_{\rm R}(\varphi_{\rm L}^+, \varphi_{\rm R}^+) + \xi_{\rm S} \qquad (12.98)$$

Equations (12.98) define the two Langevin equations, which account for the balance of currents in the left and right conductors. The physical meanings of different terms in these equations are illustrated by means of the equivalent circuit in Fig. 12.10. Here the current values in the left and right conductors $I_{\rm L,R}$ in the presence of arbitrary time-dependent voltages $\dot\varphi_{\rm L,R}^{+}/e$ are defined as

$$I_r(\varphi_{\rm L}^+, \varphi_{\rm R}^+) = -e\frac{\delta \tilde S_{\rm R}}{\delta\varphi_r^-(t)}\bigg|_{\zeta=0} = -e\int d\tau_1 d\tau_3\, {\rm tr}\Big\{\big[\delta(\tau_1 - t)\hat\Lambda_r\delta(t-\tau_3)$$
$$- \hat{\boldsymbol S}^\dagger(\tau_1 - t)\hat\Lambda_r\hat{\boldsymbol S}(t-\tau_3)\big]e^{i\hat\varphi^+(\tau_3)}\check f(\tau_3-\tau_1)e^{-i\hat\varphi^+(\tau_1)}\Big\}, \quad r = L,\ R,$$
$$(12.99)$$

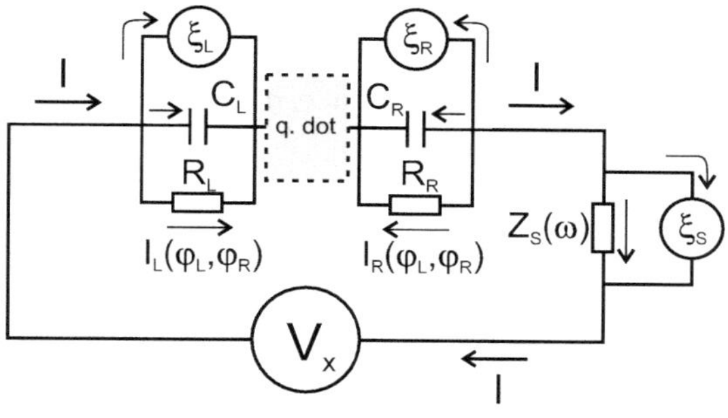

Figure 12.10 Equivalent electric circuit illustrating different contributions to the Langevin equations (12.98). Here $C_{\rm L}$, $C_{\rm R}$ are the capacitances and $R_{\rm L}$, $R_{\rm R}$ are the resistances of the conductors; I is the total current flowing through the external circuit and measured in the experiment; $I_{\rm L}(\varphi_{\rm L}^+, \varphi_{\rm R}^+)$ and $I_{\rm R}(\varphi_{\rm L}^+, \varphi_{\rm R}^+)$ are the deterministic currents flowing through the left and the right junctions, respectively; V_x is the total bias voltage; $Z_{\rm S}(\omega)$ is the environment impedance; $\xi_{\rm L}$, $\xi_{\rm R}$ describe the current noise in the junctions; $\xi_{\rm S}$ defines the current noise of the impedance $Z_{\rm S}(\omega)$; arrows indicate the directions of flowing currents; shadowed area in the center indicates the quantum dot.

while the stochastic variables $\xi_{\rm L,R}$ describing the current noise read

$$\xi_r(\varphi_{\rm L}^+, \varphi_{\rm R}^+) = -e\frac{\delta \tilde{S}_{\rm R}}{\delta \varphi_r^-(t)}\bigg|_{\tilde{f}=0} = -e\int d\tau_1 d\tau_3 \,{\rm tr}\left\{\left[\delta(\tau_1 - t)\hat{\Lambda}_r\delta(t-\tau_3)\right.\right.$$
$$\left.\left. - \hat{\boldsymbol{S}}^\dagger(\tau_1 - t)\hat{\Lambda}_r\hat{\boldsymbol{S}}(t-\tau_3)\right]e^{i\hat{\varphi}^+(\tau_3)}\hat{\varsigma}(\tau_3, \tau_1)e^{-i\hat{\varphi}^+(\tau_1)}\right\}, \quad r = L, R. \qquad (12.100)$$

In both these equations, we employed the matrices

$$\hat{\Lambda}_{\rm L} = \begin{pmatrix} \hat{1}_L & 0 \\ 0 & 0 \end{pmatrix}, \quad \hat{\Lambda}_{\rm R} = \begin{pmatrix} 0 & 0 \\ 0 & \hat{1}_R \end{pmatrix} \qquad (12.101)$$

We stress that, as compared to short coherent scatterers discussed in Chapter 11, both the current and noise terms in Eqs. (12.99) and (12.100) explicitly depend on the time-dependent scattering matrix $\hat{\boldsymbol{S}}(t)$. This dependence reflects complicated dynamics of electrons inside the dot and, in particular, accounts for retardation effects in electron scattering. As a result, there appear nonzero cross-correlations between the noise terms $\xi_{\rm L}$ and $\xi_{\rm R}$.

It is also worth pointing out that, while the general form of the Langevin equations (12.98) is insensitive to the relation between the electron phase/energy relaxation times τ_φ and $\tau_{\rm in}$ and the dwell time $\tau_{\rm D}$, Eqs. (12.99) and (12.100) essentially depend on the relation between these timescales. For instance, in the inelastic limit $\tau_{\rm in} \ll \tau_{\rm D}$, the scattering matrix $\hat{\boldsymbol{S}}(t)$ in the above equations should be replaced by its disorder averaged value,

$$\hat{\boldsymbol{S}}(t) \to \langle \hat{\boldsymbol{S}}(t) \rangle_{\rm dis} = \begin{pmatrix} \hat{r}_{\rm L} & 0 \\ 0 & \hat{r}'_{\rm R} \end{pmatrix} \delta(t) \qquad (12.102)$$

where $\hat{r}_{\rm L}$ and $\hat{r}'_{\rm R}$ are the reflection matrices for the left and right conductors. In this case, one finds $I_{\rm L} = \dot{\varphi}^+_{\rm L}/eR_{\rm L}$ and $I_{\rm R} = -\dot{\varphi}^+_{\rm R}/eR_{\rm R}$. No correlation between the noise terms $\xi_{\rm L}$, $\xi_{\rm R}$ occurs. In this limit, the solution to Eqs. (12.98) reduces to the expression for the current (12.59) derived in Section 12.3.1.

Turning back to the coherent regime (12.84), let us define the dc current I through the quantum dot equal to the average value of the current in the left conductor:

$$I = \left\langle I_{\rm L}(\varphi^+_{\rm L}, \varphi^+_{\rm R}) \right\rangle + \left\langle \xi_{\rm L}(\varphi^+_{\rm L}, \varphi^+_{\rm R}) \right\rangle \qquad (12.103)$$

Making use of the current conservation, one also finds

$$I = -\left\langle I_{\rm R}(\varphi^+_{\rm L}, \varphi^+_{\rm R}) \right\rangle - \left\langle \xi_{\rm R}(\varphi^+_{\rm L}, \varphi^+_{\rm R}) \right\rangle \qquad (12.104)$$

Note that the minus sign in the latter formula is due to the adopted convention according to which the electric potentials of the two leads have the same sign and equal, respectively, to $\dot{\varphi}^+_{\rm L}/e$ and $\dot{\varphi}^+_{\rm R}/e$.

Similar to our previous considerations, in order to derive the interaction correction to the current, we will proceed perturbatively in the noise terms $\xi_{\rm L}$ and $\xi_{\rm R}$. In the non-interacting limit in Eq. (12.98), we set $\xi_{\rm L} = \xi_{\rm R} = 0$ and obtain $\varphi^+_{\rm L}(t) = eV_{\rm L}t$, $\varphi^+_{\rm R} = eV_{\rm R}t$ and $V_{\rm L} + V_{\rm R} = V$. Then from (12.103), we easily recover the standard Landauer formula

$$\begin{aligned} I &= I_0(V_{\rm L}, V_{\rm R}) \\ &= \frac{e}{2\pi} \int dE \,\mathrm{tr}\left[\hat{t}^\dagger(E)\hat{t}(E)\right]\left[f(E + eV_{\rm R}) - f(E + eV_{\rm L})\right] \end{aligned} \qquad (12.105)$$

In the presence of interactions, the phase variables fluctuate and, hence, $\varphi^+_{\rm L,R} = eV_{\rm L,R}t + \delta\varphi^+_{\rm L,R}$ and

$$\left\langle I_{\rm L}(\varphi^+_{\rm L}, \varphi^+_{\rm R}) \right\rangle = V_{\rm L}/R_{\rm L} + \delta I_{\rm L}.$$

Expanding both the current and the noise terms in small phase fluctuations $\delta\varphi^{+}_{\mathrm{L,R}}$ and keeping only the leading order terms, from Eqs. (12.99) and (12.100), we get

$$\delta I_L = \frac{e}{2}\int d\tau_1 d\tau_3\, \tilde{f}(\tau_3-\tau_1)\,\mathrm{tr}\Big\{\big[\delta(\tau_1-t)\hat{\Lambda}_L\delta(t-\tau_3) - \hat{\boldsymbol{S}}^{\dagger}(\tau_1-t)\hat{\Lambda}_L\hat{\boldsymbol{S}}(t-\tau_3)\big]\langle[\delta\hat{\varphi}^{+}(\tau_3)-\delta\hat{\varphi}^{+}(\tau_1)]^2\rangle e^{\mathrm{i}e\hat{V}(\tau_3-\tau_1)}\Big\}, \quad (12.106)$$

$$\langle\xi_{\mathrm{L}}\rangle = -\mathrm{i}e\int d\tau_1 d\tau_3\,\mathrm{tr}\Big\{\big[\delta(\tau_1-t)\hat{\Lambda}_r\delta(t-\tau_3) - \hat{\boldsymbol{S}}^{\dagger}(\tau_1-t)\hat{\Lambda}_r\hat{\boldsymbol{S}}(t-\tau_3)\big] \times e^{\mathrm{i}e\hat{V}\tau_3}\big\langle\delta\hat{\varphi}^{+}(\tau_3)\hat{\boldsymbol{\xi}}(\tau_3,\tau_1)-\hat{\boldsymbol{\xi}}(\tau_3,\tau_1)\delta\hat{\varphi}^{+}(\tau_1)\big\rangle e^{-\mathrm{i}e\hat{V}\tau_1}\Big\} \quad (12.107)$$

Now let us express the phase fluctuations $\delta\varphi^{+}_{\mathrm{L,R}}$ via $\zeta_{\alpha\beta}$. This goal can be achieved with the aid of Eqs. (12.98). In the limit $Z_{\mathrm{S}}(\omega)\to 0$, we have $\dot{\varphi}^{+}_{\mathrm{L}}/e - \dot{\varphi}^{+}_{\mathrm{R}}/e = V_x$ and, hence, $\delta\varphi^{+}_{\mathrm{L}} = \delta\varphi^{+}_{\mathrm{R}} = \delta\varphi$, where $\delta\varphi$ is related to the fluctuating electric potential of the dot. Combining the two equations (12.98), we find

$$C_\Sigma\frac{\delta\ddot{\varphi}(t)}{e} - \mathrm{i}e\int d\tau_1 d\tau_3\,\tilde{f}(\tau_3-\tau_1)[\delta\varphi(\tau_3)-\delta\varphi(\tau_1)] \times \mathrm{tr}\Big\{\big[\delta(\tau_1-t)\hat{1}\delta(t-\tau_3) - \hat{\boldsymbol{S}}^{\dagger}(\tau_1-t)\hat{\boldsymbol{S}}(t-\tau_3)\big]e^{\mathrm{i}e\hat{V}(\tau_3-\tau_1)}\Big\} = -\xi_d(t) \quad (12.108)$$

where the stochastic variable $\xi_{\mathrm{d}}(t) = \xi_{\mathrm{L}}(eV_{\mathrm{L}}t, eV_{\mathrm{R}}t) + \xi_{\mathrm{R}}(eV_{\mathrm{L}}t, eV_{\mathrm{R}}t)$ accounts for fluctuations of the quantum dot charge. The solution to this equation reads

$$\delta\varphi(t) = -\frac{1}{e}\int dt' K(t-t')\xi_d(t') \quad (12.109)$$

where

$$K(t) = \int\frac{d\omega}{2\pi}\,e^{-\mathrm{i}\omega t}\,K_\omega \quad (12.110)$$

is the response function with the Fourier transform

$$K_\omega = \frac{-1}{\frac{C_\Sigma\omega^2}{e^2} + \mathrm{i}\int\frac{dE}{2\pi}\,\mathrm{tr}\Big\{\big[\hat{1}-\hat{S}^{\dagger}(E)\hat{S}(E+\omega)\big]\big[\hat{f}(E)-\hat{f}(E+\omega)\big]\Big\}} \quad (12.111)$$

Here we defined the matrix of the distribution functions

$$\hat{f}(E) = \begin{pmatrix} \hat{1}_L\, f(E+eV_{\mathrm{L}}) & 0 \\ 0 & \hat{1}_R\, f(E+eV_{\mathrm{R}}) \end{pmatrix} \quad (12.112)$$

With the aid of the above equations, one readily finds

$$\left\langle \left[\delta\varphi(t) - \delta\varphi(0)\right]^2 \right\rangle = 2 \int \frac{dE\,d\omega}{(2\pi)^2} |K(\omega)|^2 \left[1 - \cos\omega t\right]$$
$$\times \operatorname{tr}\left[\hat{h}(E)(\hat{1} - \hat{S}^\dagger(E)\hat{S}(E+\omega))\,\hat{f}(E+\omega)(\hat{1} - \hat{S}^\dagger(E+\omega)\hat{S}(E))\right] \tag{12.113}$$

Substituting now the results (12.109), (12.113) into Eqs. (12.106), (12.107) and performing averaging over the noise variables $\zeta_{\alpha\beta}$, after a straightforward though somewhat tedious algebra, we obtain

$$\delta I_{\rm L} \equiv \langle I_{\rm L}(\varphi_{\rm L}^+, \varphi_{\rm R}^+)\rangle - V/(R_{\rm L} + R_{\rm R}) = \frac{e}{4} \int \frac{dE\,dE'\,d\omega}{(2\pi)^3} |K(\omega)|^2$$
$$\times \operatorname{tr}\left[\hat{h}(E')(1 - \hat{S}^\dagger(E')\hat{S}(E'+\omega))\,\hat{f}(E'+\omega)(1 - \hat{S}^\dagger(E'+\omega)\hat{S}(E'))\right]$$
$$\times \operatorname{tr}\left[\left(\hat{\Lambda} - \hat{S}^\dagger(E)\hat{\Lambda}\hat{S}(E)\right)\left(2\hat{f}(E) - \hat{f}(E-\omega) - \hat{f}(E+\omega)\right)\right] \tag{12.114}$$

$$\langle \xi_L \rangle = -e\,{\rm Im} \int \frac{dE\,d\omega}{(2\pi)^2} K_\omega \operatorname{tr}\left[(\hat{1} - \hat{S}^\dagger(E)\hat{S}(E+\omega))\hat{h}(E+\omega)\right.$$
$$\left.\times\ (\hat{S}^\dagger(E+\omega)\hat{\Lambda}\hat{S}(E+\omega) - \hat{S}^\dagger(E)\hat{\Lambda}\hat{S}(E))\,\hat{f}(E)\right] \tag{12.115}$$

where we defined

$$\hat{\Lambda} = \begin{pmatrix} \hat{1}_L & 0 \\ 0 & -\hat{1}_R \end{pmatrix}, \quad \hat{h}(E) = \hat{1} - \hat{f}(E) \tag{12.116}$$

Combining now Eqs. (12.114) and (12.115) with Eq. (12.103), we recover the I–V curve of our quantum dot in the presence of electron–electron interactions. Note that the interaction correction to the standard Ohmic term $V/(R_{\rm L} + R_{\rm R})$ is defined by the sum $\delta I_{\rm L} + \langle \xi_{\rm L} \rangle$ expressing this correction in terms of an *arbitrary* scattering matrix $\hat{S}(E)$ for a non-interacting quantum dot. As before, the only restriction for the above results is set by the requirement $\delta I_{\rm L} + \langle \xi_{\rm L}\rangle \ll V/(R_{\rm L} + R_{\rm R})$ satisfied either provided at least one of the resistors $R_{\rm L}$ or $R_{\rm R}$ remains much smaller than $R_{\rm q}$ or at sufficiently high temperatures and/or voltages.

12.4.2 Large Metallic Dots

Let us further specify the above general results for the case of large metallic quantum dots with very small level spacing δ. In this regime, one can neglect mesoscopic fluctuations and perform disorder averaging in both contributions to the interaction correction (12.114) and (12.115). This averaging will be carried out with the aid of the random matrix theory describing disorder configurations in terms of different realizations of the scattering matrix $\hat{S}(E)$.

Without loss of generality, we will assume that each of the two conductors has $N \gg 1$ conducting channels being described by one of the two $2N \times 2N$ scattering matrices having the usual block structure

$$\hat{S}_{\mathrm{L}} = \begin{pmatrix} \hat{r}_{\mathrm{L}} & \hat{t}'_{\mathrm{L}} \\ \hat{t}_{\mathrm{L}} & \hat{r}'_{\mathrm{L}} \end{pmatrix}, \quad \hat{S}_{\mathrm{R}} = \begin{pmatrix} \hat{r}_{\mathrm{R}} & \hat{t}'_{\mathrm{R}} \\ \hat{t}_{\mathrm{R}} & \hat{r}'_{\mathrm{R}} \end{pmatrix} \tag{12.117}$$

Then the scattering matrix of the quantum dot takes the form [69]

$$\hat{S}(E) = \hat{R} + \hat{T}'\left[\hat{1} - \hat{S}_0(E)\hat{R}'\right]^{-1}\hat{S}_0(E)\hat{T}, \tag{12.118}$$

where we defined

$$\hat{R} = \begin{pmatrix} \hat{r}_{\mathrm{L}} & 0 \\ 0 & \hat{r}'_{\mathrm{R}} \end{pmatrix}, \quad \hat{R}' = \begin{pmatrix} \hat{r}'_{\mathrm{L}} & 0 \\ 0 & \hat{r}_{\mathrm{R}} \end{pmatrix} \tag{12.119}$$

$$\hat{T} = \begin{pmatrix} \hat{t}_{\mathrm{L}} & 0 \\ 0 & \hat{t}'_{\mathrm{R}} \end{pmatrix}, \quad \hat{T}' = \begin{pmatrix} \hat{t}'_{\mathrm{L}} & 0 \\ 0 & \hat{t}_{\mathrm{R}} \end{pmatrix} \tag{12.120}$$

and introduced the $2N \times 2N$ matrix $\hat{S}_0(E)$ for an open quantum dot (i.e., the dot with fully open N conducting channels in both left and right conductors: $T_n = 1$ for $n \leq N$ and $T_n = 0$ for $n > N$). The scattering matrices $\hat{S}_{\mathrm{L}}$ and $\hat{S}_{\mathrm{R}}$ are not affected by disorder in the quantum dot, while the matrix $\hat{S}_0(E)$ fluctuates. The random matrix theory [69, 333] provides the rules of averaging for the matrix products involving the scattering matrix $\hat{S}_0(E)$. In particular, in the limit $N \gg 1$, we have

$$\left\langle \mathrm{tr}\left[\hat{A}\hat{S}_0^{\dagger}(E_1)\hat{B}\hat{S}_0(E_2)\right]\right\rangle = \frac{2\,\mathrm{tr}\left[\hat{A}\right]\mathrm{tr}\left[\hat{B}\right]}{2N - \frac{4\pi \mathrm{i} E_{21}}{\delta}} \tag{12.121}$$

$$\left\langle \mathrm{tr}\left[\hat{A}\hat{S}_0^\dagger(E_1)\hat{B}\hat{S}_0(E_2)\hat{C}\hat{S}_0^\dagger(E_3)\hat{D}\hat{S}_0(E_4)\right]\right\rangle = -\frac{8\,\mathrm{tr}\left[\hat{A}\right]\mathrm{tr}\left[\hat{B}\right]\mathrm{tr}\left[\hat{C}\right]\mathrm{tr}\left[\hat{D}\right]\left(2N-\frac{4\pi\mathrm{i}(E_{23}+E_{41})}{\delta}\right)}{\left(2N-\frac{4\pi\mathrm{i}E_{21}}{\delta}\right)\left(2N-\frac{4\pi\mathrm{i}E_{43}}{\delta}\right)\left(2N-\frac{4\pi\mathrm{i}E_{41}}{\delta}\right)\left(2N-\frac{4\pi\mathrm{i}E_{23}}{\delta}\right)} \tag{12.122}$$

where $\hat{A}$, $\hat{B}$, $\hat{C}$, and $\hat{D}$ are arbitrary $2N \times 2N$ square matrices; and $E_{ij} = E_i - E_j$.

Employing the above expressions, one can derive similar rules for the full scattering matrix (12.118). For example, the corresponding equation analogous to Eq. (12.121) reads

$$\left\langle \mathrm{tr}\left[\hat{A}\hat{S}^\dagger(E_1)\hat{B}\hat{S}(E_2)\right]\right\rangle = \mathrm{tr}\left[\hat{A}\hat{R}^\dagger\hat{B}\hat{R}\right] + \frac{2\,\mathrm{tr}\left[\hat{T}\hat{A}\hat{T}^\dagger\right]\mathrm{tr}\left[\hat{T}'^\dagger\hat{B}\hat{T}'\right]}{(g_\mathrm{L}+g_\mathrm{R})(1-\mathrm{i}E_{21}\tau_D)} \tag{12.123}$$

where we defined the average electron dwell time in the quantum dot

$$\tau_\mathrm{D} = \frac{4\pi}{(g_\mathrm{L}+g_\mathrm{R})\delta_\mathrm{d}} \tag{12.124}$$

One can also construct the average of the product of four matrices analogous to Eq. (12.122). The corresponding, rather complicated, expression can be found in the paper [163].

Employing the above rules of averaging in Eqs. (12.114) and (12.115), we observe that the term δI_L in the general expression for the current

$$I = \frac{V}{R_\mathrm{L}+R_\mathrm{R}} + \delta I_\mathrm{L} + \langle \xi_\mathrm{L}\rangle \tag{12.125}$$

vanishes identically after averaging,

$$\delta I_\mathrm{L} \equiv 0 \tag{12.126}$$

since the total average transmission of the quantum dot $\langle \hat{t}^\dagger(E)\hat{t}(E)\rangle$ does not depend on energy and, hence, the current I_L remains insensitive to voltage fluctuations across the left and right conductors.

It follows immediately that in this case, the interaction correction to the current is given only by the term $\langle \xi_\mathrm{L}\rangle$, thereby emphasizing again a direct relation between shot noise and interaction effects in mesoscopic conductors already established in Chapter 11. We obtain

$$\langle \xi_{\rm L} \rangle = -\frac{e}{2}\frac{g_{\rm L}g_{\rm R}(g_{\rm L}\beta_{\rm R}+g_{\rm R}\beta_{\rm L})}{(g_{\rm L}+g_{\rm R})^3}\int \frac{d\omega}{2\pi}\frac{1}{\omega}\left(\frac{1}{1+\omega^2\tau^2}-\frac{1}{1+\omega^2\tau_D^2}\right)$$
$$\times\left((\omega+eV)\coth\frac{\omega+eV}{2T}-(\omega-eV)\coth\frac{\omega-eV}{2T}\right) \tag{12.127}$$

where
$$\frac{1}{\tau}=\frac{1}{\tau_0}+\frac{1}{\tau_{\rm D}} \tag{12.128}$$

and the times τ_0 and $\tau_{\rm D}$ are defined, respectively, in Eqs. (12.57) and (12.124). Performing the frequency integration in Eq. (12.127), we arrive at the final result for the $I-V$ curve of a big metallic dot

$$I = \frac{V}{R_{\rm L}+R_{\rm R}}$$
$$-\frac{e}{\pi}\frac{g_{\rm L}g_{\rm R}(g_{\rm L}\beta_{\rm R}+g_{\rm R}\beta_{\rm L})}{(g_{\rm L}+g_{\rm R})^3}\,{\rm Im}\left[\left(\frac{1}{\tau}+{\rm i}eV\right)\Psi\left(1+\frac{1}{2\pi T\tau}+{\rm i}\frac{eV}{2\pi T}\right)\right.$$
$$\left.-\left(\frac{1}{\tau_{\rm D}}+{\rm i}eV\right)\Psi\left(1+\frac{1}{2\pi T\tau_{\rm D}}+{\rm i}\frac{eV}{2\pi T}\right)\right] \tag{12.129}$$

In the limit $V \to 0$, this formula yields the expression for the zero-bias conductance in the form

$$G_0 = \frac{1}{R_{\rm L}+R_{\rm R}}-\frac{2}{R_{\rm q}}\frac{g_{\rm L}g_{\rm R}(g_{\rm L}\beta_{\rm R}+g_{\rm R}\beta_{\rm L})}{(g_{\rm L}+g_{\rm R})^3}[\mathcal{P}(T\tau)-\mathcal{P}(T\tau_D)] \tag{12.130}$$

where we defined

$$\mathcal{P}(x)=\Psi\left(1+\frac{1}{2\pi x}\right)+\frac{1}{2\pi x}\Psi'\left(1+\frac{1}{2\pi x}\right) \tag{12.131}$$

Let us compare the $I-V$ curve for fully coherent quantum dots (12.129) with that derived in the regime of strong inelastic relaxation, Eqs. (12.59) and (12.56). The corresponding differential conductances dI/dV are depicted in Fig. 12.11. Since in the coherent regime the current (12.129) depends only on the ratio between the total bias voltage V and temperature T, the Coulomb dip in the differential conductance is narrower in this case. In the high-temperature limit, its half-width coincides with that of a single junction (cf. Eq. (7.37)). In contrast, in the inelastic regime, the current (12.59), (12.56) depend on the two voltages $V_{\rm L}$ and $V_{\rm R}$

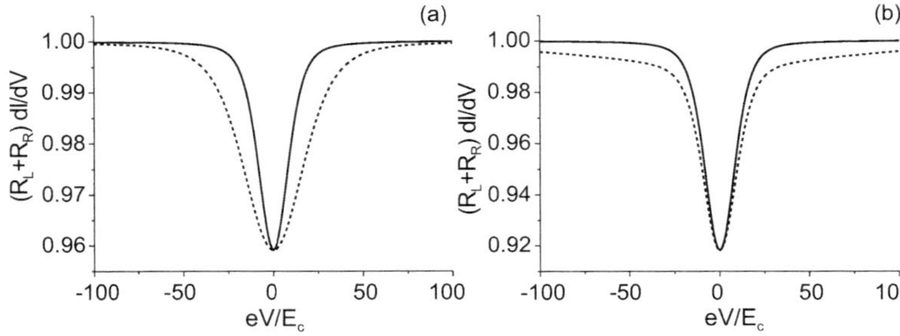

Figure 12.11 Differential conductance of a quantum dot in the regime of coherent transport, as given by Eq. (12.129), (solid line) and in the regime of complete inelastic relaxation, Eq. (12.59) (dashed line). (a) Symmetric quantum dot with the parameters $g_L = g_R = 10$, $\beta_L = \beta_R = 1$, $\delta = 0.01 E_C$, $T = 3E_C$; (b) asymmetric quantum dot with the parameters $g_L = 2$, $g_R = 18$, $\beta_L = \beta_R = 1$, $\delta = 0.01 E_C$, $T = 3E_C$.

separately, and the Coulomb dip becomes wider. This difference is particularly pronounced in the case of symmetric quantum dots in which case strong inelastic relaxation yields doubling of the dip width; see Fig. 12.11(a). In asymmetric dots, the effect of the inelastic relaxation on the $I-V$ curve also exists but turns out to be somewhat weaker; see Fig. 12.11(b).

Comparing zero-bias conductances (12.60), (12.61) and (12.130), (12.131), we note that they essentially coincide at $T\tau_D \gg 1$, implying that fluctuations in the left and right conductors become independent provided temperature exceeds the Thouless energy of the quantum dot $1/\tau_D$. In this regime, Eqs. (12.130) and (12.131) yield

$$G_0 = \frac{1}{R_L + R_R} + \frac{2}{R_q}\frac{g_L g_R(g_L\beta_R + g_R\beta_L)}{(g_L + g_R)^3}\ln(T\tau_0) \tag{12.132}$$

for $T\tau_0 \ll 1$ and

$$G_0 = \frac{g}{R_L + R_R} - \frac{1}{R_q}\frac{g_L g_R(g_L\beta_R + g_R\beta_L)}{(g_L + g_R)^3}\left[\frac{E_C}{3T} - \frac{3\zeta(3)g}{2\pi^4}\left(\frac{E_C}{T}\right)^2\right] \tag{12.133}$$

at even higher temperatures $T\tau_0 \gg 1$.

In the opposite limit $T\tau_D < 1$, the quantum dot behaves as a single coherent conductor and its conductance (12.130) saturates to the value

$$G_{\min} = \frac{1}{R_L + R_R}\left[1 - 2\frac{g_L\beta_R + g_R\beta_L}{(g_L + g_R)^2}\ln\frac{\tau_D}{\tau}\right] \qquad (12.134)$$

This formula may be interpreted as an extension of the elastic cotunneling result (12.49) to the high conductance regime $g_{L,R} \gg 1$.

In contrast, in the regime of strong inelastic relaxation, the conductance (12.61) keeps decreasing down to very low temperatures. It saturates as soon as the inelastic relaxation time $\tau_{in}(T)$ (which usually grows with decreasing T) becomes comparable to the electron dwell time τ_D and the quantum dot crosses over to the coherent regime.

Turning to the differential conductance, in the low-temperature limit $T \ll eV$, we obtain from Eq. (12.129)

$$\frac{dI}{dV} = \frac{1}{R_L + R_R}\left[1 - \frac{g_L\beta_R + g_R\beta_L}{(g_L + g_R)^2}\ln\frac{e^2V^2 + 1/\tau^2}{e^2V^2 + 1/\tau_D^2}\right] \qquad (12.135)$$

At voltages below the Thouless energy $eV \lesssim 1/\tau_D$, it again saturates to the value (12.134). At higher temperatures $T/E_C \gg g_L + g_R$, one can expand Eq. (12.129) in powers of $1/T$ with the result

$$\frac{dI}{dV} = \frac{1}{R_L + R_R}\left[1 - \frac{E_C}{T}\frac{g_L\beta_R + g_R\beta_L}{g_L + g_R}\frac{\frac{eV}{2T}\cosh\frac{eV}{2T} - \sinh\frac{eV}{2T}}{\sinh^3\frac{eV}{2T}}\right] \qquad (12.136)$$

Finally, let us outline the validity conditions of the above results. Note that in big metallic dots, one has $\tau_D/\tau_0 = 4E_C/\delta_d \gg 1$, i.e., the electron dwell time τ_D strongly exceeds the charge relaxation time τ_0 and, hence, $\ln(\tau_D/\tau)$ may also be large. Exactly as in the case of a short coherent scatterer analyzed in Chapter 11, we conclude that Eq. (12.129) holds provided the interaction correction to the current remains small as compared to $V/(R_L + R_R)$. Then from Eq. (12.134), we get

$$2\frac{g_L\beta_R + g_R\beta_L}{(g_L + g_R)^2}\ln\frac{4E_C}{\delta_d} \ll 1 \qquad (12.137)$$

As long as this inequality is satisfied, Eq. (12.129) applies at all temperatures T and bias voltages V. If, however, Eq. (12.137) is violated, the validity range of our result (12.129) is restricted by the conditions

$$\max\{eV, T\} \gg E_C, \quad \text{at } g_L + g_R \lesssim 1,$$
$$\max\{eV, T\} \gg (1/\tau_0)e^{-(g_L+g_R)/2} \quad \text{at } g_L + g_R \gtrsim 1 \qquad (12.138)$$

which are fully analogous to those already encountered, e.g., in Chapters 7 and 11 in the case of tunnel barriers and short coherent conductors.

12.4.3 Effect of External Impedance

So far we merely addressed the case of ideally conducting external leads $Z_S(\omega) \to 0$. In many experimental situations, however, this voltage-biased model is not appropriate since the impedance of external leads remains nonzero in the relevant frequency range. For this reason, it is important to find out how the above results for the $I-V$ curve are modified in the case $Z_S(\omega) \neq 0$.

The corresponding analysis [163] is straightforward, but the general results are somewhat cumbersome and, therefore, are not presented here. In order to understand the effect of an external impedance here, we restrict our consideration to a simple limiting case of symmetric quantum dots with $g_L = g_R = g/2$, $\beta_L = \beta_R = \beta$, $R_L = R_R = R$ and $C_L = C_R = C$. We will also assume the external impedance to be Ohmic $Z_S(\omega) = R_S$ at all relevant frequencies. Then we get

$$\delta I_{\text{tot}}(V) = \delta I(V) + \delta I_S(V) \tag{12.139}$$

where $\delta I(V)$ is the interaction correction evaluated above for $Z_S = 0$ and $\delta I_S(V)$ represents an additional contribution due to the external shunt. It has the form [163]

$$\delta I_S(V) = -\frac{e}{4\pi}\frac{g}{g+4g_S}\int_0^\infty dt \frac{\pi^2 T^2}{\sinh^2 \pi T t}\sin eVt$$
$$\times \left[\beta(1-\mathrm{e}^{-t/\tau_S}) + 1 - \frac{\tau_D \mathrm{e}^{-t/\tau_D} - \tau_S \mathrm{e}^{-t/\tau_S}}{\tau_D - \tau_S}\right] \tag{12.140}$$

where $\tau_S = RR_SC/(R_S+2R)$. As before, this integral can be expressed in terms of the Ψ-functions.

Let us concentrate on the linear in voltage regime. Making use of the above general results, one can easily determine the total linear conductance G of the dot in the presence of an external Ohmic shunt.

It reads

$$G = \frac{1}{2R} - \frac{1}{2R_q}\left\{\beta\big(\mathcal{P}(T\tau) - \mathcal{P}(T\tau_D)\big) + \frac{g}{g + 4g_S}\right.$$
$$\left.\times\left[\frac{\tau_D\big(\mathcal{P}(T\tau_D) - \mathcal{P}(T\tau_S)\big)}{\tau_D - \tau_S} + (\beta + 1)(\mathcal{P}(T\tau_S) + \gamma)\right]\right\} \quad (12.141)$$

where the function $\mathcal{P}(x)$ was already defined in Eq. (12.131).

For large quantum dots, one typically has $\tau_D \gg \tau_0 > \tau_S$. In this case at sufficiently low temperatures $T\tau_D \ll 1$, the whole quantum dot can be considered a single scatterer, and we find

$$G_0 = \tilde{G}_0 - \frac{2\tilde{\beta}}{R_q}\frac{g}{g + 4g_S}\left(\gamma + 1 + \ln\frac{1}{2\pi T\tau_D}\right) \quad (12.142)$$

where

$$\tilde{G}_0 = \frac{1}{2R} - \frac{\beta}{2R_q}\left(\ln\frac{\tau_D}{\tau} + \frac{g}{g + 4g_S}\ln\frac{\tau_D}{\tau_S}\right) \quad (12.143)$$

and $\tilde{\beta} = (\beta + 1)/4$ is the effective total Fano factor of a symmetric chaotic quantum dot, which follows from a more general expression

$$\tilde{\beta} = \frac{g_L g_R}{(g_L + g_R)^2} + \frac{g_R^3\beta_L + g_L^3\beta_R}{(g_L + g_R)^3} \quad (12.144)$$

to be derived in Chapter 13 within a somewhat broader context; see Eq. (13.86).

In the opposite high-temperature limit, $T\tau_D \gg 1$, albeit $T\tau_0 \ll 1$, we obtain

$$G \approx \frac{1}{2R} - \frac{\beta}{2R_q}\frac{g}{g + 4g_S}\ln\frac{\tau_0}{\tau_S} - \frac{\beta}{R_q}\frac{g + 2g_S}{g + 4g_S}\left(\gamma + 1 + \ln\frac{1}{2\pi T\tau_0}\right) \quad (12.145)$$

Let us compare the results (12.142) and (12.145). In both temperature regimes, we observe that the interaction correction depends logarithmically on temperature, though with different prefactors in front of the logarithm. At high temperatures $T\tau_D \gg 1$, this prefactor is determined by the sum of two different contributions originating from the terms δI and δI_S. In this regime, the two barriers of the quantum dot behave as independent ones. In the other regime $T\tau_D < 1$, the first logarithmic term saturates and does not depend on T anymore providing effective renormalization of the non-interacting conductance (12.143). On the contrary, the

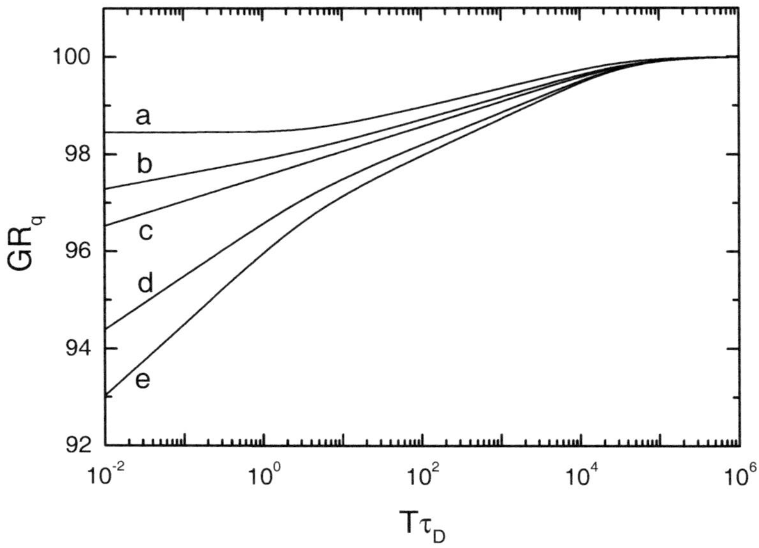

Figure 12.12 Conductance of a symmetric quantum dot, as given by Eq. (12.141). The Fano factor of each barrier is chosen to be $\beta = 1/3$, $R = R_q/200$, and $\tau_D/\tau_0 = 10^4$. Different curves correspond to different values of the shunt resistor: (a) $R_S = 2.5 \times 10^{-5}\, R_q$, (b) $R_S = 0.0025\, R_q$, (c) $R_S = 0.005\, R_q$, (d) $R_S = 0.025 R_q$, and (e) $R_S = 0.25 R_q$. A crossover between the two logarithmic regimes is clearly observed at $T\tau_D \sim 1$.

logarithmic temperature dependence (12.142) in the second term survives down to exponentially small temperatures. In this regime, the quantum dot behaves as a single coherent scatterer. Its internal structure becomes insignificant in this case, and the interaction correction scales with the total Fano factor $\tilde{\beta}$. A crossover between the two logarithmic regimes is clearly observed in Fig. 12.12. It is also important to emphasize that, while the logarithmic in T correction (12.142) vanishes for $R_S \to 0$, in the opposite limit $R_S \gg R$, it becomes practically independent of the shunt resistance.

The presence of two different logarithmic regimes can also be understood bearing in mind a close relation between shot noise and interaction effects in mesoscopic conductors. At high temperatures $T\tau_D \gg 1$, i.e., in the regime of independent barriers, the noise at each of them determines the corresponding contribution to the interaction correction. Hence, the latter is proportional to the parameter β. In the opposite limit $T\tau_{\mathrm{D}} \ll 1$, the barriers are not anymore independent. In this regime, the temperature-dependent part of the interaction correction should be related to the total shot noise of the quantum dot, which depends on the parameter $\tilde{\beta}$. The high-frequency noise $\omega\tau_{\mathrm{D}} \gtrsim 1$ now only provides the conductance renormalization (12.143), which is again proportional to β.

It follows from the above analysis that at higher temperatures $T\tau_{\mathrm{D}} \gg 1$, the interaction correction to the conductance of the dot vanishes completely for fully transparent barriers $\beta \to 0$. By contrast, at lower temperatures $T\tau_{\mathrm{D}} \lesssim 1$, only the T-independent renormalization in Eq. (12.143) disappears in this limit, while the interaction correction in Eq. (12.142) scales with $\tilde{\beta}$ and remains finite even for $\beta = 0$. We also observe that in the case $\beta \to 0$, the conductance (12.142) becomes completely independent of the charging energy E_{C}.

12.4.4 Boltzmann–Langevin Approach

Our previous analysis of electron transport in coherent quantum dots in the presence of electron–electron interactions was based on the scattering matrix formalism. Averaging over disorder was performed with the aid of the random matrix theory. Although technically the whole procedure appears rather involved, it eventually yields a relatively simple result for the current, Eq. (12.129). It is, therefore, tempting to construct a simpler and, perhaps, physically more transparent derivation of this result. This goal is accomplished within the framework of the so-called Boltzmann–Langevin approach. A detailed description of this approach in the case of non-interacting electrons can be found in the review by Blanter and Büttiker [60]. In the remaining part of this chapter, we will include interaction effects and work out the corresponding generalization of the Boltzmann–Langevin approach.

To begin with, let us recall that in the limit of strong inelastic relaxation, $\tau_{\mathrm{in}} \ll \tau_{\mathrm{D}}$, the electron distribution function $f_{\mathrm{d}}(E)$ inside the dot has the universal form and just coincides with the Fermi function $f_0(E) = 1/[1+e^{E/T}]$. In the opposite case of weak inelastic relaxation, $\tau_{\mathrm{in}}, \tau_{\varphi} \gg \tau_{\mathrm{D}}$, this distribution function is not anymore universal and it should be determined self-consistently from the kinetic equation. Let us first consider a simple model and write down this equation assuming that the voltages V_{L}, V_{R} across the left and the right contacts do not fluctuate in time. The random matrix theory predicts that the electron escape rates through these two contacts are, respectively,

$$\Gamma_{\mathrm{L}} = \frac{g_{\mathrm{L}}\delta_{\mathrm{d}}}{4\pi}, \quad \Gamma_{\mathrm{R}} = \frac{g_{\mathrm{R}}\delta_{\mathrm{d}}}{4\pi} \tag{12.146}$$

Accordingly, the kinetic equation for the distribution function $f_{\mathrm{d}}(E)$ reads

$$\frac{\partial f_{\mathrm{d}}(E)}{\partial t} = -(\Gamma_{\mathrm{L}} + \Gamma_{\mathrm{R}}) f_{\mathrm{d}}(E) + \Gamma_{\mathrm{L}} f_0(E + eV_{\mathrm{L}}) + \Gamma_{\mathrm{R}} f_0(E + eV_{\mathrm{R}}) \tag{12.147}$$

In the stationary case $\partial f_{\mathrm{d}}(E)/\partial t = 0$, one readily finds

$$f_{\mathrm{d}}(E) = \frac{g_{\mathrm{L}} f_{\mathrm{L}}(E + eV_{\mathrm{L}}) + g_{\mathrm{R}} f_{\mathrm{R}}(E + eV_{\mathrm{R}})}{g_{\mathrm{L}} + g_{\mathrm{R}}} \tag{12.148}$$

Combining the general expressions for the current through the left and the right contacts

$$\begin{aligned} I_{\mathrm{L}} &= -\frac{2e\Gamma_{\mathrm{L}}}{\delta_{\mathrm{d}}} \int dE \left[f_0(E + eV_{\mathrm{L}}) - f_{\mathrm{d}}(E) \right], \\ I_{\mathrm{R}} &= -\frac{2e\Gamma_{\mathrm{R}}}{\delta_{\mathrm{d}}} \int dE \left[f_{\mathrm{d}}(E) - f_0(E + eV_{\mathrm{R}}) \right] \end{aligned} \tag{12.149}$$

(the factor 2 in front of these expressions accounts for spin degeneracy) with Eq. (12.148), we immediately recover the standard result for the current through the quantum dot

$$I = I_{\mathrm{L}} = I_{\mathrm{R}} = \frac{e^2}{2\pi} \frac{g_{\mathrm{L}} g_{\mathrm{R}}}{g_{\mathrm{L}} + g_{\mathrm{R}}} (V_{\mathrm{L}} - V_{\mathrm{R}}) \tag{12.150}$$

The kinetic equation (12.147) does not yet include the effect of the current noise. In order to account for this effect, let us generalize this equation by including additional stochastic terms $\eta_{\mathrm{L}}(E, t)$ and $\eta_{\mathrm{R}}(E, t)$:

$$\begin{aligned} \frac{\partial f_{\mathrm{d}}(E, t)}{\partial t} = {} & -(\Gamma_{\mathrm{L}} + \Gamma_{\mathrm{R}}) f_{\mathrm{d}}(E, t) + \Gamma_L f_0(E + eV_{\mathrm{L}}) + \Gamma_{\mathrm{R}} f_0(E + eV_{\mathrm{R}}) \\ & + \eta_{\mathrm{L}}(E, t) + \eta_{\mathrm{R}}(E, t) \end{aligned} \tag{12.151}$$

where

$$\langle\eta_{\mathrm{L}}(E_1,t_1)\eta_{\mathrm{L}}(E_2,t_2)\rangle = \frac{\delta_d^2}{4e^2R_{\mathrm{L}}}\delta(E_1-E_2)\delta(t_1-t_2)$$
$$\times\left\{\beta_{\mathrm{L}}\left[f_{\mathrm{d}}(E,t)\left(1-f_0(E+eV_{\mathrm{L}})\right)+f_0(E+eV_{\mathrm{L}})\left(1-f_{\mathrm{d}}(E,t)\right)\right]\right.$$
$$\left.+(1-\beta_{\mathrm{L}})\left[f_{\mathrm{d}}(E,t)\left(1-f_{\mathrm{d}}(E,t)\right)+f_0(E+eV_{\mathrm{L}})\left(1-f_0(E+eV_{\mathrm{L}})\right)\right]\right\},$$
$$\langle\eta_R(E_1,t_1)\eta_R(E_2,t_2)\rangle = \frac{\delta_d^2}{4e^2R_{\mathrm{R}}}\delta(E_1-E_2)\delta(t_1-t_2)$$
$$\times\left\{\beta_{\mathrm{R}}\left[f_{\mathrm{d}}(E,t)\left(1-f_0(E+eV_{\mathrm{R}})\right)+f_0(E+eV_{\mathrm{R}})\left(1-f_{\mathrm{d}}(E,t)\right)\right]\right.$$
$$\left.+(1-\beta_{\mathrm{R}})\left[f_{\mathrm{d}}(E,t)\left(1-f_{\mathrm{d}}(E,t)\right)+f_0(E+eV_{\mathrm{R}})\left(1-f_0(E+eV_{\mathrm{R}})\right)\right]\right\},$$
$$\langle\eta_{\mathrm{L}}(E_1,t_1)\eta_{\mathrm{R}}(E_2,t_2)\rangle = 0 \qquad (12.152)$$

The stochastic variables $\eta_{\mathrm{L,R}}$ may be interpreted as terms describing current noise at a fixed electron energy. They can be related to fluctuations of the distribution functions of the incoming electrons $\zeta(t_1,t_2)$ introduced in the previous section. However, in contrast to $\zeta(t_1,t_2)$, the variables $\eta_{\mathrm{L,R}}$ describe fluctuations of the distribution function inside the quantum dot. The physical meaning of Eq. (12.151) is transparent: Due to the finite size of the quantum dot, fluctuations of the number of electrons passing through the contacts, within a given time interval and at a given energy, induce fluctuations of the distribution function $f_{\mathrm{d}}(E,t)$.

It is important to emphasize that Eq. (12.151) correctly describes only low-frequency fluctuations and becomes inaccurate at higher frequencies. As our main goal here is to account for interaction effects, it is necessary to further generalize our approach in a way that allows to include fluctuations of voltages V_{L} and V_{R} with arbitrary frequencies.

A rigorous derivation of the corresponding kinetic equation is rather complicated. For its detailed description, we refer the reader to the paper [164]. Here we only point out that the correct form of this kinetic equation may be guessed already from Eq. (12.151) if one simply rewrites the latter in terms of the Keldysh Green functions for both the dot

$$G_{\mathrm{d}}(t_1,t_2) = \int\frac{dE}{2\pi}\,e^{-\mathrm{i}E(t_1-t_2)}\left[1-2f_{\mathrm{d}}\left(E,\frac{t_1+t_2}{2}\right)\right] \qquad (12.153)$$

and the leads

$$G_j = \frac{-\mathrm{i}T}{\sinh[\pi T(t_1-t_2)]}e^{\mathrm{i}[\varphi_j^+(t_1)-\varphi_j^+(t_2)]},\quad j=\mathrm{L,R}. \qquad (12.154)$$

In terms of these functions, the kinetic equation (12.151) transforms into the following one:

$$\left(\frac{\partial}{\partial t_1}+\frac{\partial}{\partial t_2}\right)G_{\rm d}(t_1,t_2)=\Gamma_{\rm L}G_{\rm L}(t_1,t_2)+\Gamma_{\rm R}G_{\rm R}(t_1,t_2)$$
$$-(\Gamma_{\rm L}+\Gamma_{\rm R})G_{\rm d}(t_1,t_2)+\eta_{\rm L}(t_1,t_2)+\eta_{\rm R}(t_1,t_2) \qquad (12.155)$$

where the stochastic variables $\eta_{\rm L}(t_1, t_2)$, $\eta_{\rm R}(t_1, t_2)$ are characterized by the pair correlators

$$\langle\eta_{\rm L}(t_1,t_2)\eta_{\rm L}(t_3,t_4)\rangle=2\frac{\delta_{\rm d}^2 g_{\rm L}}{16\pi^2}\delta(t_3-t_2)\delta(t_1-t_4)$$
$$-\frac{\delta_{\rm d}^2 g_{\rm L}}{16\pi^2}\beta_{\rm L}\left[G_{\rm L}(t_3,t_2)G_{\rm d}(t_1,t_4)+G_{\rm d}(t_3,t_2)G_{\rm L}(t_1,t_4)\right]$$
$$-\frac{\delta_{\rm d}^2 g_{\rm L}}{16\pi^2}(1-\beta_{\rm L})\left[G_{\rm d}(t_3,t_2)G_{\rm d}(t_1,t_4)+G_{\rm L}(t_3,t_2)G_{\rm L}(t_1,t_4)\right] \qquad (12.156)$$

$$\langle\eta_{\rm R}(t_1,t_2)\eta_{\rm R}(t_3,t_4)\rangle=2\frac{\delta_{\rm d}^2 g_{\rm R}}{16\pi^2}\delta(t_3-t_2)\delta(t_1-t_4)$$
$$-\frac{\delta_{\rm d}^2 g_{\rm R}}{16\pi^2}\beta_{\rm R}\left[G_{\rm d}(t_3,t_2)G_{\rm R}(t_1,t_4)+G_{\rm R}(t_3,t_2)G_{\rm d}(t_1,t_4)\right]$$
$$-g_{\rm R}(1-\beta_{\rm R})\left[G_{\rm R}(t_3,t_2)G_{\rm R}(t_1,t_4)+G_{\rm d}(t_3,t_2)G_{\rm d}(t_1,t_4)\right] \qquad (12.157)$$

$$\langle\zeta_{\rm L}(t_1,t_2)\zeta_{\rm R}(t_3,t_4)\rangle=0 \qquad (12.158)$$

One can demonstrate [164] that the kinetic equation (12.155), together with Eqs. (12.156)–(12.158), already applies at all frequencies and will serve as a basis for our subsequent analysis.

Let us define the electric charge Q accumulated in the quantum dot. In the stationary case, i.e., provided the distribution function $f_{\rm d}(E)$ does not depend on time, one has

$$Q=-\frac{2e}{\delta_{\rm d}}\int dE\left[f_{\rm d}(E)-\frac{1}{1+e^{E/T}}\right] \qquad (12.159)$$

In a non-stationary situation, this formula has to be replaced by a more general one which involves the Green function $G_{\rm d}(t_1, t_2)$ and reads

$$Q(t)=\frac{2\pi e}{\delta_{\rm d}}\lim_{t'\to t}\left[G_{\rm d}(t,t')+\frac{iT}{\sinh[\pi T(t-t')]}\right] \qquad (12.160)$$

With this in mind, from Eq. (12.155) we obtain

$$\dot Q=-(\Gamma_{\rm L}+\Gamma_{\rm R})Q+\frac{\dot\varphi_{\rm L}^+}{eR_{\rm L}}+\frac{\dot\varphi_{\rm R}^+}{eR_{\rm R}}+\frac{2\pi e}{\delta_{\rm d}}\eta_{\rm L}(t,t)+\frac{2\pi e}{\delta_{\rm d}}\eta_{\rm R}(t,t) \qquad (12.161)$$

This equation can also be written in the form

$$\dot{Q} = \tilde{I}_{\rm L} + \tilde{I}_{\rm R} \tag{12.162}$$

where the currents

$$\begin{aligned} \tilde{I}_{\rm L} &= \frac{\dot{\varphi}_{\rm L}^{+}}{eR_{\rm L}} - \Gamma_{\rm L} Q + \frac{2\pi e}{\delta_{\rm d}} \eta_{\rm L}(t, t), \\ \tilde{I}_{\rm R} &= \frac{\dot{\varphi}_{\rm R}^{+}}{eR_{\rm R}} - \Gamma_{\rm R} Q + \frac{2\pi e}{\delta_{\rm d}} \eta_{\rm R}(t, t) \end{aligned} \tag{12.163}$$

account for the charge transfer through the contacts. Adding the displacement currents that describe recharging of the capacitors, we get

$$\begin{aligned} C_{\rm L}\frac{\ddot{\varphi}_{\rm L}^{+}}{e} + \frac{\dot{\varphi}_{\rm L}^{+}}{eR_{\rm L}} - \Gamma_{\rm L} Q + \frac{2\pi e}{\delta_{\rm d}} \eta_{\rm L}(t, t) &= I, \\ C_{\rm R}\frac{\ddot{\varphi}_{\rm R}^{+}}{e} + \frac{\dot{\varphi}_{\rm R}^{+}}{eR_{\rm R}} - \Gamma_{\rm R} Q + \frac{2\pi e}{\delta_{\rm d}} \eta_{\rm R}(t, t) &= -I, \\ \frac{V_x}{Z_{\rm S}(0)} - \int_{-\infty}^{t} dt' k_{\rm S}(t-t') \left(\frac{\dot{\varphi}_{\rm L}^{+}}{e} - \frac{\dot{\varphi}_{\rm R}^{+}}{e} \right) - \xi_{\rm S}(t) &= I \end{aligned} \tag{12.164}$$

Here we defined the total current I flowing in the external circuit; see Fig. 12.10. The accumulated charge Q can be found from Eq. (12.161):

$$\begin{aligned} Q(t) = \int_{-\infty}^{t} dt'\, e^{-(t-t')/\tau_{\rm D}} &\left[\frac{\dot{\varphi}_{\rm L}^{+}(t')}{eR_{\rm L}} + \frac{\dot{\varphi}_{\rm R}^{+}(t')}{eR_{\rm R}} \right. \\ &\left. + \frac{2\pi e}{\delta_{\rm d}} \eta_{\rm L}(t', t') + \frac{2\pi e}{\delta_{\rm d}} \eta_{\rm R}(t', t') \right] \end{aligned} \tag{12.165}$$

Combining Eqs. (12.164) and (12.165), we obtain

$$\begin{aligned} C_{\rm L}\frac{\ddot{\varphi}_{\rm L}^{+}}{e} + I_{\rm L}^{(0)}(\varphi_{\rm L}^{+}, \varphi_{\rm R}^{+}) + \int_{-\infty}^{t} dt' k_{\rm S}(t-t') \left(\frac{\dot{\varphi}_{\rm L}^{+}}{e} - \frac{\dot{\varphi}_{\rm R}^{+}}{e} \right) \\ = \frac{V_x}{Z_{\rm S}(0)} - \xi_{\rm L}(\varphi_{\rm L}^{+}, \varphi_{\rm R}^{+}) - \xi_{\rm S}, \\ C_{\rm R}\frac{\ddot{\varphi}_{\rm R}^{+}}{e} + I_{\rm R}^{(0)}(\varphi_{\rm L}^{+}, \varphi_{\rm R}^{+}) - \int_{-\infty}^{t} dt' k_{\rm S}(t-t') \left(\frac{\dot{\varphi}_{\rm L}^{+}}{e} - \frac{\dot{\varphi}_{\rm R}^{+}}{e} \right) \\ = -\frac{V_x}{Z_{\rm S}(0)} - \xi_{\rm R}(\varphi_{\rm L}^{+}, \varphi_{\rm R}^{+}) + \xi_{\rm S} \end{aligned} \tag{12.166}$$

These equations have the same form as Eqs. (12.98); however, both the currents $I^{(0)}_{\mathrm{L,R}}(\varphi^+_{\mathrm{L}}, \varphi^+_{\mathrm{R}})$ and the noise variables $\xi_{\mathrm{L,R}}(\varphi^+_{\mathrm{L}}, \varphi^+_{\mathrm{R}})$ differ from the corresponding terms in Eqs. (12.98). Namely, here we find

$$I^{(0)}_{\mathrm{L}}(\varphi^+_{\mathrm{L}}, \varphi^+_{\mathrm{R}}) = \frac{\dot{\varphi}^+_{\mathrm{L}}}{eR_{\mathrm{L}}} - \Gamma_{\mathrm{L}} \int_{-\infty}^{t} dt'\, e^{-\frac{t-t'}{\tau_{\mathrm{D}}}} \left[\frac{\dot{\varphi}^+_{\mathrm{L}}(t')}{eR_{\mathrm{L}}} + \frac{\dot{\varphi}^+_{\mathrm{R}}(t')}{eR_{\mathrm{R}}}\right],$$
$$I^{(0)}_{\mathrm{R}}(\varphi^+_{\mathrm{L}}, \varphi^+_{\mathrm{R}}) = \frac{\dot{\varphi}^+_{\mathrm{R}}}{eR_{\mathrm{R}}} - \Gamma_{\mathrm{R}} \int_{-\infty}^{t} dt'\, e^{-\frac{t-t'}{\tau_{\mathrm{D}}}} \left[\frac{\dot{\varphi}^+_{\mathrm{L}}(t')}{eR_{\mathrm{L}}} + \frac{\dot{\varphi}^+_{\mathrm{R}}(t')}{eR_{\mathrm{R}}}\right] \quad (12.167)$$

and, similarly,

$$\xi_{\mathrm{L}} = \frac{2\pi e}{\delta_{\mathrm{d}}} \left(\eta_{\mathrm{L}}(t,t) - \Gamma_{\mathrm{L}} \int_{-\infty}^{t} dt'\, e^{-\frac{t-t'}{\tau_{\mathrm{D}}}} \left[\eta_{\mathrm{L}}(t',t') + \eta_{\mathrm{R}}(t',t')\right]\right),$$
$$\xi_{\mathrm{R}} = \frac{2\pi e}{\delta_{\mathrm{d}}} \left(\eta_{\mathrm{R}}(t,t) - \Gamma_{\mathrm{R}} \int_{-\infty}^{t} dt'\, e^{-\frac{t-t'}{\tau_{\mathrm{D}}}} \left[\eta_{\mathrm{L}}(t',t') + \eta_{\mathrm{R}}(t',t')\right]\right) \quad (12.168)$$

We emphasize that Eqs. (12.167) and (12.168) do not require any disorder averaging since such averaging has already been carried out in the course of the derivation of Eq. (12.155). The internal dynamics of electrons inside the quantum dot is characterized by a single timescale τ_{D}. These are the main differences of our present approach as compared to the scattering matrix formalism discussed above. Of course, Eqs. (12.167) and (12.168) can also be reconstructed within the latter formalism employing averaging over realizations of the random $S-$matrices with the aid of Eq. (12.123).

Let us now rederive the $I-V$ curve for our system making use of the Bolzmann–Langevin approach. The total current I can be found from the relation

$$I = \langle I^{(0)}_{\mathrm{L}}(\varphi^+_{\mathrm{L}}, \varphi^+_{\mathrm{R}})\rangle + \langle \xi_{\mathrm{L}}(\varphi^+_{\mathrm{L}}, \varphi^+_{\mathrm{R}})\rangle \quad (12.169)$$

For simplicity, we again consider a voltage-biased quantum dot and set $Z_{\mathrm{S}}(\omega) = 0$. Then we have $\dot{\varphi}^+_{\mathrm{L}}/e - \dot{\varphi}^+_{\mathrm{R}}/e = V$ and

$$\langle I^{(0)}_{\mathrm{L}}(\varphi^+_{\mathrm{L}}, \varphi^+_{\mathrm{R}})\rangle = \frac{1}{eR_{\mathrm{L}}}\frac{\Gamma_{\mathrm{R}}\langle\dot{\varphi}^+_{\mathrm{L}}\rangle}{\Gamma_{\mathrm{L}}+\Gamma_{\mathrm{R}}} - \frac{1}{eR_{\mathrm{R}}}\frac{\Gamma_{\mathrm{L}}\langle\dot{\varphi}^+_{\mathrm{R}}\rangle}{\Gamma_{\mathrm{L}}+\Gamma_{\mathrm{R}}} = \frac{V}{R_{\mathrm{L}}+R_{\mathrm{R}}} \quad (12.170)$$

The average value of the noise variable ξ_{L} may be expressed in the form

$$\langle\xi_{\mathrm{L}}\rangle = \frac{2\pi e}{\delta}\frac{\Gamma_{R}\langle\eta_{\mathrm{L}}\rangle - \Gamma_{\mathrm{L}}\langle\eta_{\mathrm{R}}\rangle}{\Gamma_{\mathrm{L}}+\Gamma_{\mathrm{R}}} \quad (12.171)$$

As we have already demonstrated before, the average values $\langle\eta_{\mathrm{L,R}}\rangle$ differ from zero due to the dependence of these variables on the phase fluctuations $\delta\varphi = \delta\varphi_{\mathrm{L}}^{+} = \delta\varphi_{\mathrm{R}}^{+}$. Assuming that these fluctuations remain small, we obtain

$$\begin{aligned}\langle\eta_{\mathrm{L}}\rangle &= \langle\eta_{\mathrm{L}}\big(eV_{\mathrm{L}}t+\delta\varphi(t), eV_{\mathrm{R}}t+\delta\varphi(t)\big)\rangle \\ &= \int dt' \left\langle\left(\frac{\delta\eta_{\mathrm{L}}(t,t)}{\delta\varphi_{\mathrm{L}}(t')}+\frac{\delta\eta_{\mathrm{L}}(t,t)}{\delta\varphi_{\mathrm{R}}(t')}\right)\delta\varphi(t')\right\rangle + \text{subleading corrections}\end{aligned}$$

As before, the phase fluctuations are determined by the noise variables $\xi_{\mathrm{L,R}}$ as

$$\delta\varphi(t') = -\frac{1}{e}\int_{-\infty}^{t'} dt'' K(t'-t'')\big(\xi_{\mathrm{L}}(t'')+\xi_{\mathrm{R}}(t'')\big) \quad (12.172)$$

where $K(t)=\int d\omega e^{-\mathrm{i}\omega t}K_{\omega}/2\pi$ and

$$K_{\omega} = \frac{e^2}{-\omega^2 C_{\Sigma} - \mathrm{i}\omega\left(\frac{1}{R_{\mathrm{L}}}+\frac{1}{R_{\mathrm{R}}}\right)\frac{-\mathrm{i}\omega\tau_{\mathrm{D}}}{1-\mathrm{i}\omega\tau_{\mathrm{D}}}} \quad (12.173)$$

Employing Eqs. (12.168), we obtain the relation

$$\xi_{\mathrm{L}}(t'')+\xi_{\mathrm{R}}(t'') = \frac{2\pi e}{\delta_{\mathrm{d}}}\int dt''' P(t''-t''')\big(\eta_{\mathrm{L}}(t''',t''')+\eta_{\mathrm{R}}(t''',t''')\big) \quad (12.174)$$

where we defined

$$P(t) = \int\frac{d\omega}{2\pi}e^{-\mathrm{i}\omega t}\frac{-\mathrm{i}\omega\tau_{\mathrm{D}}}{1-\mathrm{i}\omega\tau_{\mathrm{D}}} \quad (12.175)$$

Combining now Eqs. (12.172)–(12.175), we find

$$\delta\varphi(t') = -\frac{2\pi}{\delta}\int_{-\infty}^{t'} dt''\tilde{K}(t'-t'')\big(\eta_{L}(t'',t'')+\eta_{\mathrm{R}}(t'',t'')\big) \quad (12.176)$$

where the Fourier transform of the function $\tilde{K}(t)$ reads

$$\tilde{K}_{\omega} = \frac{e^2}{-\omega^2 C_{\Sigma} - \mathrm{i}\omega\left(\frac{1}{R_{\mathrm{L}}}+\frac{1}{R_{\mathrm{R}}}\right)\frac{-\mathrm{i}\omega\tau_{\mathrm{D}}}{1-\mathrm{i}\omega\tau_{\mathrm{D}}}}\frac{-\mathrm{i}\omega\tau_{\mathrm{D}}}{1-\mathrm{i}\omega\tau_{\mathrm{D}}} \quad (12.177)$$

Making use of the fact that the noise variables η_{L} and η_{R} are uncorrelated, we have

$$\begin{aligned}\langle\eta_{\mathrm{L}}\rangle = -\frac{2\pi}{\delta_{\mathrm{d}}}\int dt'\int_{-\infty}^{t'} dt''\tilde{K}(t'-t'') \\ \left\langle\left(\frac{\delta\eta_{\mathrm{L}}(t,t)}{\delta\varphi_{\mathrm{L}}^{+}(t')}+\frac{\delta\eta_{\mathrm{L}}(t,t)}{\delta\varphi_{\mathrm{R}}^{+}(t')}\right)\eta_{\mathrm{L}}(t'',t')\right\rangle\end{aligned}$$

or, equivalently,

$$\langle \eta_{\mathrm{L}} \rangle = -\frac{2\pi}{\delta} \int dt' \int_{-\infty}^{t'} dt'' \tilde{K}(t'-t'') \\ \times \left(\frac{\delta}{\delta \varphi_{\mathrm{L}}^{+}(t')} + \frac{\delta}{\delta \varphi_{\mathrm{R}}^{+}(t')} \right) \left\langle \eta_{\mathrm{L}}(t,t) \eta_{\mathrm{L}}(t'',t'') \right\rangle \quad (12.178)$$

The correlator $\left\langle \eta_{\mathrm{L}}(t,t) \eta_{\mathrm{L}}(t'',t'') \right\rangle$ is defined in Eq. (12.156) and depends on the phase φ_{L}^{+} via the Green functions G_{L} and G_{d}. The latter has to be found from the kinetic equation (12.155). To the lowest order in $\eta_{\mathrm{L,R}}$, it reads

$$G_{\mathrm{d}}(t_1, t_2) = \frac{-\mathrm{i}T}{\sinh[\pi T(t_1 - t_2)]} \int_{-\infty}^{(t_1+t_2)/2} d\tau \exp\left(-\frac{t_1 + t_2 - 2\tau}{2\tau_{\mathrm{D}}} \right) \\ \left\{ \Gamma_{\mathrm{L}} \exp\left(\mathrm{i} \left[\varphi_{\mathrm{L}}^{+}\left(\tau + \frac{t_1 - t_2}{2} \right) - \varphi_{\mathrm{L}}^{+}\left(\tau - \frac{t_1 - t_2}{2} \right) \right] \right) \right. \\ \left. + \Gamma_{\mathrm{R}} \exp\left(\mathrm{i} \left[\varphi_{\mathrm{R}}^{+}\left(\tau + \frac{t_1 - t_2}{2} \right) - \varphi_{\mathrm{R}}^{+}\left(\tau - \frac{t_1 - t_2}{2} \right) \right] \right) \right\} \quad (12.179)$$

Combining these results and evaluating the functional derivative in Eq. (12.178), we find

$$\left(\frac{\delta}{\delta \varphi_{\mathrm{L}}^{+}(t')} + \frac{\delta}{\delta \varphi_{\mathrm{R}}^{+}(t')} \right) \left\langle \eta_{\mathrm{L}}(t,t) \eta_{\mathrm{L}}(t'',t'') \right\rangle = \frac{\delta_d^3}{32\pi^3} g_{\mathrm{L}} g_{\mathrm{R}} \beta_{\mathrm{L}} \\ \times \frac{T^2 \sin[eV(t-t'')]}{\sinh^2[\pi T(t-t'')]} \left[e^{-\frac{t-t'}{\tau_{\mathrm{D}}}} - \tau_{\mathrm{D}} \delta(t-t') - e^{-\frac{t''-t'}{\tau_{\mathrm{D}}}} + \tau_{\mathrm{D}} \delta(t''-t') \right]$$

Substituting this result into Eq. (12.178) and changing to the frequency domain, we arrive at the result

$$\langle \eta_{\mathrm{L}} \rangle = -\frac{\delta_{\mathrm{d}}}{4\pi} \frac{g_{\mathrm{L}} g_{\mathrm{R}} \beta_{\mathrm{L}}}{(g_{\mathrm{L}} + g_{\mathrm{R}})^2} \int \frac{d\omega}{2\pi} \frac{1}{\omega} \left(\frac{1}{1+\omega^2\tau^2} - \frac{1}{1+\omega^2\tau_{\mathrm{D}}^2} \right) \\ \times \left((\omega + eV) \coth \frac{\omega + eV}{2T} - (\omega - eV) \coth \frac{\omega - eV}{2T} \right) \quad (12.180)$$

The expression for the average $\langle \eta_{\mathrm{R}} \rangle$ is derived analogously. It is obtained from Eq. (12.180) by replacing $\beta_{\mathrm{L}} \to \beta_{\mathrm{R}}$ and by inverting the sign in front of the whole expression. Finally, with the aid of Eq. (12.171), we establish the average value of the current noise $\langle \xi_{\mathrm{L}} \rangle$, which, as expected, coincides with our result (12.127) derived in Section 12.4.2.

Chapter 13

Coulomb Blockade in Quantum Dot Chains and Metallic Wires

13.1 Introduction and Basic Model

In Chapter 11, we elucidated a close relation between interaction effects and shot noise in short coherent conductors. In particular, the leading interaction correction to the current through such conductors is determined by the second cumulant of the current operator, i.e., by the power spectrum of the shot noise. Moreover, the interaction correction to the shot noise was, in turn, found proportional to the third cumulant of the current operator. Even more generally, the RG analysis developed in Section 11.3.1 allows to conclude that the lowest-order interaction correction to the n-th current cumulant is determined by $1 + n$-th current cumulant for all values of n. Since the current cumulants, in the absence of interactions, can be conveniently analyzed within the FCS formalism [269], the above observations provide a great deal of information about the effect of electron–electron interactions as well.

In Chapter 12, we extended this analysis to metallic quantum dots. A close relation between shot noise and interaction effects exists there too, although the corresponding physics becomes

Dissipative Quantum Mechanics of Nanostructures: Electron Transport, Fluctuations, and Interactions
Andrei D. Zaikin and Dmitry S. Golubev

ISBN 978-981-4774-50-5 (Hardcover), 978-0-429-29823-3 (eBook)
www.jennystanford.com

somewhat more complex, as it now essentially depends on the relation between the electron dwell time τ_{D} and such parameters as temperature, applied voltage as well as both electron energy and phase relaxation times τ_{in} and τ_{φ} inside the dot. In this chapter, we will relax a size restriction for our conductors and further extend our analysis to virtually arbitrary metallic systems where interacting electrons may propagate at a large distance and suffer numerous scattering events at an arbitrary number of scatterers.

Our previous analysis demonstrated that the formalism of quasiclassical Langevin equations appears to be the most adequate tool enabling one to analyze the relation between shot noise and interaction effects in a physically transparent manner. For instance, the interaction correction to the current can be derived within this formalism simply by taking into account the noise contribution in the equation describing the balance of currents across the scatterer. In previous chapters, this program was accomplished for relatively compact scatterers, in which case the description of interaction effects with the aid of a uniform in space fluctuating field φ is sufficient. For spatially extended conductors to be addressed here, the coordinate dependence of this field cannot anymore be disregarded and the whole analysis needs to be modified. As we will demonstrate below in this chapter, such modification is simple if one models the system, e.g., by an array of scatterers connected between each other via relatively big metallic islands where the electron distribution function remains in equilibrium. The latter assumption is appropriate provided the electron dwell time τ_{D} inside each island remains much longer than the inelastic relaxation time τ_{in}. If this condition is violated, the electron distribution function is driven out of equilibrium and the whole consideration becomes much more involved. Nevertheless, also in this case, it is possible to formulate the generalized Langevin equation approach, which allows to describe shot noise in arrays (or chains) of metallic quantum dots and to derive a complete formula for the interaction correction to the I–V curve of disordered metallic conductors.

Below we will model a metallic conductor by a chain of $N - 1$ chaotic quantum dots, which is schematically depicted in Fig. 13.1. Each of these quantum dots can be viewed as an island in between two scatterers/barriers, which in turn connect adjacent quantum

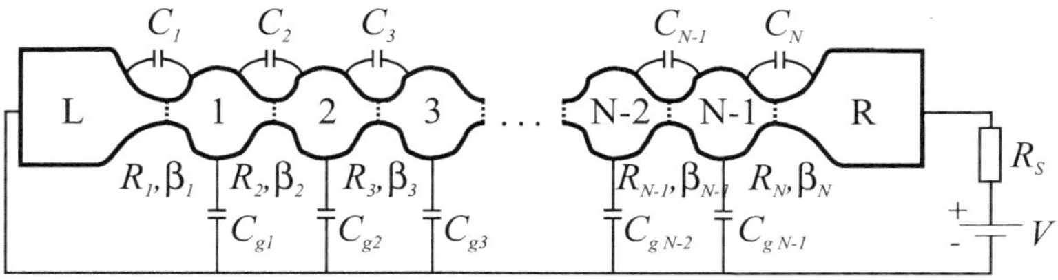

Figure 13.1 A chain of $N - 1$ chaotic quantum dots and N barriers. The n-th quantum dot is characterized by mean level spacing δ_n and gate capacitance $C_{\mathrm{g}n}$. The n-th barrier is described by its Landauer conductance $1/R_n$, capacitance C_n, and Fano factor β_n. The array is placed in between two big metallic reservoirs, which are connected to the voltage source via Ohmic resistor R_{S}.

dots. Electrons can enter the dot through one of the barriers, spend some time there, possibly being scattered at the barriers, outer walls or otherwise, and finally leave the dot through another barrier.

As before, electron scattering at each of the N barriers will be accounted for within the most general scattering matrix formalism. This formalism was already described in details in Chapter 10. Introducing again effective transmissions $T_k^{(n)}$ of the conducting modes in the n-th barrier, we define its Landauer conductance $1/R_n = (e^2/\pi)\sum_k T_k^{(n)}$, dimensionless conductance $g_n = 2\pi/e^2 R_n$, and the Fano factor $\beta_n = \sum_k T_k^{(n)}(1-T_k^{(n)})/\sum_k T_k^{(n)}$. The capacitance of the n-th barrier will be denoted by C_n. The n-th dot will be characterized by the mean level spacing $\delta_n = 1/N_0\mathcal{V}_n$, where $\mathcal{V}_n$ is the dot volume and N_0 is the density of states at the Fermi level. For the sake of generality, we will also assume that each dot has an additional capacitance to the ground $C_{\mathrm{g}n}$. Finally, the first and the last scatterers are connected to two big metallic reservoirs, which in turn are connected to the voltage source via external leads with an Ohmic resistance R_{S}.

The above model embraces practically all types of disordered conductors ranging from tunnel junction arrays to diffusive metallic wires. In the latter case, each scatterer can account, e.g., for a point-like impurity or a boundary imperfection, thereby enabling one to microscopically treat every scattering event of electrons propagating inside the wire. As we will see below in this chapter and also in Chapter 15, this model proves very useful in describing electron–electron interaction effects in disordered metallic conductors.

13.2 Strong Inelastic Relaxation Limit

We begin our analysis from the most simple strong inelastic relaxation limit. Specifically, we will require the dwell time τ_{D} in a single quantum dot to be much longer than the inelastic relaxation time τ_{in}:

$$\tau_{\text{D}} \gg \tau_{\text{in}} \tag{13.1}$$

In this case, the electron distribution function is locally maintained in its equilibrium form inside each quantum dot (or metallic island) depicted in Fig. 13.1.

13.2.1 Highly Resistive Tunnel Barriers

Let us first address the limit of highly resistive tunnel barriers with $g_n \ll 1$. Since $\tau_{\text{D}} \sim e^2 R_n/\delta_n$, the condition (13.1) is usually well satisfied in this limit provided the level spacing δ_n remains sufficiently small, which is typical for metallic islands. In this case, at sufficiently high temperatures $T > e^2/2C_n$, the differential conductance of the array is dominated by sequential tunneling. For identical junction resistances $R_1 = R_2 = \ldots = R_N = R$ and capacitances $C_1 = C_2 = \ldots = C_N = C$, it can be evaluated easily and reads [137, 325]

$$\frac{dI}{dV} = \frac{1}{NR}\left(1 - \frac{N-1}{N}\frac{e^2}{2CT}\frac{\frac{eV}{2NT}\cosh\frac{eV}{2NT} - \sinh\frac{eV}{2NT}}{\sinh^3\frac{eV}{2NT}}\right) \tag{13.2}$$

At lower temperatures, the system enters the regime of strong Coulomb blockade. In this case, the sequential tunneling contribution G_0^{seq} to the zero-bias conductance of our system is exponentially suppressed as

$$G_0^{\text{seq}} \sim e^{-U/T} \tag{13.3}$$

where the activation barrier U depends on the details of the model. Here U denotes the electrostatic energy of the charge configuration with one electron located in the middle of the array. For example, in the weak screening limit $G_g \to 0$, the potentials of the dots only linearly decay with the distance from the charged dot. As a result, the activation barrier grows with the number of the junctions in the chain, $U = Ne^2/8C$. In the regime of stronger screening, $\sqrt{C/C_g} \ll N$ (but still $C_g \ll C$), the potentials decay exponentially, U becomes N-independent, and one finds $U \approx e^2/2\sqrt{C_g C}$. Finally, at $C = 0$ one simply has $U = e^2/2C_g$.

At T, $eV/N \lesssim E_C$, the current is also sensitive to the gate voltage. Unfortunately, no simple theory describing the gate modulation of the current in chains of tunnel barriers can be constructed. As we already observed in Chapter 12, the stability diagram of a quantum dot (i.e., the color plot representing the differential conductance $dI(V, V_g)/dV$ as a function of both the gate and transport voltages) reveals the characteristic diamond structure; see Fig. 12.2. The stability diagram of a realistic tunnel junction chain usually looks rather messy since the islands may have different sizes and gate capacitances generating different periods in the diamond structure. Then the resulting picture represents a nonlinear overlap of the diamond structures from individual quantum dots. Quite generally, one can show that the borders between various parts of this diagram are formed by N sets of parallel lines, i.e., every tunnel junction in the chain generates its own set of lines [96].

In the long chain with $N \gg 1$, the gate-voltage-induced current oscillations have different periods and phases and, hence, almost cancel each other. In this case, the observed value of the current may be obtained by performing an average over the random fluctuations of the gate charges $Q_{gn} = C_{gn} V_{gn}$ under the assumption that they are independent and homogeneously distributed in the interval $-0.5 < Q_{gn} < 0.5$. Monte Carlo analysis of this problem was carried out [302] for a chain of quantum dots with approximately equal sizes and, hence, with equal junction and gate capacitances, $C_n = C$, $G_{gn} = C_g$ for $n = 1, 2, \ldots N$. In this case, it was demonstrated that at $T = 0$ the transport current is fully suppressed (i.e., the chain exhibits strong Coulomb blockade) at bias voltages below a certain threshold value

$$V_{th} = \frac{eN}{C_g} \alpha \left(\frac{C}{C_g} \right) \tag{13.4}$$

Here $\alpha(C/C_g)$ is the dimensionless function, which equals to $\alpha(C/C_g) = 0.5$ at $C \ll C_g$ and $\alpha(C/C_g) = 0.1C_g/C$ at $C \gg C_g$. For $V > V_{th}$, the current grows linearly as

$$I \approx \frac{e}{2RC_g}\left(\frac{V}{V_{th}} - 1\right) \tag{13.5}$$

Thus, the shape of the I–V curve of the quantum dot chain resembles that of the I–V curve of a single tunnel junction embedded in a high Ohmic environment; see Eq. (7.18). Zero-bias conductance of a long chain was studied numerically in the limit of weak screening $C_g \ll C$ with the result [217]

$$G_0^{\rm seq} \approx e^{-0.35e^2/CT}/NR \tag{13.6}$$

which indicates that the activation barrier is much smaller than $U = Ne^2/8C$ found for an ideal array with identical junction parameters, see above. The origin of this reduction is transparent: Owing to the random distribution of the gate charges in the array, some of the dots turn out to be almost open and their conductances are suppressed only weakly.

As in the case of single quantum dots, at sufficiently low temperatures, the sequential tunneling contribution to the linear conductance (13.3) becomes exponentially small, and the cotunneling contribution gains importance. For single quantum dots, this electron transport mechanism was already analyzed in details in Section 12.2.3. Here we extend our analysis to a linear chain of quantum dots. For the sake of convenience and also for pedagogical purposes, here we will illustrate an alternative approach [148] that employs path integrals in combination with the ImF-method based on Eq. (4.10).

For simplicity, here we will ignore the gate effect and set all capacitances to the ground equal to zero, i.e., $C_{gn} = 0$. Then the grand partition function of our chain can be expressed in the form

$$\mathcal{Z} = \sum_{m_n} \int \prod_{n=1}^{N} d\varphi_n \int_{\varphi_n}^{\varphi_n+2\pi m_n} \mathcal{D}\varphi_n \int \mathcal{D}q \exp\left(-\sum_{n=1}^{N} S_n[\varphi_n] - W[\varphi_n, q]\right) \tag{13.7}$$

where

$$S_n[\varphi_n] = \int_0^\beta d\tau \frac{C_n}{2}\left(\frac{\dot{\varphi}_n}{e}\right)^2 - \frac{g_n T^2}{4}\int_0^\beta d\tau \int_0^\beta d\tau' \frac{\cos[\varphi_n(\tau) - \varphi_n(\tau')]}{\sin^2[\pi T(\tau - \tau')]} \tag{13.8}$$

is the action of the n-th tunnel barrier (cf. Eq. (7.87)), which depends on the phase variable φ_n related to the fluctuating voltage V_n across this barrier as $V_n = \dot{\varphi}_n/e$. The term $W[\varphi_n, q]$ accounts for the effect of an external circuit. It reads

$$W[\varphi_n, q] = \mathrm{i}\int_0^\beta d\tau \left(\sum_{n=1}^{N} \dot{\varphi}_n\right)\frac{q}{e} + \frac{T}{2}\sum_l |\omega_l| Z_S(-\mathrm{i}\omega_l)|q_{\omega_l}|^2 - \int_0^\beta d\tau V q(\tau) \tag{13.9}$$

Here $\omega_l = 2\pi l T$ and the variable $q(\tau)$ defines the fluctuating current in the external circuit $I(\tau) = \dot{q}$; see Section 6.5 for more details.

Provided the voltage bias V differs from zero, the problem becomes unstable implying that the free energy of our system F formally diverges and acquires an imaginary part. This situation was repeatedly encountered, e.g., in Chapters 4–7. In our case, it simply means that electrons tunnel across the barriers between different islands and, hence, the nonzero net current I flows through our system. As in the low-temperature limit this current is predominantly determined by electron cotunneling $I = e\Gamma_{\mathrm{cot}}$, the task at hand is to evaluate the cotunneling rate Γ_{cot} by means of the formula

$$\Gamma_{\mathrm{cot}} = -2\mathrm{Im}\delta F_{\mathrm{cot}} \tag{13.10}$$

where δF_{cot} is the contribution to the free energy that accounts for the cotunneling process. In order to extract this contribution, we have to bear in mind that cotunneling in our chain of tunnel junctions just constitutes of (almost simultaneous) single-electron tunneling events across each of the N junctions. Hence, in order to identify the term δF_{cot}, it is necessary to expand the expression for the grand partition function (13.7)–(13.9) in powers of g_n keeping only the terms proportional to the product $\prod_{n=1}^{N} g_n$ and describing only the forward tunneling of electrons, which effectively implies the replacement

$$\cos[\varphi_n(\tau) - \varphi_n(\tau')] \to \exp\{\mathrm{i}[\varphi_n(\tau) - \varphi_n(\tau')]\}/2.$$

Performing the path integrals over both phase and charge variables in these terms, after simple algebra we obtain

$$\delta F_{\text{cot}} = (-1)^{N+1} T \int_0^{\beta} \dots \int_0^{\beta} \prod_{n=1}^{N} \left[d\tau_n d\tau_n' \frac{g_n T^2}{4\sin^2[\pi T(\tau_n - \tau_n')]} \right]$$
$$\times \exp\left[eV \sum_{n=1}^{N} \frac{C_0}{C_n}(\tau_n' - \tau_n) \right] \exp\left[- \sum_{n=1}^{N} \sum_{p=1}^{N} \frac{E_{np}}{2} \right.$$
$$\left. \times \int_0^{\beta} d\tau [\theta(\tau - \tau_n) - \theta(\tau - \tau_p) - \theta(\tau - \tau_n') + \theta(\tau - \tau_p')]^2 \right]$$
$$\times \exp\left[\sum_{n=1}^{N} \sum_{p=1}^{N} E_{np} [F(\tau_n - \tau_p) + F(\tau_n' - \tau_p') - F(\tau_n - \tau_p') - F(\tau_n' - \tau_p)] \right] \tag{13.11}$$

where we defined $E_{np} = e^2 C_0/(2C_n C_p)$,

$$C_0 = \left(\sum_{n=1}^{N} \frac{1}{C_n} \right)^{-1}, \quad F(\tau) = T \sum_l \frac{|\omega_l| Z_S(-i\omega_l)}{1/C_0 + |\omega_l| Z_S(-i\omega_l)} \frac{1 - \cos(\omega_l \tau)}{\omega_l^2} \tag{13.12}$$

Note that the integrals in Eq. (13.11) run over the time moments τ_n and τ_n' at which single-electron tunneling events occur in the n-th junction. It is easy to observe that the main contribution to δF_{cot} (13.11) comes from the integration interval with $\tau_1 \approx \tau_2 \approx \dots \approx \tau_N$, $\tau_1' \approx \tau_2' \approx \dots \approx \tau_N'$ (implying that single electrons tunnel across all N junctions approximately at the same times) and large values $\tau_n - \tau_n'$. With this in mind, we make the following shift in the integration parameters: $\tau_1 \to \tau_1$, $\tau_1' \to \tau + \tau_1$, while for $n \geq 2$, we replace $\tau_n \to \tau_1 + \tau_n$, $\tau_n' \to \tau + \tau_1 + \tau_n'$. Besides that we assume $\tau_n, \tau_n' \ll \tau$ for $n \geq 2$. Then we observe that the integrand in Eq. (13.11) no longer depends on τ_1. Hence, the integral over this time just produces the factor $\beta = 1/T$, which cancels the temperature T standing in front of the whole expression. Next we replace all the differences $\tau_n - \tau_p'$ by τ. Considering the limit of sufficiently low temperatures, we finally arrive at the result

$$\delta F_{\text{cot}} = (-1)^{N+1} \prod_{n=1}^{N} g_n \int_0^{\beta} d\tau \frac{A^2 (T/2)^{2N}}{\sin^{2N}[\pi T \tau]} \exp\left[-\frac{e^2 F(\tau)}{C_0} + eV\tau \right] \tag{13.13}$$

where we have explicitly indicated the integral over τ and combined all the remaining integrals over τ_n, τ_n', $n \geq 2$, into a constant prefactor A^2 with

$$A = \int_{-\infty}^{\infty} \prod_{n=2}^{N} d\tau_n \exp\left[-\sum_{p=2}^{N} \frac{e^2 C_0}{C_1 C_p} G(\tau_p) - \sum_{n=2}^{N-1} \sum_{p=n+1}^{N} \frac{e^2 C_0}{C_n C_p} G(\tau_n - \tau_p)\right] \tag{13.14}$$

and

$$G(\tau) = \frac{|\tau|}{2} - F(\tau) = \int_0^{\infty} \frac{d\omega}{\pi} \frac{1 - \cos\omega\tau}{\omega^2[1 + \omega Z_S(-i\omega) C_0]} \tag{13.15}$$

In order to evaluate an imaginary part of δF_{cot}, we perform an analytic continuation of the expression (13.13). Proceeding in exactly the same way as, e.g., in Chapter 5 (cf. Eqs. (5.42)–(5.45)), we arrive at the result for $\text{Im}\delta F_{\text{cot}}$ and, with the aid of Eq. (13.10), find both the cotunneling rate and the current $I = e\Gamma_{\text{cot}}$. In the zero-temperature limit and in case of Ohmic dissipation, i.e., for $Z_S(\omega) = R_S$, we obtain [148]

$$I = \frac{2\pi e A^2}{\Gamma(2N + 2/g_S)} \left[\prod_{n=1}^{N} \left(\frac{g_n}{4\pi^2}\right)\right] (eV)^{2N-1} \left(e^{-\gamma} eV \tau_S\right)^{2/g_S} \tag{13.16}$$

where we denoted $g_S = 2\pi/e^2 R_S$ and $\tau_S = R_S C_0$. At higher temperatures $T \gg eV$, the same analysis yields

$$I = V \frac{2\pi e^2 A^2 \Gamma^2(2N + 1/g_S)}{\Gamma(2N + 2/g_S)} \left[\prod_{n=1}^{N} \left(\frac{g_n}{4\pi^2}\right)\right] (2\pi T)^{2N-2} \left(2\pi e^{-\gamma} T \tau_S\right)^{2/g_S} \tag{13.17}$$

Equations (13.16) and (13.17) determine the current through a chain of weakly conducting tunnel barriers with $g_n \ll 1$ in the Coulomb blockade regime

$$\max(eV, T) \ll \min(e^2/C_n, 1/\tau_S) \tag{13.18}$$

The value of the parameter A can be determined analytically in several limiting cases. In the limit of very weak dissipation $g_S \ll 1$, we obtain

$$G(\tau) \approx -\frac{\tau^2}{2\pi\tau_S} \left(\ln\frac{|\tau|}{\tau_S} + \gamma - \frac{3}{2}\right)$$

and

$$A \approx \left(\frac{2\pi g_S}{\ln\frac{1}{g_S} - 2\gamma + 3}\right)^{(N-1)/2} \left[\prod_{n=1}^{N} \left(\frac{C_n}{C_0}\right)\right]^{1/2} \tau_S^{N-1} \tag{13.19}$$

In the opposite limit of strong dissipation in the external circuit $g_S \gg 1$ and for identical capacitances $C_n = C$, we get $G(\tau) = |\tau|/2$ and

$$A = \frac{N^N}{E_C^{N-1}(N-1)!}, \qquad E_C = \frac{e^2}{2C} \tag{13.20}$$

The result (13.16) demonstrates that at $T \to 0$, the I–V curve of our system is highly nonlinear and also essentially depends on the external impedance Z_S. While each junction is almost decoupled from the external circuit due to the presence of other junctions and, hence, single-electron tunneling in any of the junctions is insensitive to the external impedance, the electron cotunneling process turns out to be much more sensitive to the particular value of $Z_S(\omega)$. Indeed, for small $Z_S(\omega)$, charge relaxation in the external circuit is fast and thus effectively destroys Coulomb blockade of electron cotunneling in the whole chain of junctions. Accordingly, in the limit $Z_S \to 0$, the current $I(V)$ reaches its maximum and Eqs. (13.16) and (13.20) reduce to a simple result [35]

$$I = eE_C \left(\prod_{n=1}^{N} g_n\right) \frac{N^{2N}}{((N-1)!)^2\,(2N-1)!} \left(\frac{CV}{\pi e}\right)^{2N-1} \tag{13.21}$$

On the other hand, for large $Z_S(\omega)$, electron cotunneling is strongly suppressed by Coulomb effects, thus making charge transfer across the system highly improbable at low temperatures and voltages.

According to Eqs. (13.16)–(13.21), the cotunneling probability quickly decays with growing number of junctions in the chain. As a result, at sufficiently large N, one should also account for the process of so-called multiple cotunneling: An electron crosses the chain by several consecutive cotunneling jumps staying on quantum dots between them. It turns out that this transport mechanism may yield the dominating contribution to the current at large values of N.

The exact treatment of this complicated regime is problematic, and one can only perform a rough estimate for the corresponding contribution to the current [48, 49]. To this end let us assume that the charge Q_n is accumulated in the n-th quantum dot of our chain. The dynamics of the charge is governed by the equation

$$\frac{\partial Q_n}{\partial t} = \sum_j \Gamma_{\text{cot}}(j) Q_{n-j} - \sum_j \Gamma_{\text{cot}}(j) Q_n \tag{13.22}$$

where $\Gamma_{\rm cot}(j)$ is the probability of the electron cotunneling through j junctions. The charges Q_n may switch between three allowed values: $-e$, 0, e. Averaging Eq. (13.22) over time and making use of the charge conservation, one can approximately express the current through the n-th junction in the form

$$I_n = \sum_j \Gamma_{\rm cot}(j)(\langle Q_{n-j}\rangle - \langle Q_n\rangle) \tag{13.23}$$

The average charge values $\langle Q_n\rangle$ may be related to the average potentials of the quantum dots V_n as $\langle Q_n\rangle \approx d_0 V(N-n)/N$, where d_0 is an average diameter of a metallic island in our array. Hence, we obtain $\langle Q_{n-j}\rangle - \langle Q_n\rangle \approx d_0 jV/N$. Assuming further that for positive V the forward cotunneling rate strongly exceeds the backward one, we get

$$I \approx \frac{d_0 V}{N}\sum_{j=1}^{\infty} j\Gamma_{\rm cot}(j) \tag{13.24}$$

This expression defines the current as the sum of partial contributions from cotunneling processes at different lengths of our chain. The rate of such processes $\Gamma_{\rm cot}(j)$ can be derived from (13.13) if one replaces N by j and eV by the electrostatic energy difference $-\Delta E(j)$ associated with the transfer of one electron via j junctions. In addition, one should set $g_{\rm S} \to \infty$. Then with the aid of Eq. (5.45), one finds

$$\begin{aligned}\Gamma_{\rm cot}(j) &= \frac{g^j E_{\rm C}}{2\pi}\left(\frac{T}{E_{\rm C}}\right)^{2j-1}\frac{j^{2j}e^{-\Delta E(j)/2T}}{\Gamma^2(j)\Gamma(2j)}\left|\Gamma\left(j - {\rm i}\frac{\Delta E(j)}{2\pi T}\right)\right|^2 \\ &\approx \frac{E_C}{2\pi}\left(\frac{ge^2}{4}\right)^j\sqrt{\frac{j}{\pi}}\left(\frac{T^2}{E_{\rm C}^2} + \frac{\Delta E^2(j)}{4\pi^2 E_{\rm C}^2 j^2}\right)^{j-1/2} {\rm e}^{-\frac{\Delta E(j)}{2T}\left(1+\frac{2}{\pi}\arctan\frac{\Delta E(j)}{2\pi T j}\right)}\end{aligned} \tag{13.25}$$

Here the charging energy may roughly be estimated as $E_{\rm C} \approx e^2/d_0$. In order to evaluate $\Delta E(j)$, i.e., the electrostatic energy of our chain with an electron-hole pair separated by j junctions, it is necessary to know an explicit expression for the capacitance matrix of the chain. The simple model introduced in the beginning of this section neglects the mutual capacitances between distant islands. While such approximation remains applicable in the case of sufficiently large metallic islands, it may become inadequate in granular materials with a typical grain size of a few nanometers. In the latter case, screening of the electric field is weak and we may set

$$\Delta E(j) \approx \frac{E_C}{j} - \frac{eV}{N} j \tag{13.26}$$

Here the first term comes from the unscreened Coulomb interaction between an electron and a hole separated by j junctions, $E_C/j = e^2/jd_0$, and the second term describes the reduction in the potential barrier by an applied electric field. The expressions (13.24)–(13.26) hold at $eV/N,\ T \lesssim E_C$ and provide a sufficiently accurate estimate of the current, which applies for metallic granular arrays of any dimension as long as the current remains exponentially suppressed.

The above results for the current may be simplified further in certain limiting cases. Notice that the sum in Eq. (13.24) is dominated by one term with the biggest rate corresponding to the optimal cotunneling distance $j_{\rm opt}$. The latter can be estimated by finding the maximum value for the cotunneling rate (13.25). We obtain

$$j_{\rm opt} \approx \sqrt{\frac{E_C}{T \ln[E_C^2/gT^2]}}, \quad \text{at } \frac{eV}{N} \lesssim T\sqrt{\frac{T \ln[E_C^2/gT^2]}{E_C}},$$
$$j_{\rm opt} \approx \sqrt{\frac{NE_C}{eV}}, \quad \text{at } T \ln\frac{E_C^2}{gT^2} \lesssim \frac{eV}{N} \lesssim E_C \tag{13.27}$$

Accordingly, in the linear response regime

$$eV/N \lesssim T\sqrt{T \ln[E_C^2/gT^2]/E_C}$$

with the exponential accuracy, we find

$$I \sim V\frac{d_0 E_C}{N}\left(\frac{E_C}{T}\right)^{5/4} \exp\left[-\sqrt{\frac{4E_C}{T}\ln\frac{E_C^2}{gT^2}}\right] \tag{13.28}$$

We observe that in the multiple cotunneling regime, the linear conductance of our metallic granular arrays is proportional to the exponent $\propto \exp\left[-\sqrt{T^*/T}\right]$, which is also encountered as a characteristic feature of the so-called variable range hopping transport mechanism in semiconductors with impurities [109]. At sufficiently high bias, $eV/N \gtrsim T \ln[E_C^2/gT^2]$, one finds

$$I \sim V\frac{d_0 E_C}{N}\frac{E_C}{T}\left(\frac{NE_C}{eV}\right)^{3/4} \exp\left[-\sqrt{\frac{NE_C}{eV}\ln\frac{E_C^2}{gT^2}}\right] \tag{13.29}$$

The cotunneling effects discussed above have been observed in a number of different experiments. For instance, the power-law dependence $I \propto V^{2N-1}$ (13.21) was demonstrated by measuring the transport current in the vertical direction through two-dimensional granular arrays with N monolayers [378]. The regime of multiple cotunneling characterized by the temperature dependence of the linear conductance $G_0 \sim \exp\left[-\sqrt{T^*/T}\right]$ was observed in the experiments [377], while the exponential dependence of the current on the bias voltage (13.29) was detected in the experiment [97].

13.2.2 Low Resistive Barriers

Now let us turn to the case of low resistive barriers with $g_n \gg 1$. As we already demonstrated in Chapters 11 and 12, in this limit fluctuations of the phase variables φ_n remain weak and, hence, the system dynamics can be conveniently treated within the formalism of quasiclassical Langevin equations adopted to a chain of scatterers as depicted in Fig. 13.1. Below we will essentially follow the analysis [114, 148] generalizing it to the case of scatterers with arbitrary transmission distributions $T_k^{(n)}$.

The corresponding Langevin equation for the n-th scatterer can be written in the form

$$C_n \frac{\hbar \ddot{\varphi}_n}{e} + \frac{1}{R_n} \frac{\dot{\varphi}_n}{e} = \dot{q} + \xi_n \tag{13.30}$$

where, as before, the two terms in the left-hand side represent the displacement and Ohmic currents through this scatterer, the term $\dot{q}$ defines the instantaneous value of the current in our circuit and the stochastic variable ξ_n with the correlator

$$\begin{aligned} \langle \xi_n(t_1)\xi_n(t_2)\rangle = &\left(\frac{1-\beta_n}{R_n} + \frac{\beta_n}{R_n}\cos\left[\varphi(t_1) - \varphi(t_2)\right]\right) \\ &\times \left(-\frac{\pi T^2}{\sinh^2[\pi T(t_1 - t_2)]}\right) \end{aligned} \tag{13.31}$$

represents the current shot noise (cf., e.g., Eq. (11.10)). Setting $n = 1, 2, \ldots, N$ in Eq. (13.30), we get N equations for N phase variables φ_n and one (quasi)charge variable q. The remaining equation is obtained taking into account the balance of voltages in our circuit. Assuming for simplicity that the external impedance is purely Ohmic $Z_S(\omega) = R_S$, we have

$$V = \dot{q} R_S + \sum_{n=1}^{N} \frac{\dot{\varphi}_n}{e} - \xi_S R_S \tag{13.32}$$

where the stochastic variable ξ_S with the correlator

$$\langle \xi_S(t_1)\xi_S(t_2)\rangle = \frac{1}{R_S}\int \frac{d\omega}{2\pi}\omega \coth(\frac{\omega}{2T})e^{i\omega t} \tag{13.33}$$

accounts for the current noise in the resistor R_S. Equations (13.30)–(13.33) represent a closed set of equations that controls the system dynamics within the quasiclassical approximation. For $N = 1$, this set of equations reduces back to Eqs. (11.7)–(11.10) analyzed in Chapter 11.

The I–V curve for our chain of scatterers is obtained by averaging Eqs. (13.30) and (13.32) over the noise realizations. In this way, we get

$$I(V) = \langle \dot{q} \rangle = \frac{1}{R_\Sigma}\left\langle \sum_{n=1}^{N} \frac{\dot{\varphi}_n}{e} \right\rangle - \sum_{n=1}^{N} \frac{R_n}{R_\Sigma}\langle \xi_j \rangle \tag{13.34}$$

where

$$R_\Sigma \equiv \sum_{n=1}^{N} R_n \tag{13.35}$$

is the total resistance of the chain.

Let us denote small deviations of the phases and of the shot noise from their average values, respectively, as $\delta\varphi_n \equiv \varphi_n - \langle \varphi_n \rangle$ and $\delta\xi_n \equiv \xi_n - \langle \xi_n \rangle$. Excluding the current $\dot{q}$ from Eqs. (13.30) and (13.32), we obtain

$$C_n \frac{\delta\ddot{\varphi}_n}{e} + \frac{1}{R_n}\frac{\delta\dot{\varphi}_n}{e} + \frac{1}{R_S}\sum_{p=1}^{N} \frac{\delta\dot{\varphi}_p}{e} = \xi_S + \delta\xi_n \tag{13.36}$$

These equations can be formally solved with the result

$$\left.\frac{\delta\dot{\varphi}_n}{e}\right|_\omega = Z_n(\omega)\delta\xi_{n,\omega} + \sum_{p\neq n} a_{np}(\omega)\delta\xi_{p,\omega} \tag{13.37}$$

where $Z_n(\omega)$ is the total impedance "seen" by the j-th scatterer and the functions $a_{np}(\omega)$ describe the mutual influence of the scatterers on each other.

Note that Eq. (13.37) applies to an arbitrary array of scatterers of any dimensionality. In the particular case of one-dimensional chains from Eq. (13.36), one has

$$Z_n(\omega) = \frac{R_n}{1 - \mathrm{i}\omega R_n C_n} \frac{R_S + \sum\limits_{p \neq n} \frac{R_p}{1 - \mathrm{i}\omega R_p C_p}}{R_S + \sum\limits_{p} \frac{R_p}{1 - \mathrm{i}\omega R_p C_p}} \tag{13.38}$$

From Eq. (13.37), one readily finds

$$\delta\varphi_n / e = \frac{1}{e^2} \int\limits_{-\infty}^{t} dt' [K_n(t - t')\delta\xi_n(t') A_{np}(t - t')\delta\xi_p(t')] \tag{13.39}$$

where the response function $K_n(t)$ is defined as

$$K_n(t) = e^2 \int \frac{d\omega}{2\pi} \frac{Z_j(\omega)}{-\mathrm{i}\omega + 0} \, \mathrm{e}^{-\mathrm{i}\omega t} \tag{13.40}$$

The functions $A_{np}(t)$ are defined analogously via $a_{np}(\omega)$. We do not specify these functions here because the contribution of the corresponding terms turns out to vanish since there exists no correlation between the current noise across different scatterers [114].

Equations (13.37) and (13.39) are the integral equations that can be solved by iterations. Here we restrict ourselves to the first iteration and put $\delta\varphi_n = 0$ in the right-hand side of Eq. (13.39). Evaluating the averages $\langle \xi_n \rangle$ in exactly the same way as in Chapter 11 (see Eqs. (11.14)–(11.16)), from Eq. (13.34) we eventually arrive at the expression for the I–V curve

$$I = \frac{V}{R_\Sigma} - \frac{1}{eR_\Sigma} \int\limits_0^\infty dt \, \frac{\pi T^2}{\sinh^2[\pi T t]} \left[\sum_{k=1}^{N} \beta_n K_n(t) \sin(eV_n t) \right] \tag{13.41}$$

The last term in this formula defines the interaction correction to the current flowing through our chain of scatterers.

In order to evaluate this interaction correction in a closed form, we restrict our further analysis to chains composed of identical scatterers with $R_n \equiv R$, $C_n \equiv C$, and $\beta_n \equiv \beta$. Then we get

$$Z_n(\omega) = \frac{R_S + (N-1)\frac{R}{1 - \mathrm{i}\omega RC}}{R_S\left(\frac{1}{R} - \mathrm{i}\omega C\right) + N} \tag{13.42}$$

$$K_n(t) = \frac{N-1}{N} e^2 R \left(1 - \mathrm{e}^{-t/RC}\right) + \frac{e^2 R_\mathrm{S} R}{N(R_\mathrm{S} + NR)} \left(1 - \mathrm{e}^{-\frac{R_\mathrm{S}+NR}{R_\mathrm{S}RC}t}\right) \tag{13.43}$$

and the current (13.41) becomes

$$I = \frac{Tv}{eR} - \frac{eT\beta}{\pi} \left[\frac{N-1}{N} F(v, u) + \frac{F(v, u_\mathrm{S})}{N(1 + N\frac{R}{R_\mathrm{S}})}\right] \tag{13.44}$$

Here we introduced the notations $v = eV/NT$, $u = 1/(2\pi T RC)$, $u_\mathrm{S} = u(1 + NR/R_\mathrm{S})$, and

$$\begin{aligned} F(v, u) = v\mathrm{Re}\left[\Psi\left(1 + u - \mathrm{i}\frac{v}{2\pi}\right) - \Psi\left(1 - \mathrm{i}\frac{v}{2\pi}\right)\right] \\ - 2\pi u\, \mathrm{Im}\Psi\left(1 + u - \mathrm{i}\frac{v}{2\pi}\right) \end{aligned} \tag{13.45}$$

Equations (13.44)–(13.45) define the I–V curve for a chain of identical highly conducting scatterers in the presence of Coulomb effects. As in Chapters 11 and 12, we again observe that the magnitude of the interaction correction to the current scales with the Fano factor β, indicating a close relation between shot noise and interaction effects also in such systems. In the limit $\beta \to 1$, Eqs. (13.44)–(13.45) reduce to the corresponding results [114, 148] derived for arrays of low resistive tunnel barriers.

The temperature and voltage dependence of the interaction correction is determined by the same combination of Ψ-functions as in the case of single coherent scatterers and quantum dots. Hence, at low enough temperatures and voltages, Eqs. (13.44)–(13.45) again yield the logarithmic dependence of the interaction correction on T and V.

Now let us put $R_\mathrm{S} = 0$ and consider the high-temperature limit $u \ll 1$. Then in the first order in u, we find

$$I = \frac{Tv}{eR} - \frac{N-1}{2N} \frac{e\beta}{RC} \left[\coth \frac{v}{2} - \frac{v}{2\sinh^2 \frac{v}{2}}\right] \tag{13.46}$$

The expression for the differential conductance dI/dV derived from Eq. (13.46) in the limit $\beta \to 0$ exactly coincides with the result (13.2) obtained for high-resistance tunnel barriers. As we already discussed, such kind of dependencies can be conveniently used, e.g., for the Coulomb blockade thermometry.

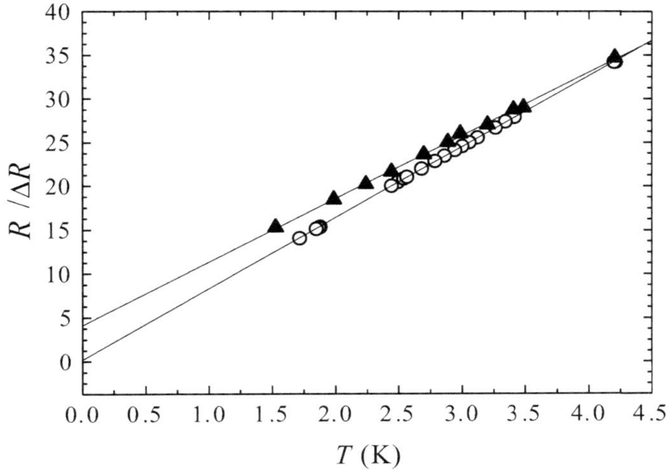

Figure 13.2 The temperature dependence of the zero-bias voltage anomaly, $R/\Delta R$, for two tunnel junction chains with per-junction resistances of 1.2 kΩ and 23 kΩ (solid triangles and open circles, respectively). The solid lines are linear fits to the experimental data [114].

The zero-bias conductance can be recovered also from Eqs. (13.44) and (13.45), which yield

$$R_\Sigma G_0 = 1 - \frac{N-1}{N}\frac{e^2\beta R}{\pi}\left\{\Psi(1+u) + \gamma + u\,\Psi'(1+u)\right\} \tag{13.47}$$

At high temperatures (small u), we find from (13.47)

$$R_\Sigma G_0 = 1 - \frac{N-1}{N}\frac{\beta E_C}{3T} + A\frac{N-1}{N}\frac{3\zeta(3)\beta}{2\pi^4}\frac{R_q}{R}\frac{E_C^2}{T^2} + \cdots \tag{13.48}$$

Within the first order in $u = \hbar/2\pi k_B T\,RC$, the inverse resistance enhancement at zero-bias voltage is obtained from Eq. (13.48) in the form

$$\frac{R}{\Delta R} = \frac{3N}{\beta(N-1)}\frac{T}{E_C} + a\frac{R_q}{R} + b \tag{13.49}$$

Here $\Delta R \equiv R(V=0) - R$, and a, b are known constants.

The temperature dependence $R/\Delta R$ measured for two tunnel junction arrays [114] is displayed in Fig. 13.2. In agreement with Eq. (13.49), one observes an almost perfect linear dependence of the value $R/\Delta R$ on temperature. This dependence provides an efficient way for an accurate experimental determination of per-junction capacitances in arrays of normal metal junctions.

13.3 Weak Inelastic Relaxation Limit

Now let us turn to the opposite limit of weak inelastic relaxation inside each quantum dot, i.e., similar to the analysis in Section 12.4, we will assume that both the electron inelastic relaxation time $\tau_{\rm in}^{(n)}$ and its dephasing time $\tau_{\varphi}^{(n)}$ inside the n-th quantum dot strongly exceed the electron dwell time $\tau_{\rm D}^{(n)}$ of the electron in this dot; see Eq. (12.84). Moreover, we will assume that the average time $\tilde{\tau}_{\rm D}$, which an electron spends in the whole array of quantum dots, is much shorter than the inelastic relaxation time $\tau_{\rm in}$, i.e., $\tilde{\tau}_{\rm D} \ll \tau_{\rm in}$. In this case, the electron distribution function can be driven out of equilibrium and, furthermore, the electron motion inside the dot remains coherent. As a result, the whole analysis of electron transport in the presence of electron–electron interactions becomes substantially more complicated as compared to that outlined above. Fortunately, in Section 12.4.4, we already developed a convenient Boltzmann–Langevin formalism, which appears to be the most adequate for such kind of situations. Below we will proceed within the same theoretical framework.

13.3.1 Kinetic Equation Analysis

The starting point of our analysis is the kinetic equation for the Keldysh Green function of the n-th quantum dot $G_n(t_1, t_2)$. It reads [164] (see also Eqs. (12.155)–(12.158))

$$\frac{4\pi}{\delta_n}\left(\frac{\partial}{\partial t_1}+\frac{\partial}{\partial t_2}+\mathrm{i}\dot{\Phi}_n(t_1)-\mathrm{i}\dot{\Phi}_n(t_2)\right)G_n$$
$$=g_nG_{n-1}+g_{n+1}G_{n+1}-(g_n+g_{n+1})G_n+\tilde{\eta}_n-\tilde{\eta}_{n+1} \qquad (13.50)$$

Here the phases Φ_n are related to the voltages V_n across the scatterers as

$$\dot{\Phi}_n(t)=\sum_{j=1}^{n}eV_j(t) \qquad (13.51)$$

and the stochastic variables $\tilde{\eta}_n$, which account for the current noise in the scatterers, are characterized by the correlators

$$\begin{aligned}&\langle\tilde{\eta}_n(t_1, t_2)\tilde{\eta}_n(t_3, t_4)\rangle = 2g_n\delta(t_3 - t_2)\delta(t_1 - t_4)\\ &- g_n\beta_n\left[G_{n-1}(t_3, t_2)G_n(t_1, t_4) + G_n(t_3, t_2)G_{n-1}(t_1, t_4)\right]\\ &- g_n(1 - \beta_n)\left[G_n(t_3, t_2)G_n(t_1, t_4) + G_{n-1}(t_3, t_2)G_{n-1}(t_1, t_4)\right] \quad (13.52)\end{aligned}$$

We also add that current noise in different scatterers is uncorrelated, i.e., $\langle\tilde{\eta}_n\tilde{\eta}_m\rangle = 0$ for $n \neq m$.

The boundary conditions for Eq. (13.50) take the form

$$\begin{aligned}G_0(t_1, t_2) &= \frac{-\mathrm{i}T}{\sinh[\pi T(t_1 - t_2)]},\\ G_N(t_1, t_2) &= \frac{-\mathrm{i}T e^{-eV(t_1-t_2)}}{\sinh[\pi T(t_1 - t_2)]}\end{aligned} \quad (13.53)$$

where G_0 and G_N are the Green functions of the leads. For simplicity, at this stage we ignore the effect of an external circuit and set $Z_S(\omega) \to 0$ (see Fig. 13.1). Hence, our system is biased directly by the external voltage V.

In full analogy with Section 12.4.4, let us employ the current conservation condition and express the dc current across our chain of scatterers as the average of the current through one of the scatterers

$$I = \left\langle \frac{V_n}{R_n} + \xi_n(t) \right\rangle \quad (13.54)$$

where the current noise $\xi_n(t)$ is defined as $\xi_n(t) = e\tilde{\eta}(t, t)/2$. Next we note that at any time and for any noise realizations, the sum of the voltages across the scatterers equals to the bias voltage V, i.e.,

$$\sum_j V_j(t) = V \quad (13.55)$$

With this in mind, Eq. (13.54) can be conveniently rewritten in the form

$$I = \frac{V}{R_\Sigma} + \sum_{n=1}^{N} \frac{R_n}{R_\Sigma}\langle\xi_n(t)\rangle \quad (13.56)$$

where R_Σ again denotes the total resistance of our chain (13.35).

As before, the last term in Eq. (13.56) defines the Coulomb interaction correction to the leading Ohmic term V/R_Σ. Below we will again restrict our attention to the metallic limit $g_n \gg 1$, in which case this interaction correction remains small and, hence, one can proceed perturbatively in the noise keeping only the lowest non-vanishing order terms in $\tilde{\eta}_n$.

As we already demonstrated in Chapters 11 and 12, the average $\langle \xi_n(t) \rangle$ generally differs from zero because of its dependence on the fluctuating voltages $V_n(t)$ or phases

$$\varphi_n(t) = \int_{-\infty}^{t} dt' \, eV_n(t') \tag{13.57}$$

These averages vanish $\langle \xi_n(t) \rangle \equiv 0$ only provided the voltages V_n do not depend on time, which is the case in the absence of electron–electron interactions. Obviously in this case, we just reproduce the standard Ohm's law $I = V/R_\Sigma$. In the presence of Coulomb interaction, both voltages V_n and phases φ_n fluctuate in time. Introducing $\delta V_n(t) = V_n(t) - \langle V_n(t) \rangle$ and $\delta\varphi_n(t) = \int_{-\infty}^{t} dt' \, e\delta V_n(t')$ and proceeding perturbatively in the phase fluctuations, we obtain

$$\langle \xi_n(t) \rangle = \sum_{m=1}^{N} \int dt' \left\langle \frac{\delta \xi_n(t)}{\delta \varphi_m(t')} \delta\varphi_m(t') \right\rangle . \tag{13.58}$$

The phase fluctuations $\delta\varphi_m(t)$ can, in turn, be formally expressed via the current noise variables as

$$\delta\varphi_m(t') = -\frac{1}{e} \sum_{k=1}^{N} \int dt'' K_{mk}(t' - t'') \xi_k(t'') \tag{13.59}$$

Employing these relations and bearing in mind that noise variables for different scatterers are uncorrelated, we get

$$\langle \xi_n(t) \rangle = -\frac{1}{e} \sum_{m=1}^{N} \int dt' dt'' K_{mn}(t' - t'') \frac{\delta}{\delta\varphi_m(t')} \langle \xi_n(t) \xi_n(t'') \rangle \tag{13.60}$$

The task at hand is to evaluate both the response function $K_{mn}(t)$ and the noise correlator $\langle \xi_n(t) \xi_n(t') \rangle$.

We begin with the response function $K_{mn}(t)$, which accounts for propagation of electric signals through our chain of scatterers. As in the case of a single quantum dot analyzed in Chapter 12, let us define the electric charges Q_n accumulated in the quantum dots:

$$Q_n(t) = \frac{2\pi e}{\delta_n} \lim_{t' \to t} \left[G_n(t, t') + \frac{\mathrm{i}T}{\sinh[\pi T(t - t')]} \right] + \frac{2e}{\delta_n} \dot{\Phi}_n(t) \tag{13.61}$$

With the aid of the kinetic equation (13.50), we find

$$\dot{Q}_n = \tilde{I}_n - \tilde{I}_{n+1} \tag{13.62}$$

where the current $\tilde{I}_n$ through the n-th scatterer reads

$$\tilde{I}_n = \frac{V_n}{R_n} + \frac{g_n\delta_{n-1}}{4\pi}Q_{n-1} - \frac{g_n\delta_n}{4\pi}Q_n + \xi_n \tag{13.63}$$

Adding the displacement currents, which recharge the capacitors, we arrive at the set of n equations describing the balance of currents in our chain

$$C_n\dot{V}_n + C_{gn}\sum_{j=1}^{n}\dot{V}_j + \frac{V_n}{R_n} + \frac{g_n\delta_{n-1}}{4\pi}Q_{n-1} - \frac{g_n\delta_n}{4\pi}Q_n = I - \xi_n \tag{13.64}$$

Since the charges do not accumulate in big metallic electrodes attached on both sides of our chain, the boundary conditions $Q_0 = Q_N = 0$ should be imposed. On the other hand, excess charges can accumulate across the capacitances implying that

$$Q_n = C_{n+1}V_{n+1} - C_nV_n - C_{gn}\sum_{j=1}^{n}V_j \tag{13.65}$$

Linearizing these equations and treating ξ_n as small variables, one can establish the relation between $\delta\varphi_m$ and ξ_k, which fixes the response function $K_{mn}(t)$. For instance, in an important special case of chains composed of identical scatterers and quantum dots with $C_n = C, C_{gn} = C_g, g_n = g, \tau_{\mathrm{D}}^{(n)} = \tau_{\mathrm{D}}, \beta_n = \beta$, and $R_n = R$ we obtain

$$K_{mn}(t) = \frac{2e^2}{N}\sum_{q=1}^{N-1}\int\frac{d\omega}{2\pi}\frac{Z_{\omega q}}{-\mathrm{i}\omega+0}e^{-\mathrm{i}\omega t} \times \cos\left[\frac{\pi q}{N}\left(m-\frac{1}{2}\right)\right]\cos\left[\frac{\pi q}{N}\left(n-\frac{1}{2}\right)\right], \tag{13.66}$$

where we defined an effective impedance

$$Z_{\omega q} = \frac{1}{\left(-\mathrm{i}\omega + \frac{1-\cos[\pi q/N]}{\tau_D}\right)\left(C + \frac{C_g}{2(1-\cos[\pi q/N])}\right) + \frac{1}{R}} \tag{13.67}$$

Now let us evaluate the functional derivative

$$\delta\langle\xi_n(t)\xi_n(t'')\rangle/\delta\varphi_m(t'),$$

which also enters in Eq. (13.60) for the noise average. Bearing in mind that $\xi_n(t) = e\tilde{\eta}(t, t)/2$, from Eq. (13.52) we find

$$\langle \xi_n(t)\xi_n(t'')\rangle = \frac{e^2}{2} g_n \delta(t''-t)\delta(t-t'')$$
$$-\frac{e^2}{4} g_n \beta_n \left[G_{n-1}(t'', t)G_n(t, t'') + G_n(t'', t)G_{n-1}(t, t'')\right]$$
$$-\frac{e^2}{4} g_n (1-\beta_n)\left[G_n(t'', t)G_n(t, t'') + G_{n-1}(t'', t)G_{n-1}(t, t'')\right] \quad (13.68)$$

The Green functions G_n can be derived from Eq. (13.50) perturbatively in the interaction. Setting $\tilde{\eta}_n = 0$ and $\Phi_n = \sum_{j=1}^{n} eV_j$ and solving Eq. (13.50), we recover the non-perturbed Green functions

$$G_n^{(0)}(t_1, t_2) = \frac{-\mathrm{i}T\left(1 - a_n + a_n e^{-\mathrm{i}eV(t_1-t_2)}\right)}{\sinh[\pi T(t_1-t_2)]}, \quad a_n = \sum_{j=1}^{n} \frac{R_j}{R_\Sigma} \quad (13.69)$$

The first-order corrections to the Green functions $\delta G_n \propto \delta\varphi_n$ obey the equation

$$\frac{4\pi}{\delta_n}\left(\frac{\partial}{\partial t_1} + \frac{\partial}{\partial t_2}\right)\delta G_n - g_n \delta G_{n-1} - g_{n+1}\delta G_{n+1} + (g_n + g_{n+1})\delta G_n$$
$$= -\mathrm{i}\frac{4\pi}{\delta_n}\left(\delta\dot{\Phi}_n(t_1) - \delta\dot{\Phi}_n(t_2)\right) G_n^{(0)}$$

Its solution can be expressed via the diffuson $D_{nm}(t)$, which just constitutes the fundamental solution to the discrete diffusion equation

$$\frac{\partial D_{n,m}}{\partial t} - \frac{\delta_n g_n}{4\pi} D_{n-1,m} - \frac{\delta_n g_{n+1}}{4\pi} D_{n+1,m} + \frac{\delta_n (g_n + g_{n+1})}{4\pi} D_{n,m} = \delta_{nm}\delta(t) \quad (13.70)$$

with the boundary conditions

$$D_{0m} = D_{Nm} = D_{n0} = D_{nN} = 0 \quad (13.71)$$

For example, in the case of homogeneous quantum dot chains, one finds

$$D_{nm}(t) = \frac{2}{N}\sum_{q=1}^{N-1}\int \frac{d\omega}{2\pi} e^{-\mathrm{i}\omega t} D_{\omega q} \sin\frac{\pi q n}{N} \sin\frac{\pi q m}{N} \quad (13.72)$$

where

$$D_{\omega q} = \frac{1}{-\mathrm{i}\omega + \frac{1}{\tau_{\mathrm{D}}}\left(1 - \cos\frac{\pi q}{N}\right)} \quad (13.73)$$

The leading correction to the Green function can now be expressed in the form

$$\delta G_n(t_1, t_2) = -\mathrm{i}\sum_m \int dt' D_{nm}\left(\frac{t_1+t_2}{2} - t'\right)\left(\delta\dot{\Phi}_m\left(t' + \frac{t_1-t_2}{2}\right)\right.$$
$$\left. - \delta\dot{\Phi}_m\left(t' - \frac{t_1-t_2}{2}\right)\right) G_m^{(0)}\left(t' + \frac{t_1-t_2}{2}, t' - \frac{t_1-t_2}{2}\right) \tag{13.74}$$

and we easily obtain

$$\frac{\delta G_n(t_1, t_2)}{\delta\Phi_m(t_3)} = \mathrm{i}G_m^{(0)}(t_1 - t_2)\frac{\partial}{\partial t_3}\left[D_{nm}(t_1 - t_3) - D_{nm}(t_2 - t_3)\right] \tag{13.75}$$

Employing the identity $\Phi_m = \sum_{j=1}^{m}\varphi_j$, one may also rewrite the above equation in the form

$$\frac{\delta G_n(t_1, t_2)}{\delta\varphi_m(t_3)} = -\sum_{k=m}^{N}\mathrm{i}G_k^{(0)}(t_1 - t_2)\left[\dot{D}_{nk}(t_1 - t_3) - \dot{D}_{nk}(t_2 - t_3)\right] \tag{13.76}$$

This formula can be further transformed with the aid of the diffusion equation (13.70), which may symbolically be expressed in the operator form as $D^{-1}D = 1$. Inverting this equation, we get $DD^{-1} = 1$ or, explicitly,

$$\frac{\partial D_{n,m}}{\partial t} - \frac{\delta_{m-1}g_m}{4\pi}D_{n,m-1} - \frac{\delta_{m+1}g_{m+1}}{4\pi}D_{n,m+1} + \frac{\delta_m(g_m + g_{m+1})}{4\pi}D_{n,m}$$
$$= \delta_{nm}\delta(t)$$

Furthermore, from the stationary limit of the kinetic equation for $G_k^{(0)}$, one trivially finds

$$(g_k + g_{k+1})G_k^{(0)} = g_k G_{k-1}^{(0)} + g_{k+1}G_{k+1}^{(0)} \tag{13.77}$$

With the aid of these two identities, one can transform the functional derivative of the Green function to the form

$$\frac{\delta G_n(t_1, t_2)}{\delta\varphi_m(t_3)} = -\mathrm{i}\left[\delta(t_1 - t_3) - \delta(t_2 - t_3)\right]G_n^{(0)}(t_1 - t_2)\theta(n - m + 0)$$
$$-\mathrm{i}\frac{\delta_{m-1}g_m}{4\pi}\left[D_{n,m-1}(t_1 - t_3) - D_{n,m-1}(t_2 - t_3)\right]G_m^{(0)}(t_1 - t_2)$$
$$+\mathrm{i}\frac{\delta_m g_m}{4\pi}\left[D_{n,m}(t_1 - t_3) - D_{n,m}(t_2 - t_3)\right]G_{m-1}^{(0)}(t_1 - t_2) \tag{13.78}$$

which, in contrast to Eq. (13.76), does not contain the summation over the quantum dot number k.

Making use of Eq. (13.78), one can evaluate the derivative of the noise correlator $\delta\langle\xi_n(t)\xi_n(t'')\rangle/\delta\varphi_m(t')$ and eventually the correction to the current (13.60). Collecting all relevant contributions, we arrive at the general expression for the current in the chain [164]:

$$I = \frac{V}{R_\Sigma} + \delta I_1 + \delta I_2 \tag{13.79}$$

where

$$\begin{aligned}\delta I_1 = \sum_{n,m=1}^{N} \frac{g_m}{4\pi^2 e R_\Sigma} \int dt dt' \frac{\pi^2 T^2 \sin eVt}{\sinh^2 \pi T t} K_{mn}(t-t') \\ \times \big[\delta_m(a_{m-1} - a_{n-1}) D_{n-1,m}(t') - \delta_{m-1}(a_m - a_{n-1}) D_{n-1,m-1}(t') \\ + \delta_m(a_{m-1} - a_n) D_{n,m}(t') - \delta_{m-1}(a_m - a_n) D_{n,m-1}(t')\big]\end{aligned} \tag{13.80}$$

and

$$\begin{aligned}\delta I_2 = -\sum_{n,m=1}^{N} \frac{\beta_n R_n}{\pi e R_\Sigma^2} \int dt dt' \frac{\pi^2 T^2 \sin eVt}{\sinh^2 \pi T t} K_{mn}(t-t') \bigg[\delta_{nm}\delta(t') \\ - \frac{g_m}{4\pi}\left(\delta_{m-1} D_{n-1,m-1}(t') - \delta_m D_{n-1,m}(t')\right) \\ - \frac{g_m}{4\pi}\left(\delta_m D_{nm}(t') - \delta_{m-1} D_{n,m-1}(t')\right)\bigg]\end{aligned} \tag{13.81}$$

define the leading Coulomb interaction corrections to the I–V curve for an arbitrary chain of quantum dots.

Equations (13.79)–(13.81) allow to make several important observations. For instance, we notice that the interaction corrections (13.80), (13.81) depend on two different response functions $K_{nm}(t)$ and $D_{nm}(t)$ describing, respectively, electric signal propagation along the chain and electron diffusion between different quantum dots. In addition, we observe that the term δI_1 does not depend of the Fano factors of the scatterers; therefore, it is not sensitive to the distribution of channel transmissions. In contrast, the contribution δI_2 is proportional to the Fano factors β_n and, hence, achieves its maximum value in the tunneling limit $T_n^{(n)} \ll 1$ corresponding to $\beta_n \to 1$. In the particular case of a single quantum dot, the term δI_1 vanishes, while the correction δI_2 reduces to the form (12.127). Several other specific examples will be considered below in this chapter.

13.3.2 Composite Fano Factor

Before turning to the analysis of the general expressions for the interaction correction (13.80) and (13.81), let us evaluate the shot noise power in chains of quantum dots in the absence of Coulomb interaction. This information may be of significance in various contexts, in particular due the established relation between shot noise and interaction effects. It turns out that our Boltzmann–Langevin kinetic equation approach is well suited for this purpose as well and, hence, we will again stick to this approach here. As before in this section, we will restrict our consideration to the limit of vanishingly small inelastic relaxation.

To begin with, let us recall that zero-frequency noise spectral power for the n-th scatterer $\mathcal{S}_n$ just coincides with the Fourier transform of the noise correlator (13.68). Evaluating this expression in the non-interacting limit, we will employ the solution to the kinetic equation (13.50) for the Green functions G_n in the form (13.69). Then after a straightforward calculation, we arrive at the following result [164]:

$$\mathcal{S}_n(0) = \left[1 - a_n(1-a_n) - a_{n-1}(1-a_{n-1}) - \beta_n(a_n - a_{n-1})^2\right]\frac{2T}{R_n} + \left[a_{n-1}(1-a_{n-1}) + a_n(1-a_n) + \beta_n(a_n - a_{n-1})^2\right]\frac{eV}{R_n}\coth\frac{eV}{2T}. \tag{13.82}$$

As our goal here is to analyze fluctuations δI of the total current I in our system, it is useful to relate such fluctuations to the noise variables ξ_n. In the interesting-for-us low-frequency limit, we can disregard the contribution from displacement currents $C_n\delta\dot{V}_n$. Then we obtain

$$\delta I = \frac{\delta V_n}{R_n} + \xi_n \tag{13.83}$$

Having in mind that in the voltage bias limit considered here, voltage fluctuations vanish, $\sum_{n=1}^{N}\delta V_n = 0$, we have

$$\delta I = \sum_{n=1}^{N}\frac{R_n\xi_n}{R_\Sigma} \tag{13.84}$$

Then for the noise spectral power of the whole chain of scatterers, one readily finds

$$\mathcal{S}(0) = \sum_{n=1}^{N} \frac{R_n^2}{R_\Sigma^2} \mathcal{S}_n = (1 - \tilde{\beta}) \frac{2T}{R_\Sigma} + \tilde{\beta} \frac{eV}{R_\Sigma} \coth \frac{eV}{2T} \tag{13.85}$$

where

$$\tilde{\beta} = \frac{1}{3} + \sum_{n=1}^{N} \frac{R_n^3}{R_\Sigma^3} \left(\beta_n - \frac{1}{3} \right) \tag{13.86}$$

represents the composite Fano factor for our chain.

Equation (13.86) constitutes an important result, which allows to reconstruct the Fano factor virtually for any configuration of scatterers with known individual resistances R_n and Fano factors β_n. For instance, in the case of arrays composed of diffusive scatterers with $\beta_n = 1/3$, one obviously gets $\tilde{\beta} = 1/3$ for all values R_n. This result can be easily expected since any chain of N diffusive conductors is also a diffusive conductor. For homogeneous arrays with $R_n = R$ and $\beta_n = \beta$, Eq. (13.86) yields

$$\tilde{\beta} = \frac{1}{3} + \frac{1}{N^2} \left(\beta - \frac{1}{3} \right) \tag{13.87}$$

Equations (13.86) and (13.87) demonstrate that in the limit of large $N \to \infty$, an array of arbitrary—not necessarily diffusive—scatterers should behave as a diffusive conductor with $\tilde{\beta} \to 1/3$. In the case of identical transmissions for all conducting channels, Eq. (13.87) reduces to the result [318]. In the case of two arbitrary scatterers ($N = 2$), Eq. (13.86) yields the result (12.144), which we already announced in Chapter 12.

13.3.3 Two Quantum Dots in Series

Now let us return to the analysis of the interaction correction to the I–V curve defined in Eqs. (13.80) and (13.81). Our first specific example is a system of two quantum dots connected in series. In this case, the total number of barriers obviously equals to $N = 3$ (see Fig. 13.1), and the problem allows for an explicit analytic solution to be worked out below.

Let us evaluate the response function $K_{nm}(t)$ for the system of two quantum dots. This function can be represented in the form

$$K_{nm}(t) = e^2 \int \frac{d\omega}{2\pi} \, \mathrm{e}^{-\mathrm{i}\omega t} \, \frac{Z_{nm}(\omega)}{-\mathrm{i}\omega + 0} \tag{13.88}$$

where $Z_{nm}(\omega)$ is an effective impedance that allows to express the Fourier-transformed voltage fluctuations via the current noise in the barriers as $\delta V_{\omega,n} = -\sum_m Z_{nm}(\omega)\xi_{\omega,m}$. This effective impedance obeys the following set of equations:

$$\begin{aligned}
\delta q_n &= C_{n+1} Z_{n+1,m} - C_n Z_{nm} - C_{gn} \sum_{j=1}^{n} Z_{jm}, \\
-\mathrm{i}\omega \delta q_n &= \frac{Z_{nm}}{R_n} - \frac{Z_{n+1,m}}{R_{n+1}} + \frac{g_n \delta_{n-1}}{4\pi} \delta q_{n-1} + \frac{g_{n+1}\delta_{n+1}}{4\pi} \delta q_{n+1} \\
&\quad - \frac{(g_n + g_{n+1})\delta_n}{4\pi} \delta q_n - \delta_{nm} + \delta_{n+1,m}, \\
\sum_{n=1}^{N} Z_{nm} &= 0
\end{aligned} \tag{13.89}$$

Resolving these equations for $N = 3$, we find

$$\begin{pmatrix} Z_{11} & Z_{12} & Z_{13} \\ Z_{21} & Z_{22} & Z_{23} \\ Z_{31} & Z_{32} & Z_{33} \end{pmatrix} = \frac{1}{A} \left(\begin{array}{c|c|c} a_{11} & -a_{11} - a_{12} & a_{12} \\ \hline -a_{11} - a_{21} & a_{11} + a_{12} + a_{21} + a_{22} & -a_{12} - a_{22} \\ \hline a_{21} & -a_{21} - a_{22} & a_{22} \end{array} \right)$$

where

$$\begin{aligned}
a_{11}(\omega) &= \left(-\mathrm{i}\omega + \frac{(g_2 + g_3)\delta_2}{4\pi} \right) C_{\Sigma 2} + \frac{1}{R_2} + \frac{1}{R_3} + \frac{g_2 \delta_1}{4\pi} C_2, \\
a_{12}(\omega) &= -\left(\left(-\mathrm{i}\omega + \frac{(g_1 + g_2)\delta_1}{4\pi} \right) C_2 + \frac{1}{R_2} + \frac{g_2 \delta_2}{4\pi} C_{\Sigma 2} \right), \\
a_{21}(\omega) &= -\left(\left(-\mathrm{i}\omega + \frac{(g_2 + g_3)\delta_2}{4\pi} \right) C_2 + \frac{1}{R_2} + \frac{g_2 \delta_1}{4\pi} C_{\Sigma 1} \right), \\
a_{22}(\omega) &= \left(-\mathrm{i}\omega + \frac{(g_1 + g_2)\delta_1}{4\pi} \right) C_{\Sigma 1} + \frac{1}{R_1} + \frac{1}{R_2} + \frac{g_2 \delta_2}{4\pi} C_2,
\end{aligned} \tag{13.90}$$

$C_{\Sigma 1} = C_1 + C_2 + C_{g1}$, $C_{\Sigma 2} = C_2 + C_3 + C_{g2}$, and $A(\omega) = a_{11}(\omega)a_{22}(\omega) - a_{12}(\omega)a_{21}(\omega)$.

For $N = 3$, the diffuson $D_{nm}(t)$ becomes a 2×2 matrix. Performing the Fourier transformation in the diffusion equation (13.70), we obtain

$$\begin{pmatrix} -\mathrm{i}\omega + 1/\tau_1 & -\delta_1 g_2/4\pi \\ -\delta_2 g_2/4\pi & -\mathrm{i}\omega + 1/\tau_2 \end{pmatrix} \begin{pmatrix} D_{\omega 11} & D_{\omega 12} \\ D_{\omega 21} & D_{\omega 22} \end{pmatrix} = \begin{pmatrix} 1 & 0 \\ 0 & 1 \end{pmatrix} \tag{13.91}$$

Here we defined $D_{\omega nm} = \int e^{\mathrm{i}\omega t} D_{nm}(t)$ and the inverse electron dwell times in the two quantum dots $1/\tau_1 = (g_1 + g_2)\delta_1/4\pi$ and $1/\tau_2 = (g_2 + g_3)\delta_2/4\pi$.

Solving the above equation, we get

$$\begin{pmatrix} D_{\omega 11} & D_{\omega 12} \\ D_{\omega 21} & D_{\omega 22} \end{pmatrix} = \frac{1}{d(\omega)} \begin{pmatrix} -\mathrm{i}\omega + 1/\tau_2 & \delta_1 g_2/4\pi \\ \delta_2 g_2/4\pi & -\mathrm{i}\omega + 1/\tau_1 \end{pmatrix} \tag{13.92}$$

where $d(\omega) = (-\mathrm{i}\omega + 1/\tau_1)(-\mathrm{i}\omega + 1/\tau_2) - \delta_1\delta_2 g_2^2/16\pi^2$.

What remains is to define the two charge relaxation rates κ_1, κ_2 and the two diffusion rates γ_1, γ_2. These rates, respectively, coincide with the roots of the equations $A(-\mathrm{i}\kappa) = 0$ and $d(-\mathrm{i}\gamma) = 0$. In what follows, we will set $\gamma_1 < \gamma_2$, $\kappa_1 < \kappa_2$ and assume that the charge relaxation is a much faster process than the electron diffusion, i.e., $\kappa_{1,2} \gg \gamma_{1,2}$. Then the interaction corrections to the current (13.80), (13.81) are evaluated analytically in a closed form and we obtain [415]

$$\delta I_1 = -2eT \frac{R_1 R_2 R_3}{R_\Sigma^3} \frac{W\left(\frac{\gamma_2 + \mathrm{i}eV}{2\pi T}\right) - W\left(\frac{\gamma_1 + \mathrm{i}eV}{2\pi T}\right)}{\sqrt{1 - \frac{4\delta_1\delta_2 R_1 R_2 R_3 R_\Sigma}{[\delta_1 R_2 R_3 + \delta_2 R_1 R_2 + (\delta_1 + \delta_2) R_1 R_3]^2}}} \tag{13.93}$$

$$\begin{aligned} \delta I_2 = & -\frac{2eT}{R_\Sigma^2 (C_{\Sigma 1} C_{\Sigma 2} - C_2^2)} \left\{ \mathcal{B}_1 \frac{W\left(\frac{\kappa_2 + \mathrm{i}eV}{2\pi T}\right) - W\left(\frac{\kappa_1 + \mathrm{i}eV}{2\pi T}\right)}{\kappa_2 - \kappa_1} \right. \\ & + \mathcal{B}_2 \left[\frac{\kappa_2 W\left(\frac{\kappa_1 + \mathrm{i}eV}{2\pi T}\right) - \kappa_1 W\left(\frac{\kappa_2 + \mathrm{i}eV}{2\pi T}\right)}{\kappa_1\kappa_2(\kappa_2 - \kappa_1)} - \frac{\gamma_2 W\left(\frac{\gamma_2 + \mathrm{i}eV}{2\pi T}\right) - \gamma_1 W\left(\frac{\gamma_1 + \mathrm{i}eV}{2\pi T}\right)}{\kappa_1\kappa_2(\gamma_2 - \gamma_1)} \right] \\ & \left. + \mathcal{B}_3 \frac{W\left(\frac{\gamma_2 + \mathrm{i}eV}{2\pi T}\right) - W\left(\frac{\gamma_1 + \mathrm{i}eV}{2\pi T}\right)}{\kappa_1\kappa_2(\gamma_2 - \gamma_1)} \right\}, \quad W(x) = \mathrm{Im}[x\Psi(1 + x)] \end{aligned} \tag{13.94}$$

Here we defined the system resistance $R_\Sigma = R_1 + R_2 + R_3$ and introduced the parameters

$$\mathcal{B}_1 = C_{\Sigma 1}\beta_1 R_1 + C_{\Sigma 2}\beta_3 R_3 + (C_{\Sigma 1} + C_{\Sigma 2} - 2C_2)\beta_2 R_2,$$
$$\mathcal{B}_2 = \beta_1 \frac{R_1(R_2+R_3)}{R_2 R_3} + \beta_2 \frac{R_2(R_1+R_3)}{R_1 R_3} + \beta_3 \frac{R_3(R_1+R_2)}{R_1 R_2},$$
$$\mathcal{B}_3 = \left(\left(\frac{1}{R_2} + \frac{1}{R_3}\right)\frac{1}{\tau_2} + \frac{\delta_1 g_2}{4\pi}\frac{1}{R_2}\right)\beta_1 R_1$$
$$+ \left(\left(\frac{1}{R_1} + \frac{1}{R_2}\right)\frac{1}{\tau_1} + \frac{\delta_2 g_2}{4\pi}\frac{1}{R_2}\right)\beta_3 R_3 + \left(\frac{1}{R_1}\frac{\delta_1 g_1}{4\pi} + \frac{1}{R_3}\frac{\delta_2 g_3}{4\pi}\right)\beta_2 R_2$$

The above equations fully determine the leading interaction correction to the current in a system of two coupled quantum dots.

These equations can now be analyzed in various physical limits. For instance, it is easy to verify that in the limit $R_2 = 0$, the problem reduces to that of a single quantum dot, in which case we have $\delta I_1 \equiv 0$ and δI_2 coincides with Eq. (12.127). Of interest is also the limit of fully open quantum dots $\beta_{1,2,3} = 0$. In this case, the term (13.94) vanishes identically, $\delta I_2 \equiv 0$, and the effect of electron–electron interactions on the I–V curve is described only by the β-independent term δI_1 (13.93). At $T \to 0$ for the differential conductance, we obtain

$$\frac{dI}{dV} = \frac{1}{R_\Sigma} - \frac{e^2 R_1 R_2 R_3}{2\pi R_\Sigma^3 \sqrt{1 - \frac{4\delta_1\delta_2 R_1 R_2 R_3 R_\Sigma}{[\delta_1 R_2 R_3 + \delta_2 R_1 R_2 + (\delta_1+\delta_2) R_1 R_3]^2}}}$$
$$\times \left[\ln\left(1 + \frac{\gamma_2^2}{e^2 V^2}\right) - \ln\left(1 + \frac{\gamma_1^2}{e^2 V^2}\right)\right] \tag{13.95}$$

This equation demonstrates that at large voltages $eV \gg \gamma_2$, the interaction correction remains small, while for $eV < \gamma_1$, it saturates to a finite voltage-independent value, i.e., at such voltages, the I–V curve of our system is Ohmic. In the limit $\gamma_1 \ll \gamma_2$, the interaction correction depends logarithmically on voltage, provided the condition $\gamma_1 \ll eV \ll \gamma_2$ is fulfilled.

Assume now that one of the two dots (e.g., the second one) is very large. In this case, both δ_2 and $1/\tau_2$ remain small. Setting $\delta_2 \to 0$, for $eV\tau_1 \ll 1$ from (13.95), one finds

$$\frac{dI}{dV} = \frac{1}{R_\Sigma}\left(1 + \frac{2}{g_\Sigma}\frac{R_1 R_2 R_3}{R_\Sigma^3}\ln(eV\tau_1)\right) \tag{13.96}$$

where $g_\Sigma = 2\pi/e^2 R_\Sigma$. For $\delta_2 = 0$, this result remains valid down to exponentially low voltages.

Finally, we consider the limit of identical outer barriers, $R_1 = R_3 = R_0$ and $\beta_1 = \beta_3 = \beta$, and assume that the second (internal) barrier has a much smaller resistance $R_2 \gg R_0$. For simplicity, we also put $C_1 = C_3 = C$, $C_{g1} = C_{g2} = C_g$, $g_1\delta_1/4\pi = g_3\delta_2/4\pi = 1/\tau_\mathrm{D}$ and assume $C_2 \ll C$. In this particular case, we find $\delta I_1 = 0$ and

$$\delta I_2 = -\frac{4eT\beta_2 R_0}{R_2}\left[W\left(\frac{1}{2\pi T\tau} + \frac{ieV}{2\pi T}\right) - W\left(\frac{1}{2\pi T\tau_\mathrm{D}} + \frac{ieV}{2\pi T}\right)\right] \tag{13.97}$$

where

$$\frac{1}{\tau} = \frac{1}{\tau_\mathrm{D}} + \frac{1}{R_0(C + C_\mathrm{g})} \tag{13.98}$$

is the effective charge relaxation time. The differential conductance in the limit $eV \gg T$ reads

$$\frac{dI}{dV} = \frac{1}{R_2} - \beta_2 \frac{e^2}{\pi}\frac{R_0}{R_2}\left[\ln\left(1 + \frac{1}{e^2V^2\tau^2}\right) - \ln\left(1 + \frac{1}{e^2V^2\tau_\mathrm{D}^2}\right)\right] \tag{13.99}$$

Comparing this result with the differential conductance of a single scatterer embedded in the Ohmic environment (11.20), we conclude that in the regime $1/\tau_\mathrm{D} \ll eV \ll 1/\tau$, the outer quantum dots act as an effective Ohmic shunt resistor $R_\mathrm{S} = 2R_0$. At lower voltages, $eV < 1/\tau_\mathrm{D}$, the differential conductance saturates to a finite value in the same way as the conductance of a single quantum dot does.

13.3.4 Chains of Identical Quantum Dots

Another important specific example is a chain of quantum dots and scatterers with identical parameters $\tau_n = \tau_\mathrm{D}$, $R_n = R$, $C_n = C$, $G_{gn} = C_g$, etc. In this case, the functions $K_{nm}(t)$ and $D_{nm}(t)$ are defined, respectively, in Eqs. (13.66) and (13.72). The interaction corrections to the current (13.80), (13.81) then take the form [164]

$$\delta I_1 = -e\sum_{q=1}^{N-1}\frac{1+\cos\frac{\pi q}{N}}{2\tau_D R N^2}\int\frac{d\omega}{2\pi}\,\mathrm{Im}(Z_{\omega q}D_{\omega q}^2)\,B(\omega, V, T)$$
$$+e\sum_{q=1}^{N-1}\frac{1+\cos\frac{\pi q}{N}}{\tau_D^2 R N^3}\mathrm{Im}\int\frac{d\omega}{2\pi}Z_{\omega q}D_{\omega q}^3\,B(\omega, V, T)$$
$$\times\frac{1-(-1)^q u^N(\omega)}{1+(-1)^q u^N(\omega)}\sqrt{-\mathrm{i}\omega\tau_\mathrm{D}(2-\mathrm{i}\omega\tau_\mathrm{D})} \tag{13.100}$$

and

$$\delta I_2 = -\frac{e\beta}{2N^2R}\sum_{q=1}^{N-1}\int\frac{d\omega}{2\pi}\,\mathrm{Im}(Z_{\omega q}D_{\omega q})B(\omega, V, T), \tag{13.101}$$

where $u(\omega) = 1-\mathrm{i}\omega\tau_\mathrm{D}-\sqrt{(1-\mathrm{i}\omega\tau_\mathrm{D})^2-1}$ and

$$B(\omega, V, T) = (\omega+eV)\coth\frac{\omega+eV}{2T}-(\omega-eV)\coth\frac{\omega-eV}{2T}. \tag{13.102}$$

The integrals in Eqs. (13.100) and (13.101) can be evaluated analytically. As before, assuming for simplicity that relaxation of the electric charge is much faster than electron diffusion, $R(C+C_\mathrm{g})\ll\tau_\mathrm{D}$, one finds

$$I = \frac{V}{NR}-\frac{2Te}{N^2}\sum_{q=1}^{N-1}\left[\beta W\left(\frac{2\sin^2\frac{\pi q}{2N}}{\pi T R\left(4C\sin^2\frac{\pi q}{2N}+C_\mathrm{g}\right)}+\frac{\mathrm{i}eV}{2\pi T}\right)\right.$$
$$\left.-\left[\beta-\frac{1-\frac{2(1-(-1)^q)}{N^2}\cot^2\frac{\pi q}{2N}}{1-\cos\frac{\pi q}{N}}\right]W\left(\frac{\sin^2\frac{\pi q}{2N}}{\pi T\tau_\mathrm{D}}+\frac{\mathrm{i}eV}{2\pi T}\right)\right] \tag{13.103}$$

where we again employed the function $W(x)$ defined in Eq. (13.94). The corresponding zero-bias conductance then reads

$$G_0 = \frac{1}{NR}-\frac{e^2}{\pi N^2}\sum_{q=1}^{N-1}\left[\beta\mathcal{P}\left(\frac{TR\left(C\sin^2\frac{\pi q}{2N}+C_\mathrm{g}/4\right)}{\sin^2\frac{\pi q}{2N}}\right)\right.$$
$$\left.-\left(\beta-\frac{1-\frac{2(1-(-1)^q)}{N^2}\cot^2\frac{\pi q}{2N}}{1-\cos\frac{\pi q}{N}}\right)\mathcal{P}\left(\frac{T\tau_\mathrm{D}}{2\sin^2\frac{\pi q}{2N}}\right)\right] \tag{13.104}$$

where the function $\mathcal{P}(x)$ was already defined in Eq. (12.131). In the tunneling limit $\beta \to 1$ and for very long chains $N \to \infty$, Eq. (13.104) matches with the result [47] derived by diagrammatic means.

In the limit $N \gg 1$, the summation over q in Eqs. (13.103) and (13.104) may be replaced by the integration and these expressions can be specified further. We begin with the zero-bias conductance (13.104). In the high-temperature limit $T R(C + C_{\rm g}) \gg 1$, we obtain

$$G_0 = \frac{1}{NR} - \frac{e^2\beta}{6NTRC}\left(1 - \sqrt{\frac{C_{\rm g}}{C_{\rm g} + 4C}}\right) \tag{13.105}$$

At intermediate temperatures $1/RC,\ 1/RC_{\rm g} \gg T \gg 1/\tau_{\rm D}$, Eq. (13.104) yields

$$G_0 \approx \frac{1}{NR} - \frac{e^2\beta}{\pi N}\left[\ln\frac{1}{2\pi TRC^*} + 1\right] \tag{13.106}$$

where $C^* = \left(\sqrt{C_{\rm g}} + \sqrt{C_{\rm g} + 4C}\right)^2/4$. In the interval $\pi^2/2N^2\tau_{\rm D} \ll T \ll 1/\tau_{\rm D}$, we find

$$\begin{aligned} G_0 \approx\ & \frac{1}{NR} - \frac{e^2\beta}{\pi N}\ln\frac{2\tau_{\rm D}}{RC^*} - \frac{e^2}{\pi N}\frac{3\zeta(3/2)}{4\sqrt{\pi T\tau_{\rm D}}} \\ & + \frac{e^2}{\pi N}\left[1 + \left(\beta - \frac{1}{4}\right)\left(1 - \frac{\gamma}{2}\right)\sqrt{\pi T\tau_{\rm D}} + \frac{\pi}{3NT\tau_{\rm D}}\right] \end{aligned} \tag{13.107}$$

Finally, in the limit of very low temperatures, $T \ll \pi^2/2N^2\tau_{\rm D}$, the interaction correction to the conductance saturates, and we get

$$G_0 \approx \frac{1}{NR} - \frac{e^2\beta}{\pi N}\ln\frac{2\tau_{\rm D}}{RC^*} - \left(0.368 - \frac{1}{N}\right)\frac{e^2}{\pi} \tag{13.108}$$

This formula remains applicable, provided the non-interacting value of the total resistance NR is much smaller than the quantum resistance unit $R_{\rm q}$ or, equivalently, for $N \ll g = R_{\rm q}/R$.

Consider now the nonlinear regime $e|V| \gg T$ in which case the I–V curve is determined by Eq. (13.103). At very small voltages and temperatures, $e|V| \ll \pi^2/2N^2\tau_{\rm D}$, one finds $dI/dV = G_0$, where G_0 is again defined by Eq. (13.108). At higher voltages, $\pi^2/2N^2\tau_{\rm D} \ll e|V| \ll 1/\tau_{\rm D}$, we obtain

$$\frac{dI}{dV} = \frac{1}{NR} - \frac{e^2\beta}{\pi N}\ln\frac{2\tau_{\rm D}}{RC^*} - \frac{e^2}{\pi N}\frac{1}{\sqrt{e|V|\tau_{\rm D}}} \tag{13.109}$$

At even higher voltage, $1/\tau_D \ll e|V| \ll 1/RC^*$, the differential conductance takes the form

$$\frac{dI}{dV} = \frac{1}{NR} - \frac{e^2\beta}{\pi N}\ln\frac{1}{e|V|RC^*} \qquad (13.110)$$

Finally, at high temperature $T \gg 1/RC,\ 1/RC_g$, one finds

$$\frac{dI}{dV} = \frac{1}{NR} - \beta\frac{N-1}{N^2}\frac{e^2}{2RCT}\left(1-\sqrt{\frac{C_g}{C_g+4C}}\right)\frac{\frac{eV}{2T}\cosh\frac{eV}{2T}-\sinh\frac{eV}{2T}}{\sinh^3\frac{eV}{2T}} \qquad (13.111)$$

We observe that the width of the interaction-induced dip in the differential conductance of the chain of quantum dots at high temperatures is the same as in the case of a single scatterer embedded in the high Ohmic environment. It does not increase with the number of the junctions N in the chain. As in the case of a single quantum dot, the reason for such a behavior is the absence of inelastic relaxation in our system. In contrast, strong inelastic relaxation results in the conductance dip with the width $\propto NT$; cf. Eq. (13.2).

The linear conductance G_0 as a function of temperature and the differential conductance dI/dV at $T \to 0$ as a function of the applied voltage are depicted in Fig. 13.3 for different number of scatterers N in the system. One observes that both quantities (as functions of the corresponding variables) demonstrate a very similar behavior. In both cases, four different regimes can be distinguished: (i) low-energy regime $eV, T \ll E_{\text{Th}}$, where $E_{\text{Th}} = \pi^2/2N^2\tau_D$ is the Thouless energy of the array; the array in this case essentially behaves as a single scatterer with the conductance (13.108) renormalized by interactions; (ii) diffusive regime $E_{\text{Th}} < eV, T < 1/\tau_D$ where the array of quantum dots behaves like a metallic diffusive wire; (iii) regime of approximately independent junctions $1/\tau_D < eV, T < 1/RC, 1/RC_g$, where the interaction correction to the conductance has the logarithmic dependence on eV, T; (iv) high-energy regime $eV, T > 1/RC, 1/RC_g$.

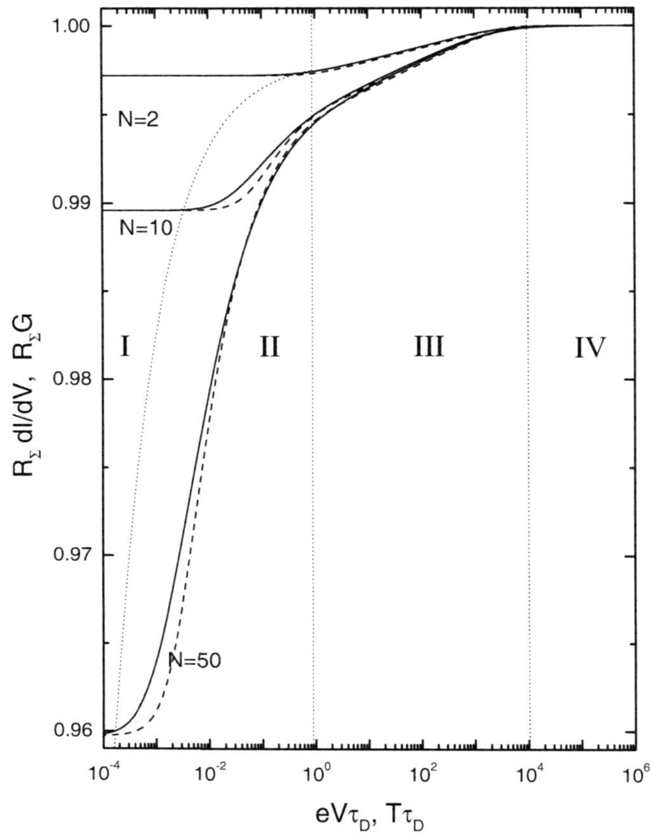

Figure 13.3 The linear conductance $G \equiv G_0$ (solid lines) as a function of temperature T together with the differential conductance dI/dV (dashed lines) as a function of the applied voltage V at $T = 0$. The results are obtained from Eqs. (13.100) and (13.101) for $g = 1000$, $\beta = 1/3$, $\tau_D/RC = 10^4$, $C_g/C = 2.5$ and for three different numbers of barriers: $N = 2$, 10, and 50. We identify four different regimes (the boundaries between them are shown by dotted lines): (I) saturation regime eV, $T \lesssim \pi^2/2N^2\tau_D$, Eq. (13.108); (II) diffusive regime $\pi^2/2N^2\tau_D \lesssim eV$, $T \lesssim 1/\tau_D$, Eqs. (13.107) and (13.109); (III) logarithmic regime of almost independent barriers $1/\tau_D \lesssim eV$, $T \lesssim 1/RC$, Eqs. (13.106) and (13.110); and (IV) high-temperature (classical) regime eV, $T \gtrsim 1/RC$, Eq. (13.105).

13.3.5 Effect of External Environment

In order to complete the above analysis, let us also include the effect of external leads into our consideration. For simplicity, we will assume the leads impedance R_S to be purely Ohmic; see Fig. 13.1. As V is now the voltage applied to the whole system "chain+leads," the condition (13.55) has to be replaced by

$$\sum_{j=1}^{N} V_j + I R_S = V \tag{13.112}$$

Equation (13.112) is solved together with Eqs. (13.50), (13.64), (13.65), and (13.70) in exactly the same way as it was done above in the absence of the shunt. As a result, for a chain composed of identical quantum dots, the effect of R_S is accounted for by means of a simple replacement

$$K_{nm}(t) \longrightarrow K_{nm}(t) + K_S(t) \tag{13.113}$$

where $K_{nm}(t)$ is defined in Eq. (13.66) and

$$K_S(t) = \frac{e^2}{N} \int \frac{d\omega}{2\pi} \frac{\mathrm{e}^{-\mathrm{i}\omega t}}{(-\mathrm{i}\omega + 0)\left(-\mathrm{i}\omega + \frac{1}{R} + \frac{N}{R_S}\right)} \tag{13.114}$$

Note that the function $K_S(t)$ does not depend on n and m. After such a replacement, the new expression for $K_{nm}(t)$ should be substituted into Eqs. (13.80) and (13.81), which yield the desired result for the interaction correction to the I–V curve. This general expression turns out to be rather cumbersome, and for this reason, we will avoid presenting it here. Below we will only address the effect of external leads on the linear conductance G_0 in the low-temperature limit $T < \pi^2/2N^2\tau_D$. It turns out that for nonzero R_S, the conductance saturation is lifted and the result (13.108) becomes incomplete. Taking into account the effect of an external shunt, one finds

$$G_0 \approx \frac{1}{NR} - \frac{e^2\beta}{\pi N} \ln \frac{2\tau_D}{RC*} - \left(0.368 - \frac{1}{N}\right)\frac{e^2}{\pi} - \frac{e^2\tilde{\beta}}{\pi} \frac{R_S}{R_S + NR} \ln \frac{\pi^2}{2N^2 T \tau_D} \tag{13.115}$$

where

$$\tilde{\beta} = \frac{\beta}{N^2} + \frac{1}{N^4} \sum_{q=1}^{N-1} \frac{(1 - (-1)^q)\cos^2 \frac{\pi q}{2N}}{\sin^4 \frac{\pi q}{2N}} \tag{13.116}$$

Equation (13.115) is applicable as long as the last logarithmic term remains much smaller than $1/NR$.

The sum in Eq. (13.116) is evaluated exactly and just yields the Fano factor of the array (13.87). Thus, in the presence of an external shunt, the conductance keeps decreasing logarithmically with T even at very low temperatures. As before, this logarithmic contribution scales linearly with the total Fano factor of the array $\tilde{\beta}$, which tends to the universal value 1/3 in the limit of large N. This result is in agreement with our previous findings for a single short coherent scatterer, and it once again emphasizes a direct relation between shot noise and interaction effects in disordered conductors.

13.3.6 Weakly Disordered Metallic Wires

Let us take the limit of a very long chain $N \to \infty$ and assume that the dwell time in a quantum dot $\tau_{\mathrm{D}} \to 0$ is sufficiently small. Then one can effectively replace the summation over q in Eqs. (13.100) and (13.101) by the integration. Denoting a typical quantum dot size as d, one can also rewrite the diffuson (13.73) in the form

$$\mathcal{D}_{\omega,k} = \frac{1}{-\mathrm{i}\omega + D_{\mathrm{eff}}k^2} \tag{13.117}$$

where we introduced the wave vector $k = \pi q/X$, the system length $X \simeq Nd$, and the effective diffusion coefficient $D_{\mathrm{eff}} = d^2/2\tau_{\mathrm{D}}$. This form of the diffuson is standard and is typically employed in extended conductors, such as diffusive metallic wires.

Let us consider such a wire with length X and cross section $\mathcal{A}$ and assume that the electron elastic mean free path ℓ is not too short, $k_{\mathrm{F}}\ell \gg 1$, i.e., the wire is weakly disordered. At scales exceeding ℓ, the electron motion is diffusive, i.e., at such scales, it is described by essentially the same diffusion equation as the one employed above for a chain of chaotic quantum dots. Hence, one can model a diffusive metallic wire by such a chain, which consists of quantum dots with the longitudinal size $d \sim \ell$.

Employing the Drude formula for the total wire conductance $1/R_{\mathrm{Drude}} = 2e^2\mathcal{A}N_0D/X$ (where, as usual, $N_0 = mp_{\mathrm{F}}/2\pi^2$ is the electron density of states at the Fermi level and $D = v_{\mathrm{F}}\ell/3$ is the diffusion coefficient) and identifying R_{Drude} with $NR = (X/d)R$, we immediately observe that for $d \sim \ell$, the conductance of a single scatterer becomes $1/R \approx N_{ch}/R_q$, where $N_{ch} = k_F^2\mathcal{A}/2\pi \gg 1$ is the total number of conducting channels in the wire. Combining this estimate with the formula for the diffusion coefficient $D_{eff} = d^2/2\tau_D = \mathcal{A}\delta_d/2e^2R$ (where $\delta_d = 2\pi^2 v_F/k_F^2\mathcal{A}^{3/2}$ is the energy level spacing in the dot), we obtain $D_{eff} \sim v_F\sqrt{\mathcal{A}}$. Bearing in mind that the electron elastic mean free path in quasi-one-dimensional metallic wires is typically $\ell \sim \sqrt{\mathcal{A}}$, we find $D_{eff} \sim D = v_F\ell/3$, i.e., the diffusion coefficients in the two models indeed coincide up to an unimportant numerical constant of order one.

It follows immediately that without loss of generality, one can model a weakly disordered metallic wire with point-like impurities by a chain of fully open quantum dots with scatterer resistances $R = R_q/N_{ch} \ll R_q$. In this case, N just coincides with the total number of impurities in the wire. Accordingly the Fano factor of the scatterers can be set equal to zero $\beta \to 0$. In this limit, the capacitances drop out from the expression for the current and we obtain

$$I = \frac{V}{R_{\mathrm{Drude}}} - \frac{8Te}{L}\int_{\pi/L}^{\infty}\frac{dk}{2\pi}\frac{1}{k^2}\left(1 - \frac{3\pi^2}{k^2L^2}\right)W\left(\frac{D_{\mathrm{eff}}k^2}{2\pi T} + \frac{ieV}{2\pi T}\right) \tag{13.118}$$

In the high bias regime $eV \gg T$ and $eV \gg \pi^2 D_{eff}/X^2$, Eq. (13.118) yields

$$\frac{dI}{dV} = \frac{1}{R_{\mathrm{Drude}}} - \sqrt{\frac{2D_{\mathrm{eff}}}{X^2e|V|}}. \tag{13.119}$$

This result agrees with Eq. (13.109) and applies to wires shorter than the inelastic relaxation length $X < L_{in}$. Equation (13.118) also provides the expression for the zero-bias conductance, which reads

$$G_0 = \frac{1}{R_{\mathrm{Drude}}} - \frac{3}{2}\zeta\left(\frac{3}{2}\right)\frac{e^2}{\pi X}\sqrt{\frac{D_{\mathrm{eff}}}{2\pi T}} \tag{13.120}$$

where, as before, $\zeta(x)$ is the Riemann zeta-function. The last term in Eq. (13.120) is just the well-known Altshuler–Aronov interaction correction to the conductance of a diffusive metallic wire, which was previously derived diagrammatically [12]; see also [120].

In the limit $T, eV \to 0$, the zero-bias conductance dI/dV reaches its minimal value

$$G_{\min} = \frac{1}{R_{\mathrm{Drude}}} - \frac{16}{3\pi^2}\frac{e^2}{\pi} \tag{13.121}$$

Note that the last term in Eq. (13.121) is slightly different from that in Eq. (13.108) due to a different short range cutoff employed here. And again, Eq. (13.121) remains valid as long as $R_{\text{Drude}} \ll R_q$. For very long wires, this condition is inevitably violated, in which case our perturbative in the interaction approach breaks down at low enough energies and a non-perturbative analysis in the spirit of Section 11.3 needs to be worked out.

13.3.7 Discussion

For clarity, let us briefly summarize again our key observations made in this section.

In Chapter 13, we extended our analysis of electron–electron interaction effects to more complicated structures as compared to those studied in Chapters 11 and 12. In particular, we demonstrated that virtually any type of disordered conductors can be modeled by an array of metallic quantum dots. This model allows us to treat both electron scattering and Coulomb interaction effects universally and on most general footing. Here we focused our attention on quasi-one-dimensional conductors; however, after trivial modifications, one can apply the whole analysis also to two- and three-dimensional disordered conductors.

We observed that a fundamental relation between shot noise and interaction effects established in Chapter 11 for short coherent conductors can be conveniently extended to arbitrary systems. The key idea of the Boltzmann–Langevin approach employed in this section is to properly account for non-equilibrium effects, which trigger modifications of the shot noise and to self-consistently describe these effects and their impact on fluctuating charges and voltages inside the conductor. Implementing this program, we derived a general expression for the current in the presence of (weak) Coulomb effects; see Eqs. (13.79)–(13.81).

These results demonstrate that at sufficiently high energies (exceeding the inverse dwell time of a single quantum dot, $1/\tau_{\text{D}}$), the individual scatterers in our system behave as effectively independent ones, and one can identify two different regimes (regimes III and IV in Fig. 13.3) described, respectively, by Eqs. (13.106) and (13.110) and by Eq. (13.105). At such energies, the interaction correction scales with the Fano factor β of individual scatterers and in a wide interval of energies depends logarithmically on temperature or voltage. At energies below $1/\tau_{\rm D}$ (regime II), scatterers located sufficiently close to each other become effectively correlated. The number of such scatterers $N_{\rm eff}$ in one "correlated" segment of the array grows with decreasing temperature (or voltage) as $N_{\rm eff} \sim 1/\sqrt{T\tau_{\rm D}}$ (or $N_{\rm eff} \sim 1/\sqrt{eV\tau_{\rm D}}$). In this regime, the system can be viewed as a chain of $\sim N/N_{\rm eff}$ segments, each of them now playing the role of a "new" independent scatterer with an effective conductance $g_{\rm eff} \sim g/N_{\rm eff} \sim g\sqrt{T\tau_{\rm D}}$ (or $\sim g\sqrt{eV\tau_{\rm D}}$). Then the results can be applied again. In the corresponding expression for the interaction correction δG, one should only substitute $g_{\rm eff}$ instead of g. In this case, the logarithmic dependence of the interaction correction on temperature/voltage drops out and, e.g., for the linear conductance, one finds $\delta G/G_0 \sim -\beta_{\rm eff}/g_{\rm eff}$, where $\beta_{\rm eff}$ is the Fano factor of a segment with $N_{\rm eff}$ scatterers. According to Eq. (13.87), for sufficiently large $N_{\rm eff} \gg 1$, the factor $\beta_{\rm eff}$ approaches the universal value 1/3, and we obtain $\delta G/G_0 \sim -1/g\sqrt{T\tau_{\rm D}}$; cf. Eq. (13.107) which, in addition, contains a temperature-independent contribution $\propto \beta$ coming from high-energy modes. Finally, as $N_{\rm eff}$ approaches N, the system conductance either saturates (for $R_{\rm S} \to 0$, regime I) or crosses over to the low-energy logarithmic regime (13.115) caused by additional voltage fluctuations across the array due to nonzero external shunt resistance $R_{\rm S}$.

In this section, we also demonstrated that in the limit $\beta \to 0$ and at distances exceeding the quantum dot size $d \sim \ell$, our array of quantum dots is equivalent to a diffusive normal metal with the effective diffusion constant $D_{\rm eff} = d^2/2\tau_{\rm D} \sim D = v_F\ell/3$. In this case, the term (13.81) vanishes identically, and the whole interaction correction to the current is defined by Eq. (13.80), which then yields the result (13.118). In the linear in-voltage regime, this result in turn reduces to the standard one (13.120) derived diagrammatically for weakly disordered metallic wires [12].

Finally, we would like to emphasize that in this chapter, we merely dealt with Coulomb blockade of electron transport and did not yet address another fundamentally important effect: weak localization. This effect has to do with quantum interference of electrons in disordered conductors, which may yield significant modifications of the system conductance at low enough temperatures. While in the absence of interactions propagating electrons remain fully coherent, Coulomb interaction causes phase relaxation processes and, hence, limits the ability of electrons to interfere. The next two chapters of this book are devoted to a detailed analysis of a nontrivial interplay between quantum interference and interaction-induced quantum decoherence of electrons in disordered metallic conductors as well as their impact on transport properties of such systems.

Chapter 14

Weak Localization and Electron Dephasing in Disordered Conductors I: Metallic Limit

14.1 Introduction

Conduction electrons propagating in disordered metallic conductors suffer elastic scattering at impurities, boundary imperfections, and possibly some other obstacles. As a result of such scattering, electron motion becomes diffusive at length scales exceeding the elastic mean free path ℓ. The value of ℓ in typical metals varies significantly ranging from ~ 100 nm in comparatively clean samples to just a few nanometers in dirtier ones. As compared to these mean free path values, the electron Fermi wavelength $\lambda_{\rm F}$ usually remains very short, typically of the order of 1 Å. Hence, the condition

$$k_{\rm F}\ell \gg 1 \tag{14.1}$$

(where $k_{\rm F} \equiv 2\pi/\lambda_{\rm F}$ is the Fermi wave vector) is normally well satisfied in most types of metallic conductors. This condition allows to describe the electron motion in such conductors quasiclassically.

Dissipative Quantum Mechanics of Nanostructures: Electron Transport, Fluctuations, and Interactions
Andrei D. Zaikin and Dmitry S. Golubev

ISBN 978-981-4774-50-5 (Hardcover), 978-0-429-29823-3 (eBook)
www.jennystanford.com

At the same time, even in the limit (14.1), quantum mechanical interference effects may become crucially important at sufficiently low temperatures. For instance, it is well known that in the absence of any interaction between electrons in low-dimensional conductors, such as disordered metallic wires and films, quantum interference should yield *localization* of their wave functions at any value of the product $k_F\ell$ provided the sample size exceeds the localization length ξ_{loc}. This Anderson localization [17] is an essentially quantum phenomenon, which was predicted to drastically change both ground state and low energy transport properties of disordered metallic conductors. For example, the theory predicts that the resistance of a disordered metallic wire should grow exponentially (not linearly!) with its length L as soon as this length exceeds $\xi_{loc} \sim N_{ch}\ell$ where, as before, N_{ch} is the number of conducting channels in the wire.

In practice, however, Anderson localization turns out to be a pretty fragile phenomenon because it requires complete phase coherence of the electron wave functions. While the electron coherence is not damaged by elastic scattering (e.g., on impurities or otherwise), it can be destroyed by processes such as electron–electron and electron–phonon interactions, spin-flip electron scattering on magnetic impurities, and so on. As a result, quantum coherence of electrons propagating in disordered conductors can be maintained only during some finite time τ_φ and within some finite length $L_\varphi = \sqrt{D\tau_\varphi}$, where $D = v_F\ell/3$ is the electron diffusion coefficient. The so-called dephasing time τ_φ and dephasing length L_φ were already encountered before in Chapters 3 and 8. These parameters will also play a crucial role in our subsequent considerations. For instance, we will demonstrate that in generic metallic conductors of a reduced dimension (and described by the condition (14.1)), electron–electron interactions restrict the electron dephasing length L_φ to always be *parametrically shorter* than the localization length ξ_{loc}. Under this condition, strong (Anderson) localization cannot be observed and turns into *weak localization* (WL), which will be the main subject of both this and the next chapters.

14.2 Weak Localization of Non-interacting Electrons

The phenomenon of WL in disordered conductors was intensively studied for a long time. The results of these studies are summarized and detailed in a number of review papers; see, e.g., [12, 54, 87]. In this section, we will provide a brief overview of the WL theory in the case of non-interacting electrons.

Our starting point is the Kubo formula for the conductivity

$$\sigma = \lim_{\omega\to 0}\frac{1}{3}\int d^3\boldsymbol{r}_2\int_{-\infty}^{t_1} dt_2\frac{e^{\mathrm{i}\omega(t_1-t_2)}}{\omega}\left\langle \hat{\boldsymbol{j}}(X_1)\hat{\boldsymbol{j}}(X_2)-\hat{\boldsymbol{j}}(X_2)\hat{\boldsymbol{j}}(X_1)\right\rangle, \tag{14.2}$$

where the four dimensional vector $X = (t, \boldsymbol{r})$ combines time and spatial coordinates,

$$\hat{\boldsymbol{j}}(X) = \frac{\mathrm{i}e}{2m}\sum_{\alpha=\uparrow,\downarrow}\left[\hat{\Psi}^\dagger_\alpha(X)\,\nabla\hat{\Psi}_\alpha(X)-\nabla\hat{\Psi}^\dagger_\alpha(X)\,\hat{\Psi}_\alpha(X)\right] - \frac{e^2}{mc}\boldsymbol{A}(X)\,\hat{\Psi}^\dagger_\alpha(X)\hat{\Psi}_\alpha(X) \tag{14.3}$$

is the current density operator expressed in terms of the electron creation and annihilation operators $\hat{\Psi}^\dagger_\alpha$ and $\hat{\Psi}_\alpha$, and the angular brackets imply a combination of quantum averaging with averaging over disorder potential realizations. As before, m is the electron mass and c denotes the speed of light.

For simplicity, in what follows let us set the vector potential equal to zero $\boldsymbol{A} = 0$, assume twofold spin degeneracy of all electron energy levels, and ignore all interaction effects. Defining the retarded and advanced electron Green functions

$$\begin{aligned} G^{\mathrm{R}}(E,\boldsymbol{r}_1,\boldsymbol{r}_2) &= -\mathrm{i}\int dt_1\, e^{\mathrm{i}E(t_1-t_2)}\theta(t_1-t_2) \\ &\quad\times\langle\hat{\Psi}_\uparrow(X_1)\hat{\Psi}^\dagger_\uparrow(X_2)+\hat{\Psi}^\dagger_\uparrow(X_2)\hat{\Psi}_\uparrow(X_1)\rangle, \\ G^{\mathrm{A}}(E,\boldsymbol{r}_1,\boldsymbol{r}_2) &= \mathrm{i}\int dt_1\, e^{\mathrm{i}E(t_1-t_2)}\theta(t_2-t_1) \\ &\quad\times\langle\hat{\Psi}_\uparrow(X_1)\hat{\Psi}^\dagger_\uparrow(X_2)+\hat{\Psi}^\dagger_\uparrow(X_2)\hat{\Psi}_\uparrow(X_1)\rangle \end{aligned} \tag{14.4}$$

and combining Eqs. (14.2) and (14.3), we rewrite the conductivity in the form

$$\sigma = \frac{e^2}{24m^2}\int d^3\boldsymbol{r}_2 \int \frac{dE}{2\pi}\frac{1}{4T\cosh^2\frac{E}{2T}} \lim_{\boldsymbol{r}_4\to\boldsymbol{r}_1}\lim_{\boldsymbol{r}_3\to\boldsymbol{r}_2}$$
$$\times\Big\langle \left[\nabla_{\boldsymbol{r}_4}-\nabla_{\boldsymbol{r}_1}\right]\left[\nabla_{\boldsymbol{r}_2}-\nabla_{\boldsymbol{r}_3}\right]\left[G^{\mathrm{R}}(E,\boldsymbol{r}_1,\boldsymbol{r}_2)-G^{\mathrm{A}}(E,\boldsymbol{r}_1,\boldsymbol{r}_2)\right]$$
$$\times\left[G^{\mathrm{R}}(E,\boldsymbol{r}_3,\boldsymbol{r}_4)-G^{\mathrm{A}}(E,\boldsymbol{r}_3,\boldsymbol{r}_4)\right]\Big\rangle_{\mathrm{dis}} \qquad (14.5)$$

where the brackets $\langle\ldots\rangle_{\mathrm{dis}}$ indicate averaging over disorder. Such averaging can be performed, e.g., by means of the diagrammatic methods [1]. Here we briefly summarize the idea and the results of this procedure.

We will employ the standard Hamiltonian describing single electrons (9.78) in a normal metal with the chemical potential μ

$$\hat{H} = \hat{H}_0 - \mu = \hat{H}_{00} + U(\boldsymbol{r}), \quad \hat{H}_{00} = -\frac{\nabla^2}{2m} - \mu, \qquad (14.6)$$

where the Hamiltonian $\hat{H}_0$ is also defined in Eq. (9.78) and the potential $U(\boldsymbol{r})$ accounts for the presence of impurities. In order to proceed, we will treat the impurity potential $U(\boldsymbol{r})$ as a Gaussian random variable with zero average and the pair correlator

$$\langle U(\boldsymbol{r}_1)U(\boldsymbol{r}_2)\rangle_{\mathrm{dis}} = \frac{1}{2\pi N_0\tau_e}\delta(\boldsymbol{r}_1-\boldsymbol{r}_2) \qquad (14.7)$$

Here, as usual, $N_0 = mk_{\mathrm{F}}/2\pi^2$ is the density of states per unit volume and per one spin at the Fermi surface, and $\tau_e = \ell/v_{\mathrm{F}}$ is the elastic electron scattering time.

In the case of non-interacting electrons considered here, the Fourier-transformed retarded and advanced Green functions (14.4) can be formally expressed as

$$G^{\mathrm{R}}(E) = [E - H_{00} - U + \mathrm{i}0]^{-1}, \quad G^{\mathrm{A}}(E) = [E - H_{00} - U - \mathrm{i}0]^{-1} \qquad (14.8)$$

Expanding these expressions in powers of the disorder potential U, we obtain

$$G^{\mathrm{R}}(E) = [E - H_{00} + i0]^{-1} + [E - H_{00} - U + \mathrm{i}0]^{-1}U\,[E - H_{00} - U + \mathrm{i}0]^{-1}$$
$$+[E - H_{00} + \mathrm{i}0]^{-1}U\,[E - H_{00} + \mathrm{i}0]^{-1}U\,[E - H_{00} + \mathrm{i}0]^{-1} + \ldots$$

and similarly for $G^{\mathrm{A}}(E)$. In this way, we effectively construct the perturbation theory in the parameter $1/k_{\mathrm{F}}\ell$. Applying Eq. (14.7) to every term of this perturbation theory and identifying the leading set of diagrams, one arrives at the so-called non-crossing diagrams, which dominate the result of averaging in this case.

At sufficiently small energies $E \ll 1/\tau_e$, one finds [1]

$$\left\langle G^{\mathrm{R}}(E, \boldsymbol{r}_1, \boldsymbol{r}_2)\right\rangle_{\mathrm{dis}} = \int \frac{d^3k}{(2\pi)^3} \frac{e^{\mathrm{i}\boldsymbol{k}(\boldsymbol{r}_1-\boldsymbol{r}_2)}}{E - \frac{k^2}{2m} + \mu + \frac{\mathrm{i}}{2\tau_e}},$$
$$\left\langle G^{\mathrm{A}}(E, \boldsymbol{r}_1, \boldsymbol{r}_2)\right\rangle_{\mathrm{dis}} = \int \frac{d^3k}{(2\pi)^3} \frac{e^{\mathrm{i}\boldsymbol{k}(\boldsymbol{r}_1-\boldsymbol{r}_2)}}{E - \frac{k^2}{2m} + \mu - \frac{\mathrm{i}}{2\tau_e}} \tag{14.9}$$

Averaging of the products of two Green functions can be performed in a similar manner, although the corresponding diagrams turn out to be somewhat more complicated. For more details, we refer the reader, e.g., to the review [3]. Here we only quote the final result:

$$\left\langle G^{\mathrm{R}}(E_1, \boldsymbol{r}_1, \boldsymbol{r}_2) G^{\mathrm{R}}(E_2, \boldsymbol{r}_3, \boldsymbol{r}_4)\right\rangle_{\mathrm{dis}} = \left\langle G^{\mathrm{R}}(E_1, \boldsymbol{r}_1, \boldsymbol{r}_2)\right\rangle_{\mathrm{dis}} \times \left\langle G^{\mathrm{R}}(E_2, \boldsymbol{r}_3, \boldsymbol{r}_4)\right\rangle_{\mathrm{dis}}, \tag{14.10}$$

$$\left\langle G^{\mathrm{A}}(E_1, \boldsymbol{r}_1, \boldsymbol{r}_2) G^{\mathrm{A}}(E_2, \boldsymbol{r}_3, \boldsymbol{r}_4)\right\rangle_{\mathrm{dis}} = \left\langle G^{\mathrm{A}}(E_1, \boldsymbol{r}_1, \boldsymbol{r}_2)\right\rangle_{\mathrm{dis}} \times \left\langle G^{\mathrm{A}}(E_2, \boldsymbol{r}_3, \boldsymbol{r}_4)\right\rangle_{\mathrm{dis}}, \tag{14.11}$$

$$\begin{aligned}\left\langle G^{\mathrm{R}}(E_1, \boldsymbol{r}_1, \boldsymbol{r}_2) G^{\mathrm{A}}(E_2, \boldsymbol{r}_3, \boldsymbol{r}_4)\right\rangle_{\mathrm{dis}} &= \left\langle G^{\mathrm{R}}(E_1, \boldsymbol{r}_1, \boldsymbol{r}_2)\right\rangle_{\mathrm{dis}} \left\langle G^{\mathrm{A}}(E_2, \boldsymbol{r}_3, \boldsymbol{r}_4)\right\rangle_{\mathrm{dis}} \\ &+ 2\pi N_0\, w(|\boldsymbol{r}_1 - \boldsymbol{r}_4|) w(|\boldsymbol{r}_2 - \boldsymbol{r}_3|) \mathcal{D}\left(E_1 - E_2, \frac{\boldsymbol{r}_1 + \boldsymbol{r}_4}{2}, \frac{\boldsymbol{r}_2 + \boldsymbol{r}_3}{2}\right) \\ &+ 2\pi N_0\, w(|\boldsymbol{r}_1 - \boldsymbol{r}_3|) w(|\boldsymbol{r}_2 - \boldsymbol{r}_4|) \mathcal{C}\left(E_1 - E_2, \frac{\boldsymbol{r}_1 + \boldsymbol{r}_3}{2}, \frac{\boldsymbol{r}_2 + \boldsymbol{r}_4}{2}\right),\end{aligned} \tag{14.12}$$

where we defined $w(r) = e^{-r/2\ell} \sin k_F r / k_F r$.

Equation (14.12) formally includes all effects of disorder on the electron transport in metallic conductors in the lowest and next-to-the lowest orders in the small parameter $1/k_{\mathrm{F}}\ell$. The right-hand side in the first line of this equation containing the product of two independent averages $\left\langle G^R\right\rangle_{\mathrm{dis}}$ and $\left\langle G^A\right\rangle_{\mathrm{dis}}$, being combined with Eq. (14.5), yields the standard expression for the classical Drude conductivity of a disordered metal (9.108).

Technically the second line of Eq. (14.12) originates from the summation of the so-called ladder diagrams [1]. It includes *the diffuson* $\mathcal{D}(E_1 - E_2, \boldsymbol{r}_1, \boldsymbol{r}_2)$, which obeys the following diffusion equation:

$$\left(-\mathrm{i}\omega - D\nabla^2_{\boldsymbol{r}_1}\right)\mathcal{D}(\omega, \boldsymbol{r}_1, \boldsymbol{r}_2) = \delta(\boldsymbol{r}_1 - \boldsymbol{r}_2) \quad (14.13)$$

and, hence, describes the classical diffusion of electrons. Finally, the third line of Eq. (14.12) formally comes from the so-called maximally crossed diagrams of the perturbation theory [1], which yields *the Cooperon* $\mathcal{C}(E_1 - E_2, \boldsymbol{r}_1, \boldsymbol{r}_2)$ obeying the diffusion equation

$$\left(-\mathrm{i}\omega - D\nabla^2_{\boldsymbol{r}_1} + \frac{1}{\tau_{\mathrm{H}}} + \frac{1}{\tau_{\varphi}}\right)\mathcal{C}(\omega, \boldsymbol{r}_1, \boldsymbol{r}_2) = \delta(\boldsymbol{r}_1 - \boldsymbol{r}_2) \quad (14.14)$$

Below we will demonstrate that the Cooperon essentially accounts for the interference of the electron wave functions in a disordered metal, hence, being responsible for the WL correction to the conductivity σ_{WL}. This WL correction has the form [173]

$$\sigma_{\mathrm{WL}} = -\frac{2e^2}{\pi} D\,\mathcal{C}(0, \boldsymbol{r}_1, \boldsymbol{r}_1) \quad (14.15)$$

Note that both Eqs. (14.13) and (14.14) contain the same diffusion constant D and, in addition, Eq. (14.14) for the Cooperon also includes two different phase relaxation rates $1/\tau_{\mathrm{H}}$ and $1/\tau_{\varphi}$, which account, respectively, for the effect of an external magnetic field H and for all types of interactions violating the time-reversal symmetry. For non-interacting electrons addressed in this section and in the absence of the magnetic field, both these dephasing rates tend to zero, in which case Eqs. (14.13) and (14.14) become identical. For our later purposes, we will, however, keep both these rates in Eq. (14.14). At this stage, both τ_{H} and τ_{φ} can be understood as phenomenological parameters to be specified in our subsequent considerations.

In the case of non-interacting electrons, the final expression for the conductivity of a disordered metal σ is given by the sum of classical and quantum contributions defined, respectively, in Eqs. (9.108) and (14.15), i.e.,

$$\sigma = \sigma_{\mathrm{D}} + \sigma_{\mathrm{WL}} \quad (14.16)$$

This result can be derived by means of a direct calculation from Eq. (14.5) combined with Eqs. (14.9)–(14.12). Here we will proceed differently and reconstruct Eq. (14.16) with the aid of the path integral analysis already developed in Chapter 9. This will be convenient for us for a number of reasons. As we will see, this analysis allows for a clear physical interpretation of both the diffuson and the Cooperon in terms of quasiclassical electron paths. Furthermore, the path integral technique of Chapter 9 fully accounts for the effect of electron–electron interactions, which will be the central subject of our analysis below in this chapter.

Let us make use of Eq. (9.103) for the linear conductivity of our system, which—after averaging over disorder configurations—becomes formally identical to Eq. (14.5). In the absence of electron–electron interactions, the kernel J in Eq. (9.103) takes a particularly simple form (cf. Eq. (9.99)):

$$J(t,t';\boldsymbol{r}_1,\boldsymbol{r}_2;\boldsymbol{r}_{1i},\boldsymbol{r}_{2i}) = \int_{\boldsymbol{r}_F(t')=\boldsymbol{r}_{1i}}^{\boldsymbol{r}_F(t)=\boldsymbol{r}_1}\mathcal{D}\boldsymbol{r}_F\int_{\boldsymbol{r}_B(t')=\boldsymbol{r}_{2i}}^{\boldsymbol{r}_B(t)=\boldsymbol{r}_2}\mathcal{D}\boldsymbol{r}_B\int\mathcal{D}\boldsymbol{p}_F\mathcal{D}\boldsymbol{p}_B$$
$$\times\exp\{\mathrm{i}S_0[\boldsymbol{p}_F,\boldsymbol{r}_F]-\mathrm{i}S_0[\boldsymbol{p}_B,\boldsymbol{r}_B]\} \tag{14.17}$$

where the free electron action S_0 is defined in Eq. (9.97). Note that the momentum integrals in the above expression for J can be carried out exactly, and this kernel can be defined in the coordinate representation only. We, however, chose to keep these momentum integrals here because, as it was already explained in Chapter 9, they will be needed in order to properly account for the electron–electron interaction effects to be considered below in Section 14.4.

The double path integral (14.17) can also be rewritten as a product of two matrix elements

$$J(t,t';\boldsymbol{r}_1,\boldsymbol{r}_2;\boldsymbol{r}_{1i},\boldsymbol{r}_{2i}) = \langle\boldsymbol{r}_1|\hat{\boldsymbol{u}}_0(t,t')|\boldsymbol{r}_{1i}\rangle\langle\boldsymbol{r}_{2i}|\hat{\boldsymbol{u}}_0(t',t)|\boldsymbol{r}_2\rangle \tag{14.18}$$

where $\hat{\boldsymbol{u}}_0(t,t') = e^{-\mathrm{i}\hat{H}(t-t')}$ is the evolution operator pertaining to the free electron Hamiltonian (14.6).

In the quasiclassical limit (14.1), the path integral (14.17) is dominated by the saddle point trajectories for the action S_0, which are just classical electron paths determined by the Hamilton equations

$$\dot{\boldsymbol{p}} = -\frac{\partial H_0(\boldsymbol{p},\boldsymbol{r})}{\partial\boldsymbol{r}},\quad \dot{\boldsymbol{r}} = \frac{\partial H_0(\boldsymbol{p},\boldsymbol{r})}{\partial\boldsymbol{p}} \tag{14.19}$$

with obvious boundary conditions $\boldsymbol{r}_F(t') = \boldsymbol{r}_{1i}$, $\boldsymbol{r}_F(t) = \boldsymbol{r}_1$ for the action $S_0[\boldsymbol{r}_F,\boldsymbol{p}_F]$ and $\boldsymbol{r}_B(t') = \boldsymbol{r}_{2i}$, $\boldsymbol{r}_B(t) = \boldsymbol{r}_2$ for the action $S_0[\boldsymbol{r}_B,\boldsymbol{p}_B]$. Substituting these saddle point trajectories into Eq. (14.17) and integrating out small fluctuations around them, for each of the matrix elements in Eq. (14.18), one arrives at the well-known van Vleck formula

$$\langle \boldsymbol{r}_1|\hat{\boldsymbol{u}}_0(t,t')|\boldsymbol{r}_{1i}\rangle = \sum_{r_{\rm F}} A[\boldsymbol{r}_{\rm F}]e^{iS_0(t,t';\boldsymbol{r}_1,\boldsymbol{r}_{1i})},$$
$$A[\boldsymbol{r}_{\rm F}] = \sqrt{\frac{i^3}{8\pi^3}\left|\det\frac{\partial^2 S_0(\boldsymbol{r}_1,\boldsymbol{r}_{1i})}{\partial\boldsymbol{r}_1\partial\boldsymbol{r}_{1i}}\right|} \tag{14.20}$$

and

$$\langle \boldsymbol{r}_{2i}|\hat{\boldsymbol{u}}_0(t',t)|\boldsymbol{r}_2\rangle = \sum_{r_{\rm B}} A^*[\boldsymbol{r}_{\rm B}]e^{-iS_0(t,t';\boldsymbol{r}_2,\boldsymbol{r}_{2i})},$$
$$A^*[\boldsymbol{r}_{\rm B}] = \sqrt{\frac{-i^3}{8\pi^3}\left|\det\frac{\partial^2 S_0(\boldsymbol{r}_2,\boldsymbol{r}_{2i})}{\partial\boldsymbol{r}_2\partial\boldsymbol{r}_{2i}}\right|}, \tag{14.21}$$

where the actions $S_0(t,t';\boldsymbol{r}_{1,2},\boldsymbol{r}_{1i,2i})$ are taken on all possible classical paths $\boldsymbol{r}_{\rm F,B}(t)$ satisfying Eqs. (14.19) together with the above boundary conditions. Then for the kernel J, we obtain

$$J(t,t';\boldsymbol{r}_1,\boldsymbol{r}_2;\boldsymbol{r}_{1i},\boldsymbol{r}_{2i}) = \sum_{r_{\rm F}} A[\boldsymbol{r}_{\rm F}]\sum_{r_{\rm B}} A^*[\boldsymbol{r}_{\rm B}]e^{iS_0(t,t';\boldsymbol{r}_1,\boldsymbol{r}_{1i})-iS_0(t,t';\boldsymbol{r}_2,\boldsymbol{r}_{2i})} \tag{14.22}$$

Since in a random impurity potential $U(\boldsymbol{r})$, there exist in general no correlations between different classical paths $\boldsymbol{r}_{\rm F}(t)$ and $\boldsymbol{r}_{\rm B}(t)$, these paths give no contribution to the double sum (14.22). Hence, the difference between the two actions in the exponent of Eq. (14.22) may have an arbitrary value and the result simply averages out after the summation over the paths $\boldsymbol{r}_{\rm F}(t)$ and $\boldsymbol{r}_{\rm B}(t)$. Thus, only the paths for which $S_0[\boldsymbol{r}_{\rm F},\boldsymbol{p}_{\rm F}] \simeq S_0[\boldsymbol{r}_{\rm B},\boldsymbol{p}_{\rm B}]$ may provide a non-vanishing contribution to (14.22). Two different classes of such paths can be distinguished (cf., e.g., [87]):

(i) The two classical paths are almost the same: $\boldsymbol{r}_{\rm F}(t'') \simeq \boldsymbol{r}_{\rm B}(t'')$, $\boldsymbol{p}_{\rm F}(t'') \simeq \boldsymbol{p}_{\rm B}(t'')$ (see Fig. 14.1a). For such pairs, we obviously have $\boldsymbol{r}_{1i} \simeq \boldsymbol{r}_{2i} \simeq \boldsymbol{r}_i$ and $\boldsymbol{r}_1 \simeq \boldsymbol{r}_2 \simeq \boldsymbol{r}_f$. In this case in Eq. (14.17), one integrates only over trajectories with $|\boldsymbol{r}_{\rm F}(t'') - \boldsymbol{r}_{\rm B}(t'')| \lesssim 1/p_{\rm F}$. Physically, this corresponds to the picture of electrons propagating as almost classical particles described by the diagonal elements of the density matrix only. The corresponding contribution to the kernel J (14.17) from such kind of trajectories reads

$$W_1(t-t', \boldsymbol{r}_f, \boldsymbol{r}_i) = \sum_{\boldsymbol{r}(t'')} |A(\boldsymbol{r}(t''))|^2, \qquad (14.23)$$

where the sum is taken over all classical trajectories $\boldsymbol{r}(t'')$, which start at $\boldsymbol{r}(t') = \boldsymbol{r}_i$ and end at $\boldsymbol{r}(t) = \boldsymbol{r}_f$. The function $W_1(t-t', \boldsymbol{r}_f, \boldsymbol{r}_i)$ is, therefore, nothing but the classical transition probability for an electron from the point $\boldsymbol{r} = \boldsymbol{r}_i$ to the point $\boldsymbol{r} = \boldsymbol{r}_f$ during the time interval $t - t'$.

(ii) The pairs of time-reversed paths. In this case, $\boldsymbol{r}_{1i} \simeq \boldsymbol{r}_2, \boldsymbol{r}_1 \simeq \boldsymbol{r}_{2i}$ (Fig. 14.1b). In the path integral (1.16), the trajectories $\boldsymbol{r}_{\rm F}$ and $\boldsymbol{r}_{\rm B}$ are related as $\boldsymbol{r}_{\rm B}(t'') \simeq \boldsymbol{r}_{\rm F}(t+t'-t'')$ and $\boldsymbol{p}_{\rm B}(t'') \simeq -\boldsymbol{p}_{\rm F}(t+t'-t'')$. In other words, in Eq. (14.17), one integrates over paths with $|\boldsymbol{r}_{\rm B}(t'') - \boldsymbol{r}_{\rm F}(t+t'-t'')| \lesssim 1/p_{\rm F}$; however, the difference $|\boldsymbol{r}_{\rm F}(t'') - \boldsymbol{r}_{\rm B}(t'')|$ may be arbitrarily large in this case. These paths essentially determine the evolution of the off-diagonal elements of the electron density matrix and yield the contribution to the kernel J (14.17) in the form

$$W_2(t-t', \boldsymbol{r}_f, \boldsymbol{r}_i) = \sum_{\boldsymbol{r}(t'')} A(\boldsymbol{r}(t''))A^*(\boldsymbol{r}(t+t'-t'')) \qquad (14.24)$$

where $\boldsymbol{r}(t'')$ are the same classical paths as in Eq. (14.23). The quantity $W_2(t-t', \boldsymbol{r}_f, \boldsymbol{r}_i)$ cannot be interpreted as a classical probability, as it is defined on time-reversed electron paths. However, in the absence of any kind of interactions breaking the time-reversal symmetry, the quantity (14.24) exactly coincides with the classic probability, i.e.,

$$W_2(t-t', \boldsymbol{r}_f, \boldsymbol{r}_i) = W_1(t-t', \boldsymbol{r}_f, \boldsymbol{r}_i) \qquad (14.25)$$

If the summation in Eqs. (14.23) and (14.24) is carried out over diffusive paths, both these quantities obey the same diffusion equation

$$\left(\frac{\partial}{\partial t} - D\nabla^2_{\boldsymbol{r}_1}\right) W_{1,2}(t-t', \boldsymbol{r}_1, \boldsymbol{r}_2) = \delta(t-t')\delta(\boldsymbol{r}_1 - \boldsymbol{r}_2) \quad (14.26)$$

In this case, the probability W_1 coincides with the diffuson $\mathcal{D}$ (cf. Eq. (14.13)), and the quantity W_2 determines the Cooperon $\mathcal{C}$.

It follows from the above analysis that in order to evaluate the conductivity of a disordered conductor, it is necessary to substitute

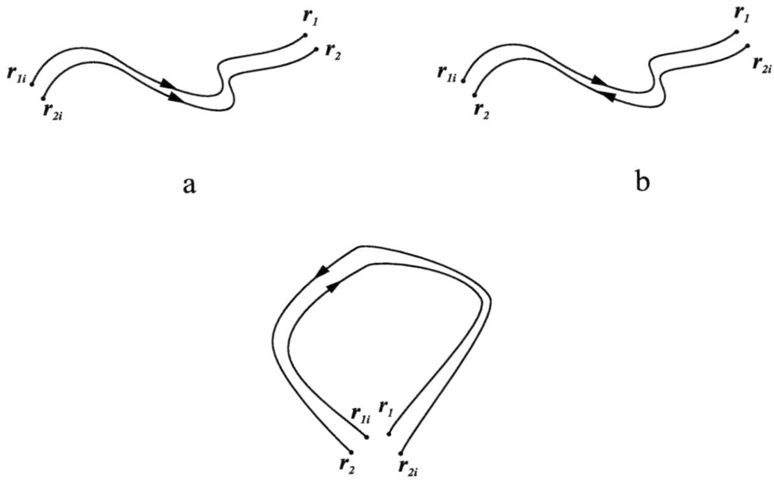

Figure 14.1 Different quasiclassical trajectories: (a) $\boldsymbol{r}_{\rm F}$ and $\boldsymbol{r}_{\rm B}$ are close to each other; (b) time-reversed paths; (c) time-reversed paths returning to the vicinity of the initial point, these paths determine the weak localization correction to conductivity.

the sum of the two contributions (14.23) and (14.24) to the path integral (14.17),

$$J = W_1 + W_2, \tag{14.27}$$

into Eq. (9.103). As a result, one recovers two different contributions to the conductivity (14.16). The first one originates from the diffuson (or, equivalently, from the term W_1) and, as we already demonstrated in Section 9.8.1, yields the standard Drude conductivity $\sigma_{\rm D}$ (9.108). The second term represents the WL correction $\sigma_{\rm WL}$, which comes from W_2.

In order to derive this quantum correction to the conductivity, we note that according to Eq. (9.103), both coordinates $\boldsymbol{r}_{\rm F}$ and $\boldsymbol{r}_{\rm B}$ in the kernel J (14.17) should be kept close to each other, i.e., $\boldsymbol{r}_{\rm F} \approx \boldsymbol{r}_{\rm B}$. For the term W_2, this observation implies that both diffusive time-reversed paths should start and end approximately at the same point, as it is illustrated in Fig. 14.1c. At this point, it is worthwhile to remind the reader that in Chapter 3 (see Section 3.2.1), we already dealt with the return quasiprobability W_2 defined on pairs of time-reversed paths. This quantity is exactly analogous to the one defined here, cf., e.g., the second Eq. (3.31) and Fig. 3.1c.

Keeping the coordinates $\boldsymbol{r}_{\rm F}$ and $\boldsymbol{r}_{\rm B}$ in the expression for W_2 close to each other, we arrive at the result

$$\sigma_{\rm WL} = -\frac{2e^2 D}{\pi}\int_{\tau_e}^{\infty} dt W_2(t, \boldsymbol{r}, \boldsymbol{r}) = -\frac{2e^2 D}{\pi}\int_{-\infty}^{t-\tau_e} dt' W_2(t-t', \boldsymbol{r}, \boldsymbol{r}) \tag{14.28}$$

In this expression, we introduced a short time cutoff τ_e indicating that the electron propagation in a disordered metal becomes diffusive only at time scales exceeding the elastic relaxation time.

Note that Eq. (14.28) is essentially identical to Eq. (14.15). Furthermore, due to Eq. (14.25), the quantity $W_2(t, \boldsymbol{r}, \boldsymbol{r})$ should coincide with the classical probability $W_1(t, \boldsymbol{r}, \boldsymbol{r})$ for diffusive paths to return to the same point $\boldsymbol{r}$ after the time t; see also Chapter 3. This return probability can be easily evaluated from Eq. (14.26). It does not depend on the particular point $\boldsymbol{r}$ and is given by the formula

$$W_1(t, \boldsymbol{r}, \boldsymbol{r}) \equiv W(t) = (4\pi Dt)^{-d/2} a^{-(3-d)} \tag{14.29}$$

where d is the system dimension and a is the transverse sample size (the film thickness for $d = 2$ and the square root of the wire cross section $a = \sqrt{\mathcal{A}}$ for $d = 1$).

Making use of Eqs. (14.25) and (14.29) and employing the formula (14.28), one can evaluate the magnitude of the WL correction to the Drude conductivity. Substituting the function $W(t)$ (14.29) instead of $W_2(t)$ into Eq. (14.28), it is easy to see that the time integral formally diverges for $d \leq 2$. In the fully coherent limit, this divergence is naturally cut off at the scale determined by the total electron diffusion time through the sample or, equivalently, by the inverse Thouless energy.

Here we present an explicit expression for the conductance G of a diffusive metallic wire with cross section $\mathcal{A}$ and length L. In the limit $T \to 0$, it reads [299]

$$G = \frac{\sigma_{\rm D}\mathcal{A}}{L} - \frac{e^2}{3\pi} \tag{14.30}$$

We observe that the WL correction to the total wire conductance is of the order of the quantum conductance unit e^2/h. This result demonstrates that the quantum correction to the Drude term remains small provided the wire is not very long, i.e., for $L \ll \xi_{\rm loc} \sim k_{\rm F}^2 \mathcal{A}\ell$. For longer wires with $L > \xi_{\rm loc}$ at $H = 0$ and in the case of non-interacting electrons, one can expect to observe a crossover to the regime of strong Anderson localization in which case the conductance would decrease exponentially with the increasing wire length as $G \sim \exp[-L/\xi_{\rm loc}]$. A detailed description of this crossover can be found, e.g., in the review [44]. Unfortunately, it turns out to be problematic to realize this scenario in any generic metallic wire. The main reason for that is the effect of electron dephasing. This effect plays a crucial role in mesoscopic conductors and will be addressed below in this and next chapters.

14.3 Electron Dephasing by Classical Fields

As we already indicated above, on a phenomenological level, the effect of electron decoherence in disordered conductors can be accounted for by including the two different phase relaxation rates $1/\tau_{\rm H}$ and $1/\tau_\varphi$ in Eq. (14.14) for the Cooperon. The dephasing rate $1/\tau_{\rm H}$ describes the effect of an externally applied magnetic field, while the rate $1/\tau_\varphi$ results from interactions of an electron with other degrees of freedom (such as phonons, electrons, magnetic impurities, etc.) as well as with fluctuating electromagnetic fields.

Let us first consider a quasi-one-dimensional metallic wire of length L and cross section $\mathcal{A}$ coupled to two bulk metallic leads on both sides. Provided the wire length exceeds the dephasing length, $L \gg L_\varphi = \sqrt{D\tau_\varphi}$, from Eq. (14.14) one readily finds

$$\mathcal{C}(0, \boldsymbol{r}_1, \boldsymbol{r}_1) = \frac{1}{\mathcal{A}} \int \frac{dq}{2\pi} \frac{1}{Dq^2 + \frac{1}{\tau_{\rm H}} + \frac{1}{\tau_\varphi}} = \frac{1}{2\mathcal{A}} \sqrt{\frac{\tau_\varphi \tau_{\rm H}}{D(\tau_\varphi + \tau_{\rm H})}} \qquad (14.31)$$

Hence, in this case, the WL correction (14.15) takes the form

$$\sigma_{\rm WL} = -\frac{e^2}{\pi \mathcal{A}} \sqrt{\frac{D\tau_\varphi \tau_{\rm H}}{\tau_\varphi + \tau_{\rm H}}}, \qquad d = 1. \qquad (14.32)$$

Below we will demonstrate that in the case of quasi-one-dimensional metallic wires, the phase relaxation time $\tau_{\rm H}$ is defined

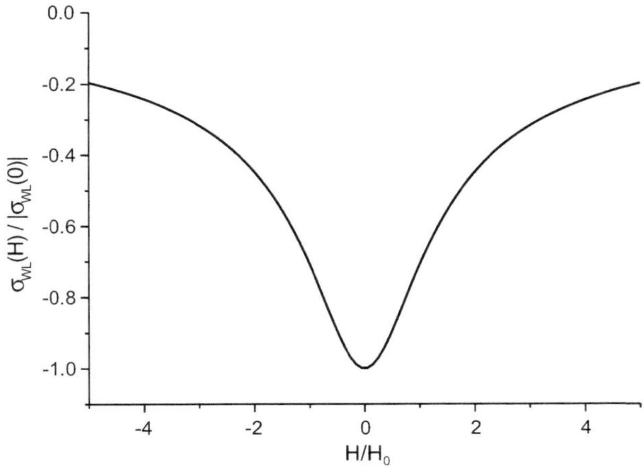

Figure 14.2 Weak localization correction to the conductivity of a diffusive metallic wire (14.32) as a function of an externally applied magnetic field H. The characteristic magnetic field H_0 is set by the formula $H_0 = (c/e)\sqrt{3/D\tau_\varphi \mathcal{A}}$.

by the formula

$$\tau_{\mathrm{H}} = \frac{3c^2}{De^2 H^2 \mathcal{A}} \tag{14.33}$$

Hence, in the limit of high magnetic fields, electron dephasing is dominated by τ_{H} and the WL correction tends to zero as $\sigma_{\mathrm{WL}} \sim 1/H$. On the contrary, at weak magnetic fields $H \to 0$, the electron phase relaxation time is set by τ_φ, and one finds

$$\sigma_{\mathrm{WL}} = -\frac{e^2}{\pi \mathcal{A}} L_\varphi \tag{14.34}$$

The dependence of the WL correction on the magnetic field is illustrated in Fig. 14.2. It demonstrates that within the frames of our analysis, the WL correction always remains negative. At this stage, it is worth mentioning that σ_{WL} may also turn positive provided a metal is characterized by sufficiently strong spin-orbit coupling. For further details, we refer the reader, e.g., to the review [54]. For simplicity, here and below we will ignore spin-orbit effects and, hence, disregard the regime of positive WL correction to the conductance.

In the case of diffusive conductors of higher dimensions ($d = 2, 3$), the WL correction σ_{WL} is evaluated analogously. For disordered metallic films of thickness a, we obtain

$$\sigma_{\mathrm{WL}} = -\frac{e^2}{2\pi^2}\ln\left(\frac{\tau_\varphi \tau_{\mathrm{H}}}{\tau_e(\tau_\varphi + \tau_{\mathrm{H}})}\right), \quad d = 2, \tag{14.35}$$

and in the case of bulk metallic samples, one finds

$$\sigma_{\mathrm{WL}} = -\frac{\sqrt{3}e^2}{2\pi^{3/2}\ell}, \quad d = 3 \tag{14.36}$$

We observe that in the case of 2d films, the WL correction is still sensitive to the effect of electron dephasing. In contrast, in 3d samples, σ_{WL} remains essentially independent of both τ_{H} and τ_φ. The reason for this difference was already indicated above: While for $d \leq 2$, the time integral in Eq. (14.28) (with W_2 substituted by $W(t)$ (14.29)) diverges at long times and should effectively be cut at the shorter of the two time scales τ_{H} and τ_φ, for $d = 3$, the magnitude of this time integral is determined by short times $t \sim \tau_e$ at which electron dephasing is still negligible.

Let us now derive Eq. (14.33). In the presence of a non-vanishing vector potential $\boldsymbol{A}$, the action of an electron moving along any path $\boldsymbol{r}(t)$ acquires an additional term proportional to $\boldsymbol{A}$. In order to account for this extra contribution in Eq. (14.22), it is necessary to make a replacement

$$S_0 \to S_0 + \int_0^t d\tau\, \frac{e}{c}\dot{\boldsymbol{r}}(\tau)\boldsymbol{A}(\boldsymbol{r}(\tau)) \tag{14.37}$$

The WL correction to conductivity (14.28) then takes the form

$$\begin{aligned}\sigma_{\mathrm{WL}} &= -\frac{2e^2 D}{\pi}\int_{\tau_e}^{\infty} dt\, W_2(t, \boldsymbol{r}, \boldsymbol{r}) \\ &\times \left\langle \exp\left\{ i\frac{e}{c}\int_0^t d\tau\, [\dot{\boldsymbol{r}}_{\mathrm{cl}}(\tau)\boldsymbol{A}(\boldsymbol{r}_{\mathrm{cl}}(\tau)) + \dot{\boldsymbol{r}}_{\mathrm{cl}}(t-\tau)\boldsymbol{A}(\boldsymbol{r}_{\mathrm{cl}}(t-\tau))]\right\}\right\rangle \\ &= -\frac{2e^2 D}{\pi}\int_{\tau_e}^{\infty} dt\, W_2(t, \boldsymbol{r}, \boldsymbol{r})\left\langle \exp\left\{ i\frac{2e}{c}\int_0^t d\tau\, \dot{\boldsymbol{r}}_{\mathrm{cl}}(\tau)\boldsymbol{A}(\boldsymbol{r}_{\mathrm{cl}}(\tau))\right\}\right\rangle,\end{aligned} \tag{14.38}$$

where we already replaced the paths $\boldsymbol{r}_{\mathrm{F}}$, $\boldsymbol{r}_{\mathrm{B}}$ by a pair of classical time-reversed paths $\boldsymbol{r}_{\mathrm{cl}}(\tau)$ and $\boldsymbol{r}_{\mathrm{cl}}(t-\tau)$.

The averaging in Eq. (14.38) is taken over all realizations of classical diffusive paths with the same initial and final point $\boldsymbol{r}$. This averaging can be performed by means of the path integral technique. For further details of this approach, we refer the reader to the review [12]. Below we will adopt a somewhat different averaging procedure, which yields the same final result (14.33).

In order to proceed, we will consider the limit of a constant in space magnetic field described by the vector potential $\boldsymbol{A}(\boldsymbol{r}(\tau)) = \boldsymbol{H} \times \boldsymbol{r}(\tau)/2$ and assume that the corresponding $\boldsymbol{A}$-dependent correction to the action is a Gaussian stochastic value. By adding a constant shift to the vector potential and choosing it in the form $\boldsymbol{A}(\boldsymbol{r}(\tau)) = \boldsymbol{H} \times (\boldsymbol{r}(\tau) - \boldsymbol{r})/2$, it is possible to relax the restriction for the electron trajectory to return to the initial point $\boldsymbol{r}$ after time t. Then we obtain

$$\left\langle \exp\left\{ \mathrm{i}\frac{2e}{c}\int_0^t d\tau\, \dot{\boldsymbol{r}}_{\mathrm{cl}}(\tau)\boldsymbol{A}(\boldsymbol{r}_{\mathrm{cl}}(\tau))\right\}\right\rangle$$
$$= \exp\left\{ -\frac{e^2}{2c^2}\int_0^t d\tau d\tau' \left\langle \left(\dot{\boldsymbol{r}}_{\mathrm{cl}}(\tau)\left(\boldsymbol{H} \times (\boldsymbol{r}_{\mathrm{cl}}(\tau) - \boldsymbol{r})\right)\right)\right.\right.$$
$$\left.\left.\times \left(\dot{\boldsymbol{r}}_{\mathrm{cl}}(\tau')\left(\boldsymbol{H} \times (\boldsymbol{r}_{\mathrm{cl}}(\tau') - \boldsymbol{r})\right)\right)\right\rangle \right\} \tag{14.39}$$

While the absolute value of the electron velocity in a metal does not change in time being equal to the Fermi velocity v_{F}, the direction of the vector $\boldsymbol{v}_{\mathrm{F}}$ gets quickly randomized in the course of the electron motion because of its frequent scattering on impurities. In the diffusive limit, we have

$$\dot{r}_i(\tau)\dot{r}_j(\tau') = \frac{v_{\mathrm{F}}^2}{3}e^{-|\tau-\tau'|/\tau_e}\,\delta_{ij} \tag{14.40}$$

Replacing the exponent $e^{-|\tau-\tau'|/\tau_e}$ by the δ-function $2\tau_e\delta(\tau-\tau')$ and averaging over the velocity directions in Eq. (14.39), we find

$$\left\langle \exp\left\{ \mathrm{i}\frac{2e}{c}\int_0^t d\tau\, \dot{\boldsymbol{r}}_{\mathrm{cl}}(\tau)\boldsymbol{A}(\boldsymbol{r}_{\mathrm{cl}}(\tau))\right\}\right\rangle$$
$$= \exp\left\{ -\frac{e^2 D}{c^2}\int_0^t d\tau \left\langle \left(\boldsymbol{H} \times (\boldsymbol{r}_{\mathrm{cl}}(\tau) - \boldsymbol{r})\right)^2\right\rangle \right\} \tag{14.41}$$

For the sake of definiteness, let us consider a wire with rectangular cross section denoting its width and thickness, respectively, as

a_x and a_y. Assuming that the magnetic field $\boldsymbol{H}$ is parallel to the wire and points to the z-direction, we get

$$\left\langle \left(\boldsymbol{H} \times (\boldsymbol{r}_{\rm cl}(\tau) - \boldsymbol{r})\right)^2 \right\rangle = \int_0^{a_x} \frac{dx dx_0}{a_x^2} \int_0^{a_y} \frac{dy dy_0}{a_y^2} \times \left[(x - x_0)^2 + (y - y_0)^2\right] \quad (14.42)$$

where we defined the vector components as $\boldsymbol{r}_{\rm cl}(\tau) = (x, y)$ and $\boldsymbol{r} = (x_0, y_0)$. Here we assumed that (i) the path $\boldsymbol{r}_{\rm cl}(\tau)$ quickly covers the wire cross section and (ii) initial coordinates of the paths are randomly and homogeneously distributed across the wire. Evaluating the integrals in Eq. (14.42), we arrive at the result

$$\left\langle \exp\left\{ \mathrm{i}\frac{2e}{c} \int_0^t d\tau\, \dot{\boldsymbol{r}}_{\rm cl}(\tau) \boldsymbol{A}(\boldsymbol{r}_{\rm cl}(\tau)) \right\} \right\rangle = e^{-t/\tau_{\rm H}} \quad (14.43)$$

with

$$\tau_{\rm H} = \frac{6c^2}{e^2 D(w_x^2 + w_y^2) H^2} \quad (14.44)$$

For $a_x = a_y = \sqrt{\mathcal{A}}$, this expression reduces to Eq. (14.33).

It is also straightforward to extend our analysis to the case of two-dimensional metallic films and to generalize it in order to describe the effect of an arbitrarily oriented external magnetic field. The corresponding expressions for $\tau_{\rm H}$ can be found, e.g., in the review [12].

Let us now assume that our sample is placed in an external electromagnetic field, which is described by a classic scalar potential $V_{\rm cl}(\tau, \boldsymbol{r})$ that changes both in space and time in an arbitrary manner. As usual, interaction between electrons and this electromagnetic field at the point $\boldsymbol{r}$ is accounted for by the term $eV_{\rm cl}(\tau, \boldsymbol{r})$, which now should be added to the electron Hamiltonian H (14.6). Accordingly, the single electron action S_0 (9.97) should be modified as

$$S_0 \to S_0 + e \int_0^t d\tau\, V_{\rm cl}(\tau, \boldsymbol{r}(\tau)) \quad (14.45)$$

For simplicity, we also assume that the electromagnetic field is not too strong and only weakly perturbs the electron paths, implying that the last term in Eq. (14.45) can be treated as a small perturbation to S_0.

Making use of the replacement (14.45), one again repeats the same analysis as in Section 14.2 taking into account the two classes of paths illustrated in Fig. 14.1. On each of such paths $\boldsymbol{r}_{\mathrm{F}}(\tau)$ and $\boldsymbol{r}_B(\tau)$, an electron picks up an extra phase

$$\delta\varphi[\boldsymbol{r}_{\mathrm{F,B}}(\tau)] = e\int_0^t d\tau\, V_{\mathrm{cl}}(\tau, \boldsymbol{r}_{\mathrm{F,B}}(\tau)) \tag{14.46}$$

For almost the same classical paths $\boldsymbol{r}_{\mathrm{F}}(\tau) \simeq \boldsymbol{r}_B(\tau)$, these phases are equal and, hence, cancel each other in the exponent of the expression for the kernel J (14.22). This cancelation implies that both the diffuson $\mathcal{D}$ and the Drude conductance σ_{D} remain practically unaffected by a sufficiently weak external electromagnetic field. In the case of the Cooperon defined on pairs of time-reversed paths $\boldsymbol{r}_{\mathrm{F}}(\tau) \simeq \boldsymbol{r}_{\mathrm{B}}(t-\tau)$, the two phases (14.46) are different and do not cancel out in the expression for J (14.22). Accordingly, the expression for the WL correction now reads

$$\sigma_{\mathrm{WL}} = -\frac{2e^2}{\pi}D\int_0^\infty dt\, W(t)\left\langle e^{\mathrm{i}\int_0^t d\tau\{eV_{\mathrm{cl}}(\tau,\boldsymbol{r}(\tau))-eV_{\mathrm{cl}}(\tau,\boldsymbol{r}(t-\tau))\}}\right\rangle_{\boldsymbol{r},V_{\mathrm{cl}}} \tag{14.47}$$

where the classical return probability $W(t)$ is defined in Eq. (14.29).

On the other hand, it follows, e.g., from Eqs. (14.14) and (14.15) that in the limit $H \to 0$, the same quantity can also be expressed in the form

$$\sigma_{\mathrm{WL}} = -\frac{2e^2}{\pi}D\int_0^\infty dt\, W(t)\, e^{-t/\tau_\varphi} \tag{14.48}$$

Comparing Eqs. (14.47) and (14.48), we immediately identify

$$e^{-t/\tau_\varphi} = \left\langle e^{\mathrm{i}\int_0^t d\tau\{eV_{\mathrm{cl}}(\tau,\boldsymbol{r}(\tau))-eV_{\mathrm{cl}}(\tau,\boldsymbol{r}(t-\tau))\}}\right\rangle_{\boldsymbol{r},V_{\mathrm{cl}}} \tag{14.49}$$

The average in the right-hand side of this equation can be estimated as

$$\begin{aligned}
&\left\langle e^{\mathrm{i}\int_0^t d\tau\{eV_{\mathrm{cl}}(\tau,\boldsymbol{r}(\tau))-eV_{\mathrm{cl}}(\tau,\boldsymbol{r}(t-\tau))\}}\right\rangle_{\boldsymbol{r},V_{\mathrm{cl}}}\\
&\simeq e^{-\frac{e^2}{2}\int_0^t d\tau d\tau'\langle\{eV_{\mathrm{cl}}(\tau,\boldsymbol{r}(\tau))-eV_{\mathrm{cl}}(\tau,\boldsymbol{r}(t-\tau))\}\{eV_{\mathrm{cl}}(\tau,\boldsymbol{r}(\tau'))-eV_{\mathrm{cl}}(\tau,\boldsymbol{r}(t-\tau'))\}\rangle_{\boldsymbol{r},V_{\mathrm{cl}}}}\\
&= \exp\left[-\frac{e^2}{2}\int_0^t d\tau d\tau' e^{-\mathrm{i}\omega(\tau-\tau')}\int\frac{d\omega d^d\boldsymbol{q}}{(2\pi)^{d+1}a^{3-d}}\left\langle|V_{\mathrm{cl}}|^2\right\rangle_{\omega,\boldsymbol{q}}\right.\\
&\left.\times\left\langle e^{\mathrm{i}\boldsymbol{q}(\boldsymbol{r}(\tau)-\boldsymbol{r}(\tau'))}+e^{\mathrm{i}\boldsymbol{q}(\boldsymbol{r}(t-\tau)-\boldsymbol{r}(t-\tau'))}-e^{\mathrm{i}\boldsymbol{q}(\boldsymbol{r}(\tau)-\boldsymbol{r}(t-\tau'))}-e^{\mathrm{i}\boldsymbol{q}(\boldsymbol{r}(t-\tau)-\boldsymbol{r}(\tau'))}\right\rangle_{\boldsymbol{r}}\right].
\end{aligned}$$

Here we defined the spectral density for the voltage fluctuations

$$\left\langle|V_{\mathrm{cl}}|^2\right\rangle_{\omega,\boldsymbol{q}} = \int dt d^d\boldsymbol{r}\,\langle V_{\mathrm{cl}}(t,\boldsymbol{r})V_{\mathrm{cl}}(0,0)\rangle\, e^{\mathrm{i}(\omega t-\boldsymbol{q}\boldsymbol{r})} \tag{14.50}$$

For diffusive classical electron paths $\boldsymbol{r}(\tau)$, one has

$$\left\langle e^{\mathrm{i}\boldsymbol{q}(\boldsymbol{r}(\tau)-\boldsymbol{r}(\tau'))}\right\rangle_{\boldsymbol{r}} = e^{-Dq^2|\tau-\tau'|} \tag{14.51}$$

Then the dephasing rate becomes equal to

$$\frac{1}{\tau_\varphi} \simeq \lim_{t\to\infty}\frac{e^2}{t}\int_0^t d\tau d\tau' e^{-\mathrm{i}\omega(\tau-\tau')}\int\frac{d\omega d^d\boldsymbol{q}}{(2\pi)^{d+1}a^{3-d}}\left\langle|V_{\mathrm{cl}}|^2\right\rangle_{\omega,\boldsymbol{q}} \times\left(e^{-Dq^2|\tau-\tau'|}-e^{-Dq^2|t-\tau-\tau'|}\right) \tag{14.52}$$

and we arrive at an important result [11]

$$\frac{1}{\tau_\varphi} \simeq e^2\int\frac{d\omega d^d\boldsymbol{q}}{(2\pi)^{d+1}a^{3-d}}\left\langle|V_{\mathrm{cl}}|^2\right\rangle_{\omega,\boldsymbol{q}}\frac{Dq^2}{\omega^2+D^2q^4} \tag{14.53}$$

This expression establishes the relation between the electron dephasing rate in a d-dimensional diffusive conductor and the correlator for an external fluctuating electromagnetic field.

It follows immediately from this relation that no dephasing can be produced by a uniform space field ($\boldsymbol{q} \to 0$) fluctuating only in time. Likewise, electron dephasing cannot be caused by a constant in time electric field. In fact, these two important conclusions follow already from Eqs. (14.47) or (14.49): Neglecting coordinate and/or time dependence of the fluctuating fields $V_{\mathrm{cl}}(\tau, \boldsymbol{r}(\tau))$, one observes identical cancelation of these fields in the above expressions, in which case the quantum coherence of the electron wave functions is fully retained. From a fundamental viewpoint, this result can be understood easily, since constant in space and/or time electric potentials do not violate time-reversal symmetry of our problem and, hence, cannot cause any dephasing. In contrast, an external magnetic field H violates this symmetry and, hence, yields non-vanishing electron phase relaxation in disordered conductors; cf. Eq. (14.33).

14.4 Dephasing by Electron–Electron Interactions

Let us now turn to the dephasing effect caused by electron–electron interactions. At the first glance, this effect can also be described with the aid of Eq. (14.53). Indeed, in this case physically it might appear reasonable to simply identify an external scalar potential

$V_{\rm cl}(t, \boldsymbol{r})$ with the fluctuating potential $V(t, \boldsymbol{r})$ produced by electrons in a disordered conductor. This is because any single electron propagating in such a conductor would "feel" the effect of both these fields in a similar way. In other words, along the lines with our analysis in Chapter 8, one could again be tempted to reduce the complicated many-body problem with disorder and electron–electron interactions to that of a single electron propagating in an effective fluctuating field produced by all other electrons. If so, the problem could be resolved easily, since the expression for the correlator for this field $\langle VV\rangle$ in a diffusive metal is well known. It reads (cf. Eq. (8.1)):

$$\left\langle |V|^2 \right\rangle_{\omega,q} = \mathrm{Im}\left(\frac{-4\pi}{q^2\epsilon(\omega, \boldsymbol{q})}\right)\coth\frac{\omega}{2T} \tag{14.54}$$

where $\epsilon(\omega, \boldsymbol{q})$ is the dielectric susceptibility of such a metal defined in Eq. (8.15). Substituting the correlator (14.54) into Eq. (14.53) instead of $\left\langle |V_{\rm cl}|^2 \right\rangle_{\omega,\boldsymbol{q}}$ and performing the integrals, one would recover the electron dephasing rate $1/\tau_\varphi$ caused by electron–electron interactions. The dephasing rate derived in this manner decreases with temperature but remains nonzero at any temperature, including $T \to 0$.

Although the above scenario looks intuitively appealing and, furthermore, is consistent with experimental observations [304] demonstrating saturation of τ_φ at low T, it nevertheless leaves a number of important questions without any answer. For instance, it is easy to observe that the corresponding integrals in the expression for $1/\tau_\varphi$ contain both ultraviolet (in all dimensions) and infrared (for $d = 1, 2$) divergencies. Even more importantly, the above analysis completely misses the Pauli principle. In Chapter 8, we deliberately ignored Fermi statistics by distinguishing a particle on a ring from electrons forming an effective environment. Here, however, all electrons are clearly indistinguishable and, hence, Pauli principle needs to be properly accounted for.

On a phenomenological level, an attempt to do so [11] amounts to keeping only the "thermal part" of the voltage noise corresponding to frequencies lower than temperature,

$$\left\langle |V_{\rm cl}|^2 \right\rangle_{\omega,q} = \mathrm{Im}\left(\frac{-4\pi}{q^2\epsilon(\omega, \boldsymbol{q})}\right)\frac{2T}{\omega}\theta(T - |\omega|) \tag{14.55}$$

and simply ignoring all modes with $\omega > T$. In this way one, of course, immediately gets rid of all ultraviolet divergencies. Infrared divergencies, on the other hand, are frequently cured by cutting out also the low-frequency modes with $\omega < 1/\tau_\varphi$. This phenomenological low-frequency cutoff is routinely explained by saying that such modes of the fluctuating field can contribute to electron dephasing only at long times $t > \tau_\varphi$ by which quantum coherence of the electron wave function should have already been destroyed anyway. At this point, we only remark that more rigorous analysis to be presented below in Section 14.4.1 does not require any artificial low-frequency cutoffs.

Collecting all the above arguments and combining Eqs. (14.55) and (8.15) with Eq. (14.53), in the case of quasi-one-dimensional diffusive metallic wires with cross section $\mathcal{A}$, one eventually arrives at the result

$$\frac{1}{\tau_\varphi} = \frac{2Te^2\sqrt{2D}}{\sigma_D \mathcal{A}} \int_{1/\tau_\varphi}^{T} \frac{d\omega}{2\pi} \frac{1}{\omega^{3/2}} \tag{14.56}$$

which obviously remains meaningful only at not-too-low temperatures $T\tau_\varphi \gg 1$. Performing the frequency integral and resolving the resulting equation with respect to τ_φ, one readily finds [11]

$$\frac{1}{\tau_\varphi(T)} \approx \left(\frac{2e^2T\sqrt{2D}}{\pi\sigma_\mathrm{D}\mathcal{A}}\right)^{2/3} \tag{14.57}$$

This formula indicates that the electron dephasing rate decreases with temperature and should eventually tend to zero in the limit $T \to 0$. A similar conclusion also follows from the above consideration in the case of two- and three-dimensional disordered conductors.

Thus, on a phenomenological level, we have two competing approaches to the problem of electron dephasing by electron–electron interactions. The first one suggests to employ the full voltage–voltage correlator (14.54) and yields non-vanishing dephasing rate at $T \to 0$, while the second one prescribes to include only the "classical" part of this correlator (14.55), leading to the conclusion about vanishing electron dephasing rate in the zero-temperature limit.

It is clear that the latter conclusion, if correct, would require a much more solid justification than the above procedure, which amounts to cutting out environmental modes with $\omega > T$ "by hand." While this procedure can be sufficient at high-enough temperatures, in the limit $T \to 0$, it is by no means satisfactory since *all* modes of the fluctuating field are effectively eliminated in this case. In other words, according to Eq. (14.55) at $T = 0$, electrons would not "feel" Coulomb interaction (mediated by the fluctuating field $V(t, \mathbf{r})$) at all!

What could be the arguments in favor of such a procedure? Usually two such arguments (or a combination of those) are discussed in this context. One of them is the golden rule type of argument stating that at low temperatures, the Pauli principle "restricts the scattering space" for interacting electrons to the energy interval of order T. If so, at $T \to 0$ electrons at the Fermi surface can neither lose nor gain energy; hence, they cannot decohere.

Within this line of reasoning, the Pauli principle is viewed merely as an energy constraint that prohibits any energy exchange between electrons at $T = 0$. Therefore, this argument is by no means specific to Fermi systems and can be tested for any low energy quantum particle interacting with a dissipative quantum environment in the limit $T \to 0$. In fact, this test was already carried out, e.g., for the exactly solvable Caldeira–Leggett model in Chapter 3 of this book where we demonstrated that the golden rule approximation yields qualitatively incorrect results in the low-temperature limit. Exactly the same conclusion also applies to the problem of interacting electrons. As before, the main drawback of the golden rule type of arguments is that they essentially resort to the language of a scattering problem employing *non-interacting* in- and out-scattering states. This language is irrelevant here since long-range Coulomb interaction can never be "turned off" and, hence, such states for interacting electrons simply cannot be constructed. One can also add that by effectively setting time equal to infinity and, hence, fixing the electron energy and leaving no room for energy exchanges at $T = 0$, the golden rule approximation violates the quantum mechanical energy-time uncertainty relation, which is crucially important to keep track of here. Further discussion of these and related issues is presented in Section 3.1.2; the same arguments hold for the case of interacting electrons as well.

Another argument that is frequently used in order to support the replacement of the full voltage–voltage correlation function (14.54) by its classical part (14.55) in Eq. (14.53) is based on the results of the first-order perturbation theory in the interaction. Within this approach, one can rigorously evaluate the first-order interaction correction to the Cooperon or to $\sigma_{\rm WL}$. A complicated diagrammatic calculation [4] allows to express this correction in terms of a sum of the so-called coth and tanh terms, which cancel each other (although not completely, see below) in the limit $T \to 0$. If to identify this perturbative result with the first-order term in the formal expansion of Eq. (14.48) in powers of the dephasing rate $1/\tau_\varphi$, one would again arrive at the expression of the form (14.53) with the correlator $\left\langle |V_{\rm cl}|^2 \right\rangle_{\omega,\boldsymbol{q}}$ substituted by [4]

$$\left\langle |V|^2 \right\rangle_{\omega,q} = \operatorname{Im}\left(\frac{-4\pi}{q^2\epsilon(\omega,\boldsymbol{q})}\right)\left(\coth\frac{\omega}{2T} - \frac{\partial}{\partial\omega}\omega\coth\frac{\omega}{2T}\right) \quad (14.58)$$

This expression decays quickly at $|\omega| > T$, thus being essentially equivalent to Eq. (14.55).

The crucial point here is an *assumption* that the phase relaxation process remains strictly exponential at all times, as suggested, e.g., by Eq. (14.48). Unfortunately, this assumption is violated. A detailed calculation to be presented below demonstrates that in the presence of Coulomb interaction, the Cooperon decays in time not purely exponentially but in a more complicated manner. Hence, in general not much can be learned from the first-order perturbation theory in the interaction, which can only probe the Cooperon dynamics *at short times* being unable to access the interesting time interval of longer times $t \gtrsim \tau_\varphi$. The Cooperon decay at such longer times can be adequately described only within a truly non-perturbative interaction analysis.

Thus, we can conclude that none of the reviewed arguments can actually justify the replacement of the full correlator (14.54) by its classical part (14.55). On the other hand, as we already explained above, including the quantum part of this correlator (i.e., fluctuating modes with $\omega > T$) in the expression for the dephasing rate also requires a detailed justification, which can only be obtained by means of a microscopic theory that non-perturbatively accounts for long-range Coulomb interaction, disorder, and Fermi statistics. Fortunately, in Chapter 9 we already derived the Feynman–Vernon influence functional for a degenerate gas of interacting electrons, which properly takes care of all these ingredients. Below in this chapter, we will apply this influence functional formalism to the problem of interaction-induced decoherence of electrons in disordered conductors.

14.4.1 Exponent

Following our path integral analysis in Section 14.2, in order to evaluate the conductivity σ of a disordered metal in the quasiclassical limit (14.1) and in the presence of electron–electron interactions, we will again employ the same classes of electron paths depicted in Fig. 14.1, which produce two different contributions W_1 and W_2 to the kernel J; Eq. (14.27). According to the results of Chapter 9, this kernel now has the form specified in Eqs. (9.99)–(9.101), where the free electron action S_0 is again defined by Eq. (9.97) and the terms $\mathrm{i}S_{\mathrm{R}}$ and S_{I} in the exponent of Eq. (9.99) determine the influence functional for an electron in a disordered conductor. As we already demonstrated in Chapter 9, within RPA this influence functional fully accounts for the effect of electron–electron interactions in our system.

In the case of a disordered metal, it is easy to verify that in the limit (14.1) considered here, the classical electron paths determined by Eqs. (14.19) are only weakly disturbed by the influence functional describing electron–electron interactions. Actually, the corresponding corrections to the classical trajectories are controlled by the same parameter $1/k_{\mathrm{F}}\ell \ll 1$ as the accuracy of the whole calculation of σ_{WL} and, hence, can be safely neglected in our future analysis.

As before, identical classical paths $\boldsymbol{r}_{\mathrm{F}}(\tau) = \boldsymbol{r}_{\mathrm{B}}(\tau)$, $\boldsymbol{p}_{\mathrm{F}}(\tau) = \boldsymbol{p}_{\mathrm{B}}(\tau)$ of Fig. 14.1a control the probability W_1 or, equivalently, the diffuson $\mathcal{D}$. According to general properties of the influence functionals (see Chapter 1) for any pair of such paths, we have $S_{\mathrm{R}} = S_{\mathrm{I}} = 0$ and, hence, within the saddle point approximation $W_1(t)$ just coincides with the classical return probability $W(t)$ (14.29), thus being insensitive to electron–electron interactions.

Let us now turn to the contribution produced by pairs of time-reversed diffusive paths $\boldsymbol{r}_{\mathrm{F}}(\tau) \equiv \boldsymbol{r}(\tau) \simeq \boldsymbol{r}_{\mathrm{B}}(t-\tau)$ $\boldsymbol{p}_{\mathrm{B}}(\tau) \equiv \boldsymbol{p}(\tau) \simeq -\boldsymbol{p}_{\mathrm{F}}(t-\tau)$ (Fig. 14.1c), which define the return quasiprobability $W_2(t)$ (or the Cooperon $\mathcal{C}$) and the WL correction σ_{WL}. The latter quantity now reads

$$\sigma_{\mathrm{WL}} = -\frac{2e^2}{\pi} D \int_0^{\infty} dt\, W(t) \left\langle e^{-\mathrm{i}S_{\mathrm{R}}[\boldsymbol{r}(\tau),\, \boldsymbol{p}(\tau); \boldsymbol{r}(t-\tau),\, \boldsymbol{p}(t-\tau)] - S_{\mathrm{I}}[\boldsymbol{r}(\tau), \boldsymbol{r}(t-\tau)]} \right\rangle_{\boldsymbol{r}} \tag{14.59}$$

where the average is taken over all diffusive paths $\boldsymbol{r}(\tau)$ returning to the initial point after the time t and the classical return probability $W(t)$ is again defined in Eq. (14.29).

Within the accuracy of our calculation, it suffices to perform averaging over diffusive paths in the exponent, in which case Eq. (14.59) can be rewritten in a somewhat simpler form

$$\sigma_{\mathrm{WL}} = -\frac{2e^2}{\pi} D \int_{\tau_e}^{\infty} dt\, W(t) \exp(-f_d(t)) \tag{14.60}$$

where

$$f_d(t) \simeq \langle \mathrm{i}S_{\mathrm{R}}[\boldsymbol{r}(\tau),\, \boldsymbol{p}(\tau); \boldsymbol{r}(t-\tau),\, \boldsymbol{p}(t-\tau)] - S_{\mathrm{I}}[\boldsymbol{r}(\tau), \boldsymbol{r}(t-\tau)] \rangle_{\boldsymbol{r}} \tag{14.61}$$

It turns out that the imaginary contribution to $f_{\mathrm{d}}(t)$ vanishes identically on any pair of time-reversed diffusive paths [155], i.e.,

$$S_{\mathrm{R}}[\boldsymbol{r}(\tau),\, \boldsymbol{p}(\tau); \boldsymbol{r}(t-\tau),\, \boldsymbol{p}(t-\tau)] = 0 \tag{14.62}$$

In order to verify Eq. (14.62), let us specify the expression for S_{R} (9.100) for a pair of time-reversed paths $\boldsymbol{r}_{\mathrm{F}}(\tau) = \boldsymbol{r}(\tau)$, $\boldsymbol{r}_{\mathrm{B}}(\tau) = \boldsymbol{r}(t-\tau)$. We obtain

$$\begin{aligned} S_{\mathrm{R}} = \frac{e^2}{2} \int_0^t dt_1 \int_0^t dt_2 \big\{ & R(t_1-t_2, \boldsymbol{r}(t_1) - \boldsymbol{r}(t_2))[1 - 2n_0(\boldsymbol{p}(t_2), \boldsymbol{r}(t_2))] \\ & - R(t_1-t_2, \boldsymbol{r}(t-t_1) - \boldsymbol{r}(t-t_2))[1 - 2n_0(-\boldsymbol{p}(t-t_2), \boldsymbol{r}(t-t_2))] \\ & + R(t_1-t_2, \boldsymbol{r}(t_1) - \boldsymbol{r}(t-t_2))[1 - 2n_0(-\boldsymbol{p}(t-t_2), \boldsymbol{r}(t-t_2))] \\ & - R(t_1-t_2, \boldsymbol{r}(t-t_1) - \boldsymbol{r}(t_2))[1 - 2n_0(\boldsymbol{p}(t_2), \boldsymbol{r}(t_2))] \big\} \end{aligned} \tag{14.63}$$

Bearing in mind that any trajectory $\boldsymbol{r}(t)$ is nothing but a solution to the classical Hamilton equations of motion (14.19), we conclude that the energy of an electron propagating along this trajectory remains conserved being equal to its initial value E_0. Since the equilibrium distribution function n_0 depends only on the electron energy (and not, e.g., on its coordinate or momentum direction), it stays constant along this trajectory too. Accordingly, the action S_R (14.63) may be cast to the form

$$\begin{aligned} S_R = \frac{e^2}{2}[1 - 2n_0(E_0)] \int_0^t dt_1 \int_0^t dt_2 \\ \times \left\{ R(t_1 - t_2, \boldsymbol{r}(t_1) - \boldsymbol{r}(t_2)) - R(t_1 - t_2, \boldsymbol{r}(t - t_1) - \boldsymbol{r}(t - t_2)) \right. \\ \left. + R(t_1 - t_2, \boldsymbol{r}(t_1) - \boldsymbol{r}(t - t_2)) - R(t_1 - t_2, \boldsymbol{r}(t - t_1) - \boldsymbol{r}(t_2)) \right\} \end{aligned} \quad (14.64)$$

It is easy to observe that the first two terms in the curly brackets in Eq. (14.64) cancel each other identically. This fact becomes obvious after we interchange time coordinates as $t_1 \to t - t_1$, $t_2 \to t - t_2$, and $t_1 \leftrightarrow t_2$. Hence, Eq. (14.64) reduces to

$$\begin{aligned} S_R = \frac{e^2}{2}[1 - 2n_0(E_0)] \int_0^t dt_1 \int_0^t dt_2 \\ \times \left\{ R(t_1 - t_2, \boldsymbol{r}(t_1) - \boldsymbol{r}(t - t_2)) - R(t_1 - t_2, \boldsymbol{r}(t - t_1) - \boldsymbol{r}(t_2)) \right\} \end{aligned} \quad (14.65)$$

This remaining contribution to S_R vanishes for diffusive trajectories $\boldsymbol{r}(t)$, i.e., at sufficiently long times $t \gg \tau_e$, which is the only relevant limit here. Indeed, after an electron experiences many scattering events, the information about its initial coordinate gets entirely lost, enabling one to perform arbitrary shifts in the time argument in $\boldsymbol{r}(t)$. Thus, replacing

$$\begin{aligned} \boldsymbol{r}(t_1) - \boldsymbol{r}(t - t_2) \to \boldsymbol{r}(0) - \boldsymbol{r}(t - t_1 - t_2), \\ \boldsymbol{r}(t - t_1) - \boldsymbol{r}(t_2) \to \boldsymbol{r}(t - t_1 - t_2) - \boldsymbol{r}(0) \end{aligned} \quad (14.66)$$

in Eq. (14.65) and having in mind that $R(t, \boldsymbol{r})$ is an even function of $\boldsymbol{r}$, we arrive at Eq. (14.62) valid for any pair of diffusive time-reversed paths.

In fact, the property (14.62) is not specific for the problem considered here; rather, it constitutes a generic feature of a wide class of influence functionals describing dissipative environments; see, e.g., Chapter 3. Equation (14.62) also demonstrates that the Fermi statistics, although fully accounted for by our analysis, turns out not to affect the function $f_d(t)$ (and, hence, the dephasing time τ_φ) by any means, since the electron distribution function (equal to the Fermi function $n_0(\boldsymbol{p}, \boldsymbol{r})$) enters only into the dissipative part of the action S_R and not to S_I.

It follows immediately from Eqs. (14.61) and (14.62) that the function $f_\mathrm{d}(t)$ is purely real, being determined solely by the imaginary part of the action S_I (9.101), which accounts for the Johnson–Nyquist noise in our conductor. We get

$$f_\mathrm{d}(t) = e^2 \int_0^t dt_1 \int_0^t dt_2 \Big\langle I(t_1 - t_2, \boldsymbol{r}(t_1) - \boldsymbol{r}(t_2)) - I(t_1 + t_2 - t, \boldsymbol{r}(t_1) - \boldsymbol{r}(t_2)) \Big\rangle_{\boldsymbol{r}} \qquad (14.67)$$

where the function I is defined in Eq. (8.10) with the Drude dielectric function (8.15).

In order to derive the expression for the function $f_\mathrm{d}(t)$ (14.67), it is convenient to perform the Fourier transformation of the function $I(t, \boldsymbol{r})$ with respect to both time and coordinates and then evaluate the average over the diffusive trajectories with the aid of a standard replacement (14.51). After the momentum integration, we obtain [157]

$$f_\mathrm{d}(t) = \frac{4e^2 D^{1-d/2}}{\sigma_d (2\pi)^d} \left(\int \frac{d^d x}{1 + x^4} \right) \int \frac{d\omega\, d\omega'}{(2\pi)^2} \left[\frac{|\omega'|^{d/2-2}(\omega - \omega') \coth \frac{\omega-\omega'}{2T}}{\omega^2} + \frac{|\omega'|^{d/2-2} \omega \coth \frac{\omega}{2T} + |\omega|^{d/2-2} \omega' \coth \frac{\omega'}{2T}}{\omega'^2 - \omega^2} \right] (1 - \cos \omega t) \qquad (14.68)$$

where we defined the Drude conductivity for a d-dimensional sample $\sigma_d = \sigma_\mathrm{D} a^{3-d}$.

In lower dimensions $d = 1, 2$, the integral over ω' in the first term diverges at $\omega' \to 0$. It is easy to check, however, that this divergence is exactly canceled by the second term; the whole integral is finite in any dimension and does not require artificial infrared cutoffs. It is also useful to note that at $T \to 0$, the leading contribution to $f_\mathrm{d}(t)$ in the long time limit is insensitive to the infrared divergence contained in the first term in the square brackets (14.68) and can be derived only from this term.

Ultraviolet divergencies in the integrals (14.68) require a separate discussion. To begin with, the result (14.68) demonstrates that high-frequency modes of our effective electronic bath all contribute to the electron dephasing. This is a pretty standard feature encountered for a large variety of quantum dissipative environments, including, e.g., the Caldeira–Leggett model (see Chapters 1 and 3) as well as many other models analyzed in details elsewhere [400]. In all these models, it is usually assumed that the environment does not contain oscillators with frequencies above a certain value ω_c determined on physical grounds for the corresponding model. In the case of a diffusive electronic bath considered here, one naturally chooses this maximum frequency to be of order of the inverse elastic time $\omega_c \sim 1/\tau_e$; see also Section 8.1. This is because the electron motion in a metal is diffusive only at time and length scales longer than, respectively, τ_e and ℓ. Accordingly, the Drude form of the dielectric function (8.15) used in deriving Eq. (14.68) remains applicable only at frequencies $\omega \lesssim 1/\tau_e$ and wave vectors shorter than $1/\ell$.

Hence, the frequency integrals in Eq. (14.68) should be performed up to $\omega \sim 1/\tau_e$ and there exists no room for any artificial cutoffs at lower frequencies, e.g., of order temperature or otherwise. What remains is only to verify that no significant contribution to $f_{\rm d}(t)$ (and, hence, to τ_φ) comes from the time/length scales shorter than the elastic time/length where the electron motion is ballistic and Eq. (8.15) does not apply. We will postpone this important task to Chapter 15 and now proceed with our calculation.

The integrals in Eq. (14.68) can be handled in a straightforward manner. For technical details of the corresponding calculation, we refer the reader to the paper [157]. Here we only present the final results. In the case of quasi-one-dimensional conductors from Eq. (14.68), we obtain [157]

$$f_1(t) = \frac{e^2}{\pi\sigma_1}\sqrt{\frac{2D}{\tau_e}}t + \frac{2e^2}{\pi\sigma_1}\sqrt{\frac{Dt}{\pi}}\left(\ln\frac{2\pi t}{\tau_e} - 6\right), \quad \pi T t \ll 1 \quad (14.69)$$

$$f_1(t) = \frac{2e^2}{\pi\sigma_1}\sqrt{\frac{D}{\pi}}\left\{\sqrt{\frac{\pi}{2\tau_e}}t + \frac{2\pi}{3}T t^{3/2} + \frac{\pi\zeta(1/2)}{\sqrt{2}}t\sqrt{T} - \frac{3\zeta(3/2)}{4\sqrt{2}}\frac{1}{\sqrt{T}}\right.$$
$$\left. + \sqrt{t}\ln\left(\frac{1}{4T\tau_e}\right) + 2\pi T t^{3/2} e^{-2\pi T t}\right\}, \quad \pi T t \gg 1, \quad (14.70)$$

where $\zeta(x)$ is the dzeta-function. We observe that in both low- and high-temperature limits (14.69) and (14.70), the function $f_1(t)$ monotonously grows with time and tends to infinity in the long time limit. Accordingly, the characteristic decay time of the Cooperon or, equivalently, the electron dephasing time can be naturally found from the condition $f_1(\tau_\varphi) \sim 1$.

In the low-temperature limit (14.69), the expression for $f_1(t)$ is dominated by a linear-in-time temperature-independent contribution to $f_1(t)$, which determines the dephasing rate at $T \to 0$ in the form [153, 155]

$$\frac{1}{\tau_\varphi(0)} \equiv \frac{1}{\tau_{\varphi 0}} \approx \frac{e^2}{\pi \sigma_{\rm D} \mathcal{A}} \sqrt{\frac{2D}{\tau_e}} \tag{14.71}$$

In the high-temperature limit (14.70), there appears an extra term $\propto T t^{3/2}$ that yields dominating contribution to τ_φ at high temperatures $T \gtrsim T_q^{(1)} \sim 1/\sqrt{\tau_\varphi \tau_e}$, where the result (14.57) $\tau_\varphi(T) \propto T^{-2/3}$ is recovered.

In addition to the abovementioned contributions, there exist extra terms in both Eqs. (14.69) and (14.70), which also grow with time, albeit slower than the main terms. These subleading terms also contribute to dephasing even at $T = 0$; however, this contribution is always smaller than that of the leading terms, typically in the parameter $\sqrt{\tau_e/\tau_\varphi}$.

A similar calculation can be performed in two and three dimensions. In any dimension, the result can be expressed in the form

$$f_{\rm d}(t) = \frac{t}{\tau_{\varphi 0}} + \delta f_{\rm d}(T, t) \tag{14.72}$$

where we defined the zero-temperature dephasing rate

$$\frac{1}{\tau_{\varphi 0}} = \frac{\kappa_d e^2 (2D)^{1-d/2}}{\pi \sigma_d \tau_e^{d/2}} \tag{14.73}$$

A numerical prefactor κ_d in (14.73) (determined for a sharp high-frequency cutoff at $\omega = 1/\tau_e$) is $\kappa_1 = 1$ for 1d, $\kappa_2 = 1/4$ for 2d, and $\kappa_3 = 1/3\pi$ for 3d. The expression for the function $\delta f_1(T, t)$ is already specified in Eqs. (14.69) and (14.70). For two- and three-dimensional systems, we will only present the leading order contributions to $\delta f_{\rm d}(T, t)$. In the low-temperature limit $Tt \ll 1$ (though $t \gg \tau_e$), we find

$$\delta f_2(t) \simeq 0.37 \frac{\tau_e}{\tau_{\varphi 0}} \ln \frac{t}{\tau_e}, \quad \delta f_3(t) \simeq 7.8 \frac{\tau_e}{\tau_{\varphi 0}} \tag{14.74}$$

while at higher temperatures $Tt \gg 1$, we obtain

$$\delta f_2(t) = \frac{2\tau_e}{\tau_{\varphi 0}} Tt \ln(Tt), \quad \delta f_3(t) \simeq 3\frac{(T\tau_e)^{3/2}}{\tau_{\varphi 0}} t \qquad (14.75)$$

Obviously, at low T, the terms $\delta f_{\mathrm{d}}(t)$ (14.74) always remain much smaller than the main contribution $t/\tau_{\varphi 0}$ to the function $f_{\mathrm{d}}(t)$ (14.72) and, hence, cannot provide any significant impact on the dephasing rate (14.73). On the other hand, at high-enough temperatures, $\delta f_{\mathrm{d}}(t)$ (14.75) already exceeds the temperature-independent term $t/\tau_{\varphi 0}$ and, hence, determines the electron dephasing rate to be (cf. [11])

$$\frac{1}{\tau_\varphi(T)} \sim \begin{cases} (T\tau_e/\tau_{\varphi 0})\ln(\tau_{\varphi 0}/\tau_e), & d = 2, \\ (T\tau_e)^{3/2}/\tau_{\varphi 0}, & d = 3, \end{cases} \qquad (14.76)$$

where, as before, $\tau_{\varphi 0}$ is defined in Eq. (14.73). It follows from the above results that in two and three dimensions, the temperature-dependent contribution (14.76) to τ_φ can dominate over the temperature-independent one (14.73) only at sufficiently high temperatures $T \simeq T_q^{(d)}$, where $T_q^{(2)} \sim 1/(\tau_e \ln(\tau_{\varphi 0}/\tau_e))$ and $T_q^{(3)} \sim 1/\tau_e$. In contrast, the classical-to-quantum crossover temperature $T_q^{(1)} \sim 1/\sqrt{\tau_\varphi \tau_e}$ turns out to be much lower and, hence, the saturation of the temperature dependence of τ_φ in metallic wires usually can be expected only at sufficiently low T.

Note that while deriving the above expressions for τ_φ in lower dimensions $d = 1, 2$, we implicitly assumed $a \lesssim \ell$. In this case, our results remain applicable for the whole temperature range. At sufficiently high temperatures $T \simeq T_q^{(d)}$, this condition can be softened because in this case τ_φ is mainly determined by low-frequency fluctuations of the environment. Then the system can be regarded as a quasi-one(two)-dimensional one if its transversal dimension is smaller than the corresponding phase breaking length $a \ll L_\varphi = \sqrt{D\tau_\varphi}$. However, at low temperatures $T \ll T_q^{(d)}$, high-frequency environmental modes become important and the situation changes. For instance, in the limit $a \gg \ell$, the diffusion process has two (and sometimes even three) stages: at short times (i.e., at frequencies higher than D/a^2), diffusion is obviously a three-dimensional process, whereas for longer times, it can effectively turn into one- or two-dimensional ones.

For large $a \gg L_\varphi$, the system should obviously be treated as three-dimensional at all T. In the intermediate case $\ell \ll a \ll L_\varphi$, it is necessary to use the corresponding one- or two-dimensional expressions for τ_φ at high temperatures $T \gg T_{qa}^{(d)}$ and the three-dimensional result (14.73),

$$\tau_{\varphi 0} \sim m p_F \ell^3, \tag{14.77}$$

in the low-temperature limit $T \ll T_{qa}^{(d)}$. The crossover temperature $T_{qa}^{(d)}$ can be determined either directly from the integral (14.68) or just by comparison of the corresponding expressions for τ_φ. We get $T_{qa}^{(d)} \approx T_q^{(d)}(a/\pi\ell)^{3-d}$. Thus, in the limit $a \gg \ell$, the low-temperature saturation of τ_φ takes place already at somewhat higher temperatures than in the case of small a. For $d = 1, 2$, the saturation value $\tau_{\varphi 0}$ is relatively smaller for large a, since it is defined by the three-dimensional result (14.77) rather than by the corresponding one- and two-dimensional results.

While such tiny details may be significant for a quantitative comparison with experimental results, of much greater importance is the very fact of the low-temperature saturation of τ_φ to a finite value $\tau_{\varphi 0}$. This saturation implies a fundamental limitation imposed by nature on the observation of quantum coherent effects in disordered conductors. In other words, electron–electron interactions significantly restrict the ability of electrons in such conductors to interfere. Even at $T \to 0$, quantum interference effects cannot extend beyond the length scale set by the electron dephasing length $L_{\varphi 0} = \sqrt{D\tau_{\varphi 0}}$. For example, in the case of quasi-one-dimensional disordered wires with the aid of Eq. (14.71), one readily finds

$$L_{\varphi 0} \sim \sqrt{N_{\mathrm{ch}}}\,\ell \tag{14.78}$$

Provided the number of conducting channels is large $N_{\mathrm{ch}} \gg 1$ (which is always the case in generic metallic wires), the zero-temperature dephasing length (14.78) turns out to be parametrically shorter than the localization length, $L_{\varphi 0}/\xi_{\mathrm{loc}} \sim 1/\sqrt{N_{\mathrm{ch}}} \ll 1$. Hence, quantum interference of electrons in metallic wires should be suppressed by electron–electron interactions already at length scales much shorter than ξ_{loc}, thereby leaving no room for Anderson localization of electrons in such wires at $T \to 0$. Accordingly, the WL correction

$\sigma_{\rm WL}$ (14.34) in such systems always remains much smaller than the Drude conductivity $\sigma_{\rm D}$. At $T \to 0$, we estimate

$$\frac{\sigma_{\rm WL}}{\sigma_{\rm D}} \sim -\frac{1}{\sqrt{N_{\rm ch}}} \tag{14.79}$$

Qualitatively similar conclusions also apply to disordered metallic films.

At this stage, it is worth pointing out that the conclusion about zero-temperature saturation of the electron decoherence time to a finite value $\tau_{\varphi 0}$ was debated in the literature [5, 156]. A formal reason for that was the observation [4, 5] that the above results of the quasiclassical path integral analysis [153, 155] allegedly do not match with those of the diagrammatic first-order perturbation theory in the interaction [4]. A closer inspection of this issue [157] demonstrated, however, that all diagrams of the perturbation theory are fully reproduced within the path integral influence functional approach developed in Chapter 9. Furthermore, the results of the perturbative first-order [4] and the non-perturbative saddle point [155] calculations cannot be directly compared since they account for two different (and not overlapping!) time domains of our problem and, in addition, are performed with a different accuracy.

This issue is quite important and, hence, requires a discussion. For clarity, from now on and till the end of this subsection, we will restore Planck constant $\hbar$, which was previously set equal to unity. Evaluating the influence functional $F(t) = \exp(-{\rm i}S_{\rm R}(t)/\hbar - S_{\rm I}(t)/\hbar)$ for the Cooperon, within the above analysis we only included the contributions from all pairs of *classical* time-reversed paths. As we demonstrated, this quasiclassical approximation (controlled by the inequality (14.1)) yields $F(t) \approx \exp(-S_{\rm I}(t)/\hbar)$ and remains accurate *to all orders in the interaction* provided $S_{\rm I}(t) \gg \hbar$, i.e., in the long time limit $t \gg \tau_\varphi$. In contrast, the first-order perturbation theory (derived by expanding $F(t)$ to the first order in ${\rm i}S_{\rm R}/\hbar - S_{\rm I}/\hbar$) probes only short times as it requires the condition $S_{\rm R,I}(t) \ll \hbar$ to be fulfilled for all relevant paths. The latter condition also indicates that in this short time limit (contrary to the long time one), it is not sufficient to only consider the classical paths, and also fluctuations around these paths need to be taken into account. As a result, the contribution of the $S_{\rm R}$-term in the influence functional is not anymore equal to zero and contributes to the first-order perturbation theory on equal footing with the S_{I}-term.

In general, this short time perturbative expansion is unsuitable for the calculation of τ_φ, since this would unavoidably require an extrapolation to long times, for instance, by exponentiation of the first-order perturbative result. Such a step is ambiguous and may yield erroneous results. The latter point is well illustrated, e.g., by the problem of quantum decay of a metastable state in the presence of dissipation. This problem was treated quasiclassically in Chapter 4 of this book, and the decay rate $\Gamma = B\exp(-A)$ (4.41) was derived, where the exponential factor A is set by the action value on the relevant saddle point path, while the pre-exponent B is determined by fluctuations around this path. In the weak coupling limit $\alpha \ll 1$ (4.56), one finds $A = A_0 + \alpha A_1$, $B \simeq B_0 + \alpha B_1$ where $A_{0,1}$ and $B_{0,1}$ are identified from Eqs. (4.54) and (4.78) and the dimensionless friction parameter α is defined in Eq. (4.55). An expansion of Γ to the first order in α yields $\Gamma = B_0\exp(-A_0)[1 - \alpha(A_1 - B_1)]$, i.e., both exponent A and pre-exponent B contribute to the first-order term. It is now obvious that the specific separation into exponent and prefactor cannot be derived unambiguously from a purely perturbative α calculation. For instance, simple exponentiation of the first-order term would, in general, be inaccurate in the interesting parameter domain (4.57).

Similarly, for the calculation of the dephasing rate, it is important to separate the pre-exponent, which is of little importance in the relevant limit $t \gg \tau_\varphi$, from the contributions to the exponent. In contrast to the first-order perturbation theory in the interaction, our non-perturbative saddle point analysis provides a definite prescription for this separation and, hence, allows to unambiguously determine τ_φ.

The task at hand is now to complete our calculation of the Cooperon dynamics in the presence of electron–electron interactions by considering fluctuations around the classical time-reversed paths. Such fluctuations—just as in the above simple example with the decay rate Γ—give rise to the pre-exponent in the expression for C. This task will be accomplished below. We will then perform a short time expansion of our final non-perturbative result and explicitly recover the first order perturbation theory in the interaction [4, 157].

14.4.2 Pre-exponent

At the first glance it might appear that the path integral representation of the kernel J, Eqs. (9.99)–(9.101), is most appropriate for calculation of the pre-exponent in the expression for the WL correction $\sigma_{\rm WL}$. Indeed, evaluating this kernel in the quasiclassical limit (14.1), one could try to proceed in a standard manner expanding the full action in the exponent of Eq. (9.99) up to the second order in small fluctuations around classical time-reversed paths and subsequently performing Gaussian path integrals over these fluctuations. In this way, one would normally recover the expression for the pre-exponent where the contribution generated by the terms $S_{\rm R}$ and $S_{\rm I}$ properly accounts for the effect of electron–electron interactions. As usually, the accuracy of this semiclassical calculation would be controlled by the inequality (14.1).

A closer inspection of the influence functional in Eqs. (9.99)–(9.101) demonstrates, however, that this straightforward approach turns out to be somewhat problematic in the case of a degenerate interacting electron gas considered here. The problem arises due to the presence of the Fermi function $n_0(\boldsymbol{p}(t), \boldsymbol{r}(t))$ in the term $S_{\rm R}$ (9.100) describing dissipation by the effective electronic environment. As we already pointed out, keeping this Fermi function in the influence functional is essential, as it allows to include the Pauli principle into our consideration. In the interesting-for-us low-temperature limit $n_0(\boldsymbol{p}(t), \boldsymbol{r}(t))$ (9.98) obviously tends to the step function of the electron energy, i.e., this function changes rapidly in the vicinity of the Fermi energy. Hence, at $T \to 0$ and such energies the term $S_{\rm R}$ (9.100) in our effective action can vary significantly even for tiny deviations from the classical paths. In other words, already small path fluctuations can be strongly non-Gaussian, and the above semiclassical pre-exponent evaluation procedure becomes obsolete.

Fortunately, our semiclassical analysis can be repaired. This is because under the condition (14.1), the electron dynamics in a disordered metal remains almost classical also in the presence of Coulomb interactions with other electrons. Therefore, the basic idea is to reformulate our theory in a way that would allow to take the main source of difficulties—the electron density matrix $\boldsymbol{\rho}$—out of the path integral and to describe only the electron motion quasiclassically, while keeping $\boldsymbol{\rho}$ in its fully quantum form throughout the whole calculation. Technically, this idea can be accomplished in the following manner [162]:

Instead of dealing with the double path integral (9.99)–(9.101) for the kernel J of the Keldysh evolution operator, where averaging over the fluctuating fields $V^{\pm}$ has already been performed, let us employ the forward/backward operators $\hat{\boldsymbol{U}}_{\mathrm{F,B}}$ (9.80), (9.81) and the related general expression for the conductivity σ defined in Eqs. (9.88) and (9.89). In what follows we will also exploit the exact relations (9.87) between the operators $\hat{\boldsymbol{U}}_{\mathrm{F,B}}$ depending on the electron density matrix $\hat{\boldsymbol{\rho}}_V$ and the operators $\hat{\boldsymbol{U}}_{\mathrm{F,B}}$ (9.83), (9.84), which do not depend on this matrix. To begin with, let us recall two simple approximations, both related to the fact that in the quasiclassical limit (14.1), one can also treat Coulomb interaction between electrons as a weak perturbation. One of these approximations essentially amounts to replacing the exact single-electron density matrix by its non-interacting value

$$\hat{\boldsymbol{\rho}}_V(t') \longrightarrow \hat{\boldsymbol{\rho}}_0 = [1 + \mathrm{e}^{(\hat{H}_0 - \mu)/T}]^{-1} \tag{14.80}$$

in Eqs. (9.88) and (9.87). A similar approximation (9.98) was already employed while deriving the influence functional in Eqs. (9.99)–(9.101). Another approximation directly follows from the observation that under the condition (14.1), quasiclassical electron trajectories in a metal remain practically non-disturbed by electron–electron interactions, and the contributions of the fluctuating fields $V^{\pm}$ add up independently. The same approximation was also involved, e.g., in the derivation of the function $f_{\mathrm{d}}(t)$ (14.68) in Section 14.4.1. In terms of the operators $\hat{\boldsymbol{u}}_{\mathrm{F,B}}$, it amounts to splitting them as

$$\hat{\boldsymbol{u}}_{\mathrm{F,B}}(t, t') \simeq \hat{\boldsymbol{u}}_0(t, t')\hat{s}(t, t', V^+)\hat{s}(t, t', \pm V^-/2) \tag{14.81}$$

where, as before, $\hat{\boldsymbol{u}}_0(t, t') = e^{-\mathrm{i}\hat{H}(t-t')}$ is the free electron evolution operator pertaining to the Hamiltonian (14.6) and we defined

$$\hat{s}(t, t', V) = \hat{\boldsymbol{u}}_0(t', t)\mathcal{T} \exp\left[-\frac{\mathrm{i}}{\hbar}\int_{t'}^{t} d\tau(\hat{H}_0 - e\hat{V}(\tau))\right] \tag{14.82}$$

Within the same accuracy, we can replace

$$\hat{\boldsymbol{u}}_0(t, t')\hat{s}(t, t', V^+) \simeq \hat{\boldsymbol{u}}_+(t, t') = \mathcal{T}\exp\left(-\frac{\mathrm{i}}{\hbar}\int_{t'}^{t} d\tau[\hat{H}_0 - e\hat{V}^+(\tau)]\right) \tag{14.83}$$

Combining Eqs. (14.81)–(14.83) with Eq. (9.87), we obtain

$$\begin{aligned}\hat{\boldsymbol{U}}_{\mathrm{F}}(t, t') &\simeq \hat{\boldsymbol{u}}_+(t, t')\left\{(1 - \hat{\rho}_0)\hat{s}^{-1}(t, t', V^-/2) + \hat{\rho}_0\hat{s}^{-1}(t, t', -V^-/2)\right\}^{-1},\\ \hat{\boldsymbol{U}}_{\mathrm{B}}(t', t) &\simeq \left\{\hat{s}(t, t', -V^-/2)(1 - \hat{\rho}_0) + \hat{s}(t, t', V^-/2)\hat{\rho}_0\right\}^{-1}\hat{\boldsymbol{u}}_+(t', t)\end{aligned} \tag{14.84}$$

Let us now proceed with our calculation of the conductivity σ. For that purpose, we will make use of Eq. (9.88), according to which it is necessary to evaluate the matrix element $\langle \boldsymbol{r}_1|\hat{\boldsymbol{U}}_{\mathrm{F}}(t, t')[\hat{\boldsymbol{r}}, \hat{\rho}_{\mathrm{V}}(t')]\hat{\boldsymbol{U}}_{\mathrm{B}}(t', t)|\boldsymbol{r}_2\rangle$ or, equivalently,

$$\int d\boldsymbol{r}_{1i}\int d\boldsymbol{r}_{2i}\langle \boldsymbol{r}_1|\hat{\boldsymbol{U}}_{\mathrm{F}}(t, t')|\boldsymbol{r}_{1i}\rangle\langle \boldsymbol{r}_{2i}|\hat{\boldsymbol{U}}_{\mathrm{B}}(t', t)|\boldsymbol{r}_2\rangle\langle \boldsymbol{r}_{1i}|[\hat{\boldsymbol{r}}, \hat{\rho}_{\mathrm{V}}(t')]|\boldsymbol{r}_{2i}\rangle \tag{14.85}$$

Under the approximation (14.80), the last matrix element in (14.85) becomes simply $(\boldsymbol{r}_{1i} - \boldsymbol{r}_{2i})\rho_0(\boldsymbol{r}_{1i}, \boldsymbol{r}_{2i})$; cf. also Eq. (9.103). The two other matrix elements will be transformed with the aid of Eqs. (14.84). The matrix element of the operator $\hat{\boldsymbol{U}}_{\mathrm{F}}$ can be decomposed as

$$\langle \boldsymbol{r}_1|\hat{\boldsymbol{U}}_{\mathrm{F}}(t, t')|\boldsymbol{r}_{1i}\rangle = \int d\boldsymbol{r}_1'\langle \boldsymbol{r}_1|\hat{\boldsymbol{u}}_+(t, t')|\boldsymbol{r}_1'\rangle L_{\mathrm{F}}(\boldsymbol{r}_1', \boldsymbol{r}_{1i}, V^-) \tag{14.86}$$

where

$$\begin{aligned}L_{\mathrm{F}}(\boldsymbol{r}_1', \boldsymbol{r}_{1i}, V^-) = \langle \boldsymbol{r}_1'|\{&(1 - \hat{\rho}_0)\hat{s}^{-1}(t, t', V^-/2)\\ &+ \hat{\rho}_0\hat{s}^{-1}(t, t', -V^-/2)\}^{-1}|\boldsymbol{r}_{1i}\rangle\end{aligned} \tag{14.87}$$

Likewise, for the remaining matrix element, we obtain

$$\langle \boldsymbol{r}_{2i}|\hat{\boldsymbol{U}}_{\mathrm{B}}(t', t)|\boldsymbol{r}_2\rangle = \int d\boldsymbol{r}_2' L_{\mathrm{B}}(\boldsymbol{r}_2', \boldsymbol{r}_{2i}, V^-)\langle \boldsymbol{r}_2'|\hat{\boldsymbol{u}}_+(t', t)|\boldsymbol{r}_2\rangle \tag{14.88}$$

with

$$\begin{aligned}L_{\mathrm{B}}(\boldsymbol{r}_2', \boldsymbol{r}_{2i}, V^-) = \langle \boldsymbol{r}_{2i}|\{&\hat{s}(t, t', -V^-/2)(1 - \hat{\rho}_0)\\ &+ \hat{s}(t, t', V^-/2)\hat{\rho}_0\}^{-1}|\boldsymbol{r}_2'\rangle\end{aligned} \tag{14.89}$$

Let us now express the matrix elements of the operators $\hat{\boldsymbol{u}}_{+}$ via the path integrals

$$\langle \boldsymbol{r}|\hat{\boldsymbol{u}}_{+}(t, t')|\boldsymbol{r}'\rangle = \int_{\boldsymbol{r}'}^{\boldsymbol{r}} \mathcal{D}\boldsymbol{r}_{\mathrm{F}}(\tau) e^{\frac{\mathrm{i}}{\hbar} S_{+}[\boldsymbol{r}_{\mathrm{F}}(\tau)]} \tag{14.90}$$

where

$$S_{+}[\boldsymbol{r}(\tau)] = S_0 + \int_{t'}^{t} d\tau e V^{+}(\tau, \boldsymbol{r}(\tau)),$$
$$S_0 = \int_{t'}^{t} d\tau \left(\frac{m\dot{\boldsymbol{r}}^2}{2} - U(\boldsymbol{r}(\tau)) \right) \tag{14.91}$$

As these matrix elements do not contain the electron density matrix $\boldsymbol{\rho}$, they can be treated quasiclassically exactly as it was already done in Eqs. (14.20) and (14.21).

Employing again the van Vleck formula, we get

$$\langle \boldsymbol{r}_1|\hat{\boldsymbol{u}}_{+}(t, t')|\boldsymbol{r}_1'\rangle = \sum_{\boldsymbol{r}_{\mathrm{F}}} A[\boldsymbol{r}_{\mathrm{F}}] e^{\mathrm{i}S_0(t,t';\boldsymbol{r}_1,\boldsymbol{r}_1') + \mathrm{i}\int_{t'}^{t} d\tau e V^{+}(\tau, \boldsymbol{r}_{\mathrm{F}}(\tau))} \tag{14.92}$$

and

$$\langle \boldsymbol{r}_2'|\hat{\boldsymbol{u}}_{+}(t', t)|\boldsymbol{r}_2\rangle = \sum_{\boldsymbol{r}_{\mathrm{B}}} A^*[\boldsymbol{r}_{\mathrm{B}}] e^{-\mathrm{i}S_0(t,t';\boldsymbol{r}_2,\boldsymbol{r}_2') - \mathrm{i}\int_{t'}^{t} d\tau e V^{+}(\tau, \boldsymbol{r}(\tau))} \tag{14.93}$$

where, as before, the actions $S_0(t, t'; \boldsymbol{r}_{1,2}, \boldsymbol{r}_{1,2}')$ are taken on all possible classical paths $\boldsymbol{r}_{\mathrm{F,B}}(t)$ obeying the saddle point equation

$$\delta S_{+}[\boldsymbol{r}_{\mathrm{F,B}}, V^{+}]/\delta \boldsymbol{r}_{\mathrm{F,B}} = 0 \tag{14.94}$$

which—bearing in mind that quasiclassical electron trajectories are weakly affected by electron–electron interactions—in the leading approximation reduces simply to

$$\delta S_0[\boldsymbol{r}_{\mathrm{F,B}}]/\delta \boldsymbol{r}_{\mathrm{F,B}} = 0 \tag{14.95}$$

Thus, Eqs. (14.92) and (14.93) differ, respectively, from Eqs. (14.20) and (14.21) only by the presence of an extra phase accumulated by an electron propagating in the fluctuating potential V^{+}; cf. also Eqs. (14.45) and (14.46).

Collecting now all the above terms, we obtain for the kernel J

$$J(t, t'; \boldsymbol{r}_1, \boldsymbol{r}_2; \boldsymbol{r}_{1i}, \boldsymbol{r}_{2i}) = \int d\boldsymbol{r}_1' \int d\boldsymbol{r}_2' \sum_{\boldsymbol{r}_{\mathrm{F}}} A[\boldsymbol{r}_{\mathrm{F}}] \sum_{\boldsymbol{r}_{\mathrm{B}}} A^*[\boldsymbol{r}_{\mathrm{B}}]$$
$$\times \langle L_{\mathrm{F}}(\boldsymbol{r}_1', \boldsymbol{r}_{1i}, V^{-}) L_{\mathrm{B}}(\boldsymbol{r}_2', \boldsymbol{r}_{2i}, V^{-}) e^{\mathrm{i}S_0(t,t';\boldsymbol{r}_1,\boldsymbol{r}_1') - \mathrm{i}S_0(t,t';\boldsymbol{r}_2,\boldsymbol{r}_2')}$$
$$\times e^{\mathrm{i}\int_{t'}^{t} d\tau (eV^{+}(\tau, \boldsymbol{r}_{\mathrm{F}}(\tau)) - eV^{+}(\tau, \boldsymbol{r}_{\mathrm{B}}(\tau)))} \rangle_{V^{\pm}} \tag{14.96}$$

where averaging over the fluctuating fields is routinely defined as

$$\langle\ldots\rangle_{V^{\pm}} = \int \mathcal{D}V^{\pm}(\ldots)\exp\left\{\mathrm{i}S_{\text{field}}[V^{+}, V^{-}]\right\} \tag{14.97}$$

with the action $S_{\text{field}}[V^{+}, V^{-}]$ specified in Eq. (9.57). As this action is Gaussian in both fields $V^{\pm}$, the averaging in (14.97) can be accomplished exactly. In order to do so, it is convenient to first integrate over V^{+}. Observing that the action $S_{\text{field}}[V^{+}, V^{-}]$ (9.57) is linear in V^{+}, this integration simply yields the $\delta-$function

$$\delta(V^{-}(\tau, \boldsymbol{r}) - V_0(\tau, \boldsymbol{r}, \boldsymbol{r}_{\text{F}}(s), \boldsymbol{r}_{\text{B}}(s)),$$

where

$$V_0(\tau, \boldsymbol{r}, \boldsymbol{r}_{\text{F}}(s), \boldsymbol{r}_{\text{B}}(s)) = -e\int_{t'}^{t} ds[R(s-\tau, \boldsymbol{r}_{\text{F}}(s)-\boldsymbol{r}) \\ - R(s-\tau, \boldsymbol{r}_{\text{B}}(s)-\boldsymbol{r})] \tag{14.98}$$

and the function $R(t, \boldsymbol{r})$ is defined in Eq. (9.65). Due to this δ-function, the subsequent integration over V^{-} also becomes trivial and we get

$$J(t, t'; \boldsymbol{r}_1, \boldsymbol{r}_2; \boldsymbol{r}_{1i}, \boldsymbol{r}_{2i}) = \int d\boldsymbol{r}_1' \int d\boldsymbol{r}_2' \sum_{\boldsymbol{r}_{\text{F}}} A[\boldsymbol{r}_{\text{F}}] \sum_{\boldsymbol{r}_{\text{B}}} A^*[\boldsymbol{r}_{\text{B}}] \\ \times L_{\text{F}}(\boldsymbol{r}_1', \boldsymbol{r}_{1i}, V_0) L_{\text{B}}(\boldsymbol{r}_2', \boldsymbol{r}_{2i}, V_0) e^{\mathrm{i}S_0(t,t';\boldsymbol{r}_1,\boldsymbol{r}_1')-\mathrm{i}S_0(t,t';\boldsymbol{r}_2,\boldsymbol{r}_2')-S_{\text{I}}[\boldsymbol{r}_{\text{F}},\boldsymbol{r}_{\text{B}}]} \tag{14.99}$$

Combining Eqs. (14.99) and (9.103), we arrive at a general expression for the linear conductivity of an arbitrary disordered conductor in the presence of electron–electron interactions [162]

$$\sigma = \frac{e^2}{3m}\int_{-\infty}^{t} dt' \int d\boldsymbol{r}_{i1} d\boldsymbol{r}_{i2} \int d\boldsymbol{r}_1' \int d\boldsymbol{r}_2' \left(\nabla_{r_1} - \nabla_{r_2}\right)\Big|_{\boldsymbol{r}_1=\boldsymbol{r}_2} \\ \times \sum_{\boldsymbol{r}_{\text{F}}} A[\boldsymbol{r}_{\text{F}}] \sum_{\boldsymbol{r}_{\text{B}}} A^*[\boldsymbol{r}_{\text{B}}] e^{\mathrm{i}S_0(t,t';\boldsymbol{r}_1,\boldsymbol{r}_1')-\mathrm{i}S_0(t,t';\boldsymbol{r}_2,\boldsymbol{r}_2')-S_{\text{I}}[\boldsymbol{r}_{\text{F}},\boldsymbol{r}_{\text{B}}]} \\ \times L_{\text{F}}(\boldsymbol{r}_1', \boldsymbol{r}_{1i}, V_0)(\boldsymbol{r}_{1i} - \boldsymbol{r}_{2i})\rho_0(\boldsymbol{r}_{1i}, \boldsymbol{r}_{2i}) L_{\text{B}}(\boldsymbol{r}_2', \boldsymbol{r}_{2i}, V_0) \tag{14.100}$$

This expression applies *to all orders* in the interaction and remains valid in the quasiclassical limit $k_{\text{F}}l \gg 1$. It is also important to stress again that no quasiclassical approximation for the electron density matrix $\boldsymbol{\rho}$ was employed during our derivation and no averaging over impurities was performed at all. Of course, the latter averaging can be accomplished at a later stage of the calculation, but we intentionally avoid doing it here, since a purely technical issue of disorder averaging has nothing to do with the main subject of our present discussion.

Let us briefly analyze our results. According to the arguments already employed above in this chapter, there exist two types of pairs of classical paths ($\boldsymbol{r}_{\rm F}$ and $\boldsymbol{r}_{\rm B}$), identical and time-reversed ones, which play an important role in a disordered quasiclassical limit $k_{\rm F}l \gg 1$. For a pair of identical paths $\boldsymbol{r}_{\rm F}(s) = \boldsymbol{r}_{\rm B}(s)$, as usually, the two actions S_0 in the exponent in Eqs. (14.99) and (14.100) cancel each other identically, the term $S_{\rm I}$ vanishes, the function $V_0(\tau, \boldsymbol{r}, \boldsymbol{r}_F(s), \boldsymbol{r}_B(s))$ vanishes too and, hence, the matrix elements $L_{\rm F,B}$ (14.87), (14.89) reduce to the δ-functions, i.e., $L_{\rm F} = \delta(\boldsymbol{r}_1' - \boldsymbol{r}_{1i})$ and $L_{\rm B} = \delta(\boldsymbol{r}_2' - \boldsymbol{r}_{2i})$. In this way from Eq. (14.99), we recover the well-known property that the diffuson does not decay in time even in the presence of interactions. With the aid of Eq. (14.100), we then reproduce the standard Drude contribution to the conductivity $\sigma_{\rm D}$; cf. Section 9.8.1.

Here we are interested in the quantum correction to conductivity $\sigma_{\rm WL}$ arising from the time-reversed paths $\boldsymbol{r}_{\rm F}(s)(s) = \boldsymbol{r}_{\rm B}(t + t' - s)$. For any pair of such paths, the actions S_0 cancel again, but the interaction term $S_{\rm I}$ is now positive; it grows with time and yields (exponential) decay of the kernel J (14.99) in the long time limit. This is nothing but the Cooperon decay in time that manifests quantum decoherence, as we already discussed in details in Section 14.4.1. The new element here is the expression for the pre-exponent containing the matrix elements $L_{\rm F,B}$ [Eqs. (14.87), (14.89)], which in turn depend on the exact electron density matrix. Unlike in the above example of equivalent paths, in the case of time-reversed paths, the matrix elements $L_{\rm F,B}$ explicitly depend on both interaction and time. It is obvious, however, that $L_{\rm F,B}$ *cannot grow* at long times because the function V_0 (14.98) is purely real and, hence, $\hat{s}(t, t', \pm V_0/2)$ are the unitary operators. The matrix elements of such operators can only oscillate (provided the function V_0 changes in time), thereby further reducing the kernel $J \propto \exp(-S_{\rm I}(t - t'))$ decaying at sufficiently large $t - t'$. Hence, in full agreement with our previous considerations, the whole expression under the integral over t' in Eq. (14.100) decays exponentially for any pair of time-reversed paths and identically reproduces our results for τ_φ derived in Section 14.4.1.

Although in principle one can proceed further and under certain approximations specify the values of the matrix elements (14.87), (14.89) for pairs of time-reversed paths, we avoid doing it here simply because particular values of $L_{\mathrm{F,B}}$ are irrelevant for dephasing. For our purposes, it suffices just to observe that these matrix elements do not grow at long times implying that τ_φ remains essentially insensitive to the electron density matrix $\boldsymbol{\rho}$. For more discussions on this and related subjects, we refer the reader to the paper [162].

14.4.3 Relation to Perturbation Theory

Although the matrix elements $L_{\mathrm{F,B}}$ cannot grow at long times, they can well do so in the short lime limit probed by means of the perturbation theory in the interaction. In order to recover the first-order terms of this perturbation theory, let us expand the exact expressions for the evolution operators $\hat{\boldsymbol{U}}_{\mathrm{F,B}}$ (9.87) to the first order in the interaction (which implies the second-order expansion in the fluctuating fields $V^{\pm}$). Substituting the result of this expansion into Eq. (9.88) and also performing the replacement (14.80), we obtain

$$\begin{aligned}\sigma = &-\frac{ie}{3}\int_{-\infty}^{t} dt' \left\langle \mathrm{tr}\left(\hat{\boldsymbol{j}}(\boldsymbol{r})\hat{\boldsymbol{u}}_0(t,t')[\hat{\boldsymbol{r}},\hat{\boldsymbol{\rho}}_0]\hat{\boldsymbol{u}}_0(t',t)\right)\right\rangle_V \\ &-\frac{2e^3}{3}\int_{-\infty}^{t} dt'\int_{t'}^{t} d\tau_1 \int_{t'}^{\tau_1} d\tau_2 \mathrm{Im}\Bigg[\Bigg\langle \mathrm{tr}\Bigg(\hat{\boldsymbol{j}}(\boldsymbol{r})\hat{\boldsymbol{u}}_0(t,\tau_1)\hat{V}^+(\tau_1)\hat{\boldsymbol{u}}_0(\tau_1,\tau_2) \\ &\times \left\{\hat{V}^+(\tau_2)+\frac{1}{2}(1-2\hat{\boldsymbol{\rho}}_0)\hat{V}^-(\tau_2)\right\}\hat{\boldsymbol{u}}_0(\tau_2,t')[\hat{\boldsymbol{r}},\hat{\boldsymbol{\rho}}_0]\hat{\boldsymbol{u}}_0(t',t)\Bigg) \\ &+\mathrm{tr}\Bigg(\hat{\boldsymbol{j}}(\boldsymbol{r})\hat{\boldsymbol{u}}_0(t,\tau_2)\left\{\hat{V}^+(\tau_2)+\frac{1}{2}(1-2\hat{\boldsymbol{\rho}}_0)\hat{V}^-(\tau_2)\right\} \\ &\hat{\boldsymbol{u}}_0(\tau_2,t')[\hat{x},\hat{\boldsymbol{\rho}}_0]\hat{\boldsymbol{u}}_0(t',\tau_1)\hat{V}^+(\tau_1)\hat{\boldsymbol{u}}_0(\tau_1,t)\Bigg)\Bigg\rangle_V\Bigg] \end{aligned} \quad (14.101)$$

This result can be expressed in the form

$$\sigma = \sigma^{(0)} + \delta\sigma_{\mathrm{I}}^{(1)} + \delta\sigma_{\mathrm{R}}^{(1)} \quad (14.102)$$

where $\sigma^{(0)}$ is the non-interacting contribution to the conductance defined by the term in the first line in Eq. (14.101), $\delta\sigma_{\mathrm{I}}^{(1)}$ corresponds to the terms in Eq. (14.101), which contain the product $\hat{V}^{+}\hat{V}^{+}$, while $\delta\sigma_{\mathrm{R}}^{(1)}$ is given by the terms containing $\hat{V}^{+}(1-2\hat{\boldsymbol{\rho}}_0)\hat{V}^{-}$ and $(1-2\hat{\boldsymbol{\rho}}_0)\hat{V}^{-}\hat{V}^{+}$.

After averaging over $V^{\pm}$, Eq. (14.101) is exactly equivalent to one derived by means of diagrammatic methods [4]; see also [157] for more details on this issue. In particular, it contains the combination $\coth\frac{\omega}{2T}+\tanh\frac{\epsilon-\omega}{2T}$ leading to (partial) cancelation of the leading contributions to the first-order terms $\delta\sigma_{\mathrm{I}}^{(1)}$ and $\delta\sigma_{\mathrm{R}}^{(1)}$ at $T\to 0$. The term proportional to $\coth\frac{\omega}{2T}$ emerges from the average $\langle\hat{V}^{+}(\tau_1)\hat{V}^{+}(\tau_2)\rangle$, while $\tanh\frac{\xi-\omega}{2T}$ appears from the combination $1-2\hat{\boldsymbol{\rho}}_0$.

Note that Eq. (14.101) represents the exact first-order expression obtained without any evaluation of the path integrals. Let us compare it with the result of our path integral analysis defined in Eq. (14.100). The latter equation remains valid to all orders in the interaction, but it was derived by evaluating the path integrals in the quasiclassical limit $k_{\mathrm{F}}l\gg 1$. In order to extract the first-order terms in the interaction, one needs to expand the exponent in Eq. (14.100) to the first order in S_{I} and the matrix elements $L_{\mathrm{F,B}}$ to the first order in V_0. The structure of this first-order quasiclassical result turns out to be identical to that of Eq. (14.101); in particular, the first-order term in S_{I} generates the correction $\delta\sigma_{\mathrm{I}}^{(1)}$ proportional to $\coth\frac{\omega}{2T}$ (S_{I}-terms), while the terms originating from the expansion of $L_{\mathrm{F,B}}$ in V_0 give rise to the correction $\delta\sigma_{\mathrm{R}}^{(1)}$ containing $\tanh\frac{\xi-\omega}{2T}$ (S_{R}-terms). The only (totally unimportant) difference between the exact and quasiclassical expressions for $\delta\sigma_{\mathrm{I,R}}^{(1)}$ is that in the latter one, the quasiclassical form (14.20), (14.21) for the matrix elements of the operator $\hat{\boldsymbol{u}}_0$ should be employed and the coordinates in the arguments of the fields V^{+} should be replaced by the classical paths, $V^{+}(\tau,\boldsymbol{r})\to V^{+}(\tau,\boldsymbol{r}_{\mathrm{F,B}}(\tau))$. Note that such a replacement should be performed neither for the field V^{-} nor for the electron density matrix $\hat{\boldsymbol{\rho}}_0$ because no quasiclassical approximation was employed with this matrix.

A clear advantage of the quasiclassical path integral formulation over the standard first-order perturbation theory in the interaction is that the former allows to unambiguously trace out the origin of both terms $\delta\sigma_{\rm I}^{(1)}$ and $\delta\sigma_{\rm R}^{(1)}$ (coming, respectively, from the exponent and the pre-exponent in Eq. (14.100)), whereas the latter leaves no chance for correct identification of their origin. Obviously, by further expanding Eq. (14.100)), one can also reconstruct the second and higher orders of the perturbation theory in the interaction, where no complete cancelation of the leading terms should occur even at $T \to 0$.

For the sake of completeness of our discussion, it can also be instructive to identically transform our first-order quasiclassical results for $\delta\sigma_{\rm I,R}^{(1)}$ to a somewhat different form. For this purpose, let us express the electron density matrix $\boldsymbol{\rho}_0$ as

$$\langle \boldsymbol{r}_1|\hat{\boldsymbol{\rho}}_0|\boldsymbol{r}_2\rangle \equiv \boldsymbol{\rho}_0(\boldsymbol{r}_1, \boldsymbol{r}_2) = \frac{1}{2}\int\limits_{-\infty}^{+\infty} ds_2 \left[\delta(s_2) + \frac{{\rm i}T}{\sinh(\pi T s_2)}\right] u_0(-s, \boldsymbol{r}_1, \boldsymbol{r}_2) \quad (14.103)$$

where $u_0(-s, \boldsymbol{r}_1, \boldsymbol{r}_2) = \langle \boldsymbol{r}_1|\hat{\boldsymbol{u}}_0(0, s)|\boldsymbol{r}_2\rangle$ is the matrix element of the evolution operator for non-interacting electrons. In addition, we note that the functions $I(t, \boldsymbol{r})$ (9.64) and $R(t, \boldsymbol{r})$ (9.65) are related to each other by means of the identity

$$I(t, \boldsymbol{r}) = \frac{T}{2}\int ds_1 \coth(\pi T s_1)[R(t - s_1, \boldsymbol{r}) + R(-t - s_1, \boldsymbol{r})].$$

Making use of the above identities, we arrive at the following expressions for $\delta\sigma_{\rm I,R}^{(1)}$:

$$\begin{aligned}\delta\sigma_{\rm I}^{(1)} = {} & \frac{e^4}{12m}\int_{-\infty}^{t} dt' \int_{-\infty}^{\infty} ds_1 ds_2 \int d\boldsymbol{r}_1' d\boldsymbol{r}_2' \frac{-{\rm i}T^2\coth(\pi T s_1)}{\sinh(\pi T s_2)} \\ & \times \left(\nabla_{r_1} - \nabla_{r_2}\right)\Big|_{\boldsymbol{r}_1=\boldsymbol{r}_2} \sum_{\boldsymbol{r}_{\rm F}} A[\boldsymbol{r}_{\rm F}] \sum_{\boldsymbol{r}_{\rm B}} A^*[\boldsymbol{r}_{\rm B}] e^{{\rm i}S_0(t,t';\boldsymbol{r}_1,\boldsymbol{r}_1') - {\rm i}S_0(t,t';\boldsymbol{r}_2,\boldsymbol{r}_2')} \\ & \times \left\{\int_{t'}^{t} d\tau_1 \int_{t'}^{t} d\tau_2 \Big[R(\tau_1 - \tau_2 - s_1, \boldsymbol{r}_{\rm F}(\tau_1) - \boldsymbol{r}_{\rm F}(\tau_2)) \right. \\ & + R(\tau_1 - \tau_2 - s_1, \boldsymbol{r}_{\rm B}(\tau_1) - \boldsymbol{r}_{\rm B}(\tau_2)) \\ & - R(\tau_1 - \tau_2 - s_1, \boldsymbol{r}_{\rm F}(\tau_1) - \boldsymbol{r}_{\rm B}(\tau_2)) \\ & \left. - R(\tau_1 - \tau_2 - s_1, \boldsymbol{r}_{\rm B}(\tau_1) - \boldsymbol{r}_{\rm F}(\tau_2)) \Big] \right\} (\boldsymbol{r}_1' - \boldsymbol{r}_2') u_0(-s_2, \boldsymbol{r}_1', \boldsymbol{r}_2')\end{aligned} \quad (14.104)$$

and

$$
\begin{aligned}
\delta\sigma_{\mathrm{R}}^{(1)} &= \frac{e^4}{12m}\int_{-\infty}^{t} dt' \int_{-\infty}^{\infty} ds_1 ds_2 \int d\boldsymbol{r}_1' d\boldsymbol{r}_2' d\boldsymbol{r} d\boldsymbol{r}' \, \frac{T}{\sinh(\pi T s_1)} \frac{-\mathrm{i}T}{\sinh(\pi T s_2)} \\
&\times \left(\nabla_{r_1} - \nabla_{r_2}\right)\big|_{\boldsymbol{r}_1=\boldsymbol{r}_2} \sum_{\boldsymbol{r}_{\mathrm{F}}} A[\boldsymbol{r}_{\mathrm{F}}] \sum_{\boldsymbol{r}_{\mathrm{B}}} A^*[\boldsymbol{r}_{\mathrm{B}}] e^{\mathrm{i}S_0(t,t';\boldsymbol{r}_1,\boldsymbol{r}_1') - \mathrm{i}S_0(t,t';\boldsymbol{r}_2,\boldsymbol{r}_2')} \\
&\times \left\{ \int_{t'}^{t} d\tau_1 \int_{t'}^{t} d\tau_2 \left[u_0(t' - \tau_2 - s_1, \boldsymbol{r}_1', \boldsymbol{r}) \big[R(\tau_1 - \tau_2, \boldsymbol{r}_{\mathrm{F}}(\tau_1) - \boldsymbol{r}) \right.\right. \\
&- R(\tau_1 - \tau_2, \boldsymbol{r}_{\mathrm{B}}(\tau_1) - \boldsymbol{r}) \big] u_0(\tau_2 - t', \boldsymbol{r}, \boldsymbol{r}')(\boldsymbol{r}' - \boldsymbol{r}_2') u_0(-s_2, \boldsymbol{r}', \boldsymbol{r}_2') \\
&- (\boldsymbol{r}_1' - \boldsymbol{r}') u_0(-s_2, \boldsymbol{r}_1', \boldsymbol{r}') u_0(t' - \tau_2, \boldsymbol{r}', \boldsymbol{r}) \big[R(\tau_1 - \tau_2, \boldsymbol{r}_{\mathrm{F}}(\tau_1) - \boldsymbol{r}) \\
&\left.\left. - R(\tau_1 - \tau_2, \boldsymbol{r}_{\mathrm{B}}(\tau_1) - \boldsymbol{r}) \big] u_0(\tau_2 - t' - s_1, \boldsymbol{r}, \boldsymbol{r}_2') \right] \right\}.
\end{aligned}
\tag{14.105}
$$

Equations (14.104) and (14.105) have a very similar structure. These two expressions are, however, not fully identical even at $T \to 0$, and hence in the first-order result (14.102), they can cancel each other only approximately but not exactly; see also the corresponding discussion in [157].

Further evaluation of Eqs. (14.104) and (14.105) makes little sense because, as it was already explained above, the first-order perturbation theory cannot provide any useful information about the electron dephasing time at low temperatures. Nevertheless, the above expressions are of a certain interest, since they help to illustrate the relation between perturbative and non-perturbative results as soon as the quasiclassical approximation has already been performed. Apart from already discussed "coth" and "tanh" terms fully reproduced within the path integral analysis, another useful observation concerns the relation between the quasiclassical paths emerging from the path integrals and those entering the first-order results for $\delta\sigma_{\mathrm{R}}^{(1)}$. The WL correction to the conductivity is defined on pairs of time-reversed path, and only such paths (plus fluctuations around them) are relevant for the path integral analysis of this quantity. Of course, the same paths enter if the general result is expanded to the first order in the interaction before the transformation (14.103). However, *after* this transformation, there appear additional matrix elements u_0 of the electron evolution operator. Proceeding quasiclassically, one can evaluate these matrix elements by means of the van Vleck formula, i.e., to write

$$u_0(t' - \tau_2 - s_1, \boldsymbol{r}'_1, \boldsymbol{r}) \propto e^{iS_0(t'-\tau_2-s_1,0,\boldsymbol{r}'_1,\boldsymbol{r})} \tag{14.106}$$

and similarly for other matrix elements. Substituting u_0 in the form (14.106) into Eq. (14.105), one can interpret the result in terms of the electron motion along artificial classical paths $\boldsymbol{r}_{\mathrm{cl}}(s)$ going, say, first from $\boldsymbol{r}'_1$ to $\boldsymbol{r}$ and then from $\boldsymbol{r}$ to $\boldsymbol{r}'$ (note that some of these paths violate the requirement of causality; see Fig. 3 in [157] and related discussion there). Observing this fact, one could naively conclude that these artificial paths are missing in the path integral formulation. This is obviously not the case, as demonstrated by the above analysis.

14.5 Relation to *P*(*E*)-Theory of Electron Tunneling

In order to complete our discussion, let us draw the attention of the reader to yet one more aspect of the obtained results, namely to a close relation between our theory of quantum decoherence in disordered conductors and the $P(E)$-theory of electron tunneling in the presence of an electromagnetic environment introduced in Chapter 7.

As we already emphasized, the fundamental physical reason behind electron dephasing by electron–electron interactions is the presence of energy exchange between any electron propagating in a disordered conductor and modes of an intrinsic electromagnetic environment produced by all other electrons. Due to this exchange, the electron energy E is smeared and fluctuates even at $T = 0$, which, among other consequences, yields non-vanishing electron decoherence down to zero temperature. Quantitatively, this smearing can be described as a result of the electron interaction with the quantum fluctuating electromagnetic field V, as it was expressed, e.g., in Eq. (9.95) for the kernel J of the evolution operator for propagating electrons. The part of this equation responsible for the electron energy smearing (which is, in essence, just the influence functional) can be defined as

$$\mathcal{P}(t) = \left\langle e^{\mathrm{i}\int_0^t dt' \int d\boldsymbol{r}\left(\rho^- V^+ + \rho^+ V^-\right)} \right\rangle_{V^+, V^-} \qquad (14.107)$$

where the "charge densities" $\rho^{\pm}(t, \boldsymbol{r})$ are specified in Eqs. (9.96). The expression in the exponent (14.107) has a simple physical meaning: It accounts for an extra phase picked up by an electron propagating in the quantum fluctuating field V. The Fourier transform of Eq. (14.107)

$$P(E) = \frac{1}{2\pi} \int_{-\infty}^{\infty} dt e^{\mathrm{i}Et} \mathcal{P}(t) \qquad (14.108)$$

plays the role of the energy probability distribution.

At this stage, let us recall that in Chapter 7 devoted to Coulomb effects in metallic tunnel junctions, we have already introduced the function $P(E)$ (7.71), which has exactly the same physical meaning of the energy distribution for tunneling electrons interacting with an external electromagnetic environment. In that case, $\mathcal{P}(t)$ is again defined by Eq. (14.107) with

$$\int d\boldsymbol{r} \rho^-(t, \boldsymbol{r}) = e, \qquad \int d\boldsymbol{r} \rho^+(t, \boldsymbol{r}) = e/2 \qquad (14.109)$$

Energy exchange between tunneling electrons and environmental modes causes dissipation and influences the probability of electron tunneling. Making use of the results derived in Section 7.3, we recover a direct relation between the conductance of the tunnel junction G and the function $\mathcal{P}(t)$:

$$G(V) = \frac{1}{R_{\mathrm{N}}} - \frac{2}{\pi R_{\mathrm{N}}} \int_0^{\infty} dt \; t \left(\frac{\pi T}{\sinh \pi T t} \right)^2 \mathrm{Im}\left[\mathcal{P}(t)\right] \cos(eVt) \qquad (14.110)$$

Here, as before, R_{N} is the junction resistance in the absence of interactions and V is the applied voltage.

For comparison, employing the results for the WL correction obtained here and disregarding for simplicity the interaction correction (as we always do in this chapter), we may express the total conductance σ of a d-dimensional disordered conductor in the form

$$\sigma = \sigma_{\mathrm{D}} - \frac{2e^2 D}{\pi} \int_{\tau_e}^{\infty} dt \; \frac{\langle \mathcal{P}(t) \rangle_{\mathrm{diff}}}{(4\pi Dt)^{d/2} \, a^{3-d}} \tag{14.111}$$

where $\langle \ldots \rangle_{\mathrm{diff}}$ implies averaging over diffusive trajectories.

We observe an obvious similarity between the results (14.110) and (14.111). While electron trajectories differ significantly in these two cases (they are confined to the junction area in the case of tunneling electrons and they are extended in space for electrons propagating in a disordered conductor), in both cases the quantum corrections to the conductance are expressed in terms of the function $\mathcal{P}(t)$, which has exactly the same physical meaning.

Also the form of the effective energy distribution function $P(E)$ is similar in both these cases. For a tunnel junction interacting with an Ohmic electromagnetic environment, the function $P(E)$ is defined by Eq. (7.78). In the limit $T \to 0$, the latter equation yields

$$P(E) \propto E^{-1+2/g_{\mathrm{S}}}, \quad E \geq 0 \tag{14.112}$$

where g_{S} is a dimensionless conductance of an Ohmic shunt. In the case of disordered conductors, we have $\langle \mathcal{P}(t) \rangle_{\mathrm{diff}} \simeq A_{\mathrm{d}}(t) \exp(-f_{\mathrm{d}}(t))$, where the function $f_{\mathrm{d}}(t)$ in the exponent is given by Eq. (14.68), and the pre-exponent $A_{\mathrm{d}}(t)$ was identified in Section 14.4.2. Substituting this averaged function $\langle \mathcal{P}(t) \rangle_{\mathrm{diff}}$ into Eq. (14.108) instead of $\mathcal{P}(t)$, in the limit $T \to 0$ we obtain

$$P(E) \sim \frac{\tau_{\varphi 0}}{1 + E^2 \tau_{\varphi 0}^2}, \quad E \geq 0 \tag{14.113}$$

Note that in the absence of electron–electron interactions, the function $P(E)$ reduces to $P(E) = \delta(E)$ in both cases (14.112) and (14.113).

Finally, it is worth pointing out that a non-perturbative analysis of exactly the same type was used in order to account for the interaction effects and to recover the correct form of $P(E)$. One should also bear in mind that these effects can be easily mistreated or even missed completely by insufficient approximations. For instance, a simple perturbation theory in the interaction (essentially equivalent to the short time expansion or the expansion in powers of $1/g_{\mathrm{S}}$) clearly fails because of the non-analytic behavior of $P(E)$ at low energies; cf. Eqs. (7.78) and (14.112). On the other hand, the long time expansion (basically equivalent to the golden-rule-type approximation) also yields incorrect results at low T, being unable to capture smearing of $P(E)$ (14.112) at $T \to 0$; see also Chapter 3. The same approximations do not work also in the case of the WL correction in the presence of electron–electron interactions, as we demonstrated above in this chapter. Further discussion of this and related subjects can be found in the paper [158].

Chapter 15

Weak Localization and Electron Dephasing in Disordered Conductors II: Beyond Quasiclassics

15.1 Introduction

In Chapter 14, we worked out a theoretical approach that allows us to analyze electron interference effects in the presence of disorder and electron–electron interactions at any temperature, including the most interesting limit $T \to 0$. This formalism directly employs the generalization of the Feynman–Vernon–Caldeira–Leggett path integral influence functional technique (Chapter 1) to fermionic systems with disorder and interactions elaborated in Chapter 9 of this book. Within this method, one conveniently extends the Chakravarty–Schmid description of weak localization [87] formulating the problem in terms of quasiclassical electron paths (Section 14.2), evaluating the weak localization correction to the system conductance and extracting the electron decoherence time τ_φ in the presence of Coulomb interactions between electrons. The analysis performed in Chapter 14 demonstrates that in the low-

Dissipative Quantum Mechanics of Nanostructures: Electron Transport, Fluctuations, and Interactions
Andrei D. Zaikin and Dmitry S. Golubev

ISBN 978-981-4774-50-5 (Hardcover), 978-0-429-29823-3 (eBook)
www.jennystanford.com

temperature limit, τ_φ saturates to a finite value due to the effect of electron–electron interactions.

Non-vanishing electron decoherence in disordered conductors at $T \to 0$ is a fundamentally important phenomenon, which is expected to occur in all types of such conductors, including both weakly and strongly disordered ones, spatially extended and spatially restricted structures, etc. At this stage, it is worth pointing out that the path integral technique developed in Chapter 9 is an exact procedure that should formally cover all these situations. However, for some types of disordered conductors, such as quantum dots and granular metals, it can be rather difficult to directly evaluate the weak localization correction within this technique for the following reasons.

First of all, our description in terms of quasiclassical electron trajectories adopted in Chapter 14 may turn insufficient in both spatially restricted and strongly disordered conductors where electron scattering on disorder should be treated on more general footing. In other words, in the above cases, it might be necessary to go beyond quasiclassics. Another (purely technical) point is averaging over disorder. Within the path integral approach of Chapter 14, it was convenient for us to postpone disorder averaging until the last stage of the calculation. In some cases—like ones studied below—it might be, on the contrary, more appropriate to perform disorder averaging already in the beginning of the whole consideration. In addition, it is highly desirable to deal with the model that would embrace various types of conductors with well-defined properties both in the long and short wavelength limits. This feature will help to construct a fully self-contained theory free of any divergencies and cutoff parameters; cf. our discussion below Eq. (14.68).

In order to develop a unified theoretical description that would cover essentially all types of disordered conductors, we will adopt the model already introduced in Chapter 13. Namely, we will model a disordered conductor by an array of (metallic) quantum dots connected via junctions (scatterers) with arbitrary transmission distribution of their conducting channels (cf. Fig. 13.1). As desired, this model allows us to easily crossover between the limits of a granular metal and that with point-like impurities as well as to treat

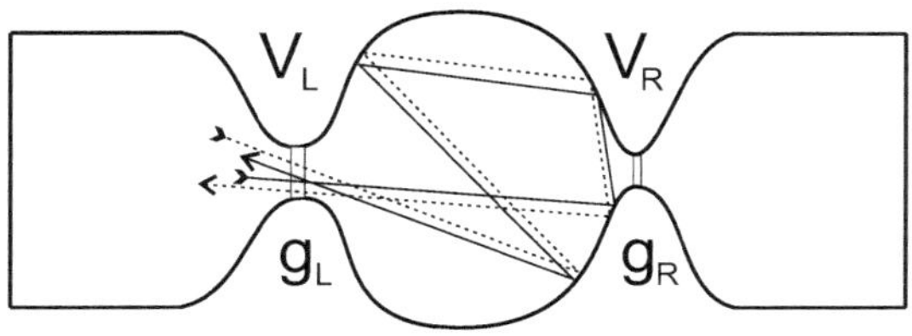

Figure 15.1 Single quantum dot and a pair of time-reversed electron paths. Fluctuating voltages V_{L} and V_{R} are assumed to drop only across the barriers and not inside the dot.

spatially restricted and spatially extended conductors within the same theoretical framework.

15.2 Dephasing in Quantum Dots: Qualitative Arguments

Before turning to a detailed calculation, it is instructive to discuss a simple qualitative picture that demonstrates under which conditions electron dephasing by interaction is expected to occur in spatially restricted conductors, such as metallic quantum dots.

Consider first the simplest system of two scatterers separated by a cavity (quantum dot, Fig. 15.1). As we already learned in Chapter 14, the weak localization correction to conductance of a disordered system G_{WL} is known to arise from interference of pairs of time-reversed electron paths. In the absence of interactions for a single quantum dot of Fig. 15.1, this correction can be evaluated in a general form (see below). Our primary goal here is to understand how electron–electron interactions can affect the value of G_{WL} in quantum dots. As we have repeatedly demonstrated in previous chapters, the effect of such interactions can be described in terms of fluctuating electric potentials mediating these interactions. Let us assume that the voltage can drop only across the barriers and consider two time-reversed electron paths that cross the left barrier (with fluctuating voltage $V_{\mathrm{L}}(t)$) twice at times t_i and t_f, as shown in Fig. 15.1. It is easy to see that the voltage-dependent random phase factor $\exp(\mathrm{i}\int_{t_i}^{t_f} V_{\mathrm{L}}(t)dt)$ acquired by the electron wave function Ψ

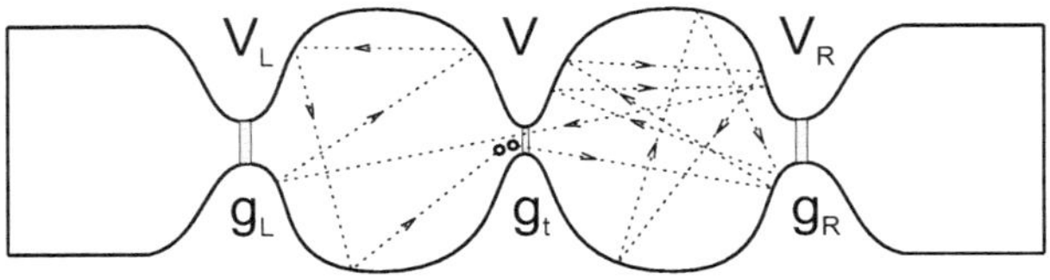

Figure 15.2 Two quantum dots and a typical electron path. Fluctuating voltages V_L, V, and V_R are again assumed to drop only across the barriers.

along any path turns out to be exactly the same as that for its time-reversed counterpart. Hence, in the product $\Psi\Psi^*$, these random phases cancel each other and quantum coherence of electrons remains fully preserved. This observation implies that for the system of Fig. 15.1, fluctuating voltages or, equivalently, electron–electron interactions *do not cause any dephasing*.

This qualitative conclusion can be verified by means of more rigorous considerations. For instance, one can demonstrate [163] that the scattering matrix of the system remains unitary in the presence of electron–electron interactions, which implies that the only effect of such interactions is transmission renormalization but not electron decoherence. A similar conclusion can also be reached by directly evaluating the weak localization correction to the system conductance [70]. Thus, for the system of two scatterers of Fig. 15.1, electron–electron interactions can only yield energy-dependent (logarithmic at sufficiently low energies; cf. Eq. (11.70)) renormalization of the dot channel transmissions but not electron dephasing.

Let us now add one more scatterer and consider the system of two quantum dots depicted in Fig. 15.2. We again assume that fluctuating voltages are concentrated at the barriers and not inside the cavities. The phase factor accumulated along the path that crosses the central barrier twice (at times t_i and $t > t_i$) and returns to the initial point (at a time t_f) is $e^{i[\varphi(t_i)-\varphi(t)]}$, where $\dot{\varphi}/e = V(t)$ is the fluctuating voltage across the central barrier. Similarly, the phase factor picked up along the time-reversed path reads $e^{i[\varphi(t_f+t_i-t)-\varphi(t_f)]}$. Hence, the overall phase factor acquired by the product $\Psi\Psi^*$ for a pair of time-reversed paths is $\exp(i\Phi_{\text{tot}})$, where

$$\Phi_{\text{tot}}(t_i, t_f, t) = \varphi(t_i) - \varphi(t) - \varphi(t_f + t_i - t) + \varphi(t_f).$$

Averaging over phase fluctuations, which for simplicity are assumed Gaussian, we obtain

$$\left\langle e^{i\Phi_{tot}(t_i,t_f,t)}\right\rangle = e^{-\frac{1}{2}\left\langle \Phi^2_{tot}(t_i,t_f,t)\right\rangle} = e^{-2F(t-t_i)-2F(t_f-t)+F(t_f-t_i)+F(t_f+t_i-2t)} \quad (15.1)$$

where we defined the phase correlation function

$$F(t) = \langle(\varphi(t)-\varphi(0))^2\rangle/2 \quad (15.2)$$

Should this function grow with time, the electron phase coherence decays and, hence, $G_{\rm WL}$ is suppressed below its non-interacting value due to the interaction-induced electron decoherence.

The above arguments are, of course, not specific to systems with three barriers only. They can also be applied to any system with larger number of scatterers, i.e., virtually to any disordered conductor where—exactly for the same reasons—one also expects non-vanishing interaction-induced electron decoherence at any temperature, including $T = 0$. Below we are going to confirm and extend our qualitative physical picture by elaborating a complete quantum mechanical analysis of the problem, which fully accounts for Fermi statistics of electrons and treats electron–electron interactions in terms quantum fields produced internally by fluctuating electrons.

15.3 Weak Localization in a System with a Tunnel Barrier

15.3.1 Effective Action and Weak Localization

We begin our analysis by considering a simple example of a system containing a single tunnel barrier. In this case, both weak localization and Coulomb effects can be treated exactly. As in Chapter 2, we will describe electron tunneling across the barrier within the frames of the tunneling Hamiltonian approach. The electronic part of the effective action $S_{\rm el}$ related to the tunnel barrier is defined by Eq. (2.52), where the inverse Green–Keldysh function $\check{G}^{-1}$ is expressed as a combination of the four 2×2 sub-blocks,

$$\check{G}^{-1} = \begin{pmatrix} \check{G}_1^{-1} & \check{t} \\ \check{t}^{\dagger} & \check{G}_2^{-1} \end{pmatrix} \quad (15.3)$$

Here, as before, the inverse Green functions

$$\check{G}_j^{-1} = \begin{pmatrix} \mathrm{i}\frac{\partial}{\partial t} + \frac{\nabla_r^2}{2m} + \mu & 0 \\ 0 & -\mathrm{i}\frac{\partial}{\partial t} - \frac{\nabla_r^2}{2m} - \mu \end{pmatrix}, \quad j = L, R \qquad (15.4)$$

represent the inverse Green functions on both sides of the barrier and the matrix

$$\check{t} = \begin{pmatrix} -t(\boldsymbol{r})e^{-\mathrm{i}\varphi_\mathrm{F}} & 0 \\ 0 & t(\boldsymbol{r})e^{-\mathrm{i}\varphi_\mathrm{B}} \end{pmatrix} \qquad (15.5)$$

(with coordinate $\boldsymbol{r}$ restricted to the junction area $\mathcal{A}$) describes electron tunneling across the barrier. A simpler form of these equations as compared to Eqs. (2.46) and (2.47) is naturally related to the fact that in this chapter, we consider the case of normal (rather than superconducting) electrodes. Following our analysis in Chapters 10–13, here we stick to the definition of the fluctuating phases on the two parts of the Keldysh contour

$$\varphi_{\mathrm{F,B}} = \int^t dt' eV_{\mathrm{F,B}}(t') \qquad (15.6)$$

(cf., e.g., Eq. (10.37)), where $V_{\mathrm{F,B}}$ are the corresponding fluctuating voltages across the barrier. Note that the phases (15.6) are the factor of two smaller than the analogous variables employed, e.g., in Chapters 2 and 7.

The Keldysh Green function in the left metallic electrode reads

$$\begin{aligned} \check{G}_\mathrm{L}(t, \boldsymbol{r}_1, \boldsymbol{r}_2) = \frac{1}{2}\int \frac{dE}{2\pi} e^{-\mathrm{i}Et}\big[& G_\mathrm{L}^R(E, \boldsymbol{r}_1, \boldsymbol{r}_2)\big(\check{1} + \check{Q}_L(E)\big)\hat{\sigma}_z \\ & + G_\mathrm{L}^A(E, \boldsymbol{r}_1, \boldsymbol{r}_2)\big(\check{1} - \check{Q}_L(E)\big)\hat{\sigma}_z\big] \end{aligned} \qquad (15.7)$$

where the matrix $\check{Q}_\mathrm{L}(E)$ is defined in Eq. (2.68). The Green function of the right electrode is defined analogously.

As before, in the tunneling limit, it is convenient to expand the exact effective action S_el (2.52) in powers of the small tunneling amplitude $\check{t}$. In Chapter 2, we carried out a detailed analysis of the lowest non-vanishing order terms (second order in $\check{t}$) of this expansion. In order to capture the weak localization effects, it is necessary to proceed further and to expand the action up to the next (fourth) order in the tunneling amplitude. Then we obtain

$$S_\mathrm{el} = S^{(2)} + S^{(4)} \qquad (15.8)$$

where

$$\mathrm{i}S^{(2)} = -2\int dX_1 dX_2 \times \mathrm{tr}\left[\check{G}_{\mathrm{L}}(t_1-t_2,\boldsymbol{r}_1,\boldsymbol{r}_2)\check{t}(t_2,\boldsymbol{r}_2)\check{G}_{\mathrm{R}}(t_2-t_1,\boldsymbol{r}_2,\boldsymbol{r}_1)\check{t}^{\dagger}(t_1,\boldsymbol{r}_1)\right] \tag{15.9}$$

and

$$\begin{aligned}\mathrm{i}S^{(4)} = -\int dX_1 dX_2 dX_3 dX_4 &\times \mathrm{tr}\left[\check{G}_{\mathrm{L}}(t_1-t_2,\boldsymbol{r}_1,\boldsymbol{r}_2)\check{t}(t_2,\boldsymbol{r}_2)\check{G}_{\mathrm{R}}(t_2-t_3,\boldsymbol{r}_2,\boldsymbol{r}_3)\check{t}^{\dagger}(t_3,\boldsymbol{r}_3)\right.\\ &\left.\times \check{G}_{\mathrm{L}}(t_3-t_4,\boldsymbol{r}_3,\boldsymbol{r}_4)\check{t}(t_4,\boldsymbol{r}_4)\check{G}_{R}(t_4-t_1,\boldsymbol{r}_4,\boldsymbol{r}_1)\check{t}^{\dagger}(t_1,\boldsymbol{r}_1)\right].\end{aligned} \tag{15.10}$$

Obviously Eq. (15.10) can also be reconstructed by keeping the second-order terms in the general expansion of the action (11.86) in the transmissions of the barrier conducting channels T_n and making use of the standard relation (see, e.g., [44])

$$T_n = \frac{4\pi^2|\tilde{t}_n|^2/\delta_{\mathrm{L}}\delta_{\mathrm{R}}}{(1+\pi^2|\tilde{t}_k|^2/\delta_{\mathrm{L}}\delta_{\mathrm{R}})^2} \tag{15.11}$$

where $\delta_{\mathrm{L,R}}$ represent the mean level spacing in the left and right electrodes and $\tilde{t}_n$ are the eigenvalues of the matrix composed by the matrix elements of the tunneling Hamiltonian (2.28) with the electron eigenstates on both sides of the barrier.

Substituting the Green functions $\check{G}_{\mathrm{L,R}}$ (15.7) into the expression (15.9), we again recover the action (2.113). Of particular interest for us here is the fourth-order correction to the action (15.10), which we are now going to evaluate.

Unlike in Chapter 14 (where disorder averaging was postponed till the last stage of the calculation), here it will be convenient for us to perform averaging of the Green functions over disorder in both metallic electrodes already in the beginning of our analysis. A nontrivial part of this procedure corresponds to averaging of the products $G^R G^A$ in each electrode to be carried out with the aid of the formula (14.12). We can identify several contributions as follows.

The first term in the right-hand side of Eq. (14.12) (where the two Green functions are averaged separately) formally provides the leading contribution to the action $S^{(4)}$ (15.10). This term describes the correction to the barrier transmission $\propto |t|^4$, i.e., it is already included in the action of the barrier expressed via its scattering matrix (10.63) addressed earlier in Chapter 10. The combination in the second line of Eq. (14.12) accounts for the process of classical diffusion of electrons in the metallic electrodes and determines their resistance values. In order to illustrate this statement, let us recall the standard Ohm's law $I = V/(R_t + r_l)$, where R_t is the barrier resistance while r_l is the series one of the leads. In the limit $r_l \ll R_t$, one obviously finds $I \simeq V/R_t - r_l V/R_t^2$. It is straightforward to demonstrate that the last term in this formula originates from the second line of Eq. (14.12). Finally, the combination in the third line of Eq. (14.12) containing the Cooperon accounts for quantum interference, thus being responsible for the weak localization correction to the conductance (14.15). Combining the last line of Eq. (14.12) with Eq. (15.10), after a straightforward though rather lengthy algebra, we arrive at the weak localization correction to the action of our tunnel barrier [167]

$$\begin{aligned}
S_{\rm WL}^{(4)} = &- \int dt_1 \dots dt_4 \int d\tau_1 d\tau_2 \int_{\mathcal{A}} d^2\boldsymbol{r}\, d^2\boldsymbol{r}' \frac{g_t(\boldsymbol{r}) g_t(\boldsymbol{r}')}{4\pi^2 \nu_L \nu_R} \\
&\times \tilde{C}_L(t_1 - \tau_1, \boldsymbol{r}', \boldsymbol{r}) \tilde{C}_R(t_2 - \tau_2, \boldsymbol{r}, \boldsymbol{r}')\, e^{i\tilde{\Phi}(t_1, t_2, t_3, t_4)} \sin\frac{\varphi^-(t_1)}{2} \\
&\times \left[\tilde{h}(\tau_1 - t_2) e^{-i\varphi^-(t_2)/2} + \tilde{f}(\tau_1 - t_2) e^{i\varphi^-(t_2)/2}\right] \\
&\times \left[\tilde{h}(\tau_2 - t_3) e^{i\varphi^-(t_3)/2} \tilde{f}(t_1 + t_3 - t_4 - \tau_1)\right. \\
&\left. - \tilde{f}(\tau_2 - t_3) e^{-i\varphi^-(t_3)/2} \tilde{h}(t_1 + t_3 - t_4 - \tau_1)\right] \\
&\times \left[e^{-i\varphi^-(t_4)/2} \tilde{f}(-t_1 + t_2 + t_4 - \tau_2)\right. \\
&\left. + e^{i\varphi^-(t_4)/2} \tilde{h}(-t_1 + t_2 + t_4 - \tau_2)\right] + \left\{C_L \leftrightarrow C_R, \varphi^\pm \to -\varphi^\pm\right\}
\end{aligned} \tag{15.12}$$

Here we routinely defined the classical $\varphi^+ = (\varphi^F + \varphi^B)/2$ and quantum $\varphi^- = \varphi^F - \varphi^B$ phases as well as the Fourier-transformed distribution functions

$$\tilde{f}(\tau) = \int \frac{dE}{2\pi} \frac{e^{-\mathrm{i}E\tau}}{1+e^{E/T}} = \frac{1}{2}\left(\delta(\tau) + \frac{\mathrm{i}T}{\sinh \pi T \tau}\right),$$
$$\tilde{h}(\tau) = \int \frac{dE}{2\pi}\left(1 - \frac{1}{1+e^{E/T}}\right) e^{-\mathrm{i}E\tau} = \frac{1}{2}\left(\delta(\tau) - \frac{\mathrm{i}T}{\sinh \pi T \tau}\right) \tag{15.13}$$

The action (15.12) also contains the densities of states in the metallic electrodes ν_{L} and ν_{R}, the combination of the classical phases

$$\tilde{\Phi}(t_1, t_2, t_3, t_4) = \varphi^+(t_1) - \varphi^+(t_2) + \varphi^+(t_3) - \varphi^+(t_4) \tag{15.14}$$

and the dimensionless barrier conductance per unit area $g_t(\boldsymbol{r})$ related to the total conductance G_t as

$$G_t \equiv 1/R_t = \frac{e^2}{2\pi} \int_{\mathcal{A}} d^2\boldsymbol{r}\, g_t(\boldsymbol{r}) \tag{15.15}$$

The corresponding quantum interference contribution to the current across the barrier I_{WL} can be expressed as a path integral

$$I_{\mathrm{WL}}(t) = -e \int \mathcal{D}\varphi^{\pm} \frac{\delta}{\delta\varphi^-(t)} S_{\mathrm{WL}}^{(4)}[\varphi^{\pm}]$$
$$\times \frac{\exp\left\{\mathrm{i}S[\varphi^{\pm}] + \mathrm{i}\int dt' \, \frac{V_x \varphi^-(t')}{Z(0)e}\right\}}{\int \mathcal{D}\varphi^{\pm} \exp\left\{\mathrm{i}S[\varphi^{\pm}] + \mathrm{i}\int dt' \, \frac{V_x\varphi^-(t')}{Z(0)e}\right\}} \tag{15.16}$$

where V_x is the external bias voltage, $S[\varphi^{\pm}]$ is the total action of our system for $V_x = 0$, and $Z(\omega)$ is an effective environment impedance "seen" by the tunnel barrier. Note that for the sake of simplicity, here we stick to the limit $R_t \gg Z(\omega)$, in which case one can safely drop the barrier contributions (15.9) and (15.10) from the total action S in the exponent of Eq. (15.16).

15.3.2 Non-interacting Limit

To begin with, let us consider the low external impedance limit $Z(\omega) \to 0$, in which case the voltage drop across the barrier V does not fluctuate being equal to the external bias, i.e., $V \equiv V_x$. Coulomb interaction effects become negligible in this limit, and we can simply replace $\varphi^+(t) \to eVt$ and $\varphi^- \to \chi$, where, as before, χ denotes the so-called counting field. Employing Eq. (10.96) together with

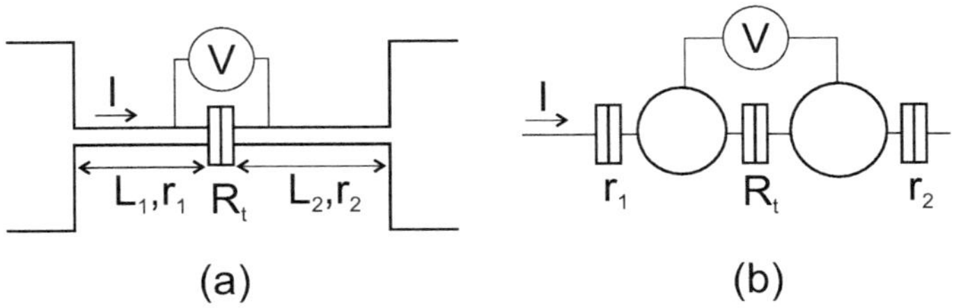

Figure 15.3 Two setups under consideration: (a) Tunnel barrier with resistance R_t connects two wires with lengths L_1 and L_2 and resistances r_1 and r_2. (b) Tunnel barrier placed in between two quantum dots coupled to bulk reservoirs via junctions with resistances r_1 and r_2; cf. also Fig. 15.2. In both cases (a) and (b), we assume r_1, $r_2 \ll R_t$.

(15.12), we identify the weak localization correction $\mathcal{F}_{\text{WL}}(t, \chi)$ to the cumulant generating function for our tunnel barrier $\mathcal{F}(t, \chi)$. For $V > 0$ and $eV \gg T$, we obtain

$$\mathcal{F}(t, \chi) = \frac{Vt}{e} G_t \left(e^{-\mathrm{i}\chi} - 1\right) + \mathcal{F}_{WL}(t, \chi) \tag{15.17}$$

where

$$\mathcal{F}_{\text{WL}}(t, \chi) = \frac{Vt}{e} G_{\text{WL}}^{(0)} \left[e^{-2\mathrm{i}\chi} - e^{-\mathrm{i}\chi}\right] \tag{15.18}$$

and

$$G_{\text{WL}}^{(0)} = -\int_{\mathcal{A}} d^2\boldsymbol{r}\, d^2\boldsymbol{r}' \frac{e^2 g_t(\boldsymbol{r}) g_t(\boldsymbol{r}')}{8\pi^3 \nu_{\text{L}} \nu_{\text{R}}} \mathcal{C}_{\text{L}}(0, \boldsymbol{r}', \boldsymbol{r}) \mathcal{C}_{\text{R}}(0, \boldsymbol{r}, \boldsymbol{r}') \tag{15.19}$$

is the absolute value of the weak localization correction to the conductance in the non-interacting limit. In this case, the current trough the barrier reads

$$I = \frac{\mathrm{i}e}{t} \frac{\partial \mathcal{F}}{\partial \chi}\bigg|_{\chi=0} = \left(G_t + G_{\text{WL}}^{(0)}\right) V \tag{15.20}$$

while the shot noise takes the form

$$\mathcal{S} = -\frac{e^2}{t} \frac{\partial^2 \mathcal{F}}{\partial \chi^2}\bigg|_{\chi=0} = \left(G_t + 3G_{\text{WL}}^{(0)}\right) eV \tag{15.21}$$

Further analysis essentially depends on the system under consideration. For instance, one can consider the wires with cross section $\mathcal{A}$ and lengths L_1 and L_2 attached to the tunnel barrier and to two bulk metallic reservoirs, as shown in Fig. 15.3a. In this geometry, the equation for the Cooperon (14.14) in the first wire acquires the form

$$\left(-\mathrm{i}\omega - D_1\frac{\partial^2}{\partial x^2} + \frac{1}{\tau_{H1}} + \frac{1}{\tau_{\varphi 1}}\right)\mathcal{C}_{\mathrm{L}}(\omega, x, x') = \frac{1}{\mathcal{A}}\delta(x - x') \tag{15.22}$$

This equation should be supplemented by the boundary conditions

$$\frac{\partial}{\partial x}\mathcal{C}_{\mathrm{L}}(\omega, 0, x') = 0, \quad \mathcal{C}_{\mathrm{L}}(\omega, L_1, x') = 0 \tag{15.23}$$

implying that, on one hand, the tunnel barrier is a poorly conducting object and, on the other hand, electrons quickly diffuse out of the wire as soon as they reach the edge of the bulk reservoir. Resolving Eqs. (15.22) and (15.23), we get

$$\mathcal{C}_{\mathrm{L}}(\omega, \boldsymbol{r}, \boldsymbol{r}') = \frac{\theta(x' - x)\cosh[\kappa_1(\omega)x]\sinh[\kappa_1(\omega)(L_1 - x')]}{\kappa_1(\omega)\mathcal{A}D_1\cosh[\kappa_1(\omega)L_1]} + \frac{\theta(x - x')\sinh[\kappa_1(\omega)(L_1 - x)]\cosh[\kappa_1(\omega)x']}{\kappa_1(\omega)\mathcal{A}D_1\cosh[\kappa_1(\omega)L_1]} \tag{15.24}$$

where

$$\kappa_1(\omega) = \sqrt{\frac{-\mathrm{i}\omega}{D_1} + \frac{1}{D_1\tau_{H1}} + \frac{1}{D_1\tau_{\varphi 1}}} \tag{15.25}$$

The Cooperon in the right wire is defined analogously. Since both Cooperon coordinates $\boldsymbol{r}$ and $\boldsymbol{r}'$ in Eq. (15.19) belong to the barrier surface, it is necessary to set $x = x' = 0$ in Eq. (15.24). Then the weak localization correction to the conductance (15.19) takes the form

$$G_{\mathrm{WL}}^{(0)} = -\frac{2e^2}{\pi}G_t^2 r_1 r_2 \frac{\tanh[\kappa_1(0)L_1]}{\kappa_1(0)L_1}\frac{\tanh[\kappa_2(0)L_2]}{\kappa_2(0)L_2} \tag{15.26}$$

where

$$r_1 = L_1/2e^2\nu_{\mathrm{L}}D_{\mathrm{L}}\mathcal{A}, \quad r_2 = L_2/2e^2\nu_{\mathrm{R}}D_{\mathrm{R}}\mathcal{A} \tag{15.27}$$

are the wire resistances. In the absence of any source of electron decoherence, one should set $1/\tau_{\mathrm{H}} = 1/\tau_\varphi = 0$. Then Eq. (15.26) reduces to

$$G_{\mathrm{WL}}^{(0)} = -\frac{2e^2}{\pi}G_t^2 r_1 r_2 \tag{15.28}$$

We remind the reader that Eqs. (15.26) and (15.28) remain applicable in the limit $G_t^2 r_1 r_2 \ll 1$.

Let us now consider a different geometry. Namely, let us assume that the metallic wires on both sides of the junction are replaced by quantum dots, which are coupled to the bulk electrodes via two additional junctions with the resistances r_1, r_2, respectively. In this case, the Cooperons do not depend on coordinates. For example, for the left dot, we obtain

$$\mathcal{C}_{\rm L} = \frac{1}{\mathcal{V}_{\rm L}\left(-{\rm i}\omega + \frac{1}{\tau_{\rm DL}} + \frac{1}{\tau_{\rm H1}} + \frac{1}{\tau_{\varphi 1}}\right)} \tag{15.29}$$

where $\mathcal{V}_{\rm L}$ is the volume of the left dot and $\tau_{\rm DL} = 2e^2\nu_{\rm L}\mathcal{V}_{\rm L}r_1$ is the electron dwell time in this dot evaluated for $G_t r_1, G_t r_2 \ll 1$. Substituting Eq. (15.29) together with the analogous expression for $\mathcal{C}_{\rm R}$ into Eq. (15.19), we find

$$G_{\rm WL}^{(0)} = -\frac{2e^2}{\pi}\frac{G_t^2 r_1 r_2}{\left(1 + \frac{\tau_{\rm DL}}{\tau_{\rm H1}} + \frac{\tau_{\rm DL}}{\tau_{\varphi 1}}\right)\left(1 + \frac{\tau_{\rm DR}}{\tau_{\rm H2}} + \frac{\tau_{\rm DR}}{\tau_{\varphi 2}}\right)} \tag{15.30}$$

In the fully coherent limit $1/\tau_{\varphi 1,2} = 1/\tau_{\rm H1,2} = 0$, we again recover Eq. (15.28).

15.3.3 Effect of Electron–Electron Interactions

Let us now include electron–electron interactions into our consideration. In order to evaluate the path integral over the phases $\varphi^\pm$ in Eq. (15.16), we should specify the environmental contribution to the action S. Below we will consider a general Gaussian environment described by an effective impedance $Z(\omega)$ and the action $S_{\rm env}$ defined by Eqs. (10.105) and (10.107). Note that in general our external circuit does not necessarily consist of linear elements. For instance, the contributions of the outer barriers with dimensionless conductances $g_{\rm L,R}$ (see Fig. 15.2) are, strictly speaking, non-Gaussian. However, in the interesting-for-us metallic limit $g_{\rm L,R} \gg 1$, phase fluctuations can be considered small down to exponentially low energies (see Chapters 7 and 11), in which case it suffices to expand both contributions up to the second-order $\varphi^\pm$. Moreover, this Gaussian approximation becomes *exact* in the limit of fully open outer barriers; see Section 11.2.1. Thus, in the metallic limit, the integral (15.16) remains Gaussian at all relevant energies and can be easily performed.

The path integral (15.16) is handled in exactly the same manner as in the case of a tunnel barrier embedded in the electromagnetic environment with linear impedance $Z(\omega)$; see, e.g., Section 7.3. This task can be accomplished with the aid of the following correlation functions:

$$\langle(\varphi^+(t) - \varphi^+(0))\varphi^+(0)\rangle = -F(t) \quad (15.31)$$

$$\langle\varphi^+(t)\varphi^-(0) + \varphi^-(t)\varphi^+(0)\rangle = \mathrm{i}\tilde{K}(t) \quad (15.32)$$

$$\langle\varphi^+(t)\varphi^-(0) - \varphi^-(t)\varphi^+(0)\rangle = \mathrm{i}\tilde{K}(|t|) \quad (15.33)$$

$$\langle\varphi^-(t)\varphi^-(0)\rangle = 0 \quad (15.34)$$

where the last relation, as usual, follows directly from the causality principle (cf. Eq. (9.66)). Here

$$F(t) = e^2 \int \frac{d\omega}{2\pi} \coth\frac{\omega}{2T} \mathrm{Re}\left[Z(\omega)\right] \frac{1 - \cos\omega t}{\omega} \quad (15.35)$$

is the correlation function for the classical phases φ^+ (cf. Eqs. (7.66) and (15.2)) and

$$\tilde{K}(t) = e^2 \int \frac{d\omega}{2\pi} \mathrm{Re}\left[Z(\omega)\right] \frac{\sin\omega t}{\omega} \quad (15.36)$$

is the response function.

As a result, we obtain [167]

$$\begin{aligned} I_{\mathrm{WL}} = & \frac{e}{8\pi^3\nu_{\mathrm{L}}\nu_{\mathrm{R}}} \mathrm{Re} \int_{\mathcal{A}} d^2\mathbf{r}\, d^2\mathbf{r}'\, g_t(\mathbf{r}) g_t(\mathbf{r}') \int dE\, d\omega_1 d\omega_2 d\omega_3 \\ & \times \mathcal{C}_{\mathrm{L}}(-\omega_3, \mathbf{r}', \mathbf{r}) \mathcal{C}_{\mathrm{R}}(-\omega_2, \mathbf{r}, \mathbf{r}')\, f_{\mathrm{L}}(E + \omega_3 - \omega_1) h_{\mathrm{R}}(E - eV - \omega_2) \\ & \times \left[f_{\mathrm{L}}(E - \omega_1) h_{\mathrm{R}}(E - eV) P_1(\omega_1, \omega_2, \omega_3) \right. \\ & + f_{\mathrm{L}}(E - \omega_1) f_{\mathrm{R}}(E - eV) P_2(\omega_1, \omega_2, \omega_3) \\ & + h_{\mathrm{L}}(E - \omega_1) h_{\mathrm{R}}(E - eV) P_2(\omega_1, \omega_3, \omega_2) \\ & \left. + h_{\mathrm{L}}(E - \omega_1) f_{\mathrm{R}}(E - eV) P_3(\omega_1, \omega_2, \omega_3) \right] - \{V \to -V\} \end{aligned} \quad (15.37)$$

Here $f_{\mathrm{L,R}}(E)$ and $h_{\mathrm{L,R}}(E) = 1 - f_{\mathrm{L,R}}(E)$ are, respectively, the electron and hole distribution functions in the left and right leads. On a qualitative level, the functions $P_j(\omega_1, \omega_2, \omega_3)$ can be interpreted as probabilities to excite or absorb photons in the environment. In this respect, they are similar to the probability function $P(E)$ (7.71) introduced in Chapter 7. Note, however, that—in contrast to the latter—the functions $P_j(\omega_1, \omega_2, \omega_3)$ are complex valued and, hence, they cannot be defined as probabilities in the strict mathematical sense. These functions read

$$P_j(\omega_1, \omega_2, \omega_3) = \int \frac{dt_1 dt_2 dt_3}{(2\pi)^3} e^{i[\omega_1 t_1 + \omega_2 t_2 + \omega_3 t_3]} \times e^{-\tilde{F}(t_1, t_2, t_3)} \mathcal{Q}_j(t_1, t_2, t_3) \quad (15.38)$$

where

$$\tilde{F}(t_1, t_2, t_3) = F(t_1 + t_3) + F(t_3) + F(t_1 + t_2) + F(t_2) - F(t_1 + t_2 + t_3) - F(t_2 - t_3) \quad (15.39)$$

and

$$\mathcal{Q}_1 = e^{-i[\tilde{K}(t_2)+\tilde{K}(t_3)+\tilde{K}(|t_2-t_3|)]} \left\{ 2e^{i[\tilde{K}(|t_1+t_2+t_3|)+\tilde{K}(t_1+t_3)+\tilde{K}(t_1+t_2)]} - e^{i[\tilde{K}(t_1+t_2+t_3)+\tilde{K}(|t_1+t_3|)+\tilde{K}(|t_1+t_2|)]} \right\},$$

$$\mathcal{Q}_2 = e^{i[\tilde{K}(|t_1+t_2+t_3|)-\tilde{K}(t_2)-\tilde{K}(|t_3|)+\tilde{K}(t_1+t_3)-\tilde{K}(|t_1+t_2|)-\tilde{K}(t_3-t_2)]},$$

$$\mathcal{Q}_3 = e^{i[\tilde{K}(t_1+t_2+t_3)-\tilde{K}(|t_2|)-\tilde{K}(|t_3|)-\tilde{K}(t_1+t_3)-\tilde{K}(t_1+t_2)+\tilde{K}(|t_3-t_2|)]} \quad (15.40)$$

One can demonstrate that the functions $P_j(\omega_1, \omega_2, \omega_3)$ obey the detailed balance condition. For this purpose, let us introduce the functions

$$U(\omega_1, \omega_2, \omega_3) = \int \frac{dt_1 dt_2 dt_3}{(2\pi)^3} e^{i[\omega_1 t_1 + \omega_2 t_2 + \omega_2 t_3]} e^{i\left[-\tilde{K}(|t_2-t_3|)+\tilde{K}(|t_2+t_3|)\right]} \times e^{-J(t_1-t_3)-J(t_1+t_3)-J(t_1-t_2)-J(t_1+t_2)+F(t_2+t_3)+F(t_2-t_3)}$$

and

$$\tilde{W}(\omega_1, \omega_2, \omega_3) = \int \frac{dt_1 dt_2 dt_3}{(2\pi)^3} e^{i[\omega_1 t_1 + \omega_2 t_2 + \omega_2 t_3]} e^{i\left[\tilde{K}(|t_2-t_3|)-\tilde{K}(|t_3|)-\tilde{K}(|t_2|)\right]} \times e^{-J(t_1+t_3)-F(t_3)-F(t_2)-J(t_2+t_1)+F(t_2-t_3)+J(t_2+t_3+t_1)}$$

where $J(t) = F(t) + i\tilde{K}(t)$. Employing the above definitions, one can express the functions P_j in the form

$$P_1(\omega_1, \omega_2, \omega_3) = 4U(\omega_2 + \omega_3 - 2\omega_1, \omega_2, \omega_3) - \tilde{W}^*(\omega_2 + \omega_3 - \omega_1, \omega_2, \omega_3),$$

$$P_2(\omega_1, \omega_2, \omega_3) = \tilde{W}(\omega_2 - \omega_1, \omega_2, -\omega_3),$$

$$P_3(\omega_1, \omega_2, \omega_3) = \tilde{W}(\omega_1, \omega_2, \omega_3) \quad (15.41)$$

With the aid of the identity $J(t - \mathrm{i}/T) = J(-t)$, one finds

$$U(\omega_1, \omega_2, \omega_3) = e^{\omega_1/T} U(-\omega_1, \omega_2, \omega_3),$$
$$\tilde{W}(\omega_1, \omega_2, \omega_3) = e^{\omega_1/T} \tilde{W}(-\omega_1, \omega_2, \omega_3) \tag{15.42}$$

and, hence,

$$\begin{aligned} P_1(\omega_1, \omega_2, \omega_3) &= 4e^{(\omega_2+\omega_3-2\omega_1)/T} U(\omega_2 + \omega_3 - 2\omega_1, \omega_2, \omega_3) \\ &\quad - e^{(\omega_2+\omega_3-\omega_1)/T} \tilde{W}^*(\omega_2 + \omega_3 - \omega_1, \omega_2, \omega_3), \\ P_2(\omega_1, \omega_2, \omega_3) &= e^{(\omega_2-\omega_1)/T} P_2(2\omega_2 - \omega_1, \omega_2, \omega_3), \\ P_3(\omega_1, \omega_2, \omega_3) &= e^{\omega_1/T} P_3(-\omega_1, \omega_2, \omega_3) \end{aligned} \tag{15.43}$$

which are the announced detailed balance conditions. In particular, in the limit $T = 0$, these conditions yield

$$\begin{aligned} P_1 &= 0 \;\text{ at }\; \omega_1 > \max\left\{\omega_2 + \omega_3, \frac{\omega_2 + \omega_3}{2}\right\}, \\ P_2 &= 0 \;\text{ at }\; \omega_1 > \omega_2, \\ P_3 &= 0 \;\text{ at }\; \omega_1 < 0 \end{aligned} \tag{15.44}$$

Evaluating the integral over energy E in Eq. (15.37), we arrive at the result

$$\begin{aligned} I_{\rm WL} = &-\frac{e}{16\pi^3 \nu_{\rm L}\nu_R} \,\mathrm{Re} \int_{\mathcal{A}} d^2\boldsymbol{r}\, d^2\boldsymbol{r}'\, g_t(\boldsymbol{r}) g_t(\boldsymbol{r}') \\ &\times \int d\omega_1 d\omega_2 d\omega_3\, \mathcal{C}_{\rm L}(-\omega_3, \boldsymbol{r}', \boldsymbol{r}) \mathcal{C}_{\rm R}(-\omega_2, \boldsymbol{r}, \boldsymbol{r}') \\ &\times \big\{2eV\, P_1 + P_1 B(\omega_1 - \omega_2 - \omega_3, V, T) \\ &+ (P_1 - P_2') N(\omega_3) \big[B(\omega_1 - \omega_2 - \omega_3, V, T) - B(\omega_1 - \omega_2, V, T)\big] \\ &+ (P_1 - P_2) N(\omega_2) \big[B(\omega_1 - \omega_2 - \omega_3, V, T) - B(\omega_1 - \omega_3, V, T)\big] \\ &+ (P_1 - P_2 - P_2' + P_3) N(\omega_2) N(\omega_3) \big[B(\omega_1, V, T) - B(\omega_1 - \omega_2, V, T) \\ &- B(\omega_1 - \omega_3, V, T) + B(\omega_1 - \omega_2 - \omega_3, V, T)\big]\big\}. \end{aligned} \tag{15.45}$$

Here $N(\omega) = 1/(e^{\omega/T} - 1) = \frac{1}{2}\left(\coth \frac{\omega}{2T} - 1\right)$ is the Bose distribution function, $P_2' = P_2(\omega_1, \omega_3, \omega_2)$, and the function $B(\omega, V, T)$ was defined in Eq. (13.102). For brevity in Eq. (15.45), we drop the arguments of the functions $P_j(\omega_1, \omega_2, \omega_3) \equiv P_j$.

We observe that the weak localization correction to the current (15.45) consists of a linear Ohmic term (which dominates at large voltages) and the nonlinear correction due to Coulomb blockade. This correction tends to a constant offset value at $V \to \infty$. Thus, the overall behavior of the quantum correction I_{WL} resembles that of the "classical" current (7.67).

With the aid of the above results, one can also derive the weak localization contribution to the differential conductance $G_{\mathrm{WL}} = dI_{\mathrm{WL}}/dV$. In the zero voltage bias limit $V \to 0$ from Eq. (15.45), one gets

$$
\begin{aligned}
G_{\mathrm{WL}}(T) = &-\frac{e^2}{8\pi^3 \nu_{\mathrm{L}} \nu_{\mathrm{R}}} \operatorname{Re} \int_{\mathcal{A}} d^2\mathbf{r}\, d^2\mathbf{r}'\, g_t(\mathbf{r}) g_t(\mathbf{r}') \\
&\times \int d\omega_1 d\omega_2 d\omega_3\, \mathcal{C}_{\mathrm{L}}(-\omega_3, \mathbf{r}', \mathbf{r}) \mathcal{C}_{\mathrm{R}}(-\omega_2, \mathbf{r}, \mathbf{r}') \\
&\times \big\{ P_1 + P_1 w(\omega_1 - \omega_2 - \omega_3) \\
&+ \big(P_1 - P_2'\big) N(\omega_3) \big[w(\omega_1 - \omega_2 - \omega_3) - w(\omega_1 - \omega_2)\big] \\
&+ \big(P_1 - P_2\big) N(\omega_2) \big[w(\omega_1 - \omega_2 - \omega_3) - w(\omega_1 - \omega_3)\big] \\
&+ (P_1 - P_2 - P_2' + P_3) N(\omega_2) N(\omega_3) \big[w(\omega_1) \\
&- w(\omega_1 - \omega_2) - w(\omega_1 - \omega_3) + w(\omega_1 - \omega_2 - \omega_3)\big] \big\},
\end{aligned}
\tag{15.46}
$$

where we introduced the function $w(\omega) = \frac{\partial}{\partial \omega} \omega \coth \frac{\omega}{2T}$. The result (15.46) contains valuable information, which we will elaborate on further below.

15.3.4 Dephasing

Turning to the analysis of the above general results, we first notice that both functions (15.35) and (15.36) are purely real and, hence, $|Q_j| \leq 1$. Furthermore, at sufficiently long times $|t| > \tau_{\mathrm{RC}}$ (where τ_{RC} is an effective RC-time of our system to be defined later), we have

$$
F(t) = \frac{2}{g_{\mathrm{Z}}} \left[\ln \frac{\sinh[\pi T t]}{\pi T \tau_{\mathrm{RC}}} + \gamma \right], \quad \tilde{K}(t) = \frac{\pi}{g_{\mathrm{Z}}} \operatorname{sign} t, \tag{15.47}
$$

where $g_{\mathrm{Z}} = 2\pi/e^2 Z(0)$, and $\gamma \simeq 0.577$ is the Euler constant. The correlation function $F(t)$ grows with time at any temperature, including $T \to 0$. In contrast, the response function $\tilde{K}(t)$ saturates to a constant $\sim 1/g_{\mathrm{Z}}$ at sufficiently long times. In order to proceed, we will consider the limit $g_{\mathrm{Z}} \gg 1$ in which case the correlator $\tilde{K}(t)$ always remains much smaller than unity and, hence, it can be safely ignored in the leading order in $1/g_{\mathrm{Z}}$. Within this approximation, one has $Q_j \equiv 1$, i.e., the functions P_j (15.38) become identical to each other. As a result, the Fermi functions $f_{L,R}(E)$ drop out from the general formula (15.37) and we get $I_{WL}(V) = G_{WL}V$, where

$$G_{WL} = -\frac{e^2}{8\pi^3 \nu_L \nu_R} \int_{\mathcal{A}} d^2\mathbf{r}\, d^2\mathbf{r}'\, g_t(\mathbf{r}) g_t(\mathbf{r}') \int dt_1 dt_2$$
$$\times\, \mathcal{C}_L(t_1, \mathbf{r}', \mathbf{r}) \mathcal{C}_R(t_2, \mathbf{r}, \mathbf{r}')\, e^{-2F(t_1)-2F(t_2)+F(t_1+t_2)+F(t_1-t_2)} \tag{15.48}$$

Identifying $t_1 = t_f - t$ and $t_2 = t - t_i$ in this integral, we observe that the exponent in Eq. (15.48) exactly coincides with the expression (15.1) derived from simple considerations involving electrons propagating along time-reversed paths in the external fluctuating field.

Thus, employing no approximations except for setting $g_Z \gg 1$, we arrive at an important conclusion: In the leading order in $1/g_Z$, the weak localization correction G_{WL} is affected by electron–electron interactions mainly due to dephasing produced *only* by the "classical" component φ^+ of the fluctuating field that mediates such interactions. This effect is described by the phase correlation function $F(t)$ (15.35). At the same time, fluctuations of the "quantum" field φ^- turn out to be irrelevant for dephasing. As we will demonstrate below in Section 15.3.5, such fluctuations are responsible for the Coulomb blockade correction to G_{WL} described by the response function $\tilde{K}(t)$ (15.36). In the limit $g_Z \gg 1$, this correction remains weak except in the immediate vicinity of $T = 0$.

Let us now explicitly evaluate the weak localization correction G_{WL} in Eq. (15.48). For simplicity, we will stick to the symmetric situation, i.e., we set $\tau_{DL} = \tau_{DR} = \tau_D$ and

$$\mathcal{C}_L = \mathcal{C}_R = \theta(t) \frac{e^{-t/\tau_D}}{\mathcal{V}} \tag{15.49}$$

where $\mathcal{V}$ is the volume of each of the quantum dots on both left and right sides of the tunnel barrier (see Figs. 15.2 and 15.3b). Substituting the first Eq. (15.47) together with Eq. (15.49) into Eq. (15.48) and performing the integrals, we obtain

$$G_{WL} = G_{WL}^{(0)} \left(\frac{e^{-\gamma} \tau_{RC}}{\tau_D} \right)^{4/g_Z} \frac{g_Z}{2} (2\pi T \tau_D)^{\frac{4}{g_Z}-1} \tag{15.50}$$

at sufficiently high temperatures $T \gg g_Z/\tau_D$ and

$$G_{WL} = G_{WL}^{(0)} \left(\frac{4e^{-\gamma} \tau_{RC}}{\tau_D} \right)^{4/g_Z} u(g_Z) \tag{15.51}$$

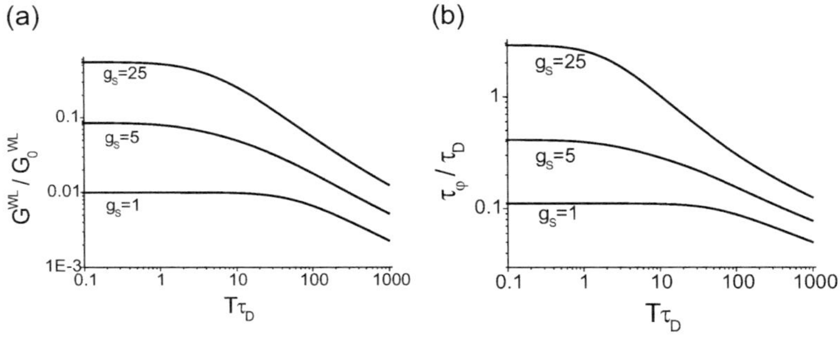

Figure 15.4 (a) Temperature dependence of the weak localization correction $G_{\rm WL}$ defined by Eq. (15.48). (b) Temperature dependence of the dephasing time τ_φ (15.53). In both cases, we have fixed $\tau_{\rm D}/\tau_{\rm RC} = 50$.

in the low-temperature limit $T \ll 1/\tau_{\rm D}$. Here $G_{\rm WL}^{(0)}$ is the non-interacting weak localization correction (15.28) and

$$u(g_{\rm Z}) = \frac{\Gamma\left(2 - \frac{4}{g_{\rm Z}}\right)\Gamma\left(1 - \frac{4}{g_{\rm Z}}\right)\Gamma\left(\frac{1}{2} + \frac{1}{g_{\rm Z}}\right)}{2\Gamma\left(\frac{3}{2} - \frac{3}{g_{\rm Z}}\right)} \tag{15.52}$$

is the dimensionless factor.

The temperature dependence of the weak localization correction $G_{\rm WL}(T)$ (15.48) in the symmetric case (15.49) is also illustrated in Fig. 15.4a. We observe that in accordance with the above analytic results, $G_{\rm WL}(T)$ increases with decreasing T and eventually saturates to Eq. (15.51) at temperatures $T \simeq 1/\pi\tau_{\rm D}$.

The above analysis also allows us to derive the electron dephasing time τ_φ due to electron–electron interactions. To this end, we again employ Eqs. (15.29) and (15.30) where the dephasing times were introduced phenomenologically. As before, we stick to the symmetric situation with $C_1 = C_2$, $\tau_{\varphi 1} = \tau_{\varphi 2} = \tau_\varphi$ and $\tau_{\rm D1} = \tau_{\rm D2} = \tau_{\rm D}$. Setting $1/\tau_{\rm H1,2} = 0$ in Eq. (15.30) and identifying this expression with the weak localization correction $G_{\rm WL}(T)$ (15.48), one arrives at the interaction-induced dephasing time τ_φ defined as

$$\tau_\varphi = \frac{\tau_{\rm D}}{\sqrt{G_{\rm WL}^{(0)}/G_{\rm WL} - 1}}, \tag{15.53}$$

where $G_{\rm WL}^{(0)}$ is again determined by Eq. (15.28).

Combining Eqs. (15.53) and (15.50), we obtain

$$\tau_\varphi(T) \approx \tau_{\rm D}\left(\frac{g_{\rm Z}}{4\pi T\tau_{\rm D}}\right)^{\frac{1}{2}-\frac{2}{g_{\rm Z}}} \tag{15.54}$$

This result remains valid at sufficiently high temperatures $T > g_{\rm Z}/\tau_{\rm D}$ and for $g_{\rm Z} \gg 1$. It demonstrates that at such values of T, the electron dephasing time τ_φ increases with decreasing temperature, i.e., it demonstrates qualitatively the same behavior as in the case of spatially extended disordered conductors analyzed in Chapter 14.

It also follows from Eq. (15.53) and the above results for $G_{\rm WL}(T)$ (see Eq. (15.51) and Fig. 15.4a) that, as temperature gets below $T \sim 1/\pi\tau_{\rm D}$, the dephasing time stops increasing and saturates to the value

$$\tau_{\varphi 0} = \frac{\tau_{\rm D}}{\left(\frac{\tau_{\rm D}}{4e^{-\gamma}\tau_{\rm RC}}\right)^{2/g_{\rm Z}}\frac{1}{\sqrt{u(g_{\rm Z})}} - 1} \tag{15.55}$$

which in the limit $g_{\rm Z} \gg 1$ reduces further to

$$\tau_{\varphi 0} = \frac{g_{\rm Z}\tau_{\rm D}}{2\left(\ln\frac{\tau_{\rm D}}{4\tau_{\rm RC}} + \gamma\right)} \tag{15.56}$$

This result demonstrates that similar to the case of diffusive conductors studied in Chapter 14, the electron decoherence time in quantum dots remains finite down to very low temperatures due to electron–electron interactions. Moreover, in Section 15.4, we will establish a direct relation between the result (15.56) and Eq. (14.71) derived within a completely different path integral technique.

The expression (15.56) can also be rewritten in the following manner. If we identify the external impedance as $Z(0) = r_1 + r_2$ and consider a symmetric situation with $r_1 = r_2$, the electron dwell time becomes $\tau_{\rm D} = e^2 Z(0)/\delta$, where $\delta = 1/\nu\mathcal{V}$ is the level spacing in a quantum dot. Then Eq. (15.56) reduces to

$$\tau_{\varphi 0} = \frac{\pi}{\delta\,(\ln[E_{\rm C}/2\delta] + \gamma)} \tag{15.57}$$

i.e., we arrive at an important conclusion that the low-temperature dephasing time $\tau_{\varphi 0}$ for electrons in a quantum dot is simply proportional to the inverse level spacing $1/\delta$ in this dot and is practically insensitive to other parameters.

15.3.5 Coulomb Blockade of Weak Localization Correction

Let us now investigate the applicability range of the above results for the weak localization correction $G_{\rm WL}(T)$. To begin with, we notice that with the aid of Eqs. (15.44), it is straightforward to verify that the weak localization correction (15.46) vanishes identically in the zero temperature limit, i.e., $G_{\rm WL}(0) \equiv 0$. The physical meaning of this result is transparent: It is just an obvious consequence of complete Coulomb blockade of electron tunneling across the barrier, which we already addressed in details in Chapter 7. At the same time, this result implies that, e.g., Eq. (15.51) becomes inaccurate at very low temperatures and should be modified accordingly in this limit.

The reason for this inaccuracy is clear: In order to derive Eq. (15.48) from the exact result (15.46), we neglected the correlator $\tilde{K}(t)$ (15.47) assuming $g_{\rm Z} \gg 1$. This approximation can be improved if we expand the exact result for the current (15.46) in powers of the "quantum" phase correlator $\tilde{K}(t)$ in the same way as it was done earlier; see Eqs. (7.67) and (7.50), which define the lowest-order weak Coulomb blockade correction to the Ohmic current. If one adopts the model of an Ohmic environment and chooses $Z(\omega) = Z(0)/(1 - {\rm i}\omega\tau_{\rm RC})$, then the expansion in $\tilde{K}(t)$ becomes equivalent to that in powers of the inverse dimensionless conductance $1/g_{\rm Z} \ll 1$. Assuming also that the charge relaxation time $\tau_{\rm RC}$ is much shorter than the electron dwell times in both quantum dots and the wires on both sides of the barriers (cf. Fig. 15.3), we recover the leading order correction to the result (15.48), i.e., we find [167]

$$G_{\rm WL}(T) \to G^{\rm WL}(T)\left\{1 - \frac{2}{g_{\rm Z}}\int_0^\infty \frac{d\omega}{\omega}\left(\frac{\partial}{\partial\omega}\omega\coth\frac{\omega}{2T}\right)\right.$$
$$\left.\times\,{\rm Re}\left[2\frac{\mathcal{C}_{\rm L}(\omega)}{\mathcal{C}_{\rm L}(0)} + 2\frac{\mathcal{C}_{\rm R}(\omega)}{\mathcal{C}_{\rm R}(0)} - \frac{\mathcal{C}_{\rm L}(-\omega)\mathcal{C}_{\rm R}(\omega)}{\mathcal{C}_{\rm L}(0)\mathcal{C}_{\rm R}(0)}\right]\right\} \quad (15.58)$$

For simplicity, here we omitted the coordinate dependence of the Cooperons $\mathcal{C}_1$ and $\mathcal{C}_2$ assuming that the central barrier is sufficiently small. One can observe that the Cooperons $\mathcal{C}_j(\omega)$ decay at higher frequencies $\omega \gtrsim 1/\tau_{{\rm D}j}$ and saturate at lower ones $\omega \lesssim 1/\tau_{{\rm D}j}$, where, as before, $\tau_{\mathrm{D}j}$ denote the electron dwell times in the leads. Hence, at higher temperatures $T \gtrsim 1/\tau_{\mathrm{D}j}$, the first correction $\propto 1/g_{\mathrm{Z}}$ in Eq. (15.58) can be safely ignored, implying that Eq. (15.48) remains accurate. Estimating the integral in Eq. (15.58) at lower temperatures $T \lesssim 1/\tau_{\mathrm{D}j}$, one can rewrite this equation as

$$G_{\mathrm{WL}}(T) \to G^{\mathrm{WL}}(T)\left\{1 + \frac{6}{g_{\mathrm{Z}}}\ln(T\tilde{\tau}_{\mathrm{D}})\right\} \tag{15.59}$$

where $\tilde{\tau}_{\mathrm{D}} = (\tau_{\mathrm{D}1}^2\tau_{\mathrm{D}2}^2/\min\{\tau_{\mathrm{D}1},\tau_{\mathrm{D}2}\})^{1/3}$. Hence, for $g_{\mathrm{Z}} \gg 1$, the Coulomb correction to the result (15.48) remains small and can be neglected provided the temperature exceeds an exponentially small value

$$T^* \sim \frac{1}{\tilde{\tau}_{\mathrm{D}}}e^{-g_{\mathrm{Z}}/6} \tag{15.60}$$

In other words, for $g_{\mathrm{Z}} \gg 1$, Eq. (15.51) remains applicable within the temperature interval $T^* \lesssim T \lesssim 1/\tau_{\mathrm{D}}$. In the limit of very low temperatures $T \lesssim T^*$, the Coulomb correction to G_{WL} becomes appreciable and it is necessary to take higher orders in $1/g_{\mathrm{Z}}$ into account. One can demonstrate that in the limit $T \to 0$, the exact expression for the weak localization correction (15.46) tends to zero as a power law,

$$G_{\mathrm{WL}} \propto (T\tau_{\mathrm{D}})^{6/g_{\mathrm{Z}}} \tag{15.61}$$

which is consistent with the expansion (15.59).

To supplement our analytic analysis, one can also evaluate the weak localization correction (15.46) numerically. The result of this calculation for the system of symmetric quantum dots is displayed in Fig. 15.5. This figure illustrates our main observations, which can be briefly summarized as follows. In the presence of electron–electron interactions, the weak localization correction $G_{\mathrm{WL}}(T)$ to the conductance of a tunnel barrier is suppressed below its non-interacting value $G_{\mathrm{WL}}^{(0)}$ by two physically different mechanisms—electron dephasing and Coulomb blockade. At sufficiently high temperatures $T\tau_{\mathrm{D}} \gtrsim 1$, only the first of these mechanisms is important, while at lower temperatures, both dephasing and Coulomb effects contribute on equal footing. The relative contributions from these two mechanisms can be clearly identified in Fig. 15.5.

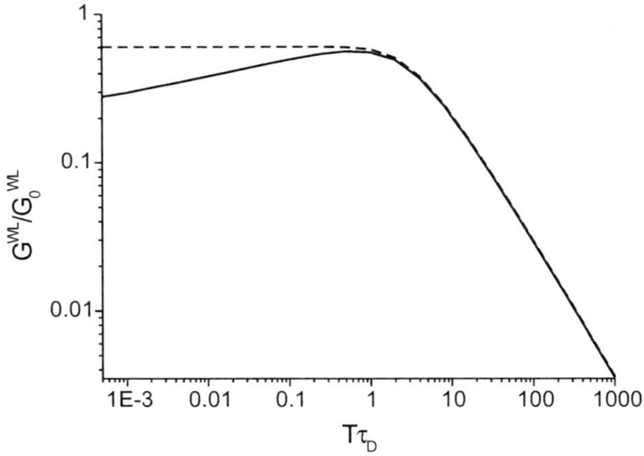

Figure 15.5 Temperature dependence of the weak localization correction $G_{\mathrm{WL}}(T)$ for a tunnel barrier placed in between two quantum dots, as shown in Fig. 15.3b. The parameters are chosen as follows: $\tau_{\mathrm{D1}} = \tau_{\mathrm{D2}} = \tau_{\mathrm{D}}$, $\tau_{\mathrm{RC}} = 0.001\tau_{\mathrm{D}}$ and $g_{\mathrm{Z}} = 50$. The exact expression (15.46) is indicated by the solid line, while the approximate expression (15.48) (which only includes dephasing and ignores Coulomb blockade) is shown by the dashed line.

To conclude our analysis, we have explicitly demonstrated that electron dephasing in quantum dots is produced by the "classical" part of the fluctuating field φ^+, which mediates electron–electron interactions and is accounted for by the correlation function $F(t)$ (15.35). Fluctuations of the "quantum" field φ^- (described by the response function $\tilde{K}(t)$ (15.36)) are, in contrast, totally irrelevant for dephasing being responsible only for the Coulomb blockade correction to G_{WL}. Note that this important conclusion fully matches with that already obtained in Chapter 14 for spatially extended disordered conductors within an entirely different theoretical framework. There it was also demonstrated that only the S_{I}-term in the effective action describing the "classical" part of the fluctuating scalar potential V^+ contributes to electron dephasing, while the S_{R}-term involving the "quantum" field V^- is completely irrelevant for this effect. In this respect, Eq. (15.48) derived here for a tunnel barrier is directly analogous to the results (14.60) and (14.67) obtained in Chapter 14.

15.4 Weak Localization in Arrays of Quantum Dots

Let us now extend our analysis to more complicated disordered structures, which—without any loss of generality—can be modeled by arrays of quantum dots. Similar to Chapter 13, in what follows we will merely restrict our consideration to the case of one-dimensional arrays of quantum dots, as shown in Fig. 13.1. For simplicity, we will assume that our array consists of $N - 1$ identical quantum dots separated by N identical barriers. Each of these barriers has Ohmic resistance R, dimensionless conductance $g = 2\pi/e^2R$, and Fano factor β. Likewise, each quantum dot is characterized by its mean level spacing δ, dwell time $\tau_{\rm D} = 2\pi/g\delta$, and two electron dephasing times τ_φ and $\tau_{\rm H}$. As before, the latter parameter defines the dephasing time due to an externally applied magnetic field H [44]

$$\tau_{\rm H} = b\frac{c^2}{e^2 v_{\rm F} d^2 \min\{d, \ell\}\, H^2} \tag{15.62}$$

where b is a geometry-dependent numerical prefactor, c is the speed of light, $v_{\rm F}$ is the Fermi velocity, d is the quantum dot size, and ℓ is the elastic mean free path. Similar to Section 15.3.2, at this stage the dephasing time τ_φ is introduced as a phenomenological parameter. Later on it will be attributed to the effect of electron–electron interactions and explicitly evaluated.

15.4.1 Non-interacting Limit

For a moment, however, let us ignore the effect of electron–electron interactions. In order to proceed, we will again analyze the Cooperon $\mathcal{C}_{nm}(\omega)$, whose spatial dependence is now accounted for by integer numbers n and m enumerating quantum dots in the array. Accordingly, the Cooperon satisfies the discrete version of the diffusion equation (14.14), which reads

$$\left(-\mathrm{i}\omega + \frac{1}{\tau_\varphi} + \frac{1}{\tau_{\rm H}}\right)\mathcal{C}_{nm} + \frac{\delta g}{4\pi}\left[2\mathcal{C}_{nm} - \mathcal{C}_{n-1,m} - \mathcal{C}_{n+1,m}\right] = \delta_{nm} \tag{15.63}$$

The solution to Eq. (15.63) corresponding to boundary conditions $\mathcal{C}_{0m} = \mathcal{C}_{n0} = \mathcal{C}_{Nm} = \mathcal{C}_{nN} = 0$ has the form

$$\mathcal{C}_{nm}(\omega) = \frac{2}{N}\sum_{q=1}^{N-1} \frac{\sin\frac{\pi q n}{N}\sin\frac{\pi q m}{N}}{-\mathrm{i}\omega + \frac{1}{\tau_\varphi} + \frac{1}{\tau_\mathrm{H}} + \frac{1-\cos[\pi q/N]}{\tau_\mathrm{D}}} \tag{15.64}$$

An important step in the whole analysis is to derive the weak localization correction $G^{(0)}_{\mathrm{WL}n}$ to the conductance of a single barrier. This goal can be accomplished in several ways. One of them is to employ the perturbation theory in the tunneling Hamiltonian, as we already did in Section 15.3, and to extend it to higher orders. Alternatively, one can apply the scattering matrix approach combined with the so-called σ−model formalism [165]. Technically both these methods are rather involved; therefore, here we only present the final result, which, however, turns out to be quite simple. The weak localization correction to the conductance of the n-th barrier in our array reads [165]

$$G^{(0)}_{\mathrm{WL}n} = -\frac{e^2}{2\pi\tau_\mathrm{D}}\{\beta\left[\mathcal{C}_{n-1,n}(0) + \mathcal{C}_{n,n-1}(0)\right] \\ + (1-\beta)\left[\mathcal{C}_{n,n}(0) + \mathcal{C}_{n-1,n-1}(0)\right]\} \tag{15.65}$$

This expression can be considered as a direct generalization of Eq. (15.19) derived in the previous section for a tunnel barrier between two metallic leads. In order to recover Eq. (15.19) from the formula (15.65), it is necessary to put $\beta = 1$, in which case the second line of Eq. (15.65) vanishes and the whole result reduces to that derived in the paper [83]. In addition, one should bear in mind that in Section 15.3, the barrier conductance was assumed to be much smaller as compared to that of the leads. For the structure considered here, this regime is reached if one assumes that the n-th barrier conductance g_n is much smaller than those of all other barriers, i.e., $g_n \ll g$. Taking into account g_n in Eq. (15.64) and solving this equation perturbatively in g_n, one readily finds $\mathcal{C}_{n-1,n}(0) = \mathcal{C}_{n,n-1}(0) \propto g_n^2\mathcal{C}_{n-1,n-1}(0)\mathcal{C}_{n,n}(0)$, and Eq. (15.65) indeed reduces to Eq. (15.19).

Let us now determine the weak localization correction $G^{(0)}_{\mathrm{WL}}$ to the conductance of the whole array. This correction is defined as

$$G^{(0)}_{\mathrm{WL}} = \frac{1}{NR + \sum_{n=1}^{N}\delta R_{\mathrm{WL}n}} - \frac{1}{NR} \approx -\frac{1}{N^2R^2}\sum_{n=1}^{N}\delta R_{\mathrm{WL}n}, \tag{15.66}$$

where $\delta R_{\mathrm{WL}n} = -R^2 G^{(0)}_{\mathrm{WL}n}$ is the weak localization correction to the resistance of the n-th barrier. Hence, we obtain

$$G^{(0)}_{\mathrm{WL}} = \frac{1}{N^2}\sum_{n=1}^{N} G^{(0)}_{\mathrm{WL}n} \tag{15.67}$$

Combining Eqs. (15.67), (15.65), and (15.64), we arrive at the result

$$G^{(0)}_{\mathrm{WL}} = -\frac{e^2}{\pi N^2}\sum_{q=1}^{N-1} \frac{1-\beta+\beta\cos\frac{\pi q}{N}}{1-\cos\frac{\pi q}{N}+\frac{\tau_{\mathrm{D}}}{\tau_{\mathrm{H}}}+\frac{\tau_{\mathrm{D}}}{\tau_{\varphi}}} \tag{15.68}$$

The sum in Eq. (15.68) can be evaluated exactly, and we obtain

$$G^{(0)}_{\mathrm{WL}} = -\frac{e^2}{\pi N^2}\left[\left(N\frac{1+w^{2N}}{1-w^{2N}} - \frac{1+w^2}{1-w^2}\right)\frac{\beta(1+w^2)+2(1-\beta)w}{1-w^2} - (N-1)\beta\right], \tag{15.69}$$

where

$$w = 1+\frac{\tau_{\mathrm{D}}}{\tau_{\mathrm{H}}}+\frac{\tau_{\mathrm{D}}}{\tau_{\varphi}} - \sqrt{\left(1+\frac{\tau_{\mathrm{D}}}{\tau_{\mathrm{H}}}+\frac{\tau_{\mathrm{D}}}{\tau_{\varphi}}\right)^2 - 1} \tag{15.70}$$

Let us consider several specific limits. In the fully coherent regime $1/\tau_{\mathrm{H}} = 1/\tau_{\varphi} = 0$, one has $w = 1$ and

$$G^{(0)}_{\mathrm{WL}} = -\frac{e^2}{\pi}\left[\frac{1}{3} - \frac{\beta}{N} + \frac{1}{N^2}\left(\beta - \frac{1}{3}\right)\right] \tag{15.71}$$

In the limit $N \to \infty$, this correction approaches the result for a long diffusive wire (14.30).

In the diffusive limit $\tau_{\mathrm{D}} \ll \tau_{\mathrm{H}}$ and $\tau_{\varphi} \ll 2N^2\tau_{\mathrm{D}}/\pi^2$, Eq. (15.69) reduces to the form

$$G^{(0)}_{\mathrm{WL}} = -\frac{e^2}{\pi L}\sqrt{\frac{\tilde{D}\tau_{\varphi}\tau_{\mathrm{H}}}{\tau_{\varphi}+\tau_{\mathrm{H}}}} \tag{15.72}$$

where $L = Nd$ is the total system length and $\tilde{D} = d^2/2\tau_{\mathrm{D}}$ is the effective diffusion constant. Again we observe that the result (15.72) is equivalent to that derived for a metallic wire (14.32).

Finally, by setting $N = 2$ in Eq. (15.69), we immediately recover the weak localization correction to the conductance of a single quantum dot [69]

$$G^{(0)}_{\mathrm{WL}} = -\frac{e^2(1-\beta)}{4\pi}\frac{1}{1+\frac{\tau_{\mathrm{D}}}{\tau_{\mathrm{H}}}+\frac{\tau_{\mathrm{D}}}{\tau_{\varphi}}} \tag{15.73}$$

Note that this correction vanishes in the tunneling limit $\beta = 1$.

15.4.2 Electron Dephasing by Coulomb Interactions

One of the main conclusions reached in the previous section is that the electron decoherence time is fully determined by fluctuations of the phase fields φ^+ (and the corresponding correlation function $F(t)$), whereas the phases φ^- (and the response function $\tilde{K}(t)$) are irrelevant for τ_φ, causing only a weak Coulomb correction to $G_{\rm WL}$. This conclusion is general, being independent of a number of scatterers in our system. Thus, in order to evaluate the decoherence time for interacting electrons in arrays of quantum dots, it is sufficient to account for the fluctuating voltages V^+ ignoring the fields V^-.

In order to proceed, it is convenient to introduce the time-dependent Cooperon $C^{(0)}_{nm}(t)$, which is defined by means of the Fourier transformation of the Cooperon (15.64), i.e.,

$$C_{nm}(t) = \int \frac{d\omega}{2\pi}\, e^{-i\omega t} C_{nm}(\omega) \tag{15.74}$$

The Cooperon $C_{nm}(t)$ can be represented in the form $C_{nm}(t) = C^{\rm bulk}_{n-m}(t) - C^{\rm bulk}_{n+m}(t)$ where

$$C^{\rm bulk}_{n-m}(t) = \frac{1}{N}\sum_{q=1}^{N-1} \int \frac{d\omega}{2\pi}\, e^{-i\omega t} \frac{\cos\frac{\pi q(n-m)}{N}}{-i\omega + \frac{1-\cos\frac{\pi q}{N}}{\tau_{\rm D}}} \tag{15.75}$$

is the Cooperon in a very long array far from its edges. Provided the total number of barriers in our array is large $N \gg 1$, the term $C^{\rm bulk}_{n+m}(t)$ can be ignored and we obtain $C_{nm}(t) \approx C^{\rm bulk}_{n-m}(t)$.

Let us represent the Cooperon $C^{\rm bulk}_{n-m}(t)$ as a sum over all possible integer-valued paths $\nu(\tau)$ starting and ending at, respectively, the m–th and the n–th dots (i.e., $\nu(0) = m$, $\nu(t) = n$) and describing successive jumps from one dot to another at times t_j. The corresponding expression can be easily recovered if one expands Eq. (15.75) in powers of $\tau_{\rm D}^{-1}\cos[\pi q/N]$ with subsequent summation over q in every order of this expansion. Bearing in mind that electrons acquire additional phase factors in the presence of a fluctuating potential V^+_ν, we obtain

$$C_{nm}(t) = \sum_{k=|n-m|}^{\infty} \sum_{\nu(\tau)} \Bigg|_{\nu(0)=m}^{\nu(t)=n} \frac{1}{(2\tau_{\rm D})^k} \int_0^t dt_k \int_0^{t_k} dt_{k-1} \ldots \int_0^{t_3} dt_2 \int_0^{t_2} dt_1$$
$$\times\, e^{-\frac{t-t_k}{\tau_{\rm D}}} e^{-\frac{t_k-t_{k-1}}{\tau_{\rm D}}} \ldots e^{-\frac{t_2-t_2}{\tau_{\rm D}}} e^{-\frac{t_2-t_1}{\tau_{\rm D}}} e^{-\frac{t_1}{\tau_{\rm D}}} e^{i\int_0^t d\tau \left[eV^+_{\nu(\tau)}(\tau) - eV^+_{\nu(t-\tau)}(\tau)\right]} \tag{15.76}$$

Averaging this formula over Gaussian fluctuations of voltages V^+ and utilizing the symmetry of the voltage correlator $\langle V^+_{\nu_1}(\tau_1)V^+_{\nu_2}(\tau_2)\rangle = \langle V^+_{\nu_2}(\tau_1)V^+_{\nu_1}(\tau_2)\rangle$, we get

$$C_{nm}(t) = \sum_{k=|n-m|}^{\infty} \sum_{\nu(\tau)} \Bigg|_{\nu(0)=m}^{\nu(t)=n} \frac{e^{-t/\tau_{\rm D}}}{(2\tau_{\rm D})^k} \int_0^t dt_k \int_0^{t_k} dt_{k-1} \dots \int_0^{t_3} dt_2 \int_0^{t_2} dt_1$$

$$\times\, e^{-e^2\int_0^t d\tau_1 \int_0^t d\tau_2 \left[\langle V^+_{\nu(\tau_1)}(\tau_1)V^+_{\nu(\tau_2)}(\tau_2)\rangle - \langle V^+_{\nu(\tau_1)}(\tau_1)V^+_{\nu(t-\tau_2)}(\tau_2)\rangle\right]} \quad (15.77)$$

The expression for the voltage–voltage correlator in Eq. (15.77) can be easily recovered from the action for the fluctuating fields $V^\pm$ [166]

$$S = \frac{2}{N}\sum_{q=1}^{N}\int \frac{d\omega}{2\pi}\Bigg[4C\left(1-\cos\frac{\pi q}{N}\right) + C_g$$

$$+ \frac{g\tau_D e^2}{\pi}\frac{1-\cos\frac{\pi q}{N}}{-i\omega\tau_{\rm D}+1-\cos\frac{\pi q}{N}}\Bigg] V_q^+(\omega)V_q^-(-\omega)$$

$$+ \frac{2i}{N}\sum_{q=1}^{N}\int\frac{d\omega}{2\pi}\frac{g\tau_{\rm D}^2 e^2}{\pi}\frac{\left(1-\cos\frac{\pi q}{N}\right)\omega\coth\frac{\omega}{2T}}{\omega^2\tau_{\rm D}^2+\left(1-\cos\frac{\pi q}{N}\right)^2} V_q^-(\omega)V_q^-(-\omega), \quad (15.78)$$

where we defined

$$V_q^\pm(\omega) = \sum_{n=1}^{N-1}\int dt\, \sin\frac{\pi q}{N}\, e^{i\omega t}\, V_n^\pm(t) \quad (15.79)$$

The action (15.78) represents a discrete version of the standard quadratic action (9.57), which determines quantum dynamics of the fluctuating electromagnetic fields in a metallic wire. The latter action was already derived in Chapter 9 within the random phase approximation; therefore, there is no need to go into further details here. Both correlators $\langle V^+V^+\rangle$ (F-function) and $\langle V^+V^-\rangle$ ($\tilde{K}$-function) responsible, respectively, for electron decoherence and Coulomb blockade can be determined from Eq. (15.78); cf., e.g., Eqs. (9.64) and (9.65). As we already explained, only the first of these two correlation functions is relevant in the context of our present analysis of electron dephasing. It reads

$$\langle V_n^+(t_1)V_m^+(t_2)\rangle = \frac{2}{N}\sum_{q=1}^{N-1}\int\frac{d\omega}{2\pi}e^{-i\omega(t_1-t_2)}\frac{\tau_{\rm D}^2\,\omega\coth\frac{\omega}{2T}}{\omega^2\tau_{\rm D}^2+\left(1-\cos\frac{\pi q}{N}\right)^2}$$

$$\times \frac{\frac{ge^2}{\pi}\left(1-\cos\frac{\pi q}{N}\right)\sin\frac{\pi q n}{N}\sin\frac{\pi q m}{N}}{\left|4C\left(1-\cos\frac{\pi q}{N}\right)+C_g+\frac{g\tau_{\rm D}e^2}{\pi}\frac{1-\cos\frac{\pi q}{N}}{-i\omega\tau_{\rm D}+1-\cos\frac{\pi q}{N}}\right|^2} \quad (15.80)$$

It is straightforward to verify that in the continuous limit $N \to \infty$ and for sufficiently low frequencies $\omega \ll 1/\tau_D$, the correlator (15.80) reduces to that of a diffusive metal determined by Eq. (9.64) with the Drude dielectric function (8.15).

In what follows, we restrict our analysis to diffusive paths $\nu(\tau)$, in which case one can approximate

$$\langle V^+_{\nu(\tau_1)}(\tau_1)V^+_{\nu(\tau_2)}(\tau_2)\rangle \approx \sum_{n,m=1}^{N-1} \frac{\langle V_n^+(\tau_1)V_m^+(\tau_2)\rangle D_{nm}(|\tau_1-\tau_2|)}{N-1} \tag{15.81}$$

where $D_{nm}(\tau)$ is the diffuson, which exactly coincides with the Cooperon for non-interacting electrons $C^{(0)}_{n,m}(t)$ in the absence of the magnetic field $H \to 0$, i.e.,

$$D_{nm}(t) = C^{(0)}_{nm}(t) = \frac{2}{N}\sum_{q=1}^{N-1}\int \frac{d\omega}{2\pi}\, e^{-i\omega t}\, \frac{\sin\frac{\pi q n}{N}\sin\frac{\pi q m}{N}}{-i\omega + \frac{1-\cos\frac{\pi q}{N}}{\tau_D}} \tag{15.82}$$

Substituting Eq. (15.81) into Eq. (15.77), we obtain

$$C_{nm}(t) \approx C^{(0)}_{nm}(t)\, e^{-\mathcal{F}(t)} \tag{15.83}$$

where the function

$$\mathcal{F}(t) = \frac{e^2}{N-1}\sum_{n,m=1}^{N-1}\int_0^t dt_1 dt_2 \langle V_n^+(t_1)V_m^+(t_2)\rangle$$
$$\times \left[D_{nm}(|t_1-t_2|) - D_{nm}(|t-t_1-t_2|)\right] \tag{15.84}$$

controls the Cooperon decay in time, i.e., describes electron decoherence for our one-dimensional array of quantum dots. The weak localization correction G_{WL} in the presence of electron–electron interactions is then recovered by substituting the result (15.83) into Eqs. (15.67) and (15.65).

The electron dephasing time τ_φ is naturally determined from the equation $\mathcal{F}(\tau_\varphi) \simeq 1$. From Eq. (15.84) with a good accuracy, we obtain

$$\frac{1}{\tau_\varphi} = \frac{e^2}{N-1}\sum_{n,m=1}^{N-1}\int d\tau \langle V_n^+(\tau)V_m^+(0)\rangle D_{nm}(\tau) \tag{15.85}$$

Combining this formula with Eqs. (15.80) and (15.82), in the most interesting limit $T \to 0$ and for $\tau_D \gg R(4C + C_g)$, we find

$$\frac{1}{\tau_{\varphi 0}} = \frac{1}{2g\tau_D(N-1)}\sum_{q=1}^{N-1}\ln\frac{2e^2}{\delta\left(4C\left(1-\cos\frac{\pi q}{N}\right)+C_g\right)}$$

which yields

$$\tau_{\varphi 0} = \frac{2g\tau_{\mathrm{D}}}{\ln(4\tilde{E}_{\mathrm{C}}/\delta)} = \frac{4\pi}{\delta \ln(4\tilde{E}_{\mathrm{C}}/\delta)} \qquad (15.86)$$

where $\tilde{E}_{\mathrm{C}}$ is the effective charging energy of a single quantum dot. In the limit $C_g \gg C$, it reads $\tilde{E}_{\mathrm{C}} = e^2/2C_g$ for $C_g \gg C$, while for $C_g \ll C$, one has $\tilde{E}_{\mathrm{C}} = e^2/4C$.

We observe that apart from an unimportant numerical factor of order one, the result for $\tau_{\varphi 0}$ (15.86) in one-dimensional array of quantum dots coincides with the exact expression (15.57) derived in Section 15.3 for the case of two quantum dots. Thus, we arrive at an important conclusion: In the limit of low temperatures, the electron decoherence time $\tau_{\varphi 0}$ is practically independent of the number of scatterers in the conductor (provided the latter number exceeds two) and is essentially determined by *local* properties of the system.

The electron dephasing length $L_\varphi = \sqrt{\tilde{D}\tau_\varphi}$ in our array can be recovered if we define the effective diffusion coefficient

$$\tilde{D} = \frac{d^2}{2\tau_{\mathrm{D}}} = \frac{d^2 g\delta}{4\pi} \qquad (15.87)$$

Combining Eqs. (15.86) and (15.87), at $T = 0$ we obtain

$$L_{\varphi 0} = \sqrt{\tilde{D}\tau_{\varphi 0}} = d\sqrt{g/\ln(4\tilde{E}_{\mathrm{C}}/\delta)} \qquad (15.88)$$

At nonzero temperatures, thermal fluctuations provide an additional contribution to the dephasing rate $1/\tau_\varphi$. Again substituting Eqs. (15.80) and (15.82) into Eq. (15.85), we get

$$\frac{1}{\tau_\varphi(T)} \simeq \frac{1}{\tau_{\varphi 0}} + \frac{T}{2\pi g}\min\{N, N_\varphi\}, \qquad (15.89)$$

where $N_\varphi = L_\varphi/d \sim \sqrt{\tau_\varphi/\tau_{\mathrm{D}}}$ is the number of quantum dots within the length L_φ. We observe that for sufficiently small $N < N_\varphi$ (but still $N \gg 1$), the dephasing rate increases linearly both with temperature and the number of N. At larger $N > \sqrt{g/\ln[4\tilde{E}_{\mathrm{C}}/\delta]}$ and/or at high-enough temperatures, N_φ becomes smaller than N and Eq. (15.89) for τ_φ should be resolved self-consistently. In this case, we obtain

$$\tau_\varphi \simeq (2\pi g\sqrt{\tau_{\mathrm{D}}}/T)^{2/3}, \qquad (15.90)$$

cf. Eq. (14.57). Equation (15.89) also allows to estimate the temperature $T^* \simeq 2\pi g/[\tau_{\varphi 0}\min\{N, N_\varphi\}]$ at which the crossover to the temperature-independent regime (15.86) occurs. We find

$$T^* \simeq \frac{\pi \ln[4\tilde{E}_C/\delta]}{N\tau_D}, \quad N \lesssim \sqrt{\frac{g}{\ln[4\tilde{E}_C/\delta]}},$$
$$T^* \simeq \frac{\pi \ln^{3/2}[4\tilde{E}_C/\delta]}{\tau_D\sqrt{2g}}, \quad N \gtrsim \sqrt{\frac{g}{\ln[4\tilde{E}_C/\delta]}} \tag{15.91}$$

The above analysis and conclusions can be generalized further to the case of two- and three-dimensional structures. This generalization is straightforward; therefore, we will not discuss it here. At $T \to 0$, one again arrives at the same result for the dephasing time $\tau_{\varphi 0}$ (15.86). This is, of course, by no means surprising since, as we already established, $\tau_{\varphi 0}$ is determined only by the local properties of the system.

15.4.3 Good Metals and Strongly Disordered Conductors

Let us discuss the relation between our present results and those derived in Chapter 14 for the case of good metals by means of a different approach; see, e.g., Eqs. (14.71) and (14.73). Expressing the dot mean level spacing via the average dot size d as $\delta = 1/\nu_0 d^3$ (where $\nu_0 = mp_F/2\pi^2$ is the electron density of states at the Fermi level), we arrive at the diffusion constant (15.87) in the form

$$\tilde{D} = \frac{g}{4\pi \nu_0 d} \tag{15.92}$$

Below we consider two different physical limits of (a) good metals and (b) strongly disordered (granular) conductors. For the model (a), we assume that quantum dots are in a very good contact with each other. In this case, g scales linearly with the contact area $\mathcal{A} = \gamma d^2$, where γ is a numerical factor of order (typically smaller than) one, whose particular value depends on geometry. For good metals, most conducting channels in such contacts can be considered open. Hence, we set $g = p_F^2 \mathcal{A}/2\pi$ and

$$\tilde{D} = \gamma v_F d/4, \tag{15.93}$$

i.e., $\tilde{D} \propto d$. If most channels are not fully transparent, then the factor γ in Eq. (15.93) also accounts for their transmission values. Comparing Eq. (15.93) with the standard definition of D for a bulk diffusive conductor, $D = v_F \ell_e/3$, we immediately observe that within our model the average dot size is comparable to the elastic mean free path, $\ell_e \sim \gamma d$ or simply $\ell_e \sim d$, as it should be for weakly disordered metals.

Expressing $\tau_{\varphi 0}$ (15.86) via D, in this limit we get

$$\tau_{\varphi 0} = \frac{64}{\pi \gamma^3} \frac{m^2}{v_F^2} \frac{\tilde{D}^3}{\ln(\tilde{D}/D_{c1})} \tag{15.94}$$

where m is the electron mass and D_{c1} is constant, which depends on $\tilde{E}_C$. Setting, e.g., $\tilde{E}_C = e^2/2d$, one obtains $D_{c1}^{-1} = 4\pi\sqrt{2}ev_0^{3/2}$.

Note that apart from an unimportant numerical prefactor and the logarithm in the denominator, Eq. (15.94) coincides with that derived in Chapter 14 for a bulk diffusive metal within the framework of a completely different approach; cf. Eq. (14.73) with $d = 3$. Within that quasiclassical approach, local properties of the model remained not fully defined. For this reason, as we already explained in Chapter 14, in the corresponding integrals in the expressions for $\tau_{\varphi 0}$, we could not avoid using an effective high-frequency cutoff procedure that correctly captures the leading dependence $\tau_{\varphi 0} \propto D^3$ and misses only a weak (logarithmic) extra dependence on D in the denominator of Eq. (15.94). Our present approach is divergence-free and, hence, it does not require any cutoffs.

Note that Eq. (15.94) is consistent with the quasiclassical results derived in Chapter 14 not only for bulk metals but also for quasi-one-dimensional and quasi-two-dimensional metallic conductors. In order to establish the correspondence between these results and Eq. (15.94), instead of $\mathcal{A} \sim d^2$, one should simply set $\mathcal{A} \sim da$ for quasi-two-dimensional and $\mathcal{A} \sim a^2$ for quasi-one-dimensional conductors, where a denotes the transversal size a of our array, which now remains smaller than d. Then Eq. (15.86) yields $\tau_{\varphi 0} \propto D^2/\ln D$ and $\tau_{\varphi 0} \propto D/\ln D$, respectively, in two-dimensional and one-dimensional cases. Again, up to the term $\ln D$, these dependencies coincide with the ones derived in Chapter 14; cf. Eq. (14.73) with $d = 2$ and $d = 1$ as well as Eq. (14.71).

Thus, the agreement between the results obtained here and in Chapter 14 allows us to formally justify the high-frequency cutoff procedure employed in the course of our derivation of Eqs. (14.71) and (14.73). Furthermore, the analysis developed in this chapter sheds light on deep physical grounds behind this procedure. Indeed, as we have already demonstrated in Section 15.2, no dephasing by electron–electron interactions occurs for electrons propagating in a single quantum dot provided fluctuating voltages drop only at the barriers, not in the interior parts of the dot. In other words, no contribution to dephasing can be expected from frequencies above the inverse dwell time $1/\tau_{\rm D}$ and wave vectors above the inverse dot size $1/d$. Since $d \sim \ell_e$ and, hence, the dot dwell time is of the order of the elastic time in the Drude model for a weakly disordered metal, $\tau_{\rm D} \sim \tau_e$, the corresponding frequency integrals, e.g., in Eq. (14.68) should indeed be performed up to $\omega \sim 1/\tau_e$, just as we did in Chapter 14.

Now let us turn to the model (b) of strongly disordered or granular conductors. In contrast to the situation (a), we will assume that the contact between dots (grains) is rather poor, and inter-grain electron transport may occur only via a limited number of conducting channels. In this case, the dimensionless conductance g can be approximated by some $\mathcal{A}$-independent constant $g = g_c$. Substituting g_c instead of g into Eq. (15.92), we observe that in the case of strongly disordered structures, one can expect $\tilde{D} \propto 1/d$. Accordingly, for $\tau_{\varphi 0}$ (15.86), one finds

$$\tau_{\varphi 0} = \frac{g_c^3}{32\pi^2 \nu_0^2 \tilde{D}^3 \ln(D_{c2}/D)} \tag{15.95}$$

where D_{c2} again depends on $\tilde{E}_{\rm C}$. For $\tilde{E}_{\rm C} \approx e^2/2d$, we have $D_{c2}^{-1} = 2\pi\sqrt{2\nu_0}/\alpha$. Hence, the dependence of $\tau_{\varphi 0}$ on D for strongly disordered or granular conductors (15.95) is qualitatively different from that for sufficiently clean metals (15.94).

One can also roughly estimate the crossover between the regimes (a) and (b) by requiring the values of $\tilde{D} = \gamma v_{\rm F} d/4$ (15.93) and $\tilde{D} = g_c/4\pi\nu_0 d$ to be of the same order. This condition yields $(p_{\rm F} d)^2 \sim 2\pi g_c/\gamma$, and we arrive at the estimate for $\tilde{D}$ at the crossover

$$\tilde{D} \approx \frac{0.6\hbar}{m}\sqrt{\frac{g_c}{\gamma}} \tag{15.96}$$

For clarity, in Eq. (15.96) we restored the Planck constant $\hbar$ set equal to unity elsewhere in this chapter.

Let us try to qualitatively understand and interpret seemingly contradicting dependencies of $\tau_{\varphi 0}$ on D predicted above for weakly and strongly disordered (or granular) conductors. While the trend "less disorder–less decoherence" (15.94) in weakly disordered conductors looks quite natural, the opposite trend "more disorder–less decoherence" (15.95) in strongly disordered structures may, at the first sight, appear counterintuitive and requires a comment. The latter dependence may indicate that with increasing disorder, electrons spend more time in the areas with fluctuating in time but spatially uniform potentials. According to our analysis in Section 15.2, such fluctuating potentials do not dephase and thus $\tau_{\varphi 0}$ gets effectively increased. In other words, in this case the corresponding effective dwell time τ_{D} in Eq. (15.86) becomes longer with increasing disorder and, hence, the electron decoherence time $\tau_{\varphi 0}$ does so, too. We will continue with these arguments below in the next section.

15.4.4 Comparison with Experiments

Having in mind the fundamental importance of the effect of low-temperature saturation of the electron dephasing time in disordered conductors, it is instructive to present a brief overview of the experimental situation and to compare the above results for τ_φ with the available experimental data.

Numerous experiments investigating the low-temperature behavior of the weak localization correction to the conductance were performed over the past decades. These experiments embrace a variety of disordered conductors, including both weakly and strongly disordered metals of different dimensions, semiconductors, quantum dots, carbon nanotubes, etc. In *all* these experiments, the effect of low-temperature saturation of the electron decoherence time has been observed and, hence, it is presently considered firmly established and indisputably existing phenomenon. At the same time, both the temperature region where saturation of $\tau_\varphi(T)$ is observed and the saturated value $\tau_{\varphi 0}$ may vary by several decades from experiment to experiment and from sample to sample. Typical dependence of τ_φ on temperature extracted from the weak localization data is presented in Fig. 15.6.

Various authors proposed different explanations of the effect, most of which are *extrinsic* in nature as opposed to the *intrinsic* mechanism of electron–electron interactions studied above. Commonly discussed extrinsic mechanisms of low-temperature

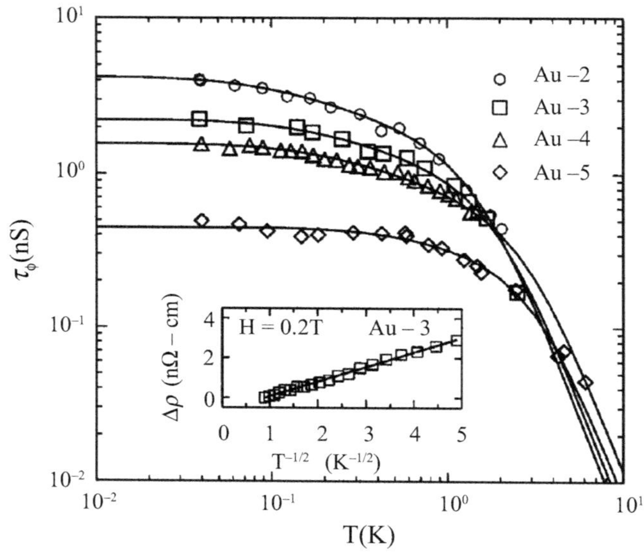

Figure 15.6 Typical dependence of the electron dephasing time τ_φ on temperature extracted from weak localization measurements for four Au wires. The inset shows the temperature dependence of the simultaneously measured interaction correction to the wire conductance. Reprinted with permission from Ref. [304], Copyright 1997, by the American Physical Society.

electron decoherence include scattering on magnetic impurities and/or two-level systems possibly available in the sample, effect of external electromagnetic radiation, unavoidable sample overheating at low-enough T, etc. Although the above extrinsic mechanisms can, under certain conditions, indeed contribute to electron dephasing, none of them is universal. Furthermore, in a vast majority of experiments, these mechanisms can be ruled out on purely experimental grounds. For instance, sample overheating can be probed, e.g., by simultaneous measuring of the interaction correction to the conductance in the same sample, as it was done by Mohanty et al. [304, 396]. Provided the latter correction keeps depending on T in the temperature interval where τ_φ already saturates (see the inset in Fig. 15.6), overheating can obviously be ruled out as a cause of this saturation. Likewise, the dephasing effect of magnetic impurities can be probed by applying a magnetic field sufficiently strong in order to fully polarize any magnetic moments should they exist in the sample [305] or by deliberately doping the sample with extra magnetic impurities [304, 396]. In many cases, the effect of magnetic impurities can be ruled out already by observing that the measured values of $\tau_{\varphi 0}$ would require unrealistically high concentrations of magnetic impurities in a given sample. A possible effect of spurious electromagnetic noise, if not fully eliminated by appropriate room shielding and other precautions, can be probed, e.g., by re-measuring the same samples in different laboratories, as it was done by Lin et al. [281]. For more detailed analysis of experimental checks of various extrinsic dephasing mechanisms, we refer the reader, e.g., to the review [303].

In contrast to all extrinsic dephasing mechanisms, the effect of electron–electron interactions is universal, i.e., it should inevitably show up in all experiments without any exception. Below we will compare the available experimental data for $\tau_{\varphi 0}$ with theoretical predictions and demonstrate that the observed τ_{φ}-saturation is indeed well explained by the effect of electron–electron interactions.

Let us first specifically address the experiments with quantum dots. Hackens et al. [186] have analyzed the values of $\tau_{\varphi 0}$ observed in their experiments with open quantum dots as well in earlier experiments by different groups [59, 91, 200, 332]. For all 14 samples reported by these five experimental groups, the values $\tau_{\varphi 0}$ were found to closely follow a simple dependence

$$\tau_{\varphi 0} \approx \tau_{\mathrm{D}} \tag{15.97}$$

This approximate scaling holds within the interval of dwell times τ_{D} of at least 3 decades, as it is shown in Fig. 15.7.

Let us now recall that our results for $\tau_{\varphi 0}$ [(15.56), (15.86)] scale practically linearly with the dwell time τ_{D}, which is essentially the scaling (15.97). Thus, electron–electron interactions are indeed responsible for the low-temperature saturation of the electron dephasing time observed in the experiments with quantum dots [59, 91, 186, 200, 332].

Note that naively one could have problems to reconcile this conclusion, e.g., with our qualitative arguments in Section 15.2, according to which no dephasing by electron–electron interactions

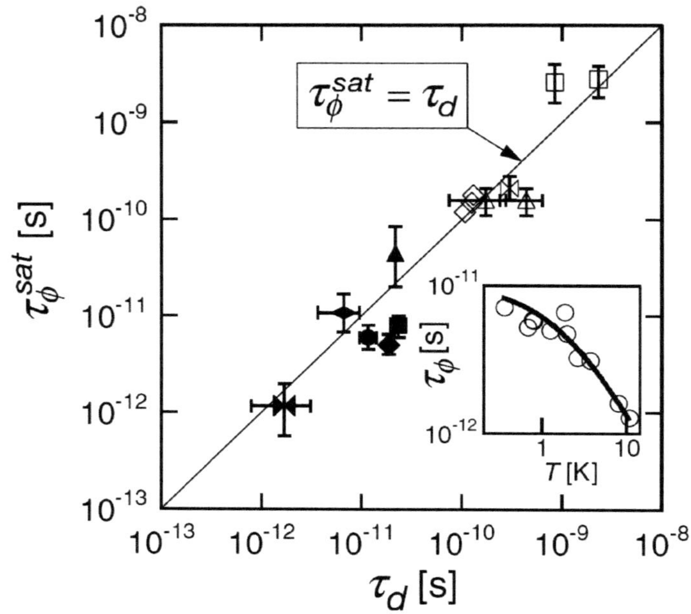

Figure 15.7 The saturated values of the electron dephasing time ($\tau_\varphi^{sat} \equiv \tau_{\varphi 0}$) in InGaAs and GaAs quantum dots plotted versus the dwell time ($\tau_d \equiv \tau_D$) [186]. The data points are taken from the experiments [59, 91, 186, 200, 332] and the solid line corresponds to Eq. (15.97). The inset shows the dependence of τ_φ on temperature for one of the samples [186]. Adapted with permission from Ref. [186], Copyright 2005, by the American Physical Society.

can occur in single quantum dots at any T. There exists no contradiction, however, since these arguments as well as the subsequent rigorous analysis remain applicable to single quantum dots only provided fluctuating voltages drop strictly across the two outer barriers. In realistic quantum dots studied in experiments [59, 91, 186, 200, 332], fluctuating voltages also penetrate inside the dot rather than just drop at its edges. One can easily mimic this situation by introducing additional scatterers inside the dot, in which case fluctuating voltages already do dephase.

For illustration, let us consider a strongly asymmetric double dot system, i.e., we replace, say, the left barrier in Fig. 15.1 by a small quantum dot with the electron flight time $\tau_{\rm fl} \ll \tau_{\rm D}$. The left dot then models a barrier of a finite length in a single dot configuration of Fig. 15.1. Employing Eq. (15.48), we evaluate the weak localization correction and again arrive at Eq. (15.51) with $\tau_{\rm D}$ being replaced by $\tau_{\rm fl}$. Extracting then the decoherence time $\tau_{\varphi 0}$ with the aid of Eq. (15.53), we obtain

$$\tau_{\varphi 0} \approx \frac{\tau_{\rm D}}{\left(\frac{\tau_{\rm fl}}{4e^{-\gamma}\tau_{\rm RC}}\right)^{2/g_{\rm Z}} \frac{1}{\sqrt{u(g_{\rm Z})}} - 1} \tag{15.98}$$

for $\tau_{\rm fl} \gtrsim \tau_{\rm RC}$ and vanishing decoherence rate in the opposite limit $\tau_{\rm fl} \ll \tau_{\rm RC}$. Provided the denominator in Eq. (15.98) is not very large (typically of order one), we arrive at Eq. (15.97). We also note that $\tau_{\varphi 0}$ in a single dot (15.98) is always longer than that in the system of two dots (15.55).

Alternatively, one can model a dot by a chain of several (N) scatterers with $\tau_\varphi(T)$ defined in Eqs. (15.86), (15.89), and (15.90). Then at higher temperatures, we obtain $\tau_\varphi \propto T^{-\nu}$ with ν ranging from 2/3 to 1, as observed in a number of dots [59, 91, 186, 200, 332]. Substituting $g\tau_{\rm D}$ by $g_{\rm tot}\tau_{\rm D}^{\rm tot}/N$ (where $g_{\rm tot}$ and $\tau_{\rm D}^{\rm tot}$ are, respectively, the dimensionless conductance and the dwell time of a composite dot), bearing in mind that $g_{\rm tot} \approx 1 \div 12$ [186] and assuming N to be not very large, at low T from Eq. (15.86), one finds $\tau_{\varphi 0} \sim \tau_{\rm D}^{\rm tot}$ in agreement with experimental results displayed in Fig. 15.7. More complicated configurations of scatterers can also be considered with essentially the same results.

We conclude that our theory of electron dephasing by electron–electron interactions is in a good quantitative agreement with the weak localization experiments [59, 91, 186, 200, 332] demonstrating low-temperature saturation of τ_φ in quantum dots.

Now let us turn to experiments with spatially extended conductors. In a number of publications [153–155, 171], a good quantitative agreement was demonstrated between quasiclassical results for the electron dephasing time derived in Chapter 14 and the available data for $\tau_{\varphi 0}$ obtained by many experimental groups in numerous metallic wires and quasi-one-dimensional semiconductors. Our quasiclassical theory predicts a pronounced increase in the electron dephasing time with the diffusion coefficient $\tau_{\varphi 0} \propto D^d$; cf. Eq. (14.73). Such a trend was indeed clearly observed in not very strongly disordered conductors, e.g., in the experiments on high mobility GaAs/AlGaAs quantum wires [317] and in a number of other experiments.

On the other hand, Lin et al. [198, 278, 281, 282] analyzed numerous experimental data for $\tau_{\varphi 0}$ obtained by various groups in rather strongly disordered conductors with $D \lesssim 10$ cm^2/s and observed systematic *decrease* in $\tau_{\varphi 0}$ with increasing D. The data could be fitted by the dependence $\tau_{\varphi 0} \propto D^{-\alpha}$ with $\alpha \gtrsim 1.5 \div 2$. This trend is clearly just the *opposite* to one observed in less disordered conductors with $D \gtrsim 10$ cm^2/s.

In order to perform a representative comparison between theory and experiment, it is necessary to select the available data for the samples with similar parameters (such as the density of states, the Fermi velocity, etc.) in which only the diffusion coefficient D (and perhaps also geometry) would vary from sample to sample. Metallic samples are well suited for this purpose.

In Fig. 15.8, we collected the experimental data for $\tau_{\varphi 0}$ obtained in about 130 metallic samples with diffusion coefficients varying by $\sim$4 decades, from $D \approx 0.3$ cm^2/s to $D \approx 350$ cm^2/s. The data were taken from about 30 different experiments listed in the figure caption. We see that the measured values of $\tau_{\varphi 0}$ strongly depend on D. Furthermore, this dependence turns out to be non-monotonous: For relatively weakly disordered structures with $D \gtrsim 10$ cm^2/s, $\tau_{\varphi 0}$ increases with increasing D, while for strongly disordered conductors with $D \lesssim 10$ cm^2/s, the opposite trend takes place. In addition to the data points in Fig. 15.8, we indicate the dependencies $\tau_{\varphi 0}(D)$ (15.94) and (15.95) for the two models (a) and (b) discussed in Section 15.4.3. The three-dimensional dependence (15.94) (rather than one-dimensional and two-dimensional ones) is appropriate because in all the samples, the elastic mean free path values do not exceed their transversal dimensions.

We observe that for $D \gtrsim 10$ cm^2/s, the data points clearly follow the scaling (15.94). Practically all data points remain within the strip between the two lines corresponding to Eq. (15.94) with $\gamma = 1$ (dashed line) and $\gamma = 0.2$ (solid line). On the other hand, for more disordered conductors with $D \lesssim 10$ cm^2/s, the data are consistent with the scaling (15.95) obtained within the model (b). We stress that theoretical curves (15.94) and (15.95) are presented in Fig. 15.8 with *no fit parameters* except for a geometry factor γ for the first

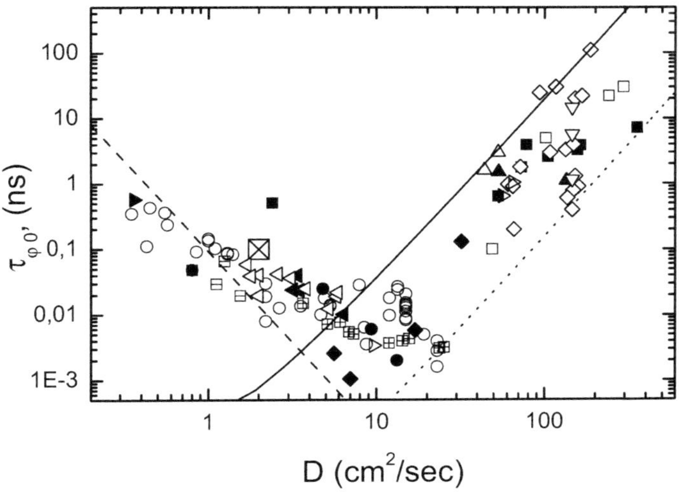

Figure 15.8 The low-temperature dephasing times observed in various experiments for the following samples: Au-1 to Au-6 [304], Au-7 [396], Au-8 and Au-10 [Mohanty (2003), unpublished] (■); 15 samples (Au, Ag, and Cu) from Table I [330] and AgMI6N [Birge (2003), unpublished) (◇); CF-1 and CF-2 [305] (△); A, B (Au) [347], Au1 [41], Ag1 [342] and Ag2 [292] (□); S, M, and L (Pt) [279, 280] (•); D and F [379] (▷); Ag, AgFe1, and AgFe2 [13] (▽); [339] (⊠, 10 samples within the box); 2 (Au) [105] and 1 (Au) [106] (▲); 44 samples (AuPd and AgPd) from 10 different papers collected in Table I [282] (○); 18 samples [281]: Au_2Al (◀), Sb (◁), $Sc_{85}Ag_{15}$ (▶), V_3Al (⊟); 9 samples (CuGeAu) [198] (⊞); al-1 [9] (♦). Our Eq. (15.94) for $\gamma = 0.2$ and 1 is indicated, respectively, by solid and dotted lines, while Eq. (15.95) for $g_c = 150$ is depicted by dashed line.

dependence and the value $g_c \approx 150$ for the second one. This value of g_c was estimated from the crossover condition (15.96) with $D \sim 10$ cm^2/s and $\gamma \sim 1$.

Note that some data points with $D \gtrsim 10$ cm^2/sek correspond to the samples contaminated by magnetic impurities. Remarkably, these data points also demonstrate—though with somewhat larger scatter—*systematic* increase in $\tau_{\varphi 0}$ with increasing D. At the same time, for similar values of D, the samples with higher concentration of magnetic impurities have systematically lower $\tau_{\varphi 0}$ than samples with a few or no magnetic impurities. Hence, for samples with relatively high concentration of magnetic impurities,

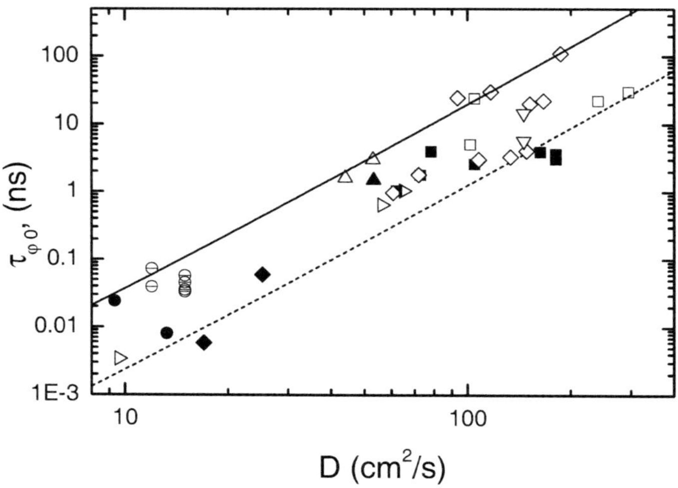

Figure 15.9 The same as in Fig. 15.6 for 9 cm^2/s $< D <$ 300 cm^2/s. Only the data for the following samples with no or a few magnetic impurities have been selected: Au-1, Au-3, Au-4, Au-6 [304], Au-7 [396], Au-8 and Au-10 [Mohanty (2003), unpublished] (■); A to F (AuPd) [313] (⊖); Ag(6N)a,b,c,d, Au(6N), Ag(5N)a,b, Cu(6N)b, Cu(5N)a [330] and AgMI6N [Birge (2003), unpublished] (◊); CF-1, CF-2 [305] (△); B (Au) [347], Au1 [41], Ag1 [342] and Ag2 [292] (□); M and L (Pt) [279, 280] (•); D (AuPd), F and H (Ag) [379] (▷); Ag and AgFe1 [13] (▽); 2 (Au) [105] (▲); au-3 and al-1 [9](♦). Our Eq. (15.94) for $\gamma = 0.2$ and 0.5 is indicated, respectively, by solid and dashed lines.

both mechanisms of electron–electron interactions and spin-flip scattering independently provide substantial contributions to τ_φ, being responsible, respectively, for the scaling $\tau_{\varphi 0} \propto D^3/\ln D$ and for additional non-universal shift in the data points downwards.

In order to carry out more accurate comparison with our theory of electron–electron interactions, let us now leave out the data points for the samples with allegedly high concentration of magnetic impurities. In Fig. 15.9, we selected the data for about 40 different metallic wires with no or a few magnetic impurities and with diffusion coefficients in the range 9 cm^2/s $< D <$ 300 cm^2/s from 17 experiments of various groups (see figure caption).

The data in Fig. 15.9 again clearly support the scaling of $\tau_{\varphi 0}$ with D (15.94) due to electron–electron interactions. Practically all data points are now located in between the solid and dashed lines, which indicate the dependence (15.94) with, respectively, $\gamma = 0.2$ and $\gamma = 0.5$. We cannot exclude that the remaining scatter among the data points with similar values of D may be, to a certain extent, due to the relatively small amount of magnetic impurities possibly residing in some samples. Also minor differences in metal parameters (e.g., in Fermi velocities) may slightly contribute to this effect. It appears, however, that sample-to-sample fluctuations of the geometry parameter γ play the major role here. This parameter is defined as a ratio between the square root of the inter-dot (inter-grain) contact area $\mathcal{A}^{1/2}$ and the average dot (grain) size $\mathcal{V}^{1/3}$. It also includes information about the transmission of inter-grain barriers. In other words, γ accounts for *local* properties of the sample that can be (and usually are) highly non-universal.

To illustrate this statement, let us recall that electrons in metallic wires get scattered both in the bulk and on the surface. Hence, the surface quality and/or details of the wire geometry—along with other factors—may significantly impact the value of γ. In fact, it is remarkable that all the data points in Fig. 15.9 are described by the values of γ varying only within the factor ~ 2.5. Given all the uncertainties, no better agreement could actually be expected in principle. At the same time, bearing in mind a strong dependence of $\tau_{\varphi 0}$ (15.94) on γ, it is by no means surprising to observe that different wires with similar values of D may have $\tau_{\varphi 0}$ differing by a few times.

Thus, Eq. (15.94) is demonstrated to be in a quantitative agreement with the available experimental data for quasi-one-dimensional metallic wires with diffusion coefficients in the range $9\ \text{cm}^2/\text{s} < D < 300\ \text{cm}^2/\text{s}$. A similarly good agreement is observed between the dependencies $\tau_{\varphi 0} \propto D/\ln D$ (1d) and $\tau_{\varphi 0} \propto D^2/\ln D$ (2d) and the experimental results on high mobility GaAs/AlGaAs quantum wires [317].

Now let us return to Fig. 15.8 and consider the data for strongly disordered conductors with $D < 10\ \text{cm}^2/\text{s}$. As we already pointed out, there exists a reasonably good agreement between the data points and the dependence (15.95) predicted within our simple model (b), in particular for samples with $D < 3\ \text{cm}^2/\text{s}$. At higher diffusion coefficients, most of the data points indicate a somewhat weaker dependence of $\tau_{\varphi 0}$ on D, which appears natural in the vicinity of the crossover to the dependence (15.94).

Our model (b) not only correctly reproduces the trend "more disorder–less decoherence" in strongly disordered conductors but also allows to understand the basic physics behind this trend. While for cleaner conductors, D usually increases with increasing average dot (or grain) size d (model (a)), for strongly disordered structures, the tendency can be just the opposite (model (b)): Increasing resistivity with increasing grain size is routinely observed in many experiments. Furthermore, since local conductance fluctuations increase with increasing disorder, several grains can form a cluster with internal inter-grain conductances strongly exceeding those at its edges. In this case, fluctuating potentials remain almost uniform inside the whole cluster, which then plays the role of an effective (bigger) grain/dot. Accordingly, the average volume of such "composite dots" $\mathcal{V} \propto 1/\delta$ may grow with increasing disorder, electrons spend more time in these bigger dots, and hence the electron decoherence time (15.86) increases in agreement with numerous experimental observations.

On the basis of the above comparison, we conclude that the theory of weak localization in the presence of electron–electron interactions outlined in this chapter universally and quantitatively describes the low-temperature electron decoherence observed in (i) quantum dots as well as in (ii) weakly and (iii) strongly disordered spatially extended conductors. In addition, our comparison allows to rule out the extrinsic mechanisms as a universal cause of low-temperature saturation of τ_{φ}. None of these mechanisms can explain both nontrivial scaling of the electron decoherence time in the cases (i), (ii), and (iii) (see Figs. 15.7–15.9) and—in a vast majority of cases—even the magnitude of the dephasing effect observed in experiments. For more details on the comparison between theory and experiment, we refer the reader to the papers [166, 171].

15.5 Further Extensions: Aharonov–Bohm Effect

Above in this chapter, we demonstrated that modeling of a disordered conductor by an array of quantum dots can substantially advance our understanding of weak localization effects in such systems, in particular in the presence of electron–electron interactions. It is tempting to extend this description to other quantum coherent effects in disordered conductors. In this section, we employ the same model in order to analyze the effect of electron–electron interactions on Aharonov–Bohm (AB) oscillations of conductance in metallic nanorings pierced by an external magnetic flux. Such oscillations serve as yet another manifestation of quantum nature of electrons, which can pick up different phases while propagating in different parts of the ring and demonstrate flux-sensitive quantum interference. This interference affects the system conductance, which oscillates as a function of the magnetic flux.

It is obvious that these AB conductance oscillations are only possible provided the ring perimeter does not exceed the electron dephasing length L_φ, which is in turn sensitive to electron–electron interactions. On the contrary, in the limit of short L_φ, quantum coherence of electrons inside the ring is essentially suppressed. As a result, electrons lose their ability to interfere and the amplitude of AB oscillations tends to zero in this case. Thus, investigations of AB effect in metallic nanorings can provide a great deal of information concerning both quantum coherent behavior of electrons and interaction-induced decoherence in such structures.

15.5.1 Ring with Two Quantum Dots

We will start our analysis by considering a simple system depicted in Fig. 15.10. The structure consists of two chaotic quantum dots (L and R) characterized by mean level spacing $\delta_{\rm L}$ and $\delta_{\rm R}$. Here we will restrict our attention to the case of metallic quantum dots with $\delta_{\rm L,R}$ being the lowest energy parameters in the problem. These dots are interconnected via two tunnel junctions $\rm J_1$ and $\rm J_2$ with conductances G_{t1} and G_{t2} forming a ring-shaped configuration, as shown in Fig. 15.10. The left and right dots are also attached to the leads (LL and RL), respectively, via the barriers $\rm J_L$ and $\rm J_R$ with conductances $G_{\rm L}$ and $G_{\rm R}$.

The whole structure is pierced by the magnetic flux Φ through the hole between two central barriers in such a way that electrons passing from left to right through different junctions acquire

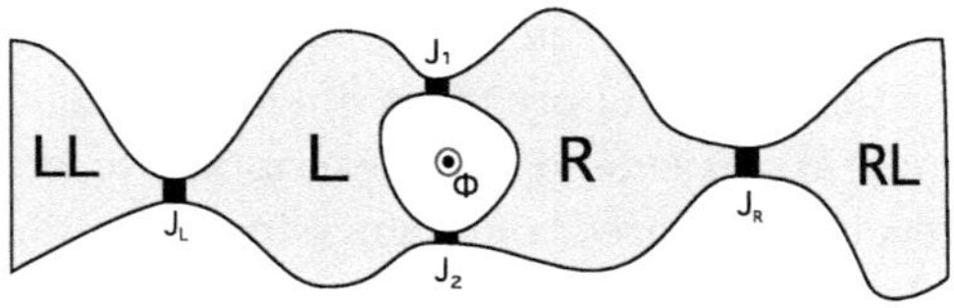

Figure 15.10 The ring-shaped quantum dot structure under consideration.

different geometric phases

$$\varphi_g^{(1,2)} = \frac{e}{c} \int_L^R dx_\mu A_\mu(x) \tag{15.99}$$

where the integration contour starts in the left dot, crosses the first ($\varphi_g^{(1)}$) or the second ($\varphi_g^{(2)}$) junction, and ends in the right dot. The difference between these two geometric phases equals to

$$\varphi_g^{(1)} - \varphi_g^{(2)} = 2\pi\,\Phi/\Phi_0 \tag{15.100}$$

where Φ_0 is the flux quantum. In the zero flux limit $\Phi \to 0$, the structure in Fig. 15.10 obviously reduces to that of two quantum dots depicted in Fig. 15.2. In this case, we simply identify $G_{t1} + G_{t2} = G_t$.

Below we will essentially follow the analysis [351] and assume that dimensionless conductances of the outer barriers $g_{\mathrm{L,R}} \equiv G_{\mathrm{L,R}} R_q$ are much larger than unity, while the conductances $g_{t1} \equiv G_{t1} R_q$ and $g_{t2} \equiv G_{t1,2} R_q$ are small as compared to $g_{\mathrm{L,R}}$, i.e.,

$$g_{\mathrm{L}}, g_{\mathrm{R}} \gg 1, g_{t1}, g_{t2} \tag{15.101}$$

As in Section 15.3.1, let us expand the exact action S_{el} (2.52) in powers of the tunneling amplitude characterizing both barriers J_1 and J_2. Keeping the terms up to the fourth order in this amplitude, we again arrive at Eqs. (15.8)–(15.10), where we should now include extra information about the geometric phases (15.99), (15.100). This expansion can also be illustrated by the diagrams in Fig. 15.11.

Averaging over realizations of transmission amplitudes yields two types of terms presented in Fig. 15.11b,c. It is straightforward to verify that only the contribution generated by the diagram (c) depends on the external magnetic flux, while the diagram (b) does not depend on Φ and, hence, is irrelevant for our present analysis. On top of that, the terms originating from the diagram (b) turn out to be parametrically small in the metallic limit.

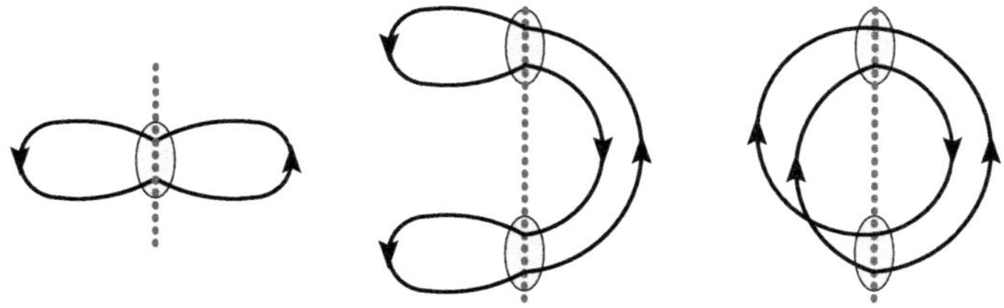

Figure 15.11 Diagrammatic representation of different contributions originating from expansion of the effective action in powers of the central barrier transmissions: second-order terms (a) corresponding to Eq. (15.9) and two kinds of fourth-order terms (b,c), which account for Eq. (15.10).

It follows immediately that only the diagram in Fig. 15.11c is responsible for the AB effect in our system. Its contribution to the action $S_{\rm WL}^{(4)}(\Phi)$ is again defined by Eq. (15.12) where one should now replace

$$\int_{\mathcal{A}} d^2\mathbf{r}\, d^2\mathbf{r}' \frac{g_t(\mathbf{r})g_t(\mathbf{r}')}{4\pi^2\nu_{\rm L}\nu_{\rm R}} \to \sum_{m,n=1,2} e^{2i(\varphi_g^{(n)}-\varphi_g^{(m)})} \int_{J_n} d^2\mathbf{r} \int_{J_m} d^2\mathbf{r}' \frac{g_t(\mathbf{r})g_t(\mathbf{r}')}{4\pi^2\nu_L^2\nu_R^2} \tag{15.102}$$

After this replacement, the action $S_{\rm WL}^{(4)}(\Phi)$ fully accounts for coherent oscillations of the system conductance in the lowest non-vanishing order in tunneling. It is worth reminding the reader that no additional approximations were employed during the derivation and, in particular, the fluctuating phases are exactly accounted for.

The total current I flowing through our system can be routinely split into two terms $I = I_0 + \delta I(\Phi)$, where I_0 is the leading flux-independent contribution and $\delta I(\Phi)$ is the quantum correction to the current sensitive to the magnetic flux Φ. As before, this correction is determined by the formula (15.16).

The whole calculation of $\delta I(\Phi)$ carried out with the aid of Eqs. (15.16), (15.12), and (15.12) is fully analogous to that already described in Section 15.3. Therefore, there is no need to go into further details here. Averaging over fluctuating phases is again accomplished with the aid of the correlation functions (15.31)–(15.36) and yields an exact expression for $\delta I(\Phi)$, which is similar to Eq. (15.37) but now also depends on the magnetic flux Φ.

In the limit $g_Z \gg 1$, one can again drop the small correlation function $\tilde{K}(t) \ll 1$ (exactly as we did in Section 15.3.4), in which case our result reduces to

$$\delta I(\Phi) = -\sum_{m,n=1,2} \frac{e^2 V e^{2i(\varphi_g^{(n)}-\varphi_g^{(m)})}}{8\pi^3 \nu_L \nu_R} \int d\tau_1 d\tau_2 \int_{J_n} d\mathbf{r} \int_{J_m} d\mathbf{r}' g_t(\mathbf{r}) g_t(\mathbf{r}')$$
$$\times \mathcal{C}_L(\tau_1; \mathbf{r}', \mathbf{r}) \mathcal{C}_R(\tau_2; \mathbf{r}, \mathbf{r}') e^{-2F(\tau_1)-2F(\tau_2)+F(\tau_1-\tau_2)+F(\tau_1+\tau_2)} \tag{15.103}$$

cf. Eq. (15.48). As before, this approximation effectively implies dropping the interaction correction to $\delta I(\Phi)$, which remains small except at exponentially small temperatures; see Section 15.3.5.

The result (15.103) can also be rewritten as

$$\delta I(\Phi) = -I_{AB} \cos(4\pi \Phi/\Phi_0) - I_{WL1} - I_{WL2} \tag{15.104}$$

where the first term in the right-hand side explicitly accounts for AB oscillations, while the two other terms $I_{WL1,2}$ represent the remaining part of the quantum correction to the current independent of Φ.

For simplicity, let us neglect the contribution of external leads (which can be trivially restored if needed), in which case in the limit (15.101), we can identify $g_Z \simeq g_L g_R/(g_L + g_R)$. As an important example, below we will evaluate the amplitude of AB oscillations I_{AB} for the symmetric system with two identical quantum dots characterized by the dwell time τ_D, volume $\mathcal{V}$, mean level spacing δ, and dimensionless conductances $g_L = g_R \equiv 2g_Z = 4\pi/\delta\tau_D$. In this case, the Cooperons take the form (15.49). Identifying dimensionless conductances of the central barriers as $g_{t1,2} = \int_{J_{1,2}} g_t(\mathbf{r}) d\mathbf{r}$, we obtain [351]

$$I_{AB} = \frac{e^2 g_{t1} g_{t2} \delta^2 V}{4\pi^3} \int_0^\infty d\tau_1 d\tau_2 e^{-\frac{\tau_1+\tau_2}{\tau_D} - 2F(\tau_1) - 2F(\tau_2) + F(\tau_1-\tau_2) + F(\tau_1+\tau_2)} \tag{15.105}$$

In the absence of electron–electron interactions ($F(\tau) \to 0$), this formula yields

$$I_{AB}^{(0)} = \frac{e^2 g_{t1} g_{t2} V}{\pi g_Z^2} \tag{15.106}$$

Including now electron–electron interactions, from Eq. (15.105) at high-enough temperatures, we get

$$\frac{I_{\rm AB}}{I_{\rm AB}^{(0)}} = \begin{cases} e^{-\frac{4\gamma}{g_{\rm Z}}} \frac{(2\pi T \tau_{\rm RC})^{4/g_{\rm Z}}}{1+2\pi T \tau_{\rm D}/g_{\rm Z}}, & \tau_{\rm D}^{-1} \lesssim T \lesssim \tau_{\rm RC}^{-1}, \\ \frac{1}{\tau_{\rm D}} \left(\frac{g_{\rm Z}\tau_{\rm RC}}{2T}\right)^{1/2}, & \tau_{\rm RC}^{-1} \lesssim T, \end{cases} \tag{15.107}$$

while in the low-temperature limit, we find

$$\frac{I_{\rm AB}}{I_{\rm AB}^{(0)}} = e^{-\frac{4\gamma}{g_{\rm Z}}} \left(\frac{2\tau_{\rm RC}}{\tau_{\rm D}}\right)^{4/g_{\rm Z}}, \qquad T \lesssim \tau_D^{-1} \tag{15.108}$$

The latter result demonstrates that interaction-induced suppression of AB oscillations in metallic dots with $\tau_{\rm RC} \ll \tau_{\rm D}$ persists down to $T = 0$. As in the case of two quantum dots considered above (cf. Section 15.3.4), this suppression is a direct result of electron dephasing produced by electron–electron interactions.

Let us also briefly address additional quantum corrections to the current $I_{\rm WL1,2}$ in Eq. (15.104). These terms do not depend on Φ but need to be accounted for in order to establish a direct relation between the AB effect and the phenomenon of weak localization addressed in Section 15.3. With the aid of Eq. (15.103), one easily finds

$$\frac{I_{\rm WL1}}{I_{\rm AB}} = \frac{g_{t1}}{2g_{t2}}, \qquad \frac{I_{\rm WL2}}{I_{\rm AB}} = \frac{g_{t2}}{2g_{t1}} \tag{15.109}$$

Combining this equation with the above results for $I_{\rm AB}$, we immediately identify the terms $I_{\rm WL1}$ and $I_{\rm WL2}$ as weak localization corrections to the current originating from the two central barriers in our structure. In the absence of the magnetic field $\Phi = 0$, the total quantum correction to the current $\delta I(0)$ (15.104) exactly coincides with the weak localization correction $I_{\rm WL}$ provided the two central barriers in Fig. 15.10 are viewed as a composite tunnel barrier with total dimensionless conductance $g_{t1} + g_{t2}$.

15.5.2 Ring Composed of a Quantum Dot Chain

Let us now turn to the analysis of AB oscillations in nanorings composed of a chain of quantum dots, as shown in Fig. 15.12. Employing exactly the same arguments as in Section 15.4, we may conclude that the dominant effect of electron–electron interactions

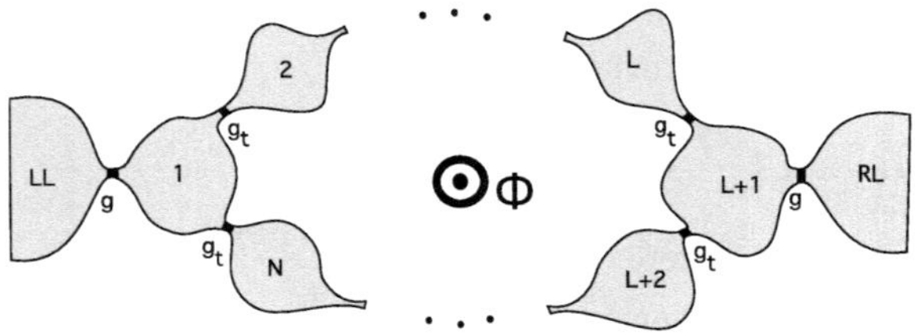

Figure 15.12 Ring composed of N quantum dots.

is electron dephasing fully determined by fluctuations of the phase φ^+. At the same time, fluctuations of the phase φ^- (which control the interaction correction to I_{AB}) turn out to be essentially irrelevant for the whole issue down to exponentially small temperatures. Hence, in order to address the problem in the many-dot configuration of Fig. 15.12, it suffices to ignore the fluctuating field φ^- and account only for the phase φ^+. As in Section 15.4, this observation yields significant simplifications in our calculation to be presented below.

For simplicity, we will consider the case of identical quantum dots (with mean level spacing δ and dwell time $\tau_{\text{D}} = 2\pi/(g_t\delta)$) coupled by barriers with conductances g_t and the Fano factor β_t. Leads are coupled to the ring at the dots with numbers 1 and $L+1$ via junctions with conductance g. The whole calculation is in many respects similar to that carried out in Section 15.4 for a chain of quantum dots. Here, however, we need to account both for the ring geometry and for the presence of the magnetic flux Φ. Below we will essentially follow the analysis [354].

The leading interference correction δG_1 to the conductance of the n-th barrier is derived analogously to Eq. (15.65). We obtain

$$\delta G_1 = -\frac{e^2 g_t \delta}{4\pi^2} \int_0^\infty dt [\beta_t C_{n,n+1}(t) e^{\frac{4\pi i\Phi}{N\Phi_0}} + (1-\beta_t)(C_{n,n}(t) + C_{n+1,n+1}(t)) + \beta_t C_{n+1,n}(t) e^{-\frac{4\pi i\Phi}{N\Phi_0}}] \tag{15.110}$$

where $C_{m,n}(t)$, as before, denotes the Cooperon. The quantum correction to conductance of the whole system δG can be obtained with the aid of the Kirchhoff law. For the case $Ng \ll g_t$ considered here, one finds

$$\delta G \approx \frac{L(N-L)g^2}{4Ng_t^2}\delta G_1 \tag{15.111}$$

In the absence of electron–electron interactions, the Cooperon $C_{m,n}(t)$ satisfies the diffusion-like equation

$$\frac{\partial C_{n,m}(t)}{\partial t} + \frac{2C_{n,m}(t) - C_{n+1,m}(t)e^{-\frac{4\pi i\Phi}{N\Phi_0}}}{2\tau_{\mathrm{D}}} - \frac{C_{n-1,m}(t)e^{\frac{4\pi i\Phi}{N\Phi_0}}}{2\tau_{\mathrm{D}}} = \delta_{n,m}\delta(t) \tag{15.112}$$

in the case $n \neq 1,\ L+1$ and

$$\frac{\partial C_{n,m}(t)}{\partial t} + \frac{2C_{n,m}(t) - C_{n+1,m}(t)e^{-\frac{4\pi i\Phi}{N\Phi_0}}}{2\tau_{\mathrm{D}}} - \frac{C_{n-1,m}(t)e^{\frac{4\pi i\Phi}{N\Phi_0}}}{2\tau_{\mathrm{D}}} + \frac{g\delta}{4\pi}C_{n,m}(t) = \delta_{n,m}\delta(t) \tag{15.113}$$

for $n = 1$ or $n = L + 1$. The solution to the above diffusion equation can be conveniently expressed in the form

$$C^{(0)}_{n,m}(t) = \sum_{k=|n-m|}^{\infty} \sum_{\nu(0)=n}^{\nu(t)=m} e^{\frac{4\pi i(n-m+NW[\nu(t)])\Phi}{N\Phi_0}} \int_0^t dt_k \int_0^{t_k} dt_{k-1} \ldots \int_0^{t_2} dt_1 \frac{e^{-\frac{t}{\tau_{\mathrm{D}}}}}{(2\tau_{\mathrm{D}})^k} \tag{15.114}$$

Here the summation is performed over all possible discrete trajectories with fixed endpoints, and $W[\nu(t)]$ denotes the winding number for a given trajectory.

Let us now include electron–electron interactions. Taking into account only the V^+-component of the fluctuating field, one can easily incorporate the effect of interactions into the above expression for the Cooperon. One finds

$$C_{n,m}(t) = \sum_{k=|n-m|}^{\infty} \sum_{\nu(0)=n}^{\nu(t)=m} e^{\frac{4\pi i(n-m+NW[\nu(t)])\Phi}{N\Phi_0}} \times \int_0^t dt_k \ldots \int_0^{t_2} dt_1 \frac{e^{-\frac{t}{\tau_{\mathrm{D}}} + ie\int_0^t d\tau (V^+_{\nu(\tau)}(\tau) - V^+_{\nu(\tau)}(t-\tau))}}{(2\tau_{\mathrm{D}})^k} \tag{15.115}$$

i.e., as expected, the fluctuating field V^+ just modifies the phases of the electron wave functions. Averaging over Gaussian fluctuations of V^+, we get

$$\left\langle \exp\left[ie \int_0^t d\tau (V_{\nu(\tau)}(\tau) - V_{\nu(\tau)}(t-\tau)) \right] \right\rangle_{V^+}$$

$$= \exp\left[-e^2 \int_0^t d\tau_1 d\tau_2 (F_{\nu(\tau_1),\nu(\tau_2)}(\tau_1 - \tau_2) - F_{\nu(\tau_1),\nu(\tau_2)}(t - \tau_1 - \tau_2)) \right] \tag{15.116}$$

Here $F_{m,n}(t) = \langle V_m^+(t) V_n^+(0) \rangle_{V^+}$ defines the correlator for fluctuating voltages.

In order to evaluate the Cooperon in the presence of interactions, we will routinely use an approximate procedure, which amounts to substituting $\langle\langle e^F \rangle\rangle \to e^{\langle\langle F \rangle\rangle}$, i.e., we first expand the exponent in Eq. (15.116) in Taylor series, then perform the summation over all trajectories, and after that re-exponentiate the result. Averaging over diffusive paths is performed with the aid of the diffuson $D_{m,n}(t)$:

$$\langle\langle F_{\nu(\tau_1),\nu(\tau_2)}(\tau_1 - \tau_2) \rangle\rangle = \frac{1}{N} \sum_{m,n=1}^{N} F_{m,n}(\tau_1 - \tau_2) D_{m,n}(|\tau_1 - \tau_2|) \tag{15.117}$$

As a result, we again arrive at Eq. (15.83) with

$$\mathcal{F}(t) = \frac{e^2}{N} \sum_{n,m=1}^{N} \int_0^t d\tau_1 d\tau_2 F_{m,n}(\tau_1 - \tau_2) \times (D_{m,n}(|\tau_1 - \tau_2|) - D_{m,n}(|t - \tau_1 - \tau_2|)) \tag{15.118}$$

The correlator for fluctuating voltages can be derived, e.g., by means of the nonlinear sigma model [165]. Analogous to Eq. (15.80), we obtain

$$F_{m,n}(t) = \frac{\tau_{\mathrm{D}}}{N} \sum_{q=1}^{N} \int \frac{d\omega}{2\pi} e^{-i\omega t} \omega \coth\frac{\omega}{2T} \frac{f(q) e^{\frac{2\pi i q}{N}(m-n)}}{\omega^2 \tau_{\mathrm{D}}^2 + \varepsilon^2(q)} \tag{15.119}$$

where

$$f(q) = \frac{g_t \tau_{\mathrm{D}} e^2}{\pi} \frac{\epsilon(q)}{(4C\epsilon(q) + C_g)^2}, \quad \varepsilon(q) = \epsilon(q) + \frac{g_t \tau_{\mathrm{D}} e^2}{\pi} \frac{\epsilon(q)}{4C\epsilon(q) + C_g} \tag{15.120}$$

and $\epsilon(q) = 1 - \cos\frac{2\pi q}{N}$. As above, here C and C_g denote, respectively, the junction and the dot capacitances.

Finally, we specify the expressions for the diffuson and the Cooperon in the absence of electron–electron interactions. They read

$$D_{m,n}(t) = \frac{\tau_{\rm D}}{N}\sum_{q=1}^{N}\int\frac{d\omega}{2\pi}\frac{e^{-{\rm i}\omega t+\frac{2\pi{\rm i}q}{N}(m-n)}}{-{\rm i}\omega\tau_{\rm D}+\epsilon(q)} \tag{15.121}$$

$$C^{(0)}_{m,n}(t) = \frac{\tau_{\rm D}}{N}\sum_{q=1}^{N}\int\frac{d\omega}{2\pi}\frac{e^{-{\rm i}\omega t+\frac{2\pi{\rm i}q}{N}(m-n)}}{-{\rm i}\omega\tau_{\rm D}+\epsilon(q-2\Phi/\Phi_0)} \tag{15.122}$$

The above equations are sufficient to evaluate the function $\mathcal{F}(t)$ in a general form. Here we are primarily interested in AB oscillations and, hence, we only need to account for the flux-dependent contributions determined by the electron trajectories fully encircling the ring at least once. Obviously, one such traverse around the ring takes time $t \geq N^2\tau_{\rm D}$. Hence, the behavior of the function $\mathcal{F}(t)$ only at such time scales needs to be studied for our present purpose. In this long-time limit, $\mathcal{F}(t)$ is a linear function of time with the slope

$$\mathcal{F}'(t \geq N^2\tau_{\rm D}) \approx \frac{2e^2\tau_{\rm D}^2}{N}\sum_{q=1}^{N-1}\int\frac{d\omega}{2\pi}\frac{f(q)\epsilon(q)\omega\coth\frac{\omega}{2T}}{(\omega^2\tau_{\rm D}^2+\epsilon^2(q))(\omega^2\tau_{\rm D}^2+\varepsilon^2(q))} \tag{15.123}$$

This observation implies that at such time scales, electron–electron interactions yield exponential decay of the Cooperon in time

$$C_{m,n}(t) \approx C^{(0)}_{m,n}(t)e^{-\frac{t}{\tau_\varphi}} \tag{15.124}$$

where we identify

$$\frac{1}{\tau_\varphi} = \mathcal{F}'(t \geq N^2\tau_{\rm D}) \tag{15.125}$$

as an effective electron dephasing time for our problem. In the case $C_g \gg C$ and $\tau_{\rm D} \gg \tau_{\rm RC} \equiv 2\pi C_g/(e^2 g_t)$ from Eqs. (15.123) and (15.125), we obtain

$$\frac{1}{\tau_\varphi} = \begin{cases} \frac{\delta}{\pi}\ln\frac{4E_{\rm C}}{\delta} & T \ll 1/N\tau_{\rm D}, \\ \frac{\pi NT}{3g_t} & T \gg 1/N\tau_{\rm D}, \end{cases} \tag{15.126}$$

where $E_C = e^2/(2C_g)$. These expressions are fully consistent with the corresponding results derived for in Section 15.4 for chains of quantum dots.

Let us emphasize again that the above results for $\mathcal{F}(t)$ apply at sufficiently long times relevant for AB conductance oscillations. At the same time, other physical quantities, such as weak localization correction to conductance, can be determined by the function $\mathcal{F}(t)$ at shorter time scales. Our general results allow us to easily recover the corresponding behavior as well. For instance, at $T \gg \tau_D$ and $t \ll N^2\tau_D$, we get

$$\mathcal{F}(t) \approx \frac{4T}{3g_t}\left(\frac{2\pi}{\tau_D}\right)^{1/2} t^{3/2} + \dots \tag{15.127}$$

in agreement with the results of Section 15.4. This expression yields the well-known dependence $\tau_\phi \propto T^{-2/3}$, which—in contrast to Eq. (15.126)—does not depend on N and remains applicable in the high-temperature limit.

To proceed further, let us integrate the expression for the Cooperon over time. We obtain

$$\int_0^\infty C_{m,n}(t)dt = \frac{\tau_D}{N}\sum_{q=1}^{N} \frac{e^{\frac{2\pi i q}{N}(m-n)}}{\epsilon(q - 2\Phi/\Phi_0) + \tau_D/\tau_\phi + g/(g_t N)} \tag{15.128}$$

where the term $g/(g_t N)$ in the denominator accounts for the effect of external leads and remains applicable as long as $Ng \ll g_t$. Combining Eqs. (15.110), (15.111), and (15.128), after the summation over q, we arrive at the final result for the AB conductance [354]

$$\delta G_{AB} = \frac{e^2 L(N-L)g^2}{2\pi N g_t^2} \frac{(\beta_t\alpha + 1 - \beta_t)(z^{-N} - \cos(4\pi\Phi/\Phi_0))}{\sqrt{\alpha^2 - 1}(z^N + z^{-N} - 2\cos(4\pi\Phi/\Phi_0))} \tag{15.129}$$

where $\alpha = 1 + \frac{\tau_D}{\tau_\phi} + \frac{g}{g_t N}$ and $z = \alpha + \sqrt{\alpha^2 - 1}$. Together with Eq. (15.126), this result fully determines AB oscillations of conductance in nanorings composed of metallic quantum dots in the presence of electron–electron interactions.

Expanding Eq. (15.129) in Fourier series, we obtain

$$\delta G^{AB} = \sum_{k=1}^{\infty} \delta G^{(k)} \cos(4\pi k\Phi/\Phi_0) \tag{15.130}$$

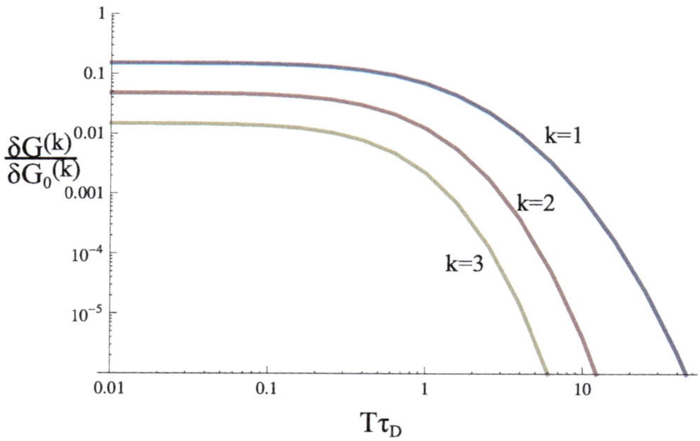

Figure 15.13 Temperature dependence of the first three harmonics of AB conductance for $g_t = 500$, $g = 30$, $N = 10$, $\beta_t = 1$, and $\tau_D/\tau_{RC} = 120$.

where

$$\delta G^{(k)} = -\frac{e^2 L(N-L)g^2(\beta_t\alpha + 1 - \beta_t)}{2\pi N g_t^2 \sqrt{\alpha^2 - 1}} z^{-N|k|} \tag{15.131}$$

In the limit $\tau_\varphi \gg \tau_D$, we have $z \approx 1 + \sqrt{2\tau_D/\tau_\varphi} + \ldots$, hence $\delta G^{(k)}$ behaves as

$$\delta G^{(k)} \propto e^{-N|k|\sqrt{\frac{2\tau_D}{\tau_\phi}}} \tag{15.132}$$

i.e., at high temperatures, $\log|\delta G|$ scales with N as $N^{3/2}$, while at low temperatures, it scales as N. The temperature dependence of the first three harmonics of AB conductance in the presence of electron–electron interactions is depicted in Fig. 15.13.

15.5.3 Diffusive Nanorings

The above results allow us to formulate quantitative predictions regarding the effect of electron–electron interactions on AB oscillations of conductance for a wide class of disordered nanorings embraced by our model. Of particular interest is the situation of a large number of dots $N \gg 1$, which essentially mimics the behavior of diffusive nanorings. In order to establish a direct relation to this important limit, we again introduce an effective electron diffusion coefficient $\tilde{D} = d^2/(2\tau_D)$ (15.87) and define the electron density of states $\nu = 1/(d^3\delta)$, where d is a linear dot size. Then we obtain with exponential accuracy:

$$\delta G^{(k)} \sim \begin{cases} e^{-|k|(\mathcal{L}/L_\varphi)}, & T \ll \tilde{D}/(\mathcal{L}d), \\ e^{-|k|(\mathcal{L}/L_\varphi)^{3/2}} & T \gg \tilde{D}/(\mathcal{L}d) \end{cases}$$

Here we introduced the ring perimeter $\mathcal{L} = Nd$ and the effective decoherence length

$$L_\varphi = \begin{cases} \left(\frac{\pi \nu d^3 \tilde{D}}{\ln \frac{4E_C}{\delta}}\right)^{1/2} & T \ll \tilde{D}/(\mathcal{L}d), \\ \left(\frac{12\nu d^2 \tilde{D}^2}{T}\right)^{1/3} & T \gg \tilde{D}/(\mathcal{L}d). \end{cases} \tag{15.133}$$

In the high-temperature limit, the dependence $L_\varphi \propto T^{-1/3}$ (15.133) was also derived for diffusive metallic nanorings [287, 374] by means of different approaches taking into account only classical component of the fluctuating electromagnetic fields. At low temperatures $T \to 0$, Eq. (15.133) for L_φ agrees with Eq. (15.88) established by weak localization arguments. In the quasiclassical limit, the latter in turn matches with L_φ derived in Chapter 14 for weakly disordered metals. A nontrivial feature demonstrated by our analysis is that in contrast to weak localization, the crossover from thermal to quantum dephasing is controlled by the ring perimeter $\mathcal{L}$. This is because only sufficiently long electron paths fully encircling the ring are sensitive to the magnetic flux and may contribute to AB conductance oscillations.

15.6 Short Summary

For clarity and having in mind fundamental importance of the issue addressed in this chapter, let us briefly recapitulate our key observations.

In the course of our analysis, we employed a model of an array of quantum dots, which was also extensively exploited in Chapter 13 (see Fig. 13.1) in order to study Coulomb blockade effects in disordered conductors. This model embraces various types of such conductors and allows us to investigate electron transport in the presence of interactions within a very general theoretical framework. In this chapter, we merely concentrated on the impact of electron–electron interactions on weak localization with the emphasis put on the interaction-induced decoherence of electrons at low temperatures. We formulated a fully self-contained theory free of any divergencies and cutoffs, which allows to conveniently handle disorder averaging for both spatially restricted and spatially extended conductors and treat electron scattering without employing any quasiclassics. At the same time, in the quasiclassical limit of weak disorder, our general results essentially reduce to those of Chapter 14 derived within a different theoretical framework. Furthermore, the analysis developed in the present chapter allowed to formally justify the phenomenological high-frequency cutoff procedure performed in Chapter 14 and also to understand the basic physical reason behind this procedure explaining why the electron dephasing rate is controlled only by fluctuations with frequencies and wave vectors below, respectively, the inverse electron elastic time and length. In addition, we illustrated a transparent relation between electron dephasing by electron–electron interactions and the so-called $P(E)$-theory describing electron tunneling in the presence of an external environment.

We constructed a unified description of electron dephasing by Coulomb interaction in different structures, including (i) metallic quantum dots, (ii) weakly disordered conductors, and (iii) strongly disordered conductors. Remarkably, in all these cases at $T \to 0$, the electron decoherence time $\tau_{\varphi 0}$ is determined by a simple formula

$$\tau_{\varphi 0} \sim \frac{g\tau_{\mathrm{D}}}{\ln(E_{\mathrm{C}}/\delta)}$$

demonstrating that low-temperature saturation of τ_φ is universally caused by electron–electron interactions.

In the case of quantum dots, this formula establishes a simple dependence between $\tau_{\varphi 0}$ and the electron dwell time τ_{D} in the dot, which agrees well with experimental observations (cf. Fig. 15.7). For weakly disordered conductors, the above formula yields $\tau_{\varphi 0} \propto D^3/\ln D$ and matches with the quasiclassical results of Chapter 14, while in the case of strongly disordered conductors, it helps to explain the trend "more disorder–less decoherence" (see Fig. 15.8) observed in an number of experiments. A detailed analysis of the data between theory and numerous experiments for the whole scope of disordered structures (i), (ii), and (iii) demonstrates a good agreement between theory and experiment, thus providing additional support for the results and conclusions of this and previous chapters.

Finally, the employed model of quantum dot arrays also proved to be very useful for the analysis of other quantum coherent effects in disordered conductors with electron–electron interactions, such as the AB conductance oscillations in metallic nanorings. Similar to the weak localization correction to conductance, such oscillations are also affected by electron–electron interactions, primarily due to quantum dephasing of electrons. Zero temperature dephasing time and length derived from the analysis of AB oscillations are fully consistent with analogous parameters extracted from the theory of weak localization.

Chapter 16

Electron Transport, Fluctuations, and Coulomb Effects in Normal-Superconducting Hybrids

16.1 Introduction

In Chapters 10 and 11, we discussed a general theory describing electron transport and fluctuation effects in short coherent conductors placed in between two normal metallic reservoirs and elucidated a fundamental relation between these effects and electron–electron interactions. In this chapter, we further extend these ideas by considering hybrid metallic structures containing both normal and superconducting electrodes.

It is well known that low-energy electron transport across the interface between a normal metal and a superconductor (NS) is provided by the mechanism of Andreev reflection [19]. This mechanism involves conversion of a subgap quasiparticle entering the superconductor from the normal metal into a Cooper pair together with simultaneous creation of a hole that goes back into the normal metal. This process is illustrated in Fig. 16.1. Each such act of electron–hole reflection corresponds to transferring twice the

Dissipative Quantum Mechanics of Nanostructures: Electron Transport, Fluctuations, and Interactions
Andrei D. Zaikin and Dmitry S. Golubev

ISBN 978-981-4774-50-5 (Hardcover), 978-0-429-29823-3 (eBook)
www.jennystanford.com

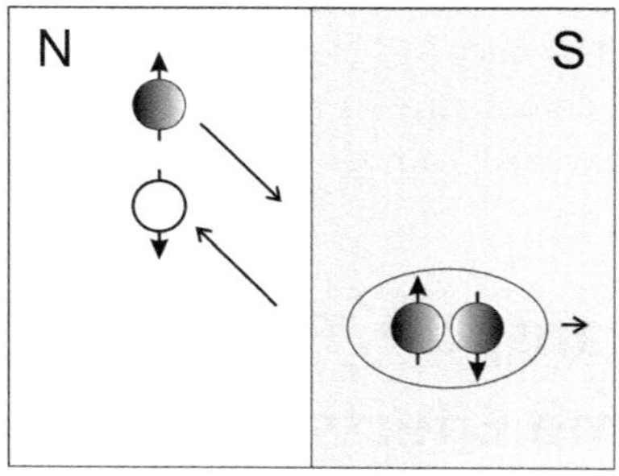

Figure 16.1 Andreev reflection process in NS hybrid structures.

electron charge $e^* = 2e$ across the NS interface. As a result, the NS structure acquires nonzero conductance at subgap energies $E < \Delta_0$ [61].

Let us assume that two bulk metallic electrodes, one normal and one superconducting, are connected by an arbitrary—though sufficiently short—coherent conductor as it is schematically shown in Fig. 16.2. The model for this conductor is identical to that already employed in Chapter 10. In the spin-degenerate case to be addressed below, this conductor is characterized by the normal state conductance

$$G_{\rm N} = \frac{2}{R_{\rm q}} \sum_n T_n, \tag{16.1}$$

cf. Eq. (10.4). As before, T_n define transmissions of all conducting channels, $R_{\rm q} = h/e^2$ is the normal quantum resistance unit, and the factor 2 accounts for spin degeneracy. Evaluating the conductance $G_{\rm A}$ of the NS structure in Fig. 16.2, at temperatures/voltages well below the superconducting gap Δ_0, one finds [61]

$$G_{\rm A} = \frac{1}{R_{\rm Q}} \sum_n \mathcal{T}_n, \tag{16.2}$$

where $R_{\rm Q} = h/(2e)^2 \equiv R_{\rm q}/4$ is the "superconducting" quantum resistance unit and $\mathcal{T}_n$ are "Andreev transmissions" related to T_n as

$$\mathcal{T}_n = T_n^2/(2 - T_n)^2 \tag{16.3}$$

Comparing Eqs. (16.2) and (16.3) with the Landauer formula (16.1), one immediately observes that Andreev conductance $G_{\rm A}$ formally

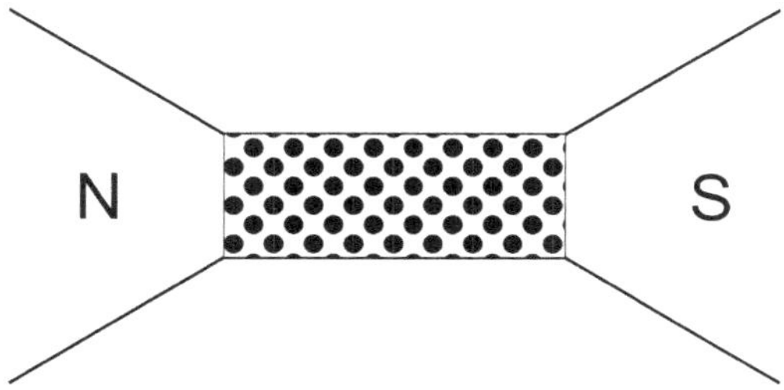

Figure 16.2 Short coherent conductor connecting normal and superconducting reservoirs.

describes "normal" transport of *spinless* quasiparticles (hence, no extra factor 2 in front of the sum) with charge $e^* = 2e$ across an effective coherent scatterer with channel transmissions $\mathcal{T}_n$ (16.3).

Later it was realized that this formal analogy applies not only to electron transport but also to low-frequency shot noise [98, 310] and eventually to full counting statistics [50]. Consider, for instance, current fluctuations $\delta I(t) = I(t) - I$ around its average value $I \equiv \langle \hat{I}(t) \rangle$. As we already discussed in Chapter 10, in normal conductors at $T \to 0$ and in the zero-frequency limit, the correlator for such fluctuations is determined by Eq. (10.94), i.e.,

$$\langle |\delta I|^2 \rangle = e|V|G_{\mathrm{N}}\beta_{\mathrm{N}}, \quad \beta_{\mathrm{N}} = \frac{\sum_n T_n(1 - T_n)}{\sum_n T_n} \tag{16.4}$$

where V is the average voltage across the conductor and β_{N} is the normal Fano factor (10.136). In NS systems, Andreev reflection also leads to the current shot noise at energies below the superconducting gap. In this case in the zero energy/frequency limit and at $T \to 0$, one obtains [98]

$$\langle |\delta I|^2 \rangle = 2e|V|G_{\mathrm{A}}\beta_{\mathrm{A}}, \quad \beta_{\mathrm{A}} = \frac{\sum_n \mathcal{T}_n(1 - \mathcal{T}_n)}{\sum_n \mathcal{T}_n} \tag{16.5}$$

where $\mathcal{T}_n$ is defined in Eq. (16.3). Again, a close similarity between Eqs. (16.4) and (16.5) is obvious: The result (16.5) just describes shot noise produced by carriers with effective charge $e^* = 2e$ in a coherent scatterer with conductance G_{A} and Fano factor β_{A}.

In the important case of diffusive NS structures, doubling of the carrier charge also implies doubling of the shot noise [98].

Evaluating the sums over transmission channels in this particular case, one finds

$$G_{\rm N} = G_{\rm A}, \quad \beta_{\rm N} = \beta_{\rm A} = 1/3 \tag{16.6}$$

which yield $\langle|\delta I|^2\rangle = 2e|V|G_{\rm N}/3$ for NS structures and the two times smaller result in the normal case. This doubling of the shot noise in diffusive NS systems was also observed experimentally [207, 238].

In this chapter, we develop a theory describing an interplay between Coulomb blockade, shot noise, and Andreev reflection.

16.2 Shot Noise and Coulomb Blockade of Andreev Reflection

Let us consider big normal and superconducting reservoirs connected by a rather short normal bridge (conductor) with arbitrary transmission distribution T_n of its conducting modes (for each n, the value T_n is the same for spin-up and spin-down electrons). This system is depicted in Fig. 16.2. Both phase and energy relaxation of electrons may occur only in the reservoirs and not inside the conductor, length of which is assumed to be shorter than dephasing and inelastic relaxation lengths. The Thouless energy of the normal conductor will not play any role in our analysis as it is supposed to be higher than any other relevant energy scale in the problem. Coulomb interaction between electrons in the conductor area is accounted for by some effective capacitance C.

16.2.1 Effective Action

In order to analyze electron transport in the presence of interactions, we will again make use of the effective action formalism combined with the scattering matrix technique. As we already demonstrated in Chapters 10–13, this approach proved to be very successful in the case of normal conductors and it can be conveniently generalized to the superconducting case. In fact, as it was verified in Chapter 2, the structure of the effective action remains the same also in the latter case, one should only replace normal propagators by 2×2 matrix Green functions, which account for superconductivity; cf., e.g., Eqs. (2.173) and (2.175). Making use of this observation together with the result (10.74), we immediately recover the Keldysh effective action for a coherent scatterer placed in between normal and superconducting reservoirs. This action accounts for electron transfer between these reservoirs and it can be expressed in the form

$$S_{\mathrm{sc}}[\varphi] = -\frac{\mathrm{i}}{2}\sum_n \mathrm{Tr}\ln\left[1+\frac{T_n}{4}\left(\{\check{G}_{\mathrm{N}},\check{G}_{\mathrm{S}}\}-2\right)\right] \tag{16.7}$$

where $\check{G}_{\mathrm{N}}$ and $\check{G}_{\mathrm{S}}$ are 4×4 Green–Keldysh matrices of normal and superconducting electrodes, product of which implies time convolution and whose anticommutator is denoted by curly brackets. The phase variable φ is related to the fluctuating voltage across the scatterer, $V(t)$, in a standard manner as $\dot{\varphi}(t) = eV$. As usual, this phase variable is defined on the forward (φ_{F}) and the backward (φ_{B}) branches of the Keldysh contour, which can be routinely combined giving rise to the "classical" and "quantum" parts of the phase, respectively $\varphi^+ = (\varphi_{\mathrm{F}}+\varphi_{\mathrm{B}})/2$ and $\varphi^- = \varphi_{\mathrm{F}}-\varphi_{\mathrm{B}}$.

The total effective action for our structure $S[\varphi]$ is obtained by adding the standard charging contribution

$$S_c[\varphi] = \frac{C}{2e^2}\int_0^t dt'(\dot{\varphi}_{\mathrm{F}}^2-\dot{\varphi}_{\mathrm{B}}^2) \equiv \frac{C}{e^2}\int_0^t dt\dot{\varphi}^+\dot{\varphi}^- \tag{16.8}$$

to the term (16.7), i.e., $S[\varphi] = S_{\mathrm{sc}}[\varphi]+S_c[\varphi]$.

For the sake of convenience and without loss of generality in what follows, we set the electric potential and fluctuating phases of the superconducting terminal equal to zero. Then the Green–Keldysh matrix of this electrode can be written in a simple form $\check{G}_{\mathrm{S}} = \begin{pmatrix}\hat{g}^{\mathrm{R}} & \hat{g}^{\mathrm{K}}\\ 0 & \hat{g}^{\mathrm{A}}\end{pmatrix}$ with

$$\hat{g}^{\mathrm{R/A}}(t) = \pm\delta(t)\hat{\tau}_z - \theta(\pm t)\left(\Delta_0\hat{\tau}_z J_1(\Delta_0 t)+\mathrm{i}\hat{\Delta}J_0(\Delta_0 t)\right) \tag{16.9}$$

and $\hat{g}^{\mathrm{K}} = \hat{g}^{\mathrm{R}}F - F\hat{g}^{\mathrm{A}}$, where $F(t) = -\mathrm{i}T/\sinh[\pi T t]$ is the Fourier transform of $1-2n(\varepsilon)$ and $n(\varepsilon) = 1/(1+e^{\varepsilon/T})$ is the Fermi function. Here $J_{0,1}$ are the Bessel functions, $\hat{\tau}_{x,y,z}$ are the Pauli matrices, $\theta(t)$ is the Heaviside step function, and $\hat{\Delta} = \mathrm{i}\Delta_0\hat{\tau}_y$, where Δ_0 is chosen real.

The Green–Keldysh matrix of the normal terminal is defined as

$$\check{G}_{\rm N}(t,t') = \frac{1}{2}\begin{pmatrix} \hat{1} & \hat{1} \\ \hat{1} & -\hat{1} \end{pmatrix} \check{Q}_{\rm N}(t,t') \begin{pmatrix} \hat{1} & \hat{1} \\ \hat{1} & -\hat{1} \end{pmatrix} \tag{16.10}$$

where

$$\check{Q}_{\rm N}(t,t') = \int \frac{d\varepsilon}{2\pi} e^{-i\varepsilon(t-t')} \begin{pmatrix} e^{i\varphi_F(t)\hat{\tau}_z} & 0 \\ 0 & e^{i\varphi_B(t)\hat{\tau}_z} \end{pmatrix}$$
$$\times \begin{pmatrix} (1-2n(\varepsilon))\hat{\tau}_z & 2n(\varepsilon)\hat{\tau}_z \\ 2(1-n(\varepsilon))\hat{\tau}_z & (2n(\varepsilon)-1)\hat{\tau}_z \end{pmatrix} \begin{pmatrix} e^{-i\varphi_F(t')\hat{\tau}_z} & 0 \\ 0 & e^{-i\varphi_B(t')\hat{\tau}_z} \end{pmatrix} \tag{16.11}$$

Substituting the above expressions for $\check{G}_{\rm S}$ and $\check{G}_{\rm N}$ into Eq. (16.7), we arrive at the action that fully describes transfer of electrons between N- and S-terminals to all orders in T_n.

In the limit of small channel transmissions, one can expand $S_{\rm sc}$ in powers of T_n. Keeping the terms up to $\sim T_n^2$, one immediately recovers the contribution from Andreev reflection. It is straightforward to verify that at low energies, this part of the action takes the same form as that derived in Chapter 2 for normal tunnel barriers; cf., e.g., Eq. (2.113) in which one should only substitute e by $2e$ and the conductance $G_{\rm N}$ by $G_{\rm A}$ (see also [411]). Here, however, we are aiming at a more general description, which includes arbitrary transmission values T_n. For this reason, we will proceed differently and employ the trick that allows us to perform an effective re-summation of the action to all orders in T_n [132].

Let us define the matrix

$$\check{X}_0[\varphi^+] = 1 - T_n/2 + (T_n/4)\left\{\check{G}_{\rm N}, \check{G}_{\rm S}\right\}|_{\varphi^-=0} \tag{16.12}$$

As the action $S_{\rm sc}$ (16.7) vanishes for $\varphi^-(t) = 0$, one has $\mathrm{Tr}\ln \check{X}_0 = 0$. Making use of this property, we can identically transform the action (16.7) to

$$S_{\rm sc} = -\frac{i}{2}\sum_n \mathrm{Tr}\ln\left[1 + \check{X}_0^{-1} \circ \check{X}'\right] \tag{16.13}$$

where we introduce the matrix

$$\check{X}' = 1 + (T_n/4)\left(\left\{\check{G}_{\rm N}, \check{G}_{\rm S}\right\} - 2\right) - \check{X}_0 \tag{16.14}$$

At temperatures and voltages well below the superconducting gap, Andreev contribution to the action dominates. Hence, it suffices to consider the limit of low energies $\varepsilon \ll \Delta$, in which case one can set $\check{G}_S \to \begin{pmatrix} \hat{\tau}_y & 0 \\ 0 & \hat{\tau}_y \end{pmatrix}$. Then we obtain

$$\check{X}_0^{-1}(t, t') = \frac{2}{2 - T_n} \left(\begin{array}{c|c} \delta(t-t')\hat{1} & \frac{2T_n}{T_n - 2} \sin\left[\varphi^+(t) - \varphi^+(t')\right] F(t-t')\mathrm{i}\hat{\tau}_y \\ \hline 0 & \delta(t-t')\hat{1} \end{array} \right) \tag{16.15}$$

and

$$\check{X}'(t, t') = \frac{T_n}{2}\delta(t-t') \left(\begin{array}{c|c} 0 & -\sin\varphi^-(t)\mathrm{i}\hat{\tau}_y \\ \hline \sin\varphi^-(t)\mathrm{i}\hat{\tau}_y & 0 \end{array} \right)$$
$$+ T_n F(t-t') \left(\begin{array}{c|c} a(t,t')\mathrm{i}\hat{\tau}_y & b(t,t')\mathrm{i}\hat{\tau}_y \\ \hline c(t,t')\mathrm{i}\hat{\tau}_y & d(t,t')\mathrm{i}\hat{\tau}_y \end{array} \right) \tag{16.16}$$

where

$$a(t, t') = -\cos\frac{\varphi^-(t)}{2}\sin\frac{\varphi^-(t')}{2}\cos(\varphi^+(t) - \varphi^+(t')),$$
$$b(t, t') = \left[\cos\frac{\varphi^-(t)}{2}\cos\frac{\varphi^-(t')}{2} - 1\right]\sin(\varphi^+(t) - \varphi^+(t')),$$
$$c(t, t') = \sin\frac{\varphi^-(t)}{2}\sin\frac{\varphi^-(t')}{2}\sin(\varphi^+(t) - \varphi^+(t')),$$
$$d(t, t') = \sin\frac{\varphi^-(t)}{2}\cos\frac{\varphi^-(t')}{2}\cos(\varphi^+(t) - \varphi^+(t')). \tag{16.17}$$

Now let us assume that either the dimensionless Andreev conductance $g_A = 4\sum_n \mathcal{T}_n$ is large, $g_A \gg 1$, or temperature is sufficiently high (though still smaller than Δ_0). In either case—just as we already did, e.g., in Chapters 10 and 11 in the case of normal conductors—one can describe quantum dynamics of the phase variable φ within the quasiclassical approximation, which amounts to expanding S_{sc} in powers of (small) "quantum" part of the phase $\varphi^-(t)$. Employing Eqs. (16.13)–(16.17) and expanding S_{sc} up to terms $\sim (\varphi^-)^2$, we recover the effective action [132]

$$\mathrm{i}S_{sc} = \mathrm{i}S_R - S_I \tag{16.18}$$

where

$$S_R = -\frac{g_A}{2\pi}\int_0^t dt'\, \varphi^-(t')\dot{\varphi}^+(t'), \tag{16.19}$$

$$S_I = -\frac{g_A}{4}\int_0^t dt' \int_0^t dt'' \frac{T^2}{\sinh^2[\pi T(t' - t'')]}\varphi^-(t')\varphi^-(t'')$$
$$\times[\beta_A \cos(2\varphi^+(t') - 2\varphi^+(t'')) + 1 - \beta_A] \tag{16.20}$$

Comparing these results with those derived in Chapters 10 and 11, we observe the action S_{sc} (16.18)–(16.20) is expressed in exactly the same form as that for normal conductors derived within the same quasiclassical approximation for the phase variable $\varphi(t)$. In order to observe the correspondence between the action (11.3) and that defined in Eqs. (16.18)–(16.20), one only needs to ignore all cubic terms $\sim (\varphi^-)^3$ in Eq. (11.3), to interchange

$$G_N \leftrightarrow G_A, \quad \beta_N \leftrightarrow \beta_A \tag{16.21}$$

and to account for an extra factor 2 in front of the phase φ^+ under cos in Eq. (16.20). This factor just implies effective charge doubling $e \to 2e$ in the course of Andreev reflection.

16.2.2 Shot Noise and Interaction Correction

Further analysis is formally similar to that developed in Chapter 11 for normal conductors. Hence, we can immediately proceed to our final results. Let us define the average current and the current–current correlator as

$$\langle \hat{I}(t)\rangle = \mathrm{i}e \int \mathcal{D}\varphi^{\pm} \frac{\delta}{\delta\varphi^-(t)} e^{\mathrm{i}S[\varphi]}, \tag{16.22}$$

$$\frac{1}{2}\langle \hat{I}\hat{I}\rangle_+ = -e^2 \int \mathcal{D}\varphi^{\pm} \frac{\delta^2}{\delta\varphi^-(t)\delta\varphi^-(t')} e^{\mathrm{i}S[\varphi]} \tag{16.23}$$

where $\langle \hat{I}\hat{I}\rangle_+ = \langle \hat{I}(t)\hat{I}(t') + \hat{I}(t')\hat{I}(t)\rangle$. In the absence of interactions, we set $\dot{\varphi}^+ = eV$ and immediately recover the standard result $I = G_A V$ with the Andreev conductance defined in Eqs. (16.2) and (16.3). For the current fluctuations $\delta I(t)$ from Eqs. (16.18)–(16.23), we obtain [132]

$$\frac{\langle |\delta I|^2_\omega\rangle}{G_A} = (1-\beta_A)\omega \coth\frac{\omega}{2T} + \frac{\beta_A}{2}\sum_{\pm}(\omega \pm 2eV)\coth\frac{\omega \pm 2eV}{2T} \tag{16.24}$$

This equation fully describes current noise in NS structures at energies well below the superconducting gap. For $eV \gg T, \omega$ Eq. (16.24) trivially reduces to the result (16.5), while in the diffusive regime, the correlator (16.24)—combined with Eqs. (16.6)—matches with the semiclassical expression [311].

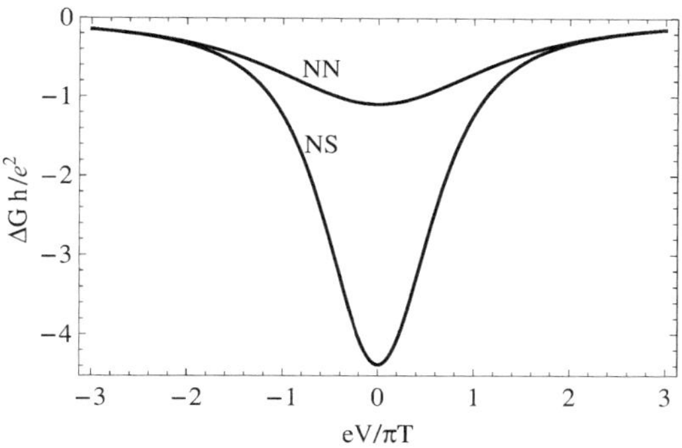

Figure 16.3 The interaction correction $\delta G = dI/dV - G_N$ for short diffusive conductors at $T = G_N/2\pi C$. The upper and lower curves correspond to normal and NS structures, respectively.

Let us now turn on interactions. Technically, this amounts to considering the full action $S[\varphi]$, including the charging term (16.8) and to allowing for arbitrary phase fluctuations. Proceeding along the same lines as in Chapter 11, for $g_A \gg 1$ or $\max(T, eV) \gg E_C = e^2/2C$, we get [132]

$$I = G_A V - 2e\beta_A T \,\mathrm{Im}\left[w\Psi\left(1+\frac{w}{2}\right) - \mathrm{i}v\Psi\left(1+\frac{\mathrm{i}v}{2}\right)\right] \quad (16.25)$$

where $\Psi(x)$ is the digamma function, $w = g_A E_C/\pi^2 T + \mathrm{i}v$, and $v = 2eV/\pi T$. This result is displayed in Fig. 16.3.

The last term in Eq. (16.25) is the interaction correction to the I–V curve, which scales with the Andreev Fano factor β_A in exactly the same way as the shot noise (16.5). Thus, we arrive at an important conclusion: *Interaction correction to Andreev conductance of NS structures is proportional to the shot noise power in such structures.* This fundamental relation between interaction effects and shot noise goes along with that established in Chapter 11 for normal conductors directly extending the latter to superconducting systems. In both cases, the fundamental physical reason behind this relation lies in discrete nature of the charge carriers passing through the conductor, respectively, single electrons and Cooper pairs.

It is remarkable that the interaction correction to Andreev conductance defined in Eq. (16.25) has exactly the same functional form as that for normal conductors; cf. Eq. (11.18). For diffusive systems, one can make use of Eqs. (16.6) and conclude that the only difference between the interaction corrections to the $I-V$ curve in normal and NS systems is the charge doubling in the latter case. As a result, the Coulomb dip on the $I-V$ curve of a diffusive NS system at any given T is exactly *two times narrower* than that in the normal case. This effect was experimentally detected in normal wires attached to superconducting electrodes [63].

Thus, seemingly different experiments [207, 238] and [63] are actually closely related: Doubling of the shot noise found in NS structures [207, 238] corresponds to the narrowing of the $I-V$ curve (otherwise pertaining its form exactly as in the normal case) observed in [63]. The absolute value of the interaction correction to the conductance is proportional to (effective charge) $\times$ (shot noise power), i.e., doubling of the shot noise in diffusive NS structures implies *four times bigger* interaction correction to conductance than in the normal case, as illustrated in Fig. 16.3.

16.3 Combined Effect of Disorder and Interactions

In the above analysis, the effect of disorder was taken into account only inside a short coherent conductor (scatterer) connecting normal and superconducting reservoirs (see Fig. 16.2) but not in these reservoirs. Now let us also include electron scattering on disorder in the bulk metallic electrodes into our consideration. As we already learned in previous chapters, scattering on disorder turns out to be crucially important for low-temperature transport properties of normal metals giving rise, e.g., to the phenomenon of weak localization as well as drastically increasing Coulomb interaction effects. Below we will demonstrate that qualitatively similar features are also observed in hybrid NS structures.

Actually the effect of disorder can even be much bigger in the latter case. Indeed, the magnitudes of both weak localization and interaction corrections to the Drude conductance of a normal metal are typically small, being proportional to the inverse dimensionless conductance of such a metal (see, e.g., Chapters 11–15). For

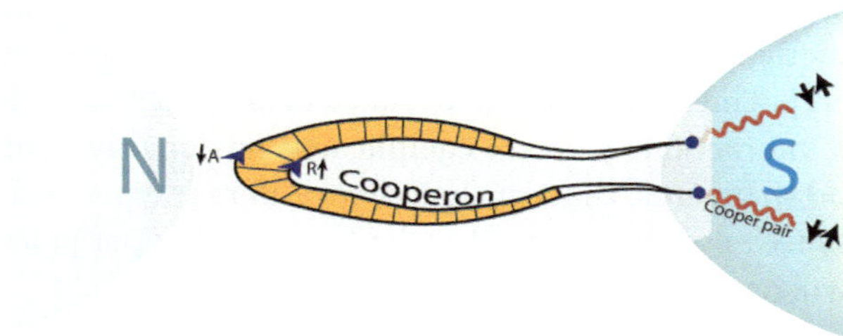

Figure 16.4 Hybrid NS structure under consideration. A normal metal piece of length d_n is connected on one side (at $x = 0$) to a superconducting reservoir via a tunnel barrier with normal state conductance G_{N} and on the opposite side (at $x = -d_n$) to a normal reservoir via another barrier with normal state conductance $\tilde{G}_t$. The diagram illustrates conversion of a Cooper pair into a pair of electrons propagating diffusively inside the normal metal (the Cooperon).

comparison, in NS hybrid structures, electron transport at low energies is dominated by quantum coherent mechanism of Andreev reflection, which is sensitive to both disorder and interactions. Hence, one can expect that the effect of disorder and interactions on the low-energy conductance of NS structures may be large (of order unity or even bigger) rather than $\sim 10^{-3} \div 10^{-4}$ as in the case of generic normal metals.

The effect of disorder on the subgap Andreev conductance G_{A} of NS hybrids is schematically illustrated in Fig. 16.4. Two electrons forming a Cooper pair penetrate into a normal metal from a superconductor where they propagate and suffer scattering at disorder. As a result of multiple scattering events, electron trajectories in the normal metal may form loops (cf. Fig. 14c), so that time-reversed paths account for quantum interference, effectively enhancing Andreev conductance of NS structures and resulting in the so-called zero-bias anomaly of G_{A} in the limit of low temperatures and voltages [218]. Technically, the whole process is described by the Cooperon that—apart from its spin structure—is similar to that encountered in the weak localization theory. It follows immediately that Cooper pairs penetrating a normal metal can—just like single electrons addressed in Chapters 14 and 15—be the subject of *dephasing* by electron–electron interactions [199, 358, 364]. Another effect of electron–electron interactions is

Coulomb blockade of Andreev reflection [411], which may result in yet another type of zero-bias anomaly of Andreev conductance [131, 199]. Thus, the combined effect of disorder and electron–electron interactions yields a complicated low-energy behavior of NS hybrid structures, which includes two different types of zero-bias conductance anomalies as well as dephasing of Cooper pairs. These effects will be the subject of our subsequent analysis.

16.3.1 Andreev Effective Action

The effect of disorder on subgap electron transport in normal-superconducting hybrid structures was investigated theoretically by means of different techniques, including real-time Keldysh–Usadel approach [388], tunneling Hamiltonian method [190, 191], scattering matrix theory combined with scaling arguments [45], and effective action analysis [411]. Below we will follow the latter approach [411], which appears most adequate for describing a combined effect of disorder and electron–electron interactions.

Let us consider a piece of a normal metal of length d_n and conductance G_n attached to a massive superconducting electrode via a tunnel barrier located in the plane $x = 0$, as it is shown in Fig. 16.4. For simplicity, we will assume that normal and superconducting metals are characterized by the same electron mean free path ℓ and diffusion coefficient $D = v_{\mathrm{F}}\ell/3$. We will also assume that the Thouless energy of the normal conductor $\epsilon_{\mathrm{Th}} = D/d_n^2$ remains much smaller than the superconducting gap, i.e., $\epsilon_{\mathrm{Th}} \ll \Delta_0$.

Electron dynamics in both metals can be described within the standard quasiclassical theory based on the Eilenberger equations (2.155) for the energy-integrated Green function matrix $\hat{g}$. In the diffusive limit considered here, this matrix becomes almost isotropic in the momentum space and one can apply the standard approximation [383]

$$\hat{g}(\boldsymbol{v}_{\mathrm{F}}, \boldsymbol{r}) = \hat{G}(\boldsymbol{r}) - \frac{\boldsymbol{v}_{\mathrm{F}}}{|\boldsymbol{v}_{\mathrm{F}}|}\hat{G}(\boldsymbol{r})\nabla_r\hat{G}(\boldsymbol{r}) \tag{16.26}$$

where the matrix $\hat{G}(\boldsymbol{r}) = G(\boldsymbol{r})\hat{\tau}_z + F(\boldsymbol{r})\mathrm{i}\tau_y$ does not depend on $\boldsymbol{v}_{\mathrm{F}}$. The expression for the current density (2.158) then reads

$$j(\boldsymbol{r}, \tau) = -\frac{\mathrm{i}ep_{\mathrm{F}}^2\ell}{6\pi}\int d\tau_1 \mathrm{Sp}[\hat{\tau}_z\hat{G}(\boldsymbol{r}; \tau, \tau_1)\nabla_r\hat{G}(\boldsymbol{r}; \tau_1, \tau)] \tag{16.27}$$

The boundary conditions (2.167) describing electron transfer between two metals through a tunnel barrier at the NS interface take the form [244]

$$\hat{G}(x=+0)\nabla_x\hat{G}(x=+0) = \hat{G}(x=-0)\nabla_x\hat{G}(x=-0)$$
$$= \frac{G_N R_q}{N_{ch}\ell}[\hat{G}(x=+0)\hat{G}(x=-0) - \hat{G}(x=-0)\hat{G}(x=+0)] \quad (16.28)$$

where, as before, G_N is the normal state conductance of the tunnel barrier defined in Eq. (16.1) with all $T_n \ll 1$, $N_{ch} = p_F^2\mathcal{A}/4\pi$ is the number of conducting channels per spin and $\mathcal{A}$ is the barrier cross section.

In order to describe Andreev processes and to establish the corresponding contribution to the effective action, we will assume that the normal state conductance of the tunnel barrier G_N remains much smaller than that of the normal conductor $1/R$, i.e.,

$$G_N R \ll 1 \quad (16.29)$$

In this limit, one can proceed perturbatively and evaluate the Green function matrix $\hat{G}$ keeping the terms proportional to G_N. Inverting the differential operators in Eq. (16.28) and making use of the normalization condition (2.157) $\hat{G}^2 = \hat{1}$, we arrive at the formal expressions for the Green function matrices on both sides of the tunnel barrier:

$$\hat{G}(x=-0,\tau,\tau') = \hat{G}_n(\tau,\tau') - \frac{3g_N}{16N_{ch}\ell}\left[\int d\tau_1 \nabla^{-1}_{x=-0}(\tau,\tau_1)\hat{G}_s(\tau_1,\tau')\right.$$
$$\left. - \int d\tau_1 \int d\tau_2 \int d\tau_3 \nabla^{-1}_{x=-0}(\tau,\tau_1)\hat{G}_n(\tau_1,\tau_2)\hat{G}_s(\tau_2,\tau_3)\hat{G}_n(\tau_3,\tau')\right]$$
$$(16.30)$$

and

$$\hat{G}(x=+0,\tau,\tau') = \hat{G}_s(\tau,\tau') + \frac{3g_N}{16N_{ch}\ell}\left[\int d\tau_1 \nabla^{-1}_{x=+0}(\tau,\tau_1)\hat{G}_n(\tau_1,\tau')\right.$$
$$\left. - \int d\tau_1 \int d\tau_2 \int d\tau_3 \nabla^{-1}_{x=+0}(\tau,\tau_1)\hat{G}_s(\tau_1,\tau_2)\hat{G}_n(\tau_2,\tau_3)\hat{G}_s(\tau_3,\tau')\right],$$
$$(16.31)$$

where $g_N = G_N R_q$ is the dimensionless barrier conductance, $\hat{G}_n(\tau) = G_n(\tau)\hat{\tau}_z$ and $\hat{G}_s(\tau) = G_s(\tau)\hat{\tau}_z + F_s(\tau)i\hat{\tau}_y$ are the Green function matrices in the bulk normal and superconducting electrodes.

Combining Eqs. (16.30) and (16.31) with Eq. (16.27) and employing the formula for the action (2.159), we recover the Andreev contribution to the effective action

$$S^{(A)}[\varphi] = -\int d\tau_1 \int d\tau_2 \int d\tau_3 \int d\tau_4 \gamma(\tau_1, \tau_2, \tau_3, \tau_4) \\ \times \cos(\varphi(\tau_1) - \varphi(\tau_2) - \varphi(\tau_3) + \varphi(\tau_4)), \qquad (16.32)$$

where

$$\gamma(\tau_1, \tau_2, \tau_3, \tau_4) = \frac{3g_{\rm N}^2}{128 N_{\rm ch}\ell}[\mathcal{G}_n(\tau_1 - \tau_2)F_s(\tau_2 - \tau_3)G_n(\tau_3 - \tau_4)F_s(\tau_4 - \tau_1) \\ -G_n(\tau_1 - \tau_2)\mathcal{F}_s(\tau_2 - \tau_3)G_n(\tau_3 - \tau_4)F_s(\tau_4 - \tau_1)] \qquad (16.33)$$

and we defined

$$\mathcal{G}_n(\tau - \tau') = \int d\tau_1 \nabla^{-1}_{x=-0}(\tau, \tau_1)G_n(\tau_1, \tau'), \\ \mathcal{F}_s(\tau - \tau') = \int d\tau_1 \nabla^{-1}_{x=+0}(\tau, \tau_1)F_s(\tau_1, \tau') \qquad (16.34)$$

Making use of the well-known expressions for $G_n(\tau) = -\mathrm{i}T/\sin(\pi T\tau)$ and

$$G_s(\tau) = T\sum_\omega e^{-\mathrm{i}\omega\tau}\frac{\omega}{\Omega}, \quad F_s(\tau) = T\sum_\omega e^{-\mathrm{i}\omega\tau}\frac{\Delta_0}{\Omega}, \quad \Omega = \sqrt{\omega^2 + \Delta_0^2}, \qquad (16.35)$$

cf. Eq. (2.162), in the adiabatic limit (i.e., at frequencies $\omega \ll \Delta_0$), we obtain

$$S^{(\mathrm{A})}[\varphi] = -\int d\tau \int d\tau' \gamma(\tau - \tau')\cos(2\varphi(\tau) - 2\varphi(\tau')), \quad (16.36)$$

where

$$\gamma(\tau) = \frac{3g_{\rm N}^2}{128 N_{\rm ch}\ell}[\mathcal{G}_n(\tau) - \tilde{\mathcal{G}}_n(\tau)]G_n(-\tau) \qquad (16.37)$$

and we defined $\tilde{\mathcal{G}}_n = G_n\mathcal{F}_sF_s^{-1}$.

What remains is to evaluate the inverse operators $\nabla^{-1}_{x=\pm 0}$ in Eqs. (16.34). For this purpose, it suffices to solve the Usadel equations [383] in both normal and superconducting electrodes. Assuming the planar geometry (in which case the Green functions depend only on the coordinate x normal to the tunnel barrier plane) and making use of a substitution $F_\omega = \sin\chi_\omega$, one can reduce the Usadel equation to a very simple form [427]

$$\frac{d^2\chi}{dx^2} = \frac{2|\omega|}{D}\sin\chi - \frac{2\Delta_0}{D}\cos\chi \qquad (16.38)$$

In the superconducting electrode ($x > 0$), this equation can be trivially solved bearing in mind that due to the presence of a tunnel barrier at the NS interface, the deviation of the F-function from its bulk value F_s (16.35) remains small, i.e., $\delta\chi = \chi - \arcsin(\Delta_0/\Omega) \ll 1$. Linearizing Eq. (16.38), one gets $d^2\delta\chi/dx^2 = (2\Omega/D)\delta\chi$, which immediately yields

$$\nabla^{-1}_{x=+0} = -\sqrt{\frac{D}{2\Omega}} \qquad (16.39)$$

An analogous procedure can be worked out for the normal electrode, i.e., for $-d_n < x < 0$. We again solve the Usadel equation (16.38), now setting $\Delta_0 \equiv 0$ and imposing a proper boundary condition for χ at $x = -d_n$. Note that at not too small frequencies $\omega \gg \epsilon_{\mathrm{Th}}$, the form of the F-function is actually insensitive to this boundary condition. At such frequencies, the Cooper pair amplitude penetrating into the normal metal from the superconductor through the tunnel barrier remains small $\chi \ll 1$ and from Eq. (16.38) with $\Delta_0 = 0$, one readily gets $\chi \sim \exp((2|\omega|/D)^{1/2}x)$ implying

$$\nabla^{-1}_{x=-0} = \sqrt{\frac{D}{2|\omega|}} \qquad (16.40)$$

In the low-frequency limit $\omega \ll \epsilon_{\mathrm{Th}}$, the choice of the boundary condition at $x = -d_n$ becomes significant. Below we will specify the results for two different physical situations. One of them corresponds to the presence of a massive metallic (non-superconducting) reservoir directly attached to the normal metal at $x = -d_n$. In this case, the Cooper pair amplitude at $x = -d_n$ is suppressed, i.e., we have $\chi(-d_n) = 0$. Another situation is realized if a weakly transparent tunnel barrier is located at $\chi(-d_n) = 0$. Then with the probability close to one, incident electrons get reflected at this barrier and we arrive at the boundary condition $\chi'(-d_n) = 0$. One can demonstrate that these two different boundary conditions are realized, respectively, for $\tilde{G}_t R \ll 1$ and $\tilde{G}_t R \gg 1$, where $\tilde{G}_t$ denotes the conductance of a tunnel barrier situated at $x = -d_n$.

Employing the general solution to the Usadel equation,

$$\chi(x) = \chi(0) + \chi'(0)x + \frac{x^2|\omega|}{D}\sin\chi(0), \quad \nabla^{-1}_{x=-0} = \frac{\chi(0)}{\chi'(0)} \tag{16.41}$$

for these two physical situations, one readily finds

$$\nabla^{-1}_{x=-0} = \begin{cases} d_n, & \chi(-d_n) = 0, \\ D/2|\omega|d_n, & \chi'(-d_n) = 0, \end{cases} \tag{16.42}$$

where the second equation remains valid as long as $\chi(0) \ll 1$. Note that with the aid of Eq. (16.28), the latter inequality can be rewritten as

$$|\omega| \gg \epsilon_{\rm Th} G_{\rm N} R \tag{16.43}$$

where for the sake of definiteness, we set $1/R = 2e^2 N_0 D\mathcal{A}/d_n$ with N_0 being the electron density of states at the Fermi level. For even smaller frequencies (and provided $\chi'(-d_n) = 0$), the Cooper pair amplitude induced in the normal metal due to the proximity effect is of order one even for weakly transparent tunnel barrier at the NS interface. Hence, provided the inequality (16.43) is violated, our perturbative in tunneling approach ceases to be valid.

Having established the expressions for the inverse operators $\nabla^{-1}_{x=\pm 0}$ in all relevant limits, we are now ready to reconstruct the kernel (16.37). Substituting Eq. (16.39) into the second Eq. (16.34), for the term originating from the superconducting electrode, we find

$$-\tilde{\mathcal{G}}_n(\tau)G_n(-\tau) = \sqrt{\frac{D}{2\Delta_0}}\frac{1}{(\pi\tau)^2} \tag{16.44}$$

Likewise, combining Eq. (16.40) with the first Eq. (16.34), within the frequency interval $\epsilon_{\rm Th} \ll |\omega| \ll \Delta$, we obtain

$$\mathcal{G}_n(\tau)G_n(-\tau) = \frac{D^{1/2}}{2(\pi|\tau|)^{3/2}} \tag{16.45}$$

At lower frequencies $|\omega| \ll \epsilon_{\rm Th}$, from Eqs. (16.42) and (16.34), we get

$$\mathcal{G}_n(\tau)G_n(-\tau) = \begin{cases} d_n/(\pi\tau)^2, & F(-d_n) = 0, \\ D/4\pi d_n|\tau|, & F'(-d_n) = 0 \end{cases} \tag{16.46}$$

Equations (16.36), (16.37), and (16.44)–(16.46) fully describe the contribution of Andreev processes to the effective action of

NS hybrid structures at subgap frequencies/energies and in the presence of disorder in the electrodes.

As we already pointed out, the applicability of the second equation (16.46) is restricted to the frequency interval $\epsilon_{\mathrm{Th}} G_{\mathrm{N}} R \ll |\omega| \ll \epsilon_{\mathrm{Th}}$. One can demonstrate that for the case $F'(-d_n) = 0$ and at even lower frequencies $|\omega| \ll \epsilon_{\mathrm{Th}} G_{\mathrm{N}} R$, the kernel $\gamma(\tau)$ in Eq. (16.36) is determined by a non-perturbative expression $\gamma(\tau) \propto G_{\mathrm{N}}/\tau^2$. We also note that in the interesting-for-us subgap limit $|\omega| \ll \Delta_0$, the contribution (16.44) always remains smaller than that defined in Eqs. (16.45) and (16.46). Nevertheless, we will keep all these contributions for the sake of completeness.

16.3.2 Andreev Conductance Affected by Disorder and Interactions

Let us now make use of the above results and evaluate the Andreev conductance G_{A} of our NS structure in the presence of both disorder and electron–electron interactions.

For the sake of future comparison, we first present a simple formula for G_{A} not affected by disorder and interactions. Combining Eqs. (16.1)–(16.3), in the tunneling limit $T_n \ll 1$ and assuming for simplicity that all transmission values T_n are equal to each other, for the subgap (Andreev) current, one readily finds

$$I = G_{\mathrm{A}} V, \quad G_{\mathrm{A}} = \frac{G_{\mathrm{N}}^2 R_{\mathrm{Q}}}{N_{\mathrm{ch}}} \tag{16.47}$$

Now let us include disorder and interactions into our consideration. This task can be conveniently accomplished within the framework of the effective action approach. As usual, taking into account charging effects by introducing the capacitance C of a tunnel barrier at the NS interface, we arrive at the effective action

$$S[\varphi] = \frac{C}{2e^2} \int d\tau \dot{\varphi}^2 + S^{(A)}[\varphi] + S_D[\varphi] \tag{16.48}$$

where $S^{(A)}[\varphi]$ (16.36) describes Andreev processes in the presence of disorder and the term $S_{\mathrm{D}}[\varphi]$ accounts for dissipation and will be specified later.

Let us, for a moment, ignore dissipation and disregard the corresponding contribution $S_{\mathrm{D}}[\varphi]$ to the effective action. In order to proceed, we first evaluate the Andreev tunneling rate Γ_{2e}, i.e., the tunneling rate of two subgap electrons across the NS interface; see Fig. 16.1. This rate can be easily evaluated by means of the ImF technique described in Chapter 4 and extensively employed, e.g., in Chapter 5 and subsequent chapters.

In the absence of tunneling, an applied voltage V induces an external charge $Q = VC$ at the tunnel barrier. Electron tunneling makes this charge state unstable. Its decay rate Γ_{2e} at low temperatures can be evaluated by means of the general formula (4.10). Proceeding along the same lines as, e.g., in Chapters 5 and 7 (cf. Sections 5.1.3 and 7.4) and treating Andreev processes perturbatively, one immediately arrives at the partition function

$$\mathcal{Z} \sim e^{-(Q^2/2C+F_{2e})/T}, \quad F_{2e} = -\sum_{\pm}\int_0^\infty d\tau\gamma(\tau)\exp[(\pm 2eV - 4E_{\rm C})\tau] \tag{16.49}$$

where F_{2e} defines the contribution of Andreev processes to the free energy, $E_{\rm C} = e^2/2C$ is the charging energy, and the expression for the kernel $\gamma(\tau)$ is set by Eqs. (16.37), (16.44)–(16.46). The Andreev tunneling rate is then derived from the formula $\Gamma_{2e} = -2{\rm Im}F_{2e}$.

From Eq. (16.49), it is easy to observe that for $|V| < e/C$ and $T \to 0$, the value F_{2e} contains no imaginary part and, hence, $\Gamma_{2e} = 0$. This is Coulomb blockade of Andreev reflection. For $|V| > e$, the integral in the expression for F_{2e} (16.49) formally diverges and the free energy acquires an imaginary part, which can be established by means of the standard analytic continuation procedure, as we repeatedly demonstrated in previous chapters. As a result, we obtain

$$\Gamma_{2e} = \Gamma_{2e}^N + \Gamma_{2e}^S \tag{16.50}$$

where the two terms in the right-hand side originate, respectively, from the two contributions to the kernel $\gamma(\tau)$ (16.37). They can be interpreted as disorder-induced quantum interference contributions of normal (Γ_{2e}^N) and superconducting (Γ_{2e}^S) electrodes. Making use of Eqs. (16.49), (16.37), and (16.44), after an analytic continuation for the latter contribution, one gets

$$\Gamma_{2e}^S = \frac{G_{\rm N}^2 R_{\rm Q}}{4eN_{\rm ch}}\frac{V - e/C}{(2\Delta_0\tau_e/3)^{1/2}} \tag{16.51}$$

where, as before, $\tau_e = \ell/v_{\rm F}$ is the electron elastic time. Likewise, employing Eqs. (16.49), (16.37), (16.45), and (16.46), one finds

$$\Gamma_{2e}^{N} = \frac{G_{\rm N}^2 R_{\rm Q}}{4eN_{\rm ch}}\sqrt{\frac{3(V - e/C)}{2e\tau_e}} \tag{16.52}$$

for $eV - 2E_{\rm C} \gg \epsilon_{\rm Th}$ and

$$\Gamma_{2e}^{N} = \begin{cases} \frac{3G_{\rm N}^2 R_{\rm Q}}{4eN_{\rm ch}}\frac{d_n}{\ell}(V - e/C), & F(-d_n) = 0, \\ \frac{\pi G_{\rm N}^2 R_{\rm Q}}{32e^2 N_{\rm ch}}\frac{v_{\rm F}}{d_n}, & F'(-d_n) = 0 \end{cases} \tag{16.53}$$

in the opposite limit $eV - 2E_{\rm C} \ll \epsilon_{\rm Th}$. As before, the validity range of the second Eq. (16.53) is additionally restricted by the condition $eV - 2E_{\rm C} \gg \epsilon_{\rm Th} G_{\rm N} R$.

Combining these results for Γ_{2e} with the equation

$$I = 2e\Gamma_{2e} \tag{16.54}$$

one arrives at the expression for the subgap (Andreev) current across our NS structure. In accordance with the above considerations, this current is blocked by Coulomb interaction $I = 0$ at $|V| < e/C$ and $T \to 0$. At higher subgap voltages, Andreev reflection remains the dominant charge transfer mechanism, and the I–V curve is determined by Eqs. (16.51)–(16.54).

Let us, for a moment, ignore Coulomb effects by setting $E_{\rm C} \to 0$ and focus our attention on the effect of disorder. Then, e.g., at not too low voltages $eV \gg \epsilon_{\rm Th}$, one arrives at the following expression for the differential Andreev conductance:

$$G_{\rm A} = \frac{dI}{dV} = \frac{G_{\rm N}^2 R_{\rm Q}}{4N_{\rm ch}}\left(\sqrt{\frac{3}{2eV\tau_e}} + \sqrt{\frac{6}{\Delta_0 \tau_e}}\right) \tag{16.55}$$

where the two terms in the brackets represent the contributions from, respectively, the normal and superconducting electrodes. Having in mind that $\Delta_0 \tau_e \ll 1$ and comparing the result (16.55) with that in the ballistic limit (16.47), we immediately conclude that disorder-enhanced interference effects on both sides of the tunnel barrier yield a strong increase in the Andreev conductance $G_{\rm A}$. In the limit $eV \ll \Delta$ considered here, the interference effects in the normal metal provide the dominant contribution and we have $G_{\rm A} \propto 1/\sqrt{V}$.

For large values of d_n, this nontrivial behavior holds down to very low voltages and is frequently called *zero-bias anomaly* (ZBA) of Andreev conductance. At $eV \sim \epsilon_{\rm Th}$, the dependence (16.55) crosses over to a different one, which can easily be recovered from Eqs. (16.53) and (16.54).

It is also instructive to express the above results for $G_{\rm A}$ in a slightly different form. For instance, in the physical situation corresponding to the "open" boundary condition $F(-d_n) = 0$ and disregarding the small extra contribution generated by the superconducting reservoir (last term in Eq. (16.55)), one can express the Andreev conductance in a remarkably simple form

$$G_{\rm A}(V) = \begin{cases} G_{\rm N}^2 R(\epsilon_{\rm Th}/8eV)^{1/2} & eV \gg \epsilon_{\rm Th}, \\ G_{\rm N}^2 R, & eV \ll \epsilon_{\rm Th} \end{cases} \qquad (16.56)$$

demonstrating that, e.g., in the low-voltage limit, one has $G_{\rm A}/G_{\rm N} = G_{\rm N} R$.

Let us now return to Coulomb effects. As we have already learned in the previous chapters, dynamical effects related to electron–electron interactions may play a crucial role in disordered conductors at low temperatures leading to both quantum dissipation and quantum dephasing. The simplest way to account for dissipation is to phenomenologically consider an extra Ohmic resistance $R_{\rm S}$ shunting our NS hybrid structure. This extra resistance is accounted for by the standard Caldeira–Leggett term (2.184)

$$S_{\rm D} = \frac{\pi T^2}{16e^2 R_{\rm S}} \int_0^{1/T} d\tau \int_0^{1/T} d\tau' \frac{[\varphi(\tau) - \varphi(\tau')]^2}{\sin^2[\pi T(\tau - \tau')]} \qquad (16.57)$$

which should now be included into the effective action (16.48). Proceeding then in much the same way as above, we again arrive at the partition function in the form (16.49) with

$$F_{2e} = -\sum_{\pm} \int_{\tau_{\rm RC}}^{\infty} d\tau \gamma(\tau) \exp\left[\pm 2eV\tau - \frac{4}{g_{\rm S}} \ln\left(\frac{\sin^2(\pi T\tau)}{(\pi T \tau_{\rm RC})^2}\right)\right] \qquad (16.58)$$

where $\tau_{\rm RC} = R_{\rm S}C$ and $g_{\rm S} = R_q/R_{\rm S}$. This expression accounts for the contribution of Andreev processes to the free energy in the presence of dissipation. It strongly resembles the one that accounts for single-electron tunneling in normal tunnel barriers considered in Chapter 7 (cf., e.g., Eq. (7.108)).

At nonzero voltages, the expression (16.58) formally diverges, thus signaling the instability with respect to tunneling of two electrons across the NS boundary. Employing again the ImF-method, we evaluate the corresponding tunneling rate $\Gamma_{2e} = -2\text{Im}F_{2e}$ and arrive at the expression for the I–V curve (16.54). For example, at low temperatures and not too small voltages $eV \gg \epsilon_{\text{Th}}$, one finds [199]

$$G_{\text{A}}(V) \propto V^{8/g_{\text{S}}-1/2} \tag{16.59}$$

This result illustrates a nontrivial tradeoff between two different types of ZBA. As already discussed above, one ZBA is caused by disorder-enhanced quantum interference tending to increase G_{A} with decreasing V (or energy). Another ZBA is due to Coulomb effects that, on the contrary, suppress the Andreev conductance in the low-energy limit, similar to the case of single-electron tunneling considered in Chapter 7 (cf. Eq. (7.69)). As it is clear from Eq. (16.59), at higher shunt conductances $g_{\text{S}} > 16$, the former tendency wins, whereas for $g_{\text{S}} < 16$, Coulomb blockade of G_{A} dominates and the opposite trend is realized.

Note that within the above phenomenological model of an Ohmic shunt R_{S}, one can exactly account for Coulomb interaction effects by explicitly evaluating the $P(E)$-function for the effective environment at all energies [199]. While this model correctly captures certain features of the problem such as the competition between two types of ZBA (16.59), it is, in general, insufficient to properly account for the effect of electron–electron interactions in a disordered normal metal attached to a superconductor. Treating a normal resistor as a zero-dimensional object, this phenomenological model obviously fails to account for the sample geometry, which turns out to be crucially important in the context of Coulomb effects. As a result, the above model yields qualitatively incorrect results for G_{A} in the low-voltage limit $eV \ll \epsilon_{\text{Th}}$, and it also completely misses the effect of quantum dephasing of Cooper pairs by electron–electron interactions inside a disordered normal metal.

An adequate microscopic description of the problem can be worked out, e.g., within the framework of the nonlinear sigma-model technique [131, 358, 364]. As the corresponding calculations are relatively complex, we will avoid going into technical details here and immediately proceed to the final results.

One can demonstrate that both the form and the magnitude of the interaction correction to the Andreev conductance G_A of diffusive NS structures essentially depend on the dimensionality of the sample as well as on the relation between temperature and applied voltage to the relevant Thouless energy of the system. The interaction effects turn out to be virtually negligible in bulk metallic electrodes and may become significant in quasi-one-dimensional normal metals. In the latter case (i.e., for the structure depicted in Fig. 16.4), at not too low voltages $\epsilon_{\rm Th} \ll eV \ll 1/\tau_{\rm RC}$, one finds [131]

$$G_{\rm A}(V) = \frac{G_{\rm N}^2 R}{2}\sqrt{\frac{\epsilon_{\rm Th}}{|2eV|}}\left[e^{-\kappa} + \frac{\sqrt{\pi}e^{-8\gamma_0/g_n}}{\Gamma(\frac{8}{g_n}+\frac{1}{2})}\left((2|eV|\tau_{\rm RC})^{8/g_n} - (\epsilon_{\rm Th}\tau_{\rm RC})^{8/g_n}\right)\right] \quad (16.60)$$

where $g_n = R_q/R$, $\tau_{\rm RC} = RC_n$, $C_n \simeq d_n/\ln(d_n^2/\mathcal{A})$ is the geometric capacitance of the normal metal, $\gamma_0 \simeq 0.577\ldots$ is the Euler constant and

$$\kappa \simeq \frac{8}{g_n}\ln\frac{1}{\tau_{\rm RC}\epsilon_{\rm Th}} = \frac{8}{g_n}\ln\frac{e^2 p_{\rm F}^2\mathcal{A}\ln(d_n^2/\mathcal{A})}{\pi^2 v_F} \quad (16.61)$$

Here for simplicity, we neglected the capacitance of the tunnel barrier at the NS interface, which can be easily restored, if necessary.

The result (16.60) is consistent with that in Eq. (16.59), provided one identifies $g_s = g_n$. In the most relevant large conductance limit $g_n \gg 1$, the expression (16.60) can also be rewritten in a simpler form

$$G_{\rm A}(V) = \frac{G_{\rm N}^2 R}{2}\sqrt{\frac{\epsilon_{\rm Th}}{|2eV|}}e^{-\kappa}\left[1 + \frac{8}{g_n}\ln\frac{|eV|}{\epsilon_{\rm Th}}\right]. \quad (16.62)$$

At lower voltages $eV \lesssim \epsilon_{\rm Th}$, the Andreev conductance saturates to a constant value [131]

$$G_{\rm A} = G_{\rm N}^2 R e^{-\kappa} \quad (16.63)$$

Combining Eq. (16.63) with Eq. (16.61) and bearing in mind that for typical values of the Fermi velocity in metals, one has $e^2/v_{\rm F} \sim 1$, we rewrite the result (16.63) as

$$G_{\rm A} \sim G_{\rm N}^2 R N_{\rm Ch}^{-8/g_n} \quad (16.64)$$

Equation (16.64) demonstrates that at low voltages, Coulomb suppression of the Andreev conductance in quasi-one-dimensional

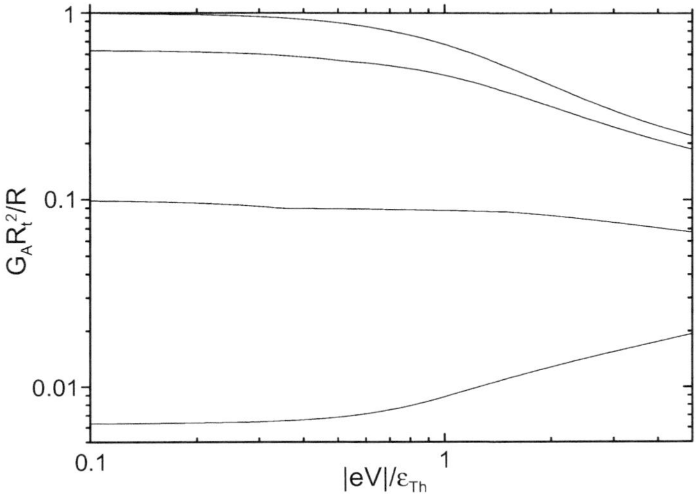

Figure 16.5 The zero-temperature differential Andreev conductance $G_{\rm A} = dI/dV$ of a diffusive NS system depicted in Fig. 16.4. We have set $RC\epsilon_{\rm Th} = 0.01$ and denoted $G_{\rm N} \equiv 1/R_t$. The dimensionless parameter $8/g_n$ equals to 0, 0.1, 0.5, 1.1 (from top to bottom).

diffusive samples is determined by the number of conducting channels to the power $8/g_n$. Note that this result cannot be recovered within the standard $P(E)$-theory employing the dissipative environment with the action (16.57), which does not account for the sample geometry. Instead of Eq. (16.64), the latter approach would then yield $G_{\rm A}$ vanishing as $V^{8/g_{\rm S}}$ in the limit $V \to 0$.

Let us also point out that all the above results remain applicable provided $eV \gg T$. In order to recover the Andreev conductance in the voltage range $eV \ll T$, it is sufficient to simply substitute T instead of eV in the above expressions for $G_{\rm A}$. This procedure yields the correct form of the interaction correction up to an unimportant numerical prefactor of order one.

The Andreev conductance of quasi-one-dimensional diffusive NS structures in the presence of electron–electron interactions is displayed in Fig. 16.5 for several values of g_n. At larger voltages, one observes a competition between two types of ZBA. As a result of this competition, $G_{\rm A}(V)$—depending on the value of g_n—either grows or decreases as V gets lower and eventually saturates to the voltage-independent value [Eqs. (16.63) and (16.64)] for all values of g_n at voltages $eV < \epsilon_{\mathrm{Th}}$. In this regime, G_{A} is always suppressed by electron–electron interactions, and this suppression becomes stronger with decreasing g_n.

The predicted behavior of the Andreev conductance displayed in Fig. 16.5—both the power-law dependencies of G_{A} on voltage and temperature and its saturation at lower voltages $eV < \epsilon_{\mathrm{Th}}$—was observed in experiments with NS structures composed of a superconductor and multiwalled carbon nanotubes [371]. The results of these experiments agree well with the above theoretical analysis.

It is important to emphasize that the above analysis remains applicable as long as quantum coherence of Cooper pairs penetrating into the normal metal remains unaffected by electron–electron interactions, i.e., for $d_n \ll L_\varphi$, where L_φ denotes the dephasing length for Cooper pairs [358, 364]. One can anticipate that the length scale L_φ should be identical to that already evaluated in previous chapters for single electrons. This is because quantum coherence of both single electrons and Cooper pairs in disordered normal metals gets destroyed due to fluctuations of the same random electromagnetic field mediating electron–electron interactions. Somewhat more formally, the subgap Andreev current across a disordered NS structure can be expressed via the Cooperon in the normal metal [190, 191, 199, 364]. The Cooperon was already encountered in the problem of weak localization in Chapters 14 and 15 and at low T was shown to decay at a typical time scale τ_φ (14.73) in the presence of electron–electron interactions. Hence, one can expect that the same time scale (as well as the length scale $L_\varphi = \sqrt{D\tau_\varphi}$) would show up in diffusive NS structures as well.

Despite these arguments, a closer inspection of the problem [364] demonstrates that the Cooperon analyzed in Chapters 14 and 15 is in many respects different, both qualitatively and quantitatively, from that encountered here. The most important difference between them is that the spin structure of the Cooperon responsible for the Andreev current in NS structures (antisymmetric combination of spin-singlets) corresponds to that of a Cooper pair and is entirely different from that for the Cooperon in the weak localization problem. In addition, the Cooperon describing propagating Cooper pairs in the normal metal is naturally bound to the NS interface, which is obviously not the case in weak localization. As a result, these two Cooperons are defined by formally different diagrammatic series and, hence, no a priori conclusions could possibly be drawn for our present problem from the analysis in Chapters 14 and 15.

As we already indicated above, an adequate microscopic analysis of the problem can be worked out within the real-time nonlinear sigma-model technique. Assuming that our diffusive NS structure is placed in a uniform magnetic field H, after a relatively cumbersome calculation, one arrives at the final result for the Andreev conductance in the form [358]:

$$G_{\mathrm{A}} = \frac{\pi T G_{\mathrm{N}}^2 R \epsilon_{\mathrm{Th}}}{4} \int_0^\infty dt^2 \frac{\cos(eVt)\cos(\omega_{\mathrm{Z}}t/2)}{\sinh(\pi T t)} \vartheta_2(0, e^{-t/\tau_{\mathrm{D}}}) e^{-f(t)} \tag{16.65}$$

where ϑ_2 is the second Jacobi theta function, $\tau_{\mathrm{D}} = 2/(\pi^2 \epsilon_{\mathrm{Th}})$ is the Thouless time, $\omega_{\mathrm{Z}} = g\mu_{\mathrm{B}} H$ (with $g\mu_{\mathrm{B}}$ being the product of a g-factor and the Bohr magneton) describes Zeeman splitting, and the function $f(\tau)$ accounts for dephasing of Cooper pairs due to both the magnetic field and the electron–electron interactions.

For $H = 0$, $T \ll eV$ and in the absence of electron–electron interactions (in which case $f(t) \to 0$), Eq. (16.65) naturally reduces to our previous result for G_{A} (16.56). Turning on both the magnetic field and electron–electron interactions, at sufficiently low temperatures $\pi T t \ll 1$, one finds [358]

$$f(t) \simeq \frac{t}{2\tau_{\mathrm{H}}} + \frac{8}{g_n} \ln\left(\frac{t}{\tau_{\mathrm{RC}}}\right) + \frac{t}{\tau_\varphi} + \sqrt{\frac{\pi t \tau_c}{4\tau_\varphi^2}} \ln\left(\frac{\tau_c}{t}\right) \tag{16.66}$$

Here τ_{H} (already defined in Eq. (14.33)) represents the electron decoherence time in a disordered conductor due to the presence of an external magnetic field H. As follows from Eqs. (16.65) and (16.66), the same time scale also controls decay of superconducting correlations penetrating into a normal metal from a superconductor. The remaining contributions to $f(t)$ in Eq. (16.66) are caused by electron–electron interactions. The first of these contributions (generated by spatially uniform fluctuations of the scalar potential in the normal metal) was already recovered by our previous analysis (cf. Eqs. (16.60)–(16.64)), while the last two terms in Eq. (16.66) originate from non-uniform in space fluctuations in the normal metal and define the new scales in our problem—Cooper pair decoherence time

$$\tau_\varphi = 2\pi N_0 \mathcal{A}\sqrt{2D\tau_c} \tag{16.67}$$

and decoherence length $L_\varphi = \sqrt{D\tau_\varphi}$. In Eqs. (16.66) and (16.67), we introduced an effective short time cutoff $\tau_c \sim \tau_e$, which is fully analogous to that employed in the weak localization problem; cf. Eq. (14.69).

We observe that the expression (16.67), which determines the dephasing time for Cooper pairs in diffusive normal metals—up to an unimportant prefactor of order one—agrees with the corresponding result (14.71) established for the weak localization problem. As we anticipated above, this agreement is, by no means, a pure coincidence; rather, it reflects fundamental and universal nature of interaction-induced low-temperature quantum decoherence in various types of disordered conductors. At the same time, there exist several important differences between the two problems, such as: (i) Unlike single electrons in normal metals, Cooper pairs in NS structures get dephased already by spatially uniform fluctuations of the scalar potential in the normal metal, as described by the second term in Eq. (16.66); (ii) unlike in the case of the weak localization problem, non-vanishing dephasing of Cooper pairs at $T = 0$ occurs already within the first-order perturbation theory in the electron–electron interactions [364]; (iii) already at $T = 0$, the Cooperon studied here decays somewhat differently as compared to the Cooperon studied in Chapter 14; cf., e.g., the last terms in Eqs. (16.66) and (14.69); and (iv) at not very low T, the temperature-dependent decay time for the Cooperon in NS systems is entirely different from that in the weak localization problem [364].

Turning to concrete results, we first set $H = 0$ and consider the low-voltage limit $eV \ll T$. According to Eqs. (16.65) and (16.66), at high temperatures $T \gg 1/\tau_\varphi$, the penetration length of Cooper pairs into the normal metal is defined by the temperature length $\sim\sqrt{D/T}$, while L_φ is irrelevant and dephasing is only due to spatially uniform fluctuations described by the second term in Eq. (16.66). In this case,

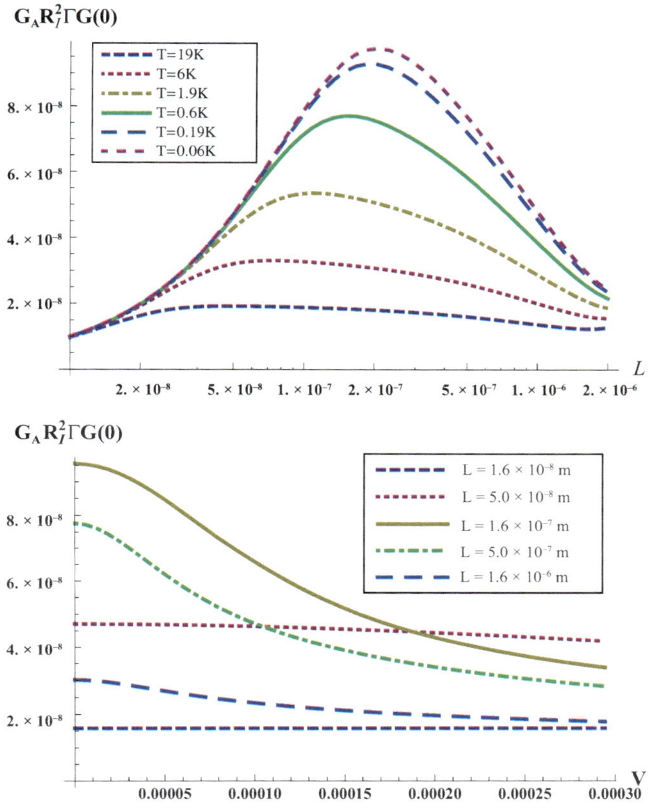

Figure 16.6 The subgap Andreev conductance $G_{\rm A}$ plotted as a function of $d_n \equiv L$ for $T \gg eV$ (left panel) and as a function of V at $T \ll eV$ (right panel). In both plots, we set $\mathcal{A} = 10$ nm and $D = 21$ cm^2/s. For these parameter values, one finds $1/\tau_\varphi \sim 0.6$ K and $L_\varphi \sim 0.2$ μm.

one naturally finds

$$G_{\rm A} \propto T^{8/g_n - 1/2} \tag{16.68}$$

for $T \gg \epsilon_{\rm Th}$ and Eq. (16.63) in the opposite limit $T \ll \epsilon_{\rm Th}$.

At low temperatures $T \ll 1/\tau_\varphi$, on the contrary, the temperature length becomes irrelevant and the penetration length of superconducting correlations into the normal metal is set by L_φ. For $d_n \ll L_\varphi$, the dephasing effect of electron–electron interactions remains weak and we again recover Eqs. (16.68) and (16.63). In the opposite limit

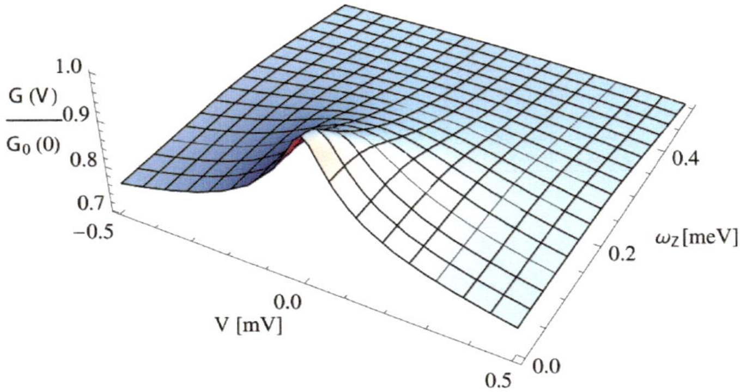

Figure 16.7 Andreev conductance G_A of an NS structure as a function of the external voltage V and the magnetic field (in units of ω_Z) at $T = 0$. The parameters are the same as in Fig. 16.6.

$d_n \gg L_\varphi$, the linear subgap conductance becomes [364]

$$G_\mathrm{A}(0) \simeq G_\mathrm{N}^2 R \frac{L_\varphi}{d_n} \left(\frac{4\tau_\mathrm{RC}}{\tau_\varphi} \right)^{8/g_n} \frac{\Gamma\left(1/2 - 8/g_n\right)}{\sqrt{2\pi}} \tag{16.69}$$

where $\Gamma(x)$ is Euler gamma function and $\zeta(x)$ is Riemann zeta-function. The dependence of the linear Andreev conductance G_A on d_n at different temperatures is displayed in the left panel of Fig. 16.6. At low T, it shows a pronounced maximum at $d_n \sim L_\varphi$, which can be conveniently used for experimental analysis of low-temperature dephasing of Cooper pairs in NS systems.

The same information can also be extracted from the nonlinear subgap conductance $G_\mathrm{A}(V)$, which shows the ZBA peak at low voltages. Again the dephasing effect of electron–electron interactions is only at $L \gtrsim L_\varphi$, in which case (and in the limit $T \to 0$) the width of this peak is roughly determined by $\sim 1/\tau_\varphi$ (see the right panel of Fig. 16.6). In particular, for $L \gg L_\varphi$ and $T = 0$, we get [364]

$$G_\mathrm{A}(V) \simeq G_\mathrm{A}(0) \mathrm{Re} \frac{1}{(1 + \mathrm{i}eV\tau_\varphi)^{1/2 - 8/g_n}} \tag{16.70}$$

where $G_\mathrm{A}(0)$ is defined in Eq. (16.69).

Including now the magnetic field H into our consideration, from Eq. (16.65) we observe that its effect is twofold. First, the magnetic field causes extra dephasing of Cooper pairs. This effect can be accounted for in Eqs. (16.69) and (16.70) by the substitution

$$\tau_\varphi \to \frac{2\tau_{\rm H}\tau_\varphi}{\tau_\varphi + 2\tau_{\rm H}} \tag{16.71}$$

Second, Zeeman splitting between states with opposite spins also influences the subgap conductance. From Eq. (16.65), we obtain

$$G_{\rm A}(V, H) = \frac{1}{2}\left(G_{\rm A}\left(V + \frac{\omega_{\rm Z}}{2e}\right) + G_{\rm A}\left(V - \frac{\omega_{\rm Z}}{2e}\right)\right) \tag{16.72}$$

This result implies that in the presence of the magnetic field, the ZBA peak gets additionally smeared due to Zeeman splitting. The behavior of the nonlinear conductance $G_{\rm A}(V)$ in the presence of an external magnetic field H and for $L > L_\varphi$ is displayed in Fig. 16.7.

It is important to emphasize that our analysis also allows to determine the subgap conductance for other geometries. For example, in the case of bulk (three-dimensional) normal metallic electrode, the decoherence effect from spatially uniform fluctuations is negligible [131] and at $T \ll 1/\tau_\varphi$, the dephasing of Cooper pairs in the normal metal is controlled by the second term in Eq. (16.66) with $\tau_\varphi \propto D^3$ (14.73).

We finally note that the above results for subgap Andreev conductance of NS hybrid structures are consistent with experimental observations [101] demonstrating that the low-temperature magnetoconductance of such structures is determined by phase coherent electron paths with a typical size restricted by the temperature-independent dephasing length L_φ rather than by the thermal length $\sim\sqrt{D/T}$ diverging in the low-temperature limit.

16.3.3 Shot Noise

The effective action analysis also allows us to study fluctuations of the Andreev current in diffusive NS structures in the presence of electron–electron interactions. Here we only briefly remark about the Andreev current noise restricting ourselves to the most interesting coherent limit $d_n \ll L_\varphi$.

The current noise spectrum is expressed in terms of the current operators $\hat{I}$ in a standard way as

$$\mathcal{S}(t, t') = \frac{1}{2}\left\langle \hat{I}(t)\hat{I}(t') + \hat{I}(t')\hat{I}(t)\right\rangle - \left\langle \hat{I}\right\rangle^2 \tag{16.73}$$

Making use of the Andreev action together with the definitions of current and noise [Eqs. (16.22), (16.23)], one can derive the following relationship [131]

$$\mathcal{S}(\omega, V, T) = e\sum_{\pm} \coth\left(\frac{\omega \pm 2eV}{2T}\right) I\left(\frac{\omega}{2e} \pm V, T\right) \qquad (16.74)$$

which remains valid irrespective of the dimensionality at frequencies and voltages smaller than both the inverse charge relaxation time $1/\tau_{\mathrm{RC}}$ and the inverse elastic scattering time $1/\tau_e$. Equation (16.74) remains applicable in the presence of electron–electron interactions and allows us to conveniently recover the Andreev current noise spectrum in diffusive NS structures from the above expressions for the current. In the zero-frequency limit, Eq. (16.74) reduces to the generalized Schottky relation

$$\mathcal{S}(\omega, V, T) = 2e \coth(eV/T) I(V, T) \qquad (16.75)$$

previously derived in the non-interacting limit [331].

It is worth pointing out that similar relations between the current and the noise spectrum hold also for normal tunnel barriers. In that case, the relation between $\mathcal{S}(\omega, V, T)$ and the current expectation value has exactly the same form as Eq. (16.74) [262, 369]; one should only replace $2e$ by e standing for the effective charge of the carriers.

Combining Eq. (16.74) with the above results for the subgap Andreev current, one arrives at a complete description of the effect of Coulomb interaction on the Andreev current noise in diffusive NS structures. For instance, at voltages $eV > \epsilon_{\mathrm{Th}}$ and in the low-frequency limit, the shot noise for quasi-one-dimensional NS structures is predicted to scale as

$$\mathcal{S} \propto |V|^{8/g_n - 1/2} \qquad (16.76)$$

whereas at lower voltages $eV < \epsilon_{\mathrm{Th}}$, one expects suppression of the shot noise spectrum by the factor $\exp(-\kappa) \sim N_{\mathrm{Ch}}^{8/g_n}$. Simultaneous current and shot noise measurements would help to experimentally verify the general relation (16.74) in hybrid NS structures.

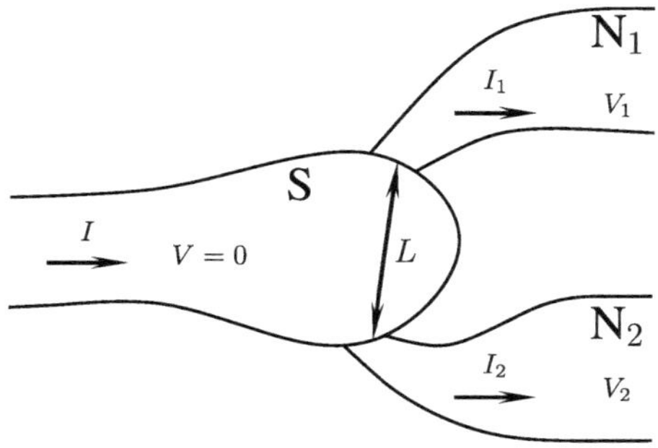

Figure 16.8 Schematics of our NSN device.

16.4 Non-local Electron Transport, Shot Noise, and Interactions

16.4.1 Crossed Andreev Reflection

Above in this chapter, we considered the processes related to Andreev reflection, which dominates charge transfer through an interface between a normal metal and a superconductor at sufficiently low temperatures and subgap energies. The physical mechanism of this *local* Andreev reflection is illustrated in Fig. 16.1 and can again be summarized as follows: A subgap electron propagating from the normal metal enters the superconductor at a length of order of the superconducting coherence length ξ, forms a Cooper pair together with another electron, while a hole goes back into the normal metal. As a result, the net charge $2e$ is transferred through the NS interface.

Let us now consider hybrid NSN structures with two N-terminals as shown in Fig. 16.8. In this case, electrons may suffer Andreev reflection and penetrate a superconductor through both NS interfaces. Provided the superconductor size (distance between two NS interfaces) L strongly exceeds ξ, Andreev reflection processes at these interfaces are independent. If, however, the distance L is smaller than or comparable with ξ, two additional non-local processes come into play. These processes are illustrated in Fig. 16.9. First, an electron with subgap energy propagating from one

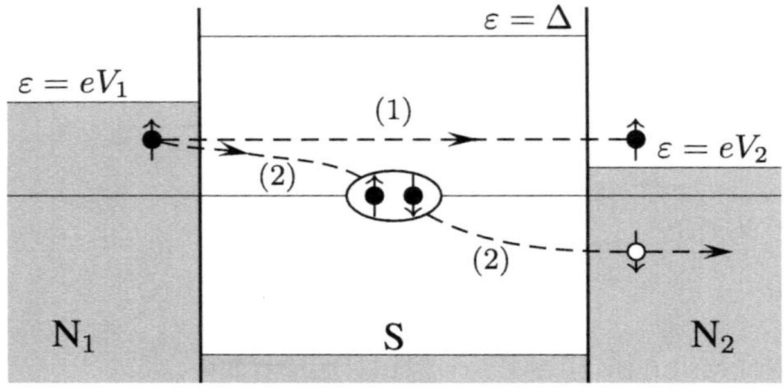

Figure 16.9 Two elementary processes contributing to non-local conductance of an NSN device: (1) direct electron transfer and (2) crossed Andreev reflection.

N-metal can penetrate through the superconductor into another N-electrode with the probability $\sim\exp(-L/\xi)$. In the tunneling limit (i.e., at small interface transmissions), this process is usually called elastic cotunneling (EC) [32]. Second, an electron penetrating the superconductor from the first N-terminal may form a Cooper pair by "picking up" another electron from the second N-terminal. In this case, a hole will go into the second (not the first!) N-metal and, hence, Andreev reflection turns into a *non-local* effect. The probability of this process—usually called non-local or crossed Andreev reflection (CAR) [78, 99]—also decays as $\sim\exp(-L/\xi)$. An interplay between these two processes essentially determines charge transfer in hybrid NSN at subgap energies.

Let us apply the voltages V_1 and V_2 to two normal electrodes and for convenience set the electric potential of the S-electrode equal to zero, as shown in Fig. 16.8. Then the currents I_1 and I_2 flowing through SN_1 and SN_2 interfaces can be expressed in the form

$$I_1 = I_{11}(V_1) + I_{12}(V_2), \tag{16.77}$$

$$I_2 = I_{21}(V_1) + I_{22}(V_2) \tag{16.78}$$

We observe that, e.g., the current I_1 is controlled not only by the local voltage V_1, but also by the voltage V_2 applied to the opposite normal terminal. In order to quantify this non-local effect, let us introduce the non-local conductance G_{12}, which can be determined from the equation $I_{12} = -G_{12}V_2$. In the interesting-for-us limit T, $V_{1,2} \ll \Delta_0$, only subgap quasiparticles contribute to the charge transfer across both NS interfaces. Then in the case of ballistic terminals, one finds [213, 214]

$$\frac{G_{12}}{G_{N_{12}}} = \frac{T_1 T_2 (1 - \tanh^2 L\Delta_0/v_F)}{[1 + (1 - T_1)(1 - T_2) + (2 - T_1 - T_2)\tanh L\Delta_0/v_F]^2} \tag{16.79}$$

where T_1 and T_2 are effective normal transmissions of two NS interfaces and $G_{N_{12}} \propto T_1 T_2/(p_F L)^2$ is the non-local conductance of our system in the normal state.

As follows from Eq. (16.79), the value G_{12} gets strongly suppressed with decreasing $T_{1,2}$ and increasing L. The result (16.79) also demonstrates that relative contributions to G_{12} from the two different charge transfer processes illustrated in Fig. 16.9 are essentially determined by the interface transmissions. At small transmissions $T_{1,2} \ll 1$ and at $T \to 0$, the non-local conductance of our device vanishes identically, $G_{12} = 0$, in the lowest order in tunneling $\propto T_1 T_2$, implying that EC and CAR contributions cancel each other exactly in this limit [112]. This cancelation of EC and CAR terms is lifted only in the next order in tunneling, in which case one has $G_{12} \propto T_1^2 T_2^2 \exp(-2L\Delta_0/v_F)$. It was also argued that CAR could prevail over EC already in the lowest order in tunneling in the presence of Coulomb interactions [272] or an external ac field [169].

For a given L, the non-local conductance G_{12} increases with increasing transmissions, reaching its maximum in the case of reflectionless interfaces with $T_1 = T_2 = 1$. In this limit, it reduces to

$$G_{12} = G_{N_{12}}(1 - \tanh^2 L\Delta_0/v_F) \tag{16.80}$$

Interestingly, in this case for small $L \ll v_F/\Delta$, the conductance G_{12} identically coincides with its normal state value $G_{N_{12}}$. This result implies that CAR contribution to the non-local conductance *vanishes for fully open barriers* [213, 214]. Actually the latter conclusion is general and applies for any value of L.

The reason for the absence of CAR contribution to G_{12} at $T_1 = T_2 = 1$ and for ballistic terminals is simple. Clearly, this contribution is associated with the flow of Cooper pairs out of the NS boundaries into the S-terminal (see Fig. 16.8). This flow is formally accounted for by electron trajectories that end deep inside the superconductor. It is obvious, on the other hand, that CAR requires "mixing" of these trajectories with those going straight between two normal terminals. Provided there exists no normal electron reflection at both NS interfaces (i.e., at $T_1 = T_2 = 1$), such mixing does not occur and, hence, CAR contribution to G_{12} vanishes identically. One can thus conclude that this effect is a simple consequence of the quasiparticle momentum conservation.

The situation changes provided at least one of the transmissions T_1 or T_2 is smaller than one. In this case, the quasiparticle momentum is not anymore conserved and scattering at SN interfaces mixes up trajectories connecting N_1 and N_2 terminals with ones going deep into the superconductor. As a result, CAR contribution to G_{12} does not anymore vanish in this case.

In realistic systems, electron scattering usually occurs not only at NS interfaces but also at disorder inside metallic terminals. As a result, the electron motion becomes diffusive rather than ballistic, and the expression for G_{12} needs to be modified. In the diffusive limit at low temperatures and any interface transmissions $T_{1,2}$, one finds [144]

$$G_{12} = G_{11}G_{22}R_{12}, \quad R_{12} = \frac{R_\xi}{2}\, e^{-L/\xi} \tag{16.81}$$

where G_{11} and G_{22} are local Andreev conductances of two NS interfaces defined in Eqs. (16.2) and (16.3), and R_{12} is the non-local resistance. It is remarkable that R_{12} does not depend on the interface transmissions at all being proportional to the normal state resistance R_ξ of the superconducting wire segment of length ξ (here we assume a quasi-one-dimensional geometry of a superconducting terminal).

Non-local electron transport in multiterminal NSN structures was intensively studied in numerous experiments [6, 42, 65, 79, 227, 228, 337], where the presence of CAR and, hence, entanglement between electrons in different normal terminals was convincingly demonstrated, a number of theoretical predictions were verified, and an interplay between non-local transport and additional effects (such as charge imbalance) was analyzed.

In this section, we will focus our attention on both fluctuation and interaction effects in NSN systems. In particular, we further develop our ideas and extend the relation between shot noise and Coulomb interaction to non-local effects. We will demonstrate that

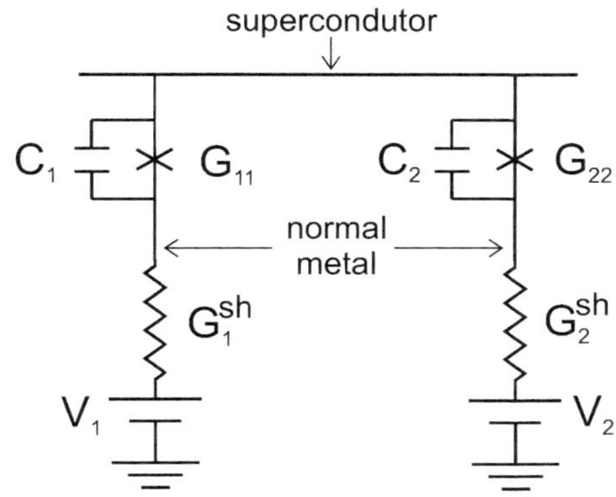

Figure 16.10 Schematics of the system under consideration.

this relation turns out to be much more subtle than in the local case merely because of (a) a variety of different processes contributing to non-local shot noise and (b) positive cross-correlations, which may occur in normal-superconducting hybrids in contrast to normal conductors where cross-correlations of fluctuating currents are known to be negative [60]. In the tunneling limit $T_{1,2} \ll 1$, EC and CAR processes provide, respectively, negative and positive contributions to non-local shot noise in NSN systems [58]. Below we will analyze non-local shot noise beyond the tunneling limit and find that at higher transmissions also direct electron transfer can yield positive cross-correlations in addition to CAR. At full transmissions only positive cross-correlations due to CAR survive and yield *Coulomb anti-blockade* of non-local electron transport. Our analysis of non-local fluctuation and interaction effects will essentially be based on the work [170].

16.4.2 The Model and Effective Action

In what follows, we will consider a hybrid structure consisting of two normal electrodes coupled to a superconductor via NS barriers with local subgap conductances G_{11} and G_{22} and capacitances C_1 and C_2 (Fig. 16.10). External voltages V_1 and V_2 are applied to normal electrodes via Ohmic shunts with conductances $G_1^{\rm sh}$ and $G_2^{\rm sh}$.

The Hamiltonian of the system reads

$$H = H_1 + H_2 + H_S + H_{T,1} + H_{T,2} \tag{16.82}$$

where

$$H_r = \sum_{\alpha=\uparrow,\downarrow} \int d\mathbf{x} \hat{\psi}^\dagger_{r,\alpha} \left(-\frac{\nabla^2}{2m} - \mu\right) \hat{\psi}_{r,\alpha}, \quad r = 1, 2,$$

are the Hamiltonians of the normal metals, m is electron mass, μ is the chemical potential,

$$H_S = \int d\mathbf{x} \left[\sum_\alpha \hat{\chi}^\dagger_\alpha \left(-\frac{\nabla^2}{2m} - \mu\right) \hat{\chi}_\alpha + \Delta \hat{\chi}^\dagger_\uparrow \hat{\chi}^\dagger_\downarrow + \Delta^* \hat{\chi}_\downarrow \hat{\chi}_\uparrow\right]$$

is the Hamiltonian of the superconductor with order parameter Δ and

$$H_{T,r} = \mathcal{A}_r \sum_{\alpha,\beta=\uparrow,\downarrow} \left[t_r e^{i\varphi_r} \hat{\psi}^\dagger_\beta \hat{\chi}_{r,\alpha} + t_r^* e^{-i\varphi_r} \hat{\chi}^\dagger_{r,\alpha} \hat{\psi}_\beta\right] \tag{16.83}$$

are tunneling Hamiltonians describing transfer of electrons across the contacts with area $\mathcal{A}_r$ and tunneling amplitude t_r.

For the sake of simplicity, we will assume that both NS barriers are uniform, implying that all $N_r = p_F^2 \mathcal{A}_r/4\pi$ conducting channels in the r-th barrier are characterized by equal normal transmission values T_r. We partially relax this assumption toward the end of this subsection. Making use of Eq. (10.32), we express T_r in terms of the transmission amplitudes t_r as

$$T_r = 4\pi^2 \nu_r \nu_S |t_r|^2 / \left(1 + \pi^2 \nu_r \nu_S |t_r|^2\right)^2, \quad r = 1, 2 \tag{16.84}$$

where ν_j ($j = r, S$) is the density of states in the corresponding terminal. Accordingly, in the low-energy limit to be considered below, local subgap conductances (16.2) reduce to $G_{rr} = N_r \mathcal{T}_r / R_Q$, where $\mathcal{T}_r = T_r^2/(2 - T_r)^2$ are effective Andreev transmissions of the two NS barriers as defined in Eq. (16.3). As usual, fluctuating phases φ_r introduced in Eq. (16.83) are linked to the voltage drops across the barriers v_r by means of the relation $\dot{\varphi}_r = ev_r$ and are treated as quantum operators.

We routinely express the kernel J of the Keldysh evolution operator via the path integral over the phase fields

$$J = \int \prod_{r=1,2} \mathcal{D}\varphi_r^F \mathcal{D}\varphi_r^B \exp(iS_{\text{env}}[\varphi] + iS_T[\varphi]) \tag{16.85}$$

where $\varphi_r^{\rm F}$ and $\varphi_r^{\rm B}$ are fluctuating phases defined, respectively, on the forward and backward branches of the Keldysh contour, $S_{\rm env}$ is the action of electromagnetic environment, and the term $S_{\rm T}$ accounts for electron transfer between the terminals. In the case of linear Ohmic environment considered here, one has

$$\mathrm{i}S_{\rm env} = \sum_{r=1,2}\left[\mathrm{i}\int dt\frac{(eV_r-\dot{\varphi}_r)(-C_r\dot{\varphi}_r^- + G_r^{\rm sh}\varphi_r^-)}{e^2}\right.$$
$$\left.-\frac{G_r^{\rm sh}}{2e^2}\int dtdt'\varphi_r^-(t)K(t-t')\varphi_r^-(t')\right] \qquad (16.86)$$

where $\varphi_r = (\varphi_r^{\rm F}+\varphi_r^{\rm B})/2$, $\varphi_r^- = \varphi_r^{\rm F}-\varphi_r^{\rm B}$ and $K(t) = -\pi T^2/\sinh^2(\pi T t)$. The action $S_{\rm T}$ reads

$$\mathrm{i}S_{\rm T} = \mathrm{tr}\ln\mathcal{G}^{-1}, \quad \mathcal{G}^{-1} = \begin{pmatrix} \check{G}_1^{-1} & \check{t}_1 & 0 \\ \check{t}_1^\dagger & \check{G}_S^{-1} & \check{t}_2 \\ 0 & \check{t}_2^\dagger & \check{G}_2^{-1} \end{pmatrix}, \qquad (16.87)$$

where 4×4 matrices $\check{G}_j^{-1}$ represent the inverse Keldysh Green functions of isolated normal ($j = 1, 2$) and superconducting ($j = S$) terminals and $\check{t}_r$ is diagonal 4×4 matrix

$$\check{t}_r = \begin{pmatrix} -t_r e^{-\mathrm{i}\varphi_r^{\rm F}} & 0 & 0 & 0 \\ 0 & t_r e^{-\mathrm{i}\varphi_r^{\rm B}} & 0 & 0 \\ 0 & 0 & t_r e^{\mathrm{i}\varphi_r^{\rm F}} & 0 \\ 0 & 0 & 0 & -t_r e^{\mathrm{i}\varphi_r^{\rm B}} \end{pmatrix} \qquad (16.88)$$

After some exact manipulations, we obtain

$$\mathrm{i}S_{\rm T} = \mathrm{tr}\ \ln\left[\check{1} - \check{t}_1^\dagger\check{G}_1\check{t}_1\check{G}_{\rm S} - \check{t}_2^\dagger\check{G}_2\check{t}_2\check{G}_{\rm S}\right] \qquad (16.89)$$

While the expression (16.89) for the action remains formally exact, it is still too complicated to be directly employed in our calculations. In order to proceed, we will make several additional steps, which yield necessary simplifications.

As the first step, we restrict ourselves to the limit of high conductances

$$g_r = (G_r^{\rm sh} + G_{rr})R_q \gg 1 \qquad (16.90)$$

in which case, phase fluctuations are weak and it suffices to expand the action (16.89) to the second order in φ_r^-, just as we already did, e.g., in previous sections of this chapter. Technically, we expand the matrices $\check{t}_r$ to the second order in the quantum phases φ_r^-, i.e., we make a replacement

$$\check{t}_r \to \check{t}_r(\varphi_r)\left(\check{1} - \mathrm{i}\frac{\varphi_r^-}{2}\check{\Lambda} - \frac{\varphi_r^{-2}}{4}\check{1}\right) \tag{16.91}$$

Here $\check{t}_r(\varphi_r)$ is defined by Eq. (16.88) with $\varphi_r^{\mathrm{F,B}}$ being replaced by the classical phase φ_r, and $\check{\Lambda}$ is a diagonal matrix with nonzero elements $\Lambda_{11} = 1$, $\Lambda_{22} = -1$, $\Lambda_{33} = -1$, and $\Lambda_{44} = 1$. Accordingly, we can write the product $\check{t}_r^\dagger \check{G}_r \check{t}_r$ in the form

$$\check{t}_r^\dagger \check{G}_r \check{t}_r = \check{\Sigma}_r + \check{\Sigma}_r^{(1)} + \check{\Sigma}_r^{(2)} + \mathcal{O}(\varphi_r^{-3}), \tag{16.92}$$

where we defined the self-energies

$$\begin{aligned}
\check{\Sigma}_r &= \check{t}_r^\dagger(\varphi_r)\check{G}_r\check{t}_r(\varphi_r),\\
\check{\Sigma}_r^{(1)} &= \frac{\mathrm{i}}{2}\left[\varphi_r^-\check{\Lambda},\, \check{\Sigma}_r\right],\\
\check{\Sigma}_r^{(2)} &= -\frac{1}{4}\{\varphi_r^{-2}\check{1},\, \check{\Sigma}_r\} + \frac{1}{4}\varphi_r^-\check{\Lambda}\check{\Sigma}_r\varphi_r^-\check{\Lambda}.
\end{aligned}$$

In order to evaluate $\check{\Sigma}_r$, we employ the Keldysh Green functions of the normal leads

$$\check{G}_r = (\hat{G}_r^{\mathrm{R}} + \hat{G}_r^{\mathrm{A}}) \otimes \hat{\tau}_z/2 + (\hat{G}_r^{\mathrm{R}} - \hat{G}_r^{\mathrm{A}}) \otimes \hat{Q}_r\hat{\tau}_z/2 \tag{16.93}$$

Here, as before, $\hat{\tau}_z$ is the Pauli matrix and the 2×2 matrices $\hat{G}_r^{\mathrm{R,A}}$ and $\hat{Q}_r$ are defined as

$$\hat{G}_r^{\mathrm{R,A}} = \begin{pmatrix} G_r^{\mathrm{R,A}} & 0 \\ 0 & G_r^{\mathrm{R,A}+} \end{pmatrix}, \quad \hat{Q}_r = \begin{pmatrix} 1 - 2n_r & 2n_r \\ 2 - 2n_r & 2n_r - 1 \end{pmatrix}$$

where n_r is the quasiparticle distribution function in the r-th normal lead. In equilibrium, it coincides with the Fermi function $n_r = 1/(1 + \exp(E/T))$. Neglecting the proximity effect in the normal leads and performing the summation over the corresponding electron states, we express the zeroth order self-energies (16.93) in the form

$$\check{\Sigma}_r(\varphi_r, \boldsymbol{r}) = \frac{\Gamma_r}{2\mathrm{i}}h_r(\boldsymbol{r})\begin{pmatrix} \hat{\tau}_z e^{\mathrm{i}\varphi_r}\hat{Q}_r e^{-\mathrm{i}\varphi_r} & 0 \\ 0 & \hat{\tau}_z e^{-\mathrm{i}\varphi_r}\hat{Q}_r e^{\mathrm{i}\varphi_r} \end{pmatrix} \tag{16.94}$$

with $\Gamma_r = 2\pi\nu_r|t_r|^2$. The function $h_r(\boldsymbol{r})$ in this expression differs from zero only at the interface of the r-th junction, and it obeys the following normalization condition

$$\int d^3\boldsymbol{r}\, h_r(\boldsymbol{r}) = N_r \tag{16.95}$$

We now assume that the barrier cross sections remain sufficiently small and put $h_r(\boldsymbol{r}) = N_r\delta(\boldsymbol{r} - \boldsymbol{r}_r)$. This assumption just implies that each of the barriers has N_r conducting channels with identical transmissions T_r (16.84), as we already indicated above. In this case, we can reduce the full coordinate dependence of the Green functions to that on the two indices i and j, which label the barriers and, hence, can take only two values 1 and 2. Accordingly, e.g., the Green function $\check{G}_{\mathrm{S}}(\boldsymbol{r}, \boldsymbol{r}')$ reduces to the 2×2 matrix in the "junction space" $\check{G}_{\mathrm{S}}^{ij}$. In addition, we should bear in mind that $\check{G}_{\mathrm{S}}^{ij}$ are the matrices in the space of conducting channels.

We note that the above assumptions are not really restrictive since they do not affect the general structure of the effective action to be derived below. At the same time, they allow to establish relatively simple expressions for the parameters entering the action. Expanding the action (16.89) in powers of φ_r^-, we arrive at the following expression:

$$\begin{aligned} \mathrm{i}S_{\mathrm{T}} = &-\mathrm{tr}\left[\check{K}\left(\check{\Sigma}_1^{(1)} + \check{\Sigma}_2^{(1)}\right)\check{G}_{\mathrm{S}}\right] - \mathrm{tr}\left[\check{K}\left(\check{\Sigma}_1^{(2)} + \check{\Sigma}_2^{(2)}\right)\check{G}_{\mathrm{S}}\right] \\ &- \frac{1}{2}\,\mathrm{tr}\left[\left(\check{K}\left(\check{\Sigma}_1^{(1)} + \check{\Sigma}_2^{(1)}\right)\check{G}_{\mathrm{S}}\right)^2\right] \end{aligned} \tag{16.96}$$

where we define the operator

$$\check{K} = \left[1 - \check{\Sigma}_1\check{G}_{\mathrm{S}} - \check{\Sigma}_2\check{G}_{\mathrm{S}}\right]^{-1} \tag{16.97}$$

Our second step allows to establish an explicit expression for the operator (16.97). Namely, in the interesting-for-us low-energy limit

$$T,\, eV_r \ll \Delta_0 \tag{16.98}$$

we can set the energy argument E in the superconductor Green function $\check{G}_{\mathrm{S}}(E)$ equal to zero. After that, $\check{G}_{\mathrm{S}}$ reduces to the time/energy-independent matrix

$$\check{G}_{\mathrm{S}}(E = 0) = \left(G_{\mathrm{S}}^{\mathrm{R}} + G_{\mathrm{S}}^{\mathrm{A}}\right) \otimes (\hat{\tau}_z/2) \tag{16.99}$$

where we introduced the retarded and advanced Green functions of the superconductor $G_{\mathrm{S}}^{\mathrm{R}}$ and $G_{\mathrm{S}}^{\mathrm{A}}$. In the limit $E \to 0$, they are equal to each other both being 4×4 matrices in the Nambu $\otimes$ "junction" space

$$G_{\mathrm{S}}^{\mathrm{R}} = G_{\mathrm{S}}^{\mathrm{A}} = \begin{pmatrix} G_{11}^{\mathrm{R}} & F_{11}^{\mathrm{R}} & G_{12}^{\mathrm{R}} & F_{12}^{\mathrm{R}} \\ F_{11}^{\mathrm{R}} & -G_{11}^{\mathrm{R}} & F_{12} & -G_{12}^{\mathrm{R}} \\ G_{21}^{\mathrm{R}} & F_{21}^{\mathrm{R}} & G_{22}^{\mathrm{R}} & F_{22}^{\mathrm{R}} \\ F_{21}^{\mathrm{R}} & -G_{21}^{\mathrm{R}} & F_{22}^{\mathrm{R}} & -G_{22}^{\mathrm{R}} \end{pmatrix} \tag{16.100}$$

In addition, each of the matrix elements in Eq. (16.100) is itself a matrix in the channel space. For instance, G^{R}_{11} and G^{R}_{12} are, respectively, $N_1 \times N_1$ and $N_1 \times N_2$ matrices, while matrices G^{R}_{21} and G^{R}_{22} have dimensions $N_2 \times N_1$ and $N_2 \times N_2$, respectively.

As the Keldysh Green function (16.99) depends neither on time nor on the quasiparticle distribution function, it commutes with the phase factors $e^{\pm \mathrm{i}\varphi_r(t)}$ entering the self-energies. This observation, combined with the multiplication rule for the Q–matrices $\hat{Q}_1\hat{Q}_2 = \hat{1} - \hat{Q}_1 + \hat{Q}_2$, allows us to express the operator $\check{K}$ (16.97) as a linear combination of these matrices:

$$\begin{aligned}\check{K}(t,t') = {} & [(K_{\mathrm{R}} + K_{\mathrm{A}})/2] \otimes \hat{1}\,\delta(t-t') \\ & - \mathrm{i}(\Gamma_1/2)K_{\mathrm{R}}X_1K_{\mathrm{A}} \otimes e^{\mathrm{i}\varphi_1(t)}\hat{\tau}_z\hat{Q}_1(t-t')\hat{\tau}_z e^{-\mathrm{i}\varphi_1(t')} \\ & - \mathrm{i}(\Gamma_1/2)K_{\mathrm{R}}X_2K_{\mathrm{A}} \otimes e^{-\mathrm{i}\varphi_1(t)}\hat{\tau}_z\hat{Q}_1(t-t')\hat{\tau}_z e^{\mathrm{i}\varphi_1(t')} \\ & - \mathrm{i}(\Gamma_2/2)K_{\mathrm{R}}X_3K_{\mathrm{A}} \otimes e^{\mathrm{i}\varphi_2(t)}\hat{\tau}_z\hat{Q}_2(t-t')\hat{\tau}_z e^{-\mathrm{i}\varphi_2(t')} \\ & - \mathrm{i}(\Gamma_2/2)K_{\mathrm{R}}X_4K_{\mathrm{A}} \otimes e^{-\mathrm{i}\varphi_2(t)}\hat{\tau}_z\hat{Q}_2(t-t')\hat{\tau}_z e^{-\mathrm{i}\varphi_2(t')}\end{aligned} \tag{16.101}$$

Here 4×4 matrices X_j are defined as follows:

$$X_1 = \begin{pmatrix} G^{\mathrm{R}}_{11} & F^{\mathrm{R}}_{11} & G^{\mathrm{R}}_{12} & F^{\mathrm{R}}_{12} \\ 0 & 0 & 0 & 0 \\ 0 & 0 & 0 & 0 \\ 0 & 0 & 0 & 0 \end{pmatrix}, \quad X_2 = \begin{pmatrix} 0 & 0 & 0 & 0 \\ F^{\mathrm{R}}_{11} & -G^{\mathrm{R}}_{11} & F^{\mathrm{R}}_{12} & -G^{\mathrm{R}}_{12} \\ 0 & 0 & 0 & 0 \\ 0 & 0 & 0 & 0 \end{pmatrix}$$

$$X_3 = \begin{pmatrix} 0 & 0 & 0 & 0 \\ 0 & 0 & 0 & 0 \\ G^{\mathrm{R}}_{21} & F^{\mathrm{R}}_{21} & G^{\mathrm{R}}_{22} & F^{\mathrm{R}}_{22} \\ 0 & 0 & 0 & 0 \end{pmatrix}, \quad X_4 = \begin{pmatrix} 0 & 0 & 0 & 0 \\ 0 & 0 & 0 & 0 \\ 0 & 0 & 0 & 0 \\ F^{\mathrm{R}}_{21} & -G^{\mathrm{R}}_{21} & F^{\mathrm{R}}_{22} & -G^{\mathrm{R}}_{22} \end{pmatrix}$$

while 4×4 matrices $K_{\mathrm{R,A}}$ read

$$K_{\mathrm{R}} = K^*_{\mathrm{A}} = \left[1 + \mathrm{i}\frac{\Gamma_1}{2}(X_1 + X_2) + \mathrm{i}\frac{\Gamma_2}{2}(X_3 + X_4)\right]^{-1} \tag{16.102}$$

Having established the expression for the operator $\check{K}$, we can now substitute it into the action (16.96). Proceeding along these lines and going through a straightforward but rather tedious calculation, we arrive at the result that still turns out to be too complicated for our present purposes. Further simplification amounts to neglecting local interference terms involving the products of the Green functions G^{R}_{11}, F^{R}_{11}, G^{R}_{22} and F^{R}_{22}. Technically, this step is equivalent to replacing these Green functions by their averaged-over-disorder values, i.e.,

$$\langle G_{11}^{\mathrm{R}}\rangle = \langle G_{22}^{\mathrm{R}}\rangle = 0, \quad \langle F_{11}^{\mathrm{R}}\rangle = \langle F_{22}^{\mathrm{R}}\rangle = -\pi\nu_{\mathrm{S}} \tag{16.103}$$

After this step, we arrive at the following result [170]:

$$S_{\mathrm{T}} = S_{11} + S_{22} + S_{12} \tag{16.104}$$

where

$$\mathrm{i}S_{11} = -\mathrm{i}\frac{G_{11}}{e^2}\int dt\dot{\varphi}_1\varphi_1^- - \int dtdt'\frac{\varphi_1^-(t)\,\tilde{\mathcal{S}}_{11}^{tt'}\,\varphi_1^-(t')}{2e^2} \tag{16.105}$$

$$\mathrm{i}S_{12} = \mathrm{i}\frac{G_{12}}{e^2}\int dt\left(\dot{\varphi}_1\varphi_2^- + \dot{\varphi}_2\varphi_1^-\right) - \int dtdt'\frac{\varphi_1^-(t)\,\tilde{\mathcal{S}}_{12}^{tt'}\,\varphi_2^-(t')}{e^2} \tag{16.106}$$

and the term S_{22} is obtained by interchanging the indices $1 \leftrightarrow 2$ in Eq. (16.105). The functions $\tilde{\mathcal{S}}_{rl}^{tt'}$ read

$$\begin{aligned}\tilde{\mathcal{S}}_{11}^{tt'} &= G_{11}K(t-t')\big(1-\beta_1+\beta_1\cos[2\varphi_1^{tt'}]\big)\\ &+2G_{12}K(t-t')\big(\alpha_1-\eta_1\cos[2\varphi_1^{tt'}]\big)\\ &+(G_{12}/2)K(t-t')\big(\kappa_1^+\cos[\varphi_1^{tt'}+\varphi_2^{tt'}]+\kappa_1^-\cos[\varphi_1^{tt'}-\varphi_2^{tt'}]\big),\end{aligned} \tag{16.107}$$

$$\begin{aligned}\tilde{\mathcal{S}}_{12}^{tt'} &= -G_{12}K(t-t')\big(1-\beta_1+\beta_1\cos[2\varphi_1^{tt'}]\big)\\ &-G_{12}K(t-t')\big(1-\beta_2+\beta_2\cos[2\varphi_2^{tt'}]\big)\\ &+(G_{12}/2)K(t-t')\big(\gamma_+\cos[\varphi_1^{tt'}+\varphi_2^{tt'}]-\gamma_-\cos[\varphi_1^{tt'}-\varphi_2^{tt'}]\big)\end{aligned} \tag{16.108}$$

Here we denoted $\varphi_r^{tt'} = \varphi_r(t) - \varphi_r(t')$ and introduced Andreev Fano factors $\beta_r = 1 - \mathcal{T}_r$.

The zero-bias non-local subgap conductance G_{12} and the parameters α_r, η_r, κ_r, γ are expressed as traces of certain combinations of the matrices K_{R}, K_{A}, and X_j. Namely, one finds

$$G_{12} = \frac{e^2}{2\pi}\,\mathrm{tr}\left[(X_3-X_4)K_{\mathrm{R}}(X_1-X_2)K_{\mathrm{A}}\right] \tag{16.109}$$

$$\begin{aligned}\kappa_1^+ = \frac{e^2\Gamma_1\Gamma_2}{\pi G_{12}}\,\mathrm{tr}\big[&2K_{\mathrm{R}}X_4K_{\mathrm{A}}X_1 + 2\Gamma_1^2K_{\mathrm{R}}X_1K_{\mathrm{A}}(X_1-X_2)K_{\mathrm{R}}X_4K_{\mathrm{A}}(X_1-X_2)\\ &-2\mathrm{i}\Gamma_1(X_1-X_2)K_{\mathrm{R}}X_1K_{\mathrm{R}}X_4K_{\mathrm{A}} + 2\mathrm{i}\Gamma_1X_1K_{\mathrm{A}}(X_1-X_2)K_{\mathrm{R}}X_4K_{\mathrm{A}}\big]\end{aligned} \tag{16.110}$$

and

$$\gamma_+ = \frac{e^2\Gamma_1\Gamma_2}{\pi G_{12}} \mathrm{tr}\left[K_\mathrm{R}X_4K_\mathrm{R}X_1 + K_\mathrm{A}X_4K_\mathrm{A}X_1\right. \tag{16.111}$$
$$- \Gamma_1\Gamma_2K_\mathrm{R}X_1K_\mathrm{A}(X_1 - X_2)K_\mathrm{R}X_4K_\mathrm{A}(X_4 - X_3)$$
$$- \Gamma_1\Gamma_2K_\mathrm{R}X_1K_\mathrm{A}(X_4 - X_3)K_\mathrm{R}X_4K_\mathrm{A}(X_1 - X_2)$$
$$+ i\Gamma_2(X_4 - X_3)K_\mathrm{R}X_1K_\mathrm{R}X_4K_\mathrm{A} + i\Gamma_1(X_1 - X_2)K_\mathrm{R}X_4K_\mathrm{R}X_1K_\mathrm{A}$$
$$\left. - i\Gamma_2X_1K_\mathrm{A}(X_4 - X_3)K_\mathrm{R}X_4K_\mathrm{A} - i\Gamma_1X_4K_\mathrm{A}(X_1 - X_2)K_\mathrm{R}X_1K_\mathrm{A}\right]$$

The parameters κ_1^- and γ_- are defined by Eqs. (16.110) and (16.111) with interchanged matrices $X_3 \leftrightarrow X_4$. The remaining parameters α_r and η_r in Eqs. (16.107) and (16.108) are defined in a similar manner.

In order to reduce the expressions (16.109)–(16.111) to a tractable form, we will make use of the condition

$$R_\xi \ll R_q/(2N_r\mathcal{T}_r) \tag{16.112}$$

which is usually well satisfied for generic systems. This condition simply states that the normal state resistance R_ξ of the superconducting wire segment of length ξ remains much smaller than those of the barriers and enables one to treat the Green functions connecting the two barriers as small parameters expanding the traces in Eqs. (16.109)–(16.111) in powers of G_{12}^R, G_{21}^R, F_{12}^R, and F_{21}^R. Keeping the leading corrections $\propto G_{12}^\mathrm{R}G_{21}^\mathrm{R}$ and $\propto F_{12}^\mathrm{R}F_{21}^\mathrm{R}$ and making use of the fact that $G_{12}^\mathrm{R}G_{21}^\mathrm{R}$ becomes equal to $F_{12}^\mathrm{R}F_{21}^\mathrm{R}$ after averaging over disorder, we arrive at explicit expressions for the parameters

$$\alpha_r = \frac{\mathcal{T}_r(1-2\mathcal{T}_r)}{\sqrt{\mathcal{T}_1\mathcal{T}_2}}, \quad \eta_r = \frac{2\mathcal{T}_r(1-\mathcal{T}_r)}{\sqrt{\mathcal{T}_1\mathcal{T}_2}},$$
$$\kappa_r^\pm = \pm(4\mathcal{T}_r - 3) + \frac{1}{\sqrt{\mathcal{T}_1\mathcal{T}_2}},$$
$$\gamma_\pm = \pm 1 + \frac{1 - 2\mathcal{T}_1 - 2\mathcal{T}_2 + 4\mathcal{T}_1\mathcal{T}_2}{\sqrt{\mathcal{T}_1\mathcal{T}_2}} \tag{16.113}$$

and for zero-bias non-local conductance

$$G_{12} = \frac{2e^2N_1N_2\mathcal{T}_1\mathcal{T}_2}{\pi^3\nu_S^2}G_{12}^\mathrm{R}G_{21}^\mathrm{R} \tag{16.114}$$

In the important case of diffusive superconductors, one has to further average Eq. (16.114) over disorder. In a simple quasi-one-dimensional geometry considered here, this averaging yields the result (16.81). Let us also note that in the diffusive limit, the inequality (16.112) helps to relax our assumption of identical normal transmissions for all conducting channels in a given barrier. Due to the diffusive nature of electron transport and full randomization of electron trajectories inside the superconductor, the average $\langle G^{\rm R}_{12}G^{\rm R}_{21}\rangle$ takes the same value for all possible pairs of channels in the junctions 1 and 2. Employing this observation and introducing channel transmissions $T_{1,m}$ and $T_{2,n}$ (which now may differ from channel to channel) instead of T_1 and T_2, one can rewrite the non-local conductance (16.114) in the form:

$$G_{12} = \frac{2e^2}{\pi^3\nu_{\rm S}^3}\left\langle G^{\rm R}_{12}G^{\rm R}_{21}\right\rangle \sum_m \mathcal{T}_{1,m}\sum_n \mathcal{T}_{2,n} \tag{16.115}$$

where $\mathcal{T}_{r,n} = T_{r,n}^2/(2 - T_{r,n})^2$ and the sums are running over the channel indices m and n. Analogously, we have

$$\kappa_r^{\pm} = \frac{\sum_{m,n}\mathcal{T}_{r,m}\mathcal{T}_{3-r,n}\left(\pm(4\mathcal{T}_{r,m}-3)+\frac{1}{\sqrt{\mathcal{T}_{r,m}\mathcal{T}_{3-r,n}}}\right)}{\sum_m \mathcal{T}_{1,m}\sum_n \mathcal{T}_{2,n}}, \tag{16.116}$$

$$\gamma_{\pm} = \frac{\sum_{m,n}\mathcal{T}_{1,m}\mathcal{T}_{2,n}\left(\pm 1+\frac{1-2\mathcal{T}_{1,m}-2\mathcal{T}_{2,n}+4\mathcal{T}_{1,m}\mathcal{T}_{2,n}}{\sqrt{\mathcal{T}_{1,m}\mathcal{T}_{2,n}}}\right)}{\sum_m \mathcal{T}_{1,m}\sum_n \mathcal{T}_{2,n}} \tag{16.117}$$

and similarly for α_r and η_r.

Let us specify the above expressions in the important limit of diffusive barriers, which can be, e.g., relatively short diffusive normal metallic conductors attached directly to N- and S-terminals. As we already discussed in Chapter 11, in this particular limit, the transmission distribution probabilities $P_r(T_r)$ are described by the universal formula [102]

$$P_r(T_r) = \pi G_{rr}/2e^2 T_r\sqrt{1-T_r} \tag{16.118}$$

(cf. Eq. 11.77). One can verify that Andreev transmission values $\mathcal{T}_r$ are distributed in exactly the same way, i.e., the probability $P_r(\mathcal{T}_r)$ is again defined by Eq. (16.118) where one should only substitute $T_r \to \mathcal{T}_r$. Making use of this remarkable property of diffusive barriers, we immediately recover the standard result for Andreev Fano factors $\beta_r = 1/3$ (cf. Eq. (16.6)). In the diffusive limit, we also find

$$\alpha_r = -\pi^2/4, \quad \eta_r = \pi^2/4, \quad \kappa_r^{\pm} = \pi^2 \mp 4/3, \quad \gamma_{\pm} = \pm 1 \tag{16.119}$$

16.5 Langevin Equations and Shot Noise

As we already learned in Chapter 2 (see Section 2.3), the quadratic in φ^- (in our case in $\varphi^-_{1,2}$) action can be rewritten in the form of an equivalent Langevin equation. With this in mind, we can express our effective action $S_{\rm env} + S_{\rm T}$ in terms of the following equations:

$$C_1\dot{v}_1 + (G_1^{\rm sh} + G_{11})v_1 - G_{12}v_2 = G_1^{\rm sh}V_1 + \xi_1^{\rm sh} + \xi_1,$$
$$C_2\dot{v}_2 + (G_2^{\rm sh} + G_{22})v_2 - G_{12}v_1 = G_2^{\rm sh}V_2 + \xi_2^{\rm sh} + \xi_2 \quad (16.120)$$

which describe the current balance in our system. Here $\xi_r^{\rm sh}$ are stochastic variables with pair correlators

$$\langle \xi_r^{\rm sh}(t)\xi_r^{\rm sh}(0)\rangle = G_r^{\rm sh}K(t) \quad (16.121)$$

describing Gaussian current noise in the shunt resistors, while the variables ξ_r with the correlators

$$\langle \xi_r(t)\xi_l(t')\rangle = \tilde{\mathcal{S}}_{rl}^{tt'} \quad (16.122)$$

describe shot noise in NS barriers.

Let us first "turn off" electron–electron interactions by taking the limit $1/G_{1,2}^{\rm sh} \to 0$ and, hence, setting $\varphi_r = eV_rt$. Defining the non-local noise spectrum as $\mathcal{S}_{rl}(\omega) = 2\int dt\,\tilde{\mathcal{S}}_{rl}^{tt'}\exp(-i\omega(t-t'))$, from Eq. (16.108) we obtain [170]

$$\begin{aligned}\mathcal{S}_{12}(\omega) = &-2G_{12}(2-\beta_1-\beta_2)W(\omega, 0) \quad (16.123)\\ &- 2G_{12}\beta_1 W(\omega, 2V_1) - 2G_{12}\beta_2 W(\omega, 2V_2)\\ &+ G_{12}\gamma_+ W(\omega, V_1+V_2) - G_{12}\gamma_- W(\omega, V_1 - V_2)\end{aligned}$$

where

$$W(\omega, V) = \frac{1}{2}\sum_{\pm}(\omega \pm eV)\coth\frac{\omega \pm eV}{2T} \quad (16.124)$$

Equations (16.123) and (16.124) fully determine non-local shot noise correlations in NSN structures at subgap energies $T, \omega, eV_r \ll \Delta_0$.

In clean NSN systems and in the specific limit $T_1 = T_2$, $V_1 = V_2$, $T = \omega = 0$, Eq. (16.123) was also derived by means of a different technique [123]. In the tunneling limit $T_{1,2} \ll 1$, we have

$\gamma_\pm = 4/\mathcal{T}_1\mathcal{T}_2 \gg \beta_{1,2} \simeq 1$ and Eq. (16.123) is dominated by the last two terms, which further reduce to the result [58]

$$\mathcal{S}_{12}(0) = eG_{12}\left[\gamma_+(V_1+V_2)\coth\frac{e(V_1+V_2)}{2T} - \gamma_-(V_1-V_2)\coth\frac{e(V_1-V_2)}{2T}\right] \quad (16.125)$$

in the limit $\omega \to 0$. One of these contributions $\propto \gamma_-$ is due to EC and describes negative shot noise correlations, while the other one $\propto \gamma_+$ comes from CAR and accounts for positive cross-correlations of fluctuating currents. Provided one of the voltages, say V_1, equals to zero, these EC and CAR terms exactly cancel each other for any V_2 (since $\gamma_+ = \gamma_-$ in the leading order in tunneling), i.e., $\mathcal{S}_{12}(0)$ tends to zero in the tunneling limit.

At higher transmissions $\mathcal{T}_{1,2}$, the value γ_- becomes negative, thus implying *positive* cross-correlations produced by direct electron transfer at such values of $\tau_{1,2}$ in addition to CAR. In the limit of fully transparent barriers $\tau_{1,2} = 1$, one has $\beta_{1,2} = \gamma_- = 0$. Then Eq. (16.123) yields

$$\mathcal{S}_{12}(0) = -8TG_{12} + 2eG_{12}(V_1+V_2)\coth\frac{e(V_1+V_2)}{2T} \quad (16.126)$$

At $T \to 0$, only *positive* cross-correlations due to CAR survive, whereas no direct electron transfer contribution to shot noise occurs for fully open barriers. Accordingly, no non-local shot noise is expected in this case at $V_1 = -V_2$ and $T \to 0$.

16.6 Interaction Correction to the Current

Now we again "turn on" electron–electron interactions and evaluate the current I_1 across the first barrier. Solving Eqs. (16.120) perturbatively in $1/g_r \ll 1$, in the lowest nontrivial order in this parameter, we get

$$I_1 = G_{11}V_1 - G_{12}V_2 - \langle\xi_1\rangle \quad (16.127)$$

In the same way as for normal junctions discussed in Chapter 11, the average $\langle\xi_1\rangle$ does not vanish since according to Eqs. (16.107) and (16.108), the noise ξ_1 depends on the phases $\varphi_{1,2}$, which in turn depend on $\xi_{1,2}$ by virtue of Eqs. (16.120). Hence, we obtain

$$\langle \xi_1 \rangle = \langle \delta\varphi_1 \; \partial\xi_1/\partial\varphi_1 \rangle + \langle \delta\varphi_2 \; \partial\xi_1/\partial\varphi_2 \rangle \tag{16.128}$$

where the phase fluctuations $\delta\varphi_r$ are found from Eqs. (16.120) and read

$$\delta\varphi_r(t) = e \int_{-\infty}^{t} dt' \frac{1 - e^{-(t-t')/\tau_{\rm RC}}}{G_r^{\rm sh}} \xi_r(t') \tag{16.129}$$

Here we have assumed $G_{12} \ll G_{rr} \ll G_r^{\rm sh}$ and introduced the RC-time $\tau_{\rm RC} = C_r/G_r^{\rm sh}$. Substituting this expression into Eq. (16.128), we find

$$\frac{\langle \xi_1 \rangle}{e} = \sum_{r=1,2} \int_{-\infty}^{t} dt' \frac{1 - e^{-\frac{t-t'}{\tau_{\rm RC}}}}{G_r^{\rm sh}} \left. \frac{\partial \langle \xi_1(t) \xi_r(t') \rangle}{\partial \varphi_r(t)} \right|_{\varphi_r = eV_r t} \tag{16.130}$$

Making use of Eqs. (16.107) and (16.108) and performing the time integral in Eq. (16.130), we arrive at the following expression for the current through the first NS interface [170]:

$$\begin{aligned} I_1 = G_{11}V_1 - G_{12}V_2 - \frac{2G_{11}\beta_1 - 4G_{12}\eta_1}{g_1} F_0(2V_1) + \frac{2G_{12}\beta_2}{g_2} F_0(2V_2) \\ -\delta G_+ F_0(V_1 + V_2) - \delta G_- F_0(V_1 - V_2) \end{aligned} \tag{16.131}$$

where $\delta G_\pm = G_{12}\left(\kappa_1^\pm/g_1 + \gamma_\pm/g_2\right)$ and

$$\begin{aligned} F_0(V) = \operatorname{Re} \Bigg[-V\Psi\left(1 + \mathrm{i}\frac{eV}{2\pi T}\right) \\ + \left(V - \frac{\mathrm{i}}{e\tau_{\rm RC}}\right) \Psi\left(1 + \frac{1}{2\pi T \tau_{\rm RC}} + \mathrm{i}\frac{eV}{2\pi T}\right) \Bigg] \end{aligned} \tag{16.132}$$

Accordingly the local and non-local differential conductances read

$$\begin{aligned} \frac{\partial I_1}{\partial V_1} = G_{11} - \frac{4G_{11}\beta_1 - 8G_{12}\eta_1}{g_1} F(2V_1) \\ - \delta G_+ F(V_1 + V_2) - \delta G_- F(V_1 - V_2) \end{aligned} \tag{16.133}$$

and

$$\begin{aligned} \frac{\partial I_1}{\partial V_2} = -G_{12}\left[1 - \frac{4\beta_2}{g_2} F(2V_2)\right] \\ - \delta G_+ F(V_1 + V_2) + \delta G_- F(V_1 - V_2) \end{aligned} \tag{16.134}$$

where

$$F(V) = \mathrm{Re}\left[\Psi\left(1+\frac{1}{2\pi T\tau_{\mathrm{RC}}}+\mathrm{i}\frac{eV}{2\pi T}\right)\right.$$
$$+\left(\frac{1}{2\pi T\tau_{\mathrm{RC}}}+\mathrm{i}\frac{eV}{2\pi T}\right)\Psi'\left(1+\frac{1}{2\pi T\tau_{\mathrm{RC}}}+\mathrm{i}\frac{eV}{2\pi T}\right)$$
$$\left.-\Psi\left(1+\mathrm{i}\frac{eV}{2\pi T}\right)-\mathrm{i}\frac{eV}{2\pi T}\Psi'\left(1+\mathrm{i}\frac{eV}{2\pi T}\right)\right] \quad (16.135)$$

and $\Psi(x)$ is the digamma function. Equations (16.133) and (16.134) together with Eq. (16.135) define the conductance matrix of our NSN device.

In the interaction correction to the local conductance in Eq. (16.133), we recover the standard Coulomb blockade term $\propto \beta_1$ (cf. Eq. (16.25)) and, in addition, three non-local contributions. The first of them, $\propto \eta_1$, enhances the conductance, while the second one, $\propto \delta G_+$, provides additional Coulomb suppression of $\partial I_1/\partial V_1$. The last term, $\propto \delta G_-$, can be both positive (at $\mathcal{T}_{1,2} \ll 1$) and negative (at bigger $\mathcal{T}_{1,2}$), implying the tendency to Coulomb anti-blockade in the latter case. The first term $\propto \beta_2$ in Eq. (16.134) has an opposite sign as compared to G_{12} (thus implying Coulomb blockade), while the second one $\propto \delta G_+$ yields Coulomb anti-blockade. Finally, the third $\propto \delta G_-$ tends to suppress or enhance the absolute value of the non-local conductance, respectively, for $\delta G_- > 0$ and $\delta G_- < 0$.

The origin of each of the terms in both interaction corrections can be easily identified from the corresponding shot noise correlators (16.107), (16.108), and (16.123), illustrating again a fundamental relation between shot noise and Coulomb effects in electron transport. This relation turns out to be considerably more complicated than in the local case. In the tunneling limit $T_{1,2} \ll 1$ in Eq. (16.133), the non-local terms add up to the local one and $\partial I_1/\partial V_1$ evolves from a typical Coulomb blockade V-like dependence at small V_2 toward a non-standard W-like one (with extra minima at $V_1 = \pm V_2$) at higher V_2 (Fig. 16.11a). In Eq. (16.134), in contrast, the last two terms exactly cancel each other for $V_1 \to 0$ and any V_2 since $\delta G_+ = \delta G_-$. This cancelation has the same origin as that of EC and CAR contributions to shot noise discussed above. For nonzero V_1 and V_2, the last two terms in Eq. (16.134) do not cancel anymore and the curve $\partial I_1/\partial V_2$ approaches the S-like shape with maximum at

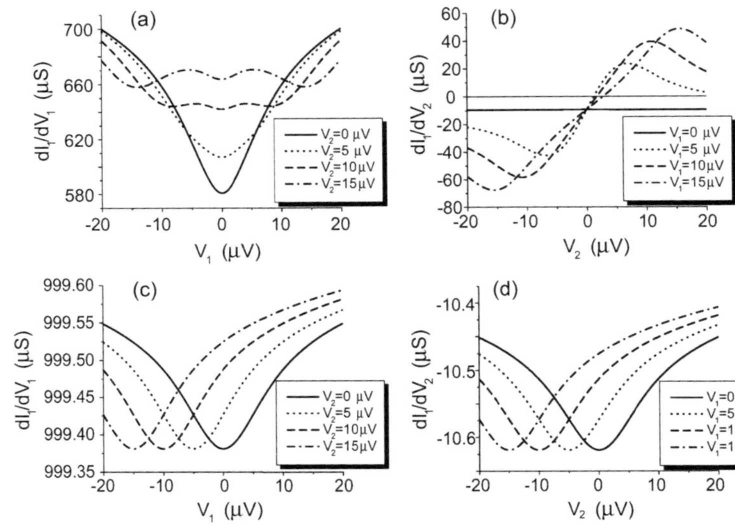

Figure 16.11 Local (a,c) and non-local (b,d) differential conductances defined, respectively, in Eqs. (16.133) and (16.134). The parameters of the system are: $T = 20$ mK, $G_{11} = 1$ mS, $G_{12} = 10\ \mu$S, $g_1 = g_2 = 516$. The dependencies (a) and (b) are plotted at small transmissions $T_1 = 0.063$, $T_2 = 0.11$, while the graphs (c) and (d) correspond to fully open barriers $T_1 = T_2 = 1$.

$V_1 = V_2$ and minimum at $V_1 = -V_2$ (Fig. 16.11b). In this case, the interaction term $\propto \delta G_{\pm} \propto T_1 T_2 (1/g_1 + 1/g_2)$ can exceed $G_{12} \propto T_1^2 T_2^2$ and, hence, $\partial I_1/\partial V_2$ can change its sign. For fully open barriers with $T_{1,2} = 1$, we get $\beta_{1,2} = 0$ and $\delta G_- = 0$, i.e., only CAR terms containing $\delta G_+ = 2G_{12}(1/g_1 + 1/g_2)$ survive in Eqs. (16.133) and (16.134), implying Coulomb blockade for local conductance $\partial I_1/\partial V_1$ and anti-blockade for non-local conductance $|\partial I_1/\partial V_2|$ in this limit (Figs. 16.11c,d).

16.7 Brief Summary

For clarity, let us briefly summarize our main observations.

In this section, we have evaluated non-local current–current correlators for an NSN structure at arbitrary transmissions of NS interfaces and arbitrary frequencies, Eqs. (16.123) and (16.124). This result demonstrates that positive cross-correlations in shot noise increase with increasing interface transmissions and dominate the result for fully open barriers in which case only CAR contribution survives. Positive noise cross-correlations in NSN structures were observed, e.g., in experiments [398].

Turning to the effect of electron–electron interactions on non-local electron transport, we would like to emphasize several important new features demonstrated within our analysis. One of them is that in the tunneling limit, almost no effect of Coulomb interaction on non-local conductance is expected if one of the applied voltages, V_1 or V_2, equals to zero. This effect is directly related to the cancelation between EC and CAR contributions to shot noise in the corresponding limit. For nonzero V_1 and V_2, no such cancelation exists anymore and the non-local conductance $-\partial I_1/\partial V_2$ approaches the S-like shape being enhanced at $V_1 \approx V_2$ and partially suppressed at $V_1 \approx -V_2$; see Fig. 16.11b. Both these features have a clear physical interpretation. Indeed, at $V_1 \approx -V_2$ negative cross-correlations due to EC dominate non-local shot noise leading to Coulomb blockade of non-local conductance, while at $V_1 \approx V_2$ positive cross-correlations due to CAR prevail and Coulomb anti-blockade of non-local transport is observed. At higher interface transmissions, only Coulomb anti-blockade of non-local conductance remains (Fig. 16.11d), which is again related to CAR-induced positive cross-correlations in shot noise.

As for experiments, the S-shaped non-local signal was observed, e.g., in [65]. Some of the features similar to those predicted above were also observed in the experimental work [398]. At the same time, it is worth pointing out that in realistic experiments—in particular at higher bias voltages—Cooper pair splitting and Coulomb interaction features can be essentially mimicked by heating effects [230]. Therefore, sufficient care is needed in order to appropriately interpret the experimental data on non-local electron transport in NSN structures.

Chapter 17

Superconducting Contacts beyond the Tunneling Limit

17.1 Introduction

In Chapter 16, we demonstrated the efficiency of path integral effective action techniques in describing Coulomb effects and quantum fluctuations in hybrid normal-superconducting (NS) structures. In this chapter, we will continue developing these ideas in the case of superconducting weak links, i.e., for the systems consisting of two large superconducting reservoirs connected via possibly a non-superconducting barrier. In Chapter 2, we already worked out an effective action description for such structures in the case of (usually very thin) insulating tunnel barriers. The presence of the Josephson effect, i.e., the possibility for a non-vanishing supercurrent to flow through a tunnel barrier between two superconductors is an important property of such systems, implications of which for a number of macroscopic quantum phenomena were explored in Chapters 4–6.

At this stage, it is worth pointing out that dc Josephson effect can be realized not only in the case of tunnel barriers but also in many other types of weak links, such as superconducting quantum point contacts [43, 185, 242] as well as both ballistic [130, 203,

Dissipative Quantum Mechanics of Nanostructures: Electron Transport, Fluctuations, and Interactions
Andrei D. Zaikin and Dmitry S. Golubev

ISBN 978-981-4774-50-5 (Hardcover), 978-0-429-29823-3 (eBook)
www.jennystanford.com

241] and diffusive [103, 406, 427] superconductor-normal-metal-superconductor (SNS) junctions. In contrast to tunnel junctions, in SNS systems, an appreciable supercurrent can flow at sufficiently low temperatures even through a normal layer as thick as a few microns.

In such systems, the Josephson effect is directly related to the phenomenon of Andreev reflection, which we already encountered in Chapter 16 in the context of NS hybrids. Suffering Andreev reflections at both NS interfaces, quasiparticles with energies below the superconducting gap are effectively "trapped" inside the junction, forming a discrete set of levels [20], which can be tuned by passing the supercurrent across the system [241]. At the same time, these subgap Andreev levels themselves contribute to the supercurrent, thus making the behavior of superconducting point contacts and SNS junctions in many respects different from that of tunnel barriers. Various features of the Josephson effect in different types of superconducting weak links are summarized in review papers [51, 172].

While the analysis of these review papers is mainly focused on larger junctions where fluctuation and interaction effects can be disregarded, here we will specifically address the behavior of small-size superconducting contacts, in which case quantum fluctuations and Coulomb effects may be important and, hence, should be properly accounted for. Unlike in the first part of this book (Chapters 2, 4–6), in this chapter we will examine the properties of superconducting contacts going beyond the tunneling limit. We will derive a tractable expression for the effective action for such contacts with arbitrary transmission distribution of conducting channels. Provided fluctuations of the Josephson phase remain sufficiently small, this effective action allows to fully describe a nontrivial interplay between fluctuation and charging effects in such systems. Our analysis allows us to discover a number of qualitatively new features of non-tunnel superconducting contacts, which do not occur at all in the tunneling limit. The key reason for that is the presence of subgap Andreev bound states inside such contacts. These low-energy bound states become particularly important at higher barrier transmissions and, as we shall see, may strongly influence the behavior of the systems under consideration.

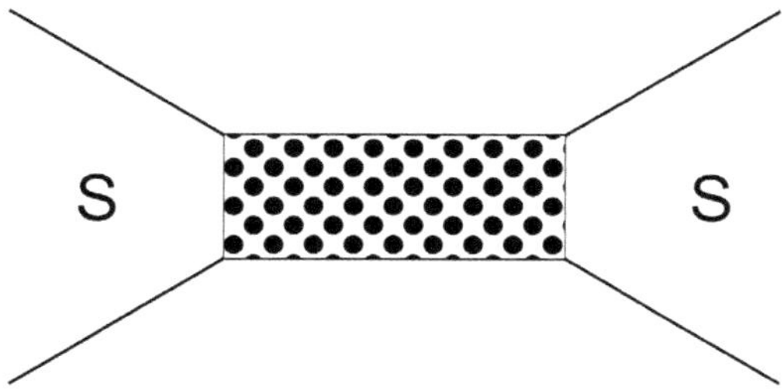

Figure 17.1 Short coherent conductor between two superconducting reservoirs.

17.2 Effective Action and Phase Fluctuations

Let us employ the standard model of a superconducting contact and consider two big superconductors connected to each other via a normal conductor (see Fig. 17.1) characterized by arbitrary transmission distribution T_n of its spin-degenerate conducting channels. Here we will merely restrict our attention to sufficiently short normal conductors with effective Thouless energy $\varepsilon_{\rm Th}$ strongly exceeding the superconducting gap Δ_0 in both reservoirs, $\varepsilon_{\rm Th} \gg \Delta_0$. In addition, we will assume the normal conductor length to be much shorter than both dephasing and inelastic relaxation lengths. Coulomb interaction between electrons in the contact area is routinely described by an effective capacitance C.

We will assume that the contact is biased by external current I, which does not exceed the critical one $I_{\rm C}$ and, hence, can flow through the contact without any dissipation, i.e., $I \equiv I_{\rm S}$. In the absence of fluctuations, this external current sets the value of the order parameter phase difference χ between two superconductors. The corresponding implicit dependence of χ on the supercurrent $I_{\rm S}$ has the form [242]

$$I_{\rm S}(\chi) = \frac{e\Delta_0 \sin\chi}{2} \sum_n \frac{T_n}{\sqrt{1 - T_n \sin^2(\chi/2)}} \tanh \frac{\Delta\sqrt{1 - T_n \sin^2(\chi/2)}}{2T} \tag{17.1}$$

In what follows, we will allow for fluctuations of the superconducting phase difference around its average value χ and employ the effective action formalism combined with the scattering matrix technique in the same spirit as it was already done in the case of normal conductors (Chapter 10) and NS hybrid structures (Chapter 16). Following the same procedure, we express the kernel J of the evolution operator on the Keldysh contour in terms of the path integral

$$J = \int \mathcal{D}\varphi_1 \mathcal{D}\varphi_2 \exp(\mathrm{i}S_{\mathrm{c}}[\varphi] + \mathrm{i}S_{\mathrm{sc}}[\varphi]), \tag{17.2}$$

where the terms $S_{\mathrm{c}}[\varphi]$ and $S_{\mathrm{sc}}[\varphi]$ account, respectively, for charging effects and for the transfer of electrons and Cooper pairs through a coherent scatterer between two superconducting reservoirs. As usual, both these terms represent the functionals of the fluctuating phase variables $\varphi_{\mathrm{F,B}}(t)$ defined on the forward and backward parts of the Keldysh contour and related to fluctuating voltages $V_{\mathrm{F,B}}$ across the conductor as $\dot{\varphi}_{\mathrm{F,B}}(t) = eV_{\mathrm{F,B}}$. With the aid of the Josephson relation, one trivially identifies the superconducting phase difference on two branches of the Keldysh contour as $\chi + 2\varphi_{\mathrm{F,B}}(t)$.

The charging term is taken in the standard form (16.8). The effective action $S_{\mathrm{sc}}[\varphi]$ for a general short coherent conductor placed in between two superconducting reservoirs was already derived in Chapter 2 within the Matsubara technique (see Eqs. (2.173) and (2.175)). An analogous expression can also be recovered within the Keldysh technique (cf. Eq. (10.74) in the normal case and Eq. (16.7) for NS structures), i.e., we have

$$S_{\mathrm{sc}}[\varphi] = -\frac{\mathrm{i}}{2}\sum_n \mathrm{Tr}\ln\left[1 + \frac{T_n}{4}\left(\{\check{G}_{\mathrm{L}}(\varphi), \check{G}_{\mathrm{R}}\} - 2\right)\right], \tag{17.3}$$

where $\check{G}_{\mathrm{L,R}}$ are 4×4 Green–Keldysh matrices of the left and right superconducting electrodes. Just like in Eq. (16.7), the product of these matrices implies time convolution and curly brackets denote anticommutation.

Without loss of generality, we can set the electric potential (and, hence, fluctuating phases) of the right superconducting terminal equal to zero. Then the Green–Keldysh matrix of this electrode can be written in a simple form

$$\check{G}_{\mathrm{R}} = \begin{pmatrix} \hat{g}^{\mathrm{R}} & \hat{g}^{\mathrm{K}} \\ 0 & \hat{g}^{\mathrm{A}} \end{pmatrix} \tag{17.4}$$

where $\hat{g}^{\mathrm{R,A}}$ are retarded and advanced 2×2 matrix functions

$$\hat{g}^{\mathrm{R,A}} = \frac{\epsilon \hat{\tau}_z + \hat{\Delta}}{\xi^{\mathrm{R,A}}}, \quad \xi^{\mathrm{R,A}} = \pm\sqrt{(\epsilon \pm \mathrm{i}0)^2 - \Delta_0^2} \tag{17.5}$$

(cf. also Eq. (16.9)) and $\hat{g}^{\mathrm{K}} = \hat{g}^{\mathrm{R}} F - F \hat{g}^{\mathrm{A}}$ is the Keldysh matrix, where $F(\epsilon) = \tanh(\epsilon/2T)$ is the Fourier transform of $F(t) = -\mathrm{i}T/\sinh[\pi T t]$. For simplicity, we choose the order parameter Δ_0 of the right superconductor real and, hence, we can set $\hat{\Delta} = \mathrm{i}\Delta_0 \hat{\tau}_y$. In order to properly account for analytic properties of the functions $\xi^{\mathrm{R,A}}$, here we keep an infinitesimally small imaginary part i0, which allows to define $\xi^{\mathrm{R,A}} = \pm \mathrm{sgn}\,\epsilon \sqrt{\epsilon^2 - \Delta_0^2}$ for $|\epsilon| > \Delta_0$ and $\xi^{\mathrm{R,A}} = \mathrm{i}\sqrt{\Delta_0^2 - \epsilon^2}$ for $|\epsilon| < \Delta_0$.

The Green–Keldysh matrix $\check{G}_{\mathrm{L}}$ of the left superconducting electrode reads

$$\check{G}_{\mathrm{L}}(\varphi)(t, t') = \check{\mathcal{L}} \check{\mathcal{M}}_+(t) \check{\mathcal{L}} \check{G}_{\mathrm{R}}(t, t') \check{\mathcal{L}} \check{\mathcal{M}}_-(t') \check{\mathcal{L}} \tag{17.6}$$

where we introduced the matrices

$$\check{\mathcal{L}} = \frac{1}{\sqrt{2}} \begin{pmatrix} \hat{1} & \hat{1} \\ \hat{1} & -\hat{1} \end{pmatrix}$$

and

$$\check{\mathcal{M}}_\pm = \begin{pmatrix} \exp\left[\pm\mathrm{i}\left(\frac{\chi}{2} + \varphi_{\mathrm{F}}(t)\right)\hat{\tau}_z\right] & 0 \\ 0 & \exp\left[\pm\mathrm{i}\left(\frac{\chi}{2} + \varphi_{\mathrm{B}}(t)\right)\hat{\tau}_z\right] \end{pmatrix} \tag{17.7}$$

Substituting the above expressions for $\check{G}_{\mathrm{L}}$ and $\check{G}_{\mathrm{R}}$ into Eq. (17.3), we arrive at the action that fully describes transfer of electrons and Cooper pairs to all orders in T_n.

In the case of tunnel barriers, all the channel transmissions remain small and one can expand S_{sc} in powers of T_n. Keeping the lowest-order terms $\sim T_n$ of this expansion one, of course, immediately recovers the AES action (2.76)–(2.78), which we already derived in Chapter 2. Here, however, we need to go beyond the tunneling limit and analyze fluctuation effects at arbitrary transmission values T_n. One can go on with the formal expansion of the logarithm in Eq. (17.3) and reconstruct the contributions to all orders in T_n (cf. Eq. (2.173)). However, the resulting series representation of the action turns out to be rather cumbersome, and, in particular, it does not allow us to easily extract the information about subgap Andreev energy states.

It order to avoid these complications and to effectively include all orders in T_n, we will again make use of the approach already employed in Chapter 16 and introduce the matrix

$$\check{X}_0 = 1 - T_n/2 + (T_n/4)\left\{\check{G}_\mathrm{L}, \check{G}_\mathrm{R}\right\}|_{\varphi^\pm=0} \tag{17.8}$$

With the aid of the identity $\mathrm{Tr}\ln\check{X}_0 = 0$, we rewrite the action (17.3) in the form (cf. Eq. (16.13))

$$S_\mathrm{sc} = -\frac{\mathrm{i}}{2}\sum_n \mathrm{Tr}\ln\left[1 + \check{X}_0^{-1}\circ\check{X}'\right], \tag{17.9}$$

where

$$\check{X}' = 1 + (T_n/4)\left(\left\{\check{G}_\mathrm{L}, \check{G}_\mathrm{R}\right\} - 2\right) - \check{X}_0 \tag{17.10}$$

With the aid of the above expressions for $\check{G}_\mathrm{L,R}$, we obtain

$$\check{X}_0^{-1} = \begin{pmatrix} \frac{\xi^2\hat{1}}{(\epsilon+i\delta)^2-\epsilon_n^2} & \frac{-\mathrm{i}\pi\xi^2\hat{1}}{\epsilon_n}\tanh\frac{\epsilon_n}{2T}\sum_\pm\delta(\epsilon\pm\epsilon_n) \\ 0 & \frac{\xi^2\hat{1}}{(\epsilon-\mathrm{i}0)^2-\epsilon_n^2} \end{pmatrix}, \tag{17.11}$$

where $\xi^2 = \epsilon^2 - \Delta_0^2$, and the subgap Andreev level inside the contact with energies $\pm\epsilon_n(\chi)$ is defined as

$$\epsilon_n(\chi) = \Delta_0\sqrt{1 - T_n\sin^2(\chi/2)} \tag{17.12}$$

Now let us assume that fluctuating phases (or fluctuating voltages) across the contact are sufficiently small $\varphi^\pm(t) \ll 1$ and perform regular expansion of the exact effective action in powers of these phases. Expanding $\check{G}_\mathrm{L}(\varphi)$ up to the second order in $\varphi^\pm$ (thus finding the matrix $\check{X}'$), from Eq. (17.9) we obtain [133]

$$\mathrm{i}S_\mathrm{sc} = -\frac{\mathrm{i}}{e}\int_0^t dt' I_\mathrm{S}(\chi)\varphi_-(t') + \mathrm{i}S_\mathrm{R} - S_\mathrm{I}, \tag{17.13}$$

where the supercurrent $I_\mathrm{S}(\chi)$ is defined in Eq. (17.1) and

$$S_\mathrm{R} = \int_0^t dt' \int_0^t dt'' \mathcal{R}(t'-t'')\varphi^-(t')\varphi^+(t''), \tag{17.14}$$

$$S_\mathrm{I} = \int_0^t dt' \int_0^t dt'' \mathcal{I}(t'-t'')\varphi^-(t')\varphi^-(t''). \tag{17.15}$$

Here both kernels $\mathcal{R}(t)$ and $\mathcal{I}(t)$ are real functions. Consider first the kernel $\mathcal{I}(t)$. It can be split into three contributions of different physical origin:

$$\mathcal{I}(t) = \mathcal{I}_1(t) + \mathcal{I}_2(t) + \mathcal{I}_3(t) \tag{17.16}$$

The first of these terms, $\mathcal{I}_1(t)$, represents the subgap contribution due to discrete Andreev states. The Fourier transform of this term reads

$$\mathcal{I}_{1\omega} = \frac{\pi\Delta_0^4}{4}\sum_n \left\{ \frac{T_n^2 \sin^2\chi}{2\epsilon_n^2(\chi)\cosh^2(\epsilon_n(\chi)/2T)}\delta(\omega) \right. \tag{17.17}$$
$$+\frac{T_n^2(1-T_n)\sin^4(\chi/2)}{\epsilon_n^2(\chi)}\left[1+\tanh^2\left(\frac{\epsilon_n(\chi)}{2T}\right)\right]$$
$$\left.\times\left[\delta\left(\omega-2\epsilon_n(\chi)\right)+\delta\left(\omega+2\epsilon_n(\chi)\right)\right]\right\}$$

It is obvious that this contribution is not contained in the AES action (2.76)–(2.78) at all.

The general expression for the Fourier transform of the second term $\mathcal{I}_2(t)$ is defined as

$$\mathcal{I}_{2\omega} = \sum_n T_n^{3/2}\sum_{\pm}\left[1\mp F\left(\epsilon_n(\chi)\right)F\left(\omega\pm\epsilon_n(\chi)\right)\right]$$
$$\times\,\theta(\omega-\Delta_0\pm\epsilon_n(\chi))\left|\sin\frac{\chi}{2}\right|$$
$$\times\frac{\Delta_0\left(\omega\epsilon_n(\chi)\pm\Delta_0^2(1+\cos\chi)\right)}{4\epsilon_n(\chi)\left((\omega\pm\epsilon_n(\chi))^2-\epsilon_n^2(\chi)\right)}\sqrt{(\omega\pm\epsilon_n(\chi))^2-\Delta_0^2} \tag{17.18}$$

where $\theta(t)$ is the Heaviside step function. In the limit of small barrier transmissions, this term scales as $\mathcal{I}_2 \propto T_n^{3/2}$ and, hence, is not contained in the AES action (2.76)–(2.78) either. This contribution can be interpreted as the "interference term" between subgap Andreev levels and quasiparticle states above the gap. In the low-temperature limit $T \to 0$, the Fourier transform of this term $\mathcal{I}_{2\omega}$ differs from zero only at sufficiently high frequencies $|\omega| > \Delta_0 + \epsilon_n(\chi)$. At higher temperatures $T \gtrsim \epsilon_n(\chi)$, however, $\mathcal{I}_{2\omega}$ vanishes only for $|\omega| < \Delta_0 - \epsilon_n(\chi)$ and remains nonzero otherwise.

Finally, the third term $\mathcal{I}_3(t)$ accounts for the contribution of quasiparticles with energies above the gap. The Fourier transform of this term is represented by the formula

$$\mathcal{I}_{3\omega} = \sum_n T_n \left[\int_{\Delta_0}^{\infty} \frac{d\epsilon}{2\pi} L_{n+}(\epsilon) + \frac{1}{2}\theta(\omega - 2\Delta_0) \int_{\Delta_0}^{\omega-\Delta_0} \frac{d\epsilon}{2\pi} L_{n-}(\epsilon) \right] \tag{17.19}$$

where

$$L_{n\pm}(\epsilon) = \frac{\sqrt{\epsilon^2 - \Delta_0^2}\sqrt{(\epsilon \pm \omega)^2 - \Delta_0^2}}{(\epsilon^2 - \epsilon_n^2(\chi))((\epsilon \pm \omega)^2 - \epsilon_n^2(\chi))} (1 \mp F(\epsilon)F(\omega \pm \epsilon))$$
$$\times \left(\epsilon(\omega \pm \epsilon) \pm \Delta_0^2 \cos\chi \pm T_n \Delta_0^2 \sin^2\frac{\chi}{2}\right) \tag{17.20}$$

In the high-frequency limit $\omega \gg \Delta_0$, Eqs. (17.19)–(17.20) reduce to the standard expression for a normal conductor

$$\mathcal{I}_{3\omega} \to \frac{\omega}{2e^2 R_{\mathrm{N}}} \coth\frac{\omega}{2T} \tag{17.21}$$

where $R_{\mathrm{N}} \equiv 1/G_{\mathrm{N}}$ is the normal contact resistance determined by the Landauer formula (16.1).

Turning now to the function $\mathcal{R}(t)$ in Eq. (17.14), we note that it can be decomposed as

$$\mathcal{R}(t) = \mathcal{R}_1(t) + \mathcal{R}_2(t) + \mathcal{R}_3(t) \tag{17.22}$$

analogously to Eq. (17.16). Furthermore, its Fourier transform can be represented as $\mathcal{R}_\omega = \mathcal{R}'_\omega + \mathrm{i}\mathcal{R}''_\omega$, where both $\mathcal{R}'_\omega$ and $\mathcal{R}''_\omega$ are real functions. The function $\mathcal{R}'_\omega$ is even in ω, while $\mathcal{R}''_\omega$ is an odd function of ω, thus implying that the function $\mathcal{R}(t)$ is indeed real as it was already indicated above.

The kernels $\mathcal{R}'_\omega$, $\mathcal{R}''_\omega$, and $\mathcal{I}_\omega$ are not independent. The function $\mathcal{R}''_\omega$ is related to $\mathcal{I}_\omega$ by means of the FDT as

$$\mathcal{R}''_\omega = 2\mathcal{I}_\omega \tanh\frac{\omega}{2T} \tag{17.23}$$

and the two functions $\mathcal{R}'_\omega$ and $\mathcal{R}''_\omega$ are in turn linked to each other by the causality principle, implying that the kernel $\mathcal{R}(t)$ should vanish at $t < 0$. The general expression for $\mathcal{R}'_\omega$ at $\omega > 0$ reads

$$\mathcal{R}'_\omega = \sum_n \Bigg\{ T_n(1-T_n)\frac{F(\epsilon_n(\chi))}{\epsilon_n(\chi)}\Delta_0^2\sin^2\frac{\chi}{2} - 2\frac{T_n^2(1-T_n)F(\epsilon_n(\chi))}{(\omega^2-4\epsilon_n^2(\chi))\epsilon_n(\chi)}\Delta_0^4\sin^4\frac{\chi}{2}$$
$$+2T_n \int_{\Delta_0-\min\{\omega,2\Delta_0\}}^{\Delta_0} \frac{d\epsilon}{2\pi}\frac{F(\epsilon+\omega)\left(\epsilon(\omega+\epsilon)+\Delta_0^2\cos\chi+T_n\Delta_0^2\sin^2(\chi/2)\right)}{(\epsilon^2-\epsilon_n^2(\chi))((\epsilon+\omega)^2-\epsilon_n^2(\chi))}$$
$$\times\sqrt{\Delta_0^2-\epsilon^2}\sqrt{(\epsilon+\omega)^2-\Delta_0^2}+\frac{T_n^{3/2}\Delta_0|\sin(\chi/2)|F(\epsilon_n(\chi))}{2\epsilon_n(\chi)}$$
$$\times\left[\sum_\pm \theta(\Delta_0\pm\epsilon_n(\chi)-\omega)\frac{\Delta_0^2(1+\cos\chi)\mp\omega\epsilon_n(\chi)}{(\omega\mp\epsilon_n(\chi))^2-\epsilon_n^2(\chi)}\sqrt{\Delta_0^2-(\omega\mp\epsilon_n(\chi))^2}\right]\Bigg\}. \tag{17.24}$$

Collecting all the above contributions to $\mathcal{R}$, one can also write

$$\mathcal{R}(t) = \sum_n \Bigg\{ T_n(1-T_n)\frac{F(\epsilon_n(\chi))}{\epsilon_n(\chi)}\Delta_0^2\sin^2\frac{\chi}{2}\delta(t) - \frac{T_n}{\pi}\delta'(t)$$
$$+2T_n^2(1-T_n)\frac{F(\epsilon_n(\chi))}{\epsilon_n^2(\chi)}\Delta_0^4\sin^4\frac{\chi}{2}\theta(t)\sin(2\epsilon_n(\chi)t) \tag{17.25}$$
$$-\frac{2}{\pi}\theta(t)\int_0^\infty d\omega\sin(\omega t)\left(f_\omega+\frac{T_n}{\pi}\omega\right)\Bigg\}.$$

Here the first and the second terms (first line in Eq. (17.25)) describe, respectively, the supercurrent correction (which is proportional to the factor $T_n(1-T_n)$) and Ohmic damping, while the third term (second line in Eq. (17.25)) accounts for the effect of discrete Andreev levels. The expression for f_ω in the last term of Eq. (17.25) can be directly reconstructed with the aid of Eqs. (17.16)–(17.24); see also Galaktionov and Zaikin [133]. At sufficiently short times $\epsilon_m t \ll 1$ with the logarithmic accuracy, we obtain

$$\int_0^\infty d\omega\sin(\omega t)\left(f_\omega+\frac{T_n}{\pi}\omega\right)\simeq T_n\left(\Delta_0^2\cos\chi+T_n\Delta_0^2\sin^2\frac{\chi}{2}\right)\ln\frac{1}{t\epsilon_m} \tag{17.26}$$

where we denoted $\epsilon_m = \max\{\Delta_0, T\}$.

With the aid of the above effective action, one can easily evaluate the phase–phase correlation functions. Combining Eqs. (17.13)–

(17.15) with Eq. (16.8), one readily finds

$$\langle\varphi^+(t_1)\varphi^+(t_2)\rangle = -\int\limits_{-\infty}^{\infty}\frac{d\omega}{2\pi}\mathrm{Im}\left(\frac{1}{C\omega^2/e^2+\mathcal{R}_\omega}\right)\coth\frac{\omega}{2T}e^{-\mathrm{i}\omega(t_1-t_2)} \tag{17.27}$$

$$\langle\varphi^+(t_1)\varphi^-(t_2)\rangle = \mathrm{i}\int\limits_{-\infty}^{\infty}\frac{d\omega}{2\pi}\left(\frac{1}{C\omega^2/e^2+\mathcal{R}_\omega}\right)e^{-\mathrm{i}\omega(t_1-t_2)}, \tag{17.28}$$

$$\langle\varphi^-(t_1)\varphi^-(t_2)\rangle = 0 \tag{17.29}$$

17.3 Equilibrium Supercurrent Noise

We first employ our effective action formalism in order to describe current fluctuations in superconducting contacts in the absence of electron–electron interactions. Let us define the current noise spectrum $\mathcal{S}_\omega$ as

$$\frac{1}{2}\langle\hat{I}(t_1)\hat{I}(t_2)+\hat{I}(t_2)\hat{I}(t_1)\rangle - I_S^2 = \int\limits_{-\infty}^{\infty}\frac{d\omega}{2\pi}\mathcal{S}_\omega e^{-\mathrm{i}\omega(t_1-t_2)} \tag{17.30}$$

The current–current correlation function can be evaluated with the aid of Eq. (16.23) where after performing functional derivatives with respect to $\varphi^-(t)$, one should formally set $\varphi^- = 0$. Substituting Eqs. (17.13)–(17.15) into Eq. (16.23), for $\omega \ll 1/R_\mathrm{N}C$ one easily finds

$$\mathcal{S}_\omega = 2e^2\mathcal{I}_\omega \tag{17.31}$$

Together with Eqs. (17.16)–(17.20), this result fully describes the equilibrium noise power spectrum in superconducting contacts with an arbitrary distribution of channel transmissions T_n.

In the low-temperature limit $T \to 0$ and at subgap frequencies $0 \le \omega < 2\Delta_0$, the above general result reduces to the following expression [133]:

$$\mathcal{S}_\omega = e^2\sum_n\left\{\frac{\pi\Delta_0^2T_n^2(1-T_n)\sin^4(\chi/2)}{1-T_n\sin^2(\chi/2)}\delta\left(\omega-2\epsilon_n(\chi)\right)\right.$$
$$+T_n^{3/2}\left|\sin\frac{\chi}{2}\right|\frac{\Delta_0\left(\omega\epsilon_n(\chi)-\Delta_0^2(1+\cos\chi)\right)}{\epsilon_n(\chi)\left((\omega-\epsilon_n(\chi))^2-\epsilon_n^2(\chi)\right)}$$
$$\left.\times\sqrt{(\omega-\epsilon_n(\chi))^2-\Delta_0^2}\,\theta(\omega-\Delta_0-\epsilon_n(\chi))\right\} \tag{17.32}$$

The contribution in the first line of Eq. (17.32) was also derived by Martin-Rodero et al. [293].

Equation (17.32) demonstrates that the contribution of each transmission channel to the noise spectrum has a narrow peak at $\omega = 2\epsilon_n(\chi)$, while at higher frequencies $\omega > \Delta_0 + \epsilon_n(\chi)$, continuous noise spectrum sets in. For even higher frequencies $\omega > 2\Delta_0$, also quasiparticles with energies above the gap contribute to the noise spectrum and in the high-frequency limit $\omega \gg \Delta_0$, Eqs. (17.31) and (17.16)–(17.20) reduce to the standard Nyquist expression for normal conductors

$$\mathcal{S}_\omega \simeq \frac{\omega}{R_\mathrm{N}} \coth \frac{\omega}{2T} \tag{17.33}$$

Let us discuss some properties of the quantum low-frequency current noise (17.32) in detail. We observe that due to the presence of subgap bound states, this noise spectrum essentially depends on the Josephson phase χ and remains nonzero even in the zero-temperature limit and at subgap frequencies. In addition, $\mathcal{S}_\omega$ essentially depends on the channel transmission values T_n. The amplitude of the peak at $\omega = 2\epsilon_n(\chi)$ increases with T_n at small transmissions and decreases at higher T_n, vanishing in the limit of perfect channel transmission $T_n \to 1$, except for a special point $\chi = \pi$ in which case the contribution of a fully open channel reduces to the universal peak at zero frequency. Combining this peak with the continuous spectrum contribution, for a fully open single channel at $T = 0$ and $\chi = \pi$, we obtain

$$\mathcal{S}_\omega = \pi e^2 \Delta_0^2 \delta(\omega) + e^2 \Delta_0 \sqrt{1 - \frac{\Delta_0^2}{\omega^2}} \theta(\omega - \Delta_0) \tag{17.34}$$

In the case of many conducting channels with different T_n, narrow peaks originating from different channels occur at different frequencies and a smoother noise spectrum is observed. An important example is a diffusive conductor characterized by the transmission distribution

$$P(T_n) \propto \frac{1}{T_n\sqrt{1 - T_n}} \tag{17.35}$$

Averaging the result (17.32) with this transmission distribution, we arrive at the equilibrium zero-temperature noise spectrum of diffusive superconducting contacts. The corresponding results are

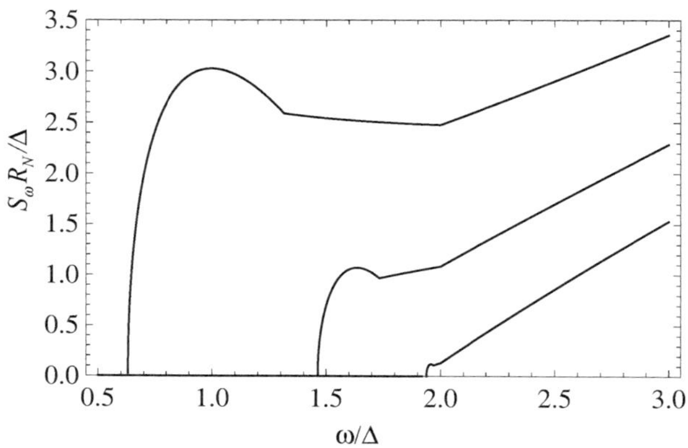

Figure 17.2 Low-temperature noise spectrum for diffusive superconducting contacts at $\chi = 2.5, 1.5, 0.5$ (from top to bottom).

displayed in Fig. 17.2 for different values of the phase difference χ. The noise spectrum is zero for $\omega < 2\Delta_0|\cos(\chi/2)|$; it increases with ω at $\omega > 2\Delta_0|\cos(\chi/2)|$, reaching the maximum at

$$\frac{\omega}{\Delta_0} = \frac{\sqrt{1+\cos\chi+\sqrt{(1+\cos\chi)(17+\cos\chi)}}}{\sqrt{2}} \tag{17.36}$$

and showing cusps at $\omega = \Delta_0(1+|\cos(\chi/2)|)$ and $\omega = 2\Delta_0$. In the limit $\omega \gg 2\Delta_0$, the Nyquist noise (17.33) is recovered.

At nonzero temperatures, there appear additional contributions to the noise spectrum [28, 133, 293]. In particular, an extra peak at zero frequency emerges with the amplitude, which depends on temperature as $\propto \cosh^{-2}(\epsilon_n(\chi)/2T)$; cf. Eq. (2.94).

In order to complete our analysis, it is instructive to compare the above results for the supercurrent noise in superconducting contacts with those for persistent current noise in normal nanorings, which is already discussed in Chapter 8. Similar to our present findings, the analysis of Chapter 8 demonstrates that persistent current noise spectrum has the form of sharp peaks, which occur at zero frequency and at frequencies determined by the interlevel distances for quantum states with nonzero transition matrix elements. In the low-temperature limit, the zero-frequency peak disappears, while the peaks at nonzero ω persist down to

$T = 0$. Essentially, the same situation is observed in superconducting contacts analyzed here.

17.4 Capacitance Renormalization

Let us now take into account small voltage fluctuations $V(t)$ across the contact. Since our present consideration is restricted to small fluctuations of the phase $\varphi^+(t) = e\int_0^t dt' V(t') \ll 1$, the constant in time part of the voltage should be equal to zero and the Fourier amplitude of its fluctuating part should obey the condition $eV_\omega \ll \omega$.

Under these conditions, the total current I across the superconducting contact takes the form

$$I = I_S(\chi) + \frac{C}{e}\ddot{\varphi}^+(t) - e\int dt' \mathcal{R}(t-t')\varphi^+(t') + \delta I(t) \tag{17.37}$$

where $I_S(\chi)$ is the supercurrent (17.1), the term involving C represents the displacement current, the $\mathcal{R}$-dependent term accounts for the retarded current response on the fluctuating voltage $V(t)$, and $\delta I(t)$ is the stochastic contribution to the current with the correlator $\langle \delta I \delta I \rangle_\omega = \mathcal{S}_\omega$ studied in the previous section. Equation (17.37) represents the quasiclassical Langevin equation describing small fluctuations of the Josephson phase in superconducting contacts.

Let us analyze the Fourier amplitude I_ω of the current in the limit of small temperatures and frequencies

$$T, \omega \ll 2\epsilon_n(\chi) \tag{17.38}$$

We remark that the condition (17.38) may yield parametrically different restrictions for weakly and highly transparent channels; cf. Eq. (17.12). We will return to this point in Section 17.7. In the limit (17.38), the noise term δI in Eq. (17.37) vanishes, while the kernel $\mathcal{R}$ can be expanded in ω up to $\sim \omega^2$ terms. Then at nonzero frequencies, we obtain

$$I_\omega = e\Delta_0\varphi_\omega^+ \sum_n \frac{T_n\left(\cos\chi + T_n\sin^4(\chi/2)\right)}{\left(1 - T_n\sin^2(\chi/2)\right)^{3/2}} - \frac{C^*(\chi)}{e}\omega^2\varphi_\omega^+ \tag{17.39}$$

The first term in this expression accounts for a trivial shift in χ by $2\varphi_\omega^+$ in Eq. (17.1). The renormalized capacitance $C^*(\chi)$ involved in the second term of Eq. (17.39) is defined as

$$C^*(\chi) = C + \delta C(\chi) \tag{17.40}$$

where in the limit $T \to 0$, we have [133]

$$\delta C(\chi) = \frac{e^2}{4\Delta_0}\sum_n \left\{\frac{2-(2-T_n)\sin^2(\chi/2)}{T_n \sin^4(\chi/2)} - \left(1-T_n\sin^2(\chi/2)\right)^{-5/2}\left[2T_n(T_n-2)\sin^2(\chi/2) + 5 + T_n + \frac{2-2(1+2T_n)\sin^2(\chi/2)}{T_n\sin^4(\chi/2)}\right]\right\}. \tag{17.41}$$

This expression fully describes the effect of capacitance renormalization in superconducting weak links with arbitrary distribution of transmission values T_n. While geometric capacitance of superconducting contacts C can be very small, our analysis demonstrates that retardation effects yield additional "capacitance-like" contributions (17.41), which depend on both the transmission distribution T_n and the Josephson phase χ and can well exceed the geometric capacitance term.

Let us specify Eq. (17.41) in some important limits. In the tunneling limit $T_n \ll 1$, this result reduces to

$$\delta C(\chi) = \frac{3\pi}{32\Delta_0 R_N}\left(1 - \frac{\cos\chi}{3}\right) \tag{17.42}$$

in agreement with Eq. (2.110) already derived in Chapter 2. In the case of small Josephson phases $\chi \ll \pi$, Eq. (17.41) yields the universal expression

$$\delta C \simeq \frac{\pi}{16\Delta_0 R_N} \tag{17.43}$$

which remains applicable for any distribution of channel transmissions. Provided all channels are transparent, i.e., $T_n \simeq 1$, Eq. (17.41) yields

$$\delta C \simeq \frac{\pi}{16\Delta_0 R_N \cos^4(\chi/4)} \tag{17.44}$$

in the region $0 < \chi < \pi$ and for χ not too close to π. At $\pi - \chi \lesssim (1-T_n)^{1/5}$, the function $\delta C(\chi)$ deviates from Eq. (17.44) and tends to

$$\delta C \simeq \frac{e^2}{4\Delta_0}\sum_n \frac{1}{(1-T_n)^{3/2}} \tag{17.45}$$

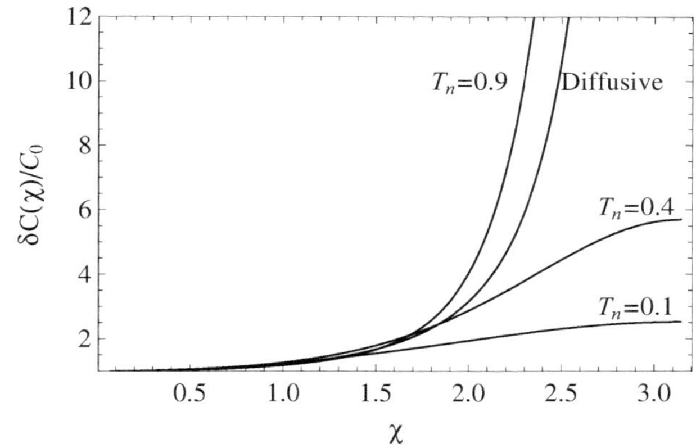

Figure 17.3 The capacitance $\delta C(\chi)$ normalized by $C_0 = \pi/16\Delta_0 R_N$. As indicated in the plot, three curves correspond to uniform barriers with channel transmissions $T_n = 0.1, 0.4, 0.9$, while the fourth curve corresponds to diffusive contacts.

for $\chi = \pi$. In the important case of diffusive contacts, averaging of Eq. (17.41) with the bimodal transmission distribution (17.35) yields

$$\delta C \simeq \frac{1.05}{\Delta_0 R_N(\pi - \chi)^2} \tag{17.46}$$

for $\pi - \chi \ll \pi$, while for small values of χ, we again reproduce Eq. (17.43). We observe that the renormalized capacitance (17.46) for diffusive superconducting contacts diverges as the phase difference χ approaches π. This divergence naturally occurs since (i) the contribution of almost fully open channels (17.45) becomes large in this limit and (ii) many such channels are available in diffusive barriers. The behavior of the renormalized capacitance is also illustrated in Fig. 17.3.

Equation (17.39) also allows us to determine the low-temperature Josephson plasma frequency ω_J of oscillations near the bottom of the Josephson potential well. We obtain

$$\omega_J^2 = \frac{e^2\Delta_0}{C^*(\chi)} \sum_n \frac{T_n \cos\chi + T_n^2 \sin^4(\chi/2)}{(1 - T_n \sin^2(\chi/2))^{3/2}} \tag{17.47}$$

where $C^*(\chi)$ is defined in Eqs. (17.40) and (17.41). Strictly speaking, this expression applies only provided the condition (17.38) is fulfilled. However, qualitatively it remains valid also at $\omega_J \sim 2\epsilon_n(\chi)$ up to a prefactor of order one. With this in mind, below we will employ the above expressions also in this case.

In the limit of large geometric capacitance of the junction $C \gg \delta C$, the capacitance renormalization can be neglected. In this case, we have $\omega_J \propto 1/\sqrt{C}$. In many cases, however, geometric capacitance turns out to be negligibly small so that $C^* \simeq \delta C$. In such cases, the combination ω_J/Δ_0 depends only on χ and on the barrier transmissions. This situation will be considered below.

For small $\chi \ll \pi$ and $T \to 0$, the Josephson current (17.1) reduces to $I_S(\chi) = \pi\Delta_0\chi/(2eR_N)$ for any transmission distribution. Combining this expression with Eq. (17.43), we get $\omega_J/\Delta_0 = 4$. This result universally holds for small values of the Josephson phase. For higher values of χ, the Josephson plasma frequency becomes smaller. For instance, for tunnel barriers with $T_n \ll 1$ at $-\pi/2 < \chi < \pi/2$, one trivially finds

$$\frac{\omega_J}{\Delta_0} = 4\sqrt{\frac{2\cos\chi}{3-\cos\chi}} \qquad (17.48)$$

In the case of highly transparent contacts with several conducting modes with close transmissions $T_n \simeq 1$, the Josephson plasma frequency can be written as

$$\frac{\omega_J}{\Delta_0} = 4\cos^2\frac{\chi}{4}\sqrt{\cos\frac{\chi}{2}} \qquad (17.49)$$

where we assume that $-\pi < \chi < \pi$ and $\pi - |\chi| > (1-T_n)^{1/5}$. As in this case the critical current is achieved at $\chi_c \simeq \pi - 2(1-T_n)^{1/4}$, for χ close to χ_c, we obtain

$$\frac{\omega_J}{\Delta_0} = 2\sqrt{2}(1-T_n)^{1/8}\sqrt{\chi_c - \chi} \qquad (17.50)$$

Provided the current is close to the critical one, the Josephson plasma frequency tends to zero as

$$\frac{\omega_J}{\Delta_0} = b\left(1 - \frac{I_S^2}{I_C^2}\right)^{1/4} \qquad (17.51)$$

where $b = 4\sqrt{2/3}$ for tunnel barriers, $b \simeq 2.53$ in the diffusive limit, and $b = 2\sqrt{2}(1-T_n)^{1/8}$ for highly transparent junctions. The behavior of $\omega_J(\chi)$ for tunnel, diffusive, and highly transparent barriers is also illustrated in Fig. 17.4.

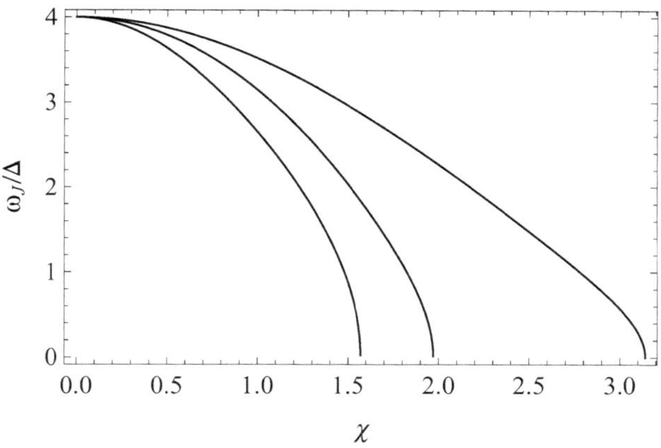

Figure 17.4 The Josephson plasma frequencies of highly transparent ($T_n = 0.9$), diffusive and tunnel barriers (from top to bottom) in the limit $\delta C \gg C$.

Finally, let us display a simple adiabatic form of effective action for a superconducting contact, which remains applicable for small phase fluctuations under the condition (17.38). It reads

$$S_{\rm c} + S_{\rm sc} \simeq \int_0^t dt' \left[\frac{C^*(\chi)}{e^2} \dot{\varphi}^+ \dot{\varphi}^- - \frac{I_{\rm S}(\chi + 2\varphi^+(t'))}{e} \varphi^-(t') \right] \tag{17.52}$$

where $I_{\rm S}(\chi)$ is defined in Eq. (17.1).

17.5 Andreev Level Spectroscopy

As we already demonstrated in previous sections, our effective action analysis is well suited for describing quantum properties of superconducting contacts involving subgap Andreev states. Here we consider one more example of the same kind, i.e., we will develop an effective action theory for Andreev level spectroscopy in superconducting contacts with arbitrary distribution of transmission coefficients T_n. First experiments aimed at directly detecting Andreev levels by means of microwave spectroscopy of non-tunnel superconducting atomic contacts have already been performed [67, 68] and their main idea can be illustrated as follows.

Let us consider a small superconducting tunnel junction with the Josephson coupling energy E_{JS} coupled to a voltage source V via an external impedance $Z(\omega)$. Assuming that the value of E_{JS} is sufficiently small, one can evaluate the inelastic Cooper pair rate $\gamma(V)$ across the junction perturbatively in E_{JS}. In fact, we have already solved this problem in Chapter 6 with the result (6.152). There we merely addressed the case of an RC circuit (cf. Eq. (6.147)); however, Eq. (6.152) remains applicable for a general impedance $Z(\omega)$.

The same result can also be expressed via the $P(E)$-function [202]. Making use of Eq. (6.152) and evaluating the Cooper pair current $I = 2e\gamma(V)$ at subgap voltages, one readily finds

$$I = \frac{eE_{\mathrm{JS}}^2}{2}\left(P(2eV) - P(-2eV)\right) \tag{17.53}$$

where

$$P(E) = \int_{-\infty}^{\infty} dt e^{\mathrm{i}Et} \exp\left\{\frac{4e^2}{\pi}\int_0^{\infty}\frac{d\omega}{\omega}\mathrm{Re}\left[Z(\omega)\right]\right. \\ \left. \times \left[\coth\frac{\omega}{2T}\left(\cos(\omega t) - 1\right) - \mathrm{i}\sin(\omega t)\right]\right\} \tag{17.54}$$

is the function describing energy smearing of tunneling Cooper pairs due to their interaction with the electromagnetic environment at a temperature T. As the reader remembers, we already encountered this so-called $P(E)$-function, e.g., in Chapter 7 while describing single-electron tunneling and Coulomb blockade effects in normal tunnel junctions. The same $P(E)$-theory applies to the process of Cooper pair tunneling.

According to Eqs. (17.53) and (17.54), the form of the I–V curve essentially depends on the environmental impedance $Z(\omega)$. In Chapter 6, we already considered one important example and performed a detailed analysis of Cooper pair tunneling in small Josephson junctions interacting with a dissipative environment described by a linear Ohmic impedance $Z(\omega) = R_{\mathrm{S}}$. One can also consider a number of other examples. For instance, provided $P(E)$ has the form of a delta function $P(E) \propto \delta(E - E_0)$, the current will be peaked as $I(V) \propto \delta(2eV - E_0)$. More generally, one can construct an environment containing a single quantum mode, e.g., produced by an *LC* circuit. In this case, the environmental impedance takes a simple form

$$Z(\omega) = \frac{i\omega}{C((\omega + i0)^2 - \omega_0^2)}. \tag{17.55}$$

Here C is an effective capacitance of the *LC* circuit and ω_0 is the oscillation frequency. Employing Eq. (17.54), in the limit of low temperatures, one finds [202]

$$P(E) = 2\pi e^{-\rho} \sum_{k=0}^{\infty} \frac{\rho^k}{k!} \delta(E - k\omega_0), \quad \rho = \frac{4E_C}{\omega_0} \tag{17.56}$$

As usual, here and below, $E_C = e^2/2C$ is the effective charging energy. Combining Eqs. (17.56) and (17.53), we arrive at the I–V curve consisting of narrow current peaks at voltages

$$2eV = k\omega_0, \quad k = 1, 2, \ldots \tag{17.57}$$

The physics behind this result is transparent: A Cooper pair with energy $2eV$ that tunnels across the junction releases this energy by exciting the environmental modes. In the case of an environment with a single harmonic quantum mode considered above, this process can occur only at discrete set of voltages (17.57).

Now, one can make one more step and consider a somewhat more complex environment consisting of an *LC* circuit and a superconducting contact containing a discrete set of Andreev levels $\epsilon_n(\chi)$ (17.12). In this case, tunneling of a Cooper pair can also be accompanied by upward transitions between these states and—along with the current peaks at voltages (17.57)—one can now expect the appearance of extra peaks at

$$2eV = k\omega_0 + 2\epsilon_n(\chi), \quad k = 0, 1, 2, \ldots \tag{17.58}$$

This simple consideration serves as a basic principle for the Andreev spectroscopy experiments [67, 68]. Following these authors, we will consider the circuit depicted in Fig. 17.5. This circuit can be divided into two parts. The part to the right of the vertical dashed line represents a superconducting loop pierced by an external magnetic flux Φ. This loop includes a Josephson tunnel junction with normal state resistance R_N and Josephson

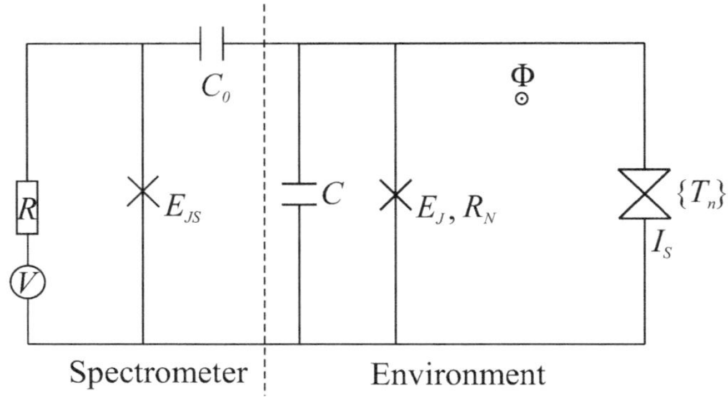

Figure 17.5 The circuit under consideration. The measured system, shown to the right of the dashed line, represents an asymmetric SQUID comprising a Josephson tunnel junction with resistance R_N and Josephson coupling energy E_J and a non-tunnel superconducting contact, characterized by an arbitrary set of transmission coefficients T_n of its conducting channels. The total capacitance C consists of a sum of geometric capacitances of both superconducting junctions C_Σ and also includes the renormalization term from the Josephson element. The superconducting loop is pierced by the magnetic flux Φ. The measuring device (the spectrometer) is shown to the left of the dashed line. It incorporates a voltage-biased tunnel junction with Josephson coupling energy E_{JS} connected to the measured system via a low resistance R and a large capacitance C_0.

coupling energy E_J connected to a non-tunnel superconducting contact, thereby forming an asymmetric SQUID. The latter contact is characterized by an arbitrary set of transmission coefficients T_n of their transport channels and, provided the superconducting phase difference χ is imposed, may conduct the supercurrent (17.1).

In what follows, we will assume that temperature T is sufficiently low $T \ll \Delta_0$ and we will stick to the limit

$$R_N \ll R_{Nc} \tag{17.59}$$

where $1/R_{Nc} = (e^2/\pi)\sum_n T_n$ is the normal state resistance of a non-tunnel contact. In this case, the critical current of the Josephson tunnel junction $\propto 1/R_N$ strongly exceeds that of the non-tunnel superconducting contact $\propto 1/R_{Nc}$. In this limit, the phase jump across the Josephson junction is close to zero, while such a jump across the non-tunnel contact is $\chi \approx 2\pi\Phi/\Phi_0$. The remaining part of the circuit in Fig. 17.5 (one to the left of the vertical dashed line) serves as measuring device called a spectrometer [68]. It consists of a voltage-biased superconducting tunnel junction with Josephson coupling energy E_{JS} connected to the asymmetric SQUID via a large capacitance C_0.

One observes a clear similarity between the system depicted in Fig. 17.5 and the above example of the LC circuit with a superconducting contact. Indeed, the asymmetric SQUID configuration on the right of Fig. 17.5 plays the role of an effective inelastic environment for the spectrometer junction E_{JS}. Bearing in mind an effective capacitance of the system

$$C = C_\Sigma + \frac{\pi}{16\Delta_0 R_N} \tag{17.60}$$

(cf. Eq. (17.43)) as well as kinetic inductances of both the Josephson element and the non-tunnel superconducting contact, to a certain approximation this environment can also be viewed as an effective LC circuit. The superconducting contact also contains extra quantum states—discrete Andreev levels (17.12). Hence, according to the above phenomenological arguments, one should expect to observe the I–V curve of the spectrometer in the form of current peaks at voltage values (17.57) and (17.58).

The effective action formalism developed in this chapter allows us to set up a fully microscopic analysis of this problem. In order to proceed, let us again denote the total phase difference across the non-tunnel superconducting contact as $\chi + 2\varphi(t)$, where χ is the constant part determined by the magnetic flux Φ and $2\varphi(t)$ is the fluctuating part of the superconducting phase. Assuming that the Josephson coupling energy of a tunnel junction, E_J, is sufficiently large, $E_J \gg E_C$, one can restrict further analysis to small phase fluctuations $2\varphi(t) \ll 1$ in both tunnel and non-tunnel contacts forming our asymmetric SQUID.

The total action S_{env}, which accounts for our effective environment (see Fig. 17.5), consists of three terms:

$$S_{env} = S_{Ch} + S_J + S_{sc} \tag{17.61}$$

describing, respectively, the charging energy, the Josephson tunnel junction, and the non-tunnel superconducting contact. At subgap frequencies, the sum of the first two terms in Eq. (17.61) reads

$$S_{Ch} + S_J = -\int dt\varphi_-(t)[\ddot{\varphi}_+(t)/(2E_C) + 4E_J\varphi_+(t)] \tag{17.62}$$

whereas the term S_{sc} is defined in Eqs. (17.13)–(17.15), where one can only include interesting for us subgap terms $\mathcal{I}_{1\omega}$ (17.17) and

$$\mathcal{R}_{1\omega} = \sum_n \frac{\gamma_n}{4\epsilon_n^2(\chi) - (\omega + \mathrm{i}0)^2} \tag{17.63}$$

ignoring quasiparticle contributions $\mathcal{I}_{2,3}$ and $\mathcal{R}_{2,3}$ in Eqs. (17.16) and (17.22). In Eq. (17.63) and also for later purposes, we defined

$$\gamma_n = 4T_n^2(1 - T_n)\frac{\Delta_0^4}{\epsilon_n(\chi)} \sin^4 \frac{\chi}{2} \tanh \frac{\epsilon_n(\chi)}{2T}. \tag{17.64}$$

Now we are in a position to evaluate the current through the spectrometer in the second order in E_{JS}. Under the approximations adopted here, the average is Gaussian, and it can be handled in a straightforward manner. As a result, we again arrive at Eqs. (17.53) and (17.54), where the inverse impedance of our effective environment takes the form [135]

$$\frac{1}{Z(\omega)} = \frac{C\left(\omega^2 - \omega_0^2\right)}{\mathrm{i}\omega} + \sum_n \frac{e^2\gamma_n}{\mathrm{i}\omega\left[4\epsilon_n^2(\chi) - \omega^2\right]}. \tag{17.65}$$

Here and below, $\omega_0 = \omega_{\mathrm{J}} \equiv \sqrt{8E_{\mathrm{J}}E_{\mathrm{C}}}$ is the Josephson plasma frequency.

It is obvious from Eqs. (17.53) and (17.54) that the positions of the current peaks are determined by zeroes of the inverse impedance (17.65). Equation (17.65), combined with Eqs. (17.12) and (17.64), allows to explicitly evaluate the $P(E)$-function and to establish both the positions and relative heights of these peaks.

For the sake of simplicity, let us assume that only one transport channel with transmission coefficient T_n in our superconducting contact is important, while all others do not exist or are irrelevant for some reason. In this case, under the condition $\gamma_n \ll E_{\mathrm{C}}\max(\omega_0^2, \epsilon_n^2(\chi))$ (which is typically well satisfied for the parameters under consideration) and outside an immediate vicinity of the "level intersection" point $\omega_0 = 2\epsilon_n$ from Eq. (17.65), we obtain

$$\mathrm{Re}\,[Z(\omega)] = \frac{\pi}{2C}\Bigg[\delta(\omega - \omega_0) + \delta(\omega + \omega_0) + \frac{2E_{\mathrm{C}}\gamma_n}{\left(\omega_0^2 - 4\epsilon_n^2\right)^2}\left(\delta(\omega - 2\epsilon_n) + \delta(\omega + 2\epsilon_n)\right)\Bigg]. \tag{17.66}$$

Introducing dimensionless parameters

$$\kappa_n = \frac{E_C \omega_0 \gamma_n}{\epsilon_n \left(\omega_0^2 - 4\epsilon_n^2\right)^2}, \qquad (17.67)$$

and retaining terms up to the first order in κ_n, we get [135]

$$P(E) = 2\pi e^{-\rho(1+\kappa_n)} \sum_{k=0}^{\infty} \frac{\rho^k}{k!} \left[\delta(E - k\omega_0) + \kappa_n \rho \delta(E - k\omega_0 - 2\epsilon_n)\right]. \qquad (17.68)$$

Substituting this result into Eq. (17.53), we recover the I–V curve of our device at subgap voltages, which fully determines the heights of all current peaks. For instance, Eq. (17.68) in combination with Eq. (17.53) yields the following ratio for the intensities of the two principal (voltage-integrated) current peaks occurring at the points $2eV = 2\epsilon_n$ and $2eV = \omega_0$:

$$\frac{\int_{eV\approx\epsilon_n} I(V)dV}{\int_{eV\approx\omega_0/2} I(V)dV} = \kappa_n \propto \frac{\sin^4 \frac{\chi}{2}}{\epsilon_n^2(\chi)\left(\omega_0^2 - 4\epsilon_n^2(\chi)\right)^2} \qquad (17.69)$$

Here the integration is performed over the voltage values in the immediate vicinity of the points $V = \epsilon_n/e$ and $V = \omega_0/2e$, respectively, in the numerator and in the denominator.

Equation (17.69) determines relative intensities of the principal spectral lines as a function of the phase χ or, equivalently, of the applied magnetic flux Φ. It applies not too close to the point $\omega_0 = 2\epsilon_n$ and holds irrespective of the fact that in any realistic experiment, the δ-function current peaks can be somewhat broadened by inelastic effects. The reasons for such broadening in realistic structures can be diverse. In the absence of extra sources of dissipation and under the assumption of thermalization, our effective action approach allows us to demonstrate that at low-enough temperatures, thermal broadening of the current peaks is determined by the factor $\sim\exp(-\epsilon_n(\chi)/T)$ [135]. Hence, broadening can become significant at values of χ close to π (i.e., at $\Phi \approx \Phi_0/2$) and provided at least one of the transmission values T_n is close to unity.

17.6 Supercurrent Affected by Electron–Electron Interactions

Let us now briefly analyze the effect of electron–electron interactions on the equilibrium Josephson current $I_S(\chi)$ (17.1). In Chapters 4 and 6, we already discussed the interaction correction to the supercurrent in Josephson tunnel barriers with linear Ohmic dissipation; cf. Eqs. (4.92) and (6.9). Here we will consider a more general case of superconducting contacts with arbitrary distribution of the channel transmission values T_n.

In order to evaluate the interaction correction to $I_S(\chi)$, it is necessary to go beyond the Gaussian effective action (17.13)–(17.15) and to evaluate the higher-order contribution to the action $\delta S_{sc} \sim \varphi^3$. It is straightforward to verify that the leading order interaction correction to the supercurrent is provided by the following terms:

$$\mathrm{i}\delta S_{sc} = \int\int\int dt_1 dt_2 dt_3 Y(t_1, t_2, t_3)\varphi^-(t_1)\varphi^+(t_2)\varphi^+(t_3)$$
$$+\int\int\int dt_1 dt_2 dt_3 Z(t_1, t_2, t_3)\varphi^+(t_1)\varphi^-(t_2)\varphi^-(t_3) \quad (17.70)$$

where

$$Y(t_1, t_2, t_3) = \int\int \frac{d\omega_1}{2\pi}\frac{d\omega_2}{2\pi} Y(\omega_1, \omega_2) e^{-\mathrm{i}\omega_1(t_1-t_2)} e^{-\mathrm{i}\omega_2(t_1-t_3)},$$

with $Y(\omega_1, \omega_2) = Y(\omega_2, \omega_1)$ and similarly for the function $Z(t_1, t_2, t_3)$.

Adding the non-Gaussian terms (17.70) to the action and employing Eq. (16.22), we arrive at the following expression for the interaction correction:

$$\delta I_S(\chi) = \mathrm{i}e \int_{-\infty}^{\infty} \frac{d\omega}{2\pi} \left[Y(\omega, -\omega)\langle\varphi^+\varphi^+\rangle_\omega + 2Z(0, -\omega)\langle\varphi^+\varphi^-\rangle_\omega\right] \quad (17.71)$$

where the phase–phase correlators are defined in Eqs. (17.27) and (17.28).

Let us consider the first term in the right-hand side of Eq. (17.71). It is easy to see that in the low-temperature limit, only frequencies

$|\omega| > \Delta_0 + \epsilon_n(\chi)$ contribute to the integral in Eq. (17.27) for $\langle\varphi_+\varphi_+\rangle$, while the contribution from the frequency interval $|\omega| < \Delta_0 + \epsilon_n(\chi)$ vanishes. Furthermore, the leading contribution from the first term in Eq. (17.71) is picked up logarithmically from the interval $2\Delta_0 \ll |\omega| \ll 1/R_N C$, where

$$\langle\varphi^+\varphi^+\rangle_\omega \simeq \frac{e^2 R_N}{|\omega|} \tag{17.72}$$

R_N is again the normal state contact resistance, and the function $Y(\omega, -\omega)$ tends to a frequency-independent value.

After a straightforward but tedious calculation in the interesting frequency range $\omega \gg \Delta_0$, from Eq. (17.3) one derives the following expression for the function $Y(\omega, -\omega)$ [133]:

$$Y(\omega, -\omega) = \frac{i\Delta_0 \sin\chi}{4} \sum_n \frac{T_n(1-T_n)(2 - T_n \sin^2(\chi/2))}{(1 - T_n \sin^2(\chi/2))^{3/2}} F(\epsilon_n(\chi)) \tag{17.73}$$

Similar to the first term in Eq. (17.25), this high-frequency term involves the factor $1 - T_n$, i.e., it vanishes identically for fully open conducting channels. Combining Eqs. (17.72) and (17.73) with Eq. (17.71), we arrive at the expression for the supercurrent

$$I(\chi) = I_S(\chi) + \delta I_S(\chi) \tag{17.74}$$

where in the low-temperature limit, the interaction correction reads

$$\delta I_S(\chi) = -\frac{e\Delta_0}{2g_N} \ln\left(\frac{1}{2\Delta_0 R_N C}\right) \sin\chi \times \sum_n \frac{T_n(1-T_n)}{(1 - T_n \sin^2(\chi/2))^{3/2}} \left(2 - T_n \sin^2\frac{\chi}{2}\right) \tag{17.75}$$

As usual, the leading interaction correction (17.75) turns out to be inversely proportional to the dimensionless normal state conductance of the contact $g_N = R_q/R_N$. This result is justified as long as the Coulomb correction $\delta I_S(\chi)$ remains much smaller than the non-interacting term $I_S(\chi)$ (17.1). Typically, this condition requires the dimensionless conductance to obey the condition

$$g_N \gg \ln(1/2\Delta_0 R_N C) \tag{17.76}$$

As the reader surely noticed, Eq. (17.75) was derived only from the first term in Eq. (17.71). The second term in this equation involving the function $Z(0, -\omega)$ and the correlator $\langle\varphi^+\varphi^-\rangle$ is treated analogously. The corresponding result turns out to be smaller than the contribution from the first term by the logarithmic factor $\sim\ln(1/2\Delta_0 R_N C)$. Hence, the second term in Eq. (17.71) can be safely neglected for our purposes.

It is interesting to point out that the result (17.75) can also be reproduced if one just substitutes $T_n \to T_n + \delta T_n$ into Eq. (17.1), where

$$\delta T_n = -\frac{2}{g_N} \ln\left(\frac{1}{2\Delta_0 R_N C}\right) T_n(1 - T_n) \tag{17.77}$$

and then expands the result to the first order in δT_n. Let us recall that the same transmission renormalization (17.77) follows from the RG equations (11.72) derived in Chapter 11 for normal conductors. In order to arrive at Eq. (17.77) after setting $g_S \to 0$ in Eq. (11.72), one should just start the corresponding RG flow at $\epsilon = 1/R_N C$ and stop it at $\epsilon = 2\Delta_0$. Thus, the result (17.75) can be interpreted in a very simple manner: Coulomb interaction provides high-frequency renormalization of channel transmissions $T_n + \delta T_n$ (17.77) of the barrier transmissions, which should be substituted into the expression for the supercurrent (17.1) instead T_n. It should be stressed, however, that the validity of the last step remains, by no means, obvious without our rigorous derivation since the Coulomb correction to the Josephson current originates from the term $\sim\varphi^-(\varphi^+)^2$ in the effective action, which is, of course, totally absent in the normal case.

One can further specify the expression for the interaction correction (17.75) in the important special case of short diffusive contacts. In the absence of interactions and at $T \to 0$, the Josephson current in such contacts (17.1) and takes the form (4.50)

$$I_S(\chi) = \frac{\pi\Delta_0}{2eR_N} \cos\frac{\chi}{2} \ln\frac{1 + \sin\frac{\chi}{2}}{1 - \sin\frac{\chi}{2}} \tag{17.78}$$

Including interactions and averaging (17.75) with the bimodal transmission distribution (17.35), one finds

$$\delta I_S(\chi) = -\frac{e}{8}\Delta_0 \ln\left(\frac{1}{2\Delta_0 R_N C}\right) \cot(\chi/2) \times \left[\left(\sin\frac{\chi}{2} + \sin^{-1}\frac{\chi}{2}\right) \ln\frac{1 + \sin(\chi/2)}{1 - \sin(\chi/2)} - 2\right] \tag{17.79}$$

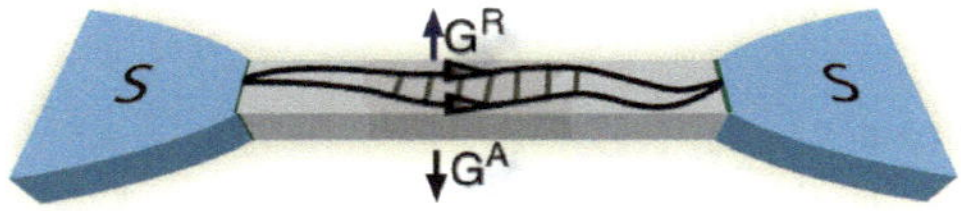

Figure 17.6 Diffusive SNS junction. The figure also illustrates the Cooperon and its spin structure corresponding to that of a Cooper pair.

To conclude this part of our analysis, under the condition (17.76), the Coulomb correction δI_C to the Josephson critical current I_C of superconducting weak links takes the universal form

$$\frac{\delta I_C}{I_C} = -\frac{\alpha}{g_N} \ln\left(\frac{1}{2\Delta_0 R_N C}\right), \quad (17.80)$$

where the numerical prefactor α depends on the transmission distribution. This prefactor reaches its maximum value $\alpha = 2$ in the case of tunnel barriers and becomes smaller for higher transmissions, e.g., $\alpha \simeq 0.72$ for diffusive contacts. In the case of open barriers $T_n \to 1$, the prefactor α tends to zero, $\alpha \to 0$, implying that *no Coulomb blockade* of the Josephson current is expected in fully transparent superconducting contacts.

Let us emphasize again that up to now in this chapter, we only considered the limit of relatively short contacts in which case the characteristic Thouless energy of the contact, ϵ_{Th}, exceeds the superconducting order parameter Δ_0. In the opposite case of long junctions $\epsilon_{\mathrm{Th}} \ll \Delta_0$, the interaction correction to the supercurrent turns out to have the same structure (17.80) with $2\Delta_0$ substituted by ϵ_{Th}. More importantly, in addition to (comparatively weak) Coulomb blockade effects in long SNS junctions, electron–electron interactions also cause dephasing of Cooper pairs, which may yield exponential suppression of the Josephson critical current down to $T = 0$ [359].

Let us consider a long diffusive SNS junction, as shown in Fig. 17.6. As before, the length and cross section of the normal wire connecting two superconductors will be denoted, respectively, as d_n and $\mathcal{A}$. For the sake of definiteness, we will also assume that at both NS interfaces there exist tunnel barriers with resistances $R_{\mathrm{L,R}}$ strongly exceeding that of a normal metal wire $R = d_n/(2e^2 N_0 D\mathcal{A})$.

In this limit and in the absence of electron–electron interactions, the Josephson current across the structure takes the form [25]

$$I \simeq \frac{\sin\chi}{4e^3 N_0 R_\mathrm{L} R_\mathrm{R} d_n \mathcal{A}} \times \begin{cases} \frac{4d_n}{L_\mathrm{T}} e^{-d_n/L_\mathrm{T}}, & d_n \gg L_\mathrm{T}, \\ \ln\left(\frac{4\epsilon_\mathrm{Th}}{\pi T}\right) + \gamma_0 + \frac{7\pi^2 T^2}{540\epsilon_\mathrm{Th}^2} + \ldots, & d_n \ll L_\mathrm{T} \end{cases} \quad (17.81)$$

where $\gamma_0 = 0.577\ldots$ is the Euler constant, $\epsilon_\mathrm{Th} = D/d_n^2$ is the Thouless energy of the normal wire (assumed to be well below the superconducting gap $\epsilon_\mathrm{Th} \ll \Delta_0$), and $L_\mathrm{T} = \sqrt{D/2\pi T}$ is the thermal length. Note that the current (17.81) formally diverges in the zero-temperature limit $T \to 0$. In the absence of interactions, this divergence can be cured only by taking into account higher-order tunneling terms.

Alternatively, one can naturally regularize the non-interacting result (17.81) in the limit $T \to 0$ by including Coulomb blockade effects [72]. Within the model adopted here, a detailed nonlinear sigma-model calculation yields [359]

$$I \simeq \frac{\mathcal{B}\sin\chi}{4e^3 N_0 R_\mathrm{L} R_\mathrm{R} d_n \mathcal{A}} \times \begin{cases} \ln(2\epsilon_\mathrm{Th}/(\pi^2\epsilon_\mathrm{C})), & \epsilon_\mathrm{Th} \gg \epsilon_\mathrm{C}, \\ 0.271\epsilon_\mathrm{Th}/\epsilon_\mathrm{C}, & \epsilon_\mathrm{Th} \ll \epsilon_\mathrm{C} \end{cases} \quad (17.82)$$

where we defined

$$\mathcal{B} \approx \begin{cases} (\epsilon_{Th}\tau_\mathrm{RC})^{\frac{8}{3g_n}} e^{-\frac{6.105}{g_n} - \frac{8\pi T}{3g_n\epsilon_\mathrm{Th}}}, & T \ll \epsilon_\mathrm{Th}, \\ (T\tau_\mathrm{RC})^{\frac{8}{3g_n}} e^{-\frac{2.892}{g_n} - \frac{8\pi T}{3g\epsilon_\mathrm{Th}}}, & \epsilon_\mathrm{Th} \ll T \end{cases} \quad (17.83)$$

and introduced the dimensionless conductance of the normal wire $g_n = R_q/R_n$ as well as the corresponding RC-time $\tau_\mathrm{RC} = RC_n$ and the Coulomb energy $\epsilon_\mathrm{C} = e^2/C_n$ with $C_n \simeq d_n/\ln(d_n^2/\mathcal{A})$ being geometric capacitance of the normal wire.

In the limit $\epsilon_\mathrm{Th} \gg \epsilon_\mathrm{C}$, Eq. (17.82) holds within the logarithmic accuracy and demonstrates that Coulomb blockade naturally eliminates the divergence of the non-interacting result (17.81) at $T \to 0$. A similar result is obtained within the model taking into account the gate capacitance and/or those of the tunnel barriers [72].

Finally, let us include the effect of Cooper pair dephasing into our consideration. The underlying physics of this effect is essentially the same as that already discussed in Chapter 16 for NS structures: Two electrons initially forming a Cooper pair propagate in the normal metal between two superconductors, pick up random phases while interacting with the fluctuating electromagnetic field produced by other electrons, and eventually become incoherent at length scales exceeding $L_\varphi = \sqrt{D\tau_\varphi}$. Technically, the analysis is also similar and again requires a detailed nonlinear sigma-model calculation. The supercurrent can be expressed via the Cooperon (as illustrated in Fig. 17.6), which decays in time in the presence of Cooper pair dephasing in the normal metal. As a result, for the Josephson current, one finds [359]

$$I = \frac{\pi T \mathcal{B} \sin\chi}{2e^3 \nu R_\mathrm{L} R_\mathrm{R} d_n \mathcal{A}} \int_0^\infty dt \frac{\vartheta_3\left(\frac{1}{2}, e^{-\pi^2 \epsilon_\mathrm{Th} t}\right) e^{-\frac{t}{\tau_\varphi}}}{\sinh(2\pi T t)} \tag{17.84}$$

where $\vartheta_3(z, q)$ is the third Jacobi theta-function and

$$\frac{1}{\tau_\varphi} \simeq \frac{1}{4\pi N_0 \mathcal{A}\sqrt{2D\tau_e}} - \frac{\ln\left(\frac{2}{\epsilon_\mathrm{Th}\tau_e}\right)}{4\pi N_0 d_n \mathcal{A}} + \frac{\pi T^2}{270 N_0 \epsilon_\mathrm{Th}^2 d_n \mathcal{A}} + \ldots \tag{17.85}$$

is the inverse dephasing time for Cooper pairs due to electron–electron interactions. One observes that—as compared to the non-interacting limit (17.81)—the Josephson current essentially depends on the extra energy scale, which is the Cooper pair dephasing rate $1/\tau_\varphi$. By formally setting this rate equal to zero in Eq. (17.84), one immediately recovers the result (17.81).

Provided the temperature is sufficiently high $L_\mathrm{T} \lesssim d_n$, the Josephson current (17.84) reduces to an exponentially small value

$$I = \frac{2\pi T \mathcal{B} L_c \sin\chi}{e^3 N_0 D R_\mathrm{L} R_\mathrm{R} \mathcal{A}} e^{-\frac{d_n}{L_c}} \tag{17.86}$$

where

$$L_c = \sqrt{\frac{D\tau_\varphi}{1 + 2\pi T \tau_\varphi}} = \frac{L_\mathrm{T} L_\varphi}{\sqrt{L_\mathrm{T}^2 + L_\varphi^2}} \tag{17.87}$$

defines the critical length, which—unlike in the non-interacting case—now depends on both temperature and the dephasing time τ_φ.

In the opposite low-temperature limit $d_n, L_\varphi \ll L_\mathrm{T}$, one finds

$$I \simeq \frac{\mathcal{B} \sin\chi}{2e^3 N_0 R_\mathrm{L} R_\mathrm{R} d_n \mathcal{A}} \ln\left(\coth\left(\frac{d_n}{2L_\varphi}\right)\right) \tag{17.88}$$

Hence, non-vanishing Cooper pair dephasing also provides a natural cutoff of the divergence in Eq. (17.81) at $T \to 0$.

The above results clearly demonstrate that dephasing of Cooper pairs by electron–electron interactions may strongly influence the Josephson current in long diffusive SNS junctions at low temperatures. The corresponding dephasing length L_φ can be detected experimentally by measuring this current as a function of the normal wire length d_n.

Note that an additional interaction-induced suppression of the Josephson current is encoded in the parameter $\mathcal{B}$ (17.83). This is a specific contribution to the dephasing of Cooper pairs provided by uniform-in-space fluctuations of the electromagnetic field. The magnitude of this effect is controlled by the dimensionless conductance g_n of the normal wire. As this parameter typically remains large for generic metallic junctions, the corresponding reduction of the supercurrent may be less significant than that caused by non-uniform-in-space electromagnetic fluctuations giving rise to the parameter τ_φ (17.85).

Finally, we observe that Eq. (17.85), up to a numerical factor of order one, coincides with the corresponding results for τ_φ already derived in Chapters 14–16 from the analysis of the Andreev conductance of NS structures and of the weak localization correction to the conductance of normal metals. This agreement again illustrates the universality of the phenomenon of low-temperature quantum decoherence by electron–electron interactions, which can be observed in a variety of normal and hybrid NS structures. An important peculiar feature of our present situation is that unlike in Chapters 14–16, here we address *non-dissipative* electron transport demonstrating that quantum dephasing of Cooper pairs may occur exactly in the equilibrium ground state.

17.7 Andreev Levels as an Effective Environment

Let us now return to short superconducting junctions with arbitrary distribution of normal transmissions T_n. In Section 17.4, we already demonstrated that at sufficiently low energies/frequencies (17.38), the system dynamics can be described in a simple manner by means of an effective capacitance renormalization in Eqs. (17.40) and (17.41). It is easy to observe that provided either the phase χ is not very close to π or T_n is not too close to unity, the energies of Andreev bound states $\epsilon_n(\chi)$ (17.12) remain of the order of the superconducting gap $\epsilon_n(\chi) \sim \Delta_0$, in which case the condition (17.38) can be simply rewritten as

$$T, \omega \ll 2\Delta_0 \tag{17.89}$$

On the other hand, for $1 - T_n \ll 1$ and $\pi - \chi \ll \pi$, we have $\epsilon_n(\chi) \ll \Delta_0$ and, hence, the conditions (17.38) and (17.89) may differ significantly in this limit. This observation implies that highly transparent superconducting contacts should be treated somewhat more carefully.

Let us again make use of the influence functional (17.14), (17.15) and decompose each of the kernels $\mathcal{R}(t)$ and $\mathcal{I}(t)$ into three different contributions, according to Eqs. (17.22) and (17.16). In the adiabatic limit (17.89), the terms $\mathcal{I}_2$ and $\mathcal{I}_3$ (which have to do with quasiparticles above the gap) vanish, while the analogous terms $\mathcal{R}_2$ and $\mathcal{R}_3$ can again be expanded in powers of ω up to terms $\sim \omega^2$, which yields an effective capacitance renormalization defined by the first line in Eq. (17.41). In the limit $1 - T_n \ll 1$ and $\pi - \chi \ll \pi$, this result reduces to a simple form

$$\tilde{C} \simeq C + \frac{e^2 \mathcal{N}}{4\Delta_0} \tag{17.90}$$

where, as before, C is the geometric capacitance and $\mathcal{N}$ is the total number of conducting channels in our contact. Comparing this partially renormalized capacitance $\tilde{C}$ with our previous result for C^* (17.41), we observe that the difference between these two expressions is given by the second and third lines in Eq. (17.41) originating from the expansion of the kernel $\mathcal{R}_1$ in powers of $\omega/\epsilon_n(\chi)$. As the inequality (17.38) is now violated, this parameter is not anymore small and we need to retain the full expressions for $\mathcal{R}_1(t)$ and $\mathcal{I}_1(t)$. Making use of Eqs. (17.17) and (17.63), we get [129]

$$\mathcal{R}_1(t) = \sum_n \frac{\gamma_n}{2\epsilon_n(\chi)} \theta(t) \sin(2\epsilon_n(\chi)t) \tag{17.91}$$

and

$$\mathcal{I}_1(t) = \sum_n \frac{\gamma_n}{8\epsilon_n(\chi)} \coth\left(\frac{\epsilon_n(\chi)}{T}\right) \cos\left(2\epsilon_n(\chi)t\right) \tag{17.92}$$

where the parameter γ_n is defined in Eq. (17.64). Collecting now all contributions to the effective action in Eq. (17.2), in the case of high contact transmissions and in the low-frequency limit (17.89), we obtain

$$\mathrm{i}S_{\mathrm{c}}+\mathrm{i}S_{\mathrm{sc}}=\mathrm{i}\tilde{S}_0[\varphi^+ +\varphi^-/2]-\mathrm{i}\tilde{S}_0[\varphi^+ -\varphi^-/2]+\mathrm{i}S_{\mathrm{R}}^{(1)}-S_{\mathrm{I}}^{(1)} \quad (17.93)$$

where

$$\tilde{S}_0[\varphi]=\int_0^t dt'\left[\frac{\tilde{C}\dot{\varphi}^2}{2e^2}-U\left(\chi+2\varphi(t')\right)\right] \quad (17.94)$$

is the "non-interacting" part of the action for the fluctuating phase φ,

$$U(\chi)=-2T\sum_{n=1}^{\mathcal{N}}\ln\left[\cosh\frac{\epsilon_n(\chi)}{2T}\right]-\frac{I\chi}{2e} \quad (17.95)$$

is the effective potential where the first term is recovered by integrating the supercurrent (17.1) over χ and the second term, as usually, accounts for the bias current I. Note that in Eq. (17.93), we refrain from expanding the action $\tilde{S}_0$ in φ^- and retain the exact expression for the potential energy U (17.95). Finally, the terms

$$S_{\mathrm{R}}^{(1)}=\int_0^t dt'\int_0^t dt''\mathcal{R}_1(t'-t'')\varphi^-(t')\varphi^+(t''), \quad (17.96)$$

$$S_{\mathrm{I}}^{(1)}=\int_0^t dt'\int_0^t dt''\mathcal{I}_1(t'-t'')\varphi^-(t')\varphi^-(t'') \quad (17.97)$$

represent the two contributions to the influence functional quadratic in the phase fluctuations.

Comparing these results with those of Chapter 1 (see Section 1.3 and 1.4), we immediately observe that our influence functional $\mathrm{i}S_{\mathrm{R}}^{(1)}-S_{\mathrm{I}}^{(1)}$ for the fluctuating phase $\varphi(t)$ is *exactly equivalent* to that of $\mathcal{N}$ harmonic oscillators with frequencies

$$\omega_n=2\epsilon_n(\chi) \quad (17.98)$$

cf., e.g., Eqs. (17.91), (17.92), (17.96), (17.97) and Eqs. (1.44), (1.45), (1.52), (1.53). In other words, the influence of subgap Andreev levels on quantum dynamics of the phase variable φ is identical to that of an effective bath of oscillators. Hence, just like in Chapter 1, our system can be described by the Hamiltonian (1.8), where now

$$\hat{H}_0 = -\frac{e^2}{2\tilde{C}}\frac{\partial^2}{\partial\varphi^2} + U(\chi + 2\varphi) \tag{17.99}$$

is the Hamiltonian of our "Josephson particle" (cf. Eq. (17.94)) and

$$\hat{H}_{\text{env}} + \hat{H}_{\text{int}} = \sum_{n=1}^{\mathcal{N}} \left[\frac{\hat{P}_n^2}{2M_n} + \frac{M_n\omega_n^2}{2}\left(Q_n - \frac{c_n\varphi}{M_n\omega_n^2}\right)^2 \right] \tag{17.100}$$

describe the Hamiltonian of an effective environment, which consists of $\mathcal{N}$ harmonic oscillators with frequencies (17.98) and their interaction with the "particle coordinate" φ. The values of the corresponding coupling constants c_n are identified by the condition

$$\frac{c_n^2}{M_n} = \gamma_n \tag{17.101}$$

It follows immediately from Eq. (17.101) combined with the expression for γ_n (17.64) that coupling between the phase φ and the n-th oscillator of our effective environment decreases with increasing transmission T_n and eventually vanishes in the limit $T_n = 1$. Thus, environmental modes corresponding to fully open transport channels are totally decoupled from the phase variable φ and, hence, cannot influence its quantum dynamics.

On the other hand, channels with $T_n < 1$ do affect the behavior of φ. As long as the characteristic frequency of the process ω_φ remains much smaller than ω_n (cf. Eqs. (17.38), (17.98)), this n-th environmental mode may only yield extra capacitance renormalization. Provided the inequality $\omega_\varphi \ll \omega_n$ is fulfilled for all n, we again recover the result (17.41), i.e., in this limit the term $S_{\text{R}}^{(1)}$ (17.96) just accounts for the difference between C^* and $\tilde{C}$. In the opposite limit $\omega_\varphi \gg \omega_n$, however, our harmonic oscillations may already act as a *quantum dissipative environment*. Dissipation for the fluctuating phase variable φ appears when the oscillators get excited to higher energy states as a result of their interaction with this variable. Note that since phase fluctuations remain small $|\varphi(t)| \ll \pi$, the dual charge variable may fluctuate strongly, implying that multiple electron charge transfer is possible through each conducting channel. Accordingly, many of such electrons can, in general, get excited to the higher of the two Andreev levels while passing through the n-th channel. These processes correspond to the excitation of harmonic oscillators to higher levels, illustrating the physical reason why the effect of Andreev levels is well accounted for by an effective environment formed by such oscillators.

The frequency spectrum of this quantum dissipative environment depends on the particular distribution of normal transmissions T_n in our contact and, hence, can be diverse. This spectrum can be defined in the same way as it is already done, e.g, in Chapter 1 for the Caldeira–Leggett model (cf. Eq. (1.54)), i.e.,

$$J(\omega, \chi) = \frac{\pi}{2}\sum_n \frac{c_n^2}{M_n\omega_n}\delta(\omega - \omega_n)$$
$$= \pi\sum_n T_n^2(1 - T_n)\frac{\Delta_0^4}{\epsilon_n^2(\chi)}\sin^4\frac{\chi}{2}\tanh\frac{\epsilon_n(\chi)}{2T}\delta(\omega - 2\epsilon_n(\chi)) \tag{17.102}$$

If one assumes that all the transmissions have the same value $T_n = \mathcal{T}$, one immediately recovers the model of a single mode environment. For a junction with a large number of channels, and with transmissions densely distributed in the interval $0 < T_n < 1$, the spectral density takes the form [129]

$$J(\omega, \chi) = \frac{\pi(\Delta^2 - \omega^2/4)}{2\omega}P(T_\omega)T_\omega(1 - T_\omega)\tanh\frac{\omega}{4T} \tag{17.103}$$

where $P(T_n)$ is the transmission distribution and

$$T_\omega = \frac{1 - \omega^2/4\Delta_0^2}{\sin^2(\chi/2)} \tag{17.104}$$

The expression (17.103) holds at frequencies $\omega_{\min}(\chi) < \omega < \omega_{\max}$ and vanishes outside this frequency interval. The cutoff frequencies are defined as

$$\omega_{\min}(\chi) = 2\Delta_0|\cos(\chi/2)|, \quad \omega_{\max} = 2\Delta_0 \tag{17.105}$$

Dissipation due to an effective environment formed by Andreev levels may occur, provided the Josephson plasma frequency (17.47) obeys the condition

$$\omega_{\mathrm{J}}(\chi) \gtrsim \omega_{\min}(\chi) \tag{17.106}$$

Let us define the phase value χ_c at which the critical value of the supercurrent I_{S} (17.1) flowing through our contact at a given temperature T is reached,

$$I_{\mathrm{S}}(\chi_c) = I_c(T) \tag{17.107}$$

and consider the phase interval $-\chi_c < \chi < \chi_c$ corresponding to the stable part of $I_S(\chi)$. In the limit of large geometric capacitance C, one typically has $\omega_J(0) \ll 2\Delta_0$, while the condition (17.106) may only be satisfied at bias current values I very close to I_c and provided $\pi - \chi_c < \omega_J^2(0)/8\Delta_0^2$. The latter condition can be satisfied only if the current-phase relation is strongly non-sinusoidal and, hence, most of the channels are nearly open. In this case, Andreev levels may make quantum dynamics of the phase dissipative at subgap energies.

For smaller values of C, the plasma frequency becomes higher, and one can even reach the limit $\omega_J(0) > \omega_{\max}$ in which case damping due to quasiparticles above the gap sets in. With increasing I, the plasma frequency decreases, first reaching the value $\omega_J(\chi_1) = \omega_{\max}$ (17.105) at a phase value $\chi = \chi_1$ and then approaching $\omega_{\min}$ at $\chi = \chi_2 > \chi_1$. Within this scenario, quasiparticles are responsible for dissipation for $0 < I < I_S(\chi_1)$, while at higher bias values $I_S(\chi_1) < I < I_S(\chi_2)$, quantum dissipation is solely due to Andreev levels. At even bigger currents $I > I_S(\chi_2)$, no dissipation occurs, and the capacitance renormalization (17.41) remains the only effect of the environment.

Consider, e.g., a qubit formed by the two lowest energy levels close to the bottom of the Josephson potential well. It follows from our analysis that for $I_S(\chi_1) < I < I_S(\chi_2)$, the energy relaxation time of this qubit $\tau_{\text{in}}(\chi)$ still remains finite due to dissipation produced by Andreev levels. In this case, we obtain

$$\frac{1}{\tau_{\text{in}}} = \frac{e^2 J\left(\omega_J(\chi), \chi\right)}{2\,\omega_J(\chi)\, C_{\text{eff}}} \coth \frac{\omega_J(\chi)}{2T} \tag{17.108}$$

where the effective capacitance C_{eff} coincides with the geometric one C, provided ω_J is not much smaller than $2\Delta_0$. Only in the limit $\omega_J \ll 2\Delta_0$, one should also include renormalization effects and $C_{\text{eff}} = \tilde{C}$.

The phase values $\chi_{1,2}$ depend on temperature, capacitance C, and the transmission distribution $P(T_n)$. For example, in diffusive contacts described by the zero-temperature current–phase relation (17.78), we have $\chi_c = 1.25(\pi/2)$ and, hence, at $T \to 0$ the low-frequency cutoff is never smaller than $\omega_{\min} \simeq 1.1\Delta_0$. Only provided χ_c is sufficiently close to π and, in addition, there exist highly transmitting channels with $T \approx 1$ the cutoff frequency $\omega_{\min}$ can be much smaller than $2\Delta_0$. An example of that kind is discussed in the next section.

17.8 Supercurrent Decay in Highly Transparent Contacts

One important application of the above analysis is the problem of macroscopic tunneling (MQT) of the superconducting phase in contacts with high values of their normal transmissions T_n. An extended theory of this phenomenon in the presence of linear Ohmic dissipation is detailed in Chapter 4 for the case of standard Josephson tunnel barriers. In that chapter, we also discussed the properties of superconducting junctions with higher normal transmissions, though without making use of any microscopic model for dissipation caused by subgap Andreev states. Below we will fill in this gap.

As we already learned in the previous chapters, the problem of quantum tunneling is conveniently described within the imaginary time Matsubara formalism. Therefore, our first step is to reformulate our results for the influence functional within the framework of the Matsubara technique. This task is accomplished in exactly the same manner as it was already done, e.g., in Chapter 1 (see Section 1.6). We again define the grand partition function for our system

$$\mathcal{Z} = \mathrm{Sp} e^{-\hat{H}/T} \tag{17.109}$$

where now the effective Hamiltonian $\hat{H}$ is defined in Eqs. (1.8), (17.99)–(17.101). Rewriting Eq. (17.109) in the form of a path integral over both φ and the oscillator coordinates Q_n and performing Gaussian integrals over all Q_n-variables, we obtain

$$\mathcal{Z} = \int \mathrm{D}\varphi \exp\left(-S_{\mathrm{eff}}[\varphi(\tau)]\right) \tag{17.110}$$

where

$$S_{\mathrm{eff}} = \int_0^{1/T} d\tau \left[\frac{\tilde{C}\dot{\phi}^2}{8e^2} + U(\chi + \phi(\tau))\right] + \int_0^{1/T} d\tau_1 \int_0^{1/T} d\tau_2 Y(\tau_1 - \tau_2)\phi(\tau_1)\phi(\tau_2) \tag{17.111}$$

is the imaginary time effective action for our superconducting contact. For convenience, here and below we re-denote the small fluctuating part of the superconducting phase as $\phi = 2\varphi$. As usual, the last (non-local in time) term defines the imaginary time version of the influence functional with [129]

$$Y(\tau) = \sum_n \frac{\gamma_n}{32\epsilon_n} \left(\frac{\delta(\tau)}{\epsilon_n} - \frac{\cosh\left[2\epsilon_n \left(|\tau| - \frac{1}{2T}\right)\right]}{\sinh[\epsilon_n/T]} \right) \quad (17.112)$$

Note that the same expression can also be derived by means of direct analytic continuation of the influence functional in Eqs. (17.91), (17.92), (17.96), and (17.97) from the Keldysh contour to the imaginary time one.

Expanding the kernel $Y(\tau)$ (17.112) in the Fourier series

$$Y(\tau) = T \sum_{\nu=-\infty}^{\infty} Y_{\omega_\nu} e^{-i\omega_\nu \tau}, \quad \omega_\nu = 2\pi\nu T \quad (17.113)$$

and combining the result with Eq. (17.64), we get

$$Y_{\omega_\nu} = \sum_n \frac{T_n^2(1-T_n)\Delta_0^4 \sin^4(\chi/2)}{8\epsilon_n^3(\chi)} \frac{\omega_\nu^2}{4\epsilon_n^2(\chi) + \omega_\nu^2} \tanh\left(\frac{\epsilon(\chi)}{2T}\right) \quad (17.114)$$

Turning now to the MQT problem, let us first fix the value of the constant part of the superconducting phase χ. In order to do so, we will bear in mind that, as we already discussed in Chapter 4, MQT is most pronounced at the bias current values I close to the critical one I_c for our superconducting contact. Thus, from now on we can set the value $\chi = \chi_c$; see Eq. (17.107). Phase fluctuations around this value are then fully described by the variable $\phi(\tau)$. Introducing the parameter

$$q(T) = 1 - I/I_c(T) \ll 1 \quad (17.115)$$

expanding the potential energy $U(\chi_c + \phi)$ in powers of $\phi \ll 1$ up to the cubic terms, and disregarding irrelevant for us ϕ-independent term, one readily finds

$$U = \frac{I_c(T)q(T)}{2e}\phi + \frac{I_S''(\chi_c)}{12e}\phi^3 \quad (17.116)$$

where $I_S''(\chi_c)$ is the second derivative of the supercurrent with respect to χ. Note that this expansion of the effective potential energy U remains applicable either at not-too-low temperatures or provided χ_c is not too close to the value π (see below).

Let us first stick to the low-temperature limit $T \to 0$. In this case, Eq. (17.116) reduces to

$$U = \frac{\Delta_0 \mathcal{N}}{2}\left[q\phi - \frac{\phi^3}{6}\right] \quad (17.117)$$

Note that in Eq. (17.117), we made use of the condition $1 - T_n \ll 1$ and neglected the terms $\sim \sqrt{1 - T_n}$ as compared to 1. It is easy to estimate the plasma oscillation frequency ω_ϕ near the bottom of the potential well (17.117). Combining the latter equation with Eq. (17.111) and neglecting for a moment the influence functional by formally setting $Y(\tau) \to 0$, one readily finds

$$\omega_\phi^2 = 2^{3/2} e^2 \Delta_0 \mathcal{N} q^{1/2} / \tilde{C} \quad (17.118)$$

As we already discussed, provided the frequency ω_ϕ obeys the adiabaticity condition (17.38), i.e.,

$$\omega_\phi \ll 2\Delta_0 \sqrt{1 - T_n \sin^2(\chi_c/2)} \quad (17.119)$$

for all n, the influence of Andreev levels on quantum dynamics of the phase variable ϕ is reduced to an effective extra renormalization of the capacitance $\tilde{C}$ defined by the second and third lines of Eq. (17.41). Then we get back our previous result for C^* (17.40). Accordingly, the plasma oscillation mode becomes softened as

$$\omega_0 = \omega_\phi \sqrt{\tilde{C}/C^*} \quad (17.120)$$

Quantum decay of the supercurrent is described in a standard manner as quantum tunneling of the Josephson particle with effective mass $C^*/4e^2$ in the potential (17.117), as illustrated in Fig. 4.1 (process (a)). The corresponding decay rate is then defined in Eq. (4.41) with the exponent A given by Eq. (4.52), where now

$$U_0 = \frac{\Delta_0 \mathcal{N} (2q)^{3/2}}{3} \quad (17.121)$$

and the frequency ω_0 is set by Eqs. (17.120) and (17.118).

Let us specify this result for the model of a single mode environment with $T_n = \mathcal{T}$, in which case we have

$$\chi_c = \pi - \arccos\left[\frac{(1 - \sqrt{1 - \mathcal{T}})^2}{\mathcal{T}}\right] \quad (17.122)$$

$$\epsilon_n(\chi_c) \equiv \epsilon(\chi_c) = \Delta_0(1 - \mathcal{T})^{1/4} \tag{17.123}$$

and

$$C^* = \tilde{C} + \frac{e^2\mathcal{N}}{4\Delta_0(1 - \mathcal{T})^{1/4}} \tag{17.124}$$

We may further assume that geometric capacitance C of our contact is not large and the expression for C^* is dominated by the last term in Eq. (17.124). Then we get $\omega_0 = 2^{7/4}\Delta_0 q^{1/4}(1 - \mathcal{T})^{1/8}$, and Eq. (4.52) yields

$$A = 2^{7/4}\frac{3\mathcal{N}q^{5/4}}{5(1 - \mathcal{T})^{1/8}} \tag{17.125}$$

This result remains applicable in the adiabatic limit (17.119). As soon as this adiabaticity condition is violated, quantum tunneling becomes dissipative (process (b) in Fig. 4.1) because the oscillators in our effective environment can now get excited to higher levels. In Chapter 4, we already learned that quantum tunneling can be described in terms of classical dynamics of the particle ϕ propagating in the inverted potential $-U$ along the bounce trajectory. Identifying the characteristic bounce frequency with that of small oscillations $\tilde{\omega}_0$ near the bottom of this potential at $\phi = \sqrt{2q}$ and setting again $\tilde{C} \simeq e^2\mathcal{N}/(4\Delta_0)$, after trivial algebra we get

$$\tilde{\omega}_0 = \Omega(q, r^{1/4}) \tag{17.126}$$

where we defined $r = 1 - \mathcal{T}$ and introduced the function

$$\Omega(q, y) = 2\Delta\sqrt{\frac{Z}{2} + \sqrt{\frac{Z^2}{4} + y^2\sqrt{8q}}}, \quad Z = \sqrt{8q} - y^2 - r/y^3 \tag{17.127}$$

For the potential (17.117), we find

$$A \simeq 36U_0/(5\tilde{\omega}_0) \tag{17.128}$$

This formula together with Eqs. (17.126) and (17.127) accounts for a trade-off between two different tunneling regimes. Let us define the value q_c from the equation

$$\Omega(q_c, r^{1/4}) = 2\epsilon(\chi_c) = 2\Delta r^{1/4} \tag{17.129}$$

which has a solution

$$q_c = \sqrt{r}/32 + (r^{3/4} + r)/8 \tag{17.130}$$

In the adiabatic limit $\omega_\phi \ll 2\Delta r^{1/4}$ (or $q < q_c$), Andreev oscillators are "fast" and can only yield strong capacitance renormalization (17.124). In this regime, the particle ϕ becomes heavier, but its energy is conserved during tunneling and we get back to Eq. (17.125). In the opposite antiadiabatic limit $\omega_\phi \gg 2\Delta r^{1/4}$ (or $q > q_c$), Andreev oscillators become "slow," generating effective potential renormalization $U \to U + \Delta\mathcal{N} r^{1/4}\phi^2/8$. Such oscillators can get excited to higher levels taking energy from the tunneling particle ϕ. Hence, for $q_c < q < \sqrt{r}$, Eq. (17.128) describes the regime of strong dissipation.

Turning now to the regime $q > \sqrt{r}$, we observe that in this limit the value χ_c is very close to π and, hence, the approximation (17.117) is not anymore sufficient. Instead, it is necessary to employ the exact non-quasiclassical potential (17.95), i.e.,

$$U(\chi) = -I\chi/2e - \mathcal{N}\Delta_0\sqrt{1 - \mathcal{T}\sin^2(\chi/2)} \tag{17.131}$$

cf. also Eq. (4.51). As we already discussed, for fully open contacts with $\mathcal{T} = 1$, the oscillators are decoupled from the phase ϕ and the MQT problem is again reduced to that of quantum tunneling of the particle with mass $\tilde{C}/4e^2$ through the potential $U(\chi + \phi)$. Employing the standard WKB formula (4.53), we obtain

$$A = \frac{24\sqrt{2}(3 - \sqrt{6})U_0}{5\omega_\phi} \tag{17.132}$$

Here the height of the potential barrier U_0 (despite its form change) is again given by Eq. (17.121), and the frequency ω_ϕ is defined in Eq. (17.118). We note that the numerical prefactor in Eqs. (17.132) and (4.52) is about two times smaller than that in Eq. (4.52). Also bearing in mind that ω_0 is never bigger than ω_ϕ and that $A \gg 1$, we conclude that for $1 - \mathcal{T} \ll 1$, the quantum decay rate $\Gamma \propto \exp(-A)$ can increase by orders of magnitude while the transmission value $\mathcal{T}$ changes only by a tiny amount. This behavior is illustrated in Fig. 17.7. The physical reason for it is simple: The potential barrier in the limit $q \gg \sqrt{1 - \mathcal{T}}$ is substantially "thinner" than in the opposite limit $q \ll \sqrt{1 - \mathcal{T}}$, while the barrier height U_0 remains the same in both limits. Accordingly, the probability for the system to tunnel can be much bigger in the former limit.

At $\mathcal{T} < 1$ and as long as the condition $\sqrt{r} \ll q$ is fulfilled, the effect of dissipation due to Andreev levels can be treated perturbatively

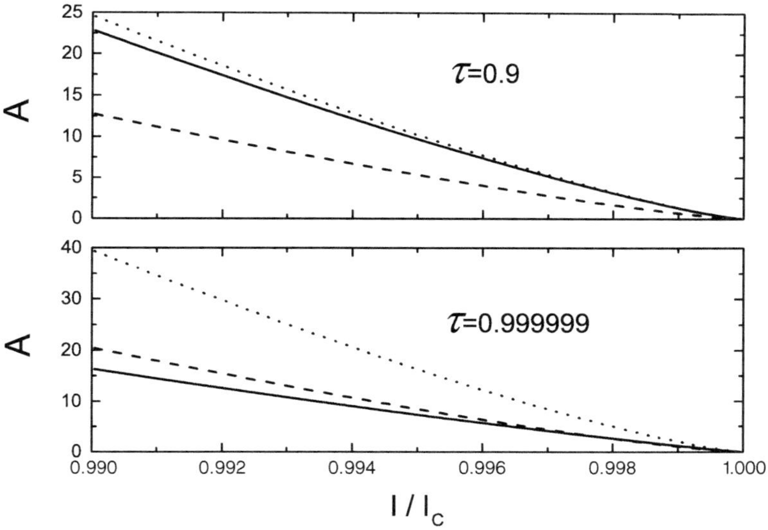

Figure 17.7 The exponential factor in the expression for the decay rate, $A = -\ln(\Gamma/B)$, in the vicinity of the critical current $I_c - I \ll I_c$ for two values of the contact transmission $\mathcal{T} = 0.9$ and $\mathcal{T} = 0.999999$. The contact geometric capacitance is supposed to be large $C \gg \delta C(\chi)$. Here we set $\mathcal{N}\Delta_0 C/e^2 = 5 \times 10^6$. Solid lines denote the integral (4.53) evaluated numerically, while dashed and dotted lines correspond to approximate results, respectively, in Eqs. (17.132) and (4.52).

and yields only a small correction to the result (17.132) in the parameter $\sqrt{r}/q$. Provided the renormalized capacitance $\tilde{C}$ is dominated by the last term in Eq. (17.90), we obtain

$$A = 2^{17/4}\frac{3-\sqrt{6}}{5}\mathcal{N}q^{5/4}\left[1+\alpha\left(\frac{\sqrt{r}}{q}\right)\right] \tag{17.133}$$

Here the function $\alpha \ll 1$ accounts for dissipation. In order to estimate α in Eq. (17.133), we employ Eq. (17.114) with χ corresponding to the local minimum of the potential (17.131). Then we obtain $\alpha(z) \sim z^2 \ll 1$. Choosing $\chi = \chi_c$ would give $\alpha(z) \sim \sqrt{z} \ll 1$; however, this choice would be inaccurate for $\sqrt{r} \ll q$ and $T \to 0$ since in this limit typical phase fluctuations $\delta\phi \sim \sqrt{q}$ exceed $\pi - \chi_c \sim r^{1/4}$. As long as $\sqrt{r} \lesssim q$, the value A grows with increasing r merely due to the potential profile change (as it is illustrated, e.g., in Fig. 17.7) rather than due to dissipation. Hence, for $\sqrt{r} \ll q$, the function α in Eq. (17.133) can be safely neglected. At larger $\sqrt{r} \sim q$, however, this function becomes of order unity $\alpha(1) \sim 1$ and the result (17.133) matches smoothly with Eq. (17.128).

To conclude, as long as the geometric capacitance C in Eq. (17.90) is small and can be neglected, one distinguishes three different MQT regimes. In the regime $q \gtrsim \sqrt{r}$, dissipation due to the effective environment is weak and the system tunnels through a non-quasiclassical potential illustrated in Fig. 4.3. For $q < \sqrt{r}$, quantum decay occurs in a smooth effective potential (17.117) and is essentially affected by Andreev levels. In the regime $q_c(r) < q < \sqrt{r}$, the environment formed by Andreev levels yields strong dissipation, whereas at $q < q_c(r)$, it produces strong capacitance renormalization. If, on the other hand, the geometric capacitance C is large enough to neglect the effect of Andreev levels, there are only two regimes $q \ll \sqrt{r}$ and $q \gg \sqrt{r}$ for which the exponent A is defined, respectively, in Eqs. (4.52) and (17.132).

In order to complete our analysis of MQT in highly transparent superconducting contacts, let us briefly address the case of nonzero temperatures. As we have already learned in Chapter 4, quantum decay can only occur at not very high temperatures $T < T_0$, whereas at $T > T_0$ thermally activated decay takes over with $\Gamma \propto \exp(-U_0/T)$. It follows from the analysis in Chapter 4 (see the discussion above Eq. (4.107)) that T_0 is formally determined as the lowest possible temperature at which the bounce still retains its trivial form $\phi(\tau) = \phi_m$, where ϕ_m is the phase value at which the maximum of the potential U (17.95) is reached. In our case, the corresponding condition takes the form

$$\pi^2 T^2 \tilde{C}/e^2 + 2Y_{2\pi T} = -U''(\phi_m) \tag{17.134}$$

Provided we approximate U by the cubic parabola (17.117), we have $\phi_m = \sqrt{2q}$ and, hence, $-U''(\phi_m) = \mathcal{N}\Delta_0\sqrt{q/2}$. Assuming that geometric capacitance C is small and $\tilde{C} \simeq e^2\mathcal{N}/(4\Delta_0)$, we resolve Eq. (17.134) with respect to T and obtain

$$T_0 = \Omega(q, y)/(2\pi), \quad y = \epsilon(\chi_c(T_0))/\Delta_0 \tag{17.135}$$

In the limit $T_0 \ll \Delta_0 r^{1/4}/\pi$, we have $y \simeq r^{1/4}$ and Eq. (17.135) reduces to $T_0 = \tilde{\omega}_0/(2\pi)$. In particular, at small enough q, we get

$$T_0 = \frac{2^{3/4}}{\pi}\Delta_0 q^{1/4}(1-\mathcal{T})^{1/8} \tag{17.136}$$

For $T_0 \gtrsim \Delta_0 r^{1/4}/\pi$, Eq. (17.135) approaches the $\mathcal{T}$-independent result

$$T_0 = \Delta_0(8q)^{1/4}/\pi \tag{17.137}$$

In order to describe the crossover between these two expressions for T_0, let us first proceed analytically and consider the temperature interval

$$\Delta_0 r^{1/4} \ll T \ll \Delta_0 \tag{17.138}$$

which obviously requires a strong inequality $r^{1/4} \ll 1$ to be fulfilled. In this case, the expression (17.1) reduces to

$$I_S(\chi) = e\mathcal{N}\Delta_0 \sin\frac{\chi}{2}\tanh\left(\frac{\Delta_0}{2T}\left|\cos\frac{\chi}{2}\right|\right) \tag{17.139}$$

and reaches its maximum at $\chi = \chi_c$, where

$$\chi_c = \pi - \frac{2T}{\Delta_0} W\left(\frac{2\Delta_0^2}{T^2}\right) \tag{17.140}$$

and $W(z)$ is the Lambert W-function providing the solution to the equation $W\exp(W) = z$. For $z \gg 1$, one has $W(z) = \ln z - \ln\ln z + o(1)$. Hence, in the temperature interval (17.138) within the logarithmic accuracy, we get

$$\epsilon(\chi_c) \simeq 2T\ln(\Delta_0/T) \tag{17.141}$$

Under the condition

$$q \ll T^2/\Delta_0^2 \tag{17.142}$$

we can again expand the potential energy U (17.95) in the phase fluctuations around the point $\chi = \chi_c$. In this way, we again reproduce Eq. (17.117). Rewriting now Eq. (17.134) in the temperature interval (17.138), with the aid of Eqs. (17.114) and (17.141), we obtain

$$\pi^2 T^2 \frac{\tilde{C}}{e^2} + \frac{\pi^2\mathcal{N}\Delta_0^4(1-\mathcal{T})}{2^7 T^3 \ln^5(\Delta_0/T)} = \frac{\mathcal{N}\Delta_0\sqrt{q}}{\sqrt{2}} \tag{17.143}$$

Resolving this equation, we determine the crossover temperature between the regimes of thermal activation and quantum tunneling. It is easy to observe that Eq. (17.143) has a solution consistent with the condition (17.138) only provided geometric capacitance C remains sufficiently small. In what follows we will neglect C and substitute the value $\tilde{C} = \mathcal{N}e^2/(4\Delta_0)$ into Eq. (17.143). If, to begin with, we simply ignore the effect of Andreev levels by dropping the second term in the left-hand side of Eq. (17.143), we immediately arrive at the result (17.137). This temperature obviously obeys the required condition (17.142) as well as the condition (17.138) in the limit $q \gg r$. Also it is easy to verify that at $T \sim T_0$ (17.137), the first term in the left-hand side of Eq. (17.143) strongly exceeds the second one for

$$q \gg r^{4/5} \tag{17.144}$$

Combining all these observations, we conclude that the result (17.137) applies under the condition (17.144).

Let us investigate Eq. (17.143) in more detail. For a qualitative analysis, it suffices to set all numerical and logarithmic factors equal to unity and refrain from writing equal signs and strong inequalities. Minimizing the left-hand side of this equation with

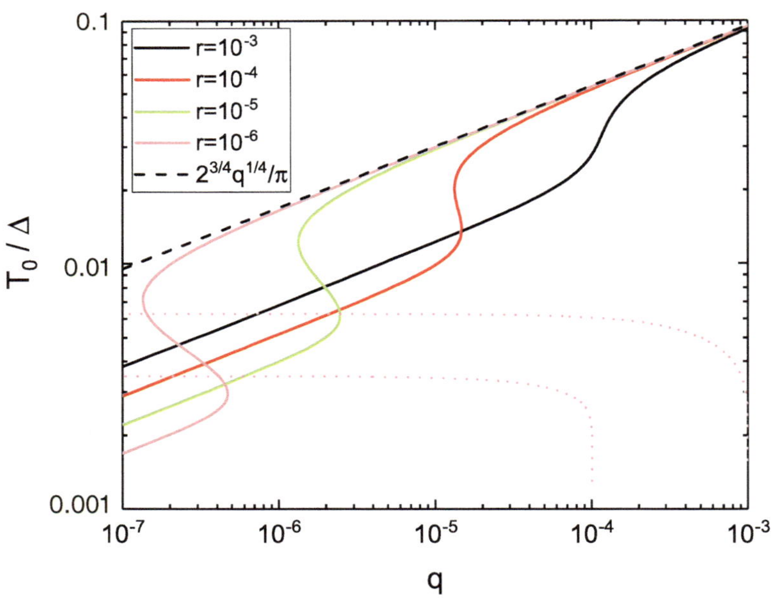

Figure 17.8 The function $T_0(q)$ (17.135) for different values of r (solid lines) together with the dependencies $q(T)$ for $r = 10^{-6}$ (dotted lines). The crossover temperature $T_0(I)$ is determined as the intersection point of $T_0(q)$ and $q(T)$.

respect to temperature, we obtain

$$T_{\min} \propto \Delta_0 r^{1/5} \tag{17.145}$$

Substituting this temperature back into Eq. (17.143), we estimate the minimum of the expression in its left-hand side to be $\sim\Delta_0 r^{2/5}$. Thus, Eq. (17.143) has a solution only provided $q > q_{\min} \propto r^{4/5}$. Under this condition, Eq. (17.143) has two solutions. One of them is given by Eq. (17.137); another one is easily recovered if we assume that in the relevant temperature interval, the first term in the left-hand side of Eq. (17.143) remains small as compared to the second one. Then for the second solution, we estimate

$$T_0 \sim \Delta_0 r^{1/3} q^{-1/6} \tag{17.146}$$

This simple qualitative analysis can be supplemented by a rigorous numerical calculation of the function $T_0(q)$ defined by Eq. (17.135). The corresponding results [129] are displayed in Figs. 17.8 and 17.9.

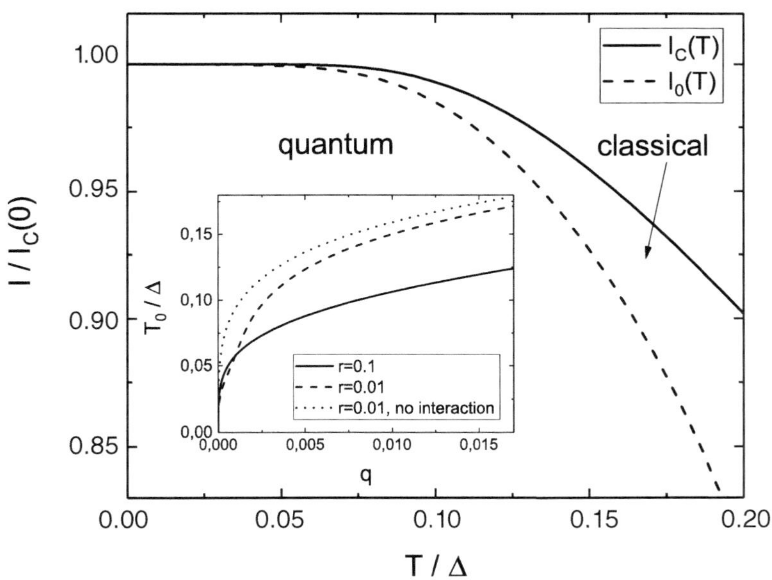

Figure 17.9 The bias current I_0 separating the regimes of quantum tunneling and thermal activation as a function of temperature plotted together with $I_C(T)$ for $r = 0.1$. The inset shows the dependence $T_0(q)$ (17.135) for $r = 0.1$ and $r = 0.01$.

As we already anticipated, for very small $r \lesssim 10^{-4}$, the function $T_0(q)$ is multivalued for some values of q. This behavior, however, does not imply the presence of several crossover temperatures for a given bias I because the critical current I_C also depends on temperature and, hence, $q = q(T)$. For each value I, the crossover temperature T_0 should be obtained from the equation $T_0 = T_0(q(T_0))$, as it is also illustrated in Fig. 17.8. As a result, we arrive at the bias current value $I = I_0(T)$ at which quantum-to-classical crossover occurs at a given temperature. The function $I_0(T)$ is plotted in Fig. 17.9 for $r = 0.1$ together with $I_C(T)$. With decreasing r, classical activation region shrinks and $I_0(T)$ rapidly approaches $I_C(T)$.

We conclude that the presence of low lying Andreev levels makes quantum decay of supercurrent in highly transparent superconducting contacts qualitatively different in many respects from that in more conventional Josephson tunnel junctions addressed in Chapter 4.

Chapter 18

Effective Action and Superconducting Fluctuations

18.1 Introduction

In Chapters 2, 16, and 17, we elaborated a detailed microscopic derivation of the effective action for various types of superconducting junctions. In doing so, we followed the basic idea of the Feynman–Vernon influence functional technique and integrated out all the electronic degrees of freedom as well as fluctuating electromagnetic potentials expressing the effective action of the system in terms of only one collective variable—the superconducting phase difference $\varphi(t)$ across the junction. In this case, we also neglected fluctuations of the absolute value of the superconducting order parameter $|\Delta|$ in bulk superconducting electrodes forming the junction.

In this chapter, we make use of the same effective action approach but apply it to spatially extended superconductors, which can be considered uniform at scales exceeding the electron elastic mean free path ℓ and do not contain, e.g., tunnel junctions or other types of inter-metallic contacts. We will again integrate out the electron variables that play the role of an effective bath in

Dissipative Quantum Mechanics of Nanostructures: Electron Transport, Fluctuations, and Interactions
Andrei D. Zaikin and Dmitry S. Golubev

ISBN 978-981-4774-50-5 (Hardcover), 978-0-429-29823-3 (eBook)
www.jennystanford.com

our problem. Unlike in the case of superconducting structures considered in Chapters 2, 16, and 17, however, it is not anymore possible to express the effective action in terms of only one quantum variable $\varphi(t)$ because (i) the superconducting order parameter phase now varies not only in time but also in space, i.e., $\varphi \equiv \varphi(\boldsymbol{r}, t)$ represents a quantum field rather than just one quantum degree of freedom; (ii) fluctuations of the absolute value of the order parameter $|\Delta(\boldsymbol{r}, t)|$ both in space and in time cannot, in general, be neglected; (iii) fluctuating electromagnetic potentials cannot, in general, be eliminated (or integrated out); and (iv) superconducting fluctuations and other dynamical processes may drive the quasiparticle distribution function far from equilibrium. All these features make our present task by far more complicated as compared to previously considered examples.

As it was demonstrated by a number of authors [24, 160, 385, 416], a significant progress can be achieved if one explicitly accounts for gauge invariance of the problem and employs the corresponding Ward identities. Although perturbation expansions up to a definite order may break gauge invariance, the Ward identities allow one to construct a manifestly gauge-invariant effective action within perturbation theory. As a result, e.g., the superconducting phase φ appears in the effective action only via gauge-invariant combinations representing the superfluid velocity $\boldsymbol{v}_S = \frac{1}{2m}(\nabla\varphi + \frac{2e}{c}\boldsymbol{A})$ and the so-called chemical potential of Cooper pairs $\Phi = (V - \dot{\varphi}/2e)$. Note that by setting $\Phi = 0$, we immediately recover the Josephson relation (2.29), which was extensively used, e.g., in Chapter 2 and elsewhere. As we already discussed, in many cases this relation is obeyed with a high accuracy. On the other hand, in non-equilibrium situations the chemical potential for Cooper pairs Φ may differ from zero and, hence, the Josephson relation (2.29) can be violated.

The real-time Keldysh effective action of a superconductor to be constructed below allows to account for diverse dynamical and fluctuation effects at any temperature, including $T \to 0$. In addition, one can demonstrate that under certain conditions, this effective action reduces to the system of Langevin equations, which fully describe quantum dynamics of the order parameter field and electromagnetic fields within the Gaussian approximation. Below we will also discuss a relation between our effective action formalism and a simplified approach based on the so-called time-dependent Ginzburg–Landau (TDGL) equations frequently used in order to describe dynamics of the superconducting order parameter close to the critical temperature T_{C}.

As an important example, we will employ our approach in order to explicitly analyze Gaussian fluctuations of the order parameter in superconducting nanowires. Non-Gaussian fluctuations in such wires—thermal and quantum phase slips (TAPS and QPS)—will be addressed in Chapter 19. A convenient theoretical approach to treat QPS effects or, equivalently, macroscopic quantum tunneling of the phase in superconducting wires is the Matsubara technique; see, e.g., Chapters 4–6. For this reason toward the end of this chapter, we will also derive an imaginary time version of the effective action and further specify it for an important case of superconducting nanowires.

18.2 Effective Action

As before, in what follows we will describe our superconductor by means of the Hamiltonian (2.17), which includes both Coulomb and BCS types of interactions between electrons. Our starting point is the formal expression for the quantum evolution operator on the Keldysh contour or the so-called Keldysh "partition function." As we have already done it in previous chapters, let us express the kernel of this evolution operator J (1.13) in the form of a path integral over quantum fields defined both at the forward and the backward branches of the Keldysh contour. We have

$$J = \int \mathcal{D}\Psi^{\alpha}_{\mathrm{F/B}}\, \mathcal{D}\bar{\Psi}^{\alpha}_{\mathrm{F/B}} \mathcal{D}\Delta_{\mathrm{F/B}}\, \mathcal{D}V_{\mathrm{F/B}}\, \mathcal{D}\boldsymbol{A}_{\mathrm{F/B}} e^{\mathrm{i}S[\Psi^{\alpha}_{\mathrm{F}},\Delta_{\mathrm{F}},V_{\mathrm{F}},\boldsymbol{A}_{\mathrm{F}}]-\mathrm{i}S[\Psi^{\alpha}_{\mathrm{B}},\Delta_{\mathrm{B}},V_{\mathrm{B}},\boldsymbol{A}_{\mathrm{B}}]} \tag{18.1}$$

Here, as before, $\Psi^{\alpha}_{\mathrm{F/B}}$, $\bar{\Psi}^{\alpha}_{\mathrm{F/B}}$ are electron Grasmann fields; $\alpha = \uparrow, \downarrow$ is the spin index; $\Delta_{\mathrm{F/B}}$ are superconducting complex order parameter fields that emerge as a result of the standard Hubbard–Stratonovich decoupling of the BCS coupling term in Eq. (2.17); $V_{\mathrm{F/B}}$ and $\boldsymbol{A}_{\mathrm{F/B}}$ are, respectively, the scalar and vector potentials. The action S is defined as follows:

$$\mathrm{i}S = \sum_{\alpha}\int d^4X \left[\bar{\Psi}^{\alpha}\left(\mathrm{i}\frac{\partial}{\partial t} + \frac{1}{2m}\left(\nabla + \mathrm{i}\frac{e}{c}\boldsymbol{A}\right)^2 + \mu - U(\boldsymbol{r}) + eV\right)\Psi^{\alpha} - \Delta\bar{\Psi}^{\uparrow}\bar{\Psi}^{\downarrow} - \Delta^{*}\Psi^{\downarrow}\Psi^{\uparrow}\right] + \int d^4X \left(\frac{\boldsymbol{E}^2 - \boldsymbol{H}^2}{8\pi} - \frac{N_0|\Delta|^2}{\lambda}\right) \tag{18.2}$$

We again define ϵ_{F} as the Fermi energy, $U(\boldsymbol{r})$ as a disorder potential, N_0 as the density of states per unit spin at the Fermi level, and λ as the BCS coupling constant. In our notations, the electron charge is $e^* = -e$, i.e., we define $e > 0$. Here and below, we also employ the notation $d^4X = dt\, d^3\boldsymbol{r}$.

As the above action is quadratic in the electron fields (cf. also Eq. (2.22)), one can integrate them out exactly, as we already did in Chapter 2. After that one arrives at the path integral

$$J = \int \mathcal{D}\Delta_{\mathrm{F/B}}\, \mathcal{D}V_{\mathrm{F/B}}\, \mathcal{D}\boldsymbol{A}_{\mathrm{F/B}}\, e^{\mathrm{i}S_{\mathrm{eff}}[\Delta_{\mathrm{F/B}}, V_{\mathrm{F/B}}, \boldsymbol{A}_{\mathrm{F/B}}]} \tag{18.3}$$

where the effective action reads

$$\mathrm{i}S_{\mathrm{eff}} = \mathrm{i}\int d^4X \left(\frac{\boldsymbol{E}^2 - \boldsymbol{H}^2}{8\pi} - \frac{N_0|\Delta|^2}{\lambda}\right) + \mathrm{Tr}\,\ln \check{\mathcal{G}}^{-1} \tag{18.4}$$

The inverse 4×4 matrix Green–Keldysh function $\check{\mathcal{G}}^{-1}$ can be split into 2×2 sub-blocks (indicated by a hat):

$$\check{\mathcal{G}}^{-1} = \begin{pmatrix} \hat{\mathcal{G}}_{11}^{-1} & -\hat{\sigma}_z\hat{\Delta} \\ -\hat{\sigma}_z\hat{\Delta} & \hat{\mathcal{G}}_{22}^{-1} \end{pmatrix}. \tag{18.5}$$

Similar to our analysis in Chapter 2, let us perform a gauge transformation under the trace, which leaves the action (18.4) invariant but allows to remove the phase φ from the superconducting order parameter field; cf., e.g., Eqs. (2.147) and (2.148). Then we obtain

$$\hat{\mathcal{G}}_{11}^{-1} = \hat{\sigma}_z\left(\mathrm{i}\frac{\partial}{\partial t} - \hat{H}_0 + \frac{\mathrm{i}}{2}\{\nabla, \hat{\boldsymbol{v}}_{\mathrm{S}}\} - \frac{m\hat{\boldsymbol{v}}_{\mathrm{S}}^2}{2} + e\hat{\Phi}\right)$$
$$\hat{\mathcal{G}}_{22}^{-1} = \hat{\sigma}_z\left(\mathrm{i}\frac{\partial}{\partial t} + \hat{H}_0 + \frac{\mathrm{i}}{2}\{\nabla, \hat{\boldsymbol{v}}_{\mathrm{S}}\} + \frac{m\hat{\boldsymbol{v}}_{\mathrm{S}}^2}{2} - e\hat{\Phi}\right) \tag{18.6}$$

and

$$\hat{\Delta} = \begin{pmatrix} \Delta_{\mathrm{F}} & 0 \\ 0 & \Delta_{\mathrm{B}} \end{pmatrix}. \tag{18.7}$$

Note that both fields $\Delta_{\rm F}$ and $\Delta_{\rm B}$ are now real since we have already eliminated the phase factors $\varphi_{\rm F/B}$ by the gauge transformation.

Here $\hat{H}_0 = -\nabla^2/2m + U(\boldsymbol{r}) - \mu$ is the single-electron Hamiltonian,

$$\hat{\boldsymbol{v}}_{\rm S} = \frac{1}{2m}\begin{pmatrix} \nabla\varphi_{\rm F} + \frac{2e}{c}\boldsymbol{A}_{\rm F} & 0 \\ 0 & \nabla\varphi_{\rm B} + \frac{2e}{c}\boldsymbol{A}_{\rm B} \end{pmatrix}, \tag{18.8}$$

$$\hat{\Phi} = \begin{pmatrix} V_{\rm F} - \frac{\dot{\varphi}_{\rm F}}{2e} & 0 \\ 0 & V_{\rm B} - \frac{\dot{\varphi}_{\rm B}}{2e} \end{pmatrix} \tag{18.9}$$

define the gauge-invariant combinations of the superconducting order parameter phase φ and the electromagnetic potentials.

18.2.1 Perturbation Theory

Although the above expression for the effective action $S_{\rm eff}$ (18.4) is formally exact, it remains too complicated for practical calculations. In order to bring this action to a tractable form, we will employ several approximations. To begin with, we will perform a perturbative expansion of the action up to the second order in $\hat{\boldsymbol{v}}_{\rm S}$ and $\hat{\Phi}$. This perturbation theory is sufficient for nearly all practical purposes, because nonlinear electromagnetic effects described by higher-order terms are usually very small in the systems under consideration. In addition, we will assume that deviations of the amplitude of the order parameter field from its equilibrium value Δ_0 are relatively small and expand the effective action to the second-order terms in these deviations. For this purpose, we split the order parameter into the BCS mean field term Δ_0 and the fluctuating part. Performing a shift $\Delta_{\rm F/B} \to \Delta_0 + \Delta_{\rm F/B}$, we re-define the order parameter field in a way to describe fluctuations by the fields $\Delta_{\rm F/B}$. Expanding the effective action (18.4) in powers of $\hat{\Delta}$, $\hat{\boldsymbol{v}}_{\rm S}$ and $\hat{\Phi}$ to the second order, we obtain the action describing quadratic fluctuations in our superconducting system

$$S_{\rm eff}^{(2)} = S_1 + S_\Delta + S_{\rm em} + S_{\varphi\Delta} \tag{18.10}$$

where

$$\mathrm{i}S_1 = \mathrm{Tr}\left[\hat{\sigma}_z\frac{\mathrm{i}}{2}\{\nabla, \hat{v}_{\rm S}\}(\hat{G} + \hat{\bar{G}}) + \hat{\sigma}_z e\hat{\Phi}(\hat{G} - \hat{\bar{G}}) - \hat{\sigma}_z\hat{\Delta}(\hat{F} + \hat{\bar{F}})\right] \tag{18.11}$$

defines the first-order (in $\hat{\Phi}$, $\hat{\boldsymbol{v}}_S$, and $\hat{\Delta}$) contribution to the action

$$\mathrm{i}S_\Delta = -\mathrm{i}\frac{N_0}{\lambda}\int d^4X\left(|\Delta_0+\Delta_\mathrm{F}|^2 - |\Delta_0+\Delta_\mathrm{B}|^2\right)$$
$$-\frac{1}{2}\,\mathrm{Tr}\left\{(\hat{F}\hat{\sigma}_z\hat{\Delta})^2 + (\hat{\hat{F}}\hat{\sigma}_z\hat{\Delta})^2 + 2\hat{\hat{G}}\hat{\sigma}_z\hat{\Delta}\hat{G}\hat{\sigma}_z\hat{\Delta}\right\} \quad (18.12)$$

accounts for fluctuations of the absolute value of the order parameter field,

$$\mathrm{i}S_\mathrm{em} = \mathrm{i}\int d^4X\,\frac{\boldsymbol{E}_\mathrm{F}^2 - \boldsymbol{E}_\mathrm{B}^2 - \boldsymbol{H}_\mathrm{F}^2 + \boldsymbol{H}_\mathrm{B}^2}{8\pi}$$
$$-\frac{1}{2}\,\mathrm{Tr}\left\{m\hat{\boldsymbol{v}}_\mathrm{S}^2(\hat{G}-\hat{\hat{G}}) + \left[\hat{\sigma}_z\left(\frac{\mathrm{i}}{2}\{\nabla,\hat{\boldsymbol{v}}_\mathrm{S}\}+e\hat{\Phi}\right)\hat{G}\right]^2\right.$$
$$+\hat{\sigma}_z\left(\frac{\mathrm{i}}{2}\{\nabla,\hat{\boldsymbol{v}}_\mathrm{S}\}+e\hat{\Phi}\right)\hat{F}\hat{\sigma}_z\left(\frac{\mathrm{i}}{2}\{\nabla,\hat{\boldsymbol{v}}_\mathrm{S}\}-e\hat{\Phi}\right)\hat{\hat{F}}$$
$$\left.+\left[\hat{\sigma}_z\left(\frac{\mathrm{i}}{2}\{\nabla,\hat{\boldsymbol{v}}_\mathrm{S}\}-e\hat{\Phi}\right)\hat{\hat{G}}\right]^2\right\} \quad (18.13)$$

describes electromagnetic fields and their coupling to the phase of the order parameter field and

$$\mathrm{i}S_{\varphi\Delta} = \mathrm{Tr}\left\{\hat{\sigma}_z\left(\frac{\mathrm{i}}{2}\{\nabla,\hat{\boldsymbol{v}}_\mathrm{S}\}+e\hat{\Phi}\right)\hat{F}\hat{\sigma}_z\hat{\Delta}\hat{G} + \hat{\sigma}_z\left(\frac{\mathrm{i}}{2}\{\nabla,\hat{\boldsymbol{v}}_\mathrm{S}\}+e\hat{\Phi}\right)\hat{G}\hat{\sigma}_z\hat{\Delta}\hat{\hat{F}}\right.$$
$$\left.+\hat{\sigma}_z\left(\frac{\mathrm{i}}{2}\{\nabla,\hat{\boldsymbol{v}}_S\}-e\hat{\Phi}\right)\hat{\hat{G}}\hat{\sigma}_z\hat{\Delta}\hat{F} + \hat{\sigma}_z\left(\frac{\mathrm{i}}{2}\{\nabla,\hat{\boldsymbol{v}}_\mathrm{S}\}-e\hat{\Phi}\right)\hat{\hat{F}}\hat{\sigma}_z\hat{\Delta}\hat{\hat{G}}\right\} \quad (18.14)$$

is responsible for coupling of electromagnetic and phase fluctuations to the absolute value of the order parameter field.

All the above contributions to the action are expressed in terms of equilibrium non-perturbed normal and anomalous 2×2 Green–Keldysh matrices, which enter as sub-blocks $\hat{G}$, $\hat{F}$, $\hat{\hat{G}}$, $\hat{\hat{F}}$ into the 4×4 matrix

$$\check{\mathcal{G}}_0 = \begin{pmatrix}\hat{G} & \hat{F}\\ \hat{\hat{F}} & \hat{\hat{G}}\end{pmatrix}, \quad (18.15)$$

obtained by inverting the expression (18.5) at $\hat{\Phi} = 0$, $\hat{\boldsymbol{A}} = 0$, and $\Delta_{\mathrm{F}} = \Delta_{\mathrm{B}} = \Delta_0$. These sub-blocks can be expressed in the form

$$
\begin{aligned}
\mathrm{i}\hat{G}(t, \boldsymbol{r}_1, \boldsymbol{r}_2) &= \sum_n \psi_n(\boldsymbol{r}_1)\psi_n(\boldsymbol{r}_2)\left[u_n^2 \hat{A}_n(t) + v_n^2 \hat{B}_n(t)\right],\\
\mathrm{i}\hat{\bar{G}}(t, \boldsymbol{r}_1, \boldsymbol{r}_2) &= \sum_n \psi_n(\boldsymbol{r}_1)\psi_n(\boldsymbol{r}_2)\left[v_n^2 \hat{A}_n(t) + u_n^2 \hat{B}_n(t)\right],\\
\mathrm{i}\hat{F}(t, \boldsymbol{r}_1, \boldsymbol{r}_2) &= \mathrm{i}\hat{\bar{F}}(t, \boldsymbol{r}_1, \boldsymbol{r}_2)\\
&= \sum_n \psi_n(\boldsymbol{r}_1)\psi_n(\boldsymbol{r}_2) u_n v_n \left[\hat{A}_n(t) - \hat{B}_n(t)\right],
\end{aligned}
\tag{18.16}
$$

where $\psi_n(\boldsymbol{r})$ obey the Schrödinger equation

$$\hat{H}_0 \psi_n(\boldsymbol{r}) = \xi_n \psi_n(\boldsymbol{r}) \tag{18.17}$$

and form a complete basis of the eigenfunctions of our problem. Here ξ_n denote the electron energies in the normal state. We also define the BCS coherence factors

$$u_n^2 = \frac{1}{2}\left(1 + \frac{\xi_n}{E_n}\right), \quad v_n^2 = \frac{1}{2}\left(1 - \frac{\xi_n}{E_n}\right) \tag{18.18}$$

where $E_n = \sqrt{\xi_n^2 + \Delta_0^2}$ are the quasiparticle energies in the superconducting state and introduce the combinations

$$
\begin{aligned}
\hat{A}_n(t) &= e^{-\mathrm{i}E_n t}\begin{pmatrix} \theta(t) - f_n & -f_n \\ 1 - f_n & \theta(-t) - f_n \end{pmatrix},\\
\hat{B}_n(t) &= e^{\mathrm{i}E_n t}\begin{pmatrix} f_n - \theta(-t) & -(1 - f_n) \\ f_n & f_n - \theta(t) \end{pmatrix}
\end{aligned}
\tag{18.19}
$$

where f_n are the occupation probabilities of the quasiparticle states. In thermodynamic equilibrium, these filling factors just coincide with the Fermi function $f_n = 1/(1 + e^{E_n/T})$.

One can further demonstrate that the matrix Green–Keldysh functions satisfy two different types of Ward identities [24, 160]:

$$
\begin{aligned}
\mathrm{i}\hat{F}\hat{\beta} + \mathrm{i}\hat{\beta}\hat{F} &= \hat{G}\hat{\sigma}_z\left(-\hat{\bar{\beta}} + \frac{\mathrm{i}}{2m}\{\nabla, \hat{\beta}\}\right)\hat{F}\\
&\quad + \hat{F}\hat{\sigma}_z\left(\hat{\bar{\beta}} + \frac{\mathrm{i}}{2m}\{\nabla, \hat{\beta}\}\right)\hat{\bar{G}} + 2\mathrm{i}\Delta_0\hat{G}\hat{\sigma}_z\hat{\beta}\hat{\bar{G}} - 2\mathrm{i}\Delta_0\hat{F}\hat{\sigma}_z\hat{\beta}\hat{F},\\
-\mathrm{i}\hat{\bar{F}}\hat{\beta} - \mathrm{i}\hat{\beta}\hat{\bar{F}} &= \hat{\bar{F}}\hat{\sigma}_z\left(-\hat{\bar{\beta}} + \frac{\mathrm{i}}{2m}\{\nabla, \hat{\beta}\}\right)\hat{G}\\
&\quad + \hat{\bar{G}}\hat{\sigma}_z\left(\hat{\bar{\beta}} + \frac{\mathrm{i}}{2m}\{\nabla, \hat{\beta}\}\right)\hat{\bar{F}} + 2\mathrm{i}\Delta_0\hat{\bar{F}}\hat{\sigma}_z\hat{\beta}\hat{\bar{F}} - 2\mathrm{i}\Delta_0\hat{\bar{G}}\hat{\sigma}_z\hat{\beta}\hat{G},
\end{aligned}
\tag{18.20}
$$

and

$$\hat{G}\hat{\beta} - \hat{\beta}\hat{G} = \mathrm{i}\hat{G}\hat{\sigma}_z\left(-\hat{\hat{\beta}} + \frac{\mathrm{i}}{2m}\{\nabla, \hat{\beta}\}\right)\hat{G} - \mathrm{i}\hat{F}\hat{\sigma}_z\left(\hat{\hat{\beta}} + \frac{\mathrm{i}}{2m}\{\nabla, \hat{\beta}\}\right)\hat{\hat{F}},$$
$$\hat{\hat{G}}\hat{\beta} - \hat{\beta}\hat{\hat{G}} = \mathrm{i}\hat{\hat{F}}\hat{\sigma}_z\left(-\hat{\hat{\beta}} + \frac{\mathrm{i}}{2m}\{\nabla, \hat{\beta}\}\right)\hat{F} - \mathrm{i}\hat{\hat{F}}\hat{\sigma}_z\left(\hat{\hat{\beta}} + \frac{\mathrm{i}}{2m}\{\nabla, \hat{\beta}\}\right)\hat{\hat{F}},$$
(18.21)

where

$$\hat{\beta} = \begin{pmatrix} \beta_{\mathrm{F}}(t, \boldsymbol{r}) & 0 \\ 0 & \beta_{\mathrm{B}}(t, \boldsymbol{r}) \end{pmatrix} \tag{18.22}$$

is an arbitrary diagonal 2×2 matrix.

The identities (18.20) and (18.21) hold for any distribution of impurities in our system as well as for any dependence of the order parameter field on both space and time coordinates. The Ward identities (18.20) directly follow from the gauge invariance of our problem. On the contrary, the Ward identities (18.21) are not the result of gauge invariance and remain valid even for uncharged particles. It is also interesting that Eqs. (18.20) and (18.21) coincide identically in the absence of superconductivity $\Delta_0 \to 0$.

Making use of Eq. (18.20), we can rewrite the first-order contribution to the action in the form

$$\mathrm{i}S_1 = -\,\mathrm{Tr}\left[\hat{\sigma}_z\hat{\Delta}(\hat{F} + \hat{\hat{F}})\right] + \mathrm{i}\Delta_0\,\mathrm{Tr}\left[\hat{\sigma}_z\hat{\varphi}(\hat{F} - \hat{\hat{F}})\right]$$
$$+ \frac{\mathrm{i}e}{2mc}\,\mathrm{Tr}\left[\hat{\sigma}_z\{\nabla, \hat{\boldsymbol{A}}\}(\hat{G} + \hat{\hat{G}})\right] + e\,\mathrm{Tr}\left[\hat{\sigma}_z\hat{V}(\hat{G} - \hat{\hat{G}})\right].$$

Observing that $\hat{F} = \hat{\hat{F}}$ and introducing the current $\boldsymbol{j}_0$ and the charge density ρ_0 in the non-perturbed state, we further transform the action S_1 and get

$$\mathrm{i}S_1 = -2\,\mathrm{Tr}\left[\hat{\sigma}_z\hat{\Delta}\hat{F}\right] - \frac{\mathrm{i}}{c}\int d^4X\, \boldsymbol{j}_0(\boldsymbol{A}_{\mathrm{F}} - \boldsymbol{A}_{\mathrm{B}}) - \mathrm{i}\int d^4X\, \rho_0(V_{\mathrm{F}} - V_{\mathrm{B}}) \tag{18.23}$$

In addition, assuming that the non-perturbed system remains in thermodynamic equilibrium, we set both the current and the charge density equal to zero. Then we obtain

$$\mathrm{i}S_1 = -2\,\mathrm{Tr}\left[\hat{\sigma}_z\hat{\Delta}\hat{F}\right] \tag{18.24}$$

Finally, we assume that the equilibrium value of the order parameter $\Delta_0(T)$ satisfies the standard BCS gap equation

$$\frac{1}{\lambda} = \int_0^{\omega_D} d\xi_p \frac{\tanh \frac{\sqrt{\xi_p^2+\Delta_0^2}}{2T}}{\sqrt{\xi_p^2 + \Delta_0^2}} \tag{18.25}$$

where ω_D is the Debye frequency. In this case, S_1 is exactly canceled by the first-order contribution from the term S_Δ and the action does not anymore contain the first-order terms in $\hat{\Delta}$.

The Ward identities (18.20) and (18.21) also allow one to transform the contribution S_{em} (18.13) and cast it to the form

$$S_{em} = S_J + S_L + S_D,$$

where S_J, S_L, and S_D define the terms of a different physical origin, which we will identify, respectively, as Josephson, London, and Drude contributions to the effective action. They read

$$iS_J = 2e^2 \operatorname{Tr} \left[\hat{\sigma}_z \hat{\Phi} \hat{F} \hat{\sigma}_z \hat{\Phi} \hat{\bar{F}}\right], \tag{18.26}$$

$$\begin{aligned} iS_L &= -i \int d^4X \, \frac{\boldsymbol{H}_F^2 - \boldsymbol{H}_B^2}{8\pi} + 2 \operatorname{Tr} \left[\hat{\sigma}_z \{\nabla, \hat{\boldsymbol{v}}_S\} \hat{F} \hat{\sigma}_z \{\nabla, \hat{\boldsymbol{v}}_S\} \hat{\bar{F}}\right], \\ iS_D &= i \int d^4X \, \frac{\boldsymbol{E}_F^2 - \boldsymbol{E}_B^2}{8\pi} + \frac{1}{2} \operatorname{Tr} \Big[\left(\hat{\sigma}_z \{\nabla, \hat{\boldsymbol{u}}\} \hat{G}\right)^2 + \left(\hat{\sigma}_z \{\nabla, \hat{\boldsymbol{u}}\} \hat{\bar{G}}\right)^2 \\ &\quad - 2\hat{\sigma}_z \{\nabla, \hat{\boldsymbol{u}}\} \hat{F} \hat{\sigma}_z \{\nabla, \hat{\boldsymbol{u}}\} \hat{\bar{F}} - 4m \hat{\sigma}_z \hat{\boldsymbol{u}}^2 (\hat{G} - \hat{\bar{G}}) \Big], \end{aligned} \tag{18.27}$$

where we defined

$$\hat{\boldsymbol{u}} = \frac{e}{2m} \left(\int_{t_0}^{t} dt' \, \nabla \hat{V} + \frac{1}{c} \hat{\boldsymbol{A}} \right) \tag{18.28}$$

At low frequencies and temperatures, the Josephson contribution S_J can be large, thus suppressing fluctuations of the gauge-invariant potential $\hat{\Phi}$. In this case, one can set $\hat{\Phi} = 0$ and get the Josephson relation (2.29) between the phase and the electric potential. Note that, e.g., in the case of ultrathin superconducting wires, this Josephson relation can be violated, as it will be demonstrated in the next chapter. The London contribution S_L is responsible for screening of the magnetic field penetrating inside the superconductor. Finally, the Drude contribution S_D remains nonzero in the normal state where it accounts for Ohmic dissipation due to the flowing electric currents.

Since the correction to the magnetic susceptibility in normal metals is usually small, it is appropriate to drop the vector potential in the expression for $\hat{\boldsymbol{u}}$ (18.28). Afterwards one can again apply the Ward identities (18.21) and rewrite S_{D} as

$$\mathrm{i}S_{\mathrm{D}} = \mathrm{i}\int d^4X\, \frac{\boldsymbol{E}_{\mathrm{F}}^2 - \boldsymbol{E}_{\mathrm{B}}^2}{8\pi} - \frac{e^2}{2}\,\mathrm{Tr}\left[\left(\hat{\sigma}_z\hat{V}\hat{G}\right)^2 + \left(\hat{\sigma}_z\hat{V}\hat{\hat{G}}\right)^2 + 2\hat{\sigma}_z\hat{V}\hat{F}\hat{\sigma}_z\hat{V}\hat{\hat{F}}\right]. \tag{18.29}$$

At last, let us consider the cross term $S_{\varphi\Delta}$. Again applying the identity (18.20), we cast this term to the form similar to Eq. (18.23):

$$\mathrm{i}S_{\varphi\Delta} = -\,\mathrm{Tr}\left[\hat{\sigma}_z\hat{\Delta}\left(\delta\hat{F} - \delta\hat{\hat{F}}\right)\right] - \frac{\mathrm{i}}{c}\int d^4X\,\delta\boldsymbol{j}_0(\boldsymbol{A}_{\mathrm{F}} - \boldsymbol{A}_{\mathrm{B}})$$
$$-\,\mathrm{i}\int d^4X\,\delta\rho_0(V_{\mathrm{F}} - V_{\mathrm{B}})$$

where $\delta\boldsymbol{j}$, $\delta\rho_0 \propto \Delta_{\mathrm{F}}, \Delta_{\mathrm{B}}$ define the first-order corrections, respectively, to the current and to the charge density due to fluctuating order parameter fields. One can verify that in the absence of both particle–hole asymmetry and charge imbalance, these corrections vanish. Likewise, in this case we have $\delta\hat{F} = \delta\hat{\hat{F}}$. Thus, we get

$$S_{\varphi\Delta} = 0 \tag{18.30}$$

Summarizing the above results, we conclude that the full effective action describing quadratic fluctuations in a superconductor reads

$$S_{\mathrm{eff}}^{(2)} = S_1 + S_\Delta + S_{\mathrm{J}} + S_{\mathrm{L}} + S_{\mathrm{D}} \tag{18.31}$$

where the terms S_1, S_Δ, S_{J}, S_{L}, and S_{D} are defined, respectively, in Eqs. (18.24), (18.12), (18.26), (18.27), and (18.29).

18.2.2 Gaussian Fluctuations in Dirty Superconductors

Below we will be mainly interested in the most generic case of so-called dirty superconductors, i.e., we assume that the concentration of defects in the system is sufficiently high and the electron motion is diffusive. In the case of ultrathin superconducting wires, only this limit appears to be of practical interest, since usually the electron elastic mean free path ℓ does not exceed the diameter of the wire. For such wires, we typically have $\ell \ll \xi_0 \sim v_{\mathrm{F}}/\Delta_0$ and $\ell \ll \lambda_{\mathrm{L}}$.

In order to account for processes with characteristic length scales exceeding the electron mean free path ℓ, it will be convenient for us to perform disorder averaging directly in the effective action. To this end, we substitute explicit expressions for the Green functions into the effective action derived in the previous section and then apply the standard rules of averaging for the electron wave functions. In the diffusion approximation, we have

$$\sum_{mn} \langle \delta(\xi_n - \xi_1)\delta(\xi_m - \xi_2)\psi_n(\boldsymbol{r})\psi_n(\boldsymbol{r}')\psi_m(\boldsymbol{r}')\psi_m(\boldsymbol{r})\rangle$$

$$= \frac{N_0}{\pi} \operatorname{Re} \mathcal{D}(\xi_1 - \xi_2, \boldsymbol{r}, \boldsymbol{r}'),$$

$$\sum_{mn} \langle \delta(\xi_n - \xi_1)\delta(\xi_m - \xi_2) \left(\nabla^\alpha_{\boldsymbol{r}_1} - \nabla^\alpha_{\boldsymbol{r}_4}\right) \left(\nabla^\beta_{\boldsymbol{r}_3} - \nabla^\beta_{\boldsymbol{r}_2}\right)$$

$$\times \psi_n(\boldsymbol{r}_1)\psi_n(\boldsymbol{r}_2)\psi_m(\boldsymbol{r}_3)\psi_m(\boldsymbol{r}_4)\rangle \,\big|_{\boldsymbol{r}_4=\boldsymbol{r}_1=\boldsymbol{r},\ \boldsymbol{r}_3=\boldsymbol{r}_4=\boldsymbol{r}'}$$

$$= \frac{4m^2 D}{\pi} \delta_{\alpha\beta} \operatorname{Re} \left[i(\xi_1 - \xi_2)\mathcal{D}(\xi_1 - \xi_2, \boldsymbol{r}, \boldsymbol{r}')\right]. \quad (18.32)$$

As before, $D = v_F\ell/3$ is the diffusion constant and $\mathcal{D}(\omega, \boldsymbol{r}, \boldsymbol{r}')$ is the diffuson defined as a solution to the diffusion equation

$$\left(-\mathrm{i}\omega - D\nabla^2_{\boldsymbol{r}}\right) \mathcal{D}(\omega, \boldsymbol{r}, \boldsymbol{r}') = \delta(\boldsymbol{r} - \boldsymbol{r}') \quad (18.33)$$

In the following, we will mainly consider spatially extended systems in which case one has

$$\mathcal{D}(\omega, \boldsymbol{r}, \boldsymbol{r}') = \int \frac{d^3\boldsymbol{q}}{(2\pi)^3} \frac{e^{\mathrm{i}\boldsymbol{q}(\boldsymbol{r}-\boldsymbol{r}')}}{-\mathrm{i}\omega + Dq^2} \quad (18.34)$$

Employing Eqs. (18.32), we arrive at the following expression for the effective action:

$$\mathrm{i}S^{(2)}_{\mathrm{eff}} = \mathrm{i}\int d^4X \, \frac{\boldsymbol{E}^-\boldsymbol{E}^+ - \boldsymbol{H}^-\boldsymbol{H}^+}{4\pi} + \mathrm{i}\int d^4X d^4X' \Big\{ \Delta^-(X)K^{X,X'}_{\Delta}\Delta^+(X')$$

$$+ \Phi^-(X)K^{X,X'}_{\mathrm{J}}\Phi^+(X') - \boldsymbol{v}^-_{\mathrm{S}}(X)K^{X,X'}_{\mathrm{L}}\boldsymbol{v}^+_{\mathrm{S}}(X') + \boldsymbol{E}^-(X)K^{X,X'}_{\mathrm{D}}\boldsymbol{E}^+(X')\Big\}$$

$$- \int d^4Xd^4X' \Big\{ \Delta^-(X)\tilde{K}^{X,X'}_{\Delta}\Delta^-(X') + \Phi^-(X)\tilde{K}^{X,X'}_{\mathrm{J}}\Phi^-(X')$$

$$- \boldsymbol{v}^-_{\mathrm{S}}(X)\tilde{K}^{X,X'}_{\mathrm{L}}\boldsymbol{v}^-_{\mathrm{S}}(X') + \boldsymbol{E}^-(X)\tilde{K}^{X,X'}_{\mathrm{D}}\boldsymbol{E}^-(X')\Big\}, \quad (18.35)$$

where, as before, we routinely introduced "classical" $\Delta^+ = (\Delta_F + \Delta_B)/2$ and "quantum" $\Delta^- = \Delta_F - \Delta_B$ components of the order parameter field and used analogous definitions for other fluctuating variables $\mathbf{v}_S^{\pm}$, $\Phi^{\pm}$ and $\boldsymbol{E}^{\pm}$. The four kernels $K_j^{X,X'}$ $(j = \Delta, L, J, D)$ are defined as follows:

$$K_j^{X,X'} = \int \frac{d^4Q}{(2\pi)^4} e^{-i\omega(t-t')+i\boldsymbol{q}(\boldsymbol{r}-\boldsymbol{r}')} \chi_j(Q),$$
$$\tilde{K}_j^{X,X'} = \int \frac{d^4Q}{(2\pi)^4} e^{-i\omega(t-t')+i\boldsymbol{q}(\boldsymbol{r}-\boldsymbol{r}')} \frac{\mathrm{Im}\,[\chi_j(Q)]}{2} \coth\frac{\omega}{2T} \tag{18.36}$$

where we denote $d^4Q = d\omega d^3\boldsymbol{q}$.

General expressions for the functions χ_Δ, χ_J, χ_L, and χ_D turn out to be rather cumbersome [24, 160]. Here we only specify these functions in several important limits.

Let us first concentrate on the low-temperature limit $T \ll \Delta_0$. In this limit at small frequencies and wave vectors ω, $Dq^2 \ll \Delta_0$, we obtain

$$\chi_\Delta = -2N_0\left(1 - \frac{\omega^2}{12\Delta_0^2} + \frac{\pi Dq^2}{8\Delta_0}\right)$$
$$\chi_J = 2e^2N_0\left(1 + \frac{\omega^2}{6\Delta_0^2} - \frac{\pi Dq^2}{8\Delta_0}\right)$$
$$\chi_L = 2\pi m^2 N_0 D\Delta_0$$
$$\chi_D = \frac{e^2N_0D}{4\Delta_0} \tag{18.37}$$

while in the limit $|\omega|$, $Dq^2 \gg \Delta_0(0) \equiv \pi e^{-\gamma}T_C$ one finds

$$\chi_\Delta = -2N_0 \ln\frac{-i\omega + Dq^2}{\Delta_0(0)}$$
$$\chi_J = -\frac{8e^2N_0\Delta_0^2(T)}{\omega(\omega^2 + D^2q^4)}\left[\omega\ln\frac{-i\omega + Dq^2}{\Delta_0(0)} - iDq^2\ln\frac{-i\omega}{2\Delta_0(0)}\right]$$
$$\chi_L = \frac{8m^2N_0D\Delta_0^2(T)}{\omega^2 + D^2q^4}\left[Dq^2\ln\frac{-i\omega + Dq^2}{\Delta_0(0)} + i\omega\ln\frac{-i\omega}{2\Delta_0(0)}\right]$$
$$\chi_D = \frac{\sigma_D}{-i\omega + Dq^2} \tag{18.38}$$

where $\sigma_D = 2e^2N_0D$ is the normal state Drude conductivity. Here we explicitly indicated the temperature dependence of the superconducting gap $\Delta_0(T)$ in order to emphasize that these asymptotic expressions are valid at all temperatures without being restricted to the limit $T \ll \Delta_0$.

Let us now specifically address higher temperatures $|T - T_C| \ll T_C$. At $T > T_C$, our general expression for χ_Δ reduces to the standard result

$$\chi_\Delta = -2N_0 \left[\ln \frac{T}{T_C} + \Psi\left(\frac{1}{2} + \frac{-i\omega + Dq^2}{4\pi T}\right) - \Psi\left(\frac{1}{2}\right) \right], \quad (18.39)$$

χ_D is again defined by the Drude formula (18.38), while two other χ-functions vanish identically in this temperature interval: $\chi_J = \chi_L = 0$. The latter observation implies that phase fluctuations remain unrestricted in this case. Hence, no Taylor expansion of the action in the phase φ can be performed. In this case, it is more convenient to undo the above gauge transformation, restoring the initial dependence of the action on the complex order parameter field and then to expand the action in this field. For simplicity ignoring electromagnetic fields and expanding the action to the second order in $\Delta^{\pm}$, we find

$$iS_{\text{eff}}^{(2)} = \frac{i}{2} \int d^4X d^4X' \left\{ \left(\Delta^-(X)\right)^* K_\Delta^{X,X'} \Delta^+(X') + \Delta^-(X) K_\Delta^{X,X'} \left(\Delta^+(X')\right)^* \right\}$$
$$- \int d^4X d^4X' \left(\Delta^-(X)\right)^* \tilde{K}_\Delta^{X,X'} \Delta^-(X') \quad (18.40)$$

In the limit of small frequencies and wave vectors $|\omega|$, $Dq^2 \ll 2\pi T_C$, one recovers the standard expression

$$\chi_\Delta = -2N_0 \ln \frac{T}{T_C} - \frac{\pi N_0}{4T}(-i\omega + Dq^2) \quad (18.41)$$

which usually serves as a starting point for the derivation of the TDGL equation, which we will address shortly below.

Turning to temperatures below the critical one, $T < T_C$ and expanding the χ-functions in powers of Δ_0, we obtain

$$\chi_\Delta = -\frac{7\zeta(3)}{2\pi^2} \frac{N_0 \Delta_0^2}{T^2} - \frac{\pi N_0}{4T}(-i\omega + Dq^2) - \frac{\pi N_0}{2T} \frac{\Delta_0^2}{-i\omega + Dq^2}$$
$$\chi_J = \frac{7\zeta(3)}{2\pi^2} \frac{e^2 N_0 \Delta_0^2}{T^2} + \frac{\pi e^2 N_0 \Delta_0^2}{T} \left(\frac{1}{-i\omega + Dq^2} + \frac{\Delta_0^2}{\omega^2} \frac{-2i\omega + Dq^2}{\left(-i\omega + Dq^2\right)^2} \right)$$
$$\chi_L = \frac{\pi m^2 N_0 D \Delta_0^2}{T} \left(\frac{-2i\omega + Dq^2}{-i\omega + Dq^2} - \frac{\Delta_0^2}{\left(-i\omega + Dq^2\right)^2} \right),$$
$$\chi_D = \frac{2e^2 N_0 D}{-i\omega + Dq^2} \left(1 - \frac{7\zeta(3)}{2\pi^2} \frac{\Delta_0^2}{T^2} \right) - \frac{\pi e^2 N_0 D \Delta_0^2}{2T} \frac{1}{(-i\omega + Dq^2)^2}$$
$$(18.42)$$

These results apply for $\Delta_0 \ll |\omega|$, Dq^2 and in the limit of small wave vectors and frequencies $|\omega|$, $Dq^2 \ll 2\pi T_C$. Here $\Delta_0(T)$ obeys the standard BCS self-consistency gap equation at $T \sim T_C$:

$$\ln\frac{T_C}{T} = \frac{7\zeta(3)}{8\pi^2}\frac{\Delta_0^2}{T^2} + \mathcal{O}\left(\frac{\Delta_0^4}{T^4}\right) \tag{18.43}$$

For $|\omega|$, $Dq^2 \ll \Delta_0$, we obtain non-analytic expressions. For example, χ_J in this limit reads

$$\chi_J = \frac{e^2 N_0 \Delta_0}{2T}\left\{\frac{7\zeta(3)}{\pi^2}\frac{\Delta_0}{T} + \frac{-i\omega}{\sqrt{\omega^2 + D^2q^4}}\left[\ln\frac{\sqrt{\omega^2 + D^2q^4} + Dq^2}{\sqrt{\omega^2 + D^2q^4} - Dq^2} + \pi i\frac{\omega}{|\omega|}\right]\right\} \tag{18.44}$$

More accurate expressions for the kernels χ_Δ, χ_J, and χ_L valid at temperatures close to T_C and at any $|\omega|$, $Dq^2 \ll 2\pi T_C$ can be found elsewhere [24, 160]. Only in the limit $|\omega| \ll Dq^2 \ll \Delta_0$, those expressions match with the well-known results for the coefficients of the linearized (time-independent) Ginzburg–Landau equation:

$$\chi_\Delta = -\frac{7\zeta(3)}{2\pi^2}\frac{N_0\Delta_0^2}{T^2}, \quad \chi_J = \frac{7\zeta(3)}{2\pi^2}\frac{e^2 N_0\Delta_0^2}{T^2}, \quad \chi_L = \frac{\pi m^2 N_0 D\Delta_0^2}{T} \tag{18.45}$$

At frequencies $Dq^2 \ll |\omega| \ll \Delta_0$, the functions χ_Δ and χ_J turn out to be parametrically different, taking much higher values:

$$\chi_\Delta = -\frac{\pi N_0\Delta_0}{4T}, \quad \chi_J = \frac{\pi e^2 N_0\Delta_0}{4T}, \tag{18.46}$$

while χ_L is still given by Eq. (18.45). The standard expression (18.38) for the Drude susceptibility χ_D applies in both limits. Thus, already at frequencies well below the gap $\Delta_0 \ll T_C$, microscopic results can strongly deviate from those frequently used within the TDGL approach. At higher frequencies and/or wave vectors $|\omega| + Dq^2 \gg \Delta_0$, this difference becomes even more pronounced; cf. Eqs. (18.42).

It is also worthwhile to point out that these kernels are not independent and obey the following exact identity:

$$\chi_\Delta = -\frac{\chi_J}{e^2} + \frac{\omega^2}{4e^2\Delta_0^2}\chi_J - \frac{q^2}{4m^2\Delta_0^2}\chi_L \tag{18.47}$$

which directly follows from the Ward identities (18.21). In addition, in the diffusive limit, the kernels χ_{J} and χ_{L} are related to each other as

$$\chi_{\mathrm{L}}(\omega, \boldsymbol{q}) = \chi_{\mathrm{L}}(\omega, 0) - \frac{m^2 D^2 q^2}{e^2} \chi_{\mathrm{J}}(\omega, \boldsymbol{q}) \tag{18.48}$$

For completeness, let us also display a useful relation between the χ-kernels derived here and some other response functions analyzed in the literature. For example, one can define the complex conductivity $\sigma(Q)$ of a superconductor by means of the relation [294]

$$\boldsymbol{j}_Q = \sigma(Q)\boldsymbol{E}_Q \tag{18.49}$$

where $\boldsymbol{j}_Q$ and $\boldsymbol{E}_Q$ are the Fourier components of, respectively, the current density and the electric field. Similarly, the function $\mathcal{Q}(Q)$ can be introduced, which expresses the current via the vector potential [1]:

$$\boldsymbol{j}_Q = -\mathcal{Q}(Q)\boldsymbol{A}_Q \tag{18.50}$$

Both functions $\mathcal{Q}(Q)$ and $\sigma(Q)$ are related to the kernels χ_{L} and χ_{D} as

$$\mathcal{Q}(Q) = \frac{-\mathrm{i}\omega}{c}\sigma(Q) = \frac{1}{c}\left[\frac{e^2}{m^2}\chi_{\mathrm{L}}(Q) - \omega^2\chi_{\mathrm{D}}(Q)\right] \tag{18.51}$$

18.3 Langevin Equations

Let us now rewrite our results in a slightly different manner. The effective action $S_{\mathrm{eff}}^{(2)}$ can be equivalently defined by means of the following formula:

$$e^{\mathrm{i}S_{\mathrm{eff}}^{(2)}} = \left\langle e^{\mathrm{i}S_{\xi}[\Delta^{\pm}, \boldsymbol{v}_S^{\pm}, \Phi^{\pm}, \boldsymbol{E}^{\pm}, \boldsymbol{H}^{\pm}, \xi_{\Delta}, \xi_{\mathrm{J}}, \boldsymbol{\xi}_{\mathrm{L}}]}\right\rangle_{\xi_j}, \tag{18.52}$$

where

$$\begin{aligned} S_{\xi} = & \int d^4X\, \frac{\boldsymbol{E}^-\boldsymbol{E}^+ - \boldsymbol{H}^-\boldsymbol{H}^+}{4\pi} + \int d^4X d^4X' \Big\{ \Delta^-(X) K_{\Delta}^{X,X'} \Delta^+(X') \\ & + \Phi^-(X) K_{\mathrm{J}}^{X,X'} \Phi^+(X') + \boldsymbol{E}^-(X) K_{\mathrm{D}}^{X,X'} \boldsymbol{E}^+(X') - \boldsymbol{v}_{\mathrm{S}}^-(X) K_{\mathrm{L}}^{X,X'} \boldsymbol{v}_{\mathrm{S}}^+(X') \Big\} \\ & - \int d^4X \left[\xi_{\Delta}\Delta^- + \xi_{\mathrm{J}}\Phi^- + \boldsymbol{\xi}_{\mathrm{L}}\boldsymbol{v}_{\mathrm{S}}^-\right], \end{aligned} \tag{18.53}$$

and averaging is performed over three different stochastic variables ξ_Δ, ξ_J, and $\boldsymbol{\xi}_\mathrm{L}$ defined by the pair correlators

$$\langle \xi_\Delta(t,\boldsymbol{r})\xi_\Delta(0,0)\rangle = \int \frac{d^4Q}{(2\pi)^4}\, e^{-\mathrm{i}\omega t+\mathrm{i}\boldsymbol{q}\boldsymbol{r}} \coth\frac{\omega}{2T}\,\mathrm{Im}\left[\chi_\Delta(Q)\right],$$

$$\langle \xi_\mathrm{J}(t,\boldsymbol{r})\xi_\mathrm{J}(0,0)\rangle = \int \frac{d^4Q}{(2\pi)^4}\, e^{-\mathrm{i}\omega t+\mathrm{i}\boldsymbol{q}\boldsymbol{r}} \coth\frac{\omega}{2T}\,\mathrm{Im}\left[\chi_\mathrm{J}(Q)+q^2\chi_\mathrm{D}(Q)\right],$$

$$\langle \xi_\mathrm{L}^\alpha(t,\boldsymbol{r})\xi_\mathrm{L}^\beta(0,0)\rangle = \delta_{\alpha\beta}\int \frac{d^4Q}{(2\pi)^4}\, e^{-\mathrm{i}\omega t+\mathrm{i}\boldsymbol{q}\boldsymbol{r}}$$
$$\times \coth\frac{\omega}{2T}\,\mathrm{Im}\left[-\chi_\mathrm{L}(Q)+\frac{m^2\omega^2}{e^2}\chi_\mathrm{D}(Q)\right],$$

$$\langle \boldsymbol{\xi}_\mathrm{L}(t,\boldsymbol{r})\xi_\mathrm{J}(0,0)\rangle = -\int \frac{d^4Q}{(2\pi)^4}\, e^{-\mathrm{i}\omega t+\mathrm{i}\boldsymbol{q}\boldsymbol{r}} \coth\frac{\omega}{2T}\,\frac{m\omega\boldsymbol{q}}{e}\,\mathrm{Im}\left[\chi_\mathrm{D}(Q)\right] \tag{18.54}$$

All other cross correlators of the above stochastic variables are equal to zero.

The representation (18.52) is just the result of the standard Hubbard–Stratonovich decoupling transformation in the effective action (18.35). We have also used the identity

$$\boldsymbol{E}^- = -\nabla V^- - \frac{1}{c}\frac{\partial \boldsymbol{A}}{\partial t} = -\nabla\Phi^- - \frac{m}{e}\frac{\partial \boldsymbol{v}_\mathrm{S}}{\partial t}. \tag{18.55}$$

Let us now find the least action paths for S_ξ. Setting the variational derivatives of the action (18.53) with respect to quantum fields Δ^-, φ^-, V^-, and $\boldsymbol{A}^-$ equal to zero, we arrive at four different equations for the fields Δ^+, φ^+, V^+, and $\boldsymbol{A}^+$, which provide the minimum for the action S_ξ (18.53). The first equation describes fluctuations of the absolute value of the order parameter. It reads

$$\int d^4X'\, K_\Delta^{X,X'}\,\Delta^+(X') = \xi_\Delta(X) \tag{18.56}$$

The second one is the continuity equation for the supercurrent. We obtain

$$\frac{\partial\rho_\mathrm{S}}{\partial t} + \nabla\boldsymbol{j}_\mathrm{S} = -\frac{\partial\xi_\mathrm{J}}{\partial t} + \frac{e}{m}\nabla\boldsymbol{\xi}_\mathrm{L} \tag{18.57}$$

where we introduced the superconducting density ρ_S and the superconducting current density $\boldsymbol{j}_\mathrm{S}$

$$\rho_\mathrm{S} = -\int d^4X'\, K_\mathrm{J}^{X,X'}\,\Phi^+(X'), \quad \boldsymbol{j}_\mathrm{S} = -\frac{e}{m}\int d^4X'\, K_\mathrm{L}^{X,X'}\,\boldsymbol{v}_\mathrm{S}^+(X') \tag{18.58}$$

The remaining two saddle point equations take the form

$$\frac{\nabla \boldsymbol{E}^{+}(X)}{4\pi} + \int d^4X'\, K_{\rm D}^{X,X'}\, \nabla \boldsymbol{E}^{+}(X') - \rho_{\rm S}(X) = \xi_{\rm J}(X) \tag{18.59}$$

and

$$\frac{\nabla \times \boldsymbol{H}^{+}}{4\pi} = \frac{1}{4\pi c}\frac{\partial \boldsymbol{E}^{+}}{\partial t} + \frac{\boldsymbol{j}_{\rm S}}{c} + \frac{\boldsymbol{j}_{\rm N}}{c} - \frac{e}{m}\boldsymbol{\xi}_{\rm L} \tag{18.60}$$

where we defined the normal quasiparticle current

$$\boldsymbol{j}_{\rm N} = \int dX'\, K_{\rm D}^{X,X'} \frac{\partial \boldsymbol{E}^{+}(X')}{\partial t'} \tag{18.61}$$

Equations (18.56), (18.57), (18.59), and (18.60) together with noise correlators (18.54) represent the set of Langevin equations fully describing quantum dynamics of the order parameter and electromagnetic fields for dirty superconductors within the Gaussian approximation. As it is clear from our derivation, these equations remain valid provided the electron distribution function is not driven far from equilibrium. Generalization of our approach to non-equilibrium situations is also possible but will not be discussed here.

18.4 Time-Dependent Ginzburg–Landau Equation

Now let us establish the relation between the above results and the approach based on the TDGL equation, which is widely used to model various non-stationary effects in superconductors at temperatures close to $T_{\rm C}$. For example, above the critical temperature, this TDGL approach allows to correctly evaluate the so-called Aslamazov–Larkin fluctuation correction to the conductivity of the system [259]. Below $T_{\rm C}$, it enables one to describe the formation of phase slip centers and the resistive state of current-biased superconducting wires. Relative simplicity of the TDGL approach makes it possible to apply powerful numerical methods, thus making this technique particularly appealing. In many cases, the TDGL-based analysis was employed even far beyond its applicability range, e.g., in order to describe QPS in superconducting nanowires at $T \to 0$.

The TDGL equation is usually written in the following simple form [259]:

$$\left[\frac{\partial}{\partial t} - 2\mathrm{i}eV + \frac{1}{\tau_{\mathrm{GL}}} - D\left(\nabla + \mathrm{i}\frac{2e}{c}\boldsymbol{A}\right)^2 + \frac{7\zeta(3)}{\pi^3 T}|\Delta|^2\right]\Delta = \tilde{\xi}_\Delta, \tag{18.62}$$

where

$$\tau_{\mathrm{GL}} = \frac{\pi}{8|T - T_{\mathrm{C}}|} \tag{18.63}$$

is the so-called Ginzburg–Landau time and

$$\langle \tilde{\xi}^*_\Delta(t, \boldsymbol{r})\tilde{\xi}_\Delta(t', \boldsymbol{r}')\rangle = \frac{16T^2}{\pi N_0}\delta(t - t')\delta(\boldsymbol{r} - \boldsymbol{r}') \tag{18.64}$$

Although this form can be justified for gapless superconductors at high-enough concentration of magnetic impurities, in general no consistent microscopic derivation of Eq. (18.62) was ever performed. Nevertheless, it is sometimes believed that Eq. (18.62) is applicable at least at temperatures above T_{C} where the average value of the BCS order parameter equals to zero, $\Delta_0 = 0$. Unfortunately, the above microscopic derivation of both the effective action and the Langevin equations does not support this viewpoint. A similar conclusion also follows from the analysis based on the real-time nonlinear σ-model [268], which yields the results consistent with our Langevin equations up to terms $\sim \Delta^+, \varphi^+$.

Furthermore, it remains unclear whether it would make any sense to pursue the Ginzburg–Landau expansion in order to keep higher-order terms $\sim\Delta^3$. A regular expansion in the effective action (18.4) up to terms $\sim\Delta^4$ generates many complicated non-local (both in space and in time) contributions containing the quantum field $(\Delta^-)^n$ with n ranging from 1 to 4. In order to recover terms $\sim\Delta^3$ in the TDGL equation, all contributions in the action $\sim(\Delta^-)^n$ with $n \geq 2$ should be ignored, which is in general difficult to justify. On top of that, the remaining terms $\sim\Delta^3$ are hardly tractable, except in the zero-frequency limit. Finally, the whole approach remains restricted to temperatures $T \sim T_{\mathrm{C}}$. In view of all these problems, it appears more appealing to perform regular expansion of the effective action in superconducting fluctuations around the mean field value Δ_0. Restricting these expansion to second-order terms in $\Delta^\pm$, one arrives at the Langevin equations (18.56)–(18.57). This approach remains applicable at all temperatures down to $T = 0$ and is sufficient for practical calculations in a large number of situations. One of such situations will be considered below.

18.5 Thin Metallic Wires

Let us turn to the specific case of sufficiently long and very thin superconducting wires, which will be of particular interest for us here. For such systems, the terms describing the action of free electromagnetic field can be rewritten in the form

$$\int d^4X \, \frac{\boldsymbol{E}^-\boldsymbol{E}^+ - \boldsymbol{H}^-\boldsymbol{H}^+}{4\pi} \to \int dtdx \left(C_w V^+V^- - \frac{A^+A^-}{L_w} \right) \tag{18.65}$$

Here we have defined the coordinate along the wire x, the capacitance per unit length of the wire C_w, and the inductance times unit length L_w. The vector-potential component parallel to the wire is denoted by A. For a cylindric wire with radius r_0 embedded in a dielectric environment with susceptibility ϵ, the capacitance C and inductance L are

$$C_w \approx \frac{\epsilon}{2\ln(R_0/r_0)}, \quad L_w \approx 2\ln(R_0/r_0) \tag{18.66}$$

where R_0 is the distance from the center of the wire and the bulk metallic electrode.

In order to transform other terms, one should apply a simple rule

$$\int d^4X \to s \int dt\, dx, \quad \int \frac{d^4Q}{(2\pi)^4} \to \frac{1}{s} \int \frac{d\omega dq}{(2\pi)^2} \tag{18.67}$$

where s is the wire cross section.

18.5.1 Propagating Modes

In the low-temperature limit $T \ll \Delta_0$, all χ-functions (18.37) are real and, hence, the noise terms in all four Langevin equations (18.56)–(18.60) vanish. This enables propagation of electromagnetic modes along a quasi-one-dimensional superconducting wire.

The equations of motion for such a wire take the form

$$\frac{1}{12\Delta_0^2}\frac{\partial^2\Delta^+}{\partial t^2}-\frac{\pi D}{8\Delta_0}\frac{\partial^2\Delta^+}{\partial x^2}+\Delta^+=0, \tag{18.68}$$

$$\frac{\partial^2\varphi^+}{\partial t^2}-2e\frac{\partial V^+}{\partial t}-\pi D\Delta_0\left(\frac{\partial^2\varphi^+}{\partial x^2}+\frac{2e}{c}\frac{\partial A^+}{\partial x}\right)=0, \tag{18.69}$$

$$C_wV^+-\frac{\sigma_{\rm D}s}{8\Delta_0}\frac{\partial^2V^+}{\partial x^2}+2e^2N_0s\left(V^+-\frac{1}{2e}\frac{\partial\varphi^+}{\partial t}\right)=0, \tag{18.70}$$

$$\frac{A^+}{L_w}+\frac{\sigma_{\rm D}s}{8\Delta_0c}\left(\frac{\partial^2V}{\partial t\partial x}+\frac{1}{c}\frac{\partial^2A^+}{\partial t^2}\right)+\frac{\pi\sigma_{\rm D}\Delta_0s}{2ec}\left(\frac{\partial\varphi^+}{\partial x}+\frac{2e}{c}A^+\right)=0. \tag{18.71}$$

For dirty metallic wires with the diameter of the order of superconducting coherence length $\xi=\sqrt{D/\Delta_0}$, one finds

$$1/L_w\gg\pi\sigma_{\rm D}\Delta_0s/c^2,\quad C_w\ll 2e^2N_0s \tag{18.72}$$

In this case, Eq. (18.71) gives $A^+\to 0$, while Eqs. (18.69) and (18.70) describe the propagation of the plasmon mode [306] with dispersion

$$\omega=c_0q \tag{18.73}$$

where

$$c_0\simeq\frac{1}{\sqrt{L_{\rm kin}C_w}}=\sqrt{\frac{\pi\sigma_{\rm D}\Delta_0s}{C_w}} \tag{18.74}$$

is the velocity of this mode and

$$L_{\rm kin}=4\pi\lambda_{\rm L}^2/s=1/(2\pi e^2N_0\Delta_0Ds) \tag{18.75}$$

is the kinetic inductance of a superconducting wire.

18.5.2 Gaussian Fluctuations of the Order Parameter

The effective action (18.35) fully accounts for Gaussian fluctuations in diffusive superconducting structures. For instance, from Eqs. (18.56)–(18.60), one readily establishes the correlation functions for all fluctuating variables in our problem. For the order parameter fields $\Delta^{\pm}$ in a quasi-one-dimensional wire, we have

$$\langle \Delta^{+}(t,r)\Delta^{+}(0,0)\rangle = -\frac{1}{s}\int \frac{d\omega\, dq}{(2\pi)^2}\, e^{-i\omega t + iqr}\mathrm{Im}\left[\frac{1}{\chi_\Delta(Q)}\right]\coth\frac{\omega}{2T} \tag{18.76}$$

Correlation functions $\boldsymbol{v}_S^{\pm}$, $\Phi^{\pm}$, and $\boldsymbol{E}^{\pm}$ are defined analogously via the corresponding kernels χ_J, χ_L, and χ_D.

Consider Gaussian fluctuations of the order parameter in thin one-dimensional wires. The simplest possible average $\langle \Delta^{+}(0,0)\Delta^{+}(0,0)\rangle/\Delta_0^2$ is divergent since the function χ_Δ grows very slowly at large ω and $\boldsymbol{q}$. Let us define and analyze a slightly different object

$$\mathcal{R} = \frac{\left\langle \Delta^{+}(1/\Delta_0,\xi)\ \Delta^{+}(0,0)\right\rangle}{\Delta_0^2} \tag{18.77}$$

One can verify that, for example, the non-local kernel $K_L^{X,X'}$ significantly decays as long as $|t|$ exceeds $1/\Delta_0$, and $|\boldsymbol{r}|$ becomes bigger than the coherence length ξ. Therefore, the parameter $\mathcal{R}$ provides a qualitative measure of the ratio between the fluctuation correction to the current and its mean field value.

In order to estimate the parameter $\mathcal{R}$, we note that at low temperatures $T \ll \Delta_0$, the kernel χ_Δ can be expressed in the form

$$\chi_\Delta = -N_0 F_0\left(\omega/\Delta_0, \xi q\right)$$

where $F_0(x,y)$ is a certain dimensionless function. Then we obtain

$$\mathcal{R} \approx \frac{1}{2\pi^2 s N_0\sqrt{D\Delta_0}}\int_0^1 dx\int_{-1}^1 dy\,\mathrm{Im}\left(\frac{1}{F_0(x,y)}\right) \sim \frac{1}{sN_0\sqrt{D\Delta_0}}$$

In the next chapter we will demonstrate that the probability for quantum tunneling of the order parameter field Δ in superconducting nanowires is proportional to $\sim\exp[-S_{\mathrm{QPS}}]$, where $S_{\mathrm{QPS}} \sim 1/\mathcal{R}$ is the action of single QPS. Hence, for $\mathcal{R} \ll 1$, Gaussian fluctuations are small and QPS events are rare, i.e., the BCS mean field theory still remains adequate. On the other hand, at $\mathcal{R} \gtrsim 1$, one enters the regime of strong non-Gaussian fluctuations, which fully suppress the mean field order parameter, thus driving the wire to a normal state. The concept of QPS also becomes ill-defined in this regime of strong quantum fluctuations.

The same conclusions can be extracted from the so-called Ginzburg–Levanyuk criterion. Let us consider the Ginzburg number Gi defined as the value $(T_C - T)/T_C$ at which the fluctuation correction to the specific heat becomes equal to the specific heat jump at the phase transition point. In the case of quasi-one-dimensional wires, this number reads [259]:

$$Gi_{1D} = \frac{1.3}{(p_F^2 s)^{2/3}(T_C \tau_e)^{1/3}}, \tag{18.78}$$

where $\tau_e = \ell/v_F$ is the elastic mean free time. Typically in thick wires, one finds $Gi_{1D} \ll 1$ and fluctuations become strong only in a very narrow region close to T_C, i.e., at $|T_C - T|/T_C < Gi_{1D}$. One can also rewrite Eq. (18.78) as

$$Gi_{1D} = \frac{0.15}{(sN_0\sqrt{D\Delta_0})^{2/3}} \sim \mathcal{R}^{2/3} \sim \frac{1}{S_{QPS}^{2/3}}, \tag{18.79}$$

or simply $Gi_{1D} \sim 1/g_\xi^{2/3}$, where g_ξ is the dimensionless conductance of the wire segment of length ξ. Thus, for $g_\xi \sim 1$, the width of the fluctuation region δT is comparable to T_C and the BCS mean field approach becomes obsolete down to $T = 0$.

We also remark that by T_C, we routinely imply the mean field BCS critical temperature for a bulk superconductor. While in many cases it is appropriate, one should also bear in mind that in low-dimensional disordered systems, such as thin films and wires, the superconducting critical temperature can be substantially reduced by Coulomb interaction as compared to its mean field value. Theoretical analysis [320] demonstrates that this reduction in T_C by Coulomb effects becomes progressively more important with both decreasing the system dimension (thickness and/or width) and increasing degree of disorder. Provided this reduction is considerable, it should be included into the expression for T_C. This renormalized value of T_C should then be used instead of the bare one in our analysis both above and below in this and next two chapters.

18.6 Matsubara Effective Action

To complete our analysis, we will briefly address the imaginary time (Matsubara) version of the effective action. As we have already demonstrated, e.g., in Chapters 4–7, technically it is more convenient to deal with this form of the action, provided one needs to account for quantum tunneling processes. This is precisely what we will do in Chapter 19 when we describe QPS in superconducting nanowires. We will proceed along the same lines as in the real-time calculation carried above.

The grand partition function $\mathcal{Z}$ for any quantum system described by the Hamiltonian $\hat{H}$ reads

$$\mathcal{Z} = \mathrm{Tr}\exp(-\hat{H}/T) \tag{18.80}$$

In our case, $\hat{H}$ just coincides with the standard Hamiltonian $\hat{H}_{\mathrm{full}}$ (2.17) for electrons in a superconductor that includes a short-range attractive BCS and a long-range repulsive Coulomb interactions as well as electromagnetic fields. As we already demonstrated above, our key idea is to exactly integrate out the electronic degrees of freedom already on the level of the partition function, so that we are left with an effective theory in terms of collective fields. For this purpose, it is convenient to apply the imaginary time Matsubara technique and to express the partition function $\mathcal{Z}$ as a path integral over the anticommuting electronic fields $\bar{\Psi}$, Ψ and the commuting gauge scalar and vector-potential fields V and $\boldsymbol{A}$, with Euclidean action

$$\begin{aligned} S_{\mathrm{E}} = \int dx \Big(& \bar{\Psi}_\sigma [\partial_\tau - \mathrm{i}eV + \xi(\boldsymbol{\nabla} - \frac{\mathrm{i}e}{c}\boldsymbol{A})]\Psi_\sigma \\ & - \lambda \bar{\Psi}_\uparrow \bar{\Psi}_\downarrow \Psi_\downarrow \Psi_\uparrow + \mathrm{i}enV + [\boldsymbol{E}^2 + \boldsymbol{B}^2]/8\pi \Big) \end{aligned} \tag{18.81}$$

Here $\xi(\nabla) \equiv -\nabla^2/2m - \mu + U(x)$ describes a single conduction band with quadratic dispersion and also includes an arbitrary impurity potential $U(x)$. The combination en denotes the background charge density of the ions. In our notation, dx denotes $d^3\boldsymbol{x}d\tau$, where τ is the imaginary time. The field strengths are functions of the gauge fields through $\boldsymbol{E} = -\boldsymbol{\nabla}V + (1/c)\partial_\tau\mathbf{A}$ and $\boldsymbol{B} = \nabla \times \mathbf{A}$ in the usual way for the imaginary time formulation.

In order to decouple the BCS interaction term and to introduce the superconducting order parameter field $\Delta = |\Delta|e^{\mathrm{i}\varphi}$, we again employ the standard Hubbard–Stratonovich transformation

$$\exp\left(\lambda\int dx\bar{\Psi}_\uparrow\bar{\Psi}_\downarrow\Psi_\downarrow\Psi_\uparrow\right) = \left[\int\mathcal{D}^2\Delta e^{-\int dx\Delta^2/\lambda}\right]^{-1}$$
$$\times\int\mathcal{D}^2\Delta e^{-\int dx\left(\Delta^2/\lambda+\Delta\bar{\Psi}_\uparrow\bar{\Psi}_\downarrow+\Delta^*\Psi_\downarrow\Psi_\uparrow\right)} \quad (18.82)$$

where the first factor is for normalization and will not be important in the following. As a result, the partition function now reads

$$\mathcal{Z} = \int\mathcal{D}^2\Delta\int\mathcal{D}^3\mathbf{A}\int\mathcal{D}V\mathcal{D}^2\psi e^{\left(-S_0-\int dx\bar{\psi}\mathcal{G}^{-1}\psi\right)}, \quad (18.83)$$

$$S_0[V,\mathbf{A},\Delta] = \int dx\left(\frac{\mathbf{E}^2+\mathbf{B}^2}{8\pi}+\mathrm{i}enV+\frac{\Delta^2}{\lambda}\right)$$

where the Nambu spinor notation for the electronic fields

$$\psi = \begin{pmatrix}\Psi_\uparrow\\ \bar{\Psi}_\downarrow\end{pmatrix}, \quad \bar{\psi} = \begin{pmatrix}\bar{\Psi}_\uparrow & \Psi_\downarrow\end{pmatrix} \quad (18.84)$$

and the matrix Green function in the Nambu space

$$\tilde{\mathcal{G}}^{-1} = \begin{pmatrix}\partial_\tau - ieV + \xi(\nabla - \frac{ie}{c}\mathbf{A}) & \Delta\\ \Delta^* & \partial_\tau + \mathrm{i}eV - \xi(\nabla + \frac{\mathrm{i}e}{c}\mathbf{A})\end{pmatrix} \quad (18.85)$$

were introduced. Integrating out the electronic degrees of freedom, we obtain

$$\mathcal{Z} = \int\mathcal{D}\Delta\,\mathcal{D}V\,\mathcal{D}\boldsymbol{A}\,e^{-S_{\text{eff}}} \quad (18.86)$$

where

$$S_{\text{eff}} = -\mathrm{Tr}\ln\tilde{\mathcal{G}}^{-1} + S_0[V,\mathbf{A},\Delta] \quad (18.87)$$

is the effective action of our superconducting system. As before, the gauge invariance of the theory enables us to rewrite the action (18.87) in the form

$$S_{\text{eff}} = -\mathrm{Tr}\ln\mathcal{G}^{-1} + S_0[V,\mathbf{A},\Delta] \quad (18.88)$$

with

$$\mathcal{G}^{-1} = \begin{pmatrix}\partial_\tau + \xi(\nabla) - \mathrm{i}e\Phi + \frac{m\mathbf{v}_s^2}{2} - \frac{\mathrm{i}}{2}\{\nabla,\mathbf{v}_s\} & \Delta\\ \Delta & \partial_\tau - \xi(\nabla) + \mathrm{i}e\Phi - \frac{m\mathbf{v}_s^2}{2} - \frac{\mathrm{i}}{2}\{\nabla,\mathbf{v}_s\}\end{pmatrix} \quad (18.89)$$

where

$$\Phi = V - \frac{\dot{\varphi}}{2e}, \quad \mathbf{v}_s = \frac{1}{2m}\left(\nabla\varphi - \frac{2e}{c}\mathbf{A}\right) \quad (18.90)$$

again represent the gauge-invariant linear combinations of the electromagnetic potentials and the phase of the order parameter, which define, respectively, the chemical potential of Cooper pairs and the superconducting velocity. The curly brackets $\{A, B\}$ in Eq. (18.89) denote the anticommutator.

As the action (18.88) cannot be evaluated exactly, we will again perform a perturbative expansion of the action up to the second order in Φ and $\mathbf{v}_s$. In addition, we will assume that deviations of the amplitude of the order parameter field from its equilibrium value Δ_0 are relatively small. This assumption allows us to also expand the effective action in powers of $\delta\Delta(x,\tau) = \Delta(x,\tau) - \Delta_0$ up to the second-order terms. The next step is to average over the random potential of impurities. After that the effective action becomes translationally invariant both in space and in time.

At this stage, we turn to the specific case of sufficiently long and very thin superconducting wires, which will be of particular interest for us here. Making use of the translational invariance of the (averaged over disorder) effective action, we perform the Fourier transformation with respect to both x and τ. Then we obtain [160, 385]

$$S_{\mathrm{E}} = \frac{s}{2}\int \frac{d\omega dq}{(2\pi)^2}\left\{\frac{|A|^2}{L_w s} + \frac{C_w|V|^2}{s} + \tilde{\chi}_{\mathrm{D}}\left|qV + \frac{\omega}{c}A\right|^2 \right.$$
$$\left. + \tilde{\chi}_{\mathrm{J}}\left|V + \frac{\mathrm{i}\omega}{2e}\varphi\right|^2 + \frac{\tilde{\chi}_{\mathrm{L}}}{4m^2}\left|\mathrm{i}q\varphi + \frac{2e}{c}A\right|^2 + \tilde{\chi}_{\Delta}|\delta\Delta|^2\right\} \quad (18.91)$$

The functions $\tilde{\chi}_{\Delta}$, $\tilde{\chi}_{\mathrm{J}}$, $\tilde{\chi}_{\mathrm{L}}$, and $\tilde{\chi}_{\mathrm{D}}$, which depend both on the frequencies and the wave vectors, are expressed in terms of the averaged products of the Matsubara Green functions [160, 385]. These functions represent an imaginary time version of the analogous real-time functions χ_j already encountered above. In order to recover the expressions for $\tilde{\chi}_j(\omega, k)$, one just needs to substitute $-\mathrm{i}\omega \to |\omega|$ in the corresponding expressions for $\chi_j(\omega, k)$ evaluated in Section 18.1. The action S_{E} (18.91) represents the imaginary time analogue of the real-time effective action (18.35).

As the action S_{E} (18.91) is quadratic both in the voltage V and the vector potential A, these variables can be integrated out exactly. Performing this integration, one arrives at the effective action that only depends on φ and $\delta\Delta$. We get

$$S = \frac{s}{2}\int \frac{d\omega dq}{(2\pi)^2}\left\{\mathcal{F}(\omega, q)|\varphi|^2 + \tilde{\chi}_\Delta|\delta\Delta|^2\right\} \tag{18.92}$$

The general expression for $\mathcal{F}(\omega, q)$ and the saddle point relations between the electromagnetic potentials and the fluctuating phase φ are specified elsewhere [160]. A simplified expression for this function is obtained if we again neglect the geometric inductance L_w of our superconducting wire as compared to the kinetic one $L_{\rm kin}$. Setting $L_w \to 0$, we find

$$\mathcal{F}(\omega, q) = \frac{\left(\frac{\tilde{\chi}_{\rm J}}{4e^2}\omega^2 + \frac{\tilde{\chi}_{\rm L}}{4m^2}q^2\right)\left(\frac{C_w}{s} + \tilde{\chi}_{\rm D}q^2\right) + \frac{\tilde{\chi}_{\rm J}\tilde{\chi}_{\rm L}}{4m^2}q^2}{\frac{C_w}{s} + \tilde{\chi}_{\rm J} + \tilde{\chi}_{\rm D}q^2} \tag{18.93}$$

and

$$V = \frac{\tilde{\chi}_{\rm J}}{\frac{C_w}{s} + \tilde{\chi}_{\rm J} + \tilde{\chi}_{\rm E}q^2}\left(\frac{-i\omega}{2e}\varphi\right), \qquad A = 0. \tag{18.94}$$

It is interesting to observe that according to Eq. (18.94), the Josephson relation $V = \dot{\varphi}/2e$ is in general *not* satisfied. This relation may approximately hold only in the limit $\tilde{\chi}_{\rm J} \gg C_w/s + \tilde{\chi}_{\rm D}q^2$. In the important limit of small elastic mean free paths ℓ, the latter condition is obeyed only at low frequencies and wave vectors $\omega/\Delta_0 \ll 1$ and $Dq^2/\Delta_0 \ll 1$.

Let us now perform yet one more approximation and expand the action in powers of ω and q^2. Keeping the terms of the order q^4 and ω^2q^2, we find

$$S_{\rm E} = \frac{s}{2}\int \frac{d\omega dq}{(2\pi)^2}\left\{\left(\frac{C_w}{s}\omega^2 + \pi\sigma_{\rm D}\Delta_0 q^2 + \frac{\pi^2}{8}\sigma_{\rm D}Dq^4 + \frac{\pi\sigma_{\rm D}}{8\Delta_0}\omega^2q^2\right)\left|\frac{\varphi}{2e}\right|^2 + 2N_0\left(1 + \frac{\omega^2}{12\Delta_0^2} + \frac{\pi Dq^2}{8\Delta_0}\right)|\delta\Delta|^2\right\} \tag{18.95}$$

The term $\propto \omega^4$ turns out to be equal to zero. At even smaller wave vectors, $Dq^2/2\Delta \ll 2C_w/\pi e^2 N_0 s \ll 1$, we get

$$S_{\rm E} = \frac{1}{2}\int \frac{d\omega dq}{(2\pi)^2}\left\{\left(C_w\omega^2 + \pi\sigma_{\rm D}\Delta_0 s q^2\right)\left|\frac{\varphi}{2e}\right|^2 + s\tilde{\chi}_A|\delta\Delta|^2\right\} \tag{18.96}$$

Here, as before, we have assumed $C_w/2e^2N_0s \ll 1$. The first two terms in this action correspond to the effective Hamiltonian of the form

$$\frac{1}{4e^2}\int dx\left[\frac{C_w}{2}(\partial_\tau\varphi)^2+\frac{1}{2L_{\rm kin}}(\partial_x\varphi)^2\right] \tag{18.97}$$

which again defines the Mooij–Schön plasma modes propagating along the wire with the velocity c_0 (18.74).

The effective action (18.92) allows to directly evaluate the fluctuation correction to the order parameter in superconducting nanowires. Performing Gaussian integration over both φ and $\delta\Delta$, we arrive at the wire free energy

$$F=F_{\rm BCS}-\frac{T}{2}\sum_{\omega,q}\left[\ln\frac{\lambda\mathcal{F}(\omega,q)}{2N_0\Delta_0^2}+\ln\frac{\lambda\tilde{\chi}_\Delta(\omega,q)}{2N_0}\right] \tag{18.98}$$

where $F_{\rm BCS}$ is the standard BCS free energy. The order parameter is defined by the saddle point equation $\partial F/\partial\Delta=0$ and can be written in the form $\Delta=\Delta_0-\delta\Delta_0$, where Δ_0 is the solution to the BCS self-consistency equation $\partial F_{\rm BCS}/\partial\Delta_0 = 0$ (18.25) and the fluctuation correction $\delta\Delta_0$ has the form

$$\delta\Delta_0=-\frac{T}{2}\left(\frac{\partial^2F_{\rm BCS}}{\partial\Delta_0^2}\right)^{-1}\frac{\partial}{\partial\Delta_0}\sum_{\omega,q}\left[\ln\frac{\lambda\mathcal{F}(\omega,q)}{2N_0\Delta_0^2}+\ln\frac{\lambda\tilde{\chi}_\Delta(\omega,q)}{2N_0}\right]. \tag{18.99}$$

Making use of the above expressions for the functions $\mathcal{F}(\omega,q)$ and $\tilde{\chi}_\Delta(\omega,q)$ and bearing in mind that for a wire of length X one has $\partial F/\partial\Delta_0=2N_0sX$, at $T\to 0$ we obtain

$$\frac{\delta\Delta_0}{\Delta_0}\sim\frac{1}{g_\xi}\sim Gi_{\rm 1D}^{3/2} \tag{18.100}$$

In Eq. (18.100), fluctuations of both the phase and the absolute value of the order parameter give contributions of the same order. The estimate (18.100) again demonstrates that at low temperatures, suppression of the order parameter in superconducting nanowires due to Gaussian fluctuations remains weak as long as $g_\xi \gg 1$ and it becomes important only for extremely thin wires with $Gi_{\rm 1D}\sim 1$.

Finally, let us return to the action (18.91), which we will use to illustrate a deficiency of the TDGL approach in the imaginary time. Considering the superconducting part of the action only and assuming that temperature is close to T_C, we can identify $\delta\Delta$ with Δ and set $\Delta_0 = 0$ in all the $\tilde{\chi}$-kernels. Then, exactly as in the real-time approach, one has $\tilde{\chi}_J = \tilde{\chi}_L = 0$ and the phase fluctuations become unrestricted. For this reason, it is again desirable to undo the gauge transformation, thereby going back to the description in terms of the complex order parameter field.

For simplicity, let us ignore both the scalar and the vector potentials. The TDGL action for the wire is then usually written in the form

$$\begin{aligned} S_{\mathrm{TDGL}} = & N_0 T s \sum_{\omega_n} \int dx \frac{\pi |\omega_n|}{8T} |\Delta|^2 \\ & + N_0 s \int d\tau dx \left(\frac{\pi D}{8T} \left| \frac{\partial \Delta}{\partial x} \right|^2 + \frac{T - T_C}{T_C} |\Delta|^2 + \mathcal{O}(|\Delta|^4) \right) \end{aligned} \tag{18.101}$$

This form can be obtained from the action (18.91) by formally expanding the kernel $\tilde{\chi}_\Delta$ in Matsubara frequencies and wave vectors ω_μ, $Dq^2 \ll 4\pi T$; cf. Eqs. (18.39) and (18.41). Note, however, that since the validity of the GL expansion is restricted to temperatures $T \sim T_C$, the Matsubara frequencies $\omega_n = 2\pi n T$ are never really smaller than $4\pi T$. Hence, the expansion $\Psi(1/2 + |\omega_n|/4\pi T) - \Psi(1/2) \to \pi\omega_n/8T$, which yields Eq. (18.101) is never correct except in the stationary case $\omega_n = 0$. Already these simple arguments illustrate the failure of the TDGL action (18.101) in the Matsubara technique. Further problems with this TDGL approach arise in the presence of the electromagnetic potentials V and A [385]. In the next chapter, we will demonstrate that the TDGL approach may yield parametrically inaccurate results even at temperatures close to the critical one $T \sim T_C$.

Chapter 19

Thermal and Quantum Phase Slips in Superconducting Nanowires

19.1 Introduction

In Chapter 18, we analyzed Gaussian fluctuations of both the absolute value and the phase of the superconducting order parameter Δ in quasi-one-dimensional systems and demonstrated that such fluctuations can play an important role only in extremely thin wires with $g_\xi \lesssim 1$; see Eq. (18.100). On the other hand, it is clear that fluctuations of the superconducting phase φ can also be crucially important even if $|\Delta|$ does not fluctuate at all. For instance, in Chapter 6 we demonstrated that under certain conditions, such phase fluctuations can yield total suppression of supercurrents in Josephson junctions. In low-dimensional superconductors—according to the general theorem [195, 301]—phase fluctuations should destroy the long-range order, i.e., long-range phase coherence. Thus, it could naively be concluded that low-dimensional conductors cannot exhibit superconducting properties because of strong phase fluctuation effects.

This conclusion, however, turns out somewhat premature if one bears in mind that any generic superconducting system has

Dissipative Quantum Mechanics of Nanostructures: Electron Transport, Fluctuations, and Interactions
Andrei D. Zaikin and Dmitry S. Golubev

ISBN 978-981-4774-50-5 (Hardcover), 978-0-429-29823-3 (eBook)
www.jennystanford.com

a finite size in which case phase coherence can be preserved at least to a certain extent. For instance, two-dimensional structures undergo Berezinskii–Kosterlitz–Thouless (BKT) phase transition [53, 233, 235], as a result of which the decay of correlations in space changes from exponential at high-enough T to power law at low temperatures. This result implies that at low T, long-range phase coherence essentially survives in samples of a finite size and, hence, generic two-dimensional films can and do exhibit superconducting properties.

Can superconductivity survive also in (quasi)-one-dimensional systems of a finite length or do phase fluctuations disrupt any supercurrent in such systems? The answer to this question is of both fundamental interest and practical importance due to the rapidly progressing miniaturization of superconducting nanocircuits. This chapter is devoted to a detailed analysis of nontrivial fluctuation effects in ultrathin superconducting wires.

19.2 General Background

The superconducting state of a quasi-one-dimensional metallic wire can be described by means of a complex order parameter $\Delta(x) = |\Delta(x)|e^{i\varphi(x)}$, where x is the coordinate along such a wire. Both thermal and quantum fluctuations cause deviations of the modulus as well as the phase of this order parameter from their equilibrium values. Small (Gaussian) fluctuations of both $|\Delta|$ and φ in space–time were already addressed in Chapter 18. In addition, there exist fluctuations of a different kind, called *phase slips*. A nontrivial fluctuation of that kind corresponds to temporal suppression of $|\Delta(x)|$ down to zero in some point $x = x_0$ inside the wire, as shown in Fig. 19.1. As soon as the modulus of the order parameter $|\Delta(x_0)|$ vanishes, the phase $\varphi(x_0)$ becomes unrestricted and can jump by the value $2\pi n$, where n is any integer number. After this process, the modulus $|\Delta(x_0)|$ gets restored, the phase becomes single valued again, and the system returns to its initial state, accumulating the net phase shift $2\pi n$. Provided such phase slip events are sufficiently rare, one can restrict n by $n = \pm 1$ (as indicated in Fig. 19.1) and totally disregard fluctuations with $|n| \geq 2$.

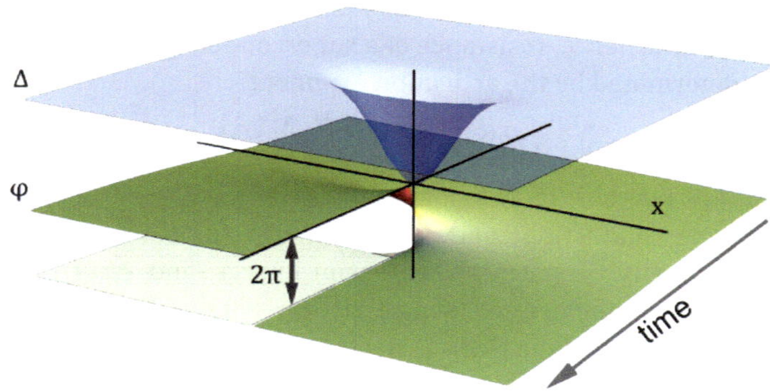

Figure 19.1 Schematics of the phase slip process in superconducting wires. The absolute value of the order parameter Δ gets locally and temporarely suppressed due to thermal and/or quantum fluctuations, while its phase φ suffers a jump by $\pm 2\pi$.

As we will see below in this chapter, phase slips may have a strong impact on the behavior of thin superconducting wires. For instance, as it was first pointed out by Little [284], quasi-one-dimensional wires made of a superconducting material can acquire a finite resistance below T_C of a bulk material due to the mechanism of thermally activated phase slips (TAPS). This mechanism works as follows.

According to the Josephson relation, each phase jump by $\delta\varphi = \pm 2\pi$ implies positive or negative voltage pulse $\delta V = \dot{\varphi}/2e$ accompanied by the transfer of one magnetic flux quantum $\Phi_0 \equiv \pi c/e = c \int |\delta V(t)| dt$ through the wire in the direction perpendicular to its axis. In the absence of any bias current, the net average numbers of "positive" ($n = +1$) and "negative" ($n = -1$) phase slips are equal; thus the net voltage drop across the wire remains zero. Applying the current $I \propto |\Delta|^2 \nabla\varphi$, one creates nonzero phase gradient along the wire and makes "positive" phase slips more likely than "negative" ones. Hence, the net voltage drop V due to TAPS differs from zero, i.e., thermal fluctuations cause nonzero resistance $R = V/I$ of superconducting wires even below T_C.

A quantitative theory of this TAPS phenomenon was first proposed by Langer and Ambegaokar [254] and then extended by McCumber and Halperin [298]. This Langer–Ambegaokar–

McCumber– Halperin (LAMH) theory predicts that the TAPS creation rate $\Gamma_{\rm TAPS}$ and, hence, resistance of a superconducting wire R below $T_{\rm C}$ are determined by the activation exponent

$$R(T) \propto \exp(-U/T), \qquad U \sim \frac{N_0 \Delta_0^2(T)}{2} s\xi(T), \tag{19.1}$$

where $U(T)$ is the effective potential barrier for TAPS proportional to the superconducting condensation energy (as before, N_0 is the metallic density of states at the Fermi energy, and $\Delta_0(T)$ is the BCS order parameter) for a part of the wire with the volume $\sim s\xi$ where superconductivity is destroyed by thermal fluctuations (s is the wire cross section and $\xi(T)$ is the superconducting coherence length). At temperatures very close to $T_{\rm C}$, Eq. (19.1) yields appreciable resistivity, which was indeed detected in experiments [288, 316] performed on small superconducting whiskers with typical diameters in the range of $\sim$0.5 μm. Close to $T_{\rm C}$, the experimental results fully confirm the activation behavior of $R(T)$ expected from Eq. (19.1). However, as the temperature is lowered further below $T_{\rm C}$, the number of TAPS inside the wire decreases exponentially and no measurable wire resistance is predicted by the LAMH theory except in the immediate vicinity of the critical temperature.

Recent progress in nanolithographic technique allowed to fabricate samples with much smaller diameters down to, and even below, 10 nm. In such systems, one can consider a possibility for phase slips to occur not only due to thermal, but also due to *quantum* fluctuations of the superconducting order parameter. The physical picture of quantum phase slippage is qualitatively similar to that of TAPS (see Fig. 19.1) except the order parameter $|\Delta(x)|$ gets virtually suppressed due to the process of quantum tunneling rather than thermal activation. As a result, the superconducting phase again suffers jumps by $\delta\varphi = \pm 2\pi$, and the magnetic flux quanta $\pm\Phi_0$ tunnel through the wire in the direction perpendicular to its axis.

Following the standard quantum mechanical arguments, one can expect that the probability of such tunneling process should be controlled by the exponent $\sim \exp(-U/\omega_0)$, i.e., one should just substitute an effective attempt frequency ω_0 instead of temperature in the activation exponent (19.1). This is because the order parameter field $\Delta(x)$ now tunnels under the barrier U rather than overcomes it by thermal activation. Since such tunneling process should obviously persist down to $T = 0$, one arrives at a fundamentally important conclusion that *in nanowires superconductivity can be destroyed by quantum fluctuations at any temperature, including* $T = 0$. Accordingly, such nanowires should demonstrate a non-vanishing resistivity down to the lowest temperatures. Assuming that $\omega_0 \sim \Delta_0(T)$, one would expect that at $\Delta_0(T) \lesssim T < T_C$, the TAPS dependence (19.1) applies, while at lower $T \lesssim \Delta_0(T)$, quantum phase slips (QPS) take over, eventually leading to saturation of the temperature dependence $R(T)$ to a nonzero value in the limit $T \ll \Delta_0$.

In this chapter, we will assume that the condition $\delta F_0 \gg T_C$ is well satisfied in our superconducting wires and, hence, the effect of Gaussian fluctuations on the order parameter $\Delta_0(T)$ can be safely neglected. This condition requires the wire to be not too thin and/or the temperature to be not too close to T_C, i.e., $(T_C - T)/T_C \gg Gi_{1D}$.

19.3 Thermally Activated Phase Slips

We will first address thermal fluctuations that are particularly important in the immediate vicinity of T_C. As we already pointed out, theoretical description of TAPS was initially worked out by Langer and Ambegaokar [254] who employed the standard Ginzburg–Landau equations and evaluated the TAPS rate Γ_{TAPS} within the exponential accuracy. More accurate analysis of the TAPS rate, including the pre-exponential factor, was then performed by McCumber and Halperin [298] who employed the time-dependent Ginzburg–Landau (TDGL) approach. More recently, it was realized that the TDGL-based approach is not sufficiently accurate to account for the effect of quantum fluctuations and, hence, to correctly determine the pre-exponent in the expression for the TAPS rate. Appropriate modifications of the LAMH theory have been worked out [168] employing the effective action approach developed in Chapter 18.

The TAPS rate Γ_{TAPS} describes the classical decay of a metastable current state in a superconducting wire. Similar to the situation already encountered in Chapter 4, this TAPS rate is defined by the standard activation dependence

$$\Gamma_{\text{TAPS}} = Be^{-\delta F/T} \tag{19.2}$$

where δF is the free energy difference or, in other words, an effective potential barrier that the system should overcome in order to create a phase slip.

19.3.1 Activation Exponent

The free energy barriers $\delta F_{\pm}$ for TAPS corresponding to overall phase jumps by $\pm 2\pi$ have been evaluated by Langer and Ambegaokar [254]. In order to obtain $\delta F_{\pm}$ entering Eq. (19.2), we make use of the standard Ginzburg–Landau free energy functional for a wire of length X:

$$F[\Delta(x)] = sN_0 \int_{-X/2}^{X/2} dx \left(\frac{\pi D}{8T} \left| \frac{\partial \Delta}{\partial x} \right|^2 + \frac{T - T_C}{T_C} |\Delta|^2 + \frac{7\zeta(3)}{16\pi^2 T^2} |\Delta|^4 \right) - \frac{I}{2e} [\varphi(X/2) - \varphi(-X/2)] \tag{19.3}$$

Here $\varphi(x)$ is the phase of the order parameter $\Delta(x)$ and I is the external current applied to the wire.

The saddle-point paths for this functional are determined by the standard Ginzburg–Landau equation

$$-\frac{\pi D}{8T} \frac{\partial^2 \Delta}{\partial x^2} + \frac{T - T_C}{T_C} \Delta + \frac{7\zeta(3)}{8\pi^2 T^2} |\Delta|^2 \Delta = 0 \tag{19.4}$$

For any given value of the bias current

$$I = \frac{\pi e N_0 D s}{2T} |\Delta|^2 \nabla \varphi \tag{19.5}$$

this equation has a number of solutions. The TAPS free energy barrier δF_{+} is determined by the two of them. The first one, $\Delta_m = |\Delta_m| \exp(i\varphi_m)$, corresponds to a metastable minimum of the free energy functional. This solution reads

$$|\Delta_m| = \Delta_0(T) \sqrt{\frac{1 + 2\cos\alpha}{3}}, \quad \varphi_m = \frac{2T}{\pi e N_0 D s} \frac{Ix}{|\Delta_m|^2} \tag{19.6}$$

Here $\Delta_0(T)$ is the equilibrium superconducting gap defined in Eq. (18.25) and the parameter

$$\alpha = \frac{\pi}{3} \theta \left(|I| - \frac{I_C}{\sqrt{2}} \right) + \frac{1}{3} \arctan \frac{2|I| \sqrt{1 - (I/I_C)^2}}{I_C \left(1 - 2(I/I_C)^2\right)} \tag{19.7}$$

accounts for the external bias current I. The Ginzburg–Landau critical current I_C is defined by the standard expression

$$I_C = \frac{16\sqrt{6}\pi^{5/2}}{63\zeta(3)} eN_0\sqrt{D}s(T_C - T)^{3/2} \tag{19.8}$$

The second (saddle-point) solution $\Delta_s(x) = |\Delta_s|\exp(i\varphi_s)$ of Eq. (19.4) has the form

$$\frac{|\Delta_s|}{\Delta_0(T)} = \sqrt{\frac{1+2\cos\alpha}{3} - \frac{2\cos\alpha - 1}{\cosh^2\left[\sqrt{2\cos\alpha - 1}\frac{x}{\xi(T)}\right]}},$$
$$\varphi_s = \frac{2TI}{\pi eN_0 Ds}\int_{-X/2}^{x}\frac{dx'}{|\Delta_s(x')|^2} \tag{19.9}$$

where $\xi(T) = \sqrt{\pi D/4(T_C - T)}$ is the superconducting coherence length in the vicinity of T_C.

The free energy barrier δF_+ in Eq. (19.2) is set by the difference

$$\delta F_+ = F[\Delta_s(x)] - F[\Delta_m(x)] = \delta F\left[\sqrt{2\cos\alpha - 1} - \sqrt{\frac{2}{3}\frac{I}{I_C}}\arctan\left(\frac{\sqrt{3}}{2}\sqrt{\frac{2\cos\alpha - 1}{1-\cos\alpha}}\right)\right] \tag{19.10}$$

where

$$\delta F = \frac{16\pi^2}{21\zeta(3)} sN_0\sqrt{\pi D}(T_C - T)^{3/2} \tag{19.11}$$

is the potential barrier for TAPS in the absence of an external current $I \to 0$. The free energy barrier δF_- for "negative" TAPS is determined analogously and is related to δF_+ as follows:

$$\delta F_- = \delta F_+ + \frac{\pi I}{e} \tag{19.12}$$

19.3.2 Pre-exponent

Now let us turn to the pre-exponent B in the expression for the TAPS rate (19.2). In order to evaluate B, one should go beyond the stationary free energy functional (19.3) and include time-dependent fluctuations of the order parameter field $\Delta(x, \tau)$. As we already mentioned above, McCumber and Halperin [298] accomplished this task within the framework of a TDGL-based analysis. Employing the

TDGL equation, it is possible to re-formulate the problem in terms of the corresponding Fokker–Planck equation [252], which can be conveniently solved for the problem in question. Since the important time scale within the TDGL approach is the Ginzburg–Landau time $\tau_{\rm GL}$ (18.63), this time also enters the expression for the pre-exponent B derived by McCumber and Halperin; see also the paper [229].

Unfortunately, as it was already pointed out in Chapter 18, the TDGL approach fails below $T_{\rm C}$. Hence, one should employ a more accurate effective action analysis. The microscopic effective action for superconducting wires derived above cannot be easily reduced to any Fokker–Planck type of equation. For this reason, below we will take a somewhat different route and combine our effective action formalism with the general formula for the decay rate of a metastable state expressed via the imaginary part of the free energy; see Eqs. (4.10) and (4.11). As we already learned in Chapter 4, this method is applicable provided the system is not driven far from equilibrium.

The decay rate Γ in the thermal activation regime can be recovered from Eq. (4.11), i.e.,

$$\Gamma = -2\frac{T^*}{T}\mathrm{Im}F(T) \tag{19.13}$$

Here T^* is an effective crossover temperature between the regimes of thermal activation and quantum tunneling; see Chapter 4 for more details. Formally T^* is defined as a temperature at which a nontrivial saddle-point solution $\Delta(\tau, x)$ describing QPS first appears upon lowering T. Within the accuracy of our calculation, it is sufficient to estimate T^* simply by setting the activation exponent equal to the exponential factor that controls the quantum tunneling of the supercconducting order parameter, i.e., the QPS action $S_{\rm QPS}(T)$. Hence, we get

$$\delta F(T^*)/T^* \simeq S_{QPS}(T^*) \tag{19.14}$$

The QPS action $S_{\rm QPS}$ will be evaluated below in Section 19.4.1. Here we only quote the corresponding result. In the limit of sufficiently small currents, one has

$$S_{\rm QPS}(T) \sim sN_0\sqrt{N_0\Delta_0(T)} \tag{19.15}$$

Then from the condition (19.14), we estimate $\Delta_0(T^*) \sim T^*$ or, equivalently, $T^* = aT_C$, where the numerical factor $a < 1$ is sufficiently close to unity, i.e., $T^* \sim T_C$. As the whole concept of TAPS is only valid at T close to T_C, one always has $T^*/T \sim 1$. Thus, with the same accuracy, one can actually use the expression for the decay rate in the quantum regime (4.10). Here we will retain the parameter T^* for the reasons that will be clear below.

In the limit $\delta F \gg T_C$, it suffices to expand the general expression for the effective action around both solutions (19.6) and (19.9) up to quadratic terms in both the phase φ and the amplitude $\delta\Delta$. One can verify that in the limit $\Delta_0(T) \ll T$, the contributions from fluctuating electromagnetic fields can be ignored and we obtain

$$S_{s/m} = F[\Delta_{s/m}] + \delta^2 S_{s/m} \tag{19.16}$$

where

$$\begin{aligned}
\delta^2 S_m = \frac{sT}{2} \sum_{\omega_n} \int dx dx' \big[\delta\Delta(\omega_n, x)\chi_\Delta^{(m)}(|\omega_n|; x - x')\delta\Delta(\omega_n, x') \\
+ \varphi(\omega_n, x) k_\varphi^{(m)}(|\omega_n|; x - x')\varphi(\omega_n, x')\big], \\
\delta^2 S_s = \frac{sT}{2} \sum_{\omega_n} \int dx dx' \big[\delta\Delta(\omega_n, x)\chi_\Delta^{(s)}(|\omega_n|; x, x')\delta\Delta(\omega_n, x') \\
+ \varphi(\omega_n, x) k_\varphi^{(s)}(|\omega_n|; x, x')\varphi(\omega_n, x')\big]
\end{aligned} \tag{19.17}$$

Here $\omega_n = 2\pi T n$ are Bose–Matsubara frequencies. The functions $\chi_\Delta^{(m)}$ and $k_\varphi^{(m)}$ are expressed in terms of the kernels $\tilde{\chi}_\Delta$, $\tilde{\chi}_J$, and $\tilde{\chi}_L$ as follows:

$$\begin{aligned}
\tilde{\chi}_\Delta^{(m)}(|\omega_n|; x - x') &= \int \frac{dq}{2\pi} e^{iq(x-x')} \tilde{\chi}_\Delta(\omega_n, q), \\
k_\varphi^{(m)}(|\omega_n|; x - x') &= \int \frac{dq}{2\pi} e^{iq(x-x')} \left(\frac{\omega_n^2}{4e^2} \tilde{\chi}_J(\omega_n, q) + \frac{q^2}{4m^2} \tilde{\chi}_L(\omega_n, q) \right)
\end{aligned} \tag{19.18}$$

The functions $\chi_\Delta^{(s)}$ and $k_\varphi^{(s)}$ describe fluctuations around the coordinate-dependent saddle point $\Delta_s(x)$ and, therefore, cannot be easily related to $\tilde{\chi}_\Delta$, $\tilde{\chi}_J$, and $\tilde{\chi}_L$. Fortunately, the explicit form of $\chi_\Delta^{(s)}$ and $k_\varphi^{(s)}$ will not be important for us here.

The pre-exponent B in Eq. (19.2) is obtained by integrating over fluctuations $\delta\Delta$ in the expression for the grand partition function. One arrives at a formally diverging expression that, as usual, signals the decay of a metastable state. After an analytic continuation of this diverging expression, one recovers an imaginary part of the free energy and finds the decay rate in the form (19.2) with

$$B = 2T^*\mathrm{Im}\prod_{\omega_n}\sqrt{\frac{\det\tilde{\chi}_\Delta^{(m)}(\omega_n)\,\det k_\varphi^{(m)}(\omega_n)}{\det\tilde{\chi}_\Delta^{(s)}(\omega_n)\,\det k_\varphi^{(s)}(\omega_n)}} \tag{19.19}$$

As before, the imaginary part of the above expression is taken since one of the eigenvalues of the operator $k_\varphi^{(s)}(0)$ is negative; see also Chapter 4.

The key point is to observe that at $T \sim T_\mathrm{C}$, all Matsubara frequencies $|\omega_n| = 2\pi T|n|$—except for one with $n = 0$—strongly exceed the order parameter, $|\omega_n| \gg \Delta_0(T)$. Hence, for all such values, the function $\chi_\Delta(\mathrm{i}|\omega_n|)$ approaches the asymptotic form (18.39), which is not sensitive to superconductivity at all. Then we have $\det\tilde{\chi}_\Delta^{(m)}(\omega_n) \simeq \det\tilde{\chi}_\Delta^{(s)}(\omega_n)$ and $\det k_\varphi^{(m)}(\omega_n) \simeq \det k_\varphi^{(s)}(\omega_n)$. The corresponding determinants in Eq. (19.19) cancel out, and only the contribution from $\omega_n = 0$ remains. It yields

$$B \simeq 2T^*\,\mathrm{Im}\sqrt{\frac{\det\tilde{\chi}_\Delta^{(m)}(0)\,\det k_\varphi^{(m)}(0)}{\det\tilde{\chi}_\Delta^{(s)}(0)\,\det k_\varphi^{(s)}(0)}}. \tag{19.20}$$

The ratio of these determinants can be evaluated at zero current with the aid of the free energy functional (19.3). Then we obtain [168]

$$\mathrm{Im}\sqrt{\frac{\det\tilde{\chi}_\Delta^{(m)}(0)\,\det k_\varphi^{(m)}(0)}{\det\tilde{\chi}_\Delta^{(s)}(0)\,\det k_\varphi^{(s)}(0)}} = \frac{2\sqrt{6}}{\sqrt{\pi}}\frac{X}{\xi(T)}\sqrt{\frac{\delta F}{T}} \tag{19.21}$$

Combining the above expressions, we arrive at the final result for the TAPS rate in the zero-bias limit:

$$\Gamma_\pm \equiv \Gamma_\mathrm{TAPS} = \frac{4\sqrt{6}}{\sqrt{\pi}}T^*\frac{X}{\xi(T)}\sqrt{\frac{\delta F}{T}}\exp\left[-\frac{\delta F}{T}\right] \tag{19.22}$$

Equation (19.22) is applicable at $T_\mathrm{C} - T \ll T_\mathrm{C}$ and as long as $\delta F \gg T$. Combining these two inequalities with Eq. (19.11), we arrive at the condition

$$Gi_{1\mathrm{D}} \ll \frac{T_\mathrm{C} - T}{T_\mathrm{C}} \ll 1, \tag{19.23}$$

where the Ginzburg number Gi_{1D} is defined in Eq. (18.78). The double inequality (19.23) is standard for the Ginzburg–Landau theory.

Turning to the case of nonzero bias, one can essentially repeat the whole calculation, which now yields two different TAPS rates $\Gamma_{\pm}$. Of practical importance is the limit of transport currents I sufficiently close to the critical one, i.e., $1 - I/I_C \ll 1$. In this regime, Γ_- is negligibly small, whereas Γ_+, on the contrary, increases since the free energy barrier

$$\delta F_+(I) = \frac{4 \times 6^{3/4}}{15} \delta F \left(1 - \frac{I}{I_C}\right)^{5/4} \tag{19.24}$$

becomes lower than that at smaller currents. Accordingly, TAPS can be detected easier in this limit [55].

The pre-exponent B_+ has essentially the same form as that defined by Eq. (19.22); one just needs to replace $\delta F \to \delta F_+(I)$ and $T^* \to T^*(I)$. In the limit $T_C - T \ll T_C$ considered here, the current dependence of the crossover temperature T^* appears insignificant in most cases, and with sufficient accuracy, one can set $T^*(I) \simeq T^*$. Indeed, very generally one can express $T^*(I) = T^* f(I/I_C(T^*))$, where $I_C(T^*)$ is the critical current at temperature T^* and $f(x)$ is some universal function with $f(x \ll 1) \simeq 1$. Having in mind the strong temperature dependence $I_C(T)$ in the temperature interval $T_C - T \ll T_C$, we find

$$I/I_C(T^*) < I_C(T)/I_C(T^*) \sim (T_C - T)^{3/2}/T_C^{3/2} \ll 1,$$

and, hence, $T^*(I) \approx T^*(0) \equiv T^*$. Thus, in the vicinity of the critical current $I_C(T) - I \ll I_C(T)$, the TAPS rate can be expressed in the form

$$\Gamma_+ \simeq 8.84 T^* \frac{X}{\xi(T)} \sqrt{\frac{\delta F}{T}} \left(1 - \frac{I}{I_C}\right)^{5/8} e^{-\delta F_+(I)/T} \tag{19.25}$$

where $\delta F_+(I)$ is defined in Eq. (19.24) and the numerical prefactor is again established from the calculation of the fluctuation determinants [168].

Collecting all the above results and substituting $T^* = aT_C$, we arrive at the final expression for the TAPS rates

$$\Gamma_{\pm}(I) = \kappa a T_C \frac{X}{\xi(T)} \sqrt{\frac{\delta F_{\pm}(I)}{T}} \exp\left[-\frac{\delta F_{\pm}(I)}{T}\right] \tag{19.26}$$

where $\kappa(I)$ is a smooth function of I varying from $\kappa(0) \simeq 5.53$ to $\kappa \simeq 8.74$ at $I_C - I \ll I_C$ and, as before, the numerical prefactor a is of order (and slightly smaller than) one. Equation (19.26) is supposed to be valid at $T_C - T \ll T_C$ and at any bias current $I < I_C$ as long as $\delta F_\pm(I) \gg T$.

19.3.3 Comparison with McCumber–Halperin Result

Let us now compare the result (19.26) with the expression for the TAPS rates derived from the TDGL-type of analysis [298]. As we expected in the beginning of our calculation, Eq. (19.26) does not contain the Ginzburg–Landau time τ_{GL} and exceeds the corresponding expression of the McCumber and Halperin theory by the factor $\sim T^*\tau_{GL} \sim (1 - T/T_C)^{-1} \gg 1$. On top of that, in the vicinity of the critical current, the pre-exponent in the TAPS rate (19.25) depends on I as $B_+ \propto (1 - I/I_C)^{5/8}$ in contrast to the result [298] $B_{TDGL} \propto (1 - I/I_C)^{15/8}$.

Since technically the approach developed by McCumber and Halperin is very different from that outlined above, for the sake of easier comparison, it is worth to briefly re-calculate the TAPS rate employing Eq. (19.13) and making use of the TDGL action (18.101). Since the whole calculation of the fluctuation determinants remains the same, one only needs to re-evaluate the crossover temperature, which we now denote as T^*_{TDGL}. To this end, we again first set $I \to 0$ and consider fluctuations of the order parameter around the saddle point $\Delta_s(x)$ along the unstable direction in the functional space [168]. We choose

$$\delta\Delta(\tau, x) = \mathrm{i}\frac{\cos\left(2\pi T^*_{TDGL}\tau\right)}{\sqrt{2}\cosh\left(x/\xi\right)}C \tag{19.27}$$

where C is a constant. Substituting this expression into the linearized TDGL equation and formally treating τ_{GL} as an independent parameter, we define the classical-to-quantum crossover temperature T^*_{TDGL} as that at which a nonzero solution ($C \neq 0$) first appears. With the aid of this definition, we obtain

$$T^*_{TDGL} = 1/4\pi\tau_{GL} \tag{19.28}$$

Substituting Eq. (19.28) into Eq. (19.22), we arrive at the expression for Γ_{TDGL} just two times bigger than that derived by McCumber and Halperin in the limit $I \to 0$.

An analogous, though slightly more complicated, analysis can be performed also at nonzero bias current I. This analysis yields [168]

$$T^*_{\mathrm{TDGL}}(I) \sim T^*_{\mathrm{TDGL}}\left(1 - \frac{I}{I_{\mathrm{C}}}\right)^{5/4} \tag{19.29}$$

Combining Eqs. (19.28) and (19.29) with the result (19.25), we arrive at the pre-exponent

$$B_{\mathrm{TDGL}} \sim \frac{1}{\tau_{\mathrm{GL}}} \frac{X}{\xi(T)} \sqrt{\frac{\delta F}{T}} \left(1 - \frac{I}{I_{\mathrm{C}}}\right)^{15/8}$$

which is again in the agreement with the result derived by McCumber and Halperin. Thus, with the aid of the general formula (19.13) describing thermal activation decay of a metastable state, we confirm that the McCumber–Halperin result for the TAPS rate is essentially correct *within the TDGL-type of formalism*. Unfortunately, however, the latter formalism is inaccurate by itself. In particular, in the expression for the TAPS rate, it does not allow to correctly obtain the classical-to-quantum crossover temperature T^*.

19.3.4 Temperature-Dependent Resistance and Noise

In order to complete our analysis, let us briefly address the relation between the above TAPS rate and physical observables, such as wire resistance and voltage noise. Every phase slip event implies changing of the superconducting phase in time in such a way that the total phase difference values along the wire before and after this event differ by $\pm 2\pi$. Since the average voltage is linked to the time derivative of the phase by means of the Josephson relation, $\langle V \rangle = \langle \dot{\varphi}/2e \rangle$, for the net voltage drop across the wire, we obtain

$$\langle V \rangle = \frac{\pi}{e}\left[\Gamma_+(I) - \Gamma_-(I)\right], \tag{19.30}$$

where $\Gamma_\pm$ are given by Eq. (19.26). In the absence of any bias current $I \to 0$, both rates are equal $\Gamma_\pm = \Gamma$ and the net voltage drop V vanishes. In the presence of a nonzero bias current, the symmetry between these two rates is lifted since, in complete analogy to the case of Josephson junctions (see Chapters 5 and 6), the potential barriers for these two processes differ. As long as the bias current I is sufficiently small, $I \ll I_C$, we obtain

$$\Gamma_{\pm}(I) = \Gamma_{\text{TAPS}} e^{\pm \Phi_0 I / 2cT} \tag{19.31}$$

Thus, at such values of I and at temperatures slightly below T_C, the I–V curve for quasi-one-dimensional superconducting wires takes a relatively simple form

$$\langle V \rangle = \frac{2\Phi_0}{c} \Gamma_{\text{TAPS}} \sinh \frac{\Phi_0 I}{2cT} \tag{19.32}$$

This important result implies that thermally activated phase slips effectively destroy long-range phase coherence in the system and the wire acquires non-vanishing resistance even below T_C. The zero-bias resistance $R(T) = (\partial V / \partial I)_{I=0}$ demonstrates exponential dependence on temperature and the wire cross section

$$R(T) = 2R_q \sqrt{6\pi} \frac{a T_C}{T} \frac{X}{\xi(T)} \sqrt{\frac{\delta F}{T}} \exp\left[-\frac{\delta F}{T}\right] \tag{19.33}$$

leading to fluctuation-induced broadening of the superconducting phase transition, which can be detected experimentally (see below).

In order to complete our description of thermal fluctuations in superconducting wires, we point out that in addition to nonzero resistance (19.33) TAPS also cause the voltage noise below T_C. Treating TAPS as independent events, one immediately concludes that they should obey Poissonian statistics. Hence, the zero-frequency voltage noise power

$$\mathcal{S}_0 = \int dt \langle \delta V(t) \delta V(0) \rangle$$

is given by the sum of the contributions of both "positive" and "negative" TAPS, i.e.,

$$\mathcal{S}_0 = (\Phi_0 / c)^2 \left[\Gamma_+(I) + \Gamma_-(I)\right] \tag{19.34}$$

At small currents $I \ll I_C$, this expression reduces to the following simple form:

$$\mathcal{S}_0 = \frac{2\Phi_0^2}{c^2} \Gamma_{\text{TAPS}} \cosh \frac{\Phi_0 I}{2cT} \tag{19.35}$$

Similar to the wire resistance, the voltage noise rapidly decreases as one lowers the temperature away from T_C. Only in the vicinity of the critical temperature, this TAPS noise remains appreciable and can be detected in modern experiments.

19.4 Theory of Quantum Phase Slips in Superconducting Nanowires

As temperature goes down, thermal fluctuations decrease and, hence, TAPS become progressively less important and eventually die out in the limit $T \to 0$. At low-enough temperature, quantum fluctuations of the order parameter field Δ take over and essentially determine the behavior of ultrathin superconducting wires. As we already discussed, the most significant non-Gaussian quantum fluctuations in such wires are QPS. Each QPS event involves suppression of the order parameter in the phase slip core and a winding of the superconducting phase around this core. This configuration describes quantum tunneling of the order parameter field through an effective potential barrier and can be conveniently described within the imaginary time effective action formalism already introduced in Chapter 18. Below we will develop a microscopic theory of QPS in superconducting nanowires. In doing so, to a large extent we will follow the papers [160, 360–362, 416, 417].

19.4.1 QPS Action

Let us denote the QPS core size as x_0 and the (imaginary time) duration of the QPS event as τ_0. At this stage, both these parameters are not yet known and remain to be determined from our subsequent analysis. It is instructive to separate the total action of a single QPS S_{QPS} into a core part S_{core}, for which the condensation energy and possibly also dissipation by normal currents inside the core are important (scales $x \leq x_0$, $\tau \leq \tau_0$), and a hydrodynamic part outside the core S_{out}, which depends on the hydrodynamics of the electromagnetic fields, i.e.,

$$S_{\text{QPS}} = S_{\text{core}} + S_{\text{out}} \tag{19.36}$$

Let us first evaluate the hydrodynamic part S_{out}. This task is simplified by the fact that outside the core, the absolute value of the order parameter field remains equal to its mean field value Δ_0, and only its phase $\varphi(x, \tau)$ changes both in space and in time. Without loss of generality, we can assume that the absolute value of the order parameter is equal to zero at $\tau = 0$ and $x = 0$. For sufficiently long wires and outside the QPS core, the saddle-point solution $\tilde{\varphi}(x, \tau)$ corresponding to a single QPS event should satisfy the identity

$$\partial_x \partial_\tau \tilde{\varphi} - \partial_\tau \partial_x \tilde{\varphi} = 2\pi \delta(\tau, x) \tag{19.37}$$

implying that after a wind around the QPS center, the phase should change by 2π. In a way, QPS is just a vortex in space–time with the phase distribution $\varphi(x, \tau)$ described by the saddle-point solution

$$\tilde{\varphi}(x, \tau) = -\arctan(x/c_0\tau) \tag{19.38}$$

Substituting the solution (19.38) into the action (18.95), we obtain

$$S_{\rm out} = \mu \ln[\min(c_0\beta, X)/\max(c_0\tau_0, x_0)] \tag{19.39}$$

where the parameter

$$\mu = \frac{\pi}{4e^2 c_0 (L_w + L_{\rm kin})} \simeq \frac{\pi}{4\alpha}\sqrt{\frac{sC_w}{4\pi\lambda_L^2}} \tag{19.40}$$

sets the scale for the hydrodynamic contribution to the QPS action. Here and below, $\alpha = e^2/c \simeq 1/137$ is the fine structure constant. We also note that at $T \to 0$, the contribution $S_{\rm out}$ (19.39) diverges logarithmically for infinitely long wires, thus making phase configurations corresponding to *single* QPS events unlikely in this limit.

Let us now turn to the core contribution to the QPS action. In order to exactly evaluate this contribution, it is necessary to explicitly find the QPS saddle point of the full nonlinear effective action. This is a formidable task, which can hardly be accomplished in practice. On the other hand, this task is greatly simplified if one is merely interested in estimating the term $S_{\rm core}$ up to a numerical prefactor of order one. Below we will recover the full microscopic expression for the core contribution $S_{\rm core}$ leaving only this numerical prefactor undetermined. In this way, we will capture all essential physics of QPS. The dimensionless prefactor can be regarded as a fit parameter, which can be extracted, e.g., from the comparison with the available experimental data.

The above strategy allows us to approximate the complex order parameter field inside the QPS core by two simple functions, which should satisfy several requirements. The absolute value of the order parameter $|\Delta(x, \tau)|$ should vanish at $x = 0$ and $\tau = 0$ and coincide with the mean field value Δ_0 outside the QPS core. The phase $\varphi(x, \tau)$ should flip at $x = 0$ and $\tau = 0$ in a way to provide the change in the net phase difference across the wire by 2π. On top of that, in a short wire and outside the QPS core, the phase φ should not depend on the spatial coordinate in the zero-bias limit. Almost any sufficiently smooth function obeying these requirements can be used to estimate S_{core}. Specifically, in what follows we will choose

$$|\delta\Delta(x, \tau)| = \Delta_0 \exp(-x^2/2x_0^2 - \tau^2/2\tau_0^2) \tag{19.41}$$

for the amplitude of the order parameter field and

$$\varphi(x, \tau) = -\frac{\pi}{2} \tanh\left(\frac{x\tau_0}{x_0\tau}\right) \tag{19.42}$$

for its phase. Rewriting the action (18.95) in the space–time domain and dropping unimportant terms $\propto (Dq^2)^2$, we obtain

$$\begin{aligned} S = \frac{s}{2} \int dx\, d\tau \Bigg\{ & \frac{C_w}{4e^2 s} \left(\frac{\partial\varphi}{\partial\tau}\right)^2 + \frac{\pi N_0 D\Delta_0}{2} \left(\frac{\partial\varphi}{\partial x}\right)^2 \\ & + \frac{\pi\sigma_D}{32e^2\Delta_0} \left(\frac{\partial^2\varphi}{\partial x\partial\tau}\right)^2 \Bigg\} \\ & + sN_0 \int dx\, d\tau \left\{ \delta\Delta^2 + \frac{1}{12\Delta_0^2} \left(\frac{\partial\delta\Delta}{\partial\tau}\right)^2 + \frac{\pi D}{8\Delta_0} \left(\frac{\partial\delta\Delta}{\partial x}\right)^2 \right\} \end{aligned} \tag{19.43}$$

Substituting the trial functions (19.41) and (19.42) into the action (19.43), we immediately recover the QPS core action in the form

$$\begin{aligned} S(x_0, \tau_0) = & \left[a_1 \frac{C_w}{e^2} + a_2 s N_0\right] \frac{x_0}{\tau_0} + a_3 s N_0 D\Delta_0 \frac{\tau_0}{x_0} \\ & + a_4 \frac{s\sigma_D}{2e^2\Delta_0} \frac{1}{x_0\tau_0} + a_5 s N_0 \Delta_0^2 x_0\tau_0 + a_6 \frac{\tilde{C}}{e^2\tau_0}, \end{aligned} \tag{19.44}$$

where a_j are numerical factors of order one, which depend on the precise form of the trial functions, $\tilde{C} = C_w X$ is the total capacitance of the wire, and X is the wire length. Note that fictitious divergencies emerging from a singular behavior of the function (19.42) at $x = x_0$ and $\tau = \tau_0$ are eliminated since the order parameter vanishes in this space–time point.

Let us first disregard capacitive effects by neglecting the last term in Eq. (19.44). Minimizing the remaining action with respect to the core parameters x_0 and τ_0, and making use of the inequality $C_w/e^2 N_0 s \ll 1$, we get

$$x_0^4 = \frac{a_3 a_4}{a_2 a_5} \frac{\sigma_D D}{2e^2 N_0 \Delta_0^2}, \qquad \tau_0^4 = \frac{a_2 a_4}{a_3 a_5} \frac{\sigma_D}{2e^2 N_0 D \Delta_0^4} \tag{19.45}$$

These values provide a minimum for the QPS core action, and we find

$$S_{\text{core}} = 2s N_0 (\sqrt{a_2 a_3 D \Delta_0} + \sqrt{a_4 a_5 \sigma_D \Delta_0 / 2e^2 N_0}) \tag{19.46}$$

Combining the Drude formula $\sigma_D = 2e^2 N_0 D$ with Eqs. (19.45) and (19.46), we arrive at the final results both for the core parameters

$$x_0 = \left(\frac{a_3 a_4}{a_2 a_5}\right)^{1/4} \sqrt{\frac{D}{\Delta_0}}, \qquad \tau_0 = \left(\frac{a_2 a_4}{a_3 a_5}\right)^{1/4} \frac{1}{\Delta_0} \tag{19.47}$$

and for the core action

$$S_{\text{core}} = \pi A N_0 s \sqrt{D \Delta_0} = A \frac{R_Q}{R_N} \frac{X}{\xi} = \frac{A}{4} g_\xi \tag{19.48}$$

Here $A = 2(\sqrt{a_2 a_3} + \sqrt{a_4 a_5})/\pi$ is the numerical prefactor, R_N is the total normal state wire resistance, $R_Q = \pi/2e^2 = 6.453$ kΩ, as before, is the "superconducting" quantum resistance unit, $\xi = \sqrt{D/\Delta_0}$ is the superconducting coherence length, and $g_\xi = 4(R_Q/R_N)(X/\xi)$ is the dimensionless normal conductance of a wire segment of length ξ.

As it was already pointed out, the results (19.47) and (19.48) hold provided the capacitive effects are small. This is the case for relatively short wires

$$X \ll \xi \frac{e^2 N_0 s}{C_w} \tag{19.49}$$

In the opposite limit, the same minimization procedure of the action (19.44) yields

$$x_0 \sim \xi, \qquad \Delta\tau_0 \sim \sqrt{X C_w / \xi e^2 N_0 s} \gg 1 \tag{19.50}$$

The QPS core action then takes a somewhat more complicated form

$$S_{\text{core}} = A' \frac{R_Q}{R_N} \left(\frac{X}{\xi}\right)^{3/2} \sqrt{\frac{C_w}{e^2 N_0 s}} \tag{19.51}$$

where A' is again a numerical prefactor. Equations (19.36), (19.39), (19.48), and (19.51) provide complete information about the QPS action in diffusive superconducting nanowires.

Let us analyze the above expressions. Introducing the total number of conducting channels in the wire, $\mathcal{N} = p_F^2 s/4\pi$, setting $C_w \sim 1$, and making use of the condition $e^2/v_F \sim 1$ satisfied for typical metals, one can rewrite the inequality (19.49) in a very simple form

$$X \ll \xi\mathcal{N} \tag{19.52}$$

Just to give an idea about the relevant length scales, for typical values $\xi \sim 10$ nm and $\mathcal{N} \sim 10^2 \div 10^3$—according to Eq. (19.52)—the wire can be considered short provided its length does not exceed $1 \div 10$ μm. This condition is met in a number of experiments; see, e.g. [55, 57, 260, 429, 430]. On the other hand, in experiments [10, 140, 141], much longer wires with lengths up to $X \sim 100$ μm were studied. Apparently such samples are already in the long wire regime $X \gg \xi\mathcal{N}$.

In the case of shorter wires, we observe a clear separation between different fluctuation effects contributing to the QPS action: Fluctuations of the order parameter field and dissipative currents determine the core part (19.48), while electromagnetic fluctuations are responsible for the hydrodynamic term (19.39). In the case of longer wires, capacitive effects also contribute to the core part (19.51). We also observe that in the short wire limit, different contributions to the QPS action depend differently on the wire thickness (or, equivalently, on the number of conducting channels $\mathcal{N}$): The core part (19.48) decreases linearly with the wire cross section, $S_{\text{core}} \propto \mathcal{N}$, whereas the hydrodynamic contribution shows a weaker dependence $\mu \propto \sqrt{\mathcal{N}}$. In the long wire limit, the dependence of the core part (19.51) on $\mathcal{N}$ also becomes weaker, $S_{\text{core}} \propto \sqrt{\mathcal{N}}$, due to capacitive effects.

Yet another important observation is that in the interesting range of wire thicknesses $r_0 \equiv \sqrt{s} \gtrsim 5 \div 10$ nm, the core part S_{core} usually exceeds the hydrodynamic term $\sim \mu$. For instance, for $C_w \sim 1$ we obtain $\mu \approx 30(r_0/\lambda_L)$. Setting $r_0 \sim 5 \div 10$ nm and estimating μ and S_{core} for typical system parameters $p_F^{-1} \sim 0.2$ nm, $l \sim 7$ nm, $\xi \sim 10$ nm, $\lambda_L \sim 100$ nm, $v_F = 10^6$ m/s, and $\Delta_0 \sim 1 \div 10$ K, we find $\mu \sim 1 \div 3$ and $S_{\text{core}} \gg \mu$. The latter inequality becomes even stronger for thicker wires. Note that the condition $S_{\text{core}} \gg \mu$ allowed us to ignore the hydrodynamic part of the QPS action while minimizing the core part with respect to x_0 and τ_0.

At the first sight, the form of the QPS action (19.48) might create an illusion that this result could be guessed without any microscopic derivation, e.g., just from the "condensation energy" term (proportional to χ_A) without taking into account dissipative effects. For instance, minimization of the contribution $\sim |\delta\Delta|^2$ (the last three terms in Eq. (19.43)) is formally sufficient to arrive at the correct estimate $S_{\rm core} \sim N_0 s\sqrt{D\Delta_0}$. At this point, one should be cautious, however, since the same equation demonstrates that not only the amplitude but also the phase fluctuations of the order parameter field provide important contributions to the QPS action. If the latter fluctuations were taken into account *without* including dissipative effects (this would correspond to formally setting $\sigma_{\rm D} \to 0$ in Eq. (19.43)), minimization of the core action would immediately yield a meaningless result $x_0 \to 0$, $\tau_0 \to 0$ (cf., Eq. (19.45)) implying that the hydrodynamic contribution $S_{\rm out}$ could not be neglected in that case. Minimization of the total action $S_{\rm QPS}$ would then yield the result parametrically different from that in Eq. (19.48). On the other hand, in the strong damping limit (approached by formally setting $\sigma \to \infty$ in Eq. (19.43)), the core size would become very large, and the action $S_{\rm core}$ (19.46) would diverge meaning that no QPS would be possible at all. These observations emphasize crucial importance of dissipative effects. Under the condition $C_w/e^2 N_0 s \ll 1$ (usually well satisfied in metallic nanowires), dissipation plays a dominant role during the phase slip event, and the correct QPS core action cannot be derived without an adequate microscopic description of dissipative currents flowing inside the wire. Only employing the Drude formula for the wire conductivity $\sigma_D = 2e^2 N_0 D$ enables one to arrive at the result (19.48), whereas for some other models of dissipation, different results for the core action would follow; cf., e.g., [416].

To complete our discussion of the QPS action, let us recall that in the course of our derivation, we employed two approximations: (i) we expanded the action up to the second order in $\delta\Delta(x,\tau) = \Delta(x,\tau) - \Delta_0$ and (ii) in Eq. (18.95), we expanded the action (18.92) in powers of ω/Δ_0 and Dq^2/Δ_0. The approximation (i) is sufficient everywhere except inside the QPS core where $\Delta(x,\tau)$ is small. In these space- and time-restricted regions, one can expand already in $\Delta(x,\tau)$ and again arrive at Eq. (18.92) with $\delta\Delta(x,\tau) \to \Delta(x,\tau)$ and with all the χ-functions taken in the limit $\Delta_0 \equiv 0$; see Chapter 18. Both expansions match smoothly at the scale of the core size $x_0 \sim \xi$, $\tau_0 \sim 1/\Delta_0$. Hence, the approximation (i) is sufficient to derive the correct QPS action up to a numerical prefactor of order one.

The approximation (ii) is sufficient within the same accuracy. One can actually avoid this approximation and substitute the trial functions (19.41) and (19.42) directly into the action (18.92). Neglecting capacitive effects in the limit (19.49), one can rewrite the QPS action as a function of the two dimensionless parameters x_0/ξ and $\Delta_0\tau_0$ only. Employing the results for the χ-functions derived in Chapter 18 and minimizing the QPS action with respect to x_0 and τ_0, one again arrives at the result (19.48) with $A \sim 1$. If the inequality (19.49) is violated, the accuracy of our expansion in powers of ω/Δ_0 can only become better (cf. Eq. (19.50)).

19.4.2 QPS Tunneling Amplitude

We now proceed further and evaluate the QPS tunneling amplitude per unit wire length γ_{QPS}. Provided the QPS action is sufficiently large $S_{\text{QPS}} \gg 1$, this amplitude can be expressed in the form

$$\gamma_{\text{QPS}} = (B/X)\exp(-S_{\text{QPS}}) \tag{19.53}$$

The results for the QPS action derived above allow to determine the amplitude γ_{QPS} with the exponential accuracy. Here we evaluate the pre-exponential factor B in Eq. (19.53). For this purpose, we will again make use of the instanton technique already employed, e.g., in Chapters 4–7 and elsewhere in this book.

Consider the grand partition function of the wire Z. As before, expressing this function via the path integral

$$\mathcal{Z} = \int \mathcal{D}\Delta\mathcal{D}\varphi \exp(-S), \tag{19.54}$$

one can perform its evaluation within the saddle-point approximation. Following this route, one constructs the least action paths

$$\delta S/|\delta\Delta| = 0, \qquad \delta S/\delta\varphi = 0, \tag{19.55}$$

which determine all possible QPS configurations in our problem. Integrating over small fluctuations around all QPS trajectories, one represents the grand partition function Z in terms of infinite series where each term describes the contribution of one particular QPS saddle point. Provided interaction between different QPS is sufficiently weak, one can perform a summation of these series in a straightforward manner with the result

$$\mathcal{Z} = \exp(-F/T) \tag{19.56}$$

where F defines the wire free energy

$$F = F_0 - T\frac{\int \mathcal{D}\delta Y \exp(-\delta^2 S_1[\delta Y])}{\int \mathcal{D}\delta Y \exp(-\delta^2 S_0[\delta Y])}\exp(-S_{\text{QPS}}) \equiv F_0 - \frac{X\gamma_{\text{QPS}}}{2} \tag{19.57}$$

Here F_0 is the free energy without quantum phase slips, $\delta Y = (\delta\Delta, \delta\varphi)$ describe fluctuations of relevant coordinates (fields), $\delta^2 S_{0,1}[\delta Y]$ are the quadratic in δY parts of the action, and the subscripts "0" and "1" denote the action, respectively, without and with one QPS.

The integrals over fluctuations in Eq. (19.57) can be evaluated exactly only in simple cases. Technically such a calculation can be quite complicated even if explicit analytical expressions for the saddle-point trajectories are available. In our case, such expressions for the QPS trajectories are not even known. Hence, an exact evaluation of the path integrals in Eq. (19.57) is not possible. Furthermore, any attempt to find an explicit value for such a prefactor would make little sense simply because the numerical value of A in Eq. (19.48) is not known exactly.

What to do in this situation? Fortunately one can suggest a simple approach, which allows to recover the correct expression for the pre-exponent B up to an unimportant numerical prefactor. This approach may be useful not only in the case of superconducting wires, but also for various other situations, since numerical prefactors in the pre-exponent are usually of little importance.

In order to evaluate the ratio of the path integrals in Eq. (19.57), it will be convenient for us to introduce the basis in the functional space $\Psi_k(z)$, in which the second variation of the action around the instanton $\delta^2 S_1[\delta Y]$ is diagonal. Here the basis functions depend on a general vector coordinate z, which is simply $z = (\tau, x)$ in our case. The first N functions Ψ_k correspond to (already familiar to us) "zero modes" (see Chapter 4), which follow from the instanton action invariance under shifts in certain directions in the functional space.

In our case, the problem has two zero modes corresponding to shifts in QPS along the wire and in imaginary time, i.e., $N = 2$. Obviously, such shifts do not cause any changes in the instanton energy. The eigenfunctions corresponding to these zero modes are $\Psi_k(X) = \partial\tilde{Y}/\partial z_k$, where $\tilde{Y}(z)$ is the instanton (or QPS) trajectory, $k \leq N$, and the number of zero modes N coincides with the dimension of the vector z. An arbitrary fluctuation $\delta Y(z)$ can be represented in terms of the Fourier expansion

$$\delta Y(z) = \sum_{k=1}^{N} \delta z_k \frac{\partial \tilde{Y}(z)}{\partial z_k} + \sum_{k=N+1}^{\infty} u_k \Psi_k(z). \tag{19.58}$$

Then we get

$$\delta^2 S_0[\delta Y] = \frac{1}{2} \sum_{k,n=1}^{\infty} A_{kn} u_k u_n^*$$
$$\delta^2 S_1[\delta Y] = \frac{1}{2} \sum_{k=N+1}^{\infty} \lambda_k |u_k|^2 \tag{19.59}$$

where for $k \leq N$, the Fourier coefficients $u_k \equiv \delta z_k$ are just the shifts in the instanton position along the k-th axis and λ_k are the eigenvalues of $\delta^2 S_1[\delta Y]$. Integrating over the Fourier coefficients, one again arrives at the standard formula for the ratio of determinants with excluded zero modes

$$\frac{\int \mathcal{D}\delta Y \exp(-\delta^2 S_1[\delta Y])}{\int \mathcal{D}\delta Y \exp(-\delta^2 S_0[\delta Y])} = \int_0^{L_1} d\delta x_1 \ldots \int_0^{L_N} d\delta x_N \sqrt{\frac{\det A_{kn}}{(2\pi)^N \prod_{k=N+1}^{\infty} \lambda_k}}, \tag{19.60}$$

where L_k is the system size in the k-th dimension; cf., e.g., also Eqs. (4.42) and (5.16).

Now we will argue that with a sufficient accuracy in the latter formula, one can retain only the contributions from the first N eigenvalues. Indeed, the contribution of the "fast" eigenmodes (corresponding to frequencies and wave vectors much larger than the inverse instanton size in the corresponding dimension) is insensitive to the presence of an instanton. Hence, the corresponding eigenvalues are the same for both $\delta^2 S_0$ and $\delta^2 S_1$ and just cancel out from Eq. (19.60). In addition to these fast modes, there are several eigenmodes with frequencies (wave vectors) of the order of the inverse instanton size. The ratio between the product of all such modes for $\delta^2 S_1$ and the product of eigenvalues for $\delta^2 S_0$ with the same numbers is dimensionless and may only affect a numerical prefactor, which is not interesting for us here. Dropping the contribution of all such eigenvalues, one gets

$$\frac{\int \mathcal{D}\delta Y \exp(-\delta^2 S_1[\delta Y])}{\int \mathcal{D}\delta Y \exp(-\delta^2 S_0[\delta Y])} \approx \int_0^{L_1} d\delta x_1 \ldots \int_0^{L_N} d\delta x_N \sqrt{\frac{\det A_{kn}|_{k,n\leq N}}{(2\pi)^N}} \tag{19.61}$$

What remains is to estimate the parameters A_{kk} for $k \leq N$. For this purpose, let us observe that the second variation of the action becomes approximately equal to the instanton action, $\delta^2 S_1 = \frac{1}{2} A_{kk} z_{0k}^2 \approx S_{\mathrm{QPS}}$, when the shift in the k-th direction becomes equal to the instanton size in the same direction $\delta z_k = z_{0k}$. Then we find $A_{kk} \approx 2S_{\mathrm{QPS}}/z_{0k}^2$ and

$$\det A_{kn}|_{k,n<N} \approx \prod_{k=1}^{N} A_{kk} \approx \frac{(2S_{\mathrm{QPS}})^N}{\prod_{k=1}^{N} z_{0k}^2} \tag{19.62}$$

Finally, combining Eqs. (19.53), (19.57), (19.61), and (19.62), we obtain

$$B = bT \left(\prod_{k=1}^{N} \frac{L_k}{z_{0k}} \right) \left(\frac{S_{\mathrm{QPS}}}{\pi} \right)^{N/2} \tag{19.63}$$

Here b is an unimportant numerical prefactor. This formula demonstrates that the functional dependence of the pre-exponent can be figured out practically without any calculation. It is sufficient to know just the instanton action, the number N of the zero modes, and the instanton effective size z_{0k} for each of these modes.

Turning to the interesting-for-us case of QPS in superconducting nanowires, we set $L_1 \equiv 1/T$, $L_2 \equiv X$. Then Eq. (19.63) yields

$$B = b\frac{S_{\mathrm{QPS}} X}{\tau_0 x_0} \tag{19.64}$$

This equation provides an accurate expression for the pre-exponent B up to a numerical factor b of order one. The inequality $S_{\mathrm{core}} \gg \mu$ allows us to substitute S_{core} (19.48) instead of S_{QPS} in Eq. (19.64).

Substituting also $x_0 \sim \xi$ and $\tau_0 \sim 1/\Delta_0$, for the QPS amplitude, we finally obtain

$$\gamma_{\rm QPS} \sim \Delta_0 \frac{R_{\rm Q}}{R_{\rm N}} \frac{X}{\xi^2} \exp(-S_{\rm QPS}) \sim (g_\xi \Delta_0/\xi) \exp(-A g_\xi/4) \quad (19.65)$$

19.4.3 QPS Interactions and Quantum Phase Transitions

Although usually the hydrodynamic part of the QPS action (19.39) is smaller than its core part (19.48), the former also plays an important role, since it determines interactions between different QPS. Consider the configuration of two phase slips essentially equivalent to a pair of vortices in space–time with the corresponding core coordinates (x_1, τ_1) and (x_2, τ_2). Provided the phase slip cores do not overlap, i.e., provided $|x_2 - x_1| > x_0$ and $|\tau_2 - \tau_1| > \tau_0$, the core contributions are independent and simply add up. In order to evaluate the hydrodynamic part, we substitute the superposition of two solutions $\tilde{\varphi}_1(x - x_1, \tau - \tau_1) + \tilde{\varphi}_2(x - x_2, \tau - \tau_2)$ satisfying the identities

$$\partial_x \partial_\tau \tilde{\varphi}_{1,2} - \partial_\tau \partial_x \tilde{\varphi}_{1,2} = 2\pi \nu_{1,2} \delta(\tau - \tau_{1,2}, x - x_{1,2}) \quad (19.66)$$

into the action (18.95) and obtain

$$S^{(2)}_{\rm QPS} = 2S_{\rm core} - \mu \nu_1 \nu_2 \ln \left[\frac{(x_1 - x_2)^2 + c_0^2(\tau_1 - \tau_2)^2}{\xi^2} \right], \quad (19.67)$$

i.e., different QPS interact logarithmically in space–time. Here $\nu_{1,2} = \pm 1$ are the topological charges of the two phase slips that fix the sign of the phase change after a wind around the QPS center to be equal to $\pm 2\pi$. We observe that QPS with opposite (equal) topological charges attract (repel) each other. Note that in Eq. (19.67) and below for the sake of simplicity, we assume that the QPS core has about the same size both in space and in time, i.e., $x_0 \sim c_0 \tau_0$. All our conclusions remain valid also if the QPS core is strongly asymmetric with respect to space and time; however, in that case the whole analysis would become slightly more complicated.

The next step is to consider a gas of n QPS. Again assuming that the QPS cores do not overlap, we can substitute a simple superposition of the saddle-point solutions for n QPS $\varphi = \sum_i^n \tilde{\varphi}_i(x - x_i, \tau - \tau_i)$ into the action and find

$$S^{(n)}_{\rm QPS} = nS_{\rm core} + S^{(n)}_{\rm int} \quad (19.68)$$

where

$$S_{\text{int}}^{(n)} = -\mu \sum_{i \neq j} \nu_i \nu_j \ln\left(\frac{\rho_{ij}}{x_0}\right) + \frac{\Phi_0}{c} I \sum_i \nu_i \tau_i \ . \tag{19.69}$$

Here $\rho_{ij} = (c_0^2(\tau_i - \tau_j)^2 + (x_i - x_j)^2)^{1/2}$ defines the distance between the i-th and j-th QPS in the (x, τ) plane and $\nu_i = \pm 1$ are the QPS topological charges. In Eq. (19.69), we also included an additional term that keeps track of the applied current I flowing through the wire. This term follows from the standard contribution to the action (see, e.g., Chapter 2)

$$\int d\tau \int dx (I/2e) \partial_x \varphi.$$

The grand partition function of the wire is represented as a sum over all possible configurations of quantum phase slips (topological charges):

$$\mathcal{Z} = \sum_{n=0}^{\infty} \frac{1}{2n!} \left(\frac{y}{2}\right)^{2n} \int_{x_0}^{X} \frac{dx_1}{x_0} \ldots \int_{x_0}^{X} \frac{dx_{2n}}{x_0} \int_{\tau_0}^{\beta} \frac{d\tau_1}{\tau_0} \ldots \int_{\tau_0}^{\beta} \frac{d\tau_{2n}}{\tau_0} \sum_{\nu_i = \pm 1} e^{-S_{\text{int}}^{(2n)}}, \tag{19.70}$$

where an effective fugacity y of these charges is related to the QPS amplitude as

$$y = x_0 \tau_0 \gamma_{\text{QPS}} \sim S_{\text{core}} \exp(-S_{\text{core}}) \tag{19.71}$$

We also note that only neutral QPS configurations with

$$\sum_{i=1}^{2n} \nu_i = 0 \tag{19.72}$$

contribute to the partition function (19.70). Similar to the situation already encountered in Chapter 6, this fact is a direct consequence of the boundary condition $\varphi(x, \tau) = \varphi(x, \tau + \beta)$ in the path integral for the partition function.

It is easy to observe that for $I = 0$, Eqs. (19.69) and (19.70) define the standard model for a two-dimensional gas of logarithmically interacting charges ν_i. The only specific feature of our present model as compared to the standard situation is that here the space and time coordinates are not equivalent and one can consider different limiting cases of "long" and "short" wires.

Let us first consider the limit of very long wires and assume that $T \to 0$. Making use of the standard renormalization group (RG) analysis of logarithmically interacting two-dimensional Coulomb gas [53, 233, 235], we arrive at the scaling equations for the interaction parameter μ and the charge fugacity y:

$$\partial_\Lambda \mu = -4\pi^2\mu^2 y^2, \quad \partial_\Lambda y = (2-\mu)y \tag{19.73}$$

Here $\Lambda = \ln(\rho/\xi_c)$ (with $\xi_c = \sqrt{x_0^2 + c_0^2\tau_0^2}$) is the scaling parameter.

Following the standard line of reasoning, we immediately conclude that a quantum phase transition for phase slips occurs in a long superconducting wire at $T \to 0$ and

$$\mu = \mu^* \equiv 2 + 4\pi y \approx 2 \tag{19.74}$$

This is just the celebrated Berezinskii–Kosterlitz–Thouless (BKT) phase transtion for charges ν_i in space–time. The only difference from the standard BKT transition in two-dimensional superconducting films is that in our case, the transition is driven by the wire thickness $\sqrt{s}$ (which enters μ) and not by temperature. In other words, for thicker wires with $\mu > \mu^*$, QPS with opposite topological charges are bound in pairs (dipols) and the *linear* resistance of a superconducting wire is T-dependent and tends to vanish in the limit $T \to 0$. Thus, we arrive at an important conclusion: at $T = 0$, a long quasi-one-dimensional superconducting wire remains in a superconducting state, with vanishing linear resistance, provided its thickness is sufficiently large and, hence, the electromagnetic interaction between phase slips is sufficiently strong, i.e., $\mu > \mu^*$.

On the other hand, for $\mu < \mu^*$, the density of free (unbound) QPS in the wire always remains finite; such fluctuations destroy the phase coherence (and, hence, superconductivity) and bring the wire into a non-superconducting state with non-vanishing resistance even at $T = 0$. Thus, another important conclusion is that superconductivity in sufficiently thin wires is *always* destroyed by quantum fluctuations.

The above analysis is valid for sufficiently long wires. For typical experimental parameters, however, $X < c_0/T$ (or even $X \ll c_0/T$), and the finite wire size needs to be accounted for. For this purpose, we modify the above RG treatment in the following manner. Starting from small scales $\rho \sim x_0$, we increase the scaling parameter Λ and renormalize both μ and y according to the RG equations (19.73). Solving these equations up to $\Lambda = \Lambda_X = \ln(X/\xi)$, we obtain the renormalized fugacity $\tilde{y} = y(\Lambda_X)$. For larger scales $\Lambda > \Lambda_X$, only the time coordinate matters and we arrive at the partition function fully equivalent to one already considered in Chapter 6 for Josephson junction in the presence of quantum fluctuations of the phase. The renormalized fugacity $\tilde{y}$ is equivalent to the tunneling amplitude of the Josephson phase (normalized by the Josephson plasma frequency), i.e.,

$$\tilde{y} \to \sqrt{S_0}\exp(-S_0) \tag{19.75}$$

where $S_0 = \sqrt{8E_J/E_C}$ is the action of an instanton (kink) in a Josephson junction, E_J and E_C are, respectively, the Josephson and the charging energies.

The subsequent analysis essentially depends on the presence (or absence) of additional dissipation in our system. Let us remind the reader that in the absence of dissipation at $T \to 0$, the Josephson junction always remains in the non-superconducting state since superconductivity is suppressed by quantum fluctuations of the phase and tunneling of Cooper pairs is prohibited (Coulomb blockade). Due to Eq. (19.75), exactly the same conclusion applies for sufficiently short superconducting wires studied here. In the presence of additional dissipation, however, quantum fluctuations of the phase can be suppressed and, hence, superconductivity can be restored; see also Chapter 6.

Of particular importance is the case of Ohmic dissipation, which is realized either provided the wire is shunted by a normal resistor R_S, or the wire itself has a non-vanishing normal conductivity σ_{qp} even far from the QPS core. The latter situation can occur, either at nonzero (and not too low) temperatures due to the presence of a sufficient number of quasiparticles above the gap or possibly also due to non-equilibrium effects. Bearing in mind that in a number of experiments with ultrathin wires, QPS effects were observed already at sufficiently high temperatures $T \gtrsim \Delta_0(T)$, it is instructive to briefly address the model with Ohmic dissipation here.

In this case, our consideration should be modified employing a two-stage scaling procedure [417]. At it was already explained, we first proceed with two-dimensional RG equations (19.73) up to the scale $\Lambda = \Lambda_X = \ln(X/\xi)$. For simplicity, we assume that Ohmic dissipation does not significantly affect the RG equations (19.73) at such small scales. Eventually we arrive at the renormalized fugacity $\tilde{y} = y(\tilde{\ell}_X)$. At larger scales, the space coordinate is irrelevant and the problem reduces to that of a one-dimensional Coulomb gas with logarithmic interaction. Therefore (for $\tilde{y} \ll 1$), further scaling is defined by the RG equations (6.22). In our present notations, these equations read

$$\partial_\Lambda \tilde{y} = (1 - \gamma)\tilde{y}, \quad \partial_\Lambda \gamma = 0 \tag{19.76}$$

where the interaction parameter γ depends on the dissipation strength being equal to either (i) $\gamma = R_Q/R_S$ or to (ii) $\gamma = \pi s \sigma_{qp}/2e^2 X$. For $\gamma > 1$, the fugacity scales down to zero, which again corresponds to a superconducting phase, whereas for $\gamma < 1$, it increases indicating a resistive phase in complete analogy to a single Josephson junction with Ohmic dissipation; see also Fig. 6.2. In the case (ii), the phase transition point again depends on the wire cross section s as well as on its total length X and the value σ_{qp}.

Thus, two different quantum phase transitions (QPTs) can occur in superconducting nanowires. One of them is the BKT-like phase transition controlled by the strength μ of inter-QPS electromagnetic interactions and eventually by the wire thickness. Another one is the Schmid phase transition occurring in the presence of Ohmic dissipation [73, 417] provided, e.g., by an external shunt resistance R_S. Note that this situation is somewhat reminiscent of that occurring in chains of resistively shunted Josephson junctions or granular arrays [62, 115] where, however, Schmid-like QPT is driven by local (rather than global) shunt resistance. Perhaps we can add that non-Ohmic dissipation in our system can also be provided if one couples a superconducting wire to a diffusive normal metallic plane [286]. In this case, one observes quantitative differences with the above QPT scenario, although qualitative dissipation again tends to suppress quantum fluctuations of the phase and, hence, favors superconductivity in the wire.

19.4.4 Wire Resistance at Low Temperatures

Let us now turn to the calculation of the wire resistance in the presence of quantum phase slips. We first consider the limit of long wires. At any nonzero T, such wires have a nonzero resistance $R(T, I)$ even in the "ordered" phase $\mu > \mu^*$. In order to evaluate $R(T)$ in this regime, we proceed perturbatively in the QPS fugacity y. For $\mu > \mu^*$, QPS form closed pairs (dipols) and, hence, interactions between different dipols can be neglected. For this reason, it suffices to evaluate the correction δF to the wire free energy due to one bound pair of quantum phase slips with opposite topological charges. This procedure is completely analogous to that already described in details in Chapter 5. Taking into account only logarithmic interactions within bound pairs of quantum phase slips, we easily sum up the series in Eq. (19.70) and arrive at the result

$$\delta F = \frac{Xy^2}{x_0\tau_0}\int_{\tau_0}^{\beta}\frac{d\tau}{\tau_0}\int_{x_0}^{X}\frac{dx}{x_0}e^{(\Phi_0 I\tau/c)-2\mu\ln[\rho(\tau,x)/x_0]}\,, \qquad (19.77)$$

where $\rho = (c_0^2\tau^2 + x^2)^{1/2}$. It is convenient to first integrate over the spatial coordinate x and take the wire length $X \to \infty$. For nonzero I, the expression in Eq. (19.77) is formally divergent for $\beta \to \infty$ and acquires an imaginary part Im δF after analytic continuation of the integral over the temporal coordinate τ; cf. Eqs. (5.42)–(5.45). This indicates a QPS-induced instability of the superconducting state of the wire: The state with a zero phase difference $\delta\varphi(X) = \varphi(X) - \varphi(0) = 0$ decays into a lower energy state with $\delta\varphi(X) = 2\pi$. The corresponding decay rate Γ_+ is defined by Eq. (4.10). The rate Γ_- for the opposite transition from 2π to 0 (or from 0 to -2π), which is nonzero as long as $T > 0$, is defined analogously with $I \to -I$. The average voltage drop across the wire is then given by the difference between these rates

$$\langle V\rangle = (\Phi_0/c)[\Gamma_+(I) - \Gamma_-(I)] \qquad (19.78)$$

Evaluating the imaginary part of the free energy ImδF from Eqs. (4.10) and (19.78), we finally obtain

$$\langle V\rangle = \frac{\Phi_0 X y^2}{c\tau_0 x_0}\frac{\sqrt{\pi}\,\Gamma(\mu-\frac{1}{2})(2\pi\tau_0 T)^{2\mu-2}}{\Gamma(\mu)\Gamma(2\mu-1)} \times \sinh\left(\frac{\Phi_0 I}{2cT}\right)\left|\Gamma\left(\mu - \frac{1}{2} + \frac{\mathrm{i}}{\pi}\frac{\Phi_0 I}{2cT}\right)\right|^2, \qquad (19.79)$$

where $\Gamma(x)$ is the Euler gamma function. For the wire resistance R, this expression yields

$$R(I, T) = \frac{\partial\langle V\rangle}{\partial I} \propto \begin{cases} y^2 T^{2\mu-3}, & T \gg \Phi_0 I \\ y^2 I^{2\mu-3}, & T \ll \Phi_0 I \end{cases} \quad (19.80)$$

Thus, for sufficiently thick wires with $\mu > \mu^*$, we expect a strong temperature dependence of the linear resistance, which eventually vanishes in the limit $T \to 0$ indicating the superconducting behavior of the wire.

Unfortunately for thinner wires with $\mu < \mu^*$ and at low temperatures, the above simple analysis becomes insufficient because of the presence of unbound QPS and the necessity to account for many-body effects in the gas of QPS. We expect that the temperature dependence of the wire resistivity should become linear at the transition to the disordered (i.e., non-superconducting) phase. At $T \ll \Phi_0 I/c$, we expect a strongly nonlinear I–V characteristics $V \sim I^a$ in thick wires, and the universal power $a(\mu^*) = 2$ in thin wires at the transition into the resistive state with $V \sim I$. Note that in contrast to the BKT transition in two-dimensional superconducting films, in the case of wires, the jump is expected to be from $a = 2$ to 1 rather than from $a = 3$ to 1.

In the case of short wires $X \ll c_0/T$, we again proceed in two steps. A two-dimensional scaling analysis yields the "global" fugacity $\tilde{y}$. In analogy with resistively shunted Josephson junctions analyzed in Chapter 6, the voltage drop derived from the imaginary part of the free energy reads

$$\langle V\rangle = \frac{2\Phi_0\tilde{y}^2}{\Gamma(2\gamma)c\tilde{\tau}_0}\sinh\left(\frac{\Phi_0 I}{2cT}\right)\left|\Gamma(\gamma + \frac{i\Phi_0 I}{2\pi cT})\right|^2 [2\pi\tilde{\tau}_0 T]^{2\gamma-1}, \quad (19.81)$$

where $\tilde{\tau}_0 \sim XC_w/e^2\gamma$. This equation gives $R \propto y^2 T^{2\gamma-2}$ and $R \propto y^2 I^{2\gamma-2}$, respectively, at high and low T. The above result is valid for $\gamma > 1$ and also for smaller γ at not very small T. In the low-temperature limit in the metallic phase, the linear resistance becomes

$$R = S\sigma_{qp}/X, \quad (19.82)$$

or $R = R_S$, i.e., R is just equal either to the quasiparticle resistance of the wire or to the shunt resistance. The physics of this result is exactly the same as in the case of resistively shunted Josephson junctions: The superconducting channel turns out to be completely blocked by quantum fluctuations, and the current can only flow through the normal resistance.

19.4.5 Discussion

The above results were derived for sufficiently uniform nanowires with (almost) constant cross section s. On the other hand, it is quite obvious that our QPS theory does not necessarily require strict uniformity of the system; it will also apply provided there exist diameter variations along the wire, which are practically unavoidable in any realistic experimental situation. The probability for QPS to occur will be higher in thinner parts of the wire than in thicker ones; otherwise, all the physics remains essentially the same. This situation should be contrasted to that of chains of Josephson junctions and one-dimensional granular wires in which case QPS occur inside the junctions (tunnel barriers) and do not require any suppression of the superconducting order parameter. In other words, in the latter case, QPS have no core at all. Accordingly, no condensation energy is lost during such phase slip events, and also dissipative effects remain insignificant (unless Josephson junctions are shunted by external normal resistors).

Note that, although the QPS core physics for uniform and granular quasi-one-dimensional wires is entirely different, the behavior of these systems at larger scales can be qualitatively similar. For instance, it is well known that the BKT-like quantum phase transition (driven by the ratio E_J/E_C) occurs in one-dimensional superconducting granular arrays [62, 64, 115] similar to the case of uniform wires [416]. Also, Schmid-like quantum dissipative phase transitions in arrays and chains of resistively shunted Josephson junctions were studied in great detail [62, 115]. One of the main conclusions reached there was that dissipative phase transitions yield *local* phase coherence in a long chain of junctions provided each of such junctions is shunted by an effective Ohmic resistance below some value of order the quantum resistance unit R_Q. In contrast, at $T \to 0$ BKT-like quantum phase transition

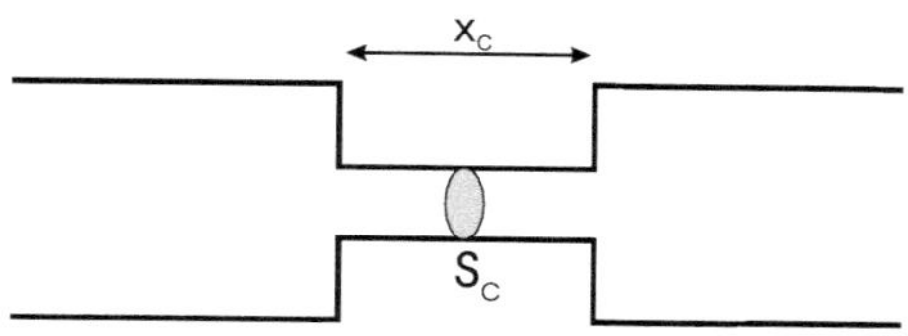

Figure 19.2 Superconducting nanowire containing a thinner part (constriction) of length x_c and cross section s_c.

yields global phase coherence in Josephson chains provided $E_{\rm J} \gtrsim E_{\rm C}$. As a result, one arrives at a nontrivial phase diagram with all possible combinations of local and global order [62, 115].

We should also stress that even if the wire diameter remains strictly constant, the system can be considered uniform only at length scales exceeding the elastic electron mean free path ℓ, whereas at the scale $\sim\ell$, this uniformity is naturally broken by the presence of impurities and defects. Scattering of electrons at such impurities and defects breaks Galilei invariance, thus preventing from momentum conservation for electrons propagating along the wire [385]. Momentum non-conservation is, in turn, necessary for QPS to occur since quantum phase slips "unwind" the supercurrent, thus generating momentum transfer. At $T = 0$ (i.e., in the absence of phonons), this extra momentum cannot be absorbed in perfectly clean and uniform wires, in which case QPS would be strictly prohibited. On a formal level, this implies that the QPS action $S_{\rm QPS}$ should become large in the clean limit [221]. In fact, this effect can be observed within our analysis too; one just needs to formally take the limit $\sigma_{\rm D} \to \infty$ in Eq. (19.46).

Although for homogeneous (at scales $\gtrsim \ell$) and uniform wires, QPT is of the BKT universality class and the temperature dependence of the resistance in the ordered (superconducting) phase is $R \propto T^{2\mu-3}$, there exist other situations in which the dependence $R \propto y^2 T^{2\gamma-2}$ may apply at not very low T. Consider, e.g., a long superconducting wire with cross section s and length X, which is uniform everywhere except for a small part (constriction) of length $x_c \ll X$ and cross section $s_c < s$, as shown in Fig. 19.2. QPS would then occur easier inside the constriction than in the rest of the wire.

Provided x_c is larger than ξ, for the corresponding QPS fugacity inside the constriction area, we obtain

$$y_c \sim g_{c\xi} \exp\left(-\frac{A}{4} g_{c\xi}\right), \tag{19.83}$$

where $g_{c\xi} = (s_c/s)g_\xi$ is the dimensionless normal state conductance of the constriction segment of length ξ.

In the opposite limit $x_c \ll \xi$, one can model the constriction as a general coherent scatterer with arbitrary distribution of (spin-degenerate) channel transmissions T_n just as it is described, e.g., in Chapter 10. In this specific limit, the condensation energy loss during a QPS event inside the constriction is small and can be neglected. Hence, y_c can be evaluated along the same lines as it was already done in Chapters 5 and 6 in the case of Josephson tunnel junctions. The fugacity is then defined by the combination (19.75) $y_c \sim \sqrt{S_0} \exp(-S_0)$, where $S_0 \sim U_0/\omega_0$ with U_0 and ω_0 being, respectively, an effective potential barrier for QPS in the constriction and an attempt frequency.

In this case, the value of the QPS action S_0 can be estimated with the aid of the results obtained in Chapter 17. The potential barrier is again set by the Josephson coupling energy $U_0 \sim E_{\rm J}$, which is easily estimated from Eq. (17.1) as $E_{\rm J} \sim g_c \Delta_0$ where $g_c = 2\sum T_n$ is the dimensionless normal state conductance of the constriction. For $E_{\rm J} \gg E_{\rm C}$, the frequency $\omega_0 \sim \sqrt{E_{\rm J} E_{\rm C}}$ can be estimated provided the effective charging energy of the constriction $E_{\rm C} = e^2/2C_c$ is known. Usually, the geometric capacitance of metallic constrictions remains small and can be neglected. Then the total effective capacitance of the constriction C_c can be recovered from Eq. (17.41). At the border of its applicability (cf. Eq. (17.38)), we find $E_{\rm C} \sim \Delta_0/g_c$. Hence, we have $\omega_0 \sim \Delta_0$ and $S_0 \sim g_c$, implying

$$y_c \sim \sqrt{g_c} \exp(-\alpha g_c) \tag{19.84}$$

where α is a numerical factor of order one, which depends on the particular distribution of T_n in the constriction. In order to evaluate this factor explicitly, it is necessary to employ the exact Matsubara effective action for superconducting constrictions (2.173), (2.175) and elaborate the instanton technique in the same spirit as it was done for Josephson junctions in Chapters 5 and 6.

In the limit $x_c \gg \ell$, the transmission distribution becomes universal being defined by Eq. (17.35). Then according to [386], one finds $\alpha = \pi^2/16$. This result remains applicable as long as the condition $\ell \ll x_c \ll \xi$ is fulfilled. For $x_c = \xi$, one has $g_c \equiv g_{c\xi}$; however, the numerical factor $A/4$ in the exponent of Eq. (19.83) cannot be identified with the aid of the results for α because, as we already explained, the latter results do not account for an extra contribution to the exponent $\sim g_{c\xi}$ originating from the condensation energy losses in the volume $\sim s_c\xi$ during a QPS event inside the constriction. In addition, at the level of the pre-exponent, there exists an obvious difference between the two limiting expressions for y_c, Eqs. (19.83) and (19.84), due to the different number of QPS zero modes in these two limits.

As before, we can express the partition function $\mathcal{Z}$ in terms of the series in powers of the QPS fugacity, which is now equal to y_c inside the constriction and to y otherwise. In other words, we should now deal with two different types of logarithmically interacting QPS. QPS in the wire are treated exactly as before, and we again arrive at the BKT-like QPT at $\mu = \mu^* \approx 2$. QPS inside the constriction interact logarithmically only in time, thus giving rise to the Schmid-like QPT at $\mu = 1$. In this case, the quasi-one-dimensional superconducting wire with Mooij–Schön plasmons as elementary excitations plays the role of an effective *dissipative* environment for the constriction. Such QPT was discussed in the literature [189] in the case of a narrow superconducting ring interrupted by a Josephson junction. In the situation considered here, this QPT appears to be of little importance since $\mu^* > 1$, i.e., at $T \to 0$, BKT-like QPT occurs when QPS inside the constriction are still bound in pairs.

On the other hand, at not very low temperatures and provided $y_c \gg y$, QPS inside the constriction may give an additional contribution to the wire resistance. In order to demonstrate that, we modify Eq. (19.77) by taking into account the difference between QPS fugacities in the wire and in the constriction. This is trivially handled by splitting the space integrals into those over the wire and constriction areas. As a result, the linear resistance of the system becomes

$$R(T) = R_w(T) + R_c(T),$$

where $R_w(T)$ is determined by QPS inside the wire (19.80), i.e., $R_w(T) \propto y^2 T^{2\mu-3}$, while the additional contribution $R_c(T)$ comes from QPS inside the constriction. This contribution has the form $R_c(T) \propto y_c^2 T^{2\mu-2}$ and can dominate over $R_w(T)$ at sufficiently high temperatures. On the contrary, at lower temperatures, $R_c(T)$ becomes irrelevant and $R(T)$ again is determined by Eq. (19.80). Obviously, the same arguments hold in the case of several constrictions or, more generally, for wires with strongly fluctuating cross section.

Finally, let us briefly address the non-superconducting phase $\mu < \mu*$. It is instructive to compare the behavior of Josephson chains and homogeneous wires. The QPS fugacities y of both systems obey RG equations (19.73) and, hence, for $\mu < \mu^*$, these fugacities grow upon renormalization in both cases. In the case of Josephson chains, this observation only implies strong fluctuations of the phases across barriers with no suppression of superconductivity inside metallic grains. In this case, inter-grain tunneling of Cooper pairs is prohibited by electron–electron interactions, and the system is in the insulating (Coulomb blockade) phase.

On the contrary, large renormalized QPS fugacity y in the case of homogeneous wires implies that the QPS gas becomes dense, the QPS cores overlap, and hence for $\mu < \mu^*$ and $T \to 0$, superconductivity gets eventually destroyed everywhere in the wire. In other words, in this regime quantum fluctuations drive homogeneous ultrathin superconducting wires *normal*. In the presence of electron–electron interactions, an exponentially small Coulomb gap develops and these disordered normal wires can again go insulating in the true zero-temperature limit, as we already discussed, e.g., in Chapters 11–13. This conclusion formally matches with the exact results derived within the sine-Gordon model [94]. However, superconductivity is essentially irrelevant in this case. Moreover, in generic metallic wires with many conducting channels and large dimensionless conductances, this insulating state can only be observed at exponentially low T, while for any experimentally attainable temperature, one expects a metallic behavior with a weak trend for the wire resistance to grow with decreasing T due to Coulomb blockade corrections; see Chapters 11, 13, and 16.

19.4.6 Phase–Charge Duality

In Chapter 6, we already established and discussed in details an important property of ultrasmall Josephson junctions, i.e., the duality between the phase and the charge variables; see Sections 6.5 and 6.6. In particular, we argued that in the limit of large Josephson energies $E_J \gg E_C$ under the duality transformation (6.76), the effective actions in the phase and in the charge representations are *exactly* transformed onto each other. Neglecting the shunting resistance, one can describe the Josephson junction in terms of the Hamiltonian (6.77) for an effective "quantum particle" in the periodic potential in the (quasi)-charge space. The charge q and the flux Φ are canonically conjugate variables exactly analogous to, respectively, the coordinate and momentum variables in quantum mechanics.

It is easy to observe [308] that all the same duality arguments remain applicable if one replaces the Josephson weak link by a *short* superconducting nanowire. For example, under the same conditions, one can again write down the Hamiltonian (6.77), where one should only replace the parameter Δ by the total QPS amplitude $\gamma_{QPS}X$. The properties of such nanowires are exactly dual to those of Josephson junctions. For this reason, a short superconducting nanowire was named a *QPS junction* [308].

The phase–charge duality arguments discussed in Chapter 6 are strictly applicable only to sufficiently short nanowires, i.e., as long as the coordinate dependence of both the phase and the charge variables can be neglected and the QPS junction can be effectively treated as a zero-dimensional object. It turns out that one can also extend these arguments further to the case of *long* superconducting nanowires. This task can be accomplished either by means a rigorous path integral analysis [357] or in terms of simple quantum mechanical operator manipulations [362] to be outlined below.

As before, let us consider a uniform superconducting wire of length X and (small) cross section s. According to our results in Sections 18.5 and 18.6, the effective Hamiltonian of the wire can be expressed in the form (18.97), i.e.,

$$\hat{H}_{\text{eff}} = \int_0^X dx \left[\frac{\hat{Q}^2(x)}{2C_w} + \frac{1}{2\mathcal{L}_{\text{kin}}} \left(\frac{\partial_x \hat{\varphi}(x)}{2e} \right)^2 \right] \tag{19.85}$$

where, as before, the wire capacitance C_w and kinetic inductance $\mathcal{L}_{\text{kin}}$ are defined in Eqs. (18.66) and (18.75); $\hat{Q}(x)$ and $\hat{\varphi}(x)$ are canonically conjugate local charge and phase operators obeying the commutation relations

$$[\hat{Q}(x), \hat{\varphi}(x')] = -2ie\delta(x - x') \tag{19.86}$$

cf. Eq. (2.14). For the sake of simplicity we will assume that the superconducting wire is isolated from any external circuit, in which case the current at its end points $x = 0$ and $x = X$ vanishes and, hence, we can define the boundary conditions for the phase in the form

$$\partial_x \hat{\varphi}(0) = \partial_x \hat{\varphi}(X) = 0 \tag{19.87}$$

Employing the Fourier series expansion, we get

$$\hat{\varphi}(x) = \hat{\varphi}_0 + \sqrt{\frac{2}{X}} \sum_{n=1}^{\infty} \hat{\varphi}_n \cos(\pi n x/X), \tag{19.88}$$

$$\hat{Q}(x) = \frac{\hat{Q}_0}{X} + \sqrt{\frac{2}{X}} \sum_{n=1}^{\infty} \hat{Q}_n \cos(\pi n x/X) \tag{19.89}$$

where

$$[\hat{Q}_0, \hat{\varphi}_0] = -2ie, \quad [\hat{Q}_m, \hat{\varphi}_n] = -2ie\delta_{mn} \tag{19.90}$$

Let us now perform the dual transformation. For this purpose, we introduce the following (dual) operators:

$$\hat{\Phi}(x) = \frac{\partial_x \hat{\varphi}(x)}{2e}, \quad \hat{\chi}(x) = -\frac{\pi}{e} \int_x^X dx' \hat{Q}(x') + \frac{\pi(X - x)}{eX} \int_0^X dx' \hat{Q}(x') \tag{19.91}$$

which can also be expressed as

$$\hat{\Phi}(x) = -\sqrt{\frac{\pi^2}{2e^2 X^3}} \sum_{n=1}^{\infty} n\hat{\varphi}_n \sin(\pi n x/X), \tag{19.92}$$

$$\hat{\chi}(x) = \sqrt{\frac{2X}{e^2}} \sum_{n=1}^{\infty} \frac{\hat{Q}_n}{n} \sin(\pi n x/X) \tag{19.93}$$

These new canonically conjugate operators obey the commutation relations

$$[\hat{\Phi}(x), \hat{\chi}(x')] = -\mathrm{i}\Phi_0\delta(x - x') \tag{19.94}$$

and obvious boundary conditions

$$\hat{\Phi}(0) = \hat{\Phi}(X) = 0, \quad \hat{\chi}(0) = \hat{\chi}(X) = 0 \tag{19.95}$$

Substituting the relations

$$\partial_x\hat{\varphi}(x) = 2e\hat{\Phi}(x), \quad \hat{Q}(x) = \frac{\hat{Q}_0}{X} + \frac{e}{\pi}\partial_x\hat{\chi}(x) \tag{19.96}$$

into Eq. (19.85), we obtain

$$\hat{H}_{\mathrm{eff}} = \frac{\hat{Q}_0^2}{2XC_w} + \hat{H}_{\mathrm{TL}} \tag{19.97}$$

where

$$\hat{H}_{\mathrm{TL}} = \int_0^X dx\left(\frac{\hat{\Phi}^2}{2L_{\mathrm{kin}}} + \frac{(\partial_x\hat{\chi})^2}{2C_w\Phi_0^2}\right) \tag{19.98}$$

is the Hamiltonian for a transmission line formed by a superconducting wire.

The above consideration does not yet account for QPS. In order to specify the QPS contribution to the wire Hamiltonian, let us first define the phase field configurations as

$$\hat{\varphi}(x)|\varphi(x)\rangle = \varphi(x)|\varphi(x)\rangle \tag{19.99}$$

and bear in mind that, as we already discussed in Chapter 6, the phase of the superconducting order parameter is a compact variable. Accordingly, the field configurations $\varphi(x)$ and $\varphi(x) + 2\pi$ correspond to the same quantum state of our system. Furthermore, in the absence of QPS, i.e., provided the absolute value of the order parameter does not fluctuate $|\Delta(x, t)| = \Delta_0$, also the states φ and $\tilde{\varphi}(x) = \varphi(x) + 2\pi\theta(x - x_1)$ (where $0 < x_1 < X$ and $\theta(x)$ is the standard Heaviside step function equal to 0 for $x \leq 0$ and to 1 for $x > 0$) are physically indistinguishable. For instance, the supercurrent operator $\hat{I}$ proportional to the combination $\Delta_0^2 \exp(-\mathrm{i}\hat{\varphi}(x))\partial_x \exp(\mathrm{i}\hat{\varphi}(x))$ remains the same in both cases.

Let us now make the step function continuous by effectively smearing it at the scale of the superconducting coherence length ξ, i.e., we substitute $\theta(x) \to \theta_\xi(x)$. The corresponding field configuration $\tilde{\varphi}_\xi(x) = \varphi(x) + 2\pi\theta_\xi(x - x_1)$, on one hand, remains very close to $\tilde{\varphi}(x)$ and, on the other hand, is already physically distinguishable from the latter. The QPS process can be viewed as quantum tunneling between these two close but physically different phase configurations.

What remains is to make use of the fact that the wire Hamiltonian does not depend on the operator $\hat{\varphi}_0$, implying that any shift by a constant phase does not change the state of our system. Hence, without loss of generality, we can set $\hat{\varphi}_0|\psi\rangle = 0$ for any system state $\psi\rangle$. This condition applies for the evolution controlled by the Hamiltonian (19.85), and it is also maintained in the presence of QPS. With this in mind, we conclude that the QPS process corresponds to quantum tunneling of the phase between the states $\varphi(x)$ and

$$\varphi'(x) = \varphi(x) + 2\pi\theta_\xi(x - x_1) - 2\pi \int_0^X dx\theta_\xi(x - x_1) \quad (19.100)$$

In the operator language, this tunneling process can be denoted as $\hat{U}_\xi(x_1)|\varphi(x)\rangle = |\varphi'(x)\rangle$, where the expression for $\hat{U}_\xi(x_1)$ just follows from the commutation relations and reads

$$\hat{U}_\xi(x_1) = \exp\left(\frac{i\pi}{e}\int_0^X dx(\hat{Q}(x) - \hat{Q}_0/X)\theta_\xi(x - x_1)\right) \quad (19.101)$$

As a result, the part of the Hamiltonian that explicitly accounts for the QPS process takes the form

$$\hat{H}_{\mathrm{QPS}} = -\int_0^X dx_1\gamma_{\mathrm{QPS}}(x_1)\cos\left(\frac{\pi}{e}\int_0^X dx(\hat{Q}(x) - \hat{Q}_0/X)\theta_\xi(x - x_1)\right) \quad (19.102)$$

where $\gamma_{\mathrm{QPS}}(x_1)$ is the QPS amplitude in the wire point $x = x_1$. Setting now $\xi \to 0$ and making use of the second Eq. (19.96), in the case of a uniform wire with $\gamma_{\mathrm{QPS}}(x) = \gamma_{\mathrm{QPS}}$, we obtain

$$\hat{H}_{\mathrm{QPS}} = -\gamma_{\mathrm{QPS}}\int_0^X dx\cos\hat{\chi} \quad (19.103)$$

The above effective Hamiltonian for an isolated superconducting wire can be easily generalized provided it is necessary to include the effect of an external circuit. For the sake of definiteness and also for the purpose of our subsequent analysis, let us consider one example

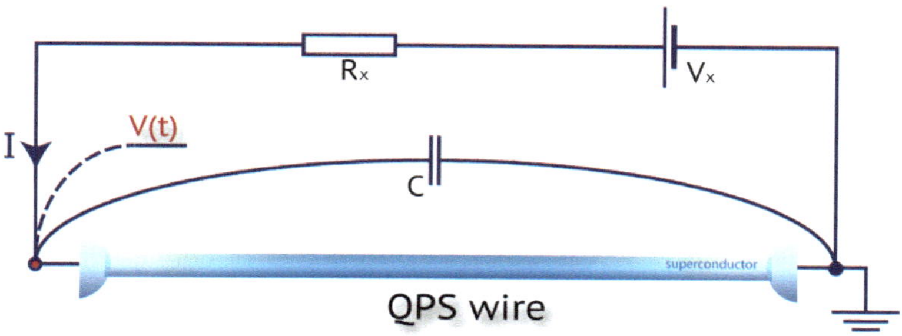

Figure 19.3 A superconducting circuit embedded in an external circuit.

of an external circuit displayed in Fig. 19.3. The system consists of an ultrathin superconducting wire and a capacitance C (which also includes the wire capacitance C_w) switched in parallel to this wire. The right end of the wire ($x = X$) is grounded as shown in the figure. The voltage $V(t)$ at its left end $x = 0$ can be measured by a detector. The whole system is biased by an external current $I = V_x/R_x$.

The system depicted in Fig. 19.3 can be described by means of the effective Hamiltonian in the mixed phase–charge representation:

$$\hat{H} = \hat{H}_{\text{wire}} + \frac{\hat{Q}_0^2}{2C} - \frac{I\hat{\varphi}}{2e} \tag{19.104}$$

As before, the term

$$\hat{H}_{\text{wire}} = \hat{H}_{\text{TL}} + \hat{H}_{\text{QPS}} \tag{19.105}$$

defined by Eqs. (19.98) and (19.103) accounts for the superconducting nanowire. The last two terms in Eq. (19.104) describe, respectively, the charging energy (which also includes the first term in the right-hand side of Eq. (19.97)) and the potential energy tilt produced by an external current I. The operator $\hat{\varphi} \equiv \hat{\varphi}(0)$ corresponds to the phase of the superconducting order parameter field $\Delta(x, t)$ at $x = 0$. Here we also set $\hat{\varphi}(X) \equiv 0$.

It is important to observe that the Hamiltonian for a superconducting wire in the presence of QPS (19.105) is exactly dual to that of a long Josephson junction [39] provided we interchange the terms describing charging and magnetic energies and map the QPS amplitude γ_{QPS} onto the Josephson coupling energy per unit length (the field $\chi(x, t)$ is dual to the Josephson phase difference across the barrier $\varphi(x, t)$ in this case). Accordingly, one model is mapped onto the other if we interchange the flux quantum and the Cooper pair charge $\Phi_0 \leftrightarrow 2e$, the Swihart velocity [370] and the Mooij–Schön velocity c_0, the Josephson penetration length λ_{J} (which controls the size of fluxons in long Josephson junctions), and the parameter $\lambda_{2e} = \sqrt{c_0/(4\pi\mu\gamma_{\mathrm{QPS}})}$ (which defines the size of $2e$ charge solitons in superconducting nanowires), and so on. Remarkably, all the phenomena known to exist in long Josephson junctions turn out to have their dual analogues in superconducting nanowires. These charge–phase duality arguments extend those for both ultrasmall Josephson junctions (outlined in Chapter 6) and QPS junctions [308] to ultrathin superconducting wires of an arbitrary length.

19.4.7 QPS Noise

In addition to non-vanishing resistance $R(I, T)$ analyzed above, QPS may cause voltage fluctuations in superconducting nanowires. Equilibrium voltage noise in such nanowires can be viewed as a direct consequence of the fluctuation-dissipation theorem (FDT): As soon as the current flow in our system becomes dissipative, equilibrium voltage fluctuations should inevitably be expected. Below we are going to demonstrate that along with such fluctuations, QPS in superconducting nanowires can also produce non-equilibrium *shot noise* of the voltage.

Note that the possibility for quantum shot noise to occur in superconducting nanowires is not immediately obvious. Indeed, as we have already discussed in the previous chapters, two key prerequisites of shot noise are (i) the presence of discrete charge carriers (e.g., electrons) in the system and (ii) scattering of such carriers at any kind of imperfections. In the absence of such scattering, the corresponding Fano factor vanishes and no shot noise may occur. Although discrete charge carriers (Cooper pairs) are certainly present in superconducting nanowires, they form a superconducting condensate flowing along the wire without any scattering. For this reason, the whole issue appears nontrivial and requires thorough investigation [360–363].

Let us again consider the system depicted in Fig. 19.3 and described by the Hamiltonian (19.104). In order to proceed, we will make use of the Keldysh path integral technique. Accordingly, our variables of interest are now defined on the forward and backward time branches of the Keldysh contour, i.e., we have $\varphi_{\mathrm{F,B}}(t)$ and $\chi_{\mathrm{F,B}}(x,t)$. As usual, it is convenient to also introduce the "classical" and "quantum" variables, respectively, $\varphi^{+}(t) = (\varphi_{\mathrm{F}}(t)+\varphi_{\mathrm{B}}(t))/2$ and $\varphi^{-}(t) = \varphi_{\mathrm{F}}(t) - \varphi_{\mathrm{B}}(t)$ (and similarly for the χ-fields). Making use of the Josephson relation between the voltage and the phase, one can formally express the expectation value of the voltage operator across the superconducting wire in the form

$$\langle V(t_1)\rangle = \frac{1}{2e}\left\langle \dot{\varphi}^{+}(t_1)e^{\mathrm{i}S_{\mathrm{QPS}}}\right\rangle_0 \tag{19.106}$$

where

$$S_{\mathrm{QPS}} = -2\gamma_{\mathrm{QPS}}\int dt \int_0^X dx \sin(\chi^{+})\sin(\chi^{-}/2) \tag{19.107}$$

and

$$\langle\ldots\rangle_0 = \int \mathcal{D}^2\varphi(t)\mathcal{D}^2\chi(x,t)(\ldots)e^{\mathrm{i}S_0[\phi,\chi]} \tag{19.108}$$

implies averaging with the Keldysh effective action S_0 corresponding to the Hamiltonian $\hat{H}_0 = \hat{H} - \hat{H}_{\mathrm{QPS}}$. Analogously, for the voltage–voltage correlator

$$\langle V(t_1)V(t_2)\rangle = \frac{1}{2}[\langle \hat{V}(t_1), \hat{V}(t_2)\rangle + \langle \hat{V}(t_2), \hat{V}(t_1)\rangle] \tag{19.109}$$

one has

$$\langle V(t_1)V(t_2)\rangle = \frac{1}{4e^2}\left\langle \dot{\varphi}^{+}(t_1)\dot{\varphi}^{+}(t_2)e^{\mathrm{i}S_{\mathrm{QPS}}}\right\rangle_0 \tag{19.110}$$

Higher voltage correlators are defined similarly.

Equations (19.106) and (19.110) are formally exact expressions, which we are now going to evaluate. For this purpose, we will employ the regular perturbation theory in γ_{QPS} regarded as a small parameter of our theory. In the zero order in γ_{QPS}, the problem is described by the quadratic (in both φ and χ) Hamiltonian $\hat{H}_0$ and all averages can be handled exactly with the aid of the Green functions

$$\begin{aligned} G^R_{ab}(X,X') &= -\mathrm{i}\langle a^{+}(X)b^{-}(X')\rangle, \\ G^K_{ab}(X,X') &= -\mathrm{i}\langle a^{+}(X)b^{+}(X')\rangle \end{aligned} \tag{19.111}$$

where $a(X)$ and $b(X)$ stand for one of the fields $\varphi(t)$ and $\chi(x, t)$. As both these fields are real, the advanced and retarded Green functions obey the condition $G^A_{ab}(\omega) = G^R_{ba}(-\omega)$. With this in mind, the Keldysh function G^K can be expressed in the form

$$G^K_{ab}(\omega) = \frac{1}{2}\coth\left(\frac{\omega}{2T}\right)\left(G^R_{ab}(\omega) - G^R_{ba}(-\omega)\right) \qquad (19.112)$$

The Green functions for the system displayed in Fig. 19.3 can be evaluated explicitly [360]. For instance, for the function $G^R_{\varphi\varphi}$, we obtain

$$G^R_{\varphi\varphi}(\omega) = \frac{1}{\frac{\omega^2}{2E_C} + \frac{i\omega}{4e^2R_x} - \frac{\omega\mu}{\pi}\cot\left(\frac{\omega X}{c_0}\right)} \qquad (19.113)$$

The corresponding expressions for $G^R_{\varphi\chi}$ and $G^R_{\chi\chi}$ turn out somewhat lengthy. In order to simplify them, it is useful to bear in mind that due to momentum conservation, plasmons in our system can only be created in pairs with the total zero momentum. All plasmons moving toward the grounded end of the wire (see Fig. 19.3) eventually disappear and never pop up again, while excitations moving in the opposite direction produce voltage fluctuations measured by a detector. Then in the long wire limit, the general expressions for $G^R_{\varphi\chi}$ and $G^R_{\chi\chi}$ reduce to

$$G^R_{\varphi\chi}(x;\omega) \simeq -\frac{2\mu e^{i\frac{\omega x}{c_0}}}{(\omega + i0)\left(\frac{\omega}{2E_C} + \frac{i\mu}{\pi}\right)}, \quad G^R_{\chi\chi}(x, x';\omega) \simeq -\frac{2\pi i\mu}{\omega + i0}e^{i\frac{\omega|x-x'|}{c_0}} \qquad (19.114)$$

In Eqs. (19.114), we also set $R_x \to \infty$ as requested in the current bias limit.

Let us now expand the general expressions (19.106) and (19.110) up to the second order in $\gamma_{\rm QPS}$. It is easy to demonstrate that linear in $\gamma_{\rm QPS}$ terms vanish identically in both expressions after averaging over the zero mode contained in the χ-field. In order to evaluate the terms $\sim\gamma^2_{\rm QPS}$, it is convenient to decompose the averages by means of the Wick theorem. Performing all necessary averages, after some manipulations, we arrive at the results expressed in terms of the Green functions (19.111) and illustrated graphically by means of the "candy" diagrams in Fig. 19.4.

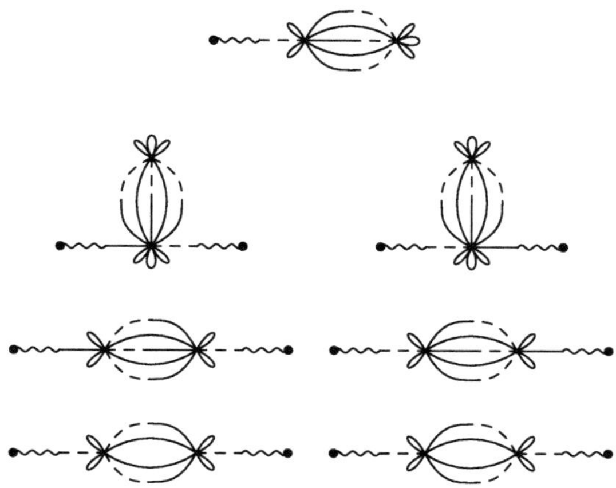

Figure 19.4 Candy-like diagrams that determine both average voltage (19.106) (upper diagram) and voltage noise (19.110) (six remaining diagrams) in the second order in γ_{QPS}. The fields φ_+, χ_+, and χ_- in the propagators (19.111) are denoted, respectively, by wavy, solid, and dashed lines.

For the average voltage, we obtain

$$\langle V \rangle = \frac{\mathrm{i}\gamma_{\text{QPS}}^2}{8e} \int_0^X dx \int_0^X dx' \left(\lim_{\omega \to 0} \omega G^R_{\varphi\chi}(x;\omega) \right) \times \left(\mathcal{P}_{x,x'}(-\Phi_0 I/c) - \mathcal{P}_{x,x'}(\Phi_0 I/c) \right) \tag{19.115}$$

where $\mathcal{P}_{x,x'}(\omega) = P_{x,x'}(\omega) + \bar{P}_{x,x'}(\omega)$

$$P_{x,x'}(\omega) = \int_0^\infty dt e^{\mathrm{i}\omega t} e^{\mathrm{i}\mathcal{G}(x,x';t,0)} \tag{19.116}$$

and

$$\mathcal{G}(x,x';t,0) = G^K_{\chi\chi}(x,x';t,0) - \frac{1}{2} G^K_{\chi\chi}(x,x;t,t) - \frac{1}{2} G^K_{\chi\chi}(x',x';0,0) + \frac{1}{2} G^R_{\chi\chi}(x,x';t,0) \tag{19.117}$$

Making use of the identity $\lim_{\omega\to 0} \omega G^R_{\varphi\chi}(x;\omega) = 2\pi\mathrm{i}$, we reduce Eq. (19.115) to the form (19.78), where we now identify $\Gamma_\pm$ as

$$\Gamma_\pm(I) = \frac{\gamma_{\text{QPS}}^2}{4} \int_0^X dx \int_0^X dx' \mathcal{P}_{x,x'}(\pm\Phi_0 I/c) \tag{19.118}$$

We immediately conclude that the combination (19.118) defines the quantum decay rate of the current state due to QPS. In Section 19.4.2, this rate was evaluated from the imaginary part of the free energy. Here we recovered the expressions for $\Gamma_{\pm}$ by means of the real-time technique without employing the ImF-method.

Combining the above results with the expressions for the Green functions (19.113) and (19.114) and keeping in mind the detailed balance condition

$$\mathcal{P}_{x,x'}(\omega) = e^{\frac{\omega}{T}}\mathcal{P}_{x,x'}(-\omega) \tag{19.119}$$

we obtain

$$\langle V\rangle = \frac{\Phi_0 X c_0 \gamma_{\rm QPS}^2}{4c}\varsigma^2\left(\frac{\Phi_0 I}{2c}\right)\sinh\left(\frac{\Phi_0 I}{2cT}\right) \tag{19.120}$$

where

$$\varsigma(\omega) = \tau_0^{\mu}(2\pi T)^{\mu-1}\frac{\Gamma\left(\frac{\mu}{2}-\frac{i\omega}{2\pi T}\right)\Gamma\left(\frac{\mu}{2}+\frac{i\omega}{2\pi T}\right)}{\Gamma(\mu)} \tag{19.121}$$

Equations (19.120) and (19.121) match with our result (19.79) already derived in Section 19.4.4 by means of the ImF-method. In particular, for the differential resistance R at $T\to 0$, we obtain

$$R = (\mu-1)\frac{\pi^2 c_0 X\gamma_{\rm QPS}^2\tau_0^{2\mu}\Phi_0^2}{2^{2\mu-2}\Gamma^2(\mu)c^2}|\Phi_0 I/c|^{2\mu-3}, \tag{19.122}$$

cf. Eq. (19.80).

Let us now turn to voltage fluctuations. Our perturbative analysis allows to recover three different contributions to the noise power spectrum [360], i.e.,

$$\mathcal{S}_{\Omega} = \int dt e^{i\Omega t}\langle V(t)V(0)\rangle = \mathcal{S}_{\Omega}^{(0)} + \mathcal{S}_{\Omega}^{r} + \mathcal{S}_{\Omega}^{a} \tag{19.123}$$

The first of these contributions $\mathcal{S}_{\Omega}^{(0)}$ has nothing to do with QPS and just defines equilibrium voltage noise for a transmission line. It reads

$$\mathcal{S}_{\Omega}^{(0)} = \frac{i\Omega^2\coth\left(\frac{\Omega}{2T}\right)}{16e^2}\left(G_{\varphi\varphi}^{R}(\Omega) - G_{\varphi\varphi}^{R}(-\Omega)\right) \tag{19.124}$$

The other two terms are due to QPS effects. The term $\mathcal{S}_{\Omega}^{r}$ is also proportional to $\coth\left(\frac{\Omega}{2T}\right)$ and contains the products of two retarded (advanced) Green functions:

$$\mathcal{S}_\Omega^r = \frac{\gamma_{\rm QPS}^2 \Omega^2 \coth\left(\frac{\Omega}{2T}\right)}{16e^2} \int_0^X dx \int_0^X dx' \text{Re}\left[G_{\varphi\chi}^R(x;\Omega)\right.$$
$$\left.\times(\mathcal{F}_{x,x'}(\Omega)G_{\varphi\chi}^R(x';\Omega) - \mathcal{F}_{x,x'}(0)G_{\varphi\chi}^R(x;\Omega))\right] \quad (19.125)$$

Here we denoted

$$\mathcal{F}_{x,x'}(\Omega) = -P_{x,x'}(\Omega + \Phi_0 I/c) - P_{x,x'}(\Omega - \Phi_0 I/c)$$
$$+ \bar{P}_{x,x'}(-\Omega + \Phi_0 I/c) + \bar{P}_{x,x'}(-\Omega - \Phi_0 I/c) \quad (19.126)$$

The remaining term $\mathcal{S}_\Omega^a$, in contrast, contains the product of one retarded and one advanced Green functions and scales with the combinations

$$\mathcal{C}_\pm = \coth\left(\frac{\Omega \pm \Phi_0 I/c}{2T}\right) - \coth\left(\frac{\Omega}{2T}\right)$$

as

$$\mathcal{S}_\Omega^a = \frac{\gamma_{\rm QPS}^2 \Omega^2}{32e^2} \int_0^X dx \int_0^X dx' G_{\varphi\chi}^R(x;\Omega) G_{\varphi\chi}^R(x';-\Omega) \quad (19.127)$$
$$\times \left[\sum_\pm \mathcal{C}_\pm \left(\mathcal{P}_{x,x'}(\Omega \pm \Phi_0 I/c) - \mathcal{P}_{x,x'}(-\Omega \mp \Phi_0 I/c)\right)\right].$$

Equations (19.123)–(19.127), together with the expressions for the Green functions (19.113) and (19.114), fully determine the voltage noise power spectrum of a superconducting nanowire in the perturbative in the QPS regime. At this point, we note that within this regime the difference between the correlator (19.109) and its irreducible version $\langle V(t)V(0)\rangle - \langle V(0)\rangle^2$ is of the order $\sim \gamma_{\rm QPS}^4$ and, hence, can be safely neglected.

In the zero-bias limit $I \to 0$, the term $\mathcal{S}_\Omega^a$ vanishes, and the equilibrium noise spectrum $\mathcal{S}_\Omega = \mathcal{S}_\Omega^{(0)} + \mathcal{S}_\Omega^r$ is determined from FDT. At nonzero bias values, the QPS noise turns non-equilibrium. In the zero-frequency limit $\Omega \to 0$, the terms $\mathcal{S}_\Omega^{(0)}$ and $\mathcal{S}_\Omega^r$ tend to zero, and the voltage noise $\mathcal{S}_{\Omega=0} \equiv \mathcal{S}_0$ is determined solely by $\mathcal{S}_\Omega^a$. Then from Eq. (19.127), we obtain

$$\mathcal{S}_0 = (\Phi_0/c)^2 (\Gamma_+(I) + \Gamma_-(I)) = (\Phi_0/c) \coth\left(\frac{\Phi_0 I}{2cT}\right) \langle V \rangle \quad (19.128)$$

where $\langle V\rangle$ is specified in Eqs. (19.78) and (19.120). Combining the result (19.128) with Eqs. (19.120) and (19.121), we find

$$\mathcal{S}_0 \propto \begin{cases} T^{2\mu-2}, & T \gg \Phi_0 I, \\ I^{2\mu-2}, & T \ll \Phi_0 I \end{cases} \tag{19.129}$$

At higher temperatures $T \gg \Phi_0 I/c$ (though still $T \ll \Delta_0$), Eq. (19.129) just describes equilibrium voltage noise $\mathcal{S}_0 = 2TR$ of a linear Ohmic resistor with $R \propto T^{2\mu-3}$. In the opposite low-temperature limit $T \ll \Phi_0 I/c$, it accounts for QPS-induced *shot noise* $\mathcal{S}_0 = (\Phi_0/c)\langle V\rangle$ obeying *Poisson statistics* with an effective "charge" equal to the flux quantum Φ_0.

This result sheds light on the physical origin of shot noise in superconducting nanowires: It is produced by coherent tunneling of magnetic flux quanta Φ_0 across the wire. In the dual picture, such flux quanta can be viewed as charged quantum particles passing through (and being scattered at) an effective spatially extended "tunnel barrier," which role is played by the nanowire.

Note that in Section 19.3.4, the result (19.34) analogous to Eq. (19.128) was derived for thermally activated phase slips. This similarity appears remarkable given a crucial physical difference between TAPS and QPS: The former can be regarded as classical (i.e., incoherent) and non-interacting objects, whereas the latter are fully coherent forming an interacting quantum gas.

Another interesting limiting case is that of sufficiently high frequencies and/or long wires $c_0/X \ll \Omega \ll \Delta_0$. In this limit, we obtain

$$S_\Omega^{(0)} = \frac{\mu}{8\pi e^2}\frac{\Omega\coth\left(\frac{\Omega}{2T}\right)}{(\Omega/2E_C)^2 + (\mu/\pi)^2} \tag{19.130}$$

This contribution is independent of the wire length X. At low T and $\Omega/\mu \gtrsim E_C = e^2/2C$, we have $\mathcal{S}_\Omega^{(0)} \propto 1/\Omega$, i.e., the wire may generate $1/f$ voltage noise. Evaluating the QPS terms $\mathcal{S}_\Omega^r$ and $\mathcal{S}_\Omega^a$, we observe that the latter scales linearly with the wire length X, whereas the former does not. Hence, the term $\mathcal{S}_\Omega^r$ can be safely neglected in the long wire limit. For the remaining QPS term $\mathcal{S}_\Omega^a$, we get

$$\mathcal{S}_\Omega^a = \frac{X\mu^2 c_0\gamma_{QPS}^2}{8e^2}\left[\varsigma\left(\frac{\Phi_0 I}{2c} - \Omega\right) - \varsigma\left(\frac{\Phi_0 I}{2c} + \Omega\right)\right] \times \frac{\sinh\left(\frac{\Phi_0 I}{2cT}\right)\varsigma\left(\frac{\Phi_0 I}{2c}\right)}{\left((\Omega/2E_C)^2 + (\mu/\pi)^2\right)\sinh\left(\frac{\Omega}{2T}\right)} \tag{19.131}$$

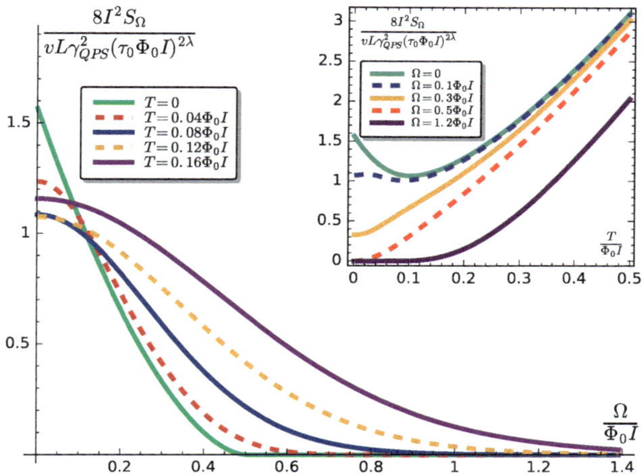

Figure 19.5 The frequency dependence of the QPS noise spectrum S_Ω (19.131) at $\mu = 2.7$, large E_C and different T in the long wire limit. The inset shows S_Ω as a function of T. In both these graphs, we identify $\lambda = \mu$, $v = c_0$ and $L = X$.

At $T \to 0$ from Eq. (19.131), we find

$$\mathcal{S}_\Omega^a \propto \begin{cases} I^{\mu-1}(I - 2\Omega c/\Phi_0)^{\mu-1}, & \Omega < \Phi_0 I/2c, \\ 0, & \Omega > \Phi_0 I/2c \end{cases} \tag{19.132}$$

This result can be interpreted as follows. At $T = 0$, each QPS event excites (at least) two plasmons with total energy $E = \Phi_0 I/c$ propagating in the opposite directions along the wire. One plasmon (with energy $E/2$) gets dissipated at the grounded end of the wire, while another one (also with energy $E/2$) reaches its opposite end causing voltage fluctuations (emits a photon) with frequency Ω measured by a detector. Clearly, at $T = 0$ this process is only possible at $\Omega < E/2$ in the agreement with Eq. (19.132).

The result (19.131) is also illustrated in Fig. 19.5. At sufficiently small Ω (we still keep $\Omega \gg c_0/X$), one observes a non-monotonous dependence of $\mathcal{S}_\Omega$ on T, which is a direct consequence of quantum coherent nature of QPS noise. Note that at nonzero T, the expression (19.131) *does not* coincide with the zero-frequency result (19.128) even in the limit $\Omega \to 0$. The point here is that before taking the zero-frequency limit in Eq. (19.131), one should formally set $X \to \infty$.

Then one gets

$$\mathcal{S}^a_{\Omega\to 0}(I) = -\frac{1}{2}c_0 X T \gamma^2_{\rm QPS}(\Phi_0/c)^2 \varsigma\left(\frac{\Phi_0 I}{2c}\right)\varsigma'\left(\frac{\Phi_0 I}{2c}\right)\sinh\left(\frac{\Phi_0 I}{2cT}\right). \tag{19.133}$$

Comparing this expression with Eq. (19.128) (combined with Eqs. (19.120) and (19.121)) obtained in the true zero-frequency limit (meaning that the limit $\Omega \to 0$ was taken prior to sending the wire length X to infinity), one can establish the following interesting identity [361]

$$\mathcal{S}_0(I, T) - \mathcal{S}^a_{\Omega\to 0}(I, T) = 2T R(I, T) \tag{19.134}$$

It follows immediately that both expressions for the zero-frequency shot noise power (19.133) and (19.128) coincide only at $T = 0$, in which case one finds

$$\mathcal{S}_0(I, 0) = \mathcal{S}^a_{\Omega\to 0}(I, 0) = \frac{\pi^2 c_0 X \gamma^2_{\rm QPS}\tau_0^{2\mu}\Phi_0^2}{2^{2\mu-1}\Gamma^2(\mu)c^2}|\Phi_0 I/c|^{2\mu-2} \tag{19.135}$$

At any nonzero T, the noise power $\mathcal{S}_0(I, T)$ (19.128) exceeds one in Eq. (19.133) and, in contrast to the latter, increases monotonously with temperature.

Finally, we point out that the perturbative in $\gamma_{\rm QPS}$ approach employed here is fully justified for not-too-thin wires with $\mu > \mu^*$ (see Eq. (19.74)). In wires with $\mu < \mu^*$ characterized by unbound QPS-anti-QPS pairs, the fugacity y (or $\gamma_{\rm QPS}$) gets effectively renormalized to higher values (cf. Section 19.4.3) and, hence, the perturbation theory eventually becomes obsolete. However, even in this case, our results may still remain applicable at sufficiently high temperature, frequency, and/or current values.

19.5 Key Experiments

As we already pointed out, soon after the development of the LAMH theory, two groups reported experiments aimed at verification of the TAPS model. In these early studies on one-dimensional superconductivity [288, 316], metallic whiskers with characteristic diameter $\sim$1 μm and lengths up to $\sim$1 mm were employed.

An example of experimental $R(T)$ dependence measured on tin whisker is shown in Fig. 19.6. One observes a very good quantitative

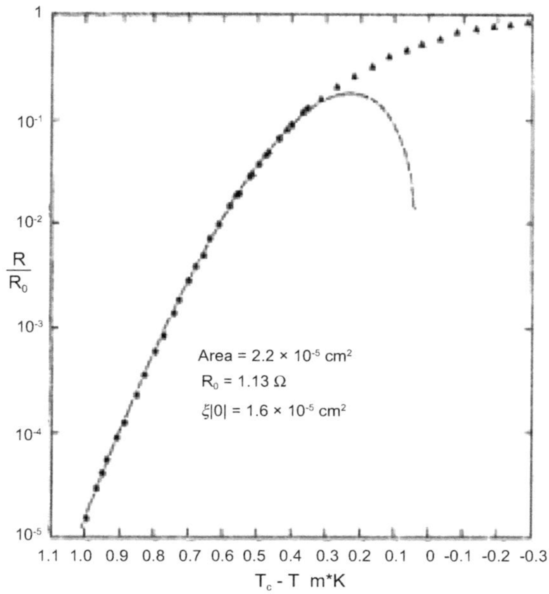

Figure 19.6 $R(T)$ dependence for a tin whisker [288]. Solid symbols are the experimental data, dashed line represents the fit to the exponential dependence defined, e.g., by Eq. (19.33). Reprinted with permission from Ref. [288], Copyright 1970, by the American Physical Society.

agreement between theory and experiment. The superconducting transition $R(T)$ observed in these experiments was very steep: The sample resistance dropped by five orders of magnitude within the temperature range $\delta T \sim 1$ mK below T_C. This sharp superconducting transition is a consequence of rather large effective diameter values $\sqrt{s} \simeq 0.5$ μm and extremely high homogeneity of single crystals employed in these experiments.

As the wire thickness decreases, the superconducting transition gets broader, exactly as it was detected, e.g., in the experiments [139–141]. These experiments clearly demonstrated a notable resistivity of ultrathin superconducting wires far below T_C. Moreover, for the thinnest nanowires, no resistance decrease at sufficiently low temperatures was observed at all. Such "foot" or "knee" structures on the I–V curves could not be adequately interpreted within the TAPS theory and could only be attributed to the effect of QPS.

It should be noted, however, that an unambiguous interpretation of the observations [139–141] in terms of QPS was questioned because of possible granularity of the samples used in those experiments. Later more detailed experiments on ultrathin superconducting wires were performed by different groups. At present there exist numerous experimental observations confirming the existence of QPS effects in superconducting nanowires. A detailed discussion of the present experimental situation in the field is provided in review papers [24, 55] and in the book [56]. Here we only mention several key experiments.

A clear manifestation of a crucial role played by quantum fluctuations in superconducting nanowires was provided in the experiments by Tinkham et al. [57, 260]. These authors developed a novel technique that allowed them to fabricate sufficiently uniform superconducting wires considerably thinner than 10 nm with lengths ranging between $\sim$100 nm and 1 μm. No sign of superconductivity was found in thinner wires [57] even well below the bulk critical temperature T_C. It turned out that all these non-superconducting wires had the normal state resistance R_N above the superconducting quantum resistance unit R_Q, while R_N for the samples showing superconducting-like behavior was lower than R_Q. This observation allowed the authors to conjecture that a dramatic difference in the behavior of these two groups of samples (otherwise having similar parameters) can be due to the dissipative Schmid-like QPT discussed in Section 19.4.3. This conjecture was later reconsidered in the light of the observations of the superconducting behavior in samples with values of R_N as high as 40 kΩ $\gg R_Q$ [260].

In Fig. 19.7, the resistance values of more than 20 different nanowires [57, 260] are presented versus the inverse normal resistance per unit length $X/R_N \propto s$. In Fig. 19.7(a), one observes a sharp crossover between normal and superconducting behavior at wire diameters $\sqrt{s} \sim 10$ nm. This crossover could be interpreted as an indication to the BKT-like QPT controlled by the parameter $\mu \propto \sqrt{s}$; see Section 19.4.3. Note, however, that all wires [57, 260] were quite short. Hence, this QPT should inevitably be significantly broadened by finite size effects. Re-plotting the same data on a semilog scale (Fig. 19.7(b)), one indeed observes a rather broad distribution of measured resistances, which—despite some

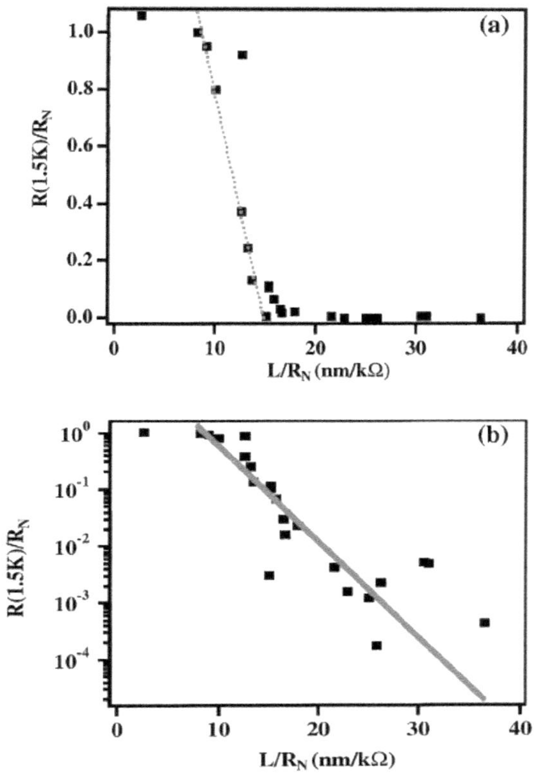

Figure 19.7 Resistance at 1.5 K (normalized to normal state resistance R_N) as a function of L/R_N. (a) Linear plot. The dotted line is a guide to the eye. (b) Semilog plot with an exponential fit. Slope of the fitted line is 0.39 kΩ/nm. Reprinted with permission from Ref. [260], Copyright 2001, by the American Physical Society.

scatter—can be fitted to the linear dependence on the wire cross section s. This fit is highly suggestive of the crossover scenario although it cannot yet rule out the BKT-like QPT either.

In order to discriminate between these two options, it is necessary to analyze the temperature dependence of the resistance $R(T)$. According to our theory, in the linear regime the wire resistance is defined by Eq. (19.80). The dependence $R \propto y^2$ comes from pairs of QPS events where the fugacity (19.71) depends on

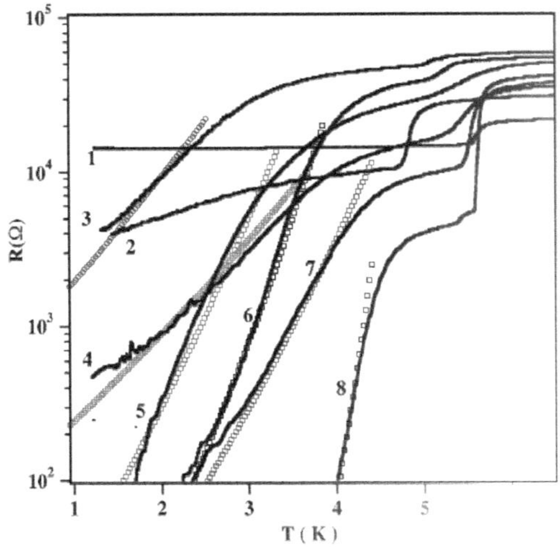

Figure 19.8 Superconducting transitions of "long" *MoGe* nanowires on top of insulating carbon nanotube used as the substrate [57, 260]. The samples' normal state resistances and lengths are 1: 14.8 kΩ, 135 nm; 2: 10.7 kΩ, 135 nm; 3: 47 kΩ, 745 nm; 4: 17.3 kΩ, 310 nm; 5: 32 kΩ, 730 nm; 6: 40 kΩ, 1050 nm; 7: 10 kΩ, 310 nm; 8: 4.5 kΩ, 165 nm. Symbols stand for calculations using Eq. (19.136) with the single numerical coefficient $A = 0.7$. The critical temperature T_C and the dirty-limit coherence length $\xi(0)$ used as fitting parameters for samples 3–8 are 3: 5.0 K, 8 nm; 4: 6.4 K, 8.5 nm; 5: 4.6 K, 8.9 nm; 6: 4.8 K, 8.9 nm; 7: 5.6 K, 11.9 nm; 8: 4.8 K, 8.5 nm [24].

temperature via $S_{\text{core}} \propto \Delta_0^{1/2}(T)$. An additional (weak) power-law dependence $\propto T^{2\mu-3}$ enters because of interactions between QPS in space–time. In order to fit the experimental data, we will ignore this power-law dependence and use the simplified formula

$$R(T) = b\frac{\Delta_0(T)S_{\text{core}}^2 X}{\xi(T)}\exp(-2S_{\text{core}}), \qquad (19.136)$$

with S_{core} defined in Eq. (19.48) and b being an unimportant constant. The results of the fit to Eq. (19.136) [24] are presented in Fig. 19.8. We observe a very good agreement between theory and experiment achieved with the same value $A \simeq 0.7$ for all six samples. Thus, the detected temperature dependence of the resistance of ultrathin *MoGe* wires [57, 260] is indeed determined by QPS effects, and the observed sharp transition between normal and superconducting behavior is most likely a thickness-governed

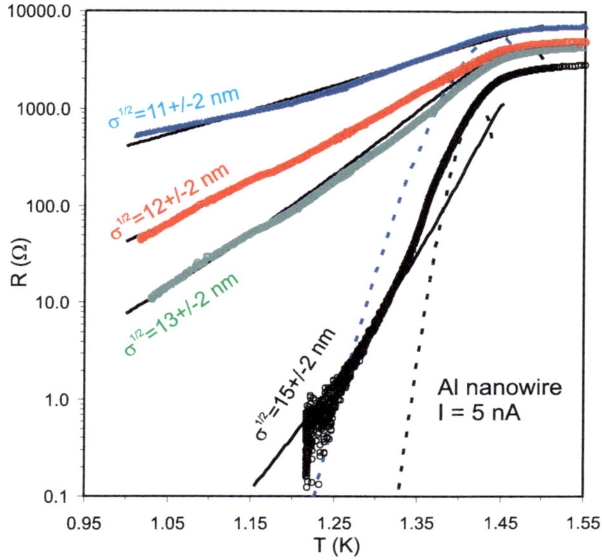

Figure 19.9 $R(T)$ curves for the thinnest samples obtained by progressive diameter reduction for the same aluminum nanowire with length $X = 10$ μm. The TAPS model fitting is shown with dashed lines for 11 and 15 nm samples with the best fit mean free path $l = 3$ and 10 nm, correspondingly, $T_C = 1.46$ K and critical magnetic field $B_c(0) = 10$ mT. Fits to Eq. (19.136) are shown by solid lines. For 11, 12, 13, and 15 nm wires, the fit parameters are $A \simeq 0.1$; T_c: 1.64 K, 1.52 K, 1.47 K, and 1.47 K; mean free path l : 7.5 nm, 8.2 nm, 9.5 nm, and 9.5 nm; normal state resistance R_N : 7200 kΩ, 5300 kΩ, 4200 kΩ, and 2700 kΩ. Reprinted with permission from Ref. [430], Copyright 2008, by the American Physical Society.

crossover between the regimes of, respectively, large and small QPS rates γ_{QPS}.

Another set of experiments [429, 430] provides an independent experimental confirmation of the QPS theory described above. These experiments were performed on several sets of aluminum nanowires with length X equal to 1, 5, and 10 μm and progressive diameter reduction in *one and the same* sample was carried out. For larger diameters $\sqrt{s} \gtrsim 20$ nm, the observed "sharp" dependence $R(T)$ can be qualitatively described by the TAPS mechanism. As the wire diameter is reduced further, deviations from the TAPS behavior become obvious; see Fig. 19.9.

A detailed comparison of the data with the QPS theory is presented in Fig. 19.9. We observe that as the sample diameter decreases, the $R(T)$ curves become progressively broader, exactly as it is predicted by the QPS theory. Here the temperature dependence is merely determined by the QPS fugacity (19.71) via $S_{\text{core}} \propto \Delta_0^{1/2}(T)$. The fits demonstrate a good agreement between theory and experiment, thus confirming the important role of QPS effects in the thinnest wires. For all aluminum nanowires, the best-fit value for the parameter A was $A \simeq 0.1$ for all curves in Fig. 19.9. This value turns out to be somewhat smaller than that extracted from the fits to the experimental data [260], which might be due to different geometry and degree of inhomogeneity of the samples used in these two experiments.

A similarly good agreement between theory and experiments was later obtained in experiments with Ti nanowires [265]. The main idea of this experiment was similar to that of the work [429, 430], and the fit of the obtained data points to Eq. (19.136) allowed to extract the value A ranging from 0.14 to 0.46. Such a scatter is again most likely due to some inhomogeneities in the samples [265]. Yet another impressive experimental confirmation of the above QPS theory was reported in the paper [40] where narrow aluminum constrictions of length $\gtrsim \xi$ and widths ranging from about 90 nm down to 5 nm were fabricated by controlled electromigration technique. In this experiment, the fit of the data to Eq. (19.136) yields $A \approx 0.55$.

To complete our brief overview, let us also mention about an experimental work [26, 326] on quantum tunneling of magnetic flux quanta demonstrating a coherent nature of QPS in superconducting nanowires and nanorings. We will return to this issue in the next chapter. Yet another experimental demonstration of quantum coherence of QPS is the observation of Coulomb blockade and Bloch oscillation features in Ti nanowires [266].

Chapter 20

Persistent Currents in Superconducting Nanorings

20.1 Persistent Currents in Bulk Superconducting Rings

It is well known that superconducting rings pierced by external magnetic flux Φ_x develop circulating persistent currents, which never vanish. This phenomenon is a fundamental consequence of macroscopic phase coherence of Cooper pair wave functions. In bulk superconducting rings (Fig. 20.1), it is responsible for Meissner screening of the magnetic field, which can penetrate inside the ring only at a typical length of order $\lambda_{\rm L}$. For sufficiently thick rings, fluctuations of the amplitude of the order parameter field are negligible, i.e., in this case, we have $|\Delta(x, \tau)| = \Delta_0$ and quantum phase slip (QPS) effects analyzed in Chapter 19 can be ignored. In addition, we will assume that temperature is sufficiently low, $T \ll \Delta_0$, in which case overgap quasiparticles are practically irrelevant and, hence, will be disregarded further below in this section.

Under these conditions, the grand partition function of a superconducting ring $\mathcal{Z}$ can be expressed via the path integral over the superconducting phase variable $\varphi(x, \tau)$. Making use of the

Dissipative Quantum Mechanics of Nanostructures: Electron Transport, Fluctuations, and Interactions
Andrei D. Zaikin and Dmitry S. Golubev

ISBN 978-981-4774-50-5 (Hardcover), 978-0-429-29823-3 (eBook)
www.jennystanford.com

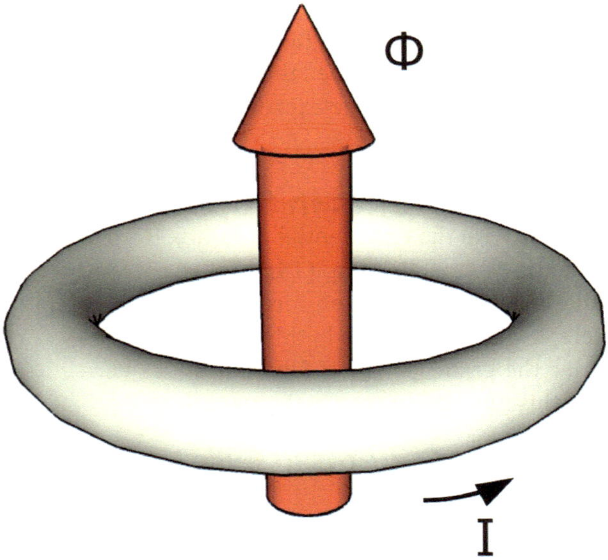

Figure 20.1 Bulk superconducting ring threaded by the magnetic flux Φ.

results derived in Chapter 18 for superconducting wires (cf., e.g., Eq. (18.97)), we obtain

$$\mathcal{Z} = \sum_{m,n} \int \mathcal{D}\varphi e^{-\frac{\mu}{2\pi} \int dx d\tau \left(c_0 (\partial_x \varphi)^2 + c_0^{-1} (\partial_\tau \varphi)^2\right)} \tag{20.1}$$

where the Mooij–Schön mode velocity c_0 and the effective coupling parameter μ are defined, respectively, in Eqs. (18.75) and (19.40).

In order to account for the ring geometry and for the magnetic flux inside the ring, it is necessary to impose proper boundary conditions for the path integral (20.1). These boundary conditions should include periodicity of the phase variable φ both in imaginary time and in space (after each full rotation around the ring). Also, it is necessary to keep in mind that the phase is defined up to $2\pi m$, where m is the winding number. Finally, the boundary conditions should be sensitive to the magnetic flux Φ_x piercing the ring. Putting all these requirements together, we get

$$\begin{aligned} \varphi(x, 0) &= \varphi(x, \beta) + 2\pi m, \\ \varphi(X, \tau) &= \varphi(0, \tau) + 2\pi(\phi_x + n) \end{aligned} \tag{20.2}$$

Here $X = 2\pi R$ is the ring perimeter, $\phi_x = \Phi/\Phi_0$, and, as before, Φ_0 is the superconducting flux quantum. Combining Eqs. (20.1) and (20.2), after a simple calculation one finds

$$\mathcal{Z}_{\phi_x} = \sum_{p=-\infty}^{\infty} \exp(-E_p(\phi_x)/T) = \sqrt{\frac{2\pi T}{E_{\rm R}}}\vartheta_3(\pi\phi_x, e^{-\frac{2\pi^2 T}{E_{\rm R}}}) \quad (20.3)$$

where

$$E_p(\phi_x) = \frac{E_{\rm R}}{2}(p+\phi_x)^2, \quad E_{\rm R} = \frac{\pi^2 N_0 D\Delta_0 s}{R} \quad (20.4)$$

and $\vartheta_3(u, q)$ is the third Jacobi theta function. It is useful to note that the partition function of the form (20.3), (20.4) was repeatedly encountered in our previous considerations: It is exactly equivalent to that of a Josephson junction in the absence of Cooper pair tunneling (cf. Eq. (6.48)), of a single-electron box in weak tunneling limit (cf. Eqs. (7.82) and (7.84)) and of a quantum particle on a ring in the absence of an external potential (cf. Eqs. (8.29) and (8.30)).

In the limit $T \to 0$, the flux-dependent supercurrent $I(\phi_x)$ in the ring is defined as a derivative of the ground-state energy $E(\Phi_x) = \min_p E_p(\phi_x)$ with respect to the flux, i.e.,

$$I(\phi_x) = c\left(\frac{\partial E(\Phi_x)}{\partial \Phi_x}\right) \quad (20.5)$$

The ground-state energy of the ring $E(\phi_x)$ is a periodic function of the magnetic flux Φ_x with the period Φ_0. Hence, the current I is also periodic in Φ_x. From Eqs. (20.4) and (20.5), we immediately recover the familiar sawtooth dependence

$$I(\phi_x) = \frac{eE_{\rm R}}{\pi}\frac{\partial}{\partial\phi_x}\min_n\frac{(n+\phi_x)^2}{2} \quad (20.6)$$

This behavior is illustrated in Fig. 20.2.

The supercurrent flowing in the ring can also be expressed via the phase variable as

$$I(\tau) = \frac{eE_{\rm R}}{2\pi^2}\left(\varphi(X,\tau) - \varphi(0,\tau)\right) \quad (20.7)$$

Combining this relation with the second Eq. (20.2) in the zero-temperature limit, we again arrive at the result (20.6) for the expectation value of the current $I(\phi_x) = \langle I(\tau)\rangle$. Furthermore, employing Eq. (20.7), it is straightforward to demonstrate that all

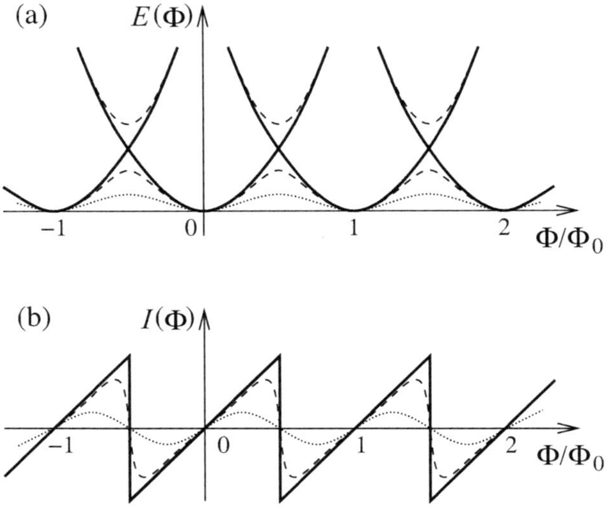

Figure 20.2 The energy (a) and persistent current (b) for a superconducting nanoring as functions of magnetic flux without fluctuations, with weak and with strong QPS effects shown, respectively, by solid, dashed, and dotted lines.

current correlators in our problem are time independent and can be expressed through the derivatives of the theta function $\vartheta_3^{(k,0)}$ as [357]

$$\langle \hat{I}(\tau_1) \dots \hat{I}(\tau_k) \rangle = \left(\frac{eE_{\mathrm{R}}}{\pi} \right)^k \frac{\sum\limits_{n=-\infty}^{\infty} (n+\phi_x)^k e^{-\beta E_n(\phi_x)}}{\sum\limits_{n=-\infty}^{\infty} e^{-\beta E_n(\phi_x)}}$$

$$= \sum_{n=0}^{2n \le k} \frac{(-eT)^k k!}{n!(k-2n)!} \left(\frac{E_{\mathrm{R}}}{2\pi^2 T} \right)^n \frac{\vartheta_3^{(k-2n,0)}(\pi\phi_x, e^{-\frac{2\pi^2 T}{E_{\mathrm{R}}}})}{\vartheta_3(\pi\phi_x, e^{-\frac{2\pi^2 T}{E_{\mathrm{R}}}})} \qquad (20.8)$$

At $T \to 0$, this expression obviously yields $\langle I^k(\tau) \rangle = \langle I(\tau) \rangle^k$ for all integer $k \ge 0$, implying that no supercurrent fluctuations can occur in the ground state of bulk superconducting rings. This conclusion is in line with our general considerations in Section 8.3, demonstrating that no persistent current fluctuations can occur at $T = 0$ provided the current operator commutes with the total Hamiltonian of the system. Bulk superconducting rings represent just one example of this physical limit.

Supercurrent fluctuations in such rings, however, do occur at nonzero temperatures. Restricting our analysis to the thermal noise of the supercurrent and introducing the noise power spectrum

$$S_{\omega} = \int dt e^{i\omega t}(\langle \hat{I}(t)\hat{I}(0) + \hat{I}(0)\hat{I}(t)\rangle - \langle \hat{I}\rangle^2), \tag{20.9}$$

where $\hat{I}(t) = e^{it\hat{H}}\hat{I}e^{-it\hat{H}}$ is the current operator in the Heisenberg picture, we proceed in exactly the same way as in Chapter 8 (see Section 8.3) and obtain

$$S_{\omega} = 4\pi\mathcal{P}\delta(\omega), \quad \mathcal{P} = \langle I^2(\tau)\rangle - \langle I(\tau)\rangle^2 \tag{20.10}$$

With the aid of Eq. (20.8), one finds [357]

$$\mathcal{P} = e^2T^2\left(\frac{E_{\rm R}}{\pi^2 T} + \frac{\vartheta_3^{(2,0)}(\pi\phi_x, e^{-\frac{2\pi^2 T}{E_{\rm R}}})}{\vartheta_3(\pi\phi_x, e^{-\frac{2\pi^2 T}{E_{\rm R}}})}\right) - e^2T^2\left(\frac{\vartheta_3^{(1,0)}(\pi\phi_x, e^{-\frac{2\pi^2 T}{E_{\rm R}}})}{\vartheta_3(\pi\phi_x, e^{-\frac{2\pi^2 T}{E_{\rm R}}})}\right)^2 \tag{20.11}$$

In the low- and high-temperature limits, this expression reduces to

$$\mathcal{P} \approx \begin{cases} \frac{2e^2E_{\rm R}^2}{\pi^2}e^{-\frac{E_{\rm R}}{2T}}\cosh\left(\frac{\pi I(\phi_x)}{eT}\right), & T \ll E_{\rm R}, \\ \frac{e^2E_{\rm R}T}{\pi^2} - 8e^2T^2e^{-\frac{2\pi^2 T}{E_{\rm R}}}\cos(2\pi\phi_x), & T \gg E_{\rm R}, \end{cases} \tag{20.12}$$

where $I(\phi_x)$ is defined in Eq. (20.6). The dependence (20.11) is depicted in Fig. 20.3 for different values of the magnetic flux. We observe that at sufficiently low temperatures, the magnitude of supercurrent fluctuations can be tuned by the external flux ϕ_x, hence indicating coherent nature of such fluctuations. At higher temperatures, quantum coherence is destroyed and $\mathcal{P}(T) \propto T$ becomes practically independent of ϕ_x.

The above picture remains applicable as long as the ring is sufficiently thick, and one can essentially ignore quantum fluctuations of the order parameter field. However, as we have already demonstrated in previous chapters, upon decreasing the thickness of a superconducting wire $\sqrt{s}$, typically down to values in the 10 nm range, one eventually reaches the new regime in

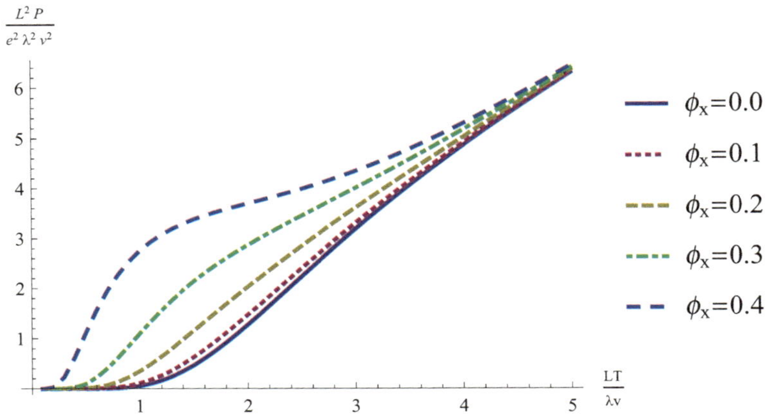

Figure 20.3 The function $\mathcal{P}(T)$ describing thermal fluctuations of the supercurrent in superconducting rings at different values of the magnetic flux ϕ_x.

which quantum fluctuations gain importance and essentially modify the low-temperature behavior of superconducting nanorings. This regime of strong quantum fluctuations of the superconducting order parameter will be considered in the next section.

20.2 Quantum Phase Slip Rings

As we have already learned in Chapter 19, at low T the most important fluctuations in superconducting nanowires are QPS. Each phase slip event implies transfer of one flux quantum Φ_0 through the wire out of or into the ring and, hence, yields the change in the total flux inside the ring by exactly the same amount. In other words, each QPS yields a jump between two neighboring energy states E_p (20.3) (i.e., between two neighboring parabolas in Fig. 20.2). As a result of such jumps, the flux inside the ring fluctuates, its average value $\langle\Phi(\Phi_x)\rangle$ decreases, and so does the persistent current I. One can also anticipate that the magnitude of this effect should increase with the ring perimeter $X = 2\pi R$. This is because the total QPS tunneling

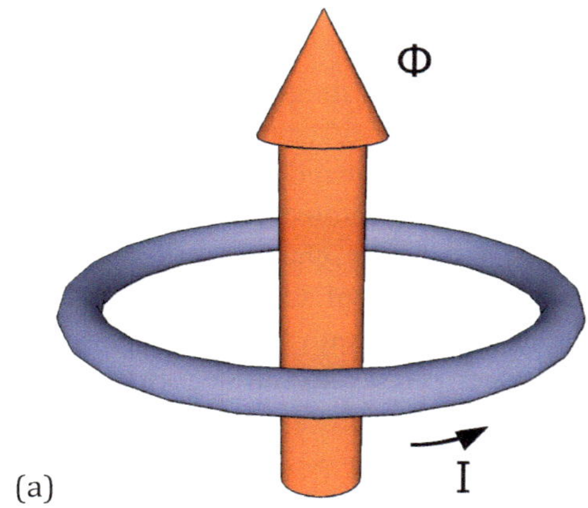

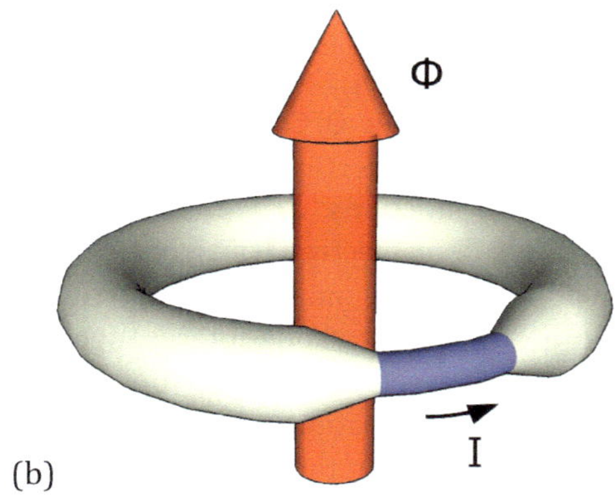

Figure 20.4 a) QPS ring, i.e., an ultrathin superconducting ring threaded by an external magnetic flux. (b) QPS junction embedded in a thick superconducting ring.

amplitude for the ring

$$\Gamma_{\text{QPS}} = 2\pi R \gamma_{\text{QPS}} \tag{20.13}$$

(cf. Eqs. (19.53) and (19.65)) increases linearly with R, i.e., the bigger the ring perimeter, the higher the probability for QPS to occur anywhere along the ring. Below in this section, we will address QPS effects in ultrathin superconducting rings (Fig. 20.4a), which from now on we will denote as *quantum phase slip rings*.

Before turning to technical details, it is instructive to point out a formal equivalence between the phenomenon discussed here and charging effects in ultrasmall Josephson junctions already analyzed in Chapter 6. Indeed, the energy states of our ring (20.3) driven by the (normalized) external magnetic flux ϕ_x (see Fig. 20.2) are fully analogous to such states of a capacitor attached to a Josephson junction driven by the (normalized) gate charge $Q_x/2e$; see Chapter 6. Furthermore, we will demonstrate that there exists a direct analogy between QPS events changing the flux inside the ring by Φ_0 and tunneling events of single Cooper pairs changing the capacitor charge by $2e$. This equivalence is reminiscent of the well-known duality between phase and charge representations, which we already discussed in Chapter 6.

Note that essentially the same behavior is expected in a somewhat different situation of rings consisting of thicker and thinner parts, as shown in Fig. 20.4b. In this case, QPS effects are negligible in a thicker part of the ring and they can only occur in its thinner part, i.e., in a QPS junction. One again observes the duality between the properties of these systems and those of mesoscopic Josephson junctions: As we already argued in Section 19.4.6, any known result on electron transport in circuits containing Josephson junctions can be mapped onto a dual result for a QPS junction in a dual circuit if one interchanges the canonically conjugate phase and charge variables describing quantum dynamics of these systems. This observation can be used in metrology, e.g., for practical implementation of the electric current standard in the above structures. It was also proposed [307] to use rings with QPS junctions for experimental realization of QPS flux qubits.

20.2.1 Non-interacting QPS: Partition Function

Let us, for a moment, turn back to the case of sufficiently thick superconducting rings of Fig. 20.1 where QPS effects can be neglected. In this case by virtue of Poisson's resummation theorem in a complete analogy to our analysis in Section 6.4., one can identically transform the ring partition function (20.3) to the following expression:

$$\mathcal{Z}_{\phi_x} = \sum_{k=-\infty}^{\infty} \exp(\mathrm{i}2\pi k\phi_x) \int d\theta_0 \int_{\theta_0}^{\theta_0+2\pi k} \mathcal{D}\theta \exp(-S_0[\theta]), \quad (20.14)$$

where

$$S_0[\theta] = \int_0^{1/T} d\tau \frac{1}{2E_R}\left(\frac{\partial\theta}{\partial\tau}\right)^2 \tag{20.15}$$

and $1/E_R$ plays the role of a "mass" for a "particle" with the coordinate θ. This coordinate represents an effective angle. It is formally analogous to the Josephson phase variable φ in which case E_R is related to the junction charging energy E_C as $E_R = 8E_C$.

Consider now thinner rings of Fig. 20.4a where QPS effects gain importance. In this case of QPS rings, it is necessary to include all transitions between different energy states (20.3) labeled by the number p. As we already discussed, these transitions are just QPS events with the rate Γ_{QPS} defined in Eq. (19.65). Formally they are fully analogous to Cooper pair tunneling events in Josephson junctions; cf. Section 6.5.

Let us fix $p = 0$ and take into account virtual transitions to all other energy states and back. Performing summation over all such contributions, we arrive at the series in powers of γ_{QPS} (or fugacity y) similar to Eq. (19.70). Assuming that both the ring perimeter X and its thickness are not too large (the latter condition restricts the parameter μ), one can neglect weak logarithmic interaction (19.67) between different phase slips. Then spatial QPS coordinates x_n can be trivially integrated out. As a result, we arrive at the following contribution to the partition function

$$\tilde{Z}(\phi_x) = \int_{\phi_x}^{\phi_x} \mathcal{D}\phi(\tau) \sum_{n=0}^{\infty} \frac{1}{2n!}\left(\frac{\Gamma_{QPS}}{2}\right)^{2n} \int_0^{1/T} d\tau_1 \ldots \int_0^{1/T} d\tau_{2n}$$

$$\times \sum_{\nu_i=\pm 1} \delta(\dot{\phi}(\tau) - \dot{\phi}_n(\tau)) \exp\left(-\int_0^{1/T} d\tau \frac{E_R\phi^2(\tau)}{2}\right) \tag{20.16}$$

(cf. Eq. (6.65)), where, as before, the summation is carried out over neutral charge configurations (19.72),

$$\phi_n(\tau) = \sum_{j=1}^{n} \nu_j \Theta(\tau - \tau_j) \tag{20.17}$$

and $\Theta(\tau)$ is the theta function. We note that the term in the exponent describes virtual energy changes occurring due to QPS events. What remains is to add up similar contributions from all other parabolas with $p \neq 0$. In this way, we arrive at the final expression for the grand partition function of the nanoring

$$\mathcal{Z}_{\phi_x} = \sum_{p=-\infty}^{\infty} \tilde{Z}(p + \phi_x) \tag{20.18}$$

Together with Eq. (20.16), this result fully accounts for all QPS events in our system.

In the absence of QPS (i.e., for $\Gamma_{\rm QPS} \rightarrow 0$), Eqs. (20.18) and (20.16) obviously reduce to Eq. (20.3). For nonzero $\Gamma_{\rm QPS}$—following the same considerations as in Section 6.5—one can rewrite the partition function (20.16), (20.18) in the equivalent form of the following path integral

$$\mathcal{Z}_{\phi_x} = \sum_{k=-\infty}^{\infty} \exp(\mathrm{i}2\pi k\phi_x) \int d\theta_0 \int_{\theta_0}^{\theta_0+2\pi k} \mathcal{D}\theta \exp(-S[\theta]) \tag{20.19}$$

where

$$S[\theta] = S_0[\theta] - \Gamma_{\rm QPS} \int_0^{1/T} d\tau \cos\theta(\tau) \tag{20.20}$$

In order to demonstrate that the partition function (20.19), (20.20) is identical to that defined in Eqs. (20.16) and (20.18), it suffices to perform the Hubbard–Stratonovich transformation of the kinetic term (20.15) (which amounts to introducing additional path integral over $\phi(\tau)$), to formally expand $\exp(-S[\theta])$ in powers of $\Gamma_{\rm QPS}$ and then to integrate out the θ-variable in all terms of these series. After these straightforward steps, we arrive back at Eqs. (20.16) and (20.18).

Equations (20.19) and (20.20) define the grand partition function for the problem of a quantum particle on a ring in a cosine periodic potential described by the Hamiltonian (8.16) with $\kappa = 1$ where we also identify $E_{\rm C} \leftrightarrow E_{\rm R}/2$ and $U_0 \leftrightarrow \Gamma_{\rm QPS}$. This problem was repeatedly encountered above in different contexts, e.g., in Chapter 8 when we analyzed the phenomenon of persistent currents (cf. Eqs. (8.17) and (8.18) with $\kappa = 1$ and $S_{\rm int} \rightarrow 0$) or in Chapter 6 when we investigated the behavior of the Josephson junctions in the presence of charging effects; cf. Eqs. (6.45) and (6.46). In the latter case, $\Gamma_{\rm QPS}$ is analogous to the Josephson energy $E_{\rm J}$.

It is also important to emphasize that the above physical picture already received a solid experimental confirmation [26, 326]. Close to the flux values $\phi_x = 1/2 + n$, the two lowest energy states of the ring are nearly degenerate; coherent QPS tunneling between these states becomes particularly intensive and the energy gap equal to $\Gamma_{\rm QPS}$ (in the limit $\Gamma_{\rm QPS} \ll E_{\rm R}$) between the lowest and the first excited energy bands develops. This is nothing but the well-known quantum mechanical effect of level repulsion, and it is this effect that was experimentally demonstrated in QPS junctions [26, 326] thereby confirming quantum coherent nature of QPS.

20.2.2 Persistent Current and Noise

In a complete analogy with Eq. (6.47), the partition function defined in Eqs. (20.19) and (20.20) can be identically rewritten as

$$\mathcal{Z}_{\phi_x} = \sum_{p=-\infty}^{\infty} \exp(-\tilde{E}_p(\phi_x)/T), \tag{20.21}$$

where $\tilde{E}_p(\phi_x)$ are the energy bands of the problem defined by the solutions to the well-known Mathieu equation. For $\Gamma_{\rm QPS} \ll E_{\rm R}$, one has

$$\tilde{E}_0(\phi) = \frac{E_{\rm R}}{2\pi^2} \arcsin^2 \left[\left(1 - \frac{\pi^2}{2} \left(\frac{\Gamma_{\rm QPS}}{E_{\rm R}} \right)^2 \right) \sin(\pi \phi_x) \right] \tag{20.22}$$

i.e., the energy bands remain nearly parabolic $\tilde{E}_p(\phi) \sim E_p(\phi)$ except in the vicinity of the crossing points where gaps open due to level repulsion (see Fig. 20.2). In this limit, the value of the gap between the lowest and the first excited energy bands just coincides with the QPS rate $\delta E_{01} = \Gamma_{\rm QPS}$. For larger $\Gamma_{\rm QPS}$, the bandwidth shrinks while the gaps become bigger. In the limit $\Gamma_{\rm QPS} \gg E_{\rm R}$, the lowest band coincides with

$$\tilde{E}_0(\phi) = \frac{8}{\sqrt{\pi}} \Gamma_{\rm QPS}^{3/4} (E_{\rm R})^{1/4} e^{-8\sqrt{\Gamma_{\rm QPS}/E_{\rm R}}} (1 - \cos(2\pi \phi_x)) \tag{20.23}$$

The gap between the two lowest bands is then $\delta E_{01} = \sqrt{\Gamma_{\rm QPS} E_{\rm R}}$.

These results are sufficient to evaluate persistent current in superconducting nanorings in the presence of QPS. As before, taking the derivative of the ground-state energy with respect to the flux Φ_x and making use of Eq. (19.65), we find that for $\Gamma_{\rm QPS} \ll E_{\rm R}$ and outside immediate vicinity of the points $\phi_x = 1/2 + p$, persistent current is again defined by Eq. (20.6). In the opposite limit $\Gamma_{\rm QPS} \gg E_{\rm R}$, we find

$$I = \tilde{I}_0 \sin(2\pi\phi_x), \quad \tilde{I}_0 = \frac{16e}{\sqrt{\pi}}\Gamma_{\rm QPS}^{3/4}(E_{\rm R})^{1/4}e^{-8\sqrt{\Gamma_{\rm QPS}/E_{\rm R}}} \tag{20.24}$$

Combining this result with Eqs. (19.65), (20.4), and (20.13), we observe that in the latter limit the supercurrent flowing in the ring is exponentially suppressed as

$$\tilde{I}_0 \propto \exp(-R/R_c) \tag{20.25}$$

where the parameter R_c is defined by a simple formula [413]

$$R_c \sim \xi \exp(S_{\rm core}/2), \quad S_{\rm core} \sim g_\xi \tag{20.26}$$

which sets a fundamental size scale beyond which persistent current in superconducting nanorings should be exponentially suppressed by quantum fluctuations of the order parameter. To give some numbers, for $\xi \sim 100$ nm and $S_{\rm core} \approx 10$ from Eq. (20.26), we arrive at an estimate $R_c \sim 1$ μm.

Following our analysis in Section 8.3, one can also investigate fluctuations of persistent current in superconducting nanorings [356]. Employing an exact equivalence between the effective action for our problem (20.15), (20.20) and that for a quantum particle in the periodic potential on a ring (8.18) (where we should set $\kappa = 1$, $1/MR^2 \to E_{\rm R} \sim g_\xi \Delta\xi/R$ and $U_0 \to \Gamma_{\rm QPS}$) and making use of our results in Section 8.3.3, in the limit $R \gg R_c$, we immediately recover the noise spectrum

$$S_\omega = \frac{e^2\omega_0^3}{\pi\Gamma_{\rm QPS}}\left(\delta(\omega - \omega_0 - \Lambda\cos(2\pi\phi_x)) + \delta(\omega + \omega_0 + \Lambda\cos(2\pi\phi_x))\right) \tag{20.27}$$

where $\omega_0 = \pi^{3/2}\sqrt{N_0 D\Delta s\gamma_{\rm QPS}}$ is the frequency of small oscillations of the "particle" near the bottom of the cosine potential and

$$\Lambda = 256\sqrt{\frac{\Gamma_{\rm QPS}^3}{\pi\Omega}}\exp(-8\Gamma_{\rm QPS}/\omega_0) \tag{20.28}$$

The result (20.27) demonstrates that in the low-temperature limit, persistent current noise power spectrum S_ω differs from zero due to the effect of QPS and has the form of two sharp peaks at frequencies well below the superconducting gap Δ_0. The exact positions of these peaks can be tuned by the flux ϕ_x piercing the ring, though only weakly, since in this case $\omega_0 \gg \Lambda$. Note that according to Eq. (20.25) in the limit of large rings $R \gg R_c$, the average persistent current is exponentially suppressed, while the peaks in the noise spectrum decrease with increasing R only as a power law $S_\omega \propto R_c/R$.

In the opposite case $R \ll R_c$, the cosine potential term in the action remains small as compared to the particle kinetic energy. Hence, in this limit the effect of QPS may be considered perturbatively in $\Gamma_{\rm QPS}$. As we have already learned in Section 20.1, in the absence of QPS, the noise power S_ω shows only one peak at zero frequency with the amplitude equal to the current dispersion $\mathcal{P}$; see Eq. (20.10). The amplitude of this peak decreases with temperature and tends to zero in the limit $T \to 0$. Employing the standard quantum mechanical perturbation theory, in the lowest non-vanishing order in $\Gamma_{\rm QPS}$, one recovers additional peaks at frequencies corresponding to the transitions between neighboring energy levels. These peaks survive even at zero temperature $T = 0$ in which case one finds

$$\begin{aligned} S_\omega = & \frac{2e^2\Gamma_{\rm QPS}^2}{\pi(1+2\phi_x)^2}\left(\delta(\omega - E_{\rm R}(1/2+\phi_x)) + \delta(\omega + E_{\rm R}(1/2+\phi_x))\right) \\ & + \frac{2e^2\Gamma_{\rm QPS}^2}{\pi(1-2\phi_x)^2}\left(\delta(\omega - E_{\rm R}(1/2-\phi_x)) + \delta(\omega + E_{\rm R}(1/2-\phi_x))\right) \end{aligned} \tag{20.29}$$

This equation is applicable as long as the condition

$$\xi g_\xi \ll R \ll \min\left(R_c, \frac{c_0}{2\pi T}\right) \tag{20.30}$$

remains fulfilled. We observe that in this case, S_ω has the form of four sharp peaks at frequencies that strongly depend on the external flux ϕ_x. By tuning the flux to the value close to one half of the superconducting flux quantum $\phi_x \approx \pm 1/2$, one should observe strong enhancement of noise peaks, which occur in the vicinity of zero frequency $\omega = 0$. The physical reason for this enhancement is the same as that already discussed in Chapter 8: The energies of two lowest levels become close to each other at such values of ϕ_x implying the possibility of rapid transitions between these states. Such intensive transitions, in turn, imply strong fluctuations of persistent current. Exactly at resonance

$\phi_x = \pm 1/2$, the second-order perturbation theory fails and more accurate treatment becomes necessary, which would properly account for level splitting and effectively regularize the divergencies emerging in Eq. (20.29) at $\phi_x \to \pm 1/2$.

20.2.3 Discussion

The above results for persistent current, Eqs. (20.25) and (20.26), are qualitatively similar to those derived within a different approach for the model of superconducting nanorings formed by Josephson chains [297]. In particular, for both granular and homogeneous rings, the dependence of persistent current on ϕ_x gradually changes from sawtooth to sinusoidal as the ring radius R increases. This change is accompanied by suppression of the supercurrent amplitude, which eventually becomes an exponentially decaying function of R (20.25) in the limit $R > R_c$. In other words, Eq. (20.26) sets the length scale R_c beyond which supercurrents in QPS rings should be essentially washed out by quantum fluctuations. In some sense, the scale $\sim R_c$ plays the role of a dephasing length in our problem demonstrating that zero-point fluctuations can destroy macroscopic phase coherence down to $T \to 0$ even in superconducting systems.

This conclusion is qualitatively consistent with the results derived in Chapters 3 and 8 for single-channel rings coupled to dissipative baths. In fact, a close similarity between the partition function (20.19), (20.20) and those in Eqs. (3.46), (3.45) or in Eqs. (8.17), (8.18), (8.22), (8.23) exists on a formal level as well: These problems are described by similar path integrals, except in the case of normal rings coupled to dissipative environments it is necessary to account for the corresponding non-local in time term into the action instead of the last term in Eq. (20.20). At sufficiently large values of R, these path integrals can also be handled in exactly the same manner: In both problems, the dominating contribution comes from instantons (kinks) describing quantum tunneling of the angle variable θ between different topological sectors of the problem. In all cases, this procedure yields the low-temperature persistent current in the form (20.24) where one finds $\tilde{I}_0 \propto \exp(-(R/L_\varphi)^2)$ in the case of Caldeira–Leggett environments (see, e.g., Eq. (3.50)) and $\tilde{I}_0 \propto \exp(-R/L_\varphi)$ in the case of so-called dirty electron gas environments (see, e.g., Eq. (8.57)).

An important qualitative difference between our present problem and those of normal nanorings with dissipation lies in the fact that in the latter case, dissipation explicitly *violates time-reversal symmetry* (thus causing genuine decoherence of electrons), while no such symmetry is violated in Eqs. (20.19) and (20.20). Hence, in our case—in contrast to the situation considered, e.g., in Chapters 3 and 8—exponential suppression of PC (20.24) can be interpreted as a nontrivial coordinate-dependent renormalization effect. This difference is essentially the same as that between *dissipativeless* Cooper pair tunneling (which, just like QPS, opens interband gaps in the spectrum of Josephson junctions, see Chapter 6) and *dissipative* single-electron tunneling (which opens no such gaps, see Chapter 7).

20.2.4 Including Interactions between QPS

Although QPS junctions and rings can indeed, under certain conditions, behave similarly to (dual) Josephson junctions, the systems depicted in Fig. 20.4 are clearly not identical to such junctions in a number of aspects. Perhaps the most important one is that a (zero-dimensional) junction is now replaced by a (quasi-one-dimensional) superconducting wire (or ring), excitation spectrum of which is rather different, e.g., due to the presence of plasmon modes lacking in small-area Josephson junctions. As a result, different QPS events start "talking" to each other and eventually become correlated due to logarithmic interaction between them both in space and in time, as we already observed in Chapter 19. Hence, the above zero-dimensional duality arguments need to be revised in order to account for interactions between different QPS.

Fortunately, the corresponding analysis can be handled in a straightforward manner. As we have already demonstrated in Section 19.4.6, the problem of logarithmically interacting gas of phase slips can be mapped onto an effective sine-Gordon model. Since the latter model is, in turn, exactly dual to that describing long Josephson barriers [39], we may conclude that the duality considerations, if properly modified, can be conveniently adapted to the case of sufficiently long QPS junctions and rings.

Alternatively, one can proceed analogously to the non-interacting case analyzed in Section 20.2.1 and now take into account inter-QPS interactions in the same spirit as it was already done in Section 19.4.3. The corresponding calculation [357] explicitly accounts for the boundary conditions (20.2) and allows to express the grand partition function for QPS rings of Fig. 20.3a in the form of the path integral over the fluctuating quantum field $\chi(x, \tau)$

$$\mathcal{Z} \sim \sum_{m,n=-\infty}^{\infty} e^{2\pi i n\phi_x} \int^{mn} \mathcal{D}\chi\, e^{-S_{\text{eff}}[\chi(x,\tau)]} \tag{20.31}$$

where

$$S_{\text{eff}} = \frac{1}{2XE_{\text{R}}} \int_0^\beta d\tau \int_0^X dx [(\partial_\tau \chi(x, \tau))^2 + c_0^2(\partial_x \chi(x, \tau))^2] - \gamma_{\text{QPS}} \int_0^\beta d\tau \int_0^X dx \cos(\chi(x, \tau)) \tag{20.32}$$

is the ring effective action, which depends on the QPS tunneling amplitude per unit length γ_{QPS} (19.65). The path integral (20.31) should be evaluated for the field configurations obeying the boundary conditions

$$\chi(x, \beta) - \chi(x, 0) = 2\pi n, \quad \chi(L, \tau) - \chi(0, \tau) = -2\pi m \tag{20.33}$$

The above equations allow to exactly map the original problem of QPS rings pierced by the magnetic flux onto an effective sine-Gordon model on torus. In full agreement with our arguments in Section 19.4.6 (and extending them to superconducting nanorings), the effective action (20.32) is exactly dual to that for spatially extended Josephson barriers. Equations (20.31)–(20.33) fully account for interactions between different QPS and serve as a convenient starting point for further analysis of the ground-state properties of QPS rings. We also note that these equations equally apply to QPS junctions in which case one should only replace the ring perimeter $X = 2\pi R$ by the effective length of the thin part of the ring in Fig. 20.3b.

At low-enough temperatures $T \ll c_0/X$ and provided the ring perimeter remains small enough, one can ignore the spatial dependence of the field χ and, hence, neglect the term containing the derivative $(\partial_x \chi)^2$ in the effective action (20.32). Then our problem reduces back to the non-interacting one already considered in Section 20.2.1. However, for not-too-small X, logarithmic interactions between QPS become significant and need to be accounted for. In order to accomplish this goal, we will make use of the effective sine-Gordon model (20.32), which can be conveniently treated by means of the Berezinskii–Kosterlitz–Thouless (BKT) scaling equations (19.73).

As usual, we start renormalization at the shortest scale $\Lambda = \ln(\rho/\xi_c) \sim 1$ (where, as before, $\xi_c = \sqrt{x_0^2 + c_0^2\tau_0^2}$ with $x_0 \sim \xi$ and $\tau_0 \sim 1/\Delta_0$) and proceed to bigger scales. Within the first-order perturbation theory in the QPS fugacity $y = \gamma_{\rm QPS} x_0 \tau_0$ (cf. Eq. (19.71), one can ignore weak renormalization of the parameter μ. Within this accuracy, the solution to Eqs. (19.73) is trivial and yields $\gamma_{\rm QPS}(\Lambda) = \gamma_{\rm QPS}(\xi_c/\rho)^{\mu}$. Our renormalization group (RG) procedure should be stopped at the maximum scale equal to the ring perimeter $\rho \sim X$. In this way, we arrive at the renormalized QPS amplitude for our system

$$\tilde{\gamma}_{\rm QPS} = \gamma_{\rm QPS}(\xi_c/X)^{\mu} \tag{20.34}$$

This result demonstrates that interaction-induced renormalization effects remain weak and can be disregarded only for very small values of $\mu \ll 1/\ln(X/\xi_c)$.

At scales $\rho > X$, the problem again becomes effectively zero-dimensional; hence in order to evaluate the supercurrent, one can employ the results derived in Section 20.2.1 in which one should only substitute the renormalized QPS amplitude instead of the bare one, i.e.,

$$\Gamma_{\rm QPS} \to \tilde{\gamma}_{\rm QPS} X \tag{20.35}$$

As before, for smaller rings with $R \ll \tilde{R}_c$ and at $T \to 0$, the current is defined by Eq. (20.6) at all values of the flux except for $\phi_x \approx 1/2 + n$ where QPS with the effective amplitude (20.34) become significant. The critical radius $\tilde{R}_c$ is now determined by the condition $E_{\rm R} \sim 2\pi\tilde{\gamma}_{\rm QPS}$, which yields

$$\tilde{R}_c \sim \xi \exp\left(\frac{S_{\rm core}}{2-\mu}\right)\left(\frac{\xi}{\xi_c}\right)^{\frac{\mu}{2-\mu}} \tag{20.36}$$

where μ is supposed not to exceed 2.

In the opposite limit of bigger rings $R \gg \tilde{R}_c$, we again reproduce the first Eq. (20.24), where now

$$\tilde{I}_0 \sim \nu^{3/4} g_\xi \Delta \left(\frac{R}{\xi}\right)^{1/2-3\mu/4} e^{-3S_{\rm core}/4-R/\tilde{R}_c}, \quad \nu = (\xi_c/\xi)^\mu \quad (20.37)$$

From Eq. (20.36), we observe that the critical radius $\tilde{R}_c$ (beyond which PC gets exponentially suppressed) increases with increasing μ and eventually diverges at the quantum BKT phase transition point $\mu = 2$, which separates a superconducting (ordered) phase $\mu > 2$ with bound QPS–anti-QPS pairs and a non-superconducting (disordered) phase $\mu < 2$ with unbound QPS; see Chapter 19. In the ordered phase $\mu > 2$, QPS effects become practically irrelevant and the supercurrent is defined by Eq. (20.6) for any value of R.

Now let us briefly analyze the effect of inter-QPS interactions on the supercurrent fluctuations. As before, in order to evaluate the supercurrent noise power spectrum S_ω, we first find the imaginary time current–current correlator and then perform its analytic continuation to real times. For further details of this procedure, we refer the reader to the paper [357]; here we only present the final result. Proceeding perturbatively in $\Gamma_{\rm QPS}$, at $T \to 0$ we obtain

$$\begin{aligned} S_\omega = {} & \frac{e^2\Gamma_{\rm QPS}^2 E_{\rm R}^2}{2\pi\Gamma^2(\mu)} \left(\frac{2\xi_c^2}{R^2}\right)^{\mu} \sum_{k=0}^{\infty} \frac{\Gamma^2(\mu+k)}{\Gamma^2(1+k)} \\ & \times \left(\frac{1}{\Omega_k^2(\phi_x)}\delta(\omega - \Omega_k(\phi_x)) + \frac{1}{\Omega_k^2(\phi_x)}\delta(\omega + \Omega_k(\phi_x)) \right. \\ & \left. + \frac{1}{\Omega_k^2(-\phi_x)}\delta(\omega - \Omega_k(-\phi_x)) + \frac{1}{\Omega_k^2(-\phi_x)}\delta(\omega + \Omega_k(-\phi_x)) \right) \end{aligned} \quad (20.38)$$

where

$$\Omega_k(\phi_x) = (1/2 + \phi_x)E_{\rm R} + 2c_0 k/R \quad (20.39)$$

denote the energy differences between the exited states and the ground state of our ring. Equation (20.39) applies for $-1/2 < \phi_x \leq 1/2$. Outside this interval, $\Omega_k(\phi_x)$ should be continued periodically with the period equal to unity.

Let us emphasize again that, owing to its coherent nature, the supercurrent noise in QPS rings can be tuned by the external

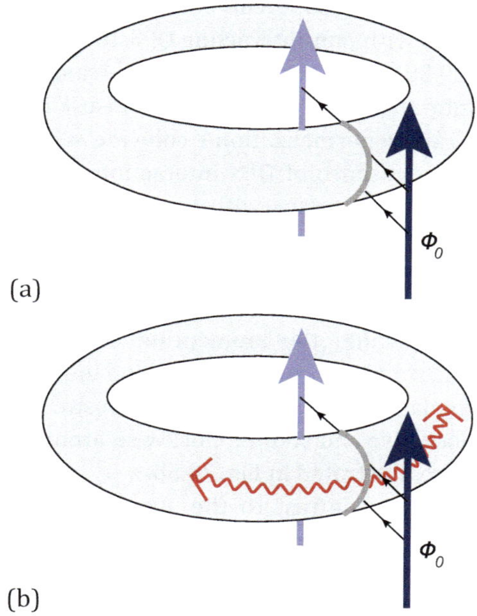

Figure 20.5 The process of coherent flux tunneling without (a) and with (b) excitation of a pair of Mooji–Schön plasma modes.

magnetic flux. Both the positions of the noise peaks at the frequencies Ω_k (corresponding to the system eigenmodes) and the magnitudes of these peaks depend periodically on ϕ_x. As follows from Eq. (20.38), in the immediate vicinity of level degeneracy points $\phi_x = \pm 1/2$, fluctuations of PC become strong and our perturbative in $\gamma_{\rm QPS}$ analysis becomes inapplicable even for $R \ll \tilde{R}_c$. In this case, it is necessary to proceed non-perturbatively in $\Gamma_{\rm QPS}$ and regularize the corresponding terms in Eq. (20.38). Roughly speaking, this regularization amounts to substituting the value $\sim\Gamma_{\rm QPS}$ instead of $\Omega_0(\phi_x)$ whenever the former exceeds the latter.

The effect of inter-QPS interactions on the supercurrent noise turns out to be somewhat richer than that for the average value of the supercurrent I. On one hand, interactions-induced renormalization of the QPS amplitude (8.57) shows up here as well: The noise power (20.38) scales with $\tilde{\gamma}^2_{\rm QPS} \propto 1/X^{2\mu}$. On the other hand, in addition to this renormalization effect, interactions also yield extra peaks in the supercurrent noise as compared to the case of small rings with non-interacting QPS. Indeed, the terms with $k = 0$ in Eq. (20.38) originating from virtual transitions of the flux quanta both into and out of the ring (see also Fig. 20.5a)—apart from the QPS rate renormalization—coincide with those already established in the absence of QPS interactions; cf. Eq. (20.29). At the same time, all extra terms with $k \neq 0$ correspond to virtual flux quanta transitions accompanied by creation of Mooij–Schön plasmon excitations. Such extra contributions to the noise power are totally absent in the zero-dimensional limit analyzed in Section 20.2.1. Keeping in mind the momentum conservation, one can conclude that the simplest process of that kind implies simultaneous creation of two plasmon excitations with opposite momenta values propagating clockwise and counterclockwise around the ring. Such processes are also illustrated in Fig. 20.5b.

To summarize, in contrast to the average supercurrent I, its fluctuations in QPS rings cannot, in general, be adequately described within the effectively zero-dimensional model of Section 20.2.1 even if the renormalization of γ_{QPS} (20.34) by interactions is taken into account. This is because virtual tunneling of flux quanta across the superconducting wire can lead to creation of plasmon modes in the system, thereby causing extra peaks in the noise power spectrum (20.38). By experimentally detecting these peaks, one can directly demonstrate the existence of Mooij–Schön plasma modes in superconducting nanorings.

Finally, let us point out a certain physical similarity between the equilibrium supercurrent noise studied here and that in point contacts between superconductors analyzed in Chapter 17. Also in the latter case, the noise power spectrum depends on the phase difference across the superconducting weak link and has the form of peaks that occur both at zero and nonzero frequencies; see Section 17.3. Similar to our problem, at $T \to 0$ the zero-frequency peak disappears, while the other peaks do not vanish except in the limit of fully transparent barriers. Unlike here, however, in the case of superconducting contacts, the noise peaks at nonzero frequencies have to do with subgap Andreev levels inside such contacts (cf., e.g., Eq. (17.32)) and are not related to QPS.

20.3 Parity Effect and Persistent Currents

In our previous analysis of QPS effects in superconducting nanorings, we followed the standard description developed for grand canonical ensembles, i.e., we implicitly assumed that the total number of electrons in the system N may fluctuate while the chemical potential is fixed. Obviously, this assumption is not correct for rings that are disconnected from any external circuit. In that case, electrons cannot enter or leave the ring and, hence, the number N is strictly fixed, but the chemical potential, on the contrary, fluctuates. It turns out that novel effects emerge in this physical situation. These effects will be discussed below.

It is well known that thermodynamic properties of isolated superconducting systems are sensitive to the parity of the total number of electrons [33, 381] even though this number is macroscopically large. This parity effect is a fundamental property of a superconducting ground-state described by the condensate of Cooper pairs. The number of electrons in the condensate is necessarily even; hence, for odd N, at least one electron remains unpaired having an extra energy equal to the superconducting energy gap Δ_0. This effect makes thermodynamic properties of the ground state with even and odd N different. Clear evidence for such parity effect was demonstrated experimentally in small superconducting islands [249, 381].

At the first sight, this parity effect should have little impact on the supercurrent because of the fundamental uncertainty relation $\delta N \delta\varphi \gtrsim 1$. Should the electron number N be fixed, fluctuations of the superconducting phase φ become large disrupting the supercurrent. On the other hand, in transport experiments with fluctuations of φ being suppressed, the parity effect cannot be observed because of large fluctuations of N.

Consider now isolated superconducting rings pierced by the magnetic flux Φ_x. In accordance with the number-phase uncertainty relation, the *global* superconducting phase of the ring fluctuates strongly in this case; however, these fluctuations are decoupled from the supercurrent and, therefore, cannot influence the latter. Below we will demonstrate that in this situation, the parity effect may substantially modify persistent current in superconducting nanorings at sufficiently low temperatures. Perhaps one of our most remarkable observations will be the presence of spontaneous supercurrent in the ground state of such rings with the odd total electron number N without any externally applied magnetic flux.

20.3.1 Parity Projection Formalism

In order to systematically investigate an interplay between the parity effect and persistent currents in superconducting nanorings, we will employ the so-called parity projection formalism. This formalism was originally proposed by a number of authors [150, 151, 206] for investigation of the parity effect in ultrasmall superconducting grains. Later it was extended to the case of superconducting nanorings [365]. The main advantage of this approach is that it allows to express the canonical partition functions and thermodynamic potentials in terms of the parity-projected grand canonical ones, thereby enormously simplifying the whole calculation. The approach works as follows.

The general grand canonical partition function

$$\mathcal{Z}(T, \tilde{\mu}) = \mathrm{Tr} e^{-\beta(\mathcal{H}-\tilde{\mu}N)} \tag{20.40}$$

is linked to the canonical one $Z(T, N)$ as

$$\mathcal{Z}(T, \tilde{\mu}) = \sum_{N=0}^{\infty} Z(T, N) \exp\left(\frac{\tilde{\mu}N}{T}\right) \tag{20.41}$$

Here and below, $\mathcal{H}$ is the system Hamiltonian. Inverting this relation and defining the canonical partition functions Z_e and Z_o, respectively, for even ($N \equiv N_e$) and odd ($N \equiv N_o$) ensembles, one gets

$$Z_{e/o}(T) = \frac{1}{2\pi} \int_{-\pi}^{\pi} du e^{-\mathrm{i}N_{e/o}u} \mathcal{Z}_{e/o}(T, \mathrm{i}Tu), \tag{20.42}$$

where

$$\begin{aligned} \mathcal{Z}_{e/o}(T, \tilde{\mu}) &= \frac{1}{2}\mathrm{Tr}\left\{\left[1 \pm (-1)^N\right] e^{-\beta(\mathcal{H}-\tilde{\mu}N)}\right\} \\ &= \frac{1}{2}\left(\mathcal{Z}(T, \tilde{\mu}) \pm \mathcal{Z}(T, \tilde{\mu} + \mathrm{i}\pi T)\right) \end{aligned} \tag{20.43}$$

are the parity-projected grand canonical partition functions. Provided the total number of particles in the system is large $N \gg 1$ (which is always the case in any generic metallic nanostructure), it is sufficient to evaluate the integral in Eq. (20.42) within the saddle point approximation

$$Z_{e/o}(T) \sim e^{-\beta(\Omega_{e/o}-\tilde{\mu}_{e/o}N_{e/o})} \tag{20.44}$$

where $\Omega_{e/o} = -T \ln \mathcal{Z}_{e/o}(T, \tilde{\mu})$ are the parity-projected thermodynamic potentials,

$$\Omega_{e/o} = \Omega_f - T \ln \left[\frac{1}{2}\left(1 \pm e^{-\beta(\Omega_b-\Omega_f)}\right)\right] \tag{20.45}$$

and

$$\Omega_{f/b} = -T \ln \left[\mathrm{Tr}\left\{(\pm 1)^N e^{-\beta(\mathcal{H}-\tilde{\mu}N)}\right\}\right] \tag{20.46}$$

Chemical potentials $\tilde{\mu}_{e/o}$ are implicitly defined by means of the saddle point condition

$$N_{e/o} = -\partial\Omega_{e/o}(T, \tilde{\mu}_{e/o})/\partial\tilde{\mu}_{e/o} \tag{20.47}$$

Note that Ω_f defines the standard grand canonical thermodynamic potential, and Ω_b represents the corresponding potential linked to the partition function $\mathcal{Z}(T, \tilde{\mu} + i\pi T)$. In order to reconstruct this function, one can make use of the following trick [150]. First one can evaluate the true grand canonical partition function $\mathcal{Z}(T, \tilde{\mu})$ and express the result as a sum over the Fermi Matsubara frequencies $\omega_f = 2\pi T(l+1/2)$ with $l = 0, \pm 1, \ldots$. Then one can just substitute the Bose–Matsubara frequencies $\omega_b = 2\pi T l$ instead of the Fermi ones ω_f and, in this way, automatically reproduce the correct expression for $\mathcal{Z}(T, \tilde{\mu} + i\pi T)$ and, hence, for Ω_b. Thus, we reduced the complicated problem of describing thermodynamic properties of canonical ensembles to a much simpler task of evaluating the grand canonical partition function.

Having found the thermodynamic potentials for the even and odd ensembles, one can easily determine the corresponding equilibrium currents I_e and I_o for isolated superconducting rings pierced by the magnetic flux Φ_x. Making use of the above expressions, one finds persistent current circulating inside the ring [365]:

$$I_{e/o} = I_f \pm \frac{I_b - I_f}{e^{\beta(\Omega_b-\Omega_f)} \pm 1}, \tag{20.48}$$

where the upper/lower sign corresponds to the even/odd ensemble and we have defined

$$I_{e/o} = c\left(\frac{\partial\Omega_{e/o}}{\partial\Phi_x}\right)_{\tilde{\mu}(\Phi_x)}, \quad I_{f/b} = c\left(\frac{\partial\Omega_{f/b}}{\partial\Phi_x}\right)_{\tilde{\mu}(\Phi_x)} \tag{20.49}$$

Equations (20.48) and (20.49) represent a direct generalization of the grand canonical formula (20.5) to canonical ensembles.

20.3.2 Homogeneous Superconducting Rings

Let us first consider homogeneous nanorings with cross section s and perimeter $X = 2\pi R$. As before, rings will be assumed sufficiently thin, $\sqrt{s} < \lambda_{\mathrm{L}}$. On the other hand, below we will neglect QPS effects, i.e., describe superconducting properties of such rings within the mean field Bardeen–Cooper–Schrieffer (BCS) theory. As we have already learned, this description is justified provided the condition $g_\xi \gg 1$ is satisfied. Hence, the ring should not be too thin, and the total number of conducting channels $\mathcal{N} = p_{\mathrm{F}}^2 s/4\pi$ should remain large $\mathcal{N} \gg 1$. In addition, the perimeter X should not exceed the critical value $2\pi R_c$ (20.26).

Let us employ the above parity projection formalism in order to analyze the properties of superconducting nanorings with even and odd total number of electrons. Assuming that the absolute value of the order parameter field $|\Delta|$ in the ring does not depend on the coordinates, evaluating the thermodynamic potentials $\Omega_{f/b}$ in a standard manner as explained above and expressing the result in terms of the quasiparticle excitation energies ε_k, one finds [150, 365]

$$\Omega_f = \tilde{\Omega} - 2T\sum_k \ln\left(2\cosh\frac{\varepsilon_k}{2T}\right), \tag{20.50}$$

$$\Omega_b = \tilde{\Omega} - 2T\sum_k \ln\left(2\sinh\frac{\varepsilon_k}{2T}\right), \tag{20.51}$$

where $\tilde{\Omega} = \mathcal{V}|\Delta|^2/\lambda + \mathrm{Tr}\{\hat{\xi}\}$

$$\hat{\xi} = \frac{1}{2m}\left(-\mathrm{i}\frac{\partial}{\partial \boldsymbol{r}} - \frac{e}{c}\boldsymbol{A}(\boldsymbol{r})\right)^2 + U(\boldsymbol{r}) - \tilde{\mu} \tag{20.52}$$

is the single-particle energy operator, $\varepsilon_k = \boldsymbol{p}_k\boldsymbol{v}_s + \sqrt{\xi_{p_k}^2 + |\Delta|^2}$, where $\boldsymbol{p}_k$ is the quasiparticle momentum, $\xi_p = (p^2 - \tilde{\mu}(\Phi_x))/2m$, $\mathcal{V} = Xs$

is the ring volume and

$$v_s = \frac{1}{4mR} \frac{\partial \min_n (n + \phi_x)^2}{\partial \phi_x} \tag{20.53}$$

is the superconducting velocity.

Let us, for a moment, turn off an external magnetic field by setting $\Phi_x = 0$ and $v_s = 0$. Then combining the relation (20.45) with Eqs. (20.50)–(20.52) within the standard BCS mean field approximation, we obtain

$$\Omega_{e/o}(T, \mu) = \min_\Delta \left\{ \Omega_{\rm BCS}(T, \mu, \Delta) - T \ln\left(\frac{1 \pm \exp(-2A(T, \Delta))}{2} \right) \right\} \tag{20.54}$$

where the upper (lower) sign corresponds to the even (odd) ensemble,

$$\Omega_{\rm BCS}(T, \mu, \Delta) = \frac{\mathcal{V}|\Delta|^2}{\lambda} - 2T \sum_k \ln\left(2\cosh \frac{\varepsilon_k}{2T} \right) + \sum_k (\xi_{p_k} - \mu) \tag{20.55}$$

defines the thermodynamic potential for the grand canonical BCS ensemble and

$$A(T, \Delta) = \sum_k \ln\left(\coth \frac{\varepsilon_k}{2T} \right) \tag{20.56}$$

Minimization of the result (20.54) with respect to the order parameter field Δ provides the equations for the equilibrium values of the order parameter for even and odd ensembles, respectively, Δ_e and Δ_o. These equations read

$$\Delta_{e/o} = \frac{\lambda}{2\mathcal{V}} \sum_k \frac{\Delta_{e/o}}{\varepsilon_k} \tanh \frac{\varepsilon_k}{2T} \pm \frac{\lambda}{2\mathcal{V}} \frac{\exp(-2A(T, \Delta_{e/o}))}{1 \pm \exp(-2A(T, \Delta_{e/o}))} \times \sum_k \frac{2\Delta_{e/o}}{\varepsilon_k \sinh \frac{\varepsilon_k}{T}} \tag{20.57}$$

The function $A(T, \Delta)$ (20.56) equals to

$$A(T \ll \Delta) \simeq N_{\rm eff}(T) e^{-\Delta/T}, \quad A(T \gg \Delta) \simeq (\pi^2/2) N_0 \mathcal{V} T \tag{20.58}$$

where $N_{\rm eff}(T) = 2N_0\mathcal{V}\sqrt{2\pi\Delta T}$ is the number of states effectively available for a quasiparticle at a given temperature. We observe that for $A(T, \Delta) \gg 1$, the parity effect is negligible and Eqs. (20.57) for both even and odd ensembles reduce to the standard BCS self-consistency equation. The difference between these two ensembles becomes important for $A(T, \Delta) \lesssim 1$. Combining this inequality with Eq. (20.58), we immediately arrive at the condition [381]

$$T \lesssim T_{\text{cr}} \approx \frac{\Delta}{\ln(N_{\text{eff}}(T))} \tag{20.59}$$

This condition restricts the temperature interval for which the difference between even and odd superconducting ensembles can be detected. For generic metallic systems, one can roughly estimate $\ln(N_{\text{eff}}(T)) \sim 10$, implying that the parity effect in isolated superconducting systems, such as nanograins or nanorings, is observable only well below T_C.

Below the crossover temperature $T < T_{\text{cr}} \ll \Delta$ from Eq. (20.57), we immediately recover the equilibrium values of the order parameter field for both even and odd ensembles [150, 151]

$$\Delta_e(T) = \Delta_0 \left(1 - 4\pi T N_0 \mathcal{V} \exp\left(-\frac{2\Delta_0}{T}\right)\right) \tag{20.60}$$

$$\Delta_o(T) = \Delta_0 \left(1 - \frac{4\pi}{3} T N_0 \mathcal{V} \exp\left(-\frac{2\Delta_0}{T}\right)\right) - \frac{1}{2N_0\mathcal{V}} \tag{20.61}$$

where Δ_0 is the standard BCS energy gap at $T = 0$.

We observe that Δ_e and Δ_o differ from each other and in general also differ from Δ_0. In particular, in the zero-temperature limit, we find [150, 206] $\Delta_e(0) = \Delta_0$ and $\Delta_o(0) = \Delta_0 - 1/2N_0\mathcal{V}$, i.e., the superconducting order parameter in the odd ensemble turns out to be smaller than in the even one by the value of order of the mean level spacing in the sample. This difference, however, may be significant only in the case of extremely small superconducting grains, which we do not consider here. Below we will always assume that the ring volume is large enough, $\mathcal{V} \gg 1/N_0\Delta_0$, in which case the difference between Δ_e, Δ_o, and Δ_0 can be neglected.

It is also instructive to define the occupation probabilities of the quasiparticle energy states ε_k:

$$f_{e/o}(\varepsilon_k) = \sum_{n_j=0,1}^{e/o} \exp(-(\varepsilon_k + \sum_{j\neq k} n_j\varepsilon_j)/T)/Z_{e/o} \tag{20.62}$$

where $\sum^{e/o}$ is taken over the states with even/odd quasiparticle numbers. With the aid of Eqs. (20.54) and (20.56), we get

$$f_e(\varepsilon_k) = \frac{1}{1+\exp(\varepsilon_k/T)} - \frac{1}{\sinh(\varepsilon_k/T)}\frac{1}{1+\exp(2A(T,\Delta_0))}$$
$$\simeq N_{\text{eff}}(T)\exp\left(-\frac{\varepsilon_k+\Delta_0}{T}\right) \tag{20.63}$$

and

$$f_o(\varepsilon_k) = \frac{1}{1+\exp(\varepsilon_k/T)} - \frac{1}{\sinh(\varepsilon_k/T)}\frac{1}{1-\exp(2A(T,\Delta_0))}$$
$$\simeq \frac{\exp\left(-\frac{\varepsilon_k-\Delta_0}{T}\right)}{N_{\text{eff}}(T)} + \frac{N_{\text{eff}}(T)}{3}\exp\left(-\frac{\varepsilon_k+\Delta_0}{T}\right) \tag{20.64}$$

It is worth pointing out that both these quasiparticle distribution functions differ from the Fermi function describing the grand canonical ensemble of electrons.

Making use of the results (20.54) and (20.56), we can also find other important thermodynamic characteristics of even and odd superconducting ensembles. For instance, defining the system entropy as $S = -\left(\frac{\partial\Omega}{\partial T}\right)_{\mathcal{V}}$, we obtain

$$S_e = 8\pi(N_0\mathcal{V}\Delta_0)^2\exp\left(-\frac{2\Delta_0}{T}\right) \tag{20.65}$$

$$S_o = \frac{8\pi}{3}(N_0\mathcal{V}\Delta_0)^2\exp\left(-\frac{2\Delta_0}{T}\right) + \ln(N_{\text{eff}}(T)) \tag{20.66}$$

Then for the specific heat $c = T\left(\frac{\partial S}{\partial T}\right)_{\mathcal{V}}$, we get

$$c_e = 16\pi(N_0\mathcal{V}\Delta_0)^2\frac{\Delta_0}{T}\exp\left(-\frac{2\Delta_0}{T}\right) \tag{20.67}$$

$$c_o = \frac{16\pi}{3}(N_0\mathcal{V}\Delta_0)^2\frac{\Delta_0}{T}\exp\left(-\frac{2\Delta_0}{T}\right) + \frac{1}{2} \tag{20.68}$$

The results (20.60)–(20.68) indicate several interesting features of BCS ensembles with fixed number of electrons. For both the even and the odd ensembles, the quantities (20.60)–(20.68) contain exponentially small terms proportional to $\exp(-2\Delta_0/T)$, whereas for the grand canonical BCS ensembles with a fixed chemical potential, the corresponding contributions are proportional to

$\exp(-\Delta_0/T)$. This difference has an obvious physical reason: In contrast to the grand canonical ensembles (in which case one can change the number of electrons one by one) for the BCS, ensembles with fixed number of electrons excitations can be created only in pairs. Thus, for the even ensemble, the energy of the first excited state (with two quasiparticles above the gap) exceeds that of the ground state by $2\Delta_0$. For the odd BCS ensemble, the gap between the ground state and the first excited state is equal to zero because at least one electron always remains unpaired and thus one quasiparticle is present above the gap even at $T = 0$. This leads to the term $\ln(N_{\rm eff}(T))$ in the expression for the entropy S_o (20.66) and to the temperature-independent contributions to the order parameter Δ_o (20.61) and the specific heat c_o (20.68) for the odd ensemble. Additional quasiparticles in this ensemble are also created in pairs leading to the terms proportional to $\exp(-2\Delta_0/T)$ in the corresponding expressions (20.61), (20.64), (20.66), and (20.68). Note that these terms for the odd ensemble are smaller by the factor 3 than the analogous terms for the even ensemble (20.60), (20.63), (20.65), and (20.67). This is a direct consequence of indistinguishability of electrons: As different number of quasiparticles (respectively, two and three) occupy the excited states of even and odd ensembles, the corresponding contributions to thermodynamic quantities differ by the factor $2!/3! = 1/3$.

Let us now again include the effect of an external magnetic flux into our consideration. With the aid of the above general relations, one can easily determine persistent currents in superconducting nanorings with even and odd number of electrons $I_{e/o}$. Here we consider the most interesting low-temperature limit $T \ll v_{\rm F}/X$. From Eqs. (20.48)–(20.51), we find that in this limit, the current I_e exactly coincides with the standard grand canonical result $I = en_s v_s s$ with $n_s \equiv n_e$, while the current I_o for the odd ensemble reads [216, 248, 365, 366]

$$I_o = en_o v_s s - e\frac{v_{\rm F}}{X}\,{\rm sgn} v_s \tag{20.69}$$

Here we introduced the electron density in the case of even/odd total number of electrons $n_{e/o} = N_{e/o}/\mathcal{V}$.

We observe that in the case of odd ensembles, there exists an additional term that modifies the flux dependence of persistent current and, as we will demonstrate below, leads to a number of fundamentally important effects. Unfortunately, in the case of homogeneous rings, the difference between persistent currents evaluated for the even and the odd ensembles turns out to be hardly detectable in any realistic experiment. Indeed, these difference is inversely proportional to the total number of conducting channels, $(I_e - I_o)/I_e \sim 1/\mathcal{N}$. For this reason, the parity effect remains vanishingly small in generic metallic rings with $\mathcal{N} \gtrsim 10^3$. On the other hand, for ultrathin nanorings with $\mathcal{N} \lesssim 10$, persistent current is essentially wiped out due to the proliferation of QPS. Estimating $g_\xi \sim \mathcal{N}\ell/\xi$, we conclude that for $g_\xi \sim \ell$ (i.e., when the QPS fugacity is already large and, hence, quantum suppression of supercurrent becomes very strong), the number of conducting channels yet remains parametrically large $\mathcal{N} \sim \xi/l \gg 1$. This estimate demonstrates that persistent current in homogeneous superconducting nanorings remains only very weakly affected by the parity effect.

20.3.3 Ballistic Superconductor–Normal–Superconductor (SNS) Rings

We now turn to the situations in which the parity effect gains importance and can be directly probed in modern experiments. Let us now slightly modify our system and consider a relatively thick superconducting ring with large $g_\xi \gg 1$ interrupted by a thin wire of length d with only a few conducting channels

$$\mathcal{N}_n \sim 1 \tag{20.70}$$

thus forming a weak link inside the superconducting ring; see Fig. 20.6. Without loss of generality, this wire can be considered normal no matter if it is made of normal or superconducting materials. In the latter case, quantum fluctuations fully suppress the superconducting order parameter inside such a wire, turning it normal. In contrast, QPS effects in superconducting parts of the ring can be neglected, thus making the mean field BCS description applicable. A clear advantage of these structures in comparison to homogeneous rings

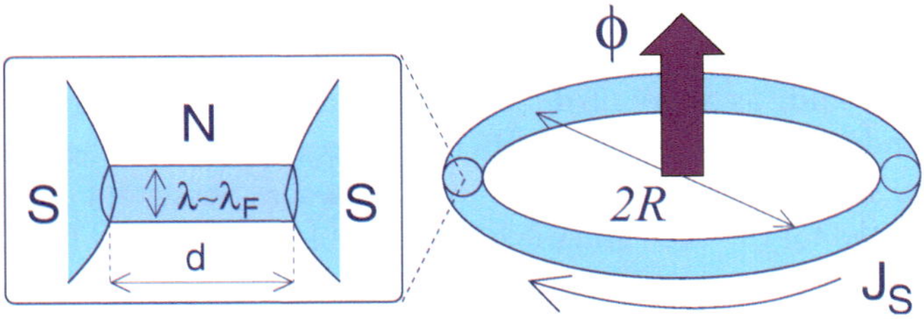

Figure 20.6 Superconducting ring with embedded SNS junction of length d.

is that in the former case, the effect of the electron parity number on persistent current can be large due to the condition (20.70).

We begin our analysis with simple physical arguments, which are already sufficient to establish a transparent relation between the supercurrent values in canonical BCS ensembles with even and odd total number of electrons in the limit of low temperatures. At $T \to 0$ and provided the total number of electrons N in the ring is even, they all form the condensate of Cooper pairs. Accordingly, all *available* electron states with energies below the Fermi energy $\tilde{\mu}_e$ are occupied, while the states with higher energies are all empty. Exactly the same situation is realized for grand canonical BCS ensembles. Hence, at $T \to 0$, the supercurrent in the ring with even number of electrons I_e should exactly coincide with the standard grand canonical result (20.6) with $n_s \equiv n_e$. If, however, the number N is odd, one electron always remains unpaired, and it occupies the lowest available energy state ε_0. An external magnetic flux Φ_x applied to the ring can shift this energy level, i.e., we have $\varepsilon_0 = \varepsilon_0(\Phi_x)$. Accordingly, the "odd" electron occupying this level should provide an extra contribution to the supercurrent I_o, which should simply be added to I_e. In other words, at $T \to 0$ we obtain [212, 365, 366]

$$I_o(\Phi) = I_e(\Phi_x) + c\frac{\partial \varepsilon_0(\Phi_x)}{\partial \Phi_x} \tag{20.71}$$

Let us apply these simple arguments to the case of superconducting rings with a short SNS weak link $d \to 0$. In practice, this limit is realized for transparent contacts with $d \ll \xi$. Even smaller values of d are required provided the contact transmission is small. In this case, the Josephson current can be expressed only

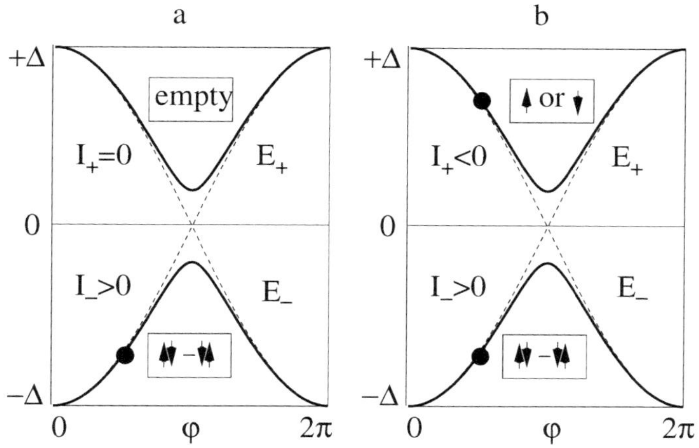

Figure 20.7 Andreev levels inside a quantum point contact and their occupation at $T = 0$ for even (a) and odd (b) ensembles.

via the contributions from discrete Andreev energy states inside this constriction $E_{i\pm} = \pm\varepsilon_i(\varphi)$ as [46]

$$I(\varphi) = 2e \sum_{i=1}^{\mathcal{N}_n} \left[\frac{\partial E_{i-}}{\partial \varphi} f_-(E_{i-}) + \frac{\partial E_{i+}}{\partial \varphi} f_+(E_{i+}) \right] \qquad (20.72)$$

where the index i labels conducting channels inside the constriction,

$$\varepsilon_i(\varphi) = \Delta_0 \sqrt{1 - \mathcal{T}_i \sin^2(\varphi/2)} \qquad (20.73)$$

are Andreev energy levels in a quantum point contact (see Fig. 20.7), $\mathcal{T}_i$ is the transmission of the i-th channel, and $\varphi = 2\pi\phi_x$ is the superconducting phase difference across the weak link.

Using the Fermi filling factors for these states $f_\pm(E_{i\pm}) = [1 + \exp(\pm\varepsilon_i(\varphi)/T)]^{-1}$, one indeed arrives at the standard grand canonical results. However, in the case of superconducting rings with fixed number of electrons N, these filling factors should be modified. As before, let us set $T \to 0$. For even N, all electrons are paired occupying available states with energies below the Fermi level (see Fig. 20.7a). In this case, one has $f_-(E_{i-}) = 1$, $f_+(E_{i+}) = 0$, the current is entirely determined by the contribution of all quasiparticle states E_{i-}, and Eq. (20.72) yields the same result as one for the grand canonical ensemble:

$$I_e = -2e \sum_{i=1}^{\mathcal{N}_n} \frac{\partial \varepsilon_i(\varphi)}{\partial \varphi} \tag{20.74}$$

By contrast, in the case of odd N, one electron always remains unpaired and occupies the lowest available energy state—in our case E_{k+} (where the index k corresponds to the conducting channel with the highest transmission value $\mathcal{T}_k$)—above the Fermi level. Hence, for odd N one has $f_-(E_{i-}) = 1$, $f_+(E_{k+}) = 1$, and $f_+(E_{i+}) = 0$ for all $i \neq k$; see Fig. 20.7b. Then from Eq. (20.72), we obtain

$$I_o = I_e + 2e\frac{\partial \varepsilon_k(\varphi)}{\partial \varphi} = -2e \sum_{i=1, i\neq k}^{\mathcal{N}_n} \frac{\partial \varepsilon_i(\varphi)}{\partial \varphi} \tag{20.75}$$

Comparing Eqs. (20.74) and (20.75), we observe that the supercurrent values in the even and odd ensembles differ because the contribution of the k-th conducting channel is totally blocked in the odd case. This is because the contributions of the two Andreev states, E_{k-} and E_{k+}, in Eq. (20.72) exactly cancel each other. For quantum point contacts with only one conducting channel, this effect implies *total blocking* of the supercurrent by the odd electron, i.e., in this case, the supercurrent remains zero for any magnetic flux Φ_x threading the ring [188, 365, 366]. In quantum point contacts with several conducting channels, the current through the most transparent channel will be blocked by the odd electron. Hence, though blocking of the supercurrent remains incomplete in this case, it may nevertheless be important also for quantum point contacts with $\mathcal{N}_n > 1$.

The above simple physical arguments are fully supported by a rigorous analysis employing the parity projection formalism. According to Eqs. (20.48) and (20.49), we need to evaluate both the currents $I_{f/b}$ and the difference between the "Fermi" and "Bose" thermodynamic potentials $\Omega_b - \Omega_f \equiv \Omega_{bf}$. The currents $I_{f/b}$ can be conveniently expressed via the phase difference across the weak link φ by means of the general formula derived for the grand canonical Fermi ensembles [130]. Employing this formula together with its extension to "Bose" ensembles, we get

$$I_{f/b} = 2e \sum_{i=1}^{\mathcal{N}_n} T \sum_{\omega_{f/b}} \frac{\sin \varphi}{\cos \varphi + W_i(\omega_{f/b})}, \tag{20.76}$$

where the sum runs over conducting channels of the normal wire and the function $W_i(\omega)$ can be derived in a straightforward manner for any type of the weak links. For instance, in the above limit of a quantum point contact with $d \to 0$, one finds [130]

$$W_i(\omega) = (2/\mathcal{T}_i)(1 + \omega^2/\Delta_0^2) - 1 \tag{20.77}$$

The difference of thermodynamic potentials Ω_{bf} is defined as a sum of the contributions from the superconducting bulk of the ring and from the weak link, $\Omega_{bf} = \Omega_{bf}^{(s)} + \Omega_{bf}^{(wl)}$. The former contribution is evaluated with the aid of Eqs. (20.50) and (20.51). Combining them with the results in Eqs. (20.54)–(20.56) and (20.58), we arrive at the standard result [381]

$$\beta\Omega_{bf}^{(s)} \sim N_{\text{eff}}(T)e^{-\frac{\Delta_0}{T}} \tag{20.78}$$

The term $\Omega_{bf}^{(wl)}$ is derived simply by integrating $I_{f/b}(\varphi)$ (20.76) over the phase φ in the spirit of the λ-trick already employed, e.g., in Chapter 2 (see Section 2.5 for details). We obtain

$$\beta\Omega_{bf}^{(wl)} = 2T\sum_j^{\mathcal{N}} \ln\coth\left(\frac{\varepsilon_j(\varphi)}{2T}\right) \tag{20.79}$$

Combining these results with Eqs. (20.48) and (20.49), we derive the final expressions for the supercurrents in both even and odd ensembles [365, 366]

$$I_{e/o} = -2e\sum_{i=1}^{N}\frac{\partial\varepsilon_i(\varphi)}{\partial\varphi}\tanh\frac{\varepsilon_i(\varphi)}{2T}\left[1 \pm \frac{(\coth\frac{\varepsilon_i(\varphi)}{2T})^2 - 1}{e^{\beta\Omega_{bf}^{(s)}}\prod_{j=1}^{N}(\coth\frac{\varepsilon_j(\varphi)}{2T})^2 \pm 1}\right] \tag{20.80}$$

where $\varepsilon_i(\varphi)$ are again defined in Eq. (20.73). The combination in front of the angular brackets in Eq. (20.80) defines the standard grand canonical result, while the term in the square brackets accounts for the parity effect in our system. For $\mathcal{N}_n = 1$ and at $T = 0$, this term reduces to unity for even ensembles and to zero for odd ones, which is just the blocking effect already discussed above.

We now turn to another important limit of superconducting rings containing a normal wire of length $d > \xi_0 \sim v_F/\Delta_0$. In contrast to the case $d \to 0$ considered above, the Josephson current in

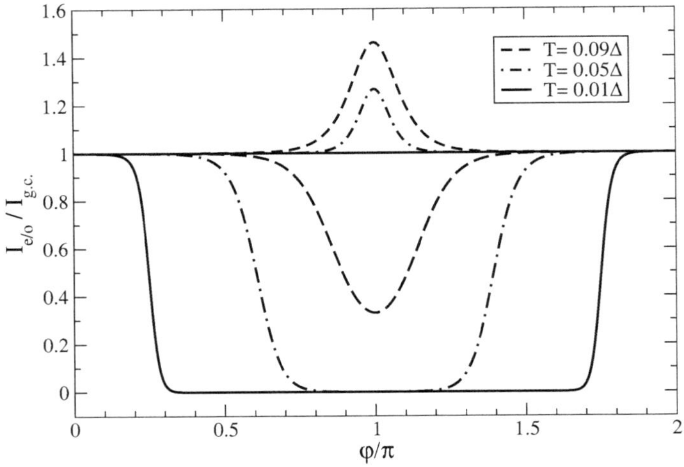

Figure 20.8 The ratio between canonical and grand canonical values of PC $I_{e/o}/I_{g.c.}$ (represented by the term in the square brackets in Eq. (20.80) versus φ in a single-mode QPC at different temperatures for even (three upper curves) and odd (three lower curves) ensembles. Here we have chosen the channel transmission $\mathcal{T} = 0.99$.

SNS structures cannot anymore be attributed only to the discrete Andreev states inside a weak link, and an additional contribution from the states in the continuum should also be taken into account. Furthermore, for any nonzero d, there are always more than two discrete Andreev levels in the system. Accordingly, significant modifications in the physical picture of the parity effect in such SNS rings can be expected.

The key difference can be understood already by comparing the typical structure of discrete Andreev levels in SNS junctions (Fig. 20.9) with that of a quantum point contact (Fig. 20.7). As before, in the limit $T \to 0$, all states below (above) the Fermi level are occupied (empty) provided the total number of electrons in the system is even (Fig. 20.9a). If, on the other hand, this number is odd, the lowest Andreev state above the Fermi energy is occupied as well (Fig. 20.9b), thus providing an additional contribution to the Josephson current. This contribution, however, cancels only that of a symmetric Andreev level below the Fermi energy, while the contributions of all other occupied Andreev levels and of the

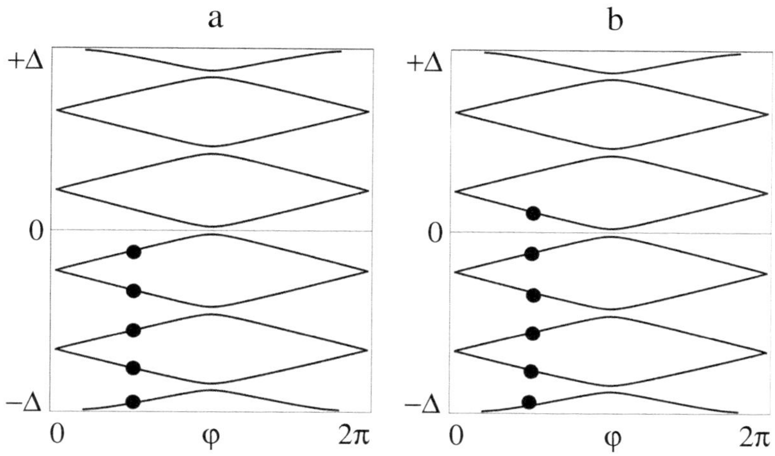

Figure 20.9 Andreev levels in a single-mode SNS junction with $d = 6\hbar v_{\rm F}/\Delta_0$ and their occupation at $T = 0$ for even (a) and odd (b) ensembles.

continuum states remain uncompensated. Hence, unlike in the case of a short $(d \to 0)$ single-channel weak link, in SNS rings one should not anymore expect the supercurrent blocking by the odd electron but rather some other nontrivial features of the parity effect.

This conclusion is fully confirmed by our quantitative analysis [365, 366]. Let us restrict our attention to transparent SNS junctions in which case the function $W_i(\omega) \equiv W(\omega)$ is the same for all transmission channels and reads [130]

$$W(\omega) = \left(\frac{2\omega^2}{\Delta_0^2} + 1\right)\cosh\left(\frac{2\omega d}{v_{\rm F}}\right) + \frac{2\omega}{\Delta_0}\sqrt{1 + \frac{\omega^2}{\Delta_0^2}}\sinh\left(\frac{2\omega d}{v_{\rm F}}\right) \quad (20.81)$$

Substituting this function into Eq. (20.76) and repeating the whole calculation as above, we arrive at the final result [365, 366]

$$I_e = \frac{ev_{\rm F}\mathcal{N}_n}{\pi d}\varphi, \qquad I_o = \frac{ev_{\rm F}\mathcal{N}_n}{\pi d}\left(\varphi - \frac{\pi\,{\rm sgn}\varphi}{\mathcal{N}_n}\right) \quad (20.82)$$

This result applies for $-\pi < \varphi < \pi$ and should be 2π-periodically continued otherwise. We observe that the current I_e again coincides with that for the grand canonical ensembles [203], while the current I_o—according to Eq. (20.71)—also contains an additional contribution from the "odd" electron occupying the lowest available

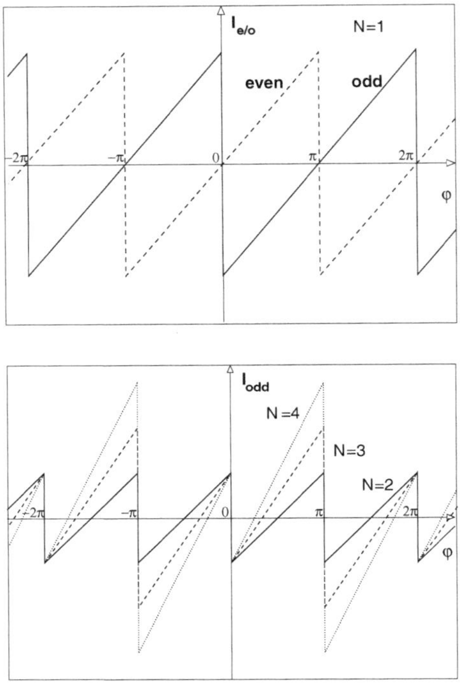

Figure 20.10 Top panel: Zero-temperature current–phase relations (20.82) $I_e(\varphi)$ (dashed line) and $I_o(\varphi)$ (solid line) for SNS rings with $\mathcal{N}_n = 1$. Bottom panel: The same relations $I_o(\varphi)$ for the odd ensembles (second Eq. (20.82)) and $\mathcal{N}_n = 2, 3, 4$.

Andreev level $E_1(\varphi) = v_{\mathrm{F}}|\pi - \varphi|/(2d)$ (where $0 < \varphi < 2\pi$); see also Fig. 20.9b. Due to this additional contribution in the case of odd ensembles, the current–phase relation is shifted by the value $\pi/\mathcal{N}_n$ as compared to that in the even case. As we have expected, this contribution indeed does not compensate for the current from other electron states. Rather it provides a possibility for a parity-induced π-junction state in our system: According to Eq. (20.82) for single-mode SNS junctions, the "sawtooth" current–phase relation will be shifted exactly by π; see Fig. 20.10. More generally, we can talk about a π/N-junction state, because in the odd case, the minimum Josephson energy (zero current) state is reached at $\varphi = \pm\pi/N$; see Fig. 20.10. For any $\mathcal{N}_n > 1$, this is a twofold degenerate state within the interval $-\pi < \varphi < \pi$. In the particular case $\mathcal{N}_n = 2$, the current–phase relation $I_o(\varphi)$ turns π-periodic.

Let us recall that the π-junction state can also be realized in SNS structures by driving the electron distribution function in the contact area out of equilibrium [387, 405, 409]. Here, in contrast, the situation of a π- or π/N-junction is achieved in thermodynamic equilibrium. Along with this important difference, there also exists a certain physical similarity between these two physical situations: In both cases, the electron distribution function in the weak link deviates substantially from the Fermi function. It is this deviation which is responsible for the appearance of the π-junction state in both physical situations.

Perhaps the most spectacular physical consequence of the parity effect in SNS rings is the presence of *spontaneous* supercurrents *in the ground state* of such rings with odd number of electrons. Similar to the case of standard π-junctions, such spontaneous supercurrents should flow even in the absence of an externally applied magnetic flux. Unlike in the usual case [74], however, here the spontaneous current state occurs for *any* inductance of the ring because of the non-sinusoidal dependence $I_o(\varphi)$.

Consider, for instance, the limit $d \gg v_F/\Delta_0$. In the case of odd number of electrons, the ground-state energy of an SNS ring can be written in a simple form

$$E = \frac{\Phi^2}{2c\mathcal{L}} + \frac{\pi v_F \mathcal{N}_n}{\Phi_0^2 d}\left(\Phi - \frac{\Phi_0 \operatorname{sgn}\Phi}{2\mathcal{N}_n}\right)^2, \tag{20.83}$$

where the first term is the magnetic energy of the ring ($\mathcal{L}$ is the ring inductance), while the second term represents the Josephson energy of an SNS junction. Minimizing Eq. (20.83) with respect to the flux Φ, one immediately concludes that the ground state of the ring is a twofold degenerate state with a non-vanishing spontaneous current

$$I = \pm\frac{ev_F}{d}\left[1 + \frac{2ev_F\mathcal{N}_n}{d}\frac{\mathcal{L}}{\Phi_0}\right]^{-1} \tag{20.84}$$

flowing either clockwise or counterclockwise. In the limit of small inductances $\mathcal{L} \to 0$, this current does not vanish and its amplitude just reduces to that of the odd electron current at $\varphi \to 0$. From the

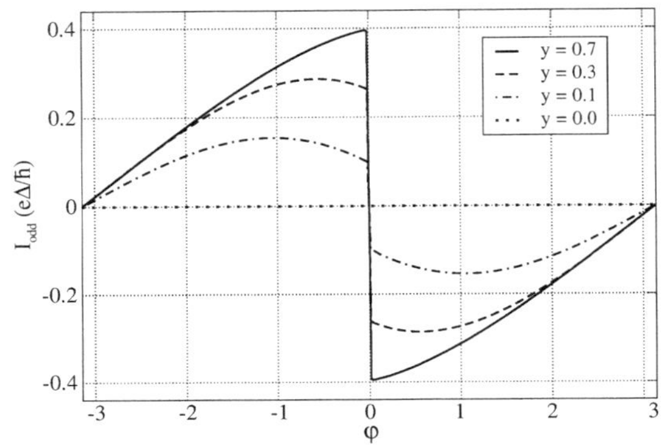

Figure 20.11 The zero-temperature current–phase relation $I_o(\varphi)$ $(-\pi < \varphi < \pi)$ for $\mathcal{N}_n = 1$ and different values of the parameter $y = d\Delta/v_{\rm F}$.

above results, one finds

$$I_{sp} = e\Delta_0^2 d/v_{\rm F}, \qquad d \ll v_{\rm F}/\Delta_0, \tag{20.85}$$

$$I_{sp} = ev_{\rm F}/\pi d, \qquad d \gg v_{\rm F}/\Delta_0. \tag{20.86}$$

At $d \sim v_{\rm F}/\Delta_0$, the amplitude of the current I_{sp} can be evaluated numerically. In agreement with Eq. (20.85), I_{sp} increases linearly with d at small d, reaches its maximum value $I_{\max} \sim 0.4e\Delta_0$ at $d \sim v_{\rm F}/\Delta_0$, and then decreases with further increase in d approaching the dependence (20.86) in the limit of large d. For generic BCS superconductors, the magnitude of this maximum current can be estimated as $I_{\max} \sim 10 \div 100$ nA. These values might be considered as surprisingly large ones having in mind that this current is associated with only one Andreev electron state.

The above analysis applies for sufficiently clean normal wires brought in a good electric contact with superconductors. Since both these assumptions can be violated in a realistic experiment, it is instructive to work out the corresponding modifications of the above results. This task will be accomplished below.

20.3.4 Resonant and Diffusive SNS Rings

Assume first that the transmissions of both NS interfaces are small $\mathcal{T}_{1,2} \ll 1$. In this limit, electron transport across the junction is mainly due to resonant tunneling through discrete energy levels inside the normal metal. For simplicity, we will restrict our analysis to a single-channel junction $\mathcal{N}_n = 1$. The most interesting physical situation is realized in the limit of short junction $d \ll v_F/\Delta_0$. In the case of a one-dimensional metal of length d, the level spacing in the vicinity of the Fermi energy is $\delta\epsilon \sim v_F/d$. Hence, the condition for the short junction regime can also be represented in the form $\delta\epsilon \gg \Delta_0$. Electron tunneling causes a nonzero linewidth of the energy levels, which is proportional to $\mathcal{T}_{1,2}\delta\epsilon$. This value is much smaller than $\delta\epsilon$; hence, the resonances remain well separated. In this situation, it suffices to take into account only the closest level to the Fermi energy inside the normal metal.

As before, making use of Eqs. (20.76) combined with the proper expression for the W-function [130], one finds [366]

$$I_{f/b} = eT \sum_{\omega_{f/b}} \frac{\Delta_0^2 \mathcal{T} \sin\varphi}{\varepsilon^2(\varphi) + \omega_{f/b}^2 (1 + 4\mathcal{D}\mathcal{T}/\mathcal{T}_{\max})}, \tag{20.87}$$

where

$$\mathcal{D} = \left(\frac{\Delta_0}{\Gamma}\right)^2 \left(1 + \frac{\omega_{f/b}^2}{\Delta_0^2}\right) + \frac{\Delta_0}{\Gamma}\sqrt{1 + \frac{\omega_{f/b}^2}{\Delta_0^2}} \tag{20.88}$$

$\varepsilon(\varphi) = \Delta_0\sqrt{1 - \mathcal{T}\sin^2(\varphi/2)}$, $\Gamma = \Gamma_1 + \Gamma_2$, $\Gamma_{1,2} = \mathcal{T}_{1,2} v_F/2d$ are the tunneling rates, $\mathcal{T}_{\max} = 4\Gamma_1\Gamma_2/\Gamma^2$, and the total transmission probability at the Fermi energy $\mathcal{T}$ is given by the Breit–Wigner formula

$$\mathcal{T} = \frac{\Gamma_1\Gamma_2}{(\epsilon_R)^2 + \frac{1}{4}\Gamma^2},$$

where ϵ_R is the energy of a resonant level.

It follows from Eqs. (20.87) and (20.88) that although the transparencies of both barriers are low, the total transmission $\mathcal{T}$ and, hence, the Josephson current shows sharp peaks provided the Fermi energy becomes close to a bound state inside the junction. On the other hand, Eqs. (20.87) and (20.88) demonstrate that even in the vicinity of resonances, the behavior of the Josephson current as a function of the phase difference φ and temperature T can substantially deviate from that for transparent SNS junctions.

In order to understand the physical reasons for such a difference, it is instructive to compare the structure of discrete Andreev

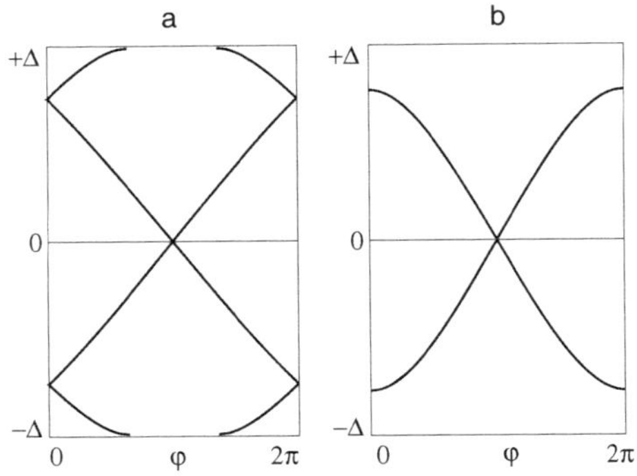

Figure 20.12 Andreev levels in a single-mode SNS junction with $d = v_F/\Delta$: (a) $\mathcal{T}_{1,2} = 1$ and (b) $\mathcal{T}_{1,2} \ll 1$ and $\Delta/\Gamma = 0.5$. In both cases, $\mathcal{T} = 1$.

levels for ballistic ($\mathcal{T}_{1,2} = 1$) SNS junctions with that for junctions with weakly transmitting NS interfaces $\mathcal{T}_{1,2} \ll 1$; see Fig. 20.12. The spectrum of the latter system consists only of a single non-degenerate state $\varepsilon_0(\varphi)$ in the interval $0 < \varepsilon_0 < \Delta_0$ (Fig. 20.12b). As a result, the behavior of $\varepsilon_0(\varphi)$ at small φ is smooth, and the derivative of ε_0 with respect to φ has no jump at $\varphi = 0$. In contrast, in the case of ballistic SNS junctions, discrete levels become split at arbitrary small values of the phase φ (Fig. 20.12a), and the derivative of the lowest Andreev level with respect to φ acquires a jump at $\varphi = 0$. As this feature is absent in resonant SNS junctions, the spontaneous current in the ground state of such systems can only develop at not very small ring inductances. The results for the supercurrent in SNS rings with resonant transmission are presented in Fig. 20.13. They clearly demonstrate that at sufficiently low temperatures, the π-junction state should be realized in the case of odd number of electrons.

Finally let us turn to the case of a disordered normal wire, which appears to be the most interesting one from the experimental point of view. As before, we will make use of Eq. (20.71), which relates the difference between the supercurrents I_o and I_e, respectively, in the odd and even ensembles at $T \to 0$ to the derivative of the lowest available energy state $\varepsilon_0(\Phi_x)$ occupied by the "odd" electron. In the case of a disordered metal connecting two superconductors, the energy $\varepsilon_0(\Phi_x)$ is just defined by the proximity-induced minigap $\varepsilon_g(\varphi)$ in the density of states of this metal [51].

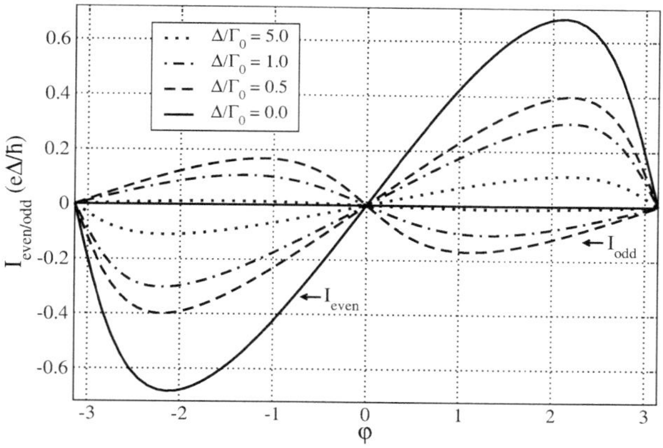

Figure 20.13 Zero-temperature current–phase relations $I_{e/o}(\varphi)$ for $\mathcal{T} = 0.9$, $\Gamma_1 = \Gamma_2 = \Gamma_0$, and different values of the parameter Δ_0/Γ_0.

At $T = 0$, the current I_e again identically coincides with one calculated for the grand canonical ensemble. Combining the result for $I_e(\varphi)$ with that for the minigap $\varepsilon_g(\varphi)$ and employing Eq. (20.71) with $\varepsilon_0 \equiv \varepsilon_g$ [212], we arrive at a typical dependence $I_o(\varphi)$ displayed in Fig. 20.14 (left panel). We observe that at sufficiently large values of $\varphi < \pi$, the absolute value of the odd electron contribution to persistent current $2e\partial\varepsilon_g/\partial\varphi$ exceeds the term $I_e(\varphi)$ and the total current I_o changes the sign. This nontrivial parity-affected current–phase relation is specific for SNS rings with disorder, and it substantially differs from the current–phase relations derived above for SNS rings with ballistic and resonant transmissions. At the same time, as in the previous cases, in the odd ensemble there exists a possibility both for a π-junction state and for spontaneous currents in the ground state of the system without any externally applied magnetic flux.

Let us evaluate the ground-state energy of the SNS junction by integrating Eq. (20.71) with respect to the phase φ. One finds

$$E_o(\varphi) = E_e(\varphi) - \varepsilon_g(0) + \varepsilon_g(\varphi), \quad E_e(\varphi) = \frac{1}{2e}\int_0^{\varphi} I_e(\varphi)d\varphi \quad (20.89)$$

where $E_{e/o}(\varphi)$ are the ground-state energies of SNS junction for even and odd number of electrons in the ring. While the energy $E_e(\varphi)$ is

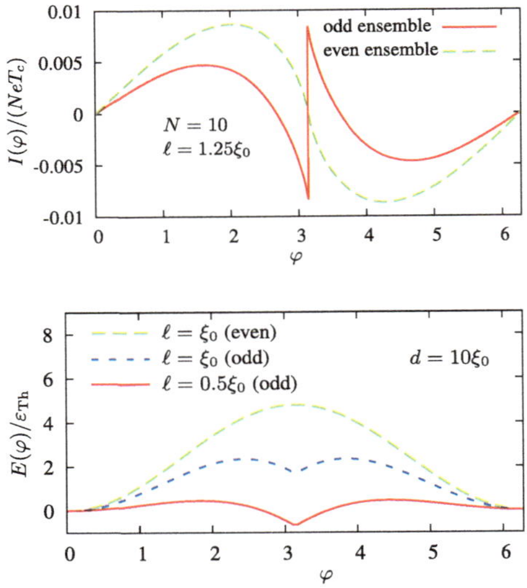

Figure 20.14 Top panel: Phase dependence of the Josephson current at $T = 0$ for the odd and even number of electrons in the ring. Bottom panel: Josephson energy $E(\varphi)$ of an SNS ring as a function of the phase difference φ for the even and odd ensembles. The solid curve corresponds to a π-junction state.

always non-negative and reaches its minimum at $\varphi = 0$, in the odd case the ground-state energy $E_o(\varphi)$ can become negative, reaching its absolute minimum at $\varphi = \pi$. This physical situation of a π-junction is illustrated in Fig. 20.14 (right panel).

It is easy to find out under which conditions the π-junction state becomes possible. For that purpose, it is sufficient to observe that for any impurity concentration, $E_e(\pi) = \gamma I_C/e$, where I_C is the grand canonical critical current at $T = 0$ and the prefactor $\gamma \sim 1$ depends on the form of the current–phase relation. The π-junction condition $E_o(\pi) < 0$ is equivalent to the inequality

$$\varepsilon_g(0) > \gamma I_C/e \tag{20.90}$$

It is obvious that in the many-channel limit, the inequality (20.90) cannot be satisfied for sufficiently large values of the mean free path ℓ. On the other hand, in the short mean free path limit, the current $I_C \propto \ell^2$ decays faster with decreasing ℓ as compared to the minigap $\varepsilon_g(0) \propto \ell$, and the π-junction state becomes possible. In particular, in the diffusive limit one has [103] $I_C \simeq 10.82\varepsilon_{Th}/eR_N = 1.53e\mathcal{N}_n v_F \ell^2/d^3$ and $\gamma \simeq 1.05$, where R_N is the Drude resistance of a normal metal.

Combining these results with the expression for the minigap $\varepsilon_g(0) \simeq 3.12\varepsilon_{Th}$ [51], from the condition (20.90) we observe that in the odd case, the π-junction state is realized provided the number of conducting channels in the junction $\mathcal{N}_n$ is smaller than

$$\mathcal{N}_n < 0.65d/\ell \tag{20.91}$$

This condition is not very restrictive, and it can certainly be achieved in experiments. For sufficiently dirty junctions, it allows for a formation of the π-junction state even in the many-channel limit. The condition (20.91) can also be rewritten as $g_N < 1.73$, where $g_N = 8\mathcal{N}_n\ell/3d$ is dimensionless conductance of a normal wire.

The condition for the presence of spontaneous currents in the ground state of SNS rings with an odd number of electrons is established analogously; one should only take into account an additional energy of the magnetic field produced by the supercurrent circulating inside the ring. The ground state with spontaneous currents is possible provided the total energy of the ring $E_{tot}(\pi)$ becomes negative, i.e.,

$$E_{tot}(\pi) = 1.8\varepsilon_{Th}\,[g_N - 1.73] + \frac{(\Phi_0/2)^2}{2\mathcal{L}} < 0 \tag{20.92}$$

This condition is more stringent than that for the π-junction state, but can also be satisfied provided $\mathcal{L}$ exceeds a certain threshold value, which can roughly be estimated as $\sim 0.1\Phi_0^2/\varepsilon_{Th}$.

We conclude that in the diffusive limit, the current–phase relation in the odd case is entirely different from that in the ballistic case. Also the restriction on the number of conducting channels $\mathcal{N}_n$ in the normal metal (20.91) is less stringent. This feature of diffusive SNS rings is rather advantageous for possible experimental observation of the effects discussed here. Still, in practice it would be necessary to fabricate SNS rings with a few conducting channels in the normal wire $\mathcal{N}_n \lesssim 10$. This condition is difficult to meet in conventional normal metals.

More promising candidates for practical realization of such structures are junctions with normal layers formed by carbon nanotubes or organic molecules where the number of channels $\mathcal{N}_n$ can be made

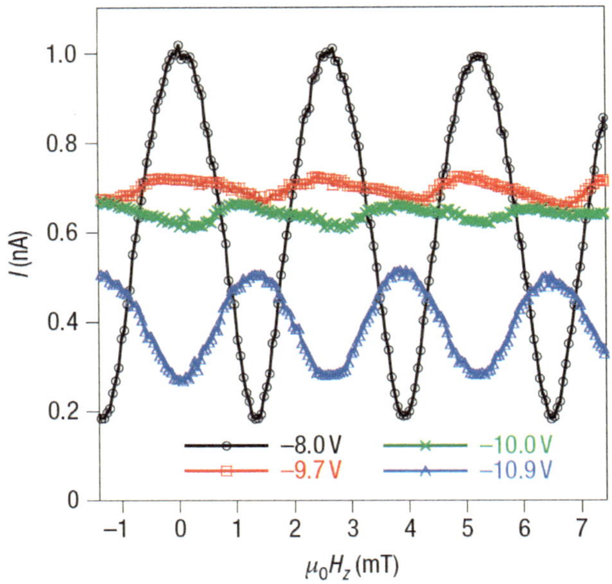

Figure 20.15 Gate-voltage-tuned π-junction behavior observed in experiments with carbon-nanotube-based SQUIDS. Reprinted by permission from Springer Nature: *Nature Nanotechnology*, Ref. [93], Copyright 2006.

sufficiently small. Observations of clear signatures of dc Josephson effect in superconducting junctions with a weak link formed by carbon nanotubes were reported by a large number of experimental groups [93, 180, 209, 210, 219, 380]. Furthermore, in one of these works [93], a gate-voltage-tuned π-junction state in SNS rings with carbon nanotubes was clearly demonstrated; see Fig. 20.15. This observation can be directly related to the superconducting parity effect discussed here. In any case, nanojunctions with carbon nanotubes are definitely among the most suitable objects, which can be employed to detect the influence of the parity effect on persistent currents in superconducting nanorings.

Bibliography

1. A. A. Abrikosov, L. P. Gor'kov, and I. Ye. Dzyaloshinskii (1975), *Methods of Quantum Field Theory in Statistical Physics*, Dover Publications, New York.
2. I. Affleck (1981), *Phys. Rev. Lett.* **46**, 388.
3. I. L. Aleiner, P. W. Brouwer, and L. I. Glazman (2002), *Phys. Rep.* **358**, 309.
4. I. L. Aleiner, B. L. Altshuler, and M. E. Gershenzon (1999a), *Waves Random Media* **9**, 201.
5. I. L. Aleiner, B. L. Altshuler, and M. E. Gershenzon (1999b), *Phys. Rev. Lett.* **82**, 3190.
6. B. Almog, S. Hakohen-Gourgy, A. Tsukernik, and G. Deutscher (2010), *Phys. Rev. B* **80**, 220512(R).
7. C. Altimiras, U. Gennser, A. Cavanna, D. Mailly, and F. Pierre (2007), *Phys. Rev. Lett.* **99**, 256805.
8. C. Altimiras, O. Parlavecchio, P. Joyez, D. Vion, P. Roche, D. Esteve, and F. Portier (2014), *Phys. Rev. Lett.* **112**, 236803.
9. F. Altomare, A. M. Chang, M. R. Melloch, Y. Hong, and C. W. Tu (2005), *Appl. Phys. Lett.* **86**, 172501.
10. F. Altomare, A. M. Chang, M. R. Melloch, Y. Yong, and C. W. Tu (2006), *Phys. Rev. Lett.* **97**, 017001.
11. B. L. Altshuler, A. G. Aronov, and D. E. Khmelnitskii (1982), *J. Phys. C* **15**, 7367.
12. B. L. Altshuler and A. G. Aronov (1985), in: *Electron–Electron Interactions in Disordered Conductors*, Elsevier Science Publishers B.V., New York.
13. G. M. Alzoubi and N. O. Birge (2006), *Phys. Rev. Lett.* **97**, 226803.
14. V. Ambegaokar and A. Baratoff (1963), *Phys. Rev. Lett.* **10**, 486.
15. V. Ambegaokar, U. Eckern, and G. Schön (1982), *Phys. Rev. Lett.* **48**, 1745.

16. V. Ambegaokar and B. I. Halperin (1969), *Phys. Rev. Lett.* **22**, 1364.
17. P. W. Anderson (1958), *Phys. Rev.* **109**, 1492.
18. P. W. Andreson and G. Yuval (1971), *J. Phys. C* **4**, 607.
19. A. F. Andreev (1964), *Sov. Phys. JETP* **19**, 1228.
20. A. F. Andreev (1966), *Sov. Phys. JETP* **22**, 455.
21. J. Ankerhold, P. Pechukas, and H. Grabert (2001), *Phys. Rev. Lett.* **87**, 086802.
22. J. Ankerhold (2004), *Europhys. Lett.* **67**, 280.
23. S. M. Apenko (1989), *Phys. Lett. A* **142**, 277.
24. K. Yu. Arutyunov, D. S. Golubev, and A. D. Zaikin (2008), *Phys. Rep.* **464**, 1.
25. L. G. Aslamazov, A. I. Larkin, and Yu. A. Ovchinnikov (1969), *Sov. Phys. JETP* **28**, 171.
26. O. V. Astafiev, L. B. Ioffe, S. Kafanov, Yu. A. Pashkin, K. Yu. Arutyunov, D. Shahar, O. Cohen, and J. S. Tsai (2012), *Nature* **484**, 355.
27. D. V. Averin (1987), *Sov. J. Low Temp. Phys.* **13**, 208.
28. D. Averin and H. T. Imam (1996), *Phys. Rev. Lett.* **76**, 3814.
29. D. V. Averin and K. K. Likharev (1986), *J. Low Temp. Phys.* **62**, 345.
30. D. V. Averin and K. K. Likharev (1991), in: *Mesoscopic Phenomena in Solids*, Elsevier, Amsterdam, p. 173.
31. D. V Averin and Yu. V. Nazarov (1990), *Phys. Rev. Lett.* **65**, 2446.
32. D. V. Averin and Yu. V. Nazarov (1992), in: *Single Charge Tunneling*, Vol. 294 of *NATO Advanced Study Institute, Series B: Physics*, Plenum, New York.
33. D. V. Averin and Yu. V. Nazarov (1992), *Phys. Rev. Lett.* **69**, 1993.
34. D. V Averin, Yu. V. Nazarov, and A. A. Odintsov (1990), *Physica B* **165/166**, 945.
35. D. V. Averin and A. A. Odintsov (1989), *Phys. Lett. A* **140**, 251.
36. D. V Averin, A. B. Zorin, and K. K. Likharev (1985), *Sov. Phys. JETP* **61**, 407.
37. D. A. Bagrets and Yu. V. Nazarov (2005), *Phys. Rev. Lett.* **94**, 056801.
38. J. Bardeen, L. N. Cooper, and J. R. Schrieffer (1957), *Phys. Rev.* **108**, 1175.
39. A. Barone and G. Paterno (1982), *Physics and Applications of the Josephson Effect*, Wiley, New York.
40. X. D. A. Baumans, D. Cerbu, O-A. Adami, V. S. Zharinov, N. Verellen, G. Papari, J. E. Scheerder, G. Zhang, V. V. Moshchalkov, A. V. Silhanek, and J. Van de Vondel (2016), *Nat. Commun.* **7**, 10560.

41. C. Bäuerle, F. Mallet, F. Shopfer, D. Mailly, G. Eska, and L. Saminadayar (2005), *Phys. Rev. Lett.* **95**, 266805.
42. D. Beckmann, H. B. Weber, and H. v. Löhneysen (2004), *Phys. Rev. Lett.* **93**, 197003.
43. C. W. J. Beenakker (1991), *Phys. Rev. Lett.* **67**, 3836.
44. C. W. J. Beenakker (1997), *Rev. Mod. Phys.* **69**, 731.
45. C. W. J. Beenakker, B. Rejaei, and J. A. Melsen (1994), *Phys. Rev. Lett.* **72**, 2470.
46. C. W. J. Beenakker and H. van Houten (1991), *Phys. Rev. Lett.* **66**, 3056.
47. I. S. Beloborodov, K. B. Efetov, A. V. Lopatin, and V. M. Vinokur (2003), *Phys. Rev. Lett.* **91**, 246801.
48. I. S. Beloborodov, A. V. Lopatin, and V. M. Vinokur (2003), *Phys. Rev. B* **72**, 125121.
49. I. S. Beloborodov, A. V. Lopatin, V. M. Vinokur, and K. B. Efetov (2007), *Rev. Mod. Phys.* **79**, 469.
50. W. Belzig (2003), in: *Quantum Noise in Mesoscopic Physics*, Kluwer, Dordrecht.
51. W. Belzig, F. K. Wilhelm, C. Bruder, G. Schön, and A. D. Zaikin (1999), *Superlatt. Microstr.* **25**, 1251.
52. E. Ben-Jacob and Y. Gefen (1985), *Phys. Lett. A* **108**, 289.
53. V. S. Berezinskii (1971), *Sov. Phys. JETP* **32** 493; **34**, 610.
54. G. Bergmann (1984), *Phys. Rep.* **107**, 1.
55. A. Bezryadin (2008), *J. Phys.: Cond. Mat.* **20**, 043202.
56. A. Bezryadin (2013), *Superconductivity in Nanowires*, Wiley-VCH, Weinheim.
57. A. Bezryadin, C. N. Lau, and M. Tinkham (2000), *Nature* **404**, 971.
58. G. Bignon, M. Houzet, F. Pistolesi, and F. W. J. Hekking (2004), *Europhys. Lett.* **67**, 110.
59. J. P. Bird, K. Ishibashi, D. K. Ferry, Y. Ochiai, Y. Aoyagi, and T. Sugano (1995), *Phys. Rev. B* **51**, 18037.
60. Ya. M. Blanter and M. Büttiker (2000), *Phys. Rep.* **336**, 1.
61. G. E. Blonder, M. Tinkham, and T. M. Klapwijk (1982), *Phys. Rev. B* **25**, 4515.
62. P. Bobbert, R. Fazio, G. Schön, and A. D. Zaikin (1992), *Phys. Rev. B* **45**, 2294.
63. A. T. Bollinger, A. Rogachev, and A. Bezryadin (2006), *Europhys. Lett.* **76**, 505.

64. R. M. Bradley and S. Doniach (1994), *Phys. Rev. B* **30**, 1138.
65. J. Brauer, F. Hübler, M. Smetanin, D. Beckmann, and H. v. Löhneysen (2010), *Phys. Rev. B* **81**, 024515.
66. A. J. Bray and M. A. Moore (1982), *Phys. Rev. Lett.* **76**, 505.
67. L. Bretheau, Ç.Ö. Girit, H. Pothier, D. Esteve, and C. Urbina (2013), *Nature* **499**, 312.
68. L. Bretheau, Ç.Ö. Girit, M. Houzet, H. Pothier, D. Esteve, and C. Urbina (2014), *Phys. Rev. B* **90**, 134506.
69. P. W. Brouwer and C. W. J. Beenakker (1996), *J. Math. Phys.* **37**, 4904.
70. P. W. Brouwer, A. Lamaraft, and C. Flensberg (2005), *Phys. Rev. B* **72**, 075316.
71. R. Brown and E. Simanek (1992), *Phys. Rev B* **45**, 6069.
72. C. Bruder, R. Fazio, and G. Schön (1994), *Physica B* **203**, 240.
73. H. P. Buchler, V. B. Geshkenbein, and G. Blatter (2004), *Phys. Rev. Lett.* **92**, 067007.
74. L. N. Bulaevskii, V. V. Kuzii, and A. A. Sobyanin (1977), *JETP Lett.* **25**, 290.
75. S. A. Bulgadaev (1984), *JETP Lett.* **39**, 315.
76. S. A. Bulgadaev (1987), *JETP Lett.* **45**, 622.
77. M. Büttiker (1992), *Phys. Rev. B* **46**, 12485.
78. J. M. Byers and M. E. Flatte (1995), *Phys. Rev. Lett.* **74**, 306.
79. P. Cadden-Zimansky and V. Chandrasekhar (2006), *Phys. Rev. Lett.* **97**, 237003.
80. A. O. Caldeira and A. J. Leggett (1981), *Phys. Rev. Lett.* **46**, 211.
81. A. O. Caldeira and A. J. Leggett (1983), *Ann. Phys. (N.Y.)* **149**, 374.
82. C. Callan and S. Coleman (1977), *Phys. Rev. D* **16**, 1762.
83. G. Campagnano and Yu. V. Nazarov (2006), *Phys. Rev. B* **74**, 125307.
84. P. Cedraschi, V. V. Ponomarenko, and M. Büttiker (2000), *Phys. Rev. Lett.* **84**, 346.
85. S. Chakravarty (1982), *Phys. Rev. Lett.* **49**, 681.
86. S. Chakravarty and A. J. Leggett (1984), *Phys. Rev. Lett.* **52**, 5.
87. S. Chakravarty and A. Schmid (1986), *Phys. Rep.* **140**, 193.
88. L. D. Chang and S. Chakravarty (1984), *Phys. Rev. B* **29**, 130; **30**, 1566 (E).
89. I. Chiorescu, Y. Nakamura, C. J. P. M. Harmans, and J. E. Mooij (2003), *Science* **299**, 1869.

90. D. Chouvaev, L. S. Kuzmin, D. S. Golubev, and A. D. Zaikin (1999), *Phys. Rev. B* **59**, 10599.

91. R. M. Clarke, I. H. Chan, C. M. Marcus, C. I. Duruoz, J. S. Harris, K. Campman, and A. C. Gossard (1995), *Phys. Rev. B* **52**, 2656.

92. A. N. Cleland, J. M. Martinis, and J. Clarke (1988), *Phys. Rev. B* **37**, 5950.

93. J.-P. Cleuziou, W. Wernsdorfer, V. Bouchiat, T. Ondarcuhu, and M. Monthioux (2006), *Nature Nanotechnol.* **1**, 53.

94. D. Controzzi, F. H. L. Essler, and A. M. Tsvelik (2001), *Phys. Rev. Lett.* **86**, 680.

95. A. J. Dahm, A. Denenstein, D. N. Langenberg, W. H. Parker, D. Rogovin, and D. J. Scalapino (1969), *Phys. Rev. Lett.* **22**, 1416.

96. A. V. Danilov, D. S. Golubev, and S. E. Kubatkin (2002), *Phys. Rev. B* **65**, 125312.

97. J.-F. Dayen, E. Devid, M. V. Kamalakar, D. S. Golubev, C. Guedon, V. Faramarzi, B. Doudin, and S. J. van der Molen (2013), *Adv. Mat.* **25**, 400.

98. M. J. M. de Jong and C. W. J. Beenakker (1994), *Phys. Rev. B* **49**, 16070.

99. G. Deutscher and D. Feinberg (2000), *Appl. Phys. Lett.* **76**, 487.

100. M. H. Devoret, J. M. Martinis, and J. Clarke (1985), *Phys. Rev. Lett.* **55**, 1908.

101. D. A. Dikin, M. J. Black, and V. Chandrasekhar (2001), *Phys. Rev. Lett.* **87**, 187003.

102. O. N. Dorokhov (1984), *Solid State Commun.* **51**, 381.

103. P. Dubos, H. Courtois, B. Pannetier, F. K. Wilhelm, A. D. Zaikin, and G. Schön (2001), *Phys. Rev. B* **63**, 064502.

104. I. E. Dzyaloshinskii and A. I. Larkin (1974), *Sov. Phys. JETP* **38**, 202.

105. P. M. Echternach, M. E. Gershenson, and H. M. Bozler (1993a), *Phys. Rev. B* **47**, 13659.

106. P. M. Echternach, M. E. Gershenson, H. M. Bozler, A. L. Bogdanov, and B. Nilsson (1993b), *Phys. Rev. B* **48**, 11516.

107. U. Eckern, G. Schön, and V. Ambegaokar (1984), *Phys. Rev. B* **30**, 6419.

108. U. Eckern and P. Schwab (2002), *J. Low Temp. Phys.* **126**, 1291.

109. A. L. Efros and B. I. Shklovskii (1975), *J. Phys. C* **8**, L49.

110. G. Eilenberger (1968), *Z. Phys.* **214**, 195.

111. K. A. Eriksen, P. Hedegard, and H. Bruus (2001), *Phys. Rev. B* **64**, 195327.

112. G. Falci, D. Feinberg, and F. W. J. Hekking (2001), *Europhys. Lett.* **54**, 255.

113. G. Falci, G. Schön, and G. Zimanyi (1995), *Phys. Rev. Lett.* **74**, 3257.
114. Sh. Farhangfar, R. S. Poikolainen, J. P. Pekola, D. S. Golubev, and A. D. Zaikin (2001), *Phys. Rev. B* **63**, 075309.
115. R. Fazio and H. van der Zant (2001), *Phys. Rep.* **355**, 235.
116. P. Fendley, A. W. W. Ludwig, and H. Saleur (1995), *Phys. Rev. B* **52**, 8934.
117. R. P. Feynman (1972), *Statistical Mechanics*, Benjamin, New York.
118. R. P. Feynman and A. R. Hibbs (1965), *Quantum Mechanics and Path Integrals*, McGraw-Hill, New York.
119. R. P. Feynman and F. L. Vernon (1963), *Ann. Phys. (N.Y.)* **24**, 118.
120. A. M. Finkelstein (1984), *Sov. Phys. JETP* **59**, 212.
121. D. S. Fisher and P. A. Lee (1981), *Phys. Rev. B* **23**, 6851.
122. M. P. A. Fisher and W. Zwerger (1985), *Phys. Rev. B* **32**, 6190.
123. A. Freyn, M. Flöser, and R. Melin (2010), *Phys. Rev. B* **82**, 014510.
124. T. A. Fulton and L. N. Dunkleberger (1974), *Phys. Rev. B* **9**, 4760.
125. A. Furusaki and K. A. Matveev (1995), *Phys. Rev. Lett.* **75**, 709; *Phys. Rev. B* **52**, 16676.
126. J. Gabelli and B. Reulet (2009), *J. Stat. Mech.*, P01049
127. A. V. Galaktionov, D. S. Golubev, and A. D. Zaikin (2003a), *Phys. Rev. B* **68**, 085317.
128. A. V. Galaktionov, D. S. Golubev, and A. D. Zaikin (2003b), *Phys. Rev. B* **68**, 235333.
129. A. V. Galaktionov, D. S. Golubev, and A. D. Zaikin (2017), *Phys. Rev. B* **96**, 134509.
130. A. V. Galaktionov and A. D. Zaikin (2002), *Phys. Rev. B* **65**, 184507.
131. A. V. Galaktionov and A. D. Zaikin (2006), *Phys. Rev. B* **73**, 184522.
132. A. V. Galaktionov and A. D. Zaikin (2009), *Phys. Rev. B* **80**, 174527.
133. A. V. Galaktionov and A. D. Zaikin (2010), *Phys. Rev. B* **82**, 184520.
134. A. V. Galaktionov and A. D. Zaikin (2011), *Phys. Rev. B* **84**, 115418.
135. A. V. Galaktionov and A. D. Zaikin (2015), *Phys. Rev. B* **92**, 214511.
136. Y. Gefen and O. Entin-Wohlman (1991), *Ann. Phys.* **206**, 68.
137. F. Giazotto, T. Heikkilä, A. Luukanen, A. M. Savin, and J. P. Pekola (2006), *Rev. Mod. Phys.* **78**, 217.
138. V. L. Ginzburg (1960), *Sov. Phys. Solid State* **2**, 1824.
139. N. Giordano (1988), *Phys. Rev. Lett.* **61**, 2137.
140. N. Giordano (1991), *Phys. Rev. B* **43**, 160.
141. N. Giordano (1994), *Physica B* **203**, 460.

142. D. S. Golubev, A. V. Galaktionov, and A. D. Zaikin (2005), *Phys. Rev. B* **72**, 205417.

143. D. S. Golubev, C. P. Herrero, and A. D. Zaikin (2003), *Europhys. Lett.* **63**, 426.

144. D. S. Golubev, M. S. Kalenkov, and A. D. Zaikin (2009), *Phys. Rev. Lett.* **103**, 067006.

145. D. S. Golubev, J. König, H. Schoeller, G. Schön, and A. D. Zaikin (1997), *Phys. Rev. B* **56**, 15782.

146. D. S. Golubev and A. D. Zaikin (1992a), *Phys. Rev. B* **46**, 10903.

147. D. S. Golubev and A. D. Zaikin (1992b), *Phys. Lett. A* **164**, 337.

148. D. S. Golubev and A. D. Zaikin (1992c), *Phys. Lett. A* **169**, 475.

149. D. S. Golubev and A. D. Zaikin (1994a), *Phys. Rev. B* **50**, 8736.

150. D. S. Golubev and A. D. Zaikin (1994b), *Phys. Lett.* **A195**, 380.

151. D. S. Golubev and A. D. Zaikin (1995), in: *Quantum Dynamics of Submicron Structures*, Vol.291 of NATO ASI Series E, p. 473, Kluwer, Dordrecht.

152. D. S. Golubev and A. D. Zaikin (1996), *JETP Lett.* **63**, 1007.

153. D. S. Golubev and A. D. Zaikin (1998a), *Phys. Rev. Lett.* **81**, 1074.

154. D. S. Golubev and A. D. Zaikin (1998b), *Physica B* **255**, 164.

155. D. S. Golubev and A. D. Zaikin (1999a), *Phys. Rev. B* **59**, 9195.

156. D. S. Golubev and A. D. Zaikin (1999b), *Phys. Rev. Lett.* **82**, 3191.

157. D. S. Golubev and A. D. Zaikin (2000a), *Phys. Rev. B* **62**, 14061.

158. D. S. Golubev and A. D. Zaikin (2000b), in: *Quantum Physics at Mesoscopic Scale*, p. 491, EDP Sciences.

159. D. S. Golubev and A. D. Zaikin (2001a), *Phys. Rev. Lett.* **86**, 4887.

160. D. S. Golubev and A. D. Zaikin (2001b), *Phys. Rev. B* **64**, 014504.

161. D. S. Golubev and A. D. Zaikin (2002), *Europhys. Lett.* **60**, 113.

162. D. S. Golubev and A. D. Zaikin (2003), *J. Low Temp. Phys.* **132**, 11.

163. D. S. Golubev and A. D. Zaikin (2004a), *Phys. Rev. B* **69**, 075318.

164. D. S. Golubev and A. D. Zaikin (2004b), *Phys. Rev. B* **70**, 165423.

165. D. S. Golubev and A. D. Zaikin (2006), *Phys. Rev. B* **74**, 245329.

166. D. S. Golubev and A. D. Zaikin (2007), *Physica E* **40**, 32.

167. D. S. Golubev and A. D. Zaikin (2008a), *New J. Phys.* **10**, 063027.

168. D. S. Golubev and A. D. Zaikin (2008b), *Phys. Rev. B* **78**, 144502

169. D. S. Golubev and A. D. Zaikin (2009), *Europhys. Lett.* **86**, 37009.

170. D. S. Golubev and A. D. Zaikin (2010), *Phys. Rev. B* **82**, 134508.

171. D. S. Golubev, A. D. Zaikin, and G. Schön (2002), *J. Low Temp. Phys.* **126**, 1355.
172. A. A. Golubov, M. Yu. Kupriyanov, and E. Il'ichev (2004), *Rev. Mod. Phys.* **76**, 411.
173. L. P. Gorkov, A. I. Larkin, and D. E. Khmelnitskii (1979), *JETP Lett.* **30**, 248.
174. G. Göppert and H. Grabert (2001), *Phys. Rev. B* **63**, 125307.
175. G. Göppert, H. Grabert, N. Prokof'ev, and B. Svistunov (1998), *Phys. Rev. Lett.* **81**, 2324.
176. H. Grabert, P. Schramm, and G. L. Ingold (1988), *Phys. Rep.* **168**, 115.
177. H. Grabert, U. Weiss, and P. Hanggi (1984), *Phys. Rev. Lett.* **52**, 2193.
178. H. Grabert and U. Weiss (1984), *Phys. Rev. Lett.* **53**, 1787.
179. H. Grabert, P. Olschowski, and U. Weiss (1987), *Phys. Rev. B* **36**, 1931.
180. K. Grove-Rasmussen, H. I. Jorgensen, and P. E. Lindelof (2007), *New J. Phys.* **9**, 124.
181. F. Guinea (2002), *Phys. Rev. B* **65**, 205317.
182. F. Guinea, V. Hakim, and A. Muramatsu (1985), *Phys. Rev. Lett.* **54**, 263.
183. F. Guinea and G. Schön (1986), *Europhys. Lett.* **1**, 585.
184. F. Guinea and G. Schön (1987), *J. Low Temp. Phys.* **69**, 219.
185. W. Haberkorn, H. Knauer, and J. Richter (1978), *Phys. Stat. Solidi (A)* **47**, K161.
186. B. Hackens, S. Faniel, C. Gustin, X. Wallart, S. Bollaert, A. Cappy, and V. Bayot (2005), *Phys. Rev. Lett.* **94**, 146802.
187. V. Hakim and V. Ambegaokar (1985), *Phys. Rev. B* **32**, 423.
188. M. Hayashi and H. Ebisawa (2003), *Phys. Rev. B* **67**, 014524.
189. F. W. J. Hekking and L. I. Glazman (1997), *Phys. Rev. B* **55**, 6551.
190. F. W. J. Hekking and Yu. V. Nazarov (1993), *Phys. Rev. Lett.* **71**, 1625.
191. F. W. J. Hekking and Yu. V. Nazarov (1994), *Phys. Rev. B* **49**, 6847.
192. C. P. Herrero, G. Schön, and A. D. Zaikin (1999), *Phys. Rev. B* **59**, 5728.
193. C. P. Herrero and A. D. Zaikin (2002), *Phys. Rev. B* **65**, 104516.
194. W. Hofstetter and W. Zwerger (1997), *Phys. Rev. Lett.* **78**, 3737.
195. P. C. Hohenberg (1967), *Phys. Rev.* **158**, 383.
196. B. Horovitz and P. le Doussal (2006), *Phys. Rev. B* **74**, 073104.
197. K. Huang (1982), *Quarks, Leptons and Gauge Fields*, World Scientific, Singapore.

198. S. M. Huang, H. Akimoto, K. Kono, and J. J. Lin (2007), *Phys. Rev. Lett.* **99**, 046601 (2007).

199. A. Huck, F. W. J. Hekking, and B. Kramer (1998), *Europhys. Lett.* **41**, 201.

200. A. G. Huibers, J. A. Folk, S. R. Patel, C. M. Marcus, C. I. Duruoz, J. S. Harris (1999), *Phys. Rev. Lett.* **83**, 5090.

201. G. L. Ingold and H. Grabert (1999), *Phys. Rev. Lett.* **83**, 3721.

202. G. L. Ingold and Yu. V. Nazarov (1992), in: *Single Charge Tunneling*, NATO ASI, Ser. B, Vol. 294, p. 21, Plenum, New York.

203. C. Ishii (1970), *Progr. Theor. Phys.* **44**, 1525.

204. Yu. M. Ivanchenko and L. A. Zil'berman (1969), *JETP* **28**, 1272.

205. L. D. Jackel, J. P. Gordon, E. L. Hu, R. E. Howard, L. A. Fetter, D. M. Tennant, R. W. Epworth, and J. Kurkijärvi (1981), *Phys. Rev. Lett.* **47**, 697.

206. B. Janko, A. Smith, and V. Ambegaokar (1994), *Phys. Rev. B* **50**, 1152.

207. X. Jehl, M. Sanquer, R. Calemczuk, and D. Mailly (2000), *Nature* (London) **405**, 50.

208. S. Jezouin, M. Albert, F. D. Parmentier, A. Anthore, U. Gennser, A. Cavanna, I. Safi, and F. Pierre (2013), *Nat. Commun.* **4**, 1802.

209. P. Jarillo-Herrero, J. A. van Dam, and L. P. Kouwenhoven (2006), *Nature* **439**, 953.

210. H. I. Jorgensen, K. Grove-Rasmussen, T. Novotny, K. Flensberg, and P. E. Lindelof (2006), *Phys. Rev. Lett.* **96**, 207003.

211. P. Joyez, V. Bouchiat, D. Esteve, C. Urbina, and M. H. Devoret (1997), *Phys. Rev. Lett.* **79**, 1349.

212. M. S. Kalenkov, H. Kloos, and A. D. Zaikin (2006), *Phys. Rev. B* **74**, 184502.

213. M. S. Kalenkov and A. D. Zaikin (2007a), *Phys. Rev. B* **75**, 172503.

214. M. S. Kalenkov and A. D. Zaikin (2007b), *Phys. Rev. B* **76**, 224506.

215. C. L. Kane and M. P. A. Fisher (1992), *Phys. Rev. B* **46**, 15233.

216. K. Kang (2000), *Europhys. Lett.* **51**, 181.

217. D. M. Kaplan, V. A. Sverdlov, and K. K. Likharev (2003), *Phys. Rev. B* **68**, 045321.

218. A. Kastalsky, A. W. Kleinsasser, L. H. Greene, R. Bhat, P. W. Milliken, and J. P. Harbison (1991), *Phys. Rev. Lett.* **67**, 3026.

219. A. Yu. Kasumov, R. Deblock, M. Kociak, B. Reulet, H. Bouchiat, I. I. Khodos, Yu. B. Gorbatov, V. T. Volkov, C. Journet, and M. Burghard (1998), *Science* **281**, 540.

220. L. V. Keldysh (1965), *Sov. Phys. JETP* **20**, 1018.

221. S. Khlebnikov (2004), *Phys. Rev. Lett.* **93**, 090403.
222. V. A. Khlus (1987), *Sov. Phys. JETP*, **66**, 1243.
223. M. Kindermann and Yu. Nazarov (2003), in: *Quantum Noise in Mesoscopic Physics*, Kluwer Academic Publishers.
224. M. Kindermann, Yu. V. Nazarov, and C. W. J. Beenakker (2003), *Phys. Rev. Lett.* **90**, 246805.
225. M. Kindermann and Yu. V. Nazarov (2003), *Phys. Rev. Lett.* **91**, 136802.
226. M. Kindermann and B. Trauzettel (2005), *Phys. Rev. Lett.* **94**, 166803.
227. A. Kleine, A. Baumgartner, J. Trbovic, and C. Schönenberger (2009), *Europhys. Lett.* **87**, 27011.
228. A. Kleine, A. Baumgartner, J. Trbovic, D. S. Golubev, A. D. Zaikin, and C. Schönenberger (2010), *Nanotechnology* **21**, 274002.
229. H. Kleinert and T. Sauer (1990), *J. Low Temp. Phys.* **81**, 123.
230. S. Kolenda, M. J. Wolf, D. S. Golubev, A. D. Zaikin, and D. Beckmann (2013), *Phys. Rev. B* **88**, 174509.
231. S. E. Korshunov (1987a), *JETP Lett.* **45**, 434.
232. S. E. Korshunov (1987b), *Sov. Phys. JETP* **65**, 1025.
233. J. M. Kosterlitz (1974), *J. Phys. C* **7**, 1046.
234. J. M. Kosterlitz (1976), *Phys. Rev. Lett.* **37**, 1577.
235. J. M. Kosterlitz and D. J. Thouless (1973), *J. Phys. C* **6**, 1181.
236. J. König and H. Schoeller (1998), *Phys. Rev. Lett.* **81**, 3511.
237. J. König, H. Schoeller, and G. Schön (1995), *Europhys. Lett.* **31**, 31.
238. A. A. Kozhevnikov, R. J. Schoelkopf, and D. E. Prober (2000), *Phys. Rev. Lett.* **84**, 3398.
239. H. Kramers (1940), *Physica (Utrecht)* **7**, 284.
240. V. A. Krupenin, A. B. Zorin, M. N. Savvateev, D. E. Presnov, and J. Niemeyer (2001), *J. Appl. Phys.* **90**, 2411.
241. I. O. Kulik (1970), *Sov. Phys. JETP* **30**, 944.
242. I. O. Kulik and A. N. Omel'yanchuk (1978), *Sov. J. Low Temp. Phys.* **4**, 142.
243. I. O. Kulik and R. I. Shekhter (1975), *Sov. Phys. JETP* **41**, 308.
244. M. Yu. Kupriyanov and V. F. Lukichev (1988), *Sov. Phys. JETP* **94**, 139.
245. L. S. Kuzmin and D. B. Haviland (1991), *Phys. Rev. Lett.* **67**, 2890.
246. L. S. Kuzmin, Yu. V. Nazarov, D. B. Haviland, P. Delsing, and T. Claeson (1991), *Phys. Rev. Lett.* **67**, 1161.
247. L. S. Kuzmin, Yu. A. Pashkin, D. S. Golubev, and A. D. Zaikin (1996), *Phys. Rev. B* **54**, 10074.

248. H.-J. Kwon and V. M. Yakovenko (2002), *Phys. Rev. Lett.* **89**, 017002.
249. P. Lafarge, P. Joyez, D. Esteve, C. Urbina, and M. H. Devoret (1993), *Phys. Rev. Lett.* **70**, 994.
250. L. D. Landau and E. M. Lifshitz, *Quantum Mechanics: Non-Relativistic Theory*, 3rd enlarged edition, Pergamon Press, Oxford.
251. J. S Langer (1967), *Ann. Phys. (N.Y.)* **41**, 108.
252. J. S. Langer (1968), *Phys. Rev. Lett.* **21**, 973.
253. J. S. Langer (1969), *Ann. Phys. (N.Y.)* **54**, 258.
254. J. S. Langer and V. Ambegaokar (1967), *Phys. Rev.* **164**, 498.
255. L. Langer, V. Bayot, E. Grivei, J.-P. Issi, J. P. Heremans, C. H. Olk, L. Stockman, C. Van Haesendonck, and Y. Bruynseraede (1996), *Phys. Rev. Lett.* **76**, 479.
256. A. I. Larkin and Yu. N. Ovchinnikov (1967), *Sov. Phys. JETP* **24**, 1035.
257. A. I. Larkin and Yu. N. Ovchinnikov (1983), *JETP Lett.* **37**, 322.
258. A. I. Larkin and Yu. N. Ovchinnikov (1984), *Sov. Phys. JETP* **59**, 420.
259. A. I. Larkin and A. Varlamov (2005), *Theory of Fluctuations in Superconductors*, Clarendon, Oxford.
260. C. N. Lau, N. Markovic, M. Bockrath, A. Bezryadin, and M. Tinkham (2001), *Phys. Rev. Lett.* **87** 217003.
261. Q. Le Masne, H. Pothier, N. O. Birge, C. Urbina, and D. Esteve (2009), *Phys. Rev. Lett.* **102**, 067002.
262. H. Lee and M. S. Levitov (1996), *Phys. Rev. B* **53**, 7383.
263. A. J. Leggett (1980), *Progr. Theor. Phys. Suppl.* **69**, 80.
264. A. J. Leggett, S. Chakravarty, A. T. Dorsey, M. P. A. Fisher, A. Garg, and W. Zwerger (1987), *Rev. Mod. Phys.* **59**, 1.
265. J. S. Lehtinen, T. Sajavaara, K. Yu. Arutyunov, M. Yu. Presnjakov, and A. L. Vasiliev (2012a), *Phys. Rev. B* **85**, 094508.
266. J. S. Lehtinen, K. Zakharov, and K. Yu. Arutyunov (2012b), *Phys. Rev. Lett.* **109**, 187001.
267. F. Lesage and H. Saleur (1998), *Phys. Rev. Lett.* **80**, 4370.
268. A. Levchenko and A. Kamenev (2007), *Phys. Rev. B* **76**, 094518.
269. L. S. Levitov, H. Lee, and G. B. Lesovik (1996), *J. Math. Phys.* **37**, 4845.
270. L. S. Levitov and M. Reznikov (2004), *Phys. Rev. B* **70**, 115305.
271. A. Levy Yeyati, A. Martin-Rodero, D. Esteve, and C. Urbina (2001), *Phys. Rev. Lett.* **87**, 046802.
272. A. Levy Yeyati, F. S. Bergeret, A. Martin-Rodero, and T. M. Klapwijk (2007), *Nat. Phys.* **3**, 455.

273. E. M. Lifshitz and L. P. Pitaevskii (1980), *Statistical Physics, Part 2*, Vol. 9 (1st ed.), Butterworth-Heinemann, Oxford.
274. E. M. Lifshitz and L. P. Pitaevskii (1981), *Physical Kinetics*, Vol. 10 (1st ed.), Butterworth-Heinemann, Oxford.
275. K. K. Likharev (1986), *Dynamics of Josephson Junctions and Circuits*, Gordon and Breach Publishers, Amsterdam.
276. K. K. Likharev and A. B. Zorin (1985), *J. Low Temp. Phys.* **59**, 347.
277. B. Limbach, P. vom Stein, C. Wallisser, and R. Scäfer (2005), *Phys. Rev. B* **72**, 045319.
278. J. J. Lin and J. P. Bird (2002), *J. Phys.: Condens. Matter* **14**, R501.
279. J. F. Lin, J. P. Bird, and L. Rotkina (2003a), *Physica E* **19**, 112.
280. J. F. Lin, J. P. Bird, L. Rotkina, and P. A. Bennett (2003b), *Appl. Phys. Lett.* **82**, 802.
281. J. J. Lin and L. Y. Kao (2001), *J. Phys.: Condens. Matter* **13**, L119.
282. J. J. Lin, T. C. Lee, and S. W. Wang (2007), *Physica E* **40**, 25.
283. J. K. Lindhard (1954), *Mat. Fys. Medd. K. Dan. Vidensk. Selsk.* **28**(8).
284. W. A. Little (1967), *Phys. Rev.* **156**, 396.
285. K. Liu, S. Roth, G. S. Düsberg, G. T. Kim, D. Popa, K. Mukhopadhyay, R. Doome, and J. B. Nagy (2000), *Phys. Rev. B* **61**, 2375.
286. A. M. Lobos, A. Iucci, M. Müller, and T. Giamarchi (2009), *Phys. Rev. B* **80**, 214515.
287. T. Ludwig and A. D. Mirlin (2004), *Phys. Rev. B* **69**, 193306.
288. J. E. Lukens, R. J. Warburton, and W. W. Webb (1970), *Phys. Rev. Lett.* **25**, 1180.
289. S. L. Lukyanov and P. Werner (2006), *J. Stat. Mech.*, P11002.
290. A. Maasen van der Brink, U. Geigenmüller, and A. D. Zaikin (1990), *Physica B* **165/166**, 939.
291. Y. Makhlin, G. Schön, and A. Shnirman (2001), *Rev. Mod. Phys.* **73**, 357.
292. F. Mallet, J. Ericsson, D. Mailly, S. Unlubayir, D. Reuter, A. Melnikov, A. D. Wieck, T. Micklitz, A. Rosch, T. A. Costi, L. Saminadayar, and C. Bäuerle (2006), *Phys. Rev. Lett.* **97**, 226804.
293. A. Martin-Rodero, A. Levy Yeyati, and F. J. Garcia-Vidal (1996), *Phys. Rev. B* **53**, R8891.
294. D. C. Mattis and J. Bardeen (1958), *Phys. Rev.* **111** 412.
295. K. A. Matveev (1991), *Sov. Phys. JETP* **72**, 892.
296. K. A. Matveev (1995), *Phys. Rev. B* **51**, 1743.

297. K. A. Matveev, A. I. Larkin, and L. I. Glazman (2002), *Phys. Rev. Lett.* **89**, 096802.
298. D. E. McCumber and B. I. Halperin (1970), *Phys. Rev. B* **1**, 1054.
299. P. A. Mello (1988), *Phys. Rev. Lett.* **60**, 1089.
300. V. I. Melnikov and S. V. Meshkov (1983), *JETP Lett.* **38**, 130.
301. N. D. Mermin and H. Wagner (1966), *Phys. Rev. Lett.* **17**, 1133.
302. A. A. Middleton and N. S. Wingreen (1993), *Phys. Rev. Lett.* **71**, 3198.
303. P. Mohanty (2001), in: *Complexity from Microscopic to Macroscopic Scales: Coherence and Large Deviations*, Kluwer, Dordrecht.
304. P. Mohanty, E. M. Q. Jariwala, and R. A. Webb (1997), *Phys. Rev. Lett.* **78**, 3366.
305. P. Mohanty and R. A. Webb (2003), *Phys. Rev. Lett.* **91**, 066604.
306. J. E. Mooij and G. Schön (1985), *Phys. Rev. Lett.* **55**, 114.
307. J. E. Mooij and C. J. P. M. Harmans (2005), *New J. Phys.* **7**, 219.
308. J. E. Mooij and Yu. V. Nazarov (2006), *Nat. Phys.* **2**, 169.
309. M. V. Moskalets (2001), *Physica B* **301**, 286.
310. B. A. Muzykantskii and D. E. Khmelnitskii (1994), *Phys. Rev. B* **50**, 3982.
311. K. E. Nagaev and M. Büttiker (2001), *Phys. Rev. B* **63**, 081301.
312. Y. Nakamura, Yu. A. Pashkin, and J. C. Tsai (1999), *Nature (London)* **398**, 786.
313. D. Natelson, R. L. Willett, K. W. West, and L. N. Pfeiffer (2001), *Phys. Rev. Lett.* **86**, 1821.
314. Yu. V. Nazarov (1999), *Superlatt. Microst.* **25**, 1221.
315. Yu. V. Nazarov (1999), *Phys. Rev. Lett.* **82**, 1245.
316. R. S. Newbower, M. R. Beasley, and M. Tinkham (1972), *Phys. Rev. B* **5**, 864.
317. M. Noguchi, T. Ikoma, T. Odagiri, H. Sakakibara, and S. N. Wang (1996), *J. Appl. Phys.* **80**, 5138.
318. S. Oberholzer, E. V. Sukhorukov, C. Strunk, and C. Schoenenberger (2002), *Phys. Rev. B* **66**, 233304.
319. A. A. Odintsov (1988), *Sov. Phys. JETP* **67**, 1265.
320. Y. Oreg and A. M. Finkelstein (1999), *Phys. Rev. Lett.* **83**, 191.
321. S. V. Panyukov and A. D. Zaikin (1988a), *J. Low Temp. Phys.* **73**, 1.
322. S. V. Panyukov and A. D. Zaikin (1988b), *Physica B* **152**, 162.
323. S. V. Panyukov and A. D. Zaikin (1991), *Phys. Rev. Lett.* **67**, 3168.

324. F. D. Parmentier, A. Anthore, S. Jezouin, H. le Sueur, U. Gennser, A. Cavanna, D. Mailly, and F. Pierre (2011), *Nat. Phys.* **7**, 935.

325. J. P. Pekola, K. P. Hirvi, J. P. Kauppinen, and M. A. Paalanen (1994), *Phys. Rev. Lett.* **73**, 2903.

326. J. T. Peltonen, O. V. Astafiev, Yu. P. Korneeva, B. M. Voronov, A. A. Korneev, I. M. Charaev, A. V. Semenov, G. N. Golt'sman, L. B. Ioffe, T. M. Klapwijk, and J. S. Tsai (2013), *Phys. Rev. B* **88**, 220506.

327. J. S. Penttila, Ü. Parts, P. J. Hakonen, M. A. Paalanen, and E. B. Sonin (1999), *Phys. Rev. Lett.* **82**, 1004.

328. J. S. Penttilä, Ü. Parts, P. J. Hakonen, M. A. Paalanen, and E. B. Sonin (2000), *Phys. Rev. B* **61**, 10890.

329. J. S. Penttila, P. J. Hakonen, E. B. Sonin, and M. A. Paalanen (2001), *J. Low Temp. Phys.* **125**, 89.

330. F. Pierre, A. B. Gougam, A. Anthore, H. Pothier, D. Esteve, and N. O. Birge (2003), *Phys. Rev. B* **68**, 085413.

331. F. Pistolesi, G. Bignon, and F. W. J. Hekking (2001), *Phys. Rev. B* **69**, 214518.

332. D. P. Pivin, A. Andresen, J. P. Bird, and D. K. Ferry (1999), *Phys. Rev. Lett.* **82**, 4687.

333. M. L. Polianski and P. W. Brouwer (2003), *J. Phys. A: Math. Gen.* **36**, 3215.

334. P. Ramond (1997), *Field Theory: A Modern Primer*, Second Edition, Westview Press, Boulder, CO.

335. B. Reulet, J. Senzier, and D. E. Prober (2003), *Phys. Rev. Lett.* **91**, 196601.

336. P. S. Riseborough, P. Hanggi, and E. Freidkin (1985), *Phys. Rev. A* **32**, 489.

337. S. Russo, M. Kroug, T. M. Klapwijk, and A. F. Morpurgo (2005), *Phys. Rev. Lett.* **95**, 027002.

338. I. Safi and H. Saleur (2004), *Phys. Rev. Lett.* **93**, 126602.

339. A. Sahnoune, J. O. Strom-Olsen, and H. E. Fisher (1992), *Phys. Rev. B* **46**, 10035.

340. H. Saleur and U. Weiss (2001), *Phys. Rev. B* **63**, 201302(R).

341. J. Salo, F. W. J. Hekking, and J. Pekola (2006), *Phys. Rev. B* **74**, 125427.

342. L. Saminadayar, P. Mohanty, R. A. Webb, P. Degiovanni, and C. Bäuerle (2007), *Physica E* **40**, 12.

343. H. Schoeller and G. Schön (1994), *Phys. Rev. B* **50**, 18436.

344. G. Schön and A. D. Zaikin (1989), *Phys. Rev. B* **40**, 5231.

345. G. Schön and A. D. Zaikin (1990), *Phys. Rep.* **198**, 237.

346. C. Schönenberger, A. Bachtold, C. Strunk, J.-P. Salvetat, and L. Forro (1999), *Appl. Phys.* **A 69**, 283.

347. F. Schopfer, C. Bäuerle, W. Rabaud, and L. Saminadayar (2003), *Phys. Rev. Lett.* **90**, 056801.

348. A. Schmid (1982), *J. Low Temp. Phys.* **49**, 609.

349. A. Schmid (1983), *Phys. Rev. Lett.* **51**, 1506.

350. D. B. Schwartz, B. Sen, C. N. Archie, and J. E. Lukens (1985), *Phys. Rev. Lett.* **55**, 1547.

351. A. G. Semenov, D. S. Golubev, and A. D. Zaikin (2009), *Phys. Rev. B* **79**, 115302.

352. A. G. Semenov and A. D. Zaikin (2009), *Phys. Rev. B* **80**, 155312.

353. A. G. Semenov and A. D. Zaikin (2010a), *J. Phys.: Condens. Mat.* **22**, 485302.

354. A. G. Semenov and A. D. Zaikin (2010b), *Physica E* **42**, 600.

355. A. G. Semenov and A. D. Zaikin (2011), *Phys. Rev. B* **84**, 045416.

356. A. G. Semenov and A. D. Zaikin (2012), *Phys. Scr.* **T151**, 014022.

357. A. G. Semenov and A. D. Zaikin (2013), *Phys. Rev. B* **88**, 054505.

358. A. G. Semenov and A. D. Zaikin (2015a), *Phys. Scr. T* **165**, 014041.

359. A. G. Semenov and A. D. Zaikin (2015b), *Phys. Rev. B* **91**, 024505.

360. A. G. Semenov and A. D. Zaikin (2016), *Phys. Rev. B* **94**, 014512.

361. A. G. Semenov and A. D. Zaikin (2017a), *Fortschr. Phys.* **65**, 1600043.

362. A. G. Semenov and A. D. Zaikin (2017b), *Low Temp. Phys. (Kharkov)* **42**, 1011.

363. A. G. Semenov and A. D. Zaikin (2017c), *J. Supercond. Nov. Magn.* **30**, 139.

364. A. G. Semenov, A. D. Zaikin, and L. S. Kuzmin (2012), *Phys. Rev. B* **86**, 144529.

365. S. V. Sharov and A. D. Zaikin (2005a), *Phys. Rev. B* **71**, 014518.

366. S. V. Sharov and A. D. Zaikin (2005b), *Physica E* **29**, 360.

367. Y. Shimazu, T. Yamagata, S. Ikehata, and S. Kobayashi (1997), *J. Phys. Soc. Jpn.* **66**, 1409.

368. I. Snyman and Yu. V. Nazarov (2008), *Phys. Rev. B* **77**, 165118.

369. E. V. Sukhorukov, G. Burkard, and D. Loss (2001), *Phys. Rev. B* **63**, 125315.

370. J. C. Swihart (1961), *J. Appl. Phys.* **32**, 461.

371. I. Takesue, T. Akazaki, S. Miyadai, N. Kobayashi, A. Tokita, M. Nomura, J. Haruyama, and H. Takayanagi (2004), *Physica E* **24**, 32.

372. R. Tarkiainen, M. Ahlskog, J. Penttilä, L. Roschier, P. Hakonen, M. Paalanen, and E. Sonin (2001), *Phys. Rev. B* **64** 195412.

373. R. Tarkiainen, M. Ahlskog, A. Zyuzin, P. Hakonen, and M. Paalanen (2004), *Phys. Rev. B* **69** 033402.

374. C. Texier and G. Montambaux (2005), *Phys. Rev. B* **72**, 115327.

375. M. Tinkham (1996), *Introduction to Supercoonductivity*, McGraw-Hill, New York.

376. A. V. Timofeev, M. Meschke, J. T. Peltonen, T. T. Heikkilä, and J. P. Pekola (2007), *Phys. Rev. Lett.* **98**, 207001.

377. T. B. Tran, I. S. Beloborodov, X. M. Lin, T. P. Bigioni, V. M. Vinokur, and H. M. Jaeger (2005), *Phys. Rev. Lett.* **95**, 076806.

378. T. B. Tran, I. S. Beloborodov, J. Hu, X. M. Lin, T. F. Rosenbaum, and H. M. Jaeger (2008), *Phys. Rev. B* **78**, 075437.

379. A. Trionfi, S. Lee, and D. Natelson (2005), *Phys. Rev. B* **72**, 035407.

380. T. Tsuneta, L. Lechner, and P. J. Hakonen (2007), *Phys. Rev. Lett.* **98**, 087002.

381. M. T. Tuominen, J. M. Hergenrother, T. S. Tighe, and M. Tinkham (1992), *Phys. Rev. Lett.* **69**, 1997.

382. N. Ubbelohde, C. Fricke, C. Flindt, F. Hohls, and R. J. Haug (2012), *Nature Commun.* **3**, 612.

383. K. D. Usadel (1970), *Phys. Rev. Lett.* **25**, 507.

384. A. M. van der Brink, U. Geigenmüller, and A. D. Zaikin (1990), *Physica B* **165/166**, 939.

385. A. van Otterlo, D. S. Golubev, A. D. Zaikin, and G. Blatter (1999), *Eur. Phys. J. B* **10**, 131.

386. M. Vanevic and Yu. V. Nazarov (2012), *Phys. Rev. Lett.* **108**, 187002.

387. A. F. Volkov (1995), *Phys. Rev. Lett.* **74**, 4730.

388. A. F. Volkov, A. V. Zaitsev, and T. M. Klapwijk (1993), *Physica C* **210**, 21.

389. R. F. Voss and R. A. Webb (1981a), *Phys. Rev. Lett.* **47**, 265.

390. R. F. Voss and R. A. Webb (1981b), *Phys. Rev. B* **24**, 7447.

391. P. Wahlgren, P. Delsing, and D. B. Haviland (1995), *Phys. Rev. B* **52**, R2293.

392. P. Wahlgren, P. Delsing, T. Claeson, and D. B. Haviland (1998), *Phys. Rev. B* **57**, 2375.

393. X. Wang, R. Egger, and H. Grabert (1997), *Europhys. Lett.* **38**, 545.

394. X. Wang and H. Grabert (1996), *Phys. Rev. B* **53**, 12621.
395. S. Washburn, R. A. Webb, R. F. Voss, and S. M. Faris (1985), *Phys. Rev. Lett.* **54**, 2712.
396. R. A. Webb, P. Mohanty, and E. M. Q. Jariwala (1998), *Fortsch. Phys.* **46**, 779.
397. H. Weber, R. Häussler, H. von Löhneysen, and J. Kroha (2001), *Phys. Rev. B* **63**, 165426.
398. J. Wei and V. Chandrasekhar (2010), *Nat. Phys.* **6**, 494.
399. U. Weiss (1996), *Solid St. Comm.* **100**, 281.
400. U. Weiss (2008) *Quantum Dissipative Systems*, 3rd enlarged edition, World Scientific, Singapore.
401. U. Weiss and H. Grabert (1985), *Phys. Lett. A* **108**, 63.
402. X. G. Wen (1990), *Phys. Rev. B* **41**, 12838.
403. P. Werner and M. Troyer (2005), J. Stat. Mech., P01003.
404. N. R. Werthamer (1966), *Phys. Rev.* **147**, 255.
405. F. Wilhelm, G. Schön, and A. D. Zaikin (1998), *Phys. Rev. Lett.* **81**, 1682.
406. F. Wilhelm, A. D. Zaikin, and G. Schön (1997), *J. Low Temp. Phys.* **106**, 305.
407. P. G. Wolynes (1981), *Phys. Rev. Lett.* **47**, 968.
408. R. Yagi, S. Kobayashi, and Y. Ootuka (1997), *J. Phys. Soc. Jpn.* **66**, 3722.
409. S. Yip (1998), *Phys. Rev. B* **58**, 5803.
410. A. D. Zaikin (1990), *J. Low Temp. Phys.* **80**, 223.
411. A. D. Zaikin (1994), *Physica B* **203**, 255.
412. A. D. Zaikin (2004), *Low Temp. Phys. (Kharkov)* **30**, 756.
413. A. D. Zaikin (2010), in *Handbook of Nanophysics: Nanotubes and Nanowires*, Taylor and Francis, p. 40–1.
414. A. D. Zaikin and D. S. Golubev (1992), *Phys. Lett. A* **164**, 337.
415. A. D. Zaikin and D. S. Golubev (2004) in: *Quantum Information and Decoherence in Nanosystems*, Elsevier Science Publishers B.V., New York.
416. A. D. Zaikin, D. S. Golubev, A. van Otterlo, and G. T. Zimanyi (1997), *Phys. Rev. Lett.* **78** 1552.
417. A. D. Zaikin, D. S. Golubev, A. van Otterlo, and G. T. Zimanyi (1998), *Physics Uspekhi* **42**, 226.
418. A. D. Zaikin and I. N. Kosarev (1988), *Phys. Lett. A* **131**, 125.

419. A. D. Zaikin, I. N. Kosarev, and S. V. Panyukov (1987), *Sov. Phys. Lebedev Inst. Reports* **N9**, 25.
420. A. D. Zaikin and S. V. Panyukov (1985a), *Sov. Phys. JETP* **62**, 137.
421. A. D. Zaikin and S. V. Panyukov (1985b), *Sov. Phys. JETP* **62**, 1091.
422. A. D. Zaikin and S. V. Panyukov (1986), *JETP Lett.* **43**, 670.
423. A. D. Zaikin and S. V. Panyukov (1987a), *Phys. Lett. A* **120**, 306.
424. A. D. Zaikin and S. V. Panyukov (1987b), *Sov. Phys. JETP* **66**, 517.
425. A. D. Zaikin and S. V. Panyukov (1988), *Sov. Phys. JETP* **67**, 2487.
426. A. D. Zaikin and S. V. Panyukov (1993), *Phys. Lett. A* **183**, 115.
427. A. D. Zaikin and G. F. Zharkov (1981), *Sov. J. Low Temp. Phys.* **7**, 184.
428. A. V. Zaitsev (1984), *Sov. Phys. JETP* **59**, 863.
429. M. Zgirski, K. P. Riikonen, V. Tuboltsev, and K. Arutyunov (2005), *Nano Lett.* **5**, 1029.
430. M. Zgirski, K. P. Riikonen, V. Tuboltsev, and K. Arutyunov (2008), *Phys. Rev. B* **77**, 054508.
431. J. König, H. Schoeller, and G. Schön (1997), *Phys. Rev. Lett.* **78**, 4482.

Index

action 10, 12, 47, 51, 53, 58, 72, 73, 77, 80, 463, 597, 598, 609, 623, 624, 680, 682
adiabaticity condition 219, 226, 762, 763
Aharonov–Bohm effect 99, 660
analytic continuation 29, 65, 116, 144, 173, 209, 295, 300, 305, 363, 541, 692, 761, 808, 828
Anderson localization 574, 602
Andreev conductance 676, 682, 683, 685, 686, 691, 693–699, 702, 703
Andreev level 733, 741, 743, 754, 755, 757–759, 762, 764, 766, 768, 770, 885, 889, 894
  discrete 745, 888
  subgap 726, 730, 731, 756, 874
Andreev oscillator 764
Andreev reflection 675, 677–683, 685, 686, 692, 693, 705, 706, 726
Andreev state 731, 886, 888
Andreev transmission 676, 710
Andreev tunneling rate 692
anharmonicity 107
array 534–538, 548, 550, 558, 565, 568, 570, 571, 641–645, 647, 649, 651, 653, 655, 657, 659
  homogeneous 558
  one-dimensional 641
Bardeen–Cooper–Schrieffer theory (BCS theory) 775, 791–793, 878, 879, 882
bath oscillator 16, 37, 86, 87, 92, 182, 184, 216
BCS theory *see* Bardeen–Cooper–Schrieffer theory
Bloch energy band 156, 210, 212
Bloch nose 240, 248, 249, 258
Bloch oscillations 223, 240, 241, 244, 245, 248–250
Bloch states 156, 210
Bloch steps 224
Bohr magneton 699
Boltzmann kinetic equation 390
Boltzmann–Langevin approach 525, 557, 570
Bose–Matsubara frequency 353, 807, 877
bounce 114, 115, 118, 119, 122, 133, 135, 137, 139, 143, 145, 163, 170, 173, 176, 766
bounce solution 124, 134, 143, 147, 148
bounce trajectory 113, 114, 117, 119, 123, 124, 127, 128, 133–136, 142, 143, 161, 163, 763
boundary conditions 76, 98, 100, 102, 122, 126, 215, 288–291, 551, 554, 687, 689, 836, 837, 856, 870
Brillouin zone 219, 220, 227, 239, 240, 243, 244

Bromwich contour 186, 187

Caldeira–Leggett bath 16, 18, 65, 70, 80, 84, 88, 89, 95, 97, 107, 113, 171, 225, 320, 333, 326, 350, 371
Caldeira–Leggett model 15–17, 19, 20, 24, 29, 60, 82, 85, 100, 323, 324, 327, 332, 349, 351, 599, 758
carbon nanotube 442, 449, 477, 651, 698, 896
causality 10, 11, 35, 54, 270, 280, 385, 404, 430, 439, 615
causality principle 55, 385, 387, 631, 732
chain 534, 535, 537, 538, 541–543, 546–548, 551, 553, 556–558, 564, 565, 567–569, 571, 655, 827, 830
  granular 449
  Josephson 831, 834, 868
  linear 538
  one-dimensional 547
  quantum dot 554
charge density 206, 218, 230, 372–376, 394, 616, 778, 780, 793
charge fluctuations 247, 250, 263, 273, 282, 481, 504
charge fugacity 168, 204, 825
charge state 211, 225, 227, 246–248, 259, 264, 267, 289, 290, 293, 294, 309, 484, 492, 692
  continuous 211, 213
  diagonal 491
  excited 306
  metastable 224
charge transfer 529, 542, 705, 706, 757
classical Drude conductivity 577
classical Drude formula 398
classical Hamilton equations 596
classical Hamilton function 36
classical Newton equation 112
classical path 93, 104, 348, 349, 580, 581, 603, 605, 610, 612
classical return probability 92, 589, 595, 596
classical transition probability 581
coherent effect 175, 235, 602, 661, 674
conductance 277, 278, 444, 446, 487, 488, 520, 521, 524, 525, 565, 568, 569, 583–585, 616, 617, 628, 629, 642, 643, 666, 670–671, 684
  Andreev 708
  gate-averaged 506
  inter-grain 660
  low-energy 685
  non-interacting 523
  nonzero 676
  normal state 676, 685, 687, 749, 832
  shunt 80, 465
  subgap 710
  temperature-dependent 282
conductivity 391, 397, 398, 575, 576, 578, 581, 582, 585, 586, 595, 606, 607, 614, 787
  complex 785
  linear 392, 579, 609
conductor 6, 329, 437, 438, 441–446, 465, 467, 471, 472, 474, 497, 505, 507, 533, 534, 620, 647, 648, 676–678
  diffusive 327, 329, 332, 350, 558, 586, 590, 637, 648, 683, 735
  granular 650
  low-dimensional 574, 799
  mesoscopic 519, 524, 584
  quasi-one-dimensional 570, 599, 649

short coherent 418, 419, 421–423, 440–442, 444, 446–452, 454, 456, 458, 460, 462–464, 466, 468, 479, 480, 508, 727, 728
single-channel 452, 472
single-mode 474
voltage-biased 426, 450
Cooperon 578, 579, 581, 584, 594, 596, 600, 603, 610, 629, 630, 644, 646, 664, 666–670, 685, 698–700, 751, 753
Cooper pair tunneling 32, 238, 239, 246, 307, 742, 826, 857, 869
correction 58, 60, 133, 134, 338, 367, 373, 594, 595, 621, 626, 633, 635, 639, 642, 663, 665
Aslamazov–Larkin fluctuation 787
Coulomb 277, 338, 444, 451, 459, 639, 644, 749–751
dissipative 123
fluctuation 791, 792, 797
flux-dependent 367
quantum 130, 151, 196–198, 236, 582, 584, 610, 617, 663–666
weak localization 397, 582, 585, 619–622, 626, 628, 629, 633, 636, 638, 639, 642, 643, 651, 655, 670, 674, 754
correlation function 209, 351, 353, 359, 374, 375, 399, 429, 450, 631, 634, 640, 644, 645, 663, 734, 790, 791
cotunneling 491, 492, 494, 539, 706
Coulomb blockade 262, 263, 271–273, 314, 485–488, 496, 497, 499–501, 503, 505, 507–509, 511, 513, 515, 638–640, 678, 679, 751, 752
Coulomb blockade regime 238, 245, 267, 272, 273, 276, 277, 281, 282, 285, 464, 466, 480, 486, 488, 490, 536, 541
Coulomb dip 488, 519, 520, 684
Coulomb effect 64, 69, 196, 238, 270, 272, 273, 308, 314, 442, 444, 445, 448, 455, 457, 459, 461–463, 465, 467, 469, 471, 693–695, 725, 726
Coulomb staircase 288, 293, 486
Coulomb suppression 282, 498, 696, 721
crossover temperature 135, 141, 601, 602, 767, 770, 806, 809–811, 880
current cumulant 356–358, 424–426, 431, 433, 435, 437, 438, 453–455, 533
current fluctuations 351, 352, 354, 356, 357, 360, 363, 364, 366, 367, 408, 437, 438, 440, 454, 734, 858
current noise 35, 53, 56, 274, 275, 431, 437, 438, 444, 458, 460, 511, 513, 526, 527, 546, 547, 551, 703, 704
current operator 43, 351, 354, 355, 359, 360, 426, 428, 429, 432, 433, 454, 533, 858, 859
current-phase relation 121, 130, 141, 197, 198, 224, 759

damping 123, 135, 141, 150, 178, 191, 390, 759
Debye frequency 779
decay rate 109–113, 115, 117–122, 124, 125, 131, 132, 134, 136, 138–142, 147, 149–153, 173–175, 225, 294, 604, 806–808
decoherence 85, 97, 186, 328, 329, 351, 352, 651, 660, 673, 869

density matrix 8, 84, 102–104, 160, 177, 178, 180, 182, 184, 228, 229, 232, 234, 235, 392, 399–402, 414–416, 502
many-body 372, 381
reduced 17, 82, 85, 192, 228, 242, 489, 501
single-electron 383, 387, 391, 400, 606
dephasing 84, 97, 157, 349, 494, 590, 600, 634, 635, 639, 640, 650, 653, 678, 685, 700, 702
Cooper pair 754
dephasing effect 590, 652, 660, 701, 702
dephasing rate 192, 578, 584, 590–592, 594, 600, 601, 604, 647
dephasing time 84, 550, 636, 641, 657, 669, 700
differential conductance 263, 272, 276–278, 281, 282, 447, 452, 485, 488, 498, 500, 519–521, 561, 562, 565, 720, 722
diffusion coefficient 80, 237, 568, 569, 647, 656, 658, 659, 686
diffusion equation 554, 555, 560, 568, 578, 581, 641, 667, 781
diffusive paths 581, 583, 596, 646, 668
diffuson 568, 578, 579, 581, 582, 589, 595, 610, 646, 668, 669, 781
disorder 406, 517, 575–577, 619, 620, 625, 651, 660, 673, 684–687, 689, 691, 693, 695, 697, 716
dissipation 5, 6, 16–18, 104, 105, 107, 108, 110–114, 122–126, 128–130, 146–148, 155, 156, 168–172, 176–178, 195–200, 257, 757–760, 764–766
Drude conductance 589, 684
Drude conductivity 327, 386, 583, 598, 603
Drude formula 386, 568, 783, 816, 818
Drude model 327, 650
Dyson equation 298, 299, 301, 362, 491

EC *see* elastic cotunneling
effective action formalism 429, 508, 678, 727, 734, 745, 773, 806, 813
effective Hamiltonian 38, 39, 41–43, 71, 428, 797, 835, 838, 839
eigenstate 20, 21, 87, 97, 109, 178, 354, 358
eigenvalue 116, 117, 125, 126, 128, 129, 134–138, 144, 146, 161, 178, 312, 413, 422, 465, 466, 625, 808, 821, 822
Eilenberger equation 51, 74, 79, 686
Eilenberger function 52, 74, 79
elastic cotunneling (EC) 205–208, 219–227, 238–241, 247–249, 252–254, 270–273, 284–288, 301–303, 310–313, 315–317, 333, 334, 339–342, 486–488, 492–496, 542–544
electron gas 7, 324–327, 329, 348, 361, 369, 370, 376, 383, 386, 396, 605
electron scattering 369, 406, 412, 424, 440, 477, 513, 535, 570, 574, 620, 673, 684, 708
electron transport 261, 264, 285, 286, 405, 406, 422, 424, 441, 443, 571, 573, 675–678, 680, 684–686, 716–718, 720–722
electron tunneling 70, 259, 262–264, 267, 271, 274, 282, 283, 288, 289, 308, 485, 486,

490, 492, 623, 624, 692, 694, 695
environment 6–12, 16, 17, 42–44, 47, 210, 211, 277, 278, 281, 282, 284, 323–329, 331, 370, 429, 430, 743–746, 754, 755, 757–759
equilibrium 49, 50, 84, 85, 324, 329, 351, 353, 383, 390, 393, 398, 399, 427, 429, 534, 772, 776
thermal 110
thermodynamic 12, 13, 16, 17, 28, 328, 385, 504, 777, 778, 891
evolution kernel 22, 24, 26, 27, 85, 394, 395
evolution operator 15, 18, 373, 375, 378–381, 391, 393, 395, 398, 400, 403, 405, 416, 613, 615
free electron 606
mechanical 7, 372
single-electron 414

Fano factor 436, 443, 444, 446, 447, 449, 452, 455, 497, 523, 524, 548, 556, 558, 568, 569, 571, 641, 666
FCS *see* full counting statistics
FDT *see* fluctuation-dissipation theorem
Fermi energy 51, 390, 396, 605, 774, 802, 884, 888, 893
Fermi function 34, 49, 50, 265, 283, 386, 395, 416, 419, 493, 525, 598, 605, 679, 881, 891
Fermi level 327, 396, 535, 568, 648, 690, 774, 885, 886, 888
Fermi statistics 327, 369–371, 390, 396, 403, 591, 595, 597, 623
Fermi velocity 74, 76, 77, 80, 327, 587, 641, 656, 659, 696
Feynman–Vernon influence functional 9, 22, 24, 27, 37, 40, 64, 70, 110, 182, 228, 369, 396, 595, 771
fluctuation-dissipation theorem (FDT) 10, 12, 17, 35, 56, 386, 430, 444, 453, 732, 840, 845
full counting statistics (FCS) 424–427, 440, 677

gate charge 293, 294, 301, 309, 313, 315, 317, 537, 862
gate effect 310, 481, 497, 500, 501, 503, 504, 507, 538
Gaussian influence functional 11, 12, 70, 92, 323
Gaussian integral 137, 198, 206, 216, 226, 230, 289, 290, 357, 458, 760
geometric capacitance 696, 738, 740, 744, 752, 755, 759, 763, 765–767, 832
Ginzburg–Landau equation 787
grand partition function 28, 98, 99, 110, 112, 164, 165, 203, 205, 206, 215, 218, 287–289, 309, 312, 330, 538, 539, 819
Grassmann field 45
Green function 23, 25, 47–50, 52, 66, 376, 377, 384, 385, 402, 554, 555, 575–577, 624, 625, 713, 715, 716, 841, 842, 844, 845
Green–Keldysh matrix 679, 680, 728, 729
ground-state energy 218, 224, 303, 306, 309, 313, 315, 333, 336, 355, 356, 468, 470, 857, 891, 895, 896

Hamiltonian 5–7, 16–18, 20, 21, 39, 40, 210–212, 214, 216–218, 325, 354–357, 371, 372, 412–414, 710, 757, 837–839, 841
harmonic oscillator 11–15, 20, 21, 23, 37, 40, 42, 92, 94, 100, 101, 103–105, 107, 158, 160, 323, 324, 756–758
Hubbard–Stratonovich transformation 18, 38, 39, 45, 62, 71, 217, 370, 372, 374, 376, 387, 511, 794, 864

impedance 33, 35, 42, 43, 60, 62, 429, 430, 446, 452, 481, 482, 522, 542, 546, 553, 559, 742
  environmental 444, 742, 743
  frequency-dependent 262, 429
  linear 33, 257, 278, 631
  Ohmic 742
inelastic relaxation 240, 246, 519–521, 525, 550, 557, 565
influence functional 6, 9–16, 37, 53, 64, 70, 81, 82, 92, 94, 95, 182, 183, 261, 264, 323–327, 329, 331, 348, 370, 371, 385, 387, 393, 396, 403, 595, 597, 603, 605, 606, 755, 760–762
instanton analysis 171, 202, 206, 221, 253, 341, 343, 474
instanton technique 133, 159, 163, 212, 221, 312, 340, 341, 343, 357, 819, 832

Josephson coupling energy 59, 197, 207, 219–221, 224, 742, 744, 745, 832, 839
Josephson effect 725, 726
Josephson element 213, 214, 224, 234, 744, 745
Josephson particle 36, 120, 121, 130, 142, 149, 156, 160, 177, 195–197, 199–201, 207, 208, 234–237, 258, 757, 762
Josephson phase 31, 35–37, 53, 64, 69, 120, 122, 123, 149, 192, 195, 199–201, 726, 735, 738, 740
Josephson relation 33, 36, 41, 42, 65, 72, 74, 79, 197, 200, 222, 223, 772, 779, 796, 801, 811

Keldysh contour 23, 27, 39, 42, 44, 179, 182, 228–231, 370, 373, 374, 394, 395, 398, 489, 490, 728, 773
Keldysh Green function 24, 48–50, 353, 387, 400, 401, 417, 527, 550, 624, 712, 714
Keldysh technique 400, 491, 728
kink 160, 163, 165, 167, 170, 173, 176, 334, 826, 868
Kirchhoff law 666

Landauer–Büttiker scattering formalism 479
Landauer formula 407, 408, 465, 514, 676, 732
Langevin equation 19, 35, 37, 62, 63, 88, 105, 222, 223, 254, 285, 286, 365, 508, 510, 512, 513, 785, 787–789
  classical 20
  quasiclassical 18, 19, 105, 254, 274, 285, 286, 364, 403, 444, 445, 534, 545, 737
linear Ohmic dissipation 19, 29, 37, 111, 140, 165, 176, 184, 196, 203, 229, 257, 258, 306, 748, 760
Luttinger model 352, 456

macroscopic quantum coherence (MQC) 32, 155–158, 160, 162, 164, 166, 168, 170, 172, 174, 176, 178, 180, 182, 192
macroscopic quantum tunneling (MQT) 32, 122, 150, 195, 224, 252, 760, 761, 766, 773
Matsubara frequency 29, 68, 115, 128, 147, 331, 353, 354, 798, 807, 808, 877
Matsubara technique 728, 760, 773, 798
metallic quantum dots 477, 478, 480–482, 484, 486, 488, 490, 492, 494, 496, 498, 500, 502, 504, 506, 532–534
metastable state 65, 107–111, 113, 114, 116, 118, 119, 131, 132, 139, 145, 148, 155, 163, 173, 806, 808, 811
MQC *see* macroscopic quantum coherence
MQT *see* macroscopic quantum tunneling
muon 7, 324, 396, 404

Nambu space 50, 51, 66, 72, 74, 77, 794
nanoring 328, 351, 354, 367, 665, 670, 736, 854, 864, 869, 880
nanowire 803, 830, 835, 840, 846, 849, 850, 853, 854
noise 12, 19, 37, 62, 286, 327, 367, 405, 429, 431, 433, 444, 445, 453, 459, 460, 524
noise correlator 63, 105, 552, 556, 557, 721, 787
noise power 360, 363, 365–367, 446, 848, 867, 873, 874
noise spectrum 354, 360, 361, 460, 461, 704, 735, 736, 866, 867
non-local conductance 706, 707, 716, 717, 722, 723

occupation probability 49, 109, 177, 190, 199, 225, 235–237, 246, 287, 381, 383, 484, 492, 777, 880
Ohmic damping 20, 733
Ohmic environment 20, 80, 262, 281–285, 295, 466, 488, 538, 562, 565, 638, 711
Ohmic shunt 70, 232, 295, 497, 522, 617, 695, 709
Ohm's law 407, 552, 626
oscillator 13, 14, 16, 17, 19, 21, 22, 81, 82, 84, 86, 87, 89, 92, 93, 95, 97, 98, 102, 104, 757, 758, 763, 764
  undamped 105
  underdamped 102

parity effect 875, 876, 879, 880, 883, 887–889, 891
partition function 98, 99, 118, 119, 159, 160, 162–167, 176, 202–204, 206, 211–219, 226, 290–292, 307, 338, 339, 793, 794, 824, 862–865
path integral 22, 25, 82, 180, 230, 251, 268, 489, 538, 540, 608, 612, 614, 820, 868
Pauli principle 23, 327, 370, 371, 378, 390, 396, 404, 424, 591, 593, 605
perturbation theory 286, 287, 289, 291, 293–296, 338, 339, 403, 462, 463, 465, 467, 469, 471, 496, 497, 577, 578, 611, 775
phase difference 33, 78, 230, 736, 739, 874, 886, 893, 896
phase relaxation 351, 508, 590
plasmon 842, 847

Poisson's resummation formula 290, 292, 335, 358
Poisson's resummation theorem 212, 297, 862

QMC analysis *see* quantum Monte Carlo analysis
QMC data 316, 317, 319, 341–343
QMC simulations 315, 318
QPS *see* quantum phase slips
QPS action 813, 814, 817–819, 823
QPS amplitude 823, 824, 838, 871, 873
QPS core 814, 815, 818, 823, 826, 834
QPS fugacity 828, 832–834, 854, 871, 883
QPS rings 863, 868, 870, 872, 874
QPT *see* quantum phase transition
quantum coherence 96, 155, 157, 195, 196, 200, 201, 328, 348, 350, 351, 361, 366, 367, 508, 509, 511, 513, 515, 698
quantum decay 69, 108, 109, 111, 113, 148, 160, 163, 199, 604, 762, 766
quantum decoherence 84, 85, 90, 95–97, 99, 351, 397, 572, 610, 615, 700, 754
quantum dephasing 88, 95, 328, 672, 674, 694
quantum dot 477–480, 482, 497, 498, 508–510, 516–518, 523–528, 536–538, 556–562, 620–622, 637, 638, 640, 641, 647–649, 653–655, 659–662, 665, 666
quantum fluctuations 121, 130, 131, 139, 140, 197, 198, 201, 725, 726, 791, 801–803, 813, 825–827, 830, 850, 859, 860, 866, 868
quantum interference 97, 348, 571, 572, 574, 602, 626, 661, 685, 695
quantum Monte Carlo analysis (QMC analysis) 208, 315, 319, 342
quantum phase 57, 62, 63, 270, 273, 279, 285, 424, 430, 434, 442, 445, 820, 824, 828, 831
quantum phase slips (QPS) 331, 773, 791, 799, 800, 802–804, 806, 808, 810, 812–850, 852, 854, 855, 860, 861, 863–867, 869–874, 883
quantum phase transition (QPT) 169, 170, 205, 208, 222, 258, 823, 827, 830, 831, 833, 850, 851
quasiclassical approximation 19, 51, 424, 509, 546, 603, 609, 612, 614, 681, 682

random phase approximation 383, 384, 456, 595, 645
RPA *see* random phase approximation

scatterer 406, 407, 423, 424, 429, 430, 447, 455, 458, 460, 461, 478, 534, 535, 545–547, 550–552, 558, 569–571, 620–623, 654, 655
  conducting 548
  diffusive 558
  short coherent 407, 429, 430, 442, 470, 479, 509, 513, 521
scattering matrix 408, 409, 411–413, 419, 464, 508, 514, 516–518, 622, 626, 727
  energy-dependent 479, 508
  unitary 407

SCHA *see* self-consistent harmonic approximation
Schrödinger equation 5, 7, 17, 37, 109, 125, 222, 242, 379, 388, 389, 391, 409, 412, 414
self-consistent harmonic approximation (SCHA) 197, 198
shot noise 444–446, 449, 451–455, 457, 459–461, 533, 534, 545, 546, 677, 678, 683, 684, 703–705, 707–709, 717–719, 721, 723, 840
sine-Gordon model 834, 869–871
Slater determinant 378–381
superconducting coherence length 705, 790, 802, 805, 816, 837
superconducting contact 726, 745, 746, 748, 761, 874
  ballistic 122
  non-tunnel 726, 744, 745
  transparent 751, 755, 766
superconducting nanowire 773, 787, 791, 793, 797, 799, 800, 802, 804, 806, 808, 810, 812–848, 850, 852, 854
superconducting phase 71, 745, 760, 761, 772, 799, 802, 811, 813, 827, 875
superconducting wire 789, 790, 796, 801–803, 806, 812, 820, 825, 827, 833, 835–837, 841, 856, 859, 869, 874
superconductor 40, 41, 44, 48–50, 73, 74, 76, 675, 698, 699, 705–710, 725, 727, 751, 753, 772, 773, 779, 780, 785
  bulk 33, 75, 792
  diffusive 716
  dirty 780, 787
  gapless 788
TAPS *see* thermally activated phase slips
TAPS rate 803, 805, 808–811
thermal activation 111, 114, 119, 120, 131, 132, 135, 140, 147, 149, 767, 769, 802, 803, 806
thermally activated phase slips (TAPS) 773, 801–805, 807–809, 811–813, 846
Thouless energy 520, 521, 565, 678, 686, 696, 751, 752
time-reversed path 92, 94, 95, 97, 348, 349, 581–583, 597, 610, 611, 614, 622, 635
trajectory 75, 82, 162, 163, 181, 183, 212, 213, 215, 309, 311, 312, 334, 335, 343, 579–581, 596, 597, 667, 668, 708
  classical 91, 96, 339, 349, 581, 595
  diffusive 597, 598, 617
  kink 161
  multi-instanton 165, 166, 170, 335, 344
  quasiclassical 405, 582
  saddle-point 820
  semiclassical 310
  straight-line 311
transition amplitude 7, 8, 81, 82, 91, 179, 180, 325
transition 10, 21, 22, 86, 181, 211, 232, 360, 364, 484, 743, 825, 828, 829, 852, 863, 867
  dissipative phase 171, 345, 830
  first-order phase 110
transmission distribution 406, 441, 466, 467, 470, 471, 497, 545, 620, 726, 735, 739, 740, 750, 751, 758, 759, 833
tunneling amplitude 41, 42, 47, 70, 73, 162, 169, 171, 173, 178, 235, 237, 334, 340, 344–348, 624

tunneling Hamiltonian 43, 56, 411, 413, 625, 642, 710

Usadel equation 688–690

van Vleck formula 580, 608, 615
voltage fluctuation 58, 437, 438, 440, 518, 571, 589, 737, 840, 842, 844, 847
voltage noise 591, 811, 812, 843, 845, 846

weak localization (WL) 177, 178, 180, 571, 573–586, 588, 590, 592, 594, 596, 618–666, 670, 672–674, 684, 685, 698–700, 887
Wick theorem 439, 842
Wigner function 102, 105, 397, 398
WL *see* weak localization

ZBA *see* zero-bias anomaly
Zener tunneling 220, 240, 241, 243–245, 249
Zener tunneling rate 220, 228, 240, 241, 245
zero-bias anomaly (ZBA) 282, 494, 685, 686, 694, 695, 697
zero-bias conductance 271, 272, 282, 480, 487, 498, 500, 501, 519, 520, 536, 538, 549, 563, 564, 569
zero mode 117–120, 129, 138, 139, 144, 145, 161–163, 312, 313, 470, 820–822, 833, 842

For Product Safety Concerns and Information please contact our EU representative GPSR@taylorandfrancis.com Taylor & Francis Verlag GmbH, Kaufingerstraße 24, 80331 München, Germany

Batch number: 10726503

Printed by Printforce, the Netherlands